The *Chlamydomonas* Sourcebook

SECOND EDITION

Volume 2: Organellar and Metabolic Processes

The *Chlamydomonas* Sourcebook

SECOND EDITION

Volume 2: Organellar and Metabolic Processes

Edited by

David B. Stern, Ph.D.
Boyce Thompson Institute
Ithaca, New York, USA

Editor in Chief

Elizabeth H. Harris, Ph.D.
Department of Biology
Duke University
Durham, North Carolina, USA

ELSEVIER

AMSTERDAM • BOSTON • HEIDELBERG • LONDON • NEW YORK
OXFORD • PARIS • SAN DIEGO • SAN FRANCISCO • SYDNEY • TOKYO

Academic Press is an imprint of Elsevier

Cover image: Backdrop is **an anti-Fus1**, **anti-tubulin labeled *Chlamydomonas* gamete**, courtesy of Mike Misamore. Overlaid images are clockwise from left, an illustration of the photorespiratory carbon oxidation cycle (see Chapter 8); regulatory circuits of the CES mechanism (see Chapter 29), and major carotenoids of vegetative cells (see Chapter 21). Figures are modified from those provided by Martin Spalding, Yves Choquet, and Martin Lohr, respectively.

Academic Press is an imprint of Elsevier
The Boulevard, Langford Lane, Kidlington, Oxford OX5 8GB
30 Corporate Drive, Suite 400, Burlington, MA 01803, USA
525 B Street, Suite 1900, San Diego, CA 92101-4495, USA

First edition 1989
Second edition 2009

Set ISBN: 978-0-12-370873-1

Volume 1: ISBN: 978-0-12-370874-8
Volume 2: ISBN: 978-0-12-370875-5
Volume 3: ISBN: 978-0-12-370876-2

Typeset by Charon Tec Ltd., A Macmillan Company. (www.macmillansolutions.com)

Printed and bound in Canada
09 10 11 12 10 9 8 7 6 5 4 3 2 1

Contents of Volume 2

Preface

As my laboratory's copy of the original *Chlamydomonas Sourcebook* became increasingly tattered, and as genome information began to turn the page on a new era in *Chlamydomonas* biology, it seemed appropriate to revisit the question of a new edition. This was around 2001, when Lib Harris and I were members of an NSF-funded and "green side oriented" *Chlamydomonas* Genome Project led by Arthur Grossman, whose initial 1999 grouping also included Pete Lefebrve, Carolyn Silflow, and John Davies. Among our goals were development of a microarray, completion of the chloroplast genome sequence, and a better molecular map. Each of these objectives was completed, but they now seem outdated and almost trivial, a fact which accentuates the pace of the *Chlamydomonas* field since that time.

Fast-forward to the months following our December, 2004 genome group meeting which at that point also included Sabeeha Merchant and Kris Niyogi. The group was enthusiastic about the concept of a new edition, and Lib took the initiative to contact Elsevier which had acquired the previous edition's publisher, Academic Press. In October, 2005 our book proposal was favorably reviewed, including the three-volume format that would permit a broader coverage of the many aspects of *Chlamydomonas* biology. I was asked to edit the "green" volume which would complement Volumes 1 and 3. I accepted the challenge because after all, I had been pushing for a new *Sourcebook*, and also because I believed it would become a bedrock for *Chlamydomonas* research for years to come.

As is obvious from the titles of the chapters in this volume, the term "green" turned out to be a misnomer. This reflects the sometimes underappreciated facts that chloroplasts do much more than photosynthesis, and that they are metabolically linked to every other compartment of the cell. Indeed, one of the images in the cover montage illustrates how the photorespiratory carbon oxidation cycle (Chapter 8) involves the movement of metabolites between chloroplasts and mitochondria. Ultimately, the purpose of Volume 2 became not just to cover photosynthesis, but more broadly chloroplast-based activities, and other metabolic activities that heavily involve the chloroplast. These include chapters on sulfur metabolism (Chapter 5), because primary sulfur assimilation takes place in the chloroplast; and starch (Chapter 1) and amino acid (Chapter 4) metabolism which are largely chloroplast-based activities.

The volume will also inform the reader who seeks insight into the nuts and bolts of photosynthesis, with detailed mechanistic and genetic descriptions of Photosystems I (Chapter 15) and II (Chapter 16), the cytochrome b_6/f complex (Chapter 17), the ATP synthase complex (Chapter 18), the light harvesting proteins (Chapter 14), heme and carotenoid biosynthesis (Chapters 20–21), glycerolipid synthesis (Chapter 2), and Rubisco (Chapter 9). The exciting advances in our understanding of state transitions are covered in Chapter 22, and the autoregulatory aspects of chloroplast translation, elucidated entirely in the past 10 years, are described in Chapter 29. Other chapters cover the organellar genomes (Chapters 12 and 24), organellar gene expression mechanisms (Chapters 25–28) and posttranslational regulation (Chapters 11 and 19), and oxidative phosphorylation in the mitochondria (Chapter 13).

Metabolic activities are often studied in the context of stress responses, reflecting conditions that *Chlamydomonas* periodically encounters in its many environmental niches. This volume covers nitrogen assimilation (Chapter 3), phosphorus deficiency (Chapter 6), transition metal nutrition (Chapter 10), high light stress (Chapter 23), and the carbon concentrating mechanism (Chapter 8). Chapter 7 is devoted to hydrogen production which increases under sulfur stress and is currently one of the high-profile biological pathways under study as a long-term contributor to our energy mix.

In all of these chapters, the influence of the genomic era is obvious. Proposed or elucidated pathways are backed up with genes and gene expression data, many long-studied mutants are now associated with specific genes, and distinctions between *Chlamydomonas* and higher plants can be discussed in terms of genome contents as well as mutant analysis. Microarray data have given insight into the cell's strategy as it encounters a stress situation, and have illuminated commonalities or overlaps between stress responses. The impact of reverse genetic tools emerges in several chapters, and is likely to be amplified as the techniques are refined in the near future. Had it been written only a year or two later, the volume would likely showcase the impact of high-throughput sequencing techniques, which will both shed unprecedented light on biological processes and help refine nuclear genome annotation. Each chapter, at its end, gives its authors' own perspectives for the future in that particular area of study, and time will tell which of those predictions are realized.

Genome information also has its limitations which are reflected in the chapters. A major frustration is the inability to predict protein localization robustly, by informatic analysis of the amino-terminal domain. As a result, whether a given protein – and by extension, its catalytic activity – is located in the chloroplast, the mitochondria, or both, is often ambiguous and awaits laborious biochemical approaches. Furthermore, with the genome still incomplete, one cannot conclude whether a gene (and its associated activity) is absent, or simply has not been found. Thus, many authors have wisely hedged their bets and indeed, this was an editorial necessity.

As mentioned earlier, comparative analyses of *Chlamydomonas* and higher plants – particularly *Arabidopsis* – are effectively used by many authors. Thus, while the volume is intended to focus on *Chlamydomonas*, I encouraged authors to use such comparisons to illustrate biological or evolutionary points. In addition to *Arabidopsis*, *Volvox*, *Ostreococcus*, and the present-day relatives of the chloroplast progenitor, *Synechococcus* and *Synechocystis*, are often cited. Where data were available, authors also mention other *Chlamydomonas* species, some of which may prove to be excellent models due to their evolutionarily selected environmental adaptations.

The chapters in this volume are truly authoritative, testifying to the thorough work and broad perspective of their writers. The viewpoints of the chapters are underpinned by tables of relevant genes and proteins, as well as illustrations to help the nonexpert understand regulatory circuits, macromolecular structures, and biosynthetic pathways. Thus, the volume is intended to serve both as a reference manual, and as a biological view of what is green in *Chlamydomonas*, along with a perhaps surprising number of processes that possess biosynthetic and/or regulatory links to photosynthesis. There is indeed much to learn about this adaptable and accessible alga and its relatives, and I hope this volume will serve you well in your intellectual and experimental journeys.

David B. Stern

Acknowledgments

I first and foremost thank Lib Harris for stimulating my interest and abilities with *Chlamydomonas* through the first *Sourcebook*, and nearly 20 years later for having the persistence and vision to get the second edition off the ground. And over the past 3 years as Volume 2 has come to fruition, Lib has shouldered more than her fair share of the burden of perfecting the mountain of manuscript pages that have graced our e-mail In-boxes, mollifying the publishers as our deadlines constantly receded, and keeping the project moving as we all juggled our many responsibilities. I also thank George Witman for his sense of humor, perfectionist instincts, and input into both style and substance. This has been an enjoyable collaboration that I will never forget.

My thanks also go to each and every author of the many chapters in this volume. Through their efforts I learned a tremendous amount of *Chlamydomonas* biology, biochemistry, physiology, and genetics, as will the reader of these chapters. I appreciate that the authors responded quickly and patiently to my endless requests for clarification and revision, right up to the last possible moment.

I am grateful to Mike Misamore for the striking image that forms the backdrop for the cover design; the other three images are modified from Chapters 8, 21, and 29. Although I designed the montage, I had a wealth of useful critique of the final version as well as its predecessors. Besides Lib and George, opinions came from numerous attendees that I cornered at the 2008 *Chlamydomonas* workshop in Hyères, France.

I thank my unflappable assistant Donna Claes for giving some sense of organization to the project and for protecting my calendar. And most importantly, my wife Karen kept me sane when the project seemed overwhelming, and distracted me with all kinds of work on our farm, Windsong. There's nothing like a country sunset to put things back in perspective.

List of Contributors

Steven G. Ball Université des Sciences et Technologies de Lille, Unité Mixte de Recherche du CNRS n 8576, Laboratoire de Chimie Biologique, Bâtiment C9, Cité Scientifique, 59655 Villeneuve d'Ascq cedex, France

Samuel I. Beale Division of Biology and Medicine, Brown University, Providence, RI 02912, USA

Christoph Benning Department of Biochemistry and Molecular Biology, Michigan State University, 215 Biochemistry Building, East Lansing, MI 48824, USA

Pierre Cardol Genetics of Microorganisms, Institute of Plant Biology B22, Bvd du rectorat, 27, University of Liège, B4000 Liège, Belgium

Yves Choquet Centre National de la Recherche Scientifique (CNRS), Université Pierre et Marie Curie (UPMC-P6) at Institut de Biologie Physico-Chimique, 13 rue Pierre et Marie Curie, 75005 Paris, France

Philippe Deschamps Université des Sciences et Technologies de Lille, Unité Mixte de Recherche du CNRS n 8576, Laboratoire de Chimie Biologique,. Bâtiment C9, Cité Scientifique, 59655 Villeneuve d'Ascq cedex, France

Catherine de Vitry Centre National de la Recherche Scientifique (CNRS), Université Pierre et Marie Curie (UPMC-P6) at Institut de Biologie Physico-Chimique, 13 rue Pierre et Marie Curie, 75005 Paris, France

Dominique Drapier Centre National de la Recherche Scientifique (CNRS), Université Pierre et Marie Curie (UPMC-P6) at Institut de Biologie Physico-Chimique, 13 rue Pierre et Marie Curie, 75005 Paris, France

Alexandra Dubini National Renewable Energy Laboratory, 1617 Cole Blvd., Golden, CO 80401, USA

Emilio Fernández Departamento de Bioquímica y Biología Molecular, Facultad de Ciencias, Universidad de Córdoba. Campus de Rabanales, Edif. Severo Ochoa, 14071-Córdoba, Spain

Francisco Figueroa Departamento de Genética Molecular, Instituto de Fisiología Celular, Universidad Nacional Autónoma de México, Circuito Exterior s/n, Colonia Copilco-Universidad, Delegación Coyoacán, 04510 México D.F., México

Giovanni Finazzi Centre National de la Recherche Scientifique (CNRS), Université Pierre et Marie Curie (UPMC-P6) at Institut de Biologie Physico-Chimique, 13 rue Pierre et Marie Curie, 75005 Paris, France

Lars-Gunnar Franzén University of Halmstad, School of Business and Engineering, Naturrum, P.O. Box 823, SE 301 18 Halmstad, Sweden

Aurora Galván Departamento de Bioquímica y Biología Molecular, Facultad de Ciencias, Universidad de Córdoba. Campus de Rabanales, Edif. Severo Ochoa, 14071-Córdoba, Spain

Maria L. Ghirardi National Renewable Energy Laboratory, Chemical and Biosciences Center, 1617 Cole Blvd., Golden, CO 80401, USA

Zoee Gokhale Donald Danforth Plant Science Center, 975 N. Warson Road, St. Louis, MO 63132, USA

Michel Goldschmidt-Clermont Departments of Molecular Biology and of Plant Biology, University of Geneva, Geneva 4, Switzerland

David Gonzalez-Ballester The Carnegie Institution, Department of Plant Biology, 260 Panama Street, Stanford, CA 94305, USA

Diego González-Halphen Departamento de Genética Molecular, Instituto de Fisiología Celular, Universidad Nacional Autónoma de México, Circuito Exterior s/n, Colonia Copilco-Universidad, Delegación Coyoacán, 04510 México D.F., Mexico

Arthur R. Grossman The Carnegie Institution, Department of Plant Biology, 260 Panama Street, Stanford, CA 94305, USA

Patrice Hamel Department of Plant Cellular and Molecular Biology and Department of Molecular and Cellular Biochemistry, The Ohio State University, 582 Aronoff Laboratory, 318 W. 12th Avenue, Columbus, OH 43210, USA

Marc Hanikenne Laboratory of Plant Cell Biology, Department of Life Sciences, Plant Biology Institute, B22, Bld du rectorat, 27, University of Liège, B4000 Liège, Belgium

Charles Hauser Bioinformatics Program, St. Edward's University, 3001 S. Congress Ave, Austin, TX 78704, USA

David L. Herrin Section of Molecular Cell and Developmental Biology, Institute for Cellular and Molecular Biology, University of Texas, Austin, TX 78712, USA

David C. Higgs Department of Biological Sciences, University of Wisconsin–Parkside, 900 Wood Road, Kenosha, WI 53141, USA

Uwe Klein Department of Molecular Biosciences, University of Oslo, Blindernveien 31, 0371 Oslo, Norway

Richard Kuras Centre National de la Recherche Scientifique (CNRS), Université Pierre et Marie Curie (UPMC-P6) at Institut de Biologie Physico-Chimique, 13 rue Pierre et Marie Curie, 75005 Paris, France

Stéphane D. Lemaire Institut de Biotechnologie des Plantes, Unité Mixte de Recherche 8618, Centre National de la Recherche Scientifique, Batiment 630, Université Paris-Sud 11, 91405 Orsay Cedex, France

Ángel Llamas Departamento de Bioquímica y Biología Molecular, Facultad de Ciencias, Universidad de Córdoba. Campus de Rabanales, Edif. Severo Ochoa, 14071-Córdoba, Spain

Martin Lohr Institut für Allgemeine Botanik, Johannes-Gutenberg Universität, 55099 Mainz, Germany

Sabeeha S. Merchant Department of Chemistry and Biochemistry, UCLA, 607 Charles E.Young Drive East, Los Angeles, CA 90095, USA

Jonathan E. Meuser Colorado School of Mines, Environmental Science and Engineering Division, Golden, CO 80401, USA

Laure Michelet Institut de Biotechnologie des Plantes, Unité Mixte de Recherche 8618, Centre National de la Recherche Scientifique, Batiment 630, Université Paris-Sud 11, 91405 Orsay Cedex, France

Jun Minagawa Institute of Low Temperature Science, Hokkaido University, N19W8, Sapporo 060-0819, Japan

Jeffrey Moseley The Carnegie Institution, 260 Panama Street, Stanford, CA 94305, USA

Krishna K. Niyogi Department of Plant and Microbial Biology, University of California, 441 Koshland Hall, Berkeley, CA 94720, USA

Matthew C. Posewitz Colorado School of Mines, Department of Chemistry and Geochemistry, Golden, CO 80401, USA

Fabrice Rappaport Centre National de la Recherche Scientifique (CNRS), Université Pierre et Marie Curie (UPMC-P6) at Institut de Biologie Physico-Chimique, 13 rue Pierre et Marie Curie, 75005 Paris, France

Kevin E. Redding Department of Chemistry & Biochemistry, Arizona State University, 1711 S Rural Rd., Tempe, AZ 85287, USA

Claire Remacle Genetics of Microorganisms, Institute of Plant Biology B22, Bvd du rectorat, 27, University of Liège, B4000 Liège, Belgium

Wayne R. Riekhof Department of Medicine, National Jewish Medical and Research Center, 1400 Jackson St., Denver, CO 80206, USA

Jean-David Rochaix Departments of Molecular Biology and Plant Biology, University of Geneva, Geneva 4, Switzerland

Richard T. Sayre Donald Danforth Plant Science Center, 975 N. Warson Road, St. Louis, MO 63132, USA

Michael Schroda Institut of Biology II / Plant Biochemistry, University of Freiburg, Schänzlestraße 1, 79104 Freiburg, Germany

Michael Seibert National Renewable Energy Laboratory, Chemical and Biosciences Center, 1617 Cole Blvd., Golden, CO 80401, USA

Martin H. Spalding Department of Genetics, Development and Cell Biology, Iowa State University, 1210 Molecular Biology, Ames, IA 50011 USA

David B. Stern Boyce Thompson Institute for Plant Research, Tower Road, Ithaca, NY 14853, USA

Olivier Vallon Centre National de la Recherche Scientifique (CNRS), Université Pierre et Marie Curie (UPMC-P6) at Institut de Biologie Physico-Chimique, 13 rue Pierre et Marie Curie, 75005 Paris, France

Francis-André Wollman Centre National de la Recherche Scientifique (CNRS), Université Pierre et Marie Curie (UPMC-P6) at Institut de Biologie Physico-Chimique, 13 rue Pierre et Marie Curie, 75005 Paris, France

Katia Wostrikoff Boyce Thompson Institute for Plant Research, Tower Road, Ithaca, NY 14853, USA

Mirko Zaffagnini Institut de Biotechnologie des Plantes, Unité Mixte de Recherche 8618, Centre National de la Recherche Scientifique, Batiment 630, Université Paris-Sud 11, 91405 Orsay Cedex, France

William Zerges Biology Department, Concordia University, 7141 Sherbrooke West, Montreal, Quebec, Canada H4B 1R6

Conventions Used

We assume that the reader is familiar with the basic vocabulary of cell and molecular biology. Abbreviations and acronyms specific to particular areas of research or to *Chlamydomonas* are defined in the list that follows. *Chlamydomonas* is assumed to refer to *C. reinhardtii* unless otherwise specified. The reference genome sequence is the DOE Joint Genome Institute (JGI) version 3.0 (http://www.genome.jgi-psf.org/Chlre3/Chlre3.home.html; Merchant et al., 2007). Supplementary material can be found on the Elsevier companion web site, http://www.elsevierdirect.com/companions/9780123708731.

Nomenclature for genes and proteins follows the guidelines, contributed by S.K. Dutcher and E.H. Harris, in the *Genetic Nomenclature Guide* published by *Trends in Genetics* in 1998, and also summarized at http://www.chlamy.org/nomenclature.html. In brief, genes encoded in the nucleus are designated by uppercase italic letters, often followed by an arabic number to distinguish different loci with the same name (*ARG7*). Three-letter names are preferred. Unless otherwise named, proteins are designated by the relevant uppercase gene symbol, but not italic (e.g., ARG7). Mutant alleles are designated in lowercase italics; different mutant alleles at the same locus are distinguished by a number separated from the gene symbol by a hyphen (e.g., *arg7-1*).

Nomenclature for organelle-encoded genes and their products is based on the system used for cyanobacteria and for plants: gene symbols consist of three letters in lowercase followed by one uppercase letter or number, in italics (*petG*, *rps12*). When a protein has no more familiar name, it may be expressed as the same letters in mixed case, not italics (PetG).

Abbreviations

αKG alpha-ketoglutarate
ACP acyl carrier protein
ALA 5-aminolevulinic acid
APS adenosine 5-phosphosulfate
ARS arylsulfatase
ASQD 2′-*O*-acyl-sulfoquinovosyldiacylglycerol
BBY particle purified photosystem II complex, named for the authors Berthold, Babcock, and Yocum
BE branching enzyme of starch metabolism
BN-PAGE blue native polyacrylamide gel electrophoresis
CA carbonic anhydrase; also used in Chapter 2 to indicate coniferonic acid
CCCP carbonylcyanide-*m*-chlorophenylhydrazone, a chemical that uncouples photosynthetic electron transport from photophosphorylation
CCM CO_2 concentrating mechanism
CES control by epistasis of synthesis
Chl chlorophyll
CoA coenzyme A
cpDNA chloroplast DNA
CPI chlorophyll protein complex I
CTE C-terminal extension
cyt cytochrome
DAG diacylglycerol
DAPI 4′,6-diamidino-2-phenylindole, used as a stain for DNA
DAS decay-associated spectrum
DBE debranching enzyme of starch metabolism
DBMIB 2,5-dibromo-3-methyl-6-isopropyl-*p*-benzoquinone, an inhibitor of photosynthetic electron transport
DCMU 3-(3,4-dichlorophenyl)-1,1-dimethylurea, herbicide that affects the photosystem II reaction center
DGDG digalactosyldiacylglycerol
DGTS diacylglyceryl-*N*,*N*,*N*-trimethylhomoserine
DVR divinyl (proto)chlorophyllide a vinyl reductase
ECP extracellular polypeptide
ELIP early light-induced protein

ENDOR	electron-nuclear double resonance
EPR	electron paramagnetic resonance
ER	endoplasmic reticulum
Etn	ethanolamine
FA	fatty acid
FAD	fatty acid desaturase
FAS	fatty acid synthase
FCCP	cyanide-*p*-trifluoromethoxyphenyl hydrazone, a chemical that uncouples photosynthetic electron transport from photophosphorylation
Fd	ferredoxin
FdUrd	5-fluorodeoxyuridine
FTIR	Fourier-transform infrared
G1P or **Glc-1-P**	glucose-1-phosphate
G3P	glyceraldehyde-3-phosphate
GBSS	granule-bound starch synthase
GFP	green fluorescent protein
GRX	glutaredoxin
GS/GOGAT	glutamine synthetase/glutamate synthase
GSA	glutamate-1-semialdehyde
GSH	reduced glutathione
GSSG	oxidized glutathione
GTF	general translation factor
HANTS	high-affinity nitrate transport system
HATS	high-affinity ammonium transport system
HL	high light
HMB	hydroxymethylbilane
HPLC	high-performance liquid chromatography
HSP	heat shock protein
IR	inverted repeat, in the context of the chloroplast genome
LANTS	low-affinity nitrate transport system
LATS	low-affinity ammonium transport system
LHC	light-harvesting complex
LL	low light
LS	large subunit of Rubisco
LSC	large single-copy region of chloroplast DNA
LSU	50S large subunit of the chloroplast ribosome
Mg–Proto	Mg–Protoporphyrin IX
MGDG	monogalactosyldiacylglycerol
Moco	molybdenum cofactor of nitrate reductase and xanthine dehydrogenase
MSX	methionine-D,L-sulfoximine
mtDNA	mitochondrial DNA
MTS	mitochondrial targeting sequence

NEP	nucleus-encoded RNA polymerase
NMR	nuclear magnetic resonance
NPQ	nonphotochemical quenching of chlorophyll fluorescence
NQNO	2-*n*-nonyl-4-hydroxyquinoline-*N*-oxide, a quinone analogue
NTE	N-terminal extension
OAA	oxaloacetate
OAS, OASTL	*O*-acetylserine, *O*-acetylserine(thiol)lyase
OEC	oxygen-evolving complex
omega (Ω)	ratio of efficiencies of Rubisco carboxylation and oxygenation reactions
OXPHOS	oxidative phosphorylation
P700	special pair of chlorophylls serving as ultimate electron donor within PS I
PABP	poly(A)-binding protein
PBG	porphobilinogen
PC	phytochelatin (Chapter 10); also appears in Figure 22.1 as an abbreviation for plastocyanin
PEP	phosphoenolpyruvate; also used in Chapter 25 for plastid-encoded RNA polymerase
PFD	photon flux density
PGA	3-phosphoglycerate
PQ, PQH2	plastoquinone, plastoquinol
PRM, PRT	phosphoribosyl-AMP, phosphoribosyl-ATP
PRP	plastid ribosomal protein
PRPP	5-phosphoribosylpyrophosphate
PS I, PS II	photosystem I, photosystem II
PTC	peptidyl transferase center of the chloroplast ribosome
PtdCho	phosphatidylcholine
PtdEtn	phosphatidylethanolamine
PtdGro	phosphatidylglycerol
PtdIns	phosphatidylinositol
PtdOH	phosphatidic acid
PtdSer	phosphatidylserine
qE	energy-dependent component of NPQ
qT	state transition component of NPQ
RBP	RNA-binding protein
RC	reaction center
RNP	ribonucleoprotein
ROS	reactive oxygen species
RP	ribosomal protein
RT-PCR	reverse transcription polymerase chain reaction
Ru5P	ribulose-5-phosphate

Rubisco	ribulose-1,5-bisphosphate carboxylase/oxygenase
RuBP	ribulose-1,5-bisphosphate
SD	Shine–Dalgarno
SDR	short dispersed repeat (in chloroplast DNA)
SEM	scanning electron microscopy
SHAM	salicylhydroxamic acid
SNP	single-nucleotide polymorphism
SQDG	sulfoquinovosyldiacylglycerol
SS	soluble starch synthase (Chapter 1), small subunit of Rubisco (Chapter 9)
SSU	30S subunit of the chloroplast ribosome
TAG	triacylglycerol
TCA	tricarboxylic acid
TDS	tridecylstigmatellin, a quinone analogue
TIC/TOC	protein translocase complex of the chloroplast envelope
TLC	thin-layer chromatography
TMD	transmembrane domain
TPP	thiamine pyrophosphate (Chapter 4); thylakoid processing peptidase (Chapter 19)
TRX	thioredoxin
TTC	2,3,5-triphenyltetrazolium chloride, used as a stain for mitochondrial function
UQ, UQH2	ubiquinone, ubiquinol
URF	unidentified reading frame
UTR	untranslated region in the 5′ or 3′ flank of a gene sequence
VHL	very high light

List of Tables

List of Figures

Contents of Volume 1

by Elizabeth H. Harris

Contents of Volume 3

edited by George B. Witman

CHAPTER 1

Starch Metabolism

Steven G. Ball and Philippe Deschamps
Université des Sciences et Technologies de Lille,
Laboratoire de Chimie Biologique,
Villeneuve d'Ascq, France

CHAPTER CONTENTS

I. STARCH STRUCTURE AND LOCALIZATION IN THE GREEN ALGAE

A. Structures of glycogen and starch

α-1,4 linked glucans branched through α-1,6 glycosidic linkages are the most widespread forms of storage polysaccharides within living cells. These polymers come in two very distinct physical states, however, defined by glycogen and starch. Glycogen is organized as small hydrosoluble spherical particles containing from 8% to 10% α-1,6 branches. Each chain is thought to support two novel chains through α-1,6 branches, leading to a concentric growth of particles with an increasing density of the peripheral glucose residues. This exponential increase is limited, however, by the physical accessibility of the chains that need to accommodate the active sites of the glycogen metabolic enzymes. Single glycogen particles therefore reach a theoretical maximal diameter that has been modeled to 40 nm. Indeed, unit glycogen particles exceeding this limit have never been observed through standard transmission electron microscopic techniques.

At variance with glycogen, starch can be considered a heterogeneous mixture of at least two distinct polysaccharide fractions: amylopectin and amylose. Amylopectin, a moderately branched (5% α-1,6 branches) polymer generates the backbone of the large insoluble granules. The relationship between the organization of sections of the starch granule and the proposed structure of amylopectin is shown in Figure 1.1 (reviewed in Buléon et al., 1998). Because of the asymmetrical distribution of the branches within amylopectin, there is no upper limit to the size reached by single molecules, and therefore by a starch granule. Amylopectin is a large biological polymer (up to 10^5 glucose residues), and adopts within the starch granule, a semi-crystalline organization that can be studied by either wide-angle powder X-ray diffraction or small-angle X-ray scattering. This insoluble semi-crystalline organization of starch has two important biological implications. First, aggregation of polysaccharide into insoluble semi-crystalline granules must define the last step of the amylopectin synthesis pathway. In fact, no enzymes of this pathway have been reported to be active on these insoluble polymers once they are packaged into starch. Second, mobilization of these glucose stores requires enzymes and biochemical steps which must be distinguishable from those of glycogen degradation. Indeed, the corresponding plant enzymes are generally unable to attack semi-crystalline polymers.

Amylose is a minor amorphous polysaccharide fraction of starch (between 10% and 30% in most wild-type starches). It consists of linear α-1,4 glucans with very few α-1,6 branches, whose sizes range from 10^2 to 10^4 glucose residues. The precise position of amylose relative to amylopectin within starch granules is unknown. The presence of this fraction is dispensable for starch granule organization. Mutants selectively lacking amylose will build

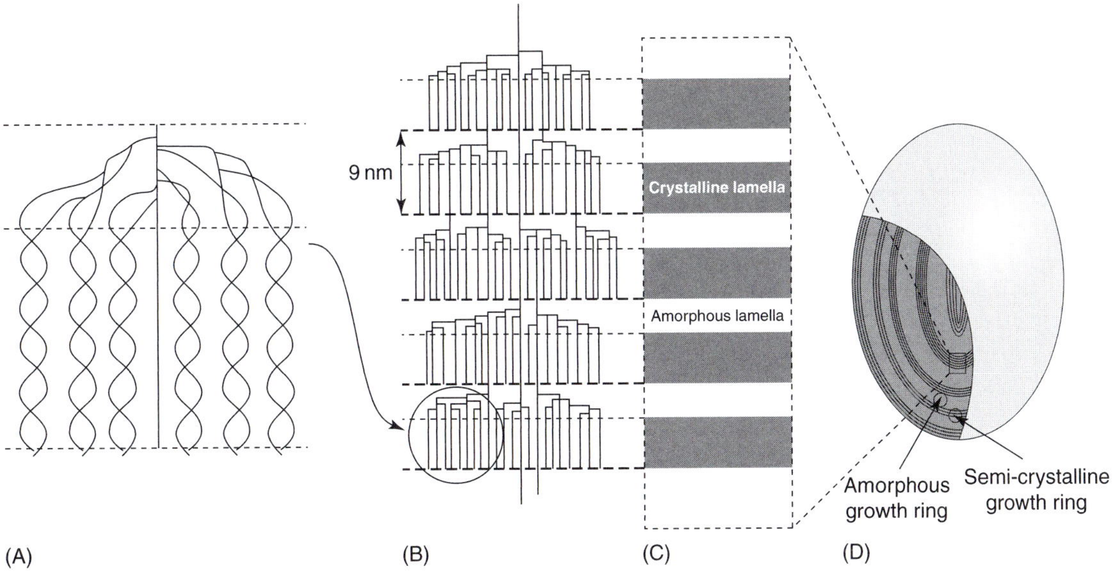

FIGURE 1.1 *Schematic structure of the amylopectin molecule and its relationship with the starch granule organization. (A) Structure of a single amylopectin cluster. The lower part is made of linear glucans organized in double helix, the upper part contains all branches. (B) Overview of the amylopectin molecule. Limitations between amorphous and crystalline lamellae corresponding to the latter's upper and lower parts are represented by the dotted lines and likened to (C) a simplified view of the succession of these lamellae. (D) Schematic overview of a semi-hydrolyzed starch granule. The concentric succession of amorphous and semi-crystalline rings is presented. The semi-crystalline rings are made of highly packed amorphous and semi-crystalline lamellae. The lamellae are more loosely organized in amorphous rings.*

wild-type amounts of starch displaying a normal semi-crystalline organization. The process of amylose synthesis involves a single elongation enzyme named granule-bound starch synthase I (GBSSI). Unlike amylopectin synthesis which relies entirely on the function of soluble enzymes, GBSSI-mediated amylose synthesis occurs within the granule polysaccharide matrix.

This chapter focuses on more recent developments concerning starch metabolism in *Chlamydomonas*. Two previous reviews have dealt with comparable issues, where the reader can find additional information dealing with compartmentalization of enzymes of carbon metabolism and the first characterizations of starch metabolic enzymes in green algae (Ball, 1998, 2002). A particular emphasis in this chapter concerns the bioinformatic analyses of relevant genes within the *Ostreococcus* and *Chlamydomonas* genomes, as compared to *Arabidopsis*. A more historical than exhaustive review (more details can be found in Ball, 2002) of genetic studies involving mutants of the starch pathway will be given in order to refine the original findings made in *Chlamydomonas*, and portray how the studies have improved our knowledge of starch metabolism in general.

B. Phylogenetic distribution and cellular localization of starch metabolism

Glycogen is found in archaea, bacteria, animals, fungi, and diverse heterotrophic protists. Starch, by contrast, is only found in lines descending either from primary endosymbiosis of the plastid or from secondary endosymbiosis of starch-accumulating red algae. In the green lineage (green algae and their land plant derivatives), starch is found within the chloroplast. However in the glaucophyte and the rhodophyte (red algae) lineages the granules are exclusively synthesized within the cytoplasm. In the secondary endosymbiotic lines, starch is either found in the cytoplasm, as in dinoflagellates and apicomplexan parasites (Coppin et al., 2005), or in the case of the cryptophytes within the periplastidial space together with the vestigial red alga nucleus of the secondary symbiont (Deschamps et al., 2006). When present, the pyrenoid of green algae is usually surrounded by closely-packed starch granules (reviewed in Chan, 1972 and Badger et al., 1998). This privileged association seems to be maintained in red algae of the Porphyridiales (Lee, 1974) and in cryptophytes (Deschamps et al., 2006), despite the fact that starch is separated from the pyrenoid in both cases by either the two plastidial membranes in the case of the rhodophytes, or the two innermost layers of the four-membrane secondary plastids of cryptophytes. The plastidial localization of starch in the green lineage is thus quite unique.

The first detailed study of starch structure in Volvocales can be traced back to the work of Bourne et al. (1950), who reported the presence of starch-like granules in the plastids of the colorless alga *Polytomella coeca*. Several other detailed studies appeared using other Volvocales including *Dunaliella bioculata* (Eddy et al., 1958), the colorless acetoflagellate *Polytoma uvella* (Manners et al., 1965), and *Haematococcus pluvialis* (Hirst et al., 1972). In all these studies, starches were demonstrated to contain both amylopectin and amylose fractions resembling those of higher plant polysaccharides. The studies also reported, however, a highly variable percentage of amylose. This variation seems to be a general rule in green algae, and to some extent has been attributed to environmental factors such as temperature (reviewed in Craigie, 1974).

The most detailed studies concerning starch structure and its variability in green algae have been performed in *Chlamydomonas reinhardtii*. Complete structural characterizations including X-ray diffraction, NMR, differential scanning calorimetry, chemical methylation, separation of starch fractions, and precise analysis of chain–length distribution have been performed over the years (Delrue et al., 1992; Fontaine et al., 1993; Libessart et al., 1995; Buléon et al., 1997). Libessart et al. (1995) found that the structure and composition of *Chlamydomonas* starch changed dramatically according to culture conditions. Unstarved cultures grown on acetate and light accumulated starch with very little ($<5\%$) amylose, whereas nutrient-starved or limited cells accumulated substantial amounts of this polysaccharide fraction (from 15% to 30%).

These fluctuations in starch composition can be traced to the kinetic properties of GBSSI, the amylose-synthesizing enzyme. Among all elongation enzymes, GBSSI displays a minimum of 10-fold higher apparent K_m for its ADP-glucose substrate. This renders amylose synthesis more sensitive to the size of the ADP-glucose pool. Any decrease of glycosyl nucleotide concentration due to modification of the environment is thus expected to have a selectively larger impact on amylose synthesis. This interpretation was strengthened by the isolation of mutants defective for either the large subunit of ADP-glucose pyrophosphorylase or for plastidial phosphoglucomutase (van den Koornhuyse et al., 1996). In both cases synthesis of ADP-glucose is substantially decreased, but not abolished. The structure of the starch isolated from nutrient-starved mutants containing low ADP-glucose could be considered a phenocopy of that which was described for unstarved wild-type cultures. These results demonstrate that the flux to ADP-glucose tightly controls the amount of amylose, and to a lesser extent the fine structure of amylopectin. The diversity of structures and amylose content reported for several of the green algae can be interpreted in the light of these results, and can thus mostly be attributed to differences in culture conditions and physiology.

II. AN OUTLINE OF THE STARCH METABOLISM PATHWAY: COMPARATIVE BIOCHEMISTRY AND GENOMICS

A. The enzymes of starch metabolism in green algae

An outline of starch metabolism is given in Figure 1.2, and details are given in subsequent sections. Briefly, biosynthesis in *Chlamydomonas* starts with glucose-1-P, which reacts with ATP to produce ADP-glucose and pyrophosphate (Figure 1.2, step 2). ADP-glucose pyrophosphorylase catalyzes this reaction and defines a major rate-controlling step of carbon flux to starch. It is typically tied to photosynthesis through the very nature of the substrates and through the nature of its effectors. As with cyanobacteria, 3-PGA, the product of CO_2 fixation, activates the enzyme of the green algae and land plants. These properties are likely to be responsible for localized synthesis of the polysaccharide around the pyrenoid in algae, where concentration gradients of these metabolites are likely to exist.

Elongation of amylopectin and amylose then proceeds through transfer of glucose residues from ADP-glucose to the non-reducing end of a growing α-1,4 linked chain through the starch (glycogen) synthases (Figure 1.2, step 3). The polymers are then branched through hydrolysis of a pre-existing α-1,4 glycosidic bond from a donor chain and transfer of a segmented chain through an α-1,6 branch to a neighboring acceptor chain (Figure 1.2, step 4).

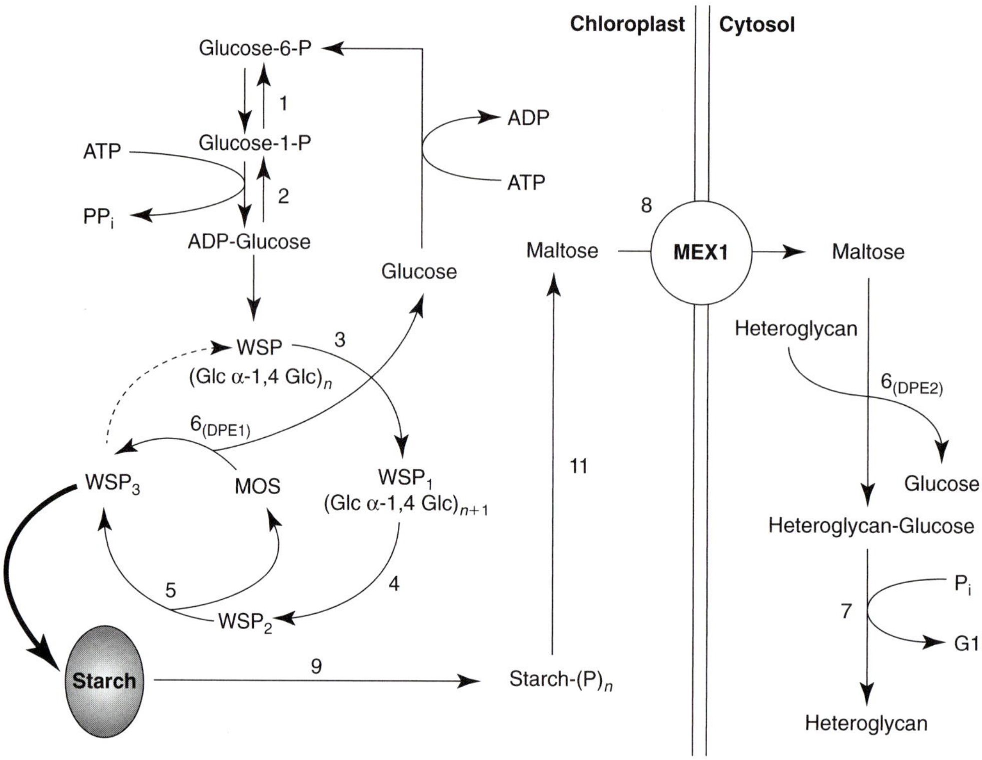

FIGURE 1.2 *Schematic representation of the major steps of the starch metabolism pathway. Numbers near arrows refer to subsections of Part II. A. in the text, where the nature of the activity is described. WSP: Water soluble polysaccharide, MOS: Malto-oligosaccharide, Starch-(P)n: Poly-phosphorylated starch. The first WSP is a pre-existing branched glucan used as primer by starch synthases. WSP_1 is the same glucan with an added glucose residue. WSP_2 is a branched glucan with higher ramifications formed by the action of a branching enzyme. WSP_3 is a less branched polysaccharide; missing branches are released by an isoamylase and become a free soluble MOS.*

Disordered branches in amylopectin are likely to produce hydrosoluble polymers that lack the asymmetrical distribution of branches that is required to generate crystallization-competent structures (see Figure 1.1). It is generally accepted that isoamylase will hydrolyze those loosely spaced branches that prevent crystallization, thereby generating the tight packing of α-1,6 branches responsible for aggregation of amylopectin into semi-crystalline granules (Figure 1.2, step 5). The processing of this transient pre-amylopectin structure results in the release of malto-oligosaccharides (MOS) generated by the cleaved misplaced chains. These MOS need to be metabolized within the plastid through a combination of α-1,4 glucanotransferases and other enzymes that are able to degrade the products generated by these transferases to glucose and glucose-1-P (Figure 1.2, step 6; labeled DPE1). The α-1,4 glucanotransferases define enzymes that transfer glucose or MOS of various length from a donor MOS to an acceptor.

Little is known about starch catabolism in green algae. It is thought that both hydrolytic (generating maltose and glucose) and phosphorolytic pathways (producing glucose-1-P) are responsible for starch mobilization. Phosphorylases are enzymes that release glucose-1-P progressively from the non-reducing end of α-1,4 linked chains in the presence of orthophosphate. Through work on dicotyledonous plants, however, it has become clear that in order to be degraded, starch first needs to be phosphorylated through a combination of glucan–water dikinases (GWD) (Figure 1.2, step 9). The phosphorylated starch is either directly or indirectly attacked by plastidial β-amylases (Figure 1.2, step 11), which generate β-maltose within the plastid. Maltose is exported to the cytoplasm through the MEX1 transporter (Figure 1.2, step 8) and hydrolyzed through another α-1,4 glucanotransferase (the product of the *DPE2* gene of *Arabidopsis*) that releases one glucose residue and transfers the other to the outer chains of a heteroglycan acceptor (Figure 1.2, step 6; labeled DPE2). These outer chains are then proposed to be recessed by a cytosolic form of phosphorylase (Figure 1.2, step 7).

In the following section the sequences identified by bioinformatics in *Chlamydomonas*, *Ostreococcus tauri* (hereafter *Ostreococcus*), and *Arabidopsis* are compared; and the *Chlamydomonas* sequences are related, when possible, to those loci identified by mutant selection. A more detailed account of functional studies is given in the subsequent section.

B. Comparative genomics of starch metabolism in the green lineage: from green algae to *Arabidopsis*

1. Plastidial phosphoglucomutase (GPM or PGM, EC 5.4.2.2)

Phosphoglucomutase interconverts glucose-6-P and glucose-1-P. Although production of glucose-1-P cannot be considered as a specific part of storage polysaccharide synthesis in eukaryotes and prokaryotes, it does define the first committed step of starch biosynthesis in green algae and land plant plastids (Figure 1.2, step 1). Phosphoglucomutase is also required in chloroplasts for the assimilation of glucose-1-P produced by the plastidial phosphorylase isoforms.

Two GPMs have been identified in *Chlamydomonas*. One of these (EDO99576 and EDO99577) displays significant homology to the three *Arabidopsis* GPMs (At5g51820, At1g70730, and At1g23190) and to the single *Ostreococcus* sequence (Ot15g02630). However there is a second gene that encodes a protein (EDP09622) homologous to a second class of enzyme found in many eukaryotes. This gene is apparently absent from other green algae and plants examined to date (with the exception of *Volvox*). Mutations have been described that lower or abolish the plastidial isoform in both *Chlamydomonas* (*sta5*; van den Koornhuyse et al., 1996) and *Arabidopsis* (*pgm-1*; Caspar et al., 1985). The *Arabidopsis* plastidial sequence (At5g51820) corresponds to the *pgm-1* mutation (Periappuram et al., 2000). Klein (1987) reported the presence of phosphoglucomutase activity in both the plastid and the cytoplasm of *Chlamydomonas*. The phylogenetic tree shown in Figure 1.3 clearly suggests that a gene has been duplicated to generate

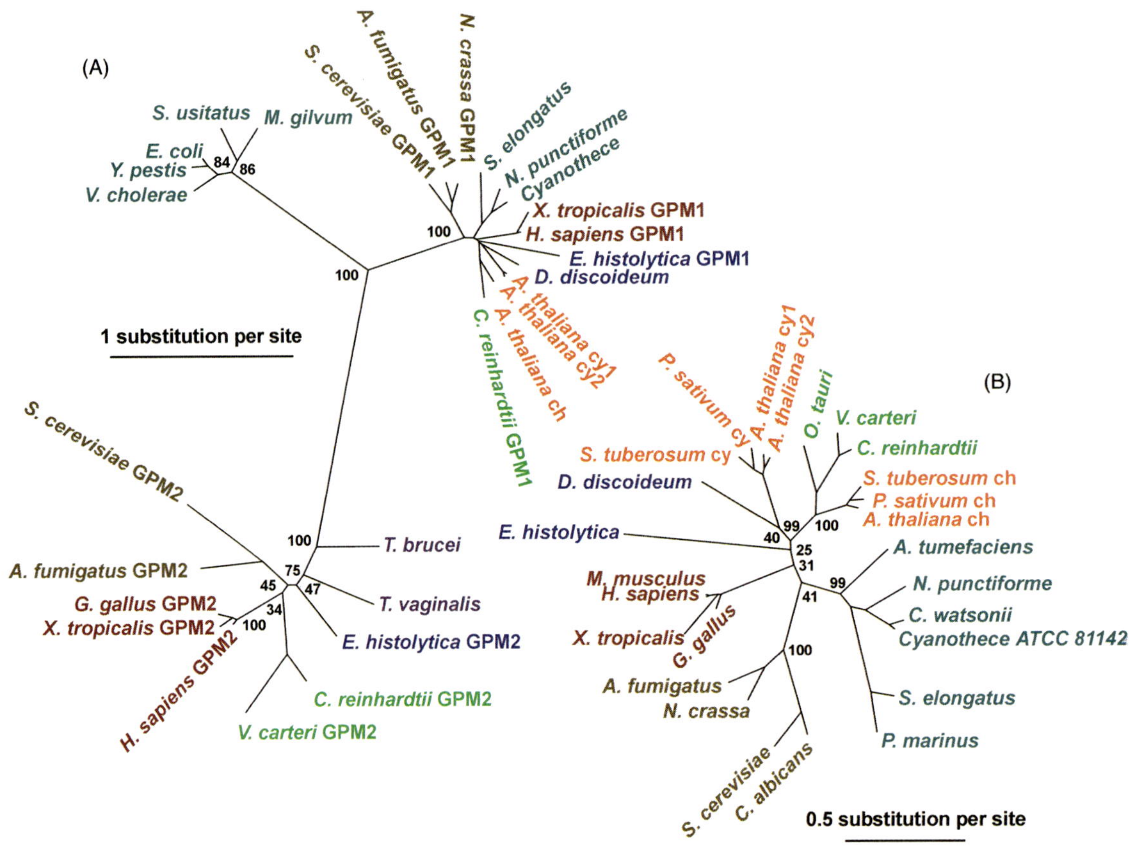

FIGURE 1.3 *(A) Neighbor-joining tree inferred for phosphoglucomutase (GPM) proteins. Protein alignment was performed using ClustalW (Thomson et al., 1994). The tree was inferred using Neighbor (Phylip package). Numbers at nodes indicate bootstrap values calculated over 500 replicates. cy: cytoplasmic isoform, pl: plastidial isoform. The tree identifies three groups of proteins. The group at the top right consists of three distinct families of homologous enzyme (plants, heterotrophic eukaryotes subgroup I and bacteria subgroup II) that are subjected to a more detailed phylogenetic analysis in panel B. Two other very distantly-related groups of sequences identify two other heterotrophic eukaryotic (heterotrophic eukaryotes subgroup II) and bacterial sequences (bacteria subgroup I). Note that Volvocales are the only plants that contain a sequence grouping with the heterotrophic eukaryotes subgroup II. (B) Maximum-likelihood tree inferred for "Plants-type" GPM. The tree was inferred using ProML (Phylip package). Numbers at nodes indicate bootstrap values calculated over 100 replicates. The top right group of Figure 1.3(A) was subjected to a more detailed analysis. The results are consistent with an eukaryotic origin for the plant enzyme. Nevertheless the close distance of the plant sequences from bacteria subgroup II and the low bootstrap values of the internal nodes both indicate that this eukaryotic origin is at best suggestive. Species names used for all trees presented here are: bacteria and cyanobacteria (light blue) – Agrobacterium tumefaciens, Bacillus subtilis, Crocosphaera watsonii, Cyanothece sp. ATCC 51142, Escherichia coli, Gloeobacter violaceus, Mycobacterium gilvum, Nostoc punctiforme, Prochlorococcus marinus, Solibacter usitatus, Synechococcus elongatus, Synechocystis sp. PCC6803, Vibrio cholerae, Yersinia pestis; fungi (brown) – Aspergillus fumigatus, Candida albicans, Neurospora crassa, Saccharomyces cerevisiae, Schizosaccharomyces pombe; animals (dark red) – Gallus gallus, Homo sapiens, Mus musculus, Xenopus tropicalis; amoebae (dark blue) – Dictyostelium discoideum, Entamoeba histolytica; green algae (green) and plants (orange) – Avena fatua, Arabidopsis thaliana, Chlamydomonas reinhardtii, Citrus reticulata, Glycine max, Hordeum vulgare, Medicago truncatula, Oryza sativa, Ostreococcus tauri, Phaseolus vulgaris, Pisum sativum, Solanum tuberosum, Triticum aestivum, Vigna angularis, Volvox carteri, Zea mays; other eukaryotes (purple) – Trichomonas vaginalis, Trypanosoma brucei.*

both plastidial and cytosolic forms of this enzyme in land plants. We do not yet know if *STA5* encodes one of the two enzyme sequences and which of the two forms is targeted to plastids in Volvocales.

2. ADP-glucose pyrophosphorylase (AGPase, EC 2.7.7.27)

ADP-glucose pyrophosphorylase is a heterotetramer of related subunits of comparable sizes (around 50 kD for most so-called "small" subunits and 2–5 kD larger for the corresponding "large" subunits) in both green algae and plants. AGPase constitutes one of the major rate-controlling steps of the starch synthesis pathway. It is typically subjected to allosteric regulation by 3-PGA and inhibition by orthophosphate, thereby coupling starch biosynthesis to photosynthesis in cyanobacteria and chloroplasts. It is generally assumed that the so-called large (regulatory) and small (catalytic) subunits of the plant heterotetramers were derived from a gene duplication of the cyanobacterial homotetrameric gene (Ballicora et al., 2003). The *Ostreococcus* (GenBank protein accessions: small, AAS88879; large 1, AAS88891; large 2, AAS88878) and *Chlamydomonas* (small, AAF75832; large 1, EDP08701; large 2, EDP04406) genomes both contain two large and one small subunit genes. *STA1* and *STA6* have been demonstrated to encode, respectively, the large 1 and the small subunit of AGPase (Ball et al., 1991; Iglesias et al., 1994; van den Koornhuyse et al., 1996; Zabawinski et al., 2001).

The *sta1* mutants lacking the protein product possess an enzyme with lower sensitivity to 3-PGA activation. The enzyme remains sensitive to orthophosphate inhibition, explaining the low starch phenotype. The *sta6* mutants lacking the small subunit were proven to be starchless as was the case for the corresponding *adg-1* mutants of *Arabidopsis*. No function has been assigned to the remaining large subunit sequence of *Chlamydomonas*. It is likely, however, that duplication of the large subunit gene has functional relevance, since both *Chlamydomonas* and *Ostreococcus* have maintained the two homologues.

In *Arabidopsis*, two small subunit (APS) genes (At5g48300 and At1g05610) and four large subunit (APL) genes (At15g19220, At1g27680, At4g39210, and At2g21590) are present (Crevillén et al., 2003). *Arabidopsis* mutants *adg-1* (with lesions in At5g48300 or *APS1*) and *adg-2* (with lesions in At15g19220 or *APL1*) behave very much like the *sta6* and *sta1* mutants, despite the greater genetic redundancy of the vascular plant genome (Lin et al., 1988a,b). Detailed expression analysis (Crevillén et al., 2003) suggested that *Arabidopsis APL3* and to a lesser extent *APL4* are the large subunit genes predominantly active in sink tissues, creating an enzymatic form less tuned to photosynthesis and thus somewhat less sensitive to 3-PGA activation.

The phylogenetic tree in Figure 1.4 shows that the L2 "orphan" large subunit of *Chlamydomonas* (which is not encoded by a *STA* gene) occurs on a long branch together with the *Ostreococcus* L1 subunit and the *Arabidopsis* S2 subunit. These long branches suggest that these genes were

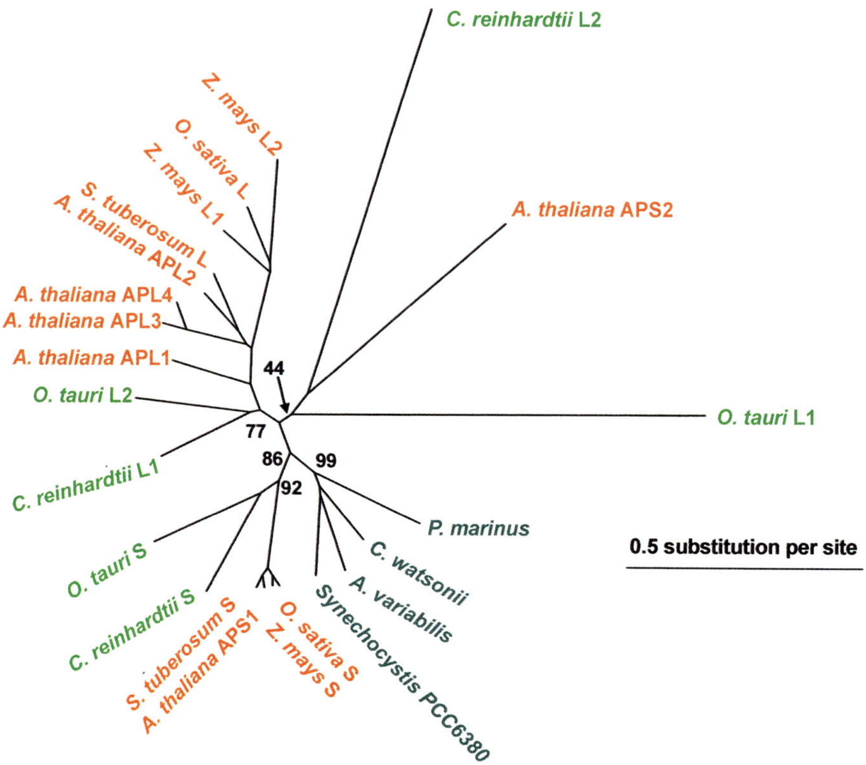

FIGURE 1.4 *Neighbor-joining tree inferred for ADP-glucose pyrophosphorylase proteins. Protein alignment was performed using ClustalW (Thomson et al., 1994). The tree was inferred using Neighbor (Phylip package). Numbers at nodes indicate bootstrap values calculated over 500 replicates. S: small subunit, L: large subunit. Species names are listed in the legend of Figure 1.3.*

not subjected to the same selective pressures as the other AGPase structural genes. Indeed, *Arabidopsis* APS2 is thought to be non-catalytic, its encoding gene is transcribed at a much lower level than the S1 subunit, and it may be on a pathway to becoming a pseudogene (Crevillén et al., 2003, 2005). Although the same could be hypothesized for the additional "large" subunits encoded in green algal genomes, the maintenance of such genes across such a large evolutionary time span argues against this view. Nevertheless, had these proteins acquired a novel common function related or not related to starch metabolism, one would have expected a lesser degree of divergence between the *Chlamydomonas* L2 and the *Ostreococcus* L1 sequences.

3. Starch synthases (SS, EC 2.4.1.21)

Six starch synthase genes have been identified in both *Ostreococcus* and *Chlamydomonas*. One of these enzymes (GenBank accession AAL28128 for *Chlamydomonas*, AAS88890 for *Ostreococcus*) corresponds to GBSSI, the only synthase that is predominantly active within the starch granule. GBSSI is the sole enzyme involved in amylose synthesis both in green algae

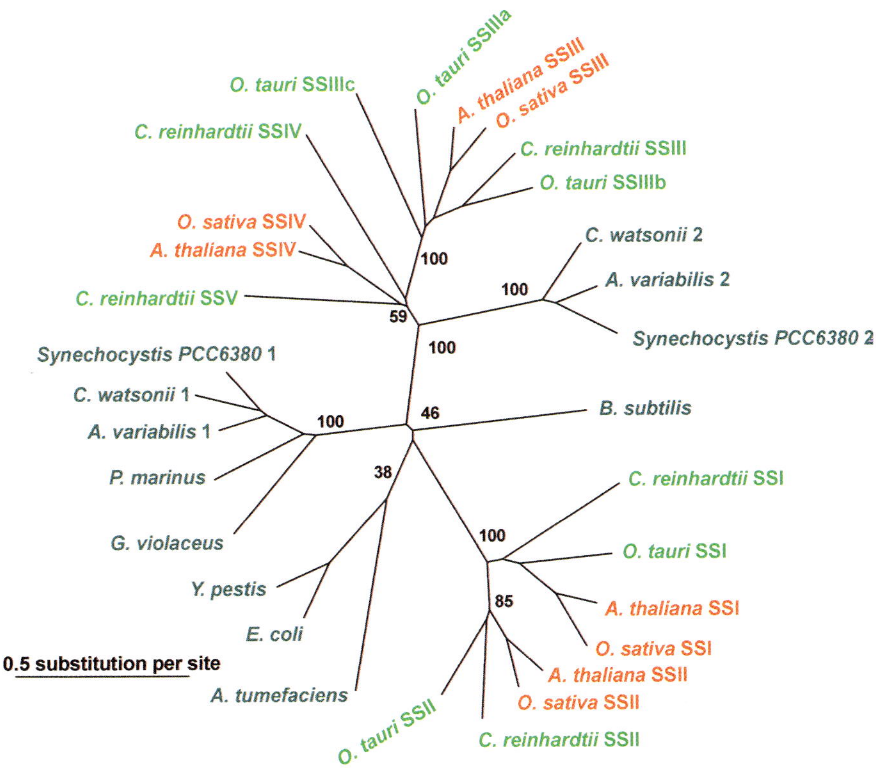

FIGURE 1.5 *Neighbor-joining tree inferred for starch synthase proteins. Protein alignment was performed using ClustalW (Thomson et al., 1994). The tree was inferred using Neighbor (Phylip package). Numbers at nodes indicate bootstrap values calculated over 500 replicates. SS: starch synthase. Species names are listed in the legend of Figure 1.3.*

and plants (for a review see Ball et al., 1998). The *Chlamydomonas STA2* locus is involved in amylose biosynthesis, and has been demonstrated to encode GBSSI (Delrue et al., 1992; Wattebled et al., 2002). Vascular plants generally contain four soluble starch synthases belonging to distinct families (SSI, SSII, SSIII, and SSIV). In addition to this, some plants may contain a fifth form generally related to SSIV.

It is perhaps surprising that *Chlamydomonas* and *Ostreococcus* both contain five genes corresponding to the same enzyme forms. Phylogenetic trees (Ral et al., 2004) clearly show the presence of SSI and SSII genes in both *Chlamydomonas* (EDO99379 and EDP04319) and *Ostreococcus* (AAS88892 and AAS88880). While *Chlamydomonas* contains a single clear SSIII-like gene sequence (AAY42381), *Ostreococcus* contains three SSIII-like sequences (AAS88893, AAS88894, and AAS88881). The *Chlamydomonas* SSIII sequence was shown (Ral et al., 2006) to correspond to the *STA3* locus, which is involved in amylopectin synthesis (see next section).

The soluble starch synthases can be traced back to two distinct subgroups of cyanobacterial glycogen synthases (Figure 1.5). SSI and SSII in this view would have a common origin, whereas SSIII and SSIV could be traced back to a single other type of cyanobacterial glycogen synthase. *Ostreococccus* seems to represent an intermediate state in the evolution of the SSIII–SSIV subfamilies. The sequence AAS88894 is clearly similar to the SSIII isoform found in *Chlamydomonas*. Even if the tree strongly supports the two last SSIII-like *Ostreococcus* sequences (AAS88893 and AAS88881) belonging to the SSIII class, these isoforms represent the base node of the group and could have evolved afterwards to become the functional equivalents of the SSIV isoforms in *Chlamydomonas* and plants.

4. Branching enzymes (BE, EC 2.4.1.18)

Three distinct types of BE belonging to two major enzyme families have been documented in most plant species. *Chlamydomonas* (EDP08206 and EDP05581) and *Ostreococcus* (AAS88883) also contain one or two copies of a BE isoform corresponding to BEII (also called the A family). Both algae contain in addition a BEI type (also called the B family) of sequence (AAS88882 for *Ostreococcus* and EDP01649 for *Chlamydomonas*). No mutants corresponding to BE structural genes have been reported to date in *Chlamydomonas*.

5. Debranching enzymes (DBE, EC 3.2.1.68)

Plant debranching enzymes fall into two major classes distinguished by their substrate specificities. Their origins can be traced back to the so-called direct debranching enzymes of bacteria. Pullulanases (also called limit dextrinases) are enzymes able to hydrolyze α-1,6 linkages provided they are tightly spaced, such as in the highly branched polymer pullulan or the root of the amylopectin cluster (see Figure 1.1). Isoamylases hydrolyze the α-1,6 linkages when they are loosely spaced, as is the case in glycogen. *Arabidopsis* encodes three isoamylase type and one pullulanase type of enzyme [At2g39930 (*AtISA1*), At1g03310 (*AtISA2*), At4g09020 (*AtISA3*), and At5g04360 (*AtPU1*); Wattebled et al., 2005]. *Chlamydomonas* and *Ostreococcus* contain, respectively, three and two isoamylase-related (accessions AAP85534, EDP04897, and EDO98700 for *Chlamydomonas* and AAS88884 and AAS88896 for *Ostreococcus*) and one pullulanase-related (EDP00234 for *Chlamydomonas* and AAS88886 for *Ostreococcus*) sequences. These sequences were used to generate the phylogenetic tree shown in Figure 1.6. Posewitz et al. (2004) demonstrated that the *Chlamydomonas STA7* locus, previously shown to be required for starch synthesis and isoamylase activity (Mouille et al., 1996), corresponds to the isoamylase subunit (AAP85534) related to *AtISA1*.

6. α-1,4 Glucanotransferases (EC 2.4.1.25)

Two distinct types of α-1,4 glucanotransferases have been described in *Arabidopsis* and most other plants. One of them, known as disproportionating enzyme (D-enzyme), encoded by *DPE1* (At5g64860), is unambiguously

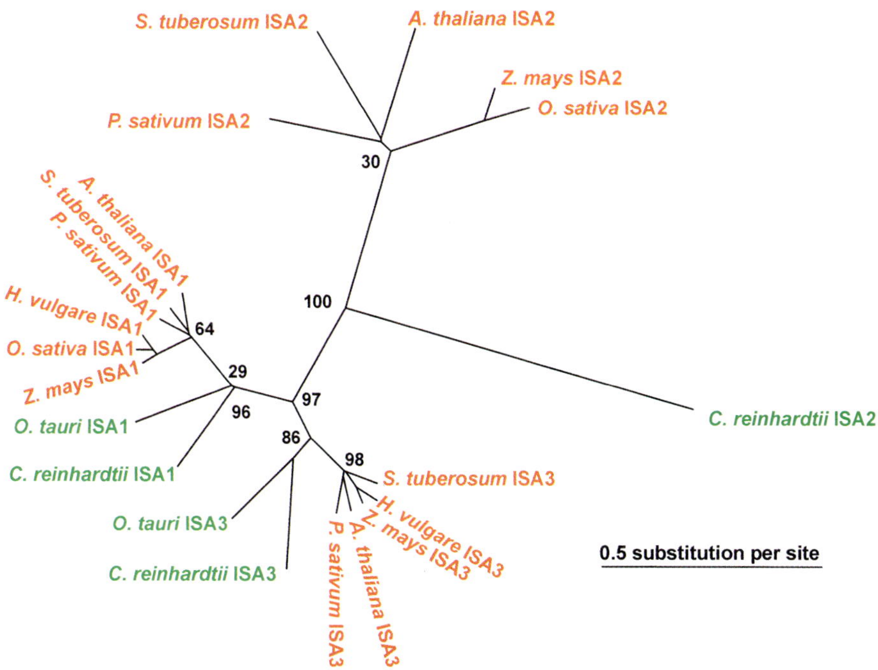

FIGURE 1.6 *Neighbor-joining tree inferred for isoamylase proteins. Protein alignment was performed using ClustalW (Thomson et al., 1994). The tree was inferred using Neighbor (Phylip package). Numbers at nodes indicate bootstrap values calculated over 500 replicates. Species names are listed in the legend of Figure 1.3.*

located within the chloroplast, whereas the other, encoded by *DPE2* (sometimes referred to as "amylomaltase," Lloyd et al., 2004; Lu and Sharkey, 2004), is thought to be located in the cytoplasm (AAL91204). Both higher plant genes, however, include candidate transit peptide sequences. corresponding genes are present in *Chlamydomonas* (*DPE1*: AAG29840; *DPE2*: EDO97689) and *Ostreococcus* (*DPE1*: AAS88889; *DPE2*: Ot11g02290). A transit peptide-like sequence is evident only in the *DPE1* sequences. Wattebled et al. (2003) demonstrated that *DPE1* corresponds to the *Chlamydomonas STA11* locus.

7. Starch phosphorylases (EC 2.4.1.1)

Two distinct types of starch phosphorylases have been documented in plants. Plastidial (Pho1 or Pho-L; in *Arabidopsis* AtPHS1: At3g29320) and cytosolic (Pho2 or Pho-H; in *Arabidopsis* AtPHS2: At3g46970) forms can be distinguished by the presence of an 80 amino acid insertion in the plastidial enzyme. In green algae the plastidial enzymes typically lack this 80 amino acid insertion. Three distinct phosphorylase sequences are present in *Ostreococcus* (AAS88885, AAS88897, and AAS88898), and one of them (AAS88885) contains a putative transit peptide and is overall more similar to the Pho-L form of plants. In *Chlamydomonas* only two genes are known, one (ABB88568) corresponding to the *STA4* locus and defining a plastidial form of phosphorylase (Dauvillée et al., 2006). Although a second phosphorylase sequence (ABB88569) was assigned to a second plastidial activity by Dauvillée et al. (2006), it is uncertain whether this enzyme is targeted only to the plastids, or to both the plastids and the cytosol. This ambiguity might be resolved if a third phosphorylase sequence emerges when the genome is ultimately completed.

8. Maltose transporter

Niittyla et al. (2004) proposed that during starch degradation (see Figure 1.2) maltose is exported to the cytoplasm in leaf cells via a plastidial transporter named MEX1. *Ostreococcus* (Ot09g03160) and *Chlamydomonas* (EDP05341) both contain the putative corresponding gene.

9. Glucan–water dikinases (GWD, EC 2.7.9.4)

GWD are enzymes that phosphorylate starch through linkage of ATP's β-phosphate to the C6 position of the glucose residues present at the root of the amylopectin clusters (Ritte et al., 2002, 2006). This phosphorylation is required to obtain significant starch degradation in plant leaves (Lorberth et al., 1998). In *Arabidopsis* three GWD-like sequences are present (Mikkelsen et al., 2004): GWD1 (*SEX1*; accession AAG47821); GWD2 (AAO42141); and GWD3 (NP_198009). GWD3 has been rechristened PWD (phosphoglucan–water dikinase) because it has been demonstrated to be only active on pre-phosphorylated amylopectin (Baunsgaard et al., 2005; Kötting et al., 2005). PWD has also been shown to be responsible for phosphorylating the C3 position of glucose residues located at the root of the amylopectin clusters (Ritte et al., 2006). Three distinct GWD-like sequences are present in *Ostreococcus* (Ot13g01510, Ot08g00790, and Ot16g02370) and *Chlamydomonas* (EDP07087, EDP05868, and EDP00979). In addition, one PWD-like sequence is encoded in *Chlamydomonas* (EDP00354), and two in *Ostreococcus* (Ot04g04170 and Ot08g01260). The phylogenetic tree shown in Figure 1.7 demonstrates the relatedness of all these sequences.

10. α-Amylases (EC 3.2.1.1)

Three *Arabidopsis* sequences have similarity to known α-amylase genes [accession numbers NP_567714 (*AtAMY1*), NP_177740 (*AtAMY2*), and NP_564977 (*AtAMY3*)]. Among these, AtAMY3 was shown to be plastidial, and contains an N-terminal starch-binding domain with significant homology to those of the GWD–PWD proteins. The other two enzymes lack putative transit peptides and one harbors an hypothetical signal peptide at the N-terminus, rendering its plastidial localization unlikely (Yu et al., 2005). *Ostreococcus* (Ot6g00380, Ot10g00260, and Ot07g02010) and *Chlamydomonas* (EDP02635, EDP00963, and EDP02737) both encode 3 α-amylase–like sequences. The green algal enzymes are more similar to

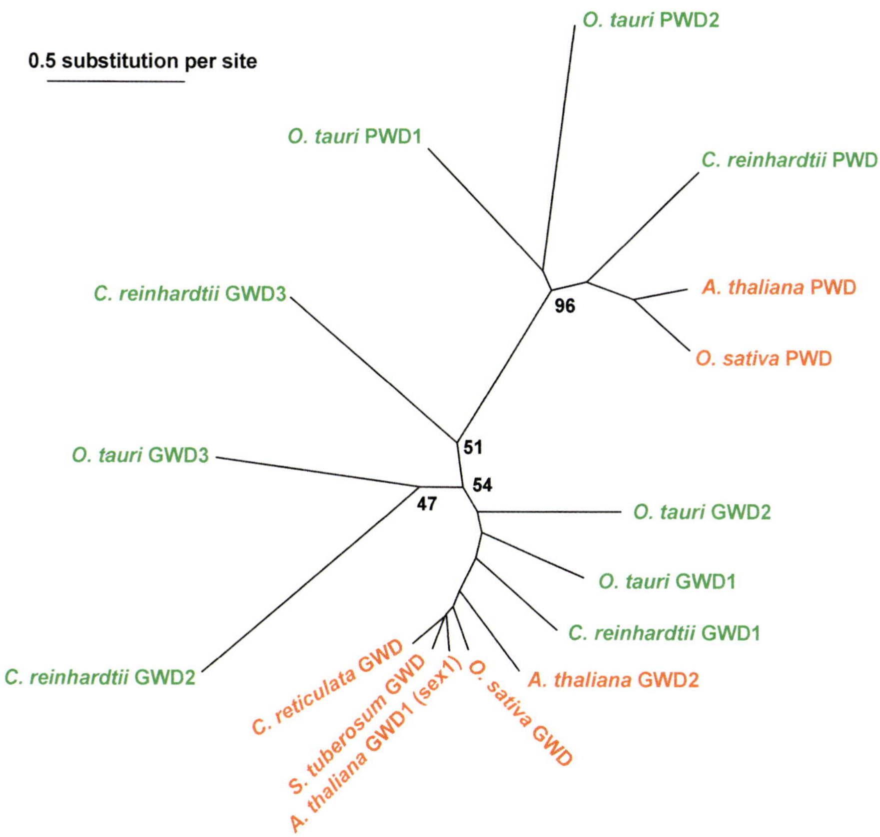

FIGURE 1.7 *Neighbor-joining tree inferred for carbohydrate dikinase proteins. Protein alignment was performed using ClustalW (Thomson et al., 1994). The tree was inferred using Neighbor (Phylip package). Numbers at nodes indicate bootstrap values calculated over 500 replicates. GWD: glucan–water dikinase, PWD: phosphoglucan water dikinase. Species names are listed in the legend of Figure 1.3.*

the higher plant enzymes than to other prokaryotic or eukaryotic sequences (Figure 1.8). Nevertheless, no clear enzyme families conserved from green algae to higher plants can be distinguished. No mutants defective for α-amylase have been documented in *Chlamydomonas*.

11. β-Amylases (EC 3.2.1.2)

Nine β-amylases are encoded by the *Arabidopsis* genome, and three of them are predicted to be chloroplast targeted (At3g23920, At4g00490, and At4g17090). Among these, the involvement of the At4g17090 product encoding a chloroplastic β-amylase (CT-BMY) in leaf starch degradation was demonstrated by analysis of *Arabidopsis* RNAi lines (Kaplan and Guy, 2005). CT-BMY is orthologous to a gene studied by antisense technology in transgenic potato (Scheidig et al., 2002). All of the transgenic lines accumulated high levels of starch in leaves. The genomes of *Chlamydomonas* and *Ostreococcus* contain, respectively, three (EDP05105, EDP08477,

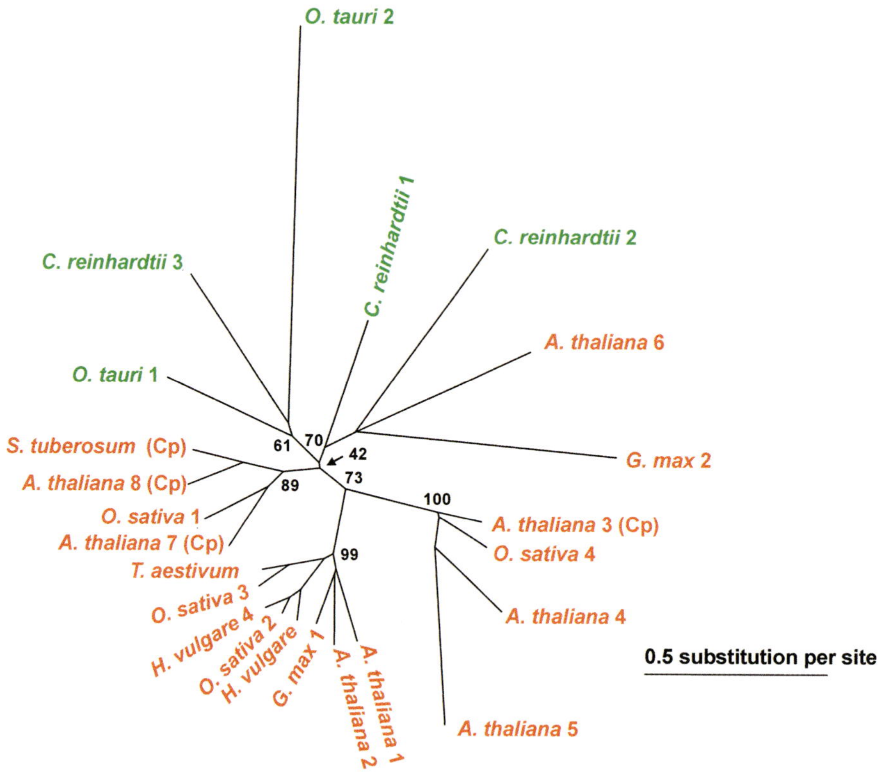

FIGURE 1.8 *Neighbor-joining tree inferred for α-amylase proteins. Protein alignment was performed using ClustalW (Thomson et al., 1994). The tree was inferred using Neighbor (Phylip package). Numbers at nodes indicate bootstrap values calculated over 500 replicates. Species names are listed in the legend of Figure 1.3.*

and EDP09926) and two (Ot03g03190 and Ot02g06980) β-amylase–like sequences. As with the α-amylases, no clear enzyme families related to the higher plant sequences can be distinguished (Figure 1.9). No mutants defective for β-amylase have been documented in *Chlamydomonas*.

C. General conclusions

Gene content analysis of the *Chlamydomonas* and *Ostreococcus* genomes demonstrates that the starch metabolism network has been highly conserved throughout evolution of green algae and land plants. Indeed, the same suite of 30–40 genes can be seen at work from *Ostreococcus* to maize. The next section concerns the genetics of starch accumulation in green algae, and demonstrates that the underlying functions of all these enzymes are substantially conserved from *Chlamydomonas* to cereals. In effect, the same degree of functional variation can be found between the cereal kernel endosperm, the potato tuber, or the *Arabidopsis* or potato leaf, as between green algae and land plants.

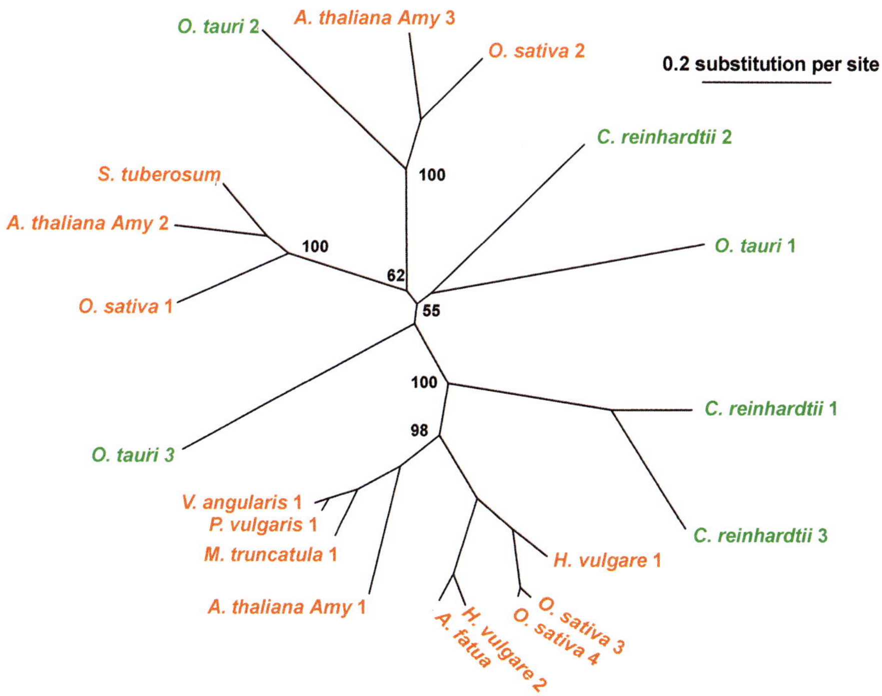

FIGURE 1.9 *Neighbor-joining tree inferred for β-amylase proteins. Protein alignment was performed using ClustalW (Thomson et al., 1994). The tree was inferred using Neighbor (Phylip package). Numbers at nodes indicate bootstrap values calculated over 500 replicates. (Cp: Chloroplast targeted isoform). Species names are listed in the legend of Figure 1.3.*

One might wonder why 30–40 genes would be devoted to the synthesis and degradation of a polymer consisting of one type of monomer (glucose) and only two types of glycosidic linkages. An explanation can be found if one compares starch to glycogen metabolism. Glycogen and starch have the same basic chemical composition and linkages but differ considerably in their structural organizations, which is much more ordered in the case of starch polymers. In the case of fungi, bacteria, or archaea, 6 to at most 12 genes can be thought of as being devoted to storage polysaccharide metabolism. One can thus speculate that to achieve the synthesis of such elaborate polymers; gene duplications followed by enzyme specialization were required. Mobilization of semi-crystalline glucose stores may also be inherently more complex than degradation of hydrosoluble glycogen particles. Very few enzymes both have access to intact starch granules, and effectively mobilize glucose from crude starch. Apparently, however, red algae or organisms derived from them through secondary endosymbiosis do not possess as much functional redundancy as green algae and vascular plants (Coppin et al., 2005; Ral et al., 2004). The apicomplexan parasite *Toxoplasma gondii*, in particular, synthesizes and degrades starch-like granules with the probable help of fewer than 10 genes.

III. THE GENETICS OF STARCH ACCUMULATION AND MOBILIZATION

A. Overview

Nine loci (*STA1* to *STA8* and *STA11*) were identified through functional screens in classical forward genetic approaches. Three of these are concerned with the supply of ADP-glucose (*STA1*, *STA6*, and *STA5*) and six with building the starch granule structure through direct or indirect effects. In addition, several mutants affected in starch biosynthesis have been isolated in other green algae. Low starch mutants were identified both in *Chlamydomonas monoica* (Rickoll et al., 1998) and in *Chlorella pyrenoidosa* (Plumed et al., 1996; Ramazanov and Ramazanov, 2006). Furthermore, a mutant defective for starch mobilization after osmotic stress was isolated in *Dunaliella* (Chitlaru and Pick, 1989). In none of these instances, however, has the enzymatic basis of the mutant phenotype been defined. The discussion below focuses on those mutants for which clear biochemical defects have been unravelled.

B. *Chlamydomonas* mutants impaired in starch synthesis

1. *Mutants impaired in the synthesis of the ADP-glucose substrate*

Ball et al. (1991) described *sta1-1*, the first green algal mutant defective for starch biosynthesis, using a simple iodine vapor spray of nitrogen-starved cell patches. The mutant accumulates less than 5% of the wild-type starch level and its AGPase is less sensitive to 3-PGA activation, but it has otherwise normal enzyme kinetics (Ball et al., 1991). While the mutant AGPase was initially thought to have lost its sensitivity to orthophosphate inhibition, purification to homogeneity of the wild-type and mutant enzymes proved that this was not the case (Iglesias et al., 1994; van den Koornhuyse et al., 1996). The decrease in starch could thus be solely attributed to the absence of 3-PGA activation, which normally counteracts the negative effect of orthophosphate inhibition. It is thus not surprising that *sta1-1* mutants synthesize anywhere from 1% to 40% residual starch, depending on the growth condition used. Molecular analyses (van den Koornhuyse et al., 1996; Zabawinski et al., 2001) subsequently showed that *STA1* encodes the large "regulatory" subunit of AGPase, and that the absence of this subunit was responsible for the *sta1-1* mutant phenotype. A starchless mutant was later found to lack both starch and AGPase activity. The corresponding *STA6* locus was demonstrated to encode the small "catalytic" subunit of the enzyme. Mutants of *STA6* together with the corresponding *adg1* mutants of *Arabidopsis* (Lin et al., 1988a) define the sole truly starchless mutants in plants. *Arabidopsis* starchless plants display a stunted growth phenotype when cultivated under a 12-hours day, 12-hours night regime. The analogous *Chlamydomonas* mutants, however, did not show any increase in generation time under laboratory conditions, and culture growth was apparently unaffected.

Characterization of the residual starch accumulated under nitrogen starvation in *sta1* mutants showed a profound modification, consisting of a change in the length distribution of amylopectin chains, and a very large decrease (or disappearance) of amylose. These changes had been previously noted when the structures of *Chlamydomonas* storage starch produced under conditions of nitrogen starvation were compared to the polysaccharide produced in nitrogen-supplied cultures (Ball et al., 1990; Libessart et al., 1995). The nitrogen-replete cells lacked amylose and displayed a similarly altered chain–length distribution within amylopectin.

The amounts of starch synthesized in cells from growing cultures are at least one order of magnitude lower than those in either growth-limited or growth-arrested nitrogen-starved cultures. The low starch amylose-less structure of wild-type cultures supplied with nitrogen mimicked the structure of the storage starch accumulated by mutants defective in ADP-glucose synthesis. This hinted that the size of the ADP-glucose pool conditions the type of starch structure produced. This assumption was confirmed by the characterization of a mutant at the *STA5* locus, which displayed an 80% decrease in total assayable phosphoglucomutase (van den Koornhuyse et al., 1996). This was accompanied by a 90% decrease in starch during nitrogen starvation, and a switch to the low amylose phenotype. Indeed, plastidial phosphoglucomutase defects are expected to lower the ADP-glucose pools in a fashion similar to the large subunit mutants of AGPase.

How does the size of the glycosyl-nucleotide pool condition starch structure? A possible answer comes from the realization that the different elongation enzymes differ significantly in their affinities for the ADP-glucose substrate. GBSSI, the enzyme responsible for amylose synthesis, possesses a 10-fold higher apparent K_m for ADP-glucose than do the soluble starch synthases devoted to amylopectin synthesis. In addition, because the different soluble starch synthases are specialized in the synthesis of distinct chain size classes, modifications of the ADP-glucose pools likely affect amylopectin structure due to differences in the kinetics of the various elongation enzymes.

2. Mutants impaired in building the starch granule architecture

The iodine spraying technique performed on nitrogen-starved cell patches gives information not only about the amount of starch present, but also on the type of polysaccharide structure. Long chains such as those present within the amylose fraction bind iodine tightly, and yield a dark olive-green color. Medium-sized chains, such as those present in the moderately branched amylopectin, yield a weaker purple stain whereas small chains, such as those present in the highly branched glycogen, do not bind iodine or bind it very weakly, giving a light orange to dark brown stain depending on the amount present. The stained mix of the purple amylopectin and dark green amylose gives the typical dark blue iodine coloration associated with starch.

"Purple" mutants at the *STA2* locus lack amylose and are defective for GBSSI. Wattebled et al. (2002) demonstrated that *STA2* encodes GBSSI, and that the enzyme contains an additional 9 kD at the C-terminal when compared to the vascular plant enzymes. GBSSI is responsible for amylose synthesis in green algae, as was demonstrated much earlier for GBSSI in maize. GBSSI also controls amylopectin structure in *Chlamydomonas*, however, based on the observations that in *sta2* mutants the long glucan content of the major amylopectin fraction (type I amylopectin) decreases significantly, while a minor second amylopectin component (type II amylopectin) disappears (Delrue et al., 1992; Maddelein et al., 1994).

The contribution of GBSSI to building amylopectin can be observed *in vitro* through incubation of purified granules with labeled ADP-glucose. In these experiments, the algal enzyme displays a specific activity at least one order of magnitude higher than the most active GBSSI from higher plants. This high specific activity seems to define a common feature of unicellular green algae and does not correlate with the C-terminal extension present in Volvocales (Oyama et al., 2006). The fact that *in vitro* synthesis of amylose from purified granules is so active has allowed the detailed mechanism of amylose synthesis to be worked out using pulse-chase experiments with purified *Chlamydomonas* starch (van de Wal et al., reviewed in Ball et al., 1998). Briefly, GBSSI progressively extends an available amylopectin chain and terminates elongation through a cleavage that generates a mature chain. GBSSI is then ready for a new round of amylose synthesis. *In vitro* experiments with pea starch correlated the production of amylose to the presence of MOS in the incubation mixture (Denyer et al., 1999). These authors concluded that MOS prime amylose synthesis because they are able to diffuse in the starch granule. The *Chlamydomonas* enzyme behaves identically to that of pea but in both *in vitro* systems, the MOS concentration required for amylose synthesis is rather high (50 mM) and possibly not physiological, and amylose synthesis through extension of amylopectin proceeds even in the complete absence of MOS. To ascertain whether MOS is indeed required for amylose synthesis, *sta2* mutants were studied *in vitro*, since they possess kinetically altered GBSSI (Ral et al., 2006). A particular mutant allele (*sta2-2*) yielded a protein that was unable to extend amylopectin but was otherwise able to extend MOS into amylose at wild-type rates (Ral et al., 2006). Because this mutant synthesizes amylose *in vivo*, the results demonstrate that at least in *Chlamydomonas*, amylose synthesis proceeds by the mechanism proposed by van de Wal et al. (1998).

The iodine spraying screening procedure yielded a number of mutants that stained olive green. These mutants, which are selectively enriched in amylose, belong to three distinct loci (*STA3*, *STA4*, and *STA8*). *STA3* mutants display an 80% decrease in starch upon nitrogen starvation, but even a modest increase in starch was observed in unstarved cultures (Ral et al., 2006). In both cases, however, the starch structure was strongly modified. All *sta3* alleles lack one of the two major soluble starch synthase activities that can be detected on zymograms.

At the time the *sta3* mutants were isolated and characterized, only one such mutant had been recorded in maize (Boyer and Preiss, 1981). In that case, however, two enzyme activities were presumed defective (SSIIa and BEIIb). Fontaine et al. (1993) were the first to characterize a particular mutant starch through enzymatic debranching and separation of individual chains by high performance anion exchange chromatography with pulsed amperometric detection, which yielded a detailed picture of amylopectin structure in *sta3* mutants. That all *sta3* mutants exhibited the same changes in the chain–length distribution led to the proposal that soluble starch synthases are responsible for the synthesis of different size classes of glucans (Fontaine et al., 1993; Maddelein et al., 1994, Delvallé et al., 2005). This proposal is in agreement with all the data subsequently produced for vascular plants (reviewed in Ball and Morell, 2003). *STA3* was noted by Buléon et al. (1997) to lack a 115-kD soluble starch synthase. Ral et al. (2006) subsequently demonstrated that the missing synthase belonged to the SSIII family (see above), and that *STA3* encoded SSIII.

Maddelein et al. (1994) first noted that *sta2* and *sta3* mutations did not display simple additive interactions, and indeed the double mutants exhibited an unexpectedly large decrease in starch content under all conditions tested. The little starch remaining in these mutants displayed a deeply modified structure and lacked chains of average degree of polymerization above 40 glucose residues. This was accompanied by a collapse of the starch granule architecture and a very significant reduction in the mass distribution of amylopectin molecules. This phenotype can be interpreted in terms of the absence of those long "spacer" chains (the so-called long B chains) which are responsible for hooking together the amylopectin clusters (see Figure 1.1).

In nitrogen-supplied medium, *sta3* mutants exhibit a spectacular increase in GBSSI mRNA abundance, correlated with a significant increase in GBSSI enzyme activity (Ral et al., 2006). This phenomenon compensates for the loss of SSIII and yields normal amounts of starch in undepleted medium. Zhang et al. (2005) also found that starch amounts were either similar or higher in SSIII mutant leaves of *Arabidopsis*. The observation was correlated with an increase in the long B chain fraction of *Arabidopsis* amylopectin. Unfortunately, no GBSSI mutant has been described or is available in *Arabidopsis* to test the likely involvement of this enzyme in the synthesis of these chains. From the studies performed in *Chlamydomonas*, it can be concluded that GBSSI, despite being the only enzyme responsible for amylose synthesis, also plays a very significant role in the synthesis of the long B spacer chains of amylopectin. SSIII, on the other hand, is the major enzyme involved in the synthesis of long B amylopectin spacers and its loss can only be compensated through a significant induction of GBSSI.

Another high amylose mutant of *Chlamydomonas* is *sta4* (Libessart et al., 1995) which like *sta3*, displayed a significant (60%) decrease in starch only under conditions of maximal starch synthesis (in nitrogen-starved cells). Despite intensive efforts, no defective activity could initially be found in the *sta4* mutants. Subsequently, it was found that all strains carrying *sta4* mutant alleles lacked one of the two phosphorylase activities associated with plastids (Dauvillée et al., 2006). In fact, *STA4* encodes the plastidial phosphorylase PHOB, and the mutants could be fully complemented by transformation with the *PHOB* gene.

The residual amylopectin synthesized in nitrogen-starved *sta4* cells is highly modified. The modifications in chain–length distribution are very different from those seen in SSIII-defective mutants. Strangely, these distributions mimic those observed in BEII mutants of vascular plants. Tetlow et al. (2004) reported the existence of a BEII-phosphorylase complex in cereal kernel amyloplasts. This complex does not operate in leaves, but is active for storage starch synthesis upon phosphorylation of its components. The fact that *sta4* leads to a selective phenotype only under nitrogen starvation for storage starch synthesis is certainly in line with the observations of Tetlow et al. (2004).

The *Chlamydomonas sta4* mutants are the first plant phosphorylase mutants with convincing defects in starch metabolism. Phosphorylases in eukaryotes and bacteria are the major enzymes of glycogen catabolism. The corresponding *E. coli* and yeast mutants accumulate excess glycogen with abnormally long outer chains (Alonso-Casajús et al., 2006). This phenotype is consistent with the supposed function of the enzyme, which is to recess the outer chains of the polysaccharides in the presence of orthophosphate, and release glucose-1-P. Defects in the plastidial form of starch phosphorylase in *Chlamydomonas* do not create a typical polysaccharide catabolism phenotype. It seems likely that its apparent function in starch biosynthesis is due to its association with BE, and the resulting mutant phenotype is predominantly due to the absence of an active BE complex, which would only be required for storage starch synthesis in cereal kernels or nitrogen-starved *Chlamydomonas* cells. Indeed, the *Chlamydomonas* starch phosphorylase activity is exquisitely sensitive to inhibition by ADP-glucose, thereby precluding a direct function of its catalytic activity in polysaccharide synthesis.

3. Mutants impaired in debranching enzyme

Mutants of the *STA7* locus stain yellow-orange when iodine is sprayed on nitrogen-starved cell patches. The yellow stain typifies mutants accumulating less than 5% residual starch, whereas the orange taint suggests the presence of minute amounts of highly branched polymers. The *sta7* mutants were shown to substitute starch completely by glycogen synthesis. In addition, *sta7* mutants synthesize minute amounts (<0.2% of wild-type starch level) of insoluble polysaccharides with an entirely novel structural organization, which consists of amylose-type chains associated with a high molecular weight network of very loosely branched long glucans. This material thus lacked the typical amylopectin clusters shown in Figure 1.1, which are responsible for the semi-crystalline insoluble nature of amylopectin. Dauvillée et al. (1999a) were able to show that the synthesis of this material was under GBSSI control, since it disappeared in *sta7 sta2* double mutants. GBSSI was not, however, bound to this insoluble polysaccharide in *sta7* mutants. This can be readily explained by assuming that binding to a semi-crystalline polysaccharide is required for normal GBSSI activity and that in its absence, GBSSI can synthesize only very low amounts of amylose-like material (Dauvillée et al., 1999a).

The substitution of starch by glycogen in *sta7* mutants is accompanied by a collapse in the amounts synthesized, which decrease by over 90%. This decrease can be best rationalized by the action of catabolic hydrosoluble enzymes which are able to degrade water-soluble polysaccharides such as glycogen, rather than by a decrease in the rate of polysaccharide synthesis. The dramatic *sta7* phenotype in *Chlamydomonas* suggested that the function encoded by *STA7* was essential for starch synthesis, and that its function defined the basic difference between starch and glycogen metabolism.

Mouille et al. (1996) initially identified the nature of the missing activity through zymogram analysis, coupled with proton NMR characterization of the affected polysaccharides. The mutants turned out to be deficient in DBE, a function which at that time was not suspected to be required during starch biosynthesis. A few months earlier James et al. (1995) had reported that the *sugary 1* (*SU1*) locus of maize encoded an isoamylase-like sequence. Maize *su-1* mutants accumulated a restricted yet significant amount of starch, and a substantial amount of glycogen. Isoamylase-type debranching enzymes can be distinguished from pullulanase-type debranching enzymes by their substrate specificities (see Section II.B.5). As with maize, *Chlamydomonas STA7* was found to encode isoamylase (Dauvillée et al., 1999b, 2000; Posewitz et al., 2004).

The situation in rice and maize turned out to be complex as both pullulanase and isoamylase decreased in *sugary* mutants, although only isoamylase is encoded at the *SU1* locus (Pan and Nelson, 1984; James et al., 1995; Nakamura et al., 1997). In *Chlamydomonas* the situation was simpler because the isoamylase-defective mutants had the same low level of pullulanase activity as the wild-type controls, and did not contain any residual starch (Mouille et al., 1996; Dauvillée et al., 2000). It was later found that the differences between maize, rice, and *Chlamydomonas* could be explained by a small overlap in function between the two types of debranching enzymes, and by the presence of a substantially higher level of pullulanase activity in cereal kernels even in the isoamylase null mutants (accounting for up to 40% in maize; Kubo et al., 1999; Dinges et al., 2003).

Because of the severity of the phenotype described in *Chlamydomonas*, it became clear that isoamylase-mediated debranching defined a new essential step of starch biosynthesis. Why might this be the case? Groups working on *Chlamydomonas* and cereals jointly proposed that amylopectin was synthesized in the form of a hydrophilic precursor named pre-amylopectin, with a disordered branching pattern that precludes crystallization and packaging into insoluble granules (Ball et al., 1996). The function of isoamylase would be to debranch and release those loosely branched chains that prevent crystallization. This process, known as glucan trimming, would be responsible for the asymmetrical distribution of branches that distinguishes amylopectin from glycogen (Ball et al., 1996; Mouille et al., 1996; Myers et al., 2000).

A second locus was subsequently demonstrated to be required for normal glucan trimming in *Chlamydomonas*. Mutants of *STA8* display a reduced but significant quantity of high amylose starch (Dauvillée et al., 1999b). In addition, *sta8* mutants synthesize glycogen in structure and amounts identical to those documented for *sta7* mutants. Dauvillée et al. (1999c) found that *sta8* mutants exhibit a reduction in size of a very large multimeric complex of isoamylase activity, which is replaced by a smaller complex consisting of two to three subunits. The isoamylase subunit encoded by *STA7* was thus present and active but misassembled, causing a threefold reduction in enzyme activity. The substrate specificity and products formed by the mutant isoamylase were apparently unchanged. Because heterozygous triploids containing one-third of the wild-type activity displayed an entirely wild-type phenotype one could conclude that reductions in enzyme activity or modified enzyme specificities did not explain the mutant phenotype. The latter seemed entirely due to the collapse of the multimeric organization of the isoamylase complex.

Vascular plants contain additional isoamylase-related isoforms termed ISA2 (Hussain et al., 2003), which is absent in *Ostreococcus* but present in *Chlamydomonas*; and ISA3, which is present in both, and involved in starch catabolism in *Arabidopsis* (Wattebled et al., 2005). It is possible that *STA8* encodes one of these subunits. Utsumi and Nakamura (2006) found that the rice isoamylase complex consisted either of five "catalytic" ISA1 subunits (the product of the *SU1* gene in maize and rice) or of one regulatory ISA2 subunit associated with five ISA1 subunits. ISA2 is believed not to harbor any catalytic activity and is suspected to play similar roles in cereals and *Chlamydomonas*.

One prediction of the glucan trimming model is that linear (unbranched) MOS would be released at the time of starch synthesis by isoamylase. Because the turnover of carbon could be substantial during trimming, it is expected that efficient systems must exist for the recovery of the energy apparently wasted through the splicing of the misplaced glucans. We believe that these systems were discovered during the characterization of the *Chlamydomonas sta11* mutants, as discussed immediately below.

4. Mutants impaired in malto-oligosaccharide metabolism

Mutants were recovered that displayed a 90% decrease in starch, a modest increase in amylose, and a substantial accumulation of unbranched MOS (*sta11*; Colleoni et al., 1999a,b). These mutants were defective for an α-1,4 glucanotransferase named D-enzyme (disproportionating enzyme) that is encoded at the *STA11* locus (Colleoni et al., 1999a,b; Wattebled et al., 2003). This enzyme transfers mostly maltosyl (consisting of two glucose residues) or longer series MOS from a donor oligosaccharide to an acceptor. The reaction involves the synthesis of a novel α-1,4 glycosidic bond on the acceptor at the expense of the cleavage of an α-1,4 glycosidic bond on the donor. This enzyme thus disproportionates (hence the name) a homogeneous solution of MOS of defined length into a series of MOS of lesser or greater degrees of polymerization. Maltose is never produced by D-enzyme, and glucose can be used as an acceptor. During characterization of the α-1,4 glucanotransferase missing in *sta11* mutants (Colleoni et al., 1999b), it was found that amylopectin outer chains serve as very good acceptors for transfer reactions involving donor unbranched MOS. Incorporation into the polysaccharide outer chains would be essentially irreversible *in vivo* because the local concentration of outer chains within polysaccharides is several orders of magnitude higher than that measured for MOS in the stroma.

Colleoni et al. (1999b) proposed that the function of D-enzyme would be to recycle efficiently the glucans sliced out by isoamylase through their transfer into the polysaccharide's outer chains before crystallization. This hypothesis was proposed to explain the small but significant modification in chain–length distribution seen in *sta11-1* mutant amylopectin. Equally possible is that D-enzyme produces a series of long MOS suitable for degradation by MOS-specific starch phosphorylase isoform(s). Both of these pathways would be much more efficient in recovering the energy contained in the misplaced glucans than standard hydrolysis, and both require D-enzyme. That D-enzyme would also be required for starch degradation when both branched and unbranched MOS are generated was proposed by Colleoni et al. (1999b) in their original discussion. However no evidence for such a role could be found in *Chlamydomonas,* and Wattebled et al. (2003)

found that MOS accumulated in *sta11* mutants during phases of net starch synthesis. Hatanaka et al. (2000) also characterized a D-enzyme from *Dunaliella parva*, but its function was not elucidated.

IV. PHYSIOLOGY OF STARCH STORAGE: CIRCADIAN CLOCK CONTROL, PYRENOIDAL STARCH SYNTHESIS AND HYDROGEN PRODUCTION

A. Clock control of starch metabolism

In the *Arabidopsis* source leaves, the onset of starch synthesis coincides with lights on, while leaf starch degradation is confined to the dark phase. This seems to suggest that at least in source tissues, starch synthesis is simply controlled by the availability of substrates and energy through photosynthesis. In sink tissues, the situation seems to be quite different. In many instances storage starch exists as granules made of a succession of layers differing in their crystallinity (see Figure 1.1D). These layers can be revealed through limited digestion with amylases followed by SEM of their shell-like structures. Buttrose (1962) showed that the layers corresponded to the number of days of starch deposition and that the deduced diurnal rhythms of starch metabolism were under circadian clock control. Starch metabolism diurnal rhythms were not under clock control in all sink tissues, however. While the layers were clearly maintained in starch from potato tubers grown under continuous light, this was not the case for cereal kernel endosperm starch.

Chlamydomonas is characterized by an atypical pattern of starch deposition in synchronized cultures. Klein (1987) and Thyssen et al. (2001) unexpectedly found that such cultures achieved minimum starch content in the middle of the light phase and that in the absence of acetate, starch amounts declined steadily after they switch to darkness. This pattern contrasted with previous results from *Chlorella*, where starch synthesis proceeded somewhat more continuously and decreased only at the time of cell division 2 hours before the dark phase (Duynstee and Schmidt, 1967). In *Chlamydomonas*, cells sampled during the dark phase displayed significantly higher starch synthesis rates than cells sampled from the middle of the light phase (Klein, 1987). Higher starch synthesis rates during darkness were also found in the presence of acetate, although under these conditions starch amounts peaked during the first half of the night and decreased during the second half, at the time of cell division (Ral et al., 2006). This study found similar patterns under continuous light or darkness, and further demonstrated that starch synthesis fulfilled all theoretical requirements defined for circadian clock-controlled rhythms. In *Chlamydomonas*, however, and possibly in most green algae, the cell cycle is gated by the circadian clock (Goto and Johnson, 1995; see also Chapter 3, Volume 1). Ral et al. (2006) found that oscillations dampened under nitrogen starvation, coincident with cell division arrest; a similar observation was made by Klein (1987). From these results it is clear that both the cell cycle and starch metabolism are under circadian clock control in *Chlamydomonas*.

Still unclear, however, is whether starch metabolism is controlled directly by the clock, or indirectly through the cell division cycle. Indeed, the latter could be the dominant factor. Oscillations of starch metabolism imply some regulation of the major rate-controlling enzyme of starch biosynthesis, as well as some control of the major triggers of starch catabolism. Zabawinski et al. (2001) noted that mRNA levels encoding the small subunit of AGPase are under circadian clock control. These observations were pursued by Ral et al. (2006), who documented the level of small subunit protein through immunoblots, the amount of enzyme activity, and mRNA levels using quantitative RT-PCR. Each of these parameters correlated with the rate of starch synthesis. This suggests that transcriptional regulation of AGPase subunit genes is sufficient to explain clock regulation of starch anabolism.

The oscillations of starch catabolism remain, however, to be explained. Unfortunately very little is known about starch degradation in *Chlamydomonas*. GWD has been found associated with starch granules, and ensuing starch phosphorylation has been observed (Ritte et al., 2004). Despite the characterization of several possible candidates, the precise nature of the predominant hydrolases and phosphorylases responsible for the initiation of starch catabolism in algae remains to be determined. Levi and Gibbs (1984) characterized amylase and phosphorylase activities that they reported to be confined to the plastid, and suspected to be involved in starch degradation in synchronously grown *Chlamydomonas* cultures. Since then, it has been demonstrated that both activities correspond to mixtures of enzyme activities encoded by distinct genes, and that a substantial portion of the phosphorylase is located in the cytosol (Mouille et al., 1996; Dauvillée et al., 2000, 2006). Despite more detailed investigations Dauvillée et al. (2006) found no clear evidence for the catabolic function of phosphorylase.

Starch mobilization in *Chlamydomonas reinhardtii* and *Chlamydomonas moewusii* is greatly stimulated in anaerobic conditions. This phenomenon, known as the Pasteur effect, is thought to be due to a selective decrease in ATP. ATP is an allosteric inhibitor of phosphofructokinase which in turn defines a major rate controlling step of glycolysis. The net effect of decreasing ATP is therefore a stimulation of glycolysis. A particular marine isolate (*Chlamydomonas* sp. MGA161) was shown to lack the Pasteur effect (Maeda et al., 2001), and this was due to higher relative concentrations of ATP under anaerobic conditions. Interest revived in anaerobic starch fermentation in green algae when breakdown products and gases such as acetate, ethanol, formate, CO_2, and H_2 were observed during starch fermentation. Of particular note was the demonstration that the starch pool is important for H_2 production (Posewitz et al., 2004). H_2 production and related topics are covered in Chapter 7.

The pattern of starch fermentation end products in *Chlamydomonas* together with the sensitivity of formate production to oxygen suggested the presence of a pyruvate formate lyase-like activity (Kreuzberg, 1984; Kreuzberg and Martin, 1984). Atteia et al. (2006) subsequently found such activity in both *Chlamydomonas* mitochondria and chloroplasts. This bears testimony to the existence of an ancient anaerobic pathway of carbon assimilation in eukaryotes, a finding with important implications for our understanding of eukaryogenesis (reviewed in Martin et al., 2001). The transition to anaerobiosis generates glycolytic oscillations in starch degradation and in the level of several metabolites generated from it (Kreuzberg and Martin, 1984). These oscillations, with a mean period of 59 minutes, provide additional evidence for a rate-controlling step catalyzed by phosphofructokinase during glycolysis. It must be stressed, however, that none of the studies on starch fermentation have yielded insights into the mechanisms involved in the initial mobilization of glucose stores from the starch particle.

B. Is starch metabolism required for normal pyrenoid function?

In green algae, the pyrenoid is in the vast majority of cases surrounded by starch (reviewed in Chan, 1972). The pyrenoid is constituted mostly if not solely of aggregated Rubisco molecules (Morita et al., 1997; Borkhsenious et al., 1998), and its formation has been correlated to particular intrinsic properties of Rubisco (Nozaki et al., 2002). Because the switch from high to low CO_2 is accompanied by the synthesis of pyrenoidal starch in a variety of green algae, including *Chlamydomonas*, it was assumed that this structure had a distinctive function in the CO_2 concentration mechanism (CCM) (Miyachi et al., 1986; Tsuzuki et al., 1986; Kuchitsu et al., 1988; reviewed in Badger et al., 1998; see also Chapter 8). Ramazanov et al. (1994) similarly found that pyrenoidal starch correlated with the absence of acetate and induction of the CCM.

On the other hand, starch synthesis was predominantly stromal in the presence of acetate, which partly inhibits induction of the CCM. Because of this, Ramazanov et al. (1994) proposed that starch metabolism *per se* might be required for normal pyrenoid function. In other words, starch would be required for full CCM function because the presence of the starch sheath would slow the diffusion of CO_2 out of the pyrenoid. That this is not the case was demonstrated through the use of starch metabolism mutants both in *Chlamydomonas* (Villarejo et al., 1996) and *Chlorella pyrenoidosa* (Plumed et al., 1996). The *Chlamydomonas sta7* mutant used in these studies carries a disruption in the gene encoding the ISA1 isoamylase subunit. It therefore accumulates a small amount of phytoglycogen but no granular starch (Mouille et al., 1996), creating an ideal situation to test this hypothesis. The *sta7* mutant showed no apparent defects in photosynthesis rates under high and low CO_2, since the half-maximal rates of photosynthesis decreased in a similar fashion to the wild type, ruling out a function of the starch sheath in CCM function (Villarejo et al., 1996). Although identical results were obtained in *Chlorella pyrenoidosa* (Plumed et al., 1996), it should be noted that the mutants were poorly characterized and no defects in starch metabolism enzymes were documented.

Despite not being required for the CCM *per se*, synthesis of pyrenoidal starch remains an important property of cell adaptation to low CO_2. Indeed among 10,368 ESTs compiled from resources built under high and low CO_2 conditions, 51 genes were shown to be induced during low CO_2 adaptation, among them those encoding GBSSI and SSIII (Miura et al., 2004). These genes were also induced in strains defective for the transcriptional regulator CCM1 (*CIA5*), which apparently defines the master regulator for the induction of genes during adaptation to low CO_2. Low CO_2 induction of GBSSI transcription could be a general response to low CO_2 adaptation in green algae. Indeed, a similar increase in GBSSI mRNA was reported in *Chlorella kessleri* (Oyama et al., 2006).

How does starch synthesis operate around the pyrenoid? An informative study was published by Süss et al. (1995) who co-localized several Calvin cycle enzymes, Rubisco, and Rubisco activase to both stromal thylakoid membranes and the inner face of the pyrenoid outer membrane. In fact, this membrane would be of the same nature as the pyrenoid tubule membranes which are generally considered as thylakoid membranes (reviewed in Badger et al., 1998). Confocal microscopy of chlorophyll fluorescence indeed revealed the presence of photosystem II in close association with the pyrenoid membranes (Gunning and Schwartz, 1999). If carbon substrates are synthesized at the inner (pyrenoid) face of the pyrenoid membrane, how would these reach the outer (stromal) face where starch synthesis occurs? Too little is known about the organization of this membrane and its function to allow any reasonable speculation. If for instance this thylakoid-like membrane were discontinuous, there might be no need to translocate carbon out of the pyrenoidal space. In cryptophyte algae, it was suggested that pyrenoidal starch synthesis occurs across the two innermost membranes separating the periplastidial starch from the chloroplast stroma (Deschamps et al., 2006). In these studies the authors discussed the relationship in red algae between the presence of GBSSI and amylose, and the close association of starch despite its extraplastidial localization with the rhodoplast inner and outer membranes which in this case surround the pyrenoid, which remains stromal. Interestingly, those red algae accumulating floridean starch devoid both of amylose and GBSSI accumulate starch in the cytoplasm, without any connections to the rhodoplast pyrenoid when present. Here again, the presence of GBSSI seems to correlate with the presence of a pyrenoid-based CCM.

V. CONCLUSIONS

A. What have we learned from *Chlamydomonas*?

A genetic screen based on the spraying of iodine vapors on nitrogen-starved cell patches has been extensively used to isolate mutants defective for various aspects of starch metabolism. Eleven loci have been identified by forward genetic approaches. Mutants of the *Chlamydomonas* heterotetrameric AGPase have been produced for both the small and large subunits (Ball et al., 1991; Zabawinski et al., 2001). Studies on the large subunit mutants contributed towards proving the *in vivo* requirement for 3-phosphoglycerate activation of AGPase in order to obtain significant starch synthesis (Ball et al., 1991; van den Koornhuyse et al., 1996). In addition, these mutants helped to establish that controlling the flux of ADP-glucose into starch had a profound consequence on the polysaccharide structure which, under low glycoside-nucleotide supply, adopts the structure of transitory starch (van den Koornhuyse et al., 1996). This conclusion was subsequently verified in pea and potato (Clarke et al., 1999; Lloyd et al., 1999).

The characterization of mutants of GBSSI, the enzyme responsible for amylose synthesis, suggested that GBSSI is an important contributor to amylopectin synthesis (Delrue et al., 1992; Maddelein et al., 1994). The isolation of a soluble starch synthase-defective mutant that was later shown to be analogous to the vascular plant SSIII enzymes, contributed to establishing that distinct soluble starch synthases had specific roles in building different size classes of amylopectin chains (Fontaine et al., 1993; Ral et al., 2006). The subsequent characterization of *Chlamydomonas* mutants defective for one plastidial form of starch phosphorylase provided the first *in vivo* evidence for a function of this enzyme in starch metabolism (Dauvillée et al., 2006).

The key breakthrough, however, came with the description of low starch mutants that completely substituted starch synthesis with a small amount of glycogen (Mouille et al., 1996). These mutants proved to lack a high mass isoamylase complex because of the absence of one of its subunits (Dauvillée et al., 2000; Posewitz et al., 2004). The phenotypes of the *Chlamydomonas* mutants appear much more severe than the orthologous ones in vascular plants, and led the authors to propose that debranching of an amylopectin precursor (pre-amylopectin) defined the step that distinguished glycogen from starch synthesis (Ball et al., 1996). This step can be considered as mandatory for obtaining semi-crystalline material which can aggregate into macrogranular insoluble structures. Again, the isolation of other *Chlamydomonas* mutants strongly suggested the presence of an active heteromultimeric isoamylase whose quaternary structure defined an important component of the polysaccharide aggregation mechanism (Dauvillée et al., 1999b,c).

Explaining the major differences between the synthesis of glycogen and starch is the "holy grail" of starch research. Because of its importance, the glucan trimming model was rightly challenged by other members of the scientific community, in particular Zeeman et al. (1998) and Burton et al. (2002). In the first case, low starch mutants of *Arabidopsis* were selected that were said to display no major alterations in chain–length distribution in the residual starch, but did accumulate glycogen. It was thus inferred that the function of isoamylase was not to trim chains to generate semi-crystalline amylopectin packaged into starch, since altered chain–length distribution should have been observed in the residual mutant amylopectin. It was proposed that the true function of isoamylase was to clear the plastid stroma of non-productive elongation and branching of MOS which would otherwise generate glycogen, and thus divert carbon from starch. This model came to be known as the WSP (water soluble polysaccharide) clearing model.

A more detailed study of *Arabidopsis* debranching enzyme mutants revealed, however, that they in fact had profoundly modified amylopectin chain–length distribution inconsistent with the WSP clearing model (Wattebled et al., 2005). Isoamylase-defective barley mutants had both now-classic phenotypes and also dramatically altered distribution of starch granule size (Burton et al., 2002). This led to the proposal that isoamylase is selectively involved in priming starch granules. This observation is inconsistent, however, with both the profound structural modification of the residual starch and the phytoglycogen accumulation of completely defective isoamylase mutants. Indeed, if isoamylase were only required for granule priming but not for continuous granule synthesis and growth, then residual granules should not be so profoundly modified. The production of abundant small granules can be rationalized by the increase in glycogen content, which provides suitable three-dimensional scaffolds for amylopectin crystallization in incompletely defective mutants.

Mutants defective for the D-enzyme (an α-1,4 glucanotransferase) established the first link between this family of enzymes and starch metabolism (Colleoni et al., 1999a,b). Because of the low starch phenotype of these mutants, the D-enzyme was inferred to play an important role in starch biosynthesis by recycling chains produced through isoamylase. The *Arabidopsis* D-enzyme was later found to be a player in leaf starch catabolism (Critchley et al., 2001). A function of *Chlamydomonas* α-1,4 glucanotransferase in both starch catabolism and starch synthesis was proposed despite the absence of a degradation phenotype in D-enzyme mutants (Colleoni et al., 1999b). It was suggested that D-enzyme metabolizes MOS whenever they are produced in the stroma and indeed MOS are expected to be produced both during both synthesis and degradation. Why do the *Chlamydomonas* and *Arabidopsis* mutants have such different phenotypes? One possibility is related to the very different energy constraints that apply to green algae and land plants, which differ in their CCM mechanisms and motility. In aqueous environments, CO_2 concentration is expected to be more costly. In addition, paralyzed mutants of *Chlamydomonas* have been reported to overaccumulate starch, suggesting that the energy consumed for motility does impact starch metabolism (Hamilton et al., 1992). In contrast, the energy supply is virtually unlimited for the *Arabidopsis* leaf cell under laboratory or greenhouse conditions. Because of these differences, the presence of a futile cycle might have no impact on starch content in *Arabidopsis*, but cause a collapse of starch synthesis in green algae.

A very useful property of *Chlamydomonas* arises from the fact that its GBSSI has a 10- to 100-fold superior specific activity as compared to those reported for plant enzymes. This property was used to establish *in vitro* synthesis of amylose with purified starch granules (van de Wal et al., 1998). These biochemical approaches established that *Chlamydomonas* GBSSI synthesizes amylose chains by extension of an amylopectin chain, followed by cleavage of mature amylose chain at the site of its synthesis. They also established that GBSSI does not contribute only negatively to the polysaccharide's crystallinity, but is also responsible for the formation of a low but significant number of B-type crystals (Wattebled et al., 2002).

B. Future prospects

At the beginning of the 1990s *Chlamydomonas* was the best system to dissect starch metabolism genetically. Since then, *Arabidopsis* has attracted more attention which mainly reflects the availability of two unique resources. First, the *Arabidopsis* genome sequence was completed 5 years before that of *Chlamydomonas*. Second and more significantly, multiple mutant libraries have been generated in *Arabidopsis* consisting of T-DNA insertions whose flanking sequences have been determined. It is now possible to order seeds heterozygous for an insertion in virtually any gene suspected to be involved in starch metabolism. One limitation is that *Arabidopsis* is mainly useful for the study of mature leaf starch synthesis, which represents only a subset of physiological constraints that have led to the selection of the starch metabolism network. The usefulness of *Chlamydomonas* will, however, very much depend on the pace with which reverse genetic tools, such as tagged mutant collection resources, are created. At the time of this writing, there is no way of telling if these resources will be as extensive and as readily available as those generated for *Arabidopsis* by a far greater number of research groups. Finally, a renewed interest in fermentative metabolism linked to the production of hydrogen gas has revived an interest in the synthesis and mobilization of the starch energy stores required for this process. This may provide the future basis for the development of approaches in green algae concerning the regulation of this important pathway.

REFERENCES

Alonso-Casajús, N., Dauvillée, D., Viale, A.M., Munoz, F.J., Baroja-Fernandez, E., Mora n-Zorzano, M.T., Eydallin, G., Ball, S., and Pozueta-Romero, J. (2006). The role of glycogen phosphorylase in glycogen breakdown and structure in *Escherichia coli*. *J. Bacteriol.* **188**, 5266–5272.

Atteia, A., van Lis, R., Gelius-Dietrich, G., Adrait, A., Garin, J., Joyard, J., Rolland, N., and Martin, W. (2006). Pyruvate: Formate lyase and a novel route of eukaryotic ATP-synthesis in anaerobic mitochondria. *J. Biol. Chem.* **281**, 9909–9918.

Badger, M.R., Andrews, T.J., Whitney, S.M., Ludwig, M., Yellowlees, D.C., Leggat, W., and Price, G.D. (1998). The diversity and coevolution of Rubisco, plastids, pyrenoids, and chloroplast-based CO_2-concentrating mechanisms in algae. *Can. J. Bot.* **76**, 1052–1071.

Ball, S. (1998). Regulation of starch biosynthesis. In: *The Molecular Biology of Chloroplasts and Mitochondria in Chlamydomonas* (J.D. Rochaix, M. Goldschmidt-Clermont, and S. Merchant, Eds.), pp. 549–567. Kluwer Academic Publishers, Dordrecht, The Netherlands.

Ball, S.G. and Morell, M.K. (2003). From bacterial glycogen to starch: Understanding the biogenesis of the plant starch granule. *Annu. Rev. Plant Biol.* **54**, 207–233.

Ball, S.G., Dirick, L., Decq, A., Martiat, J.C., and Matagne, R.F. (1990). Physiology of starch storage in the monocellular alga *Chlamydomonas reinhardtii*. *Plant Sci.* **66**, 1–9.

Ball, S., Marianne, T., Dirick, L., Fresnoy, M., Delrue, B., and Decq, A. (1991). A *Chlamydomonas reinhardtii* low-starch mutant is defective for 3-phosphoglycerate activation and orthophosphate inhibition of ADP-glucose pyrophosphorylase. *Planta* **185**, 17–26.

Ball, S., Guan, H.P., James, M., Myers, A., Keeling, P., Mouille, G., Buleon, A., Colonna, P., and Preiss, J. (1996). From glycogen to amylopectin: A model for the biogenesis of the plant starch granule. *Cell* **86**, 349–352.

Ball, S., van de Wal, M., and Visser, R. (1998). Progress in understanding the biosynthesis of amylose. *Trends Plant Sci.* **3**, 462–467.

Ballicora, M.A., Iglesias, A.A., and Preiss, J. (2003). ADP-glucose pyrophosphorylase, a regulatory enzyme for bacterial glycogen synthesis. *Microbiol. Mol. Biol. Rev.* **67**, 213–225.

Baunsgaard, L., Lutken, H., Mikkelsen, R., Glaring, M.A., Pham, T.T., and Blennow, A. (2005). A novel isoform of glucan, water dikinase phosphorylates pre-phosphorylated alpha-glucans and is involved in starch degradation in *Arabidopsis*. *Plant J.* **41**, 595–605.

Borkhsenious, O.N., Mason, C.B., and Moroney, J.V. (1998). The intracellular localization of ribulose-1,5-bisphosphate carboxylase/oxygenase in *Chlamydomonas reinhardtii*. *Plant Physiol.* **116**, 1585–1591.

Bourne, E.J., Stacey, M., and Wilkinson, I.A. (1950). The composition of the polysaccharide synthesized by *Polytomella coeca*. *J. Chem. Soc.*, 2694–2698.

Boyer, C.D. and Preiss, J. (1981). Evidence for independent genetic control of the multiple forms of maize endosperm branching enzymes and starch synthases. *Plant Physiol.* **67**, 1141–1145.

Buléon, A., Gallant, D.-J., Bouchet, B., Mouille, G., D'Hulst, C., Kossman, J., and Ball, S.G. (1997). Starches from A to C. *Chlamydomonas reinhardtii* as a model microbial system to investigate the biosynthesis of the plant amylopectin crystal. *Plant Physiol.* **115**, 949–957.

Buléon, A., Colonna, P., Planchot, V., and Ball, S. (1998). Starch granules: Structure and biosynthesis, *Int. J. Biol. Macromol.* **23**, 85–112.

Burton, R.A., Jenner, H., Carrangis, L., Fahy, B., Fincher, G.B., Hylton, C.M., Laurie, D.A., Parker, M., Waite, D., van Wegen, S., Verhoeven, T., and Denyer, K. (2002). Starch granule initiation and growth are altered in barley mutants that lack isoamylase activity. *Plant J.* **31**, 97–112.

Buttrose, M.S. (1962). The influence of environment on the shell structure of starch granules. *J. Cell Biol.* **14**, 159–167.

Caspar, T., Huber, S.C., and Somerville, C. (1985). Alterations in growth, photosynthesis, and respiration in a starchless mutant of *Arabidopsis thaliana* (L.) deficient in chloroplast phosphoglucomutase activity. *Plant Physiol.* **79**, 11–17.

Chan, K. (1972). The pyrenoids of the green algae (Chlorophyceae). *Academic Annual* **2**, 53–68.

Chitlaru, E. and Pick, U. (1989). Selection and characterization of *Dunaliella salina* mutants defective in haloadaptation. *Plant Physiol.* **91**, 788–794.

Clarke, B.R., Denyer, K., Jenner, C.F., and Smith, A.M. (1999). The relationship between the rate of starch synthesis, the adenosine 5′-diphosphoglucose concentration and the amylose content of starch in developing pea embryos. *Planta* **209**, 324–329.

Colleoni, C., Dauvillée, D., Mouille, G., Buléon, A., Gallant, D., Bouchet, B., Morell, M., Samuel, M., Delrue, B., D'Hulst, C., Bliard, C., Nuzillard, J.-M., and Ball, S. (1999a). Genetic and biochemical evidence for the involvement of α-1,4 glucanotransferases in amylopectin synthesis. *Plant Physiol.* **120**, 993–1004.

Colleoni, C., Dauvillée, D., Mouille, G., Morell, M.K., Samuel, M., Slomiany, M.C., Liénard, L., Wattebled, F., D'Hulst, C., and Ball, S. (1999b). Biochemical characterization of the *Chlamydomonas reinhardtii* α-1,4 glucanotransferase supports a direct function in amylopectin biosynthesis. *Plant Physiol.* **120**, 1005–1014.

Coppin, A., Varre, J.S., Lienard, L., Dauvillée, D., Guerardel, Y., Soyer-Gobillard, M.O., Buleon, A., Ball, S., and Tomavo, S. (2005). Evolution of plant-like crystalline storage polysaccharide in the protozoan parasite *Toxoplasma gondii* argues for a red alga ancestry. *J. Mol. Evol.* **60**, 257–267.

Craigie, J.S. (1974). Storage products. In: *Algal Physiology and Biochemistry* (W.D.P. Stewart, Ed.), pp. 206–235. Blackwell, Oxford.

Crevillén, P., Ballicora, M.A., Mérida, Á., Preiss, J., and Romero, J.M. (2003). The different large subunit isoforms of *Arabidopsis thaliana* ADP-glucose pyrophosphorylase confer distinct kinetic and regulatory properties to the heterotetrameric enzyme. *J. Biol. Chem.* **278**, 28508–28515.

Crevillén, P., Ventriglia, T., Pinto, F., Orea, A., Mérida, Á., Preiss, J., and Romero, J.M. (2005). Differential pattern of expression and sugar regulation of *Arabidopsis thaliana* ADP-glucose pyrophosphorylase-encoding genes. *J. Biol. Chem.* **280**, 8143–8149.

Critchley, J.H., Zeeman, S.C., Takaha, T., Smith, A.M., and Smith, S.M. (2001). A critical role for disproportionating enzyme in starch breakdown is revealed by a knock-out mutation in *Arabidopsis*. *Plant J.* **26**, 89–100.

Dauvillée, D., Colleoni, C., Shaw, E., Mouille, G., D'Hulst, C., Morell, M., Samuel, M.S., Bouchet, B., Gallant, D.J., Sinskey, A., and Ball, S. (1999a). Novel, starch-like polysaccharides are synthesized by an unbound form of granule-bound starch synthase in glycogen-accumulating mutants of *Chlamydomonas reinhardtii*. *Plant Physiol.* **119**, 321–329.

Dauvillée, D., Colleoni, C., Mouille, G., Buleon, A., Gallant, D.J., Bouchet, B., Morell, M.K., D'Hulst, C., Myers, A.M., and Ball, S.G. (1999b). Two loci control phytoglycogen production in the monocellular green alga *Chlamydomonas reinhardtii*. *Plant Physiol.* **125**, 1710–1722.

Dauvillée, D., Colleoni, C., Mouille, G., Morell, M.K., D'Hulst, C., Wattebled, F., Lienard, L., Delvalle, D., Ral, J.P., Myers, A.M., and Ball, S.G. (1999c). Biochemical characterization of wild-type and mutant isoamylases of *Chlamydomonas reinhardtii* supports a function of the multimeric enzyme organization in amylopectin maturation. *Plant Physiol.* **125**, 1723–1731.

Dauvillée, D., Mestre, V., Colleoni, C., Slomianny, M.-C., Mouille, G., Delrue, B., D'Hulst, C., Bliard, C., Nuzillard, J.-M., and Ball, S. (2000). The debranching enzyme complex missing in glycogen-accumulating mutants of *Chlamydomonas reinhardtii* displays an isoamylase-type specificity. *Plant Sci.* **157**, 145–156.

Dauvillée, D., Chochois, V., Steup, M., Haebel, S., Eckermann, N., Ritte, G., Ral, J.-P., Colleoni, C., Hicks, G., Wattebled, F., Deschamps, P., D'Hulst, C., Liénard, L., Buléon, A., Puteaux, J.-L., and Ball, S. (2006). Plastidial phosphorylase is required for normal starch granule biogenesis in the monocellular alga *Chlamydomonas reinhardtii*. *Plant J.* **48**, 274–285.

Delrue, B., Fontaine, T., Routier, F., Decq, A., Wieruszeski, J.M., van den Koornhuyse, N., Maddelein, M.L., Fournet, B., and Ball, S. (1992). Waxy *Chlamydomonas reinhardtii*: Monocellular algal mutants defective in amylose biosynthesis and granule-bound starch synthase activity accumulate a structurally modified amylopectin. *J. Bacteriol.* **174**, 3612–3620.

Delvallé, D., Dumez, S., Wattebled, F., Roldán, I., Planchot, V., Berbezy, P., Colonna, P., Vyas, D., Chatterjee, M., Ball, S., Mérida, A., and D'Hulst, C. (2005). Soluble starch synthase I: A major determinant for the synthesis of amylopectin in *Arabidopsis thaliana* leaves. *Plant J.* **43**, 398–412.

Denyer, K., Waite, D., Motawia, S., Moller, B.L., and Smith, A.M. (1999). Granule-bound starch synthase I in isolated starch granules elongates malto-oligosaccharides processively. *Biochem. J.* **340**, 183–191.

Deschamps, P., Haferkamp, I., Dauvillée, D., Haebel, S., Steup, M., Buléon, A., Putaux, J.-L., Colleoni, C., D'Hulst, C., Plancke, C., Gould, S., Maier, U., Neuhaus, H.E., and Ball, S. (2006). Nature of the periplastidial pathway of starch synthesis in the cryptophyte *Guillardia theta*. *Eukaryotic Cell,* **5**, 954–963.

Dinges, J.R., Colleoni, C., James, M.G., and Myers, A.M. (2003). Mutational analysis of the pullulanase-type debranching enzyme of maize indicates multiple functions in starch metabolism. *Plant Cell* **15**, 666–680.

Duynstee, E.E. and Schmidt, R.R. (1967). Total starch and amylose levels during synchronous growth of *Chlorella pyrenoidosa*. *Arch. Biochem. Biophys.* **119**, 382–386.

Eddy, B.P., Fleming, I.D., and Manners, D.J. (1958). α-1,4-glucosans. Part IX. The molecular structure of a starch-type polysaccharide from *Dunaliella bioculata*. *J. Chem. Soc.* **2**, 2827–2830.

Fontaine, T., D'Hulst, C., Maddelein, M.L., Routier, F., Pepin, T.M., Decq, A., Wieruszeski, J.M., Delrue, B., van den Koornhuyse, N., Bossu, J.P., Fournet, B., and Ball, S. (1993). Toward an understanding of the biogenesis of the starch granule. Evidence that *Chlamydomonas* soluble starch synthase II controls the synthesis of intermediate size glucans of amylopectin. *J. Biol. Chem.* **268**, 16223–16330.

Goto, K. and Johnson, C.H. (1995). Is the cell division cycle gated by a circadian clock? The case of *Chlamydomonas reinhardtii*. *J. Cell Biol.* **129**, 1061–1069.

Gunning, B.E.S. and Schwartz, O.M. (1999). Confocal microscopy of thylakoid autofluorescence in relation to origin of grana and phylogeny in the green algae. *Aust. J. Plant Physiol.* **26**, 695–708.

Hamilton, B.S., Nakamura, K., and Roncari, D.A.K. (1992). Accumulation of starch in *Chlamydomonas reinhardtii* flagellar mutants. *Biochem. Cell Biol.* **70**, 255–258.

Hatanaka, Y., Inaoka, K., Kobayashi, O., Higashihara, M., and Hiyama, K. (2000). Isolation of disproportionating enzyme from halotolerant microalga *Dunaliella parva* (Volvocales, Chlorophyceae). *Phycol. Res.* **48**, 1–8.

Hirst, E., Manners, D., and Pennie, I.R. (1972). α1,4-D-glucans Part XXI: The molecular structure of starch-type polysaccharide from *Haematococcus pluvialis* and *Tetraselmis carteriiformis*. *Carbohydr. Res.* **22**, 5–11.

Hussain, H., Mant, A., Seale, R., Zeeman, S., Hinchliffe, E., Edwards, A., Hylton, C., Bornemann, S., Smith, A.M., Martin, C., and Bustos, R. (2003). Three isoforms of isoamylase contribute different catalytic properties for the debranching of potato glucans. *Plant Cell* **15**, 133–149.

Iglesias, A.A., Charng, Y.Y., Ball, S., and Preiss, J. (1994). Characterization of the kinetic, regulatory and structural properties of ADP-glucose pyrophosphorylase from *Chlamydomonas reinhardtii*. *Plant Physiol.* **104**, 1287–1294.

James, M.G., Robertson, D.S., and Myers, A.M. (1995). Characterization of the maize gene sugary1, a determinant of starch composition in kernels. *Plant Cell* **7**, 417–429.

Kaplan, F. and Guy, C.L. (2005). RNA interference of *Arabidopsis* beta-amylase prevents maltose accumulation upon cold shock and increases sensitivity of PSII photochemical efficiency to freezing stress. *Plant J.* **44**, 730–743.

Klein, U. (1987). Intracellular carbon partitioning in *Chlamydomonas reinhardtii*. *Plant Physiol.* **85**, 892–897.

Kötting, O., Pusch, K., Tiessen, A., Geigenberger, P., Steup, M., and Ritte, G. (2005). Identification of a novel enzyme required for starch metabolism in *Arabidopsis* leaves. The phosphoglucan, water dikinase. *Plant Physiol.* **137**, 242–252.

Kubo, A., Fujita, N., Harada, K., Matsuda, T., Satoh, H., and Nakamura, Y. (1999). The starch-debranching enzymes isoamylase and pullulanase are both involved in amylopectin biosynthesis in rice endosperm. *Plant Physiol.* **121**, 399–410.

Kreuzberg, K. (1984). Starch fermentation via a formate producing pathway in *Chlamydomonas reinhardtii*, *Chlorogonium elongatum* and *Chlorella fusca*. *Physiol. Plant.* **61**, 87–94.

Kreuzberg, K. and Martin, W. (1984). Oscillatory starch degradation and fermentation in the green alga *Chlamydomonas reinhardtii*. *Biochim. Biophys. Acta,* **799**, 291–297.

Kuchitsu, K., Tsuzuki, M., and Miyachi, S. (1988). Changes of starch localization within the chloroplasts induced by changes in the CO_2 concentration during growth of *Chlamydomonas reinhardtii*: Independent regulation of pyrenoid and stroma starch. *Plant Cell Physiol.* **29**, 1269–1278.

Lee, R.E. (1974). Chloroplast structure and starch grain production as phylogenetic indicators in the lower rhodophyceae. *Br. Phycol. J.* **9**, 291–295.

Levi, C. and Gibbs, M. (1984). Starch degradation in synchronously grown *Chlamydomonas reinhardtii* and characterization of the amylase. *Plant Physiol.* **74**, 459–463.

Libessart, N., Maddelein, M.L., van den Koornhuyse, N., Decq, A., Delrue, B., Mouille, G., D'Hulst, C., and Ball, S.G. (1995). Storage, photosynthesis and growth: The conditional nature of mutations affecting starch synthesis and structure in *Chlamydomonas reinhardtii*. *Plant Cell* **7**, 1117–1127.

Lin, T.P., Caspar, T., Somerville, C., and Preiss, J. (1988a). Isolation and characterization of a starchless mutant of *Arabidopsis thaliana* (L.) Heynh lacking ADP-glucose pyrophosphorylase activity. *Plant Physiol.* **86**, 1131–1135.

Lin, T.P., Caspar, T., Somerville, C., and Preiss, J. (1988b). A starch deficient mutant of *Arabidopsis thaliana* with low ADP-glucose pyrophosphate activity lacks one of the two subunits of the enzyme. *Plant Physiol.* **88**, 1175–1181.

Lloyd, J.R., Springer, F., Buleon, A., Muller-Rober, B., Willmitzer, L., and Kossmann, J. (1999). The influence of alterations in ADP-glucose pyrophosphorylase activities on starch structure and composition in potato tubers. *Planta* **209**, 230–238.

Lloyd, J.R., Blennow, A., Burhenne, K., and Kossmann, J. (2004). Repression of a novel isoform of disproportionating enzyme (stDPE2) in potato leads to inhibition of starch degradation in leaves but not tubers stored at low temperature. *Plant Physiol.* **134**, 1347–1354.

Lorberth, R., Ritte, G., Willmitzer, L., and Kossmann, J. (1998). Inhibition of a starch–granule-bound protein leads to modified starch and repression of cold sweetening. *Nat. Biotechnol.* **16**, 473–477.

Lu, Y. and Sharkey, T.D. (2004). The role of amylomaltase in maltose metabolism in the cytosol of photosynthetic cells. *Planta* **218**, 466–473.

Maddelein, M.L., Libessart, N., Bellanger, F., Delrue, B., D'Hulst, C., van den Koornhuyse, N., Fontaine, T., Wieruszeski, J.M., Decq, A., and Ball, S. (1994). Toward an understanding of the biogenesis of the starch granule. Determination of granule-bound and soluble starch synthase functions in amylopectin synthesis. *J. Biol. Chem.* **269**, 25150–25157.

Maeda, I., Hikawa, H., Mizoguchi, T., and Yagi, K. (2001). Repression of starch degradation under anaerobic conditions by irregularly high levels of ATP in *Chlamydomonas* sp. MGA161. *Plant Sci.* **160**, 629–634.

Manners, D.J., Mercer, G.A., and Stark, J.R. (1965). Studies on the metabolism of protozoa – The molecular structure of a starch-type polysaccharide from *Polytoma uvella*. *Biochem. J.* **96**, 530–532.

Martin, W., Hoffmeister, M., Rotte, C., and Henze, K. (2001). An overview of endosymbiotic models for the origins of eukaryotes, their ATP-producing organelles (mitochondria and hydrogenosomes), and their heterotrophic lifestyle. *Biol. Chem.* **382**, 1521–1539.

Mikkelsen, R., Baunsgaard, L., and Blennow, A. (2004). Functional characterization of α-glucan, water dikinase, the starch phosphorylating enzyme. *Biochem. J.* **377**, 525–532.

Miura, K., Yamano, T., Yoshioka, S., Kohinata, T., Inoue, Y., Taniguchi, F., Asamizu, E., Nakamura, Y., Tabata, S., Yamato, K.T., Ohyama, K., and Fukuzawa, H. (2004). Expression profiling-based identification of CO_2-responsive genes regulated by CCM1 controlling a carbon-concentrating mechanism in *Chlamydomonas reinhardtii*. *Plant Physiol.* **135**, 1595–1607.

Miyachi, S., Tsuzuki, M., Maruyama, I., Gantar, M., Miyachi, S., and Matsushima, H. (1986). Effects of CO_2 concentration during growth on the intracellular structure of *Chlorella* and *Scenedesmus* (Chlorophyta). *J. Phycol.* **22**, 313–319.

Morita, E., Kuroiwa, H., Kuroiwa, T., and Nozaki, H. (1997). High localization of ribulose-1,5-bisphosphate carboxylase/oxygenase in the pyrenoids of *Chlamydomonas reinhardtii* (Chlorophyta), as revealed by immunogold electron microscopy. *J. Phycol.* **33**, 68–72.

Mouille, G., Maddelein, M.L., Libessart, N., Talaga, P., Decq, A., Delrue, B., and Ball, S. (1996). Preamylopectin processing: A mandatory step for starch biosynthesis in plants. *Plant Cell* **8**, 1353–1366.

Myers, A.M., Morell, M.K., James, M.G., and Ball, S.G. (2000). Recent progress toward understanding biosynthesis of the amylopectin crystal. *Plant Physiol.* **122**, 989–997.

Nakamura, Y., Kubo, A., Shimamune, T., Matsuda, T., Harada, K., and Satoh, H. (1997). Correlation between activities of starch de-branching enzyme and alpha-polyglucan structure in endosperms of sugary-1 mutants of rice. *Plant J.* **12**, 143–153.

Niittyla, T., Messerli, G., Trevisan, M., Chen, J., Smith, A.M., and Zeeman, S.C. (2004). A previously unknown maltose transporter essential for starch degradation in leaves. *Science* **303**, 87–89.

Nozaki, H., Onishi, K., and Morita, E. (2002). Differences in pyrenoid morphology are correlated with differences in the *rbcL* genes of members of the *Chloromonas* lineage (Volvocales, Chlorophyceae). *J. Mol. Evol.* **55**, 414–430.

Oyama, Y., Izumo, A., Fujiwara, S., Shimonaga, T., Nakamura, Y., and Tsuzuki, M. (2006). Granule-bound starch synthase cDNA in *Chlorella kessleri* 11h: Cloning and regulation of expression by CO_2 concentration. *Planta* **224**, 646–654.

Pan, D. and Nelson, O.E. (1984). A debranching enzyme deficiency in endosperms of the sugary1 mutants of maize. *Plant Physiol.* **74**, 324–328.

Periappuram, C., Steinhauer, L., Barton, D.L., Taylor, D.C., Chatson, B., and Zou, J. (2000). The plastidic phosphoglucomutase from *Arabidopsis*. A reversible enzyme reaction with an important role in metabolic control. *Plant Physiol.* **122**, 1193–1199.

Plumed, M.P., Villarejo, A., de los Rios, A., García-Reina, G., and Ramazanov, Z. (1996). The CO_2 concentrating mechanism in a starchless mutant of the green unicellular alga *Chlorella pyrenoidosa*. *Planta* **200**, 28–31.

Posewitz, M.C., Smolinski, S.L., Kanakagiri, S., Melis, A., Seibert, M., and Ghirardi, M.L. (2004). Hydrogen photoproduction is attenuated by disruption of an isoamylase gene in *Chlamydomonas reinhardtii*. *Plant Cell* **16**, 2151–2163.

Ral, J.P., Derelle, E., Ferraz, C., Wattebled, F., Farinas, B., Corellou, F., Buléon, A., Slomiany, M.C., Delvalle, D., D'Hulst, C., Rombauts, S., Moreau, H., and Ball, S. (2004). Starch division and partitioning a mechanism for granule propagation and maintenance in the picophytoplanktonic green alga *Ostreococcus tauri*. *Plant Physiol.* **136**, 3333–3340.

Ral, J.P., Colleoni, C., Wattebled, F., Dauvillee, D., Nempont, C., Deschamps, P., Li, Z., Morell, M.K., Chibbar, R., Purton, S., D'Hulst, C., and Ball, S.G. (2006). Circadian clock regulation of starch metabolism establishes GBSSI as a major contributor to amylopectin synthesis in *Chlamydomonas reinhardtii*. *Plant Physiol.* **142**, 305–317.

Ramazanov, A. and Ramazanov, Z. (2006). Isolation and characterization of a starchless mutant of *Chlorella pyrenoidosa* STL-PI with a high growth rate, and high protein and polyunsaturated fatty acid content. *Phycol. Res.* **54**, 255–259.

Ramazanov, Z., Rawat, M., Henk, M.C., Mason, C.B., Matthews, S.W., and Moroney, J.V. (1994). The induction of the CO_2-concentrating mechanism is correlated with the formation of the starch sheath around the pyrenoid of *Chlamydomonas reinhardtii*. *Planta* **195**, 210–216.

Rickoll, W., Rehkopf, D., Dunn, C., Malmberg, A., and VanWinkle-Swift, K. (1998). Note – The *sta1-1* mutation prevents assembly of starch granules in nitrogen-starved cells and serves as a useful morphological marker during sexual reproduction in *Chlamydomonas monoica* (Chlorophyceae). *J. Phycol.* **34**, 147–151.

Ritte, G., Lloyd, J.R., Eckermann, N., Rottmann, A., Kossmann, J., and Steup, M. (2002). The starch related R1 protein is an α-glucan, water dikinase. *Proc. Natl. Acad. Sci. U. S. A.* **99**, 7166–7171.

Ritte, G., Scharf, A., Eckermann, N., Haebel, S., and Steup, M. (2004). Phosphorylation of transitory starch is increased during degradation. *Plant Physiol.* **135**, 2068–2077.

Ritte, G., Heydenreich, M., Mahlow, S., Haebel, S., Kotting, O., and Steup, M. (2006). Phosphorylation of C6- and C3-positions of glucosyl residues in starch is catalysed by distinct dikinases. *FEBS Lett.* **580**, 4872–4876.

Scheidig, A., Fröhlich, A., Schulze, S., Lloyd, J.R., and Kossmann, J. (2002). Downregulation of a chloroplast-targeted β-amylase leads to a starch-excess phenotype in leaves. *Plant J.* **30**, 581–591.

Süss, K.H., Prokhorenko, I., and Adler, K. (1995). *In situ* association of Calvin cycle enzymes, ribulose-1,5-bisphosphate carboxylase/oxygenase activase, ferredoxin-NADP reductase, and nitrite reductase with thylakoid and pyrenoid membranes of *Chlamydomonas reinhardtii* chloroplasts as revealed by immunoelectron microscopy. *Plant Physiol.* **107**, 1387–1397.

Tetlow, I.J., Wait, R., Lu, Z., Akkasaeng, R., Bowsher, C.G., Esposito, S., Kosar-Hashemi, B., Morell, M.K., and Emes, M.J. (2004). Protein phosphorylation in amyloplasts regulates starch branching enzyme activity and protein–protein interactions. *Plant Cell* **16**, 694–708.

Utsumi, Y. and Nakamura, Y. (2006). Structural and enzymatic characterization of the isoamylase1 homo-oligomer and the isoamylase1–isoamylase2 hetero-oligomer from rice endosperm. *Planta* **225**, 75–87.

Thompson, J.D., Higgins, D.G., and Gibson, T.J. (1994). CLUSTAL W: Improving the sensitivity of progressive multiple sequence alignment through sequence weighting, position-specific gap penalties and weight matrix choice. *Nucl. Acids Res.* **22**, 4673–4680.

Thyssen, C., Schlichting, R., and Giersch, C. (2001). The CO_2-concentrating mechanism in the physiological context: Lowering the CO_2 supply diminishes culture growth and economises starch utilisation in *Chlamydomonas reinhardtii*. *Planta* **213**, 629–639.

Tsuzuki, M., Gantar, M., Aizawa, K., and Miyachi, S. (1986). Ultrastructure of *Dunaliella tertiolecta* cells grown under low and high CO_2 concentrations. *Plant Cell Physiol.* **27**, 737–739.

van den Koornhuyse, N., Libessart, N., Delrue, B., Zabawinski, C., Decq, A., Iglesias, A., Preiss, J., and Ball, S. (1996). Control of starch composition and structure through substrate supply in the monocellular alga *Chlamydomonas reinhardtii*. *J. Biol. Chem.* **271**, 16281–16287.

van de Wal, M., D'Hulst, C., Vincken, J.P., Buléon, A., Visser, R., and Ball, S. (1998). Amylose is synthesized *in vitro* by extension of and cleavage from amylopectin. *J. Biol. Chem.* **273**, 22232–22240.

Villarejo, A., Martinez, F., Plumed, M.D., and Ramazanov, Z. (1996). The induction of the CO_2 concentrating mechanism in a starch-less mutant of *Chlamydomonas reinhardtii*. *Physiol. Plant.* **98**, 798–802.

Wattebled, F., Buléon, A., Bouchet, B., Ral, J.P., Lienard, L., Delvalle, D., Binderup, K., Dauvillée, D., Ball, S., and D'Hulst, C. (2002). Granule-bound starch synthase I. A major enzyme involved in the biogenesis of B-crystallites in starch granules. *Eur. J. Biochem.* **269**, 3810–3820.

Wattebled, F., Ral, J.P., Dauvillée, D., Myers, A.M., James, M.G., Schlichting, R., Giersch, C., Ball, S.G., and D'Hulst, C. (2003). *STA11*, a *Chlamydomonas reinhardtii* locus required for normal starch granule biogenesis, encodes disproportionating enzyme. Further evidence for a function of alpha-1,4 glucanotransferases during starch granule biosynthesis in green algae. *Plant Physiol.* **132**, 137–145.

Wattebled, F., Dong, Y., Dumez, S., Delvallé, D., Planchot, V., Berbezy, P., Vyas, D., Colonna, P., Chatterjee, M., Ball, S., and D'Hulst, C. (2005). Mutants of *Arabidopsis* lacking a chloroplastic isoamylase accumulate phytoglycogen and an abnormal form of amylopectin. *Plant Physiol.* **138**, 184–195.

Yu, T.S., Zeeman, S.C., Thorneycroft, D., Fulton, D.C., Dunstan, H., Lue, W.L., Hegemann, B., Tung, S.Y., Umemoto, T., Chapple, A., Tsai, D.L., Wang, S.M., Smith, A.M., Chen, J., and Smith, S.M. (2005). Alpha-amylase is not required for breakdown of transitory starch in *Arabidopsis* leaves. *J. Biol. Chem.* **280**, 9773–9779.

Zabawinski, C., van den Koornhuyse, N., D'Hulst, C., Schlichting, R., Giersch, C., Delrue, B., Lacroix, J.M., Preiss, J., and Ball, S. (2001). Starchless mutants of *Chlamydomonas reinhardtii* lack the small subunit of a heterotetrameric ADP-glucose pyrophosphorylase. *J. Bacteriol.* **183**, 1069–1077.

Zeeman, S.C., Umemoto, T., Lue, W.L., Au-Yeung, P., Martin, C., Smith, A.M., and Chen, J. (1998). A mutant of *Arabidopsis* lacking a chloroplastic isoamylase accumulates both starch and phytoglycogen. *Plant Cell* **10**, 1699–1712.

Zhang, X., Myers, A.M., and James, M.G. (2005). Mutations affecting starch synthase III in *Arabidopsis* alter leaf starch structure and increase the rate of starch synthesis. *Plant Physiol.* **138**, 663–674.

CHAPTER 2

Glycerolipid Biosynthesis

Wayne R. Riekhof[1] *and Christoph Benning*[2]
[1]National Jewish Medical & Research Center, Denver, Colorado, USA
[2]Department of Biochemistry & Molecular Biology, Michigan State University, East Lansing, Michigan, USA

CHAPTER CONTENTS

I. INTRODUCTION

The importance of membranes and membrane biogenesis to cell growth and division cannot be overstated. Cellular membranes act as structural components to define the boundaries of the cell and delineate subcellular compartments. Membranes also serve as scaffolds for diverse signaling events, and participate directly in essential metabolic processes. In eukaryotic and most prokaryotic systems, diacylglycerol-based polar lipids serve as the major building blocks of membranes. In the most thoroughly studied mammalian, fungal, and bacterial model organisms, glycerophospholipids are the major structural components of membranes. In plants and algae, the non-phosphorous galactolipids and sulfolipids of the plastid membranes constitute a significant proportion of the total membrane lipid content, in addition to the phospholipids (Benning, 1998; Dörmann and Benning, 2002). Furthermore, in many bacterial and algal species, the relatively under-investigated betaine lipids are also present as non-phosphorous glycerolipid building blocks of membranes (Klug and Benning, 2001). *Chlamydomonas* spp. membranes consist of a rich complement of glycerolipids, derived from all of the classes mentioned above. In the plastid membranes, the mono- and digalactosyldiacylglycerols (MGDG and DGDG) and sulfoquinovosyldiacylglycerol (SQDG) are present, while in the extraplastidic membranes, phospholipids such as phosphatidylinositol (PtdIns) and phosphatidylethanolamine (PtdEtn), and the non-phosphorous betaine lipid diacylglyceryl-*N,N,N*-trimethylhomoserine (DGTS) are predominant. In Figure 2.1 the structures of these major membrane glycerolipids are summarized. All glycerolipids contain fatty acyl groups esterified in the *sn*-1 and *sn*-2 positions of the glycerol backbone giving rise to the diacylglycerol moiety. Fatty acyl groups are present in different combinations in individual glycerolipids, leading to a number of possible molecular species in each lipid class. The focus of this chapter will be on the predominant pathways of fatty acid biosynthesis, their modification, and on glycerolipid assembly in *Chlamydomonas*.

Much is known about the biosynthesis of glycerolipids in seed plants, and the availability of the *Chlamydomonas* genome sequence enables an informed comparison of lipid metabolism in a unicellular green alga and seed plants. From this comparison, it is apparent that glycerolipid metabolism in *Chlamydomonas* is less complex, making it an attractive model in which to study recalcitrant problems of lipid metabolism in photosynthetic organisms such as interorganelle lipid trafficking and the regulation of lipid homeostasis. A distinguishing feature of lipid metabolism in *Chlamydomonas* is the lack of phosphatidylcholine (PtdCho), whose role appears to be partially assumed by the betaine lipid DGTS. The lack of PtdCho may also explain why chloroplast lipids in *Chlamydomonas* appear to be almost exclusively synthesized directly in the plastid. In contrast, in all seed plants the endoplasmic reticulum is involved in the biosynthesis of at least a fraction of the chloroplast lipids, a process in which PtdCho is thought to act as an intermediate. The biosynthesis of nature's most abundant lipid class, the galactoglycerolipids, appears to be less complex in *Chlamydomonas* as well, because the enzymes involved are not present in multiple isoforms. Taken together, the similarities and differences

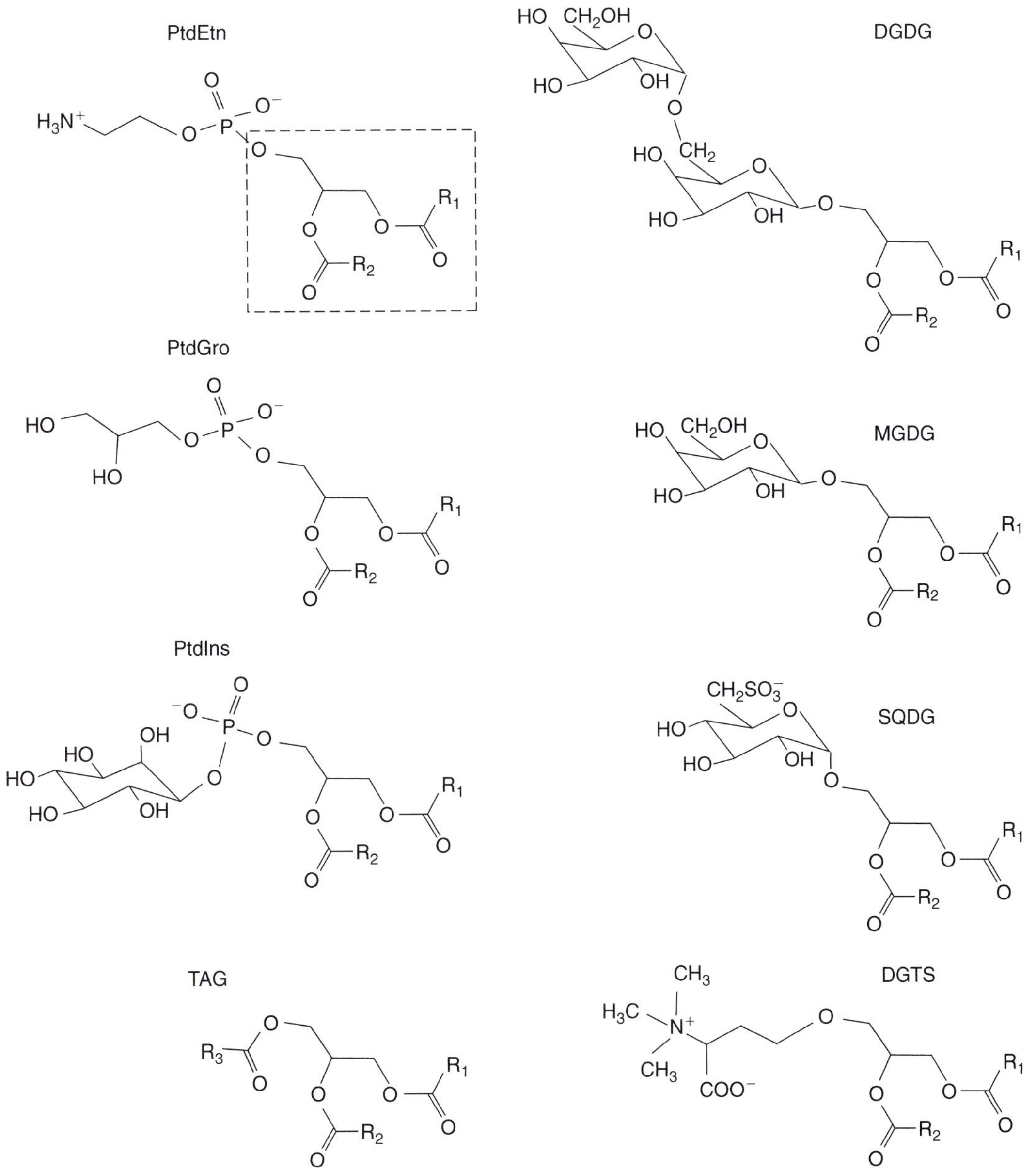

FIGURE 2.1 *Chlamydomonas glycerolipids. DGDG, digalactosyldiacylglycerol; DGTS, diacylglyceryl-N,N,N-trimethylhomoserine; MGDG, monogalactosyldiacylglycerol; PtdEtn, phosphatidylethanolamine; PtdGro, phosphatidylglycerol; PtdIns, phosphatidylinositol; SQDG, sulfoquinovosyldiacylglycerol; TAG, triacylglycerol. As an example, the diacylglycerol moiety of PtdEtn is indicated by a box.*

in lipid metabolism between *Chlamydomonas* and model seed plants such as *Arabidopsis* provide new perspectives on the understanding of lipid metabolism in photosynthetic organisms.

Recent years have seen rapid progress in the experimental elucidation of biosynthetic pathways responsible for membrane lipid biogenesis in

Chlamydomonas, and we will review these biochemical and genetic studies. In addition, the *Chlamydomonas* genome sequence makes it possible to reconstruct its lipid metabolism based on the putative presence or absence of orthologues of genes that are known to encode biosynthetic enzymes in other organisms. An initial report on the reconstruction of lipid metabolism based on sequence data from the *Chlamydomonas* genome project (Riekhof et al., 2005b) forms the basis for the genome sequence-based candidate gene predictions. We combine these *in silico* predictions with a thorough review of the experimentally verified pathways to present a comprehensive view of the identities and functions of genes involved in glycerolipid biosynthesis in *Chlamydomonas*.

II. FATTY ACID BIOSYNTHESIS AND DESATURATION

A. Overview

The common feature of glycerolipids is the presence of a *sn*-1,2-diacylglycerol moiety (Figure 2.1). The acyl chains arise from the action of two large, evolutionarily conserved enzyme complexes: acetyl-CoA carboxylase (ACCase) and fatty acid synthase (FAS). In plants, the fatty acid chain is synthesized in the plastid by a type-II (multipartite bacterial type) FAS (White et al., 2005) followed by either incorporation of the acyl chain into plastid lipids, or export and incorporation of the fatty acid into extraplastidic lipids (Ohlrogge and Browse, 1995). The FAS complex gives rise to a saturated carbon chain, after which the molecule may become a substrate for a number of fatty acid desaturase enzymes. Only a few studies on fatty acid biosynthesis in *Chlamydomonas* are available (Sirevag and Levine, 1972; Giroud and Eichenberger, 1988; Giroud et al., 1988; Grenier et al., 1991; Picaud et al., 1991), and much of our current knowledge on fatty acid biosynthetic enzymes is inferred from the genome database. Several fatty acid desaturases have been isolated and characterized at the molecular level, and the biochemistry and genetics of fatty acid chain elongation and desaturation in *Chlamydomonas* are discussed below.

B. ACCase and FAS

The plastid is known to be the major site of fatty acid synthesis in plant cells (Ohlrogge et al., 1979) and many of the genes encoding enzymes involved in plastidic fatty acid synthesis have been isolated and studied, mostly in *Arabidopsis* (Beisson et al., 2003). Fatty acid synthesis in the plastid occurs via a system very similar to that of bacteria (White et al., 2005), and represents an extant function of the original cyanobacterial ancestor of the chloroplast. As the committed step in fatty acid biosynthesis pathway, malonyl-CoA is produced by the action of ACCase. The predominant plastidic isoform of ACCase is a tetrameric bacterial-type enzyme consisting of biotin carboxylase, biotin carboxyl carrier protein, α-carboxyltransferase, and β-carboxyltransferase (Nikolau et al., 2003). ACCase is regarded as central in regulating the initiation of fatty acid biosynthesis and as such, its activity is controlled at multiple levels (Ohlrogge and Jaworski, 1997). Putative *Chlamydomonas* genes that encode the components of plastidic ACCase are presented in Table 2.1.

Table 2.1 Assignment of candidate *Chlamydomonas* genes encoding enzymes of glycerolipid metabolism[a]

Reaction	Description	Gene	Accession #
PLASTID AND MITOCHONDRIAL PATHWAYS			
1	Acetyl-CoA carboxylase components:		
1a	*Alpha-carboxyltransferase*	*ACX1*	EDP00637
1b	*Beta-carboxyltransferase*	*BCX1*	ED096563
1c	*Biotin carboxylase*	*BCR1*	ED097049
1d	Biotin carboxyl carrier protein	*BXP1 [BCC1]*	ED098131
		BXP2 [BCC2]	EDP09857
2	*Acyl carrier protein* (ACP)	*ACP1*	ED098915
		ACP2	EDP09036
3	*Malonyl-CoA: ACP transacylase*	*MCT1*	EDP09600
4	Type II fatty acid synthase components:		
4a	3-Ketoacyl-ACP synthase	*KAS1*	ED097709
		KAS2	ED098446
		KAS3	ED096631
4b	*3-Ketoacyl-ACP reductase*	*KAR1*	EDP06155
4c	*3-Hydroxyacyl-ACP dehydratase*	*HAD1*	EDP03190
4d	*Enoyl-ACP-reductase*	*ENR1*	ED097343
5	Glycerol-3-P: Acyl-ACP acyltransferase	*ATS1*	EDP02129
6	Lyso-phosphatidate: Acyl-ACP acyltransferase		no candidate
7	CDP-diacylglycerol synthetase	*CDS1 [PCT1]*	EDP04328
8	Phosphatidylglycerolphosphate synthase	*PGP1*	ED097733
		PGP2	ED098928
9	Phosphatidylglycerolphosphate phosphatase		no candidate
10	Phosphatidate phosphatase		no candidate
11	UDP-sulfoquinovose synthase	*SQD1*[b]	ED099781
12	Sulfolipid synthase	*SQD2*	EDP09400
		SQD3	ED098898
13	Sulfolipid 2′-*O*-acyltransferase		no candidate
14	Monogalactosyldiacylglycerol synthase	*MGD1*	EDP09106
15	Digalactosyldiacylglycerol synthase	*DGD1*	EDP08851
16	*Acyl-ACP thioesterase*	*FAT1*	EDP08596
CYTOSOLIC PATHWAYS			
17	Long-chain acyl-CoA synthetase	*LCS1*	ED096800
		LCS2	EDP08946
18	Glycerol-3-P: Acyl-CoA acyltransferase		no candidate
19	Lyso-phosphatidate: Acyl-CoA acyltransferase		no candidate

(Continued)

Table 2.1 *Continued*

Reaction	Description	Gene	Accession #
20	Phosphatidate phosphatase	*PAP1*	EDO97062
		PAP2	EDO99916
		PAH1	EDP05457
21	CDP-diacylglycerol synthetase	*CDS2 [PCT2]*	EDP05553
22	AdoMet synthetase	*SAS1 (METM)*	EDP08638
23	Betaine lipid synthase	*BTA1*[b]	EDP07133
24	CDP-ethanolamine: diacylglycerol ethanolamine phosphotransferase	*EPT1*[b]	EDP03425
25	Serine decarboxylase	*SDC1*	EDP09397
26	Ethanolamine kinase	*EKI1 [ETK1]*	EDP01663
27	CTP: Phosphoethanolamine cytidylyltransferase	*ECT1*[b]	EDP03532
28	Phosphatidylglycerolphosphate synthase	*PGP3*	EDP06071
29	Phosphatidylglycerolphosphate phosphatase		no candidate
30	Inositol-3-phosphate synthase	*INO1*	EDP05930
31	CDP-DAG: Inositol phosphotransferase	*PIS1*	EDP06395
FATTY ACID DESATURASES			
Plastidic isoforms (as judged by ChloroP analysis)			
32	Stearoyl-ACP-Δ9-desaturase	*FAB2*	EDP04705
33	Phosphatidylglycerol palmitate-Δ3t-desaturase		
34	Monogalactosyldiacylglycerol palmitate-Δ7-desaturase	*FAD5 [FAD5A]*	EDP02613
35	ω-6 desaturase	*FAD6a*[b] *[DES6]*	EDP03637
		FAD6b	EDP01638
36	ω-3 desaturase	*FAD7*	EDP09401
37	MGDG 16-carbon Δ4-desaturase		no candidate
Cytosolic isoforms			
38	Oleate desaturase	*FAD2a*	EDP04777
		FAD2b	
39	Linoleate desaturase	*FAD3*	
40	Δ5 Desaturase	*FAD13*[b]	EDP08027
TRIGLYCERIDE SYNTHESIS ENZYMES			
	Acyl-CoA: Diacylglycerol acyltransferase	*DGA1*	EDO96893
	Phospholipid: Diacylglycerol acyltransferase	*LRO1 [LCA1]*	EDP07444

[a]*An earlier version of this table was published by Riekhof et al. (2005b). Assignment of candidate genes was based on putative yeast, Arabidopsis or cyanobacterial orthologs and prediction of subcellular localization as previously described (Riekhof et al., 2005b). Single isoform proteins predicted to be dually targeted to mitochondria and chloroplasts are italicized. The reaction numbers correspond to the numbers in Figures 2.2 and 2.3. Gene names are as previously published (Riekhof et al., 2005b), and correspond to those of homologues from either yeast or Arabidopsis.*

[b]*Genes encoding experimentally confirmed enzymes: SQD1 (Riekhof et al., 2003); BTA1 (Riekhof et al., 2005b); EPT1 (Yang et al., 2004b); ECT1 (Yang et al., 2004a); FAD6a (Sato et al., 1997); FAD13 (Kajikawa et al., 2006).*

Following the synthesis of malonyl-CoA, malonate is transferred to a small phosphopanthetheine-containing polypeptide called acyl carrier protein (ACP). This reaction is catalyzed by malonyl-CoA:ACP trans-acylase (MCT), and the putative homologue from *Chlamydomonas* (Table 2.1) was identified in a proteomic screen for thioredoxin interacting proteins (Lemaire et al., 2004). This finding might be an indication that MCT activity is under the control of the plastid thioredoxin system (see Chapter 11), which would likely inhibit its activity through oxidation of a disulfide bridge in the dark, when energy charge is low. Conversely, it would be activated in the light during optimal growth conditions by reduction of this putative disulfide bridge. This proposed redox control of MCT, coupled with the presumed regulation of ACCase activity, could represent another layer of regulation in the initial steps of fatty acid biosynthesis.

Following the synthesis of malonyl-ACP by MCT, fatty acid synthesis in plastids continues by the action of a type-II (bacterial) heteromultimeric FAS, located in the chloroplast stroma (Ohlrogge et al., 1979; White et al., 2005). Using acyl-ACP and malonyl-ACP as substrates, the first reaction of this complex is a Claisen condensation, catalyzed by 3-ketoacyl-ACP synthase, of the malonyl- and acyl-substituents, releasing carbon dioxide as a byproduct, and forming a carbon–carbon bond to increase the chain length by two carbons. The next steps involve reduction of the 3-keto group, elimination of water from the resulting 3-hydroxy acyl species, and reduction of the enoyl-ACP species by the activities of 3-ketoacyl-ACP reductase, 3-hydroxyacyl-ACP dehydratase, and enoyl-ACP reductase, respectively. This sequence of reactions results in the addition of two fully saturated methylene carbons to the growing acyl chain, and this sequence is repeated in a cyclic fashion with addition of two carbons each cycle until the chain reaches 16 or 18 carbons. Finally, chain elongation is terminated by the action of fatty acyl-ACP thioesterases. Candidate genes for these enzymes are present in the *Chlamydomonas* genome, and are listed in Table 2.1.

Fatty acid synthesis has also been demonstrated to occur in the mitochondria of plants (Wada et al., 1997), where the medium chain fatty acid octanoate serves as the precursor of lipoic acid, a prosthetic group for certain mitochondrial enzymes. In our earlier analysis of fatty acid biosynthesis genes (Riekhof et al., 2005b), we noted that the majority of enzymatic activities for plastidic fatty acid biosynthesis in *Chlamydomonas* are represented by single genes. Given that these activities are likely to be required in both chloroplasts and mitochondria, the simplest explanation is that these "singleton" proteins are targeted to both organelles. These potentially dual-targeted proteins are italicized in Table 2.1.

C. Fatty acid desaturation

The end products of chloroplast fatty acid synthesis are 16- or 18-carbon, saturated or mono-unsaturated fatty acids (Ohlrogge and Browse, 1995). However, the predominant fatty acids in *Chlamydomonas* lipids are polyunsaturated, typically containing two to four double bonds (Giroud and Eichenberger, 1988). The fatty acid desaturases are responsible for production of these polyunsaturated acyl species, and forward and reverse genetic

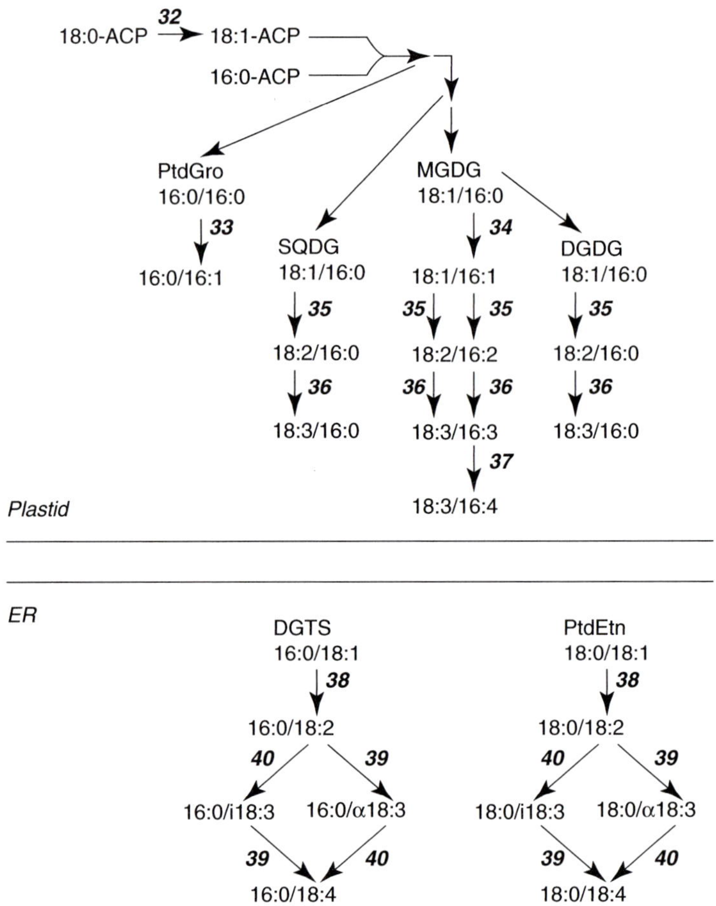

FIGURE 2.2 *Acyl-chain desaturation in Chlamydomonas. This figure is modified with permission from Figure 2 in Riekhof et al. (2005b). Numbered arrows represent desaturation steps, and the numbers correspond to the experimentally verified or predicted gene products listed in Table 2.1. Glycerolipid abbreviations are the same as those in Figure 2.1. Fatty acids are referred to by the standard abbreviation "carbon atoms:double bonds." Specifically, common names of these fatty acids include 16:0, palmitic acid; 18:0, stearic acid; 18:1, oleic acid; 16:1, palmitoleic acid; 18:2, linoleic acid; 18:3, linolenic acid; i18:3, pinolenic acid; 18:4, coniferonic acid.*

approaches have been used to identify a number of fatty acid desaturases in *Chlamydomonas* (Figure 2.2).

A mutant with high chlorophyll fluorescence (*hf-9*) was isolated in a screen for strains with photosynthetic defects, and was shown to be deficient in the plastid ω-6 fatty acid desaturase activity (Sato et al., 1995a). The defective gene was designated as *DES6*, and the wild-type *DES6* gene was cloned and shown to complement the fatty acid desaturase deficiency in the mutant (Sato et al., 1997). The complementing gene did not, however, rescue the photosynthetic defects, and photosystem II (PS II) activity was compromised in this strain regardless of the presence of a functional *DES6* gene. This led to the conclusion that other mutations, secondary to the defect in *DES6*, are responsible for the high chlorophyll fluorescence phenotype.

One of the more unusual aspects of the fatty acid composition of *Chlamydomonas* relative to seed plants is the presence of tetraene fatty acids, and the novel position of the double bonds in these species. The presence of pinolenic acid (PA, $18:3^{\Delta 5,9,12}$) and coniferonic acid (CA, $18:4^{\Delta 5,9,12,15}$) as acyl substituents of the polar lipids of *Chlamydomonas* was reported (Giroud et al., 1988). These atypical fatty acids are predominantly esterified to the extraplastidic lipids DGTS and PtdEtn. A candidate gene approach was taken to identify the Δ5 desaturase involved in PA and CA biosynthesis (Kajikawa et al., 2006). The gene encoding this enzyme was isolated on the basis of similarity to other fungal and algal Δ5 desaturases. Upon its expression in the yeast *Pichia pastoris* or in transgenic tobacco (neither of which contains PA or CA under normal circumstances) a significant amount of both CA and PA was shown to be synthesized. The 20-carbon fatty acids in these organisms were also substrates for the heterologous desaturase, however the *Chlamydomonas* enzyme was shown to introduce a Δ7 double bond in these species. Given that the Δ5 position of a C18 and the Δ7 position of a C20 fatty acid are 13 carbons from the terminal methyl unit, the authors designated the *Chlamydomonas* protein as an ω-13 fatty acid desaturase (Table 2.1, *FAD13*). This enzyme is not active on C16 fatty acids, therefore the identity of the Δ4 desaturase responsible for the production of $16:4^{\Delta 4,7,10,13}$, present in MGDG in *Chlamydomonas* (Giroud et al., 1988), remains unknown.

Several genes in the *Chlamydomonas* genome also encode proteins with similarity to characterized fatty acid desaturase enzymes from other plants. These include the stearoyl-ACP Δ9 desaturase, the plastidic ω-3 desaturase, and the extraplastidic ω-6 and ω-3 desaturases. A detailed pathway for fatty acid desaturation in *Chlamydomonas* is given in Figure 2.2, and the identities of the experimentally verified genes, as well as those based on similarity to fatty acid desaturases characterized from other organisms are given in Table 2.1.

III. GLYCEROLIPID BIOSYNTHESIS

A. Glycerophospholipids

1. Overview

Glycerophospholipids are the prevalent lipids in non-photosynthetic organisms. They are characterized by a phosphatidic acid (PtdOH) backbone. In plants approximately one-third of the organic phosphorus is found in phosphoglycerolipids (Poirier et al., 1991). This places a substantial burden on the phosphate homeostasis of a plant cell, explaining the abundance of non-phosphorous glycerolipids in photosynthetic tissues. The *de novo* synthesis of all glycerolipids involves the formation of PtdOH, from

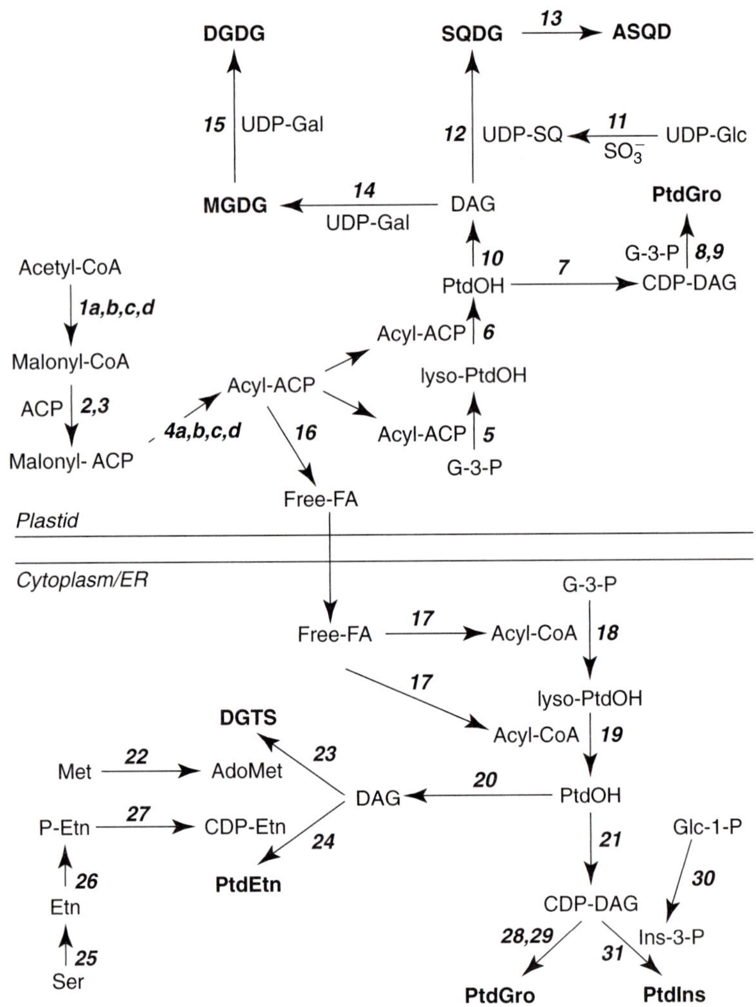

FIGURE 2.3 *Glycerolipid biosynthesis in Chlamydomonas. This figure is modified with permission from Figure 1 in Riekhof et al. (2005b). Compounds in bold correspond to the structures presented in Figure 2.1. Numbered arrows refer to enzymatic transformations except for* **2** *(acyl carrier protein), which is a structural component of the fatty acid synthesis machinery. The numbers correspond to the experimentally verified or predicted gene products listed in Table 2.1. Abbreviations: ACP, acyl carrier protein; AdoMet, S-adenosylmethionine; ASQD, 2′-O-acyl-sulfoquinovosyldiacylglycerol; CDP, cytidine-5′-diphosphate; CoA, coenzyme A; CTP, cytidine-5′-triphosphate; DAG, diacylglycerol; DGDG, digalactosyldiacylglycerol; DGTS, diacylglyceryl-N,N,N-trimethylhomoserine; Etn, ethanolamine; FA, fatty acid; G-3-P, glycerol-3-phosphate; Glc, glucose; Glc-1-P, glucose-1-phosphate; Ins, inositol; Ins-3-P, inositol-3-phosphate; Met, methionine; MGDG, monogalactosyldiacylglycerol; P-Etn, phosphoethanolamine; PtdEtn, phosphatidylethanolamine; PtdGro, phosphatidylglycerol; PtdGroP, phosphatidylglycerolphosphate; PtdIns, phosphatidylinositol; PtdOH, phosphatidic acid; Ser, serine; SQ, sulfoquinovose; SQDG, sulfoquinovosyldiacylglycerol; UDP, uridine-5′-diphosphate.*

which phosphate is lost during the synthesis of the diacylglycerol moiety of non-phosphorus glycerolipids. In fact, PtdOH is the central metabolite in the biosynthesis of all glycerolipids in both photosynthetic and non-photosynthetic organisms. The basic pathways are shown in Figure 2.3, and the corresponding enzymes and their respective genes are listed in Table 2.1.

2. Assembly of PtdOH and conversion to DAG and CDP-DAG

As described above, the acyl-ACPs produced from plastidic FAS may become substrates for the plastid glycerol 3-phosphate acyltransferase (GPAT) and lyso-PtdOH acyltransferase (LPAAT) enzymes, giving rise to the central glycerolipid precursor PtdOH. These enzymatic activities have not been studied directly in *Chlamydomonas*; however studies from other organisms may explain the *sn*-1,2 acyl chain distribution that has been observed (Giroud et al., 1988). The enzyme GPAT has been isolated from a number of plant species (Murata and Tasaka, 1997) and shown to prefer oleoyl-ACP as a substrate. Mutations in the *Arabidopsis* plastid GPAT (encoded by the *ACT1/ATS1* gene) give rise to plants that are unable to synthesize diacylglycerol moieties in the plastid, forcing an increase in flux through the so-called "eukaryotic" pathway of thylakoid lipid synthesis (Kunst et al., 1988). The second step in this sequence is catalyzed by LPAAT, and this enzyme shows higher affinity for palmitoyl-ACP (Kim and Huang, 2004). All GPAT mutant alleles are viable and grow essentially like wild type, presumably due to the increase in flux through the eukaryotic pathway (Kunst et al., 1988; Xu et al., 2006), whereas null alleles of the LPAAT gene cause an embryo-lethal phenotype (Kim and Huang, 2004; Yu et al., 2004). The combined substrate specificities of the two plastidic acyltransferases result in a DAG (diacylglycerol) *sn*-1/*sn*-2 molecular species preference of 18:1/16:0, and these are the predominant PtdOH species initially produced by the plastidic or "prokaryotic" pathway. PtdOH produced by this sequence can then be converted to CDP-DAG and used directly for phosphatidylglycerol (PtdGro) biosynthesis, or can be dephosphorylated by the action of phosphatidic acid phosphatase, forming DAG as a substrate for galactolipid and sulfolipid biosynthesis. *Chlamydomonas* orthologues of genes encoding these activities are listed in Table 2.1.

3. PtdGro synthesis

PtdGro synthesis in plants has been thoroughly studied in the context of its role in the structure and function of the thylakoid membrane (Frentzen, 2004). PtdGro is a key anionic constituent of the photosynthetic membrane, and mutants of *Arabidopsis* with reduced (Xu et al., 2002) or loss of (Hagio et al., 2002) PtdGro synthesis show either a profound decrease or a complete lack of photosynthetic activity, respectively. Mutants of *Chlamydomonas* affecting both PtdGro content and photosynthesis were also reported (Dubertret et al., 1994, 2002; El Maanni et al., 1998). These mutants, designated *mf-1* and *mf-2*, are deficient in a form of PtdGro that is only found in plastids, and show a reduced total PtdGro content. One of the thylakoid-specific molecular species of PtdGro contains a $16{:}1^{\Delta 3t}$ fatty acid, and the *mf-1* and *mf-2* mutants are specifically deficient in this form of PtdGro. This deficiency correlates with a significant reduction in thylakoid stacking and impairment of PS II function. The *mf-1* and *mf-2* mutants are allelic, indicating that a single mutant locus gives rise to both the lipid phenotype and the photosynthetic deficiencies (Pineau et al., 2004). However, given that this mutant gene has not yet been identified at the molecular level, the question of cause and effect becomes an important one; that is, does the PtdGro deficiency cause the photosynthetic defect, or does a specific mutation in some other component of the photosynthetic machinery lead to a lack of PtdGro accumulation? This question was addressed by feeding of $16{:}1^{\Delta 3t}$-PtdGro to cultures of the *mf-2* mutant (Dubertret et al., 1994). Surprisingly, this mutant can apparently transport the intact lipid into the cell, traffic it intracellularly to the chloroplast, and incorporate the exogenous lipid into the thylakoid membrane. This results in an abrogation of the severity of the photosynthetic defect, suggesting that the lack of the lipid is at least partially responsible for the PS II defects. A corresponding mutant, *fad4*, is available in *Arabidopsis* (Browse et al., 1985), but the respective gene has not been identified to date.

Aside from the plastid, PtdGro biosynthesis is thought to take place in the mitochondrion and the endoplasmic reticulum (ER). Three isoforms of the phosphatidylglycerolphosphate synthase are encoded in the genome (Table 2.1). Two of these are expressed as judged by EST analysis, and both contain a predicted plastid or mitochondrial targeting sequence (Riekhof et al., 2005b). The other is predicted to be cytosolic, but is not represented in the EST database. This may correlate with the fact that the mass of PtdGro present in the endomembrane system is predicted to be much less than that in the plastid or mitochondria (Giroud et al., 1988).

4. PtdSer, PtdEtn, and PtdCho synthesis

Aminoglycerophospholipids are basic building blocks of extraplastidic membranes in plants. Aminoglycerophospholipid metabolism in *Chlamydomonas* differs from that of seed plants by the absence of distinct pathways or branches of pathways (Figure 2.3). The most notable absence is phosphoethanolamine (phospho-Etn) *N*-methyltransferase (PEMT), which typically provides the pathway for methylation of phospho-Etn to form phospho-Cho. The PEMT enzymes from spinach and *Arabidopsis* have been functionally identified by their ability to rescue a *cho2* mutant of *Schizosaccharomyces pombe* or an *opi3* mutant of *Saccharomyces cerevisiae*, respectively (Bolognese and McGraw, 2000; Nuccio et al., 2000). The activity of PEMT generates phosphocholine (phospho-Cho) for conversion into CDP-choline followed by transfer of phosphocholine onto DAG, giving rise to PtdCho. The apparent lack of a pathway for PtdCho precursor biosynthesis in *Chlamydomonas*, as well as a lack of the PtdEtn methylation pathway, correlates with a lack of PtdCho from cellular membranes (Giroud et al., 1988). In the absence of PtdCho, the betaine lipid DGTS has been proposed as the bilayer-forming and major structural component of extraplastidic membranes (Giroud and Eichenberger, 1987; Sato and Murata, 1991). The biosynthesis and function of this lipid is discussed in detail below.

Phosphatidylserine (PtdSer) is a major precursor of PtdEtn in yeast, bacteria, and plants, and its synthesis is mediated by the action of PtdSer decarboxylase (PSD). A mitochondrial PSD enzyme has been isolated and characterized from tomato (Rontein et al., 2003). However, *Chlamydomonas* has been reported to lack PtdSer as a component of its membranes (Giroud et al., 1988). This assertion is consistent with the apparent lack of genes encoding both PtdSer synthase and phospholipid base exchange enzymes. This suggests that the biosynthesis of PtdEtn is the function of a single pathway consisting of serine decarboxylase, which is a well-described enzyme in plants (Rontein et al., 2001). This pathway generates ethanolamine (Etn), followed by phosphorylation of Etn by Etn kinase, activation of phospho-Etn with CTP to form CDP-Etn by phospho-Etn:CTP cytidylytransferase, and transfer of the phospho-Etn moiety to DAG, forming PtdEtn by the action of CDP-Etn:DAG Etn phosphotransferase. The *Chlamydomonas* enzymes CTP:phospho-Etn-cytidylyltransferase and DAG:Etn-phosphotransferase have been characterized through expression in *E. coli* and biochemical characterization of the recombinant proteins (Yang et al., 2004a,b). Additionally, putative serine decarboxylase and Etn kinase-encoding genes are predicted by genome annotation (Table 2.1).

5. PtdIns synthesis

Inositol supply and phosphatidylinositol (PtdIns) synthesis are key players in the regulation of lipid homeostasis in *S. cerevisiae* (Jesch et al., 2005), and PtdIns (and its phosphorylated forms) play important roles in signal transduction processes and membrane anchoring of proteins in plants (Stevenson et al., 2000; Mueller-Roeber and Pical, 2002; Fischer et al., 2004). Aspects of PtdIns phosphorylation and signaling during culture growth have been investigated in *Chlamydomonas eugametos* (Munnik et al., 1994) and *C. moewusii* (Munnik et al., 1994, 1998; Meijer et al., 2001) and the unicellular red alga *Galdieria sulphuraria* (Heilmann et al., 1999, 2001). In *C. reinhardtii*, mechanisms of PtdIns formation were studied in cell extracts (Blouin et al., 2003). These experiments provided evidence for the presence of a CDP-DAG:*myo*-inositol 3-phosphatidyltransferase and a manganese-dependent PtdIns:*myo*-inositol exchange reaction. Genes encoding proteins proposed to be involved in PtdIns biosynthesis (Figure 2.3) are listed in Table 2.1.

B. Non-phosphorous glycerolipids

1. Overview

Chlamydomonas contains a diverse complement of polar membrane lipids. These include the already-mentioned glycerophospholipids that are typical of other eukaryotes, glycolipids that are common to plants and cyanobacteria, as well as betaine lipids. The latter are of limited phylogenetic distribution, but nonetheless common in soil-dwelling bacteria and aquatic algae (Klug and Benning, 2001). Some of these lipids are the most common membrane building blocks found in nature. They are prevalent in photosynthetic membranes and might help photosynthetic organisms to cope with phosphate-limiting growth conditions by providing a non-phosphorous alternative to phosphoglycerolipids (Benning et al., 1993; Güler et al., 1996; Härtel et al., 1998, 2000; Yu and Benning, 2003). For example, the abundance of sulfolipid in photosynthetic picoplankton provides a distinct advantage for this abundant organism group over heterotrophic organisms in the North Pacific gyre (Van Mooy et al., 2006). Several of the pathways were directly investigated in *Chlamydomonas* and are shown in Figure 2.3. The corresponding enzymes and their respective genes are listed in Table 2.1.

2. MGDG synthesis

The most abundant membrane lipid in the biosphere is monogalactosyldiacylglycerol (MGDG; Figure 2.1), which constitutes about 50% of the polar lipids of the thylakoid membrane in most organisms with oxygenic photosynthesis (Gounaris and Barber, 1983; Joyard et al., 1998; Dörmann and Benning, 2002). The enzyme responsible for MGDG synthesis in plants was characterized as a component of the chloroplast inner envelope membrane, and a thorough biochemical characterization was carried out on partially purified fractions from spinach (Marechal et al., 1994, 1995). Cloning of the gene encoding MGDG synthase was accomplished by purifying the enzyme from expanding cucumber leaves, partial sequencing of the protein, and using this information to amplify a fragment of the gene by degenerate PCR (Shimojima et al., 1997). *Chlamydomonas* encodes a homologue of this *MGD1* gene (Table 2.1).

MGDG synthase is a member of the glycosyltransferase superfamily, catalyzing the condensation of UDP-galactose and DAG with inversion of the anomeric configuration, producing a β-galactosidic linkage (Miege et al., 1999; Awai et al., 2001). It should be noted that cyanobacteria lack this enzyme. Instead, they first synthesize monoglucosyldiacylglycerol, which is subsequently converted to MGDG by an epimerase. The responsible monoglucosyl:DAG acyltransferase was recently identified (Awai et al., 2006a). Without further biochemical data it cannot be ruled out at this time that some algae might use this pathway instead of the plant type MGDG synthase. *Arabidopsis* and other plant genomes contain two paralogues of

MGD1 (denoted *MGD2* and *MGD3*), and one paralogue of the DGDG synthase gene *DGD1* (denoted *DGD2*, see below) (Benning and Ohta, 2005). In addition, there is a processive galactosyltransferase involved in the biosynthesis of oligogalactolipids in plants. However, galactolipid biosynthesis in *Chlamydomonas* might be less complex as only single *MGD* and *DGD* genes are present (Table 2.1).

3. DGDG synthesis

In most plants, the second most abundant lipid in the plastid is derived from MGDG by addition of a second galactose moiety. This galactose is bound in a 1-α-glycosidic linkage to carbon 6 of the first galactose indicating the involvement of a retaining, non-processive galactosyltransferase. Digalactosyldiacylglycerol (DGDG; Figure 2.1) is a bilayer-forming lipid. When suspended in water, it tends to form a lamellar phase due to the tight packing of approximately cylindrical adjacent monomers (Webb and Green, 1991). In contrast, MGDG is a non-bilayer forming lipid, and its overall space filling in a conical shape allows pure preparations to attain inverted hexagonal phases in aqueous solution. The bilayer-to-non-bilayer ratio of biological membranes is typically under tight control, but structurally different non-bilayer forming lipids were found to substitute for each other functionally (Wikstrom et al., 2004; Xie et al., 2006). As the two galactolipids DGDG and MGDG are the predominant bilayer and non-bilayer forming lipids in photosynthetic membranes, respectively, the enzyme which forms DGDG from MGDG is expected to be at a control point in determining the stability of the thylakoid membrane bilayer.

To identify the plant DGDG synthase-encoding gene, a direct biochemical genetic screen was carried out involving thin-layer chromatography of lipids extracted from a mutagenized *Arabidopsis* population. This approach resulted in isolating a mutant with a 90% reduction in the amount of DGDG (Dörmann et al., 1995). Genetic mapping of the mutant locus identified the *DGD1* gene, and the activity of its product was confirmed in *E. coli* by co-expression with the *MGD1* gene from cucumber (Dörmann et al., 1999), which led to synthesis of DGDG in *E. coli*. The DGD1 protein is characterized by a galactosyltransferase motif in its C-terminal domain, implying that it uses UDP-galactose as a substrate as directly demonstrated for the DGD1 paralog DGD2 (Kelly et al., 2003). As with MGDG synthase described above, *Chlamydomonas* contains a *DGD1* gene (Table 2.1), however biochemical analysis of the encoded protein will be required to verify its role in galactolipid synthesis.

4. SQDG and ASQD synthesis

The plant sulfolipid sulfoquinovosyldiacylglycerol (SQDG, Figure 2.1), was discovered by Benson et al. (1959). As a component of the thylakoid membrane, it is one of the most abundant sulfonate compounds in the biosphere (Heinz, 1993). Genes involved in sulfolipid biosynthesis were first discovered in the purple bacterium *Rhodobacter sphaeroides* (Benning and Somerville, 1992a,b) and orthologues were subsequently isolated from cyanobacteria (Güler et al., 1996, 2000), plants (Essigmann et al., 1998; Yu et al., 2002) and *Chlamydomonas* (Sato et al., 2003b). In cyanobacteria, plants, and presumably *Chlamydomonas*, two enzymes catalyze SQDG biosynthesis. In *Arabidopsis* SQD1 produces the headgroup donor UDP-sulfoquinovose from UDP-glucose and sulfite (Mulichak et al., 1999; Sanda et al., 2001). In spinach, this protein is found in a complex with ferredoxin-dependent glutamate synthase (Shimojima and Benning, 2003; Shimojima et al., 2005), but the significance of this interaction is not yet clear. In *Arabidopsis*, the sulfoquinovosyl headgroup is transferred from UDP-sulfoquinovose onto DAG by the action of an enzyme encoded by *SQD2* (Yu et al., 2002). Simultaneous expression of *Arabidopsis* cDNAs encoding SQD1 and SQD2 led to the formation of SQDG in *E. coli*, demonstrating that these proteins are necessary and sufficient for SQDG biosynthesis (Yu et al., 2002). Both enzymes are localized in the plastid in plants and constitute the key enzymes of the nucleotide pathway of SQDG biosynthesis as diagrammed in Figure 2.3. The corresponding *Chlamydomonas* enzymes are listed in Table 2.1.

The presence of a unique sulfolipid in photosynthetic membranes suggested that it might have a specific role in photosynthesis. However sulfolipid-deficient mutants of *R. sphaeroides*, *Synechococcus*, and *Arabidopsis* are photosynthetically competent but cannot grow as well under phosphate-limited conditions (Benning et al., 1993; Güler et al., 1996; Yu et al., 2002). On the other hand, a mutant of *Synechocystis* PCC 6803 in which the *sqdB* gene is inactivated behaves as a sulfolipid auxotroph, indicating that SQDG is an essential factor for growth of this species (Aoki et al., 2004). This leads to the conclusion that there is no universal requirement for SQDG in photosynthesis, but that it is conditionally important in a species-specific manner.

In *Chlamydomonas* the *hf-2* mutant was isolated during a genetic screen for high chlorophyll fluorescence, and was later shown to be deficient in the biosynthesis of SQDG (Sato et al., 1995b). Further characterization indicated that *hf-2* has defects in the heat stability of PS II (Sato et al., 2003a), and that it is more sensitive to the herbicide DCMU, a competitive inhibitor of the Q_B plastoquinone binding site of PS II. Taken together, these data led to the conclusion that SQDG is involved in maintaining the stability of PS II. In addition, the *Chlamydomonas* cytochrome b_6f complex crystal structure revealed a sulfolipid molecule in tight association with cytochrome *f* (Stroebel et al., 2003). The sulfolipid sulfonate group forms a salt bridge with a lysine residue that was shown to be critical for the co-translational insertion of this polypeptide into the thylakoid membrane (Choquet et al., 2003), perhaps indicating that sulfolipid has a role as a membrane protein chaperone during the folding process, similar to PtdEtn acting as a chaperone for multiple-membrane spanning proteins in *E. coli* (Bogdanov and Dowhan, 1999). In short, sulfolipids have been implicated in the functions of two of the core components of the photosynthetic electron transport chain and while not necessarily essential, might have a protective function when the photosynthetic apparatus is under stress.

In addition to the functional studies on *hf-2*, which remains to be fully analyzed at the molecular level, a null mutant of the *SQD1* gene was isolated in a high-throughput genetic screen for lipid-deficient *Chlamydomonas* mutants (Riekhof et al., 2003). This strain was generated as an insertion mutant, and the disrupting cassette was shown to have concurrently caused complete deletion of the *SQD1* gene. As with SQDG-deficient mutants in other species, the *Chlamydomonas sqd1*Δ mutant was characterized by a growth defect on phosphate-limited media, as well as sensitivity to DCMU.

Surprisingly, lipid analysis after $^{35}SO_4$ labeling of the mutant and wild-type strains revealed a second labeled compound absent in *sqd1*Δ, which chromatographed with higher mobility than SQDG, indicating a likely product or precursor relationship to SQDG. Mass spectrometry and nuclear magnetic resonance spectroscopy revealed the structure to be an acylated form of SQDG, 2′-*O*-acyl-sulfoquinovosyldiacylglycerol (ASQD) (Riekhof et al., 2003). The fatty acid composition of ASQD was unique relative to SQDG and as such, this derivative of SQDG may have a specific, but as of yet undetermined function. The isolation of a mutant that is defective only in ASQD will be critical to the deconvolution of functions specific to SQDG and ASQD.

5. DGTS synthesis

Another group of polar non-phosphorous glycerolipids, the betaine lipids, consist of DAG bound in an ether linkage to a quaternary amine alcohol. These lipids are widely distributed among marine and freshwater algae, mosses, ferns, and other non-vascular plants, as well as among many fungal and bacterial species (Klug and Benning, 2001). The structure of a representative member of this class, diacylglyceryl-*N,N,N*-trimethylhomoserine (DGTS), is given in Figure 2.1.

The quaternary amine-containing, zwitterionic betaine lipids are similar in overall structure to PtdCho, and it has been postulated that betaine lipids such as DGTS can functionally replace PtdCho in membranes (Sato and Murata, 1991). Indeed, this is consistent with the fact that *Chlamydomonas* lacks PtdCho. Further correlative evidence comes from studies on *R. sphaeroides* which when deprived of phosphate, decreases the amount of PtdCho and other phospholipids in its membranes, and induces *de novo* synthesis of DGTS and other non-phosphorous lipids (Benning et al., 1995). Studies following up on the discovery of DGTS in *R. sphaeroides* focused on defining its biosynthetic pathway (Hofmann and

Eichenberger, 1996), and subsequently *btaA* and *btaB* were identified as encoding DGTS biosynthesis enzymes (Klug and Benning, 2001). Another study defined the regulatory elements that control DGTS induction in *Sinorhizobium melilotii*, and showed that PHO (phosphate utilization) regulon-mediated perception of phosphorus availability is the determining factor for the presence or absence of DGTS in this organism (Geiger et al., 1999). The *btaA* gene is proposed to encode an *S*-adenosylmethionine: DAG 3-amino-3-carboxypropyl transferase leading to the formation of diacylglycerylhomoserine. The *btaB* gene encodes an *S*-adenosylmethionine-dependent *N*-methyltransferase that converts diacylglycerylhomoserine in three consecutive methylation reactions into DGTS (Figure 2.3).

The DGTS biosynthetic pathway has also been studied in the eukaryotic algae *Ochromonas danica* (Vogel and Eichenberger, 1992) and *Chlamydomonas* (Sato, 1988; Moore et al., 2001). In *Chlamydomonas*, *BTA1* appears to encode both activities needed for the synthesis of DGTS (Riekhof et al., 2005a). The Chlamydomonas BTA1 protein has two domains, one of which is homologous to *R. sphaeroides* BtaA, and the other to BtaB (Klug and Benning, 2001; Riekhof et al., 2005b). The functions of the BtaA and BtaB proteins have been elucidated by deletion analysis in *R. sphaeroides* (Klug and Benning, 2001), and by heterologous expression in *E. coli* (Riekhof et al., 2005b), and were shown to be necessary and sufficient for DGTS synthesis. The *Chlamydomonas* BTA1 protein was also characterized by expression in *E. coli*, and was shown to be sufficient for DGTS synthesis in the heterologous host (Riekhof et al., 2005a).

The finding that a single protein contains all of the activities for synthesis of the major extraplastidic lipid in *Chlamydomonas* would seem to be an example of evolutionary parsimony, given that bacteria encode the same activity in two separate open reading frames. In *Chlamydomonas*, the fusion of activities would simplify regulation, and achieve the same end result as regulation of the bacterial operon. The regulation of DGTS biosynthesis *per se* has not been investigated in *Chlamydomonas*, but given the complex membrane composition in this organism, this process is likely to be affected by multiple inputs. Elucidation of the mechanisms of maintaining lipid homeostasis is one of the major areas in which our knowledge is severely lacking, and studying the control of *Chlamydomonas* BTA1 activity may provide clues that will lead to insights in other organisms.

6. Triacylglycerol synthesis

All glycerolipids discussed thus far are components of membranes. However, one ubiquitous non-membrane glycerolipid class is represented by triacylglycerols (TAG), neutral lipids that in general serve as storage form for acyl groups, forming oil droplets or oleosomes in the cytosol. In *Chlamydomonas* TAGs accumulate under conditions of nutrient deprivation (Weers and Gulati, 1997) or high light (Picaud et al., 1991).

Chlamydomonas can also synthesize TAGs from lipids supplied in the medium (Grenier et al., 1991). In *Arabidopsis* seeds TAG biosynthesis is catalyzed by a DAG:acyl-CoA acyltransferase encoded by *TAG1* (Routaboul et al., 1999; Hobbs et al., 1999). An orthologue of *TAG1* could not be predicted in the available *Chlamydomonas* genome. However, an homologue of the yeast DAG:acyl-CoA acyltransferase Dga1p (Sorger and Daum, 2002) is encoded (Table 2.1), as well as a gene encoding a putative phosphoglycerolipid:DAG acyl exchange enzyme (Dahlqvist et al., 2000), representing a second mechanism of TAG biosynthesis in yeast. An ortholog for this latter enzyme is also present in *Arabidopsis* (Stahl et al., 2004). A recently identified magnesium-dependent PtdOH phosphatase in yeast is encoded by *PAH1*. This orthologue of mammalian lipin is involved in providing the DAG substrate for TAG formation catalyzed by Dga1p (Han et al., 2006); mutants disrupted in *PAH1* produce less TAG. A homologue of *PAH1* encoded in *Chlamydomonas* could be involved in TAG biosynthesis.

IV. COMPARTMENTALIZATION OF LIPID BIOSYNTHESIS

In *Arabidopsis*, plastid glycoglycerolipids are assembled from DAG precursors that are either synthesized *de novo* inside the plastid or at the ER (Roughan and Slack, 1982; Frentzen, 1986). Lipid species derived from the plastid pathway contain a 16-carbon acyl chain and those derived from the ER pathway an 18-carbon acyl chain at the *sn*-2 position of the glycerol backbone (Heinz and Roughan, 1983). Because of the similarity of the acyl composition of lipids assembled in the plastid to lipids in cyanobacteria, the plastid pathway is called the "Prokaryotic Pathway," while the ER pathway is also referred to as the "Eukaryotic Pathway." Plastid lipids are exclusively derived from the ER pathway in a large number of plants (Mongrand et al., 1998). However, no seed plant is known in which plastid lipids are exclusively derived from the plastid pathway. Based on lipid analysis in *Chlamydomonas*, it appears that plastid glycoglycerolipids are exclusively assembled in the plastid (Giroud et al., 1988). As it has been proposed that PtdCho is a crucial intermediate for the formation of ER-derived plastid lipids (Roughan and Slack, 1982), the lack of PtdCho in *Chlamydomonas* would explain the lack of ER-derived plastid lipids. Apparently, DGTS present in extraplastidic membranes cannot take over this role of PtdCho.

The compartmentalization of lipid metabolism in plants requires a host of interorganelle trafficking processes, most of which are still poorly understood (Kelly and Dörmann, 2004; Benning et al., 2006). Of note are two *Arabidopsis* proteins, TGD1 and TGD2, which are components of a proposed PtdOH transport complex in the inner chloroplast envelope membrane (Xu et al., 2003, 2005; Awai et al., 2006b). *Arabidopsis* mutants deficient in either of the two proteins show a complex phenotype consistent with a disruption of the ER pathway of plastid lipid biosynthesis. Thus, in *Arabidopsis* this TGD1/TGD2 transporter is involved in providing ER-derived precursors to the lipid assembly machinery in the plastid. Consistent with a lack of the ER-derived plastid lipids (Giroud et al., 1988), no homologues of TGD1 or TGD2 are encoded in the *Chlamydomonas* genome.

V. PERSPECTIVES

Based on the discussion above, it seems clear that *Chlamydomonas* is a more simple system than *Arabidopsis* to study basic principles of lipid trafficking and regulation of membrane lipid homeostasis. For example, in the extraplastidic membranes, PtdEtn and DGTS are the predominant components. The ratio of these two lipids is probably critical to maintaining the correct lipid phase of the membrane, that is, the correct ratio of bilayer-forming (DGTS) and non-bilayer forming (PtdEtn) lipids. Given the simplicity of the biosynthesis of these lipids from DAG, catalyzed by one enzyme for each (see Figure 2.3), *Chlamydomonas* could serve as a simple model for our understanding of the control of flux from precursors to products in a tractable system. Similarly, *Chlamydomonas* could be used to identify components of the plastidic fatty acid export machinery, as present knowledge is mainly based on labeling data with intact leaves (Koo et al., 2004).

One of the most prominent factors controlling membrane lipid composition in plants and photosynthetic bacteria is the availability of phosphate (Benning et al., 1993a; Güler et al., 1996; Härtel et al., 1998, 2000; Yu et al., 2002; Yu and Benning, 2003). Similar changes in lipid composition in response to phosphate are observed for *Chlamydomonas* (Sato et al., 2000; Riekhof et al., 2003). Mechanisms of adaptation to nutrient deprivation are well studied in *Chlamydomonas* (see Chapters 5 and 6) and mutants impaired in the response to phosphorus or sulfur deprivation have been isolated (Davies et al., 1994; Shimogawara et al., 1999). These led to the isolation of central regulators (Davies et al., 1996, 1999; Wykoff et al., 1999). It might be rewarding to explore whether these regulatory factors are also involved in the regulation of membrane lipid composition in *Chlamydomonas*.

ACKNOWLEDGMENTS

Work on plant and algal lipids in the Benning lab is supported in parts by grants from the US-National Science Foundation, the US-Department of Energy, BASF-Plant Sciences, the US Airforce Office of

Scientific Research, the MSU Center of Excellence for the Structural Biology Membrane Proteins, and the Michigan Agricultural Experiment Station.

REFERENCES

Aoki, M., Sato, N., Meguro, A., and Tsuzuki, M. (2004). Differing involvement of sulfoquinovosyl diacylglycerol in photosystem II in two species of unicellular cyanobacteria. *Eur. J. Biochem.* **271**, 685–693.

Awai, K., Marechal, E., Block, M.A., Brun, D., Masuda, T., Shimada, H., Takamiya, K., Ohta, H., and Joyard, J. (2001). Two types of MGDG synthase genes, found widely in both 16:3 and 18:3 plants, differentially mediate galactolipid syntheses in photosynthetic and nonphotosynthetic tissues in *Arabidopsis thaliana*. *Proc. Natl. Acad. Sci. U. S. A.* **98**, 10960–10965.

Awai, K., Kakimoto, T., Awai, C., Kaneko, T., Nakamura, Y., Takamiya, K., Wada, H., and Ohta, H. (2006a). Comparative genomic analysis revealed a gene for monoglucosyldiacylglycerol synthase, an enzyme for photosynthetic membrane lipid synthesis in cyanobacteria. *Plant Physiol.* **141**, 1120–1127.

Awai, K., Xu, C., Tamot, B., and Benning, C. (2006b). A phosphatidic acid-binding protein of the chloroplast inner envelope membrane involved in lipid trafficking. *Proc. Natl. Acad. Sci. U. S. A.* **103**, 10817–10822.

Beisson, F., Koo, A.J., Ruuska, S., Schwender, J., Pollard, M., Thelen, J.J., Paddock, T., Salas, J.J., Savage, L., Milcamps, A., Mhaske, V.B., Cho, Y., and Ohlrogge, J.B. (2003). *Arabidopsis* genes involved in acyl lipid metabolism. A 2003 census of the candidates, a study of the distribution of expressed sequence tags in organs, and a web-based database. *Plant Physiol.* **132**, 681–697.

Benning, C. (1998). Biosynthesis and function of the sulfolipid sulfoquinovosyl diacylglycerol. *Annu. Rev. Plant Physiol. Plant Mol. Biol.* **49**, 53–75.

Benning, C. and Ohta, H. (2005). Three enzyme systems for galactoglycerolipid biosynthesis are coordinately regulated in plants. *J. Biol. Chem.* **280**, 2397–2400.

Benning, C. and Somerville, C.R. (1992a). Identification of an operon involved in sulfolipid biosynthesis in *Rhodobacter sphaeroides*. *J. Bacteriol.* **174**, 6479–6487.

Benning, C. and Somerville, C.R. (1992b). Isolation and genetic complementation of a sulfolipid-deficient mutant of *Rhodobacter sphaeroides*. *J. Bacteriol.* **174**, 2352–2360.

Benning, C., Beatty, J.T., Prince, R.C., and Somerville, C.R. (1993). The sulfolipid sulfoquinovosyldiacylglycerol is not required for photosynthetic electron transport in *Rhodobacter sphaeroides* but enhances growth under phosphate limitation. *Proc. Natl. Acad. Sci. U. S. A.* **90**, 1561–1565.

Benning, C., Huang, Z.H., and Gage, D.A. (1995). Accumulation of a novel glycolipid and a betaine lipid in cells of *Rhodobacter sphaeroides* grown under phosphate limitation. *Arch. Biochem. Biophys.* **317**, 103–111.

Benning, C., Xu, C., and Awai, K. (2006). Non-vesicular and vesicular lipid trafficking involving plastids. *Curr. Opin. Plant Biol.* **9**, 241–247.

Benson, A.A., Daniel, H., and Wiser, R. (1959). A sulfolipid in plants. *Proc. Natl. Acad. Sci. U. S. A.* **45**, 1582–1587.

Blouin, A., Lavezzi, T., and Moore, T.S. (2003). Membrane lipid biosynthesis in *Chlamydomonas reinhardtii*. Partial characterization of CDP-diacylglycerol: myo-inositol 3- phosphatidyltransferase. *Plant Physiol. Biochem.* **41**, 11–16.

Bogdanov, M. and Dowhan, W. (1999). Lipid-assisted protein folding. *J. Biol. Chem.* **274**, 36827–36830.

Bolognese, C.P. and McGraw, P. (2000). The isolation and characterization in yeast of a gene for *Arabidopsis S*-adenosylmethionine:phospho-ethanolamine *N*-methyltransferase. *Plant Physiol.* **124**, 1800–1813.

Browse, J., McCourt, P., and Somerville, C. (1985). A mutant of *Arabidopsis* lacking a chloroplast-specific lipid. *Science* **227**, 763–765.

Choquet, Y., Zito, F., Wostrikoff, K., and Wollman, F.A. (2003). Cytochrome *f* translation in *Chlamydomonas* chloroplast is autoregulated by its carboxyl-terminal domain. *Plant Cell* **15**, 1443–1454.

Dahlqvist, A., Stahl, U., Lenman, M., Banas, A., Lee, M., Sandager, L., Ronne, H., and Stymne, S. (2000). Phospholipid:diacylglycerol acyltransferase: An enzyme that catalyzes the acyl-CoA-independent formation of triacylglycerol in yeast and plants. *Proc. Natl. Acad. Sci. U. S. A.* **97**, 6487–6492.

Davies, J.P., Yildiz, F., and Grossman, A.R. (1994). Mutants of *Chlamydomonas* with aberrant responses to sulfur deprivation. *Plant Cell* **6**, 53–63.

Davies, J.P., Yildiz, F.H., and Grossman, A. (1996). Sac1, a putative regulator that is critical for survival of *Chlamydomonas reinhardtii* during sulfur deprivation. *EMBO J.* **15**, 2150–2159.

Davies, J.P., Yildiz, F.H., and Grossman, A.R. (1999). Sac3, an Snf1-like serine/threonine kinase that positively and negatively regulates the responses of *Chlamydomonas* to sulfur limitation. *Plant Cell* **11**, 1179–1190.

Dörmann, P. and Benning, C. (2002). Galactolipids rule in seed plants. *Trends Plant Sci.* **7**, 112–118.

Dörmann, P., Hoffmann-Benning, S., Balbo, I., and Benning, C. (1995). Isolation and characterization of an *Arabidopsis* mutant deficient in the thylakoid lipid digalactosyl diacylglycerol. *Plant Cell* **7**, 1801–1810.

Dörmann, P., Balbo, I., and Benning, C. (1999). *Arabidopsis* galactolipid biosynthesis and lipid trafficking mediated by DGD1. *Science* **284**, 2181–2184.

Dubertret, G., Mirshahi, A., Mirshahi, M., Gerardhirne, C., and Trémolières, A. (1994). Evidence from *in vivo* manipulations of lipid-composition in mutants that the delta(3)-trans-hexadecenoic acid-containing phosphatidylglycerol is involved in the biogenesis of the light-harvesting chlorophyll a/b-protein complex of *Chlamydomonas reinhardtii*. *Eur. J. Biochem.* **226**, 473–482.

Dubertret, G., Gerard-Hirne, C., and Trémolières, A. (2002). Importance of *trans*-delta(3)-hexadecenoic acid containing phosphatidylglycerol in the formation of the trimeric light-harvesting complex in *Chlamydomonas*. *Plant Physiol. Biochem.* **40**, 829–836.

El Maanni, A., Dubertret, G., Delrieu, M.J., Roche, O., and Trémolières, A. (1998). Mutants of *Chlamydomonas reinhardtii* affected in phosphatidylglycerol metabolism and thylakoid biogenesis. *Plant Physiol. Biochem.* **36**, 609–619.

Essigmann, B., Güler, S., Narang, R.A., Linke, D., and Benning, C. (1998). Phosphate availability affects the thylakoid lipid composition and the expression of *SQD1*, a gene required for sulfolipid biosynthesis in *Arabidopsis thaliana*. *Proc. Natl. Acad. Sci. U. S. A.* **95**, 1950–1955.

Fischer, U., Men, S., and Grebe, M. (2004). Lipid function in plant cell polarity. *Curr. Opin. Plant Biol.* **7**, 670–676.

Frentzen, M. (1986). Biosynthesis and desaturation of the different diacylglycerol moieties in higher plants. *J. Plant Physiol.* **124**, 193–209.

Frentzen, M. (2004). Phosphatidylglycerol and sulfoquinovosyldiacylglycerol: Anionic membrane lipids and phosphate regulation. *Curr. Opin. Plant Biol.* **7**, 270–276.

Geiger, O., Röhrs, V., Weissenmayer, B., Finan, T.M., and Thomas-Oates, J.E. (1999). The regulator gene *phoB* mediates phosphate stress-controlled synthesis of the membrane lipid diacylglyceryl-*N,N,N*-trimethylhomoserine in *Rhizobium (Sinorhizobium) meliloti*. *Mol. Microbiol.* **32**, 63–73.

Giroud, C. and Eichenberger, W. (1987). Lipid biosynthesis in Chlamydomonas reinhardtii. *Experientia* **43**, 660–661.

Giroud, C. and Eichenberger, W. (1988). Fatty-acids of *Chlamydomonas reinhardtii*: Structure, positional distribution and biosynthesis. *Biol Chem. Hoppe-Seyler,* **369**, 18–19.

Giroud, C., Gerber, A., and Eichenberger, W. (1988). Lipids of *Chlamydomonas reinhardtii*: Analysis of molecular species and intracellular site(s) of biosynthesis. *Plant Cell Physiol.* **29**, 587–595.

Gounaris, K. and Barber, J. (1983). Monogalactosyldiacylglycerol: The most abundant polar lipid in nature. *Trends Biochem. Sci.* **8**, 378–381.

Grenier, G., Guyon, D., Roche, O., Dubertret, G., and Trémolières, A. (1991). Modification of the membrane fatty-acid composition of *Chlamydomonas reinhardtii* cultured in the presence of liposomes. *Plant Physiol. Biochem.* **29**, 429–440.

Güler, S., Seeliger, A., Härtel, H., Renger, G., and Benning, C. (1996). A null mutant of *Synechococcus* sp. PCC7942 deficient in the sulfolipid sulfoquinovosyl diacylglycerol. *J. Biol. Chem.* **271**, 7501–7507.

Güler, S., Essigmann, B., and Benning, C. (2000). A cyanobacterial gene, *sqdX*, required for biosynthesis of the sulfolipid sulfoquinovosyldiacylglycerol. *J. Bacteriol.* **182**, 543–545.

Hagio, M., Sakurai, I., Sato, S., Kato, T., Tabata, S., and Wada, H. (2002). Phosphatidylglycerol is essential for the development of thylakoid membranes in *Arabidopsis thaliana*. *Plant Cell Physiol.* **43**, 1456–1464.

Han, G.S., Wu, W.I., and Carman, G.M. (2006). The *Saccharomyces cerevisiae* Lipin homolog is a Mg^{2+}-dependent phosphatidate phosphatase enzyme. *J. Biol. Chem.* **281**, 9210–9218.

Härtel, H., Essigmann, B., Lokstein, H., Hoffmann-Benning, S., Peters-Kottig, M., and Benning, C. (1998). The phospholipid-deficient *pho1* mutant of *Arabidopsis thaliana* is affected in the organization, but not in the light acclimation, of the thylakoid membrane. *Biochim. Biophys. Acta,* **1415**, 205–218.

Härtel, H., Dörmann, P., and Benning, C. (2000). DGD1-independent biosynthesis of extraplastidic galactolipids following phosphate deprivation in *Arabidopsis*. *Proc. Natl. Acad. Sci. U. S. A.* **97**, 10649–10654.

Heilmann, I., Perera, I.Y., Gross, W., and Boss, W.F. (1999). Changes in phosphoinositide metabolism with days in culture affect signal transduction pathways in *Galdieria sulphuraria*. *Plant Physiol.* **119**, 1331–1340.

Heilmann, I., Perera, I.Y., Gross, W., and Boss, W.F. (2001). Plasma membrane phosphatidylinositol 4,5-bisphosphate levels decrease with time in culture. *Plant Physiol.* **126**, 1507–1518.

Heinz, E. (1993). Recent investigations on the biosynthesis of the plant sulfolipid. In: *Sulfur Nutrition and Assimilation in Higher Plants* (L.J. De Kok, Ed.), pp. 163–178. SPB Academic Publishing: The Hague, The Netherlands.

Heinz, E. and Roughan, G. (1983). Similarities and differences in lipid metabolism of chloroplasts isolated from 18:3 and 16:3 plants. *Plant Physiol.* **72**, 273–279.

Hobbs, D.H., Lu, C., and Hills, M.J. (1999). Cloning of a cDNA encoding diacylglycerol acyltransferase from *Arabidopsis thaliana* and its functional expression. *FEBS Lett.* **452**, 145–149.

Hofmann, M. and Eichenberger, W. (1996). Biosynthesis of diacylglyceryl-*N,N,N*-trimethylhomoserine in *Rhodobacter sphaeroides* and evidence for lipid-linked *N*-methylation. *J. Bacteriol.* **178**, 6140–6144.

Jesch, S.A., Zhao, X., Wells, M.T., and Henry, S.A. (2005). Genome-wide analysis reveals inositol, not choline, as the major effector of Ino2p-Ino4p and unfolded protein response target gene expression in yeast. *J. Biol. Chem.* **280**, 9106–9118.

Joyard, J., Marechal, E., Miege, C., Block, M.A., Dorne, A.J., and Douce, R. (1998). Structure, distribution and biosynthesis of glycerolipids from higher plant chloroplasts. In: *Lipids in Photosynthesis: Structure, Function and Genetics* (P.-A. Siegenthaler, and N. Murata, Eds.), pp. 21–52. Kluwer Academic Publishers, Dordrecht, The Netherlands.

Kajikawa, M., Yamato, K.T., Kohzu, Y., Shoji, S., Matsui, K., Tanaka, Y., Sakai, Y., and Fukuzawa, H. (2006). A front-end desaturase from *Chlamydomonas reinhardtii* produces pinolenic and coniferonic acids by omega13 desaturation in methylotrophic yeast and tobacco. *Plant Cell Physiol.* **47**, 64–73.

Kelly, A.A., Froehlich, J.E., and Dörmann, P. (2003). Disruption of the two digalactosyldiacylglycerol synthase genes *DGD1* and *DGD2* in *Arabidopsis* reveals the existence of an additional enzyme of galactolipid synthesis. *Plant Cell* **15**, 2694–2706.

Kelly, A.A. and Dörmann, P. (2004). Green light for galactolipid trafficking. *Curr. Opin. Plant Biol.* **7**, 262–269.

Kim, H.U. and Huang, A.H.C. (2004). Plastid lysophosphatidyl acyltransferase is essential for embryo development in *Arabidopsis*. *Plant Physiol.* **134**, 1206–1216.

Klug, R.M. and Benning, C. (2001). Two enzymes of diacylglyceryl-*O*-4'-(*N,N,N*,-trimethyl)homoserine biosynthesis are encoded by *btaA* and *btaB* in the purple bacterium *Rhodobacter sphaeroides*. *Proc. Natl. Acad. Sci. U. S. A.* **98**, 5910–5915.

Koo, A.J.K., Ohlrogge, J.B., and Pollard, M. (2004). On the export of fatty acids from the chloroplast. *J. Biol. Chem.* **279**, 16101–16110.

Kunst, L., Browse, J., and Somerville, C. (1988). Altered regulation of lipid biosynthesis in a mutant *Arabidopsis* deficient in chloroplast glycerol-3-phosphate acyltransferase activity. *Proc. Natl. Acad. Sci. U. S. A.* **85**, 4143–4147.

Lemaire, S.D., Guillon, B., Le Marechal, P., Keryer, E., Miginiac-Maslow, M., and Decottignies, P. (2004). New thioredoxin targets in the unicellular photosynthetic eukaryote *Chlamydomonas reinhardtii*. *Proc. Natl. Acad. Sci. U.S.A.* **101**, 7475–7480.

Marechal, E., Block, M.A., Joyard, J., and Douce, R. (1994). Kinetic properties of monogalactosyldiacylglycerol synthase from spinach chloroplast envelope membranes. *J. Biol Chem.* **269**, 5788–5798.

Marechal, E., Miege, C., Block, M.A., Douce, R., and Joyard, J. (1995). The catalytic site of monogalactosyldiacylglycerol synthase from spinach chloroplast envelope membranes. Biochemical analysis of the structure and of the metal content. *J. Biol. Chem.* **270**, 5714–5722.

Meijer, H.J.G., Arisz, S.A., van Himbergen, J.A.J., Musgrave, A., and Munnik, T. (2001). Hyperosmotic stress rapidly generates lyso-phosphatidic acid in *Chlamydomonas*. *Plant J.* **25**, 541–548.

Miege, C., Marechal, E., Shimojima, M., Awai, K., Block, M.A., Ohta, H., Takamiya, K., Douce, R., and Joyard, J. (1999). Biochemical and topological properties of type A MGDG synthase, a spinach chloroplast envelope enzyme catalyzing the synthesis of both prokaryotic and eukaryotic MGDG. *Eur. J. Biochem.* **265**, 990–1001.

Mongrand, S., Besoule, J.-J., Cabantous, F., and Cassagne, C. (1998). The C16:3/C18:3 fatty acid balance in photosynthetic tissues from 468 plant species. *Phytochemistry* **49**, 1049–1064.

Moore, T.S., Du, Z.R., and Chen, Z. (2001). Membrane lipid biosynthesis in *Chlamydomonas reinhardtii. In vitro* biosynthesis of diacylglyceryltrimethylhomoserine. *Plant Physiol.* **125**, 423–429.

Mueller-Roeber, B. and Pical, C. (2002). Inositol phospholipid metabolism in *Arabidopsis*. Characterized and putative isoforms of inositol phospholipid kinase and phosphoinositide-specific phospholipase C. *Plant Physiol.* **130**, 22–46.

Mulichak, A.M., Theisen, M.J., Essigmann, B., Benning, C., and Garavito, R.M. (1999). Crystal structure of SQD1, an enzyme involved in the biosynthesis of the plant sulfolipid headgroup donor UDP-sulfoquinovose. *Proc. Natl. Acad. Sci. U. S. A.* **96**, 13097–13102.

Munnik, T., Irvine, R.F., and Musgrave, A. (1994). Rapid turnover of phosphatidylinositol 3-phosphate in the green alga *Chlamydomonas eugametos*: Signs of a phosphatidylinositide 3-kinase signalling pathway in lower plants? *Biochem. J.* **298** (Pt 2), 269–273.

Munnik, T., van Himbergen, J.A.J., ter Riet, B., Braun, F.J., Irvine, R.F., van den Ende, H., and Musgrave, A.R. (1998). Detailed analysis of the turnover of polyphosphoinositides and phosphatidic acid upon activation of phospholipases C and D in *Chlamydomonas* cells treated with non-permeabilizing concentrations of mastoparan. *Planta* **207**, 133–145.

Murata, N. and Tasaka, Y. (1997). Glycerol-3-phosphate acyltransferase in plants. *Biochim. Biophys. Acta,* **1348**, 10–16.

Nikolau, B.J., Ohlrogge, J.B., and Wurtele, E.S. (2003). Plant biotin-containing carboxylases. *Arch. Biochem. Biophys.* **414**, 211–222.

Nuccio, M.L., Ziemak, M.J., Henry, S.A., Weretilnyk, E.A., and Hanson, A.D. (2000). cDNA cloning of phosphoethanolamine N-methyltransferase from spinach by complementation in *Schizosaccharomyces pombe* and characterization of the recombinant enzyme. *J. Biol. Chem.* **275**, 14095–14101.

Ohlrogge, J. and Browse, J. (1995). Lipid biosynthesis. *Plant Cell* **7**, 957–970.

Ohlrogge, J.B. and Jaworski, J.G. (1997). Regulation of fatty acid synthesis. *Annu. Rev. Plant Physiol. Plant Mol. Biol.* **48**, 109–136.

Ohlrogge, J.B., Kuhn, D.N., and Stumpf, P.K. (1979). Subcellular localization of acyl carrier protein in leaf protoplasts of *Spinacia oleracea*. *Proc. Natl. Acad. Sci. U. S. A.* **76**, 1194–1198.

Picaud, A., Creach, A., and Trémolières, A. (1991). Studies on the stimulation by light of fatty-acid synthesis in *Chlamydomonas reinhardtii* whole cells. *Plant Physiol. Biochem.* **29**, 441–448.

Pineau, B., Girard-Bascou, J., Eberhard, S., Choquet, Y., Trémolières, A., Gerard-Hirne, C., Bennardo-Connan, A., Decottignies, P., Gillet, S., and Wollman, F.A. (2004). A single mutation that causes phosphatidylglycerol deficiency impairs synthesis of photosystem II cores in *Chlamydomonas reinhardtii*. *Eur. J. Biochem.* **271**, 329–338.

Poirier, Y., Thoma, S., Somerville, C., and Schiefelbein, J. (1991). A mutant of *Arabidopsis* deficient in xylem loading of phosphate. *Plant Physiol.* **97**, 1087–1093.

Riekhof, W.R., Ruckle, M.E., Lydic, T.A., Sears, B.B., and Benning, C. (2003). The sulfolipids 2′-*O*-acyl-sulfoquinovosyldiacylglycerol and sulfoquinovosyldiacylglycerol are absent from a *Chlamydomonas reinhardtii* mutant deleted in SQD1. *Plant Physiol* **133**, 864–874.

Riekhof, W.R., Andre, C., and Benning, C. (2005a). Two enzymes, BtaA and BtaB, are sufficient for betaine lipid biosynthesis in bacteria. *Arch. Biochem. Biophys.* **441**, 96–105.

Riekhof, W.R., Sears, B.B., and Benning, C. (2005b). Annotation of genes involved in glycerolipid biosynthesis in *Chlamydomonas reinhardtii*: Discovery of the betaine lipid synthase BTA1Cr. *Eukaryotic Cell*, **4**, 242–252.

Rontein, D., Nishida, I., Tashiro, G., Yoshioka, K., Wu, W.I., Voelker, D.R., Basset, G., and Hanson, A.D. (2001). Plants synthesize ethanolamine by direct decarboxylation of serine using a pyridoxal phosphate enzyme. *J. Biol. Chem.* **276**, 35523–35529.

Rontein, D., Wu, W.I., Voelker, D.R., and Hanson, A.D. (2003). Mitochondrial phosphatidylserine decarboxylase from higher plants. Functional complementation in yeast, localization in plants, and overexpression in *Arabidopsis*. *Plant Physiol.* **132**, 1678–1687.

Roughan, P.G. and Slack, C.R. (1982). Cellular organization of glycerolipid metabolism. *Annu. Rev. Plant Physiol.* **33**, 97–132.

Routaboul, J.M., Benning, C., Bechtold, N., Caboche, M., and Lepiniec, L. (1999). The TAG1 locus of *Arabidopsis* encodes for a diacylglycerol acyltransferase. *Plant Physiol. Biochem.* **37**, 831–840.

Sanda, S., Leustek, T., Theisen, M., Garavito, M., and Benning, C. (2001). Recombinant *Arabidopsis* SQD1 converts UDP-glucose and sulfite to the sulfolipid head precursor UDP-sulfoquinovose *in vitro*. *J. Biol. Chem.* **276**, 3941–3946.

Sato, N. (1988). Dual role of methionine in the biosynthesis of diacylglyceryltrimethylhomoserine in *Chlamydomonas reinhardtii*. *Plant Physiol.* **86**, 931–934.

Sato, N. and Murata, N. (1991). Transition of lipid phase in aqueous dispersions of diacylglyceryltrimethylhomoserine. *Biochim. Biophys. Acta,* **1082**, 108–111.

Sato, N., Tsuzuki, M., Matsuda, Y., Ehara, T., Osafune, T., and Kawaguchi, A. (1995a). Isolation and characterization of mutants affected in lipid metabolism of *Chlamydomonas reinhardtii*. *Eur. J. Biochem.* **230**, 987–993.

Sato, N., Sonoike, K., Tsuzuki, M., and Kawaguchi, A. (1995b). Impaired photosystem II in a mutant of *Chlamydomonas reinhardtii* defective in sulfoquinovosyl diacylglycerol. *Eur. J. Biochem.* **234**, 16–23.

Sato, N., Fujiwara, S., Kawaguchi, A., and Tsuzuki, M. (1997). Cloning of a gene for chloroplast omega 6 desaturase of a green alga, *Chlamydomonas reinhardtii*. *J. Biochem.* **122**, 1224–1232.

Sato, N., Hagio, M., Wada, H., and Tsuzuki, M. (2000). Environmental effects on acidic lipids of thylakoid membranes. *Biochem. Soc. Trans.* **28**, 912–914.

Sato, N., Aoki, M., Maru, Y., Sonoike, K., Minoda, A., and Tsuzuki, M. (2003a). Involvement of sulfoquinovosyl diacylglycerol in the structural integrity and heat-tolerance of photosystem II. *Planta* **217**, 245–251.

Sato, N., Sugimoto, K., Meguro, A., and Tsuzuki, M. (2003b). Identification of a gene for UDP-sulfoquinovose synthase of a green alga, *Chlamydomonas reinhardtii*, and its phylogeny. *DNA Res.* **10**, 229–237.

Shimogawara, K., Wykoff, D.D., Usuda, H., and Grossman, A.R. (1999). *Chlamydomonas reinhardtii* mutants abnormal in their responses to phosphorus deprivation. *Plant Physiol* **120**, 685–694.

Shimojima, M. and Benning, C. (2003). Native uridine 59-diphosphate-sulfoquinovose synthase, SQD1, from spinach purifies as a 250-kDa complex. *Arch. Biochem. Biophys.* **413**, 123–130.

Shimojima, M., Ohta, H., Iwamatsu, A., Masuda, T., Shioi, Y., and Takamiya, K. (1997). Cloning of the gene for monogalactosyldiacylglycerol synthase and its evolutionary origin. *Proc. Natl. Acad. Sci. U. S. A.* **94**, 333–337.

Shimojima, M., Hoffmann-Benning, S., Garavito, R.M., and Benning, C. (2005). Ferredoxin-dependent glutamate synthase moonlights in plant sulfolipid biosynthesis by forming a complex with SQD1. *Arch. Biochem. Biophys.* **436**, 206–214.

Sirevag, R. and Levine, R.P. (1972). Fatty acid synthetase from *Chlamydomonas reinhardi*. *J. Biol. Chem.* **247**, 2586–2591.

Sorger, D. and Daum, G. (2002). Synthesis of triacylglycerols by the acyl-coenzyme A:diacyl-glycerol acyltransferase Dga1p in lipid particles of the yeast *Saccharomyces cerevisiae*. *J. Bacteriol.* **184**, 519–524.

Stahl, U., Carlsson, A.S., Lenman, M., Dahlqvist, A., Huang, B., Banas, W., Banas, A., and Stymne, S. (2004). Cloning and functional characterization of a phospholipid:diacylglycerol acyltransferase from *Arabidopsis*. *Plant Physiol.* **135**, 1324–1335.

Stevenson, J.M., Perera, I.Y., Heilmann, I., Persson, S., and Boss, W.F. (2000). Inositol signaling and plant growth. *Trends Plant Sci.* **5**, 252–258.

Stroebel, D., Choquet, Y., Popot, J.L., and Picot, D. (2003). An atypical haem in the cytochrome b_6f complex. *Nature* **426**, 413–418.

Van Mooy, B.A., Rocap, G., Fredricks, H.F., Evans, C.T., and Devol, A.H. (2006). Sulfolipids dramatically decrease phosphorus demand by picocyanobacteria in oligotrophic marine environments. *Proc. Natl. Acad. Sci. U. S. A.* **103**, 8607–8612.

Vogel, G. and Eichenberger, W. (1992). Betaine lipids in lower plants. Biosynthesis of DGTS and DGTA in *Ochromonas danica* (Chrysophyceae) and the possible role of DGTS in lipid metabolism. *Plant Cell Physiol.* **33**, 427–436.

Wada, H., Shintani, D., and Ohlrogge, J. (1997). Why do mitochondria synthesize fatty acids? Evidence for involvement in lipoic acid production. *Proc. Natl. Acad. Sci. U. S. A.* **94**, 1591–1596.

Webb, M.S. and Green, B.R. (1991). Biochemical and biophysical properties of thylakoid acyl lipids. *Biochim. Biophys. Acta,* **1060**, 133–158.

Weers, P. and Gulati, R. (1997). Growth and reproduction of *Daphnia galeata* in response to changes in fatty acids, phosphorus, and nitrogen in *Chlamydomonas reinhardtii*. *Limnol. Oceanogr.* **42**, 1586–1589.

White, S.W., Zheng, J., Zhang, Y.M., and Rock, C.O. (2005). The structural biology of type II fatty acid biosynthesis. *Annu. Rev. Biochem.* **74**, 791–831.

Wikstrom, M., Xie, J., Bogdanov, M., Mileykovskaya, E., Heacock, P., Wieslander, A., and Dowhan, W. (2004). Monoglucosyldiacylglycerol, a foreign lipid, can substitute for phosphatidylethanolamine in essential membrane-associated functions in Escherichia coli. *J. Biol. Chem.* **279**, 10484–10493.

Wykoff, D.D., Grossman, A.R., Weeks, D.P., Usuda, H., and Shimogawara, K. (1999). Psr1, a nuclear localized protein that regulates phosphorus metabolism in *Chlamydomonas*. *Proc. Natl. Acad. Sci. U. S. A.* **96**, 15336–15341.

Xie, J., Bogdanov, M., Heacock, P., and Dowhan, W. (2006). Phosphatidylethanolamine and monoglucosyldiacylglycerol are interchangeable in supporting topogenesis and function of the polytopic membrane protein lactose permease. *J. Biol. Chem.* **281**, 19172–19178.

Xu, C., Härtel, H., Wada, H., Hagio, M., Yu, B., Eakin, C., and Benning, C. (2002). The *pgp1* locus of *Arabidopsis* encodes a phosphatidylglycerol synthase with impaired activity. *Plant Physiol.* **129**, 594–604.

Xu, C., Fan, J., Riekhof, W., Froehlich, J.E., and Benning, C. (2003). A permease-like protein involved in ER to thylakoid lipid transfer in *Arabidopsis*. *EMBO J.* **22**, 2370–2379.

Xu, C., Fan, J., Froehlich, J.E., Awai, K., and Benning, C. (2005). Mutation of the TGD1 chloroplast envelope protein affects phosphatidate metabolism in *Arabidopsis*. *Plant Cell* **17**, 3094–3110.

Xu, C., Yu, B., Cornish, A.J., Froehlich, J.E., and Benning, C. (2006). Phosphatidylglycerol biosynthesis in chloroplasts of *Arabidopsis* mutants deficient in acyl-ACP glycerol-3-phosphate acyltransferase. *Plant J.* **47**, 296–309.

Yang, W., Mason, C.B., Pollock, S.V., Lavezzi, T., Moroney, J.V., and Moore, T.S. (2004a). Membrane lipid biosynthesis in *Chlamydomonas reinhardtii*: Expression and characterization of CTP: phosphoethanolamine cytidylyltransferase. *Biochem. J.* **382**, 51–57.

Yang, W., Moroney, J.V., and Moore, T.S. (2004b). Membrane lipid biosynthesis in *Chlamydomonas reinhardtii*: Ethanolaminephosphotransferase is capable of synthesizing both phosphatidylcholine and phosphatidylethanolamine. *Arch. Biochem. Biophys.* **430**, 198–209.

Yu, B. and Benning, C. (2003). Anionic lipids are required for chloroplast structure and function in *Arabidopsis*. *Plant J.* **36**, 762–770.

Yu, B., Xu, C., and Benning, C. (2002). *Arabidopsis* disrupted in SQD2 encoding sulfolipid synthase is impaired in phosphate-limited growth. *Proc. Natl. Acad. Sci. U. S. A.* **99**, 5732–5737.

Yu, B., Wakao, S., Fan, J.L., and Benning, C. (2004). Loss of plastidic lysophosphatidic acid acyltransferase causes embryo-lethality in *Arabidopsis*. *Plant Cell Physiol.* **45**, 503–510.

CHAPTER 3

Nitrogen Assimilation and its Regulation

Emilio Fernández, Ángel Llamas and Aurora Galván
Departamento de Bioquímica y Biología Molecular,
Facultad de Ciencias, Universidad de Córdoba, Córdoba, Spain

CHAPTER CONTENTS

I. INTRODUCTION

Nitrogen (N) assimilation is part of the nitrogen cycle and has an important impact not only in ecological terms, but also economically and for animal health. Although progress in understanding organic nitrogen metabolism in *Chlamydomonas* has been limited, progress on inorganic nitrogen assimilation has been aided by molecular and genomic advances. This chapter presents nitrogen assimilation in the context of genome analysis and gene function, with particular emphasis on inorganic nitrogen transporters, molybdenum cofactor (Moco) metabolism, and regulation of the nitrate assimilation pathway. In a comparative view, if considering only nitrate assimilation genes, *Chlamydomonas* and *Arabidopsis* show similar complexity whereas *Ostreococcus*, another unicellular alga, is much simpler. Understanding the bases of these differences will provide important perspectives into the biology of this pathway.

II. USEFUL NITROGEN SOURCES FOR *CHLAMYDOMONAS*

A. Overview

Chlamydomonas preferentially uses ammonium as inorganic nitrogen for growth. When ammonium is limited, other inorganic or organic N sources can be assimilated after their transformation into ammonium, which is subsequently incorporated into carbon skeletons by the glutamine synthetase/glutamate synthase (GS/GOGAT) cycle. A few organic N compounds can be used by some *Chlamydomonas* species (Volume 1, Table 6.7) but these are considered poor sources; among them are several amino acids, the purines, urea, and acetamide. The oxidized N forms nitrate and nitrite constitute inorganic sources that are efficiently assimilated after reduction to ammonium, as mentioned above.

Little information apart from that reviewed previously (Fernández et al., 1998) has been generated on the assimilation of organic nitrogen compounds in *Chlamydomonas*. A general view of the pathways is presented here, along with the putative genes involved (Figure 3.1).

B. Organic nitrogen

1. Urea assimilation

Urea can be readily assimilated from the external medium by *Chlamydomonas*. First, it is transported by an active and energy-dependent system (Williams and Hodson, 1977) and then distributed between two pools, one large and non-metabolic, probably located in the chloroplast, and another which is metabolic. Ammonium inhibits movement

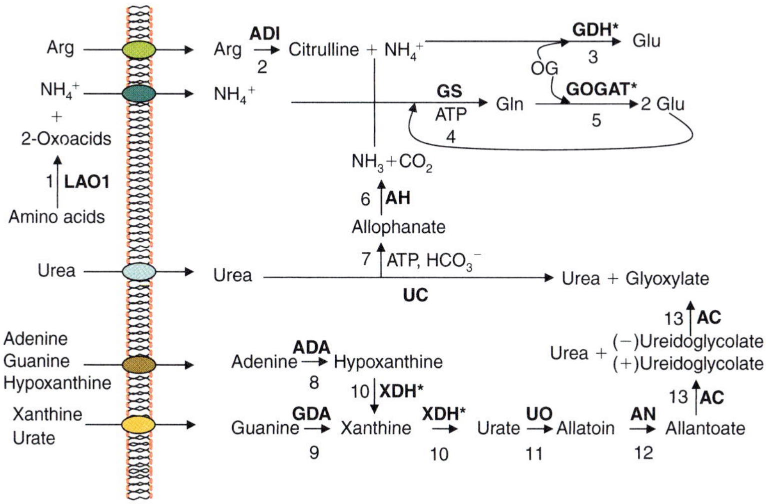

FIGURE 3.1 *Assimilation of organic nitrogen compounds in Chlamydomonas. LAO1, amino acid oxidase; ADI, arginine deiminase; GDH, glutamate dehydrogenase; GS, glutamine synthetase; GOGAT, glutamate oxoglutarate amidotransferase; AH, allophanate hydrolase; UC, urea carboxylase; ADA, adenine deaminase; GDA, guanine deaminase; XDH, xanthine dehydrogenase; UO, urate oxidase; AN, allantoinase; AC, allantoicase. Asterisks denote requirement for reducing power.*

from the non-metabolic to the metabolic pool (Dagestad et al., 1981). This compartmentation of urea into a metabolically inactive form is analogous to the storage reported for xanthine (Pérez-Vicente et al., 1995). In the absence of ammonium and the presence of urea or acetamide, *Chlamydomonas* expresses the enzyme complex ATP:urea amidolyase that catalyzes the hydrolysis of urea to produce ammonium (Leftley and Syrett, 1973; Hodson et al., 1975; Semler et al., 1975). This complex consists of two enzyme activities: urea carboxylase, which catalyzes the ATP-dependent condensation of urea and bicarbonate to yield allophanate; and allophanate hydrolase, which yields ammonia and CO_2 (Whitney and Cooper, 1973; Hodson et al., 1975). Acetamide can also be used as a nitrogen source in *Chlamydomonas* after hydrolysis catalyzed by acetamidase to produce ammonium and acetate (Hodson and Gresshoff, 1987).

2. Assimilation of amino acids

The ability of *Chlamydomonas* to use different amino acids as an N source is strongly species-dependent (Volume 1, Table 6.7). *C. reinhardtii* can use most extracellular amino acids as a sole N source, however their assimilation is strongly dependent on a supply of acetate as carbon source (Muñoz-Blanco et al., 1990). Twelve L-amino acids (asparagine, glutamine, arginine, lysine, alanine, valine, leucine, isoleucine, serine, methionine, histidine, and phenylalanine) are extracellularly deaminated *in vivo*, producing the corresponding 2-oxoacid plus ammonium. Only ammonium enters the cell readily and is subsequently incorporated; the derived oxoacids cannot be used in acetate-containing media (Muñoz-Blanco et al., 1990). The extracellular deamination of amino acids is catalyzed by L-amino acid oxidase (LAO1), which is located in the periplasm. LAO1 is induced upon N starvation in *Chlamydomonas* cultures containing a source of organic carbon (Piedras et al., 1992; Vallon et al., 1993). The purified enzyme contains FAD (flavin adenine dinucleotide) and consists of oligomeric forms of the catalytic subunit Mα with essentially similar enzymatic properties. Piedras et al. (1992) reported an active LAO protein of 450 kD with eight identical subunits of 60 kD, and Vallon et al. (1993) suggested two forms of 1.2–1.3 and 0.9–1.0 megadaltons, respectively. The lighter form is a hexadecamer of M catalytic subunits, while the heavier form contains in addition to Mα, an accessory subunit Mβ. Both polypeptides are glycosylated (Vallon et al., 1993; Vallon and Wollman, 1997). The LAO forms purified by Vallon et al. catalyze *in vitro* deamination of all L-amino acids tested except cysteine. In addition, a saturation of the enzymatic rate is obtained with 2 micromoles phenylalanine as substrate, suggesting that LAO1 may operate *in vivo* as a very efficient scavenger of ammonium from extracellular amino acids.

Arginine is reported to be the only amino acid that enters *Chlamydomonas* via a specific transport system, which is induced under N starvation and repressed by ammonium (Kirk and Kirk, 1978). Intracellular arginine can be either directly incorporated as such for biosynthetic purposes or hydrolyzed to produce citrulline and ammonium by the action of arginine deiminase (Figure 3.1).

An alternative assimilation pathway for histidine has been shown in *Chlamydomonas*. When both histidine and ammonium are in the medium, histidine is degraded only after ammonium depletion, whereupon the first two enzymes of the histidine degradation pathway, histidase and urocanase, are produced (Hellio et al., 2004). When histidine is used as an N source and acetate as a C source, the cells die rapidly (Hellio et al., 2004). The assimilation of some extracellular amino acids has also been related to the existence of specific transport systems and transaminase enzymes that could explain resistance/sensitivity to methionine-D,L-sulfoximine (MSX), a glutamine synthetase inhibitor (Franco et al., 1996a, b).

3. A gene cluster for urea and arginine metabolism

With the exception of *LAO1* (Vallon et al., 1997), no genes related to urea or arginine metabolism have been experimentally analyzed; therefore the information discussed below and presented in Table 3.1 is derived from genome-based predictions. *LAO1* encodes the catalytic subunit Mα and cDNA sequence analysis predicts a precursor of 560 residues for the periplasmic isoform (Vallon and Wollman, 1997).

Essential genes for a given N metabolic process are frequently clustered in microorganisms. Examples of such operons are those for nitrate

Table 3.1 Major elements of nitrogen assimilation in *Chlamydomonas*

	Gene	Reaction number	Linkage group; accession #	cDNA Protein # amino acids°	Characteristics	References
L-amino acid oxidase	*LAO1*	1	XII/XIII; EDP03066	U78797 AAB97101 560 aa	Periplasmic; nitrogen regulated	Vallon and Wolman (1997)
Arginine deiminase	*ADI1*	2	VIII; ED097040			
Urea carboxylase	*DUR1*	7	VIII; ED097054			
Allophanate lyase	*DUR2*	6	VIII; ED097054			
Urea active transporter	*DUR3A*		VIII; ED097038			
	DUR3B		VIII; ED097039			
	DUR3C		XIX; EDP04688			
Xanthine/uracil permeases	*UAPA1*		X; EDP05773			
	UAPA2		X; EDP05602			
	UAPA3		X; EDP06439			
	UAPA4		X; EDP06260			
	UAPA5, XUV2		X; EDP07979			
	UAPA6, XUV1		VI; EDP08303			
Urate oxidase	*UO, UOX*	11	XII/XIII; EDP05446	AF195795 AAG28465 295 aa		Merchán et al. (2001)
Xanthine dehydrogenase	*XDH1*	10	XII/XIII; EDP03026			
Allantoinase	*DAL1*	12	XII/XIII; EDP03524			
Allantoicase		13	?			
Ammonium transporter AMT1 family	*AMT1.1, AMT1A*		III; EDP00166	AF479643 AAL85345 539 aa	Pm; inducible under N-deficiency NIT2-regulated	González-Ballester et al. (2004)
	AMT1.2, AMT1B		XV; EDP03690	AF530051 AAM94623 542 aa	Chl; inducible under N-deficiency	González-Ballester et al. (2004)
	AMT1.3, AMT1C, AMT3		VI; ED097333	AF509497 AAM43910 579 aa	Pm; high expression in nitrate	González-Ballester et al. (2004); Kim et al. (2005)

(Continued)

Table 3.1 *Continued*

	Gene	Reaction number	Linkage group; accession #	cDNA Protein # amino acids°	Characteristics	References
	AMT1.4, AMT1D, AMT4		XIV; EDP08718	AY542491 AAS55466 498 aa	Chl; high expression under N-deficiency	González-Ballester et al. (2004); Kim et al. (2005)
	AMT1.5, AMT1E, AMT5		IX; EDP01915	Amt1.5a AY542492 AAS90602 579 aa Amt1.5b AY588243 AAS55467 579 aa	Pm; high expression under N-deficiency	González-Ballester et al. (2004)
	AMT1.6, AMT1F, AMT6		Unknown; EDO97275	AY548756* AAS54907	Er, Prx; high expression in ammonium	González-Ballester et al. (2004)
	AMT1.7, AMT1G, AMT7		II; XP_001699698, XP_001699699	Amt1.7a AY588244 AAS90603 411 aa *Amt1.7b AY548755 AAS54906 b: 487 aa*	a: Mit; b: Pm; high expression in ammonium	González-Ballester et al. (2004)
	AMT1.8, AMT1H, AMT8		XII-XIII; EDP03492	AY548754 AAS54905 481 aa	Pm, Chl; high expression in nitrate	González-Ballester et al. (2004)
Ammonium/CO transporter RH family	*RH, RHP1*		VI; EDP01722	AY013257 AAK14647 574 aa		
	RH2, RHP2		VI; EDP01723	AF500098 AAM19665 638 aa		
Glutamate dehydrogenase	*GDH1*	3	IX; EDP02540	AF533889 AAP83856 448 aa	Mit	
	GDH2	3	V; EDO97071	450 aa	Mit	
Glutamine synthetase	*GS1, GLN1*	4	II; EDP07598	CRU46207 AAB01817 382 aa	Cyt	Chen and Silflow (1996)
	GS2, GLN2	4	XII/XIII; EDP03611	U46208 Q42689 380 aa	Chl	Chen and Silflow (1996)
	GS3, GLN3	4	XII/XIII; EDP03496		Chl	
	GS4, GLN4	4	III; EDP03343			

(Continued)

Table 3.1 *Continued*

	Gene	Reaction number	Linkage group; accession #	cDNA Protein # amino acids°	Characteristics	References
Glutamate synthase	*NADH-GOGAT, GSN1*	5	XII/XIII; EDP03651		Chl	
	Fd-GOGAT, GSF1	5	XII-XIII; EDO96739	AAF64387A F135592 847 aa	Chl	
Nitrate transporter NRT1 family	*NRT1.1*		IV; EDP04261			
Nitrate transporter NRT2 family	*NRT2.1, NAR3*		IX; EDP00897	Z25438 CAA80925 547 aa	Pm; Nitrate inducible; NIT2-regulated	Quesada et al. (1993; 1994)
	NRT2.2, NAR4		IX; EDP00896	Z25439 CAA80926	Pm; Nitrate inducible; NIT2-regulated	Quesada et al. (1993; 1994)
	NRT2.3, NAR5		IX; EDP02491	AJ223296 CAD60538 628 aa	Pm; Nitrate inducible; NIT2-regulated	Quesada et al. (1998)
	NRT2.4-5		III	DQ363998*	Pm	
	NRT2.6		II	DQ363997	Pm	
Nitrate transporter component NAR2	*NAR2*		IX; EDP00804		Pm; Nitrate inducible; NIT2-regulated	Quesada et al. (1993; 1994)
Nitrite transporter NAR1 family	*NAR1.1*		IX; EDP00806	AF149737 AAF73173 355 aa	Chl; Nitrate inducible; NIT2-regulated	Rexach et al. (2000)
	NAR1.2, LCIA		VI; EDP04946	AY612639 AAT39454 336 aa	Chl; CO_2-deficiency inducible; CCM1-regulated	Miura et al. (2004); Mariscal et al. (2006)
	NAR1.3		Unknown; EDO97671	AY612643 AAT39458 561 aa		Mariscal et al. (2006)
	NAR1.4		VII; EDO98442	AY612640 AAT39455 406 aa		Mariscal et al. (2006)
	NAR1.5		XII-XIII; EDP03003	AY612641 AAT39456 343 aa	Chl, Mit	Mariscal et al. (2006)
	NAR1.6		I; EDP09170	AY612642 AAT39457 405 aa	Nitrate inducible; NIT2-regulated	Mariscal et al. (2006)

(Continued)

Table 3.1 *Continued*

	Gene	Reaction number	Linkage group; accession #	cDNA Protein # amino acids°	Characteristics	References
Nitrate reductase	*NIA1, NIT1*		IX; EDP00805	AF203033 AAF17595 882 aa	Cyt; Nitrate inducible; NIT2-regulated	Fernández et al. (1989)
Nitrite reductase	*NII1*		IX; EDP00895	Y08937 CAA70137 589 aa	Chl; Nitrate inducible; NIT2-regulated	Quesada et al. (1998)
Molybdenum cofactor (MoCo) biosynthesis	*CNX1G*		VI; EDP01708	DQ311645 ABC42491 210 aa		Llamas et al. (2007)
	CNX1E		X; EDP07827	DQ311646 ABC42492 450 aa		Llamas et al. (2007)
	CNX2		XIV; ED098964			Llamas et al. (2007)
	CNX3		Uncertain; EDP01035			Llamas et al. (2007)
	CNX5		IX; EDP02544			Llamas et al. (2007)
	CNX6		Unknown; EDP07307			Llamas et al. (2007)
	CNX7		Unknown; ED096035			Llamas et al. (2007)
Moco sulfurase	*ABA3*		III; EDP01442 II; EDP02004			
Moco carrier protein	*MCP1*		IX; EDP02441	AY039706 AAK77219 165 aa		Ataya et al. (2003)
Positive regulator NIT2	*NIT2*		III	DQ311647 ABC42493 1196 aa	Nuc; expression increases under N-deficiency	Schnell and Lefebvre (1993); Camargo (2006)
Alternative oxidase	*AOX1*		IX; EDP02600	AF314254 AAG33633 360 aa	Mit; component of the redox valve for nitrate	Quesada et al. (2000); Dinant et al. (2001)
$NADP^+$-malate dehydrogenase	*NADP-MDH, MDH5*		IX; EDP00894	AJ277281 CAC19083 415 aa	Chl; component of the redox valve for nitrate	Quesada et al. (2000); Gómez et al. (2002)

*Genes are named and classified into families according to their predicted function, first and second columns. Alternative names for the same gene are indicated. The reaction number in the third column is from Figure 3.1. The fourth column gives the linkage group localization and the NCBI accession number for the protein sequence derived from the genomic sequence. The fifth column gives the accession numbers if any for the full cDNA and protein sequences, and the number of amino acids; *partial cDNA sequence. Protein amino acid residues correspond to full-length cDNAs. For* AMT1.7 *two different mature mRNAs are indicated (a and b). The sixth column shows some characteristics of gene regulation and most probable protein subcellular localization: Chloroplast (Chl), mitochondria (Mit), Plasma membrane (Pm), endoplasmic reticulum (Er), peroxisome (Prx), and nucleus (Nuc).*

assimilation in the fungus *Aspergillus nidulans* (Johnstone et al., 1990), the yeast *Hansenula polymorpha* (Avila et al., 1998) and the unicellular algae *Chlamydomonas* (Quesada et al., 1993; Galván and Fernández, 2001) and *Ostreococcus*. The *Ostreococcus* genome includes four clustered genes for urea assimilation (Derelle et al., 2006), and a similar organization is observed in *Chlamydomonas* (Figure 3.2 and Table 3.1). The *Chlamydomonas* nitrate cluster is on linkage group IX and will be discussed below. The urea/arginine cluster is on linkage group VIII with the following gene organization: a putative urea carboxylase (*DUR1*), an allophanate lyase (*DUR2*), two urea active transporter-like proteins (*DUR3* and *DUR3B*), and a putative arginine deiminase (*ADI1*). Another putative urea transporter is encoded on linkage group XIX (Table 3.1).

4. Purine assimilation

Purine assimilation has been studied in a few green algae (Vogels and Van der Drift, 1976; Antia et al., 1991), particularly in *Chlamydomonas* (Pineda and Cárdenas, 1996; Piedras et al., 1998), and is summarized in Figure 3.1. *Chlamydomonas* is able to utilize adenine, guanine, hypoxanthine, urate, allantoin, or allantoate as N sources under phototrophic conditions. The standard aerobic pathway of purine degradation seems to be present in higher plants, animals, and many microorganisms (Pineda et al., 1984). However, in the step of allantoate hydrolysis, a true allantoicase has been shown to exist in *Chlamydomonas* (Piedras et al., 2000), but nothing is known of the genes that encode allantoate amidohydrolase and ureidoglycolate amidinohydrolase/urea-lyase in plants (Muñoz, et al., 2006; Todd et al., 2006).

High affinity transport systems mediating adenine, guanine, hypoxanthine, xanthine, and urate uptake have been identified in *Chlamydomonas* at the physiological level (Pineda and Cárdenas, 1996). These systems are repressed by ammonium, require *de novo* synthesis upon induction by specific purines, and are differentially inhibited (Pérez-Vicente et al., 1991; Lisa et al., 1995). At least two nucleotide transporter types seem to be involved, one specific for adenine, guanine, and hypoxanthine, and other specific for xanthine and urate (Pérez-Vicente et al., 1991; Lisa et al., 1995). Xanthine seems to be the sole N source transiently accumulated in cytoplasmic vacuoles as a solid material (Pérez-Vicente et al., 1995). The urate uptake system is expressed constitutively in N-free media (Pineda et al., 1987), as is the gene encoding the enzyme required for its metabolism, urate oxidase (UO, also called uricase; Merchán et al., 2001). Ammonium generated from purine metabolism is incorporated into carbon skeletons through the GS/GOGAT cycle. Inhibition of GS with MSX in cells utilizing urate results in ammonium excretion together with the rapid inhibition of urate uptake (Pineda et al., 1987).

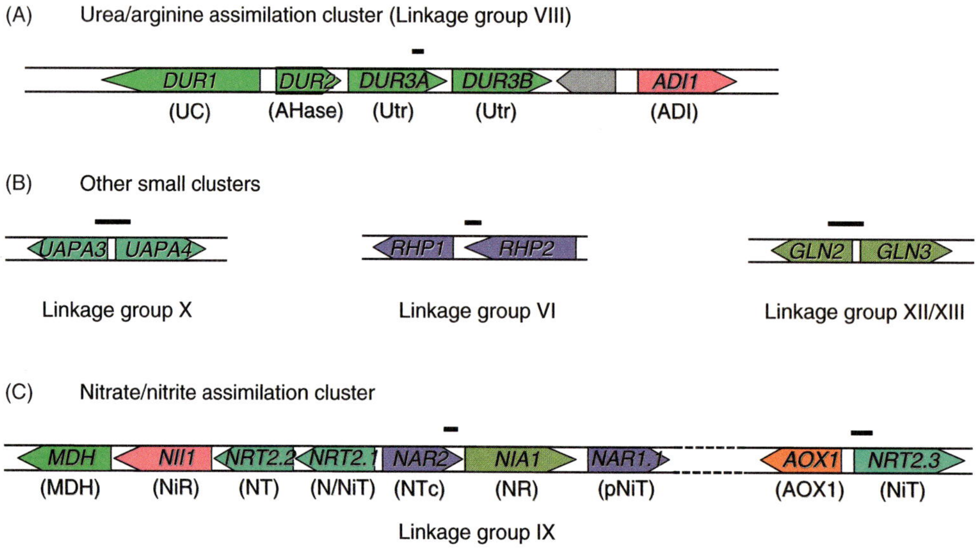

FIGURE 3.2 *Gene clusters for N assimilation in Chlamydomonas. Clusters for the assimilation of urea/arginine (A), other N-related steps (B), and nitrate/nitrite (C) are shown. Gene names are in the open arrowheads, and protein names, where known, are given below each gene. Other details are given in Table 3.1.*

Although no experimental evidence is available to assign these activities to particular nucleotide transporter genes, six putative xanthine/uracil permeases are annotated in the *Chlamydomonas* genome and are thus candidates for purine transport. Curiously, five of them are located on linkage group X, the sixth being on linkage group VI (Table 3.1). The linkage group X cluster includes two genes encoding putative purine transporters, *UAPA1* and *UAPA2*, <20 kb apart; two others, *UAPA3* and *UAPA4*, are adjacent in opposite orientation (Figure 3.2C).

Some of the enzymes for purine assimilation in *Chlamydomonas* have been purified and characterized (see details in Fernández et al., 1998). Among these are xanthine dehydrogenase (XDH), which catalyzes the oxidation of hypoxanthine to xanthine and xanthine to urate and uses NAD^+, but not oxygen, as an electron acceptor (Pérez-Vicente et al., 1988, 1992). XDH shows a diaphorase activity using NAD(P)H (Pérez-Vicente et al., 1988), and can be irreversibly inactivated by incubation with xanthine or hypoxanthine in the absence of NAD^+ (Pérez-Vicente et al., 1992). Like UO, XDH is repressed by ammonium, and is expressed in media having no N or containing poor N sources such as urea, adenine, urate, guanine, or xanthine (Pérez-Vicente et al., 1988).

Chlamydomonas UO is a homotetramer of 31 kD subunits, each containing a copper atom (Alamillo et al., 1991), and its activity appears to be modulated by ammonium (Pineda et al., 1984; Alamillo, 1992). Allantoinase and allantoicase are two separate enzymes, with allantoinase but not allantoicase being membrane-associated, suggesting that they are localized to different subcellular compartments (Piedras et al., 1995). Allantoinase seems to be a constitutive enzyme, whereas allantoicase is repressed in the presence of readily usable N sources like ammonium or nitrate (Piedras et al., 1995). Allantoicase (allantoate amidinohydrolase) is a 200-kD protein with similar-sized subunits of 34 kD, organized as two trimers of 100 kD. The enzyme catalyzes the degradation of allantoate to (−) ureidoglycolate and (+) ureidoglycolate to glyoxylate, shows hyperbolic kinetics for allantoate and ureidoglycolate, and a higher activity with allantoate than with ureidoglycolate (Piedras et al., 2000).

Of the putative genes encoding these enzymes only *UOX*, encoding a type II urate oxidase, has been characterized. This gene is on linkage group XII/XIII, is expressed under N deficiency and was previously identified as *NCG2*, a gene induced very early in gametogenesis (Merchán et al., 2001). Putative genes for xanthine dehydrogenase, *XDH1*, and allantoinase, *DAL1*, are both found on linkage group XII/XIII (Table 3.1). However, even though the allantoicase enzyme has been purified and a partial sequence is available (Piedras et al., 2000), the corresponding gene is not present in the available genomic sequence.

C. Inorganic nitrogen

Chlamydomonas efficiently uses the inorganic nitrogen forms ammonium, nitrate, and nitrite for growth. Ammonium is preferred versus the oxidized compounds, since its assimilation has less energetic cost and also because many genes involved in nitrate/nitrite assimilation are repressed in the presence of ammonium (Fernández and Cárdenas, 1982; Florencio and Vega, 1983a; Fernández et al., 1998). In spite of this, *Chlamydomonas* can utilize oxidized forms in the presence of ammonium under natural CO_2 conditions (air). In other words, ammonium repression is more efficient than nitrate induction at high CO_2, but nitrate can induce the assimilation route for oxidized N at low CO_2 even in the presence of ammonium (Rexach et al., 1999; Galván and Fernández, 2001; Llamas et al., 2002). In *Chlamydomonas* as in plants, nitrate is not only a major nutrient but also confers strong regulatory effects on both metabolic and developmental pathways and in *Chlamydomonas*, gametogenesis (Crawford, 1995; Pozuelo et al, 2000; Wang et al., 2000; Forde, 2002; Llamas et al., 2002; González-Ballester et al., 2005b).

Nitrate assimilation is an essential route for N acquisition in plants, algae, and microorganisms (Hardy and Havelka, 1975; Guerrero et al., 1981; Foyer and Ferrario, 1994; Fernández et al., 1998; Nosengo, 2003). Inorganic N assimilation has been intensively studied in a variety of organisms (Guerrero et al., 1981; Crawford, 1995; Marzluf, 1997; Crawford and

Glass, 1998; Campbell, 1999; Forde, 2000; Galván and Fernández, 2001; Williams and Miller, 2001; Siverio, 2002; Galván et al., 2006; Schwarz and Mendel, 2006). Most of the N consumed by animals is derived ultimately from the N forms assimilated by plants. The strong influence of inorganic N on plant productivity has resulted in overuse of fertilizers during the twentieth century, which has led to severe pollution, especially of aquatic ecosystems (Nosengo, 2003). In natural waters and soils, ammonium and nitrate are major forms of inorganic N but nitrite also appears in non-equilibrated biosystems as a consequence of at least three different biochemical pathways: nitrification, denitrification, and nitrate reduction to ammonium (Philips et al., 2002).

Inorganic nitrogen metabolism in *Chlamydomonas* is summarized in Figure 3.3, and consists of two transport and two reduction steps. First, nitrate enters the cell via specific transport systems. Next, nitrate is reduced to nitrite in the cytosol by nitrate reductase (NR). Nitrite is then transported to the chloroplast by specific transporters, where nitrite reductase (NiR) catalyses its reduction to ammonium. Finally, ammonium is incorporated into L-glutamate by the GS/GOGAT cycle.

Nitrite and ammonium utilization differ from nitrate assimilation in that NR or/and NiR are not required, and nitrate assimilation is dependent

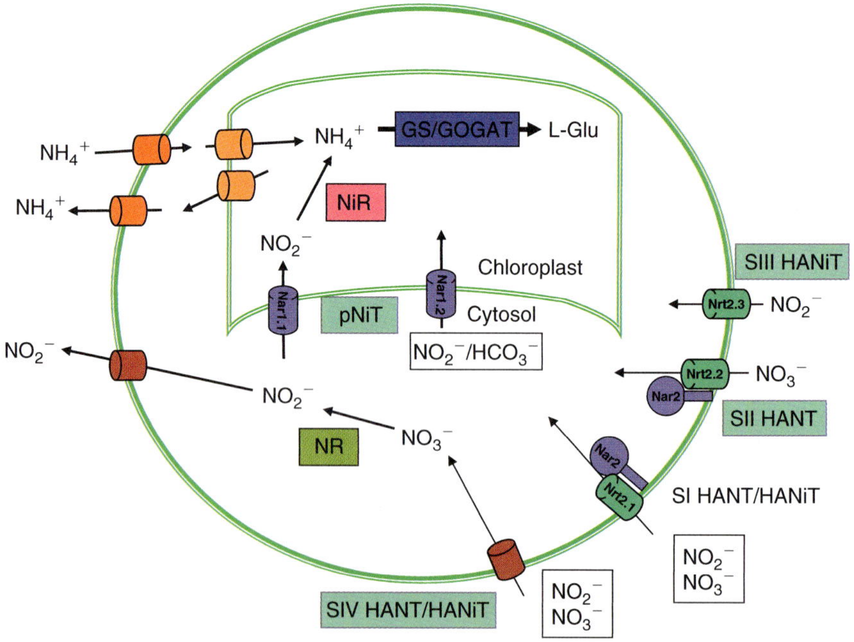

FIGURE 3.3 *The inorganic nitrogen assimilation pathway in Chlamydomonas. Transport steps for the uptake and excretion of N compounds are shown by open cylinders, and enzymatic steps in shaded squares. SI through SIV are four nitrogen transport systems discussed in the text.*

on Moco biosynthesis and its transfer to the apo-enzyme NR (Fernández et al., 1998; Schwarz and Mendel, 2006). On the other hand, nitrite and ammonium must cross two membranes to reach the chloroplast, the main site of ammonium incorporation, and thus transport systems are required. It is important to note that *Chlamydomonas* does not accumulate nitrate, nitrite, or ammonium within intracellular compartments. Instead, cells excrete nitrite or ammonium when their assimilation capacity is exceeded (Figure 3.3), thus avoiding any toxic effect of excessive intracellular amounts (Azuara and Aparicio, 1983; Navarro et al., 2000; Rexach et al., 2000; Mariscal et al., 2004). Nitrate, nitrite, and ammonium excretion activities have also been described in plants and other microorganisms (Aslam et al., 1994, 1996; Crawford and Glass, 1998). These efflux systems could play an important role in the homeostasis of inorganic nitrogen; however the molecular information on this topic is scarce, especially in plants and algae. The following parts of this chapter place *Chlamydomonas* inorganic nitrogen assimilation in the context of genome information, with emphasis on transporters, Moco biosynthesis and regulatory mechanisms.

III. AMMONIUM ASSIMILATION

A. Overview

The first step in the assimilation of extracellular ammonium is transport, with transporters expected to localize to the plasma membrane, the chloroplast, and mitochondria. Among these, the main flux should be into the chloroplast, which as the main site for ammonium assimilation contains the enzymes needed for its incorporation into carbon skeletons (Fischer and Klein, 1988). The possible distribution of transporters could be important to provide efficient metabolic exchange of ammonium among the organelles and cytosol, and the necessary efflux from the cell under limiting carbon conditions (Azuara and Aparicio, 1983; Rexach et al., 2000). Furthermore, ammonium is a strong negative regulatory signal for nitrate assimilation, and mechanisms regulating ammonium transport and incorporation could be involved in mediating its signaling effects.

Ammonium taken up from the medium or generated intracellularly from photorespiration, protein turnover, and nucleic acid catabolism, is incorporated into carbon skeletons by the GS/GOGAT cycle to produce L-glutamate (Figure 3.3; Miflin and Lea, 1975; Cullimore and Sims, 1980; Cullimore and Sims, 1981a,b; Hipkin et al., 1982; Miflin and Habash, 2002). Glutamate is a key molecule that couples the metabolic pathways of amino acids with those of carbohydrates and lipids (Torchinsky, 1987; Lam et al., 1996), with transamination reactions playing an important role in the control of internal amino acid pools (Muñoz-Blanco and Cárdenas, 1989; Lam et al., 1996). The properties and regulation of the GS/GOGAT cycle enzymes have been previously reviewed (Fernández et al., 1998), and are outlined and updated below.

B. Genome analysis for transport and metabolism of ammonium

1. *Ammonium transporters*

Although the weak base ammonia (NH_3) permeates the cell membrane at high pH, ammonium transport is required by the cells to make this process a regulated and efficient one at neutral pH. Two transport systems subject to circadian rhythms have been characterized physiologically in *Chlamydomonas*: a constitutively expressed low affinity and high capacity system (LATS) which might be related to passive K^+ channels (Ullrich et al., 1984), and a high affinity and low capacity system (HATS) which is negatively regulated by ammonium (Franco et al., 1987b, 1988; Byrne et al., 1992).

Ammonium LATS have not yet been described at the molecular level, but HATS have been characterized in plants, yeasts, bacteria, fungi, and animals, and constitute the AMT/MEP family (Marini et al., 1997; von Wirén et al., 2000; Crawford and Forde, 2002). The human rhesus (Rh) blood proteins are related to this family, and *Chlamydomonas* is one of the few organisms in which both gene families, AMT/MEP (Ammonium transporter/Ammonium permease) and RH, have been identified. A possible role of the *Chlamydomonas* RH proteins appears to be a bidirectional channel for ammonia as proposed in other systems (Marini and Andre, 2000; Westhoff et al., 2002). However, in *Chlamydomonas* there is also evidence that RH proteins are gas channels for CO_2 (Soupene et al., 2002, 2004). Tandemly duplicated *RHP* genes are found on linkage group VI (Figure 3.2B).

In *Chlamydomonas* putative ammonium transport genes encoding proteins of the AMT1 family have been isolated, painting a more complex picture than expected (González-Ballester et al., 2004; Kim et al., 2005). The *Chlamydomonas AMT1* gene family has eight members (*AMT1.1–AMT1.8*, also known as *AMT1A* through *AMT1H*) and no *AMT2* genes (Table 3.1; González-Ballester et al., 2004). The *Chlamydomonas AMT1* family is the largest described to date, and its members are scattered throughout the genome. AMT1.1–6 are phylogenetically closely related to the plant AMT proteins, whereas AMT1.7 and 1.8 are more distantly related to those of plants. The *AMT1* genes feature very small exons (some are less than 40 bp) and conserved positions of certain introns, which are probably related to ancestral gene duplications and to transcriptional regulatory mechanisms. *AMT1.5* and *AMT1.7* mRNAs are alternatively spliced in their 5′ regions, which may result in different protein localizations as suggested for isoforms deduced for AMT1.7 (González-Ballester et al., 2004). Although experimental evidence for the localizations of AMT1 proteins is unavailable, AMT1.2 and

AMT1.4 have predicted chloroplast transit peptides. The large number of AMT1 transporters might reflect the need for complementary affinities and capacities to obtain ammonium efficiently under changing environmental conditions. In addition, they may participate in a regulated intracellular flux since metabolic reactions involving ammonium take place in the cytoplasm, chloroplast, and mitochondria, and ammonium can accumulate in the vacuole or be excreted to the medium (Howitt and Udvardi, 2000; Williams and Miller, 2001; Glass, 2003). On the other hand, some AMT proteins may have partially or fully redundant functions.

Some evidence for complementary functions comes from mRNA accumulation studies for *AMT1* genes, of which *AMT1.4* and *AMT1.7* have the highest overall expression. *AMT1.1, 2, 4,* and *5* have maximal expression in N-free medium, whereas for *AMT1.3* and *8* this occurs in nitrate medium. Expression of *AMT1.6* and *7* is enhanced in ammonium medium (González-Ballester et al., 2004). Expression of *AMT1.1–4* is also strongly enhanced in media containing a poor nitrogen source such as arginine (Kim et al., 2005). *AMT1.1* exhibits complex regulation, with repression by ammonium, ammonium derivatives and nitrate. Nitrate effects are twofold, one being derived from ammonium produced during its assimilation, and the other from a negative signal generated by the *NIT2* gene product in the presence of nitrate. This observation shows that ammonium uptake and nitrate assimilation have overlapping regulation and that *NIT2* has dual roles, a positive effect on nitrate assimilation, and a negative effect on *AMT1.1* expression (Fernández et al., 1998; González-Ballester et al., 2004).

Methylammonium, an ammonium analogue, is non-metabolizable in *Chlamydomonas*. This has given insight into ammonium transport processes through the characterization of methylammonium-resistant mutants (Franco et al., 1984, 1988). Mutants defective in two systems with respectively high and low affinity for ammonium transport were described, although the complexity of the *AMT1* gene family precludes easy identification of candidate genes. Sixteen spontaneous mutants affected in methylammonium accumulation capacity and uptake were also characterized (Kim et al., 2005), and found to be mainly associated with transposable element insertions or related events (Kim et al., 2006). Twelve of them carry mutations in *AMT1.4,* and two others have lesions in different loci. These mutants have [^{14}C]methylammonium uptake activities that correlate with their respective accumulation or uptake phenotypes. The properties of the *amt4* mutants are very different from those of *RHP1* RNAi lines, supporting the idea that the physiological substrates for AMT and RH proteins are NH_3 and CO_2, respectively (Kim et al., 2005).

Ammonium is also sensed chemotactically by *Chlamydomonas* vegetative cells, which swim towards it (Ermilova et al., 1998). This behavior is not exhibited by gametes, although pregametes obtained in the dark in the absence of an N source are chemotactically competent (Ermilova et al.,

2003, 2004). These pregametes lose chemotaxis during gametogenesis in a process mediated by the blue light receptor phototropin. A connection has been proposed between ammonium transport and chemotaxis to ammonium (Ermilova et al., 2004).

2. GS/GOGAT

GS catalyzes the ATP-dependent formation of glutamine from ammonium and glutamate. Like vascular plants, *Chlamydomonas* has cytosolic (GS1; *GLN1* gene) and plastidic (GS2; *GLN2*) isoforms, both of which are homooctamers (Florencio and Vega, 1983b; Fischer and Klein, 1988). GS can be inactivated in the dark or by ammonium treatment, and reactivated *in vivo* upon reillumination (Cullimore and Sims, 1981b) or *in vitro* by thioredoxins (Florencio et al., 1993). GS has also been shown to be phosphorylated by a Ca^{2+}- and calmodulin-dependent protein kinase and to bind 14-3-3 proteins without affecting its enzyme activity, which could be related to another step of regulation (Pozuelo et al., 2001).

Expression of *GLN1* and *GLN2* transcripts and GS enzyme activity respond differentially to nitrate or ammonium (Fischer and Klein, 1988; Chen and Silflow, 1996). *GLN1* transcripts are more abundant in nitrate-grown cells than in ammonium-grown cells, while *GLN2* transcripts accumulate constitutively (Chen and Silflow, 1996). Thus, in cells transferred from nitrate to ammonium there is a decrease in total GS activity and an increase in the GS2:GS1 ratio (Fischer and Klein, 1988). It has been proposed that GS activity might control, through ammonium derivatives, the nitrate assimilation pathway (Cullimore and Sims, 1981a; Hoff et al., 1994). An increase of 30% in plastidic GS2 activity is observed upon adaptation from high to low CO_2 (Ramazanov and Cárdenas, 1994), without significant changes in transcript abundance (Chen and Silflow, 1996), suggesting a post-transcriptional regulatory mode. It is interesting that *GLN2* and a nearly identical gene, *GLN3*, are head-to-head on linkage group XII/XIII (Figure 3.2B), and not linked to *GLN1* and *GLN4*.

Glutamine oxoglutarate amidotransferase (GOGAT) catalyzes the transfer of the amide group from glutamine to 2-oxoglutarate, in a reaction which requires reducing equivalents from NADH or reduced ferredoxin (Fd) and produces two molecules of glutamate, the substrate of GS (Lea and Miflin, 1975). *Chlamydomonas* has two GOGAT isoenzymes in the chloroplast, which have been purified and characterized (Cullimore and Sims, 1981b; Galván et al., 1984; Márquez et al., 1984, 1986b; Fischer and Klein, 1988), and whose genes have been annotated (Table 3.1). Fd-GOGAT and NADH-GOGAT are probably compensatory for glutamate production depending on the supply of their specific electron donors, and whether growth conditions are photosynthetic or mixotrophic. Accordingly Fd-GOGAT activity, in contrast to that of NADH-GOGAT, increases in the light in synchronous and asynchronous cultures, and is also higher in ammonium than in nitrate media (Cullimore and Sims, 1981b; Márquez et al., 1986a; Martínez-Rivas et al., 1991).

3. Glutamate dehydrogenase

NAD(P)H-glutamate dehydrogenase (GDH) catalyzes the reversible reductive amination of 2-oxoglutarate to form glutamate, and appears to be encoded by two genes (Table 3.1). Though the *Chlamydomonas* enzyme has been purified and characterized, its role in nitrogen assimilation has been controversial (Cullimore and Sims, 1981b; Fischer and Klein, 1988; Muñoz-Blanco et al., 1989; Muñoz-Blanco and Cárdenas, 1989). Three isoenzymes named GDH1, GDH2, and GDH3 have been reported (Muñoz-Blanco et al., 1989; Moyano et al., 1992a). In contrast to Fischer and Klein (1988) who found GDH activity exclusively in the chloroplast, Moyano et al. (1992b) reported a mitochondrial localization for the three isoenzymes, which are differentially regulated depending on trophic and environmental conditions (Muñoz-Blanco et al., 1989; Moyano et al., 1995). It has been proposed that GDH1 could participate in ammonium assimilation, whereas GDH2 and GDH3 might have an anaplerotic role by producing 2-oxoglutarate (Moyano et al., 1995). The role of the glutamine synthetase-glutamate synthase pathway and glutamate dehydrogenase in the source-sink nitrogen cycle has been analyzed in tobacco (Masclaux-Daubresse et al., 2006). Regardless of leaf age, ammonium assimilation occurs via the GS/Fd- or NADH-GOGAT both in the light and darkness and not via GDH, which has a role in the deamination of Glu to provide 2-oxoglutarate and ammonium.

IV. ASSIMILATION OF NITRATE AND NITRITE

A. Overview

Chlamydomonas has single-copy genes encoding nitrate reductase (NR; *NIA1* or *NIT1*) and nitrite reductase (NiR; *NII1*), but a more complex scenario with respect to putative nitrate and nitrite transporters (N/NiT). The N/NiTs have been identified in many eukaryotes, and collectively correspond to the three families *NRT1*, *NRT2*, and *NAR1* (Crawford, 1995; Crawford and Glass, 1998; Forde, 2000; Galván and Fernández, 2001; Crawford and Forde, 2002; Siverio, 2002). It is instructive to compare the gene complement in *Arabidopsis*, *Chlamydomonas*, and *Ostreococcus* for the three transporter types: *NRT1* (4 At, 1 Cr, 1 Ot), *NRT2* (7 At, 6 Cr, 1 Ot), and *NAR1* (0 At, 6 Cr, 1 Ot). *NAR2*, which encodes an additional component for some NRT2 transporters, is also present in the three genomes (2 At, 1 Cr, 1 Ot; Derelle et al., 2006). This comparison shows that while *Ostreococcus* has a single gene for each transporter type, *Chlamydomonas* appears as complex as *Arabidopsis*, or even more so given the absence of *NAR1* genes from *Arabidopsis*. These data naturally raise the question of biological significance, especially in the unicellular *Chlamydomonas*. While some redundancy may exist, the wide adaptability of *Chlamydomonas* to nutritional and environmental changes is likely to be germane.

With respect to nitrate assimilation, *Chlamydomonas* features two gene clusters on linkage group IX (Figure 3.2C), most of which are nitrate-regulated and under control of the regulatory gene *NIT2* (Table 3.1). The main cluster is in a region of about 40 kb, and contains *NIA1*, *NII1*, *NAR1.1* (which encodes a plastidial nitrite transporter), and *NAR2*, *NRT2.1*, and *NRT2.2* (which encode nitrate transporter components). A gene encoding plastidial NADP-dependent malate dehydrogenase (*MDH*), not regulated by *NIT2*, is also present and seems to play an important role in the supply of reducing power for nitrate reduction (Quesada et al., 1993, 1994, 1998a,1998b; Gómez et al., 2002). The second cluster contains two genes related to nitrate assimilation, *NRT2.3* (which encodes a nitrite transporter) and *AOX1* (which encodes a mitochondrial alternative oxidase; Quesada et al., 1998a; Rexach et al., 1999). Both genes are nitrate-regulated, but only *NRT2.3* control depends on *NIT2*.

The clustering of nitrate assimilation genes in microorganisms is widespread, with the list including *Aspergillus nidulans* (Johnstone et al., 1990), *Hansenula polymorpha* (Pérez et al., 1997), *Chlamydomonas* (Quesada et al., 1993), and *Ostreococcus* (Derelle et al., 2006). Gene number and organization vary, but all the clusters contain the essential genes for nitrate/nitrite transport and its reduction, and in *Ostreococcus* additional genes for Moco metabolism. In plants, clustering of nitrate assimilation genes has only been reported in *Arabidopsis* where *NRT2.1* and *NRT2.2* are in a tail-to-tail configuration on Chromosome I, and *NRT2.3* and *NRT2.4* in a head-to-tail configuration on Chromosome V (Orsel et al., 2002). Thus, it appears that these clusters arose not from horizontal gene transfer, but perhaps as a cellular strategy to optimize the regulation of this pathway.

B. Nitrate and nitrite transport systems

1. NRT1 transporters

The existence of an NRT1-type transporter in *Chlamydomonas* is based on genome analysis (Table 3.1). NRT1 transporters, also called PTR (for peptide transporter) or POT (for proton-dependent oligopeptide transporters), are widely distributed in eukaryotes and prokaryotes (Paulsen and Skurray, 1994; Pao et al., 1998; Forde, 2000; Crawford and Forde, 2002). *NRT1* genes are particularly abundant in plants. *Arabidopsis*, for example, has 51 *NRT1* genes, although only four appear to function in nitrate transport (Okamoto et al., 2003). NRT1 proteins are predicted to have 12 membrane-spanning domains, separated into two sets of six by a central loop containing multiple charged residues. Both the N- and C-termini are short (Tsay et al., 1993; Forde, 2000; Crawford and Forde, 2002).

The first NRT1 characterized in relation to nitrate transport corresponds to AtNRT1.1 (CHL1). *AtNRT1.1* mutants were originally isolated based on chlorate resistance (Oostindiër-Braaksma and Feenstra, 1973), and shown to be defective in uptake of both ClO_3^- and NO_3^- (Doddema and Telkamp, 1979). Subsequently, the gene was cloned by T-DNA tagging (Tsay et al., 1993). Although initially described as a low affinity nitrate transporter, AtNRT1.1 is considered as a dual transporter that shows high affinity (HANTS) or low affinity (LANTS) depending on its phosphorylation state. The phosphorylated form (HANTS) is triggered by limiting external NO_3^- and facilitates adaptation to nutrient supply (Wang et al., 1998; Liu et al., 1999; Liu and Tsay, 2003). Further analysis showed that AtNRT1.1 is required for young organ development and stomatal opening, and contributes to drought susceptibility (Guo et al., 2001, 2003). Functional studies with other plant NRT1 transporters such AtNRT1.2 and *Brassica napus* NRT1.2 showed them as LANT (Huang et al., 1999; Zhou et al., 1998), and BnNRT1.1 was able to transport L-histidine as well as nitrate (Zhou et al., 1998). The function of the single *Chlamydomonas NRT1* gene awaits experimental analysis.

2. NRT2 transporters

NRT2 genes are also widely distributed in eukaryotes and prokaryotes (Crawford and Glass, 1998; Forde, 2000; Galván et al., 2001). NRT2 transporters (also named NNP, for nitrate–nitrite porter) have a typical carrier-type structure with 12 transmembrane domains divided into two sets. A central loop connects the sets of transmembrane domains, and the N- and C-termini are predicted to be cytoplasmic. Based on length of the central loop and N- or C-termini, NRT2 proteins have been classified into three types (Forde, 2000). Type I proteins, represented by *E. coli* NarK, have short sequences outside the transmembrane domains. Type II, represented by fungal members, have a large hydrophilic central loop between transmembrane domains 6 and 7, which is related to ubiquitination and subsequent proteolysis in the vacuole in response to glutamine (Navarro et al., 2006). Type III, represented by *Chlamydomonas* and plant members, have long C-termini.

Type III NRT2 functionality requires in some cases an auxiliary protein with a single transmembrane domain. This protein was first identified in *Chlamydomonas* and named NAR2 (for nitrate assimilation-related protein 2; Quesada et al., 1993, 1994; Galván et al., 1996), and later characterized in plants (Tong et al., 2005; Orsel et al., 2006; Okamoto et al., 2006). The requirement for NAR2 allows classification of NRT2 transporters into two categories, single (NRT2) or two-component (NRT2/NAR2) systems (Galván and Fernández, 2001; Orsel et al., 2006). The precise role of NAR2 in regulating function or affinity of NRT2 carrier proteins for nitrate remains to be elucidated, but in both *Chlamydomonas* and *Arabidopsis* it is required for HANTS (Galván et al., 1996; Zhou et al., 2000; Tong et al., 2005; Orsel et al., 2006; Okamoto et al., 2006). In addition, it was shown that AtNAR2.1 and AtNRT2.1 interact when expressed in yeast (Orsel et al., 2006). The *Chlamydomonas* NAR2 protein is required for functionality of NRT2.1 and NRT2.2, but not for the other putative NRT2 transporters (NRT2.3–6; Quesada et al., 1994; Galván et al., 1996). Also AtNAR2.1 shows specificity for AtNRT2.1 but not for the other NRT2's (Orsel et al., 2006). So, it is clear that not all NRT2 proteins require NAR2, and also that each protein requires very precise interactions to distinguish its partner.

In plants, physiological studies revealed different types of nitrate transport systems, which were named as follows: The inducible high affinity transport system (iHATS) is a saturatable system able to take up nitrate at low external concentration (1 micromoles–1 mM), and requires nitrate as an inducer. The constitutive high affinity transport system (cHATS) operates at the same low nitrate concentration range, but is not nitrate-induced. The low affinity transport system (LATS) contributes to nitrate uptake only when external nitrate concentrations are high (>1 mM; Glass and Siddiqi, 1995; Orsel et al., 2006). In *Chlamydomonas* HANT and LANT activities can also be measured, and altogether four transport systems have been characterized according to their ion specificities and inhibition characteristics (Figure 3.3). Transport system I (HANT/HANiT) is bispecific for nitrate and nitrite with K_M values of 1.6 micromoles and 1.8 micromoles for NO_3^- and NO_2^-, respectively, and requires the *NRT2.1* and *NAR2* gene products. System II (HANT) is nitrate-specific (K_M = 11 micromoles), and requires *NRT2.2* and *NAR2* (Galván et al., 1996; Zhou et al., 2000). System III (HANiT) is *NAR2*-independent and presumably corresponds to *NRT2.3*, and shows high affinity for nitrite (K_M=5 micromoles), but a low affinity for nitrate (Galván et al., 1996; Rexach et al., 1999). These three systems are nitrate-induced, under the regulatory control of *NIT2*, operate optimally at high CO_2, and are inhibited by ammonium but not by chloride. Systems I and II not only have high affinity for nitrate, but also have the highest capacity for mediating efficient growth in nitrate. Mutants lacking these two systems grow very slowly on nitrate as pale green colonies, even though other putative transport genes exist.

System IV (HANT/HANiT) is also bispecific for nitrate and nitrite, with K_M values of 40 micromoles and 33 micromoles for NO_3^- and NO_2^-, respectively. This is an orphan system since the encoding gene is unknown, and could correspond to a new *NRT2* member or another putative nitrate/nitrite transporter gene (Rexach et al., 1999; Navarro et al., 2000). System IV shows differential regulation in the sense that it operates efficiently at low CO_2, and is inhibited by chloride but not ammonium (Galván et al., 1996; Rexach et al., 1999).

3. NAR1 transporters

NAR1.1 (for nitrate assimilation-related component 1) defined the first chloroplast nitrite transporter (Quesada et al., 1993; Rexach et al., 2000), at a point when a lack of molecular evidence and proposed diffusion mechanisms had created uncertainty about the process. The NAR1.1 work both provided molecular evidence for a regulated process, and showed that nitrite transport at this level depends on carbon availability (Rexach et al., 2000; Galván, et al., 2002; Mariscal et al., 2004). That NAR1.1 is not the only chloroplast nitrite transporter in *Chlamydomonas* is supported by physiological data from *nar1.1* mutants, as well as the existence of a *NAR1* gene family (Table 3.1; Rexach et al., 2000; Galván et al., 2002; Mariscal et al., 2006).

As mentioned above, *NAR1.1* has a strategic localization in the nitrate cluster, linked to *NIA1*, and is coregulated with the other clustered genes. Thus, *NAR1.1* is expressed in nitrate but not ammonium medium and is under the control of *NIT2* (Quesada et al., 1993). NAR1.1 corresponds to an integral membrane protein with six membrane-spanning domains and chloroplast localization, based on immunoblot analysis (Rexach et al., 2000). By mediating the transport of nitrite into the chloroplast, NAR1.1 allows nitrate utilization when this nutrient is limiting. This limitation occurs at millimolar nitrate in strains lacking the HANT systems I and II, or at micromolar concentrations in strains having HANT (Rexach et al., 2000). NAR1.1 also improves efficiency of nitrate use at low CO_2. Under this condition, *nar1.1* mutants uncouple nitrate reduction from the capability to assimilate it, leading to an overproduction of ammonium from nitrate reduction and deregulated expression of enzymes and transporters for nitrate assimilation, including GS (Rexach et al., 2000; Mariscal et al., 2004).

NAR1 belongs to the FNT (formate nitrite transporters) family (Peakman et al., 1990; Suppmann and Sawers, 1994; Rexach et al., 2000). In contrast to the ubiquitous NRT1 and NRT2 transporters, FNT/NAR1 is absent from plants and animals (Mariscal et al., 2006). Most other organisms contain a single gene copy, including *Ostreococcus* as well as some fungi, algae, and protozoa. A few organisms contain several FNT/NAR1-encoding genes, such as *E. coli* which contains three (*focA*, *focB*, and *nirC*). *Chlamydomonas*, however, has the greatest number so far discovered, with six (Mariscal et al., 2006). It is remarkable that organisms such as *Saccharomyces cerevisiae* and *Plasmodium falciparum* encode NAR1-like proteins but do not assimilate nitrate or nitrite, suggesting that these proteins might fulfill important membrane influx or efflux activities for nitrite or other structurally related monovalent anions such as formate or bicarbonate. It should be noted in this context that *Chlamydomonas* has a fermentative metabolism under anaerobiosis producing, among other compounds, formate from starch degradation (Kreuzberg, 1984; Atteia et al., 2006; Grossman et al., 2007).

Analysis of the *Chlamydomonas NAR1* gene family suggests several membrane localizations for the encoded proteins. NAR1.1 and NAR1.2 are predicted to be chloroplast-targeted, whereas for NAR1.5 chloroplast and/or mitochondrial localization is predicted. In contrast, NAR1.3, NAR1.4, and NAR1.6 are predicted to be plasma membrane transporters. The NAR1 proteins fall into two classes, with NAR1.1, NAR1.2, NAR1.4, and NAR1.5 having short C-termini, and NAR1.3 and NAR1.6 long C-termini, but the significance of this characteristic is unknown.

NAR1 genes are differentially regulated by carbon and nitrogen sources, and the regulatory genes *CCM1* and *NIT2*. *NAR1.1* and *NAR1.6* respond to nitrogen source and *NIT2* (Rexach et al., 2000; Mariscal et al., 2004, 2006), whereas *NAR1.2* responds to carbon source and *CCM1*, which is the central regulatory gene for carbon assimilation (Miura et al., 2004; Mariscal et al., 2006; see Chapter 8). *NAR1.3*, *NAR1.4*, and *NAR1.5* appear to be *NIT2/CCM1*-independent. Functionality of NAR1.2 as a nitrite and bicarbonate transporter was successfully obtained in *Xenopus* oocytes (Mariscal et al., 2006). Coupled with data on NAR1.1 (Mariscal et al., 2004), one may conclude that nitrite and bicarbonate transport into the chloroplast are intertwined processes. This is consistent with the concept that *Chlamydomonas NAR1* family members have a role in integrating C and N metabolism.

In summary, a complex network of transporters for nitrate/nitrite metabolism is defined by six *NRT2*, six *NAR1*, and probably one *NRT1* gene in *Chlamydomonas*. Although the precise roles of a few of these are known, how they are regulated to maximize efficient use of nitrate, and how their activities are regulated in the context of other nutrient availability, remain to be established.

C. Nitrate and nitrite reduction

1. Nitrate reductase and its regulation

As in other photosynthetic eukaryotes, *Chlamydomonas* NR is a homodimer that uses pyridine nucleotides as electron donor for nitrate reduction to nitrite. Each monomer is a 105-kD polypeptide that contains three prosthetic groups, FAD, heme b_{557}, and Moco, which are attached to three functional domains separated by two short hinge regions (Fernández and Cárdenas, 1983; Kalakoutskii and Fernández, 1995; Campbell and Kinghorn, 1990; Moura et al., 2004). Eukaryotic NR sequences are highly conserved and differ from cyanobacterial NR, which uses ferredoxin as an electron donor, as do a number of other prokaryotic NRs (Campbell, 1999; Moreno-Vivián et al., 1999). NR exhibits two partial activities named diaphorase (from NAD(P)H to artificial electron acceptors) and terminal NR (from artificial electron donors to nitrate). These partial activities have been very useful in characterizing NR-deficient mutants (Fernández et al.,

1998). The crystal structure of a recombinant, catalytically active fragment of *Pichia angusta* molybdenum-containing NR-Mo has been obtained (Fischer et al., 2005), which identified key Arg and Trp residues as candidates for participation in nitrate binding. Four ordered water molecules in close proximity to Mo define a nitrate binding site, a penta-coordinated reaction intermediate, and product release. The sequence similarity among eukaryotic NAD(P)H:NRs has allowed elaboration of a general mechanism for catalytic nitrate reduction (Fischer et al., 2005).

One or two genes can be identified for NR in different organisms. For example, barley and *Arabidopsis* contain two structural loci (*NIA1* and *NIA2*), *NIA1* encoding the most abundant NADH-dependent isoform. In other organisms, including *Chlamydomonas*, only one gene is responsible for NR activity (Table 3.1; Fernández et al., 1989; Campbell, 1999). The *Chlamydomonas NIA1* or *NIT1* gene was successfully used to develop a transformation system in the alga by complementation of NR-deficient mutants (Fernández et al., 1989; Kindle et al., 1989; Kindle, 1990), and subsequently harnessed for approaches such as insertional mutagenesis (Tam and Lefebvre, 1993; Prieto et al., 1996; Zhang and Lefebvre, 1997; Pérez-Alegre et al., 2005).

NR activity is tightly regulated by environmental factors such as light, nitrogen, and carbon availability. Reversible inactivation of plant NR depends on phosphorylation status and binding of a 14-3-3 protein (Bachmann et al., 1996; Moorhead et al., 1996; Lillo et al., 1997; Kaiser et al., 2002; MacKintosh, 2004). The interactions of 14-3-3 proteins with NR and other metabolic enzymes are sequence-specific and in the case of *Arabidopsis* NR, operate through phosphorylation of Ser-534 in the first hinge region (Kanamaru et al., 1999). The function of 14-3-3-mediated phosphorylation is not clear and has been related to protein turnover and daily rhythms of sugar metabolism in coordination with photosynthesis, ATP production, and nitrate reduction (Huber et al., 2002; Comparot et al., 2003; MacKintosh, 2004). However, *Chlamydomonas* NR is not affected in this respect by 14-3-3 proteins, since it lacks the conserved plant sequence in the first hinge region (Pozuelo et al., 2001).

In contrast, *Chlamydomonas* NR is subject to redox regulation, becoming over-reduced and inactivated in the absence of nitrate (Guerrero et al., 1981; Fernández et al., 1998). Inactive NR can be reactivated *in vivo* by nitrate resupply and *in vitro* by ferricyanide oxidation. This interconversion takes place both *in vivo* and *in vitro*. Inactivation of *Chlamydomonas* NR has been related to enzyme turnover since enzyme half-life decreases considerably upon inactivation, and the diaphorase activity of a Moco-deficient mutant incapable of being inactivated is very stable under all conditions (Franco et al., 1987a).

Light regulates nitrate assimilation pathway genes and activities, subject to circadian rhythms (Quesada and Fernández, 1994; Pajuelo et al., 1995;

Mittag and Waltenberger, 1997), with nitrate and nitrite uptake generally peaking during the light period and reaching a minimum in the dark. The endogenous circadian clock in *Chlamydomonas* controls the RNA-binding protein Chlamy 1, an analog of the *trans*-acting factor CCTR (Clock controlled regulator) from the dinoflagellate *Gonyaulax polyedra* (Mittag and Waltenberger, 1997). Chlamy 1 is thought to act as a translational repressor, targeting mRNAs containing 3′ UTR [U + G] repeats from the end of the light phase until the end of the dark phase. Some nitrogen assimilation genes have this repeat motif and thus may be under Chlamy 1 control. These include *NRT2.3*, *NII1*, *GLN2*, and *ARG7* (for arginine biosynthesis), as well as some photosynthesis and CO_2 assimilation-related genes including *LIP36*-G, *RBCS1*, and the *POR* gene encoding NADPH-protochlorophyllide-oxidoreductase (Li and Timko, 1996; Mittag and Waltenberger, 1997; Galván et al., 2006).

Chlamydomonas NIA1 expression is also regulated by the redox state of the plastoquinone pool. A reduced plastoquinone pool is a positive signal that allows *NIA1* expression, whereas chemical inhibitors or mutations preventing loading of plastoquinone with electrons have a negative effect (Giordano, et al., 2005). In contrast, an oxidized state of the electron transport chain appears to activate expression in plants (Sherameti et al., 2002), which might reflect differences in the physiology and biochemistry of nitrate reduction in plants and algae.

2. Nitrite reductase and its regulation

Nitrite reductase (NiR) is a stromal enzyme that catalyses a six-electron step reduction of nitrite to ammonium using reduced Fd as electron donor. The holoenzyme is encoded by a single gene clustered with other nitrate assimilation genes (Figure 3.2C). The 63-kD protein contains two redox centers: a [4Fe-4S] cluster and the characteristic siroheme. The N-terminal part of the protein binds Fd, and the C-terminal moiety the [4Fe-4S] redox center (Vega and Kamin, 1977; Campbell and Kinghorn, 1990; Quesada et al., 1998b). *Chlamydomonas* NiR has been shown to share common antigenic determinants with Fd-GOGAT (Romero et al., 1988).

Eukaryotic NiR has no regulation at the activity level, and changes in enzyme amounts appear to be due to transcriptional regulation. *Chlamydomonas* NiR shows a similar regulatory pattern to NR, and is maximally expressed in the absence of ammonium and the presence of light (Martínez-Rivas et al., 1991; Quesada et al., 1998b). Reducing power is essential for nitrate and nitrite reduction and it is mainly supplied by the photosynthetic electron transfer chain, which generates the requisite reduced Fd. NADPH-ferredoxin oxidoreductase and redox valves feed the cytosol with NAD(P)H for nitrate reduction. Accordingly, glucose-6-phosphate dehydrogenase and Calvin cycle activities increase in *Chlamydomonas* grown in nitrate media (Huppe and Turpin, 1996).

3. Energy supply for nitrate and nitrite assimilation

Two genes present in the linkage group IX cluster (Figure 3.2C), *MDH* and *AOX1*, participate in the fine-tuning of redox energy and ATP supply necessary for nitrate assimilation (Quesada et al., 2000). Plastidic MDH is a component of the malate-oxalacetate shuttle, which participates in the export of reducing equivalents from the chloroplast to the cytosol. The activity of this redox valve is regulated by light, mediated by the thioredoxin-ferredoxin system (Gómez et al., 2002, Lemaire et al., 2005; see Chapter 11). Two *AOX* genes are known in *Chlamydomonas*, *AOX1* being much more strongly expressed and unlike *AOX2*, regulated by nitrate (Quesada et al., 2000; Dinant et al., 2001; Baurain et al., 2003). AOX1 is a mitochondrial terminal oxidase that drives electrons to oxygen from reduced ubiquinone, bypassing Complexes III and IV of the mitochondrial electron transfer chain (see Chapter 13). AOX1 activity leads to a lower electrochemical potential, and thus of the amount of ATP per electron pair. AOX1 could play a role as a redox valve to balance carbon metabolism and mitochondrial electron transport (Vanlerberghe and McIntosh, 1997). Nitrate specifically induces *AOX1* expression in cells growing in ammonium medium, adapting mitochondrial electron flow and ATP synthesis during assimilation of these nitrogen sources. Under conditions resulting in high cellular nitrite concentrations, AOX1 activity is critical to maintaining ATP production, since NO inhibits the mitochondrial electron transport chain. This NO might come from nitrite reduction mediated by cytosolic NR as shown in *Chlamydomonas* (Sakihama et al., 2002), or by mitochondrial AOX as shown in *Chlorella* (Tischner et al., 2004). Whether NO production is a stress response or results from a metabolic condition that triggers a signaling cascade requires further investigation.

D. Molybdenum metabolism

1. Molybdenum in molybdoenzymes

Molybdenum (Mo) is an essential micronutrient. All Mo enzymes studied, except nitrogenase, have Mo activated and chelated into a prosthetic group named molybdenum cofactor (Moco). Some of the most prominent Moco enzymes are NR, aldehyde oxidase (abscisic acid biosynthesis), xanthine dehydrogenase (purine degradation), and sulfite oxidase (sulfite detoxification) (Eilers et al., 2001; Mendel and Bittner, 2006). Moco consists of Mo covalently bound to the molybdopterin (MPT) cofactor (Rajagopalan and Johnson, 1992), which is highly conserved in eukaryotes, eubacteria, and archaebacteria.

2. Moco biosynthesis

Moco is synthesized by a conserved pathway that can be divided into several steps corresponding to the production of the intermediates cPMP

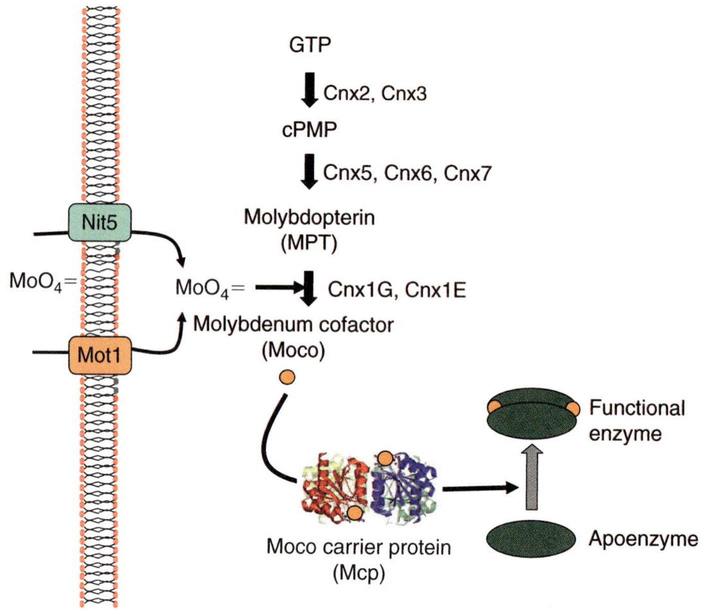

FIGURE 3.4 *Molybdenum cofactor biosynthesis in Chlamydomonas. The proposed pathway for biosynthesis of functional Moco from molybdate and GTP in Chlamydomonas, and its attachment to apoproteins, are discussed in detail in the text.*

(cyclic pyranopterin monophosphate), MPT, adenylated MPT, and finally Moco (Schwarz and Mendel, 2006). These steps are summarized according to plant information and nomenclature CNX (Cofactor for Nitrate reductase and Xanthine dehydrogenase). Figure 3.4 shows a model for Moco biogenesis in *Chlamydomonas*.

Conversion of GTP to cPMP: A complex reaction sequence starting from GTP generates the first stable intermediate, cPMP. In plants cPMP synthesis is catalyzed by CNX2 and CNX3. CNX2 belongs to the family of S-adenosyl methionine (SAM)-dependent radical enzymes (Sofia et al., 2001). The catalytic mechanism of CNX3 has not been determined.

Molybdopterin synthesis: MPT is formed by the incorporation of two sulfur atoms into cPMP in a reaction catalyzed by MPT synthase, a heterotetrameric complex of two small and two large subunits encoded by *CNX6* and *CNX7*, respectively. In a separate reaction, sulfur is transferred to the small subunit of MPT synthase by the MPT synthase sulfurtransferase, encoded by *CNX5*.

Metal insertion: MPT synthesis is followed by the incorporation of Mo, which is catalyzed by the two-domain protein CNX1 (Stallmeyer et al., 1995). The CNX1G domain catalyzes the adenylation of MPT in a Mg^{2+}- and ATP-dependent manner, and the MPT-AMP intermediate is transferred to the CNX1E domain (Llamas et al., 2004). CNX1E was shown to couple the cleavage of MPT-AMP to Mo insertion, yielding Moco. This reaction is molybdate and Zn^{2+}/Mg^{2+}-dependent (Llamas et al., 2006). Upon its formation, Moco is released from CNX1E for assemblage into apomolybdoenzymes. In plants, CNX1 is encoded by a single gene (*CNX1*) whereas in *E. coli* the E and G domains are encoded by *moeA* and *mogA*, respectively (Schwarz and Mendel, 2006).

Moco maturation: Moco biosynthesis is completed once the metal is incorporated into the pterin moiety. In plants, however, additional modifications occur before cofactor insertion into apoenzymes. Mo-apoproteins of the NR/SO family include a conserved cysteine where Moco is covalently linked. In enzymes of the XDH/AO family, however, this essential sulfur ligand is available neither from the apoprotein nor from the Moco moiety, and is added enzymatically by Moco sulfurase, encoded in plants by *ABA3* (Leon-Kloosterziel et al., 1996). Two unlinked *ABA3*-like genes have been identified in *Chlamydomonas* (Table 3.1). Regarding Moco insertion into NR or SO, the mechanism for the specific linking of Moco to the third sulfur ligand of the apoenzyme has not been elucidated.

3. Storage and transfer of molybdenum cofactor

Moco is oxygen-sensitive and very labile. It is therefore assumed that there is a negligible amount of free Moco, and that the cofactor is transferred immediately following biosynthesis to the apoenzyme, or bound to a carrier protein (MCP) for its protection and storage. MCP activity was first described in *E. coli* (Amy and Rajagopalan, 1979) and subsequently in *Chlamydomonas* (Aguilar et al., 1992); however only *Chlamydomonas* MCP has been molecularly characterized.

Purified *Chlamydomonas* MCP is a homotetramer of 16 kD subunits and its activity of Moco transfer to apoNR takes place only if the proteins are in physical contact (Witte et al., 1998; Ataya et al., 2003). *MCP1* was cloned and maps to linkage group IX, but is not part of the nitrate assimilation cluster. Its deduced amino acid sequence has a significant similarity only to bacterial proteins of unknown function (Ataya et al., 2003). The crystal structure of MCP was determined at 1.6 Å resolution. The monomer constitutes a Rossmann fold with two homodimers forming a symmetrical tetramer in solution. MCP binds Moco as well as the tungstate-substituted form of the cofactor (Wco) with high affinity, while MPT cannot be bound (Fischer et al., 2006).

4. Molybdate uptake

Molybdate transport and homeostasis are well-studied in bacteria, but poorly understood in eukaryotes. In *Chlamydomonas* genetic evidence suggests that molybdate uptake is mediated by specific transporters, and two distinct molybdate uptake activities have been reported (Llamas et al., 2000). A high-capacity system is related to *NIT5* function, and a lower-capacity system to *NIT6* (Figure 3.4; Llamas et al., 2000). The identities of these genes are currently unknown (see following section).

5. *Chlamydomonas* Moco mutants

Moco is produced constitutively in *Chlamydomonas*, with amounts being maximal in nitrate-grown cells and during exponential growth (Aguilar et al., 1991). There is no correlation between the synthesis of Moco and NR, as described for *Nicotiana tabacum* (Mendel et al., 1984; Aguilar et al., 1991). Moco species are mainly distributed in two pools, one from which Moco can be released, demonstrated by its ability to reconstitute directly NR from apoNR of the *Neurospora crassa nit-1* mutant, and another in which Moco is tightly bound to molybdoenzymes and is only releasable by heat treatment (Aguilar et al., 1991).

Biochemical and genetic characterization of *Chlamydomonas* NR-deficient mutants has allowed the identification of seven loci (*NIT3–NIT7*, *NIT10*, and *NIT11*) related to Moco biosynthesis (Fernández and Matagne, 1984; Aguilar et al., 1992; Navarro et al., 2005). Mutants that carry *nit5* (strain 21 gr) or *nit6* (CC-2901-derived) alleles are particularly interesting, since they are phenotypically wild-type with respect to nitrate utilization, and are therefore referred to as cryptic Moco mutants (Fernández and Matagne, 1984). However, the *nit5–nit6* double mutant has a nit- phenotype and lacks Moco (Fernández and Matagne, 1984; Llamas et al., 2000), leading to the proposition that *Chlamydomonas* possesses at least two molybdate transporters. The *nit5* mutant is resistant to high concentrations of molybdate and tungstate (Fernández and Aguilar, 1987; Llamas et al., 2000).

Strains carrying mutations at the *NIT4* or *NIT5/NIT6* loci lack Moco, but they recover the ability to grow in nitrate-containing media at high concentrations of molybdate. These loci are proposed to be required for molybdate processing, that is, transport, storage, and insertion of molybdate into the organic moiety of the cofactor (Fernández and Matagne, 1984; Fernández and Aguilar, 1987). Any of these mutants might correspond to the molybdate-reparable *cnx1* mutant of *Arabidopsis*.

Mutants defective at the *NIT3* or *NIT7* loci cannot be rescued with high molybdate concentrations. These loci are probably involved in molybdopterin biosynthesis (Fernández and Aguilar, 1987; Aguilar et al., 1992). It appears that the capability of NR subunits to assemble into an enzyme complex depends on the organic moiety of Moco (Johnson, 1980; Mendel and Hänsch, 2002). In agreement with this assumption, *Chlamydomonas* molybdate processing mutants (*nit4*, *nit5/nit6*) have NR subunits assembled into a 220 kD complex with only diaphorase activity. In *nit3* and *nit7* strains, NR subunits are still able to assemble to a certain extent, which suggests that these subunits might interact either by themselves in the absence of Moco, or in the presence of some molybdopterin precursor

(Fernández and Aguilar, 1987; Aguilar et al., 1992). An oxidation product of molybdopterin cofactor can be measured in cell-free *Chlamydomonas* extracts. This product was detected in extracts from *nit3*, *nit4*, *nit5/nit6*, and *nit7* mutants in an equivalent amount to wild-type cells, suggesting that feedback inhibition of Moco biosynthesis takes place even in the absence of active Moco (Aguilar et al., 1992).

The mutants described above were obtained by traditional mutagenesis and chlorate selection, and lack molecular tags. This makes it difficult to identify the corresponding affected genes. Out of 22,000 transformants in an ordered library generated by insertional mutagenesis (González-Ballester et al., 2005b), 13 were unable to grow on nitrate, and thus are potential Moco mutants. In fact, upon identification of the DNA flanking the insertion sites, five strains were found to contain the insertion within or in proximity to a gene related to Moco metabolism, including *CNX2*, which is involved in the first step of Moco biosynthesis (González-Ballester et al., 2005a). The *CNX1* gene, involved in the third step of Moco biosynthesis, is split in *Chlamydomonas* genome, a feature of some prokaryotes (e.g. *E. coli*) and eukaryotes (e.g. *C. elegans*), but not plants. The *Chlamydomonas* *CNX1E* locus (corresponding to the E domain) is on linkage group X and *CNX1G* (corresponding to G domain) is on linkage group VI. Single insertional mutants were affected at *CNX1G* and *CNX1E*, and the wild-type phenotype was restored after transformation with the corresponding DNAs. cDNAs of both genes and gene chimeras (G and E) gave functional complementation of *E. coli moeA* and *mogA* Moco mutants. It is curious that CNX1E and CNX1G domains are functional either as independent polypeptides or as chimeric proteins, independent of their orientation (Llamas et al., 2007).

Analysis of the *Ostreococcus* genome has revealed a cluster where there are essential genes for nitrate assimilation (encoding NR, NiR, and nitrate transporters) together with genes for molybdate metabolism (*CNX2*, *CNX5*, and a putative *MoT*; Derelle et al., 2006). In the *Chlamydomonas* genome, however, there are no genes related to molybdenum metabolism near the nitrate assimilation gene cluster on linkage group IX. These *CNX* genes are on different linkage groups, in contrast to other genes for nitrate, urea, purine, or ammonium metabolism (Table 3.1 and Figure 3.2).

V. REGULATION OF NITRATE ASSIMILATION

A. Nitrate signaling

As discussed above, nitrogen source, light, circadian rhythm, and availability of other nutrients such as carbon or sulfur, all impact nitrate assimilation genes. Membrane proteins are responsible for controlling the flows of nutrients and solutes across the plasma membrane and cellular organelle envelopes, and thus mediate signals which modulate responses to changing nutrient conditions.

In *Chlamydomonas*, nitrate is the main metabolic signal needed to activate genes for its assimilation. Several mechanistic conclusions were made using the arylsulfatase reporter gene under the control of the *NIA1* promoter (Llamas et al., 2002; Rexach et al., 2002). The first was that nitrate signaling occurs intracellularly, since the presence of extracellular nitrate was not sufficient to promote reporter gene activation. Therefore, mutants devoid of the major HANT systems I and II are unable to promote expression of *NIA1* at low nitrate concentrations. A second conclusion was that *NIA1* activation responds to intracellular nitrate concentrations as regulated by the nitrate transport systems. Evidence in support of this conclusion is that HANT system I can promote efficient signaling by nitrate even at submicromolar concentrations of the anion, and LANT optimally promotes signaling at millimolar nitrate concentrations. HANT system IV, which is insensitive to ammonium and active at low CO_2, allows nitrate signaling at micromolar concentrations even in the presence of ammonium.

A final conclusion was that HANT system I makes cells competent to sense the very low nitrate concentrations present in medium with no added nitrate. This fact explains what was previously termed "NR autoregulation" in different systems (Cove and Pateman, 1969; Fernández and Cárdenas, 1982). "NR autoregulation" was manifested in overexpression of nitrate assimilation genes in strains lacking NR activity, since a continuous signaling by nitrate occurs in mutants possessing active HANT, which accumulate intracellular nitrate but cannot assimilate it. Gene expression profiling using the *Arabidopsis chl1* mutant has revealed that NRT1.1 plays a direct signaling role in regulating other nitrate transporters (Muños et al., 2004). Whether an orthologous role exists for the single *Chlamydomonas NRT1.1* gene is unknown.

B. Regulatory genes

Ammonium has a negative effect on nitrate assimilation gene expression (Fernández and Quesada, 1994) that appears to be mediated by its assimilation products, for example glutamine (Florencio and Vega, 1983a). How ammonium triggers down-regulation of nitrate assimilation is an important unsolved question for photosynthetic eukaryotes. In yeasts, the nonessential ammonium transporter MEP2 has been implicated in ammonium sensing (Lorenz and Heitman, 1998; Marini and André, 2000). In *Chlamydomonas*, the methylammonium-resistant mutant 2170 is relieved from the negative effect of ammonium or methylammonium on the nitrate pathway (Franco et al., 1987b). However, this mutant has a wild-type *AMT1.4* sequence and no defects in expression of *AMT1* genes (Kim et al., 2005).

In contrast to fungi and yeast where the regulatory genes for nitrate assimilation are well-defined (Marzluf, 1997; Siverio, 2002), no regulatory genes have been identified in photosynthetic eukaryotes except *Chlamydomonas NIT2*, which is required for the co-regulated expression of nitrate assimilation genes (Quesada and Fernández, 1994; Fernández et al., 1998). *NIT2* was cloned by transposon tagging from chlorate-resistant mutants (Schnell and Lefebvre, 1993). The deduced NIT2 protein is a transcription factor unrelated to the regulatory proteins for nitrate assimilation in fungi (Camargo, 2006), containing an RWPXRK box and leucine zipper motifs present in plant proteins involved in N-control (Schauser et al., 1999; Scheible et al., 2004) and also found in the *Chlamydomonas* MID protein involved in mating type *minus* dominance (Ferris and Goodenough, 1997). *NIT2* is expressed at a basal level though its transcript levels are enhanced in N-free media, and it is itself subject to negative regulation by ammonium (Schnell and Lefebvre, 1993).

To identify new regulatory genes for nitrate assimilation, the insertional library described above (González-Ballester et al., 2005b) was used. A forward genetics approach was undertaken based on the fact that the strain used to construct the mutant library contained the *NIA1-ARS* chimeric gene described in the previous section. Four screens were used to identify regulatory genes mediating the repressive effect of ammonium or the positive effect of nitrate (Fernández and Cárdenas, 1989; Ohresser et al., 1997; Fernández et al., 1998; Llamas et al., 2002). The screens were: AI, ammonium-insensitive (showing *ARS* expression in ammonium); NI, nitrate-insensitive (lacking *ARS* expression in nitrate); ON, overexpressed in nitrate (overexpressing *ARS* in nitrate); and no growth on nitrate (*nit-*). The DNA flanking the insertions was identified using a PCR-based method (González-Ballester et al., 2005a).

All NI mutants with a *nit-* phenotype were defective at *NIT2*, showing the central role of this gene in positive signaling (González-Ballester et al., 2005b). The screen also yielded new mutants putatively defective at loci related to nitrate and ammonium signals, and able to grow on nitrate. AI mutants had defects in genes putatively encoding guanylate cyclase, protein kinase, peptidyl-prolyl isomerase, or DNA-binding proteins (González-Ballester et al., 2005b). These and previous findings (Prieto et al., 1996; Zhang and Lefebvre, 1997; Pérez-Alegre, 2005) show that negative signaling by ammonium and its derivatives in the nitrate assimilation pathway are part of a complex network. Further efforts will likely uncover the precise roles of those genes in nitrate assimilation and their connection with other pathways.

ACKNOWLEDGMENTS

We gratefully acknowledge funding from the Ministerio de Educación y Ciencia, Spain (Grant BFU2005-07521), and Junta de Andalucía, Spain (PAI, BIO-0128, and Grant number P06-CVI-1609).

REFERENCES

Aguilar, M.R., Cárdenas, J., and Fernández, E. (1991). Regulation of molybdenum cofactor species in the green alga *Chlamydomonas reinhardtii*. *Biochim. Biophys. Acta* **1073**, 463–469.

Aguilar, M., Cárdenas, J., and Fernández, E. (1992). Quantitation of molybdopterin oxidation product in wild-type and molybdenum cofactor deficient mutants of *Chlamydomonas reinhardtii*. *Biochim. Biophys. Acta* **1160**, 269–274.

Alamillo, J.M., Cárdenas, J., and Pineda, M. (1991). Purification and molecular properties of urate oxidase from *Chlamydomonas reinhardtii*. *Biochim. Biophys. Acta* **1076**, 203–208.

Alamillo, J.M., Cárdenas, J., and Pineda, M. (1992). Kinetic and catalytic characterization of urate oxidase from *Chlamydomonas reinhardtii*. *J. Mol. Catal.* **77**, 353–364.

Amy, N.K. and Rajagopalan, K.V. (1979). Characterization of molybdenum cofactor from *Escherichia coli*. *J. Bacteriol.* **140**, 114–124.

Antia, N.J., Harrison, P.J., and Oliveira, L. (1991). The role of dissolved organic nitrogen in phytoplankton nutrition, cell biology and ecology. *Phycologia* **30**, 1–89.

Aslam, M., Travis, R.L., and Huffaker, R.C. (1994). Stimulation of nitrate and nitrite efflux by ammonium in barley (*Hordeum vulgare* L.) seedlings. *Plant Physiol.* **106**, 1293–1301.

Aslam, M., Travis, R.L., Rains, D.W., and Huffaker, R.C. (1996). Effect of root perturbation and excision on nitrate influx and efflux in barley (*Hordeum vulgare*) seedlings. *Physiol. Plant.* **97**, 425–432.

Ataya, F.S., Witte, C.P., Galván, A., Igeño, M.I., and Fernández, E. (2003). Mcp1 encodes the molybdenum cofactor carrier protein in *Chlamydomonas reinhardtii* and participates in protection, binding, and storage functions of the cofactor. *J. Biol. Chem.* **278**, 10885–10890.

Atteia, A., van Lis, R., Gelius-Dietrich, G., Adrait, A., Garin, J., Joyard, J., Rolland, N., and Martin, W. (2006). Pyruvate formate-lyase and a novel route of eukaryotic ATP- synthesis in *Chlamydomonas* mitochondria. *J. Biol. Chem.* **281**, 9909–9918.

Avila, J., González, C., Brito, N., and Siverio, J.M. (1998). Clustering of the YNA1 gene encoding a Zn(II)2Cys6 transcriptional factor in the yeast *Hansenula* with the nitrate assimilation genes YNT1, YNI1 and YNR1, and its involvement in their transcriptional activation. *Biochem J.* **335**, 647–652.

Azuara, M.P. and Aparicio, P.J. (1983). *In vivo* blue-light activation of *Chlamydomonas reinhardii* nitrate reductase. *Plant Physiol.* **71**, 286–290.

Bachmann, M., Huber, J.L., Athwal, G.S., Wu, K., Ferl, R.J., and Huber, S.C. (1996). 14-3-3 proteins associate with the regulatory phosphorylation site of spinach leaf nitrate reductase in an isoform-specific manner and reduce dephosphorylation of Ser-543 by endogenous protein phosphatases. *FEBS Lett.* **398**, 26–30.

Baurain, D., Dinant, M., Coosemans, N., and Matagne, R.F. (2003). Regulation of the alternative oxidase *Aox1* gene in *Chlamydomonas reinhardtii*. Role of the nitrogen source on the expression of a reporter gene under the control of the *Aox1* promoter. *Plant Physiol.* **131**, 1418–1430.

Byrne, T.E., Wells, M.R., and Johnson, C.H. (1992). Circadian rhythms of chemotaxis to ammonium and of methylammonium uptake in *Chlamydomonas*. *Plant Physiol.* **98**, 879–886.

Camargo, A. (2006). *Regulacion positiva de la asimilacion de nitrato en algas y plantas*. Ph.D. Dissertation. Universidad de Córdoba, Spain.

Campbell, W.H. (1999). Nitrate reductase structure, function and regulation: Bridging the gap between biochemistry and physiology. *Annu. Rev. Plant Physiol. Plant Mol. Biol.* **50**, 277–303.

Campbell, W.H. and Kinghorn, J.R. (1990). Functional domains of assimilatory nitrate reductases and nitrite reductases. *Trends Biochem. Sci.* **15**, 315–319.

Chen, Q. and Silflow, C.D. (1996). Isolation and characterization of glutamine synthetase genes in *Chlamydomonas reinhardtii*. *Plant Physiol.* **112**, 987–996.

Comparot, S., Lingiah, G., and Martin, T. (2003). Function and specificity of 14-3-3 proteins in the regulation of carbohydrate and nitrogen metabolism. *J. Exp. Bot.* **54**, 595–604.

Cove, D.J. and Pateman, J. (1969). Autoregulation of the synthesis of nitrate reductase in *Aspergillus nidulans*. *J. Bacteriol.* **97**, 1374–1378.

Crawford, N.M. (1995). Nitrate: Nutrient and signal for plant growth. *Plant Cell* **7**, 859–868.

Crawford, N.M. and Forde, B.G. (2002). Molecular and developmental biology of inorganic nitrogen nutrition. In: *The Arabidopsis Book* (C.R. Somerville and E.M. Meyerowitz, Eds.), American Society of Plant Biologists, Rockville, MD, doi/10.1199/tab.0011, http://www.aspb.org/publications/Arabidopsis/.

Crawford, N.M. and Glass, A.M.D. (1998). Molecular and physiological aspects of nitrate uptake in plants. *Trends Plant Sci.* **3**, 389–395.

Cullimore, J.V. and Sims, A.P. (1980). An association between photorespiration and protein catabolism: Studies with *Chlamydomonas*. *Planta* **150**, 392–396.

Cullimore, J.V. and Sims, A.P. (1981a). Glutamine synthase of *Chlamydomonas*: Its role in the control of nitrate assimilation. *Planta* **153**, 18–24.

Cullimore, J.V. and Sims, A.P. (1981b). Pathway of ammonia assimilation in illuminated and darkened *Chlamydomonas reinhardii*. *Phytochemistry* **20**, 933–940.

Dagestad, D., Lien, T., and Knutsen, G. (1981). Degradation and compartmentalization of urea in *Chlamydomonas reinhardii*. *Arch. Microbiol.* **129**, 261–264.

Derelle, E., Ferraz, C., Rombauts, S., Rouze, P., Worden, A.Z., Robbens, S., Partensky, F., Degroeve, S., Echeynie, S., Cooke, R., Saeys, Y., Wuyts, J., Jabbari, K., Bowler, C., Panaud, O., Piegu, B., Ball, S.G., Ral, J.P., Bouget, F.Y., Piganeau, G., De Baets, B., Picard, A., Delseny, M., Demaille, J., and Van de Peer, Y. (2006). Genome analysis of the smallest free-living eukaryote *Ostreococcus tauri* unveils many unique features. *Proc. Natl. Acad. Sci. U. S. A.* **103**, 11647–11652.

Dinant, M., Baurain, D., Coosemans, N., Joris, B., and Matagne, R.F. (2001). Characterization of two genes encoding the mitochondrial alternative oxidase in *Chlamydomonas reinhardtii*. *Curr. Genet.* **39**, 101–108.

Doddema, H. and Telkamp, G.P. (1979). Uptake of nitrate by mutants of *Arabidopsis thaliana*, disturbed in uptake or reduction of nitrate: II. Kinetics. *Physiol. Plant.* **45**, 332–338.

Eilers, T., Schwarz, G., Brinkmann, H., Witt, C., Richter, T., Nieder, J., Koch, B., Hille, R., Hansch, R., and Mendel, R.R. (2001). Identification and biochemical characterization of *Arabidopsis thaliana* sulfite oxidase. A new player in plant sulfur metabolism. *J. Biol. Chem.* **276**, 46989–46994.

Ermilova, E.V., Zalutskaya, Z.M., Munnik, T., van den Ende, H., and Gromov, B.V. (1998). Calcium in the control of chemotaxis in *Chlamydomonas*. *Biologia (Bratislava)* **53**, 577–581.

Ermilova, E.V., Zalutskaya, Z.M., Lapina, T.V., and Nikitin, M.M. (2003). Chemotactic behavior of *Chlamydomonas reinhardtii* is altered during gametogenesis. *Curr. Microbiol.* **46**, 261–264.

Ermilova, E.V., Nikitin, M.M., Lapina, T.V., and Zalutskaya, Z.M. (2004). Cha1, a DNA insertional transformant of the green alga *Chlamydomonas reinhardtii* with altered chemotaxis to ammonium. *Protistology* **3**, 251–255.

Fernández, E. and Aguilar, M. (1987). Molybdate repair of molybdopterin deficient mutants from *Chlamydomonas reinhardtii*. *Curr. Genet.* **12**, 349–355.

Fernández, E. and Cárdenas, J. (1982). Regulation of the nitrate-reducing system enzymes in wild-type and mutant strains of *Chlamydomonas reinhardii*. *Mol. Gen. Genet.* **186**, 164–169.

Fernández, E. and Cárdenas, J. (1983). Nitrate reductase from a mutant strain of *Chlamydomonas reinhardtii* incapable of nitrate assimilation. *Z. Naturforsch.* **38**, 439–445.

Fernández, E. and Matagne, R.F. (1984). Genetic analysis of nitrate reductase-deficient mutants in *Chlamydomonas reinhardtii*. *Curr. Genet.* **8**, 635–640.

Fernández, E., Schnell, R., Ranum, L.P.W., Hussey, S.C., Silflow, C.D., and Lefebvre, P.A. (1989). Isolation and characterization of the nitrate reductase structural gene in *Chlamydomonas reinhardtii*. *Proc. Natl. Acad. Sci. U. S. A.* **86**, 6449–6453.

Fernández, E., Galván, A., and Quesada, A. (1998). Nitrogen assimilation and its regulation. In: *The Molecular Biology of Chloroplasts and Mitochondria in Chlamydomonas* (J.D. Rochaix, M. Goldschmidt-Clermont, and S. Merchant, Eds.), pp. 637–659. Kluwer Academic, Dordrecht.

Ferris, P.J. and Goodenough, U.W. (1997). Mating type in *Chlamydomonas* is specified by *mid*, the minus-dominance gene. *Genetics* **146**, 859–869.

Fischer, K., Barbier, G.G., Hecht, H.J., Mendel, R.R., Campbell, W.H., and Schwarz, G. (2005). Structural basis of eukaryotic nitrate reduction: Crystal structures of the nitrate reductase active site. *Plant Cell* **17**, 1167–1179.

Fischer, K., Llamas, A., Tejada-Jimenez, M., Schrader, N., Kuper, J., Ataya, F.S., Galván, A., Mendel, R.R., Fernández, E., and Schwarz, G. (2006). Function and structure of the molybdenum cofactor carrier protein from *Chlamydomonas reinhardtii*. *J. Biol. Chem.* **281**, 30186–30194.

Fischer, P. and Klein, U. (1988). Localization of nitrogen-assimilating enzymes in the chloroplast of *Chlamydomonas reinhardtii*. *Plant Physiol.* **88**, 947–952.

Florencio, F.J. and Vega, J.M. (1983a). Utilization of nitrate, nitrite and ammonium by *Chlamydomonas reinhardtii*. *Planta* **158**, 288–293.

Florencio, F.J. and Vega, J.M. (1983b). Separation, purification, and characterization of two isoforms of glutamine synthetase from *Chlamydomonas reinhardtii*. *Z. Naturforsch.* **38c**, 531–538.

Florencio, F.J., Gadal, P., and Buchanan, B.B. (1993). Thioredoxin-linked activation of the chloroplast and cytosolic forms of *Chlamydomonas reinhardtii* glutamine synthetase. *Plant Physiol. Biochem.* **31**, 649–655.

Forde, B.G. (2000). Nitrate transporters in plants: Structure, function and regulation. *Biochim. Biophys. Acta* **1465**, 219–235.

Forde, B.G. (2002). Local and long-range signaling pathways regulating plant responses to nitrate. *Annu. Rev. Plant Biol.* **53**, 203–224.

Foyer, C.H. and Ferrario, S. (1994). Modulation of carbon and nitrogen metabolism in transgenic plants with a view to improved biomass production. *Biochem. Soc. Trans.* **22**, 909–915.

Franco, A.R., Cárdenas, J., and Fernández, E. (1984). Ammonium (methylammonium) is the co-repressor of nitrate reductase in *Chlamydomonas reinhardtii*. *FEBS Lett.* **176**, 453–456.

Franco, A.R., Cárdenas, J., and Fernández, E. (1987a). Involvement of reversible inactivation in the regulation of nitrate reductase enzyme levels in *Chlamydomonas reinhardtii*. *Plant Physiol.* **84**, 665–669.

Franco, A.R., Cárdenas, J., and Fernández, E. (1987b). A mutant of *Chlamydomonas reinhardtii* altered in the transport of ammonium and methylammonium. *Mol. Gen. Genet.* **206**, 414–418.

Franco, A.R., Cárdenas, J., and Fernández, E. (1988). Two different carriers transport both ammonium and methylammonium in *Chlamydomonas reinhardtii*. *J. Biol. Chem.* **263**, 14039–14043.

Franco, A.R., López-Siles, F.J., and Cárdenas, J. (1996a). Resistance to phosphinothricin (glufosinate) and its utilization as a nitrogen source by *Chlamydomonas reinhardtii*. *Appl. Environ. Microbiol.* **62**, 3834–3839.

Franco, A.R., Díaz, M.E., Pineda, M., and Cárdenas, J. (1996b). Characterization of a mutant of *Chlamydomonas reinhardtii* that uses L-methionine-S-sulfoximine and phosphinothricin as nitrogen sources for growth. *Plant Physiol.* **110**, 1215–1222.

Galván, A. and Fernández, E. (2001). Eukaryotic nitrate and nitrite transporters. *Cell Mol. Life Sci.* **58**, 225–233.

Galván, F., Márquez, A., and Vega, J.M. (1984). Purification and molecular properties of ferredoxin-glutamate synthase from *Chlamydomonas reinhardtii*. *Planta* **162**, 180–187.

Galván, A., Quesada, A., and Fernández, E. (1996). Nitrate and nitrite are transported by different specific transport systems and by a bispecific transporter in *Chlamydomonas reinhardtii*. *J. Biol. Chem.* **271**, 2088–2092.

Galván, A., Rexach, J., Mariscal, V., and Fernández, E. (2002). Nitrite transport to the chloroplast in *Chlamydomonas*: Molecular evidences for a regulated process. *J. Exp. Bot.* **53**, 845–854.

Galván, A., Mariscal, V., González-Ballester, D., and Fernández, E. (2006). The green alga *Chlamydomonas* as a tool to study the nitrate assimilation pathway in plants. In: *Model Plants and Crop Improvement* (R.K. Varshney and R.M.D. Koebner, Eds.), pp. 125–158. CRC Taylor and Francis group, Boca Raton.

Giordano, M., Chen, Y.B., Koblizek, M., and Falkowski, P.G. (2005). Regulation of nitrate reductase in *Chlamydomonas reinhardtii* by the redox state of the plastoquinone pool. *Eur. J. Phycol.* **40**, 345–352.

Glass, A.D.M. (2003). Nitrogen use efficiency of crop plants: Physiological constraints upon nitrogen absorption. *Crit. Rev. Plant. Sci.* **22**, 453–470.

Glass, A.D.M. and Siddiqi, M.Y. (1995). Nitrogen absorption by plant roots. In: *Nitrogen Nutrition in Higher Plants* (H.S. Srivastava and R.P. Singh, Eds.), pp. 21–56. Associated Publishing, New Delhi.

Gómez, I., Merchán, F., Fernández, E., and Quesada, A. (2002). NADP-malate dehydrogenase from *Chlamydomonas*: Prediction of new structural determinants for redox regulation by homology modelling. *Plant Mol. Biol.* **48**, 211–221.

González-Ballester, D., Camargo, A., and Fernández, E. (2004). Ammonium transporter genes in *Chlamydomonas*: The nitrate-specific regulatory gene *Nit2* is involved in Amt1;1 expression. *Plant Mol. Biol.* **56**, 863–878.

González-Ballester, D., de Montaigu, A., Galván, A., and Fernández, E. (2005a). Restriction enzyme site directed amplification (RESDA)-PCR: A tool to identify regions flanking a marker DNA. *Anal. Biochem.* **340**, 330–335.

González-Ballester, D., de Montaigu, A., Higuera, J.J., Galván, A., and Fernández, E. (2005b). Functional genomics of the regulation of the nitrate assimilation pathway in *Chlamydomonas*. *Plant Physiol.* **137**, 522–533.

Grossman, A.R., Croft, M., Gladyshev, V.N., Merchant, S.S., Posewitz, M.C., Prochnik, S., and Spalding, M.H. (2007). Novel metabolism in *Chlamydomonas* through the lens of genomics. *Curr. Opin. Plant Biol.* **10**, 190–198.

Guerrero, M.G., Vega, J.M., and Losada, M. (1981). The assimilatory nitrate-reducing system and its regulation. *Annu. Rev. Plant Physiol.* **32**, 169–204.

Guo, F.Q., Wang, R., Chen, M., and Crawford, N.M. (2001). The *Arabidopsis* dual-affinity nitrate transporter gene AtNRT1.1 (CHL1) is activated and functions in nascent organ development during vegetative and reproductive growth. *Plant Cell* **13**, 1761–1767.

Guo, F.Q., Young, J., and Crawford, N.M. (2003). The nitrate transporter AtNRT1.1 (CHL1) functions in stomatal opening and contributes to drought susceptibility in *Arabidopsis*. *Plant Cell* **15**, 107–117.

Hardy, R.W.F. and Havelka, U. (1975). Nitrogen fixation research: A key to world food? *Science* **188**, 633–643.

Hellio, C., Veron, B., and Le, G.Y. (2004). Amino acid utilization by *Chlamydomonas reinhardtii*: Specific study of histidine. *Plant Physiol. Biochem.* **42**, 257–264.

Hipkin, C.R., Everest, S.A., Rees, T.A.V., and Syrett, P.J. (1982). Ammonium generation by nitrogen-starved cultures of *Chlamydomonas reinhardii*. *Planta* **154**, 587–592.

Hodson, R.C. and Gresshoff, P.M. (1987). Fluoroacetamide resistance mutations in *Chlamydomonas reinhardtii*. *Arch. Microbiol.* **148**, 8–13.

Hodson, R.C., Williams, S.K., and Davidson, W.R., Jr. (1975). Metabolic control of urea catabolism in *Chlamydomonas reinhardi* and *Chlorella pyrenoidosa*. *J. Bacteriol.* **121**, 1022–1035.

Hoff, T., Truong, H.N., and Caboche, M. (1994). The use of mutants and transgenic plants to study nitrate assimilation. *Plant Cell Environ.* **17**, 489–506.

Howitt, S.M. and Udvardi, M.K. (2000). Structure, function and regulation of ammonium transporters in plants. *Biochim. Biophys. Acta* **1465**, 152–170.

Huang, N.C., Liu, K.H., Lo, H.J., and Tsay, Y.F. (1999). Cloning and functional characterization of an *Arabidopsis* nitrate transporter gene that encodes a constitutive component of low-affinity uptake. *Plant Cell* **11**, 1381–1392.

Huber, S.C., MacKintosh, C., and Kaiser, W.M. (2002). Metabolic enzymes as targets for 14-3-3 proteins. *Plant Mol. Biol.* **50**, 1053–1063.

Huppe, H.C. and Turpin, D.H. (1996). Appearance of novel glucose-6-phosphate dehydrogenase isoforms in *Chlamydomonas reinhardtii* during growth on nitrate. *Plant Physiol.* **110**, 1431–1433.

Johnson, J.L. (1980). The molybdenum cofactor common to nitrate reductase, xanthine dehydrogenase and sulphite oxidase. In: *Molybdenum and Molybdenum-Containing Enzymes* (M.P. Coughlan, Ed.), pp. 345–383. Pergamon Press, New York.

Johnstone, I.L., McCabe, P.C., Greaves, P., Gurr, S.J., Cole, G.E., Brow, M.A., Unkles, S.E., Clutterbuck, A.J., Kinghorn, J.R., and Innis, M.A. (1990). Isolation and characterisation of the *crnA-niiA-niaD* gene cluster for nitrate assimilation in *Aspergillus nidulans*. *Gene* **90**, 181–192.

Kaiser, W.M., Weiner, H., Kandlbinder, A., Tsai, C.B., Rockel, P., Sonoda, M., and Planchet, E. (2002). Modulation of nitrate reductase: Some new insights, an unusual case and a potentially important side reaction. *J. Exp. Bot.* **53**, 875–882.

Kalakoutskii, K. and Fernández, E. (1995). *Chlamydomonas reinhardtii* nitrate reductase complex has 105 kD subunits in the wild-type strain and a structural mutant. *Plant Sci.* **105**, 195–206.

Kanamaru, K., Wang, R., Su, W., and Crawford, N.M. (1999). Ser-534 in the hinge 1 region of *Arabidopsis* nitrate reductase is conditionally required for the binding of 14-3-3 proteins and the *in vitro* inhibition of NR. *J. Biol. Chem.* **274**, 4160–4165.

Kim, K.S., Feild, E., King, N., Yaoi, T., Kustu, S., and Inwood, W. (2005). Spontaneous mutations in the ammonium transport gene *AMT4* of *Chlamydomonas reinhardtii*. *Genetics* **170**, 631–644.

Kim, K.S., Kustu, S., and Inwood, W. (2006). Natural history of transposition in the green alga *Chlamydomonas reinhardtii*: Use of the *AMT4* locus as an experimental system. *Genetics* **173**, 2005–2019.

Kindle, K.L. (1990). High-frequency nuclear transformation of *Chlamydomonas reinhardtii*. *Proc. Natl. Acad. Sci. U. S. A.* **87**, 1228–1232.

Kindle, K.L., Schnell, R.A., Fernández, E., and Lefebvre, P.A. (1989). Stable nuclear transformation of *Chlamydomonas* using the *Chlamydomonas* gene for nitrate reductase. *J. Cell Biol.* **109**, 2589–2601.

Kirk, D.L. and Kirk, M.M. (1978). Carrier-mediated uptake of arginine and urea by *Chlamydomonas reinhardtii*. *Plant Physiol.* **61**, 556–560.

Kreuzberg, K. (1984). Starch fermentation via a formate producing pathway in *Chlamydomonas reinhardii*, *Chlorogonium elongatum* and *Chlorella fusca*. *Physiol. Plant.* **61**, 87–94.

Lam, H.M., Coschigano, K., Oliveira, I.C., Melo-Oliveira, R., and Coruzzi, G. (1996). The molecular-genetics of nitrogen assimilation into amino acids in higher plants. *Annu. Rev. Plant Physiol. Plant Mol. Biol.* **47**, 569–593.

Lea, P.J. and Miflin, B.J. (1975). The occurrence of glutamate synthase in algae. *Biochem. Biophys. Res. Commun.* **64**, 856–862.

Leftley, J.W. and Syrett, P.J. (1973). Urease and ATP: Urea amidolyase activity in unicellular algae. *J. Gen. Microbiol.* **77**, 109–115.

Lemaire, S.D., Quesada, A., Merchán, F., Corral, J.M., Igeño, M.I., Keryer, E., Issakidis-Bourguet, E., Hirasawa, M., Knaff, D.B., and Miginiac-Maslow, M. (2005). NADP-malate dehydrogenase from unicellular green alga *Chlamydomonas reinhardtii*. A first step towards redox regulation? *Plant Physiol.* **137**, 514–521.

Leon-Kloosterziel, K.M., Gil, M.A., Ruijs, G.J., Jacobsen, S.E., Olszewski, N.E., Schwartz, S.H., Zeevaart, J.A.D., and Koornneef, M. (1996). Isolation and characterization of abscisic acid-deficient *Arabidopsis* mutants at two new loci. *Plant J.* **10**, 655–661.

Li, J. and Timko, M.P. (1996). The *pc-1* phenotype of *Chlamydomonas reinhardtii* results from a deletion mutation in the nuclear gene for NADPH: protochlorophyllide oxidoreductase. *Plant Mol. Biol.* **30**, 15–37.

Lillo, C., Kazazaic, S., Ruoff, P., and Meyer, C. (1997). Characterization of nitrate reductase from light- and dark-exposed leaves. Comparison of different species and effects of 14-3-3 inhibitor proteins. *Plant. Physiol.* **114**, 1377–1383.

Lisa, T.A., Piedras, P., Cárdenas, J., and Pineda, M. (1995). Utilization of adenine and guanine as nitrogen sources by *Chlamydomonas reinhardtii* cells. *Plant Cell Environ.* **18**, 583–588.

Liu, K.H. and Tsay, Y.F. (2003). Switching between the two action modes of the dual-affinity nitrate transporter CHL1 by phosphorylation. *EMBO J.* **22**, 1005–1013.

Liu, K.H., Huang, C.Y., and Tsay, Y.F. (1999). CHL1 is a dual-affinity nitrate transporter of *Arabidopsis* involved in multiple phases of nitrate uptake. *Plant Cell* **11**, 865–874.

Llamas, A., Kalakoutskii, K.L., and Fernández, E. (2000). Molybdenum cofactor amounts in *Chlamydomonas reinhardtii* depend on the *Nit5* gene function related to molybdate transport. *Plant Cell Environ.* **23**, 1247–1255.

Llamas, A., Igeño, M.I., Galván, A., and Fernández, E. (2002). Nitrate signalling on the nitrate reductase gene promoter depends directly on the activity of the nitrate transport systems in *Chlamydomonas*. *Plant J.* **30**, 261–271.

Llamas, A., Mendel, R.R., and Schwarz, G. (2004). Synthesis of adenylated molybdopterin: An essential step for molybdenum insertion. *J. Biol. Chem.* **279**, 55241–55246.

Llamas, A., Otte, T., Multhaup, G., Mendel, R.R., and Schwarz, G. (2006). The mechanism of nucleotide-assisted molybdenum insertion into molybdopterin. A novel route toward metal cofactor assembly. *J. Biol. Chem.* **281**, 18343–18350.

Llamas, A., Tejada-Jiménez, M., González-Ballester, D., Higuera, J.J., Schwarz, G., Galván, A., and Fernández, E. (2007). *Chlamydomonas reinhardtii* CNX1E reconstitutes molybdenum cofactor biosynthesis in *Escherichia coli* mutants. *Eukaryotic Cell* **6**, 1063–1067.

Lorenz, M.C. and Heitman, J. (1998). Regulators of pseudohyphal differentiation in *Saccharomyces cerevisiae* identified through multicopy suppressor analysis in ammonium permease mutant strains. *Genetics* **150**, 1443–1457.

MacKintosh, C. (2004). Dynamic interactions between 14-3-3 proteins and phosphoproteins regulate diverse cellular processes. *Biochem. J.* **381**, 329–342.

Marini, A.M. and André, B. (2000). *In vivo* N-glycosylation of the Mep2 high-affinity ammonium transporter of *Saccharomyces cerevisiae* reveals an extracytosolic N-terminus. *Mol. Microbiol.* **38**, 552–564.

Marini, A.M., Soussi-Boudekou, S., Vissers, S., and Andre, B. (1997). A family of ammonium transporters in *Saccharomyces cerevisiae*. *Mol. Cell. Biol.* **17**, 4282–4293.

Mariscal, V., Rexach, J., Fernández, E., and Galván, A. (2004). The plastidic nitrite transporter NAR1.1 improves nitrate use efficiency for growth in *Chlamydomonas*. *Plant Cell Environ.* **27**, 1321–1328.

Mariscal, V., Moulin, P., Orsel, M., Miller, A.J., Fernández, E., and Galván, A. (2006). Differential regulation of the *Chlamydomonas Nar1* gene family by carbon and nitrogen. *Protist* **157**, 421–433.

Márquez, A.G., Galván, F., and Vega, J.M. (1984). Purification and characterization of the NADH-glutamate synthase from *Chlamydomonas reinhardtii*. *Plant Sci. Lett.* **34**, 305–314.

Márquez, A.G., Galván, F., and Vega, J.M. (1986a). Utilization of ammonium by mutant and wild type *Chlamydomonas reinhardtii*. Studies on the glutamate synthase activity. *J. Plant Physiol.* **124**, 95–102.

Márquez, A.J., Gotor, C., Romero, L.C., Galván, F., and Vega, J.M. (1986b). Ferredoxin-glutamate synthase from *Chlamydomonas reinhardtii*. Prosthetic groups and preliminary studies of mechanism. *Int. J. Biochem.* **18**, 531–535.

Martínez-Rivas, J.M., Vega, J.M., and Márquez, A. (1991). Differential regulation of the nitrate-reducing and ammonium-assimilatory systems in synchronous cultures of *Chlamydomonas reinhardtii*. *FEMS Microbiol. Lett.* **78**, 85–88.

Marzluf, G.A. (1997). Molecular genetics of sulfur assimilation in filamentous fungi and yeast. *Annu. Rev. Microbiol.* **51**, 73–96.

Masclaux-Daubresse, C., Reisdorf-Cren, M., Pageau, K., Lelandais, M., Grandjean, O., Kronenberger, J., Valadier, M.H., Feraud, M., Jouglet, T., and Suzuki, A. (2006). Glutamine synthetase-glutamate synthase pathway and glutamate dehydrogenase play distinct roles in the sink-source nitrogen cycle in tobacco. *Plant Physiol.* **140**, 444–456.

Mendel, R.R. and Bittner, F. (2006). Cell biology of molybdenum. *Biochim. Biophys. Acta* **1763**, 621–635.

Mendel, R.R. and Hänsch, R. (2002). Molybdoenzymes and molybdenum cofactor in plants. *J. Exp. Bot.* **53**, 1689–1698.

Mendel, R.R., Buchanan, R.J., and Wray, J.L. (1984). Characterization of a new type of molybdenum cofactor-mutant in cell cultures of *Nicotiana tabacum*. *Mol. Gen. Genet.* **195**, 186–189.

Merchán, F., van den Ende, H., Fernández, E., and Beck, C.F. (2001). Low-expression genes induced by nitrogen starvation and subsequent sexual differentiation in *Chlamydomonas reinhardtii*, isolated by the differential display technique. *Planta* **213**, 309–317.

Miflin, B.J. and Habash, D.Z. (2002). The role of glutamine synthetase and glutamate dehydrogenase in nitrogen assimilation and possibilities for improvement in the nitrogen utilization of crops. *J. Exp. Bot.* **53**, 979–987.

Miflin, B.J. and Lea, P.J. (1975). Glutamine and asparagine as nitrogen donors for reductant-dependent glutamate synthesis in pea roots. *Biochem. J.* **149**, 403–409.

Mittag, M. and Waltenberger, H. (1997). *In vitro* mutagenesis of binding site elements for the clock-controlled proteins CCTR and Chlamy 1. *Biol. Chem.* **378**, 1167–1170.

Miura, K., Yamano, T., Yoshioka, S., Kohinata, T., Inoue, Y., Taniguchi, F., Asamizu, E., Nakamura, Y., Tabata, S., Yamato, K.T., Ohyama, K., and Fukuzawa, H. (2004). Expression profiling-based identification of CO_2-responsive genes regulated by CCM1 controlling a carbon-concentrating mechanism in *Chlamydomonas reinhardtii*. *Plant Physiol.* **135**, 1595–1607.

Moorhead, G., Douglas, P., Morrice, N., Scarabel, M., Aitken, A., and MacKintosh, C. (1996). Phosphorylated nitrate reductase from spinach leaves is inhibited by 14-3-3 proteins and activated by fusicoccin. *Curr. Biol.* **6**, 1104–1113.

Moreno- Vivián, C., Cabello, P., Martinez-Luque, M., Blasco, R., and Castillo, F. (1999). Prokaryotic nitrate reduction: Molecular properties and functional distinction among bacterial nitrate reductases. *J. Bacteriol.* **181**, 6573–6584.

Moura, J.J., Brondino, C.D., Trincao, J., and Romao, M.J. (2004). Mo and W bis-MGD enzymes: Nitrate reductases and formate dehydrogenases. *J. Biol. Inorg. Chem.* **9**, 791–799.

Moyano, E., Cárdenas, J., and Muñoz-Blanco, J. (1992a). Purification and properties of three $NAD(P)^+$ isozymes of L-glutamate dehydrogenase of *Chlamydomonas reinhardtii*. *Biochim. Biophys. Acta,* **1119**, 63–68.

Moyano, E., Ramazanov, Z., Cárdenas, J., and Muñoz-Blanco, J. (1992b). Intracellular localization of three L-glutamate dehydrogenase isozymes from *Chlamydomonas reinhardtii*. *Plant Physiol.* **100**, 1575–1579.

Moyano, E., Cárdenas, J., and Muñoz-Blanco, J. (1995). Involvement of glutamate dehydrogenase isoenzymes in carbon and nitrogen metabolism in *Chlamydomonas reinhardtii*. *Physiol. Plant.* **94**, 553–559.

Muños, S., Cazettes, C., Fizames, C., Gaymard, F., Tillard, P., Lepetit, M., Lejay, L., and Gojon, A. (2004). Transcript profiling in the *chl1-5* mutant of *Arabidopsis* reveals a role of the nitrate transporter NRT1.1 in the regulation of another nitrate transporter, NRT2.1. *Plant Cell* **16**, 2433–2447.

Muñoz, A., Raso, M.J., Pineda, M., and Piedras, P. (2006). Degradation of ureidoglycolate in French bean (*Phaseolus vulgaris*) is catalysed by a ubiquitous ureidoglycolate urea-lyase. *Planta* **224**, 175–184.

Muñoz-Blanco, J. and Cárdenas, J. (1989). Changes in glutamate dehydrogenase activity of *Chlamydomonas reinhardtii* under different trophic and stress conditions. *Plant Cell Environ.* **12**, 173–182.

Muñoz-Blanco, J., Moyano, E., and Cárdenas, J. (1989). Glutamate dehydrogenase isozymes of *Chlamydomonas reinhardtii*. *FEMS Microbiol. Lett.* **61**, 315–318.

Muñoz-Blanco, J., Hidalgo-Martínez, J., and Cárdenas, J. (1990). Extracellular deamination of L-amino acids by *Chlamydomonas reinhardtii* cells. *Planta* **182**, 194–198.

Navarro, M.T., Guerra, E., Fernández, E., and Galván, A. (2000). Nitrite reductase mutants as an approach to understanding nitrate assimilation in *Chlamydomonas reinhardtii*. *Plant Physiol.* **122**, 283–290.

Navarro, M.T., Mariscal, V., Macías, M.I., Fernández, E., and Galván, A. (2005). *Chlamydomonas reinhardtii* strains expressing nitrate reductase under control of the *cabII-1* promoter: Isolation of chlorate resistant mutants and identification of new loci for nitrate assimilation. *Photosynth. Res.* **83**, 151–161.

Navarro, F.J., Machín, F., Martín, Y., and Siverio, J.M. (2006). Down-regulation of eukaryotic nitrate transporter by nitrogen-dependent ubiquitinylation. *J. Biol. Chem.* **281**, 13268–13274.

Nosengo, N. (2003). Fertilized to death. *Nature* **425**, 894–895.

Ohresser, M., Matagne, R.F., and Loppes, R. (1997). Expression of the arylsulphatase reporter gene under the control of the *nit1* promoter in *Chlamydomonas reinhardtii*. *Curr Genet.* **31**, 264–271.

Okamoto, M., Vidmar, J.J., and Glass, A.D.M. (2003). Regulation of *NRT1* and *NRT2* gene families of *Arabidopsis thaliana*: responses to nitrate provision. *Plant Cell Physiol.* **44**, 304–317.

Okamoto, M., Kumar, A., Li, W., Wang, Y., Siddiqi, M.Y., Crawford, N.M., and Glass, A.D.M. (2006). High-affinity nitrate transport in roots of *Arabidopsis* depends on expression of the *NAR2*-like gene *AtNRT3.1*. *Plant Physiol.* **140**, 1036–1046.

Oostindiër-Braaksma, F.J. and Feenstra, W.J. (1973). Isolation and characterization of chlorate-resistant mutants of *Arabidopsis thaliana*. *Mut. Res.* **19**, 175–185.

Orsel, M., Krapp, A., and Daniel-Vedele, F. (2002). Analysis of the NRT2 nitrate transporter family in *Arabidopsis*. Structure and gene expression. *Plant Physiol.* **129**, 886–896.

Orsel, M., Chopin, F., Leleu, O., Smith, S.J., Krapp, A., Daniel-Vedele, F., and Miller, A.J. (2006). Characterization of a two-component high-affinity nitrate uptake system in *Arabidopsis*. Physiology and protein–protein interaction. *Plant Physiol.* **142**, 1304–1317.

Pajuelo, E., Pajuelo, P., Clemente, M.T., and Márquez, A.J. (1995). Regulation of the expression of ferredoxin-nitrite reductase in synchronous cultures of *Chlamydomonas reinhardtii*. *Biochim. Biophys. Acta* **1249**, 72–78.

Pao, S.S., Paulsen, I.T., and Saier, M.H. (1998). The major facilitator superfamily. *Microbiol. Mol. Biol. Rev.* **62**, 1–34.

Paulsen, I.T. and Skurray, R.A. (1994). The POT family of transport proteins. *Trends Biochem. Sci.* **19**, 404.

Peakman, T., Crouzet, J., Mayaux, J., Busby, S., Mohan, S., Harborne, N., Wooton, J., Nicolson, R., and Cole, J. (1990). Nucleotide sequence, organisation and structural analysis of the products of genes in the *nirB-cysG* region of the *Escherichia coli* chromosome. *Eur. J. Biochem.* **191**, 315–323.

Pérez, M.D., González, C., Avila, J., Brito, N., and Siverio, J.M. (1997). The *YNT1* gene encoding the nitrate transporter in the yeast *Hansenula polymorpha* is clustered with genes *YNI1* and *YNR1* encoding nitrite reductase and nitrate reductase, and its disruption causes inability to grow in nitrate. *Biochem J.* **321**, 397–403.

Pérez-Alegre, M., Dubus, A., and Fernández, E. (2005). REM1, a new type of long terminal repeat retrotransposon in *Chlamydomonas reinhardtii*. *Mol. Cell Biol.* **25**, 10628–10638.

Pérez-Vicente, R., Pineda, M., and Cárdenas, J. (1988). Isolation and characterization of xanthine dehydrogenase from *Chlamydomonas reinhardtii*. *Physiol. Plant.* **72**, 101–107.

Pérez-Vicente, R., Cárdenas, J., and Pineda, M. (1991). Distinction between hypoxanthine and xanthine transport in *Chlamydomonas reinhardtii*. *Plant Physiol.* **95**, 126–130.

Pérez-Vicente, R., Alamillo, J.M., Cárdenas, J., and Pineda, M. (1992). Purification and substrate inactivation of xanthine dehydrogenase from *Chlamydomonas reinhardtii*. *Biochim. Biophys. Acta* **1117**, 159–166.

Pérez-Vicente, R., Burón, M.I., González-Reyes, J.A., Cárdenas, J., and Pineda, M. (1995). Xanthine accumulation and vacuolization in *Chlamydomonas reinhardtii* cells. *Protoplasma* **186**, 93–98.

Philips, S., Laanbroek, H.J., and Verstraete, W. (2002). Origin, causes and effects of increased nitrite concentrations in aquatic environments. *Rev. Environ. Sci. Biotech.* **1**, 115–141.

Piedras, P., Pineda, M., Muñoz, J., and Cárdenas, J. (1992). Purification and characterization of an L-amino-acid oxidase from *Chlamydomonas reinhardtii*. *Planta* **188**, 13–18.

Piedras, P., Cárdenas, J., and Pineda, M. (1995). Solubilization and extraction of allantoinase and allantoicase from the green alga *Chlamydomonas reinhardtii*. *Phytochem. Anal.* **6**, 239–243.

Piedras, P., Aguilar, M., and Pineda, M. (1998). Uptake and metabolism of allantoin and allantoate by cells of *Chlamydomonas reinhardtii* (Chlorophyceae). *Eur. J. Phycol.* **33**, 57–64.

Piedras, P., Muñoz, A., Aguilar, M., and Pineda, M. (2000). Allantoate amidinohydrolase (Allantoicase) from *Chlamydomonas reinhardtii*: Its purification and catalytic and molecular characterization. *Arch. Biochem. Biophys.* **378**, 340–348.

Pineda, M. and Cárdenas, J. (1996). Transport and assimilation of purines in *Chlamydomonas reinhardtii*. *Sci. Mar.* **60**, 195–201.

Pineda, M., Fernández, E., and Cárdenas, J. (1984). Urate oxidase of *Chlamydomonas reinhardtii*. *Physiol. Plant.* **62**, 453–457.

Pineda, M., Cabello, P., and Cárdenas, J. (1987). Ammonium regulation of urate uptake in *Chlamydomonas reinhardtii*. *Planta* **171**, 496–500.

Pozuelo, M., Merchán, F., Macías, M.I., Beck, C.F., Galván, A., and Fernández, E. (2000). The negative effect of nitrate on gametogenesis is independent of nitrate assimilation in *Chlamydomonas reinhardtii*. *Planta* **211**, 287–292.

Pozuelo, M., MacKintosh, C., Galván, A., and Fernández, E. (2001). Cytosolic glutamine synthetase and not nitrate reductase from the green alga *Chlamydomonas reinhardtii* is phosphorylated and binds 14-3-3 proteins. *Planta* **212**, 264–269.

Prieto, R., Dubus, A., Galván, A., and Fernández, E. (1996). Isolation and characterization of two regulatory mutants for nitrate assimilation in *Chlamydomonas reinhardtii*. *Mol. Gen. Genet.* **251**, 461–471.

Quesada, A. and Fernández, E. (1994). Expression of nitrate assimilation genes in *Chlamydomonas reinhardtii*. *Plant Mol. Biol.* **24**, 185–194.

Quesada, A., Galván, A., Schnell, R.A., Lefebvre, P.A., and Fernández, E. (1993). Five nitrate assimilation-related loci are clustered in *Chlamydomonas reinhardtii*. *Mol. Gen. Genet.* **240**, 387–394.

Quesada, A., Galván, A., and Fernández, E. (1994). Identification of nitrate transporters in *Chlamydomonas reinhardtii*. *Plant J.* **5**, 407–419.

Quesada, A., Hidalgo, J., and Fernández, E. (1998a). Three *Ntr2* genes are differentially regulated in *Chlamydomonas reinhardtii*. *Mol. Gen. Genet.* **258**, 373–377.

Quesada, A., Gómez, I., and Fernández, E. (1998b). Clustering of the nitrite reductase gene and a light-regulated gene with nitrate assimilation loci in *Chlamydomonas reinhardtii*. *Planta* **206**, 259–265.

Quesada, A., Gómez-García, I., and Fernández, E. (2000). Involvement of chloroplast and mitochondria redox valves in nitrate assimilation. *Trends Plant Sci.* **5**, 463–464.

Rajagopalan, K.V. and Johnson, J.L. (1992). The pterin molybdenum cofactors. *J. Biol. Chem.* **267**, 10199–10202.

Ramazanov, Z. and Cárdenas, J. (1994). Photorespiratory ammonium assimilation in chloroplasts of *Chlamydomonas reinhardtii*. *Physiol. Plant.* **91**, 495–502.

Rexach, J., Montero, B., Fernández, E., and Galván, A. (1999). Differential regulation of the high affinity nitrite transport systems III and IV in *Chlamydomonas reinhardtii*. *J. Biol. Chem.* **274**, 27801–27806.

Rexach, J., Fernández, E., and Galván, A. (2000). The *Chlamydomonas reinhardtii Nar1* gene encodes a chloroplast membrane protein involved in nitrite transport. *Plant Cell* **12**, 1441–1453.

Rexach, J., Llamas, A., Fernández, E., and Galván, A. (2002). The activity of the high-affinity nitrate transport system I (NRT2;1, NAR2) is responsible for the efficient signalling of nitrate assimilation genes in *Chlamydomonas reinhardtii*. *Planta* **215**, 606–611.

Romero, L.C., Gotor, C., Márquez, A.J., Forde, B.G., and Vega, J.M. (1988). Antigenic similarities between ferredoxin-dependent nitrite reductase and glutamate synthase from *Chlamydomonas reinhardtii*. *Biochim. Biophys. Acta* **957**, 152–157.

Sakihama, Y., Nakamura, S., and Yamasaki, H. (2002). Nitric oxide production mediated by nitrate reductase in the green alga *Chlamydomonas reinhardtii*: An alternative NO production pathway in photosynthetic organisms. *Plant Cell Physiol.* **43**, 290–297.

Schauser, L., Roussis, A., Stiller, J., and Stougaard, J. (1999). A plant regulator controlling development of symbiotic root nodules. *Nature* **402**, 191–195.

Scheible, W.R., Morcuende, R., Czechowski, T., Fritz, C., Osuna, D., Palacios-Rojas, N., Schindelasch, D., Thimm, O., Udvardi, M.K., and Stitt, M. (2004). Genome-wide reprogramming of primary and secondary metabolism, protein synthesis, cellular growth processes, and the regulatory infrastructure of *Arabidopsis* in response to nitrogen. *Plant Physiol.* **136**, 2483–2499.

Schnell, R.A. and Lefebvre, P.A. (1993). Isolation of the *Chlamydomonas* regulatory gene *nit2* by transposon tagging. *Genetics* **134**, 737–747.

Schwarz, G. and Mendel, R.R. (2006). Molybdenum cofactor biosynthesis and molybdenum enzymes. *Annu. Rev. Plant Biol.* **57**, 623–647.

Semler, B.L., Hodson, R.C., Williams, S.K., II., and Howell, S.H. (1975). The induction of allophanate lyase during the vegetative cell cycle in light-synchronized cultures of *Chlamydomonas reinhardi*. *Biochim. Biophys. Acta* **399**, 71–78.

Sherameti, I., Sopory, S.K., Trebickas, A., Pfannschmidt, T., and Oelmüller, R. (2002). Photosynthetic electron transport determines nitrate reductase gene expression and activity in higher plants. *J. Biol. Chem.* **277**, 46594–46600.

Siverio, J.M. (2002). Assimilation of nitrate by yeasts. *FEMS Microbiol. Rev.* **26**, 277–284.

Sofia, H.J., Chen, G., Hetzler, B.G., Reyes-Spindola, J.F., and Miller, N.E. (2001). Radical SAM, a novel protein superfamily linking unresolved steps in familiar biosynthetic pathways with radical mechanisms: Functional characterization using new analysis and information visualization methods. *Nucleic Acids Res.* **29**, 1097–1106.

Soupene, E., King, N., Feild, E., Liu, P., Niyogi, K.K., Huang, C.H., and Kustu, S. (2002). Rhesus expression in a green alga is regulated by CO_2. *Proc. Natl. Acad. Sci. U. S. A.* **99**, 7769–7773.

Soupene, E., Inwood, W., and Kustu, S. (2004). Lack of the Rhesus protein Rh1 impairs growth of the green alga *Chlamydomonas reinhardtii* at high CO_2. *Proc. Natl. Acad. Sci. U. S. A.* **101**, 7787–7792.

Stallmeyer, B., Nerlich, A., Schiemann, J., Brinkmann, H., and Mendel, R.R. (1995). Molybdenum co-factor biosynthesis: The *Arabidopsis thaliana* cDNA *cnx1* encodes a multifunctional two-domain protein homologous to a mammalian neuroprotein, the insect protein Cinnamon and three *Escherichia coli* proteins. *Plant J.* **8**, 751–762.

Suppmann, B. and Sawers, G. (1994). Isolation and characterization of hypophosphite resistant mutants of *Escherichia coli*: Identification of the FocA protein, encoded by the pfl operon, as a putative formate transporter. *Mol. Microbiol.* **11**, 965–982.

Tam, L.W. and Lefebvre, P.A. (1993). Cloning of flagellar genes in *Chlamydomonas reinhardtii* by DNA insertional mutagenesis. *Genetics* **135**, 375–384.

Tischner, R., Planchet, E., and Kaiser, W.M. (2004). Mitochondrial electron transport as a source for nitric oxide in the unicellular green alga *Chlorella sorokiniana*. *FEBS Lett.* **576**, 151–155.

Todd, C.D., Tipton, P.A., Blevins, D.G., Piedras, P., Pineda, M., and Polacco, J.C. (2006). Update on ureide degradation in legumes. *J. Exp. Bot.* **57**, 5–12.

Tong, Y., Zhou, J.J., Li, Z., and Miller, A. (2005). A two-component high-affinity nitrate uptake system in barley. *Plant J.* **41**, 442–450.

Torchinsky, Y. (1987). Transamination: Its discovery, biological and chemical aspects (1937–1987). *Trends Biochem. Sci.* **12**, 115–117.

Tsay, Y.F., Schroeder, J., Feldman, K.A., and Crawford, N.M. (1993). A herbicide sensitive gene CHL1 of *Arabidopsis* encodes a nitrate-inducible nitrate transporter. *Cell* **72**, 705–713.

Ullrich, W.R., Larsson, M., Larsson, C.M., Lesch, S., and Novacky, A. (1984). Ammonium uptake in *Lemna gibba* G 1, related membrane potential changes and inhibition of anion uptake. *Physiol. Plant.* **61**, 369–376.

Vallon, O. and Wollman, F.A. (1997). cDNA sequence of Mα, the catalytic subunit of the *Chlamydomonas reinhardtii* L-amino acid oxidase (accession no. U78797). A new sequence motif shared by a wide variety of flavoproteins (PGR 97-171). *Plant Physiol.* **115**, 1729.

Vallon, O., Bulte, L., Kuras, R., Olive, J., and Wollman, F.A. (1993). Extensive accumulation of an extracellular L-amino-acid oxidase during gametogenesis of *Chlamydomonas reinhardtii*. *Eur. J. Biochem.* **215**, 351–360.

Vanlerberghe, G.C. and McIntosh, L. (1997). Alternative oxidase: From gene to function. *Annu. Rev. Plant Physiol. Plant Mol. Biol.* **48**, 703–734.

Vega, J.M. and Kamin, H. (1977). Spinach nitrite reductase, Purification and properties of a siroheme-containing iron-sulfur enzyme. *J. Biol. Chem.* **252**, 896–909.

Vogels, G.D. and Van der Drift, C. (1976). Degradation of purines and pyrimidines by microorganisms. *Bacteriol. Rev.* **40**, 403–468.

von Wirén, N., Gazzarrini, S., Gojon, A., and Frommer, W.B. (2000). The molecular physiology of ammonium uptake and retrieval. *Curr. Opin. Plant Biol.* **3**, 254–261.

Wang, R., Liu, D., and Crawford, N.M. (1998). The *Arabidopsis* CHL1 protein plays a major role in high-affinity nitrate uptake. *Proc. Natl. Acad. Sci. U. S. A.* **95**, 15134–15139.

Wang, R., Guegler, K., LaBrie, S.T., and Crawford, N.M. (2000). Genomic analysis of a nutrient response in *Arabidopsis* reveals diverse expression patterns and novel metabolic and potential regulatory genes induced by nitrate. *Plant Cell* **12**, 1491–1509.

Westhoff, C.M., Ferreri-Jacobia, M., Mak, D.O., and Foskett, J.K. (2002). Identification of the erythrocyte Rh blood group glycoprotein as a mammalian ammonium transporter. *J. Biol. Chem.* **277**, 12499–12502.

Whitney, P.A. and Cooper, T. (1973). The induction of urea carboxylase and allophanate hydrolase in *Saccharomyces cerevisiae*. *J. Biol. Chem.* **248**, 6203–6209.

Williams, L. and Miller, A. (2001). Transporters responsible for the uptake and partitioning of nitrogenous solutes. *Annu. Rev. Plant. Physiol. Plant Mol. Biol.* **52**, 659–688.

Williams, S.K. and Hodson, R.C. (1977). Transport of urea at low concentrations in *Chlamydomonas reinhardi*. *J. Bacteriol.* **130**, 266–273.

Witte, C.P., Igeño, M.I., Mendel, R., Schwarz, G., and Fernández, E. (1998). The *Chlamydomonas reinhardtii* Moco carrier protein is multimeric and stabilizes molybdopterin cofactor in a molybdate charged form. *FEBS Lett.* **431**, 205–209.

Zhang, D. and Lefebvre, P.A. (1997). *FAR1*, a new negative regulatory locus required for the repression of the nitrate reductase gene in *Chlamydomonas reinhardtii*. *Genetics* **146**, 121–133.

Zhou, J.J., Theodoulou, F.L., Muldin, I., Ingemarsson, B., and Miller, A.J. (1998). Cloning and functional characterization of a *Brassica napus* transporter that is able to transport nitrate and histidine. *J. Biol. Chem.* **273**, 12017–12023.

Zhou, J.J., Fernández, E., Galván, A., and Miller, A.J. (2000). A high affinity nitrate transport system from *Chlamydomonas* requires two gene products. *FEBS Lett.* **466**, 225–227.

CHAPTER 4

Amino Acid Metabolism

Olivier Vallon[1] and Martin H. Spalding[2]
[1]Institut de Biologie Physico-Chimique, UMR 7141
CNRS/Université Pierre et Marie Curie, Paris, France
[2]Department of Genetics, Development and Cell Biology,
Iowa State University, Ames, Iowa, USA

CHAPTER CONTENTS

I. INTRODUCTION

Like all photoautotrophs, *Chlamydomonas* can synthesize the whole range of amino acids that it requires to build proteins. Amino acids also serve as precursors of many important metabolites: pyrimidines, glutathione, heme, nucleotides, polyamines, etc. As in most microorganisms, amino acid metabolism is intimately connected with that of the carbon skeletons that it uses, in particular during photorespiration. This chapter will describe our knowledge of the biosynthesis and utilization in *Chlamydomonas* of the 21 genetically encoded amino acids (the standard 20 amino acids plus selenocysteine), and touch on some of their downstream metabolites. We will largely disregard amino acid modifications that occur after they are incorporated into proteins.

Amino acid metabolism in *Chlamydomonas* has not been studied extensively, with most work having been performed by geneticists in search of auxotrophy markers or specifically interested in nitrogen nutrition. In many instances, therefore, information is essentially limited to a description of the gene repertoire based on analysis of the genome sequence (except where noted, version 3.0 has been used). Overall, it appears that most of the amino acid biosynthetic pathways in *Chlamydomonas* are similar to those of higher plants, as expected from the phylogeny. However, the plant pathways themselves are largely inferred from microbial (bacterial, fungal) pathways and may be improperly described. Furthermore, regulatory circuits operating in plants are in general not well understood, and they might differ markedly in *Chlamydomonas* owing to its microbial lifestyle.

II. BIOSYNTHESIS OF AMINO ACIDS: PATHWAYS, ENZYMES, LOCALIZATION, INHIBITORS, MUTANTS

A. Overview

Available evidence indicates that both *Chlamydomonas* and plants carry out the bulk of their amino acid biosynthesis in the plastid, perhaps because of the ready supply of reductant and ATP from photosynthesis. Differences in gene repertoire and regulation can be anticipated between *Chlamydomonas* and plant amino acid biosynthesis if the latter uses different isozymes for tissue-dependent expression, or responds differently to changes in nutrient availability. As with most organisms, the carbon backbones of amino acids in *Chlamydomonas* are generally provided by central carbon metabolism, including glycolysis (pyruvate, pyr; phospho*enol*pyruvate, PEP; 3-phosphoglycreate, 3PGA), photosynthetic carbon metabolism (3PGA; erythrose-4-phosphate, E4P; ribose-5-phosphate, R5P), the oxidative pentose phosphate pathway (E4P, R5P), the photorespiratory glycolate pathway (glyoxylate, hydroxypyruvate), and the TCA (tricarboxylic acid) cycle (alpha-ketoglutarate, αKG; oxaloacetate). The biosynthetic pathways will be discussed in groups based on similar starting materials or related intermediates. Particular attention will be directed towards the distribution of the pathways between the organelles of endosymbiotic origin (plastid and mitochondrion) and the other compartments of the cell. Still, one must bear in mind that available prediction programs do not reliably distinguish between chloroplast and mitochondrial targeting in *Chlamydomonas*.

B. Ammonia assimilation and the biosynthesis of glutamate and glutamine

Inorganic combined nitrogen (e.g. NO_3^-) must be reduced to NH_3 before it can be incorporated into N-containing organic molecules (for a detailed discussion of N metabolism see Chapter 3). NH_3 is assimilated initially into glutamine and glutamate via reactions catalyzed by the enzymes glutamine synthetase (GS), glutamate synthase (glutamine-oxoglutarate aminotransferase; GOGAT), and glutamate dehydrogenase (GDH). Each of these enzymes typically is found in multiple isoforms, which are likely to serve specific functions within the main ammonia assimilation pathways, primary nitrogen assimilation, re-assimilation of photorespiratory NH_3, and reassimilation of recycled nitrogen.

Ammonia assimilation in *Chlamydomonas* proceeds primarily via the glutamine-glutamate cycle, as in higher plants. In the first step of this cycle, NH_3 is combined with glutamate to form the amide group of glutamine in an ATP-dependent reaction catalyzed by one of two identified isoforms of GS of which one, GS1, is localized to the cytosol and the other, GS2, to the plastid (Florencio and Vega, 1983; Chen and Silflow, 1996). Both GS isoforms have been characterized as large, octameric proteins resembling the plant type of GS, rather than the form found in photosynthetic prokaryotes (Florencio and Vega, 1983). GS1 has been found to bind a 14-3-3 protein and to be phosphorylated by a Ca^{2+}- and calmodulin-dependent kinase (Pozuelo et al., 2001), but the physiological significance is unknown because activity was not affected. *Chlamydomonas* encodes four GS genes, of which two (*GLN1* and *GLN4*) encode the cytosolic GS1 and two (*GLN2* and *GLN3*) the plastid-targeted GS2 (Table 4.1). In contrast, higher plant plastidic GS2 isoforms typically are encoded by a single gene (Lam et al., 1996). The *Chlamydomonas* *GLN2* and *GLN3* genes are highly similar in nucleotide and deduced amino acid sequence and are located head-to-head in the genome, suggesting that they result from a fairly recent gene duplication, also apparent in the *Volvox carteri* genome. All the GS2 cDNA clones identified in Chen and Silflow (1996) were derived from *GLN2*, but EST data show that *GLN3* is also expressed in a variety of conditions. ESTs also provide evidence for the expression of both *GLN1* and *GLN4*. While *GLN2* is expressed at a high level in all conditions tested, *GLN1* is induced upon transfer from darkness

Table 4.1 NH_3 assimilation and glutamate/glutamine biosynthesis enzymes

Gene	Accession #	Enzyme	EC #
GDH1	EDP02540	Glutamate dehydrogenase (GDH)	1.4.1.2
GDH2	EDO97071	Glutamate dehydrogenase (GDH)	1.4.1.2
GLN1	EDP07598	Glutamine synthetase (GS1)	6.3.1.2
GLN2	EDP03611	Glutamine synthetase (GS2)	6.3.1.2
GLN3	EDP03496	Glutamine synthetase (GS2)	6.3.1.2
GLN4	EDP03343	Glutamine synthetase (GS1)	6.3.1.2
GSF1	EDO96739	Ferredoxin-dependent glutamate synthase (Fd-GOGAT)	1.4.7.1
GSN1	EDP03651	NADH-dependent glutamate synthase (NADH-GOGAT)	1.4.1.14

to light, or from ammonium- to nitrate-containing media (Chen and Silflow, 1996). Accordingly, the ratio of cytosolic to plastidic GS activity is higher in nitrate than ammonium medium (Fischer and Klein, 1988).

The incorporation of ammonium into glutamine is followed by reductive transfer of the amide group from glutamine to αKG to form two glutamates in a reaction catalyzed by two GOGAT isoforms (Cullimore and Sims, 1981a; Galván et al., 1984; Márquez et al., 1984, 1986a): one NADH-dependent (NADH-GOGAT) and the other ferredoxin-dependent (Fd-GOGAT). These are encoded by *GSN1* (NADH-GOGAT) and *GSF1* (Fd-GOGAT), and are similar to the corresponding plant enzymes both in terms of sequence and compartmentation: both appear to be plastid-localized based on cell fractionation (Fischer and Klein, 1988) and targeting predictions.

GDH generally is thought to function primarily or only in catabolic processes and not to play a significant role in NH_3 assimilation under normal circumstances in either plants or *Chlamydomonas*. However, a possible role of GDH in NH_3 assimilation has been considered under specific circumstances (Cullimore and Sims, 1981b; Lam et al., 1996). Three GDH isozymes have been reported in *Chlamydomonas* (Moyano et al., 1992a,b), but only two genes encoding GDH have been identified (*GDH1* and *GDH2*, with no splicing variants detected; Table 4.1). This suggests that the three different isozymes represent different combinations of the two potential subunits in this hexameric enzyme. While the above studies placed GDH in the mitochondrion, Fischer and Klein (1988) had found it in the chloroplast, where it could only function in the catabolic direction. In higher plants, GDH is found mainly in mitochondria, and under some conditions in the cytosol (Kichey et al., 2005).

The *Chlamydomonas* enzymes involved in NH_3 assimilation, especially those of the glutamine-glutamate cycle, are similar in sequence, structure, and compartmentation to the corresponding enzymes of higher plants, and their functions are fairly parallel. The plastidic GS2 of plants, in combination with the plastidic Fd-GOGAT, is thought to be the predominant enzyme involved in both primary NH_3 assimilation and reassimilation of photorespiratory NH_3 in photosynthetic tissue, and the *Chlamydomonas* orthologues probably play similar roles (Márquez et al., 1986b). However, the main involvement of the plant cytosolic GS1 appears to be in primary nitrogen assimilation in roots and in phloem-specific synthesis of glutamine for long-distance transport (Lam et al., 1996), functions which have no equivalent in unicellular algae. The *Chlamydomonas* NADH-GOGAT probably also functions in assimilation of exogenous NH_3 (Márquez et al., 1986b), since its activity is increased in NH_3-grown cells (Fischer and Klein, 1988) and Fd-GOGAT is unlikely to be active in the dark. GS1 may also be involved in primary nitrogen assimilation, especially in dark-grown cells where the GS1 isoform predominates (Florencio and Vega, 1983; Fischer and Klein, 1988). Note that the *GLN1* expression data of Chen and Silflow (1996) appear to contradict this view.

NADH-GOGAT also has been implicated in recycling NH_3 released from protein degradation (Márquez et al., 1986a). However, research using the GS inhibitor L-methionine-DL-sulfoximine (MSX) suggests that GDH isozymes participate in the reassimilation of NH_3 during recycling of nitrogen, for example, when remobilized by proteolysis under nitrogen limitation (Cullimore and Sims, 1980; Hipkin et al., 1982; Peltier and Thibault, 1983). Peltier and Thibault (1983) suggested that GDH could account for about 30% of observed NH_3 assimilation when GS activity was blocked with MSX under conditions of CO_2 saturation. Even though similar experiments inhibiting GS in plants with MSX failed to show NH_3 assimilation under similar conditions (Lam et al., 1996), characteristics of an *Arabidopsis* GDH mutant indicate that GDH may play a role in NH_3 assimilation in plants under conditions of nitrogen excess (Melo-Oliveira et al., 1996).

The cytosolic and plastidic glutamate pools are equilibrated via a shuttle mechanism: αKG/malate and malate/glutamate antiporters allow export of glutamate from the chloroplast in exchange for αKG, with no net transport of malate (Weber et al., 1995). The Arabidopsis *OMT1* and *DCT1/DCT2* genes have been identified as responsible for these two reactions (Taniguchi et al., 2002). The *Chlamydomonas* genome encodes two *OMT* homologues (*OMT1* and *OMT2*), and one *DCT* homologue, *DCT1*, also known as *LCI20* because it is inducible by low CO_2 and controlled by CCM1 (Miura et al., 2004). *OMT2* has been found to be the target of a miRNA that triggers cleavage within a possibly regulatory upstream ORF found in its first exon (Zhao et al., 2007).

C. Biosynthesis of amino acids from TCA cycle, photosynthesis, and photorespiration intermediates

Following initial N assimilation into glutamine and glutamate, N is distributed to other amino acids and other N-containing compounds through the activity of aminotransferases or transaminases. Since many of these transaminations use glutamate as the amino donor and thus regenerate αKG, the precursor for ammonium assimilation, these transaminations recycle the αKG, facilitating continued NH_3 assimilation without depleting the αKG precursor and without requiring net αKG synthesis.

Aspartate and Alanine: The amino acids alanine and aspartate are synthesized by transamination of the corresponding common α-keto organic acids pyruvate and oxaloacetate, respectively, usually with glutamate as the amino donor. The reversible transamination of oxaloacetate with glutamate to yield aspartate and αKG is catalyzed by aspartate aminotransferase (aspartate transaminase; glutamic-oxaloacetic transaminase; AspAT). An AspAT purified from *Chlamydomonas* (Lain-Guelbenzu et al., 1990) was also able to use phenylalanine and tyrosine as amino donors when the acceptor was αKG, and serine, alanine, and glutamine when it was oxaloacetate. Five genes (*AST1–AST5*; Table 4.2) have been annotated in the genome as AspAT. AST1–4 are most closely related to plant AspATs, but AST5 is closer to cyanobacterial homologues and could represent another type of transaminase. In plant cells, AspAT isoforms are located in multiple compartments, including the mitochondrion, plastid and cytosol, but it is not always possible to derive unambiguous orthology relationship between the higher plant and algal genes. A proteomic survey of the *Chlamydomonas* mitochondrion (see http://www.chlamy.org/mito.html) has found many peptides from the AST1 and AST4 proteins. AST3 is most closely related to plant plastidic AspAT. *Chlamydomonas* AST2 has no targeting peptide and is probably cytosolic. The localization of AST5 is the most uncertain as its N-terminal extension is not recognized as a targeting peptide by prediction programs.

Similarly, alanine can be generated by the reversible transamination of pyruvate with glutamate to yield alanine and αKG. This reaction is catalyzed by alanine aminotransferase (glutamic-pyruvic transaminase; AAT). Two *AAT* genes have been annotated in the *Chlamydomonas* genome (Table 4.2), encoding AAT1 (predicted mitochondrial targeting) and AAT2 (targeting uncertain, no peptide evidence, no EST). The induction of *AAT1* by transfer from high to low CO_2 (Chen et al., 1996), and the effect of the transaminase inhibitor aminooxyacetic acid are consistent with a role of AAT in photorespiration. An AAT purified from *Chlamydomonas* (Lain-Guelbenzu et al., 1991) was able to use alanine and glutamate as substrates with glyoxylate, in addition to the primary activity. Because of the ability of AAT to transaminate glyoxylate to glycine using glutamate as an amino donor, AAT may function in the photorespiratory glycolate pathway.

Table 4.2 Aspartate, alanine, serine, and glycine biosynthesis enzymes

Gene	Accession #	Enzyme	EC #
AAT1	EDP08011	Alanine aminotransferase (AAT)	2.6.1.2
AAT2	EDP01608	Alanine aminotransferase (AAT)	2.6.1.2
AGT1	EDO97315	Alanine-glyoxylate transaminase	2.6.1.44
AGT2	EDO96807	Alanine-glyoxylate transaminase	2.6.1.44
AGT3	EDO99530	Alanine-glyoxylate transaminase	2.6.1.44
AST1	EDP02192	Aspartate aminotransferase (AspAT)	2.6.1.1
AST2	EDP09991	Aspartate aminotransferase (AspAT)	2.6.1.1
AST3	EDP07735	Aspartate aminotransferase (AspAT)	2.6.1.1
AST4	EDP08586	Aspartate aminotransferase (AspAT)	2.6.1.1
AST5	EDO99564	Aspartate aminotransferase (AspAT)	2.6.1.1
PGD1	EDP03909	D-3-Phosphoglycerate dehydrogenase (PGDH)	1.1.1.95
PSP1	EDO97634	Phosphoserine phosphatase (PSP)	3.1.3.3
PST1	EDP05686	Phosphoserine aminotransferase (PSAT)	2.6.1.52
SGA1	EDO97195	Serine glyoxylate aminotransferase, variant SG1a	2.6.1.45
SGA1	EDO97196	Serine glyoxylate aminotransferase, variant SG1b	2.6.1.45
SHMT1	EDO97448	Serine hydroxymethyltransferase (SHMT)	2.1.2.1
SHMT2	EDO97351	Serine hydroxymethyltransferase (SHMT)	2.1.2.1
SHMT3	EDP00905	Serine hydroxymethyltransferase (SHMT)	2.1.2.1

Pyruvate also can be transaminated with glycine to yield alanine and glyoxylate, a reaction catalyzed by alanine-glyoxylate aminotransferase. This reaction, however, is not expected to play a significant role in the net synthesis of alanine.

Serine and Glycine: Both serine and glycine are synthesized as part of the photorespiratory glycolate cycle, although it is not clear whether this is a major source of these amino acids for protein synthesis. The photorespiratory glycolate pathway is discussed in more detail in Chapter 8. Briefly, this pathway is initiated by the oxygenase activity of ribulose-1,5-bisphosphate carboxylase/oxygenase (Rubisco), which generates one molecule of 3PGA and one of 2-phosphoglycolate (P-glycolate) from the oxygenation of ribulose-1,5-bisphosphate (RuBP). P-glycolate, the initial substrate of the photorespiratory glycolate cycle, undergoes a specific dephosphorylation in the plastid to glycolate, which is transported to the mitochondrion (or possibly the peroxisome) and oxidized by glycolate dehydrogenase to glyoxylate. The glyoxylate is transaminated to glycine, probably by AAT using glutamate (or possibly alanine) as amino donor as discussed above, or by alanine-glyoxylate aminotransferase using alanine as the amino donor. Half of the glyoxylate also is transaminated to glycine via serine-glyoxylate aminotransferase. Via this pathway, serine is then synthesized from two molecules of glycine in a two-step process via the glycine decarboxylase complex and serine hydroxymethyltransferase, which yields one molecule of serine, one of carbon dioxide and one of ammonia. In a continuation of the photorespiratory glycolate pathway, the serine is converted to hydroxypyruvate in a transamination reaction also converting glyoxylate to glycine (see above) and the hydroxypyruvate reduced to glycerate. The glycerate is returned to the plastid and phosphorylated in an ATP-dependent reaction catalyzed by glycerate kinase to form 3PGA, thus returning three of the four original carbons (two molecules of P-glycolate) to the Calvin cycle.

By analogy with plants and in agreement with biosynthesis in microorganisms, net serine and glycine biosynthesis can also occur via a different, non-photorespiratory pathway (Figure 4.1), the phosphorylated pathway of serine biosynthesis (Ho and Saito, 2001). This pathway begins with 3PGA, which is oxidized to phospho-hydroxypyruvate in an NAD-dependent reaction catalyzed by D-3-phosphoglycerate dehydrogenase. P-hydroxypyruvate is then transaminated via a specific phosphoserine aminotransferase to phosphoserine, probably using glutamate as amino donor, and then subsequently dephosphorylated to serine in a reaction catalyzed by a specific phosphoserine phosphatase. All of the reactions in the phosphorylated pathway of serine biosynthesis take place in the plastid of plants, so by analogy are expected to take place in the plastid of *Chlamydomonas*. Each of the enzymes in this pathway is apparently encoded by a single gene in *Chlamydomonas* (*PGD1*, *PST1*, and *PSP1*) and the products are predicted to be organelle-targeted (Table 4.2).

D. Biosynthesis of cysteine, selenocysteine, and histidine

Cysteine: Cysteine biosynthesis represents the interface between sulfur assimilation and biosynthesis of the sulfur-containing amino acids.

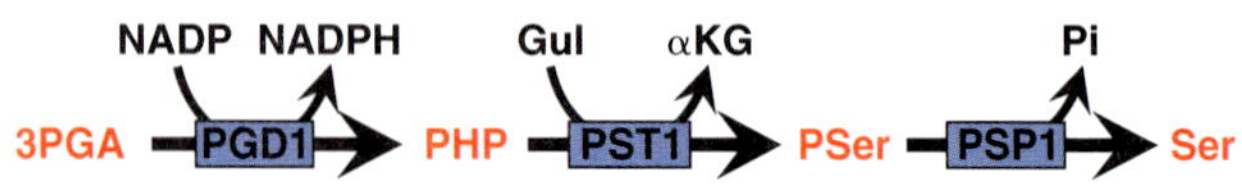

FIGURE 4.1 *Pathway for the synthesis of serine via phosphoserine. Enzymes are indicated by their gene names. Abbreviations for metabolites: 3PGA, 3-phosphoglycerate; PHP, 3-phosphohydroxypyruvate; PSer, phosphoserine.*

Cysteine itself, in addition to being an important functional protein amino acid, plays a significant role in maintaining the redox environment in cells of all organisms. Cysteine is synthesized from serine in a two step process (Figure 4.2), commencing with transfer of the acetyl group from acetyl-CoA to serine, forming *O*-acetylserine (OAS), catalyzed by serine acetyltransferase (SAT). The second step, a condensation of OAS with sulfide (or possibly with a thiosulfide carrier) to yield cysteine and acetate, is catalyzed by *O*-acetylserine(thiol)lyase (OASTL).

Chlamydomonas SAT1 was originally identified by complementation of an *Escherichia coli* SAT mutant (Ravina et al., 2002). EST evidence supports two *SAT1* mRNAs encoding identical proteins. In addition, the genome contains a putative *SAT2* gene which might be a pseudogene because its 3′ end runs into a REP2 transposon. Of the four annotated OASTL genes (*OASTL1–4*; Table 4.3), one was identified previously as a cDNA clone (*OASTL4*, named Cys1Acr by Ravina et al., 1999). Alternative transcription start sites

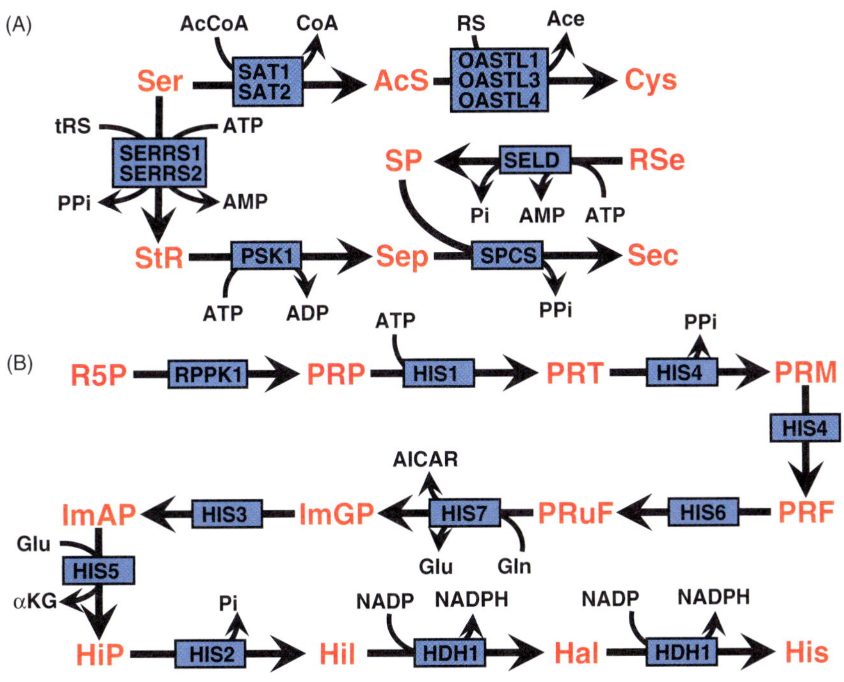

FIGURE 4.2 *Pathways for the synthesis of cysteine (A) from serine and of histidine (B) from ribose-5-phosphate. AcS, O-acetyl-L-serine; Ace, acetate; Cys, cysteine; AcCoA, acetyl-coenzyme A; CoA, coenzyme A; RS, sulfide donor; R5P, ribose-5-phosphate; PRP, 5-phosphoribosylpyrophosphate; PRT, phosphoribosyl-ATP; PPi, pyrophosphate; PRM, phosphoribosyl-AMP; PRF, phosphoribosyl-formamino-(5-amino-4-imidazole carboxamide)-phosphate; PRuF, phosphoribulosyl-formamino-(5-amino-4-imidazole carboxamide)-phosphate; ImGP, imidazole-glycerol-3-phosphate; AICAR, 5-aminoimidazole-4-carboxamide riboside; ImAP, imidazole-acetol-phosphate; HiP, histidinol-P; Pi, orthophosphate; Hil, histidinol; Hal, histidinal.*

Table 4.3 Cysteine and histidine biosynthesis enzymes

Gene	Accession #	Enzyme	EC #
HDH1	EDP06829	Histidinol dehydrogenase (HDH)	1.1.1.23
HIS1		ATP phosphoribosyl transferase (ATP-PRT)	2.4.2.17
HIS3	EDP07687	Imidazole glycerol-phosphate dehydratase (IGPD)	4.2.1.19
HIS4	EDP08791	Bifunctional phosphoribosyl-AMP cyclohydrolase (PRA-CH) and phosphoribosyl-ATP pyrophosphatase (PRA-PH)	3.5.4.19 3.6.1.31
HIS5	EDP01242	Histidinol-phosphate aminotransferase	2.6.1.9
HIS6	ED097969	Phosphoribosylformimino-5-aminoimidazole carboxamide ribonucleotide isomerase	5.3.1.16
HIS7	EDP00338	Imidazole glycerol phosphate synthase (IGPS)	2.4.2.-
OASTL1	EDP01088	*O*-acetylserine sulfhydrylase; cysteine synthase (CS)	2.5.1.47
OASTL2	ED097451	*O*-acetylserine sulfhydrylase; cysteine synthase (CS)	2.5.1.47
OASTL3	EDP05983	*O*-acetylserine sulfhydrylase; cysteine synthase (CS)	2.5.1.47
OASTL4	EDP04425	*O*-acetylserine sulfhydrylase; cysteine synthase, variant a (CS)	2.5.1.47
OASTL4	EDP04426, AAC27794	*O*-acetylserine sulfhydrylase; cysteine synthase, variant b (CS)	2.5.1.47
PSK1	EDP05206	*O*-phosphoseryl-tRNA(Sec) kinase (PSTK)	2.7.1.-
RPPK1	EDP02679	Ribose-phosphate pyrophosphokinase (RPPK)	2.7.6.1
RPPK2	EDP02478	Ribose-phosphate pyrophosphokinase (RPPK)	2.7.6.1
SAT1	EDP07949, EDP07950	Serine *O*-acetyl transferase (SAT)	2.3.1.30
SAT2	ED098094	Serine *O*-acetyl transferase (SAT)	2.3.1.30
SELD	EDP08654	Selenophosphate synthase (SelD)	2.7.9.3
SPCS	EDP02915	Selenocysteine synthase; Sep-tRNA:Sec-tRNA synthase (SepSecS)	2.9.1.1

in *OASTL4* lead to the generation of two products differing at their N-termini. Both *SAT1* and *OASTL4* have been characterized as responding to sulfur limitation, which is discussed in Chapter 5.

In higher plants, the two enzymes involved in cysteine synthesis are found in plastids, mitochondria, and the cytosol (Leustek et al., 2000) and the same appears true for *Ostreococcus*, but it is unclear whether a similar situation prevails in *Chlamydomonas*. The products of all the *Chlamydomonas OASTL* genes are predicted to be organelle-targeted, except for OASTL2 which is directed to the secretory pathway (a feature shared by its *Arabidopsis* and *Ostreococcus* orthologues At1g55880 and *OASTL2*).

Similarly, the two predicted SAT proteins in *Chlamydomonas* have putative organelle-targeting sequences. It is therefore unlikely that *Chlamydomonas* can synthesize cysteine in all three compartments, in particular in the cytosol. In the plastid, the acetyl-CoA needed for cysteine synthesis could be produced from at least two sources, pyruvate dehydrogenase or acetyl-CoA synthase (encoded by *ACS2*). The latter enzyme, which uses acetate as substrate, might be expected to be more important in mixotrophically or heterotrophically grown cells. The source of acetyl-CoA in the mitochondrion, assuming that the SAT reaction occurs in that compartment, would likely be pyruvate dehydrogenase or the mitochondrial acetyl-CoA synthase (ACS3), while ATP-citrate lyase could produce the precursor in the cytosol. In higher plants, SAT and OASTL form a reversible complex comprising a homotetramer of SAT and two dimers of OASTL (Hawkesford and De Kok, 2006). High levels of OAS bring about dissociation of the complex, which inactivates SAT and activates OASTL, thus bringing down the OAS level.

Selenocysteine: Serine serves as a precursor for an unusual amino acid of major biological significance, selenocysteine (Sec), found only in a small number of proteins. It is encoded by a UGA codon in a specific "SECIS" context, which is a particular mRNA secondary structure. The "21st" amino acid is present in all kingdoms of life, but has been lost in many branches. *Chlamydomonas* has selenoproteins (Fu et al., 2002; Novoselov et al., 2002), as do *Ostreococcus* (Palenik et al., 2007), animals and many bacteria, but land plants and fungi do not. In these organisms, selenocysteine is usually replaced in proteins by the less reactive cysteine. Selenocysteine is not found in free form, but is synthesized by modification of an aminoacyl-tRNA (Allmang and Krol, 2006). The $tRNA^{Sec}$ is first acylated with serine by an ordinary seryl-tRNA synthetase. In eukaryotes, the seryl moiety is modified in two steps. It is first phosphorylated by an *O*-phosphoseryl-$tRNA^{Sec}$ kinase (PSTK) to form *O*-phosphoserine (Sep). The product is then converted to selenocysteine by a Sep-tRNA:Sec-tRNA synthase (SepSecS) (Yuan et al., 2006). This reaction uses selenophosphate generated by a selenophosphate synthase. In *Chlamydomonas*, genes can be found for SepSecS (*SPCS*) and selenophosphate synthase (*SELD*), the latter being of clear bacterial origin. In addition, a protein with some similarity to human PSTK has been annotated as PSK1, but its identification as *O*-phosphoseryl-$tRNA^{Sec}$ kinase is still tentative.

Histidine: Histidine biosynthesis proceeds by a pathway similar to that found in plants, which is basically the same as that operating in prokaryotic microorganisms, albeit with noteworthy differences in the enzymes involved (Figure 4.2). Histidine synthesis begins with the ATP-dependent pyrophosphorylation of R5P catalyzed by ribose-phosphate pyrophosphokinase (RPPK) to form phosphoribosylpyrophosphate (PRPP). There appear to be two *RPPK* genes in *Chlamydomonas* (Table 4.3), *RPPK1* and *RPPK2*. Both proteins appear to be organelle-targeted, in contrast to higher plants which have RPPK proteins in the cytosol in addition to chloroplast and mitochondria (Krath and Hove-Jensen, 1999). Here again, intracellular localization appears more restricted in *Chlamydomonas* than in higher plants.

PRPP then serves as the first substrate for a series of 11 reactions catalyzed by 9 enzymes encoded by 8 genes (Figure 4.2 and Table 4.3). It is remarkable that different lineages have evolved different combinations of some of the catalytic activities into bi- or even tri-functional proteins. The first step in the pathway from PRPP, the condensation of ATP with PRPP to form phosphoribosyl-ATP, is catalyzed by *N′*-5′-phosphoribosyl-ATP transferase, which is a monofunctional enzyme encoded by the *HIS1* gene. The *HIS1* locus is unusual in that another gene resides in its 5th intron, but on the other DNA strand. That gene itself has an intron, encodes a phosphatase 2C, and enjoys ample support from EST and proteomics data. How the expression of the two genes might interact is not known. A similar situation has been described for other *Chlamydomonas* genes (Jain et al., 2007). The second and third steps of the histidine pathway, conversion of phosphoribosyl-ATP to phosphoribosyl formamino-5-aminoimidazole-4-carboxamide ribonucleotide (PRF), are catalyzed by phosphoribosyl-ATP phosphohydrolase and phosphoribosyl-AMP cyclohydrolase, which are contained within a single bifunctional polypeptide in plants (Stepansky and Leustek, 2006) as well as in *Chlamydomonas* (HIS4). This differs from the situation in most prokaryotic microbes, where the two catalytic activities are carried by distinct polypeptides, and in yeast, where the HIS4 protein contains not only these two catalytic activities, but also a histidinol dehydrogenase (HDH) domain. Conversion of PRF to phosphoribulosyl formamino-5-aminoimidazole-4-carboxamide ribonucleotide (PRuF) is catalyzed by PRF isomerase, encoded by *HIS6* in *Chlamydomonas*. However, the following step, conversion of PRuF to imidazoleglycerol phosphate (ImGP) and 5-aminoimidazole-4-carboxamide ribonucleotide (AICAR), requires two catalytic functions, a transamidation (using glutamine as amide donor) and a cyclization reaction. In prokaryotes, these two catalytic activities are encoded by two genes whose products form a complex but in yeast and plants, these two steps are catalyzed by a bifunctional IGP synthase (Stepansky and Leustek, 2006) which is also found in *Chlamydomonas* (*HIS7*). AICAR serves as a precursor for purine biosynthesis.

The plant and yeast IGP dehydratase, which catalyzes the dehydration of ImGP to imidazoleacteol phosphate, is a monofunctional enzyme. It lacks the histidinol phosphatase activity found in the bifunctional enzymes from prokaryotic microbes such as *E. coli* (Tada et al., 1994). Not surprisingly, the *Chlamydomonas* HIS3 enzyme also lacks a histidinol phosphatase domain. Interestingly, no histidinol phosphatase gene has been identified in either plants or *Chlamydomonas*. Histidinol phosphatase is encoded by two apparently unrelated gene families, represented by the amino-terminal domain of the bifunctional *E. coli hisB* gene and by the monofunctional *HIS2* gene of yeast. *Ostreococcus* contains a histidinol phosphatase gene of the bacterial type, but none can be found in *Chlamydomonas* or *Arabidopsis*. Thus it is likely that the conversion of histidinol phosphate to histidinol is catalyzed by a third type of enzyme in plants and Volvocales.

The enzyme catalyzing the reaction intermediate between the IGP dehydratase and histidinol phosphatase reactions, histidinol phosphate aminotransferase, is a monofunctional enzyme (HIS5). The remaining two steps in the histidine biosynthesis pathway, the sequential double oxidation of histidinol to histidine via the intermediate histidinal, are catalyzed in plants and *Chlamydomonas* by a monofunctional HDH. The plant enzyme is more closely related to the yeast histidinol dehydrogenase domain in the trifunctional HIS4 (see above) than to the monofunctional HDH enzyme in prokaryotic microbes (Stepansky and Leustek, 2006).

E. Biosynthesis of aspartate-derived amino acids

In addition to facilitating continued NH_3 assimilation by recycling the αKG precursor for glutamate and glutamine synthesis, transamination of oxalo-acetate generates aspartate, which serves as a precursor for a series of other amino acids, namely asparagine, threonine, lysine, and methionine. As described below, asparagine is derived directly from aspartate, but threonine, lysine, and methionine are synthesized in a branched pathway that originates with aspartate.

Asparagine: Asparagine is synthesized by amidation of aspartate, probably by transfer of the amide from glutamine to asparagine in an ATP-dependent reaction catalyzed by asparagine synthetase. The annotated *Chlamydomonas* asparagine synthetase (AS) gene, *ASNS* (Table 4.4), is most similar to the plant AS, which is a cytosolic enzyme. Plant AS appears to be able to also use NH_3 as a substrate for amidation, but this reaction does not appear to be physiologically significant *in vivo* (Wong et al., 2004). The *Chlamydomonas ASNS* gene product lacks an N-terminal extension, and asparagine synthesis is thus likely to proceed in the cytosol, as it does in plants. In light of this, it is tempting to suggest that the GS1 isoform of glutamine synthetase (see Section A) functions at least in part to provide a source of cytosolic glutamine for amidation of aspartate, with the resulting glutamate recycling back as substrate for GS1.

Threonine, Methionine, and Lysine: The biosynthesis of threonine, methionine, and lysine begins with aspartate and enters a branched pathway, which appears to be largely the same in *Chlamydomonas* and higher plants (Figure 4.3). The first two steps, the ATP-dependent phosphorylation of aspartate (catalyzed by aspartate kinase, AK) and the NADPH-dependent reduction of aspartate-4-phosphate into aspartate 4-semialdehyde (catalyzed by aspartate semialdehyde dehydrogenase), are common to all three amino acids. The AK reaction is the committing reaction and appears to be highly regulated in most organisms, including plants (Curien et al., 2005, 2007).

Table 4.4 Asparagine, threonine, methionine, and lysine biosynthesis enzymes

Gene	Accession #	Enzyme	EC #
AHD1	EDP01964	Bifunctional aspartate kinase/homoserine dehydrogenase (AK/HSDH)	2.7.2.4 1.1.1.3
ASK1	EDP08069	Aspartate kinase, monofunctional (AK)	2.7.2.4
ASNS	ED097202	Asparagine synthetase (AS)	6.3.5.4
ASSD	EDP02211	Aspartate semialdehyde dehydrogenase (ASADH)	1.2.1.11
CGS1	EDP01529	Cystathionine γ-synthase (CGS)	2.5.1.48
DAE1	ED099946	Diaminopimelate epimerase (DAPE)	5.1.1.7
DPA1	EDP03630	LL-diaminopimelate aminotransferase (DAP aminotransferase)	2.6.1.-
DPD1	EDP09979	Diaminopimelate decarboxylase (DAPD)	4.1.1.20
DPR1	ED099712	Dihydrodipicolinate reductase (DHDPR)	1.3.1.26
DPS1	EDP07434	Dihydrodipicolinate synthase (DHDPS)	4.2.1.520
HSD1	EDP07408	Homoserine dehydrogenase (HSDH)	1.1.1.3
HSK1	EDP06874	Homoserine kinase (HSK)	2.7.1.39
METC	ED098869	Cystathionine β-lyase (CBL)	4.4.1.8
METE	ED096787	Cobalamin-independent methionine synthase (MS)	2.1.1.14
METH1	EDP08397	Cobalamin-dependent methionine synthase (MS)	2.1.1.13
METH2	ED095867	Cobalamin-dependent methionine synthase (MS)	2.1.1.13
THS1	EDP08010	Threonine synthase (TS)	4.2.3.1

Like plants, *Chlamydomonas* appears to have two types of AK, a monofunctional aspartate kinase, encoded by *ASK1*, and a bifunctional protein containing both AK and homoserine dehydrogenase (HSDH) domains, encoded by *AHD1* (Table 4.4).

From aspartate 4-semialdehyde, the common pathway for threonine and methionine continues for two additional steps, the NADPH-dependent reduction of aspartate 4-semialdehyde to homoserine (catalyzed by homoserine dehydrogenase HSDH) and the ATP-dependent phosphorylation of homoserine to homoserine 4-phosphate (catalyzed by homoserine kinase, HSK). As mentioned above, plants and *Chlamydomonas* have a bifunctional HSDH–AK, which in plants is the only known HSDH. However, *Chlamydomonas* also contains a gene encoding a monofunctional HSDH, *HSD1*, which is more similar to the HSDH domain of the bifunctional AK–HSDH genes of plants and *E. coli* rather than the monofunctional HSDH of fungi. A similar situation is found in *Ostreococcus*. Homoserine 4-phosphate is converted to

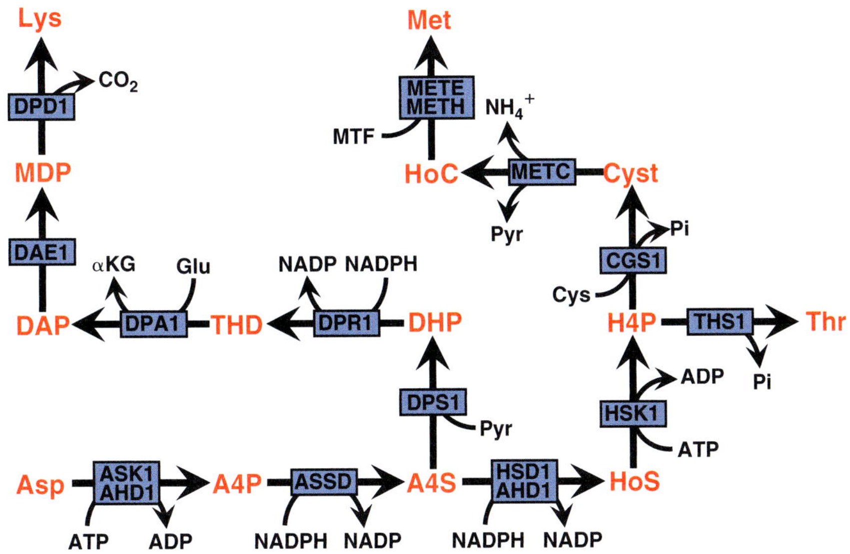

FIGURE 4.3 *Pathway for synthesis of the aspartate-derived amino acids, threonine, methionine, and lysine. A4P, aspartate-4-phosphate; A4S, aspartate-4-semialdehyde; HoS, homoserine; H4P, homoserine-4-phosphate; Pi, orthophosphate; Cyst, cystathionine; Pyr, pyruvate; HoC, homocysteine; MTF, methylene tetrahydrofolate; DHP, L-2,3-dihydropicolinate; THD, tetrahydrodipicolinate; DAP, L-diaminopimelate; MDP, meso-diaminopimelate.*

threonine by threonine synthase, which catalyzes both the dephosphorylation and the rearrangement of the hydroxyl group necessary to form threonine.

The pathway for biosynthesis of methionine from homoserine 4-phosphate in *Chlamydomonas* and in plants differs somewhat from that of prokaryotic microorganisms in that it uses homoserine 4-phosphate (*O*-phosphohomoserine) rather than *O*-succinyl-homoserine. *O*-phosphohomoserine is combined with cysteine to produce cystathionine, a reaction catalyzed by cystathionine γ-synthase (CGS). The CGS of *E. coli* and that of *Arabidopsis* are both able to utilize *O*-succinylhomoserine *in vitro*, to differing degrees, but only the *Arabidopsis* enzyme is able to use *O*-phosphohomoserine (Hacham et al., 2003). Because of the sequence similarity between the CGS gene of *Chlamydomonas*, *CGS1*, and those of higher plants, and of the apparent lack of genes encoding homoserine-*O*-succinyltransferase in *Chlamydomonas*, it appears likely that in *Chlamydomonas* cystathionine is synthesized from *O*-phosphohomoserine (Table 4.4). Higher plant CGS is feedback-regulated at the level of mRNA stability by *S*-adenosyl-methionine (Chiba et al., 2003), but the sequence responsible for this regulation (Ominato et al., 2002) is not conserved in *Chlamydomonas* or *Ostreococcus*.

Homocysteine is formed from cystathionine by elimination of the non-sulfur part of cysteine as pyruvate and NH_4^+ in a reaction catalyzed by cystathionine β-lyase and common to most organisms. The final step, the conversion of homocysteine to methionine, is catalyzed by methionine synthase (MS). *Chlamydomonas* has two types of MS enzyme (Grossman et al., 2007). One, encoded by *METE*, is a vitamin B_{12} (cobalamin)-independent enzyme similar to that of higher plants, and is found in the mitochondrial proteome. Plant MS can utilize *S*-methylmethionine or *S*-adenosylmethionine, in addition to the more usual 5-methyltetrahydropteroyltriglutamate, as methyl donors, but the *Chlamydomonas METE* gene is more closely related to cyanobacterial than to plant relatives, so it may not exhibit the methyl donor versatility of the plant enzyme. The *Chlamydomonas* genome encodes another type of MS that uses vitamin B_{12} as a cofactor and is closely related to that found in animals and bacteria. There is one complete gene of this type, *METH1*, and a very closely related but incomplete sequence, *METH2*, found on a small scaffold in version 3.0 of the genome but next to METH1 in version 4. This region still contains gaps in version 4, probably as a result of genome misassembly due to repeated sequences. Independent assembly data in fact suggest that this region holds at least three repeats in tandem of the first half of the gene, with >97% identity among them at the nucleotide level. The triplicated region contains the homocysteine *S*-methyltransferase and the pterin binding domains. In the *Volvox* genome, the *METH* region shows no sequence gaps. Here, the gene can be modeled as having three almost perfect repeats of the first three exons (but not the introns) and producing a protein with a triplication of the two N-terminal domains. The functional consequences of such a complex organization are unclear. *METH* is an ancestral gene that has been lost in many lineages, in particular in land plants. Eukaryotes are unable to synthesize vitamin B_{12}, and therefore those that have vitamin B_{12}-dependent enzymes must acquire it from bacteria that live in their environment. Interestingly, many eukaryotic algae have lost the *METE* type of MS gene, and have therefore become vitamin B_{12} auxotrophs (Croft et al., 2006). *Chlamydomonas* can live without vitamin B_{12} addition because of its *METE* gene, but uses the more efficient *METH* when vitamin B_{12} is available (Croft et al., 2005).

Six different pathways for lysine biosynthesis are recognized in various organisms, but these can be divided into two basic groups, the diaminopimelate (DAP) pathways and the α-aminoadipate pathways (Hudson et al., 2006). Lysine biosynthesis in plants and *Chlamydomonas* follows one of the four versions of the DAP pathway (Figure 4.3 and Table 4.4). The first four steps, common to all DAP pathways, include the AK and aspartate semialdehyde dehydrogenase reactions previously described. Condensation of aspartate semialdehyde and pyruvate by dihydrodipicolinate synthase (DPS1) leads to L-2,3-dihydrodipicolinate, which is then converted to tetrahydrodipicolinate by dihydrodipicolinate reductase (DPR1). The conversion of tetrahydrodipicolinate to *meso*-DAP in plants and cyanobacteria utilizes a novel transaminase, LL-DAP aminotransferase, which specifically catalyzes the conversion of tetrahydrodipicolinate to LL-DAP (Hudson et al.,

2006). DAP aminotransferase catalyzes in a single step a conversion that requires three enzymes in the DAP pathway variant found in *E. coli*, and appears to be unique to green photosynthetic organisms. The LL-DAP is then converted to *meso*-DAP by the action of DAP epimerase, encoded by *DAE1*. The last step of lysine biosynthesis, decarboxylation of *meso*-DAP to form lysine, is common to all the lysine biosynthesis pathways and is catalyzed by DAP decarboxylase encoded by *DPD1* (Table 4.4).

F. Biosynthesis of branched-chain amino acids

Based on identification of putative genes (Table 4.5), the pathway for branched-chain amino acid biosynthesis appears to be the same as that found in prokaryotic microbes and plants (Singh and Shaner, 1995). It is expected to occur in the plastid, and predicted targeting of the required gene products is indeed consistent with an organellar location.

Isoleucine and valine are synthesized in parallel pathways sharing four common enzymes, each of which utilizes dual substrates (Figure 4.4). The biosynthesis of valine starts with pyruvate, a common metabolite of intermediary metabolism, while that of isoleucine starts with 2-ketobutyrate.

Table 4.5 Isoleucine, valine, and leucine biosynthesis enzymes

Gene	Accession #	Enzyme	EC #
AAD1	EDP03205	Dihydroxyacid dehydratase (DHAD)	4.2.1.9
AAI1	EDP06428	Acetohydroxy acid isomeroreductase (AHRI)	1.1.1.86
ALSL1	EDP01876	Acetolactate synthase, large subunit (AHAS; ALS)	2.2.1.6
ALSS1	EDO98300	Acetolactate synthase, small subunit (AHAS; ALS)	2.2.1.6
BCA1	EDP06865	Branched-chain amino acid aminotransferase (BCAT)	2.6.1.42
BCA2	EDP07184	Branched-chain amino acid aminotransferase (BCAT)	2.6.1.42
BCA3	EDP08065	Branched-chain amino acid aminotransferase (BCAT)	2.6.1.42
LEU1L	EDO97224	3-Isopropylmalate dehydratase, large subunit	4.2.1.33
LEU1S	EDP08379	Isopropylmalate dehydratase; small subunit	4.2.1.33
LEU2	EDP08580	2-Isopropylmalate synthase	2.3.3.13
LEU3	EDP07327, EDP07328	3-Isopropylmalate dehydrogenase, splicing variants a and b	1.1.1.85
THD1	EDP06791	Threonine deaminase (TD)	4.3.1.19

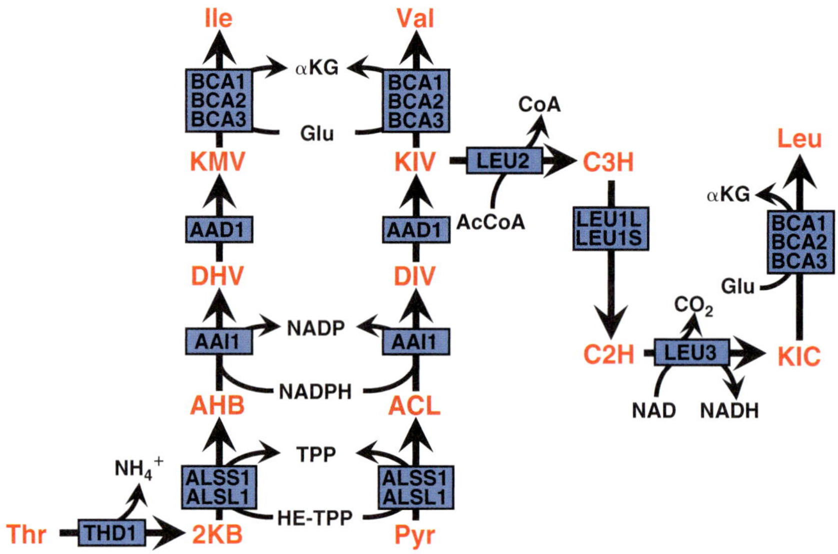

FIGURE 4.4 *Pathway for synthesis of the branched-chain amino acids, isoleucine, valine, and leucine. Pyr, pyruvate; 2KB, 2-ketobutyrate; HE-TPP, 2-hydroxyethyl-thiamine pyrophosphate; TPP, thiamine pyrophosphate; AHB, 2-acetohydroxybutyrate; ACL, 2-acetolactate; DHV, 2,3-dihydroxy-3-methylvalerate; DIV, 2,3-dihydroxyisovalerate; KMV, 3-keto-3-methylvalerate; KIV, 3-ketoisovalerate; AcCoA, acetyl-coenzyme A; CoA, coenzyme A; C3H, 3-carboxy-3-hydroxyisocaproate; C2H, 3-carboxy-2-hydroxyisocaproate; KIC, 2-ketoisocaproate.*

Synthesis of 2-ketobutyrate from threonine represents a committed step in isoleucine biosynthesis and is catalyzed by threonine deaminase (TD, threonine dehydratase). The *Chlamydomonas* enzyme (encoded by the *THD1* gene) is probably feedback regulated by isoleucine, based on similarity of its C-terminal region to the regulatory region found in TDs of plants and microbes (Samach et al., 1991). It thus belongs to the "biosynthetic" category of threonine deaminases, as opposed to the catalytic forms which help remobilize carbon skeletons and nitrogen and are not sensitive to end-product inhibition.

The first of the four parallel steps in the biosynthesis of isoleucine and valine is catalyzed by acetohydroxy acid synthase (AHAS), also known as acetolactate synthase (ALS). It is the target for three important herbicide classes (imidazolinones, sulfonylureas, and trazolopyrimidines), and AHAS mutants have been identified in plants with decreased sensitivity to each of these herbicides as the result of altered amino acid sequences (Singh and Shaner, 1995). Similarly, *Chlamydomonas* mutants with increased resistance to these herbicides have been identified and have been demonstrated to have altered amino acid sequences in the large subunit of AHAS consistent with their specific patterns of sensitivity (Hartnett et al., 1987; Winder and Spalding, 1988; Funke et al., 1999; Kovar et al., 2002). The AHAS large subunit gene (*ALSL1*) from one such mutant has been demonstrated to function as a dominant selectable marker in *Chlamydomonas* (Kovar et al., 2002). Acetohydroxy acid isomeroreductase then catalyzes the rearrangement and NADPH-mediated reduction of the intermediates, and dihydroxy-acid dehydratase (acetohydroxyacid dehydratase), a 2Fe-2S cluster protein, is responsible for the production of 2-keto-isovalerate and 2-keto-3-methylvalerate, whose transamination with glutamate yields valine and isoleucine, respectively. *Chlamydomonas* encodes three genes for branched-chain amino acid aminotransferases (BCAT).

The biosynthesis of leucine branches from the valine biosynthetic pathway. Its committing step is the addition of an acetyl group from acetyl-CoA to 2-ketoisovalerate, forming 3-carboxy-3-hydroxyisocaproate (isopropylmalate), a reaction catalyzed by 2-isopropylmalate synthase encoded by *LEU2*. Isopropylmalate isomerase (3-isopropylmalate dehydratase) is a heteromeric enzyme catalyzing the conversion of isopropylmalate to 3-carboxy-2-hydroxyisocaproate, which is then oxidatively decarboxylated by isopropylmalate dehydrogenase to form 2-ketoisocaproate. There are two splicing variants of the isopropylmalate dehydrogenase (*LEU3*) mRNA, but both products appear to be targeted to the plastid. Finally, leucine is formed by a transamination catalyzed by BCAT.

Feedback inhibition is the major regulatory mechanism in branched-chain amino acid biosynthesis. In addition to TD mentioned above, it is exerted on AHAS, which both valine and leucine inhibit, and on 2-isopropylmalate synthase, which is inhibited by micromolar concentrations of leucine (Singh and Shaner, 1995). Thus leucine and isoleucine specifically feedback inhibit their own biosynthesis, whereas inhibition of AHAS by valine and leucine can also inhibit isoleucine biosynthesis.

G. Biosynthesis of glutamate-derived amino acids

The biosynthetic pathways for arginine and proline are bound by the common precursor glutamate. In addition, ornithine, an important intermediate in the arginine biosynthetic pathway, can serve as an alternative precursor for proline biosynthesis, linking the two biosynthetic pathways in a web-like pattern.

Arginine: Arginine biosynthesis is notable both for its complexity and as one of the better studied amino acid biosynthetic pathways in *Chlamydomonas*. Exogenous arginine is transported into *Chlamydomonas* (Kirk and Kirk, 1978a), which enabled the isolation of a series of arginine auxotrophic mutants (Loppes and Heindricks, 1986; also see Volume 1, Chapter 6). Arginine biosynthesis (Figure 4.5 and Table 4.6) requires three amination reactions of the side chain. It proceeds first to the intermediate ornithine, which receives another amine by incorporation of carbamoylphosphate, and a third one from aspartate. Based on the subcellular

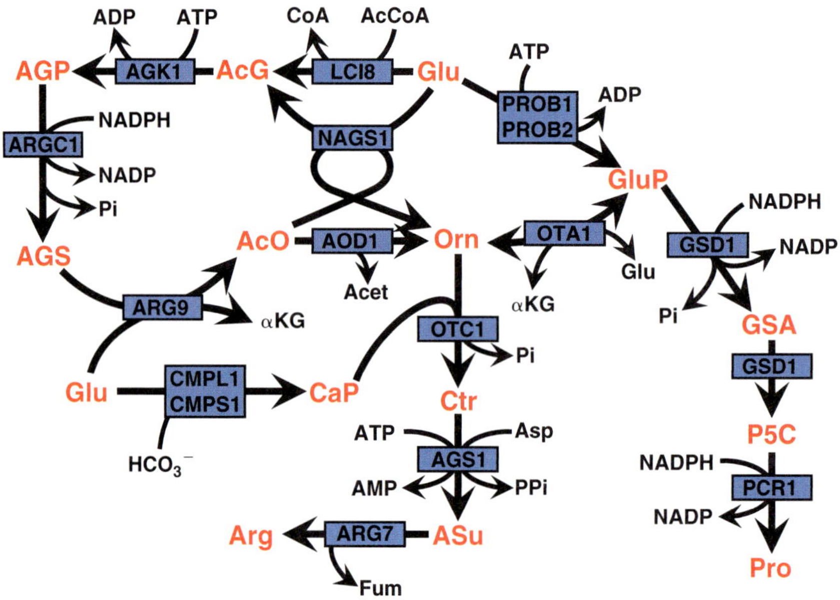

FIGURE 4.5 *Pathway for synthesis of the glutamate-derived amino acids arginine and proline. GluP, L-glutamyl-γ-phosphate; Pi, orthophosphate; GSA, glutamic γ-semialdehyde; P5C, Δ^1-pyrroline-5-carboxylate; AcCoA, acetyl-coenzyme A; CoA, coenzyme A; AcG, N-acetyl-L-glutamate; AGP, N-acetylglutamyl-phosphate; AGS, N-acetylglutamate semialdehyde; AcO, N-α-acetylornithine; Orn, L-ornithine; CaP, carbamoyl-phosphate; Ctr, citrulline; PPi, pyrophosphate; ASu, L-arginino-succinate; Fum, fumarate.*

compartmentation of the homologues in plants and on targeting predictions, most of the arginine biosynthetic pathway in plants and *Chlamydomonas* appears to be located in the plastid (Slocum, 2005).

Ornithine synthesis is initiated by the acetylation of glutamate using acetyl-CoA, catalyzed by *N*-acetylglutamate synthase. This enzyme is encoded by the *LCI8* gene, identified in a screen for low CO_2-inducible genes, and is dependent upon the CCM1 regulator for its expression (Miura et al., 2004). *N*-acetylglutamate is eventually regenerated by the last step of the ornithine cycle, where ornithine acetyltransferase/*N*-acetylglutamate synthase (NAGS1) transfers an acetyl group to glutamate without consuming acetyl-CoA. Thus, the role of the *LCI8* gene product is just to prime the cycle, which otherwise consumes only NADPH, ATP, and an amine donated by transamination from glutamate.

The second step of the ornithine pathway, phosphorylation of *N*-acetylglutamate by *N*-acetyl-L-glutamate kinase, is allosterically feedback regulated by arginine (Faragó and Dénes, 1967). *Chlamydomonas* AGK1 shows the N-terminal extension present in the regulated plant enzyme but absent in the *E. coli* enzyme that is not subject to feedback regulation. In *Arabidopsis* the enzyme is activated by binding of GLB1, the homologue of

Table 4.6 Arginine and proline biosynthesis enzymes

Gene	Accession #	Enzyme	EC #
AGK1	EDP09199	*N*-acetyl-L-glutamate kinase (NAGK)	2.7.2.8
AGS1	EDP00857	Argininosuccinate synthase (AS)	6.3.4.5
AOD1	EDP07452	*N*-acetylornithine deacetylase (NAOD)	3.5.1.16
ARG1	EDP01497	*N*-acetyl-γ-glutamyl-phosphate reductase (NAGPR)	1.2.1.38
ARG7	EDP09253	Argininosuccinate lyase (ASL)	4.3.2.1
ARG9	ED099676	*N*-acetylornithine aminotransferase (NAOAT)	2.6.1.11
CMPL1	ED096236	Carbamoyl phosphate synthase, large subunit (CPS)	6.3.5.5
CMPS1	EDP05092	Carbamoyl phosphate synthase, small subunit (CPS)	6.3.5.5
GSD1	EDP01490	Δ^1-Pyrroline-5-carboxylate synthetase	1.2.1.41
LCI8	EDP02142	*N*-acetylglutamate synthase (monofunctional) (NAGS)	2.3.1.1
NAGS1	ED098751	*N*-acetylglutamate synthase (bifunctional) (NAGS)	2.3.1.1 2.3.1.35
OTA1	ED097297	Ornithine-δ-aminotransferase (OAT)	2.6.1.13
OTC1	EDP05375	Ornithine carbamoyltransferase	2.1.3.3
PCR1	EDP00128, EDP00129	Δ^1-Pyrroline-5-carboxylate reductase (P5CR)	1.5.1.2
PROB1	EDP04671	γ-Glutamyl kinase (γGK)	2.7.2.11
PROB2	EDP04670	γ-Glutamyl kinase (γGK)	2.7.2.11

the PII protein that in bacteria plays a central role in regulation of nitrogen metabolism (Ferrario-Mery et al., 2006). *Chlamydomonas* has a GLB1 homologue that appears to be chloroplast-targeted as in higher plants. It is therefore likely that in algae as well, the PII homologue activated by high N-availability stimulates arginine biosynthesis even in the presence of high levels of the amino acid (Slocum, 2005).

N-acetylglutamyl-phosphate then undergoes an NADPH-dependent, dephosphorylating reduction catalyzed by *N*-acetylglutamyl reductase (NAGPR) to produce acetylglutamyl semialdehyde. The *arg1* mutant was reported to be defective at this step (Strijkert and Sussenbach, 1969), and the mutation maps to linkage group I, consistent with the location of the gene for NAGPR. The *arg9* mutant may lack the next enzyme, acetylornithine aminotransferase, consistent with its map position on linkage group VI

(Loppes and Heindricks, 1986). The ornithine cycle is completed by the transfer of the acetyl group from *N*-acetylornithine to glutamate by ornithine acetyltransferase/*N*-acetylglutamate synthase, as described above, a reaction probably marked by another auxotrophy mutation, *arg10* (Loppes and Heindricks, 1986). In plants and yeast, the *N*-transacetylase and NAG synthase domains are separated by a conserved PxM-ATML sequence that is the site of auto-proteolytic processing, yielding two fragments that remain associated within an $\alpha2\beta2$ heterotetramer (Slocum, 2005). The cleavage site is conserved in *Chlamydomonas* (genome version 4), *Volvox*, and *Ostreococcus*. In addition to the cyclic pathway, plants and bacteria (and *Chlamydomonas*) can synthesize ornithine by deacetylation of *N*-acetylornithine by acetylornithine deacetylase (AOD1) (Slocum, 2005). The enzyme is homologous to *E. coli* ArgE and to metallo-carboxypeptidases of the M20 family, and probably binds two Zn ions.

The downstream part of the pathway leads from ornithine to arginine in three steps. Ornithine is converted to citrulline by addition of carbamoyl-phosphate in a reaction catalyzed by ornithine carbamoyltransferase (OTC1). The *arg4* mutant is blocked in this step of the pathway (Loppes and Heindricks, 1986), and it maps on linkage group XII/XIII approximately where the *OTC1* gene lies. Strangely, the mutant is rescued neither by ornithine nor citrulline, though the latter would be expected to allow arginine synthesis in the absence of OTC. Carbamoyl phosphate is used both for arginine and for pyrimidine biosynthesis, which both occur in the chloroplast. In *Chlamydomonas* as in higher plants, each of the large and small subunits of carbamoylphosphate synthase are encoded by a single gene, so it is unclear how allocation of the substrate between the two pathways is regulated (Slocum, 2005).

Citrulline is converted to argininosuccinate by the addition of aspartate in a reaction catalyzed by argininosuccinate synthase, the gene for which, *AGS1*, is probably marked by the *arg8* mutations (three alleles, of which two showed intragenic complementation; Loppes and Heindricks, 1986). The splitting of argininosuccinate into arginine and fumarate is catalyzed by argininosuccinate lyase (ASL; sometimes misnamed arginosuccinate lyase). A number of mutations have been isolated in the corresponding *ARG7* gene, among which alleles *arg7* and *arg7–8* (formerly known as *arg2*) have been used extensively in transformation experiments (Debuchy et al., 1989). The mutation in *arg7–8* has been sequenced in a study aiming to demonstrate homologous recombination in *ARG7*, and two intragenic suppressors have been described (Mages et al., 2007). *ARG7* appears to be a hotspot for recombination, and a detailed recombination map was obtained (see Matagne, 1978). Because the enzyme is a tetramer, some alleles show intragenic (intermolecular) complementation, and *arg7* and *arg7–8* are used routinely to generate vegetative diploids thanks to the rarity of recombinants.

Proline: There are two alternative routes for proline biosynthesis in plants (Hong et al., 2000), and based on identification of genes putatively encoding proline biosynthetic enzymes (Table 4.6), the two pathways are also likely to operate in *Chlamydomonas* (Figure 4.5). In the "glutamate pathway," proline is derived directly from glutamate, whereas in the "ornithine pathway," glutamate is first converted to ornithine.

In *Chlamydomonas*, the first two steps of the glutamate pathway are catalyzed by separate enzymes, γ-glutamyl kinase (γGK) and glutamate-5-semialdehyde (GSA) dehydrogenase (GSADH; γ-glutamyl phosphate reductase), as in prokaryotic microbes. This differs from the bifunctional enzyme (Δ^1-pyrroline-5-carboxylate synthetase) found in plants (Hong et al., 2000). Interestingly, the closest homologues of the *Chlamydomonas* GSADH (encoded by *GSD1*) are GSADH domains of the plant bifunctional enzyme. However, the γGK domain of the bifunctional plant enzyme seems only distantly related to the *Chlamydomonas* γGK (*PROB1* and *PROB2* gene products). GSA spontaneously undergoes a dehydration and cyclization to form Δ^1-pyrroline-5-carboxylate, which is converted to proline by the NADPH-dependent enzyme Δ^1-pyrroline-5-carboxylate reductase, encoded by *PCR1*.

Ornithine, which also is an intermediate in the biosynthesis of arginine (see above) can serve as an alternative precursor for proline biosynthesis. The most probable entry of ornithine into the proline biosynthetic pathway is via transamination, using ornithine as amino donor and αKG as amino acceptor, to form GSA and glutamate in an ornithine-δ-aminotransferase (OTA1)-catalyzed reaction (Figure 4.5).

H. Biosynthesis of aromatic amino acids

Aromatic amino acid biosynthesis in *Chlamydomonas* appears to be fairly similar to that in higher plants (Figure 4.6), based on genes identifiable by homology (Table 4.7). The three aromatic amino acids, tryptophan, tyrosine, and phenylalanine, are derived from the common precursor chorismate in plants and *Chlamydomonas*. The seven-step synthesis of chorismate begins with the condensation of two intermediates of carbohydrate metabolism, PEP and E4P, in a reaction catalyzed by 3-deoxy-D-arabino-heptulosonate-7-phosphate (DAHP) synthase. The second enzyme in this pathway, 3-dehydroquinate synthase, catalyzes the cyclization of DAHP to form 3-dehydroquinate. The third and fourth enzymes of the chorismate biosynthesis pathway, 3-dehydroquinate dehydratase and shikimate-NADP oxidoreductase (shikimate dehydrogenase), convert 3-dehydoquinate into shikimate and are found on a bifunctional enzyme in *Chlamydomonas*, as they are in plants. Shikimate undergoes an ATP-dependent phosphorylation to form shikimate-3-phosphate, catalyzed by the fifth enzyme in the pathway, shikimate kinase. The sixth and penultimate step in the synthesis of chorismate is the reversible production of 5-*enol*pyruvylshikimate-3-phosphate (EPSP) and orthophosphate from shikimate-3-phosphate and PEP, catalyzed by EPSP synthase. EPSP synthase is the site of action of the commercially important herbicide

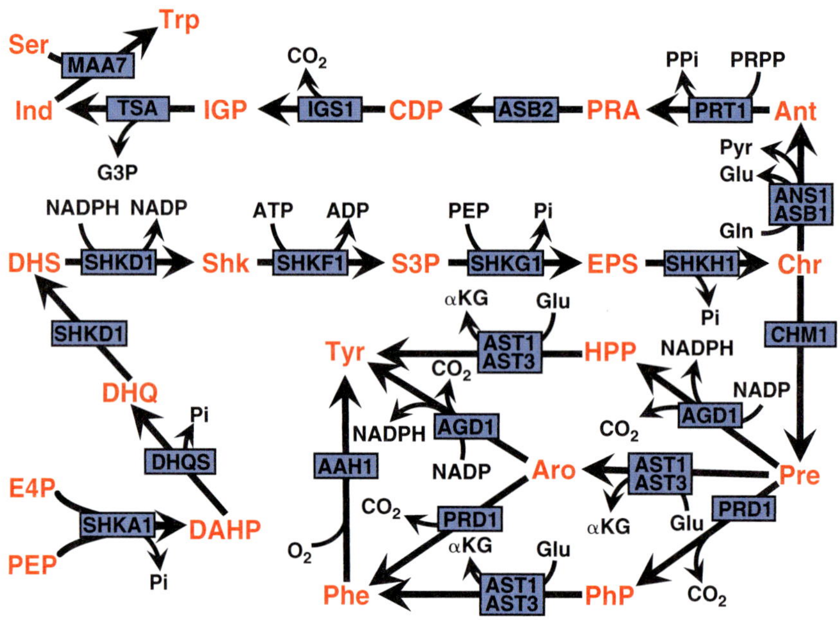

FIGURE 4.6 *Pathway for synthesis of the aromatic amino acids tyrosine, phenylalanine, and tryptophan. E4P, erythrose-4-phosphate; PEP, phosphoenolpyruvate; Pi, orthophosphate; DAHP, 3-deoxy-D-arabino-heptulosonate 7-phosphate; DHQ, 3-dehydroquinate; DHS, 3-dehydroshikimate; Shk, shikimate; S3P, shikimate 3-phosphate; EPS, 5-enolpyruvylshikimate-3-phosphate; Chr, chorismate; Pre, prephenate; HPP, 4-hydroxyphenylpyruvate; Aro, arogenate; PhP, phenylpyruvate; Pyr, pyruvate; Ant, anthranilate; PRPP, 5-phosphoribosylpyrophosphate; PPi, pyrophosphate; PRA, 5-phosphoribosyl-anthranilate; CDP, 1-(o-carboxy-phenylamino)-1-deoxyribulose 5-phosphate; IGP, indole-3-glycerol phosphate; G3P, glyceraldehyde-3-phosphate; Ind, indole.*

glyphosate, to which *Chlamydomonas* is sensitive. One of the mutations that confer resistance to 5-methylanthranilate, *maa15*, also confers resistance to glyphosate but it is not on the same linkage group as EPSP synthase (Dutcher et al., 1992, see below). Note that phenylalanine and tyrosine addition alleviate toxicity of glyphosate (Gresshoff, 1979). The final step in the synthesis of chorismate, the dephosphorylation of EPSP to form chorismate, is catalyzed by chorismate synthase.

From chorismate, biosynthesis of the three aromatic amino acids tyrosine, phenylalanine, and tryptophan diverge. The tyrosine and phenylalanine biosynthetic pathways form a network following one branch from chorismate, through prephenate, and tryptophan biosynthesis following a second branch, through anthranilate (Figure 4.6).

Tyrosine and phenylalanine biosynthesis: The first and committing step in synthesis of tyrosine and phenylalanine from chorismate is catalyzed by chorismate mutase, and involves an intermolecular rearrangement of the *enol*pyruvyl side chain of chorismate to generate prephenate. There are two

Table 4.7 Aromatic amino acid biosynthesis enzymes

Gene	Accession #	Enzyme	EC #
AAH1	EDP09772	Aromatic amino acid hydroxylase	1.14.16.-
ADT1	EDP08558	Arogenate dehydratase	4.2.1.91
AGD1	EDP01548	Arogenate/prephenate dehydrogenase	1.3.1.43
ANS1	EDP05111	Anthranilate synthase, α-subunit (ASα)	4.1.3.27
ASB1	EDP09587	Anthranilate synthase, β-subunit (ASβ)	4.1.3.27
ASB2	EDP00638	Phosphoribosylanthranilate isomerase (PAI)	5.3.1.24
AST1	EDP02192	Aspartate aminotransferase (AspAT)	2.6.1.1
AST2	EDP09991	Aspartate aminotransferase (AspAT)	2.6.1.1
AST3	EDP07735	Aspartate aminotransferase (AspAT)	2.6.1.1
AST4	EDP08586	Aspartate aminotransferase (AspAT)	2.6.1.1
AST5	EDO99564	Aspartate aminotransferase (AspAT)	2.6.1.1
CHM1	EDP00124	Chorismate mutase (CM)	5.4.99.5
DHQS	EDP01025	3-Dehydroquinate synthase (DHQ synthase)	4.2.3.4
IGS1	EDP09023	Indole-3-glycerol-phosphate synthase (MAA4) (IGPS)	4.1.1.48
MAA7	EDP06079	Tryptophan synthase, β-subunit (TSB)	4.2.1.20
PRT1	EDP06478	Anthranilate phosphoribosyltransferase (APRT)	2.4.2.18
SHKA1	EDP00515	3-Deoxy-D-arabino-heptulosonate 7-phosphate synthase (DAHP synthase)	2.5.1.54
SHKD1	EDP02779	Bifunctional dehydroquinate dehydratase-shikimate:NADP oxidoreductase (DHQase-SORase)	4.2.1.10 1.1.1.25
SHKF1	EDP05661	Shikimate kinase	2.7.1.71
SHKG1	EDO96795	5-Enolpyruvylshikimate-3-phosphate (EPSP) synthase	2.5.1.19
SHKH1	EDP01398	Chorismate synthase (CS)	4.2.3.5
TSA	EDP00665	Tryptophan synthetase, α-subunit (MAA1) (TSA)	4.2.1.20

general pathways known for the synthesis of phenylalanine and tyrosine from prephenate: branching directly from prephenate via phenylpyruvate and *p*-hydroxyphenylpyruvate, followed by transamination of each to form phenylalanine and tyrosine, respectively, or branching to the final products after transamination to the intermediate arogenate (Figure 4.6). It is not clear from the identified genes (Table 4.7) likely to encode enzymes active in these pathways whether biosynthesis of tyrosine and phenylalanine in *Chlamydomonas* proceeds by one alternative or the other, or by a combination of both general pathways. Substrate preferences of aminotransferase enzymes are not easily predicted, so the aspartate aminotransferase isozymes indicated might catalyze the transamination of prephenate to arogenate, of the intermediates phenylpyruvate and *p*-hydroxyphenylpyruvate to phenylalanine and tyrosine, respectively, or of all three.

Conversions of prephenate to *p*-hydroxyphenylpyruvate and of arogenate to tyrosine are catalyzed by prephenate dehydrogenase and arogenate dehydrogenase, respectively. The putative arogenate/prephenate dehydrogenase (AGD1) identified in *Chlamydomonas* shares greatest sequence similarity with the type 2 arogenate dehydrogenase of *Arabidopsis*, which exhibits weak prephenate dehydrogenase activity (Rippert and Matringe, 2002). However, because activity of the *Arabidopsis* type 2 arogenate dehydrogenase with prephenate was judged too weak to be physiologically relevant (Rippert and Matringe, 2002), it appears most likely that *AGD1* encodes arogenate dehydrogenase and that the synthesis of tyrosine in *Chlamydomonas* proceeds via arogenate. Nonetheless, the specificity of AGD1 for prephenate versus arogenate cannot be determined based on sequence alone, so the *p*-hydroxyphenylpyruvate pathway for tyrosine biosynthesis cannot be ruled out.

Similarly, conversions of prephenate to phenylpyruvate and of arogenate to phenylalanine are catalyzed by prephenate dehydratase and arogenate dehydratase, respectively. In plants, it has been reported that synthesis of phenylalanine occurs via the phenylpyruvate pathway, at least in the etiolated state (Warpeha et al., 2006). However, the six *Arabidopsis* genes with homology to prephenate dehydratases have been shown to encode arogenate dehydratases, with no or little activity on prephenate (Cho et al., 2007), and were renamed *ADT1-6*. This is in keeping with earlier reports (e.g. Schmid and Amrhein, 1995) where the arogenate pathway was suggested to be the only route to phenylalanine and tyrosine in plants. Because the single *Chlamydomonas* orthologue shows high similarity to the *Arabidopsis* sequences, it has been called *ADT1*. While its product is believed to act primarily on arogenate, a dual specificity cannot be completely ruled out.

Another potential pathway for synthesis of tyrosine has been suggested by identification of a putative aromatic amino acid hydroxylase (AAH1) that might be able to convert phenylalanine to tyrosine. Thus the pathways for biosynthesis of phenylalanine and tyrosine are illustrated in Figure 4.6 as a web of potential pathways, indicating our uncertainty of which is/are truly active in *Chlamydomonas*.

Tryptophan biosynthesis: Biosynthesis of tryptophan in most plants and microbes follows a single pathway (Figure 4.6), which proceeds from the branch point chorismate. Its conversion to anthranilate by elimination of the *enol*pyruvyl side chain, accompanied by an amino transfer with glutamine as donor, is catalyzed by the heteromeric enzyme anthranilate synthase (Schmid and Amrhein, 1995). The next three steps in tryptophan biosynthesis convert anthranilate to indole-3-glycerol-phosphate and are catalyzed by anthranilate phosphoribosyl transferase, phosphoribosylanthranilate isomerase, and indole-3-glycerol phosphate synthase. Genes predicted to encode these three enzymes and the α- and β-subunits of anthranilate synthase have been identified in *Chlamydomonas* (Table 4.7).

The last two steps of the tryptophan biosynthetic pathway, conversion of indole-3-glycerol-phosphate to tryptophan via an indole intermediate, are catalyzed by a single, multimeric enzyme, tryptophan synthase. Tryptophan synthase is composed of two α- and two β-subunits, each of which is responsible for and able to catalyze one of the two reactions on its own. Mutations in the gene for the β-subunit of tryptophan synthase (*MAA7*) confer resistance to 5-fluoroindole (Palombella and Dutcher, 1998), and have served as a negative selection marker in *Chlamydomonas* (Rohr et al., 2004). Other 5-fluoroindole resistance mutations mapped to two other loci, *TAR1* and *MAA2*, the latter located only 3.5 map units from *MAA7* (Palombella and Dutcher, 1998). Mutants at the *MAA2* locus had already been obtained in a screen for resistance to 5-methylanthranilate, which identified a total of 16 *MAA* genes (including *MAA7*) of which 13 have been mapped (Dutcher et al., 1992). 5-Methylanthranilate is converted to 5-methyltryptophan, which in turn represses anthranilate synthase. No tryptophan auxotrophic mutants were recovered in these studies, probably because of the absence of active tryptophan uptake. 5-Fluoroindole and 5-methylanthranilate resistance can arise from hypomorphic mutations that still permit some tryptophan biosynthesis but produce only sublethal amounts of the toxic product. Indeed, many of the resistant strains have reduced growth rates compared to the wild type (Dutcher et al., 1992). Several of the mutations map near known tryptophan biosynthesis genes (Bowers et al., 2003), which in some case have been found to contain lesions. Thus, the *maa1* mutations mark the *TSA* gene (in spite of the fact that none confer resistance to 5-fluoroindole, in contrast to mutations in *MAA7/TSB*). The *maa5* mutant excreted anthranilate and phenylalanine in the medium, and had higher anthranilate synthase and anthranilate to indole-3-glycerophosphate activities. It could be a deregulated mutant in the shikimate pathway common to the three aromatic amino acids. The *maa6* mutant was unique in that it excreted an anthranilate derivative and its slow growth phenotype was partially rescued by indole. Because it was sensitive to 5-fluoroindole, this suggested that *MAA6* encodes one of the three enzymes between anthranilate and indole-3-glycerophosphate, and indeed their combined activity was undetectable. The mutation maps on linkage group VI near the mating type locus, but this does not correspond to any of the known tryptophan biosynthesis genes. The absence of activity in *maa6* is puzzling because it does not lead to tryptophan auxotrophy. This is also true for the lack of TSB activity in the *maa7-5* mutant (Palombella and Dutcher, 1998). Either the activities were too labile to be measured or less likely, another pathway exists for tryptophan biosynthesis in *Chlamydomonas*.

III. USE OF EXOGENOUS AMINO ACIDS

A. Import of amino acids

Early studies showed that *Chlamydomonas* can use many organic compounds as a nitrogen source, including several amino acids (glutamine, ornithine, and arginine). This triggered the search for amino acid auxotrophic mutants, thought to be a valuable resource for the development of genetic tools. The harvest has been unfortunately small, yielding only arginine-requiring mutants (Gillham, 1965). Nakamura et al. (1981) linked their failure to obtain methionine auxotrophs in *Chlamydomonas* to the photodynamic action of this and other amino acids. They observed that methionine at high concentrations negates the toxic effect of MSX, and concluded that it was imported and that its intracellular photosensitization was responsible for the death of the auxotrophs. However, control experiments, using cells grown in the dark on acetate, have not been reported since then and their hypothesis has remained unproven.

Kirk and Kirk (1978a) found that among 18 amino acids tested, only arginine was efficiently taken up by *Chlamydomonas*, a situation similar to that found in *Volvox* (Kirk and Kirk, 1978b). Arginine was accumulated 150-fold against its concentration gradient, while lysine, leucine, and glutamate accumulated only 2.5-, 0.8-, and 0.8-fold, respectively. Using a mixture of 18 ^{3}H-labelled amino acids excluding arginine, they found that radioactivity uptake was completely and selectively abolished by 1 mM arginine but not by the other amino acids, implying that an arginine transporter was responsible for most of the uptake of the other amino acids, but with far lower efficiency. Compared to cultures grown on ammonium, arginine uptake was found to be enhanced several-fold by growth on urea or amino acids (including arginine itself) and even more by nitrogen starvation. Earlier reports indeed had found that ammonium inhibited arginine uptake (Strijkert et al., 1971) and the question of whether and how ammonium inhibits growth of *arg* mutants has been debated (Loppes and Strijkert, 1972). The differences observed between strains have been proposed to reflect differences in the ammonium repression of intracellular arginine breakdown, but this has not been demonstrated.

The identity of the arginine transporter(s) has not yet been determined experimentally. *Chlamydomonas* encodes a surprisingly high number of genes for the two main classes of putative amino acid transporters: seven in the AAAP (Amino Acid/Auxin Permease) family and six in the APC (Amino acid Polyamine organoCation) family. Among the latter, AOC5 and AOC6 are the closest to plant Cationic Amino acid Transporters and to the human SLC7A1–3 that form the y+ transport system. This system transports arginine, lysine, and ornithine and is subject to trans-stimulation (enhancement by prior loading with the transported amino acid). Note that the single APC-type transporter encoded in *Ostreococcus* is orthologous to AOC5/6, which suggests that this alga will also preferentially utilize arginine.

In *Chlamydomonas* AOC5 and AOC6 may also be responsible for the uptake of two other cationic amino acids that are potent antibiotics, L-canavanine and L-methionine-sulfoximine. *AOC5* is found on linkage group VII, like the *can1* mutation that confers resistance to L-canavanine (Smyth et al., 1975). This mutation is closely linked to *pf17* (no recombinants in 259 tetrads; Harris, 1989). The *PF17* gene product has been identified as *RSP9*, coding for a radial spoke protein (Volume 3, Chapter 7) which is found on the same scaffold 165 kb from *AOC5*. Similarly, *AOC6* is situated on linkage group I, between *RB47* and *CNA73*, a position compatible with the genetic mapping of the *msr1* mutation that confers resistance to MSX, and also makes it acceptable as a sole source of nitrogen (Franco et al, 1996). The *arf1* mutations described by Franco et al. may thus affect uptake of MSX, rather than its metabolization. In this respect, it should be noted that arginine protects *Chlamydomonas* against MSX toxicity: growth of wild-type strains is possible on arginine-containing plates supplemented with 100 micromoles MSX, a concentration that kills them in TAP medium (OV; http://www.chlamy.org/methods/cleanup.html). This could be explained by competition for uptake, rather than by an unusual bypass of the GS pathway.

There is no *Chlamydomonas* homologue for the animal SLC3A glycoproteins that associate with SLC7A subunits to form the y+LAT system, which in animals transports cationic and large neutral amino acids. There is also no equivalent of the SLC6A type of Na^+-dependent amino acid transporter. The other members of the APC family, of uncertain specificity, have been called AOC1–4 for amino acid carrier. Their roles cannot be deduced from sequence alone, and it is possible that some of them function in intracellular amino acid transport, in particular between the chloroplast and the cytosol. This might be also true of the *Chlamydomonas* members of the AAAP family of transporters, called AOT1–7. They are related to the SLC38A family, and generally involved in the transport of neutral amino acids. While AOT1–4 are closer to plant homologues, AOT5–7 lean toward the animal/fungal side. Their specificity and subcellular localizations await experimental determination, but AOT3 appears to be organelle-targeted, both in *Chlamydomonas* and *Ostreococcus*.

B. Intracellular catabolism of arginine and urea

There are three classical pathways for intracellular arginine degradation that produce reutilizable nitrogen. The arginase enzyme converts arginine to ornithine and urea and is the most widespread in eukaryotes. It is absent from *Chlamydomonas*, where the urea cycle is therefore not operating, in contrast to plants and the diatom *Thalassiosira* (Armbrust et al., 2004). In the agmatine pathway that converts arginine to putrescine, nitrogen is released either as urea (by agmatinase) or as ammonium. No agmatinase is encoded in *Chlamydomonas*, but a plant-type agmatin degradation pathway is probably present (see below, polyamine synthesis). In *Chlamydomonas*, the only pathway of arginine catabolism that has been documented biochemically is the bacterial-type deiminase pathway (Sussenbach and Strijkert, 1969). Arginine deiminase produces NH_3 and citrulline and thus also serves as a starting point for polyamine synthesis. It has also been described in *Giardia* (Brown et al., 1998) and in *Arabidopsis* chloroplasts (Ludwig, 1993). Arginine deiminase is probably encoded by the *ADI1* gene, whose product is closely related to the catabolic arginine deiminase of *Giardia*. The citrulline produced by arginine deiminase is then converted to ornithine and carbamoyl phosphate by ornithine carbamoyltransferase (ornithine transcarbamoylase), a chloroplast enzyme encoded by *OTC1*. Carbamoyl phosphate is probably entirely recycled into pyrimidine biosynthesis, because carbamate kinase, the enzyme that converts it to CO_2, NH_3, and ATP, appears to be missing. No orthologue can be identified in the *Chlamydomonas*, *Volvox*, or *Ostreococcus* genomes when either the *Giardia*/animal type or the diatom/bacterial type of carbamate kinase are used as queries. Carbamate kinase activity has been found in *Arabidopsis* (Ludwig, 1993), but the corresponding gene has not been identified.

Another biologically important fate of arginine is its conversion to nitric oxide and citrulline by NO synthase (NOS), the major source of NO in animals. Unfortunately, the mechanism of NO synthesis in plants is still unknown (Zemojtel et al., 2006; Wilson et al., 2008). An *Arabidopsis* protein (AtNOA1) has been described whose gene inactivation reduces NO production (Guo and Crawford, 2005), but the protein lacks NO synthase activity. AtNOA1 is rather a GTPase located in the mitochondrion, the main source of NO within the cell. *Chlamydomonas* encodes an orthologue of AtNOA1, as well as a related protein (NOA2). In addition, NO can be synthesized by *Chlamydomonas* cells fed with nitrite in the dark, in a process that is dependent upon nitrate reductase activity but independent of arginine (Sakihama et al., 2002).

Since neither arginase nor the related agmatinase appears to be encoded in *Chlamydomonas*, urea is not a product of arginine catabolism in *Chlamydomonas*. However, it is efficiently utilized as an exogenous N source. Urea degradation uses a bacterial-type mechanism combining urea carboxylase and allophanate hydrolase, rather than the urease found in plants. The two genes encoding urea carboxylase and allophanate hydrolase in *Chlamydomonas*, *DUR1* and *DUR2*, are adjacent and in divergent orientation. Other urea assimilation genes, the *DUR3A* and *DUR3B* urea transporters, are found nearby as is the arginine deiminase gene *ADI1*. This functional clustering is suggestive of a common regulation, similar to that observed for the cluster of nitrate/nitrite assimilation genes (Quesada et al., 1993; also see Chapter 3). Allophanate hydroxylase is induced by growth on urea (Semler et al., 1975), and also on acetamide. Further evidence for a link between acetamide and urea/arginine catabolism comes from mutational studies (Hodson and Gresshoff, 1987). Among a series of 5-fluoroacetamide resistant mutants, a fraction were unable to use both acetamide and urea and lacked urea transport, urea carboxylase and allophanate hydrolase. The only recognizable general amidase gene in *Chlamydomonas*, *AMI1*, is not part of the urea gene cluster, so these mutants probably have alterations in a regulatory gene necessary for the induction of acetamide and urea utilization genes, rather than a deletion of the urea cluster. *Ostreococcus*, like higher plants, uses a urease rather than urea carboxylase/allophanate lyase, but it is striking to see that the urease gene is clustered with a series of functionally related genes (involved in the assembly of the enzyme), probably also for regulatory reasons.

C. Extracellular oxidation of amino acids

Since arginine is the only amino acid that can be taken up against a concentration gradient, the ability of *Chlamydomonas* to grow in a medium where amino acids other than arginine are the only supply of nitrogen raised the question of how nitrogen was acquired. A first answer came when Muñoz-Blanco et al. (1990) described extracellular deamination of various amino acids by *Chlamydomonas* cells grown in an ammonium-free medium. The reaction produces NH_3 that can be taken up readily and a ketoacid that remains in the medium. Indeed, ketoacids had been identified in the culture medium of *Chlamydomonas* cells "grown" in the absence of nitrogen (Collins and Kalnins, 1967). Extracellular L-amino acid oxidases (LAO) have been described in a variety of marine algae, and proposed to contribute significantly to nitrogen supply in oceanic environments (Palenik and Morel, 1990). Later, a *Chlamydomonas* LAO was purified from *cw15* cells grown in a medium containing alanine as sole nitrogen source (Piedras et al., 1992). It was found as a 470-kD oligomer of 60-kD subunits, associated with cellular membranes. The enzyme contained FAD, a cofactor typical of LAOs, and had a broad specificity, but the reported K_m values were rather high (31–176 micromoles). Later, Vallon et al. (1993) identified as an LAO the mysterious protein M that had been found to accumulate to high levels in nitrogen-starved cells, up to a point where it became the most abundant protein in the cell (Bulté and Wollman, 1992). The fact that it cross-reacted with cytochrome *f* antibodies and appeared at the same time as cytochrome b_6f was degraded, had suggested a link between the two proteins, which was disproved when protein M was found to accumulate in a cytochrome *f* mutant. The cross-reactivity was traced to an EKVGN peptide shared by the two proteins (Vallon et al., 1993). Protein M oxidized all L-amino acids tested except proline and cysteine, but not D-amino acids or polyamines. L-MSX was also a substrate, which makes the use of this GS inhibitor delicate in ammonium-free medium. K_m values were in the sub-micromolar range, too low to be accurately measured, compatible with a scavenging role in an environment where amino acids are very dilute.

The enzyme was found to reside in the periplasm, in line with the presence of a signal peptide at the N-terminus of its FAD-binding subunit, Mα, encoded by the *LAO1* gene (Vallon and Wollman, 1997). The purified enzyme appeared in two high-MW forms of comparable enzymatic properties. The lighter form (900–1,000 kD) contained Mα only, while the heavier one (1,300 kD) had an additional subunit Mβ of 135 kD that could be dissociated by mild SDS treatment. Both Mα and Mβ are heavily *O*-glycosylated, which has made them useful in the characterization of the *gag1* mutants (Vallon and Wollman, 1995).

Surprisingly, homologues of LAO1 can be found neither in the available *Volvox* genome sequence, nor in the complete *Ostreococcus* genome, nor in other Viridiplantae. The closest relatives are found among animal amine oxidases, with which a shared GGRXXS/T motif has been described (Vallon, 2000). It therefore appears that *Chlamydomonas* has adapted an ancient amine oxidase to scavenge extracellular amino acids. In addition, *Chlamydomonas* contains a gene closely related to *LAO1*, called *LAO3*, whose product may also be a L-amino acid oxidase. The LAO preparations of Piedras et al. (1992) and Vallon et al. (1993) differ in several respects, so the possibility exists that *LAO3* corresponds to the former, while *LAO1* encodes the latter.

IV. ROLE OF AMINO ACIDS AS PRECURSOR OF METABOLITES

A. Polyamine metabolism

Amino acids serve as precursors for a wide variety of metabolites, from purine and pyrimidine nucleotides to heme and glutathione. Here, we will only describe the formation of polyamines and glutathione.

Polyamines are essential metabolites of the eukaryotic cell, that in plants participate in defense, cell division, development, etc. (Cona et al., 2006). They are produced by decarboxylation either of arginine (the agmatine pathway found in bacteria and plants) or of ornithine (as in most plants and animals). In the first pathway, agmatine formed by decarboxylation of arginine is either directly hydrolyzed by agmatinase to urea and putrescine, or first converted to *N*-carbamoylputrescine by agmatine iminohydrolase (AIH), then to putrescine by *N*-carbamoylputrescine amidohydrolase (CPA) (Illingworth et al., 2003). As mentioned above, agmatinase is not found in *Chlamydomonas*, but two genes each can be identified encoding AIH (*AIH1* and *AIH2*) and CPA (*CPA1* and *CPA2*). It has been proposed that the pathway is vestigial in *Chlamydomonas* (Illingworth et al., 2003), but the expression of *AIH1* is attested by EST data and that of *AIH2* by proteomics data (mitochondrial soluble proteome). The genes appear to be conserved in *Volvox*, although not in *Ostreococcus*. The identification of an arginine decarboxylase is more problematic, but in addition to ODC1 and ODC2 (Baumann et al., 2007), an OAD1 protein with an arginine/ornithine/DAP decarboxylase domain appears as a good candidate.

However, the best-documented pathway of polyamine biosynthesis in *Chlamydomonas* is the ornithine decarboxylase (ODC) pathway (Voigt et al., 2004), which after deimination of arginine to citrulline and decarboxylation to ornithine (see arginine degradation, above), yields putrescine by action of ODC. Spermidine synthase then adds to putrescine an aminopropyl group derived from decarboxylated *S*-adenosyl Met. In contrast to most eukaryotes, the pathway does not appear to proceed further to spermine, as spermine was not found in *Chlamydomonas*.

Putrescine and spermidine accumulate prior to cell division, and spermine addition induces cell cycle arrest, presumably by feedback inhibition of polyamine biosynthesis (Theiss et al., 2002). The regulation of polyamine biosynthesis thus appears to be important for the *Chlamydomonas* cell cycle. Decarboxylation of *S*-adenosyl methionine, an essential step of polyamine biosynthesis, is carried out by *S*-adenosyl-methionine decarboxylase (AdoMetDC), the product of the *DCA1* gene. Polyamines regulate this gene at the translational level, via a mechanism that is conserved between higher plants, *Chlamydomonas* (Hanfrey et al., 2005) and *Volvox*. The 5′ UTR contains a "small uORF," showing in particular a conserved C-terminal PS dipeptide (Figure 4.7). The uORF-encoded polypeptide is believed to repress translation of the downstream AdoMetDC ORF by interacting with a component of the ribosome. The small uORF is in turn negatively regulated by a 9-codon "tiny uORF" overlapping with its start codon. When cellular concentrations of polyamines are low, ribosomes would translate the tiny uORF, then reinitiate at the next AUG downstream, producing the AdoMetDC protein. In contrast, high polyamine concentrations would favor leaky scanning by the ribosome. Initiation would then occur

```
OlDCA1uORF  MMDSEDGSSDNI-RDEEEQNKVTESSFIADLRPWG-CEKYASSAYQDLTRSPS
OtDCA1uORF  -MDVEDGSSDNISRDQLELNEAFESALIADLRPWG-WEKYQDASYYDLTRPPS
CrDCA1uORF  -MPAVKRTSMRS-----VYDPIATEPIVAS-EPAR-LVVKRTACFADTGSLPS
VcDCA1uORF  -MPVS--SSMR------VFEPLA-EPVAVD-RPA--VIAKFSAAFADCVKFPS
AtDCA1uORF  -MESKGGKKKSS-SSSSLFYEAPLGYSIEDVRPNGGIKKPKSSVYSNCSKRPS
```

FIGURE 4.7 *Alignment of small ORFs upstream of AdoMetDC in Ostreococcus lucimarinus, Ostreococcus tauri, Chlamydomonas reinhardtii, Volvox carteri, and Arabidopsis thaliana. Note conservation of the C-terminal PS dipeptide.*

at the first strong AUG, that of the small uORF, leading to repression of AdoMetDC translation. Note that the small uORF but not the tiny uORF is present in *O. tauri* and *O. lucimarinus*, suggesting a distinct mechanism to regulate uORF translation in *Ostreococcus*.

In animals and fungi, ODC is another important level of regulation. The small peptide ornithine decarboxylase antizyme binds to ODC and promotes its degradation by the proteasome (Pegg, 2006). Its action is in part controlled by the antizyme inhibitor AZI, which is homologous to ODC but enzymatically inactive (Keren-Paz et al., 2007). AZI sequesters the antizyme, thus protecting ODC from degradation. Antizyme has been found in animals, fungi, and protists but not in plants, and accordingly no antizyme homologue can be found in the *Chlamydomonas* genome. AZI also appears to be lacking: there are two genes with similarity to animal and fungal ornithine decarboxylases, *ODC1* and *ODC2*, but both possess the important catalytic residues that are mutated in AZI, and are therefore presumed to be active. Compared to other ODCs, the *Chlamydomonas* proteins contain long N-terminal extensions, also found in *Volvox* and in the unique *Ostreococcus* ODC1 protein, and it cannot be excluded that some of the algal ODCs reside in the chloroplast rather than in the cytosol.

In *Chlamydomonas*, *ODC1*, the more highly expressed of the two genes, may instead be regulated post-transcriptionally. EST data show multiple alternative splicing isoforms, such as retention of introns 4 or 5, or skipping of exon 5. Of the four variants produced, only the one that skips exon 5 (variant ODC1a) has a full length ORF starting at the first AUG. The cDNA described in Voigt et al. (2004) is truncated at the 5′ end, retains exon 5 and intron 5 and shows an indel in exon 5 compared to the genomic sequence and all the available ESTs. This led to an error in the prediction of the N-terminal sequence. The other splice variants show premature stop codons that are predicted to trigger nonsense-mediated mRNA decay. Indeed, many *ODC1* ESTs show staggered 5′ ends, suggesting that they are prone to degradation. This is reminiscent of the situation encountered in the long *CTH1* mRNA produced when cells are starved for copper (Moseley et al., 2002). It is tempting to speculate that regulation of *ODC1* expression occurs at the level of splicing, with high polyamine levels favoring the formation of unproductive mRNAs.

Spermidine and spermine synthases form a small gene family in higher plants. Based on the specificity of the *Arabidopsis* proteins (Hanzawa et al., 2002), the product of the *Chlamydomonas SPD1* gene is predicted to catalyze the formation of spermidine, while that of the *SPS1* gene, which is closely related to the Arabidopsis *ACAULIS 5*, would synthesize spermine. This is surprising, because as mentioned above, no spermine has been detected in *Chlamydomonas* (Theiss et al., 2002). Both SPD1 and SPS1 have been detected in the flagellar proteome (Pazour et al., 2005), which also holds ODC1. The possible significance of a flagella-localized polyamine biosynthesis pathway is unclear.

One of the functions of spermidine is to serve as a precursor for the formation of the unusual amino acid hypusine [*N*(epsilon)-(4-amino-2-hydroxybutyl)lysine], which occurs uniquely on a single residue of eukaryotic translation initiation factor 5A (eIF5A) (Wolff et al., 2007). Deoxyhypusine synthase (DHS) first transfers the aminobutyl moiety of spermidine to a specific lysine residue of eIF5A, forming deoxyhypusine which is subsequently hydroxylated by deoxyhypusine hydroxylase (DOHH). In several systems, including *Arabidopsis*, this modification of eIF5A is important in the control of cell growth and proliferation (Duguay et al., 2007). *Chlamydomonas* encodes a single gene for DOHH, called *DOH1*, showing the HEAT repeats and His-Asp motifs characteristic of this metalloenzyme. Surprisingly, two unlinked but closely related genes *DHS1* and *DHS2* can be found for deoxyhypusine synthase. In all eukaryotes, a single gene codes for DHS, even when multiple eIF5A isoforms exist as in yeast. *Chlamydomonas* has a single eIF5A gene, so the presence of two distinct DHS is doubly puzzling. Note that only *DHS1* shows EST evidence and conservation in *Volvox*. Some higher plants possess a related enzyme, homospermidine synthase, but homospermidine has not been reported in *Chlamydomonas* and DHS1 and DHS2 are more closely related to *bona fide* DHS of plants and animals than to homospermidine synthases.

In plants, polyamines are degraded by two types of oxidases, the copper-containing and the FAD-containing amine oxidases (CuAO and PAO, respectively) (Cona et al., 2006). Their oxidation not only participates in the regulation of their accumulation, but also in the production of a variety of secondary metabolites and in the generation of H_2O_2. *Chlamydomonas* has three CuAO genes whose products, AMX1-3, can degrade putrescine into NH_3 and 4-aminobutanal. The AMX1 and AMX2 proteins have a C-terminal SKL peptide typical of peroxisome-targeted proteins. In addition, there are four genes encoding FAD-containing amine oxidases of unknown specificity, called *AOF1–3* and *CPLD35*. The latter has been identified as conserved in photosynthetic organisms, including diatoms (Merchant et al., 2007), it is predicted to be organelle-targeted, and the *Arabidopsis* homologue has been found in the chloroplast.

B. Glutathione metabolism

Glutathione (γ-L-glutamyl-L-cysteinylglycine) is an important metabolite in plants, in particular because of its role in the response to oxidative stress. It is synthesized in two steps, by the action of γ-glutamylcysteine synthetase and glutathione synthetase, encoded in *Chlamydomonas* by the *GSH1* and *GSH2* genes, respectively. While GSH1 is probably located in the chloroplast, the localization of GSH2 and hence the site of glutathione synthesis is more uncertain. In higher plants, the reported cytosolic localization (Mullineaux and Rausch, 2005) conflicts with the chloroplast targeting generally predicted from its N-terminal sequence. In addition to its role as an antioxidant and a source of reductants in enzymatic reactions, glutathione can be conjugated to proteins to affect their function or fate, including in *Chlamydomonas* (Michelet et al., 2005). Glutathione also serves as a precursor for phytochelatins, and changes in its metabolism have been investigated during cadmium treatment of *Chlamydomonas* where phytochelatins should play a protective role. Changes were observed in the levels of several proteins involved in the metabolism of glutamate, cysteine, and glycine, but not in GSH1 or GSH2 themselves (Gillet et al., 2006). Indeed, another study found no correlation between phytochelatin levels and Cd accumulation in *C. acidophila* (Nishikawa et al., 2006).

V. CONCLUSIONS

In spite of its limited ability to import amino acids, *Chlamydomonas* has proved a valuable model to explore several aspects of their metabolism in photosynthetic organisms. With the unraveling of the genome sequence and the development of reverse genetic tools, it is likely that these pathways will attract more and more attention in the coming years. But we must first find a remedy to the paucity of auxotrophs: a required first step may be the implanting of specific amino acid transport systems, which for the moment only arginine enjoys.

REFERENCES

Allmang, C. and Krol, A. (2006). Selenoprotein synthesis: UGA does not end the story. *Biochimie* **88**, 1561–1571.

Armbrust, E.V., Berges, J.A., Bowler, C., Green, B.R., Martinez, D., Putnam, N.H., Zhou, S., Allen, A.E., Apt, K.E., Bechner, M. et al. (2004). The genome of the diatom *Thalassiosira pseudonana*: Ecology, evolution, and metabolism. *Science* **306**, 79–86.

Baumann, S., Sander, A., Gurnon, J.R., Yanai-Balser, G.M., Van Etten, J.L., and Piotrowski, M. (2007). *Chlorella* viruses contain genes encoding a complete polyamine biosynthetic pathway. *Virology* **360**, 209–217.

Bowers, A.K., Keller, J.A., and Dutcher, S.K. (2003). Molecular markers for rapidly identifying candidate genes in *Chlamydomonas reinhardtii*. *ERY1* and *ERY2* encode chloroplast ribosomal proteins. *Genetics* **164**, 1345–1353.

Brown, D.M., Upcroft, J.A., Edwards, M.R., and Upcroft, P. (1998). Anaerobic bacterial metabolism in the ancient eukaryote *Giardia duodenalis*. *Int. J. Parasitol.* **28**, 149–164.

Bulté, L. and Wollman, F.A. (1992). Evidence for a selective destabilization of an integral membrane protein, the cytochrome b_6/f complex, during gametogenesis in *Chlamydomonas reinhardtii*. *Eur. J. Biochem.* **204**, 327–336.

Chen, Q. and Silflow, C.D. (1996). Isolation and characterization of glutamine synthetase genes in *Chlamydomonas reinhardtii*. *Plant Physiol.* **112**, 987–996.

Chen, Z.Y., Burow, M.D., Mason, C.B., and Moroney, J.V. (1996). A low-CO_2-inducible gene encoding an alanine: Alpha-ketoglutarate aminotransferase in *Chlamydomonas reinhardtii*. *Plant Physiol.* **112**, 677–684.

Chiba, Y., Sakurai, R., Yoshino, M., Ominato, K., Ishikawa, M., Onouchi, H., and Naito, S. (2003). S-adenosyl-L-methionine is an effector in the posttranscriptional autoregulation of the cystathionine gamma-synthase gene in *Arabidopsis*. *Proc. Natl. Acad. Sci. U. S. A.* **100**, 10225–10230.

Cho, M.H., Corea, O.R., Yang, H., Bedgar, D.L., Laskar, D.D., Anterola, A.M., Moog-Anterola, F.A., Hood, R.L., Kohalmi, S.E., Bernards, M.A., Kang, C., Davin, L.B., and Lewis, N.G. (2007). Phenylalanine biosynthesis in *Arabidopsis thaliana*. Identification and characterization of arogenate dehydratases. *J. Biol. Chem.* **282**, 30827–30835.

Collins, R.P. and Kalnins, K. (1967). Keto acids produced by *Chlamydomonas reinhardtii*. *Can. J. Microbiol.* **13**, 995–999.

Cona, A., Rea, G., Angelini, R., Federico, R., and Tavladoraki, P. (2006). Functions of amine oxidases in plant development and defence. *Trends Plant Sci.* **11**, 80–88.

Croft, M.T., Lawrence, A.D., Raux-Deery, E., Warren, M.J., and Smith, A.G. (2005). Algae acquire vitamin B12 through a symbiotic relationship with bacteria. *Nature* **438**, 90–93.

Croft, M.T., Warren, M.J., and Smith, A.G. (2006). Algae need their vitamins. *Eukaryotic Cell* **5**, 1175–1183.

Cullimore, J.V. and Sims, A.P. (1980). An association between photorespiration and protein catabolism: studies with *Chlamydomonas*. *Planta* **150**, 392–396.

Cullimore, J.V. and Sims, A.P. (1981a). Occurence of two forms of glutamate synthase in *Chlamydomonas reinhardtii*. *Phytochemistry* **20**, 597–600.

Cullimore, J.V. and Sims, A.P. (1981b). Pathway of ammonia assimilation in illuminated and darkened *Chlamydomonas reinhardtii*. *Phytochemistry* **20**, 933–940.

Curien, G., Ravanel, S., Robert, M., and Dumas, R. (2005). Identification of six novel allosteric effectors of *Arabidopsis thaliana* aspartate kinase-homoserine dehydrogenase isoforms. Physiological context sets the specificity. *J. Biol. Chem.* **280**, 41178–41183.

Curien, G., Laurencin, M., Robert-Genthon, M., and Dumas, R. (2007). Allosteric monofunctional aspartate kinases from *Arabidopsis*. *FEBS J.* **274**, 164–176.

Debuchy, R., Purton, S., and Rochaix, J.-D. (1989). The argininosuccinate lyase gene of *Chlamydomonas reinhardtii*: An important tool for nuclear transformation and for correlating the genetic and molecular maps of the *ARG7* locus. *EMBO J.* **8**, 2803–2809.

Duguay, J., Jamal, S., Liu, Z., Wang, T.W., and Thompson, J.E. (2007). Leaf-specific suppression of deoxyhypusine synthase in *Arabidopsis thaliana* enhances growth without negative pleiotropic effects. *J. Plant Physiol.* **164**, 408–420.

Dutcher, S.K., Galloway, R.E., Barclay, W.R., and Poortinga, G. (1992). Tryptophan analog resistance mutations in *Chlamydomonas reinhardtii*. *Genetics* **131**, 593–607.

Faragó, A. and Dénes, G. (1967). Mechanism of arginine biosynthesis in *Chlamydomonas reinhardtii*. II. Purification and properties of N-acetylglutamate 5-phosphotransferase, the allosteric enzyme of the pathway. *Biochim. Biophys. Acta* **136**, 6–18.

Ferrario-Mery, S., Besin, E., Pichon, O., Meyer, C., and Hodges, M. (2006). The regulatory PII protein controls arginine biosynthesis in *Arabidopsis*. *FEBS Lett.* **580**, 2015–2020.

Fischer, P. and Klein, U. (1988). Localization of nitrogen-assimilating enzymes in the chloroplast of *Chlamydomonas reinhardtii*. *Plant Physiol.* **88**, 947–952.

Florencio, F.J. and Vega, J.M. (1983). Separation, purification, and characterization of two isoforms of glutamine synthetase from *Chlamydomonas reinhardtii*. *Z. Naturforsch.* **38C**, 531–538.

Franco, A.R., Diaz, M.E., Pineda, M., and Cárdenas, J. (1996). Characterization of a mutant of *Chlamydomonas reinhardtii* that uses L-methionine-S-sulfoximine and phosphinothricin as nitrogen sources for growth. *Plant Physiol.* **110**, 1215–1222.

Fu, L.H., Wang, X.F., Eyal, Y., She, Y.M., Donald, L.J., Standing, K.G., and Ben-Hayyim, G. (2002). A selenoprotein in the plant kingdom. Mass spectrometry confirms that an opal codon (UGA) encodes selenocysteine in *Chlamydomonas reinhardtii* gluththione peroxidase. *J. Biol. Chem.* **277**, 25983–25991.

Funke, R.P., Kovar, J.L., Logsdon, J.M., Jr., Corrette-Bennett, J.C., Straus, D.R., and Weeks, D.P. (1999). Nucleus-encoded, plastid-targeted acetolactate synthase genes in two closely related chlorophytes, *Chlamydomonas reinhardtii* and *Volvox carteri*: Phylogenetic origins and recent insertion of introns. *Mol. Gen. Genet.* **262**, 12–21.

Galván, A., Márquez, A.J., and Vega, J.M. (1984). Purification and molecular properties of ferredoxin-glutamate synthase from *Chlamydomonas reinhardtii*. *Planta* **162**, 180–187.

Gillet, S., Decottignies, P., Chardonnet, S., and Le Marechal, P. (2006). Cadmium response and redoxin targets in *Chlamydomonas reinhardtii*: A proteomic approach. *Photosynth. Res.* **89**, 201–211.

Gillham, N.W. (1965). Induction of chromosomal and nonchromosomal mutations in *Chlamydomonas reinhardi* with N-methyl-N′-nitro-N-nitrosoguanidine. *Genetics* **52**, 529–537.

Gresshoff, P.M. (1979). Growth inhibition by glyphosate and reversal of its action by phenylalanine and tyrosine. *Aust. J. Plant Physiol.* **6**, 177–185.

Grossman, A.R., Croft, M., Gladyshev, V.N., Merchant, S.S., Posewitz, M.C., Prochnik, S., and Spalding, M.H. (2007). Novel metabolism in *Chlamydomonas* through the lens of genomics. *Curr. Opin. Plant Biol.* **10**, 190–198.

Guo, F.Q. and Crawford, N.M. (2005). *Arabidopsis* nitric oxide synthase1 is targeted to mitochondria and protects against oxidative damage and dark-induced senescence. *Plant Cell* **17**, 3436–3450.

Hacham, Y., Gophna, U., and Amir, R. (2003). In vivo analysis of various substrates utilized by cystathionine c-synthase and O-acetylhomoserine sulfhydrylase in methionine biosynthesis. *Mol. Biol. Evol.* **20**, 1513–1520.

Hanfrey, C., Elliott, K.A., Franceschetti, M., Mayer, M.J., Illingworth, C., and Michael, A.J. (2005). A dual upstream open reading frame-based autoregulatory circuit controlling polyamine-responsive translation. *J. Biol. Chem.* **280**, 39229–39237.

Hanzawa, Y., Imai, A., Michael, A.J., Komeda, Y., and Takahashi, T. (2002). Characterization of the spermidine synthase-related gene family in *Arabidopsis thaliana*. *FEBS Lett.* **527**, 176–180.

Harris, E.H. (1989). *The Chlamydomonas Sourcebook. A Comprehensive Guide to Biology and Laboratory Use.* Academic Press, San Diego. 780 pp.

Hartnett, M.E., Newcomb, J.R., and Hodson, R.C. (1987). Mutations in *Chlamydomonas reinhardtii* conferring resistance to the herbicide sulfometuron methyl. *Plant Physiol.* **85**, 898–901.

Hawkesford, M.J. and De Kok, L.J. (2006). Managing sulphur metabolism in plants. *Plant Cell Environ.* **29**, 382–395.

Hipkin, C.R., Everest, S.A., Rees, T.A.V., and Syrett, P.J. (1982). Ammonium generation by nitrogen-starved cultures of *Chlamydomonas reinhardtii*. *Planta* **154**, 587–592.

Ho, C.L. and Saito, K. (2001). Molecular biology of the plastidic phosphorylated serine biosynthetic pathway in *Arabidopsis thaliana*. *Amino Acids* **20**, 243–259.

Hodson, R.C. and Gresshoff, P.M. (1987). Fluoroacetamide resistance mutations in *Chlamydomonas reinhardtii*. *Arch. Microbiol.* **148**, 8–13.

Hong, Z., Lakkineni, K., Zhang, Z., and Verma, D.P.S. (2000). Removal of feedback inhibition of D1-pyrroline-5-carboxylate synthetase results in increased proline accumulation and protection of plants from osmotic stress. *Plant Physiol.* **122**, 1129–1136.

Hudson, A.O., Singh, B.K., Leustek, T., and Gilvarg, C. (2006). An LL-diaminopimelate aminotransferase defines a novel variant of the lysine biosynthesis pathway in plants. *Plant Physiol.* **140**, 292–301.

Illingworth, C., Mayer, M.J., Elliott, K., Hanfrey, C., Walton, N.J., and Michael, A.J. (2003). The diverse bacterial origins of the *Arabidopsis* polyamine biosynthetic pathway. *FEBS Lett.* **549**, 26–30.

Jain, M., Shrager, J., Harris, E.H., Halbrook, R., Grossman, A.R., Hauser, C., and Vallon, O. (2007). EST assembly supported by a draft genome sequence: An analysis of the *Chlamydomonas reinhardtii* transcriptome. *Nucleic Acids Res.* **35**, 2074–2083.

Keren-Paz, A., Bercovich, Z., and Kahana, C. (2007). Antizyme inhibitor: A defective ornithine decarboxylase or a physiological regulator of polyamine biosynthesis and cellular proliferation. *Biochem. Soc. Trans.* **35**, 311–313.

Kichey, T., Le Gouis, J., Sangwan, B., Hirel, B., and Dubois, F. (2005). Changes in the cellular and subcellular localization of glutamine synthetase and glutamate dehydrogenase during flag leaf senescence in wheat (*Triticum aestivum* L.). *Plant Cell Physiol.* **46**, 964–974.

Kirk, D.L. and Kirk, M.M. (1978a). Carrier-mediated uptake of arginine and urea by *Chlamydomonas reinhardtii*. *Plant Physiol.* **61**, 556–560.

Kirk, M.M. and Kirk, D.L. (1978b). Carrier-mediated uptake of arginine and urea by *Volvox carteri* f. *nagariensis*. *Plant Physiol.* **61**, 549–555.

Kovar, J.L., Zhang, J., Funke, R.P., and Weeks, D.P. (2002). Molecular analysis of the acetolactate synthase gene of *Chlamydomonas reinhardtii* and development of a genetically engineered gene as a dominant selectable marker for genetic transformation. *Plant J.* **29**, 109–117.

Krath, B.N. and Hove-Jensen, B. (1999). Organellar and cytosolic localization of four phosphoribosyl diphosphate synthase isozymes in spinach. *Plant Physiol.* **119**, 497–506.

Lain-Guelbenzu, B., Muñoz-Blanco, J., and Cárdenas, J. (1990). Purification and properties of L-aspartate aminotransferase of *Chlamydomonas reinhardtii*. *Eur. J. Biochem.* **188**, 529–533.

Lain-Guelbenzu, B., Cárdenas, J., and Muñoz-Blanco, J. (1991). Purification and properties of L-alanine aminotransferase from *Chlamydomonas reinhardtii*. *Eur. J. Biochem.* **202**, 881–887.

Lam, H.M., Coschigano, K.T., Oliveira, I.C., Melo-Oliveira, R., and Coruzzi, G.M. (1996). The molecular genetics of nitrogen assimilation into amino acids in higher plants. *Annu. Rev. Plant Physiol. Plant Mol. Biol.* **47**, 569–593.

Leustek, T., Martin, M.N., Bick, J.A., and Davies, J.P. (2000). Pathways and regulation of sulfur metabolism revealed through molecular and genetic studies. *Annu. Rev. Plant Physiol. Plant Mol. Biol.* **51**, 141–165.

Loppes, R. and Heindricks, R. (1986). New arginine-requiring mutants in *Chlamydomonas reinhardtii*. *Arch. Microbiol.* **143**, 348–352.

Loppes, R. and Strijkert, P.J. (1972). Arginine metabolism in *Chlamydomonas reinhardi*. Conditional expression of arginine-requiring mutants. *Mol. Gen. Genet.* **116**, 248–257.

Ludwig, R.A. (1993). *Arabidopsis* chloroplasts dissimilate L-arginine and L-citrulline for use as N source. *Plant Physiol.* **101**, 429–434.

Mages, W., Heinrich, O., Treuner, G., Vlček, D., Daubnerova, I., and Slaninova, M. (2007). Complementation of the *Chlamydomonas reinhardtii arg7-8* (*arg2*) point mutation by recombination with a truncated nonfunctional *ARG7* gene. *Protist* **158**, 435–446.

Márquez, A.J., Galván, A., and Vega, J.M. (1984). Purification and characterization of the NADH-glutamate synthase from *Chlamydomonas reinhardtii*. *Plant Sci. Lett.* **34**, 305–314.

Márquez, A.J., Gotor, C., Romero, L.C., Galván, A., and Vega, J.M. (1986a). Ferredoxin-glutamate synthase from *Chlamydomonas reinhardtii*. Prosthetic groups and preliminary studies of mechanism. *Int. J. Biochem.* **18**, 531–535.

Márquez, A.J., Galván, A., and Vega, J.M. (1986b). Utilization of ammonium by mutants and wild types of *Chlamydomonas reinhardtii*. Studies of the glutamate synthase activities. *J. Plant Physiol.* **124**, 95–102.

Matagne, R.F. (1978). Fine structure of the arg-7 cistron in *Chlamydomonas reinhardi*. *Mol. Gen. Genet.* **160**, 95–99.

Melo-Oliveira, R., Oliveira, I.C., and Coruzzi, G.M. (1996). *Arabidopsis* mutant analysis and gene regulation define a nonredundant role for glutamate dehydrogenase in nitrogen assimilation. *Proc. Natl. Acad. Sci. U. S. A.* **93**, 4718–4723.

Merchant, S.S., Prochnik, S.E., Vallon, O., Harris, E.H., Karpowicz, S.J., Witman, G.B., Terry, A., Fritz-Laylin, L.K., Maréchal-Drouard, L. et al. (2007). The *Chlamydomonas* genome reveals the evolution of key animal and plant functions. *Science* **318**, 245–250.

Michelet, L., Zaffagnini, M., Marchand, C., Collin, V., Decottignies, P., Tsan, P., Lancelin, J.M., Trost, P., Miginiac-Maslow, M., Noctor, G., and Lemaire, S.D. (2005). Glutathionylation of chloroplast thioredoxin f is a redox signaling mechanism in plants. *Proc. Natl. Acad. Sci. U. S. A.* **102**, 16478–16483.

Miura, K., Yamano, T., Yoshioka, S., Kohinata, T., Inoue, Y., Taniguchi, F., Asamizu, E., Nakamura, Y., Tabata, S., Yamato, K.T., Ohyama, K., and Fukuzawa, H. (2004). Expression profiling-based identification of CO_2-responsive genes regulated by CCM1 controlling a carbon-concentrating mechanism in *Chlamydomonas reinhardtii*. *Plant Physiol.* **135**, 1595–1607.

Moseley, J.L., Page, M.D., Alder, N.P., Eriksson, M., Quinn, J., Soto, F., Theg, S.M., Hippler, M., and Merchant, S. (2002). Reciprocal expression of two candidate di-iron enzymes affecting photosystem I and light-harvesting complex accumulation. *Plant Cell* **14**, 673–688.

Moyano, E., Cárdenas, J., and Muñoz-Blanco, J. (1992a). Purification and properties of three NAD(P)+ isozymes of L-glutamate dehydrogenase of *Chlamydomonas reinhardtii*. *Biochim. Biophys. Acta.* **1119**, 63–68.

Moyano, E., Ramazanov, Z., Cárdenas, J., and Muñoz-Blanco, J. (1992b). Intracellular localization of three L-glutamate dehydrogenase isozymes from *Chlamydomonas reinhardtii*. *Plant Physiol.* **100**, 1575–1579.

Mullineaux, P.M. and Rausch, T. (2005). Glutathione, photosynthesis and the redox regulation of stress-responsive gene expression. *Photosynth. Res.* **86**, 459–474.

Muñoz-Blanco, J., Hidalgo-Martínez, J., and Cárdenas, J. (1990). Extracellular deamination of L-amino acids by *Chlamydomonas reinhardtii* cells. *Planta* **182**, 194–198.

Nakamura, K., Lepard, S.L., and MacDonald, S.J. (1981). Is it possible to isolate methionine auxotrophs in *Chlamydomonas reinhardtii*? Consideration of photodynamic action of the amino acid. *Mol. Gen. Genet.* **181**, 292–295.

Nishikawa, K., Onodera, A., and Tominaga, N. (2006). Phytochelatins do not correlate with the level of Cd accumulation in *Chlamydomonas* spp. *Chemosphere* **63**, 1553–1559.

Novoselov, S.V., Rao, M., Onoshko, N.V., Zhi, H., Kryukov, G.V., Xiang, Y., Weeks, D.P., Hatfield, D.L., and Gladyshev, V.N. (2002). Selenoproteins and selenocysteine insertion system in the model plant cell system, *Chlamydomonas reinhardtii*. *EMBO J.* **21**, 3681–3693.

Ominato, K., Akita, H., Suzuki, A., Kijima, F., Yoshino, T., Yoshino, M., Chiba, Y., Onouchi, H., and Naito, S. (2002). Identification of a short highly conserved amino acid sequence as the functional region required for posttranscriptional autoregulation of the cystathionine gamma-synthase gene in *Arabidopsis*. *J. Biol. Chem.* **277**, 36380–36386.

Palenik, B. and Morel, F.M. (1990). Comparison of cell-surface amino acid oxidases from several marine phytoplankton. *Marine Ecology Progress Series* **59**, 195–201.

Palenik, B., Grimwood, J., Aerts, A., Rouze, P., Salamov, A., Putnam, N., Dupont, C., Jorgensen, R., Derelle, E., Rombauts, S. et al. (2007). The tiny eukaryote *Ostreococcus* provides genomic insights into the paradox of plankton speciation. *Proc. Natl. Acad. Sci. U. S. A.* **104**, 7705–7710.

Palombella, A.L. and Dutcher, S.K. (1998). Identification of the gene encoding the tryptophan synthase beta-subunit from *Chlamydomonas reinhardtii*. *Plant Physiol.* **117**, 455–464.

Pazour, G.J., Agrin, N., Leszyk, J., and Witman, G.B. (2005). Proteomic analysis of a eukaryotic cilium. *J. Cell Biol.* **170**, 103–113.

Pegg, A.E. (2006). Regulation of ornithine decarboxylase. *J. Biol. Chem.* **281**, 14529–14532.

Peltier, G. and Thibault, P. (1983). Ammonia exchange and photorespiration in *Chlamydomonas*. *Plant Physiol.* **71**, 888–892.

Piedras, P., Pineda, M., Muñoz-Blanco, J., and Cárdenas, J. (1992). Purification and characterization of an L-amino acid oxidase from *Chlamydomonas reinhardtii*. *Planta* **188**, 13–18.

Pozuelo, M., MacKintosh, C., Galván, A., and Fernández, E. (2001). Cytosolic glutamine synthetase and not nitrate reductase from the green alga *Chlamydomonas reinhardtii* is phosphorylated and binds 14-3-3 proteins. *Planta* **212**, 264–269.

Quesada, A., Galván, A., Schnell, R.A., Lefebvre, P.A., and Fernández, E. (1993). Five nitrate assimilation-related loci are clustered in *Chlamydomonas reinhardtii*. *Mol. Gen. Genet.* **240**, 387–394.

Ravina, C.G., Barroso, C., Vega, J.M., and Gotor, C., (1999). Cysteine biosynthesis in *Chlamydomonas reinhardtii*. Molecular cloning and regulation of O-acetylserine(thiol)lyase. *Eur. J. Biochem.* **264**, 848–853.

Ravina, C.G., Chang, C.I., Tsakraklides, G.P., McDermott, J.P., Vega, J.M., Leustek, T., Gotor, C., and Davies, J.P. (2002). The *sac* mutants of *Chlamydomonas reinhardtii* reveal transcriptional and posttranscriptional control of cysteine biosynthesis. *Plant Physiol.* **130**, 2076–2084.

Rippert, P. and Matringe, M. (2002). Purification and kinetic analysis of the two recombinant arogenate dehydrogenase isoforms of *Arabidopsis thaliana*. *Eur. J. Biochem.* **269**, 4753–4761.

Rohr, J., Sarkar, N., Balenger, S., Jeong, B., and Cerutti, H. (2004). Tandem inverted repeat system for selection of effective transgenic RNAi strains in *Chlamydomonas*. *Plant J.* **40**, 611–621.

Sakihama, Y., Nakamura, S., and Yamasaki, H. (2002). Nitric oxide production mediated by nitrate reductase in the green alga *Chlamydomonas reinhardtii*: An alternative NO production pathway in photosynthetic organisms. *Plant Cell Physiol.* **43**, 290–297.

Samach, A., Hareven, D., Gutfinger, T., Ken-Dror, S., and Lifschitz, E. (1991). Biosynthetic threonine deaminase gene of tomato: Isolation, structure, and upregulation in floral organs. *Proc. Natl. Acad. Sci. U. S. A.* **88**, 2678–2682.

Schmid, J. and Amrhein, N. (1995). Molecular organization of the shikimate pathway in higher plants. *Phytochemistry* **39**, 737–749.

Semler, B.L., Hodson, R.C., Williams, S.K., II., and Howell, S.H. (1975). The induction of allophanate lyase during the vegetative cell cycle in light-synchronized cultures of *Chlamydomonas reinhardi*. *Biochim. Biophys. Acta.* **399**, 71–78.

Singh, B.K. and Shaner, D.L. (1995). Biosynthesis of branched chain amino acids: From test tube to field. *Plant Cell* **7**, 935–944.

Slocum, R.D. (2005). Genes, enzymes and regulation of arginine biosynthesis in plants. *Plant Physiol. Biochem.* **43**, 729–745.

Smyth, R.D., Martinek, G.W., and Ebersold, W.T. (1975). Linkage of six genes in *Chlamydomonas reinhardtii* and the construction of linkage test strains. *J. Bacteriol.* **124**, 1615–1617.

Stepansky, A. and Leustek, T. (2006). Histidine biosynthesis in plants. *Amino Acids* **30**, 127–142.

Strijkert, P.J. and Sussenbach, J.S. (1969). Arginine metabolism in *Chlamydomonas reinhardi*. Arginine deiminase: The first enzyme of the catabolic pathway. *FEBS Lett.* **3**, 166–168.

Strijkert, P.J., Loppes, R., and Sussenbach, J.S. (1971). Arginine metabolism in *Chlamydomonas reinhardi*. Regulation of uptake and breakdown. *FEBS Lett.* **14**, 329–332.

Sussenbach, J.S. and Strijkert, P.J. (1969). Arginine metabolism in *Chlamydomonas reinhardi*. Arginine deiminase: The first enzyme of the catabolic pathway. *FEBS Lett.* **3**, 166–168.

Tada, S., Volrath, S., Guyer, D., Scheidegger, A., Ryals, J., Ohta, D., and Ward, E. (1994). Isolation and characterization of cDNAs encoding imidazoleglycerolphosphate dehydratase from *Arabidopsis thaliana*. *Plant Physiol.* **105**, 579–583.

Taniguchi, M., Taniguchi, Y., Kawasaki, M., Takeda, S., Kato, T., Sato, S., Tabata, S., Miyake, H., and Sugiyama, T. (2002). Identifying and characterizing plastidic 2-oxoglutarate/malate and dicarboxylate transporters in *Arabidopsis thaliana*. *Plant Cell Physiol.* **43**, 706–717.

Theiss, C., Bohley, P., and Voigt, J. (2002). Regulation by polyamines of ornithine decarboxylase activity and cell division in the unicellular green alga *Chlamydomonas reinhardtii*. *Plant Physiol.* **128**, 1470–1479.

Vallon, O. (2000). New sequence motifs in flavoproteins: Evidence for common ancestry and tools to predict structure. *Proteins* **38**, 95–114.

Vallon, O. and Wollman, F.-A. (1995). Mutations affecting O-glycosylation in *Chlamydomonas reinhardtii* cause delayed cell wall degradation and sex-limited sterility. *Plant Physiol.* **108**, 703–712.

Vallon, O. and Wollman, F.-A. (1997). cDNA sequence of M(α), The catalytic subunit of the *Chlamydomonas reinhardtii* L-amino acid oxidase (Accession No. U78797): A new sequence motif shared by a wide variety of flavoproteins. (PGR97-171). *Plant Physiol.* **115**, 1729.

Vallon, O., Bulté, L., Kuras, R., Olive, J., and Wollman, F.A. (1993). Extensive accumulation of an extracellular L-amino-acid oxidase during gametogenesis of *Chlamydomonas reinhardtii*. *Eur. J. Biochem.* **215**, 351–360.

Voigt, J., Fausel, M., Bohley, P., Adam, K.H., and Marquardt, O. (2004). Structure and expression of the ornithine decarboxylase gene of *Chlamydomonas reinhardtii*. *Microbiol. Res.* **159**, 403–417.

Warpeha, K.M., Lateef, S.S., Lapik, Y., Anderson, M., Lee, B.-S., and Kaufman, L.S. (2006). G-Protein-coupled receptor 1, G-protein Ga-subunit 1, and prephenate dehydratase 1 are required for blue light-induced production of phenylalanine in etiolated *Arabidopsis*. *Plant Physiol.* **140**, 844–855.

Weber, A., Menzlaff, E., Arbinger, B., Gutensohn, M., Eckerskorn, C., and Flugge, U.I. (1995). The 2-oxoglutarate/malate translocator of chloroplast envelope membranes: molecular cloning of a transporter containing a 12-helix motif and expression of the functional protein in yeast cells. *Biochemistry* **34**, 2621–2627.

Wilson, I.D., Neill, S.J., and Hancock, J.T. (2008). Nitric oxide synthesis and signalling in plants. *Plant Cell Environ.* **31**, 622–631.

Winder, T. and Spalding, M.H. (1988). Imazaquin and chlorsulfuron resistance and cross resistance in mutants of *Chlamydomonas reinhardtii*. *Mol. Gen. Genet.* **213**, 394–399.

Wolff, E.C., Kang, K.R., Kim, Y.S., and Park, M.H. (2007). Posttranslational synthesis of hypusine: Evolutionary progression and specificity of the hypusine modification. *Amino Acids.* **33**, 341–350.

Wong, H.K., Chan, H.K., Coruzzi, G.M., and Lam, H.M. (2004). Correlation of *ASN2* gene expression with ammonium metabolism in *Arabidopsis*. *Plant Physiol.* **134**, 332–338.

Yuan, J., Palioura, S., Salazar, J.C., Su, D., O'Donoghue, P., Hohn, M.J., Cardoso, A.M., Whitman, W.B., and Soll, D. (2006). RNA-dependent conversion of phosphoserine forms selenocysteine in eukaryotes and archaea. *Proc. Natl. Acad. Sci. U. S. A.* **103**, 18923–18927.

Zemojtel, T., Fröhlich, A., Palmieri, M.C., Kolanczyk, M., Mikula, I., Wyrwicz, L.S., Wanker, E.E., Mundlos, S., Vingron, M., Martasek, P., and Durner, J. (2006). Plant nitric oxide synthase: A never-ending story? *Trends Plant Sci.* **11**, 524–525.

Zhao, T., Li, G., Mi, S., Li, S., Hannon, G.J., Wang, X.J., and Qi, Y. (2007). A complex system of small RNAs in the unicellular green alga *Chlamydomonas reinhardtii*. *Genes Dev.* **21**, 1190–1203.

CHAPTER 5

Sulfur: From Acquisition to Assimilation

David Gonzalez-Ballester and Arthur R. Grossman
The Carnegie Institution, Department of Plant Biology, Stanford, California, USA

CHAPTER CONTENTS

I. INTRODUCTION

Sulfur (S) is an essential element present in proteins, lipids, and important metabolites (Meister and Anderson, 1983; Gupta et al., 1990; Grossman and Takahashi, 2001), including signaling molecules (Schultze et al., 1992). S can be limiting in the environment and also can limit plant productivity in certain agricultural settings (Mahler and Maples, 1986, 1987; Warman and Sampson, 1994). In recent history many soils have accumulated high S levels, resulting either from administration of fertilizers contaminated with SO_4^{2-} salts, or exposure to pollutants present in acid rain (Cole and Johnson, 1977; Johnson et al., 1982; David et al., 1988; MacDonald et al., 1991). However, most organisms are unable to store S, making growth and development dependent on external S sources. Therefore, with increased fertilizer purity and decreased occurrence of acid rains, low levels of available S in diverse ecosystems can limit growth and development of plants (Marschner, 1995) and microbes. This chapter describes the pathways associated with SO_4^{2-} assimilation in *Chlamydomonas*, the multiplicity of ways in which photosynthesis in eukaryotes can acclimate to limiting S conditions, and the regulatory elements associated with this acclimation process. It also uses *Chlamydomonas* genomic (Merchant et al., 2007) and experimental information to highlight the similarities and differences of assimilation and acclimation processes between *Chlamydomonas* and vascular plants.

II. SO_4^{2-} ACQUISITION AND ASSIMILATION

A. Overview

The SO_4^{2-} anion, usually the most abundant S form in soils, can be taken up by plants and microbes, activated, and used for sulfation of molecules such as lipids, polysaccharides and proteins, or incorporated into various compounds, including cysteine and methionine, by reductive assimilation (Bick and Leustek, 1998; Leustek and Saito, 1999; Saito, 2000; Smith et al., 2000). A summary of pathways used in the assimilation of SO_4^{2-}, the genes associated with assimilation, and the acclimation of *Chlamydomonas* to S-limiting conditions, are given in Figure 5.1 and Table 5.1.

B. Hydrolysis

The majority of SO_4^{2-} may not be bioavailable for most organisms in some soils. This is a result of leaching of SO_4^{2-} through the soil matrix and its tight adsorption onto the surface of soil particles. Furthermore, a large proportion of soil SO_4^{2-} may be covalently bonded to organic molecules in the form of sulfate esters and sulfonates (C–S bonds). Finally, plants and microbes release a variety of S-containing organic compounds (e.g. proteins, lipids, polysaccharides, dimethylsulfoniopropionate) when they die, which are usually rapidly cycled into the available S pool through the action of hydrolytic enzymes in the soil: these enzymes are mostly released by microorganisms.

Many microorganisms synthesize and secrete aryl- and alkyl-sulfatases and sulfonatases, which can cleave SO_4^{2-} from organic S compounds in the

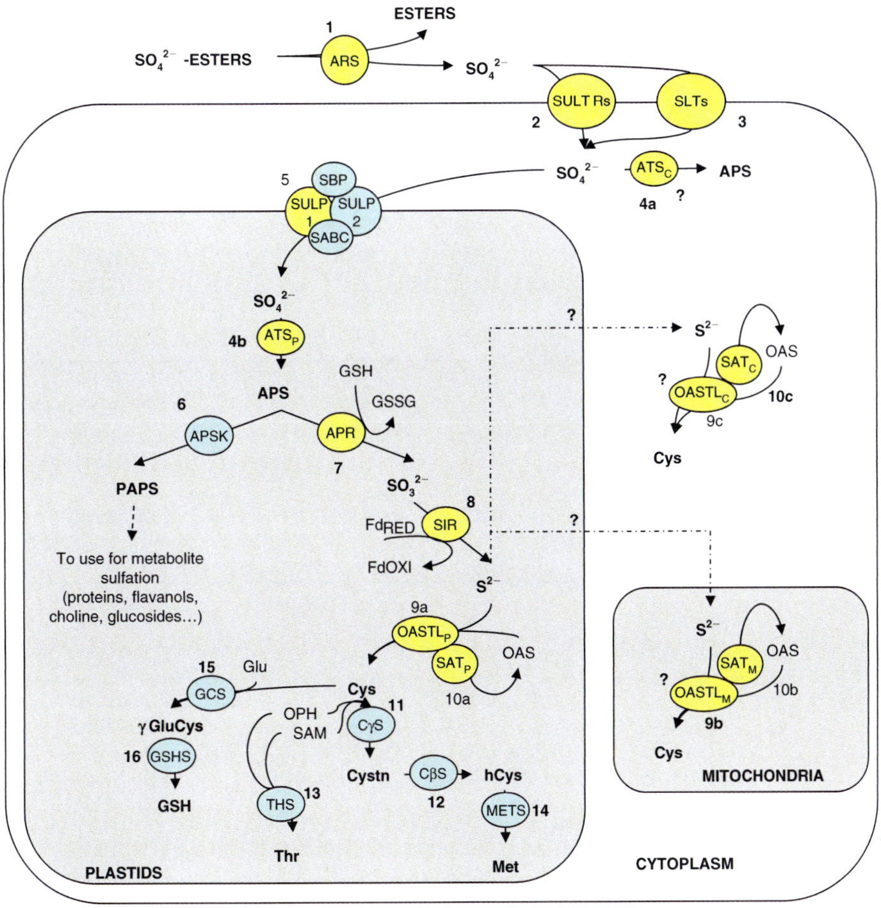

FIGURE 5.1 *SO_4^{2-} assimilation pathway.* **1**, *arylsulfatase;* **2**, *H^+/SO_4^{2-} transporter family;* **3**, *Na^+/SO_4^{2-} transporter family;* **4**, *ATP sulfurylase, with two hypothetical localizations: the cytoplasm (ATS_c) and plastid (ATS_p);* **5**, *chloroplast SO_4^{2-} transport system;* **6**, *APS kinase;* **7**, *APS reductase;* **8**, *sulfite reductase;* **9**, *acetylserine(thiol)lyase, with three possible localizations as indicated by subscripts: cytoplasm (c), plastid (p), or mitochondrion (m);* **10**, *serine acetyltransferase, with three possible localizations as indicated by subscripts: cytoplasm (c), plastid (p), or mitochondrion (m);* **11**, *Cystathionine γ-synthase;* **12**, *Cystathionine β-synthase;* **13**, *threonine synthase;* **14**, *methionine synthase;* **15**, *γ-glutamylcysteine synthetase;* **16**, *glutathione synthetase. Proteins encoded by genes that have been shown to be up-regulated under S starvation conditions in Chlamydomonas are shown in yellow. Protein designations along with their abbreviations are presented in Table 5.1. APS, adenosine 5′-phosphosulfate; PAPS, 3′-phophoadenosine 5′-phosphosulfate; GSH, reduced glutathione; GSSG, oxided glutathione; S^{-2}, sulfide; OAS, O-acetylserine; Cystn, cystathionine; hCys, homocysteine; SAM, S-adenosylmethionine; OPH, O-phosphohomoserine; γGluCys, γ-glutamylcysteine.*

soil. Sulfatases hydrolyze esterified sulfate from organic molecules, which promotes S cycling. The synthesis of sulfatases is often regulated by SO_4^{2-} availability. *Chlamydomonas* makes a prominent, extracellular, cell-wall-associated arylsulfatase (ARS) in response to S limitation (Schreiner et al.,

Table 5.1 Genes involved in S assimilation and those up-regulated during S deprivation

	Activity	Genes	Proteins	References	Accession #
1	Arylsulfatase	***ARS1*** ***ARS2***	ARS1 ARS2	[1, 2, 3, 4] [3]	EDP04612 EDP04462
2	Sulfate transporter (H^+/SO_4^{2-} family)	*SULTR1* ***SULTR2*** *SULTR3* *SULTR4*	SULTR1 SULTR2 SULTR3 SULTR4		 EDP00477 EDP00036 ED099355
3	Sulfate transporter (Na^+/SO_4^{2-} family)	***SLT1*** ***SLT2*** *SLT3*	SLT1 SLT2 SLT3	 [4] 	EDP05421 EDP05750 EDP05749
4	ATP sulfurylase	***ATS1*** ***ATS2***	ATS1 ATS2	[3, 4, 5] [4, 6]	EDP03369 EDP07432
5	Chloroplast sulfate transporter (subunits)	***SULP1*** *SULP2* *SABC* *SBP1*	SULP1 SULP2 SABC SBP1	[7, 8] [9] [9] [9]	EDP03937 EDP10015 EDP08170 EDP08595
6	APS kinase	*APK1, AKN*	APK	[4]	EDP01296
7	APS reductase	***APR, MET16***	APR	[3, 4]	EDP00776
8	Sulfite reductase	***SIR1*** ***SIR2*** *SIR3*	SIR1 SIR2 SIR3	[4] [4] [4]	ED098734 EDP05929
9	*O*-acetylserine(thiol)-lyase	*ASL1, OASTL1* *ASL2, OASTL2* *ASL3, OASTL3* ***ASL4, OASTL4, Cys1A***	OASTL1 OASTL2 OASTL3 OASTL4	 [3, 4, 10]	EDP01088 ED097451 EDP05983 EDP04425, EDP04426*
10	Serine acetyltransferase	***SERAT1, SAT1*** *SERAT2, SAT2*	SAT1 SAT2	[3, 4]	EDP07949, EDP07950* ED098094
11	Cystathionine γ-synthase	*CGS1, METB*	METB		EDP01529
12	Cystathionine β-lyase	*CBS, METC*	METC		ED098869
13	Threonine synthase	*THS*	THS		EDP08010
14	Methionine synthase	*METH* *METE*	METH METE	[11] [11]	EDP08397 ED096787
15	γ-Glutamylcysteine synthetase	*GSH1*	GSH1		EDP06622
16	GSH synthetase	*GSH2*	GSH2		EDP04651
–	Extracellular polypeptides	***ECP76*** ***ECP88***	ECP76 ECP88	[4, 12] [12]	EDP01767 EDP02984
–	Selenium binding protein	***SPD1, SBDP***	SBDP	[4]	EDP06040
–	Positive regulatory protein	*SAC1*	SAC1	[3, 4, 13]	EDP00156
–	Negative regulatory protein	*SAC3*	SAC3	[3, 14, 15]	ED096994

*The numbers in the first column correspond to those used in Figure 5.1. For those genes that have been shown to be up-regulated during S deprivation, the gene designations are in bold. The symbol * indicates that more than one transcript appears to be encoded by the gene. References are [1] de Hostos et al. (1989); [2] Davies et al. (1994); [3] Ravina et al. (2002); [4] Zhang et al. (2004); [5] Yildiz et al. (1996); [6] Allmer et al. (2006); [7] Chen et al. (2003); [8] Chen and Melis (2004); [9] Melis and Chen (2005); [10] Ravina et al. (1999); [11] Croft et al. (2006); [12] Takahashi et al. (2001); [13] Davies et al. (1996); [14] Davies et al. (1999); [15] Irihimovitch and Stern (2006).*

1975; de Hostos et al., 1988, 1989). The protein has at least three *O*-linked oligosaccharides, is stable in the extracellular space, and is translated as a precursor protein with a signal sequence that is cleaved as it is exported from the cell (de Hostos et al., 1988). The extracellular location of ARS allows it to hydrolyze soluble sulfate esters in the medium, releasing free SO_4^{2-} for assimilation. Identification of the ARS polypeptide (de Hostos et al., 1988) led to the cloning and characterization of two *ARS* genes, *ARS1* and *ARS2* (de Hostos et al., 1989; Ravina et al., 2002), that are contiguous on the genome and arranged in a tail-to-tail orientation. Transcripts from both of these genes increase in response to S deprivation (de Hostos et al., 1989; Ravina et al., 2002).

Examination of the *Chlamydomonas* genome has revealed the presence of 18 putative *ARS* genes, many of which are present in clusters (Figure 5.2A); the linked genes often showing high sequence similarity (Figure 5.2B). This organization suggests a common evolutionary origin and the potential for similar regulatory features. It is not clear if the different *ARS* genes are distinct with respect to substrate specificity, subcellular localization or catalytic properties, and it is not known how most of the individual genes (other than *ARS1* and *ARS2*) respond during exposure of the cells to S deprivation. While all of the *ARS* genes have the canonical sulfatase signature, the ARS1 and ARS2 polypeptides are the most similar (98.5% identity), which might reflect a recent gene duplication.

C. Transport

1. Overview

Initial studies with *Chlamydomonas* demonstrated that the maximum velocity (V_{max}) and the substrate concentration at which SO_4^{2-} transport was at half-maximum velocity ($K_{1/2}$) were altered when S-replete cells were transferred to medium devoid of S; the V_{max} for SO_4^{2-} increased by approximately 10-fold while the $K_{1/2}$ decreased by roughly 7-fold (Yildiz et al., 1994). Higher affinity SO_4^{2-} transport could be detected within an hour of S deprivation. The development of enhanced SO_4^{2-} transport activity following S starvation was blocked by the addition of cycloheximide (an inhibitor of cytosolic translation) to cell cultures, but not by the addition of chloramphenicol (an inhibitor of chloroplast translation). These results demonstrate the requirement for cytosolic protein synthesis during the development of new transport activities following S deprivation. Moreover, SO_4^{2-} transport in *Chlamydomonas* is an energy-dependent process, and may be driven by a proton gradient generated by a plasma membrane ATPase (Yildiz et al., 1994).

Based on whole genome analysis and EST information, seven putative eukaryotic-type SO_4^{2-} transporter genes have been identified in *Chlamydomonas*. Four of these are of the plant type (H^+/SO_4^{2-} co-transporters) and are designated *SULTR1*, *SULTR2*, *SULTR3*, and *SULTR4*. The

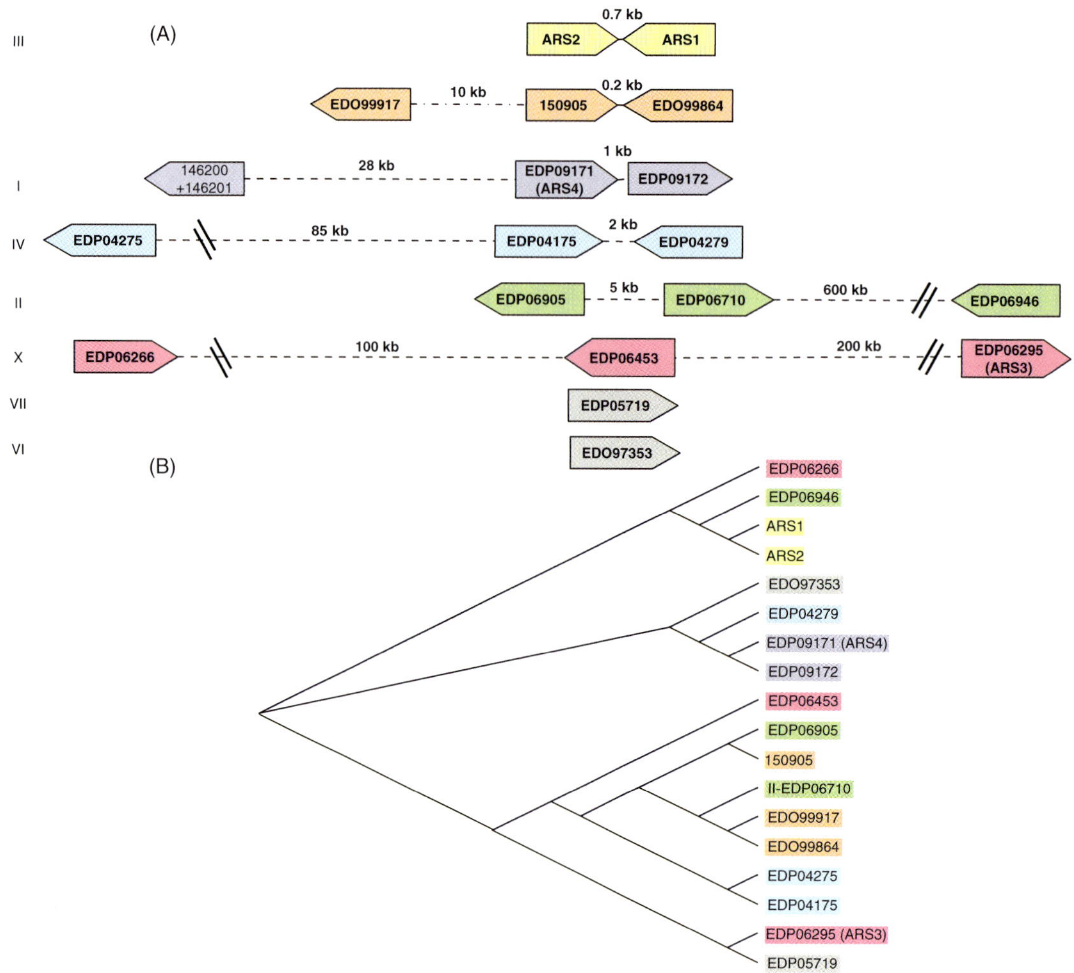

FIGURE 5.2 *The ARS gene family. (A) ARS and ARS-like genes on the Chlamydomonas genome. Blocks represent ARS genes and show their orientations. The distances between genes are indicated in kb. (B) Phylogenetic tree of the ARS-like genes (BioEdit 7.0.5.3), showing GenBank accession numbers for the proteins predicted from the genomic sequence.*

three remaining SO_4^{2-} transporters are of the animal type (Na^+/SO_4^{2-} co-transporters) and are designated *SLT1*, *SLT2*, and *SLT3*. Neither functionality nor cellular localizations for these transporters have been demonstrated using biochemical and/or genetic approaches.

2. Plant type H^+/SO_4^{2-} co-transporters

The putative plasma membrane H^+/SO_4^{2-} transporters of *Chlamydomonas* encode polypeptides with strong sequence similarity to the SLC26A anion family of polypeptides. Similar proteins are required for the transport of SO_4^{2-} across the plasma membrane and tonoplast membranes in vascular

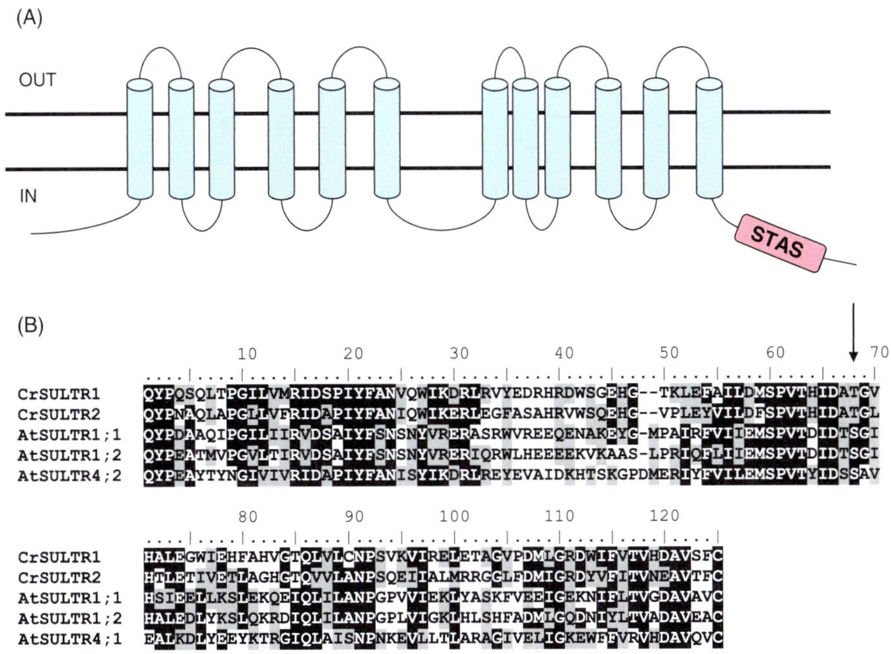

FIGURE 5.3 *SO_4^{2-} transporter structure. (A) Representation of the structure of a H^+/SO_4^{2-} transporter in the membrane, showing 12 transmembrane domains and the carboxy-terminal STAS domain that protrudes into the cytoplasm. (B) ClustalW alignment of STAS domains present in SULTR1 and SULTR2 of Chlamydomonas, and SULTR1;1, SULTR 1;2, and SULTR 4;1 of Arabidopsis. The vertical arrow indicates the putative phosphorylation site in the STAS domain. The alignment was performed using BioEdit v.7.0.5.3.*

plants, while the chloroplast appears to use a bacterial-type transport system. Genes encoding H^+/SO_4^{2-} transporters from numerous plants have been identified (Smith et al., 1997; Takahashi et al., 2000; Howarth et al., 2003). Many of these have been associated with SO_4^{2-} transport function based on rescue of the methionine-dependent phenotype of a *Saccharomyces cerevisiae* (yeast) null mutant for the major SO_4^{2-} transporters, *Sul1* and *Sul2*. Furthermore, many of these transporters have been localized to specific subcellular compartments, and show increased expression (Takahashi et al., 2006) when plant growth is limited by the SO_4^{2-} concentration.

Like most other nutrient transporters in eukaryotes, the H^+/SO_4^{2-} transporters have between 10 and 14 predicted transmembrane domains (TMD) that can be divided into two symmetrical sets of similar TMDs oriented in opposite directions along the polypeptide chain (e.g. in Figure 5.3A, TMD 1–6 have duplicated to form TMD 7–12, but they are in the inverse orientation). The H^+/SO_4^{2-} transporters also contain a cytosolic carboxy-terminal sequence that has been designated the STAS domain

(Aravind and Koonin, 2000; Figure 5.3B). This domain is attached to the catalytic moiety of the transporter through a region of low conservation and of highly variable length, designated the linker (L) domain. The STAS domain has sequence (Duncan et al., 1996) and structural (Rost et al., 1997; based on the PHD-POPITS program) similarity to the bacterial anti-sigma factor antagonist SpoIIAA, although STAS domains are generally not highly conserved. The SpoIIAA polypeptide interacts with SpoIIAB and undergoes a cycle of phosphorylation–dephosphorylation that controls the heterologous protein–protein interaction, which in turn serves to regulate sporulation in *Bacillis subtilis* (Ho et al., 2003). *Chlamydomonas* SULTR3 and SULTR4 are the most diverged from plant-type transporters, with the latter lacking the STAS domain. However, these observations should be viewed as preliminary since the gene models for SULTR3 and SULTR4 are not supported by EST data.

The STAS domain is critical for normal function of many anion transporters, and mutations that cause diastrophic dysplasia and Pendred's syndrome in humans map to STAS domains (Everett and Green, 1999). Furthermore, like the SpoIIAA, the STAS domain may bind nucleotide triphosphates (Najafi et al., 1996) and be modulated in its function via a phosphorylation–dephosphorylation cycle. These features suggest that the STAS domain serves a general regulatory function. Studies of plant SO_4^{2-} transporters also demonstrated a STAS domain requirement for transport function (Rouached et al., 2005; Shibagaki and Grossman, 2004, 2006).

Based on structural analyses, the STAS domain is thought to consist of four β-strands forming a β-sheet, surrounded by five α-helices. The β-sheet, in association with hydrophobic surfaces of the α-helices, forms a hydrophobic core that is not readily accessible to the external medium. In contrast, the externally exposed surfaces of the α-helices and the loops between α-helices and β-strands are predicted to be available for interactions with other molecules (Rouached et al., 2005). Random mutations introduced into the STAS and L domains of SULTR1;2 of *Arabidopsis* demonstrated that mutations that affect the β-strands often alter the stability and/or biogenesis of the transporter, whereas lesions introduced into specific α-helices often affect transport activity without affecting protein abundance (Shibagaki and Grossman, 2006). Furthermore, mutations in the L domain of SULTR1;2 mainly alter transporter activity, although some mutations can affect biogenesis. For the SLC26A family of transport proteins, the STAS domain appears to participate in some aspects of homeostatic regulation by influencing transport-dependent/associated processes. For example, the DRA Cl^--HCO_3^- antiporter STAS domain can directly interact with the Cystic Fibrosis Transmembrane Conductance Regulator, which is a cAMP-regulated Cl^- channel, and this interaction modulates the activity of both the transporter and the channel (Ko et al., 2004). The results discussed above, generated from studies of both plants and animal systems, would suggest that the L-STAS regions of the *Chlamydomonas* SULTR proteins are involved in transporter function and regulation and/or the responses of the alga to S deprivation.

3. Animal type Na^+/SO_4^{2-} co-transporters

Chlamydomonas contains three genes encoding proteins with strong sequence similarity to mammalian Na^+/SO_4^{2-} transporters (*SLT* genes). Two of these, *SLT2* and *SLT3*, are arranged in tandem and in the same orientation. The transcripts from *SLT1* and *SLT2* increase when the cells are deprived of SO_4^{2-} (Zhang et al., 2004). The putative SLT transporters have significant sequence similarity to a regulatory element that controls *Chlamydomonas* responses to S deprivation, designated SAC1 (Davies et al., 1996) (see below), as shown in Figure 5.4A.The Na^+/SO_4^{2-} transporters have no apparent STAS domain but do contain TrkA-C domains, which have been associated with potassium transport and which may bind NAD^+, but as yet have no defined physiological or regulatory function (Tucker and Fadool, 2002; Kraegeloh et al., 2005). The arrangement of the TrkA-C domains in the *Chlamydomonas* Na^+/SO_4^{2-} transporters and SAC1 are shown in Figure 5.4B.

4. Other SO_4^{2-} transporters

Once in the cytoplasm, SO_4^{2-} enters the plastid where it is reduced. Initially, a single *Chlamydomonas* gene encoding a subunit of a putative SO_4^{2-} transporter associated with chloroplasts was isolated and designated *SULP1* (Chen et al., 2003; Chen and Melis, 2004). The SULP1 protein is predicted to be transmembrane, with strong similarity to the hydrophobic permease-associated ABC transporter subunit CysT (characteristic of bacterial transporters). *SULP1* mRNA and protein increase upon S starvation and a *sulP* antisense strain exhibited reduced SO_4^{2-} uptake capacity, lower levels of photosynthesis (as measured by light-saturated rates of O_2 evolution), low levels of Rubisco and the Photosystem II (PS II) reaction center polypeptide D1, and signs of S deficiency, even when a moderate concentration of SO_4^{2-} was included in the medium (Chen and Melis, 2004; Chen et al., 2005). Failure to isolate antisense transformants with a large reduction in the *SULP1* transcript level suggests that a complete loss of gene function may be lethal and that this transporter defines the major path for SO_4^{2-} import into *Chlamydomonas* plastids.

Other components of the chloroplast SO_4^{2-} transporter were subsequently identified (Melis and Chen, 2005), including a second transmembrane protein (SULP2), a stromal-targeted nucleotide binding protein (SABC) and the substrate binding protein (SBP). The mature SULP1 and SULP2 polypeptides contain seven transmembrane domains and one or two large hydrophilic loops that protrude into the cytosol. The nuclear genome of the diatom *Thalassiosira pseudonana* also encodes subunits of a

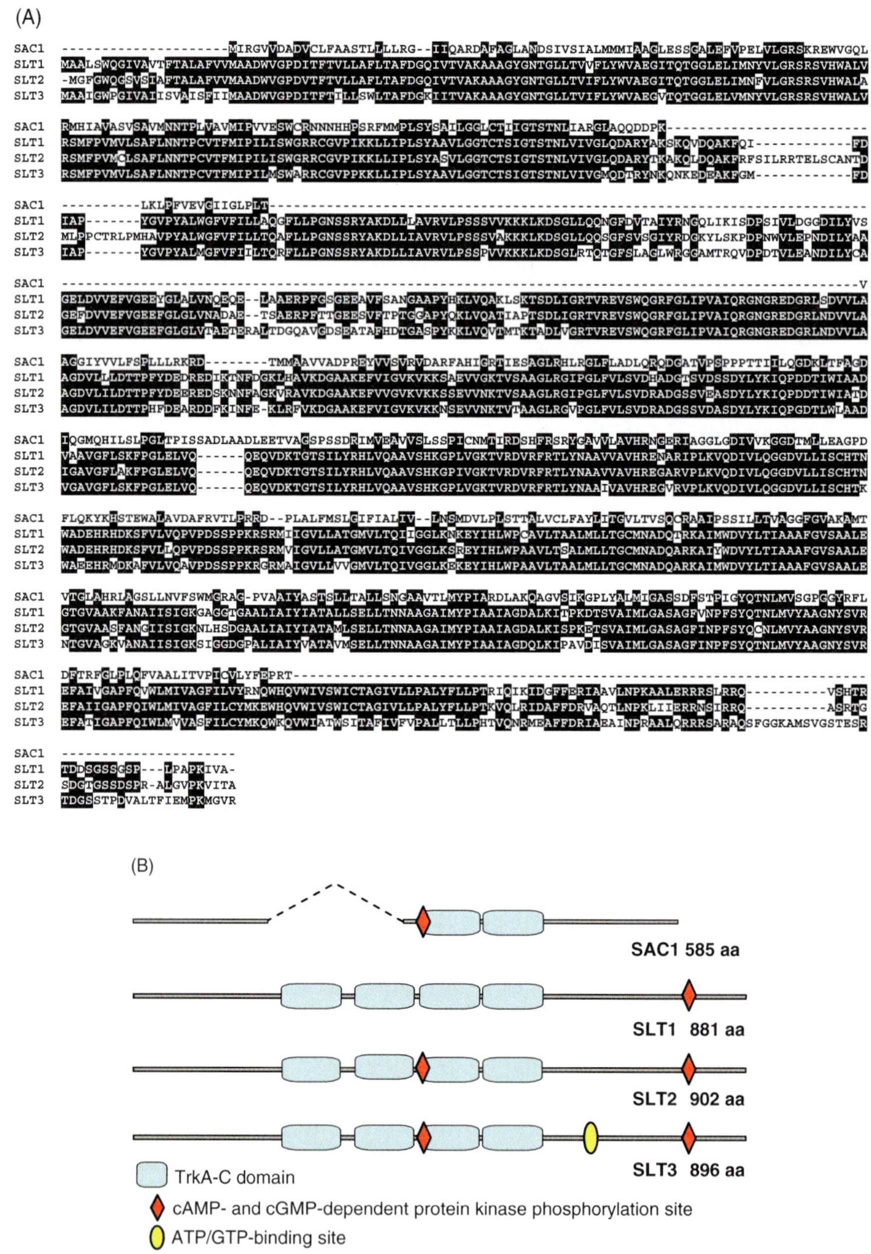

FIGURE 5.4 *Features of the SLT or SAC1-like transporters. (A) ClustalW alignment of SLT SO_4^{2-} transporters. The alignment was performed using BioEdit v.7.0.5.3. (B) Schematic representation of the TrkA-C domains in SAC1 and SLT proteins. SAC1 is physically smaller than the SLT polypeptides; it has two TrkA-C domain repeats (the broken line represents a gap in the alignment) while SLT1, SLT2, and SLT3 have four. Other domains with possible functional relevance in the activity/regulation of these proteins are also depicted.*

bacterial-type SO_4^{2-} transporter (Armbrust et al., 2004) while in the red alga *Cyanidioschyzon merolae* they are present in both the nuclear and plastid genomes (Ohta et al., 2003). Bacterial-type SO_4^{2-} transporter gene subunits have been noted to be on the chloroplast genomes of many algae. These findings suggest that the algal chloroplast SO_4^{2-} transport system resembles that of prokaryotes, and that some genes encoding these polypeptides may be transferred from the original endosymbiont genome to the nuclear genome of the host organism. The distribution of these transporter genes between the nuclear and plastid genomes varies, and studies of these genes are likely to be enlightening from biochemical and evolutionary perspectives, and reveal ways in which the expression of the different transporter subunits are coordinately regulated.

D. Incorporation into cysteine and methionine

SO_4^{2-} that enters the cell may be activated by ATP sulfurylase (ATS) to form adenosine 5′-phosphosulfate (APS). ATS isoforms are located in both the plastid and cytosol of plant cells (Rotte and Leustek, 2000). In contrast to *Arabidopsis* in which there are at least three plastidic and one putative cytosolic ATS (Rotte and Leustek, 2000), *Chlamydomonas* has two genes, *ATS1* and *ATS2* (Yildiz et al., 1996; Allmer et al. 2006), both encoding proteins with potential chloroplast transit peptides. These findings suggest that in *Chlamydomonas* all activation of SO_4^{2-} occurs in the chloroplast. However, biochemical work is required to verify subcellular locations of the ATS isozymes. *ATS1* and *ATS2* transcripts have been shown to increase during S starvation (Yildiz et al., 1996; Ravina et al., 2002; Zhang et al., 2004).

The APS generated by the ATS reaction can serve as substrate for SO_4^{2-} reduction or be further phosphorylated by APS kinase (phosphosulfate kinase), encoded in *Chlamydomonas* by a single, intronless gene, to yield 3′-phosphoadenosine 5′-phosphosulfate (PAPS) (Arz et al., 1994; Lee and Leustek, 1998). PAPS can be used by sulfotransferases to catalyze sulfation of various metabolites including flavanols, choline, and glucosides (Varin et al., 1997). APS can also be reduced to sulfite by APS reductase (gene *APR1* or *MET16*; Gutierrez-Marcos et al., 1996; Setya et al., 1996; Ravina et al., 2002). APS reductases are in the plastid (Rotte and Leustek, 2000), and the source of reductant for the enzyme is probably reduced glutathione (Bick et al., 1998; Prior et al., 1999). The mRNA and activity for APR increase during S starvation, suggesting that a key juncture for control of S assimilation occurs at the point at which APS interacts with either APS kinase or APR (Gutierrez-Marcos et al., 1996; Takahashi et al., 1997); the relative flux through the two paths would likely depend on the demand for sulfur-containing amino acids for protein synthesis. In *Chlamydomonas*, both the *APR1* transcript and APR activity increase during S starvation (Ravina et al., 2002; Zhang et al., 2004).

In plants the SO_3^{2-} generated by the APR reaction is reduced to sulfide by the plastid sulfite reductase (SIR), which uses electrons from reduced ferredoxin (Yonekura-Sakakibara et al., 2000). In *Chlamydomonas* there are two ferredoxin-type *SIR* genes (*SIR1* and *SIR2*) and one bacterial-type (*SIR3*). *SIR1* and *SIR2* expression increase in response to S deprivation (Zhang et al., 2004). The sulfide formed in the SIR reaction combines with *O*-acetylserine (OAS) to generate cysteine, a reaction catalyzed by *O*-acetylserine(thiol)lyase (OASTL). OASTL is present in the cytosol, chloroplast, and mitochondrion in *Arabidopsis* (Lunn et al., 1990). Genes encoding different OASTL isoforms (gene designation *ASL*) have been isolated from various plants (Sirko et al., 2004). *Chlamydomonas ASL4* exhibits an increase in expression during S deprivation (Ravina et al., 1999; Zhang et al., 2004), but the three additional putative *Chlamydomonas ASL* genes have not been characterized.

Serine acetyltransferase (SAT) catalyzes the formation of OAS in the cytosol, chloroplast, and mitochondrion (Ruffet et al., 1995). Cytosolic SAT is feedback inhibited at micromolar concentrations of cysteine (Saito et al., 1995; Noji et al., 1998, 2001; Noji and Saito, 2002). SAT is associated with OASTL and four molecules of pyridoxal 5′-phosphate in the cysteine synthase complex (Bogdanova and Hell, 1997), but the isolated complex is inefficient in synthesizing cysteine, and complex formation/dissociation may represent a way in which SAT and OASTL activities are regulated (Droux et al., 1998). *Arabidopsis* has five *SAT* (also known as *SERAT*) genes encoding proteins that differ in their expression levels, and changes in the abundance of the *SAT* transcripts occur in response to S starvation. Furthermore, the *Arabidopsis* SAT proteins differ in L-cysteine sensitivity and subcellular location (Kawashima et al., 2005). In *Chlamydomonas* the *SAT1* gene has been characterized, and its transcript level and protein activity increase in response to S starvation (Ravina et al., 2002; Zhang et al., 2004). The genome also encodes a second gene, *SAT2*, which like the first, has a putative chloroplast transit peptide.

Methionine is synthesized from cysteine and *O*-phosphohomoserine (OPH) through three consecutive reactions catalyzed by cystathionine γ-synthase, cystathionine β-lyase, and methionine synthase (Ravanel et al., 1998a). Cystathionine γ-synthase, localized in chloroplasts, catalyzes cystathionine formation (Wallsgrove et al., 1983; Ravanel et al., 1995a,b) and is controlled by availability of OPH and *S*-adenosylmethionine (SAM). However, OPH can serve as substrate for both cystathionine γ-synthase and threonine synthase. The relative activities of these enzymes are controlled by the S status of cells. Threonine synthase is most active during S-replete growth when levels of SAM are high (Curien et al., 1998). When SAM levels drop, threonine synthase activity declines and a greater proportion of the OPH is converted to methionine via cystathionine γ-synthase (Ravanel et al., 1998a). *Chlamydomonas* has a single cystathionine γ-synthase gene.

In *Arabidopsis*, cystathionine γ-synthase mRNA is controlled at the level of stability; destabilization of the mRNA is triggered either by methionine or a methionine metabolite (Chiba et al., 1999).

Cystathionine β-lyase generates homocysteine from cystathionine (Wallsgrove et al., 1983; Droux et al., 1995). The gene is single-copy in *Arabidopsis* and the enzyme, which has properties similar to those of the bacterial enzyme (Droux et al., 1995; Ravanel et al., 1995b, 1996, 1998b), is likely chloroplast targeted. A single gene for cystathionine β-lyase has been identified in *Chlamydomonas*. Methionine formation is catalyzed by methionine synthase via the methylation of homocysteine. *E. coli* has two methionine synthase genes, one encoding a cobalamin (vitamin B_{12})-dependent and the other a cobalamin-independent enzyme, while animals only have the cobalamin-dependent form and plants only have the cobalamin-independent form. *Chlamydomonas* has both enzyme types, and expression of the cobalamin-independent form may be controlled by a riboswitch mechanism in which cobalamin would bind the RNA and arrest translation (Croft et al., 2005; Grossman et al., 2007).

E. Glutathione synthesis

Glutathione (GSH = γ-Glu-Cys-Gly) is a dominant non-protein thiol in plants (Rennenberg, 1982); it plays a role in regulating the uptake of SO_4^{2-} by plant roots (Herschbach and Rennenberg, 1994; Lappartient and Touraine, 1996) and serves as a major antioxidant. GSH is a substrate for GSH-S-transferases, which are involved in detoxification of xenobiotics (Marrs, 1996). GSH is also the precursor of phytochelatins, peptides that help plants cope with heavy metals in the environment (Rauser, 1987; Scheller et al., 1987; Grill et al., 1989) and serve in redox buffering (Foyer and Halliwell, 1976; Law et al., 1983; Kunert and Foyer, 1993; Meister, 1994). GSH synthesis occurs in plastids and is catalyzed by γ-glutamylcysteine synthetase and GSH synthetase. *Arabidopsis* γ-glutamylcysteine synthetase and GSH synthetase cDNAs were isolated by complementation of *E. coli* mutants (May and Leaver, 1994; Rawlins et al., 1995); at least one gene encoding GSH synthetase is present in *Chlamydomonas*.

III. S STARVATION RESPONSES

A. Overview

Typically, organisms exhibit a suite of responses when their growth becomes nutrient-limited. These responses have been described as "general" and "specific." General responses are those associated with deprivation for any essential nutrient and include the cessation of cell growth and division, accumulation of storage carbohydrates and modulation of metabolic processes, including a decrease in photosynthesis. Of all of these responses, that of photosynthesis has been the most extensively studied. During nutrient limitation in *Chlamydomonas* (S and phosphorus), PS II activity declines through various mechanisms (Wykoff et al., 1998). Furthermore, mutant strains unable to shut down photosynthetic electron transport during nutrient limitation die rapidly (Davies et al., 1996), suggesting that reduction of PS II activity is critical for survival during nutrient limitation. On the other hand, a decline in photosynthetic activity (while maintaining respiratory rates) causes the cultures to become anaerobic (Wykoff et al., 1998), which can trigger fermentation metabolism and a concomitant increase in expression of two iron-only hydrogenases that can catalyze light-driven H_2 production (Forestier et al., 2003; Mus et al., 2007). This finding is being exploited by several researchers for biocatalytic H_2 production in S-depleted *Chlamydomonas* cultures (Fedorov et al., 2005; Fouchard et al., 2005). A better understanding of S deprivation responses may provide insights into metabolisms that facilitate the generation of H_2 (see Chapter 7).

The specific responses are those that are associated only with deprivation for a single nutrient. In the case of S deprivation, they are mostly associated with activities that promote scavenging of extracellular S and the recycling of intracellular S. These processes include the increased synthesis of SO_4^{2-} transporters and hydrolytic enzymes that facilitate the use of esterified sulfate, the recycling of proteins and lipids (Ferreira and Teixeira, 1992), and the replacement of proteins rich in S amino acids with functionally similar proteins that use few S amino acids (e.g. the IRL protein of maize [Petrucco et al., 1996], the ECPs of *Chlamydomonas* [Takahashi et al., 2001], and the β-conglycinin storage protein of soybean [Naito et al., 1994; Kim et al., 1999]). In general, high cysteine and glutathione levels repress SO_4^{2-} assimilation, while S starvation causes elevated activities of key enzymes in the assimilatory pathway. Administration of high concentrations of cysteine and glutathione to plant roots causes reduction in mRNA accumulation for SO_4^{2-} transporters, ATP sulfurylase and APS reductase (Bolchi et al., 1999; Lappartient et al., 1999; Lee and Leustek, 1999). Glutathione may function in repressing genes encoding key enzymes in the S assimilation pathway (Lappartient et al., 1999) while OAS, the substrate for cysteine synthesis, is a positive effector of the CYSB transcription factor in bacteria (Kredich, 1992), and also acts as a positive effector in plants. Roots subjected to exogenous OAS exhibit increased accumulation of mRNAs encoding the SO_4^{2-} transporters (Smith et al., 1997) and APS reductase (Koprivova et al., 2000).

B. Genes responsive to S deprivation

There are numerous genes in *Chlamydomonas* that respond to S deprivation (see Table 5.1 and Figure 5.1). For example, S deprivation activates genes involved in hydrolyzing SO_4^{2-} from organic compounds (*ARS1, ARS2*), transporting SO_4^{2-} across the plasma membrane (*SULTR2, SLT1,* and *SLT2)* and incorporating SO_4^{2-} into amino acids (*ATS, APR, ASL,* and *SAT)* (Davies et al., 1994; Ravina et al., 2002; Zhang et al., 2004; Eberhard et al., 2006). In addition, two prominent extracellular polypeptides, ECP76 (76 kD) and ECP88 (88 kD) are synthesized in response to S deprivation (Takahashi et al., 2001). The *ECP76* and *ECP88* genes are rapidly activated following elimination of S from the growth medium, and their mRNAs are rapidly degraded (half-life of <10 minutes) once S-deprived cells are provided with adequate levels of SO_4^{2-} (Takahashi et al., 2001). ECP76 and ECP88 have features of cell wall proteins, but the mature polypeptides together possess only a single S-containing amino acid. Hence, *Chlamydomonas* has highly regulated processes that likely tailor its protein-rich cell wall during S deprivation in ways that conserve S for other cellular activities. The new cell wall material may also accommodate conditions in which the cells are no longer elongating and dividing (e.g. changes in extensibility and elasticity).

Microarray analyses using an array representing ~3,000 unique genes demonstrated that the levels of transcripts from over 140 genes in wild-type cells increased by fourfold or more following elimination of S from the medium (Zhang et al., 2004). Approximately 40 transcripts decreased to below 25% of the level present in nutrient-replete cells. Many transcripts that increased during S deprivation were previously characterized as encoding S-deprivation–induced proteins (e.g. ARS, ECP76, proteins important for SO_4^{2-} assimilation). Others, however, encoded novel polypeptides of the light harvesting family (e.g. LHCSR2, see below), enzymes involved in scavenging reactive oxygen species, and putative regulatory elements.

C. Identification of genes controlling S deprivation responses

The responses displayed by *Chlamydomonas* during S starvation require an efficient mechanism for sensing S availability and activating a signaling pathway(s) that modulates structural, metabolic and physiological processes, leading to a new cellular homeostasis. Little is known about regulatory elements that control S acclimation processes in plants, although recently a transcription factor was isolated that appears to bind the promoters of genes that become active when S availability declines (Maruyama-Nakashita et al., 2006).

In the case of *Chlamydomonas*, the screening of thousands of insertional mutants has enabled identification of several proteins involved in the acclimation of cells to S deprivation (Davies et al., 1994; Pollock et al., 2005). Transformants of interest were identified based on their inability to synthesize ARS (Davies et al., 1996) in response to S deprivation, or because they constitutively expressed ARS activity under S-replete conditions. Potential mutants were examined for co-segregation of the mutant phenotype with the inserted "tag," survival during S-deprivation conditions and changes in the abundance of transcripts previously shown to be responsive to S levels.

The first mutant identified using this strategy was *sac1* (sulfur acclimation). The SAC1 protein plays a central role in controlling S deprivation responses (Davies et al., 1996; Zhang et al., 2004). The mutant strain is unable to synthesize ARS and exhibits abnormal SO_4^{2-} uptake in response to S deprivation. Essentially no induction of any other gene associated with SO_4^{2-} acquisition and assimilation was observed in a *sac1* mutant (Yildiz et al., 1996; Takahashi et al., 2001; Ravina et al., 2002; Zhang et al., 2004). Furthermore, transcripts from several genes associated with photosynthetic electron transport and the amelioration of damaging effects elicited by the accumulation of reactive oxygen species rise in response to S deprivation, and appear to be under SAC1 control. With respect to light harvesting and electron transport, the *LHCSR2* transcript accumulates to high levels in response to S deprivation, while transcripts encoding many other LHC polypeptides decline (Zhang et al., 2004). These changes in transcript levels are SAC1-dependent.

Transcripts from other genes show extreme accumulation in the *sac1* mutant; among them are those encoding two small chaperones thought to be associated with chloroplasts, HSP22E and HSP22F (see Chapter 19). An increase in these chaperones may reflect an extreme stress response because of the inability of mutant cells to acclimate; the cells die within 2 days of initiation of S deprivation (Davies et al., 1996). This death has been linked to electron flow out of PS II. Hence, modification of photosynthetic electron transport during S deprivation appears to be critical for cell survival. Mutants unable to perform these modifications would exhibit hyper-reduction of the electron transport chain (e.g. PQ pool), which could adversely affect cellular metabolism, leading to the accumulation of reactive oxygen species and extensive cellular damage (Davies et al., 1996).

The *SAC1* gene product is predicted to be similar to Na^+/SO_4^{2-} transporters (*SLT* genes; Davies and Grossman, 1998). Its deduced polypeptide sequence and the *sac1* mutant phenotype suggest a similarity with the yeast Snf3p system. Snf3p is a yeast "transporter-like" regulatory protein that governs expression of genes involved in hexose utilization (Ozcan et al., 1996, 1998), but also interacts with other stress-related processes (Sanz, 2003). The similarity between SAC1 and Snf3p raises the possibility that polypeptides whose original function was to bind and transport various substrates into cells may have evolved into regulatory elements that sense extracellular or intracellular nutrient concentrations. This information would then be communicated to a signal transduction pathway involved in modulating the efficiency of nutrient acquisition processes and tuning cellular metabolism to the potential for cell growth. Two TrkA-C domains present in the central region of the SAC1 protein could play a role in sensing and transduction mechanisms. The function of the TrkA-C domains (also known as RCK-C) is not well understood. They are present in many proton channels as one or two tandem repeats, and are thought to generate homodimer interactions that produce a cleft between two lobes of the transport complex (Anantharaman et al., 2001; Dong et al., 2005). Similar to SAC1, the other SAC1-like proteins (SLTs) also have TrkA-C domains, but they have four tandem repeats rather than two (Figure 5.4B). Both in SAC1 and the SLTs, the TrkA-C domains are positioned in a predicted cytosolic hydrophilic loop that separates two sets of transmembrane helices.

Another mutant, designated *sac3*, exhibits low-level constitutive ARS activity in S-replete medium, but like wild-type cells shows an increase in ARS activity following transfer of cells to S-depleted medium. Hence, SAC3 functions either directly or indirectly in maintaining repression of ARS activity when S is readily available; other S-starvation-induced genes may also be negatively regulated by SAC3 (Ravina et al., 2002). SAC3 seems to positively regulate the activity of SO_4^{2-} transporters since while their mRNAs still increase during S deprivation of the *sac3* mutant, there is little increase in the capacity of mutant cells to transport SO_4^{2-} (no change in the V_{max} for SO_4^{2-}s; Davies et al., 1999). In addition, unlike wild-type cells, *sac3* does not exhibit a decrease in chloroplast transcriptional activity (Irihimovitch and Stern, 2006) during S starvation, suggesting that SAC3 is required to inactivate the chloroplast RNA polymerase sigma factor Sig1 when the cells are starved for S. This inactivation may involve proteolytic activities that are phosphorylation-dependent, since *SAC3* encodes a putative serine-threonine kinase of the plant-specific SNRK2 family (*SAC3* has been also designated as *SNRK2.2*). The protein targets of SAC3 remain to be identified.

Subsequently we identified 49 other mutants (*ars* mutants) that do not properly acclimate to S deprivation (Pollock et al., 2005). The *Chlamydomonas* sequences flanking the insertion were determined in 80% of these strains and the genes disrupted by the insertions in some key mutants are given in Table 5.2. Not all of the mutants have been characterized for many of the responses associated with S starvation, but some strains that have been examined in some detail are described below.

Two allelic mutants, *ars11* and *ars44*, possess insertions in a gene encoding another member of the serine-threonine kinase family; like SAC3, the encoded polypeptide is similar to SNRK2 family proteins and has been designated as *SNRK2.1* (Pollock et al., 2005; Gonzalez-Ballester et al., 2008). These mutants have a *sac1*-like phenotype: no ARS activity in S-depleted medium, defective SO_4^{2-} uptake, and a faster decrease of the chlorophyll content than wild type cells during S deprivation. All of these phenotypic features are more pronounced in the *ars11* mutant than in the *sac1* mutant which suggests a more central role of SNRK2.1 in controlling the S starvation responses. Some members of the *Arabidopsis* SNRK2 kinase family appear to function in controlling S-limitation responses. *Arabidopsis* plants with T-DNA insertions in *SNRK2.3* exhibit a decreased induction of *SULTR2;2*, which encodes a low-affinity SO_4^{2-} transporter, and also accumulate OAS (Kimura et al., 2006). Overall, five of the ten *Arabidopsis SNRK2* genes show increased expression following S starvation.

Another mutant with essentially no induction of ARS activity during S starvation is interrupted for a gene encoding a putative guanylyl cyclase (*ars401*), one of more than 50 putative proteins of this family in *Chlamydomonas* (Merchant et al., 2007). It will be interesting to elucidate the specificity of these cyclases and the biological processes to which they are linked. Finally, a mutant affected in a member of the E3 ubiquitin ligase complex (SCF (Skp1/Cullin/F-box protein)-type, SGT-1-like gene), which is involved in protein degradation, also makes little active ARS when transferred to medium lacking S.

The phenotypes of all of the mutants discussed above can be rescued by introduction of a wild-type gene, although for most other mutants presented in Table 5.2, both linkage and complementation analyses are still required. In addition, double mutant analyses will help determine possible epigenetic relationships among the different genes identified by the screen. While the genetic screens performed thus far have identified a number of key proteins involved in the acclimation of *Chlamydomonas* to S deprivation, extending our knowledge of this process considerably, only a few of the mutants are represented by multiple alleles, which suggests

Table 5.2 Sulfur acclimation/arylsulfatase mutants

Mutant	ARS	Tagged	Accession #	Protein prediction
ars5	2	Yes	EDP00051	Similar to PHO81
ars11	5	Yes	EDP06815	SNRK2.1, putative Ser/Thr kinase. Similar to SAC3
ars20	10	n.a	EDP00621	PAF1: Putative RNA polymerase II regulator
ars44	10	Yes	EDP06815	SNRK2.1, putative Ser/Thr kinase. Similar to SAC3
ars53	10	n.a	EDP03402	HYDA1-iron hydrogenase precursor
ars63	2	n.a	EDP09374	SEC24-like protein
ars75	2	n.a	EDP00156	Probably affects SAC1
ars76	2	Yes	EDP05314	APPLE domain: mediates dimer formation
ars107	8	n.a	EDP09374	SEC24-like protein
ars122	2	Yes	ED096714	SGT-1-like (SCF-mediated ubiquitination)
ars124	10	n.a	ED095880	Adenylate kinase
ars401	2	Yes	EDP08005*	Guanylate cyclase-like

*ARS activity (ARS%) is as a percentage relative to the activity measured for the parental strain under S deprivation conditions. Co-segregation of the marker and phenotype (tagged mutant) was determined by genetic crosses. n.a: not analyzed. The symbol * indicates that the guanylyl cyclase JGI model is fused with a 3′-5′ exonuclease. Accession numbers are given for the wild-type protein sequences predicted to be affected by the mutations.*

that other genes that are important for normal S-deprivation responses in *Chlamydomonas* remain to be identified.

D. Model for acclimation of *Chlamydomonas* to S deprivation

In Figure 5.5 we depict a model of the regulatory processes associated with S deprivation responses. This model is not supported at this point by protein–protein interaction evidence, so its tenets rely on the analyses of mutants and how the lesions in the mutants affect ARS and SO_4^{2-} uptake activities, the expression patterns of some S deprivation-induced genes, and certain physiological responses of the mutant strains. However, the model helps put the information into a context that may guide future experimentation.

It is known that SAC1 positively enhances expression of many S deprivation-regulated genes when *Chlamydomonas* is transferred to medium devoid of S. On the other hand, the SAC3 kinase has a negative effect on some S starvation-induced genes. A double mutant *sac1 sac3*, like the *sac3* single mutant, exhibits low-level constitutive ARS activity in S-replete medium, but during S deprivation, unlike *sac1*, there is some

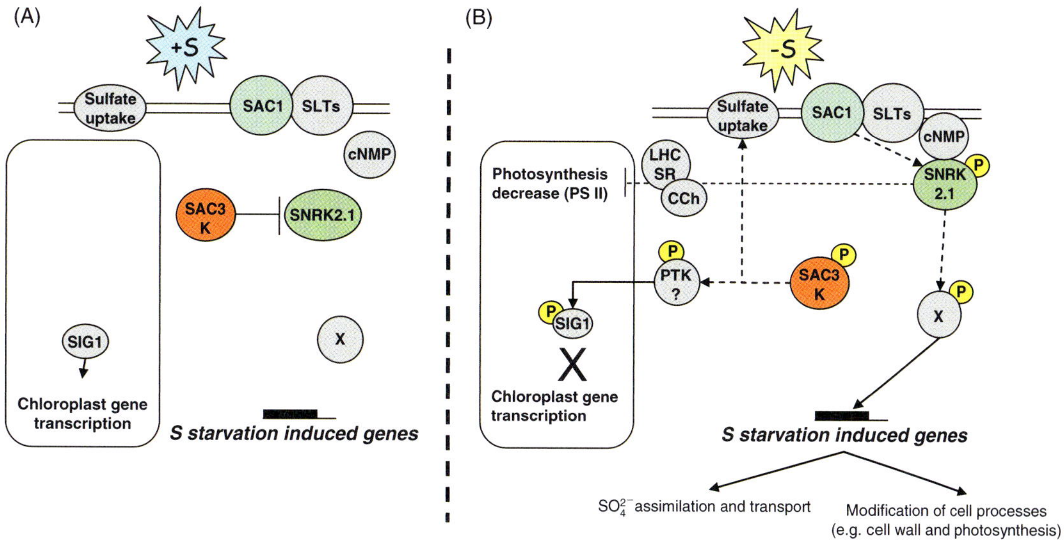

FIGURE 5.5 *Model for S-dependent signal transduction pathways in Chlamydomonas during S-replete and S deprivation conditions. (A) Under S-replete condition SNRK2.1 is not active. (B) Activation of SNRK2.1 under S deprivation is elicited by phosphorylation through SAC1 in a process involving a guanylyl cyclase (cNMP). Other proteins such as the SLT and/or SULTR transporters may interact with SAC1 in the plasma membrane to control both transport and signaling through SNRK2.1. Once SNRK2.1 is activated, it may phosphorylate transcription factor X, which then elicits elevated or depressed activity from S-responsive genes, including those encoding enzymes involved in the assimilation of SO_4^{2-}, restructuring of the cell wall and modification of the photosynthetic apparatus. However, the activity of SNRK2.1 may not be completely SAC1-dependent. SAC3 may inhibit SNRK2.1 during S-replete conditions (e.g. through a second phosphorylation of a specific site on SNRK2.1), which keeps the S-responsive genes completely off. In addition, during S-starvation conditions SAC3 might signal to a putative plastid kinase which in turn activates a system involved in proteolysis of SIG1 and also either directly or indirectly modulate SO_4^{2-} transport activity through post-transcriptional processes.*

increase in ARS expression, although lower than in either wild-type cells or the *sac3* mutant (Davies and Grossman, 1994). This finding suggests that the relationship between *SAC1* and *SAC3* is non-epistatic, and that proteins in addition to SAC1 and SAC3 may be involved in controlling *ARS* expression. SAC1, which has strong similarity to the SLT family of proteins, may act as a sensor that resides on the plasma membrane. The TrkA-C domains present in SAC1 and SLT transporters could be involved in protein–protein interactions and ligand binding. We hypothesize that SAC1 interacts with the SLT proteins and under S deprivation conditions, triggers a phosphorylation cascade that modulates expression of S-regulated genes.

The kinase encoded by *SNRK2.1* was shown to be required for ARS activity and is a key regulatory component of the S starvation responses (Pollack et al., 2005; Gonzalez-Ballester et al., 2008). SNRK2.1 is likely part of the same phosphorylation cascade as SAC1. In this scenario, SO_4^{2-} availability sensed by SAC1 initiates a phosphorylation cascade through SNRK2.1 that tunes the activity of S-responsive genes, including those encoding proteins involved in SO_4^{2-} assimilation and modification of the photosynthetic apparatus (e.g. *LHCSR* genes and some specific small chloroplast chaperones), through an undefined transcription activator designated X. This process may also be controlled by a guanylyl/adenyl cyclase via the production of cGMP/cAMP, which may serve to amplify the signal through the pathway. Severe S deprivation conditions would elicit full activation of the cascade, triggering high level expression from downstream target genes.

Nevertheless, SNRK2.1 activation is not completely SAC1-dependent since the *sac1 sac3* double mutant still possesses low ARS activity in S-replete and S-depleted conditions. We postulate that SAC3 may be critical for keeping SNRK2.1 in a completely inactive form, perhaps via a direct phosphorylation. Because there is no change in the V_{max} for SO_4^{2-} transport in S-deprived *sac3* cells (Davies and Grossman, 1994), even though SO_4^{2-} transporter mRNAs increase, it appears that SAC3 is required for full activation of SO_4^{2-} uptake, probably at the post-transcriptional level. SAC3 is also involved in controlling plastid gene expression through proteolytic degradation of SIG1, which causes down regulation of chloroplast transcription (Irihimovitch and Stern, 2006). The potential interaction between SAC3 and a putative chloroplast-localized kinase that modulates transcription in *Chlamydomonas*, remains to be established.

IV. PERSPECTIVES

Numerous polypeptides that control S deprivation responses in *Chlamydomonas* have been discovered, but it will take significantly more work to understand the exact interactions of those regulatory elements and details of the associated phosphorylation cascade. It is also critical to identify and characterize the full complement of regulators associated with S deprivation responses (e.g. Table 5.2). Furthermore, P and S metabolism in vascular plants (Wang et al., 2002), yeast (O'Connell and Baker, 1992) and *Chlamydomonas* (Moseley et al., 2006) appear to intersect, suggesting that dynamic regulatory networks integrate nutrient stress responses, and most likely responses to other conditions that limit cell growth. Establishing these links and the specific factors that facilitate this integration would be important for understanding the homeostasis of the cell as it experiences fluctuating environmental conditions.

ACKNOWLEDGMENTS

We thank the Carnegie Institution and Ministerio de Educación y Ciencia (Spain) for their financial support. This work was supported in part by NSF grant MCB 0235878 awarded to A.R.G.

REFERENCES

Allmer, J., Naumann, B., Markert, C., Zhang, M., and Hippler, M. (2006). Mass spectrometric genomic data mining: Novel insights into bioenergetic pathways in Chlamydomonas reinhardtii. *Proteomics* **6**, 6207–6220.

Anantharaman, V., Koonin, E.V., and Aravind, L. (2001). Regulatory potential, phylogenetic distribution and evolution of ancient, intracellular small-molecule-binding domains. *J. Mol. Biol.* **307**, 1271–1292.

Aravind, L. and Koonin, E.V. (2000). The STAS domain: a link between anion transporters and antisigma-factor antagonists. *Curr. Biol.* **10**, 53–55.

Armbrust, E.V., Berges, J.A., Bowler, C., Green, B.R., Martinez, D., Putnam, N.H., Zhou, S.G., Allen, A.E., Apt, K.E., Bechner, M., Brzezinski, M.A. et al. (2004). The genome of the diatom *Thalassiosira pseudonana*: Ecology, evolution, and metabolism. *Science* **306**, 79–86.

Arz, H.E., Gisselmann, G., Schiffmann, S., and Schwenn, J.D. (1994). A cDNA for adenylyl sulfate (APS)-kinase from *Arabidopsis thaliana*. *Biochim. Biophys. Acta* **1218**, 447–452.

Bick, J.A. and Leustek, T. (1998). Plant sulfur metabolism: The reduction of sulfate to sulfite. *Curr. Opin. Plant Biol.* **1**, 240–244.

Bick, J.A., Aslund, F., Chen, Y., and Leustek, T. (1998). Glutaredoxin function for the carboxyl terminal domain of the plant-type 5-adenylylsulfate (APS) reductase. *Proc. Natl. Acad. Sci. U.S.A.* **95**, 8404–8409.

Bogdanova, N. and Hell, R. (1997). Cysteine synthesis in plants: Protein–protein interactions of serine acetyltransferase from *Arabidopsis thaliana*. *Plant J.* **11**, 251–262.

Bolchi, A., Petrucco, S., Tenca, P., Foroni, C., and Ottonello, S. (1999). Coordinate modulation of maize sulfate permease and ATP sulfurylase mRNAs in response to variations in sulfur nutritional status: Stereospecific down-regulation by L-cysteine. *Plant Mol. Biol.* **39**, 527–537.

Cole, D.W. and Johnson, D.W. (1977). Atmospheric sulfate additions and cation leaching in a douglas fir ecosystem. *Water Res. Res.* **13**, 313–317.

Croft, M.T., Lawrence, A.D., Raux-Deery, E., Warren, M.J., and Smith, A.G. (2005). Algae acquire vitamin B12 through a symbiotic relationship with bacteria. *Nature* **438**, 90–93.

Curien, G., Job, D., Douce, R., and Dumas, R. (1998). Allosteric activation of *Arabidopsis* threonine synthase by S-adenosylmethionine. *Biochem.* **37**, 13212–13221.

Chen, H.C. and Melis, A. (2004). Localization and function of SulP, a nuclear-encoded chloroplast sulfate permease in *Chlamydomonas reinhardtii*. *Planta* **220**, 198–210.

Chen, H.C., Yokthongwattana, K., Newton, A.J., and Melis, A. (2003). SulP, a nuclear gene encoding a putative chloroplast-targeted sulfate permease in *Chlamydomonas reinhardtii*. *Planta* **218**, 98–106.

Chen, H.C., Newton, A.J., and Melis, A. (2005). Role of SulP, a nuclear-encoded chloroplast sulfate permease, in sulfate transport and H_2 evolution in *Chlamydomonas reinhardtii*. *Photosynth. Res.* **84**, 289–296.

Chiba, Y., Ishikawa, M., Kijima, F., Tyson, R.H., Kim, J.H., Yamamoto, A., Nambara, E., Leustek, T., Wallsgrove, R.M., and Naito, S. (1999). Evidence for autoregulation of cystathionine γ-synthase mRNA stability in *Arabidopsis*. *Science* **286**, 1371–1374.

David, M.B., Grigal, D.F., Ohmann, L.F., and Gertner, G.Z. (1988). Sulfur, carbon, and nitrogen relationships in forest soils across the northern Great Lakes states as affected by atmospheric deposition and vegetation. *Canadian J. Forest. Res.* **18**, 1386–1391.

Davies, J. and Grossman, A.R. (1994). Sequences controlling transcription of the *Chlamydomonas reinhardtii* β_2-tubulin gene after deflagellation and during the cell cycle. *Mol. Cell Biol.* **14**, 5165–5174.

Davies, J.D. and Grossman, A.R. (1998). Responses to deficiencies in macronutrients. In: *The Molecular Biology of Chlamydomonas* (J.-D. Rochaix, M. Goldschmidt-Clermont, and S. Merchant, Eds.), pp. 613–635. Kluwer Academic Publishers, Dordrecht.

Davies, J.P., Yildiz, F., and Grossman, A.R. (1994). Mutants of *Chlamydomonas reinhardtii* with aberrant responses to sulfur deprivation. *Plant Cell* **6**, 53–63.

Davies, J., Yildiz, F., and Grossman, A.R. (1996). Sac1, a putative regulator that is critical for survival of *Chlamydomonas reinhardtii* during sulfur deprivation. *EMBO J.* **15**, 2150–2159.

Davies, J.P., Yildiz, F.H., and Grossman, A.R. (1999). Sac3, a Snf1-like serine/threonine kinase that positively and negatively regulates the responses of *Chlamydomonas* to sulfur limitation. *Plant Cell* **11**, 1179–1190.

de Hostos, E.L., Togasaki, R.K., and Grossman, A.R. (1988). Purification and biosynthesis of a derepressible periplasmic arylsulfatase from *Chlamydomonas reinhardtii*. *J. Cell Biol.* **106**, 29–37.

de Hostos, E.L., Schilling, J., and Grossman, A.R. (1989). Structure and expression of the gene encoding the periplasmic arylsulfatase of *Chlamydomonas reinhardtii*. *Mol. Gen. Genet.* **218**, 229–239.

Dong, J., Shi, N., Berke, I., Chen, L., and Jiang, Y. (2005). Structures of the MthK RCK domain and the effect of Ca^{2+} on gating ring stability. *J. Biol. Chem.* **280**, 41716–41724.

Droux, M., Ravanel, S., and Douce, R. (1995). Methionine biosynthesis in higher plants. I. Purification and characterization of cystathionine β-lyase from spinach chloroplasts. *Arch. Biochem. Biophys.* **316**, 585–595.

Droux, M., Ruffet, M.-L., Douce, R., and Job, D. (1998). Interactions between serine acetyltranferase and *O*-acetylserine (thiol) lyase in higher plants. *Eur. J. Biochem.* **255**, 235–245.

Duncan, L., Alper, S., and Losick, R. (1996). SpoIIAA governs the release of the cell-type specific transcription factor sigma F from its anti-sigma factor SpoIIAB. *J. Mol. Biol.* **260**, 147–164.

Eberhard, S., Jain, M., Im, C.S., Pollock, S., Shrager, J., Lin, Y., Peek, A.S., and Grossman, A.R. (2006). Generation of an oligonucleotide array for analysis of gene expression in *Chlamydomonas reinhardtii*. *Curr. Genet.* **49**, 106–124.

Everett, L.A. and Green, E.D. (1999). A family of mammalian anion transporters and their involvement in human genetic diseases. *Hum. Mol. Genet.* **8**, 1883–1891.

Fedorov, A.S., Kosourov, S., Ghirardi, M.L., and Seibert, M. (2005). Continuous hydrogen photoproduction by *Chlamydomonas reinhardtii*: Using a novel two-stage, sulfate-limited chemostat system. *Appl. Biochem. Biotechnol.* **4**, 121–124, 403–412.

Ferreira, R.M.B. and Teixeira, A.R.N. (1992). Sulfur starvation in *Lemna* leads to degradation of ribulose-bisphosphate carboxylase without plant death. *J. Biol. Chem.* **267**, 7253–7257.

Forestier, M., King, P., Zhang, L., Posewitz, M., Schwarzer, S., Happe, T., Ghirardi, M.L., and Seibert, M. (2003). Expression of two [Fe]-hydrogenases in *Chlamydomonas reinhardtii* under anaerobic conditions. *Eur. J. Biochem.* **270**, 2750–2758.

Fouchard, S., Hemschemeier, A., Caruana, A., Pruvost, J., Legrand, J., Happe, T., Peltier, G., and Cournac, L. (2005). Autotrophic and mixotrophic hydrogen photoproduction in sulfur-deprived *Chlamydomonas* cells. *Appl. Environ. Microbiol.* **71**, 6199–6205.

Foyer, C.H. and Halliwell, B. (1976). The presence of glutathione and glutathione reductase in chloroplasts: A proposed role in ascorbic acid metabolism. *Planta* **133**, 21–25.

Gonzalez-Ballester, D., Pollock, S.V., Pootakham, W., and Grossman, A.R. (2008). The central role of a SNRK2 kinase in sulfur deprivation responses. *Plant Physiol.* **147**, 216–227.

Grill, E., Loffler, S., Winnacker, E.L., and Zenk, M.H. (1989). Phytochelatins, the heavy-metal-binding peptides of plants, are synthesized from glutathione by a specific y-glutamylcysteine dipeptidyl transpeptidase (phytochelatin synthetase). *Proc. Natl. Acad. Sci. U. S. A.* **86**, 6838–6842.

Grossman, A.R. and Takahashi, H. (2001). Macronutrient utilization by photosynthetic eukaryotes and the fabric of interactions. *Ann. Rev. Plant Physiol. Plant Mol. Biol.* **52**, 163–210.

Grossman, A.R., Croft, M., Gladyshev, V., Posewitz, M.C., Merchant, S., and Spalding, M. (2007). Novel metabolism in *Chlamydomonas* through the lens of genomics. *Curr. Opin. Plant Biol.* **10**, 190–198.

Gupta, A.S., Alscher, R.G., and McCune, D.F. (1990). Ozone exposure, glutathione levels and photosynthesis in hybrid poplar. In: *Sulfur Nutrition and Sulfur Assimilation in Higher Plants* (H. Rennenberg, C. Brunold, L.J. Dekoli, and I. Stulen, Eds.), pp. 195–197. SPB Academic Publications, The Hague, The Netherlands.

Gutierrez-Marcos, J., Roberts, M., Campbell, E., and Wray, J. (1996). Three members of a novel small gene-family from *Arabidopsis thaliana* able to complement functionally an *Escherichia coli* mutant defective in PAPS reductase activity encode proteins with a thioredoxin-like domain and "APS reductase" activity. *Proc. Natl. Acad. Sci. U. S. A.* **93**, 13377–13382.

Herschbach, C. and Rennenberg, H. (1994). Influence of glutathione (GSH) on net uptake of sulfate and sulfate transport in tobacco plants. *J. Exp. Bot.* **45**, 1069–1076.

Ho, M.S., Carniol, K., and Losick, R. (2003). Evidence in support of a docking model for the release of the transcription factor sigma F from the antisigma factor SpoIIAB in *Bacillus subtilis*. *J. Biol. Chem.* **278**, 20898–20905.

Howarth, J.R., Fourcroy, P., Davidian, J.C., Smith, F.W., and Hawkesford, M.J. (2003). Cloning of two contrasting high-affinity sulfate transporters from tomato induced by low sulfate and infection by the vascular pathogen *Verticillium dahliae*. *Planta* **218**, 58–64.

Irihimovitch, V. and Stern, D.B. (2006). The sulfur acclimation SAC3 kinase is required for chloroplast transcriptional repression under sulfur limitation in *Chlamydomonas reinhardtii*. *Proc. Natl. Acad. Sci. U. S. A.* **103**, 7911–7916.

Johnson, D.W., Turner, J., and Kelly, J.M. (1982). The effects of acid rain on forest nutrient status. *Water Res. Res.* **18**, 449–461.

Kawashima, C.G., Berkowitz, O., Hell, R., Noji, M., and Saito, K. (2005). Characterization and expression analysis of a serine acetyltransferase gene family involved in a key step of the sulfur assimilation pathway in *Arabidopsis*. *Plant Physiol.* **137**, 220–230.

Kim, H., Hirai, M.Y., Hayashi, H., Chino, M., Naito, S., and Fujiwara, T. (1999). Role of O-acetyl-L-serine in the coordinated regulation of the expression of a soybean seed storage-protein gene by sulfur and nitrogen nutrition. *Planta* **209**, 282–289.

Kimura, T., Shibagaki, N., Ohkama-Ohtsu, N., Hayashi, H., Yoneyama, T., Davies, J.P., and Fujiwara, T. (2006). *Arabidopsis* SNRK2.3 protein kinase is involved in the regulation of sulfur-responsive gene expression and O-acetyl-L-serine accumulation under limited sulfur supply. *Soil Sci. Plant Nutr.* **52**, 211–220.

Ko, S.B., Zeng, W., Dorwart, M.R., Luo, X., Kim, K.H., Millen, L., Goto, H., Naruse, S., Soyombo, A., Thomas, P.J., and Muallem, S. (2004). Gating of CFTR by the STAS domain of SLC26 transporters. *Nat. Cell Biol.* **6**, 343–350.

Koprivova, A., Suter, M., den Camp, R.O., Brunhold, C., and Kopriva, S. (2000). Regulation of sulfate assimilation by nitrogen in *Arabidopsis*. *Plant Physiol.* **122**, 737–746.

Kraegeloh, A., Amendt, B., and Kunte, H.J. (2005). Potassium transport in a halophilic member of the bacteria domain: Identification and characterization of the K^+ uptake systems TrkH and TrkI from *Halomonas elongata* DSM 2581T. *J. Bacteriol.* **187**, 1036–1043.

Kredich, N. (1992). The molecular basis for posititive regulation of *cys* promoters in *Salmonella typhimurium* and *Escherichia coli*. *Mol. Microbiol.* **6**, 2747–2753.

Kunert, K.J. and Foyer, C. (1993). Thiol/disulfide exchange in plants. In: *Sulfur Nutrition and Sulfur Assimilation in Higher Plants* (L.J. de Kok, J. Stulen, H. Rennenberg, C. Brunold, and W.E. Rauser, Eds.), pp. 139–151. SPB Academic Publishing, The Hague.

Lappartient, A.G. and Touraine, B. (1996). Demand-driven control of root ATP sulfurylase activity and SO_4^{2-} uptake in intact canola: The role of phloem-translocated glutathione. *Plant Physiol.* **111**, 147–157.

Lappartient, A.G., Vidmar, J.J., Leustek, T., Glass, A.D.M., and Touraine, B. (1999). Inter-organ signaling in plants: Regulation of ATP sulfurylase and sulfate transporter genes expression in roots mediated by phloem-translocated compound. *Plant J.* **18**, 89–95.

Law, M.Y., Charles, S.A., and Halliwell, B. (1983). Glutathione and ascorbic acid in spinach (*Spinacia oleracea*) chloroplasts: The effect of hydrogen peroxide and of paraquat. *Biochem J.* **210**, 899–903.

Lee, S. and Leustek, T. (1998). APS kinase from *Arabidopsis thaliana*: Genomic organization, expression and kinetic analysis of the recombinant enzyme. *Biochem. Biophys. Res. Commun.* **247**, 171–175.

Lee, S. and Leustek, T. (1999). The effect of cadmium on sulfate assimilation enzymes in *Brassica juncea*. *Plant Science* **141**, 201–207.

Leustek, T. and Saito, K. (1999). Sulfate transport and assimilation in plants. *Plant Physiol.* **120**, 637–643.

Lunn, J.E., Droux, M., Martin, J., and Douce, R. (1990). Localization of ATP sulfurylase and O-acetyl(thiol)lyase in spinach leaves. *Plant Physiol.* **94**, 1345–1352.

MacDonald, N.W., Burton, A.J., Jurgensen, M.F., McLaughlin, J.W., and Mroz, G.D. (1991). Variation in forest soil properties along a Great Lakes air pollution gradient. *Soil Science Soc. Amer. J.* **55**, 1709–1715.

Mahler, R.J. and Maples, R.L. (1986). Responses of wheat to sulfur fertilization. *Commun. Soil Sci. Plant Anal.* **17**, 975–988.

Mahler, R.J. and Maples, R.L. (1987). Effect of sulfur additions on soil and the nutrition of wheat. *Commun. Soil Sci. Plant Anal.* **18**, 653–673.

Marrs, K.A. (1996). The functions and regulation of glutathione S-transferases in plants. *Annu. Rev. Plant Physiol.* **47**, 127–158.

Marschner, H. (1995). *Mineral Nutrition of Higher Plants*. Academic Press Limited, London.

Maruyama-Nakashita, A., Nakamura, Y., Tohge, T., Saito, K., and Takahashi, H. (2006). *Arabidopsis* SLIM1 is a central transcriptional regulator of plant sulfur response and metabolism. *Plant Cell* **18**, 3235–3251.

May, M.J. and Leaver, C.J. (1994). *Arabidopsis thaliana* γ-glutamylcysteine synthase is structurally unrelated to mammalian, yeast, and *Escherichia coli* homologs. *Proc. Natl. Acad. Sci. U. S. A.* **91**, 10059–10063.

Meister, A. (1994). Glutathione-ascorbic acid antioxidant system in animals. *J. Biol. Chem.* **269**, 9397–9400.

Meister, A. and Anderson, M.E. (1983). Glutathione. *Annu. Rev. Biochem.* **52**, 711–760.

Melis, A. and Chen, H.C. (2005). Chloroplast sulfate transport in green algae: Genes, proteins and effects. *Photosynth. Res.* **86**, 299–307.

Merchant, S.S., Prochnik, S.E., Vallon, O., Harris, E.H., Karpowicz, S.J., Witman, G.B., Terry, A., Salamov, A., Fritz-Laylin, L.K., Maréchal-Drouard, L. et al. (2007). The *Chlamydomonas* genome reveals the evolution of key animal and plant functions. *Science* **318**, 245–250.

Moseley, J.L., Chang, C.W., and Grossman, A.R. (2006). Genome-based approaches to understanding phosphorus deprivation responses and PSR1 control in *Chlamydomonas reinhardtii*. *Eukaryotic Cell* **5**, 26–44.

Mus, F., Dubini, A., Seibert, M., Posewitz, M.C., and Grossman, A.R. (2007). Anaerobic acclimation in *Chlamydomonas reinhardtii*: Anoxic gene expression, hydrogenase induction and metabolic pathways. *J. Biol. Chem.* **282**, 25475–25486.

Naito, S., Yokota-Hirai, M., Chino, M., and Komeda, Y. (1994). Expression of soybean (*Glycine max* [L.] Merr.) seed storage protein gene in transgenic *Arabidopsis thaliana* and its response to nutritional stress and to abscisic acid mutations. *Plant Physiol.* **104**, 497–503.

Najafi, S.M., Harris, D.A., and Yudkin, M.D. (1996). The SpoIIAA protein of *Bacillus subtilis* has GTP-binding properties. *J. Bacteriol.* **178**, 6632–6634.

Noji, M. and Saito, K. (2002). Molecular and biochemical analysis of serine acetyltransferase and cysteine synthase towards sulfur metabolic engineering in plants. *Amino Acids* **22**, 231–243.

Noji, M., Inoue, K., Kimura, N., Gouda, A., and Saito, K. (1998). Isoform-dependent differences in feedback regulation and subcellular localization of serine acetyltransferase involved in cysteine biosynthesis from *Arabidopsis thaliana*. *J. Biol. Chem.* **273**, 32739–32745.

Noji, M., Takagi, Y., Kimura, N., Inoue, K., Saito, M., Horikoshi, M., Saito, F., Takahashi, H., and Saito, K. (2001). Serine acetyltransferase involved in cysteine biosynthesis from spinach: Molecular cloning, characterization and expression analysis of cDNA encoding a plastidic isoform. *Plant Cell Physiol.* **42**, 627–634.

O'Connell, K.F. and Baker, R.E. (1992). Possible cross-regulation of phosphate and sulfate metabolism in *Saccharomyces cerevisiae*. *Genetics* **132**, 63–73.

Ohta, N., Matsuzaki, M., Misumi, O., Miyagishima, S.Y., Nozaki, H., Tanaka, K., Shin, I.T., Kohara, Y., and Kuroiwa, T. (2003). Complete sequence and analysis of the plastid genome of the unicellular red alga *Cyanidioschyzon merolae*. *DNA Res.* **10**, 67–77.

Ozcan, S., Dover, J., Rosenwald, A.G., Wolf, S., and Johnston, M. (1996). Two glucose transporters in *Saccharomyces cerevisiae* are glucose sensors that generate a signal for induction of gene expression. *Proc. Natl. Acad. Sci. U. S. A.* **93**, 12428–12432.

Ozcan, S., Dover, J., and Johnston, M. (1998). Glucose sensing and signaling by two glucose receptors in the yeast *Saccharomyces cerevisiae*. *EMBO J.* **17**, 2566–2573.

Petrucco, S., Bolchi, A., Foroni, C., Percudani, R., Rossi, G.L., and Ottonello, S. (1996). A maize gene encoding as NADPH binding enzyme highly homologous to isoflavone reductase is activated in response to sulfur starvation. *Plant Cell Environ.* **8**, 69–80.

Pollock, S.V., Pootakham, W., Shibagaki, N., Moseley, J.L., and Grossman, A.R. (2005). Insights into the acclimation of *Chlamydomonas reinhardtii* to sulfur deprivation. *Photosynth. Res.* **86**, 475–489.

Prior, A., Uhrig, J.F., Heins, L., Wiesmann, A., Lilliq, C.H., Stolz, C., Soll, J., and Schwenn, J.D. (1999). Structural and kinetic properties of adenylylsulfate reductase from *Catheranthus roseus* cell cultures. *Biochem. Biophys. Res. Commun.* **1430**, 25–38.

Rauser, W.E. (1987). Changes in glutathione content of maize seedlings exposed to cadmium. *Plant Science* **51**, 171–175.

Ravanel, S., Droux, M., and Douce, R. (1995a). Methionine biosynthesis in higher plants. I. Purification and characterization of cystathionine γ-synthase from spinach chloroplasts. *Arch. Biochem. Biophys.* **316**, 572–584.

Ravanel, S., Ruffet, M.-L., and Douce, R. (1995b). Cloning of an *Arabidopsis thaliana* cDNA encoding cystathionine β-lyase by functional complementation in *Escherichia coli*. *Plant Mol. Biol.* **29**, 875–882.

Ravanel, S., Job, D., and Douce, R. (1996). Purification and properties of cystathionine β-lyase from *Arabidopsis thaliana* overexpressed in *Escherichia coli*. *Biochem. J.* **320**, 383–392.

Ravanel, S., Gakiere, B., Job, D., and Douce, R. (1998a). The specific features of methionine biosynthesis and metabolism in plants. *Proc. Natl. Acad. Sci. U. S. A.* **95**, 7805–7812.

Ravanel, S., Gakiere, B., Job, D., and Douce, R. (1998b). Cystathionine γ-synthase from *Arabidopsis thaliana*: Purification and biochemical characterization of the recombinant enzyme overexpressed in *Escherichia coli*. *Biochem. J.* **331**, 639–648.

Ravina, C.G., Barroso, C., Vega, J.M., and Gotor, C. (1999). Cysteine biosynthesis in *Chlamydomonas reinhardtii*. Molecular cloning and regulation of O-acetylserine(thiol)lyase. *Eur. J. Biochem.* **264**, 848–853.

Ravina, C.G., Chang, C.I., Tsakraklides, G.P., McDermott, J.P., Vega, J.M., Leustek, T., Gotor, C., and Davies, J.P. (2002). The *sac* mutants of *Chlamydomonas reinhardtii* reveal transcriptional and posttranscriptional control of cysteine biosynthesis. *Plant Physiol.* **130**, 2076–2084.

Rawlins, M.R., Leaver, C.J., and May, M.J. (1995). Characterization of an *Arabidopsis thaliana* cDNA encoding glutathione synthetase. *FEBS Lett.* **376**, 81–86.

Rennenberg, H. (1982). Glutathione metabolism and possible biological role in higher plants. *Phytochemistry* **21**, 2771–2781.

Rost, B., Schneider, R., and Sander, C. (1997). Protein fold recognition by prediction-based threading. *J. Mol. Biol.* **270**, 471–480.

Rotte, C. and Leustek, T. (2000). Differential subcellular localization and expression of ATP sulfurylase and 5′-adenylylsulfate reductase during ontogenesis of *Arabidopsis* leaves indicates that cytosolic and plastid forms of ATP sulfurylase may have specialized functions. *Plant Physiol.* **124**, 715–724.

Rouached, H., Berthomieu, P., El Kassis, E., Cathala, N., Catherinot, V., Labesse, G., Davidian, J.C., and Fourcroy, P. (2005). Structural and functional analysis of the C-terminal STAS (sulfate transporter and anti-sigma antagonist) domain of the *Arabidopsis thaliana* sulfate transporter SULTR1.2. *J. Biol. Chem.* **280**, 15976–15983.

Ruffet, M.-L., Lebrun, M., Droux, M., and Douce, R. (1995). Subcellular localization of serine acetyltransferase from *Pisum sativum* and characterization of an *Arabidopsis thaliana* putative cytosolic isoform. *Eur. J. Biochem.* **227**, 500–509.

Saito, K. (2000). Regulation of sulfate transport and synthesis of sulfur-containing amino acids. *Curr. Opin. Plant Biol.* **3**, 188–195.

Saito, K., Yokoyama, H., Noji, M., and Murakoshi, I. (1995). Molecular cloning and characterization of a plant serine acetyltransferase playing a regulatory role in cysteine biosynthesis from watermelon. *J. Biol. Chem.* **270**, 16321–16326.

Sanz, P. (2003). Snf1 protein kinase: A key player in the response to cellular stress in yeast. *Biochem. Soc. Trans.* **31**, 178–181.

Scheller, H.V., Huang, B., Hatch, E., and Goldsbrough, P.B. (1987). Phytochelatin synthesis and glutathione levels in response to heavy metals in tomato cells. *Plant Physiol.* **85**, 1031–1035.

Schreiner, Ø., Lien, T., and Knutsen, G. (1975). The capacity for arylsulfatase synthesis in synchronous and synchronized cultures of *Chlamydomonas reinhardtii*. *Biochim. Biophys. Acta* **384**, 180–193.

Schultze, M., Quiclet-Sire, B., Kondorosi, E., Virelizer, H., Glushka, J.N., Endre, G., Gero, S.D., and Kondorosi, A. (1992). *Rhizobium meliloti* produces a family of sulfated lipooligosaccharides exhibiting different degrees of plant host specificity. *Proc. Natl. Acad. Sci. U. S. A.* **89**, 192–196.

Setya, A., Murillo, M., and Leustek, T. (1996). Sulfate reduction in higher plants: Molecular evidence for a novel 5′-adenylylsulfate reductase. *Proc. Natl. Acad. Sci. U. S. A.* **93**, 13383–13388.

Shibagaki, N. and Grossman, A.R. (2004). Probing the function of STAS domains of the *Arabidopsis* sulfate transporters. *J. Biol. Chem.* **279**, 30791–30799.

Shibagaki, N. and Grossman, A.R. (2006). The role of the STAS domain in the function and biogenesis of a sulfate transporter as probed by random mutagenesis. *J. Biol. Chem.* **281**, 22964–22973.

Sirko, A., Blaszczyk, A., and Liszewska, F. (2004). Overproduction of SAT and/or OASTL in transgenic plants: A survey of effects. *J. Exp. Bot.* **55**, 1881–1888.

Smith, F.W., Hawkesford, M.J., Ealing, P.M., Clarkson, D.T., Vanden Berg, P.J., Belcher, A.R., and Warrilow, A.G. (1997). Regulation of expression of a cDNA from barley roots encoding a high affinity sulphate transporter. *Plant J.* **12**, 875–884.

Smith, F., Rae, A.L., and Hawkesford, M.J. (2000). Molecular mechanisms of phosphate and sulphate transports in plants. *Biochem. Biophys. Res. Commun.* **1465**, 236–245.

Takahashi, H., Yamazaki, M., Sasakura, N., Watanabe, A., Leustek, T., de Almeida Engler, J., Engler, G., Van Montagu, M., and Saito, K. (1997). Regulation of sulfur assimilation in higher plants: A sulfate transporter induced in sulfate starved roots plays a central role in *Arabidopsis thaliana*. *Proc. Natl. Acad. Sci. U. S. A.* **94**, 11102–11107.

Takahashi, H., Watanabe-Takahashi, A., Smith, F., Blake-Kalff, M., Hawkesford, M.J., and Saito, K. (2000). The roles of three functional sulphate transporters involved in uptake and translocation of sulphate in *Arabidopsis thaliana*. *Plant J.* **23**, 171–182.

Takahashi, H., Braby, C.E., and Grossman, A.R. (2001). Sulfur economy and cell wall biosynthesis during sulfur limitation of *Chlamydomonas reinhardtii*. *Plant Physiol.* **127**, 665–673.

Takahashi, H., Yoshimoto, N., and Saito, K. (2006). Anionic nutrient transport in plants: The molecular basis of the sulfate transporter gene family. *Genet. Eng.* **27**, 67–80.

Tucker, K. and Fadool, D.A. (2002). Neurotrophin modulation of voltage-gated potassium channels in rat through TrkB receptors is time and sensory experience dependent. *J. Physiol.* **542**, 413–429.

Varin, L., Marsolais, F., Richard, M., and Rouleau, M. (1997). Biochemistry and molecular biology of plant sulfotransferases. *FASEB J.* **11**, 517–525.

Wallsgrove, R.M., Lea, P.J., and Miflin, B.J. (1983). Intracellular localization of aspartate kinase and the enzymes of threonine and methionine biosynthesis in green leaves. *Plant Physiol.* **71**, 780–784.

Wang, Y.H., Garvin, D.F., and Kochian, L.V. (2002). Rapid induction of regulatory and transporter genes in response to phosphorus, potassium, and iron deficiencies in tomato roots. Evidence for cross talk and root/rhizosphere-mediated signals. *Plant Physiol.* **130**, 1361–1370.

Warman, P.R. and Sampson, H.G. (1994). Effect of sulfur additions on the yield and elemental composition of canola and spring wheat. *J. Plant Nutrition* **17**, 1817–1825.

Wykoff, D., Davies, J., and Grossman, A. (1998). The regulation of photosynthetic electron transport during nutrient deprivation in *Chlamydomonas reinhardtii*. *Plant Physiol.* **117**, 129–139.

Yildiz, F., Davies, J.P., and Grossman, A.R. (1994). Characterization of sulfate transport in *Chlamydomonas reinhardtii* during sulfur-limited and sulfur-sufficient growth. *Plant Physiol.* **104**, 981–987.

Yildiz, F.H., Davies, J.P., and Grossman, A.R. (1996). Sulfur availability and the *SAC1* gene control adenosine triphosphate sulfurylase gene expression in *Chlamydomonas reinhardtii*. *Plant Physiol.* **112**, 669–675.

Yonekura-Sakakibara, K., Onda, Y., Ashikari, T., Tanaka, Y., Kusumi, T., and Hase, T. (2000). Analysis of reductant supply systems for ferredoxin-dependent sulfite reductase in photosynthetic and nonphotosynthetic organs of maize. *Plant Physiol.* **122**, 887–894.

Zhang, Z., Shrager, J., Jain, M., Chang, C.W., Vallon, O., and Grossman, A.R. (2004). Insights into the survival of *Chlamydomonas reinhardtii* during sulfur starvation based on microarray analysis of gene expression. *Eukaryotic Cell* **3**, 1331–1348.

CHAPTER 6

Phosphate Metabolism and Responses to Phosphorus Deficiency

Jeffrey Moseley and Arthur R. Grossman
The Carnegie Institution, Department of Plant Biology, Stanford, California, USA

CHAPTER CONTENTS

I. INTRODUCTION

Phosphorus (P) is an integral component of nucleic acids and phospholipids, an essential modifier of protein function, and is fundamental to the cell's energy currency. The predominant form of available P in the environment is the inorganic phosphate anion (PO_4^{3-}, designated Pi throughout). Growth and development of microorganisms, soil-dwelling algae such as *Chlamydomonas*, and plants, in both the natural environment and agricultural settings, are often limited by P availability. To optimize crop yields, Pi is included as an abundant component of commercial fertilizers. Much of this supplementary Pi that is applied to agricultural fields leaches from the soil and flows into nearby lakes and rivers, where it can trigger algal blooms that generate anoxic conditions and cause massive fish kills (Wetzel, 1983). The sustainability of agricultural yields and the quality of aquatic ecosystems will benefit from more efficient use of Pi by crop plants, which would also decrease the dependence of farmers on rock Pi reserves, the mining of which can have serious economic and ecological consequences (Smil, 2000). The sources and cycling of P compounds in the environment are shown in Figure 6.1.

While P is an abundant element in the earth's crust, with many soils having a Pi content of between 0.5 and 1.5 mM, the majority of it may be in the form of insoluble Fe^{3+}, Al^{3+}, and Ca^{2+} salts, or esterified to organic molecules that cannot be assimilated by most organisms. The pH and anion exchange characteristics of particular soils, as well as the abundance of specific soil ligands, affect Pi availability. Therefore, plants and microbes are often in a state of P limitation and have evolved a suite of responses, extensively studied in *Chlamydomonas* and described in this chapter, that enable them to survive extended exposure to low P availability. These responses can be divided into two broad categories, the P-specific responses and the general or global stress responses (Quisel et al., 1996; Davies and Grossman, 1998; Wykoff et al., 1998). P-specific responses facilitate mobilization and acquisition of P from extracellular and intracellular stores and include the synthesis and secretion of phosphatases with broad substrate specificity, and the synthesis and accumulation of high-affinity Pi transporters in the plasma membrane (Quisel et al., 1996; Shimogawara et al., 1999). The global responses allow for long-term survival by coordinating decreased P availability with changes in cellular metabolism and reduced cell division and growth. Qualitatively similar global changes are observed when cells are starved for other nutrients, including sulfur and nitrogen (see Chapters 5 and 3, respectively).

For plants, Pi acquisition can be complex because of root system interactions with symbionts such as mycorrhizal fungi, whose proliferation can markedly enhance mobilization and uptake of soil Pi (reviewed by Karandashov and Bucher, 2005). However a number of plants, including

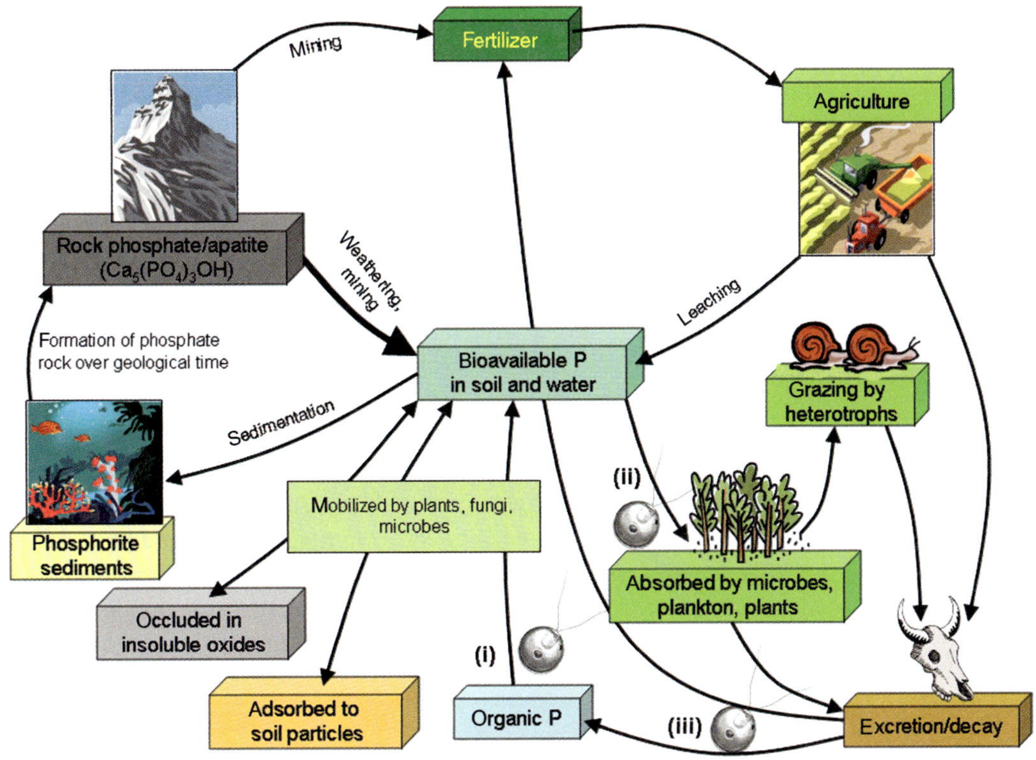

FIGURE 6.1 *P cycle in the environment. The major source of bioavailable P ($H_2PO_4^-$) comes from weathering of terrestrial, Pi-containing minerals, primarily apatite ($Ca_5(PO_4)_3OH$). Other sources include Pi that is occluded in insoluble Al and Fe oxides, adsorbed to soil particles and bound to organic molecules. Plants, microbes, and fungi secrete organic acids and enzymes to mobilize environmental P. Once Pi becomes bioavailable, it is rapidly taken up by plants, microbes, and phytoplankton, which are in turn consumed by grazers. P is returned to the environment as organic phosphates upon excretion, death, and decay of organisms. Chlamydomonas participates in the processes indicated in (i), (ii), and (iii). Pi-rich rocks are formed over millennia by sedimentation of calcium Pi salts that originate from the skeletons and shells of oceanic organisms. Tectonically uplifted phosphorite deposits are mined for use as fertilizer.*

Arabidopsis, do not form mycorrhizal associations. Work in *Arabidopsis* has centered on characterization of Pi "accessing" enzymes (i.e. phytases and phosphatases) and the elucidation of anion uptake and translocation processes (Poirier et al., 1991; Delhaize and Randall, 1995; Okumura et al., 1998; Gilbert et al., 1999; Li et al., 2002; Veljanovski et al., 2006). Since studies with vascular plants are complicated by multiple tissue types and complex developmental regulation, *Chlamydomonas* has been an attractive model for exploring nutrient deprivation responses.

Chlamydomonas induces a suite of responses upon P deprivation including the secretion of periplasmic phosphatases, which can cleave Pi from Pi esters in the immediate environment, the mobilization of internal stores of Pi from polyphosphate molecules, the activation of high-affinity Pi uptake and the substitution of glycolipids and sulfolipids for phospholipids.

The cells also cease division, down-regulate metabolic processes such as photosynthesis, and activate mechanisms that reduce the potential for generating reactive oxygen species. Under nutrient deprivation conditions the absorption of excitation energy by photosynthetic pigments could be lethal, since if the light energy absorbed by chlorophyll is not efficiently used for photochemistry, it may cause the accumulation of triplet-excited chlorophyll which could interact with oxygen to form singlet oxygen. Singlet oxygen can lead to disruption of both protein and membrane functions.

Prokaryotic and eukaryotic microorganisms as well as vascular plants exhibit similar strategies for acclimating to nutrient deprivation conditions, although the regulatory systems and the specific enzymes used for P scavenging may vary significantly. Transporters involved in low- and high-affinity Pi uptake are conserved between *Chlamydomonas* and fungi, while aspects of the regulatory system controlling transcriptional responses during P deprivation are conserved between *Chlamydomonas* and vascular plants. In this chapter, we enumerate some basic metabolic processes that are highly sensitive to cellular P levels, explore P acquisition and assimilation in *Chlamydomonas*, and summarize our current understanding of how limitation for available P alters metabolic and regulatory processes in this alga.

II. PHOSPHORUS ACQUISITION

A. Phosphatases

Chlamydomonas synthesizes a constellation of phosphatases during P limitation. Two acid phosphatases appear to be constitutively expressed, although their levels increase when the cells are limited for P (Matagne and Loppes, 1975; Matagne et al., 1976). There is also a neutral phosphatase, and possibly two alkaline phosphatases, that are derepressed upon P starvation (Lien and Knutsen, 1973; Loppes, 1976; Loppes et al., 1977; Loppes and Deltour, 1981). Mutants that exhibit aberrant expression of acid, neutral, and alkaline phosphatases have been isolated (Loppes and Matagne, 1973; Matagne and Loppes, 1975; Loppes, 1976; Loppes et al., 1977; Loppes and Deltour, 1981). The subcellular locations of some of the phosphatases have been determined; one of the acid phosphatases is vacuolar and the second may be associated with the cell wall, while the neutral and alkaline phosphatases appear to be extracellular or cell wall-associated (Matagne et al., 1976; Patni et al., 1977; Quisel et al., 1996).

Quisel et al. (1996) showed that ~90% of the extracellular phosphatase activity induced by P limitation is Ca^{2+}-dependent, strongly inhibited by low levels of Zn^{2+}, and associated with a ~190-kD monomeric polypeptide encoded by the *PHOX* gene (see Table 6.1 and section IV.C). The enzyme has a broad substrate specificity (it can use AMP) and an alkaline pH

Table 6.1 Genes with potential roles in P scavenging, transport, and homeostasis

Activity	Gene	Accession #	Expression in –P[a]
Alkaline phosphatase	***PHOX***[b]	EDO96628	+++
	PHO1	EDO97048	nt
	PHOD	EDO99918	nt
Putative metallophosphoesterase	***MPA1***	EDP01496	++
	-	EDO98607	nt
	-	EDP01249	nt
	-	EDP08805	+
	-	-	nt
	MPA2	EDP05402	nt
Phosphate transporter (H^+/PO_4^{3-} family)	***PTA1***	EDP06808	−−−
	PTA2a[c,d]	EDP04437	nc
	PTA2b	EDP04438	nc
	PTA3	EDP04436	nc
	PTA4	EDP04439	++
Phosphate transporter (Na^+/PO_4^{3-} family)	*PTB1*	EDP05393	nc
	PTB2	EDO96820	++
	PTB3	EDO96821	+
	PTB4	EDP02027	++
	PTB5	EDP02025	+++
	PTB6a	EDO99485	nt
	PTB6b	EDO99486	nt
	PTB7	EDP05556	nt
	PTB8	EDP01113	nt
	PTB9	EDP02023	nt
	PTB12	EDP02024	nt
Phosphate transporter (SLC13-carrier family)	*PTC1*	EDP08389	nc
Calcium transporter (Ca^{2+}/H^+ antiporter)	***CAX1/VCX1***	EDP00630	+
Vacuolar transport chaperone family	*VTC1*	EDP05311	+
	-	EDP00052	nt
	-	EDP00051	++
	-	EDO97213	nt
	-	-	nt
Polyribonucleotide phosphorylase	***PNP1***	EDO99387	−, nq
S-like RNase	-	-	nt
Unknown function, required for survival	*LPB1*	AAY31019	nc
Positive regulatory protein	***PSR1***	AAD55941	++

[a] *Increases in RNA abundance between 2.5- and 10-fold are indicated by (+), between 10- and 100-fold by (++) and by >100-fold by (+++). Reduction in RNA abundance between 2.5- and 10-fold is indicated by (−), and by >100-fold by (− − −). nc = < 2.5-fold change, either up or down, nt = not tested, nq = not quantified.*

[b] *Genes that have been reported to have altered expression (either up- or down-regulated) in P-deficient cells are indicate d with bold letters.*

[c] *Genes labeled with an "a" or "b" suffix are potential splice variants.*

[d] *Quantification of PTA2 expression did not distinguish between the potential alternatively spliced isoforms.*

optimum, and behaves as a 5′-nucleotidase. A second extracellular phosphatase (10% of the total) has an apparent molecular mass of 73 kD and can slowly hydrolyze IMP and AMP. This is probably the same species that is absent in a *Chlamydomonas* mutant lacking neutral phosphatase activity (Dumont et al., 1990). The third extracellular activity (<5% of the total), which can be resolved from the two others by column chromatography (Quisel et al., 1996), cannot use IMP or AMP as substrates. *PHOX* was first identified as encoding a 160 kD extracellular phosphatase in *Volvox carteri* (Hallmann, 1999).

Starvation of most plants for P results in elevated phosphatase activity. An increase in intracellular and/or extracellular acid phosphatases occurs in the aquatic macrophyte *Spirodela oligorrhiza* (Nakazato et al., 1997), *Arabidopsis* (Trull et al., 1997), and crop plants (e.g. Tadano et al., 1993). Alkaline phosphatase activity is also secreted from P-deprived suspension culture cells (Ueki and Sitiro, 1971; Ueki, 1978). These phosphatases may increase the level of available Pi by mobilizing Pi from phytic acid and inositol phosphates which can accumulate in the vacuole, facilitating the use of external organic phosphates, and recapturing Pi that leaks from cells in the form of phosphate esters. Acid phosphatases are also the predominant enzymes in *Saccharomyces cerevisiae* that scavenge Pi from Pi esters in the environment (Bostian et al., 1980; Lemire et al., 1985; Bun-ya et al., 1991). Several genes encoding acid phosphatases have been identified in plants (e.g. Baldwin et al., 2001). A number of genes encoding putative proteins with similarity to *Arabidopsis* purple acid phosphatases are found in the *Chlamydomonas* genome (see Table 6.1), but only two have been confirmed to be activated during P starvation (see section IV.C.2).

Ribonucleases (RNases) are also important for acclimation to low Pi availability. These enzymes function in the mobilization of Pi from both extracellular and intracellular RNA pools. The death and decay of living matter can make RNA an important P-containing soil component, while the turnover of intracellular RNAs (e.g. rRNA) could serve a major role during conditions of both P limitation and cell senescence (Green, 1994). In tomato at least four intracellular "S-like" RNases are induced upon P starvation; three of these are vacuolar and probably involved in recycling Pi from cellular RNA while another is targeted to the endoplasmic reticulum (Löffler et al., 1992). Nürnberger et al. (1990) identified a periplasmic RNase of tomato that was specifically synthesized during P limitation. RNase genes of *Arabidopsis* have also been characterized (Taylor and Green, 1991; Taylor et al., 1993; Bariola et al., 1994).

B. Pi transport

Pi transport has been characterized in *Chlamydomonas* cells grown in replete medium or medium devoid of P (Shimogawara et al., 1999). Upon

P starvation the V_{max} for Pi transport increases by approximately 10-fold. Two distinct kinetic components are involved in Pi uptake; the K_m for one (low affinity) is approximately 10 μM while that for the other (high affinity) is between 0.1 and 0.3 μM. Under nutrient-replete conditions the low affinity component accounts for approximately 80% of total Pi uptake, but after 24 hours of P starvation essentially all uptake occurs via the high affinity system.

P limitation can also elevate the V_{max} (but not necessarily alter the K_m) for Pi uptake into plant roots via an increase in high affinity transport (e.g. Jungk et al., 1990). The uptake of Pi by plants is a metabolically driven process and the Pi concentration in the xylem sap can be 400 times that of the soil solution. The main form of Pi transported into plants is the $H_2PO_4^-$ ion, which increases in soils as the pH drops below neutral. A number of genes encoding plant Pi transporters have been isolated and characterized (reviewed by Raghothama, 2000). However, it should be kept in mind that Pi uptake into root cortical cells of many plants is facilitated by mycorrhizal associations, as mentioned in the Introduction (Alloush et al., 2000; Yao et al., 2001).

Chlamydomonas resembles *S. cerevisiae* in containing genes encoding both H^+/PO_4^{3-} (plant-like) and the Na^+/PO_4^{3-} (animal-like) high-affinity Pi transporters (*PHO84* and *PHO89*, respectively). However, unlike the situation in *Chlamydomonas* (see section IV.C.3), in *S. cerevisiae* both gene types are derepressed during P limitation and contribute to high-affinity Pi uptake (Ueda and Oshima, 1975; Tamai et al., 1985; Bun-ya et al., 1991; Martinez and Persson, 1998; Persson et al., 1998). *Chlamydomonas* also contains a gene (*PTC1*) encoding a homologue of the yeast putative low-affinity Pi transporters PHO87, PHO90, and PHO91 (see Table 6.1; Wykoff and O'Shea, 2001). There is some evidence that the yeast proteins act as Pi sensors, since the knockout strains display constitutive activation of the PHO regulon without any noticeable defects in Pi uptake (Auesukaree et al., 2003; Pinson et al., 2004). The yeast and *Chlamydomonas* homologues all contain N-terminal SPX-domains, which may interact with the beta-subunit of G-proteins and could be involved in signal transduction. A similar domain is found in the yeast Pho81p cyclin-dependent kinase repressor (see section IV.C; Spain et al., 1995; Lenburg and O'Shea, 1996; Battini et al., 1999).

III. P STORAGE, MOBILIZATION, AND CONSERVATION

A. Polyphosphate synthesis and mobilization

Chlamydomonas cells contain very little free Pi, and most of what is present resides in the chloroplast (Hebeler et al., 1992). The major storage form of P in *Chlamydomonas* is polyphosphate, a linear polymer of orthophosphate (Hebeler et al., 1992; Siderius et al., 1996). These polymers, which can be chains containing hundreds of Pi residues, function in several cellular processes including buffering the cytosolic pH, osmoregulation (Weiss et al., 1991; Leitao et al., 1995), and modulation of translation (McInerney et al., 2006). The majority of the polyphosphate in *Chlamydomonas* is complexed with calcium, magnesium, and zinc in electron-dense bodies in the cytosol (Siderius et al., 1996; Komine et al., 2000; Ruiz et al., 2001). Polyphosphate is also found in the cell wall, and its abundance changes during the cell cycle, peaking during cytokinesis (Werner et al., 2007).

Hebeler et al. (1992) used ^{31}P NMR (Nuclear Magnetic Resonance) in a preliminary study to demonstrate that the vast majority of the Pi taken up by P-starved *Chlamydomonas* cells upon re-addition of Pi to the medium is rapidly incorporated into polyphosphate. When *Chlamydomonas* cells are starved for P, their polyphosphate levels rapidly decline to below the detection limit, and ATP, Pi, and sugar phosphate levels also significantly decline. Very little is known about the biochemistry of polyphosphate synthesis in *Chlamydomonas* beyond the observation that the alga possesses an actin-related polyphosphate kinase, with similar characteristics to the activity characterized in *Neurospora crassa* (Gomez-Garcia and Kornberg, 2004). *Chlamydomonas* also possesses homologues of genes encoding vacuolar transport chaperones; these proteins have been implicated in the maintenance of normal polyphosphate levels in yeast (Ogawa et al., 2000; Auesukaree et al., 2004). In *Chlamydomonas*, proton-pumping pyrophosphatase and a H^+/ATPase are associated with polyphosphate bodies described as "electron-dense vacuoles," and may be involved in maintaining an acidic environment within these structures (Ruiz et al., 2001).

B. Nucleic acids

Acclimation to P deprivation involves reallocation of internal P pools. Significant pools are present in nucleic acids of the cytoplasmic and chloroplast compartments. Chloroplast transcripts increase in abundance during P starvation along with down-regulation of chloroplast polynucleotide phosphorylase (PNPase; Yehudai-Resheff et al., 2007), an enzyme critical for the degradation of chloroplast RNA (cpRNA) molecules (Kudla et al., 1996; Lisitsky et al., 1997; Nishimura et al., 2004). In P-replete cells, PNPase catalyzes polymerization of poly(A) tails onto fragments of cpRNA generated by endonucleolytic cleavage. The PNPase then binds the poly(A) sequences and degrades the entire RNA fragment; the poly(A) tails destabilize cpRNA (Slomovic et al., 2006), in contrast to the situation for mRNA synthesized from the nuclear genome where 3′ polyadenylation stabilizes transcripts. However, breakdown of RNA molecules by PNPase is phosphorylytic rather than hydrolytic, and thus consumes Pi.

Under P starvation conditions the abundance of the PNPase protein is reduced fourfold, leading to increased stability and accumulation of cpRNA

(Yehudai-Resheff et al., 2007). Pi that might have been utilized for cpRNA degradation is therefore conserved for other essential processes in the chloroplast, such as ATP synthesis via cyclic photosynthetic electron transfer. The down-regulation of PNPase is controlled at least in part by the transcription factor PSR1 (see section IV.B and Figure 6.5). In the *psr1* mutant the PNPase protein levels do not decline during P starvation, and the abundance of cpRNA molecules becomes even lower than is observed in P-replete, wild-type cells (Yehudai-Resheff et al., 2007). The physiological importance of the chloroplast PNPase is demonstrated by the rapid loss of viability during P starvation of strains of *Chlamydomonas* in which RNA interference or antisense transcription was used to reduce PNPase activity (Yehudai-Resheff et al., 2007).

In contrast to the accumulation of cpRNA during P deprivation, the level of chloroplast DNA decreases (Yehudai-Resheff et al., 2007). Chloroplasts from P-replete *Chlamydomonas* cells contain 80–100 copies of the 203 kb, circular plastid genome, but only a few copies are essential to maintain gene expression (Eberhard et al., 2002). The decrease in cpDNA copy number may reflect degradation and mobilization of the DNA, or reduced synthesis as the cells divide. In either case, this would allow the cells to conserve P for essential processes that accompany P deprivation.

C. Phospholipids

In addition to activating scavenging mechanisms, P starvation of *Chlamydomonas* elicits metabolic adjustments that appear to reduce the cellular requirement for P. An extensive study of the glycerolipid composition of *Chlamydomonas* (Riekhof et al., 2003; see also Chapter 2) demonstrated that unlike other model systems, *Chlamydomonas* lacks phosphatidylcholine, and the phospholipids identified in order of relative abundance are phosphatidylglycerol, phosphatidylethanolamine, and phosphatidylinositol. The amounts of these phospholipids decrease significantly in P-starved cells, with more than a 50% reduction observed for phosphatidylglycerol. Concomitantly there is a greater than twofold increase in the level of sulfoquinovosyldiacylglycerol (SQDG). *Chlamydomonas* also synthesizes a novel sulfolipid, 2′-*O*-acyl-sulfoquinovosyldiacylglycerol (ASQD), which contains a higher ratio of unsaturated to saturated fatty acids than SQDG. Although the absolute abundance of ASQD was not reported, fractionation experiments with ^{35}S-labeled lipids demonstrated a constant ASQD:SQDG ratio of 1:20 in P-replete and P-deprived cells, suggesting that the amount of ASQD increases in concert with SQDG during P deprivation. An approximately 25% increase in digalactosyldiacylglycerol in P-starved cells is also observed.

The replacement of phospholipids with sulfolipids during P deprivation preserves the anionic lipid character of the membranes (the substitution maintains the negatively charged lipid–water interface) and enables conservation and productive use of P that does become available. The physiological relevance of this substitution is supported by the finding that the *sqd1* mutant of *Chlamydomonas* exhibits poor growth in P-deficient medium. SQD1 catalyzes the synthesis of UDP-sulfoquinovose from UDP-glucose and sulfite, and the mutant that is unable to make this protein is completely devoid of sulfolipids. Sulfolipids are also important for proper photosynthetic function in P-replete cells, as *sqd1* and another sulfolipid-deficient mutant, *hf-2*, are much more susceptible to the herbicide DCMU than wild-type cells (Minoda et al., 2002; Riekhof et al., 2003). Furthermore, the *hf-2* mutant has been shown to have impaired photoautotrophic growth and photosystem (PS) II activity even under nutrient-replete conditions (Sato et al., 1995a,b; Minoda et al., 2002).

IV. GENETIC CONTROL OF PHOSPHORUS DEPRIVATION RESPONSES

A. Mutants with altered responses to P deprivation

In spite of the identification of mutants and the considerable expenditure of effort focused on defining P deprivation responses of *Chlamydomonas*, our understanding is still limited. Mutants with aberrant control of these responses have been designated *psr* (phosphorus starvation response) and *lpb* (low phosphate bleaching) (Shimogawara et al., 1999; Chang et al., 2005). The *psr1-1* and *psr1-2* mutants exhibit very low levels of extracellular phosphatase activity upon P starvation, while the *psr2* mutant exhibits constitutive extracellular phosphatase (Shimogawara et al., 1999). The *lpb1-1* and *lpb1-2* mutants bleach much more rapidly than wild-type cells during P limitation (Chang et al., 2005).

B. PSR1

The *PSR1* locus was identified by transforming the *psr1* mutant with a *Chlamydomonas* indexed cosmid library and selecting for growth on medium in which Pi was replaced with glucose-1-phosphate (G1P). Because *psr1* is unable to accumulate extracellular phosphatases in response to P starvation, it could not access esterified Pi and therefore could not grow on G1P. In contrast, wild-type cells grow similarly whether G1P is the sole P source or present along with Pi. This discrimination facilitated identification of *PSR1* (Wykoff et al., 1999).

The *PSR1* gene encodes a protein of 762 amino acids with a number of domains characteristic of transcription regulators (Figure 6.2). The region between residues 187 and 240 is related to the DNA-binding domain of MYB proteins (Lipsick, 1996), suggesting that PSR1 is directly responsible for transcriptional activation of P starvation-responsive genes. Similar

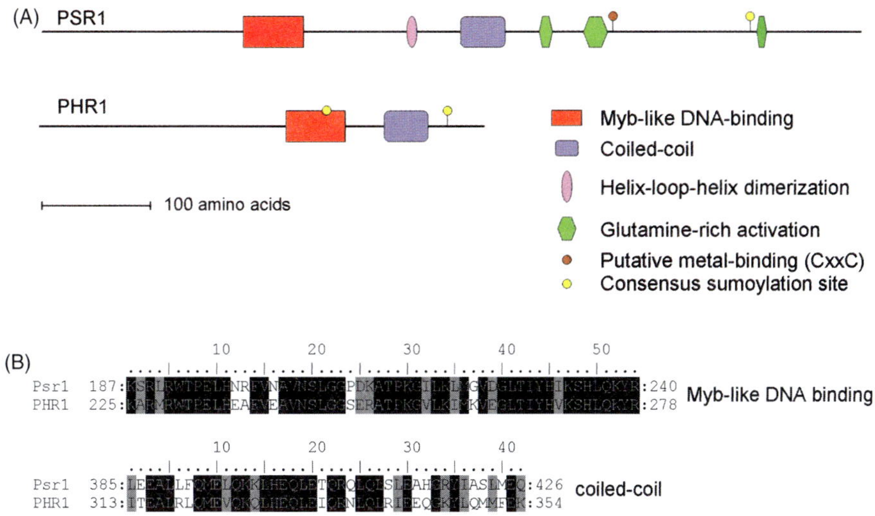

FIGURE 6.2 *The sequence of PSR1 and the domains of interest – comparison with PHR1. (A) To-scale line drawings comparing the protein domains of Chlamydomonas PSR1 (NCBI accession AAD55941) and Arabidopsis PHR1 (NCBI accession CAC59689). (B) CLUSTALW alignments of the sequences of the Myb-like DNA binding and coiled-coil domains of PSR1 and PHR1. Identical residues are shaded black with white letters while functionally similar amino acids are shaded grey with black letters.*

domains are present in CCA1, an *Arabidopsis* transcription factor that is responsible for phytochrome-mediated regulation of light harvesting genes (Wang et al., 1997), as well as in PHR1 (PHosphate starvation Response 1), the P-responsive regulator in *Arabidopsis* that was discovered in a reporter gene screen for mutants altered in the control of genes normally activated during P deprivation (Rubio et al., 2001). A *phr1* mutant cannot activate P starvation-responsive genes, including those encoding Pi transporters, phosphatases, and RNases (Rubio et al., 2001; Bari et al., 2006). Like PSR1, PHR1 is localized in the nucleus. Furthermore, PHR1 can bind to an imperfect palindromic DNA sequence that is conserved in promoters of P starvation-responsive genes.

Another domain common to PHR1, PSR1, and eukaryotic transcription factors is a coiled-coil domain; this domain is thought to promote protein dimerization and is required for sequence-specific binding of CCA1 and PHR1 to DNA (Rubio et al., 2001). Based on these features, *Chlamydomonas* PSR1 can be placed into the MYB-CC subfamily. However, in contrast to PHR1, PSR1 also contains three glutamine-rich regions, which are critical for the positive regulatory activity of eukaryotic transcription factors (Pabo and Sauer, 1992). Homologues of PHR1 in other vascular plants contain similar DNA-binding and coiled-coil domains, but all of the plant proteins lack the putative transcriptional activation domains present in the carboxy-terminus of the *Chlamydomonas* protein, raising the question of whether a co-activator protein is involved in plant P deprivation-responsive gene expression.

The *PSR1* transcript level increases when *Chlamydomonas* is exposed to 4–8 hours of P starvation; in some experiments after 24 hours of starvation, the transcript declined to just above the level observed in nutrient-replete cells. Elevation of *PSR1* transcript abundance precedes the induction of a number of other P starvation responses, including the development of high-affinity Pi transport and the secretion of phosphatase activity (Quisel et al., 1996; Shimogawara et al., 1999). PSR1 protein levels have been quantified using an epitope-tagged version, and found to increase over 10-fold after 1 day of P starvation (Wykoff et al., 1999), which correlates with elevated accumulation of the encoding transcript. Immunofluorescence analysis of cells expressing PSR1::3XHA revealed a nuclear localization regardless of P status (Wykoff et al., 1999).

While PSR1 does not contain a recognizable nuclear localization sequence, it does have a helix-loop-helix dimerization motif that has been associated with nuclear import (Nagoshi et al., 1999). Interactions of PSR1 with specific nuclear-targeted proteins may afford it passage into the nucleus. Since the same low percentage (1–5%) of PSR1 is observed in the cytoplasm of starved and nonstarved cells, nuclear translocation is unlikely to be a key regulatory event in the control of the P starvation responses. This contrasts with the situation in yeast, where under P-replete conditions the cyclin PHO80 and the cyclin-dependent kinase PHO85 form a complex that phosphorylates the PHO4 transcription factor at multiple sites, preventing nuclear import and transcription of the P deficiency-response genes (Yoshida et al., 1989; Okada and Toh-e, 1992; Komeili and O'Shea, 1999; Thomas and O'Shea, 2005).

Little is known about the regulatory molecules that function upstream of PSR1 in sensing and responding to P deficiency. Sumoylation of *Arabidopsis* PHR1 was demonstrated *in vitro*, suggesting that post-translational processes control PHR1 activity (Miura et al., 2005). Furthermore, the *Arabidopsis siz1* mutant, which lacks the SIZ1 E3 SUMO ligase, has a P starvation root morphology, exhibits increased expression of some P starvation-responsive genes in P-replete plants, and is unable to induce expression of other target genes during P deprivation (Miura et al., 2005). Nevertheless, *in vivo* sumoylation of PHR1 has not been demonstrated, and the phenotype of *siz1* does not preclude an indirect effect on PHR1-dependent gene expression (Catala et al., 2007). PSR1 lacks the consensus sumoylation sites that are present in PHR1, which themselves are poorly conserved among PHR1 plant homologues.

It is not clear how much conservation exists between the *Chlamydomonas* and vascular plant P deficiency signal transduction systems beyond the similarities noted between PSR1 and PHR1. A putative pathway for inhibition of PHR1-dependent gene expression has been demonstrated, where PHR1 is required for the induction of the microRNA transcript miR399 during

P starvation of *Arabidopsis* and other plants (Bari et al., 2006). This miRNA down-regulates the abundance of the *PHO2* mRNA by interacting with complementary sequences that are present in the *PHO2* 5′ untranslated region (Fujii et al., 2005; Chiou et al., 2006). *PHO2* encodes an E2 ubiquitin conjugase which is highly expressed under P-replete conditions and represses expression of P starvation-inducible genes. Accordingly, *pho2* mutants have elevated levels of leaf Pi and constitutively high expression of PHR1-dependent, P starvation-induced genes, including Pi transporters and phosphatases (Delhaize and Randall, 1995; Dong et al., 1998; Bari et al., 2006). Grafting experiments demonstrate that *pho2* mutant roots are necessary and sufficient to cause increased shoot Pi accumulation (Bari et al., 2006). Conversely, grafted shoots constitutively expressing miR399 down-regulate *PHO2* in wild-type roots, demonstrating that the microRNA is transported through the phloem and acts systemically to control P homeostasis (Pant et al., 2008).

Although no miRNAs with sequence similarity to miR399 have been found yet in *Chlamydomonas*, the alga expresses a wide variety of miRNAs, and at least one (candidate 161 in Zhao et al.) has potential target sequences in the coding regions of the genes specifying the putative Pi transporters PTB3 and PTA2 (Molnar et al., 2007; Zhao et al., 2007). Overall orthology between plants and *Chlamydomonas* would be to some extent surprising given the very different issues associated with Pi uptake, transport and distribution in a vascular plant and a unicellular organism.

C. PSR1-dependent, P deficiency-responsive gene expression

1. Potential PSR1 targets

Although sequence-specific DNA binding has not been demonstrated for PSR1, potential targets have been identified based on altered transcript abundance in P-deficient relative to P-replete cells, and loss of differential gene expression in *psr1* mutant strains (Moseley et al., 2006). Table 6.1 provides a list of genes that are potentially regulated by PSR1 and/or are involved in P metabolism. As expected, PSR1 is required for activation of genes involved in Pi assimilation during P starvation.

2. Phosphatases

The mRNA for the *Chlamydomonas PHOX* homologue, which encodes the major inducible extracellular phosphatase, increases by >5000-fold in cells starved for 24 hours, correlating with the accumulation of an abundant 190 kD extracellular polypeptide. Furthermore, neither the transcript nor the 190-kD protein is induced in the *psr1* mutant (Moseley et al., 2006). Another transcript encoding a putative phosphatase with some similarity to *Arabidopsis* purple acid phosphatases exhibits a marked increase in P-starved cells (Moseley et al., 2006) and may be targeted to the secretory pathway, based on TargetP analysis. However, it has not been determined whether this gene encodes either of the minor inducible extracellular phosphatases that were biochemically characterized (Quisel et al., 1996; see section II.A).

3. Transporters

High-affinity Pi uptake that develops following P deprivation in *Chlamydomonas* appears to involve expression changes for genes encoding putative Pi transporters. Ten homologues of the yeast high-affinity Na^+/Pi symporter PHO89 are encoded in *Chlamydomonas* (Figure 6.3), and the transcripts for at least four of these Pi transporter type B proteins (PTB) increase in abundance during Pi starvation and are regulated by PSR1 (Kobayashi et al., 2003; Moseley et al., 2006). The expansion of this gene family appears to be recent, as six of the *PTB* genes are found in two clusters (Moseley et al., 2006). An additional four Pi transporter type A polypeptides (PTA; Figure 6.3A) are encoded by genes homologous to *PHO84*, the yeast gene encoding the high-affinity H^+/Pi cotransporter.

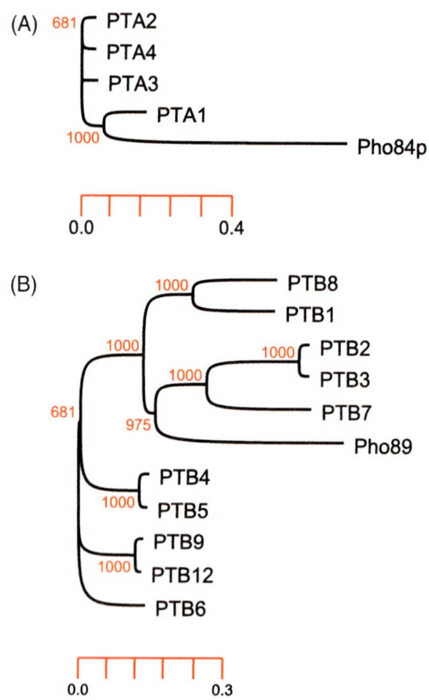

FIGURE 6.3 *Chlamydomonas Type A and Type B Pi transporters. (A) Unrooted, neighbor-joining tree showing the evolutionary relationships between the four Chlamydomonas putative H^+/Pi co-transporters (PTAs) and the S. cerevisiae homolog Pho84p. (B) Unrooted, neighbor-joining tree showing the evolutionary relationships between the 10 Chlamydomonas putative Na^+/Pi co-transporters (PTBs) and the S. cerevisiae homologue Pho89.*

Of the four *PTA* genes, only the *PTA4* transcript increases during P deprivation; this increase does not depend entirely on PSR1. The abundance of *PTA2* and *PTA3* transcripts is not affected significantly by P deprivation. Instead, *PTA3* responds to Mn starvation, with transcripts increasing 25-fold, while *PTA4* transcripts increase 5-fold, suggesting a possible role in Mn uptake (Allen et al., 2007). Transporters in this class have been shown to function as Mn^{2+}/PO_4^{3-} co-transporters (Fristedt et al., 1999; Jensen et al., 2003), so the upregulation of *PTA4* during P deprivation may reflect the reduced rate of uptake of the Mn^{2+} counter-ion. Indeed, Mn-deficient cells have reduced intracellular P content, although they show no signs of increased P deficiency-responsive gene expression (Allen et al., 2007).

The *PTA1* transcript declines by ~1000-fold following exposure of nutrient-replete cells to P starvation, and this decline does not occur in a *psr1* mutant (Moseley et al., 2006). This characteristic makes PTA1 a candidate for the low-affinity Pi uptake system that operates in cells that are P-replete. Alternatively, changing *PTA1* and *PTB* transcript levels might result from alterations in the transport of Pi into and out of specific intracellular compartments. At present nothing is known about the localization of the Pi transporter gene products; this is an area that is ripe for detailed investigation.

4. Other genes

A number of additional genes related to P metabolism are differentially regulated during P starvation and are likely to be under the control of PSR1. The *CAX1* transcript encoding a putative H^+/Ca^{2+} antiporter (Hanikenne et al., 2005) is upregulated in P-starved cells and may be involved in regulating intracellular Ca^{2+} levels when Ca^{2+} is released from polyphosphate bodies (Moseley et al., 2006; also see Chapter 10). P deficiency also activates expression of *VTC1*, which encodes a homologue of yeast vacuolar transport chaperones which have been associated with polyphosphate metabolism (Ogawa et al., 2000; Auesukaree et al., 2004). Down-regulation of the *PNP* gene encoding the chloroplast PNPase (section III.B) is likely under PSR1 control since this is not observed in *psr1* mutants (Yehudai-Resheff et al., 2007).

5. "Electron valves"

In addition to controlling cellular responses that promote the scavenging of extracellular, free and esterified Pi, PSR1 may also be involved in regulating a class of genes associated with "electron valve" function (Figure 6.4). These functions promote the dissipation of excitation energy from photosynthetic and respiratory electron transfer chains, presumably to prevent the production of harmful reactive oxygen species under conditions that limit cellular growth and division. For example, transcripts from genes encoding plastid

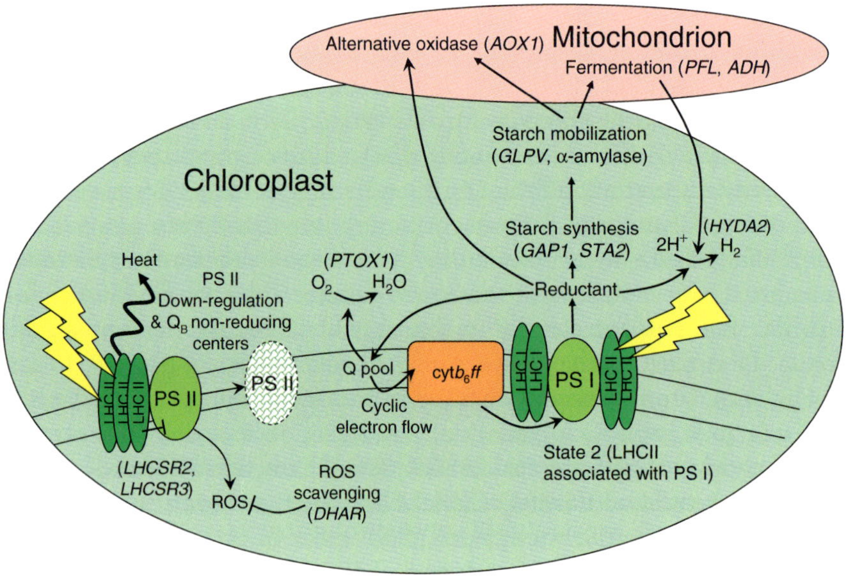

FIGURE 6.4 *Mechanisms to reduce photosynthetic excitation pressure during P starvation. Cp, chloroplast; M, mitochondrion. Genes that are upregulated in P-starved cells are shown in parentheses. LHCSR2 and LHCSR3 encode light-harvesting complex stress-related proteins that may help dissipate excess absorbed excitation energy. DHAR encodes a dehydroascorbate reductase that functions in the pathway that recycles ascorbate. PTOX1 encodes a plastoquinol terminal oxidase that is located in the chloroplast and that may function to remove excess electrons from the plastoquinol pool. GAP1 encodes a plastid localized glyceraldehyde-3-phosphate dehydrogenase. STA2 encodes a starch synthase. HYDA2 encodes an iron hydrogenase which uses electrons from PS I or from fermentation to synthesize H_2. GPLV encodes a starch phosphorylase. AOX1 encodes a mitochondrial alternative oxidase. PFL encodes pyruvate-formate lyase and ADH encodes an alcohol dehydrogenase, both of which are involved in fermentation.*

and mitochondrial alternative oxidases increase in wild-type, P-deficient cells, but not in the *psr1* mutant. Increased transcript abundance is observed for genes encoding enzymes that function in starch synthesis and degradation and fermentation metabolism (pyruvate-formate lyase and alcohol dehydrogenase), possibly reflecting a decline in the oxygen level of the cultures during P deprivation. Also, a small but reproducible elevation in the level of the *HYDA2* transcript (but not that of *HYDA1*), encoding a *Chlamydomonas* Fe-hydrogenase (see Chapter 7), is observed in P-starved cells (Moseley et al., 2006). Electrons for H_2 production can come from fermentation (Atteia et al., 2006) or from ferredoxin reduced by PS I (Benemann et al., 1973).

Activation of genes encoding electron valve functions may reflect an increased need to "bleed off" electrons from the photosynthetic electron transport chain under conditions in which the rate of linear electron flow is limited by the consumption of reducing equivalents, resulting from decreased CO_2 fixation and a decline in other anabolic activities (e.g. reduction of sulfate, phosphate, or nitrite). The inability of *psr1* mutants to increase levels of transcripts for electron valve polypeptides during P starvation may reflect a direct interaction between PSR1 and the specific gene promoters or alternatively, the mutants may develop a different metabolic homeostasis as a consequence of their inability to properly acclimate to P starvation conditions. It is clear that changes in the physiological state of mutant cells are different from those that occur in wild-type cells since a number of "stress" genes are uniquely activated in P-deprived *psr1*.

D. Low photochemical bleaching mutants

The *lpb1* mutants unmasked a component of the acclimation response to P starvation that is distinct from those controlled by PSR1. Although P deficiency-responsive gene expression occurs normally in P-starved *lpb1* cells, the cultures rapidly bleach and die, suggesting that the mutant is unable to properly acclimate to P starvation. The *lpb1* mutant could be complemented by the *LPB1* gene, which encodes a 1064 amino acid protein possessing a domain with limited similarity to nucleotide-diphospho-sugar transferases, and a "P-loop" nucleotide-binding domain common to nucleotide triphosphate hydrolases (Chang et al., 2005). LPB1 homologues are encoded in other green algae and vascular plants, where they are predicted to be chloroplast-targeted, and in acanthamoebae. The precise function of LPB1 has not been determined, although an evident possibility based on its sequence is a role in sugar or polysaccharide metabolism (Chang et al., 2005).

V. P DEFICIENCY AND PHOTOSYNTHESIS

Wykoff et al. (1998) demonstrated that *Chlamydomonas* undergoes a suite of responses during P deprivation that result in the down-regulation of photosynthesis, which would prevent photodamage under conditions in which the absorbed light energy cannot be used for growth. S starvation induces the same set of responses albeit much more rapidly (see Chapter 5), reflecting the limited availability of internal S compared to P. The efficiency of light energy utilization is severely reduced after 4 days of P starvation, and the maximal O_2 evolution rate declines to one quarter of the initial rate. Examination of partial reactions of photosynthetic electron transport demonstrated that PS II was the site at which photosynthesis was inhibited during P deprivation since electron transport through the cyt b_6f and PS I complexes was unaffected (Figure 6.4). Instead, the decline in photosynthetic yield is largely attributable to an approximately 30% decrease in the amount of functional PS II as determined by quantification of D1 protein abundance and functional Q_A levels, and an increase in the proportion of PS II centers that are not competent for electron transfer from Q_A to Q_B (Q_B non-reducing centers; Figure 6.4).

In addition to the down-regulation of PS II, P-starved cells induce non-photochemical quenching to reduce the quantum efficiency of photosynthesis, even at low light intensities. Activation of the xanthophyll cycle in P-starved cells leads to increased dissipation of absorbed light energy as heat, and

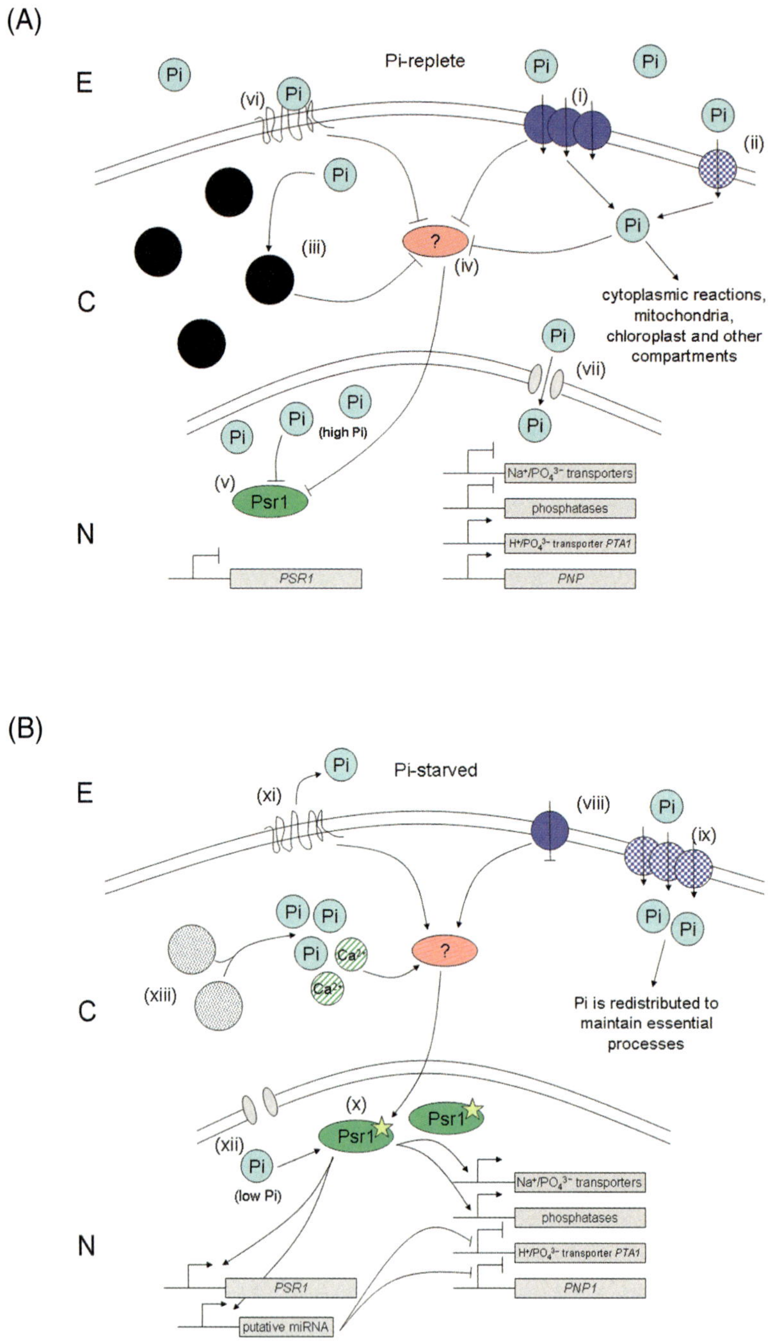

movement of the light-harvesting apparatus from PS II to PS I (state transition; cells are going from state 1 to state 2) diverts absorption away from PS II (the PS I reaction center can act as a quencher; Figure 6.4). Furthermore, cyclic electron flow around PS I appears to be promoted during P stress, presumably to maintain the ΔpH necessary for non-photochemical quenching, and to enable production of ATP for the maintenance of cellular processes (Wykoff et al., 1998). State transitions and non-photochemical quenching are discussed in more detail in Chapters 22 and 23, respectively.

VI. REGULATORY MODEL FOR P DEPRIVATION RESPONSES

A working model to explain how PSR1 regulates Pi metabolism is presented in Figure 6.5. Under nutrient-replete conditions, PSR1 is present but the protein is kept in an inactive form through a signal transduction pathway controlled by external and/or internal sensors of the cellular P status. For example, the *PTC1* gene product (Figure 6.5(A; vi) may serve as a sensor of external Pi levels as is the case with the homologous proteins in yeast, and

FIGURE 6.5 *(A) In cells growing in Pi-replete medium, Pi-uptake from the external environment into the cytoplasm occurs mainly through a low-affinity system (i) A high affinity Pi-uptake system is present in low abundance (ii) Imported Pi participates in reactions that occur in the cytoplasm or is distributed to other cellular compartments. Excess Pi is incorporated into polyphosphate-rich acidocalcisomes (iii) Cytoplasmic Pi may inhibit a cytoplasmic signal transduction component (iv) that is required for activation of PSR1 (v) in the nucleus. Alternatively (or in addition), inhibitory signals might be transmitted by an external sensor of Pi (vi), from the acidocalcisomes, or from the low-affinity transporter system itself. Pi presumably enters the nucleus through nuclear pores (vii) and might play a role in repressing PSR1. Under Pi-replete conditions the PSR1, Na^+/PO_4^{3-} transporter and inducible phosphatase genes are expressed at basal levels, whereas the PTA1 and PNP genes are fully expressed. E, cell exterior; C, cytoplasm; N, nucleus. (B) When Pi becomes limiting in the growth medium the Pi uptake activity of the low affinity uptake system declines (viii), but transport through the high-affinity system is maintained (ix). The declining Pi levels in the cell and/or transport of Pi exclusively through the high-affinity system may promote a signal that is transmitted to the nucleus and enables PSR1 to activate transcription (x). PSR1 mRNA abundance increases within 1 hour of transfer of cells to Pi-deficient medium, and this increased expression precedes the activation of the Na^+/PO_4^{3-} transporter and inducible phosphatase genes and the repression of PTA1 and PNP. Rather than acting both as an activator and a repressor of transcription, PSR1 might control PTA1 and PNP through expression of an inhibitory miRNA that reduces the abundance of those transcripts. Temporal difference in the expression of genes following the imposition of P deprivation may reflect the affinity of the individual promoters for PSR1. An alternative model is that the signal to activate PSR1 is transmitted when Pi dissociates from an external sensor (xi). PSR1 might also respond to changes in the nuclear Pi concentration (xii). Cytoplasmic poly(P) is degraded and buffers the cytoplasmic Pi concentration until the poly(P) stores are depleted (xiii). Ca^{2+} ions released from the polyphosphate bodies might provide an internal P starvation signal leading to PSR1 activation.*

might through unknown intermediates (iv) negatively regulate PSR1 activity in the presence of high external Pi. Additionally or alternatively, PSR1 might be sensitive to levels of Pi in the nucleus, and/or to the activity of Pi transporters (ii).

During P starvation PSR1 is activated by a still-undefined mechanism, and temporally regulates the transcription of specific genes. One way to achieve such regulation is if different P starvation-activated promoters bind PSR1 with different affinities. Early in the acclimation process the *PSR1* gene promoter would be activated (although it has not yet been determined if *PSR1* is autoregulated), while the promoters of genes encoding components of the low affinity Pi uptake system would be repressed. As active PSR1 protein accumulates, it would bind the lower-affinity promoters (e.g. promoters of genes encoding Na^{+}/PO_4^{3-} transporters, inducible phosphatases and electron valve functions). After 24 hours of P deprivation, cell growth rates decline along with the level of the *PSR1* transcript and transcripts from Pi transporter and phosphatase genes. These declines may reflect other regulatory processes that dominate later stages of the acclimation process. For example, energy for transcribing and translating genes that promote nutrient scavenging and growth may be redirected towards metabolic systems that promote storage, conservation and long-term survival, such as the reduction in cpRNA degradation by PNPase and the replacement of phospholipids with other lipids (discussed above). The various mechanisms which down-regulate photosynthesis also appear to become activated subsequent to 24 hours of P starvation, after the peak in Pi transporter and phosphatase gene activation. Other models involving additional regulatory elements can also be envisioned. As in plants, miRNAs and proteolysis may play a role in regulating *PSR1*-dependent gene expression, but evidence for such pathways has not yet been discovered.

VII. CONCLUDING REMARKS

Genetic and molecular characteristics coupled with genome-wide analyses have made *Chlamydomonas* an excellent model for elucidating the ways in which photosynthetic eukaryotes sense and respond to their nutrient environment. With respect to acclimation to P deprivation, it is important to integrate biochemical, physiological, genetic, molecular, and genomic information to define the interactions of the key regulator PSR1 with the promoters of target genes and potential partner proteins, to elucidate mechanisms involved in the hierarchy of responses elicited by PSR1, and to understand the activity and function of LPB1. It is also critical that we define the full range of genes controlled by PSR1 and the functions of the polypeptides that they encode if we are to understand the ways in which photosynthetic organisms, whether microbe or plant, survive conditions that severely limit their growth potential and enhance the potentially toxic effects that may be triggered by the absorption of excess excitation energy and the concomitant generation of reactive oxygen species.

ACKNOWLEDGMENTS

We thank David Gonzalez-Ballester and Wirulda Pootakham for helpful suggestions and critical reading of the manuscript. Some of this work was supported by U.S. Department of Agriculture grant 2002-35301-12178 and National Science Foundation grant MCB 0235878, both awarded to A.R.G.

REFERENCES

Allen, M.D., Kropat, J., Tottey, S., Del Campo, J.A., and Merchant, S.S. (2007). Manganese deficiency in *Chlamydomonas* results in loss of photosystem II and MnSOD function, sensitivity to peroxides, and secondary phosphorus and iron deficiency. *Plant Physiol.* **143**, 263–277.

Alloush, G.A.Z., Zeto, S.K., and Clark, R.B. (2000). Phosphorus source, organic matter, and arbuscular mycorrhiza effects on growth and mineral acquisition of chickpea grown in acidic soil. *J .Plant Nutr.* **23**, 1351–1369.

Atteia, A., van Lis, R., Gelius-Dietrich, G., Adrait, A., Garin, J., Joyard, J., Rolland, N., and Martin, W. (2006). Pyruvate formate-lyase and a novel route of eukaryotic ATP synthesis in *Chlamydomonas* mitochondria. *J. Biol. Chem.* **281**, 9909–9918.

Auesukaree, C., Homma, T., Kaneko, Y., and Harashima, S. (2003). Transcriptional regulation of phosphate-responsive genes in low-affinity phosphate-transporter-defective mutants in *Saccharomyces cerevisiae*. *Biochem. Biophys. Res. Commun.* **306**, 843–850.

Auesukaree, C., Homma, T., Tochio, H., Shirakawa, M., Kaneko, Y., and Harashima, S. (2004). Intracellular phosphate serves as a signal for the regulation of the PHO pathway in *Saccharomyces cerevisiae*. *J. Biol. Chem.* **279**, 17289–17294.

Baldwin, J.C., Karthikeyan, A.S., and Raghothama, K.G. (2001). Leps2, a phosphorus starvation-induced novel acid phosphatase from tomato. *Plant Physiol.* **125**, 728–737.

Bari, R., Datt Pant, B., Stitt, M., and Scheible, W.-R. (2006). PHO2, MicroRNA399, and PHR1 define a phosphate-signaling pathway in plants. *Plant Physiol.* **141**, 988–999.

Bariola, P.A., Howard, C.J., Taylor, C.B., Verburg, M.T., Jaglan, V.D., and Green, P.J. (1994). The *Arabidopsis* ribonuclease gene *RNS1* is tightly controlled in response to phosphate limitation. *Plant J.* **6**, 673–685.

Battini, J.-L., Rasko, J.E.J., and Miller, A.D. (1999). A human cell-surface receptor for xenotropic and polytropic murine leukemia viruses: Possible role in G protein-coupled signal transduction. *Proc. Natl. Acad. Sci. U. S. A.* **96**, 1385–1390.

Benemann, J.R., Berenson, J.A., Kaplan, N.O., and Kamen, M.D. (1973). Hydrogen evolution by a chloroplast-ferredoxin-hydrogenase system. *Proc. Natl. Acad. Sci. U. S. A.* **70**, 2317–2320.

Bostian, K.A., Lemire, J.M., Cannon, L.E., and Halvorson, H.O. (1980). *In vitro* synthesis of repressible yeast acid phosphatase: Identification of multiple mRNAs and products. *Proc. Natl. Acad. Sci. U. S. A.* **77**, 4504–4508.

Bun-ya, M., Nishimura, M., Harashima, S., and Oshima, Y. (1991). The *PHO84* gene of *Saccharomyces cerevisiae* encodes an inorganic phosphate transporter. *Mol. Cell. Biol.* **11**, 3229–3238.

Catala, R., Ouyang, J., Abreu, I.A., Hu, Y., Seo, H., Zhang, X., and Chua, N.-H. (2007). The *Arabidopsis* E3 SUMO ligase SIZ1regulates plant growth and drought responses. *Plant Cell* **19**, 2952–2966.

Chang, C.-W., Moseley, J.L., Wykoff, D., and Grossman, A.R. (2005). The *LPB1* gene is important for acclimation of *Chlamydomonas reinhardtii* to phosphorus and sulfur deprivation. *Plant Physiol.* **138**, 319–329.

Chiou, T.-J., Aung, K., Lin, S.-I., Wu, C.-C., Chiang, S.-F., and Su, C.-l. (2006). Regulation of phosphate homeostasis by microRNA in *Arabidopsis*. *Plant Cell* **18**, 412–421.

Davies, J. and Grossman, A.R. (1998). The use of *Chlamydomonas* as a model algal system for genome studies. *J. Phycol.* **34**, 907–917.

Delhaize, E. and Randall, P.J. (1995). Characterization of a phosphate-accumulator mutant of *Arabidopsis thaliana*. *Plant Physiol.* **107**, 207–213.

Dong, B., Rengel, Z., and Delhaize, E. (1998). Uptake and translocation of phosphate by *pho2* mutant and wild-type seedlings of *Arabidopsis thaliana*. *Planta* **205**, 251–256.

Dumont, F., Loppes, R., and Kremers, P. (1990). New polypeptides and *in-vitro*-translatable mRNAs are produced by phosphate-starved cells of the unicellular alga *Chlamydomonas reinhardtii*. *Planta* **182**, 610–616.

Eberhard, S., Drapier, D., and Wollman, F.-A. (2002). Searching limiting steps in the expression of chloroplast-encoded proteins: Relations between gene copy number, transcription, transcript abundance and translation rate in the chloroplast of *Chlamydomonas reinhardtii*. *Plant J.* **31**, 149–160.

Fristedt, U., van der Rest, M., Poolman, B., Konings, W.N., and Persson, B.L. (1999). Studies of cytochrome *c* oxidase-driven H^+-coupled phosphate transport catalyzed by the *Saccharomyces cerevisiae* Pho84 permease in coreconstituted vesicles. *Biochemisry* **38**, 16010–16015.

Fujii, H., Chiou, T.J., Lin, S.I., Aung, K., and Zhu, J.K. (2005). A miRNA involved in phosphate-starvation response in *Arabidopsis*. *Curr Biol* **15**, 2038–2043.

Gilbert, G.A., Knight, J.D., Vance, C.P., and Allen, D.L. (1999). Acid phosphatase activity in phosphorus-deficient white lupin roots. *Plant Cell Environ.* **22**, 801–810.

Gomez-Garcia, M.R. and Kornberg, A. (2004). Formation of an actin-like filament concurrent with the enzymatic synthesis of inorganic polyphosphate. *Proc. Natl. Acad. Sci. U. S. A.* **101**, 15876–15880.

Green, P.J. (1994). The ribonucleases of higher plants. *Annu. Rev. Plant Physiol. Plant Mol. Biol.* **45**, 421–445.

Hallmann, A. (1999). Enzymes in the extracellular matrix of *Volvox*: An inducible, calcium-dependent phosphatase with a modular composition. *J. Biol. Chem.* **274**, 1691–1697.

Hanikenne, M., Kramer, U., Demoulin, V., and Baurain, D. (2005). A comparative inventory of metal transporters in the green alga *Chlamydomonas reinhardtii* and the red alga *Cyanidioschizon merolae*. *Plant Physiol.* **137**, 428–446.

Hebeler, M., Hentrich, S., Mayer, A., Leibfritz, D., and Grimme, L.H. (1992). Phosphate regulation and compartmentation in *Chlamydomonas reinhardtii* studied by *in vivo* ^{31}P-NMR. In: *Research in Photosynthesis* (N. Murata, Ed.), Vol. 3, pp. 717–720. Kluwer Academic Publishers, Dordrecht.

Jensen, L.T., Ajua-Alemanji, M., and Culotta, V.C. (2003). The *Saccharomyces cerevisiae* high affinity phosphate transporter encoded by PHO84 also functions in manganese homeostasis. *J. Biol. Chem.* **278**, 42036–42040.

Jungk, A., Asher, C.J., Edwards, D.G., and Meyers, D. (1990). Influence of phosphate status on phosphate uptake kinetics of maize (*Zea mays*) and soybean (*Glycine max*). *Plant Soil* **124**, 175–182.

Karandashov, V. and Bucher, M. (2005). Symbiotic phosphate transport in arbuscular mycorrhizas. *Trends Plant Sci.* **10**, 22–29.

Kobayashi, I., Fujiwara, S., Shimogawara, K., Kaise, T., Usuda, H., and Tsuzuki, M. (2003). Insertional mutagenesis in a homologue of a Pi transporter gene confers arsenate resistance on *Chlamydomonas*. *Plant Cell Physiol.* **44**, 597–606.

Komeili, A. and O'Shea, E.K. (1999). Roles of phosphorylation sites in regulating activity of the transcription factor Pho4. *Science* **284**, 977–980.

Komine, Y., Eggink, L.L., Park, H., and Hoober, J.K. (2000). Vacuolar granules in *Chlamydomonas reinhardtii*: Polyphosphate and a 70-kDa polypeptide as major components. *Planta* **210**, 897–905.

Kudla, J., Hayes, R., and Gruissem, W. (1996). Polyadenylation accelerates degradation of chloroplast mRNA. *EMBO J.* **15**, 7137–7146.

Leitao, J.M., Lorenz, B., Bachinski, N., Wilhelm, C., Muller, W.E.G., and Schroder, H.C. (1995). Osmotic-stress-induced synthesis and degradation of inorganic polyphosphates in the alga *Phaeodactylum tricornutum*. *Mar. Ecol. Prog. Ser.* **121**, 279–288.

Lemire, J.M., Willcocks, T., Halvorson, H.O., and Bostian, K.A. (1985). Regulation of repressible acid phosphatase gene transcription in *Saccharomyces cerevisiae*. *Mol. Cell. Biol.* **5**, 2131–2141.

Lenburg, M.E. and O'Shea, E.K. (1996). Signaling phosphate starvation. *Trends Biochem. Sci.* **21**, 383–387.

Li, D., Zhu, H., Liu, K., Liu, X., Leggewie, G., Udvardi, M., and Wang, D. (2002). Purple acid phosphatases of *Arabidopsis thaliana*. Comparative analysis and differential regulation by phosphate deprivation. *J. Biol. Chem.* **277**, 27772–27781.

Lien, T. and Knutsen, G. (1973). Synchronous cultures of *Chlamydomonas reinhardtii*: Properties and regulation of repressible phosphatases. *Physiol. Plant.* **28**, 291–298.

Lipsick, J.S. (1996). One billion years of Myb. *Oncogene* **13**, 223–235.

Lisitsky, I., Kotler, A., and Schuster, G. (1997). The mechanism of preferential degradation of polyadenylated RNA in the chloroplast. The exoribonuclease 100 rnp/polynucleotide phosphorylase displays high binding affinity for poly(A) sequence. *J. Biol. Chem.* **272**, 17648–17653.

Löffler, A., Abel, S., Jost, W., and Glund, K. (1992). Phosphate-regulated induction of intracellular ribonucleases in cultured tomato (*Lycoperisicon esculentum*) cells. *Plant Physiol.* **98**, 1472–1478.

Loppes, R. (1976). Release of enzymes by normal and wall-free cells of *Chlamydomonas*. *J. Bacteriol.* **128**, 114–116.

Loppes, R. and Deltour, R. (1981). A pleiotropic mutant of *Chlamydomonas reinhardi* showing cell wall abnormalities and altered phosphatase activities. *Plant Sci. Lett.* **21**, 193–197.

Loppes, R. and Matagne, R.F. (1973). Acid phosphatase mutants in *Chlamydomonas*: Isolation and characterization by biochemical, electrophoretic and genetic analysis. *Genetics* **75**, 593–604.

Loppes, R., Braipson, J., Matagne, R.F., Sassen, A., and Ledoux, L. (1977). Regulation of the neutral phosphatase in *Chlamydomonas reinhardi*: An immunogenetic study of wild-type and mutant strains. *Biochem. Genet.* **15**, 1147–1157.

Martinez, P. and Persson, B.L. (1998). Identification, cloning and characterization of a derepressible Na^+-coupled phosphate transporter in *Saccharomyces cerevisiae*. *Mol. Gen. Genet.* **258**, 628–638.

Matagne, R.F. and Loppes, R. (1975). Isolation and study of mutants lacking a derepressible phosphatase in *Chlamydomonas reinhardi*. *Genetics* **80**, 239–250.

Matagne, R.F., Loppes, R., and Deltour, R. (1976). Phosphatases of *Chlamydomonas reinhardi*: Biochemical and cytochemical approach with specific mutants. *J. Bacteriol.* **126**, 937–950.

McInerney, P., Mizutani, T., and Shiba, T. (2006). Inorganic polyphosphate interacts with ribosomes and promotes translation fidelity *in vitro* and *in vivo*. *Mol. Microbiol.* **60**, 438–447.

Minoda, A., Sato, N., Nozaki, H., Okada, K., Takahashi, H., Sonoike, K., and Tsuzuki, M. (2002). Role of sulfoquinovosyl diacylglycerol for the maintenance of photosystem II in *Chlamydomonas reinhardtii*. *Eur. J. Biochem.* **269**, 2353–2358.

Miura, K., Rus, A., Sharkhuu, A., Yokoi, S., Karthikeyan, A.S., Raghothama, K. G., Baek, D., Koo, Y.D., Jin, J.B., Bressan, R.A., Yun, D-J., and Hasegawa, P.M. (2005). The *Arabidopsis* SUMO E3 ligase SIZ1 controls phosphate deficiency responses. *Proc. Natl. Acad. Sci. U. S. A.* **102**, 7760–7765.

Molnar, A., Schwach, F., Studholme, D.J., Thuenemann, E.C., and Baulcombe, D.C. (2007). miRNAs control gene expression in the single-cell alga *Chlamydomonas reinhardtii*. *Nature* **447**, 1126–1129.

Moseley, J.L., Chang, C.W., and Grossman, A.R. (2006). Genome-based approaches to understanding phosphorus deprivation responses and PSR1 control in *Chlamydomonas reinhardtii*. *Eukaryotic Cell* **5**, 26–44.

Nagoshi, E., Imamoto, N., Sato, R., and Yoneda, Y. (1999). Nuclear import of sterol regulatory element-binding protein-2, a basic helix-loop-helix-leucine zipper (bHLH-Zip)-containing transcription factor, occurs through the direct interaction of importinβ with HLH-Zip. *Mol. Biol. Cell.* **10**, 2221–2233.

Nakazato, H., Okamoto, T., Ishikawa, M., and Okuyama, H. (1997). Purification and characterization of phosphatase inducibly synthesized in *Spirodela oligorrhiza* grown under phosphate deficient conditions. *Plant Physiol. Biochem.* **35**, 437–446.

Nishimura, Y., Kikis, E.A., Zimmer, S.L., Komine, Y., and Stern, D.B. (2004). Antisense transcript and RNA processing alterations suppress instability of polyadenylated mRNA in *Chlamydomonas* chloroplasts. *Plant Cell* **16**, 2849–2869.

Nürnberger, T., Abel, S., Jost, W., and Glund, K. (1990). Induction of an extracellular ribonuclease in cultured tomato cells upon phosphate starvation. *Plant Physiol.* **92**, 970–976.

Ogawa, N., DeRisi, J., and Brown, P.O. (2000). New components of a system for phosphate accumulation and polyphosphate metabolism in *Saccharomyces cerevisiae* revealed by genomic expression analysis. *Mol. Biol. Cell* **11**, 4309–4321.

Okada, H. and Toh-e, A. (1992). A novel mutation occurring in the *PHO80* gene suppresses the *PHO4*^{c} mutations in *Saccharomyces cerevisiae*. *Curr. Genet.* **21**, 95–99.

Okumura, S., Mitsukawa, N., Shirano, Y., and Shibata, D. (1998). Phosphate transporter gene family of *Arabidopsis thaliana*. *DNA Res.* **5**, 261–269.

Pabo, C.O. and Sauer, R.T. (1992). Transcription factors: Structural families and principles of DNA recognition. *Annu. Rev. Biochem.* **61**, 1053–1095.

Pant, B.D., Buhtz, A., Kehr, J., and Scheible, W-R. (2008). MicroRNA399 is a long-distance signal for the regulation of plant phosphate homeostasis. *Plant J.* **53**, 731–738.

Patni, N.J., Dhawale, S.W., and Aaronson, S. (1977). Extracellular phosphatases of *Chlamydomonas reinhardi* and their regulation. *J. Bacteriol.* **130**, 205–211.

Persson, B.L., Berhe, A., Fristedt, U., Martinez, P., Pattison, J., Peterson, J., and Weinander, R. (1998). Phosphate permeases of *Saccharomyces cerevisiae*. *Biochem. Biophys. Res. Commun.* **1365**, 23–30.

Pinson, B., Merle, M., Franconi, J.-M., and Daignan-Fornier, B. (2004). Low affinity orthophosphate carriers regulate *PHO* gene expression independently of internal orthophosphate concentration in *Saccharomyces cerevisiae*. *J. Biol. Chem.* **279**, 35273–35280.

Poirier, Y., Thoma, S., Somerville, C., and Schiefelbein, J. (1991). Mutant of *Arabidopsis* deficient in xylem loading of phosphate. *Plant Physiol.* **97**, 1087–1093.

Quisel, J., Wykoff, D., and Grossman, A.R. (1996). Biochemical characterization of the extracellular phosphatases produced by phosphorus-deprived *Chlamydomonas reinhardtii*. *Plant Physiol.* **111**, 839–848.

Raghothama, K.G. (2000). Phosphorus acquisition: Plant in the driver's seat! *Trends Plant Sci.* **5**, 412–413.

Riekhof, W.R., Ruckle, M.E., Lydic, T.A., Sears, B.B., and Benning, C. (2003). The sulfolipids 2′-*O*-acyl-sulfoquinovosyldiacylglycerol and sulfoquinovosyldiacylglycerol are absent from a *Chlamydomonas reinhardtii* mutant deleted in *SQD1*. *Plant Physiol.* **133**, 864–874.

Rubio, V., Linhares, F., Solano, R., Martin, A.C., Iglesias, J., Leyva, A., and Paz-Ares, J. (2001). A conserved MYB transcription factor involved in phospate starvation signaling both in vascular plants and in unicellular algae. *Genes Dev.* **15**, 2122–2133.

Ruiz, F.A., Marchesini, N., Seufferheld, M., Govindjee, , and Docampo, R. (2001). The polyphosphate bodies of *Chlamydomonas reinhardtii* possess a proton-pumping pyrophosphatase and are similar to acidocalcisomes. *J. Biol. Chem.* **276**, 46196–46203.

Sato, N., Sonoike, K., Tsuzuki, M., and Kawaguchi, A. (1995a). Impaired photosystem II in a mutant of *Chlamydomonas reinhardtii* defective in sulfoquinovosyl diacylglycerol. *Eur. J. Biochem.* **234**, 16–23.

Sato, N., Tsuzuki, M., Matsuda, Y., Ehara, T., Osafune, T., and Kawaguchi, A. (1995b). Isolation and characterization of mutants affected in lipid metabolism of *Chlamydomonas reinhardtii*. *Eur. J. Biochem.* **230**, 987–993.

Shimogawara, K., Wykoff, D.D., Usuda, H., and Grossman, A.R. (1999). *Chlamydomonas reinhardtii* mutants abnormal in their responses to phosphorus deprivation. *Plant Physiol.* **120**, 1–10.

Siderius, M., Musgrave, A., van den Ende, H., Koerten, H., Cambier, P., and van der Meer, P. (1996). *Chlamydomonas eugametos* (Chlorophyta) stores phosphate in polyphosphate bodies together with calcium. *J. Phycol* **32**, 402–409.

Slomovic, S., Portnoy, V., Liveanu, V., and Schuster, G. (2006). RNA polyadenylation in prokaryotes and organelles; different tails tell different tales. *Crit. Rev. Plant Sci.* **25**, 65–77.

Smil, V. (2000). Phosphorus in the environment: Natural flows and human interferences. *Annu. Rev. Energy Environ.* **25**, 53–88.

Spain, B.H., Koo, D., Ramakrishnan, M., Dzudzor, B., and Colicelli, J. (1995). Truncated forms of a novel yeast protein suppress the lethality of a G protein alpha subunit deficiency by interacting with the beta subunit. *J. Biol. Chem.* **270**, 25435–25444.

Tadano, T., Ozawa, K., Sakai, H., Osaki, M., and Matsui, H. (1993). Secretion of acid phosphatase by the roots of crop plants under phosphorus-deficient conditions and some properties of the enzyme secreted by lupin roots. *Plant Soil* **155-156**, 95–98.

Tamai, Y., Toh-e, A., and Oshima, Y. (1985). Regulation of inorganic phosphate transport systems in *Saccharomyces cerevisiae*. *J. Bacteriol.* **164**, 964–968.

Taylor, C.B. and Green, P.J. (1991). Genes with homology to fungal and S-gene RNases are expressed in *Arabidopsis thaliana*. *Plant Physiol.* **96**, 980–984.

Taylor, C.B., Bariola, P.A., del Cardayré, S.B., Raines, R.T., and Green, P.J. (1993). RNS2: A senescence-associated RNase of *Arabidopsis* that diverged from the S-RNases before speciation. *Proc. Natl. Acad. Sci. U. S. A.* **90**, 5118–5122.

Thomas, M.R. and O'Shea, E.K. (2005). An intracellular phosphate buffer filters transient fluctuations in extracellular phosphate levels. *Proc. Natl. Acad. Sci. U. S. A.* **102**, 9565–9570.

Trull, M.C., Guiltinan, M.J., Lynch, J.P., and Deikman, J. (1997). The response of wild-type and ABA mutant *Arabidopsis thaliana* plants to phosphorous starvation. *Plant Cell Environ.* **20**, 85–92.

Ueda, Y. and Oshima, Y. (1975). A constitutive mutation, *phoT*, of the repressible acid phosphatase synthesis with inability to transport inorganic phosphate in *Saccharomyces cerevisiae*. *Mol. Gen. Genet.* **136**, 255–260.

Ueki, K. (1978). Control of phosphatase release from cultured tobacco cells. *Plant Cell Physiol.* **19**, 385–392.

Ueki, K. and Sitiro, S. (1971). Effect of inorganic phosphatase on the extracellular acid phosphatase activity of tobacco cells cultured *in vitro*. *Physiol. Plant.* **24**, 506–511.

Veljanovski, V., Vanderbeld, B., Knowles, V.L., Snedden, W.A., and Plaxton, W.C. (2006). Biochemical and molecular characterization of AtPAP26, a vacuolar purple acid phosphatase up-regulated in phosphate-deprived *Arabidopsis* suspension cells and seedlings. *Plant Physiol.* **142**, 1282–1293.

Wang, Z.Y., Kenigsbuch, D., Sun, L., Harel, E., Ong, M.S., and Tobin, E.M. (1997). A Myb-related transcription factor is involved in the phytochrome regulation of an *Arabidopsis Lhcb* gene. *Plant Cell Environ.* **9**, 491–507.

Weiss, M., Bental, M., and Pick, U. (1991). Hydrolysis of polyphosphates and permeability changes in response to osmotic shocks in cells of the halotolerant alga *Dunaliella*. *Plant Physiol.* **97**, 1241–1248.

Werner, T., Amrhein, N., and Freimoser, F. (2007). Inorganic polyphosphate occurs in the cell wall of *Chlamydomonas reinhardtii* and accumulates during cytokinesis. *BMC Plant Biol.* **7**, 51–61.

Wetzel, R.G. (1983). *Limnology*, 2nd edition. Saunders College Publishing, Philadelphia.

Wykoff, D.D. and O'Shea, E.K. (2001). Phosphate transport and sensing in *Saccharomyces cerevisiae*. *Genetics* **159**, 1491–1499.

Wykoff, D., Davies, J., and Grossman, A. (1998). The regulation of photosynthetic electron transport during nutrient deprivation in *Chlamydomonas reinhardtii*. *Plant Physiol.* **117**, 129–139.

Wykoff, D., Grossman, A., Weeks, D.P., Usuda, H., and Shimogawara, K. (1999). Psr1, a nuclear localized protein that regulates phosphorus metabolism in *Chlamydomonas*. *Proc. Natl. Acad. Sci. U. S. A.* **96**, 15336–15341.

Yao, Q., Li, X.L., Feng, G., and Christie, P. (2001). Mobilization of sparingly soluble inorganic phosphates by external mycelium of an arbuscular mycorrhizal fungus. *Plant Soil* **230**, 279–285.

Yehudai-Resheff, S., Zimmer, S.L., Komine, Y., and Stern, D.B. (2007). Integration of chloroplast nucleic acid metabolism into the phosphate deprivation response in *Chlamydomonas reinhardtii*. *Plant Cell* **19**, 1023–1038.

Yoshida, K., Ogawa, N., and Oshima, Y. (1989). Function of the *PHO* regulatory genes for repressible acid phosphatase synthesis in *Saccharomyces cerevisiae*. *Mol. Gen. Genet.* **217**, 40–46.

Zhao, T., Li, G., Mi, S., Li, S., Hannon, G.J., Wang, X.-J., and Qi, Y. (2007). A complex system of small RNAs in the unicellular green alga *Chlamydomonas reinhardtii*. *Genes Dev.* **21**, 1190–1203.

CHAPTER 7

Hydrogenases, Hydrogen Production, and Anoxia

Matthew C. Posewitz,[1] Alexandra Dubini,[1,2]
Jonathan E. Meuser,[1] Michael Seibert[2] and Maria L. Ghirardi[2]
[1]Colorado School of Mines, Department of Chemistry and Geochemistry, Golden, Colorado, USA
[2]National Renewable Energy Laboratory, Chemical and Biosciences Center, Golden, Colorado, USA

CHAPTER CONTENTS

I. BACKGROUND

Several water-oxidizing, phototrophic algae have the remarkable ability to use low-potential electrons from the photosynthetic electron transport chain to reduce protons into H_2. Consequently, renewable H_2 production can be achieved in these microorganisms using energy provided by solar irradiation and electrons derived from water. When H_2 is used as an energy carrier in fuel cell technologies only water vapor is emitted, and significant environmental benefits can be realized if sufficient quantities of H_2 are sustainably produced in a cost effective manner. Therefore, major research efforts are aimed at developing a more informed understanding of the physiological parameters facilitating algal H_2 production, with the ultimate goal of improving H_2 yields.

Hydrogen production in a green alga was first reported over 65 years ago in seminal experiments performed by Hans Gaffron and co-workers (Gaffron and Rubin, 1942; reviewed in Homann, 2003; Melis and Happe, 2004). Subsequently, several diverse phototrophic microorganisms, including numerous species of *Chlamydomonas*, have been shown to produce H_2 using the enzyme hydrogenase and low-potential electrons provided by either organic substrate fermentation or light-driven photosynthetic pathways, as illustrated in Figure 7.1 (Healey, 1970; Kessler, 1974; Weaver et al., 1980; Brand et al., 1989; Ghirardi et al., 2000, 2005, 2007; Melis and Happe, 2001; Happe et al., 2002; Boichenko et al., 2004; Kruse et al., 2005b; Seibert et al., 2008).

Chlamydomonas has emerged as the model organism for investigating H_2 metabolism in water-oxidizing, photosynthetic eukaryotes, and substantial progress has been made in elucidating the mechanisms of H_2 production, as well as in improving H_2 photoproduction activity (Weaver et al., 1980; Ghirardi et al., 2000, 2005; Melis and Happe, 2001; Kosourov et al., 2002; Fedorov et al., 2005; Kruse et al., 2005a). Detailed genetic and physiological studies describing several aspects of algal H_2 production are available, and several fascinating mutants linked to H_2 metabolism have been isolated. These research efforts, in combination with the nearly complete *Chlamydomonas* genome sequence, have dramatically enhanced our understanding of algal H_2 production, and provided remarkable new insights into the biochemistry, genetics and metabolic networks influencing hydrogenase activity. These contributions demonstrate that *Chlamydomonas* has an unprecedented repertoire of metabolic capabilities that allow adaptation to rapidly changing environmental conditions, including anoxia, which is necessary for H_2 production activity. Moreover, this work has led to the production of significant quantities of H_2 over extended periods, the determination of several algal [FeFe]-hydrogenase sequences, and identification of [FeFe]-hydrogenase maturation proteins.

The remarkable metabolic flexibility of *Chlamydomonas* likely facilitates acclimation to natural energetic fluxes arising from environmental

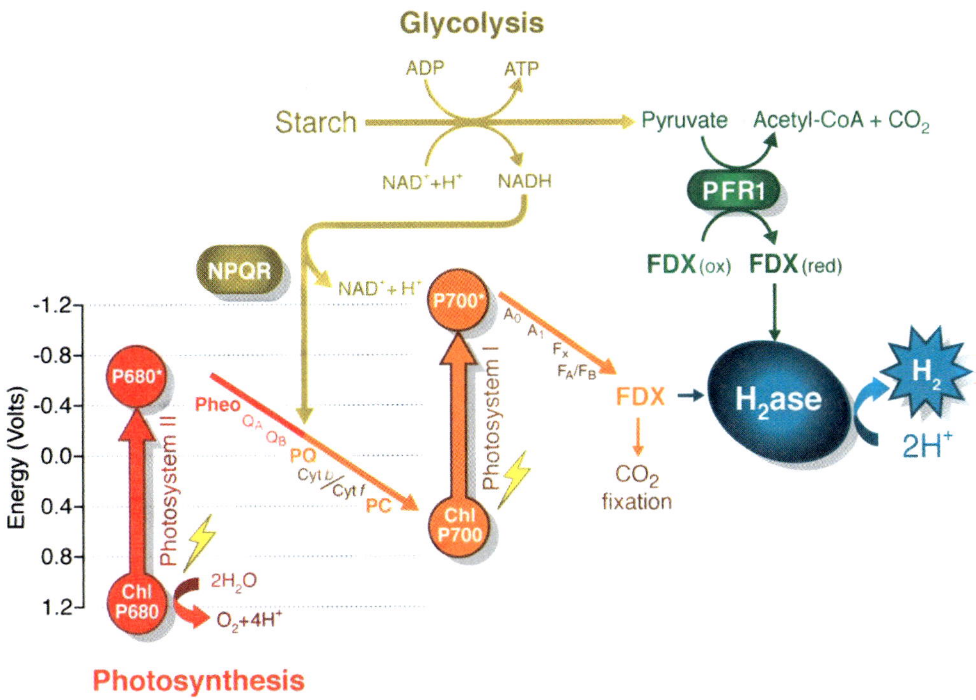

FIGURE 7.1 *Simplified illustration of the physiological pathways used for H_2 production in Chlamydomonas. The two photoproduction pathways use low-potential electrons generated from either water oxidation by PS II activity or the non-photochemical reduction of PQ by NPQR (see main text). Both the PS II-dependent (red) and NPQR-dependent (yellow) pathways require reduction of the PQ pool and PS I activity (orange). Electrons excited to higher energy (lower potential) by PS I are able to reduce ferredoxin (FDX), the physiological electron donor to hydrogenase. In the case of the PS II-independent pathway, NADH formed by the catabolism of organic substrates is used for reduction of the PQ pool. In addition, during dark fermentation the oxidation of pyruvate catalyzed by PFR (green) is used to reduce ferredoxin and putatively mediates the observed production of H_2 in the dark.*

conditions and diurnal light-dark cycles, and indicates that this alga is able to readily adapt its metabolism in response to diverse nutrient challenges. Several proteins, including the [FeFe]-hydrogenases, are typically found in strictly anaerobic organisms and are used in anoxic metabolism. The presence of fermentation pathways typically observed in anaerobic chemotrophs in an oxygenic phototroph was initially considered somewhat paradoxical. However, it is now apparent that photosynthetic microbes frequently experience extended periods of limited O_2 availability (Moseley et al., 2000; Quinn et al., 2002; Steunou et al., 2006; Grossman et al., 2007; Mus et al., 2007). Components of these fermentative pathways, such as the [FeFe]-hydrogenases, interact with the photosynthetic electron transport chain under the appropriate conditions.

Although substantial insights regarding algal H_2 production exist, critical aspects regarding hydrogenase activity remain unresolved and new advances are required to define more clearly the metabolic and enzymatic processes influencing algal H_2 production. A more detailed understanding of the physiological parameters and basic biochemistry facilitating H_2 production is necessary to improve efficiencies, and to obtain a more informed model of H_2 metabolism in *Chlamydomonas*. Fortunately, the foundation for exciting new breakthroughs has been provided by numerous advances in genetic techniques and resources, our improved understanding of *Chlamydomonas* anoxic physiology, characterization of several genes relevant to hydrogenase activity, and the genome sequence.

II. HYDROGENASES

A. Overview

Biological H_2 production in photosynthetic microorganisms is catalyzed by either the hydrogenase or nitrogenase enzymes (Weaver et al., 1980; Boichenko and Hoffmann, 1994; Tamagnini et al., 2002; Boichenko et al., 2004; Kruse et al., 2005b). As nitrogenases are not observed in eukaryotic algae, they will not be discussed further. Hydrogenases are widely distributed in diverse microorganisms (Adams, 1990; Wu and Mandrand, 1993; Vignais et al., 2001; Vignais and Colbeau, 2004; Posewitz et al., 2008), and two distinct classes of hydrogenases efficiently catalyze the reversible oxidation of H_2 ($H_2 \leftrightarrows 2H^+ + 2e^-$). These enzymes are termed the [NiFe]- and [FeFe]-hydrogenases, according to the metals at their respective active sites. The [NiFe]- and [FeFe]-hydrogenases are phylogenetically distinct, and their enzyme activities are proposed to reflect a convergent evolution of function (Vignais et al., 2001). To date, there appears to be a strict segregation with respect to the occurrence of hydrogenases in water-oxidizing phototrophs. Only the [FeFe]-hydrogenases are found in eukaryotic green algae, and only [NiFe]-hydrogenases are found in cyanobacteria (Ludwig et al., 2006). A [NiFe]-hydrogenase was proposed to be present in the eukaryotic alga *Scenedesmus obliquus* (Zinn et al., 1994); however, this finding has yet to be confirmed, and a subsequent report indicated that *S. obliquus* likely contains only [FeFe]-hydrogenases (Florin et al., 2001). It will be very interesting to better define the evolutionary and metabolic implications of these observations, which are not intuitive.

Although [NiFe]- and [FeFe]-hydrogenases are genetically unrelated, the enzyme classes are unified in that (1) their active sites contain the biologically novel ligands CO and CN^- coordinated to Fe and (2) each active site contains a binuclear metal center. In general, [NiFe]-hydrogenases catalyze H_2 uptake, whereas [FeFe]-hydrogenases catalyze H_2 evolution (Frey, 2002). Furthermore, the turnover frequency of [FeFe]-hydrogenases is typically 10–100 times higher than that of [NiFe]-hydrogenases (Frey, 2002), making the former one of the most efficient H_2 production catalysts known. However, it should be noted that several exceptions to these generalizations exist (Posewitz et al., 2008).

The [NiFe]-hydrogenases are found in a variety of anaerobic and facultative archaea, eubacteria, and cyanobacteria, and minimally consist of two distinct subunits (large and small). Because the presence of these enzymes in eukaryotic algae has not been conclusively demonstrated, they will not be discussed further; however, several excellent reviews regarding [NiFe]-hydrogenases are available (Garcin et al., 1998; Armstrong, 2004; de Lacey et al., 2005).

The [FeFe]-hydrogenases are typically found in eubacteria (e.g. *Clostridium* and *Desulfovibrio*), several anaerobic protozoa (e.g. *Trichomonas* and *Nyctotherus*), and in some green algae (e.g. *Chlamydomonas* and *Scenedesmus*) (Vignais et al., 2001). These enzymes may exist as distinct monomers or in heteromeric complexes, and X-ray crystallographic structural data are available for two representative bacterial enzymes (Peters et al., 1998; Nicolet et al., 1999; Peters, 1999). The [FeFe]-hydrogenase catalytic site is known as the H-cluster, and consists of a [4Fe-4S] cluster coordinated to four cysteines and connected via a single bridging cysteine thiolate to a unique binuclear 2Fe center (Figure 7.2). Terminal CO and CN^- ligands are bound to each Fe atom of the 2Fe center, and a third CO ligand bridges both of the Fe atoms in the enzyme's resting state. A unique non-protein dithiolate ligand, proposed to be either di(thiomethyl)amine or dithiolpropane (Nicolet et al., 2001), also bridges the 2Fe atoms of the active site in a bidentate fashion. This unusual ligand complement putatively functions to tune the unique chemistry of the

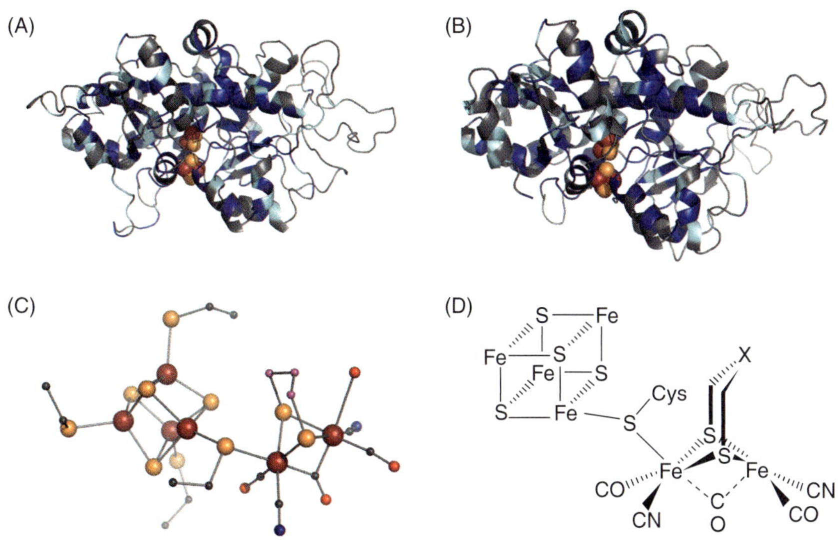

FIGURE 7.2 *Structural models of the Chlamydomonas [FeFe]-hydrogenases HYDA1 (A) and HYDA2 (B), ball and stick model of the H-cluster active site (C), and schematic of the [FeFe]-hydrogenase H-cluster (D). The atoms in (C) correspond to iron (dark red), sulfur (orange), carbon (black), nitrogen (blue), and oxygen (light red). The three light atoms of the dithiolate ligand are depicted in pink. The figures in A, B, and C were prepared by David W. Mulder and John W. Peters, Montana State University.*

H-cluster, stabilizing lower redox states of Fe that facilitate reversible H_2 oxidation chemistry.

B. [FeFe]-hydrogenases in green algae

The first published algal [FeFe]-hydrogenase gene sequences were derived from *Scenedesmus obliquus* (Florin et al., 2001; Wunschiers et al., 2001). Additional sequences have subsequently been reported for *Chlamydomonas HYDA1* (Happe and Kaminski, 2002) and *HYDA2* (Forestier et al., 2003), *Chlorella fusca* (previously known as *Scenedesmus vacuolatus*; Winkler et al., 2002a), and *Chlamydomonas moewusii* (Winkler et al., 2004). Currently, all reported algal [FeFe]-hydrogenase genes encode monomeric proteins of approximately 48 kD, which are among the smallest [FeFe]-hydrogenases characterized to date (Happe et al., 2002). Additional FeS clusters are found in most bacterial [FeFe]-hydrogenase enzymes and are putatively required for electron transport from/to redox mediators. However, the algal [FeFe]-hydrogenases lack these additional metal clusters, and the functional rationale for the evolution of these truncated [FeFe]-hydrogenases, which are found only in green algae, is unresolved. Green algae having hydrogenase activity often contain two isoforms of the [FeFe]-hydrogenase enzyme, denoted HYDA1 and HYDA2 (Forestier et al., 2003; Winkler et al., 2004). The precise contribution of each enzyme to H_2 metabolism is currently unknown. A third protein with [FeFe]-hydrogenase homology is also encoded in *Chlamydomonas* (accession number EDP03395). However, this protein is more similar to the family of Narf-like proteins that are proposed to play a role in FeS cluster assembly in some eukaryotes, and it is not believed to have hydrogenase activity (Balk et al., 2004).

The algal [FeFe]-hydrogenases are similar to the C-terminal portion of bacterial [FeFe]-hydrogenases, which coordinates the catalytic H-cluster. The algal enzymes contain the conserved [FeFe]-hydrogenase H-cluster binding motifs L1 (TSCCPxW), L2 (MPCxxKxxE), and L3 (ExMACxxGCxxGGGxP) identified by Vignais et al. (2001), which are observed in the majority of characterized [FeFe]-hydrogenases, and contain the cysteine amino acids required for H-cluster coordination. The algal [FeFe]-hydrogenases contain transit peptides of variable length at the N-terminus, which are required for translocation to the chloroplast. The algal enzymes sequenced to date also contain an insertion of 15–54 amino acids at the C-terminus, and a smaller insertion (HLEXHP), approximately 9 amino acids to the N-terminal side of L1 (Figure 7.3). The physiological significance of these insertions is currently unknown; however, they may have roles in regulating enzyme activity or protein translocation, and the large insert may participate in electrostatic interactions between the hydrogenase and physiological electron donors (Happe and Kaminski, 2002; Horner et al., 2002).

Hydrogenase activity in *Chlamydomonas* is induced during anaerobiosis. This can be achieved in the laboratory with a variety of techniques

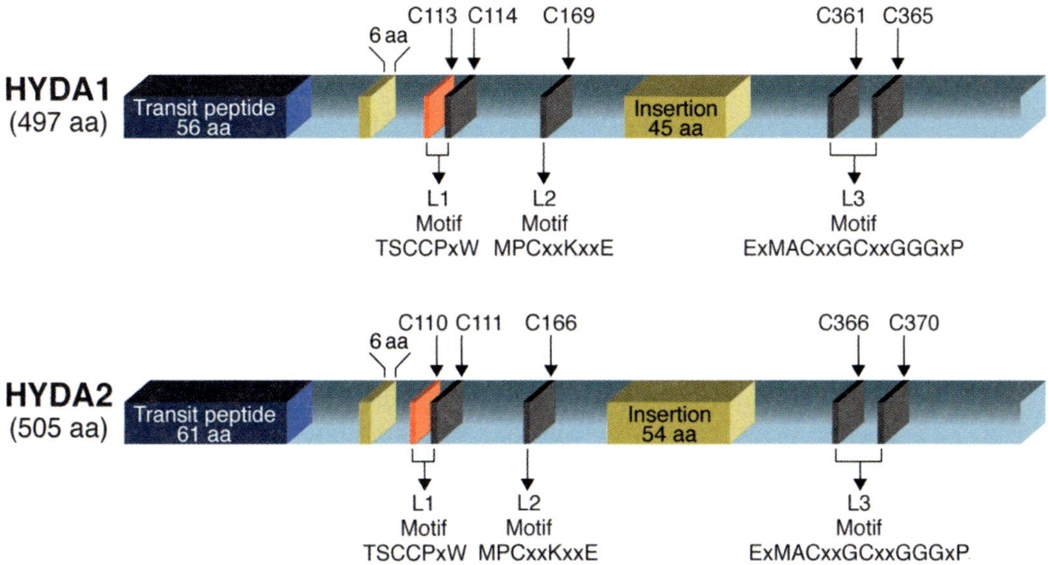

FIGURE 7.3 *Schematic representations of the two Chlamydomonas [FeFe]-hydrogenases. The mature enzymes are depicted by the long, light blue linear rectangles. Within these, the dark blue rectangles at the N-terminus represent transit peptides. Conserved cysteine residues that coordinate the catalytic H-cluster are shown in black. Cysteines putatively involved in proton transfer during H_2 evolution are shown in red. The small and large insertions, unique to algal [FeFe] hydrogenases, are shown in yellow.*

including purging with inert gas, providing exogenous reductant, and/or allowing cellular respiration to metabolize dissolved O_2. Following the establishment of anaerobiosis, cultures are sealed to prevent the introduction of O_2 from the atmosphere. Moreover, cultures grown in nutrient-replete media must be maintained in the dark or at very low light levels to prevent O_2 accumulation from endogenous photosynthetic activity. Hydrogen photoproduction is observed at high initial rates after illumination of dark, anaerobically adapted cells. However, in cultures grown in a nutrient-replete medium, these initial rates of H_2 photoproduction rapidly diminish as O_2 levels from photosynthesis increase and the *Chlamydomonas* cells adapt to an aerobic metabolism and CO_2-fixing conditions.

Anaerobiosis can also be established and maintained at higher light levels when cultures are deprived of sulfate, as described in section VI (see also Chapter 5). Sulfate-deprived cultures have attenuated rates of photosynthetic O_2 evolution (Wykoff et al., 1998), but maintain relatively high levels of respiration that exceed the rates of photosynthesis (Ghirardi et al., 2000; Melis et al., 2000). Consequently, sealed cultures become anaerobic after a 15–40 hour period of adaptation, and sustained H_2 production is observed at relatively high light intensities for several days (Ghirardi et al., 2000; Melis et al., 2000).

Chlamydomonas hydrogenases are synthesized on cytosolic ribosomes (Roessler and Lien, 1984a), and then translocated to the chloroplast stroma, where hydrogenase activity is observed (Maione and Gibbs, 1986a). Hydrogenase localization to the chloroplast was confirmed after organelle isolation, activity measurements, and immunoblotting (Happe et al., 1994). Moreover, these authors reported a ~90% decline in hydrogenase activity following the addition of cycloheximide, an inhibitor of cytosolic protein synthesis. Presently, there is no evidence that the hydrogenases are co-localized to cellular compartments outside of the chloroplast.

Immunoblot analyses failed to detect [FeFe]-hydrogenase in aerobically growing cells, suggesting that the enzyme is synthesized *de novo* upon anoxic adaptation (Happe et al., 1994), which coincides with a significant increase in the levels of *HYDA1* and *HYDA2* transcripts (Happe and Kaminski, 2002; Forestier et al., 2003; Posewitz et al., 2004a,b; Mus et al., 2007). However, these findings are in contrast with a previous report that some inactive hydrogenase was present under aerobic conditions, and is rapidly activated during anaerobiosis (Roessler and Lien, 1984a). It should be noted that elevated levels of *HYDA1* and *HYDA2* transcripts are detected in cultures that are not well illuminated or vigorously aerated, especially after centrifugation in the dark, and it is possible that this work used cells that were entering hypoxia due to cellular respiration and insufficient oxygenation, resulting in hydrogenase expression.

A partial purification of *Chlamydomonas* hydrogenase was originally reported by Erbes et al. (1979). Roessler and Lien (1984b) subsequently isolated an enzyme that was ~40% pure and calculated a specific H_2 evolution activity of 1800 U (micromole H_2/mg protein/minute) using ferredoxin as the electron mediator. Happe and Naber (1993) were able to purify the *Chlamydomonas* HYDA1 enzyme to homogeneity, allowing an N-terminal amino acid sequence to be determined, and reported specific activities of 935 U when using methyl viologen as the electron donor. It should be noted that hydrogenase activity assays are significantly influenced by several parameters, including the nature of the electron mediator (ferredoxin, methyl viologen, etc.), pH, temperature, ionic strength, and specific ions (Roessler and Lien, 1982). Care should therefore be used when comparing specific activities obtained using different assay conditions.

The metal content of purified HYDA1 suggested that the enzyme was an [FeFe]-hydrogenase (Happe and Naber, 1993), which was later confirmed by gene sequencing (Happe and Kaminski, 2002). The *HYDA2* gene product was also found to accumulate only during anaerobiosis (Forestier et al., 2003); however, its purification from *Chlamydomonas* has not been reported. The HYDA1 and HYDA2 protein sequences are highly similar (68% identity/74% similarity) and as noted previously, it appears that two homologous [FeFe]-hydrogenases are typically observed in all green algae with hydrogenase activity. Although the physiological basis for the presence of two distinct enzymes has yet to be determined, it is clear that both the *Chlamydomonas HYDA1* and *HYDA2* genes encode active [FeFe]-hydrogenases, which can be heterologously expressed in *E. coli* (King et al., 2006a).

C. Transcriptional regulation

Anaerobiosis induces significant increases in the levels of the *HYDA1* and *HYDA2* transcripts, as well as two *Chlamydomonas* [FeFe]-hydrogenase maturation genes (section II.D), *HYDEF* and *HYDG* (Posewitz et al., 2004a, 2005; Mus et al., 2007), which is consistent with anoxic enzyme accumulation (Happe et al., 1994; Happe and Kaminski, 2002). Although the precise regulatory mechanisms are currently unknown, anoxia has long been established as necessary for hydrogenase activity, and quantitative PCR data demonstrate that the levels of hydrogenase transcripts increase over 100-fold during cellular acclimation to anoxia (Mus et al., 2007). Notably, metabolic mutants have been isolated that exhibit attenuated hydrogenase gene transcription under anaerobic conditions (Posewitz et al., 2004b), suggestive of regulatory factors other than O_2. For example, two *Chlamydomonas* starchless mutants, *sta6* (Zabawinski et al., 2001) and *sta7* (Posewitz et al., 2004b; see also Chapter 1), have significantly attenuated levels of hydrogenase activity, as well as decreased levels of hydrogenase transcripts under dark anaerobic conditions (Posewitz et al., 2004b). This may be the consequence of a more oxidized plastoquinone pool during dark, anaerobic adaptation, and it is possible that the reduction states of cellular redox constituents such as plastoquinones, or other products of starch catabolism, influence signal-transduction events controlling the expression of the hydrogenase genes during anaerobiosis. The degradation of starch, which occurs under anaerobic conditions, can influence intracellular levels of NAD(P)H and/or the oxidation state of the plastoquinone pool, both of which have been demonstrated to regulate transcriptional processes (Escoubas et al., 1995; Rutter et al., 2001; Pfannschmidt et al., 2003). Further experimentation will be required to establish whether this type of regulation applies to hydrogenase gene expression in green algae.

The precise DNA regulatory elements controlling hydrogenase gene expression are also currently unknown. Experiments coupling varying lengths of the *Chlamydomonas HYDA1* promoter region to a promoterless arylsulfatase (*ARS*) reporter gene have been reported (Stirnberg and Happe, 2004). These studies demonstrated that the region between –128 and –21 relative to the *HYDA1* transcription start site is required for expression under anaerobic conditions. Further analysis is needed to fully define the promoter elements, the accompanying regulatory proteins, and the transduction mechanisms mediating hydrogenase expression. The *Chlamydomonas* genome encodes homologues of several O_2-sensing and transduction proteins found in both eubacteria and eukaryotes, which are promising candidates to probe for potential roles in regulating hydrogenase gene expression and activity.

D. Enzyme maturation

It is well established that enzymes requiring complex metalloclusters for activity (e.g. nitrogenase, cytochrome oxidase, and [NiFe]-hydrogenase) use unique maturation proteins for assembly of their respective active sites (Casalot and Rousset, 2001; Dos Santos et al., 2004; Rubio and Ludden, 2005; Bock et al., 2006). Speculation that the [FeFe]-hydrogenases also required novel maturation proteins for activation (Vignais et al., 2001), was confirmed when these proteins were identified by screening and characterizing *Chlamydomonas* insertional mutants that were unable to produce H_2 (Posewitz et al., 2004a, 2005). Hydrogen photoproduction was screened using sensitive chemochromic H_2 sensors (Seibert et al., 2001; Flynn et al., 2002), as shown in Figure 7.4. *Chlamydomonas* is especially well-suited for screening hydrogenase activity, as a functional hydrogenase is not essential for its survival under laboratory conditions, and the initial burst of H_2 produced by illuminating small colonies of dark-adapted cells is sufficient for analytical detection.

The screen discussed above led to isolation of the *hydEF-1* mutant, which is unable to produce any H_2. Transformation of the mutant with the *HYDEF* gene restored hydrogenase activity. In *Chlamydomonas HYDEF* is adjacent to another hydrogenase related gene, *HYDG*, which is transcribed in divergent orientation from the same promoter region. Both gene products are very likely required for [FeFe]-hydrogenase

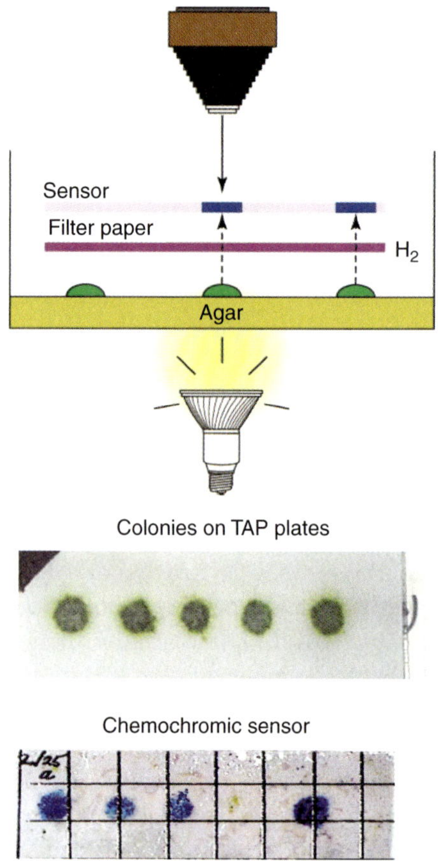

FIGURE 7.4 *Screening Chlamydomonas colonies on TAP-agar plates for hydrogenase activity. Sensitive chemochromic sensors based on thin films of Pd/WO_3 are used to detect H_2 from algal colonies and turn blue in the presence of H_2. The cells are adapted to anaerobiosis in the dark and then illuminated to stimulate H_2 production. The fourth colony from the left is unable to produce H_2 and no response is detected.*

maturation, as demonstrated by the heterologous expression of active *Chlamydomonas HYDA1* in *E. coli* (Posewitz et al., 2004a), a bacterium that lacks a native [FeFe]-hydrogenase. When the *HYDEF* and *HYDG* genes were coexpressed with *HYDA1* in *E. coli*, [FeFe]-hydrogenase activity was observed. However, expression of *HYDA1* alone or in combination with only *HYDEF* or *HYDG* resulted in a non-functional HYDA1 protein after purification. The ability to express active [FeFe]-hydrogenases heterologously, such as in *E. coli*, required the identification of the [FeFe]-hydrogenase maturases. Such systems have dramatically accelerated the pace at which native and mutant [FeFe]-hydrogenases are being characterized (Cohen et al., 2005a; Ghirardi et al., 2006; King et al., 2006a,b). Moreover, the application of biotechnological techniques such as [FeFe]-hydrogenase gene shuffling (Nagy et al., 2007) and the heterologous expression of [FeFe]-hydrogenases in other phototrophic hosts, such as cyanobacteria, are now possible.

The *HYDEF* and *HYDG* genes encode two novel radical *S*-adenosylmethionine (Radical-SAM) proteins. The HYDEF protein of *Chlamydomonas* contains two unique segments that are homologous to two distinct bacterial proteins. These strictly conserved [FeFe]-hydrogenase maturation proteins, HYDE, HYDF (or HYDEF), and HYDG are found exclusively in organisms containing [FeFe]-hydrogenase. The HYDE and HYDG proteins belong to the Radical SAM superfamily, and HYDF contains a putative GTPase domain. All three maturases coordinate FeS clusters, and the respective proteins have been shown to have the expected Radical SAM or GTPase activities (Rubach et al., 2005; Brazzolotto et al., 2006).

The precise mechanism of [FeFe]-hydrogenase assembly is unknown; however, one theory is that the maturases use common amino acids or metabolic products to synthesize a precursor to the [FeFe]-hydrogenase active site, which is then transferred to the structural enzyme (Peters et al., 2006). The Radical SAM enzymes, which have been reviewed extensively (Fontecave et al., 2001; Frey, 2001; Marsh et al., 2004; Layer et al., 2005; Walsby et al., 2005), use radical-initiated chemistry to perform a variety of anaerobic oxidations, thiol insertions, and, in the case of NifB (associated with nitrogenase maturation) and the [FeFe]-hydrogenase maturases, metallocluster assembly. Additionally, several metallocluster maturation proteins require NTPase activity, as observed for HYDF. These characteristics are consistent with the requirements necessary for the synthesis of the unique [FeFe]-hydrogenase ligands (CO, CN^-, and the dithiolate moiety) and for assembly of the [FeFe]-hydrogenase catalytic site.

Since the [FeFe]-hydrogenases are localized to the chloroplast, an intriguing and unresolved question is the compartmentalization of [FeFe]-hydrogenase maturation in *Chlamydomonas*. It is unknown whether active enzyme is transported into the chloroplast, or whether the structural enzymes and maturation proteins are all translocated to the chloroplast prior to assembly. The maturases both contain predicted transit peptides that could target them to the chloroplast.

E. Oxygen sensitivity

The *Chlamydomonas* hydrogenases are particularly sensitive to O_2 and are irreversibly inactivated within minutes after exposure to atmospheric levels of O_2 (Abeles, 1964; Erbes et al., 1979; Happe and Naber, 1993; Ghirardi et al., 1997; Cohen et al., 2005a; King et al., 2006b). The *Chlamydomonas* HYDA enzymes are among the most O_2-sensitive hydrogenases examined, and are more rapidly inactivated by O_2 than clostridial [FeFe]-hydrogenases tested under similar conditions (Cohen et al., 2005a; King et al., 2006b). It is conceivable that hydrogenases in oxygenic green algae evolved hypersensitivity to O_2 to attenuate hydrogenase activity in the presence of O_2, when the cell preferentially uses higher energy-yielding metabolic processes that require O_2 as the terminal electron acceptor. Lee and Greenbaum (2003) proposed that levels of O_2 below those required to inhibit hydrogenase activity *in vivo* (ca. 0.1%) activate an undefined metabolic pathway, perhaps related to CO_2 fixation, that effectively competes with hydrogenase for photosynthetic reductant contributing to the observed attenuation of H_2 photoproduction activity in the presence of O_2.

The examination of hydrogenase activity in whole cells and cell-free extracts of *Scenedesmus* and *Chlamydomonas* demonstrated hydrogenase inactivation within 5 minutes after exposure to O_2 in both species (Urbig et al., 1993). In whole cells, recovery of activity occurred upon repeated degassing, and re-adapted cells had H_2 photoproduction rates similar to those measured after the first anaerobic induction. However, the activity of enzymes in cell-free homogenates could not be recovered even after prolonged degassing, consistent with the well-documented irreversible inactivation of [FeFe]-hydrogenases by O_2 (Adams, 1990). When homogenates of cells with active hydrogenases were added to inactivated cell-free samples, no restoration of activity was found (Urbig et al., 1993). Since a soluble re-activation factor could not be found, *de novo* protein synthesis is the likely means for recovery of hydrogenase activity in whole cells (Roessler and Lien, 1984a; Urbig et al., 1993).

Among the different enzymes examined to date, the rates of [FeFe]-hydrogenase inactivation by O_2 are variable. For purified enzymes that were heterologously expressed in *E. coli*, the time required for 50% inhibition of hydrogenase activity after exposure to ambient levels of O_2 ranged from 2 seconds to 3 seconds for the *Chlamydomonas* HYDA1 enzyme to ~100 seconds for a *Clostridium acetobutylicum* [FeFe]-hydrogenase (Cohen et al., 2005a; King et al., 2006b). A truncated form of the *C. acetobutylicum* HydA enzyme was generated, which lacked the N-terminal accessory FeS cluster binding domains. This modified protein is structurally similar to the algal enzymes in that it contains only the H-cluster catalytic domain. The truncated *C. acetobutylicum* HydA enzyme remained less sensitive to O_2 inactivation relative to the wild-type *Chlamydomonas* HYDA1 enzyme, indicating that the presence of the N-terminal domain of the native *C. acetobutylicum* enzyme is not required for the comparatively prolonged hydrogenase activity observed after exposure to O_2 (Cohen et al., 2005a; King et al., 2006b).

To obtain sustained H_2 photoproduction in *Chlamydomonas*, an O_2-tolerant enzyme is putatively required to couple to the photosynthetic electron-transport chain under conditions of high PS II-catalyzed O_2 evolution activity. Significant research efforts are therefore aimed at engineering an [FeFe]-hydrogenase with amino acid substitutions that sterically inhibit O_2 diffusion to the enzyme active site (Ghirardi et al., 2005; Ghirardi, 2006). This in turn requires a detailed understanding of the protein cavities and channels facilitating O_2 migration to the catalytic center. Molecular dynamics simulations and solvent accessibility maps, based on the structural data available for the *C. pasteurianum* [FeFe]-hydrogenase, have demonstrated that the movement of O_2 in this protein is predominantly restricted to two well-defined pathways (Cohen et al., 2005a,b; King et al., 2006b). Although H_2 gas molecules can use the same paths, the smaller diameter of H_2 allows diffusion from the protein catalytic site through several alternative cavities and channels that are not as accessible to O_2. These results suggest the possibility of preferentially restricting the pathways available for O_2 migration to the [FeFe]-hydrogenase active site, without significantly interfering with the diffusion of H_2 gas from the enzyme active site. A restricted gas channel has been proposed as a possible mechanism for the observed O_2 tolerance exhibited by some H_2-sensing [NiFe]-hydrogenases, which have relatively low catalytic activity (Buhrke et al., 2005; Duche et al., 2005).

III. HYDROGEN METABOLISM

A. Role of H_2 metabolism in eukaryotic algae

Hydrogenase activity is not observed in all genera of green algae (Brand et al., 1989; Boichenko et al., 2004; Melis et al., 2004), and the precise role of the hydrogenases and H_2 metabolism in eukaryotic algae is not clearly resolved. The involvement of hydrogenase activity in redox balancing, photoprotection, and fermentative metabolism have all been proposed. However, it should be noted that the relative activities of the specific H_2 metabolic pathways described below are reported to vary among different green algal species. In *Chlamydomonas*, cellular redox conditions regulate the complex interactions of glycolysis, fermentation, chlororespiration, mitochondrial respiration, and photosynthesis, all of which can ultimately influence hydrogenase activity. Although many aspects of algal H_2 metabolism remain unresolved, decades of research have provided substantial insights into the fascinating role of H_2 in eukaryotic phototrophs.

B. Hydrogen production pathways

There are three distinct H_2 production pathways in *Chlamydomonas*, as illustrated in Figure 7.1. These include two photoproduction pathways, and a dark, fermentative pathway. The first of the H_2 photoproduction pathways (also termed direct biophotolysis) is dependent on the activities of both PS II and PS I. It involves light-induced water oxidation at PS II, transfer of electrons from PS II to PS I, and light-dependent excitation of electrons by PS I to reduce ferredoxin (PETF or FDX), the physiological electron donor to *Chlamydomonas* hydrogenases (Roessler and Lien, 1984b; Happe and Naber, 1993). The second photoproduction pathway (indirect biophotolysis) involves the non-photochemical reduction of plastoquinone using electrons from NAD(P)H, generated primarily from starch catabolism, followed by light-dependent reduction of ferredoxin by PS I. This pathway is independent of PS II; however, it is dependent on NAD(P)H-plastoquinone oxidoreductase (NPQR) activity (Godde and Trebst, 1980; Maione and Gibbs, 1986a; Bennoun 1998; Cournac et al., 2000; Mus et al., 2005).

In *Chlamydomonas*, the existence of H_2 photoproduction independent of PS II was confirmed by studies in which NADH oxidation was coupled to light-dependent hydrogenase activity in cell-free extracts that were treated with 3-(3′,4′-dichlorophenyl)-1,1-dimethylurea (DCMU), which inhibits electron transfer from PS II to plastoquinone (Ben-Amotz and Gibbs, 1975). This reaction was completely inhibited by 2,5-dibromo-3-methyl-6-isopropyl-*p*-benzoquinone (DBMIB), which blocks plastoquinone oxidation by the cyt b_6f complex. Early studies regarding the nature of NPQR suggested involvement of a rotenone-sensitive type I NADH dehydrogenase (Godde and Trebst, 1980). However, this interpretation may have been a consequence of mitochondrial contamination, since later inhibitor studies and genomic evidence indicate the involvement of one or more type II NAD(P)H dehydrogenases in reduction of the plastoquinone pool in *Chlamydomonas* (Mus et al., 2005). In support of this conclusion, a type II NAD(P)H dehydrogenase from *Agrobacterium tumefaciens* was shown to reduce plastoquinone in isolated *Chlamydomonas* chloroplast extracts (Bernard et al., 2006). The *Chlamydomonas* genome encodes seven proteins homologous to known NDH-2 proteins (Mus et al., 2005). However, those which are involved in reduction of the chloroplast plastoquinone pool under H_2-producing conditions remain to be identified.

The direct biophotolysis pathway contributes between 50% and 90% of the total electron flux used in H_2 photoproduction, as determined by DCMU inhibition (Kosourov et al., 2003). The impact of DCMU inhibition on H_2 photoproduction varies as a function of the specific strain, culturing and assay conditions, as well as the induction protocol used (Laurinavichene et al., 2004). Both of the H_2 photoproduction pathways minimally require PS I activity (Redding et al., 1999; Kosourov et al., 2003), and are inhibited by DBMIB. At one point it was reported that H_2 photoproduction could be directly coupled to PS II activity (Greenbaum et al., 1995; Lee et al., 1996); however, it is now believed that the mutants used in these studies maintained significant residual PS I activity (Greenbaum et al., 1997; Redding et al., 1999).

The dark, fermentative H_2 production pathway in *Chlamydomonas* is coupled to starch catabolism (Gfeller and Gibbs, 1984; Kreuzberg, 1984; Ohta et al., 1987). Although not well studied, it is likely analogous to H_2-producing heterofermentation pathways described in anaerobic bacteria (Saint-Amans et al., 2001), or in amitochondriate eukaryotes (Dyall et al., 2004). These organisms couple pyruvate oxidation to ferredoxin reduction via the enzyme pyruvate-ferredoxin-oxidoreductase (PFR1), as illustrated for *Chlamydomonas* in Figure 7.5. PFR1 oxidizes pyruvate to acetyl-CoA and CO_2, and the reduced ferredoxin is the electron donor to hydrogenase. A variety of [2Fe-2S] ferredoxins mediate electron transfer to algal hydrogenase *in vitro*, and *Chlamydomonas* encodes several [2Fe-2S] ferredoxin homologues. It is unclear if certain isoforms are the preferred physiological electron mediator for one or both of the [FeFe]-hydrogenases, or if unique ferredoxins are exclusively involved in the dark- or light-mediated H_2 production and H_2 uptake pathways (see below). Although no evidence exists at this point, the oxidation of NADH generated during glycolysis could also potentially provide reductant to hydrogenase indirectly (e.g. via NAD(P)H ferredoxin oxidoreductase activity or interacting diaphorase proteins), as proposed at high NADH concentrations in species of clostridia (Saint-Amans et al., 2001) and amitochondriate eukaryotes (Dyall et al., 2004).

C. Hydrogen utilization pathways

Hydrogen uptake activity is also observed in *Chlamydomonas*, and two distinct pathways, as initially described by Gaffron (Gaffron, 1942, 1944; Gaffron and Rubin, 1942) have been examined (Kessler, 1974; Maione and Gibbs, 1986a,b; Chen and Gibbs, 1992). The first pathway, anaerobic CO_2 photoreduction, couples H_2 oxidation and cyclic PS I activity to Rubisco-mediated CO_2 fixation. This light-dependent pathway functions in the absence of PS II activity and requires the anoxic induction of hydrogenase activity. The proposed mechanism involves hydrogenase-catalyzed electron transfer from H_2 to ferredoxin, followed by reduction of NADP to NADPH by ferredoxin-NADP oxidoreductase (FNR). This pathway provides the reductant necessary for CO_2 fixation. Cyclic electron transfer via PS I, using either reduced ferredoxin or NADPH as the mediator(s), supplies the ATP required for CO_2 fixation. Photoreduction can be disrupted by either DBMIB, which

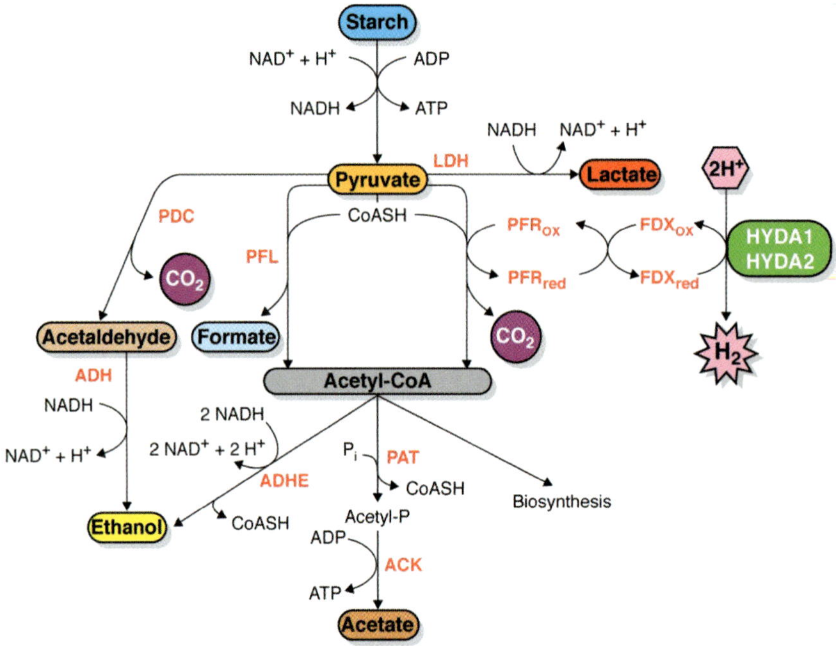

FIGURE 7.5 *Metabolic pathways contributing to the production of the fermentative products secreted during dark, anaerobic acclimation. The four predominant pathways available in Chlamydomonas for the catabolism of pyruvate during anaerobiosis are illustrated. Under these conditions, pyruvate can be metabolized to acetyl-CoA via pyruvate ferredoxin oxidoreductase (PFR1) yielding acetyl-CoA, reduced ferredoxin and CO_2. Alternatively, pyruvate can be cleaved by pyruvate formate lyase (PFL) yielding acetyl-CoA and formate. Ferredoxin (FDX or PETF) is reduced by the action of PFR. The [FeFe]-hydrogenases (HYDA1, HYDA2) can then be reduced by FDX, catalyzing H_2 production. Proteins required for activation of PFL (PFLA) and the HYDA enzymes (HYDEF and HYDG) are not shown. Pyruvate may also be decarboxylated by the putative pyruvate decarboxylase homologues (PDC) resulting in the production of acetaldehyde and CO_2. Under hypoxic conditions, NADH is reoxidized by alcohol/aldehyde dehydrogenase (ADH1) activity. Additional alcohol dehydrogenase homologues encoded in Chlamydomonas may participate in the production of ethanol from acetaldehyde as well. Oxidation of NADH may also be catalyzed by a putative D-lactate dehydrogenase (LDH). The conversion of acetyl-CoA to acetate and CoA via the sequential action of acetate kinase (ACK) and phosphotransacetylase (PAT) results in the production of ATP. The illustration depicts the primary metabolisms putatively leading to the fermentation products observed during dark, anoxic adaptation. For clarity, the illustration does not include other aspects of metabolism that could potentially influence product formation. It should also be noted that multiple homologues for some proteins are present (see Table 7.1).*

prevents plastoquinone oxidation, or cyanide-*p*-trifluoromethoxyphenyl hydrazone (FCCP), which uncouples electron transport from photophosphorylation. The effect of DBMIB demonstrates the importance of electron flow through PS I, and the effect of FCCP indicates that photophosphorylation is required for CO_2 fixation. The addition of exogenous ATP restores photoreduction in FCCP-treated *Chlamydomonas* chloroplasts (Maione and Gibbs, 1986a), indicating that under these conditions NADPH is available for CO_2 fixation.

The second H_2 uptake pathway is the oxy-hydrogen reaction, which in some cases is coupled to CO_2 reduction (Gaffron, 1942, 1944; Kessler,

1974). At O_2 levels below the inhibitory threshold of *in vivo* hydrogenase activity, the concomitant uptake of H_2 and O_2 is observed in the dark (Russell and Gibbs, 1968; Kessler, 1974; Maione and Gibbs, 1986a,b; Chen and Gibbs, 1992). The precise reactions associated with the oxy-hydrogen reaction are still not clearly understood. Maione and Gibbs proposed that electrons from H_2 oxidation, catalyzed by hydrogenase, are used to reduce ferredoxin. Reduced ferredoxin then mediates reduction of the plastoquinone pool, either directly or by the formation of NADPH followed by NAD(P)H/PQ oxidoreductase activity (Maione and Gibbs, 1986b). The chlororespiratory pathway would then reduce O_2, oxidizing the plastoquinone pool and leading to the observed uptake of both H_2 and O_2. This proposed scheme postulates that the plastoquinone pool serves as a crossover point between photosynthesis and respiration (Bennoun, 1982). The coupling of the oxy-hydrogen reaction to dark CO_2 reduction has also been reported (Gaffron, 1942, 1944; Russell and Gibbs, 1968; Maione and Gibbs, 1986b; Chen and Gibbs, 1992). Although the origin of the ATP required for CO_2 reduction under these conditions is unknown, the oxy-hydrogen reaction was proposed to require mitochondrial respiration for full activity (Maione and Gibbs, 1986a,b; Chen and Gibbs, 1992). Despite indications that chlororespiration participates in the oxy-hydrogen reaction coupled to dark CO_2 fixation (Chen and Gibbs, 1992), the chlororespiratory pathway is believed to be either non-electrogenic or poorly electrogenic under these conditions (Cournac et al., 2000; Peltier and Cournac, 2002; Mus et al., 2005). Therefore, it is likely that mitochondrial respiration plays a significant role in the oxy-hydrogen reaction coupled to CO_2 fixation, a potential sink for NADPH generated by H_2 uptake during the oxy-hydrogen reaction, by supplying the required ATP (Maione and Gibbs, 1986a; Chen and Gibbs, 1992).

D. Modulation of hydrogenase activity

Hydrogen photoproduction is significantly influenced by several metabolites including acetate, CO_2, nitrate, and nitrite. Hydrogen photoproduction is clearly enhanced when *Chlamydomonas* is cultured in the presence of acetate (Lien and Pietro, 1981; Bamberger et al., 1982; Gfeller and Gibbs, 1984; Gibbs et al., 1986; Happe et al., 1994). This is presumably a function of the accumulation of reduced organic compounds and starch derived from acetate assimilation (Gibbs et al., 1986; Willeford and Gibbs, 1989). During anaerobic catabolism, these substrates provide electrons to hydrogenase via the plastoquinone pool and PS I activity, and may also be used to respire the low levels of O_2 produced by PS II under H_2 photoproduction conditions. Moreover, Gibbs and co-workers proposed that reductant for H_2 photoproduction in the *Chlamydomonas F60* mutant may be provided by acetate photodissimilation (Gibbs et al., 1986; Willeford and Gibbs, 1989).

This process was proposed to involve acetate metabolism by both the glyoxylate and citric acid cycles. The reduced pyridines generated by these pathways would supply electrons to the photosynthetic electron transport chain for H_2 production via the non-photochemical reduction of the plastoquinone pool and PS I activity. However, caution should be used when extrapolating data obtained using the *F60* mutant, which is defective in photosynthetic CO_2 fixation, to wild-type cells.

Hydrogen photoproduction activity is also significantly affected by CO_2. Cultures grown in the presence of 3% CO_2 and then assayed in the absence of exogenous CO_2 have enhanced H_2 production rates relative to cultures grown with ambient air (Semin et al., 2003). This is likely due to the increased accumulation of intracellular stores of fixed carbon that can be oxidized to provide electrons to the photosynthetic electron transport chain.

Hydrogen photoproduction activity is significantly reduced when assayed *in vivo* in the presence of CO_2 under conditions where CO_2 fixation is able to compete with H_2 photoproduction (Cinco et al., 1993). A low level of CO_2 (0.8 ppm) enhances H_2 photoproduction relative to cultures assayed without CO_2, putatively due to bicarbonate-stimulated PS II activity. However, higher CO_2 concentrations (58 ppm and above) significantly attenuate H_2 photoproduction. This is likely due to the partitioning of photosynthetic reductant away from hydrogenase toward carbon fixation.

The reductive utilization of nitrate and nitrite also significantly suppresses H_2 photoproduction (Aparicio et al., 1985). Subsequent depletion of nitrite results in the restoration of hydrogenase activity. The mechanism by which hydrogenase activity is attenuated is unknown, but is likely the result of a process in which nitrite is reduced to ammonia by nitrite reductase activity (see also Chapter 3). This oxidizes ferredoxin in competition with hydrogenase activity (Aparicio et al., 1985). Sulfate may also provide a similar competition at the level of ferredoxin-sulfite reductase; however, detailed experiments examining this process have not been reported.

Hydrogenase activity in *Chlamydomonas* is dramatically affected by uncouplers of electrogenic phosphorylation, such as carbonylcyanide-*m*-chlorophenylhydrazone (CCCP) or FCCP, and inhibitors of ATP synthesis, such as $NaAsO_4$ (Lien and Pietro, 1981; Bamberger et al., 1982; Happe et al., 1994; Cournac et al., 2002; Lee and Greenbaum, 2003; Kruse et al., 2005a). The addition of uncouplers just prior to anaerobic adaptation suppresses hydrogenase activity presumably by attenuating energy-dependent processes such as hydrogenase expression, maturation, and/or translocation (Lien and Pietro, 1981; Happe et al., 1994). However, their addition to previously adapted cells stimulates H_2 photoproduction by removal of the inhibitory effect of the proton gradient on electron transport (Lien and Pietro, 1981; Bamberger et al., 1982; Happe et al., 1994; Cournac et al., 2002; Lee and Greenbaum, 2003). In the absence of uncouplers, and under conditions where ATP utilization by carbon fixation is limited, proton translocation is inhibited. This results in the acidification of the lumen, causing decreased photosynthetic activity (Lee and Greenbaum, 2003). This situation can be relieved by proton ionophores, resulting in increased electron flux to hydrogenase and an observed increase in H_2 photoproduction. Lee and Greenbaum (2003) proposed uncoupling proton translocation from ATP synthesis as a possible mechanism to enhance H_2 photoproduction. This approach requires introducing anaerobically inducible proton channels that span the thylakoid membrane, and that are independent of ATPase activity. Hydrogen photoproduction via the PS II-independent pathway is also prolonged by the addition of FCCP, which indicates that this pathway is likewise limited by an electrochemical gradient across the thylakoid membranes (Cournac et al., 2002).

The addition of proton ionophores stimulates anaerobic starch catabolism in both the light and dark (Bamberger et al., 1982; Gfeller and Gibbs, 1984). This process may contribute to the enhanced rates of H_2 photoproduction observed in the presence of uncouplers, as increased electron flux from NADH oxidation enters the photosynthetic electron transport chain at the plastoquinone pool via NPQR activity. The addition of FCCP to dark, anaerobically adapted cells stimulates starch degradation (Bamberger et al., 1982; Gfeller and Gibbs, 1984), but the dark, anaerobic yields of fermentative H_2 production are decreased (Bamberger et al., 1982; Cournac et al., 2002).

Although the mechanisms responsible for the dramatic effects of uncouplers on hydrogenase activity are not entirely defined, it is clear that cellular electrochemical gradients (and therefore ATP levels) significantly influence H_2 metabolism in green algae. It is also evident that numerous metabolic pathways, requiring activities in both the mitochondria and the chloroplast, are able to influence hydrogenase activities. The ability of these integrated processes to modulate H_2 production dramatically was highlighted by the isolation and characterization of the *Chlamydomonas* insertional mutant *stm6* (Schonfeld et al., 2004; Kruse et al., 2005a).

The *stm6* mutant is inhibited in cyclic electron transport during anaerobiosis, overaccumulates starch, has a decreased number of active PS II centers, exhibits increased rates of respiration, has increased H_2 production activity (up to 13-fold relative to the progenitor strain), and is capable of photoproducing H_2 at higher rates and for longer time periods compared to its progenitor under sulfur-deprived conditions (Kruse et al., 2005a,b; Rupprecht et al., 2006). The *stm6* mutant contains a disrupted *MOC1* gene. The nuclear-encoded MOC1 protein, which is targeted to the mitochondrion, is involved in the assembly of the mitochondrial respiratory chain in the light, and is homologous to a human mitochondrial transcription termination factor. Although several mitochondrial processes are significantly altered in *stm6*, a number of the phenotypes observed in this mutant, such as the inhibition of state transitions, starch accumulation, and decreased PS

II activity are manifested in the chloroplast (Schonfeld et al., 2004; Kruse et al., 2005a).

Hydrogen production in *stm6* was initially investigated under the premise that cyclic electron transport effectively competes with hydrogenase for electrons, and that mutants locked in State 1 may have increased H_2 production rates. Although this is observed in the *stm6* mutant, other state transition mutants were also examined, and did not show similarly enhanced levels of H_2 photoproduction (Kruse et al., 2005a). Therefore, the observed dramatic increase in hydrogenase activity in *stm6* is likely a consequence of the synergistic effects of attenuated cyclic electron transport, starch overaccumulation (which could provide increased levels of endogenous substrate to support H_2 production and increased rates of respiration), and lower O_2 tensions due to reduced PS II activity (Kruse et al., 2005a).

In summary, it is clear that H_2 metabolism is incorporated into several physiological processes in green algae. Additional examination of the role of hydrogenase activity in *Chlamydomonas* is necessary to deconvolute the complicated metabolic networks influencing hydrogenase activity, better define how the spatial and temporal distribution of hydrogenases within the cell influence the ability of these enzymes to utilize cellular reductant, and potentially maximize the flux of low-potential electrons to H_2 production.

IV. FERMENTATIVE METABOLISM

Hydrogenase activity is a component of *Chlamydomonas* fermentative metabolism in both the light and dark. Based on the available genome sequence and published results, it is clear that *Chlamydomonas* encodes numerous proteins associated with several diverse anoxic pathways (Hemschemeier and Happe, 2005; Atteia et al., 2006; Mus et al., 2007). Moreover, *Chlamydomonas* has several proteins usually associated with strictly anaerobic organisms, including the [FeFe]-hydrogenase enzymes. When *Chlamydomonas* is grown with limited aeration under low light intensities, O_2 consumption by respiration can exceed the rate of photosynthetic O_2 production, leading to hypoxia (Moseley et al., 2000; Quinn et al., 2002), likely a common condition in the natural environment. Pyruvate is the central metabolite of glycolysis, as illustrated in Figure 7.5, and *Chlamydomonas* encodes homologues for each of the enzymes required for the predominant pathways used in anaerobic pyruvate metabolism. This metabolic diversity is extraordinary in a single organism and indicates that *Chlamydomonas* is particularly well-suited to adapt metabolically to diverse and dynamic environmental conditions including hypoxia.

In *Chlamydomonas*, pyruvate is likely metabolized under fermentative conditions to acetyl-CoA and formate via the pyruvate formate lyase (PFL1) pathway; acetyl-CoA and CO_2 via the PFR1 pathway, which results in the reduction of ferredoxin; acetylaldehyde and CO_2 via the PDC3 pathway; or lactate coupled to NADH oxidation via lactate dehydrogenase. Reduced ferredoxin generated by PFR1 can subsequently be oxidized by a variety of cellular redox proteins including the [FeFe]-hydrogenases, resulting in the observed production of H_2 from starch catabolism. The presence of both the PFR1 and PFL1 pathways is unique in eukaryotes and highlights the diversity of metabolisms available in *Chlamydomonas*. The PFL1 protein was shown to be present in both the chloroplast and mitochondrion (Atteia et al., 2006).

Acetyl-CoA is a product of both the PFL1 and PFR1 pathways and can be converted to acetate via the phosphotransacetylase (PAT1/PAT2) and acetate kinase (ACK1/ACK2) pathway, resulting in ATP production. *Chlamydomonas* has two genes encoding each of these proteins. The PAT1 and ACK2 proteins are found in mitochondria, whereas PAT2 and ACK1, whose genes are adjacent to *HYDA2*, are putatively localized in the chloroplast (Atteia et al., 2006). The production of acetate results in substrate-level formation of ATP; however, NADH reoxidation is required to sustain glycolysis. Conversion of acetyl-CoA to ethanol via acetaldehyde oxidizes two NADH molecules and requires aldehyde and alcohol dehydrogenase activities. Both of these activities are likely catalyzed by the alcohol/aldehyde dehydrogenase (ADH1) homologue (Hemschemeier and Happe, 2005; Atteia et al., 2006; Mus et al., 2007). The *Chlamydomonas ADH1* gene model is homologous to other dual function dehydrogenases that are usually designated ADHE to distinguish the protein from other alcohol dehydrogenases without aldehyde dehydrogenase activity. The *Chlamydomonas* genome is also predicted to encode other alcohol dehydrogenase enzymes without aldehyde dehydrogenase activity (Table 7.1). Ethanol may also be formed via acetaldehyde by the PDC3 pathway, which oxidizes one NADH. *Chlamydomonas* is rare among eukaryotes in having both ethanol production pathways. The cellular locations of the enzymes responsible for ethanol formation, as well as PFR1, are unclear. Because hydrogenase activity is localized to the chloroplast, it is expected that PFR1 is similarly located, and the *PFR1* gene encodes a putative transit peptide consistent with chloroplast targeting.

Formate, acetate, and ethanol are the major dark fermentative products in *Chlamydomonas*, and H_2 and CO_2 gases are emitted as minor products (Gfeller and Gibbs, 1984; Kreuzberg, 1984; Ohta et al., 1987). Glycerol and lactate have also been reported under certain culture conditions (Kreuzberg, 1984). The accumulation of ethanol and formate and acidification of the cellular environment by organic acids can be toxic to the cells, and product ratios are likely to reflect the need to balance ATP formation, NADH oxidation, and suppress the accumulation of potentially toxic fermentation products. The respective ratios of formate:acetate:ethanol have been reported to be 2:1:1 (Gfeller and Gibbs, 1984; Kreuzberg, 1984), as well as 2:2:1 (Ohta

Table 7.1 Characterized and putative proteins proposed to be associated with the production of observed fermentation products in *Chlamydomonas*

Protein name	Accession #	Function and localization
ACK1	EDP02500	Acetate kinase; acetokinase; potential N-terminal organelle targeting sequence
ACK2	EDP04790	Acetate kinase; acetokinase; compartmentation unclear
ADH1	ED096497	Dual function alcohol dehydrogenase/acetaldehyde dehydrogenase
ADH	ED098807	Putative iron-containing aldehyde-alcohol dehydrogenase
ADH	ED098866	Putative iron-containing aldehyde-alcohol dehydrogenase
FDX2	ED099795	Fe2S2 containing redox protein (ferredoxin), predicted chloroplast localization
FDX3	EDP05114	Fe2S2 containing redox protein (ferredoxin), predicted chloroplast localization
FDX4	ED098400	Fe2S2 containing redox protein (ferredoxin), predicted chloroplast localization
FDX5	EDP04711	Fe2S2 containing redox protein (ferredoxin), predicted chloroplast localization
FDX6	ED096814	Fe2S2 containing redox protein (ferredoxin), predicted chloroplast localization
HYDA1	EDP03402	Chloroplast Fe-hydrogenase
HYDA2	EDP02498	Chloroplast Fe-hydrogenase
HYDEF	EDP05198	Iron hydrogenase assembly protein, contains domains homologous to prokaryotic HydE and HydF; radical SAM domain present in N-terminal region
HYDG	EDP05052	Related to thiazole biosynthesis protein thiH/O. Radical SAM protein required for the assembly of an active [Fe]-hydrogenase
LDH1	ABY77748	Putative lactate dehydrogenase
PAT1	EDP04895	Putative phosphate acetyltransferase phosphotransacetylase; similarity to prokaryotic phosphate acetyltransferase
PAT2	EDP02499	Putative phosphate acetyltransferase phosphotransacetylase; similarity to prokaryotic phosphate acetyltransferase
PDC3	EDP06212	Putative decarboxylase
PETF	EDP03827	Fe2S2 containing redox protein (ferredoxin), predicted chloroplast localization
PFL1	EDP09457	Pyruvate-formate lyase; may contain organelle targeting peptide
PFLA1	ED097895	Pyruvate formate lyase activating enzyme
PFR1	ED097718	Pyruvate:ferredoxin oxidoreductase

et al., 1987; Mus et al., 2007). Culture and induction conditions, and the use of different laboratory strains, likely influence the product ratios. Moreover, acetate is often present in the growth medium and this organic acid may be secreted during anaerobiosis without having originated from starch catabolism. If PFL1 activity is solely responsible for the conversion of pyruvate into acetyl-CoA, a 2:1:1 product ratio would be expected from starch-derived metabolites. A reduction in the formate stoichiometry suggests that PFL1 and PFR1 (or PDC3) are acting in concert to regulate the fermentative metabolism of pyruvate in *Chlamydomonas*.

Quantitative RT-PCR demonstrated that *PFR1* expression increases significantly during anaerobiosis, as does transcript abundance for HYDA1, HYDA2, HYDEF, and HYDG (Mus et al., 2007). The gene expression and metabolite data suggest that the PFR1 pathway is responsible for a significant amount of pyruvate catabolism during anaerobiosis under the experimental conditions used, concomitant with substantial PFL1 activity. Increasing the yields of dark, fermentative H_2 from *Chlamydomonas* will therefore likely require disruption of PFL1 to increase the flux of pyruvate to acetyl-CoA through the PFR1 pathway, to increase in turn the levels of reduced ferredoxin to drive H_2 production. Characterized and putative proteins that are relevant to the fermentation metabolisms described above, are listed in Table 7.1.

It is clear that numerous fermentative pathways are available to *Chlamydomonas* and that apparently parallel metabolic processes are occurring to some extent in both mitochondria and the chloroplast. Cellular localization of the relevant proteins, as well as metabolite analysis, will provide significant new insights into the remarkably unique anoxic metabolisms available in *Chlamydomonas*. Moreover, relatively little is known regarding the level of metabolite exchange between organelles during anoxia, the regulatory mechanisms by which *Chlamydomonas* senses and acclimates to anoxia, the repertoire of proteins required during anaerobiosis and how these proteins influence overall cellular metabolism, and the role of metabolic fluxes in sustaining H_2 production activity during anaerobiosis.

V. GENOMICS AND SYSTEMS BIOLOGY

The nearly complete *Chlamydomonas* genome sequence has provided the groundwork for experimental examination of the transcriptome, proteome, metabolome, and localisome, as well as for computational systems biology efforts. A better understanding of anaerobic metabolism in *Chlamydomonas* and metabolic fluxes associated with diurnal periods of light and dark will facilitate the development of a more comprehensive model of anoxic algal physiology. Several studies have linked fermentation and photosynthetic electron transport to H_2 production in *Chlamydomonas*. However, a more precise knowledge of the metabolic and regulatory context required for H_2 production will be necessary to understand current limitations in H_2 yields.

High-density, oligonucleotide-based microarrays have provided unique insights into the genome-wide responses initiated during dark, anoxic acclimation (Mus et al., 2007). The levels of more than 500 transcripts were observed to be significantly upregulated as a consequence of anoxia. As expected, several of these transcripts were associated with fermentative metabolism. Moreover, several genes encoding homologues of proteins involved in transcriptional/translational regulation, posttranslational modification, and stress responses also increased as the cells acclimated to anoxia. Insights from the microarray data are being integrated into metabolomic and proteomic data sets, and will soon be integrated into a larger systems biology framework focused on understanding the flexibility of whole cell metabolism under dynamic environmental conditions.

The microarray data also revealed the increased expression of genes related to oxidative stress, particularly catalase. Although intriguing, this is not entirely unexpected given that genes encoding O_2-sensitive proteins in other organisms are present in operons that also encode high-affinity oxidases. Since adaptation to anoxia requires cells to invest substantial amounts of cellular energy in the biosynthesis of O_2-labile proteins, it is conceivable that these organisms have also evolved mechanisms to protect O_2-sensitive enzymes from reactive species formed by exposure to low levels of O_2, which is likely to occur in natural settings.

The microarray data also reveal that three proline hydroxylases are significantly upregulated in response to anaerobiosis. Although proline hydroxylation is ubiquitous in algal cell wall proteins, it is unlikely that these gene products are involved in protein hydroxylation during anaerobiosis, as proline hydroxylases require O_2 as a substrate. Proline hydroxylation is used by mammalian cells to sense O_2 and subsequently target select proteins for proteolytic degradation in response to O_2 (Ivan et al., 2001; Schofield and Ratcliffe, 2005; Stiehl et al., 2006). Although highly speculative at this point, it would represent a particularly elegant mechanism to sense O_2 and target proteins used in fermentation for turnover as the cell transitions back to aerobiosis. Intriguingly, several O_2-sensitive proteins in *Chlamydomonas*, including the [FeFe]-hydrogenases, have unique amino acid insertions containing proline motifs that are not found in their bacterial counterparts, but are conserved in the proteins of other green algae.

Another observation from the microarray data is that the majority (>70%) of differentially expressed genes encode putative proteins of unknown function. *Chlamydomonas* is a metabolically versatile organism that is able to perform photosynthesis and aerobic respiration, as well as ameliorate the toxic consequences of photosynthetically generated O_2. It has several genes encoding proteins associated with anaerobic respiration, as well as one of the most extensive sets of fermentative enzymes yet observed, and *Chlamydomonas* is capable of using diverse anaerobic pathways that are unprecedented in other sequenced organisms. Moreover, under conditions of sulfate deprivation in the light, *Chlamydomonas* can balance the pathways of photosynthesis, aerobic respiration, and fermentation. It is likely that some of the proteins currently of unknown function are involved in the regulation, metabolic partitioning, and/or function of *Chlamydomonas'* extraordinary metabolic networks.

Accurate models of overall cellular metabolism will have to account for the diversity of metabolic processes available in *Chlamydomonas*, as well as consider metabolite fluxes in multiple cellular compartments and cross-talk between organelles. Additional examination of the proteins involved in anoxic adaptation, as well as accurate localization of the proteins involved, will provide a foundation for metabolic engineering and modeling, as well as facilitate a better understanding of the metabolic fluxes occurring in *Chlamydomonas*. This information may potentially be leveraged into metabolic engineering strategies designed to optimize the production of fermentative products including H_2 and/or ethanol.

VI. SULFUR DEPRIVATION

A short-term method to circumvent the O_2 sensitivity of algal H_2 production, developed by Melis et al. (2000), resulted in sustainable H_2 production for a period of approximately 4 days, an achievement that demonstrated the potential of green algae for providing large amounts of H_2 (Happe et al., 2002; Melis et al., 2004, 2007; Ghirardi, 2006; Seibert et al., 2008). Depriving cultures of sulfate results in the partial and reversible inactivation of O_2 evolution. In the absence of sulfate, protein synthesis is geared towards those products that are necessary for viability under conditions of decreased metabolic activity (Wykoff et al., 1998; see also Chapter 6). As a consequence, the D1 protein of PS II (which is essential for photosynthesis but turns over rapidly due to continuous photodamage) cannot be replaced quickly, leading to the gradual loss of O_2 evolution. Simultaneously, as a response to nutrient stress, sulfur-deprived algae start overaccumulating starch to amounts 8–20 times higher than under non-stressed conditions (Tsygankov et al., 2002; Zhang et al., 2002). When the rate of O_2 evolution decreases below the rate of respiratory O_2 consumption, the cultures undergo a rapid shift to anaerobiosis (Antal et al., 2003). The establishment of anaerobiosis in the cultures triggers a number of effects (Melis and Happe, 2001): (a) *HYDA1* and *HYDA2* are activated and the respective functional proteins can be detected after about 3–4 hours (Forestier et al., 2003), (b) measurable H_2 production is detected after another 5 hours (Antal et al., 2003), (c) starch accumulation ceases and is replaced by starch degradation (Tsygankov et al., 2002; Zhang et al., 2002), and (d) starch degradation leads to increased levels of reductant for H_2 production, as well as pyruvate, which is dissimilated into acetate, formate (Happe et al., 2002; Tsygankov et al., 2002; Kosourov et al., 2003), and, in some cases ethanol (Winkler et al., 2002b; Kosourov et al., 2003). It is clear that under these anaerobic conditions, no CO_2 fixation occurs

(Melis et al., 2000), as demonstrated by the early loss of Rubisco activity (Zhang et al., 2002).

Initially, starch accumulates during the early stages of sulfur deprivation. This energy store is degraded at later times of sulfur deprivation and provides substrate for the respiratory consumption of O_2 and the establishment of anaerobiosis under these conditions. The importance of starch overaccumulation was also demonstrated by the work of Kruse et al. (2005a) who described corresponding increases in H_2 production rates by the *Chlamydomonas* mutant *stm6*, which was discussed in section III.D. Moreover, the decreased level of H_2 production by a Rubisco-less mutant (White and Melis, 2006) supports the crucial role of starch (and respiration) in sustaining H_2 production during sulfur deprivation.

An additional role for starch degradation is related to the existence of the DCMU-insensitive pathway addressed in section III.B. Besides providing electrons to the plastoquinone pool, this pathway participates in keeping the plastoquinone pool reduced during the anaerobic stage of sulfur deprivation. A more reduced plastoquinone pool correlates with a higher level of expression of the hydrogenase genes (Posewitz et al., 2004b), although the exact mechanism underlying this effect is currently unknown.

Although elevated starch content of a culture is a *sine qua non* for high rates of H_2 photoproduction, the presence of acetate in the growth medium is not. The usual method for H_2 production by sulfur-deprived cultures involves aerobic photoheterotrophic growth conditions, followed by sulfur deprivation in tris-acetate-phosphate (TAP) medium. Several unsuccessful attempts were made to generate H_2 under photoautotrophic conditions in the past. Anaerobiosis and H_2 photoproduction were successfully observed under photoautotrophic conditions upon addition of DCMU, an inhibitor of PS II activity (Fouchard et al., 2005). Low rates of H_2 production (on the order of a few μl/hour) were reported when the green alga *Platymonas subcordiformis* was sulfur-deprived and sealed in an anaerobic photobioreactor under photoautotrophic conditions (Guan et al., 2004), and the authors reported an increase of about 40% in the rates when a carbon source was added to the reactor. Subsequently, it was shown that photoautotrophic *Chlamydomonas* cultures can produce H_2 under sulfur-deprived conditions when supplied with CO_2 gas instead of acetate during the pre-growth and the early stages of sulfur deprivation. The rates of H_2 production and the total yields of H_2, however, were still low compared to previously reported data obtained in the presence of acetate, and a special light regime was required to optimize the rates of H_2 production (Tsygankov et al., 2006). The protocol involves changing the experimental conditions such that both starch accumulation and starch degradation/O_2 consumption are optimized, and further optimizations may eventually lead to rates comparable to those obtained in the presence of acetate.

Surprisingly, not all algal species that produce H_2 under anaerobic conditions respond in the same manner to anaerobiosis induced by sulfur deprivation. For example, both *S. obliquus* and *C. fusca* contain hydrogenases whose activities are induced by purging cultures with an inert gas. However, only the former photoproduced H_2 gas under sulfur deprivation (Winkler et al., 2002b).

Production of H_2 under sulfur deprivation conditions eventually stops, possibly due to the more generalized effect of sulfur deprivation on cell metabolism. H_2 production can be sustained for longer periods, however, by operating algal cell suspensions in cycles of +S/–S (Ghirardi et al., 2000), or by physically separating the O_2 from the H_2 evolution stages in separate photobioreactors (Fedorov et al., 2005). The latter method is based on the use of a dual bioreactor system where in the first, cultures are grown photosynthetically in chemostat mode in the presence of limiting amounts of sulfate, and in the second are sulfur-deprived cultures that are already producing H_2 gas. The cells in the second bioreactor are continuously replaced with fresh cells from the first bioreactor. The system produced H_2 for a total of 6 months, albeit at low rates due to the low cell density achievable under sulfur-limited conditions.

The maximum theoretical light-conversion efficiency (percent of incident sunlight energy that is converted to energy in H_2) of green algae is about 10–13% (Ghirardi, 2006). Since this efficiency is limited by the electron transfer reactions of photosynthesis at high light intensity, the 10–13% level can only be achieved under sunlight conditions in mutants with small light-absorbing antennae (Melis et al., 1998). In sulfur-deprived cultures, the light conversion efficiency under limiting light intensity is expected to decrease to 1%, because photosynthetic capacity of the cultures is only about 10% of that of sulfur-replete cultures (Melis et al., 2000). Rao and Cammack (2001) suggested that 10% efficiency was probably required for a commercially viable system. However, an analysis by Amos (2004) indicated that applied systems might be viable at lower efficiencies as well, depending on the cost of photobioreactor material. Currently, suspension cultures yield a conversion efficiency of 0.24% at most (Ghirardi, 2006), possibly due to the low cell density used. Work focused on methods to increase cell density under sulfur-deprived conditions involved immobilizing cultures onto inert matrices such as glass fibers (Laurinavichene et al., 2006), allowing the culture medium to be changed from +S to –S without concomitant handling of the cell cultures, and yielding light conversion efficiencies of about 0.36% at low light intensities (Ghirardi, 2006).

The low light conversion efficiencies are due to the existence of competing or wasteful pathways associated with photosynthetic H_2 production under sulfur deprivation. It was suggested that in the absence of Rubisco, photosynthetically generated reductant would accumulate (Greenbaum et al., 1995), triggering a shift to State II, in which cyclic electron transport around

PS I predominates (de Vitry et al., 2004), and/or quenching of PS II-absorbed light by the activation of the xanthophyll cycle (Niyogi, 1999). Both of these responses would cause down-regulation of photosynthetic electron transport from water to the hydrogenase. The contribution of cyclic electron transport to photosynthetic electron transport in sulfur-deprived cultures has been demonstrated in the *stm6* mutant (see section III.D). On the other hand, the much smaller effect of uncouplers on H_2 production during sulfur deprivation (Zhang and Melis, 2002) suggests that the quenching of absorbed light plays a minor role in down-regulating electron transport rates under sulfur deprivation conditions.

An alternative approach that further extends the usefulness of sulfur deprivation was developed by the Melis group. Mutants affected in the *SULP* gene, which encodes a protein involved in sulfate uptake into the chloroplasts (Chen and Melis, 2004), were obtained by antisense technology (Chen et al., 2003). The rates of photosynthetic O_2 production and respiratory O_2 consumption (the P/R ratio; Melis and Melnicki, 2006) of the mutants was much closer to one, which allowed the cultures to become anaerobic even in the presence of limiting concentrations of sulfate (75 micromoles), and to photoproduce H_2 gas at levels comparable to a sulfur-deprived wild-type strain (Chen et al., 2005).

The sulfur deprivation method was applied to photosynthetic mutants that already exhibited lower rates of O_2 evolution, to test whether they would transition more quickly to anaerobiosis and thus produce more H_2. D1 mutants with different levels of attenuated PS II activity were shown to undergo faster transition to anaerobiosis upon sulfur deprivation. However, this was accompanied by low starch accumulation and reduced H_2 photoproduction (Makarova et al., 2007). The data confirm a strong correlation between the capability to accumulate starch and the subsequent ability to photoproduce H_2 at high rates.

Further improvements in the H_2 production of sulfur-deprived systems will depend on an overall understanding of the large number of metabolic and regulatory functions affected by this particular nutrient stress. Preliminary global gene expression studies were performed with cultures that were sulfur-deprived for up to 24 hours (Zhang et al., 2004). Although useful insights were obtained regarding the early response mechanisms activated upon sulfate stress (Pollock et al., 2005), these studies did not examine the role of long-term adaptation strategies such as H_2 production.

Additional improvements in the system can be achieved through a better understanding of the H_2 production process itself, which has been attempted through several approaches. Hahn et al. (2004) examined the effect of process variables such as light intensity, reactor volume, temperature, algal concentration, and culture agitation, using a factorial-analysis approach. They identified the surface area/light pathway length as being critical for future scale-up efforts. A similar factorial approach, response surface methodology, was used to examine the optimal nutrient (N, S, and P) and pH configurations that would lead to high rates of growth under sulfur-replete conditions, and subsequent high rates of H_2 production under sulfur deprivation (Jo et al., 2006). Finally, a power-law sensitivity approach yielded an initial analysis of possible biochemical limiting factors regulating H_2 production in sulfur-deprived *Chlamydomonas* (Horner and Wolinsky, 2002). Although this model did not include reductant donation through the DCMU-insensitive pathway, the authors accurately modeled a number of features of the sulfur deprivation response. Subsequent work modeled the effect of light intensity on H_2 production rates, once again demonstrating the usefulness of S-state (an analysis based on power laws that attempt to mathematically represent the synergism and saturation between different pathways) to understand the metabolic processes (Park and Moon, 2007).

VII. OUTLOOK

Hydrogenases are the focus of research efforts aimed at producing H_2 as a renewable, non-polluting energy carrier, and organisms such as *Chlamydomonas* are currently being emphasized since they have the ability to couple H_2 production to photosynthesis. However, more efficient interactions with the photosynthetic electron-transport chain under oxygenic conditions and/or an O_2-tolerant enzyme are probably required to improve yields.

The O_2 sensitivity of hydrogenase transcription, maturation, and catalytic activity are also major issues delaying the development of an applied algal photobiological H_2-producing system. Other major challenges to be solved include limitations on the rate of electron transport and the predominance of cyclic electron transfer under H_2-producing conditions, competition for reductants from ferredoxin among different pathways, and the low solar conversion efficiency of H_2 photoproduction (Ghirardi et al., 2006, 2007; Seibert et al., 2008).

Several contemporary studies examining hydrogenases and H_2 metabolism in *Chlamydomonas* have provided substantial insights into the fields of H_2 photoproduction, hydrogenase activity and maturation, and anoxic algal physiology. These include: (1) determination of genomic, cDNA, and protein sequences for several green algal [FeFe]-hydrogenases; (2) discovery of maturation proteins required for [FeFe]-hydrogenase assembly; (3) development of heterologous systems for expression of [FeFe]-hydrogenases; (4) production of bulk quantities of H_2 by depriving *Chlamydomonas* cultures of sulfate; (5) optimization of the sulfur deprivation response to prolong sustained H_2 photoproduction over a period of several months; (6) isolation of a state transition mutant with enhanced H_2 photoproduction rates and productivity; (7) examination of the partitioning of photosynthetic reductant into hydrogenase and competing pathways; (8) establishing the importance of starch in hydrogenase gene transcription and activity; (9) development of sophisticated models detailing the putative O_2 and H_2

diffusion channels in [FeFe]-hydrogenases; and (10) the first genome-wide profile of changes in gene expression that occur as cultures acclimate to anoxia. Future research into algal H_2 metabolism and the hydrogenase enzymes in *Chlamydomonas* will provide substantial and exciting new insights into the role of these fascinating enzymes in green algae, as well as potentially reveal mechanisms to utilize hydrogenases for the more efficient production of biological H_2.

ACKNOWLEDGMENTS

The authors would like to thank the U.S. Air Force Office of Scientific Research (MURI award FA9550-05-01-0365); the U.S. Department of Energy's Office of Science BES and BER programs, the Hydrogen, Fuel Cell and Infrastructure Technologies Program; the National Air and Space Administration (GSRP award NNG05GL52H); the NSF (award 0328187) for support. We also thank Dr. Paul King of NREL, and Arthur Grossman of the Carnegie Institution, for insightful comments and suggestions, and Drs. John Peters and David Mulder at Montana State University for previewing portions of this chapter and the preparation of Figure 7.2. Thanks are also extended to Andrea Manuell for assistance in the preparation of Figure 7.1.

REFERENCES

Abeles, F.B. (1964). Cell-free hydrogenase from *Chlamydomonas*. *Plant Physiol.* **39**, 169–176.

Adams, M.W.W. (1990). The structure and mechanism of iron-hydrogenases. *Biochim. Biophys. Acta,* **1020**, 115–145.

Amos, W. (2004). Updated cost analysis of photobiological hydrogen production from *Chlamydomonas reinhardtii* green algae. NREL/MP-560-35593. http://www.nrel.gov/docs/fy04osti/35593.pdf.

Antal, T.K., Krendeleva, T.E., Laurinavichene, T.V., Makarova, V.V., Ghirardi, M.L., Rubin, A.B., Tsygankov, A.A., and Seibert, M. (2003). The dependence of algal H_2 production on Photosystem II and O_2 consumption activities in sulfur-deprived *Chlamydomonas reinhardtii* cells. *Biochim. Biophys. Acta* **1607**, 153–160.

Aparicio, P.J., Azuara, M.P., Ballesteros, A., and Fernandez, V.M. (1985). Effects of light intensity and oxidized nitrogen sources on hydrogen production by *Chlamydomonas reinhardii*. *Plant Physiol.* **78**, 803–806.

Armstrong, F.A. (2004). Hydrogenases: Active site puzzles and progress. *Curr. Opin. Chem. Biol* **8**, 133–140.

Atteia, A., Van Lis, R., Gelius-Dietrich, G., Adrait, A., Garin, J., Joyard, J., Rolland, N., and Martin, W. (2006). Pyruvate formate-lyase and a novel route of eukaryotic ATP synthesis in *Chlamydomonas* mitochondria. *J. Biol. Chem.* **281**, 9909–9918.

Balk, J., Pierik, A.J., Netz, D.J. Muhenhoff, U., and Lill, R. (2004). The hydrogenase-like Nar1p is essential for maturation of cytosolic and nuclear iron-sulphur proteins. *EMBO J.* **23**, 2105–2115.

Balk, J., Pierik, A.J., Netz, D.J. Muhienhoff, U. and Lill, R. (2004). The hydrogenase-like Nar1p is essential for maturation of cytosolic and nuclear iron-sulphur proteins. *EMBO J.* **23**, 2105–2115.

Bamberger, E.S., King, D., Erbes, D.L., and Gibbs, M. (1982). H_2 and CO_2 evolution by anaerobically adapted *Chlamydomonas reinhardtii* F-60. *Plant Physiol.* **69**, 1268–1273.

Bedard, J. and Jarvis, P. (2005). Recognition and envelope translocation of chloroplast preproteins. *J. Exp. Bot.* **56**, 2287–2320.

Ben-Amotz, A. and Gibbs, M. (1975). H_2 metabolism in photosynthetic organisms II. Light-dependent H_2 evolution by preparations from *Chlamydomonas, Scenedesmus* and spinach. *Biochem. Biophys. Res. Comm.* **64**, 355–359.

Bennoun, P. (1982). Evidence for a respiratory chain in the chloroplast. *Proc. Natl. Acad. Sci. U. S. A.* **79**, 4352–4356.

Bennoun, P. (1998). Chlororespiration, sixteen years later. In: *The Molecular Biology of Chloroplasts and Mitochondria in Chlamydomonas* (J.D. Rochaix, M. Goldschmidt-Clermont, and S. Merchant, Eds.), pp. 675–683. Dordrecht, Kluwer Academic Publishers.

Bernard, L., Desplats, C., Mus, F., Cuine, S., Cournac, L., and Peltier, G. (2006). *Agrobacterium tumefaciens* type II NADH dehydrogenase. Characterization and interactions with bacterial and thylakoid membranes. *FEBS J.* **273**, 3625–3637.

Bock, A., King, P.W., Blokesch, M., and Posewitz, M.C. (2006). Maturation of hydrogenases. *Adv. Microb. Physiol.* **51**, 1–71.

Boichenko, V.A. and Hoffmann, P. (1994). Photosynthetic hydrogen production in prokaryotes and eukaryotes: Occurrence, mechanism, and functions. *Photosynthetica* **30**, 527–552.

Boichenko, V.A., Greenbaum, E., and Seibert, M. (2004). Hydrogen production by photosynthetic microorganisms. In: *Photoconversion of Solar Energy, Molecular to Global Photosynthesis*, Vol. 2 (M.D. Archer and J. Barber, Eds.), pp. 397–452. Imperial College Press, London.

Brand, J.J., Wright, J.N., and Lien, S. (1989). Hydrogen production by eukaryotic algae. *Biotech. Bioeng.* **33**, 1482–1488.

Brazzolotto, X., Rubach, J.K., Gaillard, J., Gambarelli, S., Atta, M., and Fontecave, M. (2006). The [Fe-Fe]-hydrogenase maturation protein HydF from *Thermotoga maritima* is a GTPase with an iron-sulfur cluster. *J. Biol. Chem.* **281**, 769–774.

Buhrke, T., Lenz, O., Krauss, N., and Friedrich, B. (2005). Oxygen tolerance of the H_2-sensing [NiFe] hydrogenase from *Ralstonia eutropha* H16 is based on limited access of oxygen to the active site. *J. Biol. Chem.* **280**, 23791–23796.

Casalot, L. and Rousset, M. (2001). Maturation of the [NiFe] hydrogenases. *Trends Microbiol.* **9**, 228–237.

Chen, C. and Gibbs, M. (1992). Coupling of carbon-dioxide fixation to the oxyhydrogen reaction in the isolated chloroplast of *Chlamydomonas reinhardtii*. *Plant Physiol.* **100**, 1361–1365.

Chen, H.C. and Melis, A. (2004). Localization and function of SulP, a nuclear-encoded chloroplast sulfate permease in *Chlamydomonas reinhardtii*. *Planta* **220**, 198–210.

Chen, H.C., Yokthongwattana, K., Newton, A.J., and Melis, A. (2003). SulP, a nuclear gene encoding a putative chloroplast-targeted sulfate permease in *Chlamydomonas reinhardtii*. *Planta* **218**, 98–106.

Chen, H.C., Newton, A.J., and Melis, A. (2005). Role of SulP, a nuclear-encoded chloroplast sulfate permease, in sulfate transport and H_2 evolution in *Chlamydomonas reinhardtii*. *Photosynth. Res.* **84**, 289–296.

Cinco, R., Macinnis, J., and Greenbaum, E. (1993). The role of carbon-dioxide in light-activated hydrogen-production by *Chlamydomonas reinhardtii*. *Photosynth. Res.* **38**, 27–33.

Cohen, J., Kim, K., Posewitz, M., Ghirardi, M.L., Schulten, K., Seibert, M., and King, P. (2005a). Molecular dynamics and experimental investigation of H_2 and O_2 diffusion in [Fe]-hydrogenase. *Biochem. Soc. Trans.* **33**, 80–82.

Cohen, J., Kim, K., King, P., Seibert, M., and Schulten, K. (2005b). Finding gas diffusion pathways in proteins: Application to O_2 and H_2 transport in CpI [FeFe]-hydrogenase and the role of packing defects. *Structure* **13**, 1321–1329.

Cournac, L., Redding, K., Ravenel, J., Rumeau, D., Josse, E.M., Kuntz, M., and Peltier, G. (2000). Electron flow between photosystem II and oxygen in chloroplasts of photosystem I-deficient algae is mediated by a quinol oxidase involved in chlororespiration. *J. Biol. Chem.* **275**, 17256–17262.

Cournac, L., Mus, F., Bernard, L., Guedeney, G., Vignais, P., and Peltier, G. (2002). Limiting steps of hydrogen production in *Chlamydomonas reinhardtii* and *Synechocystis* PCC 6803 as analysed by light-induced gas exchange transients. *Int. J. Hyd. Energy*, **27**, 1229–1237.

De Lacey, A., Fernandez, V., and Rousset, M. (2005). Native and mutant nickel-iron hydrogenases: Unravelling structure and function. *Coord. Chem. Rev.* **249**, 1596–1608.

de Vitry, C., Ouyang, Y., Finazzi, G., Wollman, F.A., and Kallas, T. (2004). The chloroplast Rieske iron-sulfur protein. At the crossroad of electron transport and signal transduction. *J. Biol. Chem.* **279**, 44621–44627.

Dos Santos, P., Dean, D., Hu, Y., and Ribbe, M. (2004). Formation and insertion of the nitrogenase iron-molybdenum cofactor. *Chem. Rev.* **104**, 1159–1173.

Duche, O., Elsen, S., Cournac, L., and Colbeau, A. (2005). Enlarging the gas access channel to the active site renders the regulatory hydrogenase HupUV of *Rhodobacter capsulatus* O_2 sensitive without affecting its transductory activity. *FEBS J.* **272**, 3899–3908.

Dyall, S.D., Yan, W., Delgadillo-Correa, M.G., Lunceford, A., Loo, J.A., Clarke, C.F., and Johnson, P.J. (2004). Non-mitochondrial complex I proteins in a hydrogenosomal oxidoreductase complex. *Nature* **431**, 1103–1107.

Erbes, D., King, D., and Gibbs, M. (1979). Inactivation of hydrogenase in cell-free-extracts and whole cells of *Chlamydomonas reinhardtii* by oxygen. *Plant Physiol.* **63**, 1138–1142.

Escoubas, J.M., Lomas, M., Laroche, J., and Falkowski, P.G. (1995). Light intensity regulation of cab gene transcription is signaled by the redox state of the plastoquinone pool. *Proc. Natl. Acad. Sci. U. S. A.* **92**, 10237–10241.

Fedorov, A., Kosourov, S., Ghirardi, M., and Seibert, M. (2005). Continuous hydrogen photoproduction by *Chlamydomonas reinhardtii*. *Appl. Biochem. Biotechnol.* **121**, 403–412.

Florin, L., Tsokoglou, A., and Happe, T. (2001). A novel type of iron hydrogenase in the green alga *Scenedesmus obliquus* is linked to the photosynthetic electron transport chain. *J. Biol. Chem.* **276**, 6125–6132.

Flynn, T., Ghirardi, M.L., and Seibert, M. (2002). Accumulation of O_2-tolerant phenotypes in H_2-producing strains of *Chlamydomonas reinhardtii* by sequential applications of chemical mutagenesis and selection. *Int. J. Hyd. Energy*, **27**, 1421–1430.

Fontecave, M., Mulliez, E., and Ollagnier-De-Choudens, S. (2001). Adenosylmethionine as a source of 5′-deoxyadenosyl radicals. *Curr. Opin. Chem. Biol.* **5**, 506–511.

Forestier, M., King, P., Zhang, L., Posewitz, M., Schwarzer, S., Happe, T., Ghirardi, M.L., and Seibert, M. (2003). Expression of two [Fe]-hydrogenases in *Chlamydomonas reinhardtii* under anaerobic conditions. *Eur. J. Biochem.* **270**, 2750–2758.

Fouchard, S., Hemschemeier, A., Caruana, A., Pruvost, J., Legrand, J., Happe, T., Peltier, G., and Cournac, L. (2005). Autotrophic and mixotrophic hydrogen photoproduction in sulfur-deprived *Chlamydomonas* cells. *Appl. Environ. Microbiol.* **71**, 6199–6205.

Frey, M. (2002). Hydrogenases: Hydrogen-activating enzymes. *Chembiochem.* **3**, 153–160.

Frey, P.A. (2001). The role of radicals in enzymatic processes. *Chem. Rec.* **1**, 277–289.

Gaffron, H. (1942). Reduction of carbon dioxide coupled with the oxyhydrogen reaction in algae. *J. Gen. Physiol.* **26**, 241–267.

Gaffron, H. (1944). Photosynthesis, photoreduction, and dark reduction of carbon dioxide in certain algae. *Biol. Rev.* **19**, 1–20.

Gaffron, H. and Rubin, J. (1942). Fermentative and photochemical production of hydrogen in algae. *J. Gen. Physiol.* **26**, 219–240.

Garcin, E., Montet, Y., Volbeda, A., Hatchikian, C., Frey, M., and Fontecilla-Camps, J.C. (1998). Structural bases for the catalytic mechanism of [NiFe] hydrogenases. *Biochem. Soc. Trans.* **26**, 396–401.

Gfeller, R.P. and Gibbs, M. (1984). Fermentative metabolism of *Chlamydomonas reinhardtii*, I: Analysis of fermentative products from starch in dark and light. *Plant Physiol.* **75**, 212–218.

Ghirardi, M.L. (2006). Hydrogen production by photosynthetic green algae. *Indian J. Biochem. Biophys.* **43**, 201–210.

Ghirardi, M.L., Togasaki, R.K., and Seibert, M. (1997). Oxygen sensitivity of algal H_2-production. *Appl. Biochem. Biotech.* **63**, 141–151.

Ghirardi, M.L., Zhang, L., Lee, J.W., Flynn, T., Seibert, M., Greenbaum, E., and Melis, A. (2000). Microalgae: A green source of renewable H_2. *Trends Biotechnol.* **18**, 506–511.

Ghirardi, M.L., King, P.W., Posewitz, M.C., Maness, P.C., Fedorov, A., Kim, K., Cohen, J., Schulten, K., and Seibert, M. (2005). Approaches to developing biological H_2-photoproducing organisms and processes. *Biochem. Soc. Trans.* **33**, 70–72.

Ghirardi, M.L., Cohen, J., King, P., Schulten, K., Kim, K., and Seibert, M. (2006). [FeFe]-hydrogenases and photobiological hydrogen production. Solar Hydrogen and Nanotechnology, San Diego, Proceedings, The International Society for Optical Engineering, # 63400X.

Ghirardi, M.L., Posewitz, M.C., Maness, P.C., Dubini, A., Yu, J., and Seibert, M. (2007). Hydrogenases and hydrogen photoproduction in oxygenic photosynthetic organisms. *Annu. Rev. Plant Biol.* **58**, 71–91.

Gibbs, M., Gfeller, R.P., and Chen, C. (1986). Fermentative metabolism of *Chlamydomonas reinhardtii*. III: Photoassimilation of acetate. *Plant Physiol.* **82**, 160–166.

Godde, D. and Trebst, A. (1980). NADH as electron donor for the photosynthetic membrane of *Chlamydomonas reinhardtii*. *Arch. Microbiol.* **127**, 245–252.

Greenbaum, E., Lee, J., Tevault, C., Blankinship, S., and Mets, L. (1995). CO_2 fixation and photoevolution of H_2 and O_2 in a mutant of *Chlamydomonas* lacking photosystem I. *Nature* **376**, 438–441.

Greenbaum, E., Lee, J., Tevault, C., Blankinship, S., and Mets, L. (1997). Correction: CO_2 fixation and photoevolution of H_2 and O_2 in a mutant of *Chlamydomonas* lacking photosystem I. *Nature* **388**, 808.

Grossman, A.R., Croft, M., Gladyshev, V.N., Merchant, S.S., Posewitz, M.C., Prochnik, S., and Spalding, M.H. (2007). Novel metabolism in *Chlaymdomonas* through the lens of genomics. *Curr. Opin. Plant Biol.* **10**, 190–198.

Guan, Y., Deng, M., Yu, X., and Zhang, W. (2004). Two-stage photo-biological production of hydrogen by marine green alga *Platymonas subcordiformis*. *Biochem. Eng. J.* **19**, 69–73.

Hahn, J.J., Ghiradi, M.L., and Jacoby, W.A. (2004). Effect of process variable on photosynthetic algal hydrogen production. *Biotechnol. Prog.* **20**, 989–991.

Happe, T. and Kaminski, A. (2002). Differential regulation of the Fe-hydrogenase during anaerobic adaptation in the green alga *Chlamydomonas reinhardtii*. *Eur. J. Biochem.* **269**, 1022–1032.

Happe, T. and Naber, J.D. (1993). Isolation, characterization and N-terminal amino acid sequence of hydrogenase from the green alga *Chlamydomonas reinhardtii*. *Eur. J. Biochem.* **214**, 475–481.

Happe, T., Mosler, B., and Naber, J.D. (1994). Induction, localization and metal content of hydrogenase in the green alga *Chlamydomonas reinhardtii*. *Eur. J. Biochem.* **222**, 769–774.

Happe, T., Hemschemeier, A., Winkler, M., and Kaminski, A. (2002). Hydrogenases in green algae: Do they save the algae's life and solve our energy problems? *Trends Plant Sci.* **7**, 246–250.

Healey, F. (1970). Hydrogen evolution by several algae. *Planta* **91**, 220–226.

Hemschemeier, A. and Happe, T. (2005). The exceptional photofermentative hydrogen metabolism of the green alga *Chlamydomonas reinhardtii*. *Biochem. Soc. Trans.* **33**, 39–41.

Homann, P. (2003). Hydrogen metabolism of green algae: Discovery and early research – a tribute to Hans Gaffron and his coworkers. *Photosynth. Res.* **76**, 93–103.

Horner, J. and Wolinsky, M. (2002). A power-law sensitivity analysis of the hydrogen-producing metabolic pathway in *Chlamydomonas reinhardtii*. *Int. J. Hyd. Energy,* **27**, 1251–1255.

Horner, D.S., Heil, B., Happe, T., and Embley, T.M. (2002). Iron hydrogenases – ancient enzymes in modern eukaryotes. *Trends Biochem. Sci.* **27**, 148–153.

Ivan, M., Kondo, K., Yang, H., Kim, W., Valiando, J., Ohh, M., Salic, A., Asara, J.M., Lane, W.S., and Kaelin, W.G., Jr (2001). HIFα targeted for VHL-mediated destruction by proline hydroxylation: Implications for O_2 sensing. *Science* **292**, 464–468.

Jarvis, P. and Soll, J. (2002). Toc, Tic, and chloroplast protein import. *Biochim. Biophys. Acta,* **1590**, 177–189.

Jo, J.H., Lee, D.S., and Park, J.M. (2006). Modeling and optimization of photosynthetic hydrogen gas production by green alga *Chlamydomonas reinhardtii* in sulfur-deprived circumstance. *Biotechnol. Prog.* **22**, 431–437.

Kessler, E. (1974). Hydrogenase, photoreduction and anaerobic growth. In: *Algal Physiology and Biochemistry* (W.D.P. Stewart, Ed.), pp. 456–473. Blackwell, Oxford.

King, P.W., Posewitz, M.C., Ghirardi, M.L., and Seibert, M. (2006a). Functional studies of [FeFe] hydrogenase maturation in an *Escherichia coli* biosynthetic system. *J. Bacteriol.* **188**, 2163–2172.

King, P.W., Svedruzic, D., Cohen, J., Schulten, K., Seibert, M., and Ghirardi, M.L. (2006b). Structural and functional investigations of biological catalysts for optimization of solar-driven, H_2 production systems. Solar Hydrogen and Nanotechnology, San Diego, Proceedings, The International Society for Optical Engineering, #63400Y.

Kosourov, S., Tsygankov, A., Seibert, M., and Ghirardi, M.L. (2002). Sustained hydrogen photoproduction by *Chlamydomonas reinhardtii*: Effects of culture parameters. *Biotechnol. Bioeng.* **78**, 731–740.

Kosourov, S., Seibert, M., and Ghirardi, M.L. (2003). Effects of extracellular pH on the metabolic pathways in sulfur-deprived, H_2-producing *Chlamydomonas reinhardtii* cultures. *Plant Cell Physiol.* **44**, 146–155.

Kreuzberg, K. (1984). Starch fermentation via a formate producing pathway in *Chlamydomonas reinhardtii, Chlorogonium elongatum* and *Chlorella fusca. Physiol. Plant.* **61**, 87–94.

Kruse, O., Rupprecht, J., Bader, K.P., Thomas-Hall, S., Schenk, P.M., Finazzi, G., and Hankamer, B. (2005a). Improved photobiological H_2 production in engineered green algal cells. *J. Biol. Chem.* **280**, 34170–34177.

Kruse, O., Rupprecht, J., Mussgnug, J., Dismukes, G., and Hankamer, B. (2005b). Photosynthesis: A blueprint for solar energy capture and biohydrogen production technologies. *Photochem. Photobiol. Sci.* **4**, 957–970.

Laurinavichene, T., Tolstygina, I., and Tsygankov, A. (2004). The effect of light intensity on hydrogen production by sulfur-deprived *Chlamydomonas reinhardtii. J. Biotechnol.* **114**, 143–151.

Laurinavichene, T., Fedorov, A., Ghirardi, M., Seibert, M., and Tsygankov, A. (2006). Demonstration of sustained hydrogen photoproduction by immobilized, sulfur-deprived *Chlamydomonas reinhardtii* cells. *Int. J. Hyd. Energy,* **31**, 659–667.

Layer, G., Kervio, E., Morlock, G., Heinz, D.W., Jahn, D., Retey, J., and Schubert, W.D. (2005). Structural and functional comparison of HemN to other radical SAM enzymes. *Biol. Chem.* **386**, 971–980.

Lee, J.W. and Greenbaum, E. (2003). A new oxygen sensitivity and its potential application in photosynthetic H_2 production. *Appl. Biochem. Biotechnol.* **105–108**, 303–313.

Lee, J.W., Tevault, C.V., Owens, T.G., and Greenbaum, E. (1996). Oxygenic photoautotrophic growth without photosystem I. *Science* **273**, 364–367.

Lien, S. and Pietro, A.S. (1981). Effect of uncouplers on anaerobic adaptation of hydrogenase activity in *C. reinhardtii. Biochem. Biophys. Res. Commun.* **103**, 139–147.

Ludwig, M., Schulz-Friedrich, R., and Appel, J. (2006). Occurrence of hydrogenases in cyanobacteria and anoxygenic photosynthetic bacteria: Implications for the phylogenetic origin of cyanobacterial and algal hydrogenases. *J. Mol. Evol.* **63**, 758–768.

Maione, T. and Gibbs, M. (1986a). Hydrogenase-mediated activities in isolated chloroplasts of *Chlamydomonas reinhardtii. Plant Physiol.* **80**, 360–363.

Maione, T. and Gibbs, M. (1986b). Association of the chloroplastic respiratory and photosynthetic electron transport chains of *Chlamydomonas reinhardtii* with photoreduction and the oxyhydrogen reaction. *Plant Physiol.* **80**, 364–368.

Makarova, V.V., Kosourov, S., Krendeleva, T.E., Semin, B.K., Kukarskikh, G.P., Rubin, A.B., Sayre, R.T., Ghirardi, M.L., and Seibert, M. (2007). Photoproduction of hydrogen by sulfur-deprived *C. reinhardtii* mutants with impaired Photosystem II photochemical activity. *Photosynth. Res.* **94**, 79–89.

Marsh, E.N., Patwardhan, A., and Huhta, M.S. (2004). S-adenosylmethionine radical enzymes. *Bioorg. Chem.* **32**, 326–340.

Melis, A. and Happe, T. (2001). Hydrogen production. Green algae as a source of energy. *Plant Physiol.* **127**, 740–748.

Melis, A. and Happe, T. (2004). Trails of green alga hydrogen research: From Hans Gaffron to new frontiers. *Photosynth. Res.* **80**, 401–409.

Melis, A. and Melnicki, M. (2006). Integrated biological hydrogen production. *Int. J. Hyd. Energy,* **31**, 1563–1573.

Melis, A., Neidhardt, J., and Benemann, J. (1998). *Dunaliella salina* (Chlorophyta) with small chlorophyll antenna sizes exhibit higher photosynthetic productivities and photon use efficiencies than normally pigmented cells. *J. Appl. Phycol.* **10**, 515–525.

Melis, A., Zhang, L., Forestier, M., Ghirardi, M.L., and Seibert, M. (2000). Sustained photobiological hydrogen gas production upon reversible inactivation of oxygen evolution in the green alga *Chlamydomonas reinhardtii*. *Plant Physiol.* **122**, 127–136.

Melis, A., Seibert, M., and Happe, T. (2004). Genomics of green algal hydrogen research. *Photosynth. Res.* **82**, 277–288.

Melis, A., Seibert, M., and Ghirardi, M.L. (2007). Hydrogen fuel production by transgenic microalgae. In: *Transgenic Microalgae as Green Cell Factories* (R. León, A. Galván, and E. Fernández, Eds.), pp. 110–121. Landes Bioscience, Austin.

Moseley, J., Quinn, J., Eriksson, M., and Merchant, S. (2000). The Crd1 gene encodes a putative di-iron enzyme required for photosystem I accumulation in copper deficiency and hypoxia in *Chlamydomonas reinhardtii*. *EMBO J.* **19**, 2139–2151.

Mus, F., Cournac, L., Cardettini, V., Caruana, A., and Peltier, G. (2005). Inhibitor studies on non-photochemical plastoquinone reduction and H_2 photoproduction in *Chlamydomonas reinhardtii*. *Biochim. Biophys. Acta,* **1708**, 322–332.

Mus, F., Dubini, A., Seibert, M., Posewitz, M.C., and Grossman, A.R. (2007). Anaerobic acclimation in *Chlamydomonas reinhardtii*: Anoxic gene expression, hydrogenase induction, and metabolic pathways. *J. Biol. Chem.* **282**, 25475–25486.

Nagy, L.E., Meuser, J.E., Plummer, S., Seibert, M., Ghirardi, M.L., King, P.W., Ahmann, D., and Posewitz, M.C. (2007). Application of gene-shuffling for the rapid generation of novel [FeFe]-hydrogenase libraries. *Biotechnol. Lett.* **29**, 421–430.

Nicolet, Y., Piras, C., Legrand, P., Hatchikian, C.E., and Fontecilla-Camps, J.C. (1999). *Desulfovibrio desulfuricans* iron hydrogenase: The structure shows unusual coordination to an active site Fe binuclear center. *Structure Fold Des.* **7**, 13–23.

Nicolet, Y., De Lacey, A.L., Vernede, X., Fernandez, V.M., Hatchikian, E.C., and Fontecilla-Camps, J.C. (2001). Crystallographic and FTIR spectroscopic evidence of changes in Fe coordination upon reduction of the active site of the Fe-only hydrogenase from *Desulfovibrio desulfuricans*. *J. Am. Chem. Soc.* **123**, 1596–1601.

Niyogi, K.K. (1999). Photoprotection revisited: Genetic and molecular approaches. *Annu. Rev. Plant Physiol. Plant Mol. Biol.* **50**, 333–359.

Ohta, S., Miyamoto, K., and Miura, Y. (1987). Hydrogen evolution as a consumption mode of reducing equivalents in green algal fermentation. *Plant Physiol.* **83**, 1022–1026.

Park, W. and Moon, I. (2007). A discrete multi states model for the biological production of hydrogen by phototrophic microalga. *Biochem. Eng. J.* **36**, 19–27.

Peltier, G. and Cournac, L. (2002). Chlororespiration. *Annu. Rev. Plant Biol.* **53**, 523–550.

Peters, J.W. (1999). Structure and mechanism of iron-only hydrogenases. *Curr. Opin. Struct. Biol.* **9**, 670–676.

Peters, J.W., Lanzilotta, W.N., Lemon, B.J., and Seefeldt, L.C. (1998). X-ray crystal structure of the Fe-only hydrogenase (CpI) from *Clostridium pasteurianum* to 1.8 angstrom resolution. *Science* **282**, 1853–1858.

Peters, J.W., Szilagyi, R.K., Naumov, A., and Douglas, T. (2006). A radical solution for the biosynthesis of the H-cluster of hydrogenase. *FEBS Lett.* **580**, 363–367.

Pfannschmidt, T., Schutze, K., Fey, V., Sherameti, I., and Oelmuller, R. (2003). Chloroplast redox control of nuclear gene expression: A new class of plastid signals in interorganellar communication. *Antioxid. Redox Signal,* **5**, 95–101.

Pollock, S.V., Pootakham, W., Shibagaki, N., Moseley, J.L., and Grossman, A.R. (2005). Insights into the acclimation of *Chlamydomonas reinhardtii* to sulfur deprivation. *Photosynth. Res.* **86**, 475–489.

Posewitz, M.C., King, P.W., Smolinski, S.L., Zhang, L., Seibert, M., and Ghirardi, M.L. (2004a). Discovery of two novel radical S-adenosylmethionine proteins required for the assembly of an active [Fe] hydrogenase. *J. Biol. Chem.* **279**, 25711–25720.

Posewitz, M.C., Smolinski, S.L., Kanakagiri, S., Melis, A., Seibert, M., and Ghirardi, M.L. (2004b). Hydrogen photoproduction is attenuated by disruption of an isoamylase gene in *Chlamydomonas reinhardtii*. *Plant Cell* **16**, 2151–2163.

Posewitz, M.C., King, P.W., Smolinski, S.L., Smith, R.D., Ginley, A.R., Ghirardi, M.L., and Seibert, M. (2005). Identification of genes required for hydrogenase activity in *Chlamydomonas reinhardtii*. *Biochem. Soc. Trans.* **33**, 102–104.

Posewitz, M.C., Mulder, D.W., and Peters, J.W. (2008). New frontiers in hydrogenase structure and biosynthesis. *Curr. Chem. Biol.* **2**, 178–199.

Quinn, J.M., Eriksson, M., Moseley, J.L., and Merchant, S. (2002). Oxygen deficiency responsive gene expression in *Chlamydomonas reinhardtii* through a copper-sensing signal transduction pathway. *Plant Physiol.* **128**, 463–471.

Rao, K.K. and Cammack, R. (2001). Producing hydrogen as a fuel. In: *Hydrogen as a Fuel* (R. Cammack, M. Frey, and R. Robson, Eds.), pp. 201–330. Taylor and Francis, London.

Redding, K., Cournac, L., Vassiliev, I.R., Golbeck, J.H., Peltier, G., and Rochaix, J.D. (1999). Photosystem I is indispensable for photoautotrophic growth, CO_2 fixation, and H_2 photoproduction in *Chlamydomonas reinhardtii*. *J. Biol. Chem.* **274**, 10466–10473.

Roessler, P. and Lien, S. (1982). Anionic modulation of the catalytic activity of hydrogenase from *Chlamydomonas reinhardtii*. *Arch. Biochem. Biophys.* **213**, 37–44.

Roessler, P.G. and Lien, S. (1984a). Activation and *de novo* synthesis of hydrogenase in *Chlamydomonas*. *Plant Physiol.* **76**, 1086–1089.

Roessler, P.G. and Lien, S. (1984b). Purification of hydrogenase from *Chlamydomonas reinhardtii*. *Plant Physiol.* **75**, 705–709.

Rubach, J.K., Brazzolotto, X., Gaillard, J., and Fontecave, M. (2005). Biochemical characterization of the HydE and HydG iron-only hydrogenase maturation enzymes from *Thermatoga maritima*. *FEBS Lett.* **579**, 5055–5060.

Rubio, L.M. and Ludden, P.W. (2005). Maturation of nitrogenase: A biochemical puzzle. *J. Bacteriol.* **187**, 405–414.

Rupprecht, J., Hankamer, B., Mussgnug, J.H., Ananyev, G., Dismukes, C., and Kruse, O. (2006). Perspectives and advances of biological H_2 production in microorganisms. *Appl. Microbiol. Biotechnol.* **72**, 442–449.

Russell, G.K. and Gibbs, M. (1968). Evidence for the participation of the reductive pentose phosphate cycle in photoreduction and the oxyhydrogen reaction. *Plant Physiol.* **43**, 649–652.

Rutter, J., Reick, M., Wu, L.C., and McKnight, S.L. (2001). Regulation of clock and NPAS2 DNA binding by the redox state of NAD cofactors. *Science* **293**, 510–514.

Saint-Amans, S., Girbal, L., Andrade, J., Ahrens, K., and Soucaille, P. (2001). Regulation of carbon and electron flow in *Clostridium butyricum* VPI 3266 grown on glucose-glycerol mixtures. *J. Bacteriol.* **183**, 1748–1754.

Schofield, C.J. and Ratcliffe, P.J. (2005). Signalling hypoxia by HIF hydroxylases. *Biochem. Biophys. Res. Commun.* **338**, 617–626.

Schonfeld, C., Wobbe, L., Borgstadt, R., Kienast, A., Nixon, P.J., and Kruse, O. (2004). The nucleus-encoded protein MOC1 is essential for mitochondrial light acclimation in *Chlamydomonas reinhardtii*. *J. Biol. Chem.* **279**, 50366–50374.

Seibert, M., Benson, D.K., and Flynn, T. (2001). Method and Apparatus for Rapid Biohydrogen Phenotypic Screening of Microorganisms Using a Chemochromic Sensor. U.S. Patent # 6,277,589, Midwest Research Institute.

Seibert, M., King, P.W., Posewitz, M.C., Melis, A., and Ghiradi, M.L. (2008). Photosynthetic water-splitting for hydrogen production. In: *Bioenergy* (J.D. Wall, C.S. Harwood, A. Demain, Eds.), pp. 273–291. ASM Press, Washington, D.C.

Semin, B.K., Davletshina, L.N., Novakova, A.A., Kiseleva, T.Y., Lanchinskaya, V.Y., Aleksandrov, A.Y., Seifulina, N., Ivanov, I.i., Seibert, M., and Rubin, A.B. (2003). Accumulation of ferrous iron in *Chlamydomonas reinhardtii*. Influence of CO_2 and anaerobic induction of the reversible hydrogenase. *Plant Physiol.* **131**, 1756–1764.

Steunou, A.S., Bhaya, D., Bateson, M.M., Melendrez, M.C., Ward, D.M., Brecht, E., Peters, J.W., Kuhl, M., and Grossman, A.R. (2006). *In situ* analysis of nitrogen fixation and metabolic switching in unicellular thermophilic cyanobacteria inhabiting hot spring microbial mats. *Proc. Natl. Acad. Sci. U. S. A.* **103**, 2398–2403.

Stiehl, D.P., Wirthner, R., Koditz, J., Spielmann, P., Camenisch, G., and Wenger, R.H. (2006). Increased prolyl 4-hydroxylase domain proteins compensate for decreased oxygen levels. Evidence for an autoregulatory oxygen-sensing system. *J. Biol. Chem.* **281**, 23482–23491.

Stirnberg, M. and Happe, T. (2004). Identification of a *cis*-acting element controlling anaerobic expression of *hydA* gene from *Chlamydomonas reinhardtii*. In: *Biohydrogen III* (J. Miyake, Y. Igarashi, and M. Rogner, Eds.), pp. 117–127. Elsevier, New York.

Tamagnini, P., Axelsson, R., Lindberg, P., Oxelfelt, F., Wunschiers, R., and Lindblad, P. (2002). Hydrogenases and hydrogen metabolism of cyanobacteria. *Microbiol. Mol. Biol. Rev.* **66**, 1–20.

Tsygankov, A., Kosourov, S., Seibert, M., and Ghirardi, M. (2002). Hydrogen photoproduction under continuous illumination by sulfur-deprived, synchronous *Chlamydomonas reinhardtii* cultures. *Int. J. Hyd. Energy,* **27**, 1239–1244.

Tsygankov, A., Kosourov, S., Tolstygina, I., Ghirardi, M., and Seibert, M. (2006). Hydrogen production by sulfur-deprived *Chlamydomonas reinhardtii* under photoautotrophic conditions. *Int. J. Hyd. Energy,* **31**, 1574–1584.

Urbig, T., Schulz, R., and Senger, H. (1993). Inactivation and reactivation of the hydrogenases of the green-algae *Scenedesmus obliquus* and *Chlamydomonas reinhardtii*. *Z. Naturforsch. [C]*, **48**, 41–45.

Vignais, P.M. and Colbeau, A. (2004). Molecular biology of microbial hydrogenases. *Curr. Issues Mol. Biol.* **6**, 159–188.

Vignais, P.M., Billoud, B., and Meyer, J. (2001). Classification and phylogeny of hydrogenases. *FEMS Microbiol. Rev.* **25**, 455–501.

Walsby, C., Ortillo, D., Yang, J., Nnyepi, M., Broderick, W., Hoffman, B., and Broderick, J. (2005). Spectroscopic approaches to elucidating novel iron-sulfur chemistry in the "Radical-SAM" protein superfamily. *Inorg. Chem.* **44**, 727–741.

Weaver, P.F., Lien, S., and Seibert, M. (1980). Photobiological production of hydrogen. *Solar Energy* **24**, 3–45.

White, A. and Melis, A. (2006). Biochemistry of hydrogen metabolism in *Chlamydomonas reinhardtii* wild type and a Rubisco-less mutant. *Int. J. Hyd. Energy* **31**, 455–464.

Willeford, K. and Gibbs, M. (1989). Localization of the enzymes involved in the photoevolution of H_2 from acetate in *Chlamydomonas reinhardtii*. *Plant Physiol.* **90**, 788–791.

Winkler, M., Heil, B., Heil, B., and Happe, T. (2002a). Isolation and molecular characterization of the [Fe]-hydrogenase from the unicellular green alga *Chlorella fusca*. *Biochim. Biophys. Acta,* **1576**, 330–334.

Winkler, M., Hemschemeier, A., Gotor, C., Melis, A., and Happe, T. (2002b). [Fe]-hydrogenases in green algae: Photo-fermentation and hydrogen evolution under sulfur deprivation. *Int. J. Hyd. Energy,* **27**, 1431–1439.

Winkler, M., Maeurer, C., Hemschemeier, A., and Happe, T. (2004). The isolation of green algal strains with outstanding H_2 productivity. In: *Biohydrogen III* (J. Miyake, Y. Igarashi, and M. Rogner, Eds.), pp. 103–115. Elsevier, New York.

Wu, L.F. and Mandrand, M.A. (1993). Microbial hydrogenases: Primary structure, classification, signatures and phylogeny. *FEMS Microbiol. Rev.* **10**, 243–269.

Wunschiers, R., Stangier, K., Senger, H., and Schulz, R. (2001). Molecular evidence for a Fe-hydrogenase in the green alga *Scenedesmus obliquus*. *Curr. Microbiol.* **42**, 353–360.

Wykoff, D.D., Davies, J.P., Melis, A., and Grossman, A.R. (1998). The regulation of photosynthetic electron transport during nutrient deprivation in *Chlamydomonas reinhardtii*. *Plant Physiol.* **117**, 129–139.

Zabawinski, C., Van Den Koornhuyse, N., D'Hulst, C., Schlichting, R., Giersch, C., Delrue, B., Lacroix, J.M., Preiss, J., and Ball, S. (2001). Starchless mutants of *Chlamydomonas reinhardtii* lack the small subunit of a heterotetrameric ADP-glucose pyrophosphorylase. *J. Bacteriol.* **183**, 1069–1077.

Zhang, L. and Melis, A. (2002). Probing green algal hydrogen production. *Philos. Trans. R. Soc. Lond. B. Biol. Sci.* **357**, 1499–1507.

Zhang, L., Happe, T., and Melis, A. (2002). Biochemical and morphological characterization of sulfur-deprived and H_2-producing *Chlamydomonas reinhardtii* (green alga). *Planta* **214**, 552–561.

Zhang, Z., Shrager, J., Jain, M., Chang, C.W., Vallon, O., and Grossman, A.R. (2004). Insights into the survival of *Chlamydomonas reinhardtii* during sulfur starvation based on microarray analysis of gene expression. *Eukaryotic Cell* **3**, 1331–1348.

Zinn, T., Schnackenberg, J., Haak, D., Romer, S., Schulz, R., and Senger, H. (1994). Evidence for nickel in the soluble hydrogenase from the unicellular green alga *Scenedesmus obliquus*. *Z. Naturforsch. [C]*, **49**, 33–38.

CHAPTER 8

The CO_2-Concentrating Mechanism and Carbon Assimilation

Martin H. Spalding
Department of Genetics, Development and Cell Biology,
Iowa State University, Ames, Iowa, USA

CHAPTER CONTENTS

I. THE CARBON ENVIRONMENT

A. Carbon supply

Carbon dioxide (CO_2) is present in relatively low abundance in the atmosphere, but it nonetheless is arguably among the most important gases on earth. Photosynthetic organisms absorb CO_2 as the major substrate to support photosynthesis, the beginning of energy flow into living organisms and one of the primary processes comprising the global carbon cycle. Therefore, changes in CO_2 concentrations can have a profound impact on photosynthesis and many related processes in carbon metabolism, which subsequently influence the biotic environment of earth. The overall global atmospheric CO_2 concentration remains fairly constant in the short term. However, the CO_2 concentrations in localized environments can change frequently and often do so rapidly. CO_2 limitation represents a frequently confronted and significant stress for photosynthetic organisms, not only due to low ambient CO_2 concentrations but also due to the low CO_2 affinity and very slow catalytic turnover of ribulose-1,5-bisphosphate carboxylase/oxygenase (Rubisco; EC 4.1.1.39), the key enzyme in CO_2 assimilation.

The limited atmospheric CO_2 presents challenges in all environments, but organisms, such as *Chlamydomonas*, that grow in aquatic environments face additional unique challenges regarding the efficient operation of photosynthesis. These challenges stem largely from the differences in CO_2 supply dynamics in aquatic versus terrestrial environments. These challenges include an inherently greater short-term and long-term variability in the abundance of dissolved CO_2 in aquatic systems resulting from, for example, variability in both the respiratory CO_2 output of organisms in the soil or sediment and the loss of CO_2 to the atmosphere. In addition to a variable supply from respiratory sources, equilibration of CO_2 between aqueous environments and the atmosphere is relatively slow, which can result in a depletion of the aqueous CO_2 during active photosynthesis (Talling, 1976). Adding a further challenge, the roughly 10^4-fold slower diffusion of CO_2 in water versus air greatly exacerbates the problem of CO_2 supply to Rubisco.

A characteristic of aquatic environments is that all forms of dissolved inorganic carbon (Ci), namely CO_2, HCO_3^-, and CO_3^{2-}, represent potential sources of CO_2 for photosynthesis as a result of their interconversion by a series of hydration, dehydration, and protonation reactions. Important features of this interconversion include the strong dependence of the ratio of CO_2 to HCO_3^- in solution on pH, and the very slow hydration of CO_2 to HCO_3^- relative to the photosynthetic consumption of CO_2 (Badger and Price, 1994). One important consequence of these features is that HCO_3^- becomes the dominant Ci species in solution above pH 6.3, with dissolved CO_2 representing an ever-decreasing proportion of the total Ci concentration as the pH increases. In addition, the slow conversion of HCO_3^- to CO_2, in combination with the slow dissolution of atmospheric CO_2, can cause a significant depletion of dissolved CO_2 below its expected concentration if in equilibrium with air. This is especially true in environments where photosynthetic consumption of dissolved CO_2 is relatively rapid.

As a result of the challenges to photosynthetic CO_2 assimilation and the selective pressure they exert, *Chlamydomonas* and other aquatic photosynthetic organisms have developed mechanisms which enable both CO_2 and HCO_3^- to support photosynthesis, as well as mechanisms that enable them to acclimate rapidly and specifically to changes in Ci supply. Fundamental to this acclimation is a mechanism to scavenge Ci and concentrate it internally when its supply is limited. The most extensively studied of these mechanisms are the CO_2-concentrating mechanisms (CCMs) of cyanobacteria and microalgae, especially *Chlamydomonas* (reviewed in Spalding, 1998; Kaplan and Reinhold, 1999; Badger and Spalding, 2000).

B. Multiple carbon acclimation states

An abundance of early research (reviewed by Spalding, 1989), suggested that green microalgae could exist in two distinctly different physiological states depending on the CO_2 concentration (typically, 0.03% vs. 1–5%). This phenomenon was confirmed by Berry et al. (1976), who demonstrated that *Chlamydomonas* cells grown in high (1–5%) or limiting ($\leq$0.04%; at or below air levels) CO_2 concentrations had photosynthetic characteristics very similar to terrestrial angiosperms with the C3 pathway or the C4 pathway of photosynthesis, respectively. Relative to cells grown in high CO_2 concentrations, cells acclimated to limiting CO_2 were found to have a much higher apparent affinity for CO_2 in photosynthesis. For example, cells grown in 5% CO_2 exhibited a photosynthetic $K_{½}$(Ci) similar to the $K_m(C_i)$ of Rubisco, whereas limiting CO_2-acclimated cells exhibited a $K_{½}$(Ci) value 10–100-fold lower than that of the high CO_2-grown cells or the K_m(Ci) of Rubisco (Badger and Price, 1992). So, the acclimation of *Chlamydomonas* and other microalgae to limiting CO_2 concentrations results in induction of physiological changes that improve their photosynthetic efficiency, changes that have been linked to a whole suite of acclimation responses (Spalding et al., 2002).

The physiological differences between microalgae grown at high and limiting CO_2 have been studied extensively and form an important contextual framework for research on CCMs. In most published research, high CO_2 (air enriched with 1–5% CO_2) has been used to repress CCM activity and "low" CO_2 (~0.03–0.04% CO_2; nominally, the air level of CO_2) has been used to induce CCM activity. CO_2 concentrations outside these typical ranges have been used occasionally, for example for investigating the relationship between limiting CO_2 concentrations and induction of the CCM (Matsuda et al., 1998), to assess the quantitative activity of the CCM at various limiting Ci concentrations (Karlsson et al., 1994) or to screen mutated populations for high CO_2-requiring (HCR) mutants (Colombo et al., 2002).

It has been generally assumed that limiting CO_2 concentrations below air equilibration levels only quantitatively affect acclimation and CCM function, an assumption that appears to be incorrect. Even though the preponderance of research on limiting CO_2 acclimation and the CCM in *Chlamydomonas* and other microalgae has clearly recognized only two physiological states, a more recently characterized growth phenotype of a classical *Chlamydomonas* HCR mutant, *pmp1* (Spalding et al., 1983b), demonstrated at least two distinct limiting CO_2 acclimation states (Spalding et al., 2002). The most conspicuous *pmp1* characteristic is that in addition to growing in high CO_2, it can grow in very low CO_2 (<0.01%), but it dies in low (air levels) of CO_2. These observations suggest that *pmp1* is capable of CCM under very low CO_2, but lacks a critical CCM when acclimated to atmospheric CO_2.

Further research has confirmed at least three distinct CO_2-regulated acclimation states based on growth, photosynthetic rate, and expression of limiting CO_2-inducible genes (Vance and Spalding, 2005). The defining characteristics of the high CO_2 state include lack of limiting CO_2-regulated gene expression, no growth delay following transfer from 5% CO_2, and a photosynthetic $K_{1/2}(CO_2)$ similar to the $K_m(CO_2)$ of Rubisco. The low CO_2 state can be distinguished from the high CO_2 state by an acclimation-related delay in culture growth, induction of limiting CO_2-regulated genes, smaller cell size and a markedly decreased photosynthetic $K_{1/2}(CO_2)$. The very low CO_2 state is distinguishable from the low CO_2 state by a decreased growth rate, smaller average cell size, less chlorophyll per cell, a decreased photosynthetic V_{max} and a further decreased photosynthetic $K_{1/2}(CO_2)$. The transcript abundance for a small selection of limiting CO_2-regulated genes in low CO_2 and very low CO_2 increased with decreasing CO_2 concentration without an obvious break. However, based on growth characteristics of *pmp1*, it is likely that the very low CO_2 state has unique gene expression patterns which remain to be uncovered. In summary, the differences between low CO_2 and very low CO_2 conditions affect the CCM both quantitatively and qualitatively, indicating the probable existence of multiple limiting CO_2 signals.

C. Limiting CO_2 acclimation

Relative to other limiting CO_2 acclimation responses, the *Chlamydomonas* CCM has received the most attention because of its effect, via increased intracellular CO_2 concentration, on photosynthetic and photorespiratory characteristics (Spalding, 1998). However, *Chlamydomonas* exhibits a whole suite of adaptive responses to limiting CO_2 including, as discussed below, induction (or de-repression) of the CCM, up-regulation of photorespiratory components, and changes in subcellular organization. Coinciding with these functional and structural changes are dramatic changes in gene expression, including up- and down-regulation of specific genes, as well as regulation at the transcriptional, translational, and post-translational levels.

As a cautionary note, it should be taken into account that most reports of limiting CO_2-related changes in gene expression, CCM activity, photorespiratory activity and cell structure do not distinguish between the low CO_2 and very low CO_2 states discussed above. In fact, it is difficult even to determine the likely limiting CO_2 state represented in most investigations, because the nominal CO_2 concentration used very likely does not accurately reflect the CO_2 concentration experienced by the cells or cultures (Vance and Spalding, 2005). This is because the effective CO_2 concentration often will be well below the nominal CO_2 concentration of growth because of active assimilation of CO_2 by the cells/cultures. For this reason, the term limiting CO_2 will be used in reference to low CO_2 and very low CO_2 conditions, collectively, unless the effective CO_2 environment is certain.

Coincident with induction of the CCM, expression of several genes is induced *de novo*. These genes appear to be involved in the fundamental functions of limiting CO_2 acclimation, including Ci transport systems, carbonic anhydrase (CA) and photorespiration. Moreover, *Chlamydomonas* genes involved in a wide spectrum of metabolic reactions and acclimation responses have been identified as quantitatively up- or down-regulated by changes in Ci supply. Examples of the most abundant proteins induced by limiting CO_2 include the major periplasmic CA, encoded by *CAH1* (Coleman and Grossman, 1984; Fukuzawa et al., 1990, Fujiwara et al., 1990), two nearly identical mitochondrial CAs, encoded by *CAH4* and *CAH5* (Eriksson et al., 1996; Geraghty and Spalding, 1996), and two highly similar chloroplast inner envelope proteins, encoded by *CCP1* and *CCP2* (Geraghty et al., 1990; Mason et al., 1990; Ramazanov et al., 1993; Chen et al., 1997). Several additional limiting CO_2-inducible genes were identified based on differential screening of cDNA libraries (Burow et al., 1996; Chen et al., 1996; Somanchi and Moroney, 1999), including *LCI1*, *AAT1*, and cyclophilin.

Subsequent genome-wide analyses (Im and Grossman, 2002; Im et al., 2003; Miura et al., 2004) revealed many more genes apparently regulated by limiting CO_2. Im and Grossman (2002) and Im et al. (2003) used differential display and microarray analysis to identify genes differentially expressed in high light, which also included a subset of genes apparently regulated by CO_2 limitation in high light rather than by high light *per se*. By comparing transcript abundance of cells under CO_2-rich and CO_2-limiting conditions, Miura et al. (2004) identified 51 limiting CO_2-inducible genes and 32 limiting CO_2 genes by cDNA macroarray analysis. The limiting CO_2-induced or up-regulated genes identified and in some cases confirmed in these global analyses, many of which were up-regulated only transiently (Tural and Moroney, 2005), included several photorespiration related genes, for example alanine-α-ketoglutarate aminotransferase (*AAT1*; Chen et al., 1996), phosphoglycolate phosphatase (*PGP1*; Mamedov et al., 2001), glycolate dehydrogenase (*GYD1*; Nakamura et al.,

2005), and serine hydroxymethyltransferase (*SHMT1*). Others were genes involved in starch synthesis, for example granule-bound starch synthase I (*STA2*; Wattebled et al., 2002) and soluble starch synthase III (*STA3*; Ral et al., 2006; see also Chapter 1); and multiple CA genes, including *CAH1*, *CAH4*, *CAH5*, and *CAH3*, the latter encoding an important thylakoid lumen CA often considered to be constitutively expressed but whose transcript abundance increases twofold in limiting CO_2 (Karlsson et al., 1998). Also induced were several genes whose products are candidates for Ci transport either in or across the plasma membrane, *HLA3* (*MRP1*; Im and Grossman, 2002; Miura et al., 2004) and *LCI1* (Burow et al., 1996), or the chloroplast envelope, *CCP1* and *CCP2* (Chen et al., 1997), *LCIA* (Miura et al., 2004; Mariscal et al., 2006), and *LCIB* and *LCIC* (Wang and Spalding, 2006).

Many genes also are induced or up-regulated by high CO_2. Among the genes either induced *de novo* or up-regulated in high CO_2 (or repressed in limiting CO_2) are *CAH2*, which encodes a minor periplasmic CA apparently associated with CO_2 uptake (Fujiwara et al., 1990; Fukuzawa et al., 1990; Rawat and Moroney, 1991) and *RHP1*, which encodes an Rh protein reported to act as a CO_2 channel (Soupene et al., 2002, 2004; Kustu and Inwood, 2006). Other high CO_2-induced genes include *H43*, which encodes a protein of unknown function located in the periplasm (Hanawa et al., 2007), as well as numerous genes whose products function in photosynthesis, nutrient acquisition, stress responses, and other pathways (Miura et al., 2004). Expression of at least two gene products decreases transiently in response to limiting CO_2, namely the subunits of Rubisco (*RBCS1*, *RBCS2*, and *rbcL*; Coleman and Grossman, 1984), although this is primarily regulated at the translational level (Winder et al., 1992).

Genome-wide transcriptome analyses are powerful tools, but the roles of translational regulation, as demonstrated for Rubisco, as well as post-translational modifications, such as the reported limiting CO_2-dependent phosphorylation of thylakoid proteins (Turkina et al., 2006), remain to be fully explored. In this latter research, the authors reported multiple phosphorylations of LCI5, the product of a previously identified limiting CO_2-inducible gene, *LCI5* (Im and Grossman, 2002; Miura et al., 2004). LCI5 was found to be peripherally associated to the stromal side of thylakoids, although it is not clear whether this protein plays any specific role in the CCM or in limiting CO_2 acclimation.

Because of complex interactions between the CCM and other metabolic processes, expression of some genes might be regulated by not only by Ci availability but also by other environmental conditions such as light intensity, nitrogen availability, or chloroplast redox state (Villand et al., 1997). It is also important to note that any limitation in Ci availability will affect many metabolic pathways and trigger stress responses, so some changes in gene expression will reflect responses to changing environments rather than a specific response to limiting CO_2. Thus many CO_2-responsive genes might not have any obvious link to limiting CO_2 acclimation, such as those involved in purine or cyclophilin metabolism (Somanchi and Moroney, 1999).

As suggested by the up-regulation of several genes encoding photorespiratory enzymes, substantial increases in the activity of photorespiratory enzymes are associated with limiting CO_2 acclimation, including a transient increase in phosphoglycolate phosphatase and more stable increases in glycolate dehydrogenase and glutamine synthetase (Nelson and Tolbert, 1969; Marek and Spalding, 1991; Ramazanov and Cardenas, 1992, 1994). Photorespiration and its relationship to the CCM and limiting CO_2 acclimation will be discussed in more detail in section II.B.

Acclimation to limiting CO_2 is accompanied by substantial changes in the subcellular organization in *Chlamydomonas*. One well-characterized example is the increased development of the starch sheath surrounding the chloroplast pyrenoid (Kuchitsu et al., 1988, 1991; Ramazanov et al., 1994; Geraghty and Spalding, 1996; Thyssen et al., 2001). The observed increase in pyrenoid starch is consistent with limiting CO_2 up-regulated expression of *STA2* and *STA3* (Wattebled et al., 2002; Miura et al., 2004; Ral et al., 2006), a general correlation that also was reported in *Chlorella kessleri* (Oyama et al., 2006). Once proposed to be functionally important as a diffusion barrier to minimize loss of CO_2 from the Rubisco-rich pyrenoid (Badger and Price, 1994), the absence of any detectable decrease in CCM functionality in a starchless mutant argues against this hypothesis (Villarejo et al., 1996a; see also Chapter 1).

Accompanying the changes in pyrenoid starch, the abundance of stromal starch also was reported to decrease during limiting CO_2 acclimation (Miyachi et al., 1986; Kuchitsu et al., 1988), but more extensive time courses suggested that stromal starch decreases only transiently, probably representing depletion of starch reserves in Ci-starved cells during acclimation and before the CCM becomes functional (Geraghty and Spalding, 1996; Thyssen et al., 2001).

Geraghty and Spalding (1996) also reported dramatic changes in mitochondrial distribution during acclimation to limiting CO_2, with mitochondria moving from a central position within the cup of the chloroplast to a peripheral position between the chloroplast envelope and the plasma membrane. The mechanism responsible for this organellar redistribution is unknown, but it must involve cytoskeletal elements, and the mitochondria probably pass through "gaps" observed in the cup-shaped chloroplast. This mitochondrial relocation is quite intriguing, because induction or up-regulation of several genes encoding mitochondrial proteins (Eriksson et al., 1996; Geraghty and Spalding, 1996; Im and Grossman, 2002; Miura et al., 2004; Nakamura et al., 2005) is coincident with the mitochondrial relocation, which argues that the mitochondria might play an important role in the limiting CO_2 acclimation. Discussion of possible function(s) for the mitochondria specifically in limiting CO_2 acclimation is clearly speculative, but because their move to a peripheral location coincides with increased photorespiratory pathway flux during and following limiting CO_2 acclimation (see above), the mitochondria might be involved in scavenging photorespiratory glycolate produced in the chloroplasts. Alternatively, the closer proximity of the relocated mitochondria to plasma membrane Ci transporters might reflect a mitochondrial involvement in energization of Ci transport into the cells, as reported for the microalga *Nannochloropsis gaditana* (Huertas et al., 2002).

D. Signals and signal transduction

Acclimation to limiting CO_2 allows the cells to respond to the natural variation in Ci supply in their environment. As with all acclimation responses to environmental changes, the acclimation to changes in CO_2 availability must involve recognition of a signal that is transduced through a pathway and results in changes in gene expression, and finally the biochemical and metabolic changes we recognize as the limiting CO_2 acclimation responses. The signal regulating acclimation of *Chlamydomonas* to changes in CO_2 concentration is not known definitively, nor has the signal receptor been identified. Thus it is not understood how the cells sense a change in the CO_2 supply, nor is it understood whether the multiple acclimation states represent proportionate responses to the same signal, for example CO_2 concentration. Clearly, the cells must sense the CO_2 supply change either directly as CO_2 or Ci concentration or indirectly, via the effect of Ci on a cellular process, and the different acclimation responses might include responses to both types of signals.

Not only is information still lacking regarding the identity of the "CO_2 signal(s)" and receptor(s), but also little is known about the signal transduction pathway(s) leading to the functional changes important for limiting CO_2 acclimation responses. Furthermore, although investigations in *Chlamydomonas* have focused on acclimation to limiting CO_2 concentrations, acclimation works in both directions; that is, *Chlamydomonas* also acclimates to increased CO_2 supply by changes in cell organization, and repression of CCM activity and of limiting CO_2-expressed genes.

One of the most obvious candidates for the "CO_2 signal" is the CO_2 concentration itself, and the results of acclimation experiments from several Chlorophyta species, including *Chlamydomonas*, are most consistent with the cells responding to the CO_2 concentration in the bulk medium (Matsuda and Colman, 1995, 1996; Bozzo and Colman, 2000; Bozzo et al., 2000; Giordano et al., 2005). This conclusion is based on the absolute correlation between the induction of the CCM and the external CO_2 concentration over different pH levels, regardless of the total Ci or HCO_3^- concentration in the medium. In these cases, the limiting CO_2 signal is argued to be perceived at the plasma membrane.

On the other hand, based on reports of requirements for light, and specifically for active photosynthesis, in the induction of the CCM and expression of limiting CO_2-induced genes, for example *CAH1*, in *Chlamydomonas*, others have argued that limiting CO_2 might be sensed indirectly as a metabolic consequence of CO_2 availability for photosynthesis (Spalding and Ogren, 1982; Spencer et al., 1983; Dionisio et al., 1989a,b; Dionisio-Sese et al., 1990; Fukuzawa et al., 1990; Villarejo et al., 1996b). In this case, the actual signal might be the redox potential of photosynthetic electron transport or a photosynthetic metabolite. Similarly, because *CAH1* induction was positively correlated with O_2 tension and prevented by inhibitors of the photorespiratory pathway, photorespiration or a photorespiratory metabolite have been proposed as signals for CO_2 limitation (Spalding and Ogren, 1982; Ramazanov and Semenenko, 1984; Ramazanov and Cardenas, 1992; Villarejo et al., 1996b). In addition to the obvious effects on photorespiration (e.g. draining carbon into photorespiratory pathway metabolites), O_2 might influence photosynthetic metabolism in other ways, for example by directly affecting the redox state of photosynthetic electron transport or by draining carbon away from non-photorespiratory metabolites. Therefore, if metabolic signals play a role in limiting CO_2 acclimation, O_2 might influence acclimation via those signals.

In combination with the possibility that high CO_2 might act as a repression signal, rather than limiting CO_2 as an inducing signal (Spalding et al., 2002), the metabolic signaling hypotheses draw intriguing parallels with classical metabolic signaling responses, such as sugar regulation of photosynthesis-related genes in plants (Sheen, 1994) and catabolite repression, as typified by glucose repression of alternative catabolic pathways in *S. cerevisiae* (Gancedo, 1998). The apparent repression of limiting CO_2 acclimation responses by acetate in mixotrophic growth of *Chlamydomonas* (Bailey and Coleman, 1988; Coleman et al., 1991; Fett and Coleman, 1994), supports the possibility that rich or preferred carbon sources, high CO_2 and/or acetate, obviate the need for, and thus repress, limiting CO_2 acclimation responses, and also highlight the parallels with yeast catabolite repression.

In contrast to the proposed signals in microalgae, either the external HCO_3^- or the internal Ci pool has been postulated to act as the CCM induction signal in cyanobacteria, and full induction of the CCM also reportedly requires an O_2-dependent process (Mayo et al., 1986; Miyachi et al., 2003; Woodger et al., 2003, 2005). There is little support from research on *Chlamydomonas* or other microalgae for external HCO_3^- to act as a signal, but the O_2 concentration has been implicated in the induction of genes such as *CAH1*, as discussed just above. Later work varying the concentrations of CO_2 and O_2 and the CO_2/O_2 ratio under carefully controlled conditions revealed that both CCM induction and expression of *CAH1* and *GYD1* were controlled exclusively by the external CO_2 concentration (Vance and Spalding, 2005), thus failing to support any role of O_2 or photorespiration in regulating CCM function or expression of limiting CO_2-regulated genes. However, the experiments of Vance and Spalding (2005) utilized fully acclimated cultures (i.e. under relatively steady-state conditions) rather than the process of limiting CO_2 acclimation. Thus transient O_2 effects during the acclimation process cannot be excluded.

Ci availability appears to be the primary signal, whether acting directly through CO_2 concentration or indirectly through metabolic signaling, but additional signals also appear to be involved in the regulation of limiting CO_2 acclimation. One of these is light, both photosynthetic, as mentioned previously, and non-photosynthetic. Dionisio et al. (1989a,b) reported that *CAH1* induction during limiting CO_2 acclimation was increased by a photosynthesis-independent blue light signal, as well as exhibiting a photosynthesis-dependent light requirement. How the blue light signal interacts with the regulation of CO_2-responsive genes and the CCM is not known, but it appears to have an enhancing rather than controlling effect (Dionisio et al., 1989a,b; Dionisio-Sese et al., 1990; Borodin et al., 1994). The blue light effect on *CAH1* induction is inhibited by potassium iodide, a flavin quencher, suggesting that a flavoprotein might be a receptor for this blue light signal (Dionisio et al., 1989b). This apparent flavoprotein involvement would be consistent with the action of phot, a blue light photoreceptor identified in *Chlamydomonas* and plants (Kottke et al., 2006).

Expression of limiting CO_2-regulated genes and CCM induction are complicated by the cell cycle, which itself is strongly influenced by circadian rhythms (Marcus et al., 1986; Rawat and Moroney, 1995; Eriksson et al., 1998; Spalding, 1998). *CAH1* expression and *CAH4/CAH5* expression have been reported to exhibit circadian rhythms in synchronously grown cultures (Rawat and Moroney, 1995; Fujiwara et al., 1996; Eriksson et al., 1998), which may reflect a complex regulation by both Ci and the cell cycle (Spalding, 1998). A controversial aspect of *CAH1* expression in synchronously grown cells is that Rawat and Moroney (1995) and Villarejo et al. (1996b) reported *CAH1* induction in the dark in limiting CO_2, which contradicts other reports of a light requirement for *CAH1* induction (Spalding and Ogren, 1982; Spencer et al., 1983; Dionisio et al., 1989a,b; Dionisio-Sese et al., 1990; Fukuzawa et al., 1990; Villarejo et al., 1996b). Additionally, Bailey and Coleman (1988) and Villarejo et al. (1996b) reported limiting CO_2 induction of *CAH1* in the dark in non-synchronous cultures, and Bozzo and Colman (2000) reported CCM induction and increased periplasmic CA activity induced by limiting CO_2 in the dark. However, in most cases where induction of *CAH1* in the dark has been reported, the expression level was substantially lower than that observed under similar conditions in the light, which indicates that there is at least a quantitative effect of light, and probably photosynthesis, on the acclimation process (Bailey and Coleman, 1988; Rawat and Moroney, 1995).

Part of the difficulty in rationalizing the signal(s) initiating limiting CO_2 acclimation is the apparently contradictory or conflicting information reported by different research groups, as discussed immediately above for *CAH1* induction in the dark. Clearly acclimation of *Chlamydomonas* to limiting CO_2 involves a signaling system more complex than simply a response to the bulk external CO_2 concentration, a change in photorespiratory metabolite concentration or a change in the chloroplast redox state but at present, it is difficult to incorporate all the available information on limiting CO_2 acclimation into a cogent hypothesis for the signaling mechanism. None of the single signal hypotheses can account for the documented reports of CCM regulation by bulk external CO_2 concentration, a photosynthetic light requirement for CCM and *CAH1* induction, enhanced *CAH1* induction by non-photosynthetic blue light, light-independent induction of *CAH1* in limiting CO_2, and repression of *CAH1* expression by mixotrophic growth with acetate.

A further complicating factor is the demonstration that *Chlamydomonas* has two distinct levels of acclimation to limiting CO_2, low CO_2, and very low CO_2. Most of the prior research addressing signal responses has not differentiated between the two, and the effective CO_2 concentration used in most experiments cannot be reasonably determined from the nominal CO_2 concentration because of CO_2 depletion in the cultures used. This leaves open the possibility that some apparent contradictions might arise from comparisons of acclimation to two physiological states for which the signals might differ. Nonetheless, based on the "air-dier" phenotype of the *pmp1* mutant (Spalding et al., 1983b, 2002; Wang and Spalding, 2006), it is unlikely, for example, that very low CO_2 acclimation is controlled at the level of redox, metabolite concentration, or internal Ci concentration. All of those factors should reach "inducing" thresholds even under low CO_2 conditions in the absence of LCIB-associated Ci uptake (Wang and Spalding, 2006). Thus the external Ci (probably CO_2) concentration appears the most probable controlling factor for the very low CO_2 acclimation signal. Furthermore, the results of Bozzo and Colman (2000) and Vance and Spalding (2005) both support external CO_2 as the primary signal in *Chlamydomonas* acclimation to both low CO_2 and very low CO_2. If external CO_2 concentration is the primary signal for acclimation to both low and very low CO_2, there either must be two receptors, or one receptor must detect a quantitative signal and transduce it to two effectively different transduction pathways.

It is counterintuitive that in spite of substantial evidence for multiple levels of acclimation to limiting CO_2 and the influence of multiple signals, the UV-induced *cia5* mutant, which was identified as a slow growing mutant in limiting CO_2 and is defective in only a single gene, appears to completely lack all of the recognized limiting CO_2 responses (Moroney et al., 1989; Marek and Spalding, 1991; Spalding et al., 1991). The characterization of *cia5* has therefore been extraordinarily revealing. The absence of any of the characterized limiting CO_2 responses in *cia5* argues that the *CIA5* gene product regulates either through one common signal transduction pathway, or by direct participation in multiple pathways. The *cia5* mutant serves as a prototype for identification of other mutants with defects in the CCM signal transduction pathway, but the paucity of reports of other *bona fide* signal transduction mutants suggest the existence of barriers to their identification, for example parallel, compensating elements or pathways.

The defective gene in *cia5*, *CIA5* (also known as *CCM1*), was identified independently by two research groups as encoding a predicted 698 amino acid hydrophilic protein (Fukuzawa et al., 2001; Xiang et al., 2001). The deduced *CIA5* gene product contains two putative zinc finger motifs in its N-terminal region and a glycine repeat region characteristic of transcriptional activators, suggesting that CIA5 is a transcription factor. Indeed, the point mutation in *cia5* is a His to Tyr alteration in the first of the two Zn finger motifs (Xiang et al., 2001). *CIA5* expression is independent of Ci conditions, and it is still not clear how CIA5 regulates CO_2-responsive genes nor how it interacts with other signal transduction components. Because of the apparent global regulation of limiting CO_2 acclimation responses by this "master regulator" (Xiang et al., 2001), it seems likely that a limiting CO_2 signal(s) cause modification of the constitutively accumulating CIA5 protein, which in turn either directly regulates CO_2-responsive genes or downstream signal transduction components. In this regard, several putative phosphorylation sites were identified near the C-terminus of CIA5 (Xiang et al., 2001).

Although CIA5 appears to act as a "master regulator," there is evidence that additional signal transduction pathways, either completely separate pathways or distinct branches downstream from CIA5, lead to differential regulation of limiting CO_2-responsive genes. Evidence includes the fact that while expression of *CAH1*, *CCP1/CCP2*, and *CAH4/CAH5* is correlated, they do not invariably have the same pattern. In addition, Villarejo et al. (1996b, 1997) reported that whereas *CAH1* was induced in darkness, induction of *CAH4/CAH5* and *CCP1/CCP2* required light. These authors also reported that expression of *CAH1* and *CAH4/CAH5* was repressed by inhibition of the glycolate pathway, but that expression of *CCP1/CCP2* was not. Furthermore, Chen et al. (1997) found that even the highly homologous *CCP1* and *CCP2* genes exhibited different timing in their expression during limiting CO_2 induction, indicating that the greater divergence in the promoter regions, relative to the coding regions, of these two genes is reflected in some differential expression. In addition, several genes encoding enzymes of the PCO (photorespiratory carbon oxidation) cycle are only transiently up-regulated (Tural and Moroney, 2005), and both subunits of

Rubisco are transiently down-regulated at the translational level (Winder et al., 1992). Thus it seems likely that even though the CCM and many of the limiting-CO_2 responsive genes require CIA5 function and appear to respond primarily to changing CO_2, they are often differentially regulated by multiple interacting pathways and either may require an additional signal to maintain increased expression, or may respond to a second signal reversing the increase in expression.

In the absence of additional signal transduction mutants or success in identifying proteins and/or genes that interact with CIA5, a complementary approach is the identification of *cis*-regulatory elements in the promoters of the regulated genes – from which the corresponding regulatory proteins can in principle be identified. CO_2 responsive elements in the promoters of *CAH1* and *CAH4/CAH5* have been identified using the arylsulfatase (*ARS1*) gene as a reporter (Villand et al., 1997; Kucho et al., 1999, 2003). Two regulatory regions were found in the *CAH1* promoter: a silencer region involved in the high CO_2 repression of transcription, and an enhancer region involved in the limiting CO_2 activation of transcription (Kucho et al., 1999). Two *cis*-elements, EE-1 and EE-2, were identified in the enhancer region and were found to interact with nuclear proteins (Kucho et al., 2003). In keeping with the expectation that coordinately regulated CO_2 responsive genes might share some regulatory elements, the core sequence motif present in EE-1 and EE-2, termed EEC, was also identified in the *CAH4* promoter (Kucho et al., 2003).

The identification of signal transduction mutants and the pursuit of *cis*-acting elements merged when Yoshioka et al. (2004), using an arylsulfatase gene controlled by the *CAH1* promoter, identified an insertional mutant, *lcr1*, defective in regulation of *CAH1*. *LCR1* encodes a novel Myb-type transcription factor that binds to the *CAH1* promoter. *LCR1* was demonstrated to be involved in the regulation of the limiting CO_2-regulated genes *CAH1*, *LCI1*, and *LCI6*, and its expression is induced by limiting CO_2 via CIA5 (Yoshioka et al., 2004). It is not yet clear whether CIA5 regulates *LCR1* directly or through other downstream elements of the signal transduction cascade(s).

II. METABOLISM OF CARBON ASSIMILATION

A. Photosynthetic carbon metabolism

The biochemistry and characteristics of photosynthesis and photosynthetic carbon assimilation in *Chlamydomonas* are very similar to those of higher plants, which is why *Chlamydomonas* has been used so frequently as a model to study plant-like photosynthesis. Thus the pathway for photosynthetic carbon assimilation in *Chlamydomonas* is almost certainly the same as that for C3-type higher plants, the C3 photosynthetic carbon reduction (PCR) cycle (Figure 8.1). The PCR cycle, where 3-phosphoglyceric acid

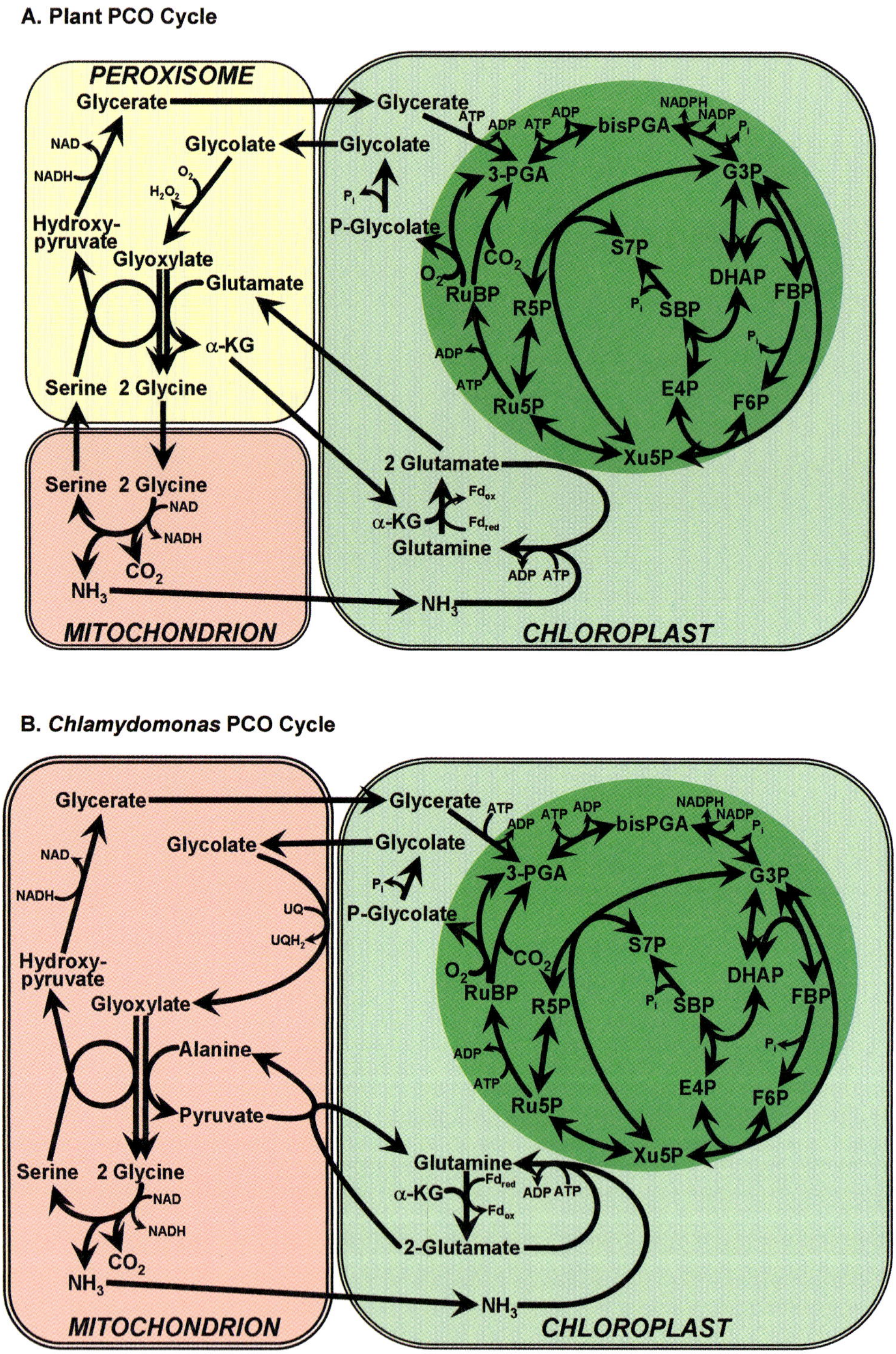

FIGURE 8.1 *Schematic illustration of the photorespiratory carbon oxidation (PCO) cycle and the closely associated photosynthetic carbon reduction (PCR) cycle (shaded pathways) in higher plants (A) and Chlamydomonas (B). Abbreviations: α-KG, α-ketoglutarate; 3-PGA, 3-phosphoglycerate; bisPGA, 1,3-bisphosphoglycerate; DHAP, dihydroxyacetone phosphate; E4P, erythrose-4-phosphate; Fd_{ox} and Fd_{red}, ferredoxin, oxidized and reduced form, respectively; F6P, fructose-6-phosphate; FBP, fructose-1,6-bisphosphate; G3P, glyceraldehyde-3-phosphate; P-glycolate, 2-phosphoglycolate; P_i, orthophosphate; R5P, ribose-5-phosphate; Ru5P, ribulose-5-phosphate; RuBP, ribulose-1,5-bisphosphate; S7P, sedoheptulose-7-phosphate; SBP, sedoheptulose-1,7-bisphosphate; Xu5P, xylulose-5-phosphate; UQ and UQH_2, ubiquinone oxidized and reduced forms, respectively.*

(3-PGA) is formed as the first stable product of photosynthesis by the Rubisco-catalyzed carboxylation of ribulose-1,5-bisphosphate (RuBP), was in fact first elucidated in the green microalga *Chlorella* (see Bassham, 1971).

Because the PCR cycle was established using microalgae grown in high CO_2 concentrations, which we now know represents a different physiological state than found in microalgae grown in air levels of CO_2, questions have been raised as to whether the "high affinity" microalgal cells grown at air CO_2 concentrations might have a different, more efficient carbon assimilation pathway. Although it seems well-documented that the C3 PCR cycle is used both in limiting CO_2 and high CO_2-acclimated microalgae (Pederson et al., 1966; Graham and Whittingham, 1968; Hogetsu and Miyachi, 1979), reports have surfaced occasionally in which substantial incorporation of radiolabeled CO_2 appears in malate and aspartate (C4 organic acids), especially during the period of limiting CO_2 acclimation and before full acclimation (Graham and Whittingham, 1968; Badour and Tan, 1983). It seems likely that as concluded by Badour and Tan (1983), the observation of substantial radiolabel incorporation into C4 organic acids might easily occur under conditions where photosynthetic carbon assimilation is greatly inhibited but anaplerotic β-carboxylation, via PEP (phosphoenolpyruvate) carboxylase is unaffected. Consistent with this conclusion, the two PEP carboxylase genes identified in *Chlamydomonas* both respond to carbon and nitrogen signals as one might expect for genes whose products are involved primarily in anaplerotic reactions (Mamedov et al., 2005).

Candidates for all the genes needed for a full PCR cycle have been identified in *Chlamydomonas* (Table 8.1). Most of these putative PCR cycle enzymes appear to be represented by single gene products targeted to the plastid, although there also are cases of apparent multiple plastid gene products, ambiguous targeting and even one gene product derived from differential splicing of an mRNA apparently encoding both cytosolic and plastidic isoforms of glyceraldehyde-3-phosphate dehydrogenase (*GAP1*).

B. Photorespiratory metabolism

All photosynthetic organisms functioning in the presence of O_2 have some level of PCO cycle activity and the fundamental pathway, studied extensively in plants, is largely the same in most of these organisms (Tolbert, 1997). The general outline of the PCO pathway involves production of phosphoglycolate (P-glycolate) as one product of the oxygenation reaction of Rubisco, for which CO_2 and O_2 are competing substrates, and the subsequent metabolism of this P-glycolate through a series of reactions apparently functioning to recover its carbon, ultimately producing one molecule of 3-PGA from two molecules of P-glycolate.

The PCO cycle in plants (Figure 8.1A) takes place in three different organelles: the chloroplast, the peroxisome, and the mitochondrion.

Table 8.1 Enzymes of photosynthetic carbon metabolism

Enzyme	EC #	Gene	Accession #
Fructose-1,6-BP aldolase[a]	4.1.2.13	*FBA1*	EDO98285
Fructose-1,6-BP aldolase	4.1.2.13	*FBA3*	EDO97897
Fructose-1,6-BPase	3.1.3.11	*FBP1*	EDP05318
Glyceraldehyde-3-P DH[b]	1.2.1.12	*GAP1*	EDO96575
Glyceraldehyde-3-P DH	1.2.1.12	*GAP3*	EDP09609
Phosphoglycerate kinase	2.7.2.3	*PGK1*	EDO98586
Phosphoribulokinase	2.7.1.19	*PRK1*	EDP02974
Ribose-5-P isomerase	5.3.1.6	*RPI1*	EDP04506
Ribulose-P-3 epimerase	2.7.1.19	*RPE1*	EDP05517
Rubisco (large subunit)	4.1.1.39	*rbcL*	P00877
Rubisco (small subunit)	4.1.1.39	*RBCS1*	EDO96904
Rubisco (small subunit)	4.1.1.39	*RBCS2*	EDO96903
Sedoheptulose-1,7-BPase	3.1.3.37	*SBP1*	EDP04487
Transketolase	2.2.1.1	*TRK1*	EDP06856
Triose-P isomerase	5.3.1.1	*TPIC*	EDP09773

[a]*Plastidic based on N-terminal extension, but greater sequence similarity to cytosolic forms.*
[b]*GAP1a splice variant, organelle targeting sequence, but not clear whether mitochondrial or plastid.*

The PCO cycle of *Chlamydomonas* (Figure 8.1B) deviates from the classical plant PCO cycle in that instead of oxidizing glycolate to glyoxylate via glycolate oxidase in the peroxisome, *Chlamydomonas* and many other microalgae apparently utilize a glycolate dehydrogenase in the mitochondrion, with other peroxisomal steps in the plant pathway thought also to be displaced to the mitochondrion (Nelson and Tolbert, 1970; Beezley et al., 1976; Spalding, 1989; Nakamura et al., 2005; Tural and Moroney, 2005). Also, because molecular O_2 is not an electron acceptor for glycolate dehydrogenase (Nelson and Tolbert, 1970; Nakamura et al., 2005), glycolate oxidation catalyzed by this enzyme does not produce H_2O_2, so catalase should not be required for the *Chlamydomonas* PCO cycle. Instead, glycolate dehydrogenase is expected to contribute electrons to the mitochondrial respiratory electron transport through reduction of the ubiquinone pool (Figure 8.1B; Nakamura et al., 2005).

An alternative to mitochondrial location of glycolate oxidation has been suggested by Goyal and Tolbert (1996), who observed rapid oxidation of glycolate to glyoxylate in *Chlamydomonas* and *Dunaliella tertiolecta* chloroplasts. They further reported that this glycolate oxidizing activity was mediated by a quinone and was associated with photosynthetic electron transport between PS II and the cytochrome b_6f complex. However, a mutant (*gdh1*, formerly HCR89) whose glycolate dehydrogenase gene (*GYD1*) is disrupted requires high CO_2 for growth, and the protein product has a predicted mitochondrial inner membrane localization (Nakamura et al., 2005).

Based on identified putative GDC genes (Table 8.2), the glycine decarboxylase complex of *Chlamydomonas* appears to be very similar to that of higher plants. Regardless of the path it takes to get there, glycine generated in the PCO cycle most likely undergoes an oxidative decarboxylation in the mitochondrion to yield (from 2xGly) serine, NH_3, and CO_2 (Figure 8.1B). Thus under normal conditions, photorespiratory CO_2 should be released, at least internally, if there is PCO cycle flux in *Chlamydomonas*.

While the PCO pathway in plants has been extensively studied and the genes encoding the plant PCO cycle enzymes have mostly been determined, neither the enzymes nor the corresponding genes in *Chlamydomonas* and other algae have been well characterized. The *Chlamydomonas* genome sequence revealed genes encoding putative products for most of the PCO cycle enzymes (Table 8.2; Chen et al., 1996; Im and Grossman, 2002; Mamedov et al., 2001; Miura et al., 2004; Nakamura et al., 2005; Tural and Moroney, 2005). Most of the photorespiratory genes are single copy, but there are apparently three genes each encoding phosphoglycolate phosphatase (PGP) and serine hydroxymethyltransferase (SHMT), although one of the SHMT genes does not appear to include an organellar targeting peptide. In addition, there are two genes encoding alanine-glyoxylate aminotransferase (AGT) and another photorespiratory gene, serine-glyoxylate aminotransferase (SGA) appears to generate an alternatively spliced transcript, but proteins from both splicing variants are predicted to be mitochondrially targeted.

Under high photorespiratory conditions, for example when high CO_2-acclimated cultures are exposed to limiting CO_2, *Chlamydomonas* cells excrete the glycolate produced rather than metabolizing it, although the glycolate excretion stops after a short time, presumably after limiting CO_2 acclimation has taken place (Tolbert et al., 1983; Moroney et al., 1986b). One reason for this apparent waste of carbon might be that *Chlamydomonas* lacks PCO cycle capacity to handle such a large flux downstream from glycolate. Because induction of the CCM should suppress much of the PCO cycle flux within a few hours, leaving only a short transition period of high photorespiratory flux, it may not be cost-effective for the cells to have more PCO cycle capacity than needed for the low flux in high CO_2. During acclimation to limiting CO_2, expression of the genes encoding most photorespiratory enzymes increases, and the expression of many increase only transiently (Tural and Moroney, 2005). This transient increase appears to correspond with the timing required for full acclimation of cells to limiting CO_2, suggesting that the CCM induced in limiting CO_2 decreases the PCO cycle flux once again to a low level (Marek and Spalding, 1991).

Table 8.2 Enzymes of photorespiratory carbon metabolism

Enzyme	EC #	Gene	Accession #	Location
Alanine aminotransferase	2.6.1.2	*AAT1*	EDP08011	Mitochondrion
Alanine-glyoxylate transaminase	2.6.1.44	*AGT1*	EDO97315	Mitochondrion[a]
Alanine-glyoxylate transaminase	2.6.1.44	*AGT2*	EDO96807	Mitochondrion[b]
Dihydrolipoyl dehydrogenase	1.8.1.4	*DLDH1*	EDP01871	Mitochondrion
Glycerate-3-kinase	2.7.1.31	*GLYK*	EDP03009	Plastid
Glycine decarboxylase, H protein		*GCSH*	EDP08614	Mitochondrion
Glycine decarboxylase, P protein	1.4.4.2	*GCSP*	EDP03562	Mitochondrion
Glycine decarboxylase, T protein	2.1.2.10	*GCST*	EDP03133	Mitochondrion
Glycolate dehydrogenase	1.1.99.14	*GYD1*	EDP01639	Mitochondrion
Hydroxypyruvate reductase	1.1.1.81	*HPR1*	EDP05213	Organelle[c]
Phosphoglycolate phosphatase	3.1.3.18	*PGP1*	EDP06184	Plastid
Phosphoglycolate phosphatase	3.1.3.18	*PGP2*	EDP05829	Plastid
Phosphoglycolate phosphatase	3.1.3.18	*PGP3*	EDP08194	Plastid
Rubisco (large subunit)	4.1.1.39	*rbcL*	P00877	Plastid
Rubisco (small subunit)	4.1.1.39	*RBCS1*	EDO96904	Plastid
Rubisco (small subunit)	4.1.1.39	*RBCS2*	EDO96903	Plastid
Serine hydroxymethyltransferase	2.1.2.1	*SHMT1*	EDO97448	Mitochondrion
Serine hydroxymethyltransferase	2.1.2.1	*SHMT2*	EDO97351	Not organellar[d]
Serine hydroxymethyltransferase	2.1.2.1	*SHMT3*	EDP00905	Mitochondrion
Serine-glyoxylate aminotransferase[e]	2.6.1.45	*SGA1a*	EDO97195	Mitochondrion
Serine-glyoxylate aminotransferase[e]	2.6.1.45	*SGA1b*	EDO97196	Mitochondrion

[a] *Predicted by TargetP as mitochondrial, but peroxisomal location supported by C-terminal PTS1 signal GKL.*
[b] *Predicted by TargetP as mitochondrial, but peroxisomal location supported by C-terminal PTS1 signal CKL.*
[c] *Ambiguous between mitochondrion and plastid.*
[d] *Predicted as neither mitochondrial nor plastid localized.*
[e] *SGA1a is the longer of two predicted splice variants, but both SGA1a and SGA1b are predicted as mitochondrial.*

The high apparent affinity for CO_2 and the lack of O_2 inhibition of photosynthesis in limiting CO_2-acclimated *Chlamydomonas* provide ample evidence that PCO cycle activity is greatly suppressed by the CCM, but there also is abundant evidence that photorespiratory flux is not completely suppressed (Moroney et al., 1986b). Some of the PCO cycle genes such as *GYD1* exhibit a non-transient increase in expression, suggesting a need for increased capacity in some steps of the PCO cycle even when the CCM is functional.

Additionally, at least two mutants with defects in PCO cycle enzymes, *pgp1* and *gdh1*, have been characterized (Suzuki et al., 1990; Mamedov et al., 2001; Nakamura et al., 2005). The very fact that these mutants require elevated CO_2 for rapid photoautotrophic growth demonstrates that the CCM does not completely suppress the oxygenase activity of Rubisco and PCO cycle activity.

C. Acetate metabolism

Many species of *Chlamydomonas* can be grown either autotrophically with CO_2 as the sole carbon source, heterotrophically by consuming acetate, or mixotrophically, using acetate as a carbon source in the light in combination with primary photosynthetic CO_2 assimilation (see Volume 1, Chapter 6 for discussion of acetate utilization by species other than *C. reinhardtii*). Although careful metabolic studies of acetate assimilation under either condition seem to be lacking, it probably is safe to assume that at least the primary metabolism of acetate occurs via the glyoxylate cycle, following incorporation into acetyl coenzyme A (acetyl-CoA; Figure 8.2). Nonetheless, details of acetate assimilation, including acetate uptake and subcellular localization of assimilation, are not at all clear.

Incorporation of acetate into acetyl-CoA can occur by one of two pathways in *Chlamydomonas*: directly into acetyl-CoA, as catalyzed by acetyl-CoA synthetase (ACS), and in two steps via acetate-phosphate, as catalyzed by acetate kinase (ACK) and phosphate acetyltransferase (PAT; Figure 8.2). Putative genes encoding all three of these enzymes have been identified in *Chlamydomonas* (Table 8.3). ACK and PAT are not thought to occur commonly in eukaryotes, although they have been identified in several recently (Ingram-Smith et al., 2006). The presence of two separate pathways for the incorporation of acetate into acetyl-CoA, as well as multiple genes encoding each enzymatic step in both pathways, underscores the importance of acetate assimilation to *Chlamydomonas* and suggests that acetate assimilation occurs in multiple compartments, although predictions for targeting of most of these gene products are inconclusive.

The three ACS genes and two ACK-PAT gene candidates (ignoring dual targeting of any gene products) makes it possible to convert acetate to acetyl-CoA in a minimum of three and a maximum of five cell compartments. It is reasonable to assume that the capability for acetate conversion to acetyl-CoA would be present in mitochondria, where the acetyl-CoA might feed directly into the TCA cycle for energy production and conversion to carbon skeletons, for example nitrogen assimilation in the plastid. In addition, a plastid location is reasonable, because of the direct use of acetyl-CoA in fatty acid biosynthesis, and a cytosolic location is similarly reasonable for fatty acid elongation and related reactions. And of course a glyoxysomal location is essential if the glyoxylate cycle is located in glyoxysomes (see discussion below), and is responsible for the majority of acetate assimilation.

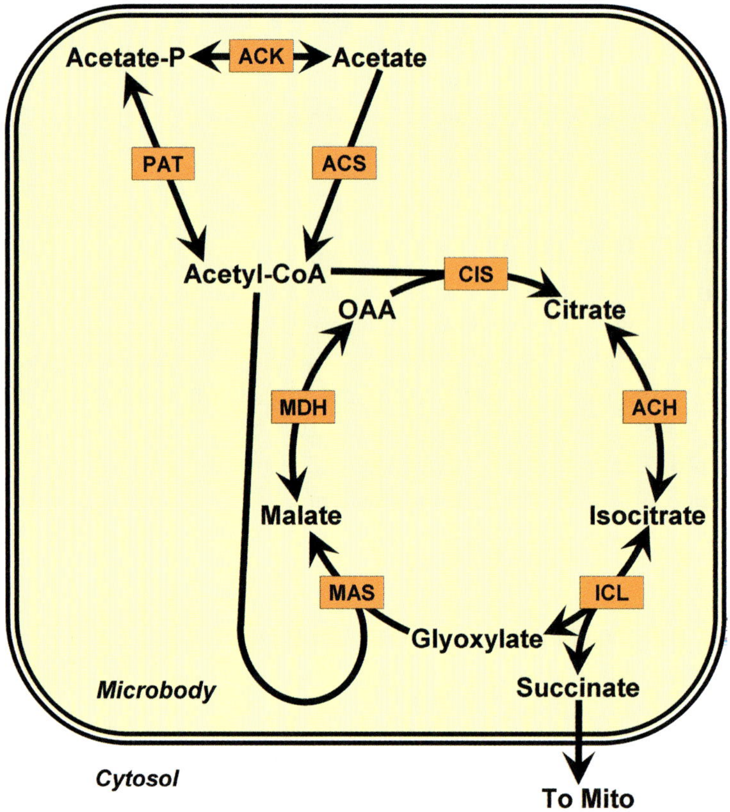

FIGURE 8.2 *Schematic illustration of proposed acetate assimilation via the glyoxylate cycle in Chlamydomonas, including the corresponding enzymes. Abbreviations: ACH, aconitase; ACK, acetate kinase; ACS, acetyl-CoA synthetase; CIS, citrate synthase; ICL, isocitrate lyase; MAS, malate synthase; MDH, malate dehydrogenase; OAA, oxaloacetic acid; PAT, phosphate acetyltransferase.*

Under the assumption that acetate assimilation occurs primarily through the glyoxylate pathway, one location of acetate assimilation into acetyl-CoA would be the compartment containing the glyoxylate cycle. Unfortunately, the compartmentation of the glyoxylate cycle itself also is uncertain in *Chlamydomonas*. Genes predicted to encode all of the enzymes of the glyoxylate cycle are present in the genome (Table 8.3), including some, such as malate synthase (Nogales et al., 2004) and isocitrate lyase (Petridou et al., 1997), that were identified previously. Malate synthase and isocitrate lyase are unique to the glyoxylate cycle, and at least malate synthase has a C-terminal amino acid sequence consistent with peroxisomal/glyoxysomal targeting (Nogales et al., 2004), but these microbodies have not been positively identified in *Chlamydomonas* (Giraud and Czanisnski, 1971). In addition, the genome encodes isoforms of citrate synthase and malate dehydrogenase that are most homologous to the

Table 8.3 Enzymes of acetate metabolism

Enzyme	EC#	Gene	Accession #	Location
Acetate kinase	2.7.2.1	*ACK1*	EDP02500	Organelle[a]
Acetate kinase	2.7.2.1	*ACK2*	EDP04790	Unknown[b]
Acetyl-CoA synthetase	6.2.1.1	*ACS1*	ED098325	Cytosol[c]
Acetyl-CoA synthetase	6.2.1.1	*ACS2*	ED098345	Organelle[a]
Acetyl-CoA synthetase	6.2.1.1	*ACS3*	ED097232	Cytosol[c]
Aconitase	4.2.1.3	*ACH1*	EDP09440	Mitochondrion[d]
Citrate synthase	2.3.3.1	*CIS2*	EDP01358	Unknown[e]
Isocitrate lyase	4.1.3.1	*ICL1*	EDP01589	Unknown[f]
Malate dehydrogenase	1.1.1.37	*MDH2*	EDP06365	Unknown[g]
Malate synthase	2.3.3.9	*MAS1*	EDP01419	Glyoxysome[h]
Phosphate acetyltransferase	2.3.1.8	*PAT1*	EDP04895	Plastid[i]
Phosphate acetyltransferase	2.3.1.8	*PAT2*	EDP02499	Organelle[a]

[a] *Predicted as targeted to mitochondria or plastid by iPSORT, TargetP, and WoLF PSORT.*

[b] *Predicted as targeted to mitochondria (iPSORT) or secreted (TargetP). WoLF PSORT is ambiguous.*

[c] *Predicted as "other" by both iPSORT and TargetP and as cytosolic by WoLF PSORT.*

[d] *Predicted as mitochondrial by iPSORT, TargetP, and WoLF PSORT.*

[e] *Predicted as "other" by iPSORT, mitochondrial by TargetP and mitochondrial or peroxisomal by WoLF PSORT.*

[f] *Predicted as "other" by both iPSORT and TargetP and as mitochondrial by WoLF PSORT.*

[g] *Predicted as "other" by both iPSORT and TargetP and as plastid or peroxisomal by WoLF PSORT.*

[h] *Predicted by WoLF PSORT and PTS1 Predictor as peroxisomal.*

[i] *Predicted by iPSORT, WoLF PSORT, and TargetP as plastid targeted.*

glyoxysomal isozymes of these enzymes in plants, although targeting of these gene products is ambiguous. On the other hand, only one gene, *ACH1*, is predicted to encode the last enzyme needed for a functional glyoxylate cycle, aconitate hydratase (aconitase), and it is most similar to plant mitochondrial aconitase genes. If aconitase is located in both mitochondria and glyoxysomes, either *ACH1* has more than one transcription start site, its transcript can be differentially spliced or the gene product can be imported into both organelles. Because of the ambiguity and for the sake of simplicity, the glyoxylate cycle of *Chlamydomonas* is assumed to be located in a microbody.

The glyoxylate cycle can generate one molecule of succinate as a net product from two molecules of acetate. Beyond the glyoxylate cycle, acetate assimilation most likely proceeds along the same pathway as gluconeogenic fatty acid conversion in oil seeds, that being conversion of succinate to malate and oxaloacetic acid (OAA) in the mitochondrion, OAA phosphorylative decarboxylation to PEP, and PEP conversion into sugar phosphates. These parts of the presumed pathway are not illustrated.

III. ACQUISITION OF INORGANIC CARBON

A. CO_2-concentrating mechanisms

CCMs represent an evolutionary adaptation acquired by microalgae and cyanobacteria in response to increased O_2 and decreased CO_2 availability, especially in aquatic environments (reviewed in Kaplan and Reinhold, 1999; Badger and Spalding, 2000; Giordano et al., 2005). It appears that cyanobacteria and microalgae probably acquired CCMs independently, since the selective pressure for acquisition of such mechanisms likely was not present until after the endosymbiotic transition of cyanobacteria-like prokaryotes to plastids in primitive photosynthetic eukaryotes (Douglas, 1999; Turmel, 1999). Nonetheless, the general function of cyanobacterial and microalgal CCMs are similar, even if the individual components may be phylogenetically unrelated. Although significant advances have been made in understanding the cyanobacterial CCMs and to a lesser extent, the microalgal CCMs, many components are still unknown or uncharacterized. The genomic era has brought new insights into the nature of the microalgal CCM and its regulation, especially in *Chlamydomonas*.

Microalgal and cyanobacterial CCMs both use active Ci uptake to increase intracellular CO_2 concentrations at the site of Rubisco, enabling high rates of photosynthetic CO_2 assimilation even when external Ci concentrations are low (Badger et al., 1980). Active Ci uptake results in the intracellular accumulation of HCO_3^-, so the CCMs also require localization of CA activity with or near Rubisco to catalyze dehydration of the accumulated HCO_3^- and provide near-saturating CO_2 concentrations for carboxylation of RuBP. Therefore, the known essential components of microalgal and cyanobacterial CCMs (reviewed in Spalding, 1998; Kaplan and Reinhold, 1999; Badger and Spalding, 2000) include at least active Ci transport and internal CA localized near Rubisco.

A classical mutational approach has proved powerful in probing the CCM in *Chlamydomonas*, especially the use of insertional mutagenesis (Fukuzawa et al., 2001; Van et al., 2001; Pollock et al., 2003; Yoshioka et al., 2004; Nakamura et al., 2005; Wang and Spalding, 2006). Identification and characterization of many high-CO_2-requiring (HCR) mutants opened the way to dissection of the CCM mechanism and the signal transduction pathway between low/very low CO_2 and gene expression.

Investigations of HCR mutants have lead to the identification of a number of gene products relevant to the limiting CO_2 acclimation, including

CIA5, the "master regulator" of limiting CO_2 acclimation and a key component of the limiting-CO_2 signal transduction pathway (Fukuzawa et al., 2001; Xiang et al., 2001); CAH3, the essential thylakoid lumen CA apparently responsible for dehydrating actively accumulated HCO_3^- to provide near-saturating substrate CO_2 concentrations for Rubisco (Spalding et al., 1983a; Moroney et al., 1986a; Hanson et al., 2003); LCIB, a plastid protein defective in the classical Ci transport mutant, *pmp1* (Wang and Spalding, 2006); LCR1, a Myb-like transcription factor involved in regulation of *CAH1*, *LCI1* and other limiting CO_2-induced genes (Yoshioka et al., 2004); RCA1 (Rubisco activase), which is required for optimal photosynthesis under limiting CO_2 conditions (Pollock et al., 2003); PGP1, a phosphoglycolate phosphatase required for dephosphorylation of P-glycolate synthesized by the oxygenase activity of Rubisco, and the first metabolite of the PCO cycle (Suzuki et al., 1990); and GYD1, a glycolate dehydrogenase required for metabolism of glycolate generated by PCO cycle activity (Nakamura et al., 2005). The precise functions of a number of these gene products were discussed in previous sections.

B. Inorganic carbon transport

The most unique aspect of the microalgal CCMs is the active uptake of Ci, which clearly distinguishes these CCMs from the C4 pathways of higher plants. Although both types of CCM concentrate CO_2 at the site of Rubisco many-fold relative to the external environment, the microalgal Ci transport systems are unique among eukaryotic photosynthetic organisms. At the whole cell level, *Chlamydomonas* can use both external CO_2 and HCO_3^- as substrates for active Ci uptake, although carbon disequilibrium studies have demonstrated that active CO_2 uptake represents the major Ci uptake flux (Marcus et al., 1984; Sültemeyer et al., 1989; Badger et al., 1994; Palmqvist et al., 1994). Active CO_2 uptake does not necessarily equate with active CO_2 transport. A CO_2 concentration gradient generated by active internal consumption of CO_2, especially an active process inside the chloroplast, could drive diffusive CO_2 uptake across both membranes in an apparently active process that is currently indistinguishable from active transport. Direct HCO_3^- uptake, presumably active HCO_3^- transport across the plasma membrane, also is well supported experimentally (Sültemeyer et al., 1989; Thielmann et al., 1990; Badger et al., 1994; Palmqvist et al., 1994; Amoroso et al., 1998). Thus *Chlamydomonas* cells can use both of the Ci forms (CO_2 and HCO_3^-) present in their habitat. It is important to note that although CO_2 was the primary Ci species actively taken up in these reports, it is not clear whether the cultures used were acclimated to low versus very low CO_2 conditions. Little is known about the differences between Ci uptake characteristics under the two distinct acclimation states, and it certainly is possible that the preferred Ci species might differ between them.

Regardless of the form taken up, the accumulated Ci must ultimately provide increased concentrations of substrate CO_2 to Rubisco, which is localized largely in the chloroplast pyrenoid (Borkhsenious et al., 1998). Thus the active Ci uptake must cross both the plasma membrane and the chloroplast envelope. Indeed, active transport of both CO_2 and HCO_3^- has been documented across both of these membranes, that is, into whole cells and into isolated chloroplasts (Marcus et al., 1984; Moroney et al., 1987; Palmqvist et al., 1988, 1994; Sültemeyer et al., 1988, 1989; Goyal and Tolbert, 1989; Badger et al., 1994; Amoroso et al., 1998; van Hunnik et al., 2002). Active uptake of HCO_3^- across both the plasma membrane and chloroplast envelope almost certainly requires a protein transporter in each case. On the other hand, active CO_2 uptake across both the plasma membrane and chloroplast envelope might occur either via a transporter or by a concentration gradient-stimulated diffusion, as discussed above. Thus the physiological evidence suggests that *Chlamydomonas* probably has a minimum of one and possibly two Ci transporters or transport complexes in each of the plasma membrane and the chloroplast envelope, which together move external Ci to the site of Rubisco. No plasma membrane or chloroplast envelope transport proteins have been unambiguously identified, although potential candidates exist.

It has long been argued that Ci must accumulate in the chloroplasts largely as HCO_3^-, which is then dehydrated by a specific chloroplast CA to provide high concentrations of substrate CO_2 for Rubisco even in the alkaline conditions of the stroma (Moroney et al., 1987; Moroney and Mason, 1991). If this is true, then the apparent active transport of both CO_2 and HCO_3^- by chloroplasts isolated from *Chlamydomonas* and other microalgae (Amoroso et al., 1998; van Hunnik et al., 2002) argues for nearly quantitative conversion of the CO_2 actively taken up into the chloroplast to HCO_3^-. CAH6, a stromal CA, could provide the mechanism for this rapid and complete conversion.

As was discussed earlier in this chapter, the *pmp1* mutant played an important role in recognition of the multi-tiered CO_2 acclimation responses in *Chlamydomonas*, and also was important in defining active Ci transport as an essential component of the CCM (Spalding et al., 1983b). Although originally identified as a conditional lethal in air levels of CO_2, later studies revealed that *pmp1* is deficient in Ci transport in low CO_2 but not in very low CO_2 conditions (Van et al., 2001; Spalding et al., 2002). Therefore, *pmp1* is defective in a component specifically controlling active Ci uptake and accumulation in low CO_2 conditions.

The gene defective in *pmp1* and an insertional allele with the same "air-dier" phenotype, *ad1*, was identified as *LCIB* (Wang and Spalding, 2006), a gene previously described as limiting CO_2-induced (Miura et al., 2004). *LCIB* is a member of a small gene family in *Chlamydomonas*, including *LCIC*, *LCID*, and *LCIE* (Table 8.4), but has no identifiable domains or motifs, and no homologies to date outside of single genes each in the green microalgae *Ostreococcus taurii*, *O. lucimarinus*, and *Volvox carteri* (Wang and Spalding, 2006; Grossman et al., 2007). *LCIB* and *LCIC*, which share 57% identity and 73% similarity in their predicted amino acid sequences, show similar CO_2 expression patterns: high expression in both low CO_2 and very low CO_2, and very low expression in high CO_2 conditions (Wang and Spalding, 2006), which is consistent with the reported phenotypic penetrance of the *pmp1* lesion (Suzuki and Spalding, 1989). *LCID* and *LCIE* also share significant deduced amino acid sequence similarity (71% identity; 78% similarity), as well as with sharing significant similarity as a pair with the *LCIB–LCIC* gene pair (40–44% identity; 62–65% similarity). The *LCID* expression pattern appears similar to that of *LCIB* and *LCIC* but at a much lower mRNA abundance, and *LCIE* expression has been detected only by recovery of a partial cDNA (Wang and Spalding, 2006).

Even though the evidence clearly argues that LCIB is required for active Ci transport in low CO_2, the lack of transmembrane regions make LCIB an unlikely candidate for a standalone Ci transporter. It is more likely that LCIB acts as a soluble subunit in a Ci transport complex or as a regulator

Table 8.4 Ci transport candidates

Gene	Accession #	Location
CCP1	EDP04238	Plastid envelope[a]
CCP2	EDP04238	Plastid envelope[a]
HLA3/MRP1	EDP07736	Plasma membrane[b]
LCI1	EDP06069	Plasma membrane[c]
LCIA	EDP04946	Plastid envelope[d]
LCIB	EDP07837	Plastid[e]
LCIC	EDP04956	Plastid[e]
LCID	EDP04142	Plastid[e]
LCIE	EDP04243	Plastid[f]
RHP1	EDP01722	Plastid envelope[g]
RHP2	EDP01723	Plastid envelope[h]
ycf10	CAA62148	Plastid envelope

[a] *Confirmed chloroplast envelope location (Ramazanov et al., 1993).*

[b] *Predicted as secretory pathway by iPSORT and TargetP, and as plasma membrane by WoLF PSORT.*

[c] *Predicted as secretory pathway by iPSORT and TargetP, and as vacuolar by WoLF PSORT.*

[d] *Predicted chloroplast envelope location (Mariscal et al., 2006).*

[e] *Predicted as mitochondrial by iPSORT, and as plastidic by TargetP and WoLF PSORT.*

[f] *No targeting prediction, because the gene model is lacking the correct N-terminus.*

[g] *Predicted as plastidic by iPSORT and TargetP, but mitochondrial by WoLF PSORT.*

[h] *Predicted as plastidic by iPSORT, TargetP, and WoLF PSORT.*

of Ci transport, as may other members of the LCIB family. The apparent non-essential nature of LCIB in very low CO_2 conditions, even though it is expressed under all limiting CO_2 conditions, raises the question of whether LCIB or any LCIB family members function in Ci transport in very low CO_2. Therefore, in addition to understanding the role of LCIB and other members of the LCIB family in the low CO_2 Ci transport system(s), it remains to be discovered what Ci transport systems are functional and essential in very low CO_2. Ci transport components other than LCIB are yet to be definitively identified. Some candidates (Table 8.4) such as *CCP1*, *CCP2*, *LCI1*, *HLA3/MRP1*, and *LCIA*, have been identified as either plastid envelope or plasma membrane transmembrane gene products induced or up-regulated by limiting CO_2 (Burow et al., 1996; Chen et al., 1997; Im and Grossman, 2002; Miura et al., 2004). These and other candidates are discussed in more detail below.

As mentioned above, there must be at least one and possibly several plasma membrane Ci transporters in *Chlamydomonas*, depending on redundancy level and whether separate carriers are involved in low versus very low CO_2 Ci transport. HLA3/MRP1 and LCI1 are the only identified Ci transport candidates predicted to be targeted to the plasma membrane. *HLA3* was identified as a high light-induced gene, but actually is induced only in limiting CO_2 (Im and Grossman, 2002) under control of CIA5 (Fukuzawa et al., 2001; Xiang et al., 2001). HLA3 also has been named MRP1 based on being a member of the multidrug resistance-related proteins (MRP) subfamily of ATP-binding cassette (ABC) type transporters (Miura et al., 2004; Hanikenne et al., 2005). Members of the MRP subfamily transport a wide variety of substrates and often are sensitive to vanadate inhibition (Rea, 2007), so in light of the reported vanadate inhibition of *Chlamydomonas* Ci transport (Palmqvist et al., 1988), the suggested Ci transport by HLA3/MRP1 is not unreasonable.

LCI1 also was identified as a limiting CO_2-induced gene regulated by both CIA5 and LCR1 (Burow et al., 1996; Miura et al., 2004; Yoshioka et al., 2004). LCI1 also has no identifiable homologues or recognizable domains, other than predicted transmembrane domains, so it appears to represent a unique membrane protein. Because it is CIA5-regulated and predicted to be targeted to the plasma membrane, it is a possible Ci transport candidate, though little else is known about this gene or its product.

The first identified Ci transport candidate in *Chlamydomonas* was the product of the plastid *ycf10* gene, which was implicated in chloroplast Ci uptake when its disruption resulted in decreased Ci uptake in chloroplasts (Rolland et al., 1997). The *ycf10* gene product is localized in the inner chloroplast envelope and thus was originally named CemA when discovered in plants (Sasaki et al., 1993); the cyanobacterial ortholog is *pxcA* (formerly *CotA*). Functionally, the *pxcA* gene product, which is involved in light-induced Na^--dependent proton extrusion and is thought to play an important role in electrical and pH homeostasis during Ci uptake (Katoh et al., 1996a,b; Sonoda et al., 1998), probably provides the greatest insight into the role of CemA in *Chlamydomonas* chloroplast Ci uptake.

The product of *LCIA* (*NAR1.2*), identified initially as a limiting CO_2-inducible gene, is a putative chloroplast envelope protein in the *Chlamydomonas NAR1* family (Galván et al., 2002; Miura et al., 2004; Mariscal et al., 2006), which encode members of the formate/nitrite transporter (FNT) family (Rexach et al., 2000) found in bacteria and, to a lesser extent in fungi, yeast, algae, and protozoa, but not in plants (see Chapter 3). LCIA was proposed as a candidate Ci transporter rather than a nitrite transporter, because its expression is regulated by CO_2 irrespective of the nitrogen source (Miura et al., 2004; Mariscal et al., 2006). LCIA also has been reported to exhibit both HCO_3^- and nitrite transport activities when expressed in *Xenopus* oocytes (Mariscal et al., 2006), although its low affinity for HCO_3^- and high affinity for nitrite leave open the question of its significance as a Ci transporter.

Chloroplast envelope proteins CCP1 and CCP2, which share 96% amino acid identity, also share sequence similarity with the mitochondrial carrier protein superfamily (Chen et al., 1997). This family includes plastid and peroxisomal as well as mitochondrial carrier proteins, whose members transport a wide spectrum of metabolites involved in carbon and energy metabolism (Kuan and Saier, 1993; Palmieri, 1994). Their limiting Ci regulation and their relationship to a superfamily of carriers with broad substrate specificity has made CCP1 and CCP2 prime candidates for chloroplast Ci transport, but RNAi knockdown of both *CCP1* and *CCP2* expression resulted in little or no effect on photosynthetic kinetics and Ci uptake characteristics (Pollock et al., 2004). These results raise doubts regarding any role of CCP1 and CCP2 in Ci transport, although it is possible that the loss of a CCP1/2-associated Ci transport system could be masked by a compensatory system.

The genomic locations of *CCP1* and *CCP2* may offer functional hints. Along with the largely uncharacterized *LCID/LCIE* gene pair, *CCP1/2* are part of a 75 kb CO_2-regulated gene cluster (Wang and Spalding, 2006; Grossman et al., 2007). In this cluster, which also includes the periplasmic CA genes *CAH1* and *CAH2*, *CCP1* and *CCP2* are aligned head to head with *LCIE* and *LCID*, respectively, in an apparent inverted repeat. The inverted regions flank *CAH1* and *CAH2*, forming a cluster of six CO_2-responsive genes, an arrangement that suggests duplication of an ancestral *CCP–LCI* gene pair flanked by an ancestral periplasmic *CAH* gene.

Although not thought to be a candidate for Ci transport under limiting CO_2 conditions, the RHP1 protein (Table 8.4) has been reported to play an important role in CO_2 uptake under high CO_2 conditions. Two genes (*RHP1* and *RHP2*) encode Rhesus (Rh) proteins, similar to the Rh proteins in the human red blood membrane, and predicted to be chloroplast envelope proteins with 12 transmembrane domains (Soupene et al., 2002). *RHP1* exhibits low level constitutive expression in limiting CO_2 but is up-regulated many-fold in high CO_2. RNAi knockdown of *RHP1* resulted in a growth defect in high CO_2 but normal growth in limiting CO_2. Since RHP1 has been concluded to function as a bidirectional CO_2 gas channel, these observations suggest that this Rh protein may function in high CO_2 to provide adequate CO_2 for photosynthesis in the absence of a CCM (Soupene et al., 2002, 2004). The role, if any, of the low level constitutive expression of Rh proteins in Ci uptake and CO_2 assimilation in limiting CO_2 has yet to be established, although the apparent lack of a limiting CO_2 phenotype in the RNAi knockdown mutants of *RHP1* argues against any substantial role.

C. Carbonic anhydrases

The first essential component of the *Chlamydomonas* CCM identified was CAH3, a thylakoid lumen CA responsible for dehydration of HCO_3^- actively accumulated internally, probably in the stroma (Funke et al., 1997; Karlsson et al., 1998). In CAH3-defective mutants, internal Ci, presumably in the form of HCO_3^-, accumulates to high levels but is not available for CO_2 assimilation (Spalding et al., 1983a; Moroney et al., 1986a; Hanson et al., 2003). Also, the first identified protein and gene induced by limiting CO_2 was *CAH1*, which encodes a periplasmic CA (Coleman and Grossman, 1984; Fukuzawa et al., 1990).

In the background of the highly expressed periplasmic CA, CAH1, significant effort has been expended to identify CA activities associated with the chloroplast. Insoluble CA activity associated with the thylakoid fraction and soluble activity, presumably localized in the stroma, have been detected (Sültemeyer et al., 1995; Amoroso et al., 1996), and a CA activity associated with the chloroplast envelope was subsequently reported (Villarejo et al., 2001). Although probable candidates for the thylakoid-associated (CAH3) and soluble (CAH6) CAs have been identified, a candidate for the chloroplast envelope CA has not yet emerged.

Twelve putative CA-encoding genes have been identified (Table 8.5), many of which apparently are regulated by CO_2 concentration and CIA5 (Mitra et al., 2005) but of them, so far only CAH3 has been shown to be essential for the CCM. The 12 putative CAs comprise members of all three major CA evolutionary lineages (Hewett-Emmett and Tashian, 1996), including three α-type CAs (CAH1, CAH2, CAH3), six β-type CAs (CAH4, CAH5, CAH6, CAH7, CAH8, CAH9) and three γ-type CAs (CAG1, CAG2, CAG3).

As mentioned earlier, *CAH1* encodes a periplasmic CA and was one of the first limiting CO_2-induced genes identified in *Chlamydomonas* (Fukuzawa et al., 1990), and represents most of the assayable CA activity under limiting CO_2 conditions. This gene has been very important as a tool for studying the regulation of gene expression during acclimation to limiting CO_2 (Spalding

Table 8.5 Carbonic anhydrases

Enzyme	EC #	Gene	Accession #	Location
α-Carbonic anhydrase	4.2.1.1	*CAH1*	EDP04241	Periplasm
α-Carbonic anhydrase	4.2.1.1	*CAH2*	EDP04240	Periplasm
α-Carbonic anhydrase	4.2.1.1	*CAH3*	EDP00852	Thylakoid lumen
β-Carbonic anhydrase	4.2.1.1	*CAH4*	EDO96058	Mitochondrion
β-Carbonic anhydrase	4.2.1.1	*CAH5*	EDP07024	Mitochondrion
β-Carbonic anhydrase	4.2.1.1	*CAH6*	EDO96552	Plastid stroma
β-Carbonic anhydrase	4.2.1.1	*CAH7*	EDO99999	Unknown[a]
β-Carbonic anhydrase	4.2.1.1	*CAH8*	EDO99999	Unknown[b]
β-Carbonic anhydrase	4.2.1.1	*CAH9*	EDP07163	Cytosolic[c]
γ-Carbonic anhydrase	4.2.1.1	*CAG1*	EDO96533	Unknown
γ-Carbonic anhydrase	4.2.1.1	*CAG2*	EDO97352	Unknown
γ-Carbonic anhydrase[d]	4.2.1.1	*CAG3*	EDP00854	Unknown[e]

[a] *Predicted by TargetP and WoLF PSORT as plastidic but cytosolic by iPSORT.*

[b] *Predicted by TargetP as plastidic, but cytosolic by iPSORT and WoLF PSORT.*

[c] *Predicted as cytosolic by iPSORT, WoLF PSORT, and Target-P.*

[d] *Predicted as a putative γ-CA based on sequence, but no CA activity detected when overexpressed in E. coli.*

[e] *Predicted as cytosolic by iPSORT and TargetP, and as plastidic by WoLF PSORT.*

et al., 2002), particularly when using its promoter in combination with reporter genes (Villand et al., 1997; Kucho et al., 1999, 2003; Yoshioka et al., 2004). The role of CAH1 in the CCM and in the three CO_2-regulated acclimation states is not entirely clear, although it has been suggested to facilitate Ci diffusion to the cell surface and uptake of CO_2 in alkaline conditions (Moroney et al., 1985; Spalding, 1998; Giordano et al., 2005; Mitra et al., 2005). However, a null *CAH1* mutant failed to exhibit any detectable growth or photosynthesis phenotype under a variety of pH and Ci conditions, arguing that CAH1 is not essential for CCM function (Van and Spalding, 1999).

While *CAH1* is expressed only in limiting CO_2, a very similar, closely linked periplasmic α-CA gene, *CAH2*, is induced only under high CO_2 conditions (Fujiwara et al., 1990; Rawat and Moroney, 1991), a situation that is consistent with the substantially greater divergence of the promoter regions of these two genes relative to their coding sequences. Little information is available regarding the potential role of CAH2 in Ci assimilation under high CO_2 conditions. Along with *CAH1* and *CAH2* being a probably recent, tandem duplication, the gene pair is clustered with other genes possibly important for CCM (Wang and Spalding, 2006; Grossman et al., 2007; see section III.B).

As mentioned previously, CAH3 is located in the thylakoid lumen and appears essential to the function of the CCM. *CAH3* mutants cannot survive under limiting CO_2 conditions and overaccumulate internal Ci in a form or location that makes it unavailable for photosynthesis (Spalding et al., 1983a; Moroney et al., 1986a; Suzuki and Spalding, 1989; Funke et al., 1997; Karlsson et al., 1995, 1998; Hanson et al., 2003). This α-CA is constitutively expressed, but modestly up-regulated in limiting CO_2 conditions, and may be associated with PS II particles (Karlsson et al., 1998; Park et al., 1999). The localization of CAH3 in the thylakoid lumen has been confirmed (Mitra et al., 2005), but its location and the possible association with PS II have led to two alternative hypotheses with regard to its primary function. One is that CAH3 supplies Rubisco with CO_2 from dehydration of the accumulated HCO_3^-. In this model, HCO_3^- accumulated in the chloroplast stroma is delivered into the thylakoid lumen, dehydrated by CAH3 in the acidic lumen to release CO_2, which diffuses back into stroma to provide near-saturating substrate for Rubisco (Pronina and Semenenko, 1992; Raven, 1997; Moroney and Somanchi, 1999; Hanson et al., 2003). This model requires a low CA activity in the vicinity of Rubisco, where the alkaline pH would favor formation of HCO_3^-. A second proposed function of CAH3 involves PS II function or proton gradient formation across the thylakoid membranes (Park et al., 1999; van Hunnik and Sültemeyer, 2002; Villarejo et al., 2002). However, a thorough re-examination of these two proposed functions using a *CAH3* mutant supported the role of CAH3 as providing CO_2 to Rubisco by dehydration of accumulated HCO_3^-, and no effect on PS II electron transport was detected (Hanson et al., 2003).

Two limiting CO_2-induced mitochondrial β-CAs, CAH4, and CAH5 are nearly identical and are encoded by recently duplicated genes arranged in an inverted repeat (Eriksson et al., 1996). Even though CAH4 and CAH5 are among the most abundant proteins induced in limiting CO_2, their roles in limiting CO_2 acclimation and/or the CCM are unclear. Various functions have been proposed, including pH buffering in the mitochondrial matrix to neutralize ammonia released during photorespiration (Eriksson et al., 1996), recycling of photorespiratory and respiratory CO_2 by hydration to HCO_3^-, which then leaks into the cytosol for transport into the chloroplast (Raven, 2001), and supplying HCO_3^- from hydration of photorespiratory and respiratory CO_2 for anaplerotic carbon recycling to provide carbon skeletons for nitrogen assimilation, and balance carbon and nitrogen assimilation (Giordano et al., 2003). Thus it may be that the functions of CAH4 and CAH5 are related to mitochondrial carbon metabolism, but not directly related to the CCM or to limiting CO_2 acclimation *per se*.

CAH6 is another β-type CA and is localized to the chloroplast stroma (Mitra et al., 2004). *CAH6* appears to be expressed in both high CO_2 and limiting CO_2 acclimated cells, but with increased expression in limiting CO_2, similar to expression of *CAH3*. Immunolocalization of CAH6 revealed a distribution throughout the stroma with a higher concentration near the starch sheath surrounding the pyrenoid, but it was not detected in the pyrenoid itself (Mitra et al., 2005). This distribution, especially the preferential concentration surrounding the pyrenoid, is consistent with the suggested function of CAH6 in trapping CO_2 that escapes Rubisco in the pyrenoid as HCO_3^- in the alkaline stroma (Mitra et al., 2004, 2005). CAH6 also may be well positioned for trapping CO_2 diffusing into the chloroplast from the cytosol, thus maintaining a steep gradient for CO_2 diffusion into the chloroplast in the light. The three additional β-CA isoforms (*CAH7*, *CAH8*, and *CAH9*) are poorly studied, apart from the confirmation of CA activity in overexpressed CAH8 (Mitra et al., 2005).

Three genes encoding γ-CA-like proteins, *CAG1*, *CAG2*, and *CAG3* (*GLP1*) are present in *Chlamydomonas*, and Mitra et al. (2004) reported the predicted amino acid sequence of CAG3 shares 30–40% similarity with the γ-CA of *M. thermophila* (Alber and Ferry, 1994) and the putative γ-CA of cyanobacteria, CcmM (Price et al., 1993). However, no CA activity was detected in when *CAG3* was overexpressed (Mitra et al., 2005). The subcellular location of CAG3 was not determined and those of CAG1 and CAG2 are also uncertain. Although these three genes encode proteins with some similarity to γ-CA, it is unclear whether any of the protein products can function as CAs.

D. Integrated model for CCM function

It is feasible to construct a speculative model of the *Chlamydomonas* CCM from the information in this chapter, as illustrated in Figure 8.3. The salient features of the model are that Ci enters the cell either by carrier as HCO_3^- or by diffusion as CO_2, that Ci is transported into the chloroplast either by carrier as HCO_3^- or by diffusion or facilitated diffusion as CO_2, accumulating in the stroma as HCO_3^- regardless of the form transported, and that the accumulated HCO_3^- enters the lumen of the pyrenoidal thylakoids where it is dehydrated by CAH3 to provide substrate CO_2 to Rubisco in the pyrenoid. This model is based on and accommodates several other models (Spalding, 1998, 2008; Grossman et al., 2007), and also preserves the Rubisco/CA co-localization from the models of Badger and Price (1992, 1994) and incorporates some elements of the Pronina and Borodin (1993) thylakoid lumen CA model. One of the key features of the model in Figure 8.3 is the inclusion of multiple, sometimes parallel, Ci transport systems, a feature based partly on characteristics of the better understood cyanobacterial Ci transport systems (Omata et al., 1999; Shibata et al., 2001, 2002; Maeda et al., 2002; Price et al., 2002, 2004). By analogy then, it is possible if not probable that multiple Ci uptake/transport systems may function at each of the two membranes the Ci must cross before reaching Rubisco.

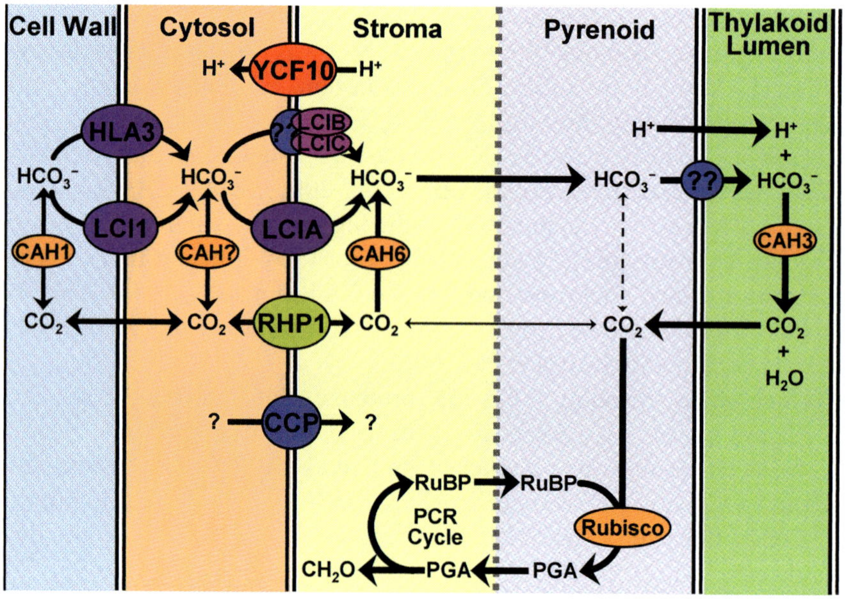

FIGURE 8.3 *Schematic illustration of the proposed Chlamydomonas CCM, including Ci transporter candidates and CA isoforms.*

In this speculative model of the *Chlamydomonas* CCM, it is assumed that active HCO_3^- transport occurs across the plasma membrane and the chloroplast inner envelope, with the prime transport candidates, as described above, being HLA3/MRP1 and LCI1 for the plasma membrane, and LCIA, an LCIB-associated transporter and possibly CCP1/2 for the chloroplast envelope. LCIB and LCIC are proposed to interact with each other and with a transporter, which could be LCIA, CCP1/2 or an unidentified transporter. Thus multiple candidate Ci transporters have been identified for Ci transport across each of the two membranes, and all are assumed to transport HCO_3^- if they transport Ci at all.

On the other hand, this Ci transport model suggests that CO_2 uptake across both the plasma membrane and the chloroplast envelope occurs by diffusion through the membrane or through a CO_2 channel (e.g. RHP1), rather than by an active, protein-mediated carrier process. This would mean that the reported active transport of CO_2 into cells and into chloroplasts reflects processes that occur after the CO_2 crosses the membrane, as apparently is the case for the active CO_2 uptake in cyanobacteria (Badger et al., 2006). For example, active CO_2 uptake into chloroplasts could occur as CAH6-mediated hydration of CO_2 to HCO_3^- driven by light-dependent stromal alkalization, effectively resulting in a near-zero stromal CO_2 concentration and thus a steep diffusion gradient for CO_2 from the cytosol into the stroma. Similarly, active uptake into cells (across the plasma membrane) might be explained as CO_2 diffusion driven by active Ci uptake into the chloroplast, either the active CO_2 uptake driven by light-dependent stromal alkalization just described or, following hydration of the CO_2 by an undiscovered cytosolic CA, active HCO_3^- transport across the chloroplast envelope.

More research will be required to determine whether this view of active CO_2 uptake is correct or whether carrier-mediated, active CO_2 transport does occur. The suggested role of CAH6 in hydration of CO_2 entering the chloroplast to HCO_3^-, thereby generating protons, might also provide an explanation for the reported requirement of CemA for chloroplast Ci uptake, which is predicted to pump protons out of the stroma (Rolland et al., 1997). Clearly, the existence or absence of a cytosolic CA isoform would have an impact on interpretation of this model, although the cytosolic CA is not an essential component of this "active CO_2 diffusion" model. With regard to potential candidates for a cytosolic CA, it is notable that CAH9 is predicted to be cytosolic (Table 8.5), and the cellular locations of both CAH7 and CAH8 are uncertain. In any case, CO_2 entering the chloroplast in the light should be rapidly hydrated to HCO_3^- by CAH6 activity under the alkaline stromal conditions.

At this point it is difficult to address the question of how Ci uptake/transport differs between the low CO_2 and very low CO_2 acclimation states, because no qualitative changes in gene expression have been documented between the two states and little else is known beyond some physiological changes and the growth characteristics of the *pmp1/ad1* mutants (Vance and Spalding, 2005; Wang and Spalding, 2006). However, we can speculate that CO_2 diffusion into cells is less likely to be a substantial component of Ci uptake or of the CCM in very low CO_2 conditions, because even with zero CO_2 in the cytosol, the gradient for diffusion may be too small to provide Ci at a physiologically significant rate. Thus, it is reasonable to predict HCO_3^- transport into the cell (e.g. possibly HLA3/MRP1 or LCI1, activity) to be a more important component of the very low CO_2 CCM. An apparently specific defect in the low CO_2 CCM resulting from lack of a functional LCIB (Wang and Spalding, 2006) argues that the differences between the acclimation states extend beyond the species of Ci entering the cell, but because the function of LCIB still is unknown, it is difficult to speculate on what may be changing.

The active Ci transport/uptake across the plasma membrane and chloroplast envelope should result in HCO_3^- accumulation in the stroma, which is dehydrated by CAH3 in the lumen to become a highly concentrated substrate for Rubisco in the pyrenoid. The necessary steps (movement of HCO_3^- into the thylakoid lumen, dehydration to CO_2 and diffusion to the pyrenoid and Rubisco), although often proposed and well supported genetically, have not been unambiguously demonstrated at the biochemical level. This is especially true for the question of how HCO_3^- moves from the stroma to the thylakoid lumen or how CAH3, which appears to be distributed in thylakoids throughout the chloroplast (Mitra et al., 2005), supplies CO_2 specifically to the pyrenoid to provide Rubisco with abundant substrate.

There is no membrane barrier between the stroma and the pyrenoid, so accumulated HCO_3^- should diffuse readily into the pyrenoid. However, almost nothing is known about the proposed HCO_3^- uptake across the thylakoid membrane, whether within the pyrenoid alone or throughout the stroma, to supply CAH3 with substrate. HCO_3^- movement into the lumen could occur via a protein carrier or an ion channel, but because of the thylakoid membrane potential in the light, it should be a passive process. Once inside, HCO_3^- would be rapidly dehydrated to CO_2 and diffuse back across the thylakoid membrane into the alkaline pyrenoid and stroma, supplying Rubisco. It is not clear whether CO_2 production by CAH3 occurs in all thylakoids, or only in or near the pyrenoid, where the CAH3 reportedly has a greater concentration (Mitra et al., 2005). This model assumes an absence of CA activity in the pyrenoid, to allow Rubisco access to the CO_2 and avoid immediate rehydration to HCO_3^-, and the absence of CAH6 from the pyrenoid (Mitra et al., 2004) supports this assumption.

The efficient operation of this model requires some type of barrier to prevent escape from the chloroplast of the CAH3-generated CO_2. Assuming RuBP is not limiting, the high concentration of Rubisco in the pyrenoid may be sufficient to consume much of the released CO_2 before it has an opportunity to escape the pyrenoid, but even if effective at minimizing CO_2 loss from the pyrenoid, it would not act as a barrier to loss of CO_2 generated outside of the pyrenoidal thylakoids. Any CO_2 released into the bulk stroma or that escapes Rubisco and leaks out of the pyrenoid should be rapidly rehydrated to HCO_3^- by CAH6 under the alkaline conditions prevalent in the light. The distribution of CAH6, which is absent from the pyrenoid and apparently concentrated in the stroma surrounding the pyrenoid (Mitra et al., 2005), suggests that trapping CO_2 leaking from the pyrenoid may be an important function for this CA. Thus CAH6 and the alkaline stroma should act as an efficient trap for CO_2 either entering the chloroplast or leaking from the thylakoid or pyrenoid.

It is clear that, in addition to further research into the identity and nature of Ci transporters, continuing research focus on subcellular and suborganellar compartmentation of CCM processes and the roles and locations of CA isoforms is essential for achieving an understanding of the operation of the *Chlamydomonas* CCM. However, the model discussed here does provide some predictions for experimental testing and appears to be consistent with most, if not all, of the available information about the function of the *Chlamydomonas* CCM.

REFERENCES

Alber, B.E. and Ferry, J.G. (1994). A carbonic anhydrase from the archaeon *Methanosarcina thermophila*. *Proc. Natl. Acad. Sci. U. S. A.* **91**, 6909–6913.

Amoroso, G., Weber, C., Sültemeyer, D.F., and Fock, H. (1996). Intracellular carbonic anhydrase activities in *Dunaliella tertiolecta* (Butcher) and *Chlamydomonas reinhardtii* (Dangeard) in relation to inorganic carbon concentration during growth: Further evidence for the existence of two distinct carbonic anhydrases associated with the chloroplasts. *Planta* **199**, 177–184.

Amoroso, G., Sültemeyer, D., Thyssen, C., and Fock, H.P. (1998). Uptake of HCO_3^- and CO_2 in cells and chloroplasts from the microalgae *Chlamydomonas reinhardtii* and *Dunaliella tertiolecta*. *Plant Physiol.* **116**, 193–201.

Badger, M.R. and Price, G.D. (1992). The CO_2 concentrating mechanism in cyanobacteria and microalgae. *Physiol. Plant.* **84**, 606–615.

Badger, M.R. and Price, G.D. (1994). The role of carbonic anhydrase in photosynthesis. *Ann. Rev. Plant Physiol. Plant Mol. Biol.* **45**, 369–392.

Badger, M.R. and Spalding, M.H. (2000). CO_2 acquisition, concentration and fixation in cyanobacteria and algae. In: *Photosynthesis: Physiology and Metabolism* (R.C. Leegood, T.D. Sharkey, and S. von Caemmerer, Eds.), pp. 369–397. Kluwer Academic Publishers, Dordrecht, The Netherlands.

Badger, M.R., Kaplan, A., and Berry, J.A. (1980). Internal inorganic carbon pool of *Chlamydomonas reinhardtii*. Evidence for a CO_2 concentrating mechanism. *Plant Physiol.* **66**, 407–413.

Badger, M.R., Palmqvist, K., and Yu, J-W. (1994). Measurement of CO_2 and HCO_3^- fluxes in cyanobacteria and microalgae during steady-state photosynthesis. *Physiol. Plant.* **90**, 529–536.

Badger, M.R., Price, G.D., Long, B.M., and Woodger, F.J. (2006). The environmental plasticity and ecological genomics of the cyanobacterial CO_2 concentrating mechanism. *J. Exp. Bot.* **57**, 249–265.

Badour, S.S. and Tan, C.K. (1983). The path of carbon in photosynthesis by *Chlamydomonas segnis* adapting to low and high carbon dioxide tension. *Z. Pflanzenphysiol.* **112**, 287–295.

Bailey, J. and Coleman, J.R. (1988). Effect of CO_2 concentration on protein biosynthesis and carbonic anhydrase expression in *Chlamydomonas reinhardtii*. *Plant Physiology* **87**, 833–840.

Bassham, J.A. (1971). Photosynthetic carbon metabolism. *Proc. Natl. Acad. Sci. U.S.A.* **68**, 2877–2882.

Beezley, B.B., Gruber, P.J., and Frederick, S.E. (1976). Cytochemical localization of glycolate dehydrogenase in mitochondria of *Chlamydomonas*. *Plant Physiol.* **58**, 315–319.

Berry, J.A., Boynton, J., Kaplan, A., and Badger, M.R. (1976). Growth and photosynthesis of *Chlamydomonas reinhardtii* as a function of CO_2 concentration. *Carnegie Institution of Washington Yearbook* **75**, 423–432.

Borkhsenious, O.N., Mason, C.B., and Moroney, J.V. (1998). The intracellular localization of ribulose-1,5-bisphosphate carboxylase/oxygenase in *Chlamydomonas reinhardtii*. *Plant Physiol.* **116**, 1585–1591.

Borodin, V., Garderström, P., and Samuelsson, G. (1994). The effect of light quality on the induction of efficient photosynthesis under low CO_2 conditions in *Chlamydomonas reinhardtii* and *Chlorella pyrenoidosa*. *Physiol. Plant.* **92**, 254–260.

Bozzo, G.G. and Colman, B. (2000). The induction of inorganic carbon and external carbonic anhydrase in *Chlamydomonas reinhardtii* is regulated by external CO_2 concentration. *Plant Cell Environ.* **23**, 1137–1144.

Bozzo, G.G., Colman, B., and Matsuda, Y. (2000). Active transport of CO_2 and bicarbonate is induced in response to external CO_2 concentration in the green alga *Chlorella kessleri*. *J. Exp. Bot.* **51**, 1341–1348.

Burow, M.D., Chen, Z.Y., Mouton, T.M., and Moroney, J.V. (1996). Isolation of cDNA clones of genes induced upon transfer of *Chlamydomonas reinhardtii* cells to low CO_2. *Plant Mol. Biol.* **31**, 443–448.

Chen, Z-Y., Burow, M.D., Mason, C.B., and Moroney, J.V. (1996). A low-CO_2-inducible gene encoding an alanine:α-ketoglutarate aminotransferase in *Chlamydomonas reinhardtii*. *Plant Physiol.* **112**, 677–684.

Chen, Z-Y., Lavigne, L.L., Mason, C.B., and Moroney, J.V. (1997). Cloning and overexpression of two cDNAs encoding the low-CO_2-inducible chloroplast envelope protein LIP-36 from *Chlamydomonas reinhardtii*. *Plant Physiol.* **114**, 265–273.

Coleman, J.R. and Grossman, A.R. (1984). Biosynthesis of carbonic anhydrase in *Chlamydomonas reinhardtii* during adaptation to low CO_2. *Proc. Natl. Acad. Sci. U.S.A.* **81**, 6049–6053.

Coleman, J.R., Luinenburg, I., Majeau, N., and Provart, N. (1991). Sequence analysis and regulation of expression of a gene coding for carbonic anhydrase in *Chlamydomonas reinhardtii*. *Can. J. Bot.* **69**, 1097–1102.

Colombo, S.L., Pollack, S.V., Eger, K.A., Mason, C.B., Adams, J.E., and Moroney, J.V. (2002). Use of bleomycin resistance to generate tagged insertional mutants of *Chlamydomonas reinhardtii* that require elevated CO_2 for optimal growth. *Funct. Plant Biol.* **29**, 231–241.

Dionisio, M.L., Tsuzuki, M., and Miyachi, S. (1989a). Light requirement for carbonic anhydrase induction in *Chlamydomonas reinhardtii*. *Plant Cell. Physiol.* **30**, 207–213.

Dionisio, M.L., Tsuzuki, M., and Miyachi, S. (1989b). Blue light induction of carbonic anhydrase activity in *Chlamydomonas reinhardtii*. *Plant Cell. Physiol.* **30**, 215–219.

Dionisio-Sese, M.L., Fukuzawa, H., and Miyachi, S. (1990). Light-induced carbonic anhydrase expression in *Chlamydomonas reinhardtii*. *Plant Physiol.* **94**, 1103–1110.

Douglas, S.E. and Penny, S.L. (1999). The plastid genome from the cryptomonad alga, *Guillardia theta*: Complete sequence and conserved synteny groups confirm its common ancestry with red algae. *J. Mol. Evol.* **48**, 236–244.

Eriksson, M., Karlsson, J., Ramazanov, Z., Gardeström, P., and Samuelsson, G. (1996). Discovery of an algal mitochondrial carbonic anhydrase: Molecular cloning and characterization of a low-CO_2-induced polypeptide in *Chlamydomonas reinhardtii*. *Proc. Natl. Acad. Sci. U.S.A.* **93**, 12031–12034.

Eriksson, M., Villand, P., Gardestrom, P., and Samuelsson, G. (1998). Induction and regulation of expression of a low-CO_2-induced mitochondrial carbonic anhydrase in *Chlamydomonas reinhardtii*. *Plant Physiol.* **116**, 637–641.

Fett, J.P. and Coleman, J.R. (1994). Regulation of periplasmic carbonic anhydrase expression in *Chlamydomonas reinhardtii* by acetate and pH. *Plant Physiol.* **106**, 103–108.

Fujiwara, S., Fukuzawa, H., Tachiki, A., and Miyachi, S. (1990). Structure and differential expression of two genes encoding carbonic anhydrase in *Chlamydomonas reinhardtii*. *Proc. Natl. Acad. Sci. U.S.A.* **87**, 9779–9783.

Fujiwara, S., Ishida, N., and Tsuzuki, M. (1996). Circadian expression of the carbonic anhydrase gene, *Cah1*, in *Chlamydomonas reinhardtii*. *Plant Mol. Biol.* **32**, 745–749.

Fukuzawa, H., Fujiwara, S., Yamamoyo, Y., Dionisio-Sese, M.L., and Miyachi, S. (1990). cDNA cloning, sequence, and expression of carbonic anhydrase in

Chlamydomonas reinhardtii: Regulation by environmental CO_2 concentration. *Proc. Natl. Acad. Sci. U.S.A.* **87**, 4383–4387.

Fukuzawa, H., Miura, K., Ishizaki, K., Kucho, K., Saito, T., Kohinata, T., and Ohyama, K. (2001). Ccm1, a regulatory gene controlling the induction of a carbon-concentrating mechanism in *Chlamydomonas reinhardtii* by sensing CO_2. *Proc. Natl. Acad. Sci. U. S. A.* **98**, 5347–5352.

Funke, R.P., Kovar, J.L., and Weeks, D.P. (1997). Intracellular carbonic anhydrase is essential to photosynthesis in *Chlamydomonas reinhardtii* at atmospheric levels of CO_2. *Plant Physiol.* **114**, 237–244.

Galván, A., Rexach, J., Mariscal, V., and Fernández, E. (2002). Nitrite transport to the chloroplast in *Chlamydomonas reinhardtii*: Molecular evidence for a regulated process. *J. Exp. Bot.* **53**, 845–853.

Gancedo, J.M. (1998). Yeast carbon catabolite repression. *Microbiol. Mol. Biol. Rev.* **62**, 334–361.

Geraghty, A.M. and Spalding, M.H. (1996). Molecular and structural changes in *Chlamydomonas* under limiting CO_2: A possible mitochondrial role in adaptation. *Plant Physiol.* **111**, 1339–1347.

Geraghty, A.M., Anderson, J.C., and Spalding, M.H. (1990). A 36 kilodalton limiting-CO_2 induced polypeptide of *Chlamydomonas* is distinct from the 37 kilodalton periplasmic carbonic anhydrase. *Plant Physiol.* **93**, 116–121.

Giordano, M., Norici, A., Forssen, M., Eriksson, M., and Raven, J.A. (2003). An anaplerotic role for mitochondrial carbonic anhydrase in *Chlamydomonas reinhardtii*. *Plant Physiol.* **132**, 2126–2134.

Giordano, M., Beardall, J., and Raven, J.A. (2005). CO_2 concentrating mechanisms in algae: Mechanisms, environmental modulation, and evolution. *Ann. Rev. Plant Biol.* **56**, 99–131.

Giraud, G. and Czanisnski, Y. (1971). Localisation ultrastructurale d'activités oxydasiques chez le *Chlamydomonas reinhardtii*. *C.R. Acad. Sci. Ser. D.* **273**, 2500–2503.

Goyal, A. and Tolbert, N.E. (1989). Uptake of inorganic carbon by isolated chloroplasts from air-adapted *Dunaliella*. *Plant Physiol.* **89**, 1264–1269.

Goyal, A. and Tolbert, N.E. (1996). Association of glycolate oxidation with photosynthetic electron transport in plant and algal chloroplasts. *Proc. Natl. Acad. Sci. U. S. A.* **93**, 3319–3324.

Graham, D. and Whittingham, C.P. (1968). The path of carbon during photosynthesis in *Chlorella pyrenoidosa* at high and low carbon dioxide concentrations. *Z. Pflanzenphysiol.* **58**, 418–427.

Grossman, A.R., Croft, M., Gladyshev, V.N., Merchant, S., Posewitz, M., Prochnik, S., and Spalding, M.H. (2007). Novel metabolism in *Chlamydomonas* through the lens of genomics. *Curr. Opin. Plant Biol.* **10**, 190–198.

Hanawa, Y., Watanabe, M., Karatsu, Y., Fukuzawa, H., and Shiraiwa, Y. (2007). Induction of a high-CO_2-inducible, periplasmic protein, H43, and its application as a high-CO_2-responsive marker for study of the high-CO_2-sensing mechanism in *Chlamydomonas reinhardtii*. *Plant Cell Physiol.* **48**, 299–309.

Hanikenne, M., Krämer, U., Demoulin, V., and Baurain, D. (2005). A comparative inventory of metal transporters in the green alga *Chlamydomonas reinhardtii* and the red alga *Cyanidioschizon merolae*. *Plant Physiol.* **137**, 428–446.

Hanson, D.T., Franklin, L.A., Samuelsson, G., and Badger, M.R. (2003). The *Chlamydomonas reinhardtii cia3* mutant lacking a thylakoid lumen-localized carbonic anhydrase is limited by CO_2 supply to Rubisco and not photosystem II function *in vivo*. *Plant Physiol.* **132**, 2267–2275.

Hewett-Emmett, D. and Tashian, R.E. (1996). Functional diversity, conservation, and convergence in the evolution of the α-, β-, and γ-carbonic anhydrase gene families. *Mol. Phylogenet. Evol.* **5**, 50–77.

Hogetsu, D. and Miyachi, S. (1979). Role of carbonic anhydrase in photosynthetic CO_2 fixation in *Chlorella*. *Plant Cell Physiol.* **20**, 747–756.

Huertas, E., Colman, B., and Espie, G.S. (2002). Mitochondrial-driven bicarbonate transport supports photosynthesis in a marine microalga. *Plant Physiol.* **130**, 284–291.

Im, C.S. and Grossman, A.R. (2002). Identification and regulation of high light-induced genes in *Chlamydomonas reinhardtii*. *Plant J.* **30**, 301–313.

Im, C.S., Zhang, Z., Shrager, J., Chang, C.W., and Grossman, A.R. (2003). Analysis of light and CO_2 regulation in *Chlamydomonas reinhardtii* using genome-wide approaches. *Photosynth. Res.* **75**, 111–125.

Ingram-Smith, C., Martin, S.R., and Smith, K.S. (2006). Acetate kinase: Not just a bacterial enzyme. *Trends Microbiol.* **14**, 249–253.

Kaplan, A. and Reinhold, L. (1999). The CO_2-concentrating mechanism of photosynthetic microorganisms. *Annu. Rev. Plant Physiol. Plant Mol. Biol.* **50**, 539–570.

Karlsson, J., Ramazanov, Z., Hiltonen, T., Garderström, P., and Samuelsson, G. (1994). Effect of vanadate on photosynthesis and the ATP/ADP ratio in low-CO_2-adapted *Chlamydomonas reinhardtii* cells. *Planta* **192**, 46–51.

Karlsson, J., Hiltonen, T., Husic, H.D., Ramazanov, Z., and Samuelsson, G. (1995). Intracellular carbonic anhydrase of *Chlamydomonas reinhardtii*. *Plant Physiol.* **109**, 533–539.

Karlsson, J., Clarke, A.K., Chen, Z-Y., Hugghins, S.Y., Park, Y.I., Husic, H.D., Moroney, J.V., and Samuelsson, G. (1998). A novel alpha-type carbonic anhydrase associated with the thylakoid membrane in *Chlamydomonas reinhardtii* is required for growth at ambient CO_2. *EMBO J.* **10**, 1208–1216.

Katoh, A., Lee, K.S., Fukuzawa, H., Ohyama, K., and Ogawa, T. (1996a). *cemA* homologue essential to CO_2 transport in the cyanobacterium in *Synechocystis* PCC 6803. *Proc. Natl. Acad. Sci. U. S. A.* **93**, 4006–4010.

Katoh, A., Sonoda, M., Katoh, H., and Ogawa, T. (1996b). Absence of light-induced proton extrusion in a *cotA*-less mutant of *Synechocystis sp.* strain PCC 6803. *J. Bacteriol.* **178**, 5452–5455.

Kottke, T., Hegemann, P., Dick, B., and Heberle, J. (2006). The photochemistry of the light-, oxygen-, and voltage-sensitive domains in the algal blue light receptor phot. *Biopolymers* **82**, 373–378.

Kuan, J. and Saier, M.H., Jr. (1993). Expansion of the mitochondrial carrier family. *Res. Microbiol.* **144**, 671–672.

Kuchitsu, K., Tsuzuki, M., and Miyachi, S. (1988). Changes in starch localization within the chloroplast induced by changes in CO_2 concentration during growth of *Chlamydomonas reinhardtii*: Independent regulation of pyrenoid starch and stromal starch. *Plant Cell Physiol.* **29**, 1269–1278.

Kuchitsu, K., Tsuzuki, M., and Miyachi, S. (1991). Polypeptide composition and enzyme activities of the pyrenoid and its regulation by CO_2 concentration in unicellular green algae. *Can. J. Bot.* **69**, 1062–1069.

Kustu, S. and Inwood, W. (2006). Biological gas channels for NH_3 and CO_2: Evidence that Rh (Rhesus) proteins are CO_2 channels. *Transfusion Cliqinue et Biologique* **13**, 103–110.

Maeda, S., Badger, M.R., and Price, G.D. (2002). Novel gene products associated with NdhD3/D4-containing NDH-1 complexes are involved in photosynthetic CO_2 hydration in the cyanobacterium *Synechococcus* sp. PCC7942. *Mol. Microbiol.* **43**, 425–435.

Mamedov, T.G., Suzuki, K., Miura, K., Kucho, K., and Fukuzawa, H. (2001). Characteristics and sequence of phosphoglycolate phosphatase from a eukaryotic green alga *Chlamydomonas reinhardtii*. *J. Biol. Chem.* **276**, 45573–45579.

Marcus, Y., Volokita, M., and Kaplan, A. (1984). The location of the transporting system for inorganic carbon and the nature of the form translocated in *Chlamydomonas reinhardtii*. *J. Exp. Bot.* **35**, 1136–1144.

Marcus, Y., Schuster, G., Michaels, A., and Kaplan, A. (1986). Adaptation to CO_2 level and changes in the phosphorylation of thylakoid proteins during the cell cycle of *Chlamydomonas reinhardtii*. *Plant Physiol.* **80**, 604–607.

Marek, L.F. and Spalding, M.H. (1991). Changes in photorespiratory enzyme activity in response to limiting CO_2 in *Chlamydomonas reinhardtii*. *Plant Physiol.* **97**, 420–425.

Mariscal, V., Moulin, P., Orsel, M., Miller, A.J., Fernández, E., and Galván, A (2006). Differential regulation of the *Chlamydomonas Nar1* gene family by carbon and nitrogen. *Protist* **157**, 421–433.

Mason, C.B., Manuel, L.J., and Moroney, J.V. (1990). A new chloroplast protein is induced by growth on low CO_2 in *Chlamydomonas reinhardtii*. *Plant Physiol.* **93**, 833–836.

Matsuda, Y. and Colman, B. (1995). Induction of CO_2 and bicarbonate transport in the green alga, *Chlorella ellipsoidea*. II. Evidence for induction in response to external CO_2 concentration. *Plant Physiol.* **108**, 253–260.

Matsuda, Y. and Colman, B. (1996). Active uptake of inorganic carbon by *Chlorella saccharophila* is not repressed by growth in high CO_2. *J. Exp. Bot.* **47**, 1951–1956.

Matsuda, Y., Bozzo, G.G., and Colman, B. (1998). Regulation of dissolved inorganic carbon transport in green algae. *Can. J. Bot.* **76**, 1072–1083.

Mayo, W.P., Williams, T.G., Birch, D.G., and Turpin, D.H. (1986). Photosynthetic adaptation by *Synechococcus leopoliensis* in response to exogenous dissolved inorganic carbon. *Plant Physiol.* **80**, 1038–1040.

Mitra, M., Lato, S.M., Ynalvez, R.A., Xiao, Y., and Moroney, J.V. (2004). Identification of a new chloroplast carbonic anhydrase in *Chlamydomonas reinhardtii*. *Plant Physiol.* **135**, 173–182.

Mitra, M., Mason, C.B., Xiao, Y., Ynalvez, R.A., Lato, S.M., and Moroney, J.V. (2005). The carbonic anhydrase gene families of *Chlamydomonas reinhardtii*. *Can. J. Bot.* **83**, 780–795.

Miura, K., Yamano, T., Yoshioka, S., Kohinata, T., Inoue, Y., Taniguchi, F., Asamizu, E., Nakamura, Y., Tabata, S., Yamato, K.T., Ohyama, K., and Fukuzawa, H. (2004). Expression profiling-based identification of CO_2-responsive genes regulated by CCM1 controlling a carbon-concentrating mechanism in *Chlamydomonas reinhardtii*. *Plant Physiol.* **135**, 1595–1607.

Miyachi, S., Tsuzuki, M., Maruyama, I., Gantar, M., and Miyachi, S. (1986). Effects of CO_2 concentration during growth on the intracellular structure of *Chlorella* and *Scenedesmus* (Chlorophyta). *J. Phycol.* **22**, 313–319.

Miyachi, S., Iwasaki, I., and Shiraiwa, Y. (2003). Historical perspective on microalgal and cyanobacterial acclimation to low- and extremely high-CO_2 conditions. *Photosynth. Res.* **77**, 139–153.

Moroney, J.V. and Mason, C.B. (1991). The role of the chloroplast in inorganic carbon acquisition by *Chlamydomonas reinhardtii*. *Can. J. Bot.* **69**, 1017–1024.

Moroney, J.V. and Somanchi, A. (1999). How do algae concentrate CO_2 to increase the efficiency of photosynthetic carbon fixation? *Plant Physiol.* **119**, 9–16.

Moroney, J.V., Husic, H.D., and Tolbert, N.E. (1985). Effects of carbonic anhydrase inhibitors on inorganic carbon accumulation by *Chlamydomonas reinhardtii*. *Plant Physiol.* **79**, 177–183.

Moroney, J.V., Tolbert, N.E., and Sears, B.B. (1986a). Complementation analysis of the inorganic carbon concentrating mechanism of *Chlamydomonas reinhardtii*. *Mol. Gen. Genet.* **204**, 199–203.

Moroney, J.V., Wilson, B.J., and Tolbert, N.E. (1986b). Glycolate metabolism and excretion by *Chlamydomonas reinhardtii*. *Plant Physiol.* **82**, 821–826.

Moroney, J.V., Kitayama, M., Togasaki, R.K., and Tolbert, N.E. (1987). Evidence for inorganic carbon transport by intact chloroplasts of *Chlamydomonas reinhardtii*. *Plant Physiol.* **83**, 460–463.

Moroney, J.V., Husic, H.D., Tolbert, N.E., Kitayama, M., Manuel, L.J., and Togasaki, R.K. (1989). Isolation and characterization of a mutant of *Chlamydomonas reinhardtii* deficient in the CO_2 concentrating mechanism. *Plant Physiol.* **89**, 897–903.

Nakamura, Y., Kanakagiri, S., Van, K., He, W., and Spalding, M.H. (2005). Disruption of the glycolate dehydrogenase gene in the high-CO_2-requiring mutant HCR89 of *Chlamydomonas reinhardtii*. *Can. J. Bot.* **83**, 820–833.

Nelson, E.B. and Tolbert, N.E. (1969). The regulation of glycolate metabolism in *Chlamydomonas reinhardtii*. *Biochim. Biophys. Acta* **184**, 263–270.

Nelson, E.B. and Tolbert, N.E. (1970). Glycolate dehydrogenase in green algae. *Arch. Biochem. Biophys.* **141**, 102–110.

Nogales, J., Guijo, M.I., Quesada, A., and Merchan, F. (2004). Functional analysis and regulation of the malate synthase from *Chlamydomonas reinhardtii*. *Planta* **219**, 325–331.

Omata, T., Price, G.D., Badger, M.R., Okamura, M., Gohta, S., and Ogawa, T. (1999). Identification of an ATP-binding cassette transporter involved in bicarbonate uptake in the cyanobacterium *Synechococcus* sp. strain PCC 7942. *Proc. Natl. Acad. Sci. U. S. A.* **96**, 13571–13576.

Oyama, Y., Izumo, A., Fujiwara, S., Shimonaga, T., Nakamura, Y., and Tsuzuki, M. (2006). Granule-bound starch synthase cDNA in *Chlorella kessleri* 11 h: cloning and regulation of expression by CO_2 concentration. *Planta* **224**, 646–654.

Palmieri, F. (1994). Mitochondrial carrier proteins. *FEBS Lett.* **346**, 48–54.

Palmqvist, K., Sjoberg, S., and Samuelsson, G. (1988). Induction of inorganic carbon accumulation in the unicellular green algae *Scenedesmus obliquus* and *Chlamydomonas reinhardtii*. *Plant Physiol.* **87**, 437–442.

Palmqvist, K., Yu, J-W., and Badger, M.R. (1994). Carbonic anhydrase activity and inorganic carbon fluxes in low and high Ci of *Chlamydomonas reinhardtii* and *Scenedesmus obliquus*. *Physiol. Plant.* **90**, 537–547.

Park, Y.I., Karlsson, J., Rojdestvenski, I., Pronina, N., Klimov, V., Oquist, G., and Samuelsson, G. (1999). Role of a novel photosystem II-associated carbonic anhydrase in photosynthetic carbon assimilation in *Chlamydomonas reinhardtii*. *FEBS Lett.* **444**, 102–105.

Pederson, T.A., Kirk, M., and Bassham, J.A. (1966). Light-dark transients in levels of intermediate compounds during photosynthesis in air adapted *Chlorella*. *Physiol. Plant.* **19**, 219–231.

Petridou, S., Foster, K., and Kindle, K. (1997). Light induces accumulation of isocitrate lyase mRNA in a carotenoid-deficient mutant of *Chlamydomonas reinhardtii*. *Plant Mol. Biol.* **33**, 381–392.

Pollock, S.V., Colombo, S.L., Prout, D.L., Jr., Godfrey, A.C., and Moroney, J.V. (2003). Rubisco activase is required for optimal photosynthesis in the green alga *Chlamydomonas reinhardtii* in a low CO_2 atmosphere. *Plant Physiol.* **133**, 1854–1861.

Pollock, S.V., Prout, D.L., Godfrey, A.C., Lemaire, S.D., and Moroney, J.V. (2004). The *Chlamydomonas reinhardtii* proteins Ccp1 and Ccp2 are required for long-term growth, but are not necessary for efficient photosynthesis, in a low CO_2 environment. *Plant Mol. Biol.* **56**, 125–132.

Price, G.D., Howitt, S.M., Harrison, K., and Badger, M.R. (1993). Analysis of a genomic DNA region from the cyanobacterium *Synechococcus* sp. strain PCC7942 involved in carboxysome assembly and function. *J. Bact.* **175**, 2871–2879.

Price, G.D., Maeda, S-I., Omata, T., and Badger, M.R. (2002). Modes of inorganic carbon uptake in the cyanobacterium *Synechococcus* sp. PCC7942. *Funct. Plant Biol.* **29**, 131–149.

Price, G.D., Woodger, F.J., Badger, M.R., Howitt, S.M., and Tucker, L. (2004). Identification of a SulP-type bicarbonate transporter in marine cyanobacteria. *Proc. Natl. Acad. Sci. U. S. A.* **101**, 18228–18233.

Pronina, N.A. and Borodin, V.V. (1993). CO_2 stress and CO_2 concentration mechanism: Investigation by means of photosystem deficient and carbonic-anhydrase-deficient mutants of *Chlamydomonas reinhardtii*. *Photosynthetica* **28**, 515–522.

Pronina, N.A. and Semenenko, V.E. (1992). Role of the pyrenoid in concentration, generation and fixation of CO_2 in the chloroplast of microalgae. *Soviet Plant Physiol.* **39**, 470–476.

Ral, J.P., Colleoni, C., Wattebled, F., Dauvillee, D., Nempont, C., Deschamps, P., Li, Z., Morell, M.K., Chibbar, R., Purton, S., d'Hulst, C., and Ball, S.G. (2006). Circadian clock regulation of starch metabolism establishes GBSSI as a major contributor to amylopectin synthesis in *Chlamydomonas reinhardtii*. *Plant Physiol.* **142**, 305–317.

Ramazanov, Z. and Cardenas, J. (1992). Involvement of photorespiration and glycolate pathway in carbonic anhydrase induction and inorganic carbon concentration in *Chlamydomonas reinhardtii*. *Physiol. Plant.* **84**, 502–508.

Ramazanov, Z. and Cardenas, J. (1994). Photorespiratory ammonium assimilation in chloroplasts of *Chlamydomonas reinhardtii*. *Physiol. Plant.* **91**, 495–502.

Ramazanov, Z. and Semenenko, V.E. (1984). Oxygen-dependent induction of synthesis of the CO_2-dependent soluble form of carbonic anhydrase in *Chlorella* cells. *Soviet Plant Physiol.* **31**, 344–349.

Ramazanov, Z., Mason, C.B., Geraghty, A.M., Spalding, M.H., and Moroney, J.V. (1993). The low CO_2-inducible 36 kDa protein is localized to the chloroplast envelope of *Chlamydomonas reinhardtii*. *Plant Physiol.* **101**, 1195–1199.

Ramazanov, Z., Rawat, M., Henk, M.C., Mason, C.B., Matthews, S.W., and Moroney, J.V. (1994). The induction of the CO_2-concentrating mechanism is correlated with the formation of the starch sheath around the pyrenoid of *Chlamydomonas reinhardtii*. *Planta* **195**, 210–216.

Raven, J.A. (1997). CO_2-concentrating mechanisms: A direct role for thylakoid lumen acidification? *Plant Cell Environ.* **20**, 147–154.

Raven, J.A. (2001). A role for mitochondrial carbonic anhydrase in limiting CO_2 leakage from low CO_2-grown cells of *Chlamydomonas reinhardtii*. *Plant Cell Environ.* **24**, 261–265.

Rawat, M. and Moroney, J.V. (1991). Partial characterization of a new isozyme of carbonic anhydrase isolated from *Chlamydomonas reinhardtii*. *J. Biol. Chem.* **266**, 9719–9723.

Rawat, M. and Moroney, J.V. (1995). The regulation of carbonic anhydrase and ribulose-1,5-bisphosphate carboxylase/oxygenase activase by light and CO_2 in *Chlamydomonas reinhardtii*. *Plant Physiol.* **109**, 937–944.

Rea, P.A. (2007). Plant ATP-binding cassette transporters. *Annu. Rev. Plant Biol.* **58**, 347–375.

Rexach, J., Fernández, E., and Galván, A. (2000). The *Chlamydomonas reinhardtii Nar1* gene encodes a chloroplast membrane protein involved in nitrite transport. *Plant Cell* **12**, 1441–1453.

Rolland, N., Dorne, A.J., Amoroso, G., Sültemeyer, D.F., Joyard, J., and Rochaix, J.D. (1997). Disruption of the plastid *ycf10* open reading frame affects uptake of inorganic carbon in the chloroplast of *Chlamydomonas*. *EMBO J.* **16**, 6713–6726.

Sasaki, Y., Sekiguchi, K., Nagano, Y., and Matsumo, R. (1993). Chloroplast envelope protein encoded by the chloroplast genome. *FEBS Lett.* **316**, 93–98.

Sheen, J. (1994). Feedback control of gene expression. *Photosynth. Res.* **39**, 427–438.

Shibata, M., Ohkawa, H., Kaneko, T., Fukuzawa, H., Tabata, S., Kaplan, A., and Ogawa, T. (2001). Distinct constitutive and low-CO_2-induced CO_2 uptake systems in cyanobacteria: Genes involved and their phylogenetic relationship with homologous genes in other organisms. *Proc. Natl. Acad. Sci. U. S. A.* **98**, 11789–11794.

Shibata, M., Katoh, H., Sonoda, M., Ohkawa, H., Shimoyama, M., Fukuzawa, H., Kaplan, A., and Ogawa, T. (2002). Genes essential to sodium-dependent bicarbonate transport in cyanobacteria. Function and phylogenetic analysis. *J. Biol. Chem.* **277**, 18658–18664.

Somanchi, A. and Moroney, J.V. (1999). As *Chlamydomonas reinhardtii* acclimates to low CO_2 conditions there is an increase in cyclophilin expression. *Plant Mol. Biol.* **40**, 1055–1062.

Sonoda, M., Katoh, H., Vermaas, W., Schmetterer, G., and Ogawa, T. (1998). Photosynthetic electron transport involved in PxcA-dependent proton extrusion in *Synechocystis* sp. strain PCC6803: Effect of *pxcA* inactivation on CO_2, HCO_3^-, and NO_3^- uptake. *J. Bact.* **180**, 3799–3803.

Soupene, E., King, N., Field, E., Liu, P., Niyogi, K.K., Huang, C.H., and Kustu, S. (2002). Rhesus expression in a green alga is regulated by CO_2. *Proc. Natl. Acad. Sci. U. S. A.* **99**, 7769–7773.

Soupene, E., Inwood, W., and Kustu, S. (2004). Lack of the Rhesus protein Rh1 impairs growth of the green alga *Chlamydomonas reinhardtii* at high CO_2. *Proc. Natl. Acad. Sci. U. S. A.* **101**, 7787–7792.

Spalding, M.H. (1989). Photosynthesis and photorespiration in freshwater green algae. *Aquat. Bot.* **34**, 181–209.

Spalding, M.H. (1998). CO_2 acquisition: Acclimation to changing carbon availability. In: *The Molecular Biology of Chloroplasts and Mitochondria in Chlamydomonas* (J.-D. Rochaix, M. Goldschmidt-Clermont, and S. Merchant, Eds.), pp. 529–547. Kluwer Academic Publishers, Dordrecht, The Netherlands.

Spalding, M.H. (2008). The CO_2-concentrating mechanism and inorganic carbon transport in *Chlamydomonas*. *J. Exp. Bot.* **59**, 1463–1473.

Spalding, M.H. and Ogren, W.L. (1982). Photosynthesis is required for induction of the CO_2-concentrating system in *Chlamydomonas reinhardii*. *FEBS Lett.* **145**, 41–44.

Spalding, M.H., Spreitzer, R.J., and Ogren, W.L. (1983a). Carbonic anhydrase deficient mutant of *Chlamydomonas* requires elevated carbon dioxide concentration for photoautotrophic growth. *Plant Physiol.* **73**, 268–272.

Spalding, M.H., Spreitzer, R.J., and Ogren, W.L. (1983b). Reduced inorganic carbon transport in a CO_2-requiring mutant of *Chlamydomonas reinhardtii*. *Plant Physiol.* **73**, 273–276.

Spalding, M.H., Winder, T.L., Anderson, J.C., Geraghty, A.M., and Marek, L.F. (1991). Changes in protein and gene expression during induction of the CO_2-concentrating mechanism in wild-type and mutant *Chlamydomonas*. *Can. J. Bot.* **69**, 1008–1016.

Spalding, M.H., Van, K., Wang, Y., and Nakamura, Y. (2002). Acclimation of *Chlamydomonas* to changing carbon availability. *Funct. Plant Biol.* **29**, 221–230.

Spencer, K.G., Kimpel, D.L., Fisher, M.L., Togasaki, R.K., and Miyachi, S. (1983). Carbonic anhydrase induction in *Chlamydomonas reinhardtii.* II. Requirements for carbonic anhydrase induction. *Plant Cell Physiol.* **24**, 301–304.

Sültemeyer, D.F., Klock, G., Kreutzberg, K., and Fock, H.P. (1988). Photosynthesis and apparent affinity for dissolved inorganic carbon by cells and chloroplasts of *Chlamydomonas reinhardtii* grown at high and low CO_2 concentrations. *Planta* **176**, 256–260.

Sültemeyer, D.F., Miller, A.G., Espie, G.S., Fock, H.P., and Canvin, D.T. (1989). Active CO_2 transport by the green alga *Chlamydomonas reinhardtii*. *Plant Physiol.* **89**, 1213–1219.

Sültemeyer, D.F., Amoroso, G., and Fock, H. (1995). Induction of intracellular carbonic anhydrases during the adaptation to low inorganic carbon concentrations in wild-type and *ca-1* mutant cells of *Chlamydomonas reinhardtii*. *Planta* **196**, 217–224.

Suzuki, K. and Spalding, M.H. (1989). Adaptation of *Chlamydomonas reinhardtii* high-CO_2-requiring mutants to limiting-CO_2. *Plant Physiol.* **90**, 1195–1200.

Suzuki, K., Marek, L.F., and Spalding, M.H. (1990). A photorespiratory mutant of *Chlamydomonas reinhardtii*. *Plant Physiol.* **93**, 231–237.

Talling, J.F. (1976). The depletion of carbon dioxide from lake water by phytoplankton. *J. Ecol.* **64**, 79–121.

Thielmann, J., Tolbert, N.E., Goyal, A., and Senger, H. (1990). Two systems for concentrating CO_2 and bicarbonate during photosynthesis by *Scenedesmus*. *Plant Physiol.* **92**, 622–629.

Thyssen, C., Schlichting, R., and Giersch, C. (2001). The CO_2-concentrating mechanism in the physiological context: Lowering the CO_2 supply diminishes culture growth and economises starch utilisation in *Chlamydomonas reinhardtii*. *Planta* **213**, 629–639.

Tolbert, N.E. (1997). The C2 oxidative photosynthetic carbon cycle. *Ann. Rev. Plant Physiol. Plant Mol. Biol.* **48**, 1–25.

Tolbert, N.E., Harrison, M., and Selph, N. (1983). Aminooxyacetate stimulation of glycolate formation and excretion by *Chlamydomonas*. *Plant Physiol.* **72**, 1075–1083.

Tural, B. and Moroney, J.V. (2005). Regulation of the expression of photorespiratory genes in *Chlamydomonas reinhardtii*. *Can. J. Bot.* **83**, 810–819.

Turkina, M.V., Blanco-Rivero, A., Vainonen, J.P., Vener, A.V., and Villarejo, A. (2006). CO_2 limitation induces specific redox-dependent protein phosphorylation in *Chlamydomonas reinhardtii*. *Proteomics* **6**, 2693–2704.

Turmel, M., Otis, C., and Lemieux, C. (1999). The complete chloroplast DNA sequence of the green alga *Nephroselmis olivacea*: Insights into the architecture of ancestral chloroplast genomes. *Proc. Natl. Acad. Sci. U. S. A.* **96**, 10248–10253.

van Hunnik, E., Amoroso, G., and Sültemeyer, D.F. (2002). Uptake of CO_2 and bicarbonate by intact cells and chloroplasts of *Tetraedron minimum* and *Chlamydomonas noctigama*. *Planta* **215**, 763–769.

Van, K. and Spalding, M.H. (1999). Periplasmic carbonic anhydrase (*Cah1*) structural gene mutant in *Chlamydomonas reinhardtii*. *Plant Physiol.* **120**, 757–764.

Van, K., Wang, Y., Nakamura, Y., and Spalding, M.H. (2001). Insertional mutants of *Chlamydomonas reinhardtii* that require elevated CO_2 for survival. *Plant Physiol.* **127**, 607–614.

Vance, P. and Spalding, M.H. (2005). Growth, photosynthesis and gene expression in *Chlamydomonas* over a range of CO_2 concentrations and CO_2/O_2 ratios: CO_2 regulates multiple acclimation states. *Can. J. Bot.* **83**, 820–833.

Villand, P., Eriksson, M., and Samuelsson, G. (1997). Carbon dioxide and light regulation of promoters controlling the expression of mitochondrial carbonic anhydrase in *Chlamydomonas reinhardtii*. *Biochem. J.* **327**, 51–57.

Villarejo, A., Martínez, F., Plumed, M.dP., and Ramazanov, Z. (1996a). The induction of the CO_2 concentrating mechanism in a starch-less mutant of *Chlamydomonas reinhardtii*. *Physiol. Plant.* **98**, 798–802.

Villarejo, A., Reina, G.G., and Ramazanov, Z. (1996b). Regulation of the low-CO_2-inducible polypeptides in *Chlamydomonas reinhardtii*. *Planta* **199**, 481–485.

Villarejo, A., Martinez, F., and Ramazanov, Z. (1997). Effect of aminooxyacetate, an inhibitor blocking the glycolate pathway, on the induction of a CO_2-concentrating mechanism and low-CO_2-inducible polypeptides in *Chlamydomonas reinhardtii* (Chlorophyta). *Eur. J. Phycol.* **32**, 141–145.

Villarejo, A., Rolland, N., Martinez, F., and Sültemeyer, D.F. (2001). A new chloroplast envelope carbonic anhydrase activity is induced during acclimation to low inorganic carbon concentrations in *Chlamydomonas reinhardtii*. *Planta* **213**, 286–295.

Villarejo, A., Shutova, T., Moskvin, O., Forssen, M., Klimov, V.V., and Samuelsson, G. (2002). A photosystem II-associated carbonic anhydrase regulates the efficiency of photosynthetic oxygen evolution. *EMBO J.* **21**, 1930–1938.

Wang, Y. and Spalding, M.H. (2006). An inorganic carbon transport system responsible for acclimation specific to air levels of CO_2 in *Chlamydomonas reinhardtii*. *Proc. Natl. Acad. Sci. U. S. A.* **103**, 10110–10115.

Wattebled, F., Buleon, A., Bouchet, B., Ral, J.P., Lienard, L., Delvalle, D., Binderup, K., Dauvillee, D., Ball, S., and D'Hulst, C. (2002). Granule-bound starch synthase I. A major enzyme involved in the biogenesis of B-crystallites in starch granules. *Eur. J. Biochem.* **269**, 3810–3820.

Winder, T.L., Anderson, J.C., and Spalding, M.H. (1992). Translational regulation of the large and small subunits of ribulose bisphosphate carboxylase/oxygenase during induction of the CO_2-concentrating mechanism in *Chlamydomonas reinhardtii*. *Plant Physiol.* **98**, 1409–1414.

Woodger, F.J., Badger, M.R., and Price, G.D. (2003). Inorganic carbon limitation induces transcripts encoding components of the CO_2-concentrating mechanism in *Synechococcus* sp. PCC7942 through a redox-independent pathway. *Plant Physiol.* **133**, 2069–2080.

Woodger, F.J., Badger, M.R., and Price, G.D. (2005). Sensing of inorganic carbon limitation in *Synechococcus* PCC7942 is correlated with the size of the internal inorganic carbon pool and involves oxygen. *Plant Physiol.* **139**, 1959–1969.

Xiang, Y.B., Zhang, J., and Weeks, D.P. (2001). The *Cia5* gene controls formation of the carbon concentrating mechanism in *Chlamydomonas reinhardtii*. *Proc. Natl. Acad. Sci. U. S. A.* **98**, 5341–5346.

Yoshioka, S., Taniguchi, F., Miura, K., Inoue, T., Yamano, T., and Fukuzawa, H. (2004). The novel Myb transcription factor LCR1 regulates the CO_2-responsive gene *Cah1*, encoding a periplasmic carbonic anhydrase in *Chlamydomonas reinhardtii*. *Plant Cell* **16**, 1466–1477.

## CHAPTER 9

Rubisco

Katia Wostrikoff and David B. Stern
Boyce Thompson Institute for Plant Research
Ithaca, New York, USA

CHAPTER CONTENTS

I. INTRODUCTION

Ribulose-1,5-bisphosphate carboxylase/oxygenase (Rubisco; EC 4.1.1.39) catalyzes the addition of gaseous carbon dioxide to ribulose-1,5-bisphosphate (RuBP), generating two molecules of 3-phosphoglyceric acid (3-PGA), and is thus the key enzyme in CO_2 assimilation. Rubisco is capable of a competing oxygenation reaction, which generates one molecule of 3-PGA and one of 2-phosphoglycolate from RuBP. In *Chlamydomonas* as in higher plants, the enzyme is a hexadecamer composed of eight large subunits (LS) and eight small subunits (SS), encoded by a single chloroplast gene (*rbcL*) and two adjacent nuclear genes (*RBCS1/2*), respectively. The number of the *RBCS* genes varies among plants but generally constitutes a small multigene family. Rubisco is mainly localized to the pyrenoid in *Chlamydomonas*.

Because of its high abundance and central metabolic role, Rubisco has been studied in many cellular contexts. Its high abundance necessitates a considerable diversion of amino acids and energy to its synthesis, and the many physiological stimuli that impinge upon the regulation of photosynthesis, require that Rubisco abundance and activity be controlled. Thus, Rubisco is discussed in several other chapters of this volume. The photorespiratory glycolate pathway, which is initiated through Rubisco oxygenase activity, is discussed in more detail in Chapter 8. This same chapter also discusses the carbon concentrating mechanism and how Rubisco may access its substrates under different growth conditions, particularly limiting CO_2, whereas Chapter 7 discusses relevance of Rubisco to hydrogen production. Chapter 29 discusses the translational regulation of LS synthesis in response to SS limitation and oxidative stress; topics which are summarized in section VII. Rubisco is also mentioned in the context of nutrient stress and transition metal deficiency, where it is subject to repression; these are referred to in section VI. Rubisco proteolysis is discussed in Chapter 19.

While Rubisco has been studied in many organisms, and thousands of *rbcL* genes have been sequenced for phylogenetic purposes (Kapralov and Filatov, 2007), *Chlamydomonas* has been particularly useful for several reasons. First, the *rbcL* gene can be readily manipulated by chloroplast transformation, leading to facile structure-function analyses. Second, the two *RBCS* genes are tightly linked and mutants exist with a deletion covering both. This is thus the only eukaryote for which there are mutants that stably and totally lack *RBCS* expression (Khrebtukova and Spreitzer, 1996; Dent et al., 2005). Third, culture conditions can be changed rapidly and uniformly. This allows a relatively uniform population to be examined for Rubisco regulation under a variety of stress conditions. This chapter focuses on studies in *Chlamydomonas*, while other aspects of Rubisco have been reviewed elsewhere. These include its potential to increase CO_2 storage in trees and diatoms (Millard et al., 2007; Roberts et al., 2007), choice of model system for studying Rubisco compartmentalization in C4 plants (Brown et al., 2005), and a series of reviews on Rubisco activity, assembly, and manipulation (Chollet and Spreitzer, 2003).

II. RUBISCO-ENCODING GENES AND ENZYME STRUCTURE

A. The *rbcL* and *RBCS* genes

The chloroplast *rbcL* gene lies within the small single-copy region, immediately upstream of *trnG2*, and is transcribed divergently from *atpA*

(Maul et al., 2002). The *rbcL* promoter and 5′ untranslated region (UTR) have been studied for their regulatory properties. The promoter contains canonical −35/−10 bacterial-type elements, as well as an enhancer that makes it one of the two most active promoters in the chloroplast genome (see Chapter 25). The 5′ UTR contains an RNA-stabilizing stem-loop structure (Suay et al., 2005), and drives reporter gene expression at a medium level when fused to the GFP coding region (Barnes et al., 2005). The *rbcL* gene is located in a region where DNA superhelicity responds to light (Salvador et al., 1998), which may have regulatory implications.

The 3′ UTR contains two stem-loop structures which stabilize the transcript against exonucleolytic decay (Blowers et al., 1993), although the 3′ UTR does not terminate transcription efficiently (Rott et al., 1996). Because of the lack of efficient termination, *Chlamydomonas* chloroplast mRNAs mature through a two-step process involving endonucleolytic cleavage and 3′ exonucleolytic resection, as modeled for *atpB* (Stern and Kindle, 1993). This pathway also occurs in *rbcL* mRNA 3′ end maturation, where the endonucleolytic cleavage destabilizes downstream sequences (Hicks et al., 2002; Rymarquis et al., 2006). This property was used in a genetic screen where the *rbcL* 3′ UTR was placed in a chimeric gene context that interrupted the *tscA* transcript, whose expression is required for splicing of *psaA* mRNA (see Chapter 26). By screening for restoration of *psaA* splicing, mutations were found which affected *cis* elements in the *rbcL* 3′ UTR as well as two nuclear loci *TAS1* and *TAS2*, whose products are apparently required for *rbcL* 3′ UTR cleavage in the chimeric gene context (Goldschmidt-Clermont et al., 2007).

The two *RBCS* genes are located on linkage group II, and were originally shown to be present in an 11-kb DNA fragment, and to respond to light and culture medium at the transcriptional level (Goldschmidt-Clermont, 1986). Their sequences and expression were determined in a subsequent study, with one finding being that the two RBCS polypeptides differ by four amino acids (Goldschmidt-Clermont and Rahire, 1986). Since a deletion mutant lacking both genes could be rescued by either gene or a chimeric version, they are functionally redundant at least under laboratory conditions (Khrebtukova and Spreitzer, 1996).

The *RBCS* promoters are strong as in all plants, and have been used in various transgene cassettes. These include the first RNAi vector for *Chlamydomonas* (Rohr et al., 2004), as well as either the standalone *RBCS2* promoter (Lumbreras et al., 1998) or a hybrid *HSP70A-RBCS2* promoter (Schroda et al., 1999) driving antisense RNA expression (reviewed in Schroda, 2006).

B. Crystal structure

The *Chlamydomonas* Rubisco crystal structure was obtained nearly simultaneously by two groups. A 1.4-Å structure was obtained from Rubisco complexed with the transition state analog 2-carboxyarabinitol-1,5-bisphosphate

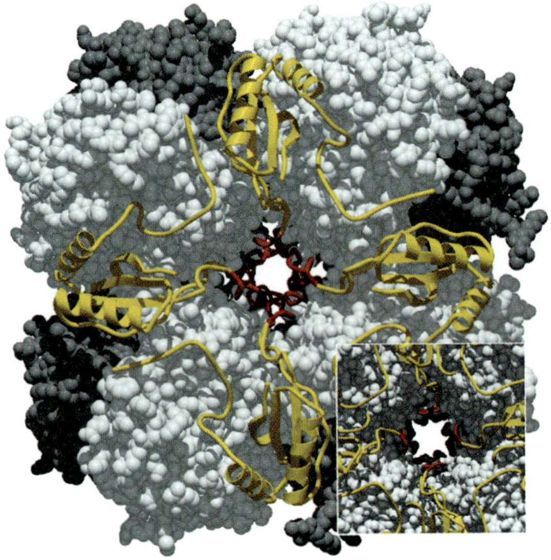

FIGURE 9.1 *The Rubisco holoenzyme from Chlamydomonas (Taylor et al., 2001). The eight LS are in dark and light gray, and the SS are in yellow. The red helices are the βA-βB loops of SS, where Chlamydomonas has an insertion that narrows the central cavity.*

(CABP), the highest-resolution Form I Rubisco holoenzyme structure obtained to date (Taylor et al., 2001). While the structure was quite similar to those obtained previously (Figure 9.1), some subtle differences were noted. These included an area of LS that interacts with Rubisco activase (section V), and both the C-terminal and an internal loop of the SS. Furthermore, several sites of probable post-translational methylation were noted. The second structure was obtained at 1.84-Å resolution, and reached similar conclusions, and identified additional post-translational modifications (Mizohata et al., 2002), which are detailed in section III.D.1. The SS insertion that affects the solvent pore is in a variable region of the polypeptide, as shown in Figure 9.2. The structures have been useful for interpreting the results from site-directed mutagenesis described in the following two sections.

III. SITE-DIRECTED MUTAGENESIS

A. Introduction

Site-directed mutagenesis of Rubisco has been carried out extensively, mostly in the context of attempting to improve its catalytic properties. These efforts have mainly focused on two parameters: the specificity factor for CO_2 ($\Omega = V_c/V_o \times K_o/K_c$), and the carboxylation efficiency. Indeed, increasing CO_2/O_2 specificity is expected to improve plant productivity by enhancing the photosynthetic rate (Ogren, 1984). Because this chapter is not devoted to the question of Rubisco improvement, the reader is referred to several excellent reviews (Spreitzer, 1999, 2003; Spreitzer and Salvucci, 2002).

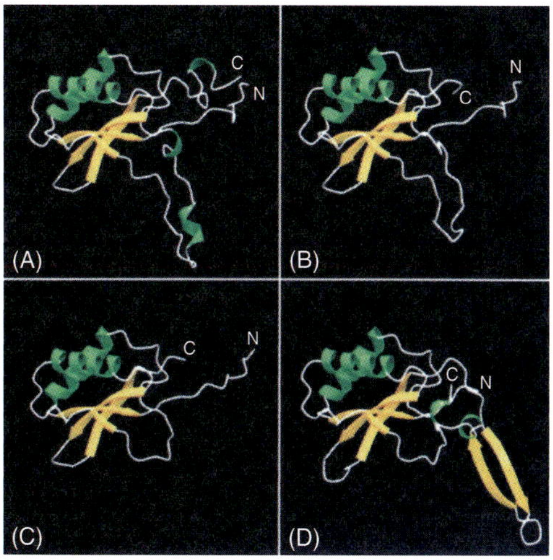

FIGURE 9.2 *SS structures (A–D) from Chlamydomonas, spinach, Synechococcus PCC6301, and the red alga Galdieria partita, as published in Mizohata et al. (2002). Helices and β-strands are colored in green and yellow, respectively.*

Chlamydomonas has proven to be an excellent model organism for Rubisco mutagenesis due to the relative ease of recovering Rubisco mutants, which can be maintained on acetate in the dark to prevent reversion, and also due to the power of forward genetics to recover revertants. Moreover, chloroplast transformation allows direct observation of mutant Rubisco properties in an homologous *in vivo* system, rather than relying on heterologous situations such as examination of *Rhodospirillum rubrum* or *Synechococcus* Rubisco in *E. coli*. Finally, as noted earlier, null mutants of both LS and SS exist for *Chlamydomonas*, providing an ideal genetic context for structure-function analyses.

Historically, mutant enzymes first allowed definition of critical regions for Rubisco holoenzyme assembly and/or activity. Recovery of revertants then identified second-site mutations that complemented the primary defect. The *Chlamydomonas* Rubisco structure (section II) facilitated elaboration of hypotheses grounded in genetic revertants, some of which have now been tested by obtaining structures of the mutant enzymes and their revertant forms. These experiments have revealed residues that are in contact, but also long-range interactions.

Phylogenetic approaches have also offered insight into how Rubisco specificity and net carboxylation differ between species. Omega (Ω) in particular is determined by the differential stabilization of the carboxylation and oxygenation transition states at the rate-determining partial reactions (Chen and Spreitzer, 1992). Although these intermediates are proposed to have similar structures, suggesting that oxygenase activity could not be separated from carboxylation activity, omega has been shown to differ among species and among mutant Rubisco versions. In the case of *Chlamydomonas*,

```
  1 mvpqtetkag agfkagvkdy rltyytpdyv vrdtdilaaf rMtpqlgvpp
                          αA                     βB

 51 eeCGaavaae sstgtwttvw tdgltsldry kgrcydiepv pgednqyiay
      αB                  310A     α0     βC             βD

101 vaypidlfee gsvtnmftsi vgNvfgfkal ralrledlri ppayvktfvg
      310B        αC        310C       βE         αD

151 pphgiqverd klnkygrgll GCTikpklgl saknygravy eClrggldft
         αE            β1                  α1            β2

201 kddenvnsqp fmrwrdRflf VAeaiYkaqa etgeVkGhyl natagtceem
                           α2               β3

251 mkravCaKel gVpiImhdyl tggftantsl aiycrdnglL lhihramhav
      α3          β4    α4-1      α4-2           β5     αF

301 idrqrnhgih frvlakalrm sggdhLhsgt Vvgklegere VTlGfvdlMr
           βF      α5          β6     loop 6          α6

351 ddyvekdrsr giyftqdwcs mpgvmpvaSg gihvwhmpal veifgddacl
      βG            βH          β7      310E   α7          β8

401 qfgggtlghp wgnapgaaan rvaleactqa rnegrdlare ggdvirsaCk
      αP                  αB                       αG

451 wspelaaaCe vwkeikfefd tiDkl
      αH               Cter
```

FIGURE 9.3 *Relevant features of the LS primary sequence and secondary structures. α-helices are highlighted in pink, 3_{10} helices in grey, and β-strands in yellow, after Mihozata et al. (2002). Loop 6 is highlighted in green, and the C-terminal in magenta. Unconserved residues are in bold underline, residues which have been studied through mutation in capitals. Where mutations strongly affect Rubisco accumulation the letters are blue; violet letters are those where catalysis was strongly affected, and red lettering indicates an effect on thermal stability. Residues mutated based on phylogenetic comparisons are highlighted in green.*

the approach has been to try to understand the critical structural determinants of the catalytic parameters observed in land plants, and to try to turn *Chlamydomonas* Rubisco into an enzyme with a specificity factor mirroring those typical of land plants, about 20% higher. From a phylogenetic point of view, one would infer that those differences can be ascribed to residues that are not conserved between species and that slightly modify Rubisco structure, allowing differential stabilization of the transition states. In *Chlamydomonas*, this has led to particular scrutiny of some of the 34 amino acids not conserved in land plants, which are underlined in Figure 9.3.

B. Mutagenesis of the large subunit

1. The N-terminus

LS mutants are summarized in Table 9.1. Rubisco's active site is formed at the interface of one LS C-terminal α/β barrel domain and the N-terminal

Table 9.1 Large subunit mutants

aa	Mutation	Location[a]	PS[b]	Holoenz acc.[c]	Thermal stability[d]	$\Omega^{e,f}$	$V_c^{f,g}$	$K_c^{f,h}$	$K_o^{f,i}$	References
WT*			+	100%	90% at 55°C 41% at 60°C 0% at 65°C	63	107	31	372	
M42	M42V	Nter	+			62	nd	nd	nd	(Du et al., 2003)
C53	C53A	Nter	+			62	nd	nd	nd	(Du et al., 2003)
	M42V-C53A	Nter	+			60	nd	nd	nd	(Du et al., 2003)
G54	G54D	Nter	–	0%						(Thow & Spreitzer, 1992)
G54	G54A	Nter	+	50%		60	100	32	360	(Spreitzer et al., 1995)
G54	G54V	Nter	+	25%		52	19	31	633	(Spreitzer et al., 1995)
N123	N123G	Nter-Cter interaction	–	100%		12	3	105	103	(Zhu and Spreitzer, 1994)
G171	G171D	β	–	100%						(Spreitzer & Mets, 1980)
C172	C172S*	β1	+	reduced	<WT	70	98	61	816	(Moreno and Spreitzer, 1999; Garcia-Murria et al., 2008)
T173	T173I	β1	–	100%						(Spreitzer et al., 1988)
C192	C192S*	α6	+			63	75	23	330	(Garcia-Murria et al., 2008)
R217	R217S	α2	–	5% est.						(Thow & Spreitzer, 1992)
R217	R217S-A242V	β3	+	25%		31	10	235	3778	(Thow et al., 1994)
V221	V221C-V235I	α2	+	100%		62	122	28	275	(Spreitzer et al., 2005)
Y226	Y226L	α2	+ +– at 35°C	80%						(Esquivel et al., 2006)

(Continued)

Table 9.1 *Continued*

aa	Mutation	Location[a]	PS[b]	Holoenz acc.[c]	Thermal stability[d]	$\Omega^{e,f}$	$V_c^{f,g}$	$K_c^{f,h}$	$K_o^{f,i}$	References
V235	V235I		+	100%		58	118	41	348	(Spreitzer et al., 2005)
L326	L326I	β6	+−	15β1%	25%	57	86	35	367	(Zhu & Spreitzer, 1996)
V331	V331A*	Loop 6	−	Near normal		40	7	166	2937	(Chen and Spreitzer, 1989)
V331	V331A-T342I*	Loop 6	+			53	35	47	829	(Chen and Spreitzer, 1989)
T342	T342I*	α6								(Karkehabadi et al., 2007)
V331	V331A-G344S*		+			46	27	73	1290	(Chen et al., 1991)
	L326I-M349L	Loop 6	+−	50%	55%	50	28	43	524	(Zhu & Spreitzer, 1996)
V341	V346I	α6	+	100%		61	103	36	390	(Zhu & Spreitzer, 1996)
M349	M349L	α6	+	75%	90%	59	97	33	379	(Zhu & Spreitzer, 1996)
S379	S379A	β7-Loop 6 interaction	−	100%		18	4	236	1673	(Zhu and Spreitzer, 1994)
C256	C256F	α3	+	100%	No effect	61	68	37	440	(Du et al., 2003)
K258	K258R	α3	+	100%	No effect	60	80	30	377	(Du et al., 2003)
I265	I265V	β4	+	100%	No effect	63	91	32	470	(Du et al., 2003)
	K258R-I265V		+−	25% est.	No effect					(Du et al., 2003)
	C256F-K258R-I265V		+−	25% est.	No effect	59	32	70	946	(Du et al., 2003)
						57	48	67	777	(Spreitzer et al., 2005)
	V221C-V235I-C256F-K258R-I265V		+ +− at 35°C	25% est.		64	48	67	1089	(Spreitzer et al., 2005)
L290	L290F*	β5	+	70%,	0% at 60°C	52	40	49	733	(Chen et al., 1988; Hong and Spreitzer, 1997)
			− at 35°C	10–17% at 35°C						(Du and Spreitzer, 2000)
						54	36	42	521	(Genkov et al., 2006)
						56	31	39	622	

L290	L290F-A222T*		+ + at 35°C	42%, 27% at 35°C	0% at 60°C	61	77	40	651	(Hong and Spreitzer, 1997)
L290	L290F-V262L		+ + at 35°C	50%, 30% at 35°C	61% at 60°C	62	72	68	1238	(Hong and Spreitzer, 1997)
A222	A222T	$\alpha 2$	+	100%	58% at 65°C	63	89	26	264	(Du and Spreitzer, 2000)
V262	V262L	Below $\beta 4$	+	100%	93% at 65°C	62	78	45	484	(Du and Spreitzer, 2000)
C449	C449S	Cter	+							(Marin-Navarro and Moreno, 2006)
C459	C459S	Cter	+							(Marin-Navarro and Moreno, 2006)
	C449S-C459S		+							(Marin-Navarro and Moreno, 2006)
D470	D470P-T471A-I472M-K474T	Cter	+	100%	No effect	69	90	36	545	(Satagopan and Spreitzer, 2008)
D473	D473A*	Cter	+	100%	No effect	52	36	77	1931	(Satagopan and Spreitzer, 2004)
	D473E	Cter	+	100%	No effect	54	34	87	1900	(Satagopan and Spreitzer, 2004)

[a] *LS regions are named according to the labeling in Figure 9.3.*

[b] *Photosynthetic growth on minimal medium at 25°C, unless otherwise indicated.*

[c] *Percent of holoenzyme accumulation compared to WT as measured, or estimated from published immunoblots (est.).*

[d] *Thermal instability was measured by the amount of activity retained by the purified enzyme after 20 minutes exposure at different temperatures. Unless otherwise indicated, the values reported correspond to activity at 55°C.*

[e] $\Omega = V_c K_O / V_O K_c$.

[f] *Values indicated are tentatively normalized to the wild-type values reported. For actual values refer to the appropriate publication.*

[g] V_c, *in* μ*mol/h/mg*

[h] K_c, *in* μM CO_2.

[i] K_O, *in* μM O_2

**Mutants for which structures are available.*

of a second LS. Mutations disrupting the hydrophobicity of this N-terminal core lead to disruption of holoenzyme stability, such as in the G54D mutant. Revertants G54A and G54V were isolated on the basis of restored growth on minimal medium, but still showed reduced holoenzyme amounts. The size of the side chain seems therefore critical in the assembly process. G54V Rubisco could be purified but showed a 17% decrease in Ω (Spreitzer et al., 1995).

The Rubisco N-terminal includes four residues that differ between *Chlamydomonas* and land plants. Replacing M42 and C53 with their land plant equivalents had no effect on holoenzyme assembly and only minor effects on catalysis (Du et al., 2003). As residues 87 and 105 are far from the interface and exposed to solvent, the tentative conclusion is that the N-terminal does not contribute substantially to differences in catalytic properties between *Chlamydomonas* and land plants.

2. Loop 6 (aa 328–338)

The α/β barrel provides a rigid framework for the active site. Residues contributing to the active site are invariant and are provided by loops of the barrel and residues from the N-terminal of another LS, to interact with the substrate (Andersson et al., 1989). Among these, loop 6 seems to have a critical role in oxygenation and carboxylation. In particular, the conserved K334 is known to orient and stabilize the transition state intermediate and is presumed to form hydrogen bonds with CO_2, as it does with its CABP analog. K334 also forms hydrogen bonds with E60 and T65 of an adjacent N-terminal (Knight et al., 1990).

Loop 6 flexibility is essential, as its movement participates in the transition from open to closed active site during catalysis. Many efforts to understand its role by site-directed mutagenesis have been deployed in *R. rubrum* and *Synechococcus* (reviewed in Spreitzer and Salvucci, 2002; Parry et al., 2003). In *Chlamydomonas*, the V331A mutant led to a thorough characterization of this region. V331A accumulates wild-type levels of holoenzyme but is nonphotosynthetic, and exhibits a strong reduction in Ω and carboxylation efficiency (Chen and Spreitzer, 1989). V331 is part of the hinge on which loop 6 moves. The reduction in the specificity factor of V331A was subsequently attributed to the enzyme being defective in stabilizing the transition state of the carboxylation step, reflected in its inability to bind CABP (Chen et al., 1991). Analysis of the V331A second-site revertants T342I and G344S showed that mutations within or close to loop 6 restore tighter binding of CABP. Structures of the V331A and V331A–T342I enzymes revealed the importance of a hydrogen bond between CABP and the active site residues T173 and S379. In V331 the bond is lengthened, creating a cavity in the loop 6 hydrophobic core, which is restored in the V331A–T342I and V331A–G344S enzymes, with a concomitant increase in the specificity factor (Karkehabadi et al., 2007). The length of this hydrogen bond therefore probably contributes to differences in kinetic constants. Similarly, mutations affecting G171 and T173 eliminate Rubisco activity (Spreitzer and Ogren, 1983; Spreitzer et al., 1988), while the S379A mutation was shown to decrease the rate of carboxylation and alter the Ω value (Zhu and Spreitzer, 1994).

3. The C-terminal (aa 463–475)

The LS C-terminal stabilizes the catalytically active closed state by packing it against loop 6, and Asp473 has been proposed to serve as a latch in this movement (Duff et al., 2000). However, this residue appears not to be critical for catalysis, since a D473E mutant showed activity, albeit with decreased Ω and carboxylation efficiency (Satagopan and Spreitzer, 2004). These results indicate that Asp473 contributes to the conformation of the active site, and substrate binding, as measured with CABP. Further evidence was provided by the crystal structure of the D473E enzyme, which displays disorder in the last six amino acids, perhaps influencing loop 6 dynamics and ultimately the catalytic site (Karkehabadi et al., 2007). The influence of this region on catalysis was also apparent from the V331A crystal structure, which showed that solvent-mediated hydrogen bonds between E338, and D473 and L475 were weakened in V331A, contributing to the decrease in specificity and carboxylation efficiency (Karkehabadi et al., 2007). Future mutagenesis of the C-terminal may be partly guided by informatic approaches, in which it is possible to perform dynamic simulations that relate the sequence of the mobile part of the C-terminal to the lifetime of the closed state (Burisch et al., 2007).

4. LS/SS interface

The LS/SS interface at the surface of the enzyme (at the bottom of the α/β barrel) has also been shown to be important for catalysis, although it is far from the active site. The L290F temperature-sensitive mutant was instrumental for its characterization. The mutant showed reduced accumulation of Rubisco at 25°C, a defect even more pronounced at 35°C, where it induced lethality on minimal medium (Chen et al., 1988). The mutation also affects catalysis, causing a 13% reduction in Ω. Its thermal instability is also associated with the enzyme being less active at higher temperature than the wild-type version.

L290F revertants growing at the restrictive temperature defined the second-site mutations A222T and V262L (Hong and Spreitzer, 1997). These intragenic suppressors exhibited a specificity factor comparable to the wild type, and increased carboxylation efficiency compared to L290F. In light of their improved growth at 35°C, the suppressors' Rubisco thermal stability was investigated. Indeed, the suppressors showed improved stability at higher temperature, with the suppressor versions carrying the wild-type L290 being even more stable than the wild-type enzyme (Hong and

Spreitzer, 1997; Du and Spreitzer, 2000). In contrast, however, the suppressor mutations had a deleterious effect on Rubisco holoenzyme accumulation at 25°C. The crystal structure showed that the highly conserved residues L290, A222, and V262 surround the βA/βB loop of Rubisco SS. One may conclude, therefore, that the LS/SS interface is critical in catalytic parameters as well as enzyme stability.

The experiments described above opened the door to a comparative mutagenesis approach in order to improve catalytic properties. In particular, three divergent residues were mutated to their land plant counterparts, by generating C256F, K258R, and I265V. The individual mutations had little effect on function, despite presumed alteration in the LS/SS interface. The double and triple mutants, however, showed decreased Rubisco accumulation, slower growth at 25°C, a 10% decrease in Ω, and a 70% decrease in carboxylation efficiency (Du et al., 2003). Thermal stability remained unaffected. By adding the substitutions V221C and V235I, creating a pentamutant, the specificity factor was restored to the *Chlamydomonas* wild type, as could be predicted by the enzyme structure (Spreitzer et al., 2005). From these studies and the subtle changes observed in many mutant versions of Rubisco (Karkehabadi et al., 2005b, 2007), it is apparent that individual substitutions are highly unlikely to improve catalysis. At the same time, they underscore the role of the SS, in terms of the function of the LS–SS interface.

C. SS mutagenesis

The SS mutants are summarized in Table 9.2. The question of how SS functions in Form I (L_8S_8) assembly and Rubisco catalysis arose through comparison to the presumed ancestral Form II Rubisco, which lacks SS. One hypothesis (Spreitzer and Salvucci, 2002) is that SS may be derived from Ccm proteins involved in assembling the carboxysome, which is a bacterial carbon-fixing structure (Cannon et al., 2001). Rubisco SS would then be required for LS assembly into octamers, thereby acting as a chaperone. One may argue that LS oligomers can be assembled without SS, as apparent from studies expressing the *Synechococcus* LS in *E. coli* (van der Vies et al., 1986). Nevertheless, it is clear that SS is required for Form I Rubisco assembly *in vivo*. Its absence leads to a concerted loss of LS, and subsequent inhibition of *rbcL* translation (Khrebtukova and Spreitzer, 1996; see Section VI). *In vitro* assembly studies performed in *E. coli* expressing a truncated *Synechococcus* SS showed that the first 20 N-terminal residues were critical for Rubisco assembly (Paul et al., 1993). The function of the *Chlamydomonas* SS C-terminal, which possesses a longer extension (11 more residues) than those of land plants and which is localized at the holoenzyme's interface, proved non-essential for function or catalysis (Taylor et al., 2001; Esquivel et al., 2002).

Unexpectedly, the βA/βB loop (aa 46–73) proved to be an important determinant of Rubisco holoenzyme stability, and was subsequently referred to as

Table 9.2 Small subunit mutants

aa	Mutation	Location	PS[a]	Holoenzyme accum[b]	Thermal stability[c]	$\Omega^{d,e}$	$V_c^{e,f}$	$K_c^{e,g}$	$K_o^{e,h}$	References
C65	C65S-L290F	βA/βB	+	100%, >WT at 35°C	Less than L290F (20%) at 55°C, 0% at 60°C	60	37	34	465	(Genkov et al., 2006)
	C65S	βA/βB	+	100%, >WT at 35°C	88% at 60°C	66	141	30	326	(Genkov et al., 2006)
	C65A	βA/βB	+		No effect	65	116	33	410	(Genkov et al., 2006)
	C65P	βA/βB	+		93% at 60°C	57	50	36	348	(Genkov et al., 2006)
Y67	Y67A	βA/βB	+, +− at 35°C	10%	WT		5% at 30°C			(Esquivel et al., 2006)
Y68	Y68A	βA/βB	+, +− at 35°C	100%	WT	62	31 60% at 30°C	22	347	(Spreitzer et al., 2001) (Esquivel et al., 2006)
Y72	Y72A	βA/βB	+,− +− at 35°C	10%	WT		5% at 30°C			(Esquivel et al., 2006)
R71	R71A	βA/βB	− at 35°C	50% est., 5% at 35°C	<WT, 0% at 55°C	58	17	40	623	(Spreitzer et al., 2001)
D69	D69A	βA/βB	+		WT	62	46	20	318	(Spreitzer et al., 2001)
R59	R59E	βA/βB	− at 35°C	20% est., 0% at 35°C						(Spreitzer et al., 2001)

(Continued)

Table 9.2 *Continued*

aa	Mutation	Location	PS[a]	Holoenzyme accum[b]	Thermal stability[c]	$\Omega^{d,e}$	$V_0^{e,f}$	$K_c^{e,g}$	$K_0^{e,h}$	References
N54	N54S-L290F	βA/βB	+ at 35°C	66%, 46% at 35°C	Comparable to L290F	63	67	32	489	(Du et al., 2000)
	N54S-RbcS1Δ	βA/βB	+		>WT, 35% at 65°C	64	77	25	331	(Du et al., 2000)
A57	A57V-L290F	βA/βB	+ at 35°C	57%, 42% at 35°C		61	71	31	461	(Du et al., 2000)
	A57V-RbcS1Δ	βA/βB	+		>WT, 88% at 65°C	62	65	25	334	(Du et al., 2000)
spinach	ABSO*	βA/βB		60% est.	85% at 60°C	64	55	21	330	(Spreitzer et al., 2005)
						62	53	23	320	(Karkehabadi et al., 2005a)
	ABSO-V221C-V235I-C256F-K258R-I265V	βA/βB	+− at 35°C	40% est.		70	59	23	328	(Spreitzer et al., 2005)
Synechococcus	ABAN*	βA/βB	+ − at 35°C	40% est.	0% at 60°C	56	58	32	260	(Karkehabadi et al., 2005a)

[a] *Photosynthetic growth on minimal medium at 25°C, unless otherwise indicated.*

[b] *Percent of holoenzyme accumulation compared to WT as measured, or estimated from published immunoblots (est.).*

[c] *Thermal instability was measured by the amount of activity retained by the purified enzyme after 20 minutes exposure at different temperatures. Unless otherwise indicated, the values reported correspond to activity at 55°C.*

[d] $\Omega = V_c K_O / V_O K_C$.

[e] *Values indicated are tentatively normalized to the wild-type values reported. In case of the Y67, Y68, and Y72 mutants, the actual data do not figure in the publication, therefore the published percentages of V_c measured at 30°C compared to WT reported were indicated. For other cases, refer to the appropriate publication for actual values.*

[f] V_c, *in μmol/h/mg.*

[g] K_c, *in μM CO_2.*

[h] K_O, *in μM O_2.*

**Mutants for which structures are available.*

an assembly domain. While SS are more divergent than LS, this divergence is mostly localized to the βA/βB loop (Taylor et al., 2001), which separates the β strands A and B (for figure see Spreitzer, 2003). This loop is located at the LS/SS interface, and shields the LS from solvent. Its length is variable: it encompasses 28 amino acids in *Chlamydomonas*, but only 22 in land plants. Its role in holoenzyme stability was first revealed by the isolation of suppressors in the *RBCS* gene from the temperature-sensitive LS mutant L290F. Unexpectedly, they clustered in the βA/βB loop. The N54S, A57V, and C65S mutations restored photosynthetic growth to L290F at the restrictive temperature, thus revealing the importance of the LS/SS interface to holoenzyme stability (Du et al., 2000; Genkov et al., 2006). When introduced in the absence of the chloroplast mutation and wild-type *RBCS* genes through transformation of mutants lacking both *RBCS* genes, the mutations did not have any obvious deleterious effect. In fact, the thermal stabilities of the N54S, A57V, and C65S mutant enzymes were not improved in the L290F mutant chloroplast backgrounds, but were in the wild-type background.

A systematic alanine-scanning mutagenesis of the βA/βB loop residues conserved between *Chlamydomonas* and land plants (R59, Y67, Y68, N69, and R71) yielded only photosynthetic strains (Spreitzer et al., 2001), which suggested that these residues do not play a critical role in assembly/stability. The fact that the R59A mutant grew on minimal medium was initially unexpected, as the phylogenetically comparable R53E substitution mutant in land plant SS failed to assemble with pea LS isolated from chloroplasts (Flachmann and Bohnert, 1992). However, R59E in *Chlamydomonas* did affect holoenzyme assembly, suggesting an importance in the nature of the substituted residue, rather than a requirement for arginine at that site. The R71A enzyme thermal stability is decreased, and is apparent by the strong reduction in Rubisco assembly at 35°C. This may result from the disruption of one of the six ionic bonds between the LS and SS subunits. Replacing the *Chlamydomonas* 28-amino acid βA/βB loop with the 22-amino acid βA/βB loop from *Synechococcus* yielded a comparable phenotype to R71A. Structure analysis shows that replacement of *Chlamydomonas* R71 with *Synechococcus* F53, and the decreased loop size, indeed removes a set of hydrogen bonds through the reorientation of K161, E223 and Y226, G261 and L266 (ABAN mutant; Karkehabadi et al., 2005a). The βA/βB loop comprises a cluster of tyrosines (Y67, Y68, Y72), whose contribution to stability through aromatic interactions was investigated. The alanine substitution mutants were photoautotrophic, but Rubisco was inactivated at a lower temperature than in the wild type. The Y67A and Y72A mutants were the most affected, as they showed a 90% decrease in holoenzyme accumulation at the restrictive temperature (Esquivel et al., 2006).

The βA/βB loop not only contributes to Rubisco stability, but also to differences in Rubisco enzymology, leading to the suggestion that species diversity in the βA/βB loop might reflect the differences observed in catalytic properties. Indeed, all the above-mentioned βA/βB loop mutations influenced the specificity factor and the carboxylation efficiency. Y67A and Y72A showed strong effects on specific activity, decreasing to 5% of the wild-type level. C65P had a deleterious effect on specificity and carboxylation efficiency. As its thermal stability property is comparable to the wild-type enzyme, this indicates that effects on catalysis may be independent of Rubisco stability.

Conversely, the N54S and A57V mutations in the L290F mutant background restored Ω to wild-type levels (Du et al., 2000), and the C65S mutation was even shown to enhance catalysis (Genkov et al., 2006). Similarly, the combination of the pentamutant mimicking spinach residues in the LS, together with the expression of a chimeric SS including the spinach βA/βB loop, caused an increase in CO_2 specificity (see mutant ABSO-V221C-V235I-C256F-K258R-I265V in Table 9.2). Its catalytic parameters became comparable to the spinach enzyme, which has a lower carboxylation efficiency. Although the decrease in V_c would result in a decrease in net CO_2 fixation at physiological concentrations of CO_2 and O_2, this study shows that mutagenesis of the LS/SS interface could lead to improvement of certain Rubisco properties. On the mechanistic side, though, it is still not completely understood how the SS βA/βB loop interaction with LS interface, at the opposite end of the α/β barrel that forms the catalytic site, has such an influence on catalytic properties. The magnitude of effect indeed does not correlate with strong structural perturbation in the active sites, as evidenced by structures of those mutant enzymes.

D. Rubisco mutagenesis and non-catalytic aspects of Rubisco biogenesis

1. Assembly

As is apparent from Table 9.1, a number of mutations in *rbcL* or *RBCS* had deleterious effects on Rubisco assembly *in vivo*, leading to strong reduction in holoenzyme accumulation, and thus pinpointing residues required for assembly and/or structural stability. Furthermore, a nuclear suppressor of the LS L290F mutation was obtained. This suppressor is called *Rbc-1*, is not allelic to *RBCS*, and leads to an increase in the amount of the holoenzyme and in its thermal stability in a wild-type chloroplast background, as well as an increase in Ω and V_c (Chen et al., 1990, 1993). Its cloning may reveal a component required for proper conformation of the βA/βB loop, thereby contributing to a better understanding of Rubisco assembly and its uniqueness compared to plants. Such a component is also suggested by the apparent absence in *Chlamydomonas* of a gene encoding the chaperone Bsd2, which is found in higher plants and required for Rubisco assembly (Brutnell et al., 1999). A candidate for a gene with *RBC1*-like function may be *RBCX* (EDP03859), which is a homologue of a Rubisco chaperone functioning at least in cyanobacteria (Saschenbrecker et al., 2007). Another gene of interest is defined by the nuclear mutant *76–5EN*, which affects both *rbcL* mRNA accumulation and its translation (Hong and Spreitzer, 1994).

Finally, the roles that post-translational modifications play in Rubisco stability/enzymology remain to be characterized. In particular, some affect residues not exposed upon assembly, indicating that they may be modified prior to assembly. *Chlamydomonas* exhibits all other known post-translational modifications, such as deformylation of LS-*N*-formyl methionine 1, followed by processing of residues M1 and S2, acetylation of P3, and carbamylation of K201 (reviewed by Houtz and Portis, 2003). Moreover, it is worth mentioning that some post-translational modifications are species-specific. In the case of *Chlamydomonas*, the specific modifications affecting Rubisco LS consist in *S*g-methylation of residues C256 and 369, as well as residues 104 and 151 being 4-hydroxyprolines (Taylor et al., 2001; Mizohata et al., 2002), which are not present for example in the spinach enzyme. These latter post-translational modifications were suggested to increase the LS folding rate following translation (Mizohata et al., 2002). Finally, the trimethylation of K14 is not observed in *Chlamydomonas* (Houtz et al., 1992), while SS residue M1 is methylated at its N-terminus, suggesting the existence of a monofunctional methyltransferase unable to methylate K14, unlike the difunctional enzyme found in pea and tobacco (Ying et al., 1999). Furthermore, this absence of methylation suggests that it is not essential for Rubisco activity.

2. Rubisco catabolism

Rubisco is a key reservoir of nitrogen, which is remobilized in senescence or stress conditions (see section V). It was shown to be oxidatively modified prior to its degradation (Mehta et al., 1992; Desimone et al., 1996), by both enzymatic and non-enzymatic reactions. This led to a characterization of the roles of cysteine residues in this process, as they may assist in transducing the redox signal into conformational changes that determine the catabolic fate of Rubisco. Indeed, the *in vitro* oxidation of cysteine residues induced conformational changes, which could be monitored by the use of proteases as structural probes, that converted Rubisco into a protease-sensitive form (Marin-Navarro and Moreno, 2003). *In vitro*, the degradation pattern is different in reducing or oxidative conditions. The degradation products result respectively from cleavage at residue K18 or within the S61–T68 loop, at the interface between the two LS that integrate the core dimers, through an apparent exposure under oxidative conditions (Marin-Navarro and Moreno, 2003). A similarly sized degradation product of 47 kD could be observed *in vivo* during senescence (Yoshida and Minamikawa, 1996).

A first hint of the involvement of cysteine residues came from the C172S mutant (Moreno and Spreitzer, 1999), which showed a delay in Rubisco turnover under oxidative conditions (10 mM H_2O_2). However, turnover was not prevented, indicating that other cysteines might contribute to redox regulation. While involvement of the nearby C192 could be envisioned through, for example, the inability to form a disulfide bond between residues 172 and C192 in the C172S mutant, structural analysis discounts this possibility (Taylor et al., 2001; Garcia-Murria et al., 2008). However, it was suggested that the C449–C459 pair may function as a redox sensor, as the distance between these two residues is variable among the available structures. Their importance in redox signaling was evidenced by mutagenesis, as the C449S–C459S double mutation prevented to some extent Rubisco oxidation and premature catabolism (Marin-Navarro and Moreno, 2006). It appears that the oxidation of this cysteine pair in the wild-type enzyme contributes to the exposure of the S61–T68 loop, rendering the double mutant more resistant to proteases *in vitro*. However the effects of the C172S and C449S–C459S mutations *in vivo* are different: while C172S mutation prevents degradation, the C449S–C459S mutant shows faster degradation, along with faster Rubisco aggregation and mobilization to the membrane fraction after saline stress (Marin-Navarro and Moreno, 2006).

Finally, the SS also contributes positively to the stabilization of Rubisco against accelerated turnover in oxidative conditions – all tyrosine mutants of the loop cluster were shown to degrade faster than the wild type under oxidative stress (Esquivel et al., 2006). Because no specific degradation products could be observed, the mutations seem to trigger a structural weakness leading to facilitated holoenzyme disassembly and increased turnover.

IV. RUBISCO ACTIVASE

A chaperone-like protein of the AAA+ family, Rubisco activase re-opens the Rubisco catalytic site and is thus required for activation and maintenance of Rubisco activity (Portis, 2003). Evidence for activase in *Chlamydomonas* first came from immunological data using an antibody raised against the purified spinach enzyme, which recognized two polypeptides (Salvucci et al., 1987). Subsequent work demonstrated, however, that *Chlamydomonas* activase is composed of a single protein, and shares approximately 65% identity with higher plant orthologues (Roesler and Ogren, 1990). Higher plants generally have two forms due to alternative splicing although cotton, for example, has two separate genes encoding proteins of slightly differing activities (Salvucci et al., 2003). The *Chlamydomonas* activase transit peptide has been modeled in solution in an effort to understand aspects of chloroplast envelope translocation signals (Krimm et al., 1999).

Using insertional mutagenesis, a *Chlamydomonas* null mutant was obtained and named *rca1* (Pollock et al., 2003), corresponding to nomenclature for the orthologous *Arabidopsis* mutant, which had been isolated some 20 years earlier (Somerville et al., 1982). While the photosynthetic impairment of the *Chlamydomonas rca1* mutant is relatively mild, it is severe when combined with the *cia5* mutation, suggesting that the carbon concentrating mechanism (see Chapter 8) can partly mitigate the inefficiency of Rubisco operating in the absence of activase.

Activase expression is under circadian control, with mRNA peaking near the end of the dark period and asynchronous with the enzyme, which accumulates to higher levels in the light period (Rawat and Moroney, 1995). Activase is also slightly induced upon transfer from high CO_2 conditions to air (Rawat and Moroney, 1995; Pollock et al., 2003), with the mRNA increasing approximately twofold based on microarray analysis (Im et al., 2003). Immunolocalization places activase mainly in the pyrenoid (McKay et al., 1991), where most Rubisco is located (Lacoste-Royal and Gibbs, 1987; Morita et al., 1997; Borkhsenious et al., 1998). This suggests that Rubisco population is largely colocalized with activase, in spite of one study that drew different conclusions (Süss et al., 1995).

The interface between Rubisco LS and activase has been studied using site-directed mutagenesis of both partners, following the observation that solanaceous activase is relatively inactive on *Chlamydomonas* Rubisco, compared to activase from spinach (Wang et al., 1992). This led to identification of LS residues that confer this species specificity (Larson et al., 1997; Ott et al., 2000), and also to the interacting region of activase (Li et al., 2005). This sets the stage for a better understanding of this critical regulatory process.

V. RUBISCO EXPRESSION IN RESPONSE TO PHYSIOLOGICAL CUES

Rubisco is principally synthesized in the light. The *rbcL* mRNA is present throughout the cell cycle in synchronous cultures, but is maximal at 4 hours of light (Herrin et al., 1986). LS synthesis also peaks in the light and is low in the dark. As LS synthesis is not responsive to rifampicin treatment, which inhibits chloroplast transcription, its primary regulation would appear to be at post-transcriptional levels (Iwanji et al., 1975). However, different results were obtained in a later study performed on cells grown phototrophically in continuous light, where LS synthesis markedly decreased upon rifampicin treatment (Eberhard et al., 2002), suggesting that *rbcL* transcripts, which are destabilized under those conditions, may be limiting.

Numerous stress-related conditions affect Rubisco. Most of them, however, do not specifically target Rubisco, but rather the photosynthetic apparatus in general. In some cases, an effect on accumulation of either *rbcL* or *RBCS* transcripts is seen, while in others, protein accumulation is affected. Copper stress induces a fivefold increase in *rbcL* mRNA accumulation at $50 \mu M$ Cu (Luis et al., 2006), and a decrease in Rubisco accumulation in the stroma, accompanied by its migration in a crosslinked state to the membrane fraction (Mehta et al., 1992). Under nitrogen-limiting conditions, Rubisco accumulates to about 60% of the normal level (Plumley and Schmidt, 1989). Under cadmium stress, a proteomic approach revealed a decrease in Rubisco holoenzyme, as well as in other proteins related to photosynthesis, the Calvin cycle and chlorophyll biosynthesis (Gillet et al., 2006).

Under sulfur stress, Rubisco was proposed to be remobilized in *Lemna* (Ferreira and Teixeira, 1992), and hence serve as an internal S source. While in *Chlamydomonas* the sulfolipid SQDG seems to be the first to be remobilized (Sugimoto et al., 2007), Rubisco indeed decreases strongly upon S starvation, declining by about 80% in the first 24 hours (Zhang et al., 2002). Sulfur stress is also not completely specific to Rubisco, and proteins such as D1, or proteins from photosystem I, are also degraded, although at a later stage following S deprivation. Microarray analysis revealed that S deprivation negatively affects *RBCS1/2* transcript accumulation (Zhang et al., 2004), and run-on experiments performed after 8 hours stress revealed a decrease in *rbcL* transcription (Irihimovitch and Stern, 2006). Thus, Rubisco is subject to both transcriptional and post-transcriptional regulation under limiting S conditions (see Chapter 5).

In most cases described above, the cascade of regulatory events leading to the observed decrease in Rubisco accumulation is poorly understood. In contrast, Rubisco's response to photooxidative stress is particularly well defined. A first response to photooxidative stress is photoacclimation. After a shift from low light (LL) to high light (HL, around 700 μmol/m^2/s), a transient decrease in LS translation, without a concomitant decrease in *rbcL* or *RBCS* mRNA, is observed as soon as 15 minutes following the shift, and rebounds in 6 hours to LL-grown levels (Shapira et al., 1997). This effect is quite specific, although D1 synthesis is also upregulated. The internal signal triggering this decrease in LS translation was shown to occur through an imbalance in the reduced/oxidized glutathione pool, a low molecular mass thiol having a regulatory role in algae and plants. This effect is dependent on the generation of reactive oxygen species upon the light stress (Irihimovitch and Shapira, 2000). Indeed, addition of the oxidant methyl viologen under LL conditions results in an increase in the free radical pool and induces a dramatic decrease in LS translation (about 80%), while ascorbate (an antioxidant) prevents the same phenomenon from occurring under HL. Furthermore, the ratio of GSS/GSSG, measured *in vivo* after the light stress, returns to its initial value in a similar time frame as the resumption of LS translation.

The translational inhibition of LS following the HL switch results in a decrease in the assembled Rubisco holoenzyme. While newly synthesized SS (which are presumably in excess) are rapidly degraded, the newly assembled holoenzyme is stable for at least an hour, as measured by pulse-chase labeling experiments (Cohen et al., 2005). LS oxidation following the shift to HL conditions is apparent from its faster migration in non-reducing gels, compared to migration of LS from LL grown cells, upon treatment of protein extracts with thiol reactive agents. A similar treatment after an *in vitro* incubation with GSSG also reveals oxidation of thiol groups of both LS and SS. Together these experiments reveal structural changes in both LS and SS upon oxidative stress. Whether these changes only affect the subunits already incorporated in the holoenzyme, or also newly synthesized LS, remains unclear.

The mechanism underlying this translational decrease upon oxidative stress has been proposed to occur through an autoregulation of the LS, which would no longer be capable of dimerization, either through structural changes upon oxidation, or through altered binding to an oxidized chaperone (Cohen et al., 2006). These redox-induced structural changes would unmask the LS N-terminal domain, which has been characterized as an RNA-binding protein, and is buried in the holoenzyme in reducing conditions at the interface of two LS domains (Yosef et al., 2004). Under oxidizing conditions, however, this motif would bind *rbcL* mRNA, resulting in translation inhibition.

Beside photoacclimation, stronger oxidative conditions (mimicked by the use of 1 mM $CuSO_4$, or 0.3 M NaCl) lead to Rubisco degradation, with a concomitant migration of covalently polymerized Rubisco to the membranes (see section III.D) (Mehta et al., 1992; Marin-Navarro and Moreno, 2006). Similarly, after 1 hour of incubation in presence of 5×10^{-5} M methyl viologen, LS degradation products appeared and were shown to exist within the holoenzyme (Knopf and Shapira, 2005). Moreover, larger aggregates composed mainly of LS, which were shown not to be membrane-associated, were detected. Although cells remained alive during this oxidative treatment, there was no survival after 24 hours, suggesting that these modifications may mark the threshold of photoacclimation, and be an early marker of cell death.

VI. REGULATION OF RUBISCO LS EXPRESSION

A. Introduction

Rubisco is like other chloroplast macromolecular complexes in that its subunits are encoded by both the nuclear and chloroplast genomes. However, it is a simple system relative to the ATPase and photosystems, because there are only two subunits with a total of three encoding genes, *RBCS1*, *RBCS2*, and *rbcL*. Mutations in all genes have been obtained, including insertional mutations inactivating both linked *RBCS* genes, and both insertional (Newman et al., 1991) and point mutations (Spreitzer et al., 1985) that inactivate *rbcL*. These provide tools and methods for modifying the genes, as described in sections III and IV.

B. LS autoregulation

LS and SS accumulate in a concerted manner, and the regulatory mechanisms underlying this phenomenon are discussed in much more detail in Chapter 29. Briefly, in the absence or limitation of LS, SS is degraded by proteolysis. In contrast, when SS is limiting, LS is subject to translational repression, either directly or indirectly involving interaction of that excess LS with *rbcL* mRNA, probably the 5′ UTR. While this has been studied in most detail in tobacco (Wostrikoff and Stern, 2007), extant data are most consistent with a similar mechanism operating in *Chlamydomonas*. One difference between plants and *Chlamydomonas*, however, is the apparent absence of the Rubisco chaperone Bsd2, first studied in maize (Brutnell et al., 1999). Bsd2 appears to intervene at the level of LS synthesis and/or initial assembly with the SS, and if this factor is indeed lacking in *Chlamydomonas*, there will be aspects unique to the alga.

In vitro evidence suggests that LS from many species, including *Chlamydomonas*, has RNA-binding properties, although under the conditions tested there was no specificity for *rbcL* mRNA (Yosef et al., 2004). A model has been proposed in which LS RNA-binding is subject to redox regulation (Cohen et al., 2005), which could also be a switch used to repress LS synthesis when SS is limiting.

While one nuclear mutant has been isolated that affects *rbcL* mRNA stability and/or translation (Hong and Spreitzer, 1994), additional mutant analysis has the potential to offer new insights into how Rubisco synthesis and assembly are regulated in *Chlamydomonas*.

VII. CONCLUSIONS AND PERSPECTIVES

Rubisco has been an intensely studied enzyme since its initial purification as "Fraction I" protein in 1947 (for an historical perspective see Portis and Parry, 2007). As is clear from the preceding text, *Chlamydomonas* is highly amenable to studies of this enzyme, and likely will continue to be far into the future. The long-term robustness of Rubisco research is assured not just by its signature role in photosynthesis, but also by the fact that its activity impinges on carbohydrate metabolism. Furthermore, its sheer abundance means that a significant amount of cellular resources must be marshaled for its synthesis and activation. Thus, many stress conditions lead to modifications of Rubisco activity or abundance, as described in this and other chapters.

While structure-function studies of Rubisco have yielded important insights into both the catalytic LS and accessory SS, the holy grail of Rubisco perfection, as a carbon-fixing enzyme, has yet to be attained. *Chlamydomonas* will likely serve as a major future platform because no other experimental system comes close in terms of amenability to high-throughput testing of new forms of the L_8S_8 enzyme. *Chlamydomonas* will also be useful in uncovering additional mechanistic details of Rubisco catalysis, modification and assembly. We also note that the LS-encoding *rbcL* gene has been a valuable reagent in its own right for studies of chloroplast gene expression, and will continue to be so.

The impact of the *Chlamydomonas* nuclear genome sequence on studies of Rubisco is likely to be substantial in the long term. On the one hand, it revealed an apparent lack of the Bsd2 chaperone which is known in higher plants, and affirmed the probable presence of RBCX. It also provides a resource for cloning genes such as those defined by the *rbc1* or *76-EN* mutants mentioned above, or future mutants obtained in suppressor screens or through forward genetics. In summary, while the structural genes and the activase have long been known, assembly and regulatory factors will undoubtedly emerge in the coming years.

ACKNOWLEDGMENTS

The authors thank Archie Portis and Jim Moroney for critical readings of the chapter. Rubisco work in the authors' laboratory was supported by National Science Foundation Award DBI-0211935 and the National Research Initiative Competitive Grants Program.

REFERENCES

Andersson, I., Knight, S., Schneider, G., Lindqvist, Y.T.L., Brändén, C.I., and Lorimer, G.H. (1989). Crystal structure of the active site of ribulose-biphosphate carboxylase. *Nature* **337**, 229–234.

Barnes, D., Franklin, S., Schultz, J., Henry, R., Brown, E., Coragliotti, A., and Mayfield, S.P. (2005). Contribution of 5′- and 3′-untranslated regions of plastid mRNAs to the expression of *Chlamydomonas reinhardtii* chloroplast genes. *Mol. Genet. Genomics* **274**, 625–636.

Blowers, A.D., Klein, U., Ellmore, G.S., and Bogorad, L. (1993). Functional *in vivo* analyses of the 3′ flanking sequences of the *Chlamydomonas* chloroplast *rbcL* and *psaB* genes. *Mol. Gen. Genet.* **238**, 339–349.

Borkhsenious, O.N., Mason, C.B., and Moroney, J.V. (1998). The intracellular localization of ribulose-1,5-bisphosphate carboxylase/oxygenase in *Chlamydomonas reinhardtii*. *Plant Physiol.* **116**, 1585–1591.

Brown, N.J., Parsley, K., and Hibberd, J.M. (2005). The future of C4 research--maize, *Flaveria* or *Cleome*? *Trends Plant Sci.* **10**, 215–221.

Brutnell, T.P., Sawers, R.J., Mant, A., and Langdale, J.A. (1999). BUNDLE SHEATH DEFECTIVE2, a novel protein required for post-translational regulation of the *rbcL* gene of maize. *Plant Cell* **11**, 849–864.

Burisch, C., Wildner, G.F., and Schlitter, J. (2007). Bioinformatic tools uncover the C-terminal strand of Rubisco's large subunit as hot-spot for specificity-enhancing mutations. *FEBS Lett.* **581**, 741–748.

Cannon, G.C., Bradburne, C.E., Aldrich, H.C., Baker, S.H., Heinhorst, S., and Shively, J.M. (2001). Microcompartments in prokaryotes: Carboxysomes and related polyhedra. *Appl. Environ. Microbiol.* **67**, 5351–5361.

Chen, Z.X. and Spreitzer, R.J. (1989). Chloroplast intragenic suppression enhances the low CO_2/O_2 specificity of mutant ribulose-bisphosphate carboxylase/oxygenase. *J. Biol. Chem.* **264**, 3051–3053.

Chen, Z. and Spreitzer, R.J. (1992). How various factors influence the CO_2/O_2 specificity of ribulose-1,5-biphosphate carboxylase/oxygenase. *Photosynth Res.* **31**, 157–164.

Chen, Z.X., Chastain, C.J., Al-Abed, S.R., Chollet, R., and Spreitzer, R.J. (1988). Reduced CO_2/O_2 specificity of ribulose-bisphosphate carboxylase/oxygenase in a temperature-sensitive chloroplast mutant of *Chlamydomonas*. *Proc. Natl. Acad. Sci. U. S. A.* **85**, 4696–4699.

Chen, Z.X., Green, D., Westhoff, C., and Spreitzer, R.J. (1990). Nuclear mutation restores the reduced CO_2/O_2 specificity of ribulose bisphosphate carboxylase/oxygenase in a temperature-conditional chloroplast mutant of *Chlamydomonas reinhardtii*. *Arch. Biochem. Biophys.* **283**, 60–67.

Chen, Z.X., Yu, W.Z., Lee, J.H., Diao, R., and Spreitzer, R.J. (1991). Complementing amino acid substitutions within loop 6 of the alpha/beta-barrel active site influence the CO_2/O_2 specificity of chloroplast ribulose-1,5-bisphosphate carboxylase/oxygenase. *Biochemistry* **30**, 8846–8850.

Chen, Z., Hong, S., and Spreitzer, R.J. (1993). Thermal instability of ribulose-1,5-bisphosphate carboxylase-oxygenase from a temperature-conditional chloroplast mutant of *Chlamydomonas reinhardtii*. *Plant Physiol.* **101**, 1189–1194.

Chollet, R. and Spreitzer, R.J. (2003). Note from the guest co-editors. *Arch. Biochem. Biophys.* **414**, 129.

Cohen, I., Knopf, J.A., Irihimovitch, V., and Shapira, M. (2005). A proposed mechanism for the inhibitory effects of oxidative stress on Rubisco assembly and its subunit expression. *Plant Physiol.* **137**, 738–746.

Cohen, I., Sapir, Y., and Shapira, M. (2006). A conserved mechanism controls translation of Rubisco large subunit in different photosynthetic organisms. *Plant Physiol.* **141**, 1089–1097.

Dent, R.M., Haglund, C.M., Chin, B.L., Kobayashi, M.C., and Niyogi, K.K. (2005). Functional genomics of eukaryotic photosynthesis using insertional mutagenesis of *Chlamydomonas reinhardtii*. *Plant Physiol.* **137**, 545–556.

Desimone, M., Henke, A., and Wagner, E. (1996). Oxidative stress induces partial degradation of the large subunit of ribulose-1,5-bisphosphate carboxylase/oxygenase in isolated chloroplasts of barley. *Plant Physiol.* **111**, 789–796.

Du, Y.C. and Spreitzer, R.J. (2000). Suppressor mutations in the chloroplast-encoded large subunit improve the thermal stability of wild-type ribulose-1,5-bisphosphate carboxylase/oxygenase. *J. Biol. Chem.* **275**, 19844–19847.

Du, Y.C., Hong, S., and Spreitzer, R.J. (2000). RbcS suppressor mutations improve the thermal stability and CO_2/O_2 specificity of *rbcL* mutant ribulose-1,5-bisphosphate carboxylase/oxygenase. *Proc. Natl. Acad. Sci. U. S. A.* **97**, 14206–14211.

Du, Y.C., Peddi, S.R., and Spreitzer, R.J. (2003). Assessment of structural and functional divergence far from the large subunit active site of ribulose-1,5-bisphosphate carboxylase/oxygenase. *J. Biol. Chem.* **278**, 49401–49405.

Duff, A.P., Andrews, T.J., and Curmi, P.M. (2000). The transition between the open and closed states of Rubisco is triggered by the inter-phosphate distance of the bound bisphosphate. *J. Mol. Biol.* **298**, 903–916.

Eberhard, S., Drapier, D., and Wollman, F.A. (2002). Searching limiting steps in the expression of chloroplast-encoded proteins: Relations between gene copy number, transcription, transcript abundance and translation rate in the chloroplast of *Chlamydomonas reinhardtii*. *Plant J.* **31**, 149–160.

Esquivel, M.G., Anwaruzzaman, M., and Spreitzer, R.J. (2002). Deletion of nine carboxy-terminal residues of the Rubisco small subunit decreases thermal stability but does not eliminate function. *FEBS Lett.* **520**, 73–76.

Esquivel, M.G., Pinto, T.S., Marin-Navarro, J., and Moreno, J. (2006). Substitution of tyrosine residues at the aromatic cluster around the betaA-betaB loop of

Rubisco small subunit affects the structural stability of the enzyme and the in vivo degradation under stress conditions. *Biochemistry* **45**, 5745–5753.

Ferreira, R.M. and Teixeira, A.R. (1992). Sulfur starvation in *Lemna* leads to degradation of ribulose-bisphosphate carboxylase without plant death. *J. Biol. Chem.* **267**, 7253–7257.

Flachmann, R. and Bohnert, H.J. (1992). Replacement of a conserved arginine in the assembly domain of ribulose-1,5-bisphosphate carboxylase/oxygenase small subunit interferes with holoenzyme formation. *J. Biol. Chem.* **267**, 10576–10582.

Garcia-Murria, M.J., Karkehabadi, S., Marin-Navarro, J., Satagopan, S., Andersson, I., Spreitzer, R.J., and Moreno, J. (2008). Structural and functional consequences of the replacement of proximal residues Cys-172 and Cys-192 in the large subunit of ribulose 1,5-bisphosphate carboxylase/oxygenase from *Chlamydomonas reinhardtii*. *Biochem. J.* **411**, 241–247.

Genkov, T., Du, Y.C., and Spreitzer, R.J. (2006). Small-subunit cysteine-65 substitutions can suppress or induce alterations in the large-subunit catalytic efficiency and holoenzyme thermal stability of ribulose-1,5-bisphosphate carboxylase/oxygenase. *Arch. Biochem. Biophys.* **451**, 167–174.

Gillet, S., Decottignies, P., Chardonnet, S., and Le Marechal, P. (2006). Cadmium response and redoxin targets in *Chlamydomonas reinhardtii*: A proteomic approach. *Photosynth. Res.* **89**, 201–211.

Goldschmidt-Clermont, M. (1986). The two genes for the small subunit of RuBP Carboxylase/oxygenase are closely linked in *Chlamydomonas reinhardtii*. *Plant Mol .Biol.* **6**, 13–21.

Goldschmidt-Clermont, M. and Rahire, M. (1986). Sequence, evolution and differential expression of the two genes encoding variant small subunits of ribulose bisphosphate carboxylase/oxygenase in *Chlamydomonas reinhardtii*. *J. Mol. Biol.* **191**, 421–432.

Goldschmidt-Clermont, M., Rahire, M., and Rochaix, J.D. (2007). Redundant *cis*-acting determinants of 3′ processing and RNA stability in the chloroplast *rbcL* mRNA of *Chlamydomonas*. *Plant J.* **53**, 566–577.

Herrin, D.L., Michaels, A.S., and Paul, A.L. (1986). Regulation of genes encoding the large subunit of ribulose-1,5-bisphosphate carboxylase and the photosystem II polypeptides D1 and D2 during the cell cycle of *Chlamydomonas reinhardtii*. *J. Cell Biol.* **103**, 1837–1846.

Hicks, A., Drager, R.G., Higgs, D.C., and Stern, D.B. (2002). An mRNA 3′ processing site targets downstream sequences for rapid degradation in *Chlamydomonas* chloroplasts. *J. Biol. Chem.* **277**, 3325–3333.

Hong, S. and Spreitzer, R.J. (1994). Nuclear mutation inhibits expression of the chloroplast gene that encodes the large subunit of ribulose-1,5-bisphosphate carboxylase-oxygenase. *Plant Physiol.* **106**, 673–678.

Hong, S. and Spreitzer, R.J. (1997). Complementing substitutions at the bottom of the barrel influence catalysis and stability of ribulose-bisphosphate carboxylase/oxygenase. *J. Biol. Chem.* **272**, 11114–11117.

Houtz, R.L., Poneleit, L., Jones, S.B., Royer, M., and Stults, J.T. (1992). Posttranslational modifications in the amino-terminal region of the large subunit of ribulose-1,5-bisphosphate carboxylase/oxygenase from several plant species. *Plant Physiol.* **98**, 1170–1174.

Houtz, R.L. and Portis, A.R., Jr. (2003). The life of ribulose 1,5-bisphosphate carboxylase/oxygenase – posttranslational facts and mysteries. *Arch. Biochem. Biophys.* **414**, 150–158.

Im, C.S., Zhang, Z., Shrager, J., Chang, C.W., and Grossman, A.R. (2003). Analysis of light and CO_2 regulation in *Chlamydomonas reinhardtii* using genome-wide approaches. *Photosynth. Res.* **75**, 111–125.

Irihimovitch, V. and Shapira, M. (2000). Glutathione redox potential modulated by reactive oxygen species regulates translation of Rubisco large subunit in the chloroplast. *J. Biol. Chem.* **275**, 16289–16295.

Irihimovitch, V. and Stern, D.B. (2006). The SAC3 kinase is required for chloroplast transcriptional repression under sulfur limitation in *Chlamydomonas reinhardtii*. *Proc. Natl. Acad. Sci. U. S. A.* **103**, 7911–7916.

Iwanij, V., Chua, N.H., and Siekevitz, P. (1975). Synthesis and turnover of ribulose biphosphate carboxylase and of its subunits during the cell cycle of *Chlamydomonas reinhardtii*. *J. Cell Biol.* **64**, 572–585.

Kapralov, M.V. and Filatov, D.A. (2007). Widespread positive selection in the photosynthetic Rubisco enzyme. *BMC Evol. Biol.* **7**, 73.

Karkehabadi, S., Peddi, S.R., Anwaruzzaman, M., Taylor, T.C., Cederlund, A., Genkov, T., Andersson, I., and Spreitzer, R.J. (2005a). Chimeric small subunits influence catalysis without causing global conformational changes in the crystal structure of ribulose-1,5-bisphosphate carboxylase/oxygenase. *Biochemistry* **44**, 9851–9861.

Karkehabadi, S., Taylor, T.C., Spreitzer, R.J., and Andersson, I. (2005b). Altered intersubunit interactions in crystal structures of catalytically compromised ribulose-1,5-bisphosphate carboxylase/oxygenase. *Biochemistry* **44**, 113–120.

Karkehabadi, S., Satagopan, S., Taylor, T.C., Spreitzer, R.J., and Andersson, I. (2007). Structural analysis of altered large-subunit loop-6/carboxy-terminus interactions that influence catalytic efficiency and CO_2/O_2 specificity of ribulose-1,5-bisphosphate carboxylase/oxygenase. *Biochemistry* **46**, 11080–11089.

Khrebtukova, I. and Spreitzer, R.J. (1996). Elimination of the *Chlamydomonas* gene family that encodes the small subunit of ribulose-1,5-bisphosphate carboxylase/oxygenase. *Proc. Natl. Acad. Sci. U. S. A.* **93**, 13689–13693.

Knight, S., Andersson, I., and Branden, C.I. (1990). Crystallographic analysis of ribulose 1,5-bisphosphate carboxylase from spinach at 2.4 Å resolution. Subunit interactions and active site. *J. Mol. Biol.* **215**, 113–160.

Knopf, J.A. and Shapira, M. (2005). Degradation of Rubisco SSU during oxidative stress triggers aggregation of Rubisco particles in *Chlamydomonas reinhardtii*. *Planta* **222**, 787–793.

Krimm, I., Gans, P., Hernandez, J.F., Arlaud, G.J., and Lancelin, J.M. (1999). A coil-helix instead of a helix-coil motif can be induced in a chloroplast transit peptide from *Chlamydomonas reinhardtii*. *Eur. J. Biochem.* **265**, 171–180.

Lacoste-Royal, G. and Gibbs, S.P. (1987). Immunocytochemical localization of ribulose-1,5-bisphosphate carboxylase in the pyrenoid and thylakoid region of the chloroplast of *Chlamydomonas reinhardtii*. *Plant Physiol.* **83**, 602–606.

Larson, E.M., O'Brien, C.M., Zhu, G., Spreitzer, R.J., and Portis, A.R., Jr. (1997). Specificity for activase is changed by a Pro-89 to Arg substitution in the large subunit of ribulose-1,5-bisphosphate carboxylase/oxygenase. *J. Biol. Chem.* **272**, 17033–17037.

Li, C., Salvucci, M.E., and Portis, A.R., Jr. (2005). Two residues of rubisco activase involved in recognition of the Rubisco substrate. *J. Biol. Chem.* **280**, 24864–24869.

Luis, P., Behnke, K., Toepel, J., and Wilhelm, C. (2006). Parallel analysis of transcript levels and physiological key parameters allows the identification of stress phase gene markers in *Chlamydomonas reinhardtii* under copper excess. *Plant Cell Environ.* **29**, 2043–2054.

Lumbreras, V., Stevens, D.R., and Purton, S. (1998). Efficient foreign gene expression in *Chlamydomonas reinhardtii* mediated by an endogenous intron. *Plant J.* **14**, 441–448.

Marin-Navarro, J. and Moreno, J. (2003). Modification of the proteolytic fragmentation pattern upon oxidation of cysteines from ribulose 1,5-bisphosphate carboxylase/oxygenase. *Biochemistry* **42**, 14930–14938.

Marin-Navarro, J. and Moreno, J. (2006). Cysteines 449 and 459 modulate the reduction-oxidation conformational changes of ribulose 1.5-bisphosphate carboxylase/oxygenase and the translocation of the enzyme to membranes during stress. *Plant Cell Environ.* **29**, 898–908.

Maul, J.E., Lilly, J.W., Cui, L., dePamphilis, C.W., Miller, W., Harris, E.H., and Stern, D.B. (2002). The *Chlamydomonas reinhardtii* plastid chromosome: Islands of genes in a sea of repeats. *Plant Cell* **14**, 2659–2679.

McKay, R.M.L., Gibbs, S.P., and Vaughn, K.C. (1991). RuBisCo activase is present in the pyrenoid of green algae. *Protoplasma* **162**, 38–45.

Mehta, R.A., Fawcett, T.W., Porath, D., and Mattoo, A.K. (1992). Oxidative stress causes rapid membrane translocation and *in vivo* degradation of ribulose-1,5-bisphosphate carboxylase/oxygenase. *J. Biol. Chem.* **267**, 2810–2816.

Millard, P., Sommerkorn, M., and Grelet, G.A. (2007). Environmental change and carbon limitation in trees: A biochemical, ecophysiological and ecosystem appraisal. *New Phytol.* **175**, 11–28.

Mizohata, E., Matsumura, H., Okano, Y., Kumei, M., Takuma, H., Onodera, J., Kato, K., Shibata, N., Inoue, T., Yokota, A., and Kai, Y. (2002). Crystal structure of activated ribulose-1,5-bisphosphate carboxylase/oxygenase from green alga *Chlamydomonas reinhardtii* complexed with 2-carboxyarabinitol-1,5-bisphosphate. *J. Mol. Biol.* **316**, 679–691.

Moreno, J. and Spreitzer, R.J. (1999). C172S substitution in the chloroplast-encoded large subunit affects stability and stress-induced turnover of ribulose-1,5-bisphosphate carboxylase/oxygenase. *J. Biol. Chem.* **274**, 26789–26793.

Morita, E., Kuroiwa, H., Kuroiwa, T., and Nozaki, H. (1997). High localization of ribulose-1,5-bisphosphate carboxylase/oxygenase in the pyrenoids of *Chlamydomonas reinhardtii* (Chlorophyta), as revealed by cryofixation and immunogold electron microscopy. *J. Phycol.* **33**, 68–72.

Newman, S.M., Gillham, N.W., Harris, E.H., Johnson, A.M., and Boynton, J.E. (1991). Targeted disruption of chloroplast genes in *Chlamydomonas reinhardtii*. *Mol .Gen. Genet.* **230**, 65–74.

Ogren, W.L. (1984). Photorespiration: Pathways, regulation, and modification. *Annu. Rev. Plant Physiol.* **35**, 415–442.

Ott, C.M., Smith, B.D., Portis, A.R., Jr., and Spreitzer, R.J. (2000). Activase region on chloroplast ribulose-1,5-bisphosphate carboxylase/oxygenase. Nonconservative substitution in the large subunit alters species specificity of protein interaction. *J. Biol. Chem* **275**, 26241–26244.

Parry, M.A., Andralojc, P.J., Mitchell, R.A., Madgwick, P.J., and Keys, A.J. (2003). Manipulation of Rubisco: The amount, activity, function and regulation. *J. Exp. Bot.* **54**, 1321–1333.

Paul, K., Morell, M.K., and Andrews, T.J. (1993). Amino-terminal truncations of the ribulose-bisphosphate carboxylase small subunit influence catalysis and subunit interactions. *Plant Physiol.* **102**, 1129–1137.

Plumley, F.G. and Schmidt, G.W. (1989). Nitrogen-dependent regulation of photosynthetic gene expression. *Proc. Natl. Acad. Sci. U. S. A.* **86**, 2678–2682.

Pollock, S.V., Colombo, S.L., Prout, D.L., Jr., Godfrey, A.C., and Moroney, J.V. (2003). Rubisco activase is required for optimal photosynthesis in the green alga *Chlamydomonas reinhardtii* in a low-CO_2 atmosphere. *Plant Physiol.* **133**, 1854–1861.

Portis, A.R., Jr. (2003). Rubisco activase – Rubisco's catalytic chaperone. *Photosynth. Res.* **75**, 11–27.

Portis, A. and Parry, M. (2007). Discoveries in Rubisco (ribulose 1,5-bisphosphate carboxylase/oxygenase): A historical perspective. *Photosynth. Res.* **94**, 121–143.

Rawat, M. and Moroney, J.V. (1995). The regulation of carbonic anhydrase and ribulose-1,5-bisphosphate carboxylase/oxygenase activase by light and CO_2 in *Chlamydomonas reinhardtii*. *Plant Physiol.* **109**, 937–944.

Roberts, K., Granum, E., Leegood, R.C., and Raven, J.A. (2007). Carbon acquisition by diatoms. *Photosynth. Res.* **93**, 79–88.

Roesler, K.R. and Ogren, W.L. (1990). Primary structure of *Chlamydomonas reinhardtii* ribulose 1,5-bisphosphate carboxylase/oxygenase activase and evidence for a single polypeptide. *Plant Physiol.* **94**, 1837–1841.

Rohr, J., Sarkar, N., Balenger, S., Jeong, B.R., and Cerutti, H. (2004). Tandem inverted repeat system for selection of effective transgenic RNAi strains in *Chlamydomonas*. *Plant J.* **40**, 611–621.

Rott, R., Drager, R.G., Stern, D.B., and Schuster, G. (1996). The 3′ untranslated regions of chloroplast genes in *Chlamydomonas reinhardtii* do not serve as efficient transcriptional terminators. *Mol. Gen. Genet.* **252**, 676–683.

Rymarquis, L.A., Webster, B.R., and Stern, D.B. (2006). The nucleus-encoded factor MCD4 participates in degradation of nonfunctional 3′ UTR sequences generated by cleavage of pre-mRNA in *Chlamydomonas* chloroplasts. *Mol. Genet. Genomics* **277**, 329–340.

Salvador, M.L., Klein, U., and Bogorad, L. (1998). Endogenous fluctuations of DNA topology in the chloroplast of *Chlamydomonas reinhardtii*. *Mol. Cell. Biol.* **18**, 7235–7242.

Salvucci, M.E., Werneke, J.M., Ogren, W.L., and Portis, A.R. (1987). Purification and species distribution of Rubisco activase. *Plant Physiol.* **84**, 930–936.

Salvucci, M.E., van de Loo, F.J., and Stecher, D. (2003). Two isoforms of Rubisco activase in cotton, the products of separate genes not alternative splicing. *Planta* **216**, 736–744.

Saschenbrecker, S., Bracher, A., Rao, K.V., Rao, B.V., Hartl, F.U., and Hayer-Hartl, M. (2007). Structure and function of RbcX, an assembly chaperone for hexadecameric Rubisco. *Cell* **129**, 1189–1200.

Satagopan, S. and Spreitzer, R.J. (2004). Substitutions at the Asp-473 latch residue of *Chlamydomonas* ribulose bisphosphate carboxylase/oxygenase cause decreases in carboxylation efficiency and CO_2/O_2 specificity. *J. Biol. Chem.* **279**, 14240–14244.

Satagopan, S. and Spreitzer, R.J. (2008). Plant-like substitutions in the large-subunit carboxy terminus of *Chlamydomonas* Rubisco increase CO_2/O_2 specificity. *BMC Plant Biol.* **8**, 85.

Schroda, M. (2006). RNA silencing in *Chlamydomonas*: Mechanisms and tools. *Curr. Genet.* **49**, 69–84.

Schroda, M., Vallon, O., Wollman, F.A., and Beck, C.F. (1999). A chloroplast-targeted heat shock protein 70 (HSP70) contributes to the photoprotection and repair of photosystem II during and after photoinhibition. *Plant Cell* **11**, 1165–1178.

Shapira, M., Lers, A., Heifetz, P.B., Irihimovitz, V., Osmond, C.B., Gillham, N.W., and Boynton, J.E. (1997). Differential regulation of chloroplast gene expression in *Chlamydomonas reinhardtii* during photoacclimation: Light stress transiently suppresses synthesis of the Rubisco LSU protein while enhancing synthesis of the PS II D1 protein. *Plant Mol. Biol.* **33**, 1001–1011.

Somerville, C.R., Portis, A.R., and Ogren, W.L. (1982). A mutant of *Arabidopsis thaliana* which lacks activation of RuBP carboxylase *in vivo*. *Plant Physiol.* **70**, 381–387.

Spreitzer, R.J. (1999). Questions about the complexity of chloroplast ribulose-1,5-biphosphate carboxylase/oxygenase. *Photosynth. Res.* **60**, 29–42.

Spreitzer, R.J. (2003). Role of the small subunit in ribulose-1,5-bisphosphate carboxylase/oxygenase. *Arch. Biochem. Biophys.* **414**, 141–149.

Spreitzer, R.J. and Mets, L.J. (1980). Non-mendelian mutation affecting ribulose-1, 5-bisphosphate carboxylase structure and activity. *Nature* **285**, 114–115.

Spreitzer, R.J. and Ogren, W.L. (1983). Rapid recovery of chloroplast mutations affecting ribulosebiphosphate carboxylase/oxygenase in *Chlamydomonas reinhardtii*. *Proc. Natl. Acad. Sci. U. S. A.* **80**, 6293–6297.

Spreitzer, R.J. and Salvucci, M.E. (2002). Rubisco: Structure, regulatory interactions, and possibilities for a better enzyme. *Annu. Rev. Plant Biol.* **53**, 449–475.

Spreitzer, R.J., Goldschmidt-Clermont, M., Rahire, M., and Rochaix, J.-D. (1985). Nonsense mutations in the *Chlamydomonas* chloroplast gene that codes for the large subunit of ribulose bisphosphate carboxylase/oxygenase. *Proc. Natl. Acad. Sci. U. S. A.* **82**, 5460–5464.

Spreitzer, R.J., Brown, T., Chen, Z., Zhang, D., and Al-Abed, S.R. (1988). Missense mutation in the *Chlamydomonas* chloroplast gene that encodes the Rubisco large subunit. *Plant Physiol.* **86**, 987–989.

Spreitzer, R.J., Thow, G., and Zhu, G. (1995). Pseudoreversion substitution at large-subunit residue 54 influences the CO_2/O_2 specificity of chloroplast ribulose-bisphosphate carboxylase/oxygenase. *Plant Physiol.* **109**, 681–685.

Spreitzer, R.J., Esquivel, M.G., Du, Y.C., and McLaughlin, P.D. (2001). Alanine-scanning mutagenesis of the small-subunit beta A-beta B loop of chloroplast ribulose-1,5-bisphosphate carboxylase/oxygenase: Substitution at Arg-71 affects thermal stability and CO_2/O_2 specificity. *Biochemistry* **40**, 5615–5621.

Spreitzer, R.J., Peddi, S.R., and Satagopan, S. (2005). Phylogenetic engineering at an interface between large and small subunits imparts land-plant kinetic properties to algal Rubisco. *Proc. Natl. Acad. Sci. U. S. A.* **102**, 17225–17230.

Stern, D.B. and Kindle, K.L. (1993). 3′ end maturation of the *Chlamydomonas reinhardtii* chloroplast *atpB* mRNA is a two-step process. *Mol. Cell. Biol.* **13**, 2277–2285.

Suay, L., Salvador, M.L., Abesha, E., and Klein, U. (2005). Specific roles of 5′ RNA secondary structures in stabilizing transcripts in chloroplasts. *Nucleic Acids Res.* **33**, 4754–4761.

Sugimoto, K., Sato, N., and Tsuzuki, M. (2007). Utilization of a chloroplast membrane sulfolipid as a major internal sulfur source for protein synthesis in the early phase of sulfur starvation in *Chlamydomonas reinhardtii*. *FEBS Lett.* **581**, 4519–4522.

Süss, K.H., Prokhorenko, I., and Adler, K. (1995). *In situ* association of Calvin cycle enzymes, ribulose-1,5-bisphosphate carboxylase/oxygenase activase, ferredoxin-NADP+ reductase, and nitrite reductase with thylakoid and pyrenoid membranes of *Chlamydomonas reinhardtii* chloroplasts as revealed by immunoelectron microscopy. *Plant Physiol.* **107**, 1387–1397.

Taylor, T.C., Backlund, A., Bjorhall, K., Spreitzer, R.J., and Andersson, I. (2001). First crystal structure of Rubisco from a green alga, *Chlamydomonas reinhardtii*. *J. Biol. Chem.* **276**, 48159–48164.

Thow, G. and Spreitzer, R.J. (1992). Missense mutations in the chloroplast *rbcL* gene that affect Rubisco holoenzyme assembly. In: *Research in Photosynthesis*, vol. 3 (N. Murata, Ed.), pp. 633–636. Kluwer Academic Publishers, Dordrecht.

Thow, G., Zhu, G., and Spreitzer, R.J. (1994). Complementing substitutions within loop regions 2 and 3 of the alpha/beta-barrel active site influence the CO_2/O_2 specificity of chloroplast ribulose-1,5-bisphosphate carboxylase/oxygenase. *Biochemistry* **33**, 5109–5114.

van der Vies, S.M., Bradley, D., and Gatenby, A.A. (1986). Assembly of cyanobacterial and higher plant ribulose bisphosphate carboxylase subunits into functional homologous and heterologous enzyme molecules in *Escherichia coli*. *EMBO J.* **5**, 2439–2444.

Wang, Z.Y., Snyder, G.W., Esau, B.D., Portis, A.R., and Ogren, W.L. (1992). Species-dependent variation in the interaction of substrate-bound ribulose-1,5-bisphosphate carboxylase/oxygenase (Rubisco) and Rubisco activase. *Plant Physiol.* **100**, 1858–1862.

Wostrikoff, K. and Stern, D. (2007). Rubisco large-subunit translation is autoregulated in response to its assembly state in tobacco chloroplasts. *Proc. Natl. Acad. Sci. U. S. A.* **104**, 6466–6471.

Ying, Z., Mulligan, R.M., Janney, N., and Houtz, R.L. (1999). Rubisco small and large subunit N-methyltransferases. Bi- and mono-functional methyltransferases that methylate the small and large subunits of Rubisco. *J. Biol. Chem.* **274**, 36750–36756.

Yosef, I., Irihimovitch, V., Knopf, J.A., Cohen, I., Orr-Dahan, I., Nahum, E., Keasar, C., and Shapira, M. (2004). RNA binding activity of the ribulose-1,5-bisphosphate carboxylase/oxygenase large subunit from *Chlamydomonas reinhardtii*. *J. Biol. Chem.* **279**, 10148–10156.

Yoshida, T. and Minamikawa, T. (1996). Successive amino-terminal proteolysis of the large subunit of ribulose 1,5-biphosphate carboxylase/oxygenase by vacuolar enzymes from French bean leaves. *Eur. J. Biochem.* **238**, 317–324.

Zhang, L., Happe, T., and Melis, A. (2002). Biochemical and morphological characterization of sulfur-deprived and H_2-producing *Chlamydomonas reinhardtii* (green alga). *Planta* **214**, 552–561.

Zhang, Z., Shrager, J., Jain, M., Chang, C.W., Vallon, O., and Grossman, A.R. (2004). Insights into the survival of *Chlamydomonas reinhardtii* during sulfur starvation based on microarray analysis of gene expression. *Eukaryotic Cell* **3**, 1331–1348.

Zhu, G. and Spreitzer, R.J. (1994). Directed mutagenesis of chloroplast ribulose-bisphosphate carboxylase/oxygenase. Substitutions at large subunit asparagine 123 and serine 379 decrease CO_2/O_2 specificity. *J. Biol. Chem.* **269**, 3952–3956.

Zhu, G. and Spreitzer, R.J. (1996). Directed mutagenesis of chloroplast ribulose-1,5-bisphosphate carboxylase/oxygenase. Loop 6 substitutions complement for structural stability but decrease catalytic efficiency. *J. Biol. Chem.* **271**, 18494–18498.

Transition Metal Nutrition: A Balance Between Deficiency and Toxicity

Marc Hanikenne[1], Sabeeha S. Merchant[2] and Patrice Hamel[3]
[1]Laboratory of Functional Genomics and Plant Molecular Imaging
University of Liège, Belgium
[2]Department of Chemistry and Biochemistry, UCLA, Los Angeles, California, USA
[3]Department of Plant Cellular and Molecular Biology and Department of Molecular and Cellular Biochemistry, The Ohio State University, Columbus, Ohio, USA

CHAPTER CONTENTS

I. INTRODUCTION

Several transition metals[1], such as zinc (Zn)[2], iron (Fe), manganese (Mn), or copper (Cu) are essential micronutrients for all organisms. Indeed, the recruitment of transition metal ions for biochemical functions is essential as evidenced by the fact that metalloproteins represent about one third of all structurally characterized proteins with a biological activity (Finney and O'Halloran, 2003). In photosynthetic organisms, transition metals act as important cofactors for many enzymes and are essential for both mitochondrial (e.g. copper in cytochrome *c* oxidase) and plastid (e.g. copper in plastocyanin) functions. Non-essential metals such as cadmium (Cd), lead (Pb), or mercury (Hg) are generally viewed as toxic at low concentrations (Warren, 1989). However, unsuspected biological functions might be unveiled for these metals as more protein structures are elucidated and analytical techniques are improved. The discovery that Cd acts as the metal cofactor of a carbonic anhydrase isoform produced under Zn deficiency in the diatom *Thalassiosira weissflogii* (Lane and Morel, 2000a) suggests that metals previously considered to be non-essential and indeed toxic may occasionally act as a prosthetic group in biologically active metalloproteins.

Their interchangeable prevalent ionic forms and affinity for functional groups occurring in proteins are unique properties of transition metals that make them useful in biochemical redox reactions. However, the same chemical properties that make metal ions useful cofactors can also be responsible for undesired reactions in a cellular environment. Indeed, because they are able to establish and maintain several stable coordinate bonds to donor atoms of organic ligands, transition metal ions can become toxic when their interaction and binding partners are not fully controlled. Metal-induced uncontrolled redox reactions or displacement of endogenous metal cofactors from their cellular binding sites can lead to cell poisoning and endanger the organism's survival (Stohs and Bagchi, 1995; Goyer, 1997). To maintain concentrations of essential metal ions within physiological limits and to cope with the toxic effects of metal overload, photosynthetic organisms have developed a tightly controlled and sophisticated metal homeostasis network that insures the balance between metal uptake, chelation, distribution and storage processes (reviewed in Clemens, 2001; Krämer and Clemens, 2005;

[1]Commonly, "transition metal" refers to the 40 elements which belong to groups 3–12 of the periodic table. In the atomic ground state, the highest-energy electron of those elements is in a d-orbital. Most of these elements have an incomplete d sub-shell, or can give rise to cations with an incomplete d sub-shell. Zinc, cadmium, and mercury possess a complete d sub-shell in most oxidation states and constitute therefore exceptions to this definition, but are generally included in the transition metals.

[2]In the chapter, the transition metal names (e.g. zinc or Zn) refer to their biologically relevant oxidation states, and not their elemental form [i.e. zinc or Zn is equivalent to Zn(II)]. For metals that can occur in different oxidation states in biological systems [e.g. Fe(II) or Fe(III), Cu(I) or Cu(II)], the oxidation state is indicated when referring explicitly to one specific oxidation state.

Colangelo and Guerinot, 2006; Grotz and Guerinot, 2006; Merchant et al., 2006; Pilon et al., 2006; Krämer et al., 2007; Puig et al., 2007).

Important progress has been achieved in our understanding of plant metal homeostasis. The completion of the *Arabidopsis* genome sequence allowed the identification of complete protein families that constitute the metal homeostasis network (with components involved mainly in metal ion transport or biosynthesis of metal-binding chelators; Axelsen and Palmgren, 2001; Mäser et al., 2001; http://www.membranetransport.org), and the roles of individual proteins in plant metal homeostasis have begun to be elucidated (Krämer, 2005; Krämer and Clemens, 2005; Williams and Mills, 2005; Colangelo and Guerinot, 2006; Grotz and Guerinot, 2006; Pilon et al., 2006; Puig et al., 2007). At the cellular level, the contribution of each member of multigene families to uptake, chelation, distribution to organelles and storage of essential metals, as well as to the detoxification of non-essential metals remains to be deciphered. As organelles of high metal ion demand due to the involvement of metal cofactors in electron transport chains, chloroplasts represent an essential component of the metal homeostasis network in photosynthetic cells (Merchant, 2006).

Investigating metal homeostasis in plants, and more generally in photosynthetic organisms, is relevant in terms of pollution management, global primary productivity, and human health. In this context, *Chlamydomonas*, as a microorganism, has served as a useful experimental model (reviewed in Merchant, 1998; Hanikenne, 2003; Merchant et al., 2006) because it is facile to manipulate the concentration of metals in the growth medium to generate metal deficient and metal overload situations. Moreover, surveys of the *Chlamydomonas* genome identified a number of transition metal transporters (Rosakis and Köster, 2004; Hanikenne et al., 2005a; Merchant et al., 2006; http://www.membranetransport.org/).

In this chapter, we describe the known components of the metal homeostasis network in *Chlamydomonas*, as well as the physiology and the molecular mechanisms of metal tolerance and metal deficiency responses.

II. COMPONENTS OF THE METAL HOMEOSTASIS NETWORK

A. Metal transporters

1. Introduction

Chlamydomonas possesses several transition metal transporters in 13 families or subfamilies of proteins (Table 10.1), potentially involved in the uptake, exclusion and/or compartmentalization of a range of essential metals such as Zn, Cu, Fe, Mn, Co, or Ni and toxic heavy metals such as Cd. Candidate transporters have been identified on the basis of sequence similarities to known transporters in *S. cerevisiae*, human and *Arabidopsis* (Table 10.2; Rosakis and Köster, 2004; Hanikenne et al., 2005a; Merchant et al., 2006;

Table 10.1 Metal transporter family or subfamily sizes in *Chlamydomonas* compared with other eukaryotes

Organism	Protein families									ABC transporters					
	CDF	ZIP	CAX	HMA	COPT CTR	FTR	NRAMP	PIC1	CCC1	MRP	ATM/ HMT	CoT	NiCoT	YSL	IREG1
S. cerevisae	5	5	4	2	2	1	3	–	1	6	1	1	–	–	–
N. crassa	9	7	8	3	2	1	2	–	1	+	2	–	1	+?	–
H. sapiens	9	14	–	2	2	–	2	–	–	12	2	–	–	–	1
C. merolae	3	4	2	2	1	4	3	1	1	2	3	–	–	–	1
Chlamydomonas	**5**	**14**	**5**	**4**	**4**	**1**	**3**	**1**	**–**	**7**	**3**	**1**	**1**	–	–
Arabidopsis	12	17	12	8	5	–	7	1	1	15	3	1–3	2	8	2–3
O. sativa	10	12	12	9	4	–	14	2	1	17	1	3	2	18	1–3

The data used to assemble this table were collected from various sources (March 2007). S. cerevisiae: PlantsT database (http://plantst.genomics.purdue.edu/); N. crassa: Blast search of genome release 3 (http://www.broad.mit.edu/annotation/genome/neurospora/Home.html) and Kiranmayi and Mohan (2006); H. sapiens: Human transporter database (http://lab.digibench.net/transporter/Family.html); **C. merolae***: the released genome annotation (http://merolae.biol.s.u-tokyo.ac.jp/) and Hanikenne et al. (2005a); Chlamydomonas: genome version 3 and Hanikenne et al. (2005a), Merchant et al. (2006), and http://www.membranetransport.org; Arabidopsis: TAIR6 annotation release (http://www.arabidopsis.org) and PlantsT database; O. sativa: Blast search of TIGR release 5 (http://www.tigr.org/tdb/e2k1/osa1/index.shtml) and Baxter et al. (2003), Jasinski et al. (2003), Klein et al. (2006) and Koike et al. (2004). +: identified by BLAST search, −: not found.*

Table 10.2 Putative metal and metalloid transporters

Protein name	Alternative name	Homologue(s)	Putative substrate(s)	Expression evidence[a]	Accession #	Homologue in *Volvox*[b]	References[c]
CDF Family (2.A.4)[d]							
MTP1	–	HsZnT4, AtMTP1	Zn(II)	qRT	EDP01521	Y	[1–3]
MTP2	–	AtMTP8, ShMtp1	Mn(II)	qRT	EDP00083	Y	[1–3]
MTP3	–	AtMTP8, ShMtp1	Mn(II)	qRT	EDP00084	Y	[1–3]
MTP4	–	AtMTP8, ShMtp1	Mn(II)	qRT	EDP00154	Y	[1–3]
MTP5	–	HsZnT9, AtMTP7	?	qRT	EDP01772	Y	[1–3]
ZIP Family (2.A.5)[d]							
ZRT1	ZIP1	ScZrt1 and 2	Zn(II)	qRT	EDP04087	N	[1, 2]
ZRT2	ZIP2	ScZrt1 and 2	Zn(II)	qRT	ED098169	N	[1, 2]
ZRT3	ZIP3	ScZrt1 and 2	Zn(II)	qRT	EDP08759	N	[1, 2]
ZRT4	ZIP9	ScZrt3	Zn(II)	qRT	ED098317	N	[1, 2]
ZRT5	ZIP4	ScZrt1 and 2	Zn(II)	qRT	ED097271	N	[1, 2]
IRT1	ZIP11	?	Fe(II)	qRT	EDP03614	Y	[1, 2]
IRT2	ZIP10	?	Fe(II)	qRT	EDP03614	Y	[1, 2]
ZIP1	ZIP5	ScZrt1 and 2	Zn(II)	qRT	ED097270	N	[1, 2]
ZIP2	ZIP13	ScZrt3	Zn(II)	qRT	EDP08783	N	[1, 2]
ZIP3	ZIP8	ScZrt3	Zn(II)	qRT	EDP04356	Y	[1, 2]
ZIP4	ZIP12	ScZrt3	Zn(II)	qRT	EDP02242	N	[1, 2]
ZIP6	ZIP6	HsZip1 and 3	Zn(II)	qRT	EDP05024	Y	[1, 2]
ZIP7	ZIL1	?	?	qRT	EDP06917	N	[1, 2, 4]
ZIP14	ZIL2	?	?	–	EDP01584	Y	[1, 2, 4]
CAX family (2.A.19)[d]							
CAX1	–	ScVcx1	?	EST	EDP00630	Y	[1]
CAX2	–	ScVcx1	?	EST	EDP00795	Y	[1]
CAX3	–	?	?	–	EDP03733	Y	[1]
CAX4	–	AtCAX7 to 9	?	–	ED099207	Y	[1]
CAX5	–	?	?	EST	ED099460	Y	[4]
HMA subfamily of P–type ATPase (3.A.3)[d]							
HMA1	–	AtHMA1	Cu(II)	EST	EDP00361	Y	[1, 2]
CTP1	HMA2	ScCcc2, Z AtRANR1	Cu(I)	qRT	ED098907	Y	[1, 2, 5]
CTP2	HMA3	AtPAA1 and 2	Cu(I)	qRT	EDP06350	Y	[1, 2]
CTP3	–	AtPAA1 and 2	Cu(I)	qRT	EDP00070	Y	[2]
Copper transporters (9.A.12)[d]							
COPT1	–	AtCOPTs	Cu(I)	qRT	EDP07016	Y	[1, 2]
CTR1	–	DdCTRs	Cu(I)	qRT	EDP08980	Y	[2]
CTR2	–	DdCTRs	Cu(I)	qRT	EDP06249	Y	[2]
CTR3	–	DdCTRs	Cu(I)	qRT	EDP06429	Y	[4]

(Continued)

Table 10.2 *Continued*							
Protein name	**Alternative name**	**Homologue(s)**	**Putative substrate(s)**	**Expression evidence**[a]	**Accession #**	**Homologue in *Volvox***[b]	**References**[c]
FTR family (9.A.10.1)[d]							
FTR1	–	ScFet3	Fe(III)	RNA blot	EDP03271		[1, 2, 5]
NRAMP family (9.A.55)[d]							
NRAMP1	DMT1	ScSmf1 and 3	Mn(II) Fe(II)?	qRT	EDP04810	N	[1–3, 6]
NRAMP2	–	ScSmf1 and 3	Mn(II) Fe(II)?	qRT	EDP07304	Y	[1–3]
RET1	NRAMP3	AtEIN2	?	qRT	EDP07047	Y	[4]
PIC1 family							
PIC1	–	AtPIC1	Fe(II)?	qRT	EDP08036	Y	[1]
ABC transporter family (3.A.1)[d]							
MRP subfamily							
MRP1	HLA3	HsABCCs	bicarbonate	RNA blot	EDP07736	Y	[1, 7]
MRP2	–	ScYcf1	GSH–Cd	qRT	EDP03509	Y	[1, 8]
MRP3	–	AtMRP11	?	EST	EDP08676	Y	[1]
MRP4	–	AtMRP11	?	–	EDP06139	Y	[1]
MRP5	–	AtMRP11	?	–	EDP07482	Y	[1]
MRP6	–	AtMRP11	?	EST	EDP01107, EDP01108	Y	[1]
MRP7	–	AtMRP11	?	–	EDP09576	Y	[1]
ATM/HMT subfamily							
CDS1	ATM/HMT1	SpHmt1, HsAbcB6	Cd(II), Fe(II)?	RNA blot	ED096584	Y	[1, 9]
ATM/ HMT2	–	SpHmt1, HsAbcB6	Cd(II), Fe(II)?	–	EDP00522	Y	[1]
ATM/ HMT3	–	AtATM3, HsAbcB7	Cd(II), Fe(II)?	–	EDP05599	Y	[1]
CoT subfamily							
COT1	–	At3g21580	Co(II)	qRT	EDP03572	Y	[4]
NiCoT family (2.A.52)[d]							
NIK1	–	AtNiCoT1 and 2	Ni(II)	qRT	EDP03691	Y	[4]
ArsAB family (3.A.4)[d]							
ARSA1	–	At3g10350	arsenite	EST	ED097076	N	[4]
ARSA2	–	At1g01910	arsenite	–	EDP03358	N	[4]
ARSB	–	Bacterial ArsB	arsenite	–	EDP02356	Y	[4]

[a] *Expression evidence: EST (Expressed Sequence Tag), qRT (quantitative real-time RT-PCR).*

[b] *Presence of homologous sequence in Volvox carteri, vista Volvox track at http://genome.jgi-psf.org/Chlre3/Chlre3.home.html.*

[c] *References: [1] Hanikenne et al. (2005a); [2] Merchant et al. (2006); [3] Allen et al. (2007a); [4] http://genome.jgi-psf.org/Chlre3/Chlre3.home.html, [5] La Fontaine et al. (2002b); [6] Rosakis and Koster (2005); [7] Im and Grossman (2002); [8] Wang and Wu (2006); [9] Hanikenne et al. (2005b).*

[d] *Transporter Classification System (Busch and Saier, 2002). Note that the annotations provided in references 1 and 2 were based on the release 2 of the genome. At, Arabidopsis thaliana; Dd, Dictyostelium discoideum, Hs, Homo sapiens; Sc, Saccharomyces cerevisiae; Sh, Stylosanthes hamata; Sp, Schizosaccharomyces pombe.*

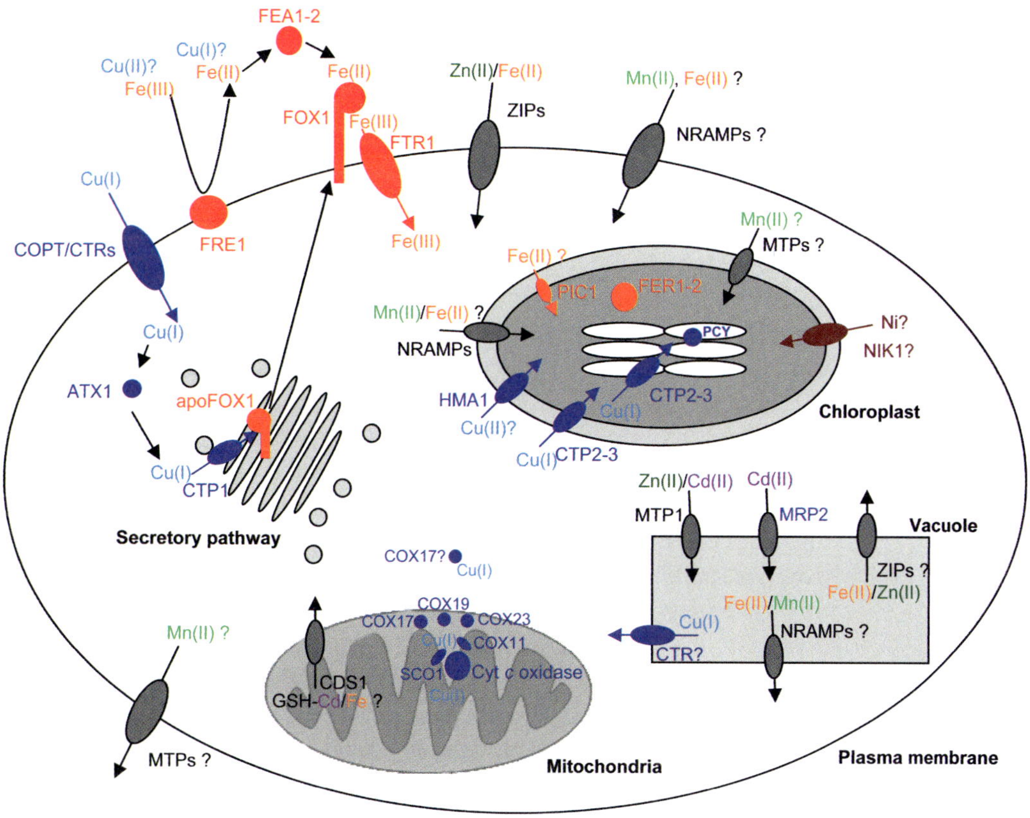

FIGURE 10.1 *Components of the metal homeostasis network. A schematic cell is represented with putative metal transporters, metal-binding proteins and chaperones (see detailed explanations in the text). Note that the nucleus is not represented. Colors are used to distinguish different metals and protein families: blue for copper, red for iron, dark green for zinc, light green for manganese, purple for cadmium, and brown for nickel. Numbers in parentheses indicate the oxidative states of the metal ions, whereas arrows show the proposed directions of transport. Uncertain components or steps are indicated with question marks. ATX1, Copper chaperone; CDS1, Cadmium-sensitive 1; COPT/CTR, Copper Transporters; COX and SCO1: chaperones for copper delivery and insertion into the cytochrome c oxidase; CTP, Copper-transporting P-type ATPases; FEA, Fe-assimilating proteins; FER, Ferritin; FOX1, Multicopper Ferroxidase 1, apoFOX1, precursor form of FOX1 prior to Cu insertion in the secretory pathway; FRE1, Ferrireductase 1; FTR1, Fe Transporter; GSH, glutathione; HMA1, Heavy Metal P-type ATPase 1; MRP2, Multidrug-Resistance associated Protein 2; MTP, Metal Tolerance Proteins; NIK, Nickel Transporter; NRAMP, Natural Resistance-Associated Macrophage Proteins; PCY, plastocyanin; PIC1, Permease in Chloroplast 1; ZIP, ZRT-, IRT-like proteins.*

http://www.membranetransport.org/). A comparison of the *Chlamydomonas* candidate transporters with the repertoire of transporters in other photosynthetic and non-photosynthetic model organisms is also given in Table 10.1. Most of the *Chlamydomonas* proteins have not been functionally characterized and at present the roles of those proteins can only be inferred from phylogenetic analysis of different families of known transporters (Hanikenne et al., 2005a), experimental data derived from functional studies with *Saccharomyces cerevisiae* and *Arabidopsis* homologues, and expression patterns in response to nutrient deficiency (Figure 10.1; La Fontaine et al., 2002b; Merchant et al., 2006; Allen et al., 2007a,b).

2. CDFs (cation diffusion facilitators)

The CDFs are a family of ubiquitous transporters that catalyze the efflux of a range of transition metals [such as Zn(II), Cd(II), Co(II), or Mn(II)] from the cytoplasm to the extracellular milieu or into subcellular compartments (e.g. vacuole or ER; Gaither and Eide, 2001). The plant members of the CDF family have been named MTPs (Metal Tolerance Proteins; Delhaize et al., 2003), and the same nomenclature was chosen to name the five *Chlamydomonas* candidate MTPs (MTP1–MTP5, Table 10.2; Hanikenne et al., 2005a).

Chlamydomonas MTP1 is closely related to a subgroup of Zn-transporting CDFs of plants (AtMTP1–AtMTP4), *S. cerevisiae* (Zrc1 and Cot1) and humans (HsZnT2–HsZnT4 and HsZnT8; Hanikenne et al., 2005a). AtMTP1 and AtMTP3 are both involved in Zn storage in the vacuole (van der Zaal et al., 1999; Kobae et al., 2004; Desbrosses-Fonrouge et al., 2005; Krämer, 2005; Arrivault et al., 2006). The *MTP1* gene is constitutively highly expressed in metal Zn/Cd hyperaccumulators such as *Arabidopsis halleri* and *Thlaspi caerulescens*, and is proposed to play an essential role in the metal hypertolerance of these species (Dräger et al., 2004; Becher et al., 2004; Craciun et al., 2006; Talke et al., 2006; van de Mortel et al., 2006). In *S. cerevisiae*, Zrc1 and Cot1 are two vacuolar transporters involved in Zn, Co and possibly Cd storage and detoxification (Gaither and Eide, 2001). Similarly to its plant and *S. cerevisiae* homologues, *Chlamydomonas* MTP1 is predicted to localize to the vacuolar membrane (Hanikenne et al., 2005a) and is likely to contribute to Zn storage in the vacuole. The three ESTs corresponding to MTP1 originate from the stress II cDNA library, which included exposure of cells to Cd among several different stress conditions, suggesting that MTP1 might also be involved in the response to Cd toxicity (Hanikenne et al., 2005a).

Chlamydomonas MTP2–MTP4 are related to a subgroup of putative plant Mn-transporting CDFs (Hanikenne et al., 2005a). The best characterized member of this subgroup is ShMTP1 that was proposed to be involved in the vacuolar storage of Mn in the Mn-tolerant legume *Stylosanthes hamata* (Delhaize et al., 2003). The *MTP11* gene has been shown to contribute to Mn tolerance in *Arabidopsis* (Peiter et al., 2007; Delhaize et al. 2007). The *Chlamydomonas MTP2–MTP4* genes arose through presumably recent gene duplications (in the ancestor common to *Volvox* and *Chlamydomonas*) and are found in a tandem organization in the genome. The corresponding proteins might be involved in vacuolar Mn storage, and perhaps in Mn delivery to the chloroplast (Figure 10.1).

Chlamydomonas MTP5 is related to *Arabidopsis* MTP7 and human ZnT9, whose functions are unknown. The *MTP5* gene was shown to be upregulated in Mn deficiency (Allen et al., 2007a). Finally, *Chlamydomonas* lacks homologs to (i) Zn-transporting CDFs involved in Zn delivery or sequestration in the secretory pathway, such as *S. cerevisiae* Msc2 and Zrg17 (Ellis et al., 2004, 2005) or human ZnT5 and 6 (Palmiter and Huang, 2004; Ellis et al., 2005); and (ii) to *S. cerevisiae* Mmt1 and Mmt2, two mitochondrial proteins that might be involved in Fe homeostasis (Li and Kaplan, 1997).

3. ZIPs (Zrt-, Irt-like proteins)

The ZIPs are another family of ubiquitous transporters that catalyze the movement of transition metals [such as Fe(II), Zn(II), Cd(II), or Mn(II)] from outside the cell or a subcellular compartment to the cytoplasm (Guerinot and Eide, 1999; Gaither and Eide, 2001). The ZIP proteins have been classified into four subfamilies based on sequence conservation: subfamily I groups most of the plant and *S. cerevisiae* ZIPs, subfamily II mainly includes mammalian proteins, the GufA subfamily contains prokaryotic and eukaryotic ZIPs of unknown function, and the Liv1 subfamily appears to be eukaryote-specific (Guerinot and Eide, 1999; Gaither and Eide, 2001). The *Chlamydomonas* ZIPs were first named based on their phylogenetic relationships (Hanikenne et al., 2005a), but later renamed according to their metal responsiveness at the transcriptional level (Merchant et al., 2006). Both nomenclatures are given in Table 10.2, but for clarity the latter has been adopted in this manuscript. *ZRT1–ZRT5* are up-regulated under Zn deficiency, whereas *IRT1* and *IRT2* are up-regulated under Fe deficiency, supporting a proposed function in Zn and Fe assimilation, respectively. Based on their sequences, the *ZIP1–6*, *ZIP7* and *ZIP14* genes belong to the ZIP family, but are not transcriptionally regulated by the metal nutritional status.

Chlamydomonas ZRT1–ZRT3, ZRT5, and ZIP1 belong to subfamily I and are related to most of the *Arabidopsis* (ZIP1–ZIP12 and IRT1–IRT3) and *S. cerevisiae* ZIPs (Zrt1 and Zrt2; Hanikenne et al., 2005a). Within this subfamily, transporters of the three species form distinct groups, suggesting that the diversification occurred independently in the subgroups through duplications of a common ancestral gene. The *ZRT5* and *ZIP1* genes are found in tandem in the *Chlamydomonas* genome. However, the two genes display different expression patterns, which is indicative of a diversification of function.

In *S. cerevisiae*, Zrt1 and Zrt2 are located in the plasma membrane and mediate high- and low-affinity Zn uptake, respectively (Zhao and Eide, 1996a,b). In *Arabidopsis*, ZIP1 to ZIP4 are involved in cellular Zn uptake and IRT1 constitutes the major Fe uptake system in roots where the gene is induced by Fe deficiency (Grotz and Guerinot, 2006). Several ZIP genes are constitutively expressed at high levels in *A. halleri* (Becher et al., 2004; Weber et al., 2004; Talke et al., 2006) and *T. caerulescens* (Lasat et al., 2000; Pence et al., 2000; Assuncão et al., 2001; van de Mortel et al., 2006) and are postulated to function in Zn hyperaccumulation. Based on the function of their homologues and their transcriptional regulation, the

Chlamydomonas subfamily I ZIPs represent likely candidates for a role in cellular Zn uptake (Figure 10.1).

Chlamydomonas ZIP6 belongs to subfamily II and is related to the mammalian ZIP1–ZIP3, which are involved in Zn uptake (Eide, 2006), whereas *Chlamydomonas* ZRT4 and ZIP2–ZIP4 cluster with members of the GufA subfamily (Hanikenne et al., 2005a). The only characterized member of this family is *S. cerevisiae* Zrt3, which remobilizes vacuolar Zn upon Zn deficiency (MacDiarmid et al., 2000). Finally, no member of the Liv1 subfamily was found in *Chlamydomonas* (Hanikenne et al., 2005a). The human LIV1 ZIPs were shown to be capable of Zn transport, and mutations in them have been implicated in various cancers, the control of the cell cycle and cell proliferation, or the control of neurodegeneration and growth (Taylor et al., 2003, 2004).

The two *Chlamydomonas* Fe-regulated ZIPs, IRT1, and IRT2, also group with the GufA proteins (Hanikenne et al., 2005a). Although they are not related to AtIRT1, they might have similar functions. Their up-regulation under Fe deficiency suggests that they might also contribute to Fe uptake. Alternatively, they might be involved in Fe remobilization from intracellular storages (e.g. vacuoles) or between one organelle (chloroplast) and another (mitochondrion). Finally, the ZIP7 and ZIP14 proteins are very distantly related to other ZIP proteins and should be considered as putative ZIPs (Table 10.2).

Compared to other unicellular organisms, *Chlamydomonas* possesses a high number of ZIP proteins with as many proteins of this family as vascular plants and humans (Table 10.1). It is noteworthy that ZIP diversification occurred in different subfamilies in different species: in subfamily I for *Arabidopsis*, Liv1 for humans, and GufA and subfamily I for *Chlamydomonas*. This evolution from a common set of ancestral genes represents a good example of convergent adaptation where the need for a similar function (i.e. cellular Zn uptake and distribution) was achieved via the diversification of different protein subfamilies.

4. CAXs (cation exchanger)

The CAX proteins are divalent cation/proton antiporters (Shigaki and Hirschi, 2006). A *vcx1 S. cerevisiae* mutant (vacuolar proton/Ca(II) exchanger) lacking a CAX protein is defective in vacuolar Ca accumulation (Cunningham and Fink, 1996). The *Arabidopsis* AtCAX1 and AtCAX2 proteins are homologues of Vcx1 and are high and low-affinity Ca(II)/proton vacuolar antiporters, respectively (Hirschi et al., 1996). AtCAX2 has also been proposed to transport Mn(II) and Cd(II) across the tonoplast of plant cells: tobacco plants overexpressing *AtCAX2* accumulate more Ca, Mn, and Cd and are more tolerant to high Mn (Hirschi et al., 2000). Root tonoplast vesicles of tobacco expressing AtCAX2 display higher transport activity of Mn and Cd (Hirschi et al., 2000). The roles of the nine other *Arabidopsis* CAXs remain unclear. Finally, the CAX protein AtMHX1 is a proton-coupled antiporter that transports Mg and Zn across the vacuolar membrane (Shaul et al., 1999).

Five CAXs (CAX1–CAX5) are encoded in the *Chlamydomonas* genome (Hanikenne et al., 2005a): CAX1 and CAX2 are highly related to *S. cerevisiae* Vcx1 and AtCAX1–AtCAX6, whereas the other CAXs are more distantly related. *Chlamydomonas CAX1* transcripts are highly represented in the cDNA libraries Stress II and III, which include cDNAs prepared from Cd-treated cells and Cu or Fe deficient cells, respectively (Shrager et al., 2003), consistent with the idea that CAX1 is involved in transition metal homeostasis or detoxification. No homologue of AtMHX1 has been found in the *Chlamydomonas* genome. The *AtMHX1* gene is mainly expressed in the vascular tissues and might be involved in Zn(II)/Mg(II) partitioning within plant organs (Shaul et al., 1999).

5. HMAs (heavy metal-transporting P-type ATPases)

Proteins of the I_B subfamily P-type ATPases (HMAs or CPx ATPases) are polytopic proteins with eight transmembrane segments that operate in transition metal transport. Their sixth transmembrane domain contains a conserved Cys-Pro-Cys/His/Ser motif (CPx motif) believed to participate in metal cation translocation across the membrane. HMAs are found in both prokaryotes and eukaryotes and are divided in two groups based on their substrate specificity: monovalent Cu(I)/Ag(I) cations or divalent Zn(II)/Cd(II)/Co(II)/Pb(II) cations (Cobbett et al., 2003; Williams and Mills, 2005).

Four HMAs are encoded in *Chlamydomonas* (Hanikenne et al., 2005a; Merchant et al., 2006). HMA1 belongs to the divalent cation-transporting subgroup that is only found in prokaryotes and plants (Williams and Mills, 2005), and is more specifically related to AtHMA1. Like AtHMA1, *Chlamydomonas* HMA1 has an uncharacteristic Ser/Pro/Cys motif in the sixth predicted transmembrane domain instead of the common Cys-Pro-Cys/His/Ser motif. This property is shared with the Co transporter CoaT of the cyanobacterium *Synechocystis* and might determine the substrate specificity of the transporter (Cobbett et al., 2003). AtHMA1 has been shown to be a plastid Cu transporter (Seigneurin-Berny et al., 2006; Puig et al., 2007).

Chlamydomonas CTP1, CTP2, and CTP3 are putative Cu-transporting proteins from the monovalent metal transporter subfamily of HMAs (Hanikenne et al., 2005a; Merchant et al., 2006). Together with the Cu chaperone ATX1, *Chlamydomonas* CTP1 was proposed to play a role in Cu delivery to the secretory pathway (Figure 10.1; section II.B.3; La Fontaine et al., 2002b), as has been functionally demonstrated for the closely related Ccc2 in *S. cerevisiae*, RAN1 in *Arabidopsis* and the Menkes and Wilson ATP7A and B in humans (Askwith et al., 1996; La Fontaine et al., 2002a,b; Lutsenko and Petris, 2003; Puig et al., 2007). Atx1 is required in *S. cerevisiae* for delivery of cytosolic Cu to apo-ferroxidase via P-type ATPase-dependent transport of Cu in a post-Golgi vesicle (Lin et al., 1997; Yuan et al., 1997;

Himelblau et al. 1998). A *Chlamydomonas* gene encoding an Atx1 ortholog has been cloned and shown to rescue high-affinity Fe uptake when expressed in an *atx1 S. cerevisiae* mutant (La Fontaine et al., 2002b). A role for CTP1 in Cu tolerance by mediating Cu efflux from the cells, as suggested for AtHMA5, should also be considered (Andrés-Colás et al., 2006).

Chlamydomonas CTP2 and CTP3 are related to the *Arabidopsis* chloroplast proteins PAA1 and PAA2 (Hanikenne et al., 2005a; Merchant et al., 2006). AtPAA1 is localized in the outer membrane of the chloroplast envelope and has been proposed to transport Cu into the stroma (Shikanai et al., 2003), whereas AtPAA2 resides in the thylakoid membrane and is postulated to drive Cu transport to the lumen for delivery to plastocyanin (Abdel-Ghany et al., 2005b). Consistent with these hypotheses, a *paa1* mutant is defective for both stromal Cu/Zn superoxide dismutase and plastocyanin, whereas a *paa2* mutant only lacks plastocyanin (Abdel-Ghany et al., 2005b). Although it is likely that CTP2 and CTP3 play similar roles in *Chlamydomonas* (Figure 10.1), their subcellular localizations and Cu delivery activities remain to be established experimentally.

6. Copper transporters (COPTs and CTRs)

Chlamydomonas encodes four putative Cu transporters (COPT1 and CTR1–CTR3) related to *S. cerevisiae* Ctr1–Ctr3 and *Arabidopsis* COPT1–COPT5 (Hanikenne et al., 2005a; Merchant et al., 2006). In *S. cerevisiae*, Cu uptake is mediated by plasma membrane reductases that reduce Cu(II) to Cu(I) (see also section IV.B.1) and Ctr1 and Ctr3-dependent high-affinity Cu(I) transport, which uses a highly specific Cu(I)/2 K(I) antiport mechanism. The role of the vacuolar *S. cerevisiae* Ctr2 remains unclear (Eide, 1998; Puig et al., 2007). Similar proteins are involved in Cu uptake in mammalian cells as well (Petris, 2004). *Chlamydomonas* COPT1 is predicted to localize to the plasma membrane, supporting a role in Cu uptake (Figure 10.1; Hanikenne et al., 2005a).

7. FTRs (Fe transporters)

In *Chlamydomonas*, the high-affinity Fe uptake system involves a plasma membrane ferric-chelate reductase (FRE1), a multicopper ferroxidase (FOX1) and an iron permease (FTR1; Herbik et al., 2002a; La Fontaine et al., 2002b; Merchant et al., 2006; Allen et al., 2007b). This system will be described in more detail in section IV.A. FTR1 is homologous to the *S. cerevisiae* high-affinity Fe permease Ftr1 (Radisky and Kaplan, 1999) and is predicted to localize to the plasma membrane (Figure 10.1; Hanikenne et al., 2005a).

8. NRAMPs (natural resistance-associated macrophage proteins)

Proteins of the NRAMP family participate in the transport of divalent metal cations. The human NRAMP1 restricts metal availability for pathogens by transporting Mn(II) and Fe(II) across the phagosomal membranes of macrophages, and is therefore required for natural resistance to intracellular bacterial infection (Mackenzie and Hediger, 2004). The second human NRAMP HsNRAMP2/HsDMT1 is involved in dietary Fe(II) uptake at the apical surface of enterocytes. It is also able to transport additional divalent metal cations [Cd(II), Co(II), Cu(II), Zn(II), Mn(II) or Pb(II)] in addition to Fe(II) (Mackenzie and Hediger, 2004). Out of six *Arabidopsis* NRAMPs, only NRAMP1, NRAMP3, and NRAMP4 have been functionally characterized. The three genes are up-regulated in Fe-deficiency and the corresponding proteins were shown to mediate Fe(II), Mn(II), and Cd(II) transport in *S. cerevisiae* (Curie et al., 2000; Thomine et al., 2000). AtNRAMP1 might contribute to Fe storage in the chloroplast, whereas AtNRAMP3 and AtNRAMP4 are involved in the remobilization of vacuolar Fe under deficiency (Thomine et al., 2003; Lanquar et al., 2005).

Arabidopsis possesses an additional protein related to NRAMPs called EIN2 (*E*thylene *In*sensitive 2), which is involved in ethylene responses and signaling (Alonso et al., 1999; Benavente and Alonso, 2006). The protein does not seem to be capable of metal transport activity. Surprisingly, *Chlamydomonas* RET1 shows homology to EIN2 (Table 10.2). As no ethylene signaling exists in *Chlamydomonas*, the role of RET1 is unclear, and it may be involved in some other signal transduction pathway.

The *Chlamydomonas* NRAMP1 and NRAMP2 proteins are related to *S. cerevisiae* Smf1–Smf3 (Hanikenne et al., 2005a) which are Mn, Cu and to a lesser extent Fe transporters. Smf1 localizes to the plasma membrane, whereas Smf2 resides in intracellular vesicle membranes, and Smf3 in the vacuolar membrane (Cohen et al., 2000). *Chlamydomonas* NRAMP1 was initially named DMT1, and complementation data in *S. cerevisiae* suggest that it is capable of Mn and possibly Fe transport (Rosakis and Köster, 2005). Both *NRAMP1* and *NRAMP2* genes are slightly induced by Mn deficiency (Allen et al., 2007a). The *Chlamydomonas* NRAMPs might be involved in Mn, and possibly Fe uptake, or in the remobilization of these metals from the vacuole under conditions of metal deficiency in *Chlamydomonas*. As in the case of Smf2 (Culotta et al., 2006), it is possible that one of the *Chlamydomonas* NRAMPs operates in the delivery of Mn to Mn superoxide dismutases.

9. PIC1 (permease in chloroplast1)

Chlamydomonas PIC1 is related to plant and cyanobacterial permeases (Duy et al., 2007). *Arabidopsis* PIC1 localizes to the inner chloroplastic membrane. *Arabidopsis pic1* mutants are dwarfed and chlorotic, and constitutively upregulate metal transport (*NRAMPs*, *IRT1*, *YSL1*) and homeostasis (*FER1*, *FER4*) genes. Moreover, PIC1 and its *Synechocystis* homologue complement the growth of metal uptake-defective *S. cerevisiae* mutants. Taken together, these data suggest that the PIC1 proteins participate in chloroplast Fe import (Figure 10.1; Duy et al., 2007).

10. ABC (ATP-binding cassette) transporters

a. Introduction

ABC transporters form a large and ubiquitous superfamily of transporters that participate in a wide range of physiological processes. The transport process is energized by ATP (reviewed in Holland et al., 2003). ABC transporters can be divided in several subfamilies based on structural similarities (Decottignies and Goffeau, 1997; Sánchez-Fernández et al., 2001; Dean et al., 2003; Rea, 2007). Here, presentation of the *Chlamydomonas* ABC transporter superfamily will be restricted to the members of the MRP and ATM/HMT subfamilies because members of these subfamilies contribute to metal transport in plants and *S. cerevisiae. Chlamydomonas* also encodes a relative of prokaryotic Co-transporting ABC proteins (Eitinger et al., 2005; Rodionov et al., 2006).

b. MRPs (multidrug-resistance associated proteins)

MRPs are full-size ABC transporters acting as glutathione-*S*-conjugate pumps (Rea et al., 1998). Two *S. cerevisiae* MRPs, namely Ycf1 (for *y*east *c*admium *f*actor) and Bpt1 (for *b*ile *p*igment *t*ransporter) are involved in heavy metal detoxification. Ycf1[3] transports glutathione (GSH)-Cd, -arsenate or -Hg complexes from the cytoplasm to the vacuole (Szczypka et al., 1994; Z.S. Li et al., 1996, 1997; Ghosh et al., 1999; Gueldry et al., 2003), and the vacuolar Bpt1 also contributes, although marginally, to Cd tolerance (Klein et al., 2002; Sharma et al., 2002). In *Arabidopsis*, MRP3 is the only MRP (out of 14) that is involved in Cd tolerance. MRP3 is upregulated by Cd and the corresponding protein complements the Cd-sensitive phenotype of a *S. cerevisiae ycf1* mutant (Bovet et al., 2003; Tommasini et al., 1998). The other characterized AtMRPs contribute to the transport of various substrates (e.g. GSH conjugates, glucuronide conjugates, or chlorophyll catabolites) and are involved in a range of physiological processes (reviewed in Klein et al., 2006; Schulz and Kolukisaoglu, 2006; Rea, 2007).

Of the seven *Chlamydomonas* MRPs, only *MRP1* and *MRP2* have been cloned and functionally characterized. MRP1 localizes to the chloroplast, and *MRP1* is regulated at the transcriptional level by light and CO_2. It is hypothesized to be involved in bicarbonate uptake (Im and Grossman, 2002; Grossman et al., 2007; see also Chapter 8). MRP2 is related to *S. cerevisiae* Ycf1 and is able to complement the Cd-sensitive phenotype of a *S. cerevisiae ycf1* mutant, which suggests a vacuolar localization. The *MRP2* gene is up-regulated by Cd. A *mrp2* mutant is Cd-sensitive and presents a modified phytochelatin (PC)-Cd[4] complex accumulation pattern, with a faster rate of low-molecular weight PC-Cd complex formation and

[3]Note that the yeast Ycf1 vacuolar protein is different from *Chlamydomonas* chloroplast Ycf1 (Boudreau et al., 1997)

[4]PC: phytochelatin. PCs are small metal binding peptides (see section II.B.1).

increased Cd sequestration by the PC-Cd complexes (Wang and Wu, 2006). These data suggest that MRP2 contributes to Cd vacuolar storage. Based on sequence similarity, it is likely that *Chlamydomonas* MRP3–MRP7 are glutathione-*S* conjugate pumps involved in xenobiotic and/or metal tolerance, but this will have to be determined experimentally.

c. ATM/HMT (ABC transporter of the mitochondria/heavy metal tolerance)

ATM/HMTs are half-size ABC transporters (Rea, 2007) that localize to the mitochondrial inner membrane or the vacuolar membrane. Mitochondrial transporters (HsABCB6 and HsABCB7, AtATM1–AtATM3 and ScAtm1) are involved in the biogenesis of iron/sulfur clusters and in Fe homeostasis (Kispal et al., 1997, 1999; Csere et al., 1998; Allikmets et al., 1999; Mitsuhashi et al., 2000; Kushnir et al., 2001; Rea, 2007), whereas vacuolar transporters (SpHMT1 and CeHMT-1) are involved in the transport of PC-Cd complexes from the cytoplasm into the vacuole and contribute to Cd tolerance (Ortiz et al., 1992, 1995; Vatamaniuk et al., 2005; Rea, 2007).

A mitochondrial ATM/HMT ortholog, named CDS1, has been studied in *Chlamydomonas* (Hanikenne et al., 2005b). The *CDS1* gene is strongly induced by Cd exposure and a *cds1* mutant is sensitive to Cd and high Fe concentrations. Two possible models can be proposed to explain the phenotype of *cds1*. First, CDS1 could be directly involved in the export of Cd out of the mitochondrial matrix, possibly as a glutathione-Cd conjugate, thereby protecting mitochondrial function from Cd toxicity. Alternatively, the Cd- and Fe-sensitive phenotype of the *cds1* mutant could be an indirect consequence of a modification to Fe homeostasis in the mitochondria. The lack of a mitochondrial CDS1 transporter could indeed result in cytosolic Fe deficiency, as suggested previously for the *S. cerevisiae* mutant *atm1*, which displays both mitochondrial Fe accumulation and cytosolic Fe deficiency phenotypes (Schueck et al., 2001). In vascular plants, Fe deficiency can often lead to an increased uptake of Cd(II) and hence sensitivity due to the induction of the Fe(II) uptake systems (Thomine et al., 2000, 2003; Connolly et al., 2002; Lombi et al., 2002).

Arabidopsis ATM2 and ATM3 are related to *Chlamydomonas* CDS1, with ATM3 being the closer homologue (Hanikenne et al., 2005a,b). D.Y. Kim et al. (2006) showed that the *Arabidopsis ATM3* gene is up-regulated by Cd and that an *atm3* mutant is Cd-sensitive. On the other hand, over-expressing plants display an increased resistance to Cd. The expression analysis of glutathione biosynthesis genes and the determination of non-protein thiol content in the mutant support a role of AtATM3 in the export of glutathione-Cd chelates from mitochondria. The ability of mitochondrial ABC transporters to act in Cd detoxification could therefore be a newly discovered property among photosynthetic organisms.

No gene encoding a vacuolar PC-Cd-transporting ABC transporter has been identified in photosynthetic organisms, despite the fact that such a transport function has been biochemically identified in plant vacuolar membranes (Vögeli-Lange and Wagner, 1990; Salt and Rauser, 1995; Rea, 2007).

d. CoT (cobalt transporter)

Chlamydomonas COT1 is related to putative prokaryotic and plant Co-transporting ABC transporters (Eitinger et al., 2005; Rodionov et al., 2006), with the closest homologues identified to date being found in the marine green alga *Ostreococcus tauri*, *Arabidopsis*, rice, cyanobacteria, and bacteria. The plant and algal proteins are predicted to localize either to the chloroplast or mitochondrion. It is worth considering whether Co may substitute for another metal in *Chlamydomonas* as it does in marine organisms (Lane and Morel, 2000b). Although there are cobalamin-dependent enzymes in *Chlamydomonas*, the cofactor is obtained from bacteria that associate with the alga (Croft et al., 2005). In plants, Co is considered a beneficial element, but no precise function has been established (Marschner, 1995; Krämer and Clemens, 2005). Further experiments will be required to establish the substrate specificity of COT1 and its plant homologues, and to determine the function of this putative Co transporter.

11. NiCoT

In anaerobic bacteria and cyanobacteria, Ni is found in hydrogenases, dehydrogenases, and methyl reductases, whereas urease is the only known Ni-requiring enzyme in vascular plants (Eitinger et al., 2005; Krämer and Clemens, 2005). Ni is an essential micronutrient in vascular plants and accordingly, Ni deficiency induces the accumulation of toxic urea concentrations in several species (Krämer and Clemens, 2005). Although no urease homologue has been identified in the annotated *Chlamydomonas* proteins, and no Ni-Fe hydrogenase activity has been found, *Chlamydomonas* does have a putative high-affinity Ni transporter, NIK1 (Table 10.2) and a putative Ni chaperone for urease or hydrogenase (accession EDP04575) similar to the urease accessory protein UreG found in bacteria and *Arabidopsis* (Krämer and Clemens, 2005). Danilov and Ekelund (2001) have shown that high Ni concentrations had stimulatory effects on photosynthesis efficiency in *Chlamydomonas*. Moreover, Quinn et al. (2003) found that Ni treatment causes a CuRE[5]-dependent upregulation of *CYC6*, *CPX1*, and *CRD1*, mimicking their induction under Cu deficiency (see also section IV.B.3.d). The authors present two models to explain the effect of Ni, either

[5]Cu-responsive element (see section IV.B.3.d).

that the target genes and Ni metabolism might be physiologically or functionally connected, or that Ni might interfere with Cu metabolism, possibly by interacting with the Cu signaling system.

12. ArsAB family

Toxic to all living forms, the metalloid arsenic is widely distributed in water, air, and soil (reviewed in Nriagu and Pacyna, 1988; Jones, 2007; Tripathi et al., 2007). In humans, chronic arsenic exposure leads to digestive disorders and cancers (Jones, 2007). Arsenate [As(V)] and arsenite [As(III)] represent the two biologically important species of arsenic, which are interconvertible depending on the redox status. In plants as in *S. cerevisiae*, arsenate enters cells via phosphate transporters, is subsequently reduced to arsenite by arsenate reductase, and is then excluded from the cell or stored in the vacuole (Rosen, 2002; Tripathi et al., 2007).

In *Chlamydomonas*, an arsenate-resistant mutant was shown to be impaired in a putative Na^+/Pi co-transporter (see section III.B, Fujiwara et al., 2000; Kobayashi et al., 2003), suggesting that the transporter is an entry route for arsenate. Once in the algal cell, arsenate does not induce phytochelatin synthesis (see section II.B.1, Kobayashi et al., 2006). However, Cd-induced GSH and/or phytochelatin accumulation was shown to increase *Chlamydomonas* arsenate tolerance, suggesting that GSH and/or phytochelatin can interact with arsenate metabolism (Kobayashi et al., 2006). No arsenate reductase homologue has been identified in the genome. However, three genes encoding putative components of a prokaryotic ATP-dependent arsenite efflux system (Rosen, 2002; Tripathi et al., 2007) are found: *ARSA1* and *ARSA2* share homologies with arsenite-translocating ATPase, whereas *ARSB* represents a putative arsenite permease (Table 10.2). Note that *ARSA* is conserved in plastid-containing organisms, suggestive of a localization in the plastid (Merchant et al., 2007). It is also possible that at least one *Chlamydomonas* MRP (Table 10.2, section II.A.10.b) contributes to arsenite storage in the vacuole, as was shown for Ycf1 in *S. cerevisiae* (Ghosh et al., 1999).

B. Metal chelation

1. Glutathione and phytochelatins

Metal chelators are key components of the metal homeostasis network. They include various organic acids (such as malate or citrate), low-molecular weight ligands, metallochaperones, and proteins that contribute to the intracellular trafficking and storage of metals. The tripeptide GSH and the derived PCs are well-known contributors to non-essential metal tolerance in a wide range of organisms (see section II.A.10.b; Z.S. Li et al., 1996, 1997; Xiang et al., 2001; Cobbett and Goldsbrough, 2002; Clemens, 2006). Whether PCs might also have a role in essential transition metal homeostasis has been discussed extensively and is still a matter of debate (see Clemens, 2006). In *Chlamydomonas*, GSH is the major compound induced after Hg exposure (Howe and Merchant, 1992), whereas PCs are the major metal-binding peptides induced by Cd (Howe and Merchant, 1992; Hu et al., 2001). Up to 70% of the Cd found in Cd-treated cells is bound to PCs (Hu et al., 2001). As in the fission yeast *S. pombe* and *Arabidopsis* (Howden et al., 1995a,b; Ortiz et al., 1992, 1995), low molecular weight PC-Cd complexes are formed and are rapidly converted to high molecular weight complexes that accumulate in the cells and contribute to a stable sequestration of Cd (Hu et al., 2001).

As shown in other algae (Mendóza-Cozatl et al., 2005), high molecular weight PC-Cd complexes are mainly found in the chloroplast in *Chlamydomonas* (Nagel et al., 1996), whereas PCs are stored in the vacuole in *S. pombe* and *Arabidopsis* (Vögeli-Lange and Wagner, 1990; Cobbett and Goldsbrough, 2002; Clemens and Simm, 2003). The analysis of four *Chlamydomonas* Cd-resistant mutants showed a correlation between Cd accumulation and PC production (Hu et al., 2001). Conversely, no correlation was observed between PC and Cd accumulation when analyzing two Cd-tolerant *Chlamydomonas acidophila* strains that exhibit contrasting Cd accumulation capacities (Nishikawa et al., 2006). In this case, Cd accumulation correlated with the ability to maintain high GSH levels via the induction of the γ-glutamylcysteine synthetase, the first enzyme and limiting step of the GSH synthesis. This enzyme and GSH synthetase, which together catalyze the two-step biosynthesis of GSH, are encoded by the *GSH1* and *GSH2* genes, respectively. A gene encoding a putative PC synthase is also found in *Chlamydomonas* (Table 10.3).

2. Metallothioneins

Metallothioneins are ubiquitous low molecular weight proteins rich in Cys residues that have high metal-binding capacities (Rauser, 1999; Cobbett and Goldsbrough, 2002). Surprisingly, the *Chlamydomonas* genome does not appear to encode metallothioneins. However, this might be an artifact caused by the difficulty of predicting small and low-complexity proteins (Merchant et al., 2006).

3. Metallochaperones

Intracellular trafficking of metals and their proper insertion into target proteins require proteins called metallochaperones (O'Halloran and Culotta, 2000). A few Cu metallochaperones are encoded in *Chlamydomonas* (Table 10.3). The putative Cu chaperone ATX1 was first identified from ESTs (La Fontaine et al., 2002b). By analogy to its *S. cerevisiae*, plant and animal homologues, ATX1 is proposed to handle the transfer of Cu from the uptake site at the plasma membrane to a Cu(I)-transporting HMA (possibly

Table 10.3 Genes encoding metal chelator synthesis proteins and putative metallochaperones

Gene name	Gene product and putative function	Expression evidence[a]	Accession #	Homologue in *Volvox*[b]	References[c]
GSH and PC synthesis					
GSH1	γ-glutamylcysteine synthetase, 1st step of GSH synthesis	EST	EDP06622	Y	[1]
GSH2	Glutathione synthetase, 2[d] step of GSH synthesis	EST	EDP04651	Y	[1]
PCS	PC synthase, Phytochelatin synthesis	-	EDP07275	Y	[1]
Copper chaperones					
ATX1	Cu delivery to the secretory pathway	RNA blot	EDP02326	Y	[1–2]
COX17	Cu delivery to the mitochondria	RNA blot	EDO97130	Y	[1–4]
COX19	Cu delivery to the mitochondria	EST	EDP08032	Y	[1, 3, 4]
COX23	Cu delivery to the mitochondria	EST	EDO98664	Y	[1, 4]
COX11	Cu delivery to Cu_B center of Cox1 subunit of Cyt *c* oxidase	EST	EDO98350	Y	[1, 3, 4]
SCO1	Cu delivery to Cu_A center of Cox2 subunit of Cyt *c* oxidase	EST	EDO97370	Y	[1, 3, 4]

[a] *Expression evidence: EST (Expressed Sequence Tag).*

[b] *Presence of homologous sequence in the genome of Volvox carteri, vista Volvox track at http://genome.jgi-psf.org/Chlre3/Chlre3.home.html*

[c] *References: [1] http://genome.jgi-psf.org/Chlre3/Chlre3.home.html; [2] La Fontaine et al. (2002b); [3] Merchant et al. (2006); [4] Cardol et al. (2005). Note that the annotations provided in reference 3 were based on the release 2 of the genome.*

CTP1) located in the secretory pathway, where Cu can be inserted into target proteins such as the multicopper ferroxidase FOX1 (see section II.A.5 and Figure 10.1, La Fontaine et al., 2002b).

The mitochondrial cytochrome *c* oxidase, a component of the electron transport chain, is a major Cu-requiring enzyme. Genes encoding Cu chaperones (COX17 and COX19) and assembly factors (COX11, COX23, and SCO1) required for Cu(I) insertion into cytochrome *c* oxidase (Cobine et al., 2006) are all present in *Chlamydomonas* (Table 10.3, Cardol et al., 2005; Merchant et al., 2006).

Finally, the Ccs Cu chaperone present in *S. cerevisiae*, plants and animals is required for the proper insertion of Cu and the activation of the Cu/Zn superoxide dismutase (Abdel-Ghany et al., 2005a; Culotta et al., 2006).

Chlamydomonas, like many other green algae, lacks a Cu/Zn superoxide dismutase (Sakurai et al., 1993) and consistently, also lacks a gene encoding a Ccs homologue.

C. Differences with other photosynthetic organisms

1. Chlamydomonas and other algae

A comparative inventory of metal transporter families in *Chlamydomonas* and the unicellular red alga *Cyanidioschyzon merolae* revealed that both algae have adapted their metal transporter complements to fit their environment (Hanikenne et al., 2005a). *Chlamydomonas* appears to be more complex than *Cyanidioschyzon* with respect to the size of the different transporter families (Table 10.1). *Chlamydomonas* is a motile organism that needs to adapt to a fluctuating environment in water and soils. It possesses a complex life cycle with distinct vegetative and sexual stages for which specialized functions may have evolved. In contrast, *Cyanidioschyzon* is an asexual alga that lives in acidic sulfur- and metal-rich (Fe, Ni, etc.) hot springs (Gross, 2000; Ciniglia et al., 2004). It is able to face stable, although extreme, conditions with a limited number of metal transporters. The metal transporter complement of *Cyanidioschyzon* can be rationalized in the context of its specific environment characterized by high Fe and low Cu availability. Among the extant genes are those encoding an IREG1-like protein for Fe efflux and a Ccc1 homologue for vacuolar Fe storage (see below in this section), whereas this alga lacks several proteins involved in Cu metabolism (the multicopper ferroxidase, a chloroplastic Cu I_B P-type ATPase, and plastocyanin; Table 10.1; Hanikenne et al., 2005a). On the other hand, most of the *Chlamydomonas* genes described here are conserved in *Volvox carteri* (Tables 10.2 and 10.3).

2. Chlamydomonas and vascular plants

The comparison of the *Chlamydomonas* and vascular plant genomes reveals that the alga possesses the basic tools to maintain cellular metal homeostasis (Table 10.1 and Figure 10.1). As such, it represents a powerful model to study the functions and the interactions of different metal transport and intracellular trafficking systems (Hanikenne, 2003; Hanikenne et al., 2005a; Merchant et al., 2006). However, vascular plants evolved adaptations of their metal homeostasis network, reflecting the need for long distance transport and for the proper distribution of metals between different organs and cell types.

Chlamydomonas indeed lacks several protein families that contribute to long-distance transport and metal distribution to plant tissues. Although widespread in prokaryotes, the occurrence of divalent cation-transporting HMAs in eukaryotes is apparently limited to plants (Hanikenne et al., 2005a; Williams and Mills, 2005). Unicellular algae possess the Cu-transporting HMAs that are involved in Cu delivery to the chloroplast or the secretory pathway (see section II.A.5), but lack divalent cation-transporting HMAs such as AtHMA2 and AtHMA4. These proteins play a central role in Zn translocation from root to shoot, possibly by loading and/or unloading the xylem (Hussain et al., 2004; Verret et al., 2004; Mills et al., 2005).

Proteins of the IREG family are also potentially involved in the long-distance transport of transition metals in plants (Curie and Briat, 2003). These plant proteins are related to the mammalian IREG1 protein (or Ferroportin1) that mediates the transport of Fe(II) from the basolateral surface of the enterocytes to the bloodstream (McKie et al., 2000). The exact functions of plant IREG proteins remains unclear. AtIREG2 was shown to have a role in Ni transport to the vacuole under Fe deficiency (Schaaf et al., 2006). Absent in *S. cerevisiae*, an *IREG1*-like gene is present in *Cyanidioschyzon* (Table 10.1; Hanikenne et al., 2005a). In *Chlamydomonas*, the presence of an *IREG1*-like gene is uncertain: BLAST searches identify a protein (EDP04441) with weak similarity to human and *Arabidopsis* IREG proteins, but experimental confirmation is required to assess its putative role in metal homeostasis. This suggests that an IREG1-like protein was present in the common ancestor of photosynthetic and non-photosynthetic eukaryotes, and was maintained in mammals and vascular plants. It is hypothesized that Fe efflux function was maintained in *Cyanidioschyzon* to cope with high Fe availability in the acidic sulfur-rich hot springs (Hanikenne et al., 2005a).

Unicellular algae such as *Chlamydomonas* do not synthesize nicotianamine (NA) and lack NAS- (nicotianamine synthase) and YSL- (Yellow Stripes-like) encoding genes (Table 10.1; Hanikenne et al., 2005a). NA is a high-affinity transition metal ligand known to contribute to long-distance transport of Fe, Cu and Zn, and metal tolerance (Pich and Scholz, 1996; Ling et al., 1999; Inoue et al., 2003; Takahashi et al., 2003; Kim et al., 2005; Pianelli et al., 2005). NA is also a precursor for the synthesis of mugineic acid phytosiderophores that are involved in the strategy II Fe uptake system in grasses (Curie and Briat, 2003; Grotz and Guerinot, 2006; section III.A.1). A functional nicotianamine synthase is present in the filamentous fungus *Neurospora crassa*, where NA is suggested to be involved in the cell-to-cell distribution of Zn and other micronutrients via the incomplete septa of the hyphae (Trampczynska et al., 2006).

The YSL proteins are members of the oligopeptide transporter (OPT) family and are involved in the transmembrane transport of metal-NA chelates. Several *Arabidopsis* and rice YSLs have been functionally characterized and contribute to the distribution of transition metals (DiDonato et al., 2004; Koike et al., 2004; Le Jean et al., 2005; Schaaf et al., 2005; Waters et al., 2006). The *S. cerevisiae* Ccc1 protein is an Fe and Mn transporter and mediates the accumulation of these metals in the vacuole (Li et al., 2001). The functional homologue of Ccc1 has been characterized in

Arabidopsis and named VIT1 (Vacuolar Iron Transporter; S.A. Kim et al., 2006). The *VIT1* gene is expressed in the developing embryo and in seeds. In young seedlings, it is mainly expressed in the vasculature. Localized at the vacuolar membrane, VIT1 is essential for Fe storage in the seeds and for seedling development under Fe-limiting conditions (S.A. Kim et al., 2006). A Ccc1 homologue is not encoded in *Chlamydomonas*, but the presence of one in *Cyanidioschyzon* suggests that a *CCC1*-like gene existed early in evolution, where it was probably maintained to provide an Fe storage system.

Finally, the sizes of the transition metal transporter families increase with the complexity of the photosynthetic organisms (Table 10.1). A similar trend is also observed in fungi: the unicellular yeast *S. cerevisiae* possesses a smaller transition metal transporter complement than does the filamentous *N. crassa* (Table 10.1; Kiranmayi and Mohan, 2006). This size expansion allows cell- and tissue-specific expressions of the transporters and the fine-tuning of the metal homeostasis network in a multicellular organism. Contradicting this general trend, some large families of metal transporters are present in unicellular organisms. A striking example is the encoding of a similar number of ZIP transporters in the *Chlamydomonas* (14, Tables 10.1 and 10.2), human (14) and *Arabidopsis* (17) genomes, whereas other unicellular organisms possess only a limited number of *ZIP* genes (Table 10.1).

III. METAL TOLERANCE

A. Tolerance and toxicity

The *Chlamydomonas* cell wall possesses a high affinity for metal cations (Collard and Matagne, 1990) and constitutes a passive protection system against excess metals. Several studies have consistently shown that strains deficient in cell walls are more sensitive to metals such as Cd, Cu, Ni, and Co (Collard and Matagne, 1990; MacFie et al., 1994; Prasad et al., 1998). It could be related to the fact that some wall-deficient strains appear to be Fe-deficient (Allen et al. 2007b), possibly because of the loss of FEA proteins (see section IV.A.1.b). It could be hypothesized that the metal sensitivity of wall-deficient strains is caused by an increased uptake due to the induction of the Fe-deficiency systems.

The toxicity of excess metals has been analyzed in *Chlamydomonas* over many years (Ben-Bassat et al., 1972; Irmer et al., 1986; Collard and Matagne, 1990; Weiss-Magasic et al., 1997; Prasad et al., 1998; Danilov and Ekelund, 2001; Devriese et al., 2001; Boswell et al., 2002; Mosulén et al., 2003; Faller et al., 2005; Morlon et al., 2005; Vega et al., 2005, 2006; Wang et al., 2005; Gillet et al., 2006; ; Jamers et al., 2006; Kobayashi et al., 2006; Luis et al., 2006). Because different media have been used and different parameters have been measured to assess metal toxicity, it is difficult to draw general conclusions from these fragmentary and mainly descriptive

Table 10.4 Metal and metalloid toxicity in *Chlamydomonas*: A summary of past studies

Chemicals	Concentrations	Described effects	References[a]
Arsenate [As(V)]	125–500 μM	Growth reduction	[1]
		Reduction of photosynthesis (measure of oxygen evolution)	[1]
Arsenite [As(III)]	4 mM	Reduction of photosynthesis (measure of PSI activity) and respiration	[2]
Cadmium [Cd(II)]	100–300 μM	Growth reduction	[3–6]
	100–300 μM	Reduction of photosynthesis (measure of PSI activity) and chlorophyll content	[2–5]
	10–100 μM	Inhibition of photoactivation	[7]
	100–300 μM	Inhibition of nitrate assimilation	[8–9]
Copper [Cu(II)]	80–300 μM	Growth reduction	[4, 10–12]
	100–300 μM	Reduction of chlorophyll content	[4, 10–12]
	2–30 μM	Reduction of photosynthesis (measure of oxygen evolution)	[13]
	7–30 μM	Inhibition of external carbonic anhydrase	[14]
	100–300 μM	Inhibition of nitrate assimilation	[8–9]
Iron [Fe(II)]	250–600 μM	Growth reduction and chlorosis	[15]
Lead [Pb(II)]	1–20 μM	Reduction of photosynthesis (measure of oxygen evolution)	[16]
	1–20 μM	Ultrastructural changes (thylakoids, mitochondria, nucleus)	[16]
	2.5–10 μM	Inhibition of external carbonic anhydrase	[14]
Nickel [Ni(II)]	3–80 μM	Stimulation of photosynthesis (measure of oxygen evolution)	[13]
Mercury [Hg(II)]	1–25 μM	Reduced growth and motility	[5, 17–18]
	1–25 μM	Reduction of chlorophyll content	[5, 17–18]
Selenite [Se(IV)]	10–500 μM	Growth reduction	[19]
	10–500 μM	Ultrastructural changes (mainly in the chloroplast)	[19]
Zinc [Zn(II)]	100–300 μM	Inhibition of nitrate assimilation	[8]
	1.5–30 μM	Reduction of photosynthesis (measure of oxygen evolution)	[13]
	8–30 μM	Inhibition of external carbonic anhydrase	[14]
Thallium [Tl(I)]	0.5–2.5 μM	Growth reduction	[20]
		Reduction of chlorophyll content	[20]

[a]*References, Chlamydomonas strains and growth conditions: [1] Kobayashi et al. (2006): strain CC-125, TAP medium 1 mM Pi, 27°C, continuous light, 80 μE/m²/s; [2] Vega et al. (2006): strain 21 gr, Sueoka medium 10 mM nitrate, 0.3 mM sulfate, 25°C, 5% CO_2, continuous light, 250 μE/m²/s; [3] Collard and Matagne (1990): strain 137c, TAP medium 1 mM glycerophosphate, 25°C, continuous light, 80 μE/m²/s; [4] Prasad et al. (1998): strain 2137, high salt medium, 22±2°C, continuous light, 8000 lux; [5] Howe and Merchant (1992): strains CC-124 or CC-1021, TAP medium, 22°C, continuous light, 125 μE/m²/s; [6] Gillet et al. (2006): strain CW15, TAP medium 1 mM glycerophosphate, 24±2°C, continuous light, 150 μE/m²/s; [7] Faller et al. (2005): strain CC-124, TAP medium; [8] Devriese et al. (2001): strain 21 gr, Sueoka medium 10 mM nitrate, 1 mM glycerophosphate, 25°C, 5% CO_2, continuous light, 250 μE/m²/s; [9] Mosulen et al. (2003): strain 21 gr, Sueoka medium 10 mM nitrate, 0.3 mM sulfate, 25°C, 5% CO_2, continuous light, 50 W m^{-2}; [10] Boswell et al. (2002): strain CC-125, TAP medium, 25°C, continuous light, 20 μE/m²/s; [11] Luis et al. (2006): strain CW15 (SAG 83.81), TAP medium, 20°C, 14–10 hours light-dark cycle, 20 μE/m²/s; [12] Jamers et al. (2006): strain SAG 11-32a, TAP medium, 25±1°C, 14–10 hours light-dark cycle, 200 μE/m²/s; [13] Danilov and Ekelund (2001): strain 137c, freshwater medium, 20°C, 16–8 hours light-dark cycle, 70 μE/m²/s; [14] Wang et al. (2005): unknown strain, artificial fresh water medium, 25±1°C, 16–8 hours light-dark cycle, 4000 lux; [15] Hanikenne et al. (2005b): strain CW15, TAP medium, 25°C, continuous light, 80 μE/m²/s; [16] Irmer et al. (1986): strain CCAP 11–32a, Chlorella medium, 25°C, 2.5% CO_2, continuous light, 140 μE/m²/s; [17] Ben-Bassat et al. (1972): strain y-1, acetate, citrate and minerals medium, unknown temperature, 5% CO_2, continuous light, 1000 foot-candles; [18] Weiss-Magasic et al. (1997): strain N.C. 27215 (Carolina Biological Supply Company), modified Mauro medium, 28°C, continuous light, unknown light intensity; [19] Morlon et al. (2005): strain CCAP 11-32b, 25±1°C, continuous light, 100 μE/m²/s; [20] Lustigman et al. (2000): strain N.C. 27215 (Carolina Biological Supply Company), modified Mauro medium, ambient temperature, continuous light, unknown light intensity. See Volume 1, Chapter 1, for more complete identification of strains.*

studies. However, Table 10.4 is an attempt to summarize these findings for As(V), As(III), Cd(II), Cu(II), Fe(II), Pb(II), Hg(II), Ni(II), Se(IV), Zn(II), and Tl(I). Most of these metals when present in excess affect growth and ultimately viability, and the photosynthetic apparatus appears as the prime target of toxicity. For example, low Cd concentrations (within the μM range) inhibit photoactivation and hence photosynthesis, most likely by competing for binding to the Ca site of Photosystem II (PS II) during the assembly of the water-splitting complex (Faller et al., 2005). A Cd-resistant mutant was shown to be impaired not in Cd uptake or detoxification but in photosynthetic activity, as revealed by reduced growth under photoautotrophic conditions, decreased – but Cd-resistant – photosynthetic oxygen evolution, reduced PS II activity and altered chlorophyll fluorescence induction kinetics in dark-adapted cells (Table 10.5; Nagel and Voigt, 1995; Voigt et al., 1998). Acetate inhibition of the water-splitting complex, which is observed in the wild type, was suppressed in the resistant mutant, suggesting that the donor side of PS II is impaired in this strain (Voigt and Nagel, 2002).

Table 10.5 Metal-resistant and metal-sensitive mutants

Metal	No of mutants	R/S[a]	Mechanisms	Genes	References[b]
Arsenate	12	R	Lower As content, higher Pi[a] content	Unknown	[1, 2]
	1	R	Lower As content Defective in a putative Pi transporter	*PTB1*	[1–3]
	2	S	Higher As content, unchanged Pi content	Unknown	[1, 2]
	1	S	Unchanged As content, impaired in the conversion of inorganic arsenic to dimethylarsinic acid	Unknown	[1, 2]
Cadmium	2	R	Lower Cd content, putative permeability mutants	Unknown	[4, 5]
	1	R	Impaired photosynthesis	Unknown	[6–9]
	2	R	Increased Cd accumulation and higher phytochelatin content	Unknown	[10]
	2	R	Reduced Cd accumulation and lower phytochelatin content	Unknown	[10]
	5	S	Unknown	Unknown	[11]
	1	S	Unknown	Unknown	[12, 13]
	5	S	Higher Cd and phytochelatin content Fe^{2+} sensitivity Defective in a mitochondrial ABC transporter	*CDS1*	[12, 13]
	1	S	Defective in a MRP ABC transporter with putative vacuolar localization	*MRP2*	[14]
Copper	5	S	Unknown	Unknown	[15]

[a]*Abbreviations: R: metal-resistant mutants, S: metal-sensitive mutants, Pi: inorganic phosphate.*

[b]*References: [1] Fujiwara et al. (2000); [2] Kobayashi et al. (2005); [3] Kobayashi et al. (2003); [4] Collard and Matagne (1990); [5] Collard and Matagne (1994); [6] Nagel and Voigt (1989); [7] Nagel and Voigt (1995); [8] Voigt et al. (1998); [9] Voigt and Nagel (2002); [10] Hu et al. (2001); [11] McHugh and Spanier (1994); [12] Hanikenne et al. (2001); [13] Hanikenne et al. (2005b); [14] Wang and Wu (2006); [15] Hanikenne (2003).*

The resistance phenotype might result from a reduced affinity for Cd of a cation-binding site, possibly for Ca or Mn, in PS II.

In *Chlamydomonas*, Cd mainly accumulates in the chloroplast (Nagel et al., 1996). The dramatic effect of Cd on photosynthesis was revealed in a proteomic study showing that about 1/3 of the soluble proteins whose abundance is decreased upon Cd treatment are involved in photosynthesis, including Rubisco, ferredoxin, ferredoxin-NADP-reductase and several enzymes of the Calvin cycle and chlorophyll biosynthesis (Gillet et al., 2006). It is possible that Cd(II) exerts its effects through interference with Fe(II) uptake and/or thiol metabolism (Howe and Merchant, 1992; Rubinelli et al., 2002).

Arsenite, Cu, and to a lesser extent Zn, were also shown to inhibit photosynthesis efficiency in *Chlamydomonas*. On the other hand, relatively high Ni and Pb concentrations had stimulatory effects (see also section II.A.11; Danilov and Ekelund, 2001, Vega et al., 2006). The effects of Cu excess on photosynthesis and the regulation of photosynthetic genes were analyzed in detail by Luis et al. (2006). High Cu causes a decrease in chlorophyll and carotenoid contents, whereas α-tocopherol synthesis is stimulated to protect the photosynthetic apparatus from the Cu-induced oxidative stress. With increasing Cu concentrations, functionality of both photosystems is gradually lost and the *psaA* and *psaB* genes are up-regulated, to cope with an increased demand for replacement of damaged PS I subunits. While the genes *FSD1* and *MSD1*, encoding Fe and Mn superoxide dismutases, are induced in response to Cu, the corresponding enzymatic activities are independent of metal status. Based on the Cu concentration at which the onset of toxicity is observed, the study suggests that stromal photosynthetic functions are more sensitive to Cu-induced reactive oxygen species than the membrane-located reactions (Luis et al., 2006). Jamers et al. (2006) performed a complementary analysis of the transcriptional response to Cu toxicity using microarrays. In presence of 125 μM Cu, the expression of genes involved in oxidative stress responses (a putative GSH-*S*-transferase, a GSH peroxidase, a mitochondrial thioredoxin and a putative peroxiredoxin), the detection and repair/proteolysis of damaged proteins (26S proteasome subunit, a DnaJ homologue, and HSP70B), and the mitochondrial electron transport chain (Rieske protein, cytochrome *c* peroxidase, and cytochrome c_1) is increased. Two thioredoxin (*TRX*) genes are also induced by Cd and Hg in *Chlamydomonas* (Lemaire et al., 1999, 2002), suggesting that Trx might play a general role in metal toxicity (Howe and Merchant, 1992). Several genes (e.g. *CRD1* and *CTH1*) involved in the Cu deficiency response (see section IV.B.3.c) are down-regulated upon exposure to high Cu concentrations (Jamers et al., 2006). Note that this study probably reflects only part of the transcriptional response to Cu overload, because only about 3000 genes were represented on the microarray.

In addition to photosynthesis, other components of cellular metabolism are affected by toxic concentrations of metals. Arsenite was shown to inhibit respiration (Vega et al., 2006). The external carbonic anhydrase of *Chlamydomonas*, which is involved in the CO_2-concentrating mechanism (see Chapter 8), is inhibited by Cu, Pb and Zn, whereas Cd has a slight stimulatory effect (Wang et al., 2005). Moreover, an excess of Cd, Cu, and Zn has also an impact on the assimilation of macronutrients such as nitrate or sulfate. These three metals inhibit nitrate uptake and synthesis of glutamine synthetase (Devriese et al., 2001; Mosulén et al., 2003; Gillet et al., 2006), whereas Cd increases the sulfate assimilation rate (Domínguez et al., 2003; Mosulén et al., 2003). Cd also induces the accumulation of enzymes of glutamate and cysteine metabolism (Vega et al., 2005; Gillet et al. 2006). The abundance of RNA encoding the *S*-adenosylmethionine (SAM) synthetase was increased by Cd, SAM being a precursor of cysteine (Rubinelli et al., 2002). Higher sulfate assimilation and increased glutamate and cysteine synthesis would support the Cd-induced accumulation of GSH and PCs (see section II.B.1).

To cope with metal-induced oxidative stress, catalase and ascorbate peroxidase activities are also induced upon Cd treatment, whereas arsenite induces catalase and GSH reductase activities (Vega et al., 2005). Cd also increases abundance of an ascorbate peroxidase and an MnSOD (product of the *MSD2* gene; Gillet et al., 2006). Similarities can be found in the response to Cd at the proteome level, and the RNA abundance changes occurring upon Cu treatment (Gillet et al., 2006; Jamers et al., 2006; see above in this section). For example, a GSH-*S*-transferase, a peroxiredoxin and HSP70 are up-regulated in both cases. Cd also has an effect on proteins of carbohydrate metabolism and fatty acid, amino acid, and protein biosynthesis (Gillet et al., 2006).

Finally, phosphates also play a role in *Chlamydomonas* metal tolerance. Cd accumulation with phosphate in the vacuole of *Chlamydomonas acidophila* is correlated with a decreased polyphosphate content (Nishikawa et al., 2003). Moreover, Cd and Cu tolerance depends on ambient phosphate concentrations (Wang and Dei, 2006), and arsenate resistance in mutants was shown to correlate with high intracellular phosphate in *C. reinhardtii* (Kobayashi et al., 2005). These mechanisms might be supported by the induction of an inorganic pyrophosphatase that catalyzes the hydrolysis of pyrophosphate into two orthophosphates (Gillet et al., 2006). It is additionally relevant that Allen et al. (2007a) described an interaction between Mn and phosphate metabolism with respect to Mn deficiency (see section IV.C).

To summarize, it appears that photosynthesis is the primary target of metal toxicity, and that the photosynthetic apparatus undergoes numerous adaptations upon metal excess in *Chlamydomonas*. Response to metal overload also involves the induction of oxidative stress protection mechanisms and of protein chaperones, and requires metabolic adaptations to support the synthesis of the metal chelators GSH and PC.

B. Tolerance mutants

To gain a better understanding of the metal tolerance mechanisms in *Chlamydomonas*, several groups have attempted to isolate metal-resistant or metal-sensitive mutants (Table 10.5; Nagel and Voigt, 1989, 1995; Collard and Matagne, 1990; Collard and Matagne, 1994; McHugh and Spanier, 1994; Voigt et al., 1998; Fujiwara et al., 2000; Hanikenne et al., 2001, 2005b; Hu et al., 2001; Kobayashi et al., 2003, 2005; Voigt and Nagel, 2002; Wang and Wu, 2006). For most of these mutants, the molecular identity of the corresponding genes is unknown. However, improving molecular and informatic tools will facilitate their identification.

A number of Cd and arsenate tolerant mutants were shown to be permeability mutants (Collard and Matagne, 1990, 1994; Fujiwara et al., 2000; Hu et al., 2001) probably resulting from an alteration of metal transport across the plasma membrane. Other Cd-resistant mutants were shown to accumulate PC-Cd high molecular weight complexes of increased size (section II.B.1; Hu et al., 2001) or to be impaired in photosynthesis (section III.A; Nagel and Voigt, 1995; Voigt et al., 1998; Voigt and Nagel, 2002).

An arsenate-resistant mutant was found to accumulate less arsenate than the corresponding wild-type strain (Fujiwara et al., 2000) and subsequently shown to be defective in a putative Na(I)/Pi co-transporter named PTB1 (Kobayashi et al., 2003). *ptb1* cells display arsenate resistance due to low arsenate uptake, suggesting that PTB1 is a major entry route for arsenate. It was also shown that *ptb1* cells accumulate higher levels of inorganic phosphate and possesses increased high-affinity inorganic phosphate transport activity relative to the wild type, which might compensate for the loss of PTB1 (Kobayashi et al., 2005).

Cd-sensitive mutants allowed the identification and the cloning of ABC transporter genes (*CDS1* and *MRP2*, Table 10.5) involved in Cd tolerance (Hanikenne et al., 2001, 2005b; Wang and Wu, 2006). CDS1 is a mitochondrial protein possibly involved in the export of Cd (section II.A.10.c), whereas MRP2 is a putative GSH-Cd vacuolar transporter (section II.A.10.b).

IV. METAL NUTRITION AND DEFICIENCY RESPONSE

A. Iron

1. Iron assimilation

a. An animal-, fungal-like iron uptake system in *Chlamydomonas*

Considering that Fe is among the most abundant elements in the earth's crust (Taylor, 1964), it is surprising that Fe deficiency is the most common nutritional deficiency in the world's population (Adamson, 2004). The poor bioavailability of Fe due to the low solubility of ferric [Fe(III)] oxides, in particular in alkaline environments, is likely to account for this paradox (Lindsay and Schwab, 1982; Marschner, 1995). However, Fe is an essential micronutrient in its role as metal cofactor for many enzymes and redox reactions. Organisms have therefore developed efficient homeostatic mechanisms for Fe assimilation and distribution within cells and organs. In photosynthetic organisms, Fe deficiency induces metabolic adaptations that allow the maintenance of both photosynthetic and respiratory functions.

Organisms possess high- and low-affinity systems for the uptake of Fe depending on its extracellular concentration and chemical state (oxidized/reduced, free or chelated, etc.). High-affinity systems are usually highly selective for Fe and operate in Fe-deficient conditions, while low-affinity systems are less selective and are active in Fe-replete conditions. Plants have developed two strategies for high-affinity Fe uptake (reviewed in Curie and Briat, 2003; Colangelo and Guerinot, 2006; Grotz and Guerinot, 2006). On the one hand, dicots and non-graminaceous monocots use strategy I (or reduction strategy) where the soil is acidified by H^+ ATPases to solubilize iron, Fe(III) is reduced by ferric chelate reductases, and Fe(II) uptake is mediated by IRT1, a protein of the ZIP family. On the other hand, grasses use strategy II (or chelation strategy) where phytosiderophores [i.e. high-affinity Fe(III) chelators] are released, and Fe(III) chelates are subsequently taken up by the YS1 transporter, a member of the OPT family (Curie et al., 2001). The chelation strategy is more efficient than the reduction strategy and allows grasses to survive on soils with very low iron availability (Curie and Briat, 2003; Grotz and Guerinot, 2006). In addition, it has been shown that in rice (a strategy II plant) iron can also be taken up by an IRT1-like protein (Bughio et al., 2002; Ishimaru et al., 2006), allowing rice to directly absorb Fe(II) which is more available in submerged conditions. These observations suggest a more general role of ZIP proteins in plant Fe uptake.

In *Chlamydomonas* the high-affinity Fe uptake system is localized to the plasma membrane and is postulated to require a ferrireductase (FRE1), a multicopper ferroxidase (FOX1), and an iron permease (FTR1), akin to the systems that operate in Fe nutrition in yeast and mammals but distinct from either of the two pathways operating in plants (Figure 10.1; Table 10.6; Herbik et al., 2002a; La Fontaine et al., 2002b; Merchant et al., 2006).

Because the complexation properties of Fe(II) and Fe(III) ions are distinct, the control of the redox state of Fe is central to its transport/mobilization in the cell. In vascular plants and algae, the involvement of ferrireductase(s) in Fe assimilation at the plasma membrane is well documented (Eckhardt and Buckhout, 1998; Lynnes et al., 1998; Schmidt, 1999; Weger et al., 2002). Ferrireductase activity is induced by Fe deficiency, and depends upon intracellular reducing power in the form of NADH or NADPH for the reduction of extracellular Fe(III) and Fe(III) chelates (Xue et al., 1998; Weger and Espie, 2000). Sequence similarities identify four ferrireductase-like proteins encoded by the *FRE1*, *CBR1*, *RBOL1*, and *RBOL2* genes in *Chlamydomonas* (Merchant et al., 2006), although their functions remain

Table 10.6 Genes encoding proteins involved in iron assimilation and storage

Gene name	Putative function	Expression evidence[a]	Accession #	Homologue in *Volvox*[b]	References[c]
FRE1	Ferrireductase, reduction of Fe(III) and Fe(III)-chelates	qRT	EDP04183	Y	[1–3]
FOX1	Multicopper ferroxidase, reoxidation of Fe(II)	RNA blot, immunoblot	EDP02580	Y	[1–3]
FOX2	Putative multicopper oxidase	RNA blot	EDP03605	Y	[1, 3]
FTR1	Iron permease, Fe(III) uptake	RNA blot	EDP03271	Y	[1–3]
FEA1	Fe-assimilating protein, Periplasmic Fe(II)-binding	RNA blot, immunoblot, purified protein	EDP02900	Y	[1, 3, 4]
FEA2	Fe-assimilating protein, Periplasmic Fe(II)-binding	RNA blot, immunoblot, purified protein	EDP02901	N	[1, 3]
FER1	Ferritin, Iron storage	RNA blot, immunoblot	EDP02528	N	[1–3]
FER2	Ferritin, Iron storage	qRT, immunoblot, RNA blot	ABW87266	Y	[1, 3]

[a] *Expression evidence: qRT (quantitative real-time RT-PCR).*

[b] *Presence of homologous sequence in the genome of Volvox carteri, vista Volvox track at http://genome.jgi-psf.org/Chlre3/Chlre3.home.html (March 2007).*

[c] *References: [1] http://genome.jgi-psf.org/Chlre3/Chlre3.home.html; [2] La Fontaine et al. (2002b); [3] Merchant et al. (2006); [4] Rubinelli et al. (2002). Note that the annotations provided in reference 3 were based on the release 2 of the genome.*

to be established experimentally. The *FRE1* gene was the most dramatically regulated by iron-deficiency (Allen et al., 2007b). The *FRE1* gene product was identified as a 100-kD plasma membrane protein in Fe-deficient cells (Reinhardt et al., 2006). This further supports a role of FRE1 as an iron reductase in Fe assimilation. *CBR1* encodes a cytochrome b_5 reductase-like protein, and the *RBOL* genes encode polypeptides with similarity to plant respiratory burst oxidase proteins, but none of these has been tested experimentally (Allen et al. 2007b).

In *Chlamydomonas* the presumed second step of Fe uptake involves a ferroxidase, that is a Cu-containing enzyme (multicopper oxidase) that catalyzes the oxidation of Fe(II) to Fe(III) ions prior to their transport by the Fe(III) permease (La Fontaine et al., 2002a). This re-oxidation step confers selectivity and specificity to high affinity Fe uptake and avoids the production of reactive oxygen species in the presence of Fe(II) (Askwith and Kaplan, 1998; Herbik et al., 2002a). A ferroxidase-like protein, originally called FLP1, was identified in a proteomic study of *Chlamydomonas* plasma membranes (Herbik et al., 2002a, b). FLP1 shares sequence similarity with *S. cerevisiae* Fet3 and mammalian multicopper oxidases such as hephaestin and ceruloplasmin, which function in Fe homeostasis (Herbik et al., 2002a, b; La Fontaine et al., 2002a). The gene encoding FLP1 was identified independently as *FOX1* on the basis of the presence of multicopper oxidase motifs (La Fontaine et al., 2002b). *FOX1* is up-regulated under Fe deficiency (La Fontaine et al., 2002b), and predicted to encode an integral membrane protein with a N-terminal hydrophobic membrane anchor. FOX1/FLP1 is believed to be glycosylated based on its electrophoretic behavior in non-denaturing gels (Herbik et al., 2002b). This is consistent with a plasma membrane localization and routing through the secretory pathway where loading of Cu into apoferroxidase also takes place (Figure 10.1; for review see Sargent et al., 2005). In *S. cerevisiae*, this copper-requiring step is essential for the complete maturation of the ferroxidase/Fe permease complex (see section IV.B.2) and its translocation to the plasma membrane (Stearman et al., 1996).

FOX1 has a paralog called FOX2 (with 55% protein sequence identity to FOX1) whose involvement in metal nutrition is unclear based on the observation that *FOX2* was both reported not to be regulated by metal availability (Merchant et al., 2006), and also described as induced upon prolonged Fe deficiency (Reinhardt et al., 2006). It is possible that the genetic background differences account for the disparity. The predicted FOX2 protein retains the copper-binding motifs found in FOX1, but lacks the hydrophobic N-terminal anchor. It is therefore likely to be localized to a different cell compartment than FOX1.

The final step of Fe uptake involves the delivery of Fe(III) across the membrane by Fe(III) permeases or transporters that are usually physically associated with multicopper oxidases. The *FTR1* gene was postulated to encode the Fe(III) permease in *Chlamydomonas* (Figure 10.1) based on the facts that the predicted FTR1 polypeptide contains characteristic Fe-binding motifs that are present in other FTR1-like proteins, and that the *FTR1* mRNA is up-regulated coordinately with *FOX1* under Fe deficiency (La Fontaine et al., 2002b). FTR1 represents the only putative Fe(III) transporter identified so far in *Chlamydomonas*.

In yeast and mammals, Fe uptake is Cu-dependent and Cu deficiency leads to Fe deficiency (Askwith et al., 1994; Askwith and Kaplan, 1998). In *Chlamydomonas*, the existence of another Fe uptake pathway that functions independently of Cu nutritional status is suspected. Indeed, Cu deficiency does not lead to Fe deficiency and even under conditions where Fe nutrition is limited, copper deficiency does not exacerbate the Fe-deficient phenotype (La Fontaine et al., 2002b). Herbik et al. (2002a) showed, however, a reduction of Fe uptake rates when Cu supply is limited. Again, this discrepancy might be attributed to strain-specific differences between laboratories or different experimental conditions. A genetic dissection of metal nutrition has demonstrated the existence of such a Cu-independent pathway through the isolation of the *crd2-1* mutant (for *c*opper *r*esponsive *d*efect), which is impaired in Fe assimilation (Eriksson et al., 2004). The Fe nutritional defect is conditional: Fe homeostasis in *crd2-1* is not affected under Cu-replete conditions, but when Cu availability decreases, *crd2* manifests a distinct chlorotic and growth inhibition phenotype. The up-regulation of the genes involved in Fe assimilation indicates that *crd2* cells experience Fe deficiency as a response to low Cu availability in the medium. As would be expected, provision of excess Fe rescues the *crd2* Cu-conditional growth phenotype. The *CRD2* gene has not been cloned yet but three alternative hypotheses can be proposed for the function of the CRD2 protein. One is that CRD2 is a Cu-independent ferroxidase that operates with FTR1 in Fe assimilation. In this context, it worth mentioning that *Cyanidioschyzon* lacks a FOX1-like multicopper oxidase (Hanikenne et al., 2005a). Because *Cyanidioschyzon* grows in an Fe- and S-rich milieu with low Cu availability, it is conceivable that this organism has evolved a Cu-independent ferroxidase associated with the FTRs. A similar situation is found in the moss *Physcomitrella patens*, which also lacks a FOX1 homologue but possesses an FTR-related protein. A second possibility is that CRD2 is a high affinity transporter of a novel type. Finally, CRD2 might be recruited to ensure that Cu is preferentially allocated to ferroxidase in situations where its availability is limited. The fact that the *crd2* mutant is unable to accumulate normal amounts of ferroxidase supports the proposal that CRD2 functions in a pathway for copper delivery to ferroxidase (Eriksson et al., 2004). The phenotype of *fox1-kd* strains, which display a growth defect in both copper-supplemented and copper-deficient cells, argues against the first two possibilities (Chen et al., 2008). The strains do not show a phenotype in iron-replete medium (20 μM), suggesting that alternative low affinity transporters (see below) may operate in this situation.

In summary, it appears that *Chlamydomonas* uses an animal/fungal-like system that is dependent on Cu availablility for high-affinity iron uptake. Multicopper oxidases have also been identified in other algae such as *Dunaliella salina*, a halotolerant unicellular green alga (see section IV.A.1.b; Paz et al., 2007), and various diatoms (Peers et al., 2005; Maldonado et al., 2006). Moreover, *Chlamydomonas* and *Cyanidioschyzon* possess Cu-independent Fe uptake mechanisms. Several genes encoding putative Fe transporters have been identified in *Chlamydomonas* including 2 NRAMPs and several ZIPs (such as IRT1 and IRT2), but there is little direct evidence for their involvement in Fe assimilation (see section II.A).

The coexistence of copper-dependent and copper-independent Fe uptake systems might be an ancestral feature of photosynthetic cells. Plants have lost the seemingly ancestral and highly specific Cu-dependent system and have adopted the two Cu-independent strategies, described above, which allow them to cope with the range of iron availabilities and oxidation states characteristic of their habitats (reviewed in Curie and Briat, 2003; Colangelo and Guerinot, 2006; Grotz and Guerinot, 2006). The actual distribution of Fe uptake systems in photosynthetic organisms might reflect past or present differences in metal ion availability in their environments.

b. Extracellular proteins involved in iron assimilation

The extracellular space of *Chlamydomonas* cells is known to harbor proteins which participate in nutrient assimilation (e.g. the arylsulfatase involved in sulfur scavenging; de Hostos et al., 1988; Fett and Coleman, 1994; Quisel et al., 1996). FEA1 and FEA2 are two related proteins that are secreted in the medium by wall-deficient cells grown under Fe deficiency (Allen et al., 2007b). The major protein, FEA1, corresponds to the previously identified *H43* gene product (Rubinelli et al., 2002), which is a homologue of the high CO_2 and Fe deficiency-inducible *Chlorococcum littorale HCR1* gene (Sasaki et al., 1998; Hanawa et al., 2007). The level of FEA1 secreted in the medium of Fe-deficient *Chlamydomonas* cultures is proportional to the level of Fe deficiency. Furthermore, *FEA1* is transcriptionally activated in response to Fe deficiency, as are *FOX1* and *FTR1* (see section IV.A.1.a). Together, these observations support a function of FEA1 in the Fe assimilation pathway (Figure 10.1). The abundance of FEA2 is lower than that of FEA1 and not as influenced by Fe availability in the medium, although *FEA2* RNA abundance increases coordinately with *FEA1* mRNA in response to Fe deficiency. *FEA2* expression is not affected by CO_2 availability. This suggests a diversification of FEA2 versus FEA1 function. The *FEA1* and *FEA2* genes are adjacent and their gene products contain conserved Cys, Asp, and Glu acid residues that could serve as Fe-binding sites.

FEA-like proteins are found in other green algae (*Scenedesmus*, *Volvox*) and also in a dinoflagellate, but are not found in vascular plants. Expression of FEA1 in *S. cerevisiae* increases the Fe content of a *fet3 fet4* mutant, which lacks ferroxidase and the low-affinity Fe transporter (Rubinelli et al., 2002), giving further credence to a role of FEA1 in Fe assimilation. Rubinelli et al. (2002) showed that *Chlamydomonas FEA1* is induced by Cd exposure, an expected finding as Cd induces Fe deficiency by competing for the same uptake systems.

A possible role for FEA1 is to increase the local concentration of Fe at the vicinity of FOX1/FTR1, to optimize uptake (Figure 10.1). It appears that algae have evolved unique mechanisms for metal acquisition as evidenced by the discovery of a transferrin-like protein in *Dunaliella salina*. A transferrin-like protein (TTF for triplicated transferrin) was found to accumulate in the plasma membrane of Fe-deficient *Dunaliella* cells (Sadka et al., 1991; Fisher et al., 1997, 1998). Previously, transferrins had been reported to be involved in Fe uptake only in animal cells (reviewed in Sargent et al., 2005), and none is apparently encoded in *Chlamydomonas*. The increased accumulation of TTF under Fe deficiency correlates with enhanced Fe uptake activity, and it has been postulated that the protein allows the cells to overcome the limited Fe availability under high salinity. The discovery that a multicopper oxidase (D-FOX) and a glycoprotein (p130B) form a complex with TTF at the plasma membrane in *Dunaliella*, and that this complex exhibits Fe binding and uptake activity, suggests that the TTF-mediated Fe assimilation pathway is of a novel type (Paz et al., 2007).

2. Metabolic adaptations to iron deficiency

a. Photosynthetic apparatus

Chlamydomonas cells are able to adapt to changes in the nutritional Fe status, and the first notable response at the molecular level is up-regulation of the genes encoding Fe assimilation components. In fact, the expression of *FOX1*, *FTR1*, and *FEA1* is significantly increased when the Fe content is reduced in the culture medium before any phenotypic symptoms of Fe deficiency are visible (La Fontaine et al., 2002b). Manifestations of Fe deficiency such as chlorosis and inhibition of cell division occur only when the Fe supply does not meet the demand for the biogenesis of Fe-requiring enzymes (i.e. Fe concentration in the sub-micromolar range; La Fontaine et al., 2002b).

Because of the central role of Fe in photosynthesis and respiration, photosynthetic organisms have evolved intracellular mechanisms to adapt to Fe deficiency in addition to the activation of the Fe assimilation pathway, which constitutes the first adaptative response to Fe deficiency. One example of such an adaptation has been described in cyanobacteria where ferredoxin, an abundant FeS protein involved in photosynthesis can be substituted by flavodoxin, a flavin mononucleotide-containing protein (Laudenbach et al., 1988; La Roche et al., 1996).

Metabolic adaptation to Fe deficiency in *Chlamydomonas* has been documented using two approaches. In one approach, the impact of Fe on the photosynthetic apparatus was examined, in particular on PS I which, with four Fe_4S_4 clusters, accounts for half of the Fe content in the photosynthetic apparatus (Moseley et al., 2002a; Naumann et al., 2005). Transition of *Chlamydomonas* cells from Fe-replete to mild and then severe Fe deficiency is accompanied by progressive modification of the interaction of light-harvesting proteins with PS I (Moseley et al., 2002a). Moseley et al. (2002a) defined several states of iron nutrition and distinguished, on the basis of the activity and composition of the photosynthetic apparatus, the iron-deficient from the iron-limited state. In particular, they noted the impact of iron deficiency on PS I and its associated light harvesting proteins, and they suggested that loss of the PSAK subunit is an early event in the remodeling of photosystem I in response to iron deficiency. The LHCA subunits of LHCI, which are involved in supplying PS I with excitation energy via chlorophyll pigments, are in tight association with PS I and form a LHCI–PSI supercomplex (Takahashi et al., 2004; see Chapter 14). In contrast to vascular plants where the LHCI–PSI complex consists of only four LHCA polypeptides (Ben-Shem et al., 2003), *Chlamydomonas* LHCI is composed of seven major and two minor LHCA polypeptides found in a complex with an estimated size of 300–400 kD (Stauber et al., 2003). Under Fe deficiency, PS I undergoes a structural modification and loses its connection to part of the LHCI pigments. PSAK, a connecting subunit between LHCI and PS I (Fromme et al., 2001; Fang and Wang, 2002; Ben-Shem et al., 2003; Kargul et al., 2003), dissociates from PS I. This disconnection results in a drop in the excitation energy transfer efficiency between PS I and LHCI and is believed to be necessary to prevent photooxidative stress resulting from the loss of FeS clusters, a consequence of Fe deficiency. Remodeling of the photosynthetic apparatus in response to Fe deficiency includes changes in the subunit composition of LHCI (and also LHCII, the LHC associated with PS II) (Moseley et al., 2002a).

In subsequent work, Naumann et al. (2005) analyzed the changes in the LHC proteins as iron-replete cells acclimate to the iron-limited state. Analysis of thylakoid membranes by 2-D separation followed by mass spectrometric identification of spots showed that Fe deficiency induces the degradation of LHCA5 and increased accumulation of LHCA4 and LHCA9 and the N-terminal processing of LHCA3, a key event that results in the dissociation of LHCI–PSI supercomplexes by physical separation of LHCI from PS I reaction centers (Naumann et al., 2005). It is conceivable that the change in LHCA subunit composition leads to the production of an antenna complex that acts as a better sink for energy dissipation. Progression to severe Fe deficiency results in the sequential proteolytic degradation of the LHCI subunits and both photosystems (Moseley et al., 2002a). Quantitative proteomics and spectroscopic data revealed that the antenna size of PS II is significantly increased under iron deficiency (Naumann et al., 2007). This enlargement in PS II antenna size results in a pronounced photoinhibition of PS II, which is believed to constitute an efficient protective mechanism against photo-oxidative damage. The photoinhibition can be explained by the fact that electron transfer is compromised under low iron because the absorbed light energy exceeds the capacity of the photosynthetic apparatus. Concomitantly with PS II antenna remodelling, stress related chloroplast polypeptides, like 2-cys peroxiredoxin and a stress-inducible light-harvesting protein, LHCSR3, as well as a novel light-harvesting protein and several proteins of unknown function were found to be induced in response to Fe deprivation (Naumann et al., 2007). The retention of PS II might also serve the purpose of chlorophyll storage so that the photosystems can be rapidly re-synthesized when iron supply is restored. Surprisingly, subunits of Fe-containing respiratory complexes in the mitochondria are not decreased suggesting that under photoheterotrophic conditions, there might be a hierarchy of iron allocation to respiratory versus photosynthetic complexes under Fe-deficient growth conditions.

In a second approach, Reinhardt et al. (2006) used electrophoretic separation and mass spectrometry to identify proteins whose synthesis and/or degradation were modified when *Chlamydomonas* cells experience Fe deficiency. Metabolic adaptation of *Chlamydomonas* cells to Fe deficiency may include the increased accumulation of PEP carboxykinase and isocitrate lyase, two cytosolic non-plastid enzymes of general metabolism that function in gluconeogenesis and the glyoxalate cycle, respectively (Reinhardt et al., 2006). The authors' interpretation of this finding, which requires validation by quantitative methods or immunoblot analysis of specific proteins, is that carbon metabolism may be redirected from lipid molecules to glucose synthesis as a result of impaired photosynthesis and respiration in Fe limited cells.

b. Ferritin

Ferritin is a multimeric molecule that binds Fe(III) (reviewed by Briat et al., 1999). Fe is mineralized within the ferritin core via oxidation of the Fe(II) form by the intrinsic ferroxidase activity of ferritin (reviewed in Chasteen and Harrison, 1999). Ferritin resides in the cytoplasm of mammalian cells (Eisenstein, 2000) but is found in the plastid stroma in plants, where it acts as a Fe supply source during plastid biogenesis (Briat and Lobréaux, 1998). Ferritin functions as an iron buffer in plastids: when the intracellular Fe is high, ferritin synthesis is increased to provide enhanced Fe chelating capacity but in situations where the intracellular demand for Fe is augmented or when facing extracellular Fe deficiency, Fe can be mobilized from ferritin-bound Fe(III). The pattern of ferritin expression in plants is consistent with these functions (reviewed in Briat et al., 1999). In *Chlamydomonas* a ferritin-like protein is encoded by *FER1*, whose expression is increased in Fe deficient cells (La Fontaine et al., 2002b). Increased abundance of ferritin was also noted in *Chlamydomonas* cells grown in Fe-deficient medium (Reinhardt et al., 2006). One explanation is that increased ferritin is needed to increase the Fe buffering capacity in the chloroplast as Fe is released following the degradation of PS I and ferredoxin in Fe-limited cells (see above). A second ferritin-encoding gene (*FER2*) is present in *Chlamydomonas* whose expression pattern differs from that of *FER1* (Table 10.6; Merchant et al., 2006). Its function remains to be elucidated.

c. Superoxide dismutases

In plants, three forms of superoxide dismutase are found which use different metal cofactors and have distinct intracellular localizations: Cu/ZnSOD is in the cytosol, plastid and peroxisome, MnSOD is in the mitochondria, and FeSOD is in the plastid (Kliebenstein et al., 1998). In *Chlamydomonas*, biochemical studies have identified one FeSOD and two MnSODs (Sakurai et al., 1993; Kitayama et al., 1999; Allen et al., 2007a). The FeSOD is encoded by *FSD1*, whereas there are five genes encoding MnSODs (*MSD1–MSD5*;

Merchant et al., 2006; Allen et al., 2007a). MnSOD activity was found not only in mitochondria as expected, but also in the chloroplast (Allen et al., 2007a). An MnSOD in diatoms is associated with PS II on the lumenal side of the thylakoids (Wolfe-Simon et al., 2005), suggesting that the occurrence of MnSOD in the chloroplast may be more common than previously suspected. The *MSD3* gene is greatly induced in Fe-deficient *Chlamydomonas* cells, presumably to compensate for the loss of FeSOD activity in the chloroplast (Allen et al., 2007a), but the correspondence between the gene products and measurable activities in isolated cell fractions has not been established. The presence of both Fe- and Mn-containing SODs in the chloroplast raises the interesting question of how specific loading of the active sites may be accomplished *in vivo* (discussed in Yang et al., 2006).

d. Iron distribution and signaling

Distributive Fe transporters to organelles such as mitochondria, plastids, and vacuoles are also expected to participate in Fe homeostasis, and candidate proteins include members of the ZIP family with a putative vacuolar localization (Figure 10.1; Hanikenne et al., 2005a; Merchant et al., 2006) and members of the ATM family with predicted organellar localization (see section II.A.10.c).

Metabolic adaptation to Fe deficiency must depend on the operation of a signaling pathway from sensing mechanism to target genes/proteins and the sensed signal is likely to be the Fe nutritional status. The involvement of a global regulator controlling Fe homeostasis is also suspected based on other regulatory circuits that control metal nutrition (Rutherford and Bird, 2004; Kropat et al., 2005). At present the molecular identities of these components are unknown, but they could be revealed through genetic approaches. The use of reporter genes established that the Fe-responsive transcriptional activity of the *FEA1*, *FOX1*, and *FTR1* genes requires Fe-responsive *cis*-elements (or FeRE) that lie in the promoter region (Allen et al. 2007b). FeRE-mediated regulation occurs via transcriptional activation of target genes whose mRNA abundance is increased when Fe is deficient. It is likely that the FeREs are the binding sites of a yet-to-be discovered transcriptional activator that stimulates the transcription of the *FEA*, *FOX1*, and *FTR1* genes (and additional target genes such as *FER1* or *MSD3*) in response to Fe deficiency. Molecular dissection of the *FOX1* promoter by mutagenesis identified two separate FeREs, FeRE1 (CACACG) and FeRE2 (CACGCG) which are both required for Fe-dependent transcriptional stimulation of the *FOX1* gene (Deng and Eriksson, 2007). A similar study using the *ATX1* promoter led to the definition of a distinct FeRE sequence (GNNGCNNTGGCATNT) (Fei and Deng, 2007). It seems that at least two types of FeREs are involved in the Fe-dependent transcriptional regulation of genes in *Chlamydomonas*. It is conceivable that the two types of FeREs are recognized by distinct transcriptional factors in response to the Fe nutritional status of the cell. Note that neither FeRE bears any similarity to *cis*-elements that were defined in Fe regulatory circuits in other organisms.

B. Copper

1. Involvement of a reductase and a transporter in Cu assimilation

Cu is an essential micronutrient because of its role as a metal cofactor in enzymes that catalyze reactions involving redox chemistry or oxygen. In *Chlamydomonas*, cytochrome *c* oxidase in the mitochondria, plastocyanin in the thylakoid lumen, and the multicopper oxidase FOX1 at the plasma membrane are the three most abundant Cu-utilizing proteins.

In *Chlamydomonas* Cu uptake is regulated in response to Cu availability (Hill et al., 1996). Investigation of Cu nutrition has led to the definition of a saturable and temperature-dependent Cu uptake pathway. This high-affinity uptake system is more active in cells that are adapted to Cu deficiency than in cells grown in Cu-replete conditions (Hill et al., 1996). *Chlamydomonas* grown in Cu-deficient medium are more sensitive to Ag(I)-induced toxicity relative to cells grown in Cu-replete conditions (Howe and Merchant, 1992). This observation was attributed to the increased capacity for Ag(I) uptake in Cu-deficient cells that are responding to Cu starvation by increasing the Cu uptake components. The identity of the Cu transporter is not established but four candidate proteins of the COPT/CTR family with selectivity for Cu(I) (Petris, 2004) are encoded in *Chlamydomonas* (see section II.A.6 and Table 10.2). One of these transporters is likely to constitute the major Cu assimilatory transporter at the plasma membrane, but this awaits experimental demonstration.

A Cu(II) reductase activity is also increased in Cu-deficient compared with Cu-replete cells (Hill et al., 1996). These findings support the involvement of a Cu(II) reductase in the Cu uptake pathway, analogous to what is observed for Fe assimilation (see section IV.A). The identity of the Cu(II) reductase is unknown and it is also unclear if the reductase is Cu-specific, or is of broad metal specificity. Biochemical studies have shown, however, that that the Cu(II) reductase activity is also induced under Fe deficiency (Weger, 1999) and Fe(II) uptake is inhibited by excess Cu(II) (Eckhardt and Buckhout, 1998), suggesting that the same enzyme might be responsible for both Cu(II) and Fe(III) reductase activities (Figure 10.1). Nevertheless it appears that Cu(II) ions need to be reduced to Cu(I) prior to uptake by the assimilatory transporter (COPT1 and/or CTR1 to 3). In summary, the Cu uptake pathway in *Chlamydomonas* is defined by a cell surface Cu(II) reductase and a Cu(I) transporter, similarly to Cu assimilation routes operating in other eukaryotes (Hassett and Kosman, 1995; Rees and Thiele, 2004).

2. Intracellular Cu distribution

Following its uptake, Cu needs to be distributed and incorporated into specific apoprotein targets such as apoplastocyanin, apoforms of Cox1 and COX2 (the two Cu-containing subunits of cytochrome *c* oxidase), and apoferroxidase. The mechanisms of Cu delivery to apoprotein substrates that reside in different compartments are controlled by distributive components such as Cu chaperones and Cu transporters. The mechanism for the incorporation of Cu into apoferroxidase is likely to be similar to the one proposed for mammalian and fungal ferroxidases, depending on the cytosolic Cu chaperone ATX1 and the Cu-transporting P-type ATPase CTP1 (see section II.A.5).

It can be predicted that two of the other candidate P-type ATPases, CTP2 and CTP3, correspond to *Arabidopsis* PAA1 and PAA2 and cyanobacterial CtaA and PacS, which distribute Cu across the plastid envelope (or cyanobacterial membrane) and thylakoid membrane, respectively (Kanamaru et al., 1994; Phung et al., 1994; Tottey et al., 2001). Definitive subcellular localization and functional dissection is required to distinguish the site of action of these P-type ATPases and confirm their involvement in Cu distribution. The recruitment of a metal chaperone in the delivery of Cu to apoplastocyanin in the thylakoid lumen has been postulated for the *Chlamydomonas PCY2* locus, whose function is required for stable holoplastocyanin accumulation (H.H. Li et al., 1996). The *pcy2-1* mutant displays a weak non-photosynthetic phenotype in Cu-supplemented medium (with apoplastocyanin accumulation), but not in Cu-deficient medium (where plastocyanin no longer participates in photosynthesis). By analogy to functions required for Cu insertion into cytochrome *c* oxidase (see below), one possible explanation is that the *PCY2* gene product acts as a Cu/apoprotein chaperone in the thylakoid lumen. The fact that metal insertion into apoplastocyanin is metal-selective *in vivo* supports the view that holoplastocyanin assembly is a catalyzed process (Hill et al., 1991), but this catalysis could occur at the level of copper delivery to the lumen via PAA1 and PAA2 homologues rather than via a metallochaperone in the lumen.

Incorporation of three Cu atoms into cytochrome *c* oxidase occurs in the mitochondrial intermembrane space and requires the activity of both Cu chaperones COX17 and COX19, and assembly factors COX11, COX23, and SCO1/2 (for reviews see Carr and Winge, 2003; Cobine et al., 2006; Fontanesi et al., 2006). While COX17 and SCO1 define a pathway for the delivery of two Cu atoms to COX2 (CuA site), COX17 and COX11 are in a pathway for CuB site formation (one Cu atom) in COX1. The actual roles of COX19 and COX23 remain unresolved. Homologues of these Cu distributing factors are encoded in *Chlamydomonas*, suggesting that the mechanisms for Cu delivery to cytochrome *c* oxidase are similar to those in fungi and mammals (Table 10.3; La Fontaine et al., 2002b; Cardol et al., 2005).

3. *Metabolic adaptations to Cu deficiency*

a. *Chlamydomonas* as a model for Cu deficiency

Studies of metabolic responses to Cu nutritional state in other organisms have historically focused on the toxicity aspect of Cu overload. Indeed, Cu deficiency is much more difficult to establish under laboratory conditions, since Cu is needed only at trace levels. Nevertheless, cases of Cu deficiency are not uncommon in natural environments and can also occur as a result of a genetic disease (reviewed by Rossi et al., 2006). *Chlamydomonas* has been a good model to dissect metabolic adaptation to Cu deficiency because the majority of the copper in the cell is found in just a few proteins and its nutritional status is easily manipulated (Quinn and Merchant, 1998). Moreover *Chlamydomonas* sustains photosynthesis under Cu deficiency when holoplastocyanin no longer accumulates, suggesting the existence of regulatory circuits and backup mechanisms (Wood, 1978; Quinn and Merchant, 1998).

b. Backup enzymes

Plastocyanin is an essential electron carrier which shuttles the electrons between cytochrome b_6f and PS I. Loss-of-function mutations in *PCY1*, the apoplastocyanin-encoding gene, result in the loss of photoautotrophic growth when *Chlamydomonas* is grown under Cu-replete conditions, but not under Cu deficiency (Gorman and Levine, 1965; Wood, 1978; H.H. Li et al., 1996; Quinn et al., 1993; Quinn and Merchant, 1998). Cells grown in Cu-deficient conditions remain photosynthetically competent because they are able to substitute plastocyanin with cytochrome c_6, a heme-containing protein that is functionally equivalent (Wood, 1978; Merchant and Bogorad, 1987b). The *CYC6* gene, which encodes cytochrome c_6, is transcriptionally responsive to Cu status (Table 10.7), with its expression being undetectable except under Cu deficiency (Merchant and Bogorad, 1987b; Quinn and Merchant, 1995). Transcriptional activation of *CYC6* is paralleled by loss of holoplastocyanin, resulting from induced proteolysis of the apoprotein in the thylakoid lumen (Li and Merchant, 1995). The replacement of plastocyanin with cytochrome c_6 under Cu deficiency is not restricted to *Chlamydomonas*, also occurring in many algae and cyanobacteria (Sandmann et al., 1983).

c. Transcriptional activation under Cu deficiency

In addition to plastocyanin loss and cytochrome c_6 gain, Cu deficiency in *Chlamydomonas* elicits a change in the transcriptional activities of three other genes: *CPX1*, *CRD1*, and *CTH1* (Table 10.7). All three genes are involved in tetrapyrrole metabolism, but the biological significance of this regulation is not clearly understood.

CPX1 encodes a soluble plastid coproporphyrinogen III oxidase whose accumulation increases in Cu-deficient medium (Hill and Merchant, 1995),

Table 10.7 Genes involved in metabolic adaptation to copper deficiency

Gene name	Gene product and function	Regulation in -Cu	Expression evidence[a]	Accession #	Homologue in *Volvox*[b]	References[c]
PCY1	Plastocyanin, electron transfer in plastid	-	RNA blot, protein purified	AAA33078	Y	[1–6]
CYC6	Cytochrome c_6, back-up of plastocyanin in Cu deficiency	Up[d]	RNA blot, purified protein	AAB00729	N	[1, 5–6, 7]
CPX1	Coproporphyrinogen III oxidase, plastid heme synthesis	Up[d]	RNA blot, protein sequence	AAD28475	Y	[1, 6, 8]
CPX2	Coproporphyrinogen III oxidase, plastid heme synthesis	-	EST	EDP06766	Y	[1, 6]
CRD1/ CHL27A	Aerobic oxidative cyclase, tetrapyrrole biosynthetic pathway	Up[d]	RNA blot, immunoblot	AAF65221	Y	[1, 6, 9–10]
CTH1/ CHL27B	Aerobic oxidative cyclase, tetrapyrrole biosynthetic pathway	Down[d]	RNA blot, immunoblot	AAK32149	Y	[1, 6, 10]
CRR1	Transcription factor, regulation of Cu deficiency response	-	RNA blot	AAZ81510	Y	[1, 6, 11–12]

[a]*Expression evidence: EST (Expressed Sequence Tag).*

[b]*Presence of homologous sequence in the genome of Volvox carteri, vista Volvox track at http://genome.jgi-psf.org/Chlre3/Chlre3.home.html (March 2007).*

[c]*References: [1] http://genome.jgi-psf.org/Chlre3/Chlre3.home.html; [2] Gorman and Levine (1966); [3] Merchant et al. (1990); [4] Quinn et al. (1993); [5] Quinn and Merchant (1998); [6] Merchant et al. (2006); [7] Merchant and Bogorad (1987); [8] Hill and Merchant (1995); [9] Moseley et al. (2000); [10] Moseley et al. (2002b); [11] Eriksson et al. (2004); [12] Kropat et al. (2005). Note that the annotations provided in reference 3 were based on the release 2 of the genome.*

[d]*Regulation mediated via CuRE* cis *elements and Crr1. Note that the Cu transporters (Copt1, Ctr1–Ctr3, Hma1 and Ctp1Ctp3) described in Table 10.2 are also likely to contribute to metabolic adaptation to copper deficiency in Chlamydomonas (see details in the text).*

and whose enzymatic activity increases accordingly (Quinn et al., 1999). A second coproporphyrinogen oxidase is encoded by *CPX2*. This isoform is closely related to the mammalian mitochondrial enzyme, whereas the *CPX1* product shows greatest sequence similarity to the plant plastid enzyme (Quinn et al., 1999). The subcellular localization of coproporphyrinogen oxidase 2 is unknown and *CPX2* mRNA abundance is not regulated by Cu nutrition (Merchant et al., 2006). In many organisms, coproporphyrinogen oxidase can become the rate-limiting step when there is a high demand for heme or under low oxygen conditions. Under Cu deficiency, cytochrome c_6 synthesis is induced in order to cope with loss of plastocyanin and there is an increased demand for its heme cofactor. The Cu deficiency-induced activation of *CPX1* might be interpreted as a response to meet this requirement.

CRD1 (*CHL27A*) and *CTH1* (*CHL27B*) encode isoforms of the aerobic oxidative cyclase (Moseley et al., 2000, 2002b). This di-iron enzyme catalyzes the oxidative cyclization of Mg-protoporphyrin IX monomethylester in aerobic chlorophyll synthesis (Tottey et al., 2003). CRD1 and CTH1 accumulation is reciprocal: CRD1 abundance is increased in Cu-depleted cells, whereas CTH1 accumulates in Cu-replete cells. In *Arabidopsis*, a single *CHL27* gene specifies a protein with dual localization to the plastid envelope and the thylakoid membrane (Tottey et al., 2003). The localizations of *Chlamydomonas* CRD1 and CTH1 have not been established. *crd1* mutants display a Cu nutrition-dependent pale/chlorotic phenotype characterized by absence of PS I and LHCI, and a reduced level of PS II and LHCII (Moseley et al., 2002a). The fact that particular chlorophyll-binding proteins are lost as a consequence of the *crd1* mutation suggests that the photosynthetic electron transfer chain undergoes remodeling to adapt to Cu deficiency. One possible explanation is that a new antenna complex with a modified subunit composition is required to accommodate the reduced catalytic efficiency of cytochrome c_6 relative to plastocyanin (Sommer et al., 2004).

The fact that Cu nutrition regulates the genes for two key oxygen-dependent steps in the tetrapyrrole pathway suggests a connection between Cu and tetrapyrrole metabolism. Indeed, *CYC6*, *CPX1*, and *CRD1* respond to oxygen deprivation as well as Cu starvation. This response is physiologically relevant and occurs through the same transcription factor (CRR1) that regulates Cu homeostasis (see section IV.B.3.d; Moseley et al., 2000; Eriksson et al., 2004; Kropat et al., 2005).

Microarray analysis confirmed the up-regulation of *CYC6*, *CPX1*, and *CRD1* in response to Cu deficiency, but also revealed induction of several genes encoding enzymes involved in photosynthesis, the pentose phosphate pathway and the mitochondrial electron transfer chain (Jamers et al., 2006). Although the results need to be validated by RNA blots or quantitative PCR, the work suggests that examination of energy metabolism in Cu-deficient cells is warranted.

d. Cu-responsive signal transduction

A *cis*-element named CuRE (Cu Responsive Element) responsible for transcriptional activation of target genes in response to Cu deficiency was defined molecularly through dissection of the *CYC6* and *CPX1* promoters (Quinn and Merchant, 1995; Quinn et al., 2000). The CuREs are able to confer Cu-dependent expression to a reporter gene and consist of a core sequence GTAC, which is absolutely necessary but not sufficient for Cu-dependent transcriptional activity. The *CYC6* promoter includes two CuREs, each of which is necessary on its own for Cu responsiveness, while a single CuRE mediates Cu-dependent transcriptional activation of *CPX1*. Putative CuREs are present in the *CRD1* promoter but remain to be experimentally tested. CuREs are required for responses to both hypoxia and Cu-deficiency (Quinn et al. 2000), although their corresponding DNA binding proteins may be different.

CRR1 is a regulatory gene whose involvement in nutritional Cu homeostasis was established genetically through a screen for mutants that display Cu-deficiency conditional phenotypes (Table 10.7; Eriksson et al., 2004). *crr1* mutations result in impaired growth under Cu deficiency that can be alleviated by addition of Cu to the medium. *CPX1*, *CYC6*, and *CRD1* transcription no longer respond to the Cu nutritional status in *crr1* mutants, indicating that *CRR1* controls their transcription, and possibly that of other genes induced by Cu deficiency. Apoplastocyanin is no longer degraded in *crr1*, suggesting that the gene encoding the protease responsible for apoplastocyanin proteolysis is one of those under *CRR1* regulation.

The CRR1 protein sequence is unique but includes a Zn-dependent DNA-binding domain called SBP (Squamosa Binding Protein; Klein et al., 1996), and a nuclear localization signal (Kropat et al., 2005). The presence of ankyrin repeats suggests that CRR1 might interact with other protein partners, and a Cys-rich domain (with similarity to metallothionein) indicates possible redox or metal sensing. The presence of metal-binding motifs such as the SBP domain and the Cys-rich segment are consistent with a role as a Cu sensor, but it is also possible that the sensing activity resides in a protein interacting with CRR1, for example via the ankyrin repeats. The CRR1 SBP domain displays Zn-dependent binding activity to a CuRE sequence *in vitro*, and competition experiments indicate that the SBP domain is 10-fold more selective toward a wild-type CuRE sequence versus a version carrying a point mutation in the GTAC core, a mutation which also confers loss of Cu responsiveness *in vivo* (Kropat et al., 2005).

Each of the known target genes of Cu deficiency is also transcriptionally activated by Ni or Co ions (Quinn et al., 2003), and excess Cu cannot reverse the effect of Ni on the transcriptional response. The response to Ni ions requires at least one CuRE and is *CRR1*-dependent, suggesting that Ni interferes with a component of the copper status signal transduction pathway. One hypothesis is that Ni and Co bind at the Cu-sensing site (CRR1 or another Cu-binding molecule) in an irreversible manner that prevents the access of Cu, and therefore the Cu-dependent regulation of CRR1 activity. In such a model, Ni or Co binding to the metal sensing site results in constitutive activation of CRR1.

The reciprocal Cu-responsive accumulation patterns of CRD1 and CTH1 are also under control of CRR1 (Moseley et al., 2002b). In Cu-replete cells, *CRD1* is expressed at a low basal level and *CTH1* produces a 2-kb mRNA, leading to the preferential accumulation of CTH1. In Cu-deficient conditions, while increased CRD1 abundance can be explained by CRR1-mediated transcriptional activation, decreased accumulation of CTH1 is due to the activity of an upstream CuRE-containing promoter in the *CTH1* transcription unit. This promoter drives the CRR1-dependent transcription of a 5′-extended 3-kb mRNA that does not lead to a translated gene product because of the presence of small ORFs in all frames in the 5′ UTR. It is likely that expression of the functional *CTH1* 2-kb transcript is blocked by occlusion of the promoter under Cu-deficient conditions.

Note that the SPB domain present in CRR1 is a plant-specific DNA binding domain that is also detected in several proteins from vascular plants. In *Arabidopsis*, some SPB domain-containing proteins also display motifs found in CRR1. It is conceivable that the activity of these proteins is controlled by Cu nutrition, similarly to *Chlamydomonas* CRR1.

C. Manganese and selenium

The study of Mn and Se nutrition is technically challenging because both metals are essential for growth in most organisms, yet required only at trace levels. Mn is an essential micronutrient because of its participation as a redox cofactor or an activator at the metal-binding sites of several enzymes (reviewed by Christianson, 1997). For instance, Mn is the redox-active cofactor of MnSOD, and it activates bound water for nucleophile-based hydrolysis of the guanidinium group of arginine in arginase, an arginine catabolic enzyme (van Loon et al., 1986; Lebovitz et al., 1996). In photosynthetic organisms, Mn occurs as a Mn_4Ca cluster involved in water splitting (reviewed by Merchant and Sawaya, 2005). *Chlamydomonas* has a higher demand for Mn as compared to non-photosynthetic organisms. Indeed, cells growing heterotrophically in Mn-deficient medium display a PS II-deficient phenotype accompanied by decreased abundance of D1, the PS II subunit that binds the Mn_4Ca cluster, and release of the oxygen evolving enhancer proteins from the thylakoid membrane (Allen et al., 2007a). Mn deficiency also impacts MnSOD activity proportionally. Mn-deficient cells experience specific peroxide-sensitive stress and the *GPX*s, *APX*, and *MSRA2* genes (encoding the anti-oxidant enzymes glutathione peroxidase, ascorbate peroxidase, and methionine sulfoxide reductase) are slightly up-regulated. The Mn, Fe, and P contents of cells grown in Mn-deficient medium are also reduced proportionately. As expected, Mn-deficient cells with reduced Fe content display a chlorotic phenotype that can be relieved by provision of extra Fe in the medium. It is conceivable that the Fe deficiency phenotype of Mn-depleted cells is the consequence of a defense mechanism aimed at reducing intracellular Fe content in order to prevent oxidative stress.

The lowered P content of Mn-deficient cells suggests that Mn starvation causes reduced P uptake. Phosphate uptake (see Chapter 6) occurs through PHO84-type transporters, which use a divalent cation as the counterion for symport (van Veen et al., 1994; Persson et al., 1999). The *PTA3* and *PTA4* genes encode such transporters and are up-regulated in response to Mn deficiency, and it is likely that Mn^{2+} is their preferred counterion. The induction of *NRAMP1* and *MTP5*, which encode candidate transporters (see section II.A.2; Table 10.2 and Figure 10.1) in Mn- but not Fe-deficient cells, suggests that the corresponding gene products may also operate in a

Mn^{2+}-selective assimilation pathway. It remains to be established which transporters function in the distribution of Mn to plastids and mitochondria.

Selenium is a rare element, present in the form of selenocysteine, a modified amino acid incorporated in proteins found in bacteria, archaea and many eukaryotes (reviewed by Rother et al., 2001; Kyriakopoulos and Behne, 2002). In bacteria, selenoproteins are recruited for catabolic reactions, whereas in eukaryotes they can have both biosynthetic and antioxidant functions. The discovery of selenoproteins and components for selenocysteine incorporation in *Chlamydomonas*, a member of the plant kingdom, came as a surprise because plants do not contain selenoproteins (Novoselov et al., 2002; Rao et al., 2003). A survey of the *Chlamydomonas* genome led to the identification of the complete selenoproteome defined by 12 selenoproteins representing 10 families (Grossman et al., 2007; Lobanov et al., 2007), among which 5 have homologues in other species that function in redox metabolism, with the functions of the remaining seven currently unknown. The algae *Ostreococcus* and *Cyanidioschyzon* also use selenoenzymes (Matsuzaki et al., 2004; Derelle et al., 2006; Palenik et al., 2007). *Chlamydomonas* growth does not appear to depend on added Se, presumably because sufficient Se is present as a trace contaminant in other media components. This is reasonable considering that the abundance of selenoproteins in *Chlamydomonas* (and other organisms) is low. However, it is conceivable that the demand for Se increases under stress conditions where redox metabolism and hence participation of selenoproteins is stimulated.

V. CONCLUSIONS AND PERSPECTIVES

For several decades, *Chlamydomonas* has been used as a model to study the physiology, biochemistry and molecular genetics of transition metal homeostasis and tolerance at the cellular level. As these data are integrated with genome information (Merchant et al., 2007), our understanding of the metal homeostasis network of a model photosynthetic cell will continue to improve.

Based on previous work (Rosakis and Köster, 2004; Hanikenne et al., 2005a; Merchant et al., 2006) and genome annotation, Table 10.2 presents a compilation of the *Chlamydomonas* metal transporter complement. Putative functions are proposed on the basis of their homology and phylogenetic relationship to functionally characterized yeast, human and plant proteins. Further experimental work will be necessary to establish their functions, metal specificities and localizations, and will probably rely in part on developing reverse genetic techniques (Rohr et al., 2004; Schroda, 2006).

The predicted metal transporter complement of *Chlamydomonas* reveals both animal- and plant-like features. *Chlamydomonas* possesses several transporters that are present in plants but absent in animals such as subfamily

I ZIPs, a putative HMA1 protein for Cu delivery to the plastid, the RET1 NRAMP protein homologous to the plant EIN2 protein and Ni/Co transporters (section II.A, Tables 10.1 and 10.2). On the other hand, high-affinity Fe uptake is driven by an animal-like Cu-dependent system, although plant-like Cu-independent systems might also contribute (section IV.A.1). Another example is the presence of selenoproteins in *Chlamydomonas*, as discussed above. Although present in prokaryotes and many eukaryotes, selenoproteins are absent in vascular plants. Thus, it appears that *Chlamydomonas* has retained various metal-related systems that have undergone specialization and/or gene loss in the animal versus the plant lineages.

Physiological studies, genetic screens and functional analyses of candidate genes have advanced our understanding of the cellular adaptations to metal deficiency or overload in *Chlamydomonas*. Transcriptomic (Eberhard et al., 2006), proteomic (Stauber and Hippler, 2004; Allmer et al., 2006; Muck et al., 2006) and metabolomic (Bölling and Fiehn, 2005) tools are starting to be applied to study metal homeostasis (Gillet et al., 2006; Jamers et al., 2006; Reinhardt et al., 2006). In the future, the use of these techniques will certainly increase our understanding of the global metabolic adaptations to metal stress in photosynthetic cells.

Chlamydomonas species live in a range of environments and geographic locations (Harris, 1989). For example, Aguilera and Amils (2005) described three *Chlamydomonas* strains isolated from an acidic- and heavy-metal rich river in Spain. This natural diversity could be exploited to analyze the genetic variation associated with the metal homeostasis network within the species and genus (Borevitz and Ecker, 2004; Koornneef et al., 2004; Weigel and Nordborg, 2005).

Future research is likely to tackle key unresolved questions, some of which are:

1. Is there a hierarchy of metal distribution to cellular compartments depending on metal availability and/or physiological needs? This question is of particular relevance in the context of metal deficiency, where it might be necessary to recycle metal from abundant metalloproteins for re-allocation to other metal-utilizing pathways. For instance, it is conceivable that Cu liberated by the degradation of plastocyanin in the thylakoid lumen is re-allocated to cytochrome *c* oxidase in the mitochondria.

2. How does deficiency or overload of one metal (e.g. Zn) influence the homeostasis of other micronutrients (e.g. Fe, Mn, or Cu) at the cellular level? Complex interactions are likely to occur between the homeostasic networks of different metals. It is well known that Fe deficiency is accompanied by secondary Cd toxicity in both plant and animal systems (Thomine et al., 2000, 2003; Connolly et al., 2002; Lombi et al., 2002; Bressler et al., 2004). Another example of such

interactions is found in the Zn hyperaccumulating plant species *A. halleri* and *T. caerulecens*, where adaptation to high Zn accumulation in shoot tissues requires the constitutive overexpression of genes associated with Fe and Cu homeostasis (Chiang et al., 2006; Filatov et al., 2006; Talke et al., 2006; van de Mortel et al., 2006).

3. What are the components involved in metal sensing, signaling and regulation, and what is the cross-talk between the homeostatic network of metals and other micro- or macronutrients? These regulatory mechanisms remain largely unknown in photosynthetic cells with a few exceptions. For example, the basic helix-loop-helix transcription factor FIT1/FER in vascular plants partly co-ordinates the iron-deficiency response in *Arabidopsis* and tomato (Colangelo and Guerinot, 2004; Brumbarova and Bauer, 2005). In *Chlamydomonas*, the central roles of the CuRE *cis* element and the CRR1 transcription factor in the regulation of the Cu deficiency response were described in section IV.B.3. Further progress could be made by comparing promoter regions of homologous genes in *Chlamydomonas* and *Volvox* to identify conserved regulatory motifs. In addition, a systematic structure-function analysis of algal and plant metal transporters will allow the identification of specific sequence motifs essential for interaction with partners, regulators, or chaperones.

4. Can we find "chaperones" for Zn, Fe, or Mn in *Chlamydomonas*? Although intracellular trafficking of metals and their proper insertion into target proteins require metallochaperones (O'Halloran and Culotta, 2000), only a few Cu metallochaperones have been identified so far in *Chlamydomonas* (Table 10.3) and more generally, in photosynthetic cells. The roles of low molecular weight metal chelators such as organic acids, GSH, or PC in metal trafficking remain to be elucidated.

5. Finally, several attempts have been made to use *Chlamydomonas* as a biotechnological tool for bioremediation. The biosorption capacities for Hg, Cd, Pb, and Cr of *Chlamydomonas* biomass or of *Chlamydomonas* immobilized on alginate beads have been studied in detail (Arica et al., 2005; Tüzün et al., 2005; Bayramoglu et al., 2006). Several genes have been expressed in *Chlamydomonas* to increase metal tolerance and metal accumulation capacities. For example, the expression of a chicken metallothionein gene enhances Cd tolerance and binding capacity (Hua et al., 1999). Moreover, the overexpression of a gene encoding the enzyme catalyzing the first step of Pro biosynthesis from Glu leads to increased Cd tolerance and accumulation (Siripornadulsil et al., 2002). The data suggested that the free Pro could act as an antioxidant in the Cd-treated cells,

resulting in a more reducing cellular environment (measured as an increased GSH/GSSH ratios) that in turn facilitates PC synthesis (Siripornadulsil et al., 2002).

The identification of genes encoding transporters for a range of transition metals provides additional targets to engineer metal accumulation. Such approaches could represent efficient and cost-effective solutions to clean-up polluted waters and soils.

ACKNOWLEDGMENTS

Research projects in the authors' laboratories are funded by the "Fonds de la Recherche Scientifique-FNRS, Belgium" (Grants No. 2.4638.05, 2.4540.06 and 2.4583.08) (M.H.), the "Fonds spéciaux pour la Recherche, University of Liège, Belgium" (M.H.), the National Institutes of Health GM42143 (SM), the Department of Energy DE-FG02-04ER15529 (SM), and the College of Biological Sciences, the College of Medicine and the Dorothy M. Davis Heart and Lung Institute at The Ohio State University (P.H.). P.H. wishes also to acknowledge the Institute of Mitochondrial Biology at The Ohio State University for intellectual input. M.H. thanks Prof. P. Motte and Dr. D. Baurain for helpful discussions. M.H. is a Postdoctoral Researcher of the "Fonds de la Recherche Scientifique-FNRS" (Belgium).

REFERENCES

Abdel-Ghany, S.E., Burkhead, J.L., Gogolin, K.A., Andrés-Colás, N., Bodecker, J.R., Puig, S., Peñarrubia, L., and Pilon, M. (2005a). AtCCS is a functional homolog of the yeast copper chaperone Ccs1/Lys7. *FEBS Lett.* **579**, 2307–2312.

Abdel-Ghany, S.E., Müller-Moulé, P., Niyogi, K.K., Pilon, M., and Shikanai, T. (2005b). Two P-type ATPases are required for copper delivery in *Arabidopsis thaliana* chloroplasts. *Plant Cell* **17**, 1233–1251.

Adamson, P. (2004). *Vitamin and mineral deficiency: A global progress report*. The Micronutient Initiative and UNICEF, Ottawa, Canada.

Aguilera, A. and Amils, R. (2005). Tolerance to cadmium in *Chlamydomonas* sp. (Chlorophyta) strains isolated from an extreme acidic environment, the Tinto River (SW, Spain). *Aquat. Toxicol.* **75**, 316–329.

Allen, M.D., Kropat, J., Tottey, S., Del Campo, J.A., and Merchant, S.S. (2007a). Manganese deficiency in *Chlamydomonas* results in loss of Photosystem II and MnSOD function, sensitivity to peroxides, and secondary phosphorus and iron deficiency. *Plant Physiol.* **143**, 263–277.

Allen, M.D., Del Campo, J.A., Kropat, J., and Merchant, S.S. (2007b). *FEA1*, *FEA2* and *FRE1*, encoding two homologous secreted proteins and a candidate ferrireductase, are expressed coordinately with *FOX1* and *FTR1* in iron-deficient *Chlamydomonas reinhardtii*. *Eukaryotic Cell* **6**, 1841–1852.

Allikmets, R., Raskind, W.H., Hutchinson, A., Schueck, N.D., Dean, M., and Koeller, D.M. (1999). Mutation of a putative mitochondrial iron transporter gene (ABC7) in X-linked sideroblastic anemia and ataxia (XLSA/A). *Hum. Mol. Genet.* **8**, 743–749.

Allmer, J., Naumann, B., Markert, C., Zhang, M., and Hippler, M. (2006). Mass spectrometric genomic data mining: Novel insights into bioenergetic pathways in *Chlamydomonas reinhardtii*. *Proteomics* **6**, 6207–6220.

Alonso, J.M., Hirayama, T., Roman, G., Nourizadeh, S., and Ecker, J.R. (1999). EIN2, a bifunctional transducer of ethylene and stress responses in *Arabidopsis*. *Science* **284**, 2148–2152.

Andrés-Colás, N., Sancenón, V., Rodríguez-Navarro, S., Mayo, S., Thiele, D.J., Ecker, J.R., Puig, S., and Peñarrubia, L. (2006). The *Arabidopsis* heavy metal P-type ATPase HMA5 interacts with metallochaperones and functions in copper detoxification of roots. *Plant J.* **45**, 225–236.

Arica, M.Y., Tüzün, I., Yalçin, E., Ince, Ö., and Bayramoglu, G. (2005). Utilisation of native, heat and acid-treated microalgae *Chlamydomonas reinhardtii* preparations for biosorption of Cr(VI) ions. *Process Biochem.* **40**, 2351–2358.

Arrivault, S., Senger, T., and Krämer, U. (2006). The *Arabidopsis* metal tolerance protein AtMTP3 maintains metal homeostasis by mediating Zn exclusion from the shoot under Fe deficiency and Zn oversupply. *Plant J.* **46**, 861 879.

Askwith, C. and Kaplan, J. (1998). Iron and copper transport in yeast and its relevance to human disease. *Trends Biochem. Sci.* **23**, 135–138.

Askwith, C., Eide, D., Van Ho, A., Bernard, P.S., Li, L., Davis-Kaplan, S., Sipe, D. M., and Kaplan, J. (1994). The *FET3* gene of *S. cerevisiae* encodes a multicopper oxidase required for ferrous iron uptake. *Cell* **76**, 403–410.

Askwith, C.C., de Silva, D., and Kaplan, J. (1996). Molecular biology of iron acquisition in *Saccharomyces cerevisiae*. *Mol. Microbiol.* **20**, 27–34.

Assuncão, A.G.L., Da Costa Martins, P., De Folter, S., Vooijs, R., Schat, H., and Aarts, M.G.M. (2001). Elevated expression of metal transporter genes in three accessions of the metal hyperaccumulator *Thlaspi caerulescens*. *Plant Cell Environ.* **24**, 217–226.

Axelsen, K.B. and Palmgren, M.G. (2001). Inventory of the superfamily of P-type ion pumps in *Arabidopsis*. *Plant Physiol.* **126**, 696–706.

Baxter, I., Tchieu, J., Sussman, M.R., Boutry, M., Palmgren, M.G., Gribskov, M., Harper, J.F., and Axelsen, K.B. (2003). Genomic comparison of P-type ATPase ion pumps in *Arabidopsis* and rice. *Plant Physiol.* **132**, 618–628.

Bayramoglu, G., Tüzün, I., Celik, G., Yilmaz, M., and Arica, M.Y. (2006). Biosorption of mercury(II), cadmium(II) and lead(II) ions from aqueous system by microalgae *Chlamydomonas reinhardtii* immobilized in alginate beads. *Int. J. Miner. Process.* **81**, 35–43.

Becher, M., Talke, I.N., Krall, L., and Krämer, U. (2004). Cross-species microarray transcript profiling reveals high constitutive expression of metal homeostasis genes in shoots of the zinc hyperaccumulator *Arabidopsis halleri*. *Plant J.* **37**, 251–268.

Ben-Bassat, D., Shelef, G., Gruner, N., and Shuval, H.I. (1972). Growth of *Chlamydomonas* in a medium containing mercury. *Nature* **240**, 43–44.

Ben-Shem, A., Frolow, F., and Nelson, N. (2003). Crystal structure of plant photosystem I. *Nature* **426**, 630–635.

Benavente, L.M. and Alonso, J.M. (2006). Molecular mechanisms of ethylene signaling in *Arabidopsis*. *Mol. BioSyst.* **2**, 165–173.

Bölling, C. and Fiehn, O. (2005). Metabolite profiling of *Chlamydomonas reinhardtii* under nutrient deprivation. *Plant Physiol.* **139**, 1995–2005.

Borevitz, J.O. and Ecker, J.R. (2004). Plant genomics: The third wave. *Annu. Rev. Genomics Hum. Genet.* **5**, 443–477.

Boswell, C., Sharma, N.C., and Sahi, S.V. (2002). Copper tolerance and accumulation potential of *Chlamydomonas reinhardtii*. *Bull. Environ. Contam. Toxicol.* **69**, 546–553.

Boudreau, E., Turmel, M., Goldschmidt-Clermont, M., Rochaix, J.D., Sivan, S., Michaels, A., and Leu, S. (1997). A large open reading frame (orf1995) in the chloroplast DNA of *Chlamydomonas reinhardtii* encodes an essential protein. *Mol. Gen. Genet.* **253**, 649–653.

Bovet, L., Eggmann, T., Meylan-Bettex, M., Polier, J.E., Kammer, P., Marin, E., Feller, U., and Martinoia, E. (2003). Transcript level of AtMRPs after cadmium treatment: induction of AtMRP3. *Plant Cell Environ.* **26**, 371–381.

Bressler, J.P., Olivi, L., Cheong, J.H., Kim, Y., and Bannona, D. (2004). Divalent metal transporter 1 in lead and cadmium transport. *Ann. N. Y. Acad. Sci.* **1012**, 142–152.

Briat, J.F. and Lobréaux, S. (1998). Iron storage and ferritin in plants. *Met. Ions Biol. Syst.* **35**, 563–584.

Briat, J.F., Lobréaux, S., Grignon, N., and Vansuyt, G. (1999). Regulation of plant ferritin synthesis: How and why. *Cell. Mol. Life Sci.* **56**, 155–166.

Brumbarova, T. and Bauer, P. (2005). Iron-mediated control of the basic helix-loop-helix protein FER, a regulator of iron uptake in tomato. *Plant Physiol.* **137**, 1018–1026.

Bughio, N., Yamaguchi, H., Nishizawa, N.K., Nakanishi, H., and Mori, S. (2002). Cloning an iron-regulated metal transporter from rice. *J. Exp. Bot.* **53**, 1677–1682.

Busch, W. and Saier, M.H., Jr (2002). The transporter classification (TC) system, 2002. *Crit. Rev. Biochem. Mol. Biol.* **37**, 287–337.

Cardol, P., González-Halphen, D., Reyes-Prieto, A., Baurain, D., Matagne, R.F., and Remacle, C. (2005). The mitochondrial oxidative phosphorylation proteome of *Chlamydomonas reinhardtii* deduced from the Genome Sequencing Project. *Plant Physiol.* **137**, 447–459.

Carr, H.S. and Winge, D.R. (2003). Assembly of cytochrome *c* oxidase within the mitochondrion. *Acc. Chem. Res.* **36**, 309–316.

Chasteen, N.D. and Harrison, P.M. (1999). Mineralization in ferritin: An efficient means of iron storage. *J. Struct. Biol.* **126**, 182–194.

Chen, J.C., Hsieh, S.I., Kropat, J., and Merchant, S.S. (2008). A ferroxidase encoded by *FOX1* contributes to iron assimilation under conditions of poor iron nutrition in *Chlamydomonas*. *Eukaryotic Cell* **7**, 541–545.

Chiang, H.C., Lo, J.C., and Yeh, K.C. (2006). Genes associated with heavy metal tolerance and accumulation in Zn/Cd hyperaccumulator *Arabidopsis halleri*: A genomic survey with cDNA microarray. *Environ. Sci. Technol.* **40**, 6792–6798.

Christianson, D.W. (1997). Structural chemistry and biology of manganese metalloenzymes. *Prog. Biophys. Mol. Biol.* **67**, 217–252.

Ciniglia, C., Yoon, H.S., Pollio, A., Pinto, G., and Bhattacharya, D. (2004). Hidden biodiversity of the extremophilic Cyanidiales red algae. *Mol. Ecol.* **13**, 1827–1838.

Clemens, S. (2001). Molecular mechanisms of plant metal tolerance and homeostasis. *Planta* **212**, 475–486.

Clemens, S. (2006). Evolution and function of phytochelatin synthases. *J. Plant Physiol.* **163**, 319–332.

Clemens, S. and Simm, C. (2003). *Schizosaccharomyces pombe* as a model for metal homeostasis in plant cells: the phytochelatin-dependent pathway is the main cadmium detoxification mechanism. *New Phytol.* **159**, 323–330.

Cobbett, C. and Goldsbrough, P. (2002). Phytochelatins and metallothioneins: Roles in heavy metaldetoxification and homeostasis. *Annu. Rev. Plant Physiol. Plant Mol. Biol.* **53**, 159–182.

Cobbett, C.S., Hussain, D., and Haydon, M.J. (2003). Structural and functional relationships between type 1B heavy metal-transporting P-type ATPases in *Arabidopsis*. *New Phytol.* **159**, 315–321.

Cobine, P.A., Pierrel, F., and Winge, D.R. (2006). Copper trafficking to the mitochondrion and assembly of copper metalloenzymes. *Biochim. Biophys. Acta* **1763**, 759–772.

Cohen, A., Nelson, H., and Nelson, N. (2000). The family of SMF metal ion transporters in yeast cells. *J. Biol. Chem.* **275**, 33388–33394.

Colangelo, E.P. and Guerinot, M.L. (2004). The essential basic helix-loop-helix protein FIT1 is required for the iron deficiency response. *Plant Cell* **16**, 3400–3412.

Colangelo, E.P. and Guerinot, M.L. (2006). Put the metal to the petal: metal uptake and transport throughout plants. *Curr. Opin. Plant Biol.* **9**, 322–330.

Collard, J.M. and Matagne, R.F. (1990). Isolation and genetic analysis of *Chlamydomonas reinhardtii* strains resistant to cadmium. *Appl. Environ. Microbiol.* **56**, 2051–2055.

Collard, J.M. and Matagne, R.F. (1994). Cd^{2+} resistance in wild-type and mutant strains of *Chlamydomonas reinhardtii*. *Environ. Exp. Bot.* **34**, 235–244.

Connolly, E.L., Fett, J.P., and Guerinot, M.L. (2002). Expression of the IRT1 metal transporter is controlled by metals at the levels of transcript and protein accumulation. *Plant Cell* **14**, 1347–1357.

Craciun, A.R., Courbot, M., Bourgis, F., Salis, P., Saumitou-Laprade, P., and Verbruggen, N. (2006). Comparative cDNA-AFLP analysis of Cd-tolerant and -sensitive genotypes derived from crosses between the Cd hyperaccumulator *Arabidopsis halleri* and *Arabidopsis lyrata* ssp. *petraea*. *J. Exp. Bot.* **57**, 2967–2983.

Croft, M.T., Lawrence, A.D., Raux-Deery, E., Warren, M.J., and Smith, A.G. (2005). Algae acquire vitamin B12 through a symbiotic relationship with bacteria. *Nature* **438**, 90–93.

Csere, P., Lill, R., and Kispal, G. (1998). Identification of a human mitochondrial ABC transporter, the functional orthologue of yeast Atm1p. *FEBS Lett.* **441**, 266–270.

Culotta, V.C., Yang, M., and O'Halloran, T.V. (2006). Activation of superoxide dismutases: putting the metal to the pedal. *Biochim. Biophys. Acta* **1763**, 747–758.

Cunningham, K.W. and Fink, G.R. (1996). Calcineurin inhibits VCX1-dependent H^+/Ca^{2+} exchange and induces Ca^{2+} ATPases in *Saccharomyces cerevisiae*. *Mol. Cell Biol.* **16**, 2226–2237.

Curie, C. and Briat, J.F. (2003). Iron transport and signaling in plants. *Annu. Rev. Plant Biol.* **54**, 183–206.

Curie, C., Alonso, J.M., Le Jean, M., Ecker, J.R., and Briat, J.F. (2000). Involvement of NRAMP1 from *Arabidopsis thaliana* in iron transport. *Biochem. J.* **347**, 749–755.

Curie, C., Panaviene, Z., Loulergue, C., Dellaporta, S.L., Briat, J.F., and Walker, E.L. (2001). Maize yellow stripe1 encodes a membrane protein directly involved in Fe(III) uptake. *Nature* **409**, 346–349.

Danilov, R.A. and Ekelund, N.G. (2001). Effects of Cu^{2+}, Ni^{2+}, Pb^{2+}, Zn^{2+} and pentachlorophenol on photosynthesis and motility in *Chlamydomonas reinhardtii* in short-term exposure experiments. *BMC Ecol.* **1**, 1.

de Hostos, E.L., Togasaki, R.K., and Grossman, A. (1988). Purification and biosynthesis of a derepressible periplasmic arylsulfatase from *Chlamydomonas reinhardtii*. *J. Cell Biol.* **106**, 29–37.

Dean, M., Rzhetsky, A., and Allikmets, R. (2003). Human and *Drosophila* ABC proteins. In: *ABC Proteins-From Bacteria to Man* (I.B. Holland, S.P.C. Cole, K. Kuchler, and C.F. Higgins, Eds.), pp. 47–61. Academic Press, London.

Decottignies, A. and Goffeau, A. (1997). Complete inventory of the yeast ABC proteins. *Nat. Genet.* **15**, 137–145.

Delhaize, E., Kataoka, T., Hebb, D.M., White, R.G., and Ryan, P.R. (2003). Genes encoding proteins of the cation diffusion facilitator family that confer manganese tolerance. *Plant Cell* **15**, 1131–1142.

Delhaize, E., Gruber, B.D., Pittman, J.K., White, R.G., Leung, H., Miao, Y., Jiang, L., Ryan, P.R., and Richardsonn, A.E. (2007). A role for the *AtMTP11* gene of *Arabidopsis* in manganese transport and tolerance. *Plant J.* **51**, 198–210.

Deng, X. and Eriksson, M. (2007). Two iron-responsive promoter elements control the expression of *FOX1* in *Chlamydomonas reinhardtii*. *Eukaryotic Cell* **6**, 2163–2167.

Derelle, E., Ferraz, C., Rombauts, S., Rouzé, P., Worden, A.Z., Robbens, S., Partensky, F., Degroeve, S., Echeynie, S., Cooke, R., Saeys, Y., Wuyts, J., Jabbari, K., Bowler, C., Panaud, O., Piegu, B., Ball, S.G., Ral, J.P., Bouget, F.Y., Piganeau, G., De Baets, B., Picard, A., Delseny, M., Demaille, J., Van de Peer, Y., and Moreau, H. (2006). Genome analysis of the smallest free-living eukaryote *Ostreococcus tauri* unveils many unique features. *Proc. Natl. Acad. Sci. U. S. A.* **103**, 11647–11652.

Desbrosses-Fonrouge, A.G., Voigt, K., Schröder, A., Arrivault, S., Thomine, S., and Krämer, U. (2005). *Arabidopsis thaliana* MTP1 is a Zn transporter in the vacuolar membrane which mediates Zn detoxification and drives leaf Zn accumulation. *FEBS Lett.* **579**, 4165–4174.

Devriese, M., Tsakaloudi, V., Garbayo, I., León, R., Vílchez, C., and Vigara, J. (2001). Effect of heavy metals on nitrate assimilation in the eukaryotic microalga *Chlamydomonas reinhardtii*. *Plant Physiol. Biochem.* **39**, 443–448.

DiDonato, R.J., Jr., Roberts, L.A., Sanderson, T., Eisley, R.B., and Walker, E.L. (2004). *Arabidopsis* Yellow Stripe-Like2 (YSL2): A metal-regulated gene encoding a plasma membrane transporter of nicotianamine-metal complexes. *Plant J.* **39**, 403–414.

Domínguez, M.J., Gutiérrez, F., León, R., Vilchez, C., Vega, J.M., and Vigara, J. (2003). Cadmium increases the activity levels of glutamate dehydrogenase and cysteine synthase in *Chlamydomonas reinhardtii*. *Plant Physiol. Biochem.* **41**, 828–832.

Dräger, D.B., Desbrosses-Fonrouge, A.G., Krach, C., Chardonnens, A.N., Meyer, R.C., Saumitou-Laprade, P., and Krämer, U. (2004). Two genes encoding *Arabidopsis halleri* MTP1 metal transport proteins co-segregate with zinc tolerance and account for high MTP1 transcript levels. *Plant J.* **39**, 425–439.

Duy, D., Wanner, G., Meda, A.R., von Wirén, N., Soll, J., and Philippar, K. (2007). PIC1, an ancient permease in *Arabidopsis* chloroplasts, mediates iron transport. *Plant Cell* **19**, 986–1006.

Eberhard, S., Jain, M., Im, C.S., Pollock, S., Shrager, J., Lin, Y., Peek, A.S., and Grossman, A.R. (2006). Generation of an oligonucleotide array for analysis of gene expression in *Chlamydomonas reinhardtii*. *Curr. Genet.* **49**, 106–124.

Eckhardt, U. and Buckhout, T.J. (1998). Iron assimilation in *Chlamydomonas reinhardtii* involves in ferric reduction and is similar to strategy I higher plants. *J. Exp. Bot.* **49**, 1219–1226.

Eide, D.J. (1998). The molecular biology of metal ion transport in *Saccharomyces cerevisiae*. *Annu. Rev. Nutr.* **18**, 441–469.

Eide, D.J. (2006). Zinc transporters and the cellular trafficking of zinc. *Biochim. Biophys. Acta* **1763**, 711–722.

Eisenstein, R.S. (2000). Iron regulatory proteins and the molecular control of mammalian iron metabolism. *Annu. Rev. Nutr.* **20**, 627–662.

Eitinger, T., Suhr, J., Moore, L., and Smith, J.A. (2005). Secondary transporters for nickel and cobalt ions: Theme and variations. *Biometals* **18**, 399–405.

Ellis, C.D., Wang, F., MacDiarmid, C.W., Clark, S., Lyons, T., and Eide, D.J. (2004). Zinc and the Msc2 zinc transporter protein are required for endoplasmic reticulum function. *J. Cell Biol.* **166**, 325–335.

Ellis, C.D., MacDiarmid, C.W., and Eide, D.J. (2005). Heteromeric protein complexes mediate zinc transport into the secretory pathway of eukaryotic cells. *J. Biol. Chem.* **280**, 28811–28818.

Eriksson, M., Moseley, J.L., Tottey, S., Del Campo, J.A., Quinn, J., Kim, Y., and Merchant, S. (2004). Genetic dissection of nutritional copper signaling in *Chlamydomonas* distinguishes regulatory and target genes. *Genetics* **168**, 795–807.

Faller, P., Kienzler, K., and Krieger-Liszkay, A. (2005). Mechanism of Cd^{2+} toxicity: Cd^{2+} inhibits photoactivation of Photosystem II by competitive binding to the essential Ca^{2+} site. *Biochim. Biophys. Acta* **1706**, 158–164.

Fang, H.M. and Wang, Y. (2002). Characterization of iron-binding motifs in *Candida albicans* high-affinity iron permease CaFtr1p by site-directed mutagenesis. *Biochem. J.* **368**, 641–647.

Fei, X. and Deng, X. (2007). A novel Fe deficiency responsive element (FeRE) regulates the expression of *atx1* in *Chlamydomonas reinharditii*. *Plant Cell Physiol.* **48**, 1496–1503.

Fett, J.P. and Coleman, J.R. (1994). Regulation of periplasmic carbonic anhydrase expression in *Chlamydomonas reinhardtii* by acetate and pH. *Plant Physiol.* **106**, 103–108.

Filatov, V., Dowdle, J., Smirnoff, N., Ford-Lloyd, B., Newbury, H.J., and Macnair, M.R. (2006). Comparison of gene expression in segregating families identifies genes and genomic regions involved in a novel adaptation, zinc hyperaccumulation. *Mol. Ecol.* **15**, 3045–3059.

Finney, L.A. and O'Halloran, T.V. (2003). Transition metal speciation in the cell: insights from the chemistry of metal ion receptors. *Science* **300**, 931–936.

Fisher, M., Gokhman, I., Pick, U., and Zamir, A. (1997). A structurally novel transferrin-like protein accumulates in the plasma membrane of the unicellular green alga *Dunaliella salina* grown in high salinities. *J. Biol. Chem.* **272**, 1565–1570.

Fisher, M., Zamir, A., and Pick, U. (1998). Iron uptake by the halotolerant alga *Dunaliella* is mediated by a plasma membrane transferrin. *J. Biol. Chem.* **273**, 17553–17558.

Fontanesi, F., Soto, I.C., Horn, D., and Barrientos, A. (2006). Assembly of mitochondrial cytochrome c-oxidase, a complicated and highly regulated cellular process. *Am. J. Physiol. Cell Physiol.* **291**, C1129–C1147.

Fromme, P., Jordan, P., and Krauss, N. (2001). Structure of photosystem I. *Biochim. Biophys. Acta* **1507**, 5–31.

Fujiwara, S., Kobayashi, I., Hoshino, S., Kaise, T., Shimogawara, K., Usuda, H., and Tsuzuki, M. (2000). Isolation and characterization of arsenate-sensitive and resistant mutants of *Chlamydomonas reinhardtii*. *Plant Cell Physiol.* **41**, 77–83.

Gaither, L.A. and Eide, D.J. (2001). Eukaryotic zinc transporters and their regulation. *Biometals* **14**, 251–270.

Ghosh, M., Shen, J., and Rosen, B.P. (1999). Pathways of As(III) detoxification in *Saccharomyces cerevisiae*. *Proc. Natl. Acad. Sci. U. S. A.* **96**, 5001–5006.

Gillet, S., Decottignies, P., Chardonnet, S., and Le Maréchal, P. (2006). Cadmium response and redoxin targets in *Chlamydomonas reinhardtii*: a proteomic approach. *Photosynth. Res.* **89**, 201–211.

Gorman, D.S. and Levine, R.P. (1965). Cytochrome f and plastocyanin: their sequence in the photosynthetic electron transport chain of *Chlamydomonas reinhardtii*. *Proc. Natl. Acad. Sci. U. S. A.* **54**, 1665–1669.

Gorman, D.S. and Levine, R.P. (1966). Photosynthetic electron transport chain of *Chlamydomonas reinhardtii*. IV. Purification and properties of plastocyanin. *Plant Physiol.* **41**, 1637–1642.

Goyer, R.A. (1997). Toxic and essential metal interactions. *Annu. Rev. Nutr.* **17**, 37–50.

Gross, W. (2000). Ecophysiology of algae living in highly acidic environments. *Hydrobiologia* **1-3**, 31–37.

Grossman, A.R., Croft, M., Gladyshev, V.N., Merchant, S.S., Posewitz, M.C., Prochnik, S., and Spalding, M.H. (2007). Novel metabolism in *Chlamydomonas* through the lens of genomics. *Curr. Opin. Plant Biol.* **10**, 190–198.

Grotz, N. and Guerinot, M.L. (2006). Molecular aspects of Cu, Fe and Zn homeostasis in plants. *Biochim. Biophys. Acta* **1763**, 595–608.

Gueldry, O., Lazard, M., Delort, F., Dauplais, M., Grigoras, I., Blanquet, S., and Plateau, P. (2003). Ycf1p-dependent Hg(II) detoxification in *Saccharomyces cerevisiae*. *Eur. J. Biochem.* **270**, 2486–2496.

Guerinot, M.L. and Eide, D. (1999). Zeroing in on zinc uptake in yeast and plants. *Curr. Opin. Plant Biol.* **2**, 244–249.

Hanawa, Y., Watanabe, M., Karatsu, Y., Fukuzawa, H., and Shiraiwa, Y. (2007). Induction of a high-CO_2-inducible, periplasmic protein, H43, and its application as a high-CO_2-responsive marker for study of the high-CO_2-sensing mechanism in *Chlamydomonas reinhardtii*. *Plant Cell Physiol.* **48**, 299–309.

Hanikenne, M. (2003). *Chlamydomonas reinhardtii* as a eukaryotic photosynthetic model for studies of heavy metal homeostasis and tolerance. *New Phytol.* **159**, 331–340.

Hanikenne, M., Matagne, R.F., and Loppes, R. (2001). Pleiotropic mutants hypersensitive to heavy metals and to oxidative stress in *Chlamydomonas reinhardtii*. *FEMS Microbiol. Lett.* **196**, 107–111.

Hanikenne, M., Krämer, U., Demoulin, V., and Baurain, D. (2005a). A comparative inventory of metal transporters in the green alga *Chlamydomonas reinhardtii* and the red alga *Cyanidioschizon merolae*. *Plant Physiol.* **137**, 428–446.

Hanikenne, M., Motte, P., Wu, M.C.S., Wang, T., Loppes, R., and Matagne, R.F. (2005b). A mitochondrial half-size ABC transporter is involved in cadmium tolerance in *Chlamydomonas reinhardtii*. *Plant Cell Environ.* **28**, 863–873.

Harris, E.H. (1989). *The Chlamydomonas Sourcebook. A Comprehensive Guide to Biology and Laboratory Use*. Academic Press, San Diego.

Hassett, R. and Kosman, D.J. (1995). Evidence for Cu(II) reduction as a component of copper uptake by *Saccharomyces cerevisiae*. *J. Biol. Chem.* **270**, 128–134.

Herbik, A., Bölling, C., and Buckhout, T.J. (2002a). The involvement of a multicopper oxidase in iron uptake by the green algae *Chlamydomonas reinhardtii*. *Plant Physiol.* **130**, 2039–2048.

Herbik, A., Haebel, S., and Buckhout, T.J. (2002b). Is a ferroxidase involved in the high-affinity iron uptake in *Chlamydomonas reinhardtii*. *Plant Soil* **241**, 1–9.

Hill, K.L. and Merchant, S. (1995). Coordinate expression of coproporphyrinogen oxidase and cytochrome c6 in the green alga *Chlamydomonas reinhardtii* in response to changes in copper availability. *EMBO J.* **14**, 857–865.

Hill, K.L., Li, H.H., Singer, J., and Merchant, S. (1991). Isolation and structural characterization of the *Chlamydomonas reinhardtii* gene for cytochrome c_6. Analysis of the kinetics and metal specificity of its copper-responsive expression. *J. Biol. Chem.* **266**, 15060–15067.

Hill, K.L., Hassett, R., Kosman, D., and Merchant, S. (1996). Regulated copper uptake in *Chlamydomonas reinhardtii* in response to copper availability. *Plant Physiol.* **112**, 697–704.

Himelblau, E., Mira, H., Lin, S.J., Culotta, V.C., Peñarrubia, L., and Amasino, R.M. (1998). Identification of a functional homolog of the yeast copper homeostasis gene ATX1 from *Arabidopsis*. *Plant Physiol.* **117**, 1227–1234.

Hirschi, K.D., Zhen, R.G., Cunningham, K.W., Rea, P.A., and Fink, G.R. (1996). CAX1, an H^+/Ca^{2+} antiporter from *Arabidopsis*. *Proc. Natl. Acad. Sci. U. S. A.* **93**, 8782–8786.

Hirschi, K.D., Korenkov, V.D., Wilganowski, N.L., and Wagner, G.J. (2000). Expression of *Arabidopsis* CAX2 in tobacco. Altered metal accumulation and increased manganese tolerance. *Plant Physiol.* **124**, 125–133.

Holland, I.B., Cole, S.P.C., Kuchler, K., and Higgins, C.F. (2003). *ABC Proteins-From Bacteria to Man*. Academic Press, London.

Howden, R., Goldsbrough, P.B., Andersen, C.R., and Cobbett, C.S. (1995a). Cadmium-sensitive, *cad1* mutants of *Arabidopsis thaliana* are phytochelatin deficient. *Plant Physiol.* **107**, 1059–1066.

Howden, R., Andersen, C.R., Goldsbrough, P.B., and Cobbett, C.S. (1995b). A cadmium-sensitive, glutathione-deficient mutant of *Arabidopsis thaliana*. *Plant Physiol.* **107**, 1067–1073.

Howe, G. and Merchant, S. (1992). Heavy metal-activated synthesis of peptides in *Chlamydomonas reinhardtii*. *Plant Physiol.* **98**, 127–136.

Hu, S., Lau, K.W.K., and Wu, M. (2001). Cadmium sequestration in *Chlamydomonas reinhartdii*. *Plant Sci.* **161**, 987–996.

Hua, C.X., Brown, C., Adhiya, J., Traina, S., Sayre, R.T., and Cai, X. (1999). Growth and heavy metal binding properties of transgenic *Chlamydomonas* expressing a foreign metallothionein gene. *Int. J. Phytorem.* **1**, 53–65.

Hussain, D., Haydon, M.J., Wang, Y., Wong, E., Sherson, S.M., Young, J., Camakaris, J., Harper, J.F., and Cobbett, C.S. (2004). P-type ATPase heavy metal transporters with roles in essential zinc homeostasis in *Arabidopsis*. *Plant Cell* **16**, 1327–1339.

Im, C.S. and Grossman, A.R. (2002). Identification and regulation of high light-induced genes in *Chlamydomonas reinhardtii*. *Plant J.* **30**, 301–313.

Inoue, H., Higuchi, K., Takahashi, M., Nakanishi, H., Mori, S., and Nishizawa, N.K. (2003). Three rice nicotianamine synthase genes, OsNAS1, OsNAS2, and

OsNAS3 are expressed in cells involved in long-distance transport of iron and differentially regulated by iron. *Plant J.* **36**, 366–381.

Irmer, U., Wachholz, I., Schäfer, H., and Lorch, D.W. (1986). Influence of lead on *Chlamydomonas reinhardtii* Dangeard (Volvocales, Chlorophyta): accumulation and ultrastructure changes. *Environ. Exp. Bot.* **26**, 97–105.

Ishimaru, Y., Suzuki, M., Tsukamoto, T., Suzuki, K., Nakazono, M., Kobayashi, T., Wada, Y., Watanabe, S., Matsuhashi, S., Takahashi, M., Nakanishi, H., Mori, S., and Nishizawa, N.K. (2006). Rice plants take up iron as an Fe^{3+}-phytosiderophore and as Fe^{2+}. *Plant J.* **45**, 335–346.

Jamers, A., Van der Ven, K., Moens, L., Robbens, J., Potters, G., Guisez, Y., Blust, R., and De Coen, W. (2006). Effect of copper exposure on gene expression profiles in *Chlamydomonas reinhardtii* based on microarray analysis. *Aquat. Toxicol.* **80**, 249–260.

Jasinski, M., Ducos, E., Martinoia, E., and Boutry, M. (2003). The ATP-binding cassette transporters: structure, function, and gene family comparison between rice and *Arabidopsis*. *Plant Physiol.* **131**, 1169–1177.

Jones, F.T. (2007). A broad view of arsenic. *Poult. Sci.* **86**, 2–14.

Kanamaru, K., Kashiwagi, S., and Mizuno, T. (1994). A copper-transporting P-type ATPase found in the thylakoid membrane of the cyanobacterium *Synechococcus* species PCC7942. *Mol. Microbiol.* **13**, 369–377.

Kargul, J., Nield, J., and Barber, J. (2003). Three-dimensional reconstruction of a light-harvesting complex I- Photosystem I (LHCI-PSI) supercomplex from the green alga *Chlamydomonas reinhardtii*. Insights into light harvesting for PSI. *J. Biol. Chem.* **278**, 16135–16141.

Kim, D.Y., Bovet, L., Kushnir, S., Noh, E.W., Martinoia, E., and Lee, Y. (2006). AtATM3 is involved in heavy metal resistance in *Arabidopsis*. *Plant Physiol.* **140**, 922–932.

Kim, S., Takahashi, M., Higuchi, K., Tsunoda, K., Nakanishi, H., Yoshimura, E., Mori, S., and Nishizawa, N.K. (2005). Increased nicotianamine biosynthesis confers enhanced tolerance of high levels of metals, in particular nickel, to plants. *Plant Cell Physiol.* **46**, 1809–1818.

Kim, S.A., Punshon, T., Lanzirotti, A., Li, L., Alonso, J.M., Ecker, J.R., Kaplan, J., and Guerinot, M.L. (2006). Localization of iron in *Arabidopsis* seed requires the vacuolar membrane transporter VIT1. *Science* **314**, 1295–1298.

Kiranmayi, P. and Mohan, P.M. (2006). Metal transportome of *Neurospora crassa*. *In Silico Biol.* **6**, 169–180.

Kispal, G., Csere, P., Guiard, B., and Lill, R. (1997). The ABC transporter Atm1p is required for mitochondrial iron homeostasis. *FEBS Lett.* **418**, 346–350.

Kispal, G., Csere, P., Prohl, C., and Lill, R. (1999). The mitochondrial proteins Atm1p and Nfs1p are essential for biogenesis of cytosolic Fe/S proteins. *EMBO J.* **18**, 3981–3989.

Kitayama, K., Kitayama, M., Osafune, T., and Togasaki, R.K. (1999). Subcellular localization of iron and manganese superoxide dismutase in *Chlamydomonas reinhardtii* (Chlorophyceae). *J. Phycol.* **35**, 136–142.

Klein, J., Saedler, H., and Huijser, P. (1996). A new family of DNA binding proteins includes putative transcriptional regulators of the *Antirrhinum majus* floral meristem identity gene *SQUAMOSA*. *Mol. Gen. Genet.* **250**, 7–16.

Klein, M., Mamnun, Y.M., Eggmann, T., Schüller, C., Wolfger, H., Martinoia, E., and Kuchler, K. (2002). The ATP-binding cassette (ABC) transporter Bpt1p mediates vacuolar sequestration of glutathione conjugates in yeast. *FEBS Lett.* **520**, 63–67.

Klein, M., Burla, B., and Martinoia, E. (2006). The multidrug resistance-associated protein (MRP/ABCC) subfamily of ATP-binding cassette transporters in plants. *FEBS Lett.* **580**, 1112–1122.

Kliebenstein, D.J., Monde, R.A., and Last, R.L. (1998). Superoxide dismutase in *Arabidopsis*: An eclectic enzyme family with disparate regulation and protein localization. *Plant Physiol.* **118**, 637–650.

Kobae, Y., Uemura, T., Sato, M.H., Ohnishi, M., Mimura, T., Nakagawa, T., and Maeshima, M. (2004). Zinc transporter of *Arabidopsis thaliana* AtMTP1 is localized to vacuolar membranes and implicated in zinc homeostasis. *Plant Cell Physiol.* **45**, 1749–1758.

Kobayashi, I., Fujiwara, S., Shimogawara, K., Kaise, T., Usuda, H., and Tsuzuki, M. (2003). Insertional mutagenesis in a homologue of a Pi transporter gene confers arsenate resistance on *Chlamydomonas*. *Plant Cell Physiol.* **44**, 597–606.

Kobayashi, I., Fujiwara, S., Shimogawara, K., Sakuma, C., Shida, Y., Kaise, T., Usuda, H., and Tsuzuki, M. (2005). High intracellular phosphorus contents exhibit a correlation with arsenate resistance in *Chlamydomonas* mutants. *Plant Cell Physiol.* **46**, 489–496.

Kobayashi, I., Fujiwara, S., Saegusa, H., Inouhe, M., Matsumoto, H., and Tsuzuki, M. (2006). Relief of arsenate toxicity by Cd-stimulated phytochelatin synthesis in the green alga *Chlamydomonas reinhardtii*. *Mar. Biotechnol. (NY)* **8**, 94–101.

Koike, S., Inoue, H., Mizuno, D., Takahashi, M., Nakanishi, H., Mori, S., and Nishizawa, N.K. (2004). OsYSL2 is a rice metal-nicotianamine transporter that is regulated by iron and expressed in the phloem. *Plant J.* **39**, 415–424.

Koornneef, M., Alonso-Blanco, C., and Vreugdenhil, D. (2004). Naturally occurring genetic variation in *Arabidopsis thaliana*. *Annu. Rev. Plant Biol.* **55**, 141–172.

Krämer, U. (2005). MTP1 mops up excess zinc in *Arabidopsis* cells. *Trends Plant Sci.* **10**, 313–315.

Krämer, U. and Clemens, S. (2005). Functions and homeostasis of zinc, copper, and nickel in plants. In: *Molecular Biology of Metal Homeostasis and Detoxification*, Vol. 14 (M. Tama´s and E. Martinoia, Eds.), pp. 216–271. Springer, Berlin.

Krämer, U., Talke, I.N., and Hanikenne, M. (2007). Transition metal transport. *FEBS Lett.* **581**, 2263–2272.

Kropat, J., Tottey, S., Birkenbihl, R.P., Depege, N., Huijser, P., and Merchant, S. (2005). A regulator of nutritional copper signaling in *Chlamydomonas* is an SBP domain protein that recognizes the GTAC core of copper response element. *Proc. Natl. Acad. Sci. U. S. A.* **102**, 18730–18735.

Kushnir, S., Babiychuk, E., Storozhenko, S., Davey, M., Papenbrock, J., De Rycke, R.R., Engler, G., Stephan, U., Lange, H., Kispal, G., Lill, R., and Van Montagu, M.M. (2001). A mutation of the mitochondrial ABC transporter Sta1 leads to dwarfism and chlorosis in the *Arabidopsis* mutant *starik*. *Plant Cell* **13**, 89–100.

Kyriakopoulos, A. and Behne, D. (2002). Selenium-containing proteins in mammals and other forms of life. *Rev. Physiol. Biochem. Pharmacol.* **145**, 1–46.

La Fontaine, S., Quinn, J., and Merchant, S. (2002a). Comparative analysis of copper and iron metabolism in photosynthetic eukaryotes vs yeast and mammals. In: *Handbook of Copper Pharmacology and Toxicology* (E.J. Massaro, Ed.), pp. 481–502. Humana Press, Inc, Totowa, N.J.

La Fontaine, S., Quinn, J.M., Nakamoto, S.S., Page, M.D., Göhre, V., Moseley, J.L., Kropat, J., and Merchant, S. (2002b). Copper-dependent iron assimilation pathway in the model photosynthetic eukaryote *Chlamydomonas reinhardtii*. *Eukaryotic Cell* **1**, 736–757.

La Roche, J., Boyd, P.W., McKay, R.M.L., and Geider, R.J. (1996). Flavodoxin as an *in situ* marker for iron stress in phytoplankton. *Nature* **382**, 802–805.

Lane, T.W. and Morel, F.M. (2000a). A biological function for cadmium in marine diatoms. *Proc. Natl. Acad. Sci. U. S. A.* **97**, 4627–4631.

Lane, T.W. and Morel, F.M. (2000b). Regulation of carbonic anhydrase expression by zinc, cobalt, and carbon dioxide in the marine diatom *Thalassiosira weissflogii*. *Plant Physiol.* **123**, 345–352.

Lanquar, V., Lelièvre, F., Bolte, S., Hamès, C., Alcon, C., Neumann, D., Vansuyt, G., Curie, C., Schröder, A., Krämer, U., Barbier-Brygoo, H., and Thomine, S. (2005). Mobilization of vacuolar iron by AtNRAMP3 and AtNRAMP4 is essential for seed germination on low iron. *EMBO J.* **24**, 4041–4051.

Lasat, M.M., Pence, N.S., Garvin, D.F., Ebbs, S.D., and Kochian, L.V. (2000). Molecular physiology of zinc transport in the Zn hyperaccumulator *Thlaspi caerulescens*. *J. Exp. Bot.* **51**, 71–79.

Laudenbach, D.E., Reith, M.E., and Straus, N.A. (1988). Isolation, sequence analysis, and transcriptional studies of the flavodoxin gene from *Anacystis nidulans* R2. *J. Bacteriol.* **170**, 258–265.

Le Jean, M., Schikora, A., Mari, S., Briat, J.F., and Curie, C. (2005). A loss-of-function mutation in AtYSL1 reveals its role in iron and nicotianamine seed loading. *Plant J.* **44**, 769–782.

Lebovitz, R.M., Zhang, H., Vogel, H., Cartwright, J. Jr., Dionne, L., Lu, N., Huang, S., and Matzuk, M.M. (1996). Neurodegeneration, myocardial injury, and perinatal death in mitochondrial superoxide dismutase-deficient mice. *Proc. Natl. Acad. Sci. U. S. A.* **93**, 9782–9787.

Lemaire, S., Keryer, E., Stein, M., Schepens, I.I., Issakidis-Bourguet, E., Gérard-Hirne, C., Miginiac-Maslow, M., and Jacquot, J.P. (1999). Heavy-metal regulation of thioredoxin gene expression in *Chlamydomonas reinhardtii*. *Plant Physiol.* **120**, 773–778.

Lemaire, S.D., Miginiac-Maslow, M., and Jacquot, J.P. (2002). Plant thioredoxin gene expression: Control by light, circadian clock, and heavy metals. *Methods Enzymol.* **347**, 412–421.

Li, H.H. and Merchant, S. (1995). Degradation of plastocyanin in copper-deficient *Chlamydomonas reinhardtii*. *J. Biol. Chem.* **270**, 23504–23510.

Li, H.H., Quinn, J., Culler, D., Girard-Bascou, J., and Merchant, S. (1996). Molecular genetic analysis of plastocyanin biosynthesis in *Chlamydomonas reinhardtii*. *J. Biol. Chem.* **271**, 31283–31289.

Li, L. and Kaplan, J. (1997). Characterization of two homologous yeast genes that encode mitochondrial iron transporters. *J. Biol. Chem.* **272**, 28485–28493.

Li, L., Chen, O.S., McVey Ward, D., and Kaplan, J. (2001). CCC1 is a transporter that mediates vacuolar iron storage in yeast. *J. Biol. Chem.* **276**, 29515–29519.

Li, Z.S., Szczypka, M., Lu, Y.P., Thiele, D.J., and Rea, P.A. (1996). The yeast cadmium factor protein (YCF1) is a vacuolar glutathione S- conjugate pump. *J. Biol. Chem.* **271**, 6509–6517.

Li, Z.S., Lu, Y.P., Zhen, R.G., Szczypka, M., Thiele, D.J., and Rea, P.A. (1997). A new pathway for vacuolar cadmium sequestration in *Saccharomyces cerevisiae*: YCF1-catalyzed transport of bis(glutathionato)cadmium. *Proc. Natl. Acad. Sci. U S A* **94**, 42–47.

Lin, S.J., Pufahl, R.A., Dancis, A., O'Halloran, T.V., and Culotta, V.C. (1997). A role for the *Saccharomyces cerevisiae* ATX1 gene in copper trafficking and iron transport. *J. Biol. Chem.* **272**, 9215–9220.

Lindsay, W.L. and Schwab, A.P. (1982). The chemistry of iron in soils and its availability to plants. *J. Plant Nutr.* **5**, 821–840.

Ling, H.Q., Koch, G., Baumlein, H., and Ganal, M.W. (1999). Map-based cloning of chloronerva, a gene involved in iron uptake of higher plants encoding nicotianamine synthase. *Proc. Natl. Acad. Sci. U. S. A.* **96**, 7098–7103.

Lobanov, A.V., Fomenko, D.E., Zhang, Y., Sengupta, A., Hatfield, D.L., and Gladyshev, V.N. (2007). Evolutionary dynamics of eukaryotic selenoproteomes: large selenoproteomes may associate with aquatic life and small with terrestrial life. *Genome Biol.* **8**, R198.

Lombi, E., Tearall, K.L., Howarth, J.R., Zhao, F.J., Hawkesford, M.J., and McGrath, S.P. (2002). Influence of iron status on cadmium and zinc uptake by different ecotypes of the hyperaccumulator *Thlaspi caerulescens*. *Plant Physiol.* **128**, 1359–1367.

Long, J.C., Sommer, F., Allen, M.D., Lu, S.F., and Merchant, S.S. (2008). *FER1* and *FER2* encoding two ferritin complexes in *Chlamydomonas reinhardtii*chloroplasts are regulated by iron. *Genetics* **179**, 137–147.

Luis, P., Behnke, K., Toepel, J., and Wilhelm, C. (2006). Parallel analysis of transcript levels and physiological key parameters allows the identification of stress phase gene markers in *Chlamydomonas reinhardtii* under copper excess. *Plant Cell Environ.* **29**, 2043–2054.

Lustigman, B., Lee, L.H., Morata, J., and Khan, F. (2000). Effect of thallium on the growth of Anacystis nidulans and *Chlamydomonas reinhardtii*. *Bull. Environ. Contam. Toxicol.* **64**, 565–573.

Lutsenko, S. and Petris, M.J. (2003). Function and regulation of the mammalian copper-transporting ATPases: Insights from biochemical and cell biological approaches. *J Membr. Biol.* **191**, 1–12.

Lynnes, J.A., Derzaph, T.L.M., and Weger, H.G. (1998). Iron limitation results in induction of ferricyanide reductase and ferric chelate reductase activities in *Chlamydomonas reinhardtii*. *Planta* **204**, 360–365.

MacDiarmid, C.W., Gaither, L.A., and Eide, D. (2000). Zinc transporters that regulate vacuolar zinc storage in *Saccharomyces cerevisiae*. *EMBO J.* **19**, 2845–2855.

MacFie, S.M., Tarmohamed, Y., and Welbourn, P.M. (1994). Effect of cadmium, cobalt, copper and nickel on growth of the green alga *Chlamydomonas reinhardtii*:The influence of the cell wall and pH. *Arch. Environ. Contam. Toxicol.* **27**, 454–458.

Mackenzie, B. and Hediger, M.A. (2004). SLC11 family of H^+-coupled metal-ion transporters NRAMP1 and DMT1. *Pflugers Arch.* **447**, 571–579.

Maldonado, M.T., Allen, A.E., Chong, J.S., Lin, K., Leus, D., Karpenko, N., and Harris, S.L. (2006). Copper-dependent iron transport in coastal and oceanic diatoms. *Limnol. Oceanog.* **51**, 1729–1743.

Marschner, H. (1995). *Mineral Nutrition of Higher Plants*, 2nd edn. Academic Press Ltd, London.

Mäser, P., Thomine, S., Schroeder, J.I., Ward, J.M., Hirschi, K., Sze, H., Talke, I.N., Amtmann, A., Maathuis, F.J., Sanders, D., Harper, J.F., Tchieu, J., Gribskov, M., Persans, M.W., Salt, D.E., Kim, S.A., and Guerinot, M.L. (2001). Phylogenetic relationships within cation transporter families of *Arabidopsis*. *Plant Physiol.* **126**, 1646–1667.

Matsuzaki, M., Misumi, O., Shin, I.T., Maruyama, S., Takahara, M., Miyagishima, S.Y., Mori, T., Nishida, K., Yagisawa, F., Yoshida, Y., Nishimura, Y., Nakao, S., Kobayashi, T., Momoyama, Y., Higashiyama, T., Minoda, A., Sano, M.,

Nomoto, H., Oishi, K., Hayashi, H., Ohta, F., Nishizaka, S., Haga, S., Miura, S., Morishita, T., Kabeya, Y., Terasawa, K., Suzuki, Y., Ishii, Y., Asakawa, S., Takano, H., Ohta, N., Kuroiwa, H., Tanaka, K., Shimizu, N., Sugano, S., Sato, N., Nozaki, H., Ogasawara, N., Kohara, Y., and Kuroiwa, T. (2004). Genome sequence of the ultrasmall unicellular red alga *Cyanidioschyzon merolae* 10D. *Nature* **428**, 653–657.

McHugh, J. and Spanier, J.G. (1994). Isolation of cadmium sensitive mutants in *Chlamydomonas reinhardtii* by transformation/insertional mutagenesis. *FEMS Microbiol. Lett.* **124**, 239–244.

McKie, A.T., Marciani, P., Rolfs, A., Brennan, K., Wehr, K., Barrow, D., Miret, S., Bomford, A., Peters, T.J., Farzaneh, F., Hediger, M.A., Hentze, M.W., and Simpson, R.J. (2000). A novel duodenal iron-regulated transporter, IREG1, implicated in the basolateral transfer of iron to the circulation. *Mol. Cell* **5**, 299–309.

Mendóza-Cozatl, D., Loza-Tavera, H., Hernández-Navarro, A., and Moreno-Sánchez, R. (2005). Sulfur assimilation and glutathione metabolism under cadmium stress in yeast, protists and plants. *FEMS Microbiol. Rev.* **29**, 653–671.

Merchant, S. (1998). Synthesis of metalloproteins involved in photosynthesis: plastoscyanin and cytochromes. In: *The Molecular Biology of Chloroplasts and Mitochondria in Chlamydomonas* (J.D. Rochaix, M. Goldschmidt-Clermont, and S. Merchant, Eds.), pp. 597–611. Kluwer Academic, Dordrecht.

Merchant, S. (2006). Trace metal utilization in chloroplasts. In: *The Structure and Function of Plastids* (R.R. Wise and J.K. Hoober, Eds.), pp. 199–218. Springer, Dordrecht.

Merchant, S. and Bogorad, L. (1987a). The Cu(II)-repressible plastidic cytochrome c. Cloning and sequence of a complementary DNA for the pre-apoprotein. *J. Biol. Chem.* **262**, 9062–9067.

Merchant, S. and Bogorad, L. (1987b). Metal ion regulated gene expression: use of a plastocyanin-less mutant of *Chlamydomonas reinhardtii* to study the Cu(II)-dependent expression of cytochrome *c*-552. *EMBO J.* **6**, 2531–2535.

Merchant, S. and Sawaya, M.R. (2005). The light reactions: A guide to recent acquisitions for the picture gallery. *Plant Cell* **17**, 648–663.

Merchant, S., Hill, K., Kim, J.H., Thompson, J., Zaitlin, D., and Bogorad, L. (1990). Isolation and characterization of a complementary DNA clone for an algal pre-apoplastocyanin. *J. Biol. Chem.* **265**, 12372–12379.

Merchant, S., Allen, M.D., Kropat, J., Moseley, J.L., Long, J.C., Tottey, S., and Terauchi, A.M. (2006). Between a rock and a hard place: Trace element nutrition in *Chlamydomonas*. *Biochim. Biophys. Acta* **1763**, 578–594.

Merchant, S.S., Prochnik, S.E., Vallon, O., Harris, E.H., Karpowicz, S.J., Witman, G. B., Terry, A., Salamov, A., Fritz-Laylin, L.K., Maréchal-Drouard, L., et al. (2007). The *Chlamydomonas* genome reveals the evolution of key animal and plant functions. *Science* **318**, 245–250.

Mills, R.F., Francini, A., Ferreira da Rocha, P.S., Baccarini, P.J., Aylett, M., Krijger, G.C., and Williams, L.E. (2005). The plant P1B-type ATPase AtHMA4 transports Zn and Cd and plays a role in detoxification of transition metals supplied at elevated levels. *FEBS Lett.* **579**, 783–791.

Mitsuhashi, N., Miki, T., Senbongi, H., Yokoi, N., Yano, H., Miyazaki, M., Nakajima, N., Iwanaga, T., Yokoyama, Y., Shibata, T., and Seino, S. (2000). MTABC3, a novel mitochondrial ATP-binding cassette protein involved in iron homeostasis. *J. Biol. Chem.* **275**, 17536–17540.

Morlon, H., Fortin, C., Floriani, M., Adam, C., Garnier-Laplace, J., and Boudou, A. (2005). Toxicity of selenite in the unicellular green alga *Chlamydomonas reinhardtii*: Comparison between effects at the population and sub-cellular level. *Aquat. Toxicol.* **73**, 65–78.

Moseley, J., Quinn, J., Eriksson, M., and Merchant, S. (2000). The Crd1 gene encodes a putative di-iron enzyme required for photosystem I accumulation in copper deficiency and hypoxia in *Chlamydomonas reinhardtii*. *EMBO J.* **19**, 2139–2151.

Moseley, J.L., Allinger, T., Herzog, S., Hoerth, P., Wehinger, E., Merchant, S., and Hippler, M. (2002a). Adaptation to Fe-deficiency requires remodeling of the photosynthetic apparatus. *EMBO J.* **21**, 6709–6720.

Moseley, J.L., Page, M.D., Alder, N.P., Eriksson, M., Quinn, J., Soto, F., Theg, S.M., Hippler, M., and Merchant, S. (2002b). Reciprocal expression of two candidate di-iron enzymes affecting photosystem I and light-harvesting complex accumulation. *Plant Cell* **14**, 673–688.

Mosulén, S., Domínguez, M.J., Vigara, J., Vílchez, C., Guiraum, A., and Vega, J.M. (2003). Metal toxicity in *Chlamydomonas reinhardtii*. Effect on sulfate and nitrate assimilation. *Biomol. Eng.* **20**, 199–203.

Muck, A., Ibanez, A.J., Stauber, E.J., Mansourova, M., and Svatos, A. (2006). Atmospheric molding of ionic copolymer MALDI-TOF/MS arrays: a new tool for protein identification/profiling. *Electrophoresis* **27**, 4952–4959.

Nagel, K. and Voigt, J. (1989). In vitro evolution and preliminary characterization of a Cadmium-resistant population of *Chlamydomonas reinhardtii*. *Appl. Environ. Microbiol.* **55**, 526–528.

Nagel, K. and Voigt, J. (1995). Impaired photosynthesis in a cadmium-tolerant *Chlamydomonas* mutant strain. *Microbiol. Res.* **150**, 105–110.

Nagel, K., Adelmeier, U., and Voigt, J. (1996). Subcellular distribution of cadmium in the unicellular green alga *Chlamydomonas reinhardtii*. *J. Plant Physiol.* **149**, 86–90.

Naumann, B., Stauber, E.J., Busch, A., Sommer, F., and Hippler, M. (2005). N-terminal processing of Lhca3 is a key step in remodeling of the photosystem I-light-harvesting complex under iron deficiency in *Chlamydomonas reinhardtii*. *J. Biol. Chem.* **280**, 20431–20441.

Naumann, B., Busch, A., Allmer, J., Ostendorf, E., Zeller, M., Kirchhoff, H., and Hippler, M. (2007). Comparative quantitative proteomics to investigate the remodeling of bioenergetic pathways under iron deficiency in *Chlamydomonas reinhardtii*. *Proteomics* **7**, 3964–3979.

Nishikawa, K., Yamakoshi, Y., Uemura, I., and Tominaga, N. (2003). Ultrastructural changes in *Chlamydomonas acidophila* (Chlorophyta) induced by heavy metals and polyphosphate metabolism. *FEMS Microbiol. Ecol.* **44**, 253–259.

Nishikawa, K., Onodera, A., and Tominaga, N. (2006). Phytochelatins do not correlate with the level of Cd accumulation in *Chlamydomonas* spp. *Chemosphere* **63**, 1553–1559.

Novoselov, S.V., Rao, M., Onoshko, N.V., Zhi, H., Kryukov, G.V., Xiang, Y., Weeks, D.P., Hatfield, D.L., and Gladyshev, V.N. (2002). Selenoproteins and selenocysteine insertion system in the model plant cell system, *Chlamydomonas reinhardtii*. *EMBO J.* **21**, 3681–3693.

Nriagu, J.O. and Pacyna, J.M. (1988). Quantitative assessment of worldwide contamination of air, water and soils by trace metals. *Nature* **333**, 134–139.

O'Halloran, T.V. and Culotta, V.C. (2000). Metallochaperones, an intracellular shuttle service for metal ions. *J. Biol. Chem.* **275**, 25057–25060.

Ortiz, D.F., Kreppel, L., Speiser, D.M., Scheel, G., McDonald, G., and Ow, D.W. (1992). Heavy metal tolerance in the fission yeast requires an ATP-binding cassette-type vacuolar membrane transporter. *EMBO J.* **11**, 3491–3499.

Ortiz, D.F., Ruscitti, T., McCue, K.F., and Ow, D.W. (1995). Transport of metal-binding peptides by HMT1, a fission yeast ABC-type vacuolar membrane protein. *J. Biol. Chem.* **270**, 4721–4728.

Palenik, B. and Morel, F.M. (1991). Amine oxidases of marine phytoplankton. *Appl. Environ. Microbiol.* **57**, 2440–2443.

Palenik, B., Grimwood, J., Aerts, A., Rouzé, P., Salamov, A., Putnam, N., Dupont, C., Jorgensen, R., Derelle, E., Rombauts, S., Zhou, K., Otillar, R., Merchant, S.S., Podell, S., Gaasterland, T., Napoli, C., Gendler, K., Manuell, A., Tai, V., Vallon, O., Piganeau, G., Jancek, S., Heijde, M., Jabbari, K., Bowler, C., Lohr, M., Robbens, S., Werner, G., Dubchak, I., Pazour, G.J.M., Ren, Q., Paulsen, I., Delwiche, C., Schmutz, J., Rokhsar, D., Van de Peer, Y., Moreau, H., and Grigoriev, I.V. (2007). The tiny eukaryote *Ostreococcus* provides genomic insights into the paradox of plankton speciation. *Proc. Natl. Acad. Sci. U. S. A.* **104**, 7705–7710.

Palmiter, R.D. and Huang, L. (2004). Efflux and compartmentalization of zinc by members of the SLC30 family of solute carriers. *Pflugers Arch.* **447**, 744–751.

Paz, Y., Katz, A., and Pick, U. (2007). A multicopper rerroxidase involved in iron binding to transferrins in *Dunaliella salina* plasma membranes. *J. Biol. Chem.* **282**, 8658–8666.

Peers, G., Quesnel, S.A., and Price, N.M. (2005). Copper requirements for iron acquisition and growth of coastal and oceanic diatoms. *Limnol. Oceanog.* **50**, 1149–1158.

Peiter, E., Montanini, B., Gobert, A., Pedas, P., Husted, S., Maathuis, F.J., Blaudez, D., Chalot, M., and Sanders, D. (2007). A secretory pathway-localized cation diffusion facilitator confers plant manganese tolerance. *Proc. Natl. Acad. Sci. U. S. A.* **104**, 8532–8537.

Pence, N.S., Larsen, P.B., Ebbs, S.D., Letham, D.L., Lasat, M.M., Garvin, D.F., Eide, D., and Kochian, L.V. (2000). The molecular physiology of heavy metal transport in the Zn/Cd hyperaccumulator *Thlaspi caerulescens*. *Proc. Natl. Acad. Sci. U. S. A.* **97**, 4956–4960.

Persson, B.L., Petersson, J., Fristedt, U., Weinander, R., Berhe, A., and Pattison, J. (1999). Phosphate permeases of *Saccharomyces cerevisiae*: Structure, function and regulation. *Biochim. Biophys. Acta* **1422**, 255–272.

Petris, M.J. (2004). The SLC31 (Ctr) copper transporter family. *Pflugers Arch.* **447**, 752–755.

Phung, L.T., Ajlani, G., and Haselkorn, R. (1994). P-type ATPase from the cyanobacterium *Synechococcus* 7942 related to the human Menkes and Wilson disease gene products. *Proc. Natl. Acad. Sci. U. S. A.* **91**, 9651–9654.

Pianelli, K., Mari, S., Marquès, L., Lebrun, M., and Czernic, P. (2005). Nicotianamine over-accumulation confers resistance to nickel in *Arabidopsis thaliana*. *Transgenic Res.* **14**, 739–748.

Pich, A. and Scholz, G. (1996). Translocation of copper and other micronutrients in tomato plants (*Lycopersicon esculentum* Mill.): nicotianamine-stimulated copper transport in the xylem. *J. Exp. Bot* **.47**, 41–47.

Pilon-Smits, E. (2005). Phytoremediation. *Annu. Rev. Plant Biol.* **56**, 15–39.

Pilon, M., Abdel-Ghany, S.E., Cohu, C.M., Gogolin, K.A., and Ye, H. (2006). Copper cofactor delivery in plant cells. *Curr. Opin. Plant Biol.* **9**, 256–263.

Prasad, M.N., Drej, K., Skawinska, A., Strzalka, K., and Stralka, K. (1998). Toxicity of cadmium and copper in *Chlamydomonas reinhardtii* wild-type (WT 2137) and cell wall deficient mutant strain (CW 15). *Bull. Environ. Contam. Toxicol.* **60**, 306–311.

Puig, S., Andrés-Colás, N., García-Molina, A., and Peñarubbia, L. (2007). Copper and iron homeostasis in *Arabidopsis*: Response to metal deficiencies, interactions and biotechnological applications. *Plant Cell Environ.* **30**, 271–290.

Quinn, J.M. and Merchant, S. (1995). Two copper-responsive elements associated with the *Chlamydomonas Cyc6* gene function as targets for transcriptional activators. *Plant Cell* **7**, 623–638.

Quinn, J.M. and Merchant, S. (1998). Copper-responsive gene expression during adaptation to copper deficiency. *Methods Enzymol.* **297**, 263–279.

Quinn, J.M., Li, H.H., Singer, J., Morimoto, B., Mets, L., Kindle, K.L., and Merchant, S. (1993). The plastocyanin-deficient phenotype of *Chlamydomonas reinhardtii ac-208* results from a frame-shift mutation in the nuclear gene encoding pre-apoplastocyanin. *J. Biol. Chem.* **268**, 7832–7841.

Quinn, J.M., Nakamoto, S.S., and Merchant, S. (1999). Induction of coproporphyrinogen oxidase in *Chlamydomonas* chloroplasts occurs via transcriptional regulation of Cpx1 mediated by copper response elements and increased translation from a copper deficiency-specific form of the transcript. *J. Biol. Chem.* **274**, 14444–14454.

Quinn, J.M., Barraco, P., Eriksson, M., and Merchant, S. (2000). Coordinate copper- and oxygen-responsive Cyc6 and Cpx1 expression in *Chlamydomonas* is mediated by the same element. *J. Biol. Chem.* **275**, 6080–6089.

Quinn, J.M., Kropat, J., and Merchant, S. (2003). Copper response element and Crr1-dependent Ni^{2+}-responsive promoter for induced, reversible gene expression in *Chlamydomonas reinhardtii*. *Eukaryotic Cell* **2**, 995–1002.

Quisel, J.D., Wykoff, D.D., and Grossman, A.R. (1996). Biochemical characterization of the extracellular phosphatases produced by phosphorus-deprived *Chlamydomonas reinhardtii*. *Plant Physiol.* **111**, 839–848.

Radisky, D. and Kaplan, J. (1999). Regulation of transition metal transport across the yeast plasma membrane. *J. Biol. Chem.* **274**, 4481–4484.

Rao, M., Carlson, B.A., Novoselov, S.V., Weeks, D.P., Gladyshev, V.N., and Hatfield, D.L. (2003). *Chlamydomonas reinhardtii* selenocysteine tRNA[Ser]Sec. *RNA* **9**, 923–930.

Rauser, W.E. (1999). Structure and function of metal chelators produced by plants: The case for organic acids, amino acids, phytin, and metallothioneins. *Cell Biochem. Biophys.* **31**, 19–48.

Rea, P.A. (2007). Plant ATP-binding cassette transporters. *Annu. Rev. Plant Biol.* **58**, 347–375.

Rea, P.A., Li, Z.S., Lu, Y.P., Drozdowicz, Y.M., and Martinoia, E. (1998). From vacuolar GS-X pumps to multispecific ABC transporters. *Annu. Rev. Plant Physiol. Plant Mol. Biol.* **49**, 727–760.

Rees, E.M. and Thiele, D.J. (2004). From aging to virulence: forging connections through the study of copper homeostasis in eukaryotic microorganisms. *Curr. Opin. Microbiol.* **7**, 175–184.

Reinhardt, I., Haebel, S., Herbik, A., and Buckhout, T.J. (2006). Proteomic studies under iron stress: Iron deficiency-induced regulation of protein synthesis in the green alga *Chlamydomonas reinhardtii*. In: *Iron Nutrition in Plants and Rhizospheric Microorganisms* (L.L. Barton and J. Abadia, Eds), pp. 371–393. Springer, Dordrecht.

Rodionov, D.A., Hebbeln, P., Gelfand, M.S., and Eitinger, T. (2006). Comparative and functional genomic analysis of prokaryotic nickel and cobalt uptake transporters: evidence for a novel group of ATP-binding cassette transporters. *J. Bacteriol.* **188**, 317–327.

Rohr, J., Sarkar, N., Balenger, S., Jeong, B.R., and Cerutti, H. (2004). Tandem inverted repeat system for selection of effective transgenic RNAi strains in *Chlamydomonas*. *Plant J.* **40**, 611–621.

Rosakis, A. and Köster, W. (2004). Transition metal transport in the green microalga *Chlamydomonas reinhardtii* – genomic sequence analysis. *Res. Microbiol.* **155**, 201–210.

Rosakis, A. and Köster, W. (2005). Divalent metal transport in the green microalga *Chlamydomonas reinhardtii* is mediated by a protein similar to prokaryotic Nramp homologues. *Biometals* **18**, 107–120.

Rosen, B.P. (2002). Transport and detoxification systems for transition metals, heavy metals and metalloids in eukaryotic and prokaryotic microbes. *Comp. Biochem. Physiol. A Mol. Integr. Physiol.* **133**, 689–693.

Rossi, L., Arciello, M., Capo, C., and Rotilio, G. (2006). Copper imbalance and oxidative stress in neurodegeneration. *Ital. J. Biochem.* **55**, 212–221.

Rother, M., Resch, A., Wilting, R., and Bock, A. (2001). Selenoprotein synthesis in archaea. *Biofactors* **14**, 75–83.

Rubinelli, P., Siripornadulsil, S., Gao-Rubinelli, F., and Sayre, R.T. (2002). Cadmium- and iron-stress-inducible gene expression in the green alga *Chlamydomonas reinhardtii*: Evidence for H43 protein function in iron assimilation. *Planta* **215**, 1–13.

Rutherford, J.C. and Bird, A.J. (2004). Metal-responsive transcription factors that regulate iron, zinc, and copper homeostasis in eukaryotic cells. *Eukaryotic Cell* **3**, 1–13.

Sadka, A., Himmelhoch, S., and Zamir, A. (1991). A 150 kilodalton cell surface protein is induced by salt in the halotolerant green alga *Dunaliella salina*. *Plant Physiol.* **95**, 822–831.

Sakurai, H., Kusumoto, N., Kitayama, K., and Togasaki, R.K. (1993). Isozymes of superoxide dismutase in *Chlamydomonas* and purification of one of the major isozymes containing Fe. *Plant Cell Physiol.* **34**, 1133–1137.

Salt, D.E. and Rauser, W.E. (1995). MgATP-dependent transport of phytochelatins across the tonoplast of oat roots. *Plant Physiol.* **107**, 1293–1301.

Sánchez-Fernández, R., Davies, T.G., Coleman, J.O., and Rea, P.A. (2001). The *Arabidopsis thaliana* ABC protein superfamily, a complete inventory. *J. Biol. Chem.* **276**, 30231–30244.

Sandmann, G., Reck, H., Kessler, E., and Boger, P. (1983). Distribution of plastocyanin and soluble plastidic cytochrome *c* in various classes of algae. *Arch. Microbiol.* **134**, 23–27.

Sargent, P.J., Farnaud, S., and Evans, R.W. (2005). Structure/function overview of proteins involved in iron storage and transport. *Curr. Med. Chem.* **12**, 2683–2693.

Sasaki, T., Kurano, N., and Miyashi, S. (1998). Cloning and characterization of high-CO_2-specific cDNAs from a marine microalga, *Chlorococcum littorale*, and effect of CO_2 concentration and iron deficiency on the gene expression. *Plant Cell Physiol.* **39**, 131–138.

Schaaf, G., Schikora, A., Häberle, J., Vert, G., Ludewig, U., Briat, J.F., Curie, C., and von Wirén, N. (2005). A putative function for the *Arabidopsis* Fe-Phytosiderophore transporter homolog AtYSL2 in Fe and Zn homeostasis. *Plant Cell Physiol.* **46**, 762–774.

Schaaf, G., Honsbein, A., Meda, A.R., Kirchner, S., Wipf, D., and von Wirén, N. (2006). AtIREG2 encodes a tonoplast transport protein involved in iron-dependent nickel detoxification in *Arabidopsis thaliana* roots. *J. Biol. Chem.* **281**, 25532–25540.

Schmidt, W. (1999). Mechanisms and regulation of reduction-based iron uptake in plants. *New Phytol.* **141**, 1–26.

Schroda, M. (2006). RNA silencing in *Chlamydomonas*: Mechanisms and tools. *Curr. Genet.* **49**, 69–84.

Schueck, N.D., Woontner, M., and Koeller, D.M. (2001). The role of the mitochondrion in cellular iron homeostasis. *Mitochondrion* **1**, 51–60.

Schulz, B. and Kolukisaoglu, H.U. (2006). Genomics of plant ABC transporters: The alphabet of photosynthetic life forms or just holes in membranes? *FEBS Lett.* **580**, 1010–1016.

Seigneurin-Berny, D., Gravot, A., Auroy, P., Mazard, C., Kraut, A., Finazzi, G., Grunwald, D., Rappaport, F., Vavasseur, A., Joyard, J., Richaud, P., and Rolland, N. (2006). HMA1, a new Cu-ATPase of the chloroplast envelope, is essential for growth under adverse light conditions. *J. Biol. Chem.* **281**, 2882–2892.

Sharma, K.G., Mason, D.L., Liu, G., Rea, P.A., Bachhawat, A.K., and Michaelis, S. (2002). Localization, regulation, and substrate transport properties of Bpt1p, a *Saccharomyces cerevisiae* MRP-type ABC transporter. *Eukaryotic Cell* **1**, 391–400.

Shaul, O., Hilgemann, D.W., de-Almeida-Engler, J., Van Montagu, M., Inzé, D., and Galili, G. (1999). Cloning and characterization of a novel Mg^{2+}/H^+ exchanger. *EMBO J.* **18**, 3973–3980.

Shigaki, T. and Hirschi, K.D. (2006). Diverse functions and molecular properties emerging for CAX cation/H^+ exchangers in plants. *Plant Biol.* **8**, 419–429.

Shikanai, T., Muller-Moule, P., Munekage, Y., Niyogi, K.K., and Pilon, M. (2003). PAA1, a P-type ATPase of *Arabidopsis*, functions in copper transport in chloroplasts. *Plant Cell* **15**, 1333–1346.

Shrager, J., Hauser, C.R., Chang, C.-W., Harris, E.H., Davies, J.P., McDermott, J., Tamse, R., Zhang, Z., and Grossman, A. (2003). *Chlamydomonas reinhardtii* genome project. A guide to the generation and use of the cDNA information. *Plant Physiol.* **131**, 401–408.

Siripornadulsil, S., Traina, S., Verma, D.P., and Sayre, R.T. (2002). Molecular mechanisms of proline-mediated tolerance to toxic heavy metals in transgenic microalgae. *Plant Cell* **14**, 2837–2847.

Sommer, F., Drepper, F., Haehnel, W., and Hippler, M. (2004). The hydrophobic recognition site formed by residues PsaA-Trp^{651} and PsaB-Trp^{627} of photosystem I in *Chlamydomonas reinhardtii* confers distinct selectivity for binding of plastocyanin and cytochrome c_6. *J. Biol. Chem.* **279**, 20009–20017.

Stauber, E.J. and Hippler, M. (2004). *Chlamydomonas reinhardtii* proteomics. *Plant Physiol. Biochem.* **42**, 989–1001.

Stauber, E.J., Fink, A., Markert, C., Kruse, O., Johanningmeier, U., and Hippler, M. (2003). Proteomics of *Chlamydomonas reinhardtii* light-harvesting proteins. *Eukaryotic Cell* **2**, 978–994.

Stearman, R., Yuan, D.S., Yamaguchi-Iwai, Y., Klausner, R.D., and Dancis, A. (1996). A permease-oxidase complex involved in high-affinity iron uptake in yeast. *Science* **271**, 1552–1557.

Stohs, S.J. and Bagchi, D. (1995). Oxidative mechanisms in the toxicity of metal ions. *Free Radic. Biol. Med.* **18**, 321–336.

Szczypka, M.S., Wemmie, J.A., Moye-Rowley, W.S., and Thiele, D.J. (1994). A yeast metal resistance protein similar to human cystic fibrosis transmembrane conductance regulator (CFTR) and multidrug resistance-associated protein. *J. Biol. Chem.* **269**, 22853–22857.

Takahashi, M., Terada, Y., Nakai, I., Nakanishi, H., Yoshimura, E., Mori, S., and Nishizawa, N.K. (2003). Role of nicotianamine in the intracellular delivery of metals and plant reproductive development. *Plant Cell* **15**, 1263–1280.

Takahashi, Y., Yasui, T.A., Stauber, E.J., and Hippler, M. (2004). Comparison of the subunit compositions of the PSI-LHCI supercomplex and the LHCI in the green alga *Chlamydomonas reinhardtii*. *Biochemistry* **43**, 7816–7823.

Talke, I.N., Hanikenne, M., and Krämer, U. (2006). Zinc-dependent global transcriptional control, transcriptional deregulation, and higher gene copy number for genes in metal homeostasis of the hyperaccumulator *Arabidopsis halleri*. *Plant Physiol.* **142**, 148–167.

Taylor, K.M., Morgan, H.E., Johnson, A., Hadley, L.J., and Nicholson, R.I. (2003). Structure-function analysis of LIV-1, the breast cancer-associated protein that belongs to a new subfamily of zinc transporters. *Biochem. J.* **375**, 51–59.

Taylor, K.M., Morgan, H.E., Johnson, A., and Nicholson, R.I. (2004). Structure-function analysis of HKE4, a member of the new LIV-1 subfamily of zinc transporters. *Biochem. J.* **377**, 131–139.

Taylor, S.R. (1964). Abundance of chemical elements in the continental crust: a new table. *Geochim. Cosmochim. Acta* **28**, 1273–1285.

Thomine, S., Wang, R., Ward, J.M., Crawford, N.M., and Schroeder, J.I. (2000). Cadmium and iron transport by members of a plant metal transporter family in *Arabidopsis* with homology to Nramp genes. *Proc. Natl. Acad. Sci. U. S. A.* **97**, 4991–4996.

Thomine, S., Lelièvre, F., Debarbieux, E., Schroeder, J.I., and Barbier-Brygoo, H. (2003). AtNRAMP3, a multispecific vacuolar metal transporter involved in plant responses to iron deficiency. *Plant J.* **34**, 685–695.

Tommasini, R., Vogt, E., Fromenteau, M., Hortensteiner, S., Matile, P., Amrhein, N., and Martinoia, E. (1998). An ABC-transporter of *Arabidopsis thaliana* has both glutathione-conjugate and chlorophyll catabolite transport activity. *Plant J.* **13**, 773–780.

Tottey, S., Rich, P.R., Rondet, S.A.M., and Robinson, N.J. (2001). Two Menkes-type ATPases supply copper for photosynthesis in *Synechocystis* sp. 6803. *J. Biol. Chem.* **276**, 19999–20004.

Tottey, S., Block, M.A., Allen, M., Westergren, T., Albrieux, C., Scheller, H.V., Merchant, S., and Jensen, P.E. (2003). *Arabidopsis* CHL27, located in both envelope and thylakoid membranes, is required for the synthesis of protochlorophyllide. *Proc. Natl. Acad. Sci. U. S. A.* **100**, 16119–16124.

Trampczynska, A., Bottcher, C., and Clemens, S. (2006). The transition metal chelator nicotianamine is synthesized by filamentous fungi. *FEBS Lett.* **580**, 3173–3178.

Tripathi, R.D., Srivastava, S., Mishra, S., Singh, N., Tuli, R., Gupta, D.K., and Maathuis, F.J. (2007). Arsenic hazards: Strategies for tolerance and remediation by plants. *Trends Biotechnol.* **25**, 158–165.

Tüzün, I., Bayramoğlu, G., Yalçin, E., Başaran, G., Çelik, G., and Arica, M.Y. (2005). Equilibrium and kinetic studies on biosorption of Hg(II), Cd(II) and Pb(II) ions onto microalgae *Chlamydomonas reinhardtii*. *J. Environ. Manag.* **77**, 85–92.

van de Mortel, J.E., Almar Villanueva, L., Schat, H., Kwekkeboom, J., Coughlan, S., Moerland, P.D., Ver Loren van Themaat, E., Koornneef, M., and Aarts, M.G. (2006). Large expression differences in genes for iron and zinc homeostasis, stress response and lignin biosynthesis distinguish *Arabidopsis thaliana* and the related metal hyperaccumulator *Thlaspi caerulescens*. *Plant Physiol.* **142**, 1127–1147.

van der Zaal, B.J., Neuteboom, L.W., Pinas, J.E., Chardonnens, A.N., Schat, H., Verkleij, J.A., and Hooykaas, P.J. (1999). Overexpression of a novel *Arabidopsis* gene related to putative zinc-transporter genes from animals can lead to enhanced zinc resistance and accumulation. *Plant Physiol.* **119**, 1047–1055.

van Loon, A.P., Pesold-Hurt, B., and Schatz, G. (1986). A yeast mutant lacking mitochondrial manganese-superoxide dismutase is hypersensitive to oxygen. *Proc. Natl. Acad. Sc.i U. S. A.* **83**, 3820–3824.

van Veen, H.W., Abee, T., Kortstee, G.J., Konings, W.N., and Zehnder, A.J. (1994). Substrate specificity of the two phosphate transport systems of *Acinetobacter johnsonii* 210A in relation to phosphate speciation in its aquatic environment. *J. Biol. Chem.* **269**, 16212–16216.

Vatamaniuk, O.K., Bucher, E.A., Sundaram, M.V., and Rea, P.A. (2005). CeHMT-1, a putative phytochelatin transporter, is required for cadmium tolerance in *Caenorhabditis elegans*. *J. Biol. Chem.* **280**, 23684–23690.

Vega, J.M., Rubiales, M.A., Vilchez, C., and Vigara, J. (2005). Effect of abiotic stress on photosynthesis, respiration and antioxidant system in *Chlamydomonas reinhardtii*. *Phyton-Annales Rei Botanicae* **45**, 97–106.

Vega, J.M., Garbayo, I., Domínguez, M.J., and Vigara, J. (2006). Effect of abiotic stress on photosynthesis and respiration in *Chlamydomonas reinhardtii* – induction of oxidative stress. *Enzyme Microb. Tech.* **40**, 163–167.

Verret, F., Gravot, A., Auroy, P., Leonhardt, N., David, P., Nussaume, L., Vavasseur, A., and Richaud, P. (2004). Overexpression of AtHMA4 enhances root-to-shoot translocation of zinc and cadmium and plant metal tolerance. *FEBS Lett.* **576**, 306–312.

Vögeli-Lange, R. and Wagner, G.J. (1990). Subcellular localization of cadmium and cadmium-binding peptides in tobacco leaves. Implication of a transport function for cadmium-binding peptides. *Plant Physiol.* **82**, 1086–1093.

Voigt, J. and Nagel, K. (2002). The donor site of photosystem II is impaired in a Cd^{2+}-tolerant mutant strain of the unicellular green alga *Chlamydomonas reinhardtii*. *J. Plant Physiol.* **159**, 941–950.

Voigt, J., Nagel, K., and Wrann, D. (1998). A cadmium-tolerant *Chlamydomonas* mutant strain impaired in photosystem II activity. *J Plant Physiol.* **153**, 566–573.

Wang, B., Liu, C.Q., and Wu, Y. (2005). Effect of heavy metals on the activity of external carbonic anhydrase of microalga *Chlamydomonas reinhardtii* and microalgae from karst lakes. *Bull. Environ. Contam. Toxicol.* **74**, 227–233.

Wang, T. and Wu, M. (2006). An ATP-binding cassette transporter related to yeast vacuolar ScYCF1 is important for Cd sequestration in *Chlamydomonas reinhardtii*. *Plant Cell Environ.* **29**, 1901–1912.

Wang, W.X. and Dei, R.C. (2006). Metal stoichiometry in predicting Cd and Cu toxicity to a freshwater green alga *Chlamydomonas reinhardtii*. *Environ. Pollut.* **142**, 303–312.

Warren, H.V. (1989). Geology, trace elements and health. *Soc. Sci. Med.* **29**, 923–926.

Waters, B.M., Chu, H.H., Didonato, R.J., Roberts, L.A., Eisley, R.B., Lahner, B., Salt, D.E., and Walker, E.L. (2006). Mutations in *Arabidopsis* yellow stripe-like1 and yellow stripe-like3 reveal their roles in metal ion homeostasis and loading of metal ions in seeds. *Plant Physiol.* **141**, 1446–1458.

Weber, M., Harada, E., Vess, C., Roepenack-Lahaye, E.V., and Clemens, S. (2004). Comparative microarray analysis of *Arabidopsis thaliana* and *Arabidopsis halleri* roots identifies nicotianamine synthase, a ZIP transporter and other genes as potential metal hyperaccumulation factors. *Plant J.* **37**, 269–281.

Weger, H.G. (1999). Ferric and cupric reductase activities in the green alga *Chlamydomonas reinhardtii*: Experiments using iron-limited chemostats. *Planta* **207**, 377–384.

Weger, H.G. and Espie, G.S. (2000). Ferric reduction by iron-limited *Chlamydomonas* cells interacts with both photosynthesis and respiration. *Planta* **210**, 775–781.

Weger, H.G., Middlemiss, J.K., and Petterson, C.D. (2002). Ferric chelate reductase activity as affected by the iron-limited growth rate in four species of unicellular green algae (Chlorophyta). *J. Phycol.* **38**, 513–519.

Weigel, D. and Nordborg, M. (2005). Natural variation in *Arabidopsis*. How do we find the causal genes? *Plant Physiol.* **138**, 567–568.

Weiss-Magasic, C., Lustigman, B., and Lee, L.H. (1997). Effect of mercury on the growth of *Chlamydomonas reinhardtii*. *Bull. Environ. Contam. Toxicol.* **59**, 828–833.

Williams, L.E. and Mills, R.F. (2005). P(1B)-ATPases: An ancient family of transition metal pumps with diverse functions in plants. *Trends Plant Sci.* **10**, 491–502.

Wolfe-Simon, F., Grzelryk, D., Schofield, O., and Falkowski, P.G. (2005). The role and evolution of superoxide dismutases in algae. *J. Phycol.* **41**, 453–465.

Wood, P.M. (1978). Interchangeable copper and iron proteins in algal photosynthesis. Studies on plastocyanin and cytochrome *c*-552 in *Chlamydomonas*. *Eur. J. Biochem.* **87**, 9–19.

Xiang, C., Werner, B.L., Christensen, E.M., and Oliver, D.J. (2001). The biological functions of glutathione revisited in *Arabidopsis* transgenic plants with altered glutathione levels. *Plant Physiol.* **126**, 564–574.

Xue, X., Collins, C.M., and Weger, H.G. (1998). The energetics of extracellular Fe(III) reduction by iron-limited *Chlamydomonas reinhardtii* (Chlorophyta). *J. Phycol.* **34**, 939–944.

Yang, M., Cobine, P.A., Molik, S., Naranuntarat, A., Lill, R., Winge, D.R., and Culotta, V.C. (2006). The effects of mitochondrial iron homeostasis on cofactor specificity of superoxide dismutase 2. *EMBO J.* **25**, 1775–1783.

Yuan, D.S., Dancis, A., and Klausner, R.D. (1997). Restriction of copper export in *Saccharomyces cerevisiae* to a late Golgi or post-Golgi compartment in the secretory pathway. *J. Biol. Chem.* **272**, 25787–25793.

Zhao, H. and Eide, D. (1996a). The yeast ZRT1 gene encodes the zinc transporter protein of a high-affinity uptake system induced by zinc limitation. *Proc. Natl. Acad. Sci. U. S. A.* **93**, 2454–2458.

Zhao, H. and Eide, D. (1996b). The ZRT2 gene encodes the low affinity zinc transporter in *Saccharomyces cerevisiae*. *J. Biol. Chem.* **271**, 23203–23210.

CHAPTER 11

Thioredoxins and Related Proteins

Laure Michelet, Mirko Zaffagnini and Stéphane D. Lemaire
Institut de Biotechnologie des Plantes, Unité Mixte de Recherche 8618,
Centre National de la Recherche Scientifique, Paris, France

CHAPTER CONTENTS

I. INTRODUCTION

Thioredoxins (TRX) are small ubiquitous redox proteins of approximately 12 kD found in nearly all organisms and playing a key role in redox signaling and oxidative stress responses. First identified in *E. coli* in the early 1960s,

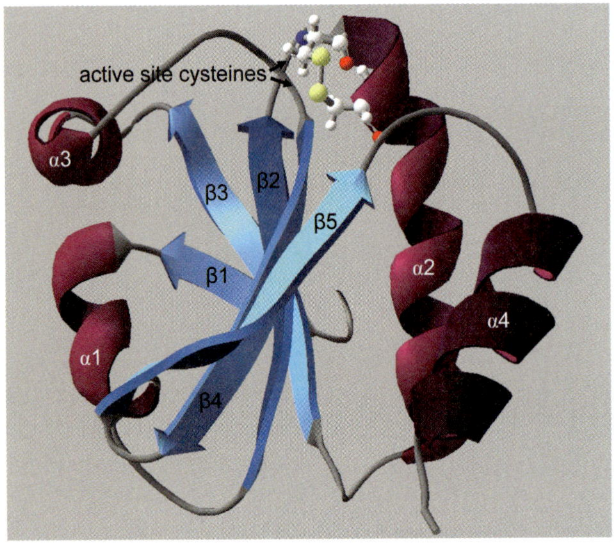

FIGURE 11.1 *Typical three-dimensional structure of a thioredoxin. Ribbon representation of a representative three-dimensional NMR structure (structure 20; Protein Database accession number 1DBY) of Chlamydomonas TRXm taken from the 28 NMR models obtained for this protein (Lancelin et al., 2000). The protein exhibits the "thioredoxin fold" characterized by the succession of the following secondary structural elements: β2, α2, β3, α3, β4, β5, α4. The backbones and side chains of the two active site cysteines are also represented.*

thioredoxins are involved in the reduction of disulfide bridges of various target proteins by thiol-disulfide exchange reactions. Thioredoxins share a common 3D structure characterized by a succession of secondary structural elements: β1, α1, β2, α2, β3, α3, β4, β5, α4 (Figure 11.1). The succession of these elements without β1 and α1 forms the "thioredoxin fold" described by Martin (1995). Additional proteins belong to the thioredoxin superfamily because they possess a thioredoxin domain. This superfamily encompasses five major groups of proteins: thioredoxins, glutaredoxins (GRX), protein disulfide isomerases (PDI), glutathione peroxidases, and glutathione-*S*-transferases.

In photosynthetic organisms, thioredoxins have been known for many years to play a major role in light signaling through regulation of carbon metabolism enzymes, including several enzymes of the Calvin cycle. *Chlamydomonas* has been an important model for the study of these proteins for more than 15 years. The availability of genome sequences has revealed that while non-photosynthetic organisms usually contain a limited number of redoxins, in photosynthetic organisms these proteins are encoded by multigene families. Proteomic studies have allowed identification of many potential targets of these redox proteins, which could constitute central regulators of the primary metabolism of photosynthetic cells. This chapter provides an overview of the functions of thioredoxins and related proteins in *Chlamydomonas*, with a focus on the TRX, GRX, and PDI families. Their functions are discussed in view of more recent results suggesting a complex cross-talk between the different redox regulatory pathways of the cell.

II. THIOREDOXINS

A. Identification of the thioredoxin system

Thioredoxin was first identified in *E. coli* as an electron donor to ribonucleotide reductase (Laurent et al., 1964). Since then, thioredoxins have been identified in all free-living organisms from bacteria to mammals, including photosynthetic organisms (for detailed aspects of thioredoxin systems in non-photosynthetic organisms, see Holmgren, 2000; Vlamis-Gardikas and Holmgren, 2002; Holmgren et al., 2005). In contrast with photosynthetic eukaryotes, non-photosynthetic eukaryotes contain a limited number of thioredoxins with usually one or two isoforms in the cytosol and one in the mitochondria. These thioredoxins are reduced by NADPH thioredoxin reductases (NTR) present in these two subcellular compartments, and are involved in several cell processes including oxidative stress defense mechanisms, by providing electrons to reactive oxygen species (ROS)-detoxifying enzymes such as the thioredoxin-dependent peroxidases peroxiredoxins (PRX), or to repair proteins such as methionine sulfoxide reductases (MSR). In non-photosynthetic eukaryotes, thioredoxins are also involved in sulfate assimilation by reducing 3′-phosphate-5′-phosphosulfate reductase (PAPS reductase), and in the control of the DNA-binding activity of several transcription factors (Vlamis-Gardikas and Holmgren, 2002). In human cells, thioredoxins appear to be involved in a multitude of processes including the modulation of the activity of several transcription factors, the regulation of apoptosis, control of the immune system, and oxidative stress defense-related mechanisms (reviewed in Pekkari and Holmgren, 2004; Burke-Gaffney et al., 2005; Nakamura, 2005; Kondo et al., 2006). Moreover, thioredoxins have also been implicated in several human disorders such as diabetes, heart disease, and cancer.

In plants, the first thioredoxin system was identified in chloroplasts in the late 1970s (Schürmann et al., 1976; Wolosiuk and Buchanan, 1977). This system was different from the NTR/TRX system previously reported in non-photosynthetic organisms, being composed of a ferredoxin, a thioredoxin, and a ferredoxin-dependent thioredoxin reductase (FTR). Chloroplast thioredoxin and FTR corresponded to the "protein factors" previously reported to be essential to ferredoxin-mediated light activation of a Calvin cycle enzyme: fructose-1,6-bisphosphatase (FBPase; Buchanan et al., 1971).

In chloroplasts, the light absorbed by photosystems triggers a photosynthetic electron transfer resulting in the reduction of ferredoxin and the production of NADPH and ATP. Photoreduced ferredoxin can in turn reduce thioredoxin through FTR, allowing thioredoxin to reduce specific disulfide bridges on target proteins, leading to the light-dependent modulation of the activity of the targets, such as FBPase. In 1979 two types of thioredoxin,

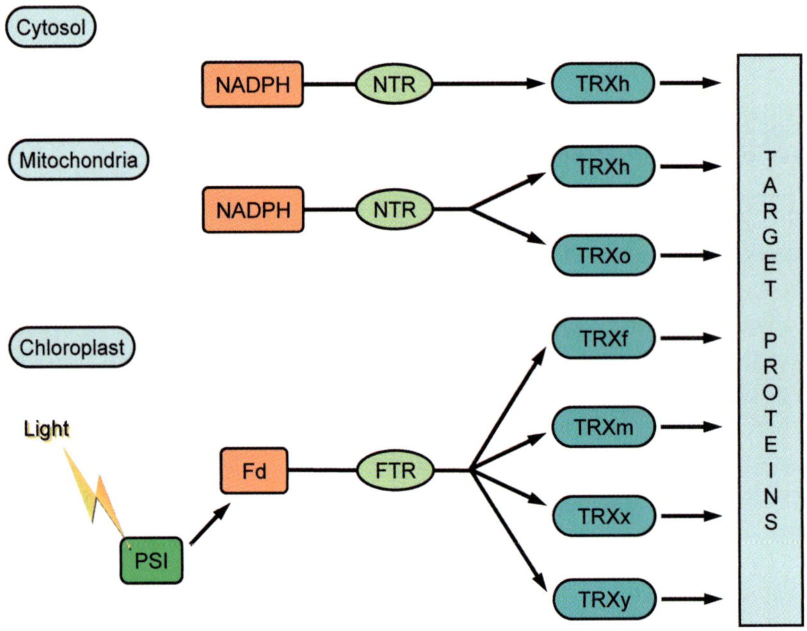

FIGURE 11.2 *Reduction of thioredoxins. In the cytosol and in mitochondria, NADPH-thioredoxin reductases (NTR) use NADPH to reduce thioredoxins (TRX). In chloroplasts, photoreduced ferredoxin reduces thioredoxins via ferredoxin thioredoxin reductase (FTR). Once reduced, thioredoxins can reduce disulfide bridges on their various target proteins.*

named f and m, were found to be present in plants (reviewed in Schürmann and Jacquot, 2000; Jacquot et al., 2002). These thioredoxins were shown to activate preferentially two chloroplast enzymes previously reported to be activated in the light: FBPase and NADP-malate dehydrogenase (NADP-MDH), respectively. TRXf and TRXm are stromal proteins and are reduced in the light by photoreduced ferredoxin and FTR (Figure 11.2). The first targets identified for these thioredoxins were key enzymes of carbon metabolism such as several Calvin cycle enzymes [FBPase, phosphoribulokinase (PRK), glyceraldehyde-3-phosphate dehydrogenase (GAPDH), and sedoheptulose-1,7-bisphosphatase (SBPase)], the γ-subunit of ATP synthase, but also NADP-MDH, which is involved in the export of reducing power out of the chloroplast. These enzymes are largely inactive in the dark, and the activated in the light by the ferredoxin/thioredoxin system. On the contrary, this system also allows inactivation in the light and activation in the dark of glucose-6-phosphate dehydrogenase, a protein of the oxidative pentose phosphate pathway.

A decade later a third type of thioredoxin, named h for "heterotrophic," was discovered (Florencio et al., 1988; reviewed in Gelhaye et al., 2004b). These thioredoxins are mainly located in the cytosol, but some isoforms can also be found in mitochondria (Gelhaye et al., 2004a; Figure 11.2). As for thioredoxins in non-photosynthetic organisms, these extra-chloroplast TRXh are reduced by NADPH via an NADPH thioredoxin reductase (Johnson et al., 1987). They are involved in the mobilization of seed reserves during germination (Kobrehel et al., 1992; Besse et al., 1996; Cho et al., 1999), in self-incompatibility mechanisms (Bower et al., 1996; Cabrillac et al., 2001), and also in ROS detoxification as electron donors to MSR (Gelhaye et al., 2003; Rouhier et al., 2007), glutathione peroxidases (GPX) (Jung et al., 2002; Herbette et al., 2002; Navrot et al., 2006a) and peroxiredoxins (Dietz, 2003; Dietz et al., 2006; Rouhier and Jacquot, 2005). Peroxiredoxins catalyze the reduction of H_2O_2, alkyl hydroperoxides or peroxynitrites using electrons provided by thioredoxins, and can be classified into four subgroups based on the number and position of their cysteine residues: 1-Cys PRX, 2-Cys PRX, type II PRX, and PRX Q. All three TRX isoforms f, m, and h, have been identified in *Chlamydomonas* (Huppe et al., 1990, 1991; Decottignies et al., 1990, 1991; Stein et al., 1995), as well as their reductases, FTR (Huppe et al., 1990) and NTR (Huppe et al., 1991).

Additional thioredoxin isoforms and types were subsequently identified upon sequencing of plant genomes, including TRXx (Mestres-Ortega and Meyer, 1999), a mitochondrial type named TRXo (Laloi et al., 2001), and in *Chlamydomonas*, TRXy, based on an EST analysis (Lemaire et al., 2003b). TRXx and TRXy are both localized to the chloroplast and seem to be involved in defense mechanisms against oxidative stress since they are efficient substrates for peroxiredoxin (Dietz, 2003; Collin et al., 2003, 2004; Rouhier et al., 2004a; Rouhier and Jacquot, 2005; Dietz et al., 2006; Figure 11.2). It was also recently shown that TRXy are the only *Arabidopsis* chloroplast thioredoxins able to reduce chloroplast GPX from poplar (Navrot et al., 2006a). Thioredoxins could also be electron donors to chloroplast MSR (Vieira dos Santos et al., 2005; Rouhier et al., 2007).

B. The thioredoxin system in *Chlamydomonas*

1. The TRX family

The completion of draft genomes for *Arabidopsis* and *Chlamydomonas* revealed a multiplicity of thioredoxin sequences (Lemaire et al., 2003b). In *Arabidopsis*, 20 thioredoxin-encoding genes have been identified, classified into six types: nine TRXh (mainly cytosolic and mitochondrial), two TRXo (mitochondrial), and nine plastidial thioredoxins: four TRXm, two TRXf, one TRXx, and two TRXy. *Chlamydomonas* encodes the same six thioredoxin types as *Arabidopsis* but with fewer isoforms for each, giving a total of eight known members (Figure 11.3). These are two TRXh, located in the cytosol; one TRXo, mitochondrial; one TRXm, two TRXf, one TRXx and one TRXy, localized in plastids (these subcellular localizations are mostly putative and deduced from those demonstrated experimentally for thioredoxins of the same type in other photosynthetic eukaryotes) (Table 11.1). Two

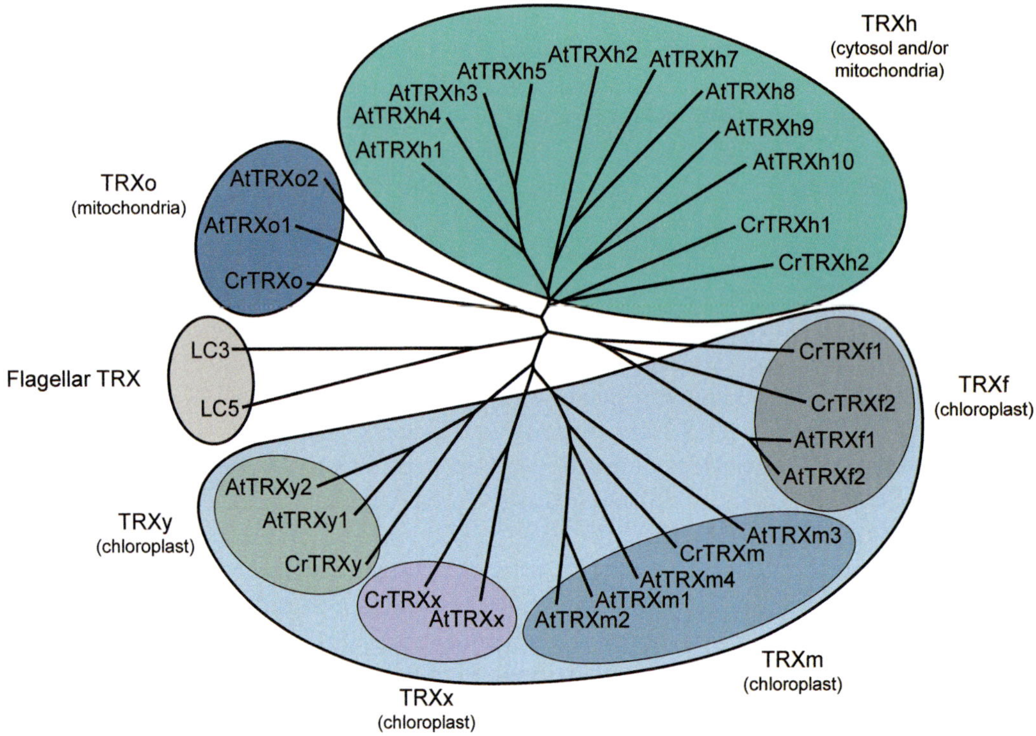

FIGURE 11. 3 *Thioredoxins in Chlamydomonas and Arabidopsis. The unrooted tree was constructed with Clustal X, and gaps were excluded. Accession numbers for Chlamydomonas thioredoxins are given in Table 11.1. Locus designations for Arabidopsis thioredoxins are: AtTRXh1: At3g51030, AtTRXh2: At5g39950, AtTRXh3: At5g42980, AtTRXh4: At1g19730, AtTRXh5: At1g45145, AtTRXh7: At1g59730, AtTRXh8: At1g69880, AtTRXh9: At3g08710, AtTRXh10: At3g56420, AtTRXo1: At2g35010, AtTRXo2: At1g3-1020, AtTRXm1: At1g03680, AtTRXm2: At4g03520, AtTRXm3: At2g15570, AtTRXm4: At3g15360, AtTRXf1: At3g02730, AtTRXf2: At5g16400, AtTRXx: At1g50320, AtTRXy1: At1g76760, AtTRXy2: At1g43560.*

additional proteins related to thioredoxin, absent in *Arabidopsis*, are located in the flagella (outer arm dynein light chains LC3 and LC5; Patel-King et al., 1996). *Chlamydomonas* and *Arabidopsis* also contain atypical thioredoxins and several thioredoxin-like proteins (Meyer et al., 2005, 2006) that will not be discussed in depth here. One of these, CITRX, has been shown to play a role in the regulation of disease resistance in tomato (Rivas et al., 2004; Nekrasov et al., 2006). The CITRX homologues encoded in *Chlamydomonas* and also *Ostreococcus*, however, must play different biological roles. In addition to the multigenic TRX family, the *Chlamydomonas* thioredoxin system also includes five thioredoxin reductases: four NTRs and one FTR.

2. Thioredoxin reductases

a. NADPH thioredoxin reductases (NTR)

Thioredoxins are reduced by different reductases, depending on their subcellular localization. In the cytosol and the mitochondria, thioredoxins are

Table 11.1 The thioredoxin superfamily in *Chlamydomonas*

Name		Putative or demonstrated subcellular localization	Accession numbers		
			V3 Genome	Gene	mRNA
TRX	*TRXh1*	Nucleus and cytosol*	EDP02569	X80887	X78822
	TRXh2	Cytosol	EDP03094	–	AY184797
	TRXo	Mitochondria	EDP07025	–	AY184798
	TRXm	Chloroplast	EDP10052	X80888	X78821
	TRXf1	Chloroplast	EDO98357	–	AY184800
	TRXf2	Chloroplast		–	–
	TRXx	Chloroplast	EDP09524	–	AY184799
	TRXy	Chloroplast	EDP07960	–	AY184796
	DLC3	Flagella*	EDP00552	–	U43610
	DLC5	Flagella*	EDO98142	–	U43609
GRX	*GRX1*	Cytosol	EDO96737	–	–
	GRX2	Cytosol	EDP02937	–	–
	GRX3	Chloroplast	EDO96837	–	–
	GRX4	Cytosol	EDO96388	–	–
	GRX5	Mitochondria	EDP07150	–	–
	GRX6	Chloroplast	EDP09482	–	–
PDI	*PDI1* (*RB60*)	Endoplasmic reticulum and chloroplast*	EDP06730	–	AF036939
	PDI2	Endoplasmic reticulum	EDP09356	–	–
	PDI3	Endoplasmic reticulum	EDO98781	–	–
	PDI4	Endoplasmic reticulum	EDP05721	–	–

Note: The subcellular localizations marked with an asterisk were confirmed experimentally for Chlamydomonas enzymes. Other TRX localizations were deduced from those demonstrated experimentally for thioredoxins of the same type in higher plants. For PDI and GRX the localizations are deduced from those predicted for homologous proteins in higher plants.

reduced by NADPH through the NADPH thioredoxin reductase present in these compartments. In the NADPH/NTR system, the reducing power of NADPH is transferred to thioredoxin via a flavoprotein disulfide oxidoreductase: NADPH thioredoxin reductase (Jacquot et al., 1994; Figure 11.2). NTRs are homodimeric enzymes and each subunit possesses two domains: a central NADPH-binding domain and a FAD-binding domain constituted by the N- and C-terminal parts of the protein, respectively. Both subunits of the dimer contain a redox-active disulfide in a Cys-Ala-Tyr-Cys motif.

NTR is universally distributed from bacteria to mammals, but two different forms have evolved. The first corresponds to a low molecular weight NTR found in bacteria, yeast, and plants. These enzymes are composed of two 35-kD subunits. Mammals contain a distinct form of homodimeric

NTR which is composed of two 55-kD subunits containing an unusual selenocysteine as the penultimate C-terminal residue. Another NTR type was later identified in rice and *Arabidopsis*, which corresponds to a bacterial type NTR fused with a thioredoxin domain in the C-terminal region (Serrato et al., 2004). This NTR has been named NTRC.

In *Chlamydomonas*, an NTR was first identified in the early 1990s (Huppe et al., 1991). This homodimer consisted of two ca. 36-kD subunits and a bound FAD molecule, but its sequence was not determined. Genome sequencing revealed the presence of four NTR-encoding genes, two of which (*NTR2* and *NTR3*) correspond to the 70-kD NTRs typically found in plants, yeast, and bacteria. In *Arabidopsis*, the genes coding the two bacterial type NTRs, *NTRA* and *NTRB*, both encode a cytosolic and a mitochondrial NTR though NTRA is mainly present in mitochondria, whereas NTRB is mainly found in the cytosol (Laloi et al., 2001; Reichheld et al., 2005). In *Chlamydomonas* the situation is different since NTR2 contains an N-terminal extension predicted to target the protein to mitochondria, whereas NTR3, which exhibits no extension, is probably cytosolic.

Another *Chlamydomonas* gene (*NTRC*) encodes an NTR-TRX fusion homologous to higher plant NTRC. Very little is known about the *Chlamydomonas* NTRC except that as in *Arabidopsis*, the NTR domain seems to be closely related to that of cyanobacteria (Meyer et al., 2006). In *Arabidopsis*, NTRC is targeted to chloroplasts and knockout mutants are hypersensitive to oxidative stress (Serrato et al., 2004; Pérez-Ruiz et al., 2006). NTRC does not appear to be able to reduce soluble thioredoxin using NADPH as electron donor (Serrato et al., 2004). Nevertheless, NTRC can use NADPH to reduce its own thioredoxin domain. NTRC was shown to be a direct and efficient reducer of 2-Cys PRX without the need of a free thioredoxin to transfer the redox flux (Moon et al., 2006; Pérez-Ruiz et al., 2006). The fourth *Chlamydomonas* NTR, NTR1, was quite unexpected since it is a mammalian type NTR containing a selenocysteine residue (Novoselov et al., 2002; Novoselov and Gladyshev, 2003). This NTR is also encoded in another alga, *Ostreococcus*, but not in higher plants (Meyer et al., 2006). The specificity of these different NTRs toward thioredoxin remains to be investigated in *Chlamydomonas*.

b. Ferredoxin and ferredoxin thioredoxin reductase (FTR)

In the chloroplast, thioredoxins are reduced by light through photosynthetically reduced ferredoxin and FTR. FTR is a heterodimeric iron-sulfur enzyme composed of a catalytic subunit of 13 kD and a variable subunit which exhibits different molecular masses from one organism to another (Schürmann, 2002; Dai et al., 2004). The catalytic subunit is highly conserved and contains a redox-active disulfide and a [4Fe-4S] cluster. *Chlamydomonas* FTR was purified by Huppe et al. (1990), and consists of a 13-kD catalytic subunit (FTRC) and a 10-kD variable subunit (FTRV). *Chlamydomonas* contains a single gene encoding each subunit. FTR classically reduces chloroplast thioredoxins using the reducing equivalents of photoreduced ferredoxin (Figure 11.2).

Ferredoxin is a soluble protein localized to the stroma, containing a [2Fe-2S] cluster of low redox potential (ca. −420 mV at pH 7.0). Ferredoxin is reduced by light through the photosynthetic electron transfer chain. It mediates electron transfer between photosystem I (PS I) and several ferredoxin-dependent enzymes including FTR. Ferredoxin interacts quite strongly with its electron acceptor proteins, forming electrostatically stabilized complexes. *Chlamydomonas* contains at least six genes encoding ferredoxin (Merchant et al., 2006). The major ferredoxin is a protein of 9.9 kD (Schmitter et al., 1988). A negatively charged ferredoxin residue specifically involved in the formation of the ferredoxin-FTR complex has been identified in spinach and in *Chlamydomonas* (E92 and E91, respectively; De Pascalis et al., 1994, Jacquot et al., 1997). The gene encoding this ferredoxin was isolated by Stein et al. (1995), and its expression was shown to be regulated by light and the circadian clock (Lemaire et al., 1999b) but not by heavy metals (Lemaire et al., 1999a). To date, this is the only ferredoxin gene isolated from photosynthetic eukaryotes that contains an intron, as do mammalian ferredoxin genes: all other plant ferredoxin genes are intronless (Elliott et al., 1989; Vorst et al., 1990; Bringloe et al., 1995).

3. Thioredoxins

a. Overview

The biochemical function of thioredoxins is fulfilled through the two vicinal cysteines of their conserved active site whose sequence is Trp-Cys-Gly/Pro-Cys. The amino terminal active-site cysteine is highly reactive due to its low pKa. This residue can thus make a nucleophilic attack on the disulfide bridge of thioredoxin target proteins (Figure 11.4A). This leads to the formation of a transient mixed disulfide between the thioredoxin and its target. This disulfide is immediately attacked by the second active-site cysteine of the thioredoxin, leading to the release of the reduced target and the oxidized thioredoxin. Thioredoxins fulfill two distinct roles towards their targets. They can function as substrates by providing electrons and protons to enzymes such as ribonucleotide reductase, MSR, PRX, or GPX. In this case, the enzymes use these electrons and protons in the reaction they catalyze. Thioredoxins can also function as modulators of the activity of target proteins by reducing regulatory disulfide bridges, as in the case of thioredoxin-regulated Calvin cycle enzymes. In the latter case, their electrons and protons are not used in the reaction catalyzed by the enzyme. The redox potential of thioredoxin is generally between −270 mV and −300 mV at pH 7.0 (Åslund et al., 1997; Hirasawa et al., 1999; Collin et al., 2003, 2004).

Chlamydomonas thioredoxin systems, including three different types of thioredoxins, h, f, and m, were identified in the early 1990s (Huppe et al.,

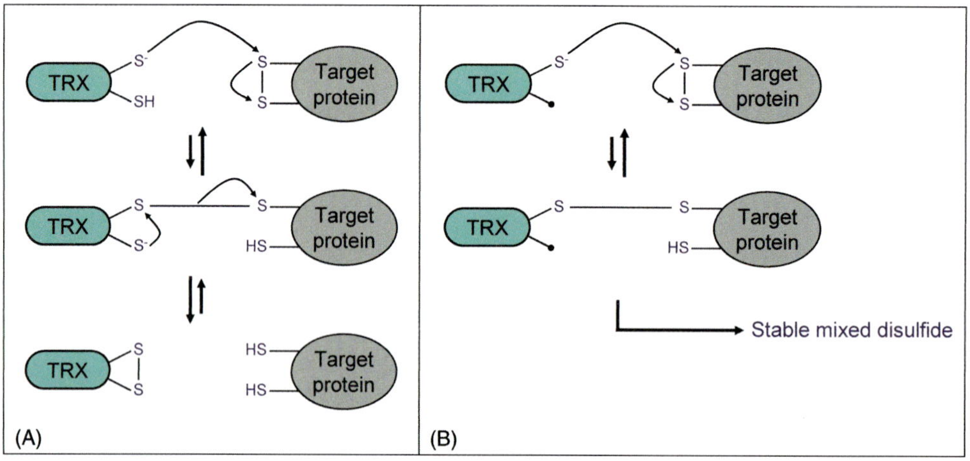

FIGURE 11.4 *(A) Catalytic mechanism for thioredoxin-mediated reduction of disulfide bridges on target proteins. After reduction by its reductase, thioredoxin (on the left) can reduce targets (on the right) by a mechanism which involves the two active site cysteines. The amino-terminal active site cysteine is the most reactive and makes a nucleophilic attack on the disulfide bridge of the target. A transient mixed disulfide between the thioredoxin and the target is formed. In an immediate second step, the other active site cysteine attacks the amino-terminal cysteine to form a disulfide bridge. At the end of the reaction, thioredoxin is oxidized, and the target protein is reduced. (B) Monocysteinic thioredoxin – principle of thioredoxin target trapping. Compared to wild-type thioredoxins, monocysteinic thioredoxin mutants lack the carboxy-terminal active site cysteine. As for wild-type thioredoxin, the first step of target reduction consists of nucleophilic attack on the target protein disulfide bridge by the amino-terminal active-site cysteine. This leads to the formation of a mixed disulfide between the target and the monocysteinic thioredoxin. However, the absence of the second active site cysteine prevents the second nucleophilic attack and thereby stabilizes the mixed disulfide between the thioredoxin and its target protein.*

1990, 1991). *Chlamydomonas* TRXh and TRXm were then purified and their primary sequences determined (Decottignies et al., 1990, 1991). TRXh and TRXm were first called Ch1 and Ch2, respectively, as the two first thioredoxins identified in *Chlamydomonas*. Upon release of the *Chlamydomonas* genome sequence and identification of the different thioredoxin isoforms, Ch1 was renamed TRXh1 (*CrTRXh1*), and Ch2, TRXm (*CrTRXm*) (Lemaire et al., 2003b). These two thioredoxins have been expressed as recombinant proteins in *E. coli*, allowing their purification in large amounts and their biochemical characterization (Jacquot et al., 1992; Stein et al., 1995). Genomic and EST data led to the identification of a second TRXh, a second TRXf, and three new thioredoxins belonging to new types: 1 TRXo, 1 TRXx, and 1 TRXy (Lemaire et al., 2003b; Lemaire and Miginiac-Maslow, 2004).

b. Thioredoxin m

Chlamydomonas contains a single gene encoding a TRXm (Figure 11.3). Phylogenetic analyses indicate that this protein, like higher plant m-type thioredoxins, is related to cyanobacterial thioredoxins (Lemaire et al., 2003b; Lemaire and Miginiac-Maslow, 2004; Meyer et al., 2006). The reactivity of TRXm with the two different reducing systems (NADPH and

NTR or photoreduced ferredoxin and FTR) was tested using a chloroplast thioredoxin target from sorghum, namely NADP-MDH. *Chlamydomonas* TRXm is preferentially reduced by ferredoxin and FTR (Huppe et al., 1990; Decottignies et al., 1990; Stein et al., 1995) and efficiently activates NADP-MDH. The three-dimensional structure of *Chlamydomonas* TRXm was solved using NMR, and was found to be typical for thioredoxin (Lancelin et al., 1993, 2000; Figure 11.1). The redox potential of TRXm is ca. −300 mV at pH 7.0, which is comparable to values for higher plant thioredoxin (Hirasawa et al., 1999).

Regulation of *Chlamydomonas TRXm* gene expression by light and heavy metals was studied (Lemaire et al., 1999a,b, 2002), as previous data suggested a possible link between thioredoxin and heavy metals (Garbisu et al., 1996; Gupta et al., 1997). It was found that the mRNA level corresponding to *TRXm* was regulated both by light and the circadian clock, with a low level during the night and a high level during the day (Lemaire et al., 1999b). Heavy metals such as cadmium or mercury also induce the expression of the *TRXm* gene, although the protein level did not change (Lemaire et al., 1999a). *In vitro*, heavy metals were shown to inhibit TRXm activity, presumably by binding to the active-site cysteines (Lemaire et al., 1999a). Later studies showed that *E. coli* thioredoxin could be inactivated by cadmium binding to an active-site Cys and a conserved Asp residue (Rollin-Genetet et al., 2004). Studies in yeast also revealed that thioredoxins play a major role in the mechanisms of cadmium resistance (Vido et al., 2001; Bae and Chen, 2004).

c. Thioredoxin f

Two genes encoding f-type thioredoxin are present in *Chlamydomonas*, *TRXf1* and *TRXf2* (Figure 11.3). Compared to TRXm and TRXh, f-type thioredoxins have been less studied in *Chlamydomonas*. Recombinant TRXf1 was purified as part of a study on the regulation of NADP-MDH (Lemaire et al., 2005). The enzyme failed to activate NADP-MDH from higher plants, a classical system to test TRXf activity, but was able to activate homologous NADP-MDH. Measurements of the ability of TRXf1 to activate mutant forms of NADP-MDH from *Sorghum vulgare* suggested that this thioredoxin has a redox potential that is less negative than most of higher plant TRXf (see section III.C.4 for details). As in *Arabidopsis*, f-type thioredoxins are probably specifically involved in the regulation of carbon metabolism enzymes in the light. Unfortunately, the poor expression of recombinant TRXf1 still prevents its detailed biochemical characterization.

TRXf sequences contain, in addition to their two active-site cysteines, a third cysteine conserved in all photosynthetic eukaryotes, including *Chlamydomonas* TRXf1 and TRXf2. In *Arabidopsis* TRXf1, this additional cysteine is the target of a reversible post-translational modification called glutathionylation, which can occur under oxidative stress (the role and importance of this modification will be discussed further in sections IV.B.1.b and IV.C). When the additional cysteine is glutathionylated, AtTRXf1 activity toward targets is impaired due to a decrease in the efficiency of the protein reduction by FTR (Michelet et al., 2006). One can assume that this modification, which specifically targets one strictly conserved cysteine residue on f-type thioredoxin, also regulates *Chlamydomonas* TRXf activity.

d. Thioredoxin x

As in the case of *Arabidopsis*, a single gene encoding TRXx is present in *Chlamydomonas* (Figure 11.3). Phylogenetic studies have shown that *Chlamydomonas* TRXx and AtTRXx are both related to cyanobacterial thioredoxin (Lemaire et al., 2003b; Meyer et al., 2006). AtTRXx has been shown to be targeted to chloroplasts (Collin et al., 2003). AtTRXx failed to activate NADP-MDH and FBPase, likely because of its less negative redox potential compared to m- and f-type thioredoxins (Collin et al., 2003). However, AtTRXx appeared to be the best substrate for chloroplast 2-Cys PRX, suggesting that this type of thioredoxin is specifically involved in oxidative stress defense.

e. Thioredoxin y

Thioredoxins belonging to the y-type were first identified in *Chlamydomonas*. A single *TRXy* gene is present in *Chlamydomonas* whereas two were identified in *Arabidopsis* (*AtTRXy1* and *AtTRXy2*; Figure 11.3). TRXy are conserved in all photosynthetic eukaryotes and are related, like TRXx and TRXm, to cyanobacterial thioredoxins (Lemaire et al., 2003b; Meyer et al., 2006). This suggested a chloroplast localization which was confirmed for the *Arabidopsis* isoforms (Collin et al., 2004). The ability of recombinant TRXy to activate the two classic chloroplast thioredoxin targets, FBPase and NADP-MDH, was investigated. Surprisingly, TRXy was found to be able to activate FBPase, a target enzyme known to be strictly dependent on TRXf for its activation. However, TRXy appeared much less efficient than *Arabidopsis* TRXf in this test (Lemaire et al., 2003a). Moreover, this unexpected property seems to be specific for *Chlamydomonas* TRXy, since neither of the two y-type thioredoxins present in *Arabidopsis* is able to activate FBPase (Collin et al., 2004). In addition, TRXy was inefficient for activation of NADP-MDH, a property shared with the *Arabidopsis* TRXy. This lack of efficiency could be linked to their redox potentials, as in the case of TRXx. Indeed, the redox potentials of AtTRXy1 and AtTRXy2 was measured to be 30 mV less negative than for classical thioredoxins. Overall, these results suggested that Chlamydomonas TRXy was not likely to be involved in the regulation of carbon assimilation, as had been established for f- and m-type thioredoxins. The identification of a new chloroplast peroxiredoxin from poplar, named PRX Q, led to the finding that TRXy is the most efficient plastidial thioredoxin for PRX Q reduction (Rouhier et al., 2004a). This result was confirmed with homologous enzymes in *Arabidopsis*, where

AtTRXy1 and AtTRXy2 were found to be much more efficient than the seven other plastidial thioredoxins in reducing *Arabidopsis* PRX Q (Collin et al., 2004). These results suggest that TRXy thioredoxins are, like TRXx, specifically involved in defense mechanisms against oxidative stress as efficient electron donors to plastidial peroxiredoxins.

f. Thioredoxin h

Reactivity of cytosolic TRXh1 towards NADP-MDH has been tested with two different reducing systems, NADPH/NTR and photoreduced Fd/FTR. Unlike TRXm, TRXh1 is preferentially reduced by NADPH and NADPH thioredoxin reductase, reinforcing the idea that this protein is cytosolic, and it also exhibits a lower reactivity towards NADP-MDH (Decottignies et al., 1991; Huppe et al., 1991; Stein et al., 1995).

Thioredoxins are generally described as heat stable proteins. For example, *E. coli* TRX1 can be heated for 1 hour at 100°C with no loss of activity (Holmgren, 1985). The analysis of the heat sensitivity of TRXh1 and TRXm revealed unexpected differences, in that TRXm was unaffected by treatments at 100°C, while TRXh1 became completely aggregated in these conditions (Stein et al., 1995). The availability of the three-dimensional NMR structures of TRXm (Lancelin et al., 1993, 2000) and TRXh1 (Mittard et al., 1995, 1997) allowed analysis of the structural determinants responsible for the differences in susceptibility to heat treatments (Lemaire et al., 2000). These differences turned out to be linked to distinct pH sensitivities rather than temperature sensitivities, and specific residues determining the pH range of aggregation could be identified.

The analysis of several mutant TRXh1 proteins has provided insight into its biochemical activity. A mutant in which the conserved active-site tryptophan (Trp35: Trp-Cys-Gly-Pro-Cys) was replaced by an alanine, was still redox active toward NTR and chemical substrates (dithiothreitol and 5,5′dithio-bis(2-nitrobenzoicacid)), but had lost its ability to activate NADP-MDH. The W35A mutation also affects the pH dependence of resonances of both active-site cysteines in ^{1}H-^{13}C NMR (Krimm et al., 1998). Comparisons of the crystal structures of wild-type and W35A TRXh1 (Menchise et al., 2000, 2001) revealed that replacement of Trp35 leads to modifications in the environment of the two catalytic cysteines, changing the active site conformation and impairing protein–protein recognition (Menchise et al., 2000). The crystal structure of D30A TRXh1 was also determined, since the conserved Asp30 was thought to be involved in the catalytic mechanism. Comparison of wild-type and mutant crystal structures confirmed the role of this aspartic acid as an acid/base in the thioredoxin oxidoreduction reactions, and also suggested the involvement of a water molecule in the catalytic mechanism (Menchise et al., 2001). Spectroscopic and calorimetric studies were undertaken on TRXh1 to investigate the unfolding of the protein induced by cold and heat at different pH values, and the reversibility or irreversibility of the phenomenon (Richardson et al., 2000). This study revealed that heat-induced unfolding of TRXh1 is reversible at acidic pH, whereas it is irreversible at neutral pH. Moreover, cold induces TRXh1 denaturation, especially at low pH (<2.5). The one-electron reduction and oxidation of TRXh1 was also studied (Lmoumene et al., 2000; Sicard-Roselli et al., 2004). TRXh1 has a redox potential of −290 mV at pH 7.0, similar to that of higher plant h-type thioredoxins (Setterdahl et al., 2003).

The regulation of *TRXh1* gene by light and heavy metals was studied (Lemaire et al., 1999a,b, 2002). Since previous data suggested a possible link between TRXh and cell division in yeast and in tobacco (Muller, 1991; Regad et al., 1995), *Chlamydomonas* cells were synchronized by light/dark cycles. These studies revealed that *TRXh1* expression was not cell cycle-regulated but like *TRXm*, was regulated both by light and the circadian clock, being expressed at a low level during the night and a higher level during the day (Lemaire et al., 1999b). Heavy metals (cadmium, mercury) induce *TRXh1* mRNA accumulation with faster kinetics than observed for *TRXm* (Lemaire et al., 1999a). For *TRXh1*, this increase was also observed at the protein level, suggesting that TRXh1 could be involved in heavy metal tolerance mechanisms.

Chlamydomonas contains two genes encoding h-type thioredoxin, *TRXh1* and *TRXh2* (Lemaire et al., 2003b; Lemaire and Miginiac-Maslow, 2004; Figure 11.3). The analysis of EST databases and *TRXh2* gene structure suggest the existence of three alternative splice variants differing in the C-terminal part of the protein. The proteins encoded by the two TRXh genes are predicted to be cytosolic.

RNAi (Rohr et al., 2004) was performed to inactivate different genes, including *TRXh1*. The study of a *TRXh1* deletion mutant called BleI, as well as of a wild-type strain in which *TRXh1* was inactivated by RNAi, implicated this thioredoxin specifically in the response to alkylation-induced DNA damage since TRXh2, the other cytosolic isoform, does not compensate for TRXh1 deficiency (Sarkar et al., 2005). The BleI strain is sensitive to DNA alkylation agents such as bleomycin and methyl methanesulfonate (MMS). Sarkar et al. also showed that a yeast strain deficient in cytosolic thioredoxin was sensitive to DNA alkylating agents, suggesting that the function of thioredoxin in DNA repair might not be restricted to *Chlamydomonas* or even photosynthetic organisms. This yeast strain exhibits a pleiotropic phenotype, including a lengthened S phase and hypersensitivity to H_2O_2. Complementation assays revealed that the deficiency in cell cycle progression and sensitivity to MMS were rescued by TRXh1 expression, whereas H_2O_2 hypersensitivity was not. Transient expression of TRXh1 in onion epidermal cells revealed a translocation of the protein from cytosol to nucleus in response to MMS treatment. All these results support a role of TRXh1 in DNA repair.

Expressed sequence tags corresponding to *TRXh2* are overrepresented in the stress-II EST library (which represents mRNA expressed during treatments with H_2O_2 among other stresses), while this is not the case for *TRXh1* (Lemaire et al., 2003b). This is consistent with the fact that TRXh1-deficient strains are not sensitive to oxidative stress and that *TRXh1* heterologous expression does not rescue the H_2O_2 sensitivity of the yeast thioredoxin mutant. These observations suggest that TRXh2 might be more specifically involved in oxidative stress responses.

g. Thioredoxin o

Chlamydomonas possesses a single gene encoding TRXo (*TRXo*) while two are present in *Arabidopsis* (*AtTRXo1* and *AtTRXo2*; Figure 11.3). AtTRXo was shown to be imported into mitochondria and to be reduced by mitochondrial NTR (Laloi et al., 2001). TRXo harbors an N-terminal extension which could confer targeting to mitochondria. The function of plant TRXo remains to be determined.

h. Thioredoxin-related proteins

Two thioredoxin-related proteins have been identified in the outer dynein arm of *Chlamydomonas* flagella, named LC3 and LC5, for dynein light chain (Patel-King et al., 1996). LC3 is a 16-kD protein with one canonical redox-active site Trp-Cys-Gly-Pro-Cys, and a second potentially redox-sensitive vicinal dithiol according to molecular modeling (Patel-King et al., 1996; Harrison et al., 2002). LC5 is a 14-kD protein with the canonical thioredoxin motif Trp-Cys-Gly-Pro-Cys. In phylogenetic analysis, LC5 clusters with animal sequences (Novoselov and Gladyshev, 2003). Although thioredoxins have been evolutionary conserved in axonemal dyneins, their role in dynein function remains to be determined. One hint may come from data suggesting that a redox-regulatory pathway impinges on outer dynein arm function and influences flagellar motility (Wakabayashi and King, 2006).

C. Thioredoxin targets

1. Fructose-1,6-bisphosphatase (FBPase)

Chloroplast FBPase was the first thioredoxin target purified from *Chlamydomonas* (Huppe and Buchanan, 1989). This Calvin cycle enzyme was only active in the presence of reduced thioredoxin, and f-type thioredoxin was the most efficient in FBPase activation. Compared to FBPase from spinach chloroplasts, the *Chlamydomonas* enzyme showed a strict requirement for a dithiol reductant irrespective of Mg^{2+} concentration, but seemed to be less specific toward reductant.

2. Glucose-6-phosphate dehydrogenase (G6PDH) and phosphoribulokinase (PRK)

The regulation of G6PDH and PRK activity has been investigated in *Chlamydomonas* (Farr et al., 1994; Huppe et al., 1994). G6PDH is a key regulatory enzyme of the chloroplast oxidative pentose phosphate pathway, whereas PRK is a Calvin cycle enzyme, involved in carbon assimilation. The activity of G6PDH was inhibited in the light and activated in the dark. On the contrary, PRK was light-activated and inhibited in the dark. These results suggested that as in higher plants, reduction by the Fd-TRX system regulates G6PDH and PRK activity so as to optimize carbon reduction and oxidation in the chloroplast in accordance with the supply of electrons, and in response to the metabolic demands for reducing power and ATP. The mRNA level for *PRK* was shown to be regulated by light and the circadian clock, as previously observed for the genes encoding ferredoxin, TRXm, and TRXh1 (Lemaire et al., 1999b).

3. PRK/CP12/Glyceraldehyde 3-phosphate dehydrogenase (GAPDH) complex

In *Chlamydomonas*, the regulation of two Calvin cycle enzymes, PRK and GAPDH, forming a supramolecular complex with a small redox polypeptide, CP12, has been extensively studied (Graciet et al., 2004; Lebreton et al., 2006). CP12 is a protein of approximately 80 amino acids universally distributed among oxygenic photosynthetic organisms but apparently absent in heterotrophs, belonging to the intrinsically unstructured proteins (IUP; Dyson and Wright, 2005), which appear to be disordered in solution. In the case of CP12, this disorder decreases when two disulfide bridges are formed under oxidizing conditions, even though the protein remains very flexible (Graciet et al., 2003). This flexibility generally allows IUP proteins to be involved in protein–protein interactions and thereby in the formation of supramolecular complexes (Dyson and Wright, 2005; Tompa et al., 2005).

Active PRK and GAPDH are dimeric and tetrameric enzymes, respectively. In photosynthetic organisms, two different GAPDH chloroplast subunits have been identified: GapA and GapB, the latter being only present in higher plants. Based on sequence homology, GapB appears to be a GapA subunit fused to the C-terminal half of CP12 at its C-terminus. In *Chlamydomonas*, where only GapA subunits are present, the only chloroplast isoform is A_4-GAPDH, whereas in higher plants two isoforms coexist: A_4-GAPDH, the minor form, and A_2B_2-GAPDH, the major form. A_4-GAPDH is considered to be redox-insensitive and is regulated via the formation of a supramolecular complex with PRK and CP12, whereas A_2B_2-GAPDH is regulated autonomously by thioredoxin and metabolites (reviewed in Trost et al., 2006). Both PRK and GAPDH exhibit a low activity in the dark, and are fully activated in the light.

In *Chlamydomonas*, the supramolecular complex is composed of two CP12 molecules associated with two PRK dimers and two A_4-GAPDH tetramers (Avilan et al., 1997, 2000; Graciet et al., 2004). Its formation initially requires oxidation of CP12 and PRK and indeed, A_4-GAPDH can bind CP12 under oxidizing conditions and only in the presence of NAD(H).

Within this complex, A_4-GAPDH is partially inactivated. Concomitantly, PRK becomes oxidized through the formation of intrasubunit disulfides, leading to a partial inactivation of the protein. Then PRK can bind the binary complex, leading to a stronger inhibition of PRK and GAPDH activity. The formation of this supramolecular complex can be reversed by different factors, including reduced thioredoxin, GAPDH substrate 1,3-bisphosphoglyceric acid (BPGA) or phosphorylated cofactor [NADP(H)].

4. NADP-malate dehydrogenase

NADP-MDH plays a major role in C4 photosynthesis, but is also present in C3 plants where it facilitates export of reducing power from the chloroplast to the cytosol. The activation mechanism of sorghum NADP-MDH by thioredoxin has been extensively studied (Miginiac-Maslow and Lancelin, 2002). The activation of this homodimer is strictly dependent on the reduction of two disulfide bridges per monomer (in the light). The N-terminal bridge of sorghum NADP-MDH has a redox potential of −290 mV (pH 7.0), whereas the C-terminal bridge has a more negative redox potential of −330 mV (pH 7.0; Hirasawa et al., 2000). Reduction of the latter bridge thus requires an excess of reduced thioredoxin. In *Chlamydomonas*, NADP-MDH contains the C-terminal cysteines but the N-terminal cysteines are absent (Gomez et al., 2002). Comparison of the activation of NADP-MDH from sorghum and *Chlamydomonas* revealed that the *Chlamydomonas* protein was more rapidly activated (Lemaire et al., 2005). Moreover, TRXf1 was able to activate *Chlamydomonas* but not sorghum NADP-MDH. Further investigations revealed that only one disulfide bridge was involved in *Chlamydomonas* NADP-MDH activation, which has a redox potential of −315 mV (pH 7.0; Lemaire et al., 2005). TRXf1 inefficiency in activating sorghum NADP-MDH could thus be due to a less negative redox potential compared to higher plant TRXf, suggesting a possible co-evolution of TRXf and NADP-MDH in each organism. Moreover, the regulation of *Chlamydomonas* NADP-MDH appears like a "first draft" of the complex regulation that has evolved in higher plants. The complexity in higher plants could be linked to their multicellular structure and indeed, in higher plants malate is a circulating form of reducing power whose concentration has to be strictly controlled (Scheibe, 2004), while in *Chlamydomonas* malate is only exchanged between the single chloroplast and the cytosol.

Comparative studies of the redox regulatory properties of enzymes identified as thioredoxin targets in photosynthetic eukaryotes in different organisms, ranging from cyanobacteria to higher plants, including diatoms and algae, can reveal how thioredoxin regulation appeared during evolution. In several cases, comparison of thioredoxin-regulated isoforms with non-regulated isoforms reveals that sequences allowing thioredoxin regulation have been introduced during evolution, suggesting a dynamism of the system (reviewed in Ruelland and Miginiac-Maslow, 1999).

5. Protein kinase STT7

Thioredoxins play a major role in the regulation of photosynthetic processes by regulating Calvin cycle enzymes, but may also be involved in the regulation of thylakoid electron transfer and especially the state transition process (see Chapter 22). The mechanism of state transition allows energy redistribution between PS I and PS II under varying illumination conditions. A mobile part of light-harvesting complex II (LHCII) associated with PS II (State I) can migrate to light-harvesting complex I (State II) when PS II receives more energy than PS I (Rochaix, 2004; Finazzi, 2005; Takahashi et al., 2006). The mechanism of state transition is based on the phosphorylation of a mobile LHCII antenna by a redox-dependent kinase, STT7, activated when the plastoquinone pool becomes overreduced. In addition to redox regulation of STT7 by the plastoquinone pool and its interaction with cytochrome b_6/f, thioredoxins have been suggested to play some role in the regulation of state transition (Rintamaki et al., 2000; Aro and Ohad, 2003). The cloning of *STT7* from *Chlamydomonas* (Depège et al., 2003) and *Arabidopsis* (Bellafiore et al., 2005) revealed the presence of two conserved cysteines in the N-terminal part of the protein, which could allow thioredoxin-dependent regulation. Moreover, several soluble or thylakoid-anchored proteins participating more directly in the photosynthetic electron transfer chain are among the new potential targets of thioredoxin identified by large scale proteomic studies in photosynthetic organisms (see section III.C.6). This suggests that thioredoxins play a major role in the regulation of photosynthetic processes not only through the Calvin cycle, but also thylakoid electron transfer.

6. Large scale identification of thioredoxin targets

The discovery of the multiplicity of thioredoxins in photosynthetic organisms prompted several groups to actively search for previously unrecognized thioredoxin targets. Indeed, the number of known thioredoxin isoforms in *Arabidopsis* exceeded the number of known targets (Buchanan and Balmer, 2005; Michelet et al., 2006). Many of these studies, including those performed in *Chlamydomonas*, were based on the use of a monocysteinic thioredoxin mutant, lacking the second active-site cysteine. According to the known reaction mechanism allowing the reduction of a target by thioredoxin (Figure 11.4A), the N-terminal cysteine remaining in the monocysteinic thioredoxin will still be able to perform a nucleophilic attack on the target disulfide. However, the absence of the second cysteine of thioredoxin active site prevents the second nucleophilic attack, thereby stabilizing the mixed disulfide formed between the thioredoxin and its target (Figure 11.4B). When bound to a column, the monocysteinic mutant protein can covalently bind thioredoxin targets present in a protein extract. After extensive washing of the column, bound targets can be eluted using a strong reductant such as DTT and subsequently separated on two-dimensional gels and identified

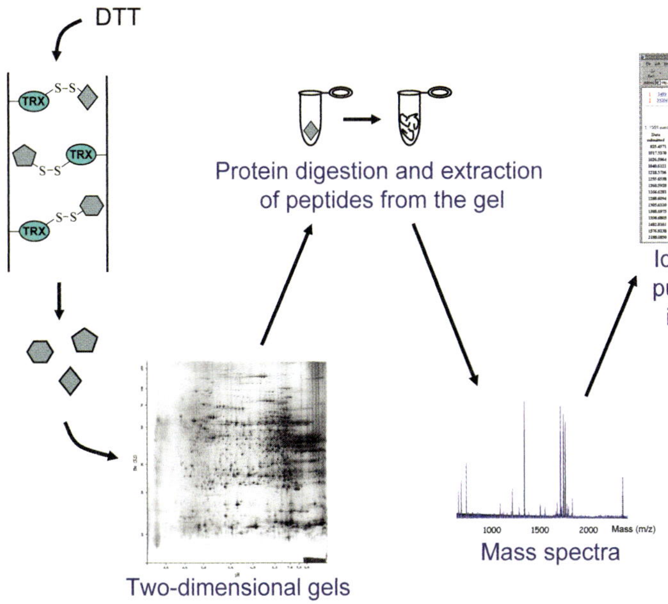

FIGURE 11.5 *Identification of thioredoxin targets by proteomics. A thioredoxin affinity column is prepared by binding of a monocysteinic thioredoxin to a matrix. After loading of a protein extract on the column and a washing step, proteins trapped by monocysteinic thioredoxins are eluted using a strong reducing agent such as dithiothreitol (DTT). The proteins are then separated on two-dimensional gels. After Coomassie blue staining, dots corresponding to the thioredoxin targets are excised, digested by a site-specific protease and peptides are extracted from the gel and analyzed by mass spectrometry. The comparison of the number of peptides and their masses observed by mass spectrometry with the peptides obtained by an in silico digestion by the same protease of all the proteins present in the database allows the identification of the putative thioredoxin targets.*

by mass spectrometry (Figure 11.5). This type of proteomic approach based on a monocysteinic column or on thiol labeling strategies allowed the identification of nearly 300 thioredoxin targets in photosynthetic organisms (reviewed in Michelet et al., 2006). In this chapter we will only describe the results obtained for *Chlamydomonas*.

A monocysteinic mutant of TRXh1 was initially used to map the cysteines targeted by thioredoxin during sorghum NADP-MDH activation (Goyer et al., 1999). When bound to a column, the monocysteinic thioredoxin allowed trapping of chloroplast 2-Cys PRX in *Chlamydomonas* (Goyer et al., 2002). A larger scale study using the same strategy subsequently identified 55 putative thioredoxin targets in *Chlamydomonas* total soluble extracts (Lemaire et al., 2004; Figure 11.6). These putative targets included many proteins whose regulation by thioredoxin had already been reported, such as Calvin cycle enzymes, demonstrating the validity of the method. In addition to known targets, many new potential targets participating in the same processes were identified. For example, among the six Calvin cycle enzymes

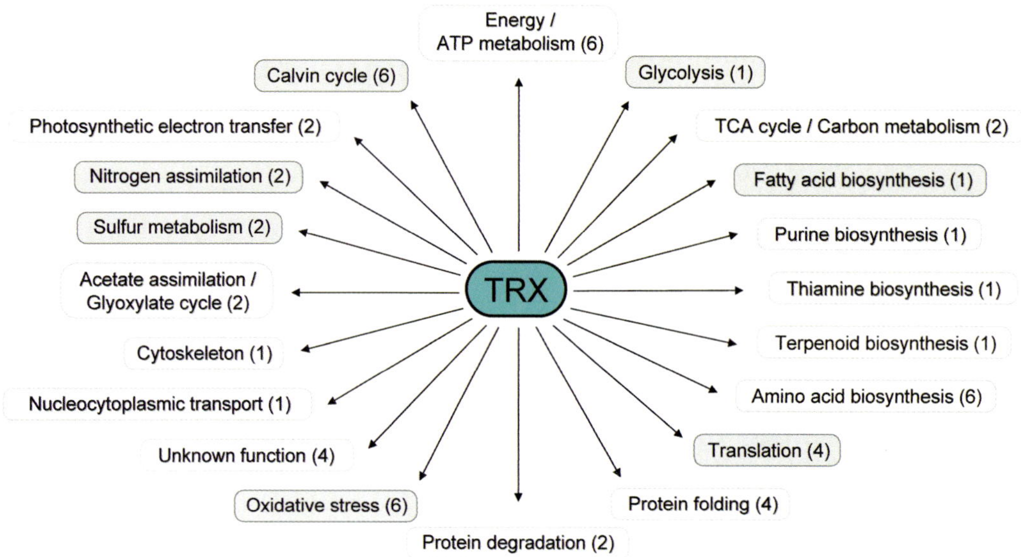

FIGURE 11.6 *Cell processes putatively regulated by thioredoxins in Chlamydomonas. The putative thioredoxin target proteins identified by proteomics participate in different cell processes (Lemaire et al., 2004). Among these targets, many are involved in metabolic processes for which regulation by thioredoxin had already been reported. These processes are listed in open rounded rectangles, with the number of targets in parentheses. Additional cell processes never reported to be regulated by thioredoxins are indicated in shaded rounded rectangles, with the number of targets in parentheses.*

bound to the thioredoxin column, only two were known targets (PRK, SBPase) while the four others (Ribose-5P-epimerase, Rubisco large and small subunits and fructose bisphosphate aldolase) were not known to be regulated by thioredoxin. Similarly, six enzymes involved in ROS detoxification were identified including 4 PRX, 1 MSR, and 1 GPX, which are all well-established targets of thioredoxin. More surprisingly, the sixth enzyme was catalase, a major hydrogen peroxide detoxification enzyme not previously known to be redox regulated. This regulation was confirmed experimentally: the reduction of catalase by thioredoxin leads to a 50% inhibition of the enzyme activity (Lemaire et al., 2004). This regulation might allow activation of catalase by oxidized thioredoxin under oxidative stress.

The proteomics study also identified 33 new possible targets involved in metabolic processes previously unrecognized to be regulated by thioredoxin (Figure 11.6). Among these, six proteins participating in several amino acid biosynthesis pathways were found. Thioredoxin-dependent regulation was confirmed for one of these enzymes, 3-isopropylmalate dehydrogenase (IPMDH), the third enzyme of leucine biosynthesis pathway. The enzyme is strongly activated by reduced thioredoxin and contains two cysteines conserved in higher plant sequences but absent in IPMDH from non-photosynthetic organisms (Lemaire et al., 2004). A unique new potential target was the Ran protein, a small Ras-like protein involved in nuclear transport, mitotic spindle formation, and nuclear envelope assembly (reviewed in Quimby and Dasso, 2003). Ran constitutes the first and only nuclear target of thioredoxin identified in photosynthetic organisms, in which Ran sequences contain 2 cysteines absent in non-photosynthetic organisms. This suggests that Ran may only be redox regulated in photosynthetic organisms.

Another possible target is isocitrate lyase (ICL), whose activation by reduced thioredoxin was confirmed in total extracts from *Chlamydomonas* (Lemaire et al., 2004). ICL catalyzes the cleavage of isocitrate to glyoxylate and succinate. In higher plants, ICL is localized in glyoxysomes and is involved in lipid conversion to carbohydrates during germination of oil-rich seeds (Eastmond and Graham, 2001). The subcellular localization of the enzyme in *Chlamydomonas* remains unknown, since glyoxysomes are considered to be absent from this organism. However, sequence alignments revealed that *Chlamydomonas* ICL is more closely related to bacterial sequences and lacks the 100 amino acid insertion characteristic of higher plant sequences. In bacteria, ICL plays a major role in carbon metabolism by allowing growth on two-carbon compounds (El-Mansi et al., 2006), and is thus a key enzyme for growth on acetate. Its identification as a putative thioredoxin target is consistent with the fact that the extracts loaded on the thioredoxin affinity column were prepared from cells grown on acetate. Moreover, acetyl-CoA synthetase, the enzyme that converts acetate into acetyl-CoA using ATP and CoA, was also retained on the column. Acetyl-CoA is involved in many different physiological processes, including the provision of carbon to the glyoxylate pathway. These results suggest that thioredoxin could be involved in the regulation of acetate assimilation in *Chlamydomonas*.

III. GLUTAREDOXINS

A. Overview

Glutaredoxins, like thioredoxins are small ubiquitous oxidoreductases of approximately 12 kD. Glutaredoxins were initially identified as an alternative hydrogen donor to ribonucleotide reductase in an *E. coli* thioredoxin mutant (Holmgren, 1976). Their redox potential is less negative than that of thioredoxins, generally between −190 mV and −230 mV (Åslund et al., 1997). Glutaredoxins are able to catalyze the reduction of disulfides, but also the reduction of mixed disulfides between a molecule of glutathione and the thiol of a protein. This latter reaction, called deglutathionylation, is specifically catalyzed by glutaredoxin, thioredoxin exhibiting a low efficiency in reducing these mixed disulfides (Jung and Thomas, 1996; Nulton-Persson et al., 2003). Glutaredoxins can reduce disulfide bridges using a dithiol mechanism similar to that of thioredoxin, and have also been reported to be able to use a monothiol mechanism requiring only one of the two active-site cysteines

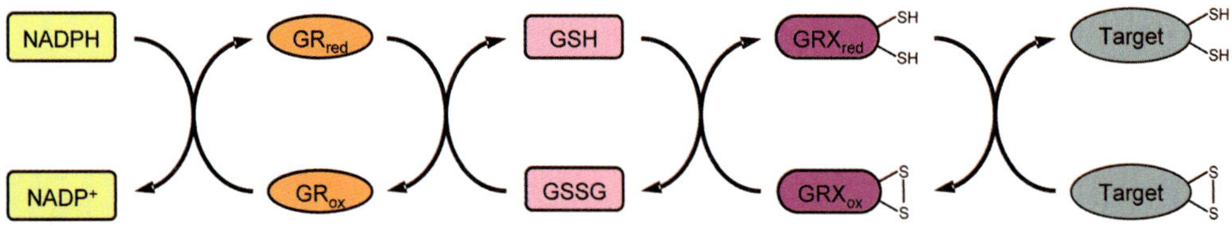

FIGURE 11.7 *Reduction of classical glutaredoxins. Glutathione reductase (GR) continuously reduces oxidized glutathione (GSSG) into reduced glutathione (GSH) using NADPH as a reductant. This GSH can reduce classical glutaredoxins (CPYC type). Reduced glutaredoxins can in turn reduce disulfide bridges on target proteins.*

(Vlamis-Gardikas and Holmgren, 2002). In contrast to thioredoxins, which are reduced by specific reductases, glutaredoxins are reduced by reduced glutathione (GSH) (Figure 11.7). Different glutaredoxins types have been defined on the basis of their active site sequences. Two types are present in *Chlamydomonas*, *E. coli* and yeast: the classical CPYC-type with a Cys-Pro-Tyr-Cys active site, and the CGFS-type with a Cys-Gly-Pro-Ser active site sequence. The third type, named CC, corresponds to glutaredoxins containing a Cys-Cys-X-Cys or Cys-Cys-X-Ser active site and seems to be specific to higher plants, since no counterparts are present in *Chlamydomonas* or other lower photosynthetic eukaryotes. An *Arabidopsis* CC-type glutaredoxin seems to be involved in petal development (Xing et al., 2005). Two additional families with high molecular weight glutaredoxin homologues have been recently defined. The first one is the GRL family and corresponds to higher plant-specific proteins loosely related to glutaredoxin and containing four Cys-X-X-Cys motifs (Navrot et al., 2006b). The second family corresponds to proteins related to glutathione-*S*-transferase and harboring a Cys-Pro-Phe-Cys or a Cys-Pro-Phe-Ser active site (Meyer et al., 2006). These two families will not be considered further.

Glutaredoxins have been much less studied than thioredoxin, and most work has been performed on glutaredoxins from non-photosynthetic organisms, especially *E. coli*, yeast, and mammals. Glutaredoxins appear to play a major role in oxidative stress responses, and glutaredoxin genes are induced in the presence of oxidants (Prieto-Alamo et al., 2000; Grant, 2001). CPYC-type glutaredoxins have been reported to function as dehydroascorbate reductase (Park and Levine, 1996; Washburn and Wells, 1999), glutathione reductase (Collinson et al., 2002), and glutathione-*S*-transferase (Collinson and Grant, 2003). Glutaredoxins are also involved in the regulation of transcription factors related to oxidative stress in mammals (Bandyopadhyay et al., 1998; Zheng et al., 1998; Nakamura et al., 1999; Hirota et al., 2000). These functions are partially redundant with those of thioredoxin. Yeast Grx5, a mitochondrial CGFS glutaredoxin, has been shown to participate in the biogenesis of Fe-S clusters (Rodriguez-Manzaneque et al., 2002; Muhlenhoff et al., 2003; Tamarit et al., 2003; Herrero et al., 2006). A similar role has been suggested for *E. coli* glutaredoxins (Achebach et al., 2004). In vertebrates, Grx5 deficiency impairs haem synthesis, suggesting a link between Fe-S cluster assembly and haem synthesis (Wingert et al., 2005). The function of glutaredoxins in Fe-S cluster biogenesis might be linked to their ability to bind Fe-S clusters (Lillig et al., 2005; Feng et al., 2006; Johansson et al., 2007). Alternatively, the binding of the Fe-S cluster could serve as a redox sensor for the activation of glutaredoxin during oxidative stress as has been proposed for human mitochondrial glutaredoxin (Berndt et al., 2007; Johansson et al., 2007). The CGFS glutaredoxins do not appear to be reduced by glutathione. Indeed, yeast Grx5 could be more efficiently reduced by thioredoxin than by glutathione (Tamarit et al., 2003). In *E. coli*, the CGFS Grx4 is coded by an essential gene and can serve as a substrate for NTR (Fernandes et al., 2005). Several reviews detail the functions of glutaredoxins (Holmgren, 2000; Fernandes and Holmgren, 2004; Lemaire, 2004; Rouhier et al., 2004b; Holmgren et al., 2005; Shelton et al., 2005; Herrero et al., 2006; Meyer et al., 2006). In *Chlamydomonas*, information on glutaredoxins is limited to the sequence analyses described in the following section.

B. The glutaredoxin system in *Chlamydomonas*

1. Glutaredoxin reducing systems

a. Glutathione

Glutathione is a tripeptide of 307 daltons composed of a glutamate, a cysteine and a glycine (γ-L-glutamyl-L-cysteinyl-glycine). This molecule can be in a reduced (GSH) or oxidized (GSSG) form. Glutathione reductase (GR) continuously regenerates GSH from GSSG using NADPH as a reductant. Glutathione is the main free soluble thiol of low molecular weight and is considered to constitute a redox buffer which maintains the intracellular environment in a reduced state. Glutathione has multiple functions in plants including ROS, heavy metal and xenobiotic detoxification but also the cell cycle G1-to-S phase transition, cell differentiation, flowering, anthocyanin accumulation, programmed cell death, and resistance against pathogens (reviewed in Ogawa, 2005; Meyer and Hell, 2005; Foyer and Noctor, 2005).

Glutathione is also involved in glutathionylation, a reversible post-translational modification, which consists of the formation of a mixed disulfide between the free thiol of a protein and a molecule of glutathione (Figure 11.8). Data from mammals and photosynthetic organisms show that glutathionylation could modulate the activity, the localization or the stability of numerous targets (reviewed in Michelet et al., 2006). Glutathionylation can occur in oxidative stress conditions, but the exact mechanism(s) leading to this modification remains to be determined. Glutaredoxins are likely to be involved in the control of protein deglutathionylation, as discussed above, but have also been suggested to be able to catalyze protein glutathionylation (Starke et al., 2003).

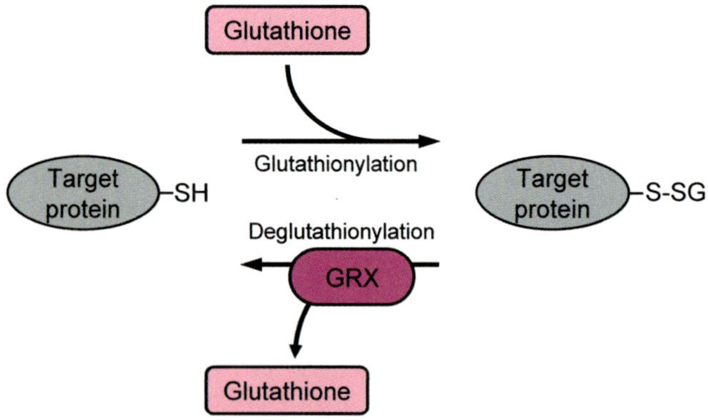

FIGURE 11.8 *Protein glutathionylation and deglutathionylation. Under oxidizing conditions, target proteins can undergo glutathionylation in the presence of glutathione. This post-translational modification is reversible and consists of the formation of a mixed disulfide between the free thiol of a protein and a molecule of glutathione. Glutaredoxins (GRX) could be involved in the removal of the glutathione adduct, that is deglutathionylation.*

b. Glutathione reductase (GR)

Glutathione reductase is an NADPH-dependent oxidoreductase. In *Chlamydomonas*, two genes encode glutathione reductases (*GSHR1* and *GSHR2*), with the *GSHR2* product being closely related to an *Arabidopsis* glutathione reductase targeted to both chloroplasts and mitochondria (Meyer et al., 2006). The presence of a transit peptide in the sequence of GSHR2, as well as the presence of several in-phase methionines, is consistent with multiple localizations of the enzyme. In *Arabidopsis*, glutathione reductases are found in cytosol, mitochondria, and chloroplasts.

2. Glutaredoxins

a. The GRX family

Genome sequencing helped to identify a multiplicity of glutaredoxins (Lemaire, 2004; Rouhier et al., 2004b). In *Arabidopsis*, thirty-one genes coding for three different types of glutaredoxins have been identified: six CPYC-type, four CGFS-type, and twenty-one CC-type (Figure 11.9). As already observed for the thioredoxin multigenic family, *Chlamydomonas* contains fewer glutaredoxin isoforms than *Arabidopsis*, but in that case, not all types are represented. Indeed, *Chlamydomonas* contains six glutaredoxins, distributed only into two different types: two CPYC-type glutaredoxins and four CGFS-type glutaredoxins (Figure 11.9 and Table 11.1).

b. CPYC GRX

The CPYC-type can be separated in two different clusters in phylogenetic analysis: the prokaryotic type, which includes bacterial and cyanobacterial members, and the eukaryotic type, which includes those of yeast, *Arabidopsis*, and *Chlamydomonas* (Lemaire, 2004; Lemaire and Miginiac-Maslow, 2004). The two CPYC glutaredoxins from *Chlamydomonas*, named

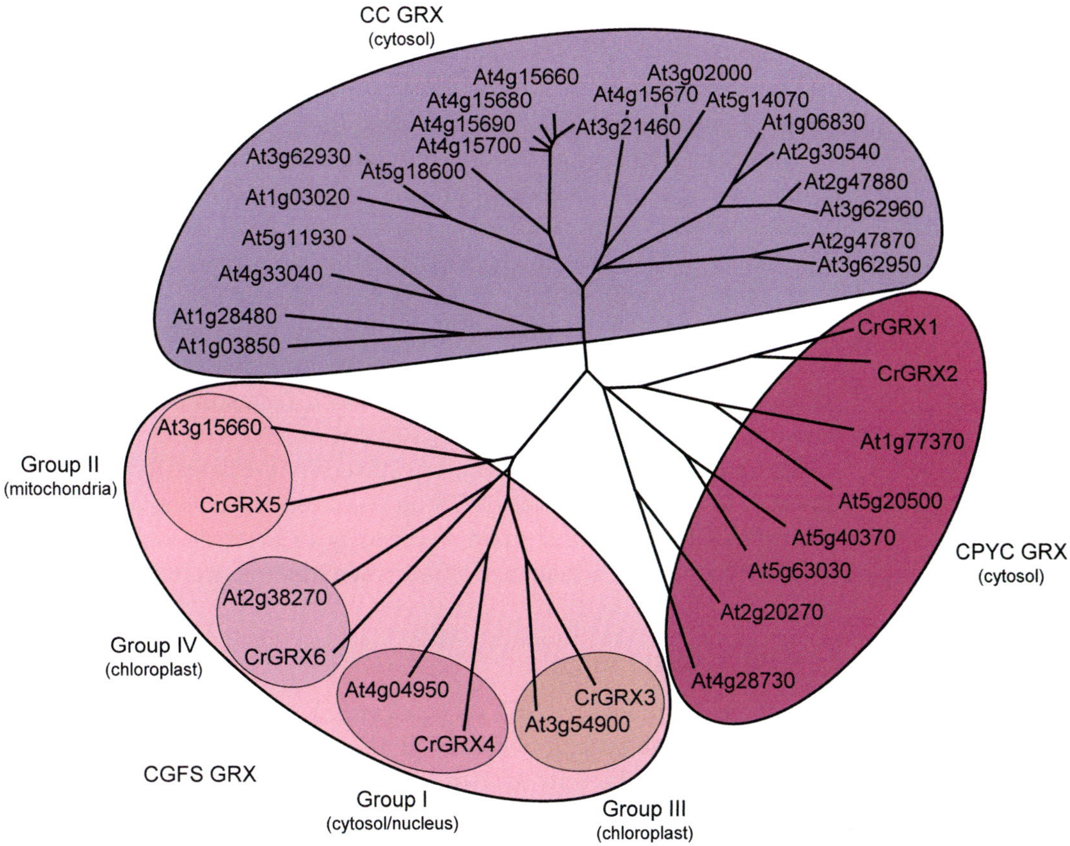

FIGURE 11.9 *Glutaredoxins in Chlamydomonas and Arabidopsis. Phylogenetic tree of the GRX family in Chlamydomonas and Arabidopsis. The unrooted tree was constructed with Clustal X, and gaps were excluded. Accession numbers for Chlamydomonas glutaredoxins are given in Table 11.1.*

GRX1 and GRX2, are predicted to be cytosolic (Figure 11.9 and Table 11.1). As in non-photosynthetic organisms, plant CPYC glutaredoxins exhibit dehydroascorbate reductase activity (Sha et al., 1997; Rouhier et al., 2003). A poplar CPYC glutaredoxin was reported to reduce a type II PRX, implicating glutaredoxins in oxidative stress responses (Rouhier et al., 2001, 2002), and a rice enzyme was able to reduce hydrogen peroxide directly with a low catalytic efficiency (Lee et al., 2002). This GRX-type has also been reported to reduce TRXh from poplar (Gelhaye et al., 2003).

c. CGFS GRX

Arabidopsis and *Chlamydomonas* each possess four genes encoding CGFS-type glutaredoxins (Figure 11.9 and Table 11.1), and each *Chlamydomonas* glutaredoxin clusters with a homologous protein in *Arabidopsis*, thus defining four subgroups (Lemaire, 2004; Lemaire and Miginiac-Maslow, 2004). Group I encompasses multipartite proteins with several thioredoxin and/or glutaredoxin domains predicted to be localized to the cytosol.

Chlamydomonas GRX4 is part of this group, along with yeast Grx3 and Grx4 which have been shown to be located in the nucleus (Molina et al., 2004). *Chlamydomonas* GRX5 belongs to Group II, which corresponds to proteins presumably targeted to mitochondria since they are clustered with yeast mitochondrial Grx5. Group II glutaredoxins might thus be involved, as in yeast, in the biogenesis of Fe-S clusters. GRX3, the most highly expressed glutaredoxin in *Chlamydomonas*, is part of Group III. The putative localizations of proteins belonging to this group are ambiguous between mitochondria or chloroplast, although the presence of cyanobacterial glutaredoxins in this group suggests that Group III proteins are more probably localized in the chloroplast. This chloroplast localization was confirmed for the *Arabidopsis* Group III glutaredoxin, which seems to be involved in oxidative stress defense (Cheng et al., 2006). *Chlamydomonas* GRX6 belongs to Group IV, whose members are unequivocally predicted to be chloroplast targeted. Compared to those of Group III, these proteins are longer, with a C-terminal glutaredoxin domain preceded by an N-terminal domain sharing high homology with a sequence of unknown function from *Synechocystis*.

C. Glutaredoxin targets

Even though glutaredoxin cDNA sequences have long been available, very little is known about glutaredoxins in photosynthetic organisms compared to thioredoxins. The only known targets were identified as a type II peroxiredoxin (Rouhier et al., 2001, 2002) and a TRXh (Gelhaye et al., 2003) from poplar. Since CPYC glutaredoxins exhibit a catalytic mechanism similar to that of thioredoxin, a monocysteinic mutant glutaredoxin column was used, leading to the identification of 94 putative targets in different higher plant tissues (Rouhier et al., 2005). However, most of these targets are also retained on monocysteinic thioredoxin columns, raising the problem of the specificity of this approach (Michelet et al., 2006).

Given the deglutathionylation activity of glutaredoxins, proteins that undergo glutathionylation are expected to be glutaredoxin targets. Even if most glutathionylated proteins identified to date are from non-photosynthetic organisms, some studies in photosynthetic organisms report the identification of glutathionylated proteins (reviewed and listed in Michelet et al., 2006). A potentially key finding is that some thioredoxins can undergo glutathionylation, suggesting a crosstalk between the thioredoxin and the glutaredoxin systems. The glutathionylation of human thioredoxin on an extra cysteine, distinct from the two active-site cysteines, leads to decreased enzyme activity, probably by decreasing its reduction by NTR (Casagrande et al., 2002). A poplar mitochondrial TRXh was shown to undergo glutathionylation with a concomitant decrease of its active-site redox potential (Gelhaye et al., 2004a). The analysis of all chloroplast thioredoxins from *Arabidopsis* and *Chlamydomonas* revealed that f-type thioredoxins are the only chloroplast thioredoxins able to undergo glutathionylation *in vitro* (Michelet et al., 2005). The glutathionylation of TRXf was shown to decrease the ability of TRXf to activate its target enzymes in the light due to a lower efficiency of reduction by FTR. Since glutathionylation of TRXf occurs on a strictly conserved additional cysteine residue shared by all TRXf, this study further suggests that TRXf glutathionylation is a general mechanism that could regulate its activity in all photosynthetic organisms.

As TRXf are specifically involved in the regulation of carbon assimilation enzymes, the modulation of TRXf activity by glutathionylation could help regulate photosynthetic metabolism under oxidative stress (Michelet et al., 2005). In addition, two Calvin cycle enzymes from *Arabidopsis* were shown to undergo glutathionylation. Fructose-1,6-bisphosphate aldolase is glutathionylated *in vivo*, but the effect of the modification on enzyme activity was not examined (Ito et al., 2003). It was also shown that the glutathionylation of *Arabidopsis* A_4-GAPDH *in vitro* leads to its reversible inactivation (Zaffagnini et al., 2007). Overall, these results suggest that under conditions of enhanced ROS production in the chloroplast, such as photoinhibitory conditions, glutathionylation of TRXf and other enzymes could lead to a slowing of the Calvin cycle. It has been suggested that this regulation might allow a redistribution of reducing power within chloroplasts by decreasing NADPH consumption by the Calvin cycle, thereby increasing NADPH availability for ROS detoxifying enzymes (Michelet et al., 2005).

IV. PROTEIN DISULFIDE ISOMERASES

A. Overview

Protein disulfide isomerases are oxidoreductases found only in eukaryotes that are usually located in the endoplasmic reticulum (ER). Protein disulfide isomerases exhibit a less negative redox potential than thioredoxins or glutaredoxins, approximately −180 mV (Lundstrom and Holmgren, 1993). These proteins of approximately 50 kD have a modular structure with different combinations of thioredoxin domains called the a and b domains. The protein disulfide isomerase active site, Trp-Cys-Gly-His-Cys, is present only in the catalytic a domains, while the b domains have thioredoxin-like folding but no sequence homology with the a domains. The b domains have an important role in the specificity of binding to substrate proteins (Freedman et al., 2002). Typical protein disulfide isomerases contain four thioredoxin domains and follow the a–b–b′–a′ architecture. Protein disulfide isomerases exhibit disulfide oxidoreductase activity allowing the reduction, the oxidation, or the isomerization of disulfides. However, their high redox potential suggests that protein disulfide isomerases could be more efficient in oxidizing than in reducing thiols. These proteins are involved in the processing and maturation of secretory proteins, acting as chaperones and catalyzing the formation and rearrangement of disulfide bonds in the lumen of the ER (reviewed in Ferrari and Söling, 1999).

B. Protein disulfide isomerases in *Chlamydomonas*

1. The PDI family

In *Chlamydomonas*, five genes encoding protein disulfide isomerases have been identified: *PDI1*, also named *RB60*; *PDI2*, *PDI3*, *PDI4*, and *PDI5* (Lemaire and Miginiac-Maslow, 2004; Figure 11.10 and Table 11.1). This chapter will discuss *PDI1–PDI4*, whose products contain at least one Trp-Cys-Gly-His-Cys active site. PDI5 has an unusual active site sequence of Trp-Cys-His-Trp-Cys, and will not be considered further. *Arabidopsis* encodes nine protein disulfide isomerases with at least one Trp-Cys-Gly-His-Cys active site. These can be subdivided into four groups based on domain distribution (Lemaire and Miginiac-Maslow, 2004), and each group contains one *Chlamydomonas* protein disulfide isomerase and one or several from *Arabidopsis*.

2. Classical PDI

The first group corresponds to classical protein disulfide isomerase with the a–b–b′–a′ architecture (Figures 11.10 and 11.11). The corresponding proteins are predicted to be localized in the ER since they possess N-terminal

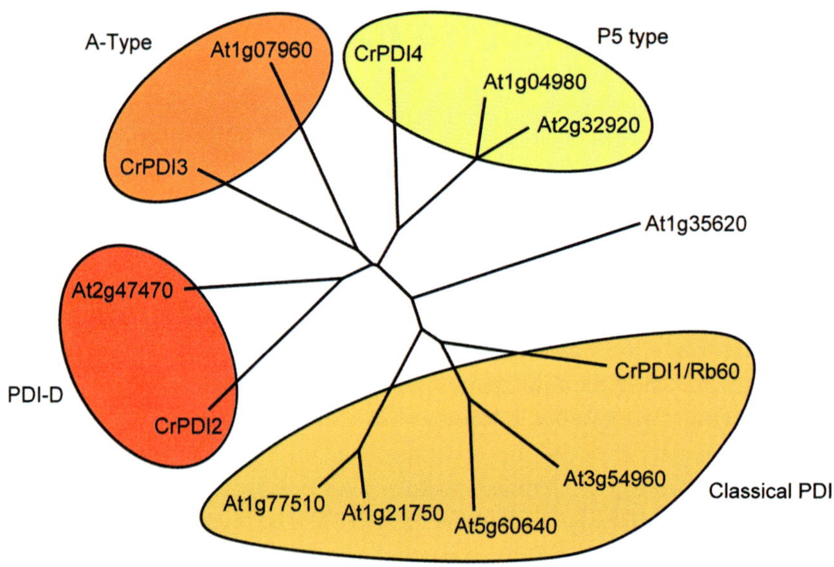

FIGURE 11.10 *Protein disulfide isomerases in Chlamydomonas and Arabidopsis. The unrooted tree was constructed with Clustal X, and gaps were excluded. Accession numbers for Chlamydomonas protein disulfide isomerases are given in Table 11.1.*

secretion signals and a C-terminal KDEL retention signal. The cloning of *Chlamydomonas PDI1* revealed that the protein belongs to this group (Kim and Mayfield, 1997). PDI1 contains an additional acidic N-terminal domain giving a domain distribution of c–a–b–b′–a′. PDI1 function suggests a chloroplast localization in *Chlamydomonas* (see section IV.C) and a plastidic localization was confirmed (Trebitsh et al., 2001). Subsequently its dual localization to the ER was demonstrated (Levitan et al., 2005).

3. P5 PDI

The second group, named P5, is composed of protein disulfide isomerases which possess an a^0–a–b–c architecture (Figure 11.11), and includes PDI4 (Figure 11.10). The role of the b domain present in all protein disulfide isomerases of this group remains to be determined. The P5-type protein disulfide isomerases found in *Arabidopsis* and *Chlamydomonas* all contain C-terminal ER retention signals, is while the presence of a putative N-terminal secretion signal predicted for *Arabidopsis* proteins but uncertain in *Chlamydomonas* PDI4. These P5 protein disulfide isomerases are thus likely to be localized to the ER.

4. PDI-D

The third group of protein disulfide isomerases is named PDI-D, and is characterized by a C-terminal α-helical domain of approximately 100 amino acids named the D domain (Ferrari and Söling, 1999). Proteins of the PDI-D group can either contain an a domain and be redox active: a^0–a–D (PDI-Dα); or contain a b domain and be redox inactive: b–D (PDI-Dβ). *Chlamydomonas* PDI2 is a member of the PDI-Dα group but lacks one thioredoxin domain, leading to an a–D architecture (Figure 11.11). The D domain might allow

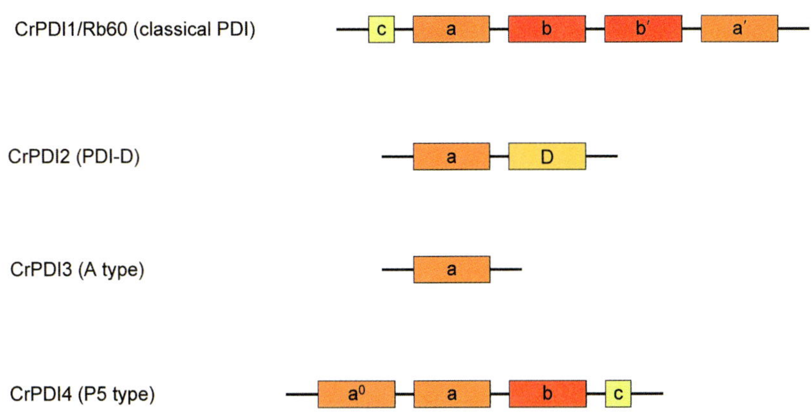

FIGURE 11.11 *Architecture of Chlamydomonas protein disulfide isomerases. The architecture of the four protein disulfide isomerases of Chlamydomonas are represented using different colors for the different domains. The scale is arbitrary, thus the size of the boxes is not proportional to the length of the corresponding amino acid sequences.*

PDI2 import into the ER, and a C-terminal EDEE sequence might function as an ER retention signal. No PDI-Dβ encoding gene has been identified in *Chlamydomonas*, but the b domain cannot be readily identified using BLAST searches (Lemaire and Miginiac-Maslow, 2004).

5. PDI-A

In contrast to the three previous groups, PDI-A seem to be specific to photosynthetic eukaryotes (Figure 11.10). The proteins of this group contain a single domain with a Trp-Cys-Gly/Lys-His-Cys active site (Figure 11.11), and include *Chlamydomonas* PDI3.

C. Roles of protein disulfide isomerases in *Chlamydomonas*

Based on EST abundance, PDI1 and PDI2 are the two most highly expressed protein disulfide isomerases in *Chlamydomonas* (Lemaire and Miginiac-Maslow, 2004). In photosynthetic organisms, the dual-targeted *Chlamydomonas* PDI1 has been the most studied protein disulfide isomerase. Its role in the ER has not been determined yet, but there it may function as a classical protein disulfide isomerase. In the chloroplast, PDI1 has been implicated as a thioredoxin-dependent translational regulator. PDI1/RB60 was discovered as a protein that bound *in vitro* to the *psbA* mRNA 5′ untranslated region along with three other proteins named RB38, RB47, and RB55, forming a complex (Danon and Mayfield, 1991). *psbA* encodes D1, a PS II core protein that is rapidly synthesized in the light, and the formation of the complex *in vitro* is greatly enhanced in extracts from light-grown cells. RB47 is a member of the poly(A)-binding protein family (Yohn et al., 1998) and light-induced changes in the stromal pH, as well as oxidoreduction of specific thiol groups on RB47 by RB60, could contribute to regulation of RB47 activity as related to D1 translation (Alergand et al., 2006). Two regulatory mechanisms have been proposed. In one, ADP-dependent phosphorylation of RB60, occurring at ADP levels only found in the dark, may inactivate the *psbA* 5′ UTR-protein complex (Danon and Mayfield, 1994b). Additionally, RB60 appears to be activated by reduction of vicinal dithiols, possibly by thioredoxins (Danon and Mayfield, 1994a) and in support of this possibility, RB60 was among putative thioredoxin targets trapped on monocysteinic thioredoxin columns (Lemaire et al., 2004). A counteracting oxidative component also appears to be involved (Trebitsh et al., 2000). However, this thiol regulation requires an initial step mediated by a light signal transduction pathway originating from the plastoquinone pool (Trebitsh and Danon, 2001).

V. CONCLUSIONS

Thioredoxins, glutaredoxins, and protein disulfide isomerases are members of the thioredoxin superfamily present in many subcellular compartments

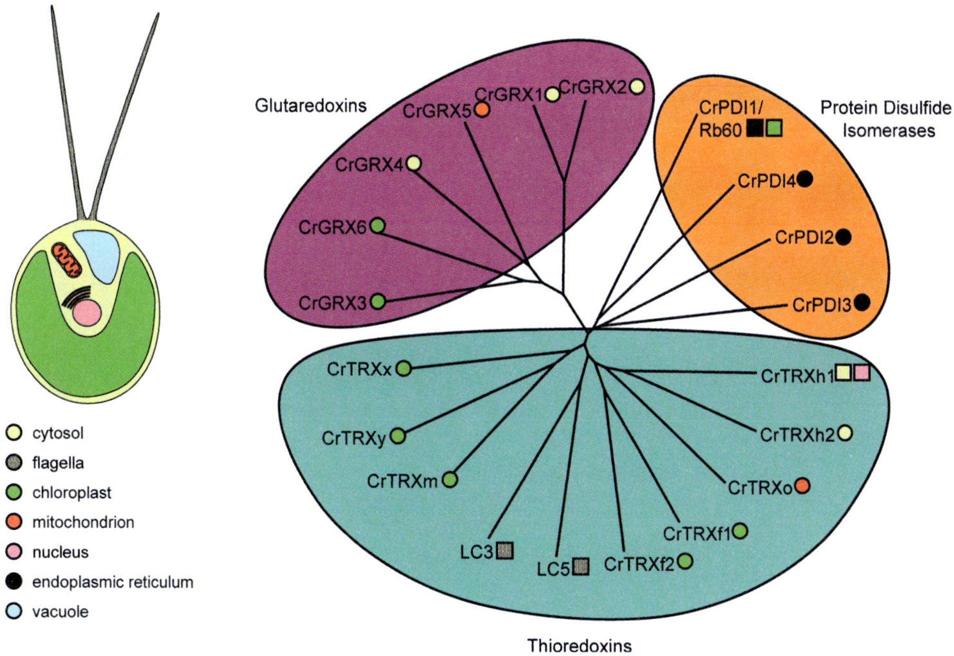

FIGURE 11.12 *Thioredoxins, glutaredoxins, and protein disulfide isomerases in Chlamydomonas. The unrooted tree was constructed with Clustal X, and gaps were excluded. Accession numbers are given in Table 11.1. Full symbols of different colors indicate the putative (circle) or established (square) subcellular localization of each protein (see Table 11.1 for details).*

(Figure 11.12). The availability of genome sequences combined with proteomic approaches have facilitated major breakthroughs, revealing thioredoxins as central metabolic regulators. New potential targets involved in numerous metabolic processes have been identified, and both the thioredoxin role and target regulation will require confirmation and characterization. By contrast, glutaredoxins and protein disulfide isomerases have been much less studied. While much attention has been paid to thioredoxins, accumulating data from photosynthetic organisms clearly suggest that other members of the thioredoxin superfamily deserve further attention. For example, glutathionylation is an emerging field related to glutaredoxins and thioredoxins. More generally, in addition to studies focusing on each family of proteins, more work is required to understand their interactions. One of the challenges of future studies will be to unravel the emerging complex interplay between the numerous redox signaling pathways of the cell.

ACKNOWLEDGMENTS

The authors would like to thank Myroslawa Miginiac-Maslow for critical reading of the manuscript and helpful suggestions. This work was supported by Agence Nationale de la Recherche Grant JC-45751.

REFERENCES

Achebach, S., Tran, Q.H., Vlamis-Gardikas, A., Mullner, M., Holmgren, A., and Unden, G. (2004). Stimulation of Fe-S cluster insertion into apoFNR by *Escherichia coli* glutaredoxins 1, 2 and 3 in vitro. *FEBS Lett.* **565**, 203–206.

Alergand, T., Peled-Zehavi, H., Katz, Y., and Danon, A. (2006). The chloroplast protein disulfide isomerase RB60 reacts with a regulatory disulfide of the RNA-binding protein RB47. *Plant Cell Physiol.* **47**, 540–548.

Aro, E.M. and Ohad, I. (2003). Redox regulation of thylakoid protein phosphorylation. *Antioxid. Redox Signal* **5**, 55–67.

Åslund, F., Berndt, K.D., and Holmgren, A. (1997). Redox potentials of glutaredoxins and other thiol-disulfide oxidoreductases of the thioredoxin superfamily determined by direct protein–protein redox equilibria. *J. Biol. Chem.* **272**, 30780–30786.

Avilan, L., Gontero, B., Lebreton, S., and Ricard, J. (1997). Memory and imprinting effects in multienzyme complexes – I. Isolation, dissociation, and reassociation of a phosphoribulokinase-glyceraldehyde-3-phosphate dehydrogenase complex from *Chlamydomonas reinhardtii* chloroplasts. *Eur. J. Biochem.* **246**, 78–84.

Avilan, L., Lebreton, S., and Gontero, B. (2000). Thioredoxin activation of phosphoribulokinase in a bi-enzyme complex from *Chlamydomonas reinhardtii* chloroplasts. *J. Biol. Chem.* **275**, 9447–9451.

Bae, W. and Chen, X. (2004). Proteomic study for the cellular responses to $Cd2^{+}$ in *Schizosaccharomyces pombe* through amino acid-coded mass tagging and liquid chromatography tandem mass spectrometry. *Mol. Cell Proteomics* **3**, 596–607.

Bandyopadhyay, S., Starke, D.W., Mieyal, J.J., and Gronostajski, R.M. (1998). Thioltransferase (glutaredoxin) reactivates the DNA-binding activity of oxidation-inactivated nuclear I. *J. Biol. Chem.* **273**, 392–397.

Bellafiore, S., Barneche, F., Peltier, G., and Rochaix, J.D. (2005). State transitions and light adaptation require chloroplast thylakoid protein kinase STN7. *Nature* **433**, 892–895.

Berndt, C., Hudemann, C., Hanschmann, E.M., Axelsson, R., Holmgren, A., and Lillig, C.H. (2007). How does iron-sulfur cluster coordination regulate the activity of human glutaredoxin 2? *Antioxid. Redox Signal* **9**, 151–157.

Besse, I., Wong, J.H., Kobrehel, K., and Buchanan, B.B. (1996). Thiocalsin: A thioredoxin-linked, substrate-specific protease dependent on calcium. *Proc. Natl. Acad. Sci. U. S. A.* **93**, 3169–3175.

Bower, M.S., Matias, D.D., Fernandes-Carvalho, E., Mazzurco, M., Gu, T., Rothstein, S.J., and Goring, D.R. (1996). Two members of the thioredoxin-h family interact with the kinase domain of a Brassica S locus receptor kinase. *Plant Cell* **8**, 1641–1650.

Bringloe, D.H., Dyer, T.A., and Gray, J.C. (1995). Developmental, circadian and light regulation of wheat ferredoxin gene expression. *Plant Mol. Biol.* **27**, 293–306.

Buchanan, B.B. and Balmer, Y. (2005). Redox regulation: A broadening horizon. *Annu. Rev. Plant Biol.* **56**, 187–220.

Buchanan, B.B., Schürmann, P., and Kalberer, P.P. (1971). Ferredoxin-activated fructose diphosphatase of spinach chloroplasts. Resolution of the system, properties of the alkaline fructose diphosphatase component, and physiological significance of the ferredoxin-linked activation. *J. Biol. Chem.* **246**, 5952–5959.

Burke-Gaffney, A., Callister, M.E., and Nakamura, H. (2005). Thioredoxin: Friend or foe in human disease? *Trends Pharmacol. Sci.* **26**, 398–404.

Cabrillac, D., Cock, J.M., Dumas, C., and Gaude, T. (2001). The S-locus receptor kinase is inhibited by thioredoxins and activated by pollen coat proteins. *Nature* **410**, 220–223.

Casagrande, S., Bonetto, V., Fratelli, M., Gianazza, E., Eberini, I., Massignan, T., Salmona, M., Chang, G., Holmgren, A., and Ghezzi, P. (2002). Glutathionylation of human thioredoxin: A possible crosstalk between the glutathione and thioredoxin systems. *Proc. Natl. Acad. Sci. U. S. A.* **99**, 9745–9749.

Cheng, N.H., Liu, J.Z., Brock, A., Nelson, R.S., and Hirschi, K.D. (2006). AtGRXcp, an *Arabidopsis* chloroplastic glutaredoxin, is critical for protection against protein oxidative damage. *J. Biol. Chem.* **281**, 26280–26288.

Cho, M.J., Wong, J.H., Marx, C., Jiang, W., Lemaux, P.G., and Buchanan, B.B. (1999). Overexpression of thioredoxin h leads to enhanced activity of starch debranching enzyme (pullulanase) in barley grain. *Proc. Natl. Acad. Sci. U. S. A.* **96**, 14641–14646.

Collin, V., Issakidis-Bourguet, E., Marchand, C., Hirasawa, M., Lancelin, J.M., Knaff, D.B., and Miginiac-Maslow, M. (2003). The *Arabidopsis* plastidial thioredoxins: New functions and new insights into specificity. *J. Biol. Chem.* **278**, 23747–23752.

Collin, V., Lamkemeyer, P., Miginiac-Maslow, M., Hirasawa, M., Knaff, D.B., Dietz, K.J., and Issakidis-Bourguet, E. (2004). Characterization of plastidial thioredoxins from *Arabidopsis* belonging to the new y-type. *Plant Physiol.* **136**, 4088–4095.

Collinson, E.J. and Grant, C.M. (2003). Role of yeast glutaredoxins as glutathione S-transferases. *J. Biol. Chem.* **278**, 22492–22497.

Collinson, E.J., Wheeler, G.L., Garrido, E.O., Avery, A.M., Avery, S.V., and Grant, C.M. (2002). The yeast glutaredoxins are active as glutathione peroxidases. *J. Biol. Chem.* **277**, 16712–16717.

Dai, S., Johansson, K., Miginiac-Maslow, M., Schürmann, P., and Eklund, H. (2004). Structural basis of redox signaling in photosynthesis: Structure and function of ferredoxin:thioredoxin reductase and target enzymes. *Photosynth. Res.* **79**, 233–248.

Danon, A. and Mayfield, S.P. (1991). Light regulated translational activators: Identification of chloroplast gene specific mRNA binding proteins. *EMBO J.* **10**, 3993–4001.

Danon, A. and Mayfield, S.P. (1994a). Light-regulated translation of chloroplast messenger RNAs through redox potential. *Science* **266**, 1717–1719.

Danon, A. and Mayfield, S.P. (1994b). ADP-dependent phosphorylation regulates RNA-binding *in vitro*: Implications in light-modulated translation. *EMBO J.* **13**, 2227–2235.

De Pascalis, A.R., Schurmann, P., and Bosshard, H.R. (1994). Comparison of the binding sites of plant ferredoxin for two ferredoxin-dependent enzymes. *FEBS Lett.* **337**, 217–220.

Decottignies, P., Schmitter, J.M., Jacquot, J.P., Dutka, S., Picaud, A., and Gadal, P. (1990). Purification, characterization, and complete amino acid sequence of a thioredoxin from a green alga, *Chlamydomonas reinhardtii*. *Arch. Biochem. Biophys.* **280**, 112–121.

Decottignies, P., Schmitter, J.M., Dutka, S., Jacquot, J.P., and Miginiac-Maslow, M. (1991). Characterization and primary structure of a second thioredoxin from the green alga, *Chlamydomonas reinhardtii*. *Eur. J. Biochem.* **198**, 505–512.

Depège, N., Bellafiore, S., and Rochaix, J.D. (2003). Role of chloroplast protein kinase Stt7 in LHCII phosphorylation and state transition in *Chlamydomonas*. *Science* **299**, 1572–1575.

Dietz, K.J. (2003). Plant peroxiredoxins. *Annu. Rev. Plant Biol.* **54**, 93–107.

Dietz, K.J., Jacob, S., Oelze, M.L., Laxa, M., Tognetti, V., de Miranda, S.M., Baier, M., and Finkemeier, I. (2006). The function of peroxiredoxins in plant organelle redox metabolism. *J. Exp. Bot.* **57**, 1697–1709.

Dixon, D.P., Skipsey, M., Grundy, N.M., and Edwards, R. (2005). Stress-induced protein S-glutathionylation in *Arabidopsis*. *Plant Physiol.* **138**, 2233–2244.

Dyson, H.J. and Wright, P.E. (2005). Intrinsically unstructured proteins and their functions. *Nat. Rev. Mol. Cell Biol.* **6**, 197–208.

Eastmond, P.J. and Graham, I.A. (2001). Re-examining the role of the glyoxylate cycle in oilseeds. *Trends Plant Sci.* **6**, 72–78.

El-Mansi, M., Cozzone, A.J., Shiloach, J., and Eikmanns, B.J. (2006). Control of carbon flux through enzymes of central and intermediary metabolism during growth of *Escherichia coli* on acetate. *Curr. Opin. Microbiol.* **9**, 173–179.

Elliott, R.C., Pedersen, T.J., Fristensky, B., White, M.J., Dickey, L.F., and Thompson, W.F. (1989). Characterization of a single copy gene encoding ferredoxin I from pea. *Plant Cell* **1**, 681–690.

Farr, T.J., Huppe, H.C., and Turpin, D.H. (1994). Coordination of chloroplastic metabolism in N-Limited *Chlamydomonas reinhardtii* by redox modulation. I. The activation of phosphoribulosekinase and glucose-6-phosphate dehydrogenase is relative to the photosynthetic supply of electrons. *Plant Physiol.* **105**, 1037–1042.

Feng, Y., Zhong, N., Rouhier, N., Hase, T., Kusunoki, M., Jacquot, J.P., Jin, C., and Xia, B. (2006). Structural insight into poplar glutaredoxin C1 with a bridging iron-sulfur cluster at the active site. *Biochemistry* **45**, 7998–8008.

Fernandes, A.P. and Holmgren, A. (2004). Glutaredoxins: Glutathione-dependent redox enzymes with functions far beyond a simple thioredoxin backup system. *Antioxid. Redox Signal,* **6**, 63–74.

Fernandes, A.P., Fladvad, M., Berndt, C., Andresen, C., Lillig, C.H., Neubauer, P., Sunnerhagen, M., Holmgren, A., and Vlamis-Gardikas, A. (2005). A novel monothiol glutaredoxin (Grx4) from *Escherichia coli* can serve as a substrate for thioredoxin reductase. *J. Biol. Chem.* **280**, 24544–24552.

Ferrari, D.M. and Söling, H.D. (1999). The protein disulphide-isomerase family: Unravelling a string of folds. *Biochem. J.* **339**, 1–10.

Finazzi, G. (2005). The central role of the green alga *Chlamydomonas reinhardtii* in revealing the mechanism of state transitions. *J. Exp. Bot.* **56**, 383–388.

Florencio, F.J., Yee, B.C., Johnson, T.C., and Buchanan, B.B. (1988). An NADP/thioredoxin system in leaves: Purification and characterization of NADP-thioredoxin reductase and thioredoxin h from spinach. *Arch. Biochem. Biophys.* **266**, 496–507.

Foyer, C.H. and Noctor, G. (2005). Redox homeostasis and antioxidant signaling: A metabolic interface between stress perception and physiological responses. *Plant Cell* **17**, 1866–1875.

Freedman, R.B., Klappa, P., and Ruddock, L.W. (2002). Protein disulfide isomerases exploit synergy between catalytic and specific binding domains. *EMBO Rep.* **3**, 136–140.

Garbisu, C., Ishii, T., Leighton, T., and Buchanan, B.B. (1996). Bacterial reduction of selenite to elemental selenium. *Chem. Geol.* **132**, 199–204.

Gelhaye, E., Rouhier, N., and Jacquot, J.P. (2003). Evidence for a subgroup of thioredoxin h that requires GSH/Grx for its reduction. *FEBS Lett.* **555**, 443–448.

Gelhaye, E., Rouhier, N., Gérard, J., Jolivet, Y., Gualberto, J., Navrot, N., Ohlsson, P.I., Wingsle, G., Hirasawa, M., Knaff, D.B., Wang, H., Dizengremel, P., Meyer, Y., and Jacquot, J.P. (2004a). A specific form of thioredoxin h occurs in plant mitochondria and regulates the alternative oxidase. *Proc. Natl. Acad. Sci. U. S. A.* **101**, 14545–14550.

Gelhaye, E., Rouhier, N., and Jacquot, J.P. (2004b). The thioredoxin h system of higher plants. *Plant Physiol. Biochem.* **42**, 265–271.

Gomez, I., Merchan, F., Fernandez, E., and Quesada, A. (2002). NADP-malate dehydrogenase from *Chlamydomonas*: Prediction of new structural determinants for redox regulation by homology modelling. *Plant Mol. Biol.* **48**, 211–221.

Goyer, A., Decottignies, P., Lemaire, S., Ruelland, E., Issakidis-Bourguet, E., Jacquot, J.P., and Miginiac-Maslow, M. (1999). The internal Cys-207 of sorghum leaf NADP-malate dehydrogenase can form mixed disulphides with thioredoxin. *FEBS Lett.* **444**, 165–169.

Goyer, A., Haslekas, C., Miginiac-Maslow, M., Klein, U., Le Maréchal, P., Jacquot, J.P., and Decottignies, P. (2002). Isolation and characterization of a thioredoxin-dependent peroxidase from *Chlamydomonas reinhardtii*. *Eur. J. Biochem.* **269**, 272–282.

Graciet, E., Gans, P., Wedel, N., Lebreton, S., Camadro, J.M., and Gontero, B. (2003). The small protein CP12: A protein linker for supramolecular complex assembly. *Biochemistry* **42**, 8163–8170.

Graciet, E., Lebreton, S., and Gontero, B. (2004). Emergence of new regulatory mechanisms in the Benson-Calvin pathway via protein-protein interactions: A glyceraldehyde-3-phosphate dehydrogenase/CP12/phosphoribulokinase complex. *J. Exp. Bot.* **55**, 1245–1254.

Grant, C.M. (2001). Role of the glutathione/glutaredoxin and thioredoxin systems in yeast growth and response to stress conditions. *Mol. Microbiol.* **39**, 533–541.

Gupta, S.D., Wu, H.C., and Rick, P.D. (1997). A *Salmonella typhimurium* genetic locus which confers copper tolerance on copper-sensitive mutants of *Escherichia coli*. *J. Bacteriol.* **179**, 4977–4984.

Harrison, A., Sakato, M., Tedford, H.W., Benashski, S.E., Patel-King, R.S., and King, S.M. (2002). Redox-based control of the gamma heavy chain ATPase from *Chlamydomonas* outer arm dynein. *Cell Motil. Cytoskeleton* **52**, 131–143.

Herbette, S., Lenne, C., Leblanc, N., Julien, J.L., Drevet, J.R., and Roeckel-Drevet, P. (2002). Two GPX-like proteins from *Lycopersicon esculentum* and *Helianthus annuus* are antioxidant enzymes with phospholipid hydroperoxide glutathione peroxidase and thioredoxin peroxidase activities. *Eur. J. Biochem.* **269**, 2414–2420.

Herrero, E., Ros, J., Tamarit, J., and Belli, G. (2006). Glutaredoxins in fungi. *Photosynth. Res.* **89**, 127–140.

Hirasawa, M., Schürmann, P., Jacquot, J.P., Manieri, W., Jacquot, P., Keryer, E., Hartman, F.C., and Knaff, D.B. (1999). Oxidation-reduction properties of chloroplast thioredoxins, ferredoxin:thioredoxin reductase, and thioredoxin f-regulated enzymes. *Biochemistry* **38**, 5200–5205.

Hirasawa, M., Ruelland, E., Schepens, I., Issakidis-Bourguet, E., Miginiac-Maslow, M., and Knaff, D.B. (2000). Oxidation-reduction properties of the regulatory disulfides of sorghum chloroplast nicotinamide adenine dinucleotide phosphate-malate dehydrogenase. *Biochemistry* **39**, 3344–3350.

Hirota, K., Matsui, M., Murata, M., Takashima, Y., Cheng, F.S., Itoh, T., Fukuda, K., and Yodoi, J. (2000). Nucleoredoxin, glutaredoxin, and thioredoxin differentially regulate NF-kappaB, AP-1, and CREB activation in HEK293 cells. *Biochem. Biophys. Res. Commun.* **274**, 177–182.

Holmgren, A. (1976). Hydrogen donor system for *Escherichia coli* ribonucleoside-diphosphate reductase dependent upon glutathione. *Proc. Natl. Acad. Sci. U. S. A.* **73**, 2275–2279.

Holmgren, A. (1985). Thioredoxin. *Annu. Rev. Biochem.* **54**, 237–271.

Holmgren, A. (2000). Antioxidant function of thioredoxin and glutaredoxin systems. *Antioxid. Redox Signal* **2**, 811–820.

Holmgren, A., Johansson, C., Berndt, C., Lönn, M.E., Hudemann, C., and Lillig, C.H. (2005). Thiol redox control via thioredoxin and glutaredoxin systems. *Biochem. Soc. Trans.* **33**, 1375–1377.

Huppe, H.C. and Buchanan, B.B. (1989). Activation of a chloroplast type of fructose bisphosphatase from *Chlamydomonas reinhardtii* by light-mediated agents. *Z. Naturforsch. [C]* **44**, 487–494.

Huppe, H.C., de Lamotte-Guéry, F., Jacquot, J.P., and Buchanan, B.B. (1990). The ferredoxin-thioredoxin system of a green alga, *Chlamydomonas reinhardtii*: Identification and characterization of thioredoxins and ferredoxin-thioredoxin reductase components. *Planta* **180**, 341–351.

Huppe, H.C., Picaud, A., Buchanan, B.B., and Miginiac-Maslow, M. (1991). Identification of an NADP/thioredoxin system in *Chlamydomonas reinhardtii*. *Planta* **186**, 115–121.

Huppe, H.C., Farr, T.J., and Turpin, D.H. (1994). Coordination of chloroplastic metabolism in N-Limited *Chlamydomonas reinhardtii* by redox modulation. II. Redox modulation activates the oxidative pentose phosphate pathway during photosynthetic nitrate assimilation. *Plant Physiol.* **105**, 1043–1048.

Ito, H., Iwabuchi, M., and Ogawa, K. (2003). The sugar-metabolic enzymes aldolase and triose-phosphate isomerase are targets of glutathionylation in *Arabidopsis thaliana*: Detection using biotinylated glutathione. *Plant Cell Physiol.* **44**, 655–660.

Jacquot, J.P., Stein, M., Hodges, M., and Miginiac-Maslow, M. (1992). PCR cloning of a nucleotidic sequence coding for the mature part of *Chlamydomonas reinhardtii* thioredoxin Ch2. *Nucleic Acids Res.* **20**, 617.

Jacquot, J.P., Rivera-Madrid, R., Marinho, P., Kollarova, M., Le Maréchal, P., Miginiac-Maslow, M., and Meyer, Y. (1994). *Arabidopsis thaliana* NAPHP thioredoxin reductase. cDNA characterization and expression of the recombinant protein in *Escherichia coli*. *J. Mol. Biol.* **235**, 1357–1363.

Jacquot, J.P., Stein, M., Suzuki, A., Liottet, S., Sandoz, G., and Miginiac-Maslow, M. (1997). Residue Glu-91 of *Chlamydomonas reinhardtii* ferredoxin is essential for electron transfer to ferredoxin-thioredoxin reductase. *FEBS Lett.* **400**, 293–296.

Jacquot, J.P., Gelhaye, E., Rouhier, N., Corbier, C., Didierjean, C., and Aubry, A. (2002). Thioredoxins and related proteins in photosynthetic organisms: Molecular basis for thiol dependent regulation. *Biochem. Pharmacol.* **64**, 1065–1069.

Johansson, C., Kavanagh, K.L., Gileadi, O., and Oppermann, U. (2007). Reversible sequestration of active site cysteines in a 2Fe-2S-bridged dimer provides a mechanism for glutaredoxin 2 regulation in human mitochondria. *J. Biol. Chem.* **282**, 3077–3082.

Johnson, T.C., Cao, R.Q., Kung, J.E., and Buchanan, B.B. (1987). Thioredoxin and NADP-thioredoxin reductase from cultured carrot cells. *Planta* **171**, 321–331.

Jung, B.G., Lee, K.O., Lee, S.S., Chi, Y.H., Jang, H.H., Kang, S.S., Lee, K., Lim, D., Yoon, S.C., Yun, D.J., Inoue, Y., Cho, M.J., and Lee, S.Y. (2002). A Chinese cabbage cDNA with high sequence identity to phospholipid hydroperoxide glutathione peroxidases encodes a novel isoform of thioredoxin-dependent peroxidase. *J. Biol. Chem.* **277**, 12572–12578.

Jung, C.H. and Thomas, J.A. (1996). S-glutathiolated hepatocyte proteins and insulin disulfides as substrates for reduction by glutaredoxin, thioredoxin, protein disulfide isomerase, and glutathione. *Arch. Biochem. Biophys.* **335**, 61–72.

Kim, J. and Mayfield, S.P. (1997). Protein disulfide isomerase as a regulator of chloroplast translational activation. *Science* **278**, 1954–1957.

Kobrehel, K., Wong, J.H., Balogh, A., Kiss, F., Yee, B.C., and Buchanan, B.B. (1992). Specific reduction of wheat storage proteins by thioredoxin h. *Plant Physiol.* **99**, 919–924.

Kondo, N., Nakamura, H., Masutani, H., and Yodoi, J. (2006). Redox regulation of human thioredoxin network. *Antioxid. Redox Signal* **8**, 1881–1890.

Krimm, I., Lemaire, S., Ruelland, E., Miginiac-Maslow, M., Jacquot, J.P., Hirasawa, M., Knaff, D.B., and Lancelin, J.M. (1998). The single mutation Trp35--> Ala in the 35-40 redox site of *Chlamydomonas reinhardtii* thioredoxin h affects its biochemical activity and the pH dependence of C36-C39 1H-13C NMR. *Eur. J. Biochem.* **255**, 185–195.

Laloi, C., Rayapuram, N., Chartier, Y., Grienenberger, J.M., Bonnard, G., and Meyer, Y. (2001). Identification and characterization of a mitochondrial thioredoxin system in plants. *Proc. Natl. Acad. Sci. U. S. A.* **98**, 14144–14149.

Lancelin, J.M., Stein, M., and Jacquot, J.P. (1993). Secondary structure and protein folding of recombinant chloroplastic thioredoxin Ch2 from the green alga *Chlamydomonas reinhardtii* as determined by 1H NMR. *J. Biochem. (Tokyo)* **114**, 421–431.

Lancelin, J.M., Guilhaudis, L., Krimm, I., Blackledge, M.J., Marion, D., and Jacquot, J.P. (2000). NMR structures of thioredoxin m from the green alga *Chlamydomonas reinhardtii*. *Proteins* **41**, 334–349.

Laurent, T.C., Moore, E.C., and Reichard, P. (1964). Enzymatic synthesis of deoxyribonucleotides. IV. Isolation and characterization of thioredoxin, the hydrogen donor from *Escherichia coli* B. *J. Biol. Chem.* **239**, 3436–3444.

Lebreton, S., Andreescu, S., Graciet, E., and Gontero, B. (2006). Mapping of the interaction site of CP12 with glyceraldehyde-3-phosphate dehydrogenase from *Chlamydomonas reinhardtii*. Functional consequences for glyceraldehyde-3-phosphate dehydrogenase. *FEBS J.* **273**, 3358–3369.

Lee, K.O., Lee, J.R., Yoo, J.Y., Jang, H.H., Moon, J.C., Jung, B.G., Chi, Y.H., Park, S.K., Lee, S.S., Lim, C.O., Yun, D.J., Cho, M.J., and Lee, S.Y. (2002). GSH-dependent peroxidase activity of the rice (*Oryza sativa*) glutaredoxin, a thioltransferase. *Biochem. Biophys. Res. Commun.* **296**, 1152–1156.

Lemaire, S.D. (2004). The glutaredoxin family in oxygenic photosynthetic organisms. *Photosynth. Res.* **79**, 305–318.

Lemaire, S.D. and Miginiac-Maslow, M. (2004). The thioredoxin superfamily in *Chlamydomonas reinhardtii*. *Photosynth. Res.* **82**, 203–220.

Lemaire, S., Keryer, E., Stein, M., Schepens, I., Issakidis-Bourguet, E., Gérard-Hirne, C., Miginiac-Maslow, M., and Jacquot, J.P. (1999a). Heavy-metal regulation of thioredoxin gene expression in *Chlamydomonas reinhardtii*. *Plant Physiol.* **120**, 773–778.

Lemaire, S.D., Stein, M., Issakidis-Bourguet, E., Keryer, E., Benoit, V., Pineau, B., Gérard-Hirne, C., Miginiac-Maslow, M., and Jacquot, J.P. (1999b). The complex regulation of ferredoxin/thioredoxin-related genes by light and the circadian clock. *Planta* **209**, 221–229.

Lemaire, S.D., Richardson, J.M., Goyer, A., Keryer, E., Lancelin, J.M., Makhatadze, G.I., and Jacquot, J.P. (2000). Primary structure determinants of the pH- and temperature-dependent aggregation of thioredoxin. *Biochim. Biophys. Acta* **1476**, 311–323.

Lemaire, S.D., Miginiac-Maslow, M., and Jacquot, J.P. (2002). Plant thioredoxin gene expression: Control by light, circadian clock, and heavy metals. *Methods Enzymol.* **347**, 412–421.

Lemaire, S.D., Collin, V., Keryer, E., Quesada, A., and Miginiac-Maslow, M. (2003a). Characterization of thioredoxin y, a new type of thioredoxin identified in the genome of *Chlamydomonas reinhardtii*. *FEBS Lett.* **543**, 87–92.

Lemaire, S.D., Collin, V., Keryer, E., Issakidis-Bourguet, E., Lavergne, D., and Miginiac-Maslow, M. (2003b). *Chlamydomonas reinhardtii*: A model organism for the study of the thioredoxin family. *Plant Physiol. Biochem.* **41**, 513–521.

Lemaire, S.D., Guillon, B., Le Maréchal, P., Keryer, E., Miginiac-Maslow, M., and Decottignies, P. (2004). New thioredoxin targets in the unicellular photosynthetic eukaryote *Chlamydomonas reinhardtii*. *Proc. Natl. Acad. Sci. U. S. A.* **101**, 7475–7480.

Lemaire, S.D., Quesada, A., Merchan, F., Corral, J.M., Igeno, M.I., Keryer, E., Issakidis-Bourguet, E., Hirasawa, M., Knaff, D.B., and Miginiac-Maslow, M. (2005). NADP-malate dehydrogenase from unicellular green alga *Chlamydomonas reinhardtii*. A first step toward redox regulation? *Plant Physiol.* **137**, 514–521.

Levitan, A., Trebitsh, T., Kiss, V., Pereg, Y., Dangoor, I., and Danon, A. (2005). Dual targeting of the protein disulfide isomerase RB60 to the chloroplast and the endoplasmic reticulum. *Proc. Natl. Acad. Sci. U. S. A.* **102**, 6225–6230.

Lillig, C.H., Berndt, C., Vergnolle, O., Lonn, M.E., Hudemann, C., Bill, E., and Holmgren, A. (2005). Characterization of human glutaredoxin 2 as iron-sulfur protein: A possible role as redox sensor. *Proc. Natl. Acad. Sci. U. S. A.* **102**, 8168–8173.

Lmoumene, C.E., Conte, D., Jacquot, J.P., and Houee-Levin, C. (2000). Redox properties of protein disulfide bond in oxidized thioredoxin and lysozyme: A pulse radiolysis study. *Biochemistry* **39**, 9295–9301.

Lundstrom, J. and Holmgren, A. (1993). Determination of the reduction-oxidation potential of the thioredoxin-like domains of protein disulfide-isomerase from the equilibrium with glutathione and thioredoxin. *Biochemistry* **32**, 6649–6655.

Martin, J.L. (1995). Thioredoxin – a fold for all reasons. *Structure* **3**, 245–250.

Menchise, V., Corbier, C., Didierjean, C., Jacquot, J.P., Benedetti, E., Saviano, M., and Aubry, A. (2000). Crystal structure of the W35A mutant thioredoxin h from *Chlamydomonas reinhardtii*: The substitution of the conserved active site Trp leads to modifications in the environment of the two catalytic cysteines. *Biopolymers* **56**, 1–7.

Menchise, V., Corbier, C., Didierjean, C., Saviano, M., Benedetti, E., Jacquot, J.P., and Aubry, A. (2001). Crystal structure of the wild-type and D30A mutant thioredoxin h of *Chlamydomonas reinhardtii* and implications for the catalytic mechanism. *Biochem. J.* **359**, 65–75.

Merchant, S.S., Allen, M.D., Kropat, J., Moseley, J.L., Long, J.C., Tottey, S., and Terauchi, A.M. (2006). Between a rock and a hard place: Trace element nutrition in *Chlamydomonas*. *Biochim. Biophys. Acta* **1763**, 578–594.

Mestres-Ortega, D. and Meyer, Y. (1999). The *Arabidopsis thaliana* genome encodes at least four thioredoxins m and a new prokaryotic-like thioredoxin. *Gene* **240**, 307–316.

Meyer, A.J. and Hell, R. (2005). Glutathione homeostasis and redox-regulation by sulfhydryl groups. *Photosynth. Res.* **86**, 435–457.

Meyer, Y., Reichheld, J.P., and Vignols, F. (2005). Thioredoxins in *Arabidopsis* and other plants. *Photosynth. Res.* **86**, 419–433.

Meyer, Y., Riondet, C., Constans, L., Abdelgawwad, M.R., Reichheld, J.P., and Vignols, F. (2006). Evolution of redoxin genes in the green lineage. *Photosynth. Res.* **89**, 1–14.

Michelet, L., Zaffagnini, M., Marchand, C., Collin, V., Decottignies, P., Tsan, P., Lancelin, J.M., Trost, P., Miginiac-Maslow, M., Noctor, G., and Lemaire, S.D. (2005). Glutathionylation of chloroplast thioredoxin f is a redox signaling mechanism in plants. *Proc. Natl. Acad. Sci. U. S. A.* **102**, 16478–16483.

Michelet, L., Zaffagnini, M., Massot, V., Keryer, E., Vanacker, H., Miginiac-Maslow, M., Issakidis-Bourguet, E., and Lemaire, S.D. (2006). Thioredoxins, glutaredoxins and glutathionylation: New crosstalks to explore. *Photosynth. Res.* **89**, 225–245.

Miginiac-Maslow, M. and Lancelin, J.M. (2002). Intrasteric inhibition in redox signalling: Light activation of NADP-malate dehydrogenase. *Photosynth. Res.* **72**, 1–12.

Mittard, V., Morelle, N., Brutscher, B., Simorre, J.P., Marion, D., Stein, M., Jacquot, J.P., Lirsac, P.N., and Lancelin, J.M. (1995). 1H, 13C, 15N-NMR resonance assignments of oxidized thioredoxin h from the eukaryotic green alga *Chlamydomonas reinhardtii* using new methods based on two-dimensional triple-resonance NMR spectroscopy and computer-assisted backbone assignment. *Eur. J. Biochem.* **229**, 473–485.

Mittard, V., Blackledge, M.J., Stein, M., Jacquot, J.P., Marion, D., and Lancelin, J.M. (1997). NMR solution structure of an oxidised thioredoxin h from the eukaryotic green alga *Chlamydomonas reinhardtii*. *Eur. J. Biochem.* **243**, 374–383.

Molina, M.M., Belli, G., de la Torre, M.A., Rodriguez-Manzaneque, M.T., and Herrero, E. (2004). Nuclear monothiol glutaredoxins of *Saccharomyces cerevisiae* can function as mitochondrial glutaredoxins. *J. Biol. Chem.* **279**, 51923–51930.

Moon, J.C., Jang, H.H., Chae, H.B., Lee, J.R., Lee, S.Y., Jung, Y.J., Shin, M.R., Lim, H.S., Chung, W.S., Yun, D.J., Lee, K.O., and Lee, S.Y. (2006). The C-type *Arabidopsis* thioredoxin reductase ANTR-C acts as an electron donor to 2-Cys peroxiredoxins in chloroplasts. *Biochem. Biophys. Res. Commun.* **348**, 478–484.

Muhlenhoff, U., Gerber, J., Richhardt, N., and Lill, R. (2003). Components involved in assembly and dislocation of iron-sulfur clusters on the scaffold protein Isu1p. *EMBO J.* **22**, 4815–4825.

Muller, E.G. (1991). Thioredoxin deficiency in yeast prolongs S phase and shortens the G1 interval of the cell cycle. *J. Biol. Chem.* **266**, 9194–9202.

Nakamura, H. (2005). Thioredoxin and its related molecules: Update 2005. *Antioxid. Redox Signal* **7**, 823–828.

Nakamura, T., Ohno, T., Hirota, K., Nishiyama, A., Nakamura, H., Wada, H., and Yodoi, J. (1999). Mouse glutaredoxin – cDNA cloning, high level expression in *E. coli* and its possible implication in redox regulation of the DNA binding activity in transcription factor PEBP2. *Free Radic. Res.* **31**, 357–365.

Navrot, N., Collin, V., Gualberto, J., Gelhaye, E., Hirasawa, M., Rey, P., Knaff, D.B., Issakidis, E., Jacquot, J.P., and Rouhier, N. (2006a). Plant glutathione peroxidases are functional peroxiredoxins distributed in several subcellular compartments and regulated during biotic and abiotic stresses. *Plant Physiol.* **142**, 1364–1379.

Navrot, N., Gelhaye, E., Jacquot, J.P., and Rouhier, N. (2006b). Identification of a new family of plant proteins loosely related to glutaredoxins with four CxxC motives. *Photosynth. Res.* **89**, 71–79.

Nekrasov, V., Ludwig, A.A., and Jones, J.D. (2006). CITRX thioredoxin is a putative adaptor protein connecting Cf-9 and the ACIK1 protein kinase during the Cf-9/Avr9- induced defence response. *FEBS Lett.* **580**, 4236–4241.

Novoselov, S.V. and Gladyshev, V.N. (2003). Non-animal origin of animal thioredoxin reductases: Implications for selenocysteine evolution and evolution of protein function through carboxy-terminal extensions. *Protein Sci.* **12**, 372–378.

Novoselov, S.V., Rao, M., Onoshko, N.V., Zhi, H., Kryukov, G.V., Xiang, Y., Weeks, D.P., Hatfield, D.L., and Gladyshev, V.N. (2002). Selenoproteins and selenocysteine insertion system in the model plant cell system, *Chlamydomonas reinhardtii*. *EMBO J.* **21**, 3681–3693.

Nulton-Persson, A.C., Starke, D.W., Mieyal, J.J., and Szweda, L.I. (2003). Reversible inactivation of alpha-ketoglutarate dehydrogenase in response to alterations in the mitochondrial glutathione status. *Biochemistry* **42**, 4235–4242.

Ogawa, K. (2005). Glutathione-associated regulation of plant growth and stress responses. *Antioxid. Redox Signal* **7**, 973–981.

Park, J.B. and Levine, M. (1996). Purification, cloning and expression of dehydroascorbic acid-reducing activity from human neutrophils: Identification as glutaredoxin. *Biochem. J.* **315**, 931–938.

Patel-King, R.S., Benashki, S.E., Harrison, A., and King, S.M. (1996). Two functional thioredoxins containing redox-sensitive vicinal dithiols from the *Chlamydomonas* outer dynein arm. *J. Biol. Chem.* **271**, 6283–6291.

Pekkari, K. and Holmgren, A. (2004). Truncated thioredoxin: Physiological functions and mechanism. *Antioxid. Redox Signal* **6**, 53–61.

Pérez-Ruiz, J.M., Spinola, M.C., Kirchsteiger, K., Moreno, J., Sahrawy, M., and Cejudo, F.J. (2006). Rice NTRC is a high-efficiency redox system for chloroplast protection against oxidative damage. *Plant Cell* **18**, 2356–2368.

Prieto-Alamo, M.J., Jurado, J., Gallardo-Madueno, R., Monje-Casas, F., Holmgren, A., and Pueyo, C. (2000). Transcriptional regulation of glutaredoxin and thioredoxin pathways and related enzymes in response to oxidative stress. *J. Biol. Chem.* **275**, 13398–13405.

Quimby, B.B. and Dasso, M. (2003). The small GTPase Ran: Interpreting the signs. *Curr. Opin. Cell Biol.* **15**, 338–344.

Regad, F., Herve, C., Marinx, O., Bergounioux, C., Tremousaygue, D., and Lescure, B. (1995). The tef1 box, a ubiquitous cis-acting element involved in the activation of plant genes that are highly expressed in cycling cells. *Mol. Gen. Genet.* **248**, 703–711.

Reichheld, J.P., Meyer, E., Khafif, M., Bonnard, G., and Meyer, Y. (2005). AtNTRB is the major mitochondrial thioredoxin reductase in *Arabidopsis thaliana*. *FEBS Lett.* **579**, 337–342.

Richardson, J.M., 3rd, Lemaire, S.D., Jacquot, J.P., and Makhatadze, G.I. (2000). Difference in the mechanisms of the cold and heat induced unfolding of thioredoxin h from *Chlamydomonas reinhardtii*: Spectroscopic and calorimetric studies. *Biochemistry* **39**, 11154–11162.

Rintamaki, E., Martinsuo, P., Pursiheimo, S., and Aro, E.M. (2000). Cooperative regulation of light-harvesting complex II phosphorylation via the plastoquinol and ferredoxin-thioredoxin system in chloroplasts. *Proc. Natl. Acad. Sci. U. S. A.* **97**, 11644–11649.

Rivas, S., Rougon-Cardoso, A., Smoker, M., Schauser, L., Yoshioka, H., and Jones, J.D. (2004). CITRX thioredoxin interacts with the tomato Cf-9 resistance protein and negatively regulates defence. *EMBO J.* **23**, 2156–2165.

Rochaix, J.D. (2004). Genetics of the biogenesis and dynamics of the photosynthetic machinery in eukaryotes. *Plant Cell* **16**, 1650–1660.

Rodriguez-Manzaneque, M.T., Tamarit, J., Belli, G., Ros, J., and Herrero, E. (2002). Grx5 is a mitochondrial glutaredoxin required for the activity of iron/sulfur enzymes. *Mol. Biol. Cell* **13**, 1109–1121.

Rohr, J., Sarkar, N., Balenger, S., Jeong, B.R., and Cerutti, H. (2004). Tandem inverted repeat system for selection of effective transgenic RNAi strains in *Chlamydomonas*. *Plant J.* **40**, 6116–6121.

Rollin-Genetet, F., Berthomieu, C., Davin, A.H., and Quemeneur, E. (2004). *Escherichia coli* thioredoxin inhibition by cadmium: Two mutually exclusive binding sites involving Cys32 and Asp26. *Eur. J. Biochem.* **271**, 1299–1309.

Rouhier, N. and Jacquot, J.P. (2005). The plant multigenic family of thiol peroxidases. *Free Radic. Biol. Med.* **38**, 1413–1421.

Rouhier, N., Gelhaye, E., Sautiere, P.E., Brun, A., Laurent, P., Tagu, D., Gérard, J., de Faÿ, E., Meyer, Y., and Jacquot, J.P. (2001). Isolation and characterization of a new peroxiredoxin from poplar sieve tubes that uses either glutaredoxin or thioredoxin as a proton donor. *Plant Physiol.* **127**, 1299–1309.

Rouhier, N., Gelhaye, E., and Jacquot, J.P. (2002). Glutaredoxin-dependent peroxiredoxin from poplar: Protein–protein interaction and catalytic mechanism. *J. Biol. Chem.* **277**, 13609–13614.

Rouhier, N., Vlamis-Gardikas, A., Lillig, C.H., Berndt, C., Schwenn, J.D., Holmgren, A., and Jacquot, J.P. (2003). Characterization of the redox properties of poplar glutaredoxin. *Antioxid. Redox Signal* **5**, 15–22.

Rouhier, N., Gelhaye, E., Gualberto, J.M., Jordy, M.N., De Fay, E., Hirasawa, M., Duplessis, S., Lemaire, S.D., Frey, P., Martin, F., Manieri, W., Knaff, D.B., and Jacquot, J.P. (2004a). Poplar peroxiredoxin Q. A thioredoxin-linked chloroplast antioxidant functional in pathogen defense. *Plant Physiol.* **134**, 1027–1038.

Rouhier, N., Gelhaye, E., and Jacquot, J.P. (2004b). Plant glutaredoxins: Still mysterious reducing systems. *Cell Mol. Life Sci.* **61**, 1266–1277.

Rouhier, N., Villarejo, A., Srivastava, M., Gelhaye, E., Keech, O., Droux, M., Finkemeier, I., Samuelsson, G., Dietz, K.J., Jacquot, J.P., and Wingsle, G. (2005). Identification of plant glutaredoxin targets. *Antioxid. Redox Signal* **7**, 919–929.

Rouhier, N., Kauffmann, B., Tete-Favier, F., Paladino, P., Gans, P., Branlant, G., Jacquot, J.P., and Boschi-Muller, S. (2007). Functional and structural aspects of poplar cytosolic and plastidial type a methionine sulfoxide reductases. *J. Biol. Chem.* **282**, 3367–3378.

Ruelland, E. and Miginiac-Maslow, M. (1999). Regulation of chloroplast enzyme activities by thioredoxins: Activation or relief from inhibition? *Trends Plant Sci.* **4**, 136–141.

Sarkar, N., Lemaire, S., Wu-Scharf, D., Issakidis-Bourguet, E., and Cerutti, H. (2005). Functional specialization of *Chlamydomonas reinhardtii* cytosolic thioredoxin h1 in the response to alkylation-induced DNA damage. *Eukaryotic Cell* **4**, 262–273.

Scheibe, R. (2004). Malate valves to balance cellular energy supply. *Physiol. Plant* **120**, 21–26.

Schmitter, J.M., Jacquot, J.P., de Lamotte-Guéry, F., Beauvallet, C., Dutka, S., Gadal, P., and Decottignies, P. (1988). Purification, properties and complete amino acid sequence of the ferredoxin from a green alga, *Chlamydomonas reinhardtii*. *Eur. J. Biochem.* **172**, 405–412.

Schurmann, P. (2002). Ferredoxin-dependent thioredoxin reductase: A unique iron-sulfur protein. *Methods Enzymol.* **347**, 403–411.

Schürmann, P. and Jacquot, J.P. (2000). Plant thioredoxin systems revisited. *Annu. Rev. Plant Physiol. Plant Mol. Biol.* **51**, 371–400.

Schürmann, P., Wolosiuk, R.A., Breazeale, V.D., and Buchanan, B.B. (1976). Two proteins function in the regulation of photosynthetic CO_2 assimilation in chloroplasts. *Nature* **263**, 257–258.

Serrato, A.J., Perez-Ruiz, J.M., Spinola, M.C., and Cejudo, F.J. (2004). A novel NADPH thioredoxin reductase, localized in the chloroplast, which deficiency causes hypersensitivity to abiotic stress in *Arabidopsis thaliana*. *J. Biol. Chem.* **279**, 43821–43827.

Setterdahl, A.T., Chivers, P.T., Hirasawa, M., Lemaire, S.D., Keryer, E., Miginiac-Maslow, M., Kim, S.K., Mason, J., Jacquot, J.P., Longbine, C.C., de Lamotte-Guéry, F., and Knaff, D.B. (2003). Effect of pH on the oxidation-reduction properties of thioredoxins. *Biochemistry* **42**, 14877–14884.

Sha, S., Yabushita, T., Minakuchi, K., Masumura, T., and Tanaka, K. (1997). Structure of the rice glutaredoxin (thioltransferase) gene. *Gene* **188**, 23–28.

Shelton, M.D., Chock, P.B., and Mieyal, J.J. (2005). Glutaredoxin: Role in reversible protein s-glutathionylation and regulation of redox signal transduction and protein translocation. *Antioxid. Redox Signal* **7**, 348–366.

Sicard-Roselli, C., Lemaire, S., Jacquot, J.P., Favaudon, V., Marchand, C., and Houee-Levin, C. (2004). Thioredoxin Ch1 of *Chlamydomonas reinhardtii* displays an unusual resistance toward one-electron oxidation. *Eur. J. Biochem.* **271**, 3481–3487.

Starke, D.W., Chock, P.B., and Mieyal, J.J. (2003). Glutathione-thiyl radical scavenging and transferase properties of human glutaredoxin (thioltransferase). Potential role in redox signal transduction. *J. Biol. Chem.* **278**, 14607–14613.

Stein, M., Jacquot, J.P., Jeannette, E., Decottignies, P., Hodges, M., Lancelin, J.M., Mittard, V., Schmitter, J.M., and Miginiac-Maslow, M. (1995). *Chlamydomonas reinhardtii* thioredoxins: Structure of the genes coding for the chloroplastic m and cytosolic h isoforms; expression in *Escherichia coli* of the recombinant proteins, purification and biochemical properties. *Plant Mol. Biol.* **28**, 487–503.

Takahashi, H., Iwai, M., Takahashi, Y., and Minagawa, J. (2006). Identification of the mobile light-harvesting complex II polypeptides for state transitions in *Chlamydomonas reinhardtii*. *Proc. Natl. Acad. Sci. U. S. A.* **103**, 477–482.

Tamarit, J., Belli, G., Cabiscol, E., Herrero, E., and Ros, J. (2003). Biochemical characterization of yeast mitochondrial Grx5 monothiol glutaredoxin. *J. Biol. Chem.* **278**, 25745–25751.

Tompa, P., Szasz, C., and Buday, L. (2005). Structural disorder throws new light on moonlighting. *Trends Biochem. Sci.* **30**, 484–489.

Trebitsh, T. and Danon, A. (2001). Translation of chloroplast *psbA* mRNA is regulated by signals initiated by both photosystems II and I. *Proc. Natl. Acad. Sci. U. S. A.* **98**, 12289–12294.

Trebitsh, T., Levitan, A., Sofer, A., and Danon, A. (2000). Translation of chloroplast *psbA* mRNA is modulated in the light by counteracting oxidizing and reducing activities. *Mol. Cell. Biol.* **20**, 1116–1123.

Trebitsh, T., Meiri, E., Ostersetzer, O., Adam, Z., and Danon, A. (2001). The protein disulfide isomerase-like RB60 is partitioned between stroma and thylakoids in *Chlamydomonas reinhardtii* chloroplasts. *J. Biol. Chem.* **276**, 4564–4569.

Trost, P., Fermani, S., Marri, L., Zaffagnini, M., Falini, G., Scagliarini, S., Pupillo, P., and Sparla, F. (2006). Thioredoxin-dependent regulation of photosynthetic glyceraldehyde-3-phosphate dehydrogenase: Autonomous vs. CP12-dependent mechanisms. *Photosynth. Res.* **89**, 1–13.

Vido, K., Spector, D., Lagniel, G., Lopez, S., Toledano, M.B., and Labarre, J. (2001). A proteome analysis of the cadmium response in *Saccharomyces cerevisiae*. *J. Biol. Chem.* **276**, 8469–8474.

Vieira Dos Santos, C., Cuine, S., Rouhier, N., and Rey, P. (2005). The *Arabidopsis* plastidic methionine sulfoxide reductase B proteins. Sequence and activity characteristics, comparison of the expression with plastidic methionine sulfoxide reductase A, and induction by photooxidative stress. *Plant Physiol.* **138**, 909–922.

Vlamis-Gardikas, A. and Holmgren, A. (2002). Thioredoxin and glutaredoxin isoforms. *Methods Enzymol.* **347**, 286–296.

Vorst, O., van Dam, F., Oosterhoff-Teertstra, R., Smeekens, S., and Weisbeek, P. (1990). Tissue-specific expression directed by an *Arabidopsis thaliana* preferredoxin promoter in transgenic tobacco plants. *Plant Mol. Biol.* **14**, 491–499.

Wakabayashi, K. and King, S.M. (2006). Modulation of *Chlamydomonas reinhardtii* flagellar motility by redox poise. *J. Cell Biol.* **173**, 743–754.

Washburn, M.P. and Wells, W.W. (1999). The catalytic mechanism of the glutathione-dependent dehydroascorbate reductase activity of thioltransferase (glutaredoxin). *Biochemistry* **38**, 268–274.

Wingert, R.A., Galloway, J.L., Barut, B., Foott, H., Fraenkel, P., Axe, J.L., Weber, G.J., Dooley, K., Davidson, A.J., Schmid, B., Paw, B.H., Shaw, G.C., Kingsley, P., Palis, J., Schubert, H., Chen, O., Kaplan, J., and Zon, L.I. (2005). Deficiency of glutaredoxin 5 reveals Fe-S clusters are required for vertebrate haem synthesis. *Nature* **436**, 1035–1039.

Wolosiuk, R.A. and Buchanan, B.B. (1977). Thioredoxin and glutathione regulate photosynthesis in chloroplasts. *Nature* **266**, 565–567.

Xing, S., Rosso, M.G., and Zachgo, S. (2005). ROXY1, a member of the plant glutaredoxin family, is required for petal development in *Arabidopsis thaliana*. *Development* **132**, 1555–1565.

Yohn, C.B., Cohen, A., Danon, A., and Mayfield, S.P. (1998). A poly(A) binding protein functions in the chloroplast as a message-specific translation factor. *Proc. Natl. Acad. Sci. U. S. A.* **95**, 2238–2243.

Zaffagnini, M., Michelet, L., Marchand, C., Sparla, F., Decottignies, P., Le Maréchal, P., Miginiac-Maslow, M., Noctor, G., Trost, P., and Lemaire, S.D. (2007). The thioredoxin independent isoform of chloroplastic glyceraldehyde-3-phosphate dehydrogenase is selectively regulated by glutathionylation. *FEBS J.* **274**, 212–226.

Zheng, M., Åslund, F., and Storz, G. (1998). Activation of the OxyR transcription factor by reversible disulfide bond formation. *Science* **279**, 1718–1721.

CHAPTER 12

The Mitochondrial Genome

Pierre Cardol,[1,2] *and Claire Remacle*[1]

[1]Genetics of Microorganisms, [2]Plant Biochemistry and Photobiology, Department of Life Sciences, Institute of Plant Biology B22, University of Liège, Belgium

CHAPTER CONTENTS

I. INTRODUCTION

Mitochondria have evolved from eubacteria-like endosymbionts whose closest known relatives are the Rickettsia group of α-proteobacteria. Over time, many genes have disappeared or have been transferred to the nuclear genome. Extant mitochondrial genomes are diverse in size, gene content, and organization. Mitochondrial DNAs (mtDNAs) from angiosperms and other land plants are the largest and the most complex genomes, comprising 187 kb (*Marchantia polymorpha*) to 2400 kb (*Cucumis melo*) (Gillham, 1994; Burger et al., 2003; Bullerwell and Gray, 2004; Knoop, 2004). In contrast, the mitochondrial genomes of vertebrates are only 16–17 kb.The size differences of mitochondrial genomes from various organisms do not reflect gene content, as the "extra" DNA in the larger genomes appears to consist primarily of non-coding sequences (Unseld et al., 1997; Kubo et al., 2000; Notsu et al., 2002; Clifton et al., 2004; Sugiyama et al., 2005). The mtDNA most resembling a primitive mitochondrial genome is that of a freshwater protozoon *Reclinomonas americana*. Its 69,034-bp genome contains the largest collection of genes (97) identified to date (Lang et al., 1997). In contrast, the 15.8 kb mitochondrial genome of *Chlamydomonas reinhardtii* has the lowest gene content (13 genes, see section II.A). For example, it lacks seven genes that encode essential components of respiratory chain complexes: *nd3*, *nd4L* (complex I), *cox2*, *cox3* (complex IV), and *atp6*, *atp8*, and *atp9* (complex V). These genes are also absent from mtDNAs of the related green algae *Chlamydomonas eugametos* (Denovan-Wright et al., 1998) and *Chlorogonium elongatum* (Kroymann and Zetsche, 1998) (see section II.A).

In higher plants, only a few mutations affecting mitochondrial genes that encode components of the respiratory chain complexes have been characterized. These mutations determine cytoplasmic male sterility in tobacco (Pla et al., 1995; Gutierres et al., 1997, 1999; Pineau et al., 2005) or a nonchromosomal stripe phenotype in maize (Karpova and Newton, 1999; Gabay-Laughnan and Newton, 2005; Yu et al., 2007). The mutants often display a decreased growth rate, have more or less severe morphological abnormalities and cannot generally be maintained as pure homoplasmic lines. In that respect, *Chlamydomonas* is a useful model to study mitochondrial function in a photosynthetic organism. Mitochondrial respiratory-deficient mutants are homoplasmic and viable if maintained under photoautotrophic conditions. This enables the isolation of complex I, III, or IV mutants that are amenable to biochemical and molecular studies. A major advance has been the establishment of biolistic transformation of the mitochondrial genome to reconstruct mutations in genes encoding complex I subunits (Remacle et al., 2006). *Chlamydomonas* thus became the second organism after the yeast *Saccharomyces cerevisiae* whose mitochondrial genome can be manipulated, opening the way to reverse genetics.

II. CHARACTERISTICS OF THE MITOCHONDRIAL GENOME

A. Structure and gene content

The *Chlamydomonas* mitochondrial genome was first characterized in terms of buoyant density (1.706 g/ml in CsCl), melting temperature (87.9°C in standard saline citrate), kinetic complexity (9.78×10^6 daltons) and physical conformation (99% linear, <1% open and closed circular molecules, all within the range of 4.0–5.4 μm) (Ryan et al., 1978). Physical mapping using restriction endonucleases and *in vitro* labeling of mtDNA molecules confirmed that the genome is linear and has unique ends (Grant and Chiang, 1980). The genome is a multicopy system organized into about 20–30 nucleoids (Nishimura et al., 1998; Hiramatsu et al., 2006). The nucleoids change in size and shape during the cell cycle and are present in mitochondria that appear as sinuous or highly branched structures extending throughout the cytoplasm, and resembling a single large mitochondrion (Ehara et al., 1995; Hiramatsu et al., 2006).

The 15.8 kb *Chlamydomonas* mitochondrial genome is fully sequenced (reviewed in Boynton and Gillham, 1996; Remacle and Matagne, 1998; Matagne and Remacle, 2002) (Figure 12.1). It is characterized by low gene content, very short intergenic sequences and no introns. Thirteen genes are present, which encode subunits of the rotenone-sensitive NADH:ubiquinone oxidoreductase or complex I (*nd1*, *nd2*, *nd4*, *nd5*, and *nd6*), apocytochrome *b* of the *bc1* complex or complex III (*cob*), subunit 1 of cytochrome *c* oxidase or complex IV (*cox1*), a reverse transcriptase-like protein (*rtl*), three tRNAs (*trnW*, *trnQ*, *trnM*) and the ribosomal RNAs (rRNAs). The *rtl* gene is thought to be a degenerate group II intron (Boer and Gray, 1988a) and putative 5′ and 3′ splice sites have been mapped downstream and upstream

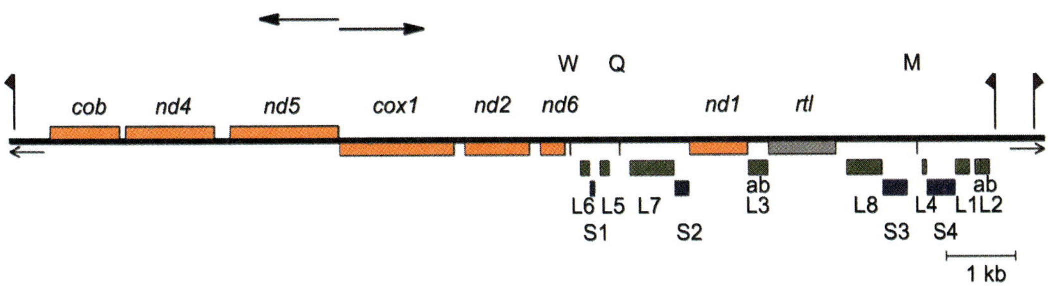

FIGURE 12.1 *Physical map of the 15.8 kb linear mitochondrial genome (GenBank accession U03843). Upper horizontal arrows indicate the bidirectional origin of transcription. The small horizontal arrows at each end of the genome denote the terminal inverted repeats, and the three copies of the 86 bp repeats are represented by flags (Vahrenholz et al., 1993). The eight genes encoding polypeptides include cob (apocytochrome b of complex III), cox1 (subunit 1 of cytochrome c oxidase or complex IV), nd1, 2, 4, 5, and 6 (subunits of NADH:ubiquinone oxidoreductase or complex I) and rtl (reverse transcriptase-like protein). L and S represent modules encoding segments of rRNA of the large and the small subunits, respectively. W, Q, and M represent tRNAs for Trp, Glu, and Met, respectively.*

of the coding sequence (Nedelcu and Lee, 1998a). The rRNA genes are discontinuous and split into four small subunit (*S*) and eight large subunit (*L*) modules, interspersed with one another and with protein and tRNA genes. The small rRNA segments of the two ribosomal subunits are thought to interact by way of extensive intermolecular pairing to form conventional rRNA molecules (Boer and Gray, 1988b).

The ends of the genome have been studied in detail (Ma et al., 1992b; Vahrenholz et al., 1993). They are characterized by terminal inverted repeats of 531–532 bp, including identical 3′ single-stranded extensions of 39–41 nt at both ends. The 86 outermost bp at each end are identical to an internal copy, which is located at the 3′ end of the *L2b* sequence, 674 bp from the right end of the molecule (Vahrenholz et al., 1993).

The mitochondrial genomes of other *Chlamydomonas* species have also been characterized. The mitochondrial genomes of the two interfertile species *C. smithii* and *C. reinhardtii* are co-linear, with the exception of a 1 kb intron located in the *cob* gene of *C. smithii* (Boynton et al., 1987). The intron (called α intron) is mobile (see section IV.B) and contains a 237-codon ORF (I-*Csm*I), which is similar to the I-*Sce*I endonuclease of *S. cerevisiae* (Colleaux et al., 1990) and to the ORF encoded by intron 2 of the *C. elongatum cob* gene (Kroymann and Zetsche, 1997). The ORF is in frame with the upstream exon and belongs to the LAGLIDADG endonuclease family.

An internal 15,637 bp fragment of the linear *C. incerta* mitochondrial genome was sequenced. This mtDNA is co-linear with its *C. reinhardtii* homologue but in addition, 1 kb introns are present in the *cob* and the *cox1* genes. Each intron sequence contains an ORF in frame with the upstream exon and having a LAGLIDADG motif (Popescu and Lee, 2007).

The mitochondrial genomes of *C. eugametos* and *C. moewusii*, which are mutually interfertile but only distantly related to *C. reinhardtii* (see Volume 1, Chapter 1), are circular and have sizes of 23 and 22 kb, respectively (Denovan-Wright and Lee, 1992). The complete sequence of *C. eugametos* mtDNA shares almost the same reduced set of coding functions and similar organization of rRNA genes with *C. reinhardtii* (Denovan-Wright et al., 1998). The corresponding small transcripts of the rRNA genes are assembled into mitochondrial ribosomes (Denovan-Wright and Lee, 1995). Nine group I introns have also been identified in the protein-coding genes or in rRNA segments, and short repetitive sequences in intergenic regions (Nedelcu and Lee, 1998b).

The mitochondrial genome of another member of the Chlamydomonadaceae, the colorless green alga *Polytomella parva*, possesses two linear mtDNA molecules of 13.5 and 3.5 kb (Fan and Lee, 2002). The 13.5 kb subgenome contains six respiratory-chain genes (*nd1*, *nd2*, *nd4*, *nd5*, *cob*, and *cox1*), one tRNA gene (*trnM*) and the scrambled rRNA genes, whereas the 3.5 kb subgenome only contains *nd6*. Both DNA components have long terminal inverted repeats which are almost identical between the 13.5- and 3.5-kb linear mtDNAs but show no similarity with the terminal repeats in the linear mtDNA of *C. reinhardtii*. The fragmented mitochondrial rRNAs have been characterized in detail (Fan et al., 2003); they are considerably shorter than those of *C. reinhardtii*. Notably, three nt are missing from the normally highly conserved GTPase center (*L5*) of the *LSU* rRNA gene.

B. Replication, transcription, and translation

How the *Chlamydomonas* mitochondrial genome replicates is not known. In each of the terminal inverted repeats, a 3′-GGCCG-5′ motif that is part of the origin of light-strand replication (O_L) in most vertebrate mtDNAs was identified (Nedelcu and Lee, 1998b). In addition, each of the two long inverted repeats contains a 5′-GGAGGGG-3′ motif complementary to the sequence recognized by the mouse mitochondrial RNA processing endoribonuclease (RNase MRP) that cleaves the RNA primer at the origin of heavy-strand replication (Nedelcu and Lee, 1998b; Martin and Li, 2007). Both a recombination-mediated model based on the presence of the three 86 nt repeat sequences and a reverse transcriptase-mediated model have been proposed for the replication of the termini of the linear mtDNA (Vahrenholz et al., 1993).

Both strands of *Chlamydomonas* mtDNA encode genetic information. Large bi-directional co-transcripts are initiated in a short intergenic region between *nd5* and *cox1* (Gray and Boer, 1988). Two potential promoter sequences for bi-directional transcription initiation presenting similarity with the consensus promoter sequence in fungal mtDNA (A/T)TA(T/G)(T/A)RR(T/G)N (Tracy and Stern, 1995) were identified. The first one, TTATTACAT, located 3 nt upstream of the ATG of *cox1*, could initiate the transcription of the rightward unit. The second one, TTATTATTT, located 3 nt upstream of the *nd5* gene initiation codon, could be the promoter of the leftward transcription unit (Duby et al., 2001). Sequences related to those described above have been proposed as potential promoters for bi-directional transcription initiation of the 13.5 kb mitochondrial subgenome of *P. parva* (Fan and Lee, 2002). The *Chlamydomonas MOC1* gene encodes a homologue of the human mitochondrial termination transcription factor (mTERF) that is targeted to mitochondria. Loss of MOC1 causes a high light sensitive phenotype and disrupts the transcription and expression profile of mitochondrial respiratory genes (Schönfeld et al., 2004).

The two primary co-transcripts are then processed to yield mature RNAs. The sizes of the mRNAs detected by RNA gel blots are close to the lengths of the coding regions, indicating that mRNAs have short 5′ and 3′ non-coding sequences. This has been confirmed in some cases by using S1 nuclease protection and primer extension to map transcript termini. The *cob*, *nd4*, *nd5*, and *cox1* mRNAs all appear to begin at the AUG initiation codon (minor species for *nd4*) while the *nd1* and *rtl* mRNAs have very short 5′ UTRs ($\leq$13 nt). The 3′ UTRs are between 35 nt and 250 nt (Boer and Gray, 1986, 1988a; Gray and Boer, 1988; Ma et al., 1989, 1990). Similarly, transcripts corresponding to the *S* and *L* modules of the rRNA genes extend only for a short distance beyond the boundaries of the modules defined by sequence comparisons (Boer and Gray, 1988b). For many of the transcripts including modules of the *rRNA* genes, the 3′ end of a given transcript is immediately followed by the 5′ end of another transcript, thus their ends abut (Boer and Gray, 1986, 1988a,b; Gray and Boer, 1988). The 3′ UTRs contain inverted repeat sequences that can be folded into hairpin structures. These structures could represent signals for transcription termination, RNA processing or RNA stability (Pratje et al., 1984; Boer and Gray, 1986, 1988a; Gray and Boer, 1988; Ma et al., 1989; Duby et al., 2001; Cardol et al., 2002). In addition, short GC-rich dispersed repeats are found in non-transcribed spacer regions, mainly located in the rRNA part of the genome (Boer and Gray, 1991). They are also capable of assuming stable hairpin structures and may play a role in mitochondrial gene expression (Boer and Gray, 1991) or be involved in fragmentation and scrambling of the rRNA coding region and ultimately reducing gene content (Nedelcu and Lee, 1998b).

As in plant mitochondria, the universal genetic code is used in *Chlamydomonas* mitochondria (Kück and Neuhaus, 1986; Michaelis et al., 1990) but in contrast to plants, no editing has been found in *Chlamydomonas*. The codon usage is highly biased. Based on GenBank accession U03843, nine codons are not used at all (or eight if the *C. smithii* intron is taken into account) (Colleaux et al., 1990; Michaelis et al., 1990). The Codon Usage Database of the Kazusa DNA Research Institute (http://www.kazusa.or.jp/codon/), which appears to be based on separately published sequences of individual genes, shows only one unused codon, however.

A minimum of 23 tRNAs is needed to translate the genetic information (Boer and Gray, 1988c). As only three tRNAs are encoded in the mitochondrial genome, most of the tRNAs used in the organelle have to be imported from the cytosol. Compilation of the eight protein-coding sequences allowed the definition of a putative ribosome-binding site ATTTTATTA or ATAATTTA, 3–7 nt upstream of the AUG codon (Colleaux et al., 1990). Ribosome binding would require an additional mechanism when mRNAs begin precisely at the AUG codon (Boer and Gray, 1986; Gray and Boer, 1988; Ma et al., 1989, 1990).

III. MUTATIONS AFFECTING THE MITOCHONDRIAL GENOME

A. Respiratory-deficient mutants

After prolonged exposure of *Chlamydomonas* cells to intercalating dyes (acriflp or ethidium bromide), the large majority of the cell population is converted into mutant cells. These cells are altered in respiratory activity and display a delayed lethality: plated in the light in the absence of dyes, the mutant cells divide 8–9 times and then stop growing, producing minute depigmented colonies (Alexander et al., 1974; Gillham et al., 1987; Matagne et al., 1989). As the two intercalating dyes cause specific elimination of mtDNA, the lethal phenotype is probably correlated with the loss of the mitochondrial genome (Gillham et al., 1987).

In addition to lethal minute colonies, acriflavine and ethidium bromide induce a low percentage of viable mutants impaired in respiratory functions. These mutants can be divided into two phenotypic classes when cultivated under heterotrophic conditions (darkness plus acetate as an exogenous carbon source). One class is composed of mutants which are unable to grow in the dark (dark$^-$ phenotype) and are thus obligate photoautotrophs, and another class is composed of mutants which grow in the dark but much more slowly than the wild-type strain (dark$^{+/-}$ phenotype). In crosses between wild-type and mutant cells, the meiotic progeny most often inherit the phenotype of the mating-type *minus* parent, a uniparental transmission pattern that is typical of mutations affecting the mitochondrial genome (see section IV.A). The respiratory-deficient mutants are thus called *dum* (for *d*ark *u*niparental transmission by the *minus* parent).

B. Mutants affected in complex III or complex IV

Mutants that have lost complex III or complex IV activity (Table 12.1) are obligate photoautotrophs deprived of the cytochrome pathway of respiration (Figure 12.2), due to mutations in *cob* (complex III mutants) or *cox1* (complex IV mutants). Their inability to grow in the dark is probably linked to the loss of the two phosphorylation sites associated with complexes III and IV. Their oxygen consumption, which is insensitive to cyanide, occurs via the activity of the salicyl hydroxamic acid (SHAM)-sensitive alternative oxidase (Figure 12.2). Under mixotrophic conditions (light + acetate), their total respiratory rates are about 50% of that of wild type cells and their growth rates are reduced. In contrast, under photoautotrophic conditions (light without acetate), their respiratory rates are similar to the wild type and their growth rates are largely unaffected (Dorthu et al., 1992).

Mutants of this type can also be identified by a staining test performed *in vivo* on colonies grown on agar plates. In anaerobic and dark conditions, wild-type colonies reduce 2,3,5-triphenyltetrazolium chloride (TTC) to red formazan and stain purple, whereas the mutant colonies remain green (Dorthu et al., 1992). The site of reduction of TTC is often considered to be at the level of cytochrome *c* oxidase (Slater et al., 1963). However, Rich et al. (2001) proposed that the reduction site of TTC is the NADH dehydrogenases (complex I and alternate NAD(P)H dehydrogenases, see Figure 12.2). Once the initial product of TTC reduction is produced by these enzymes, it would then be rapidly reoxidized by molecular oxygen, so that the generation

Table 12.1 *Chlamydomonas* mitochondrial mutants

Mutant	Growth in the dark	Biochemical defect (% of wild-type activity)	Gene mutation	Size (kD) and relative amount (%) of complex I and sub-complex	References
dum1, 2, 3, 4, 11, 14, 16	–	Complex III (0)	*cob* partially or totally deleted (0.7–1.7 deletion)	ND	Dorthu et al. (1992), Matagne et al. (1989), Randolph-Anderson et al. (1993)
dum15	–	Complex III (0)	2-bp substitution at codon 140 of *cob* TCT (Ser) → TAC (Tyr)	ND	Dorthu et al. (1992)
dum21	–	Complex III (0)	1-bp substitution at codon 166 of *cob* TGG (Trp) → TGA (Stop)	ND	Matagne and Remacle (2002)
mud2[a]	+	Resistance to myxothiazol	1-bp substitution at codon 129 of *cob*, TTC (Phe) → CTC (Leu)	ND	Bennoun et al. (1991)
bm20[a]	+	Resistance to myxothiazol	1-bp substitution at codon 132 of *cob*, TAT (Tyr) → TGT (Cys)	ND	Bennoun et al. (1992)
im1[a]	+	Resistance to myxothiazol	1-bp substitution at codon 137 of *cob*, GGC (Gly) → AGC (Ser)	ND	Bennoun et al. (1992)
dum18	–	Complex IV (0)	+1T (in a run of 3 T) at codon 145 of *cox1*	ND	Colin et al. (1995)
dum19	–	Complex IV (0)	–1T (in a run of 4T) at codon 152 of *cox1*	ND	Colin et al. (1995)
dum5	+/–	Complex I (15)	–1T (in a run of 3T) in the 3'untranslated region of *nd5*	950 (15)	Cardol et al. (2002)
dum17	+/–	Complex I (0)	–1T (in a run of 4T) at codons 143–144 of *nd6*	950 (0–2)	Colin et al. (2002)
dum20	+/–	Complex I (0)	–1T (in a run of 2 T) at codon 243 of *nd1*	950 (0–3)	Remacle et al. (2001a)
dum25	+/–	Complex I (0)	deletion of codons 199–200 of *nd1*	950 (100)	Remacle et al. (2001a)
Δ*nd4*	+/–	Complex I (0)	deletion of codons 2–24 of *nd4*	700 (15)	Remacle et al. (2006)
Δ*nd4*/*dum25*	+/–	Complex I (0)	Δ*nd4* and *dum25* mutations	700 (15)	Remacle et al. (2006)
dum22[b]	–	Complexes I (0) and III (0)	*cob* and 3'end of *nd4* deleted (3 kb deletion)	700 (15)	Remacle et al. (2001b)
dum24	–	Complexes I (0) and III (0)	*cob*, *nd4*, and 3'end of *nd5* deleted (4.35 kb deletion)	700 (15)	Duby and Matagne (1999), Duby et al. (2001)
dum19/*dum25*	–	Complexes I (0) and IV (0)	*dum19* and *dum25* mutations	950 (100)	Remacle et al. (2001a)

ND: Not determined.

[a] *The mutants retain full complex III activity.*

[b] *This mutant was isolated from a strain harboring the α intron in the* cob *gene.*

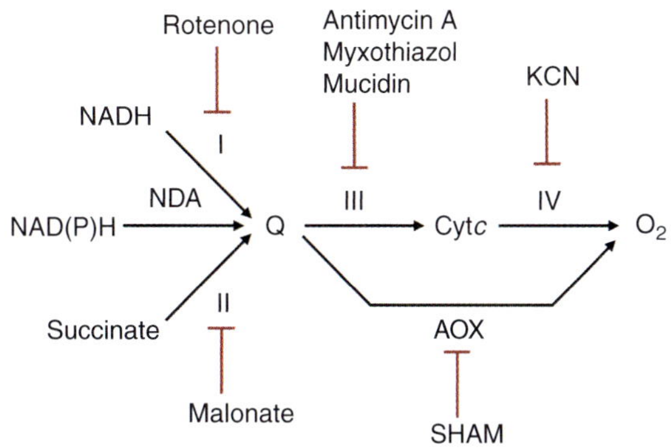

FIGURE 12.2 *Schematic view of the respiratory chain in* Chlamydomonas. *I, complex I or NADH: ubiquinone oxidoreductase; II, complex II or succinate:ubiquinone oxidoreductase; III, complex III or ubiquinol:cytochrome* c *oxidoreductase; IV, complex IV or cytochrome* c *oxidase; Cyt* c*, cytochrome* c*; Q, ubiquinone pool; NDA, alternate NAD(P)H dehydrogenases; AOX, alternative oxidase. Complexes I, III, and IV are the phosphorylating sites of the respiratory chain. Complex III, cytochrome* c *and complex IV define the "cytochrome pathway." Specific inhibitors of each complex are also indicated.*

of the stable red formazan product from this intermediate occurs only under strictly anaerobic conditions (Rich et al., 2001). Mutant colonies defective in the cytochrome pathway of respiration (complex III or complex IV) would not be able to generate sufficiently anaerobic conditions to form the stable red formazan and would thus remain green. The biochemical defect in complex III or complex IV mutants can also be determined by growth tests on agar plates containing inhibitors. The mutants are insensitive to 7.5 μM myxothiazol and 1 μM antimycin A, two inhibitors of complex III, but are sensitive to 35 μM rotenone, an inhibitor of complex I (Figure 12.2).

Many mutants (*dum1, 2, 3, 4, 11, 14,* and *16*) have a terminal deletion of the left end of the mitochondrial genome (Matagne et al., 1989; Dorthu et al., 1992; Randolph-Anderson et al., 1993). The *cob* gene is totally or partly deleted, leading to the absence of complex III activity. In mutant mitochondria, deleted monomeric genomic DNAs always coexist with dimers resulting from head-to-head fusions between deleted monomers. Two point mutations affecting the *cob* gene have also been identified: in *dum15,* a 2 bp substitution transforms TCT (Ser) into TAC (Tyr) at codon 140 (Colin et al., 1995); in *dum21,* a G → A transition changes TGG (Trp) into TGA (Stop) at codon 166 (Matagne and Remacle, 2002). Both the *dum18* and *dum19* mutants lack complex IV activity. They possess frameshift mutations at *cox1* codons 145 and 152, respectively (Colin et al., 1995).

C. Mutants resistant to myxothiazol

Myxothiazol and mucidin (strobilurin A) are complex III inhibitors which prevent electron transfer at the ubiquinone redox site Q_O located close to the outer side of the inner mitochondrial membrane. In *S. cerevisiae* and *Rhodobacter capsulatus*, mutations conferring resistance to myxothiazol or mucidin have been mapped to specific sites in the *cob* gene. In *Chlamydomonas*, several myxothiazol-resistant mutants with substitutions in the *cob* gene have also been characterized (Table 12.1; Bennoun et al., 1991, 1992). Three sites have been found: Phe (129) → Leu (*mud2*) and Tyr (132) → Cys, both located in transmembrane helix III, and Gly (137) → Ser located in the extra-membrane loop, between helices III and IV. The mutated amino acids are probably involved in the binding of myxothiazol.

D. Mutants affected in complex I

Four mutants (*dum17*, *20*, *23*, and *25*) lacking NADH:ubiquinone (complex I) activity and one mutant (*dum5*) showing reduced complex I activity have been isolated by random mutagenesis (Table 12.1). On agar plates, they grow very slowly in the dark and are insensitive to rotenone. In the light, their growth is significantly reduced when myxothiazol and antimycin A are added to the culture medium. The TTC test is positive with these mutants. Two mutations affect the *nd1* coding sequence; *dum20* being a frameshift mutation in codon 243 of *nd1*, and *dum25* corresponding to a 6 bp deletion that eliminates two residues (Glu and Ala) in a highly conserved hydrophilic segment of the protein (Remacle et al., 2001a). *dum17* is a frameshift mutation in codon 143 of *nd6* (Cardol et al., 2002). One mutation has been found in *nd5*: *dum5* is a T base deletion in the 3′ UTR region of *nd5* mRNA (Cardol et al., 2002). In *dum5* mitochondria, the *nd5* transcript is present in very low amounts, showing the role of this untranslated region in transcript stability (Cardol et al., 2002).

All the frameshift mutations mentioned above are deletions or additions of one T base in runs of two, three, or four T bases. Because all these mutations have been induced by treatment with acriflavine, it is likely that they result from replication errors favored by distortion of the double helix structure when the intercalating dye is bound to DNA.

In addition, two complex I mutants have been created by biolistic mitochondrial transformation. The Δ*nd4* mutant harbors an in-frame deletion of 23 amino acids, eliminating codons 2–24. This deletion was also introduced in the *dum25* mutant background, creating a mutant affected in both *nd1* and *nd4* (Remacle et al., 2006).

E. Mutants affected in more than one complex

A few mutants lack both complex I activity and the cytochrome pathway of respiration (Table 12.1). Contrary to the mutants described above, their growth in photoautotrophic conditions is significantly decreased. The respiratory rates of this mutant type are very low and rely solely on the activities of alternate NAD(P)H dehydrogenases, complex II and the alternative oxidase, with none of these enzymes being coupled to ATP production (Figure 12.2). One of these mutants, which lacks complex I and complex IV activities, is *dum19 dum25*, isolated after mutagenic treatment of *dum19* (Remacle et al., 2001a).

Other mutants of this type possess extensive deletions of the left part of the genome (Duby and Matagne, 1999; Remacle et al., 2001b). This is the case for the complex *dum24* mutant, which has deleted and dimeric molecules, the deletion including at least *cob*, *nd4*, and the 3′ end of *nd5* (Duby and Matagne, 1999). A detailed analysis of this mutant as well as of the *dum1* and *dum2* mutants showed that the deletion also extends to the left terminal repeat structure including the left 86 nt repeat (Duby et al., 2001). As the 86 nt repeats have been proposed to participate in replication of the terminal region (Vahrenholz et al., 1993, see above), the deletion of the left one could prevent the complete replication of the monomer and explain why dimers resulting from fusions between identical or differentially deleted monomers are always found in the mutants. Dimeric molecules that have intact terminal structures at both ends would be the only replicative forms of the genome, thus the reason for persistence of monomeric forms is unclear.

F. Complex I assembly in wild type and in *nd* mutants

In all eukaryotes, NADH:ubiquinone oxidoreductase (complex I, EC 1.6.5.3) is the largest enzyme in the respiratory chain (700–1000 kD). It is composed of two arms, one located in the inner mitochondrial membrane and the other protruding into the matrix. The membrane domain contains all the mitochondrially encoded subunits and bears the ubiquinone binding site(s) and the proton pumping machinery. The hydrophilic arm is responsible for NADH dehydrogenase activity and is composed of nucleus-encoded subunits that contain a non-covalently bound flavin mononucleotide molecule and 8–9 [Fe-S] clusters. As discussed in Chapter 13, complex I has been purified from a variety of organisms including *Chlamydomonas*. However, probably because of its large number of subunits and very large hydrophobic domain, only the peripheral arm of the corresponding complex of bacteria, which is much simpler in organization, has been crystallized (Sazanov and Hinchliffe, 2006).

Blue native polyacrylamide gel electrophoresis has been used to separate active respiratory complexes of mitochondria from various organisms (Schagger and von Jagow, 1991; Jansch et al., 1996). The technique has been successfully applied to *Chlamydomonas*, by using membrane-enriched fractions or purified mitochondria. Complex I can be detected (a) by Coomassie Blue staining of the gel, (b) by its NADH dehydrogenase activity using NADH and Nitroblue Tetrazolium as electron donor and acceptor, or (c) by immunoblotting with an antiserum against whole complex I from

Neurospora crassa. Alternatively, complex I can be separated from the other respiratory complexes by sucrose gradient centrifugation.

As shown in Table 12.1, the absence of intact ND1 (*dum20*) or ND6 (*dum17*) prevents assembly of the 950 kD complete complex I. In contrast, the loss of ND4 (*dum22*) or ND4 and ND5 (*dum24*) leads to the formation of a sub-complex of 700 kD bearing NADH dehydrogenase activity and present in reduced amounts (Duby et al., 2001; Cardol et al., 2002). In bovine complex I, ND4 and ND5 have been assigned to the distal end of the membrane domain, whereas ND1 and ND6 have been located in a connecting module between the hydrophilic and the hydrophobic arms (Sazanov et al., 2000; Carroll et al., 2003). Results with *Chlamydomonas* fit well with these locations. Loss of ND4 or ND5 would lead to the failure to assemble a hydrophobic 250-kD module, but would allow the assembly of a subcomplex loosely bound to the membrane and still retaining NADH dehydrogenase activity. In contrast, loss of a subunit of the connecting module (ND1 or ND6) would prevent the assembly of the whole complex.

In contrast, *dum25* mitochondria still contain intact complex I retaining NADH dehydrogenase activity but lacking NADH:ubiquinone activity. This indicates that the two amino acids of ND1 that are lost in this mutant are important for ubiquinone reduction but not for assembly. Finally, the *dum5* mutant exhibits reduced amounts of complex I, indicating that its assembly depends on the availability of ND5 (Cardol et al., 2002).

Mitochondrial biolistic transformation permits introduction of mutations into *nd* genes. The first mutation introduced was a deletion of *nd4* codons 2–24, which are believed to encode part of the first transmembrane segment. This Δ*nd4* mutant, like mutants with larger deletions such as *dum22*, has a subcomplex of 700 kD. This suggests that the deleted part of ND4 plays an important role in the assembly of the 250-kD module, or in its attachment to the 700-kD subcomplex. However, one cannot rule out a synthesis defect of ND4 due to a modified nucleotide context downstream of the initiation codon, even though the *nd4* mRNA level is unaffected. This deletion was also introduced in the *dum25* mutant, in which two codons of *nd1* were deleted. The *dum25*/Δ*nd4* mutant, like the Δ*nd4* mutant, forms a subcomplex of 700 kD. ND1 is thought to be a component of the 700-kD subcomplex, thus the 2-codon deletion in *nd1* which does not destabilize the intact complex I (Cardol et al., 2002), also has no effect on the stability of the 700-kD subcomplex (Remacle et al., 2006).

G. Suppressor mutations

Two spontaneous leaky revertants derived from the two frameshift mutants of the *cox1* gene, *dum18* and *dum19*, have been isolated (Remacle et al., 1998; Matagne and Baurain, 2001). An intragenic suppressor was found in the *dum18 su11* revertant, close to the original *dum18* mutation. In *dum18 su11*, the addition of one T base is maintained at codon 145 and a single

A → C substitution is present at the next codon (Remacle et al., 1998). This transversion introduces an alanine codon GCG at codon 146, instead of the glutamic acid GAG found in the *dum18* sequence. As GCG codons are very rare in the *Chlamydomonas* mitochondrial genome, it has been postulated that the partial frameshift suppression is due to the occasional abnormal translocation of the ribosome (+1 base shift) facilitated by both the run of T bases and the low level or weak interaction of alanyl-tRNA (Remacle et al., 1998). This very unusual mechanism of suppression is reminiscent of a situation encountered in the bacterium *Salmonella typhimurium* (Atkins et al., 1983).

An extragenic suppressor of the *dum19* mutant has also been characterized. The suppressor mutation is an A1090G substitution in the *L5* module of the mitochondrial LSU rRNA gene (Matagne and Baurain, 2001). The suppressor mutation affects the conserved GTPase-associated domain of the LSU rRNA. Surprisingly, the substitution A1098C in the same domain is found in both the revertant and the phenotypically dark$^-$ *dum19* mutant. It is not clear whether the change at position 1098 is also required for the expression of the suppressor phenotype. A mutation affecting an unknown nuclear gene and also involved in the suppression has been identified by genetic analysis. When this nuclear mutation is associated with the mitochondrial suppressor mutation, the capacity to grow in the dark of the *dum19* mutant is significantly increased (Matagne and Baurain, 2001).

A strain carrying both A1090G and A1098C in the *L5* gene but wild-type for the *cox1* gene sequence has been isolated (*m* mutant). The activities of complexes I, III, and IV, which contain mitochondrially encoded subunits, are reduced. In contrast, the activities of complex II and alternative oxidase, both encoded exclusively by nuclear genes, is not modified in the *m* mutant. The two substitutions do not suppress the dark$^-$ phenotype of other frameshift or UGA stop mutations, showing that the suppressor effect is dependent on the mtDNA context (Remacle et al., 2004).

Genetic data have been presented supporting the idea that two chloroplast suppressor mutations (*SUC1* and *SUC2*) act on a mitochondrial mutation (*LD1*) reducing heterotrophic growth (Bennoun and Delosme, 1999). *SUC2* also suppresses a point mutation in the chloroplast *rbcL* gene. The authors suspect that *SUC2* is probably an informational suppressor, acting in both the chloroplast and in mitochondria.

IV. MITOCHONDRIAL GENOME TRANSMISSION

A. Transmission of mitochondrial genes in meiotic zygotes

In contrast to the chloroplast genome which is preferentially transmitted to the meiotic progeny by the mt^+ or maternal parent, mtDNA is most often inherited from the mt^- or paternal parent. This was first demonstrated in crosses between *C. reinhardtii* and *C. smithii* by using RFLPs that differentiate the two species (Boynton et al., 1987; Matagne et al., 1989). Point mutations affecting mitochondrial genes follow the same rule. The percentage of meiotic products inheriting the mitochondrial marker of *mt*$^+$ origin is most often null, or lower than 3% (Bennoun et al., 1991; Dorthu et al., 1992; Colin et al., 1995). However, the uniparental paternal transmission of the mitochondrial genome can be strongly biased when the *mt*$^-$ parent contains deleted mtDNA molecules. In that case, the percentage of mtDNA of *mt*$^+$ origin can reach 16–28% (Matagne et al., 1989; Dorthu et al., 1992).

The fate of mtDNA of *mt*$^+$ and *mt*$^-$ origins was analyzed in zygotes and during maturation of the zygospore. Beckers et al. (1991) found that the disappearance of mtDNA from the *mt*$^+$ strain is rather slow. MtDNAs of both *mt*$^+$ and *mt*$^-$ origins persist in the zygote 24 hours after mating. During maturation of zygotes into zygospores in the dark, the amount of *mt*$^+$ mtDNA progressively decreases whereas *mt*$^-$ mtDNA is retained. MtDNA of *mt*$^+$ origin totally disappears 24 hours after transfer of zygospores to light to induce meiosis (Beckers et al., 1991). Nishimura et al. (1998) confirmed the slow disappearance of *mt*$^+$ mtDNA by showing that mitochondrial nucleoids of *mt*$^+$ origin are still present 3 days after mating and based on that, concluded that transmission of mtDNA is biparental in meiotic progeny. Subsequently, however, it was determined that mtDNA from *mt*$^+$ origin is eliminated between 6 and 12 hours after transfer of zygospores to light, just after nuclear DNA synthesis and before the first nuclear division (Aoyama et al., 2006).

B. Transmission of mitochondrial genes in vegetative zygotes

In *Chlamydomonas*, 1–10% of zygotes do not undergo meiosis and instead divide mitotically to give rise to stable diploid progeny (Ebersold, 1967). In crosses between strains that differ by several physical markers and that both possess or both lack the α group I intron in the *cob* gene, three types of diploid clones are produced: biparental clones that carry markers of both origins, uniparental clones with markers from the *mt*$^-$ parent only, and less frequently uniparental clones with markers from the *mt*$^+$ parent. Biparental clones show a strong bias for RFLP markers from the *mt*$^-$ parent (Remacle et al., 1990; Remacle and Matagne, 1993). A similar type of transmission is observed in crosses between wild type and *mud2* (myxothiazol resistant) (Bennoun et al., 1991). In biparental clones, recombinant genomes are frequently produced, suggesting that mitochondria fuse in vegetative zygotes. During successive mitotic divisions, the various types of DNA copies – paternal, maternal, and recombinant – segregate and after 15 divisions, most of the cells are homoplasmic, that is they contain only one type of mitochondrial genome. Recombination and segregation are also properties of diploid clones produced by artificial fusions of identical or opposite mating-type cells (Remacle and Matagne, 1993).

In crosses between *C. smithii* (α^+ strain, containing the group I intron in *cob*) and *C. reinhardtii* (α^-, intron-free *cob*), the transmission pattern is different and influenced by the mobile α group I intron (Boynton et al., 1987; Remacle et al., 1990; Remacle and Matagne, 1993). The conversion of intron-less molecules into intron-plus is very frequent and in 94% of cases, the conversion of α^- to α^+ is accompanied by the co-conversion of *mud2* located 27 bp from the insertion site of the intron. In both crosses and artificial fusions, the conversion of α^- to α^+ also influences the transmission of more distant markers, located 0.5 – 5 kb from the intron insertion site (Remacle and Matagne, 1993). The large region of co-conversion in *Chlamydomonas* mitochondria contrasts with the shorter region of co-conversion found for the ω group I intron located in the mitochondrial LSU rRNA gene of *S. cerevisiae* (Jacquier and Dujon, 1985). In that case, the co-conversion region between ω^+ and ω^- strains is limited to a few hundred bp. A long co-conversion track is also associated with the *C. eugametos* LSU.5 intron from the chloroplast genome (Bussieres et al., 1996).

The conversion of α^- into α^+ molecules is related to the action of the intronic endonuclease I-*Csm*I (Colleaux et al., 1990; Ma et al., 1992a). I-*Csm*I was purified and tested *in vitro* for cleavage of the *cob* intron-free sequence. The recognition sequence resides between 12 nt upstream and 12 nt downstream of the intron insertion site, and cleavage occurs 5 nt and 1 nt downstream of the intron insertion site on the coding and non-coding strands, respectively, creating a 4 nt 3′ overhang (Kurokawa et al., 2005). Site-directed mutagenesis has shown that the majority of substitutions that block substrate cleavage are located between −5 and +11 in relation to the insertion site. The span of critical bases is thus not centered at the intron insertion site, but is spread almost symmetrically with respect to the cleavage point of the coding and non-coding strands (Kurokawa et al., 2005). Recognition properties of I-*Csm*I were also tested *in vivo*. For that purpose, hybrid diploid cells were isolated by crossing α^+ *C. smithii* to α^- *C. reinhardtii* containing mutated target sequences. Surprisingly, the intron invasion frequency was much higher than expected from the *in vitro* cleavage frequency of the respective mutated substrates and again, markers as far as 5 kb from the insertion site were co-converted. It was proposed that the principal bottleneck for lateral intron transmission was not the sequence specificity of the homing enzyme, but instead the rare occurrence of interspecific cell fusions (Kurokawa et al., 2006).

C. Transmission of mitochondrial genes in cytoductants

By using appropriate genetic markers, haploid cytoductants that harbor the nuclear genotype of one parent and the chloroplast marker of the other parent can be readily isolated (Matagne et al., 1991). Many haploid cytoductants contain chloroplast markers from both parents and as in diploids, these DNA copies recombine frequently (Matagne et al., 1991). These observations have been extended to the mitochondrial genome by Bennoun et al. (1992), who showed that *mud2* could be efficiently transferred through cytoduction.

D. Mapping mutations by recombination analysis

More than 75% of vegetative zygotes that result from crosses between strains mutated in *cob* and *cox1* segregate recombinant wild-type cells able to grow in the dark, attesting that recombination events between mtDNAs of different origins are extremely frequent (Remacle et al., 1995). This raises the question of how the recombination frequencies are related to distances. These frequencies are calculated on the basis of the occurrence of wild-type recombinant dark$^+$ clones within a large population of vegetative diploid clones, after mitotic segregation had produced homoplasmic diploid progeny. Results involving mutations in *cob* (*dum11* and *dum15*) and *cox1* (*dum18* and *dum19*) led to an estimated recombination rate of 3% per kb (Remacle et al., 1995). This value was confirmed and used to map genetically other mutants (Remacle et al., 2001a,b; Cardol et al., 2002). The recombination rate is higher than that of the chloroplast genome (1% per kb) (Newman et al., 1990) but is much lower than that of the *S. cerevisiae* mitochondrial genome (15–25% per kb) (Dujon, 1981). This latter value is the maximum recombination rate for multicopy systems and explains why in *S. cerevisiae,* only mutation sites that are very close to each other can be mapped by recombination analysis. In contrast, the situation is much more favorable in *Chlamydomonas* since the maximal recombination frequency is reached only at 10 kb, representing two-thirds of the mitochondrial genome.

V. MITOCHONDRIAL TRANSFORMATION

In the first report of mitochondrial transformation in *Chlamydomonas*, a mutant (*dum1*) deleted for the left terminus and *cob* gene (1.5 kb deletion) was successfully transformed to respiratory competence with partially purified mtDNA from *C. reinhardtii* or *C. smithii* using a biolistic device (Randolph-Anderson et al., 1993). Some years later, biolistics were again used to transform the same kind of deletion mutant, with purified mtDNA or cloned mtDNA fragments (Yamasaki et al., 2005). In both cases, a wild-type mitochondrial genome replaced the deleted genome after selection of transformants under heterotrophic conditions (dark + acetate). Transformation efficiency was low (0.4–3 transformants/μg DNA) and did not enable any genetic manipulation of the mitochondrial genome. Moreover, transformation of point mutants failed and it was concluded that only mitochondrial genomes deleted for their left end could be transformed (Yamasaki et al., 2005).

Subsequently, biolistic transformation was optimized using cloned mtDNA or PCR fragments (Remacle et al., 2006). A mutant carrying a 1.2 kb deletion including the left terminus and part of the *cob* gene (*dum11*) could be rescued after selection in the dark, as could a double frameshift mutant in the *cox1* and *nd1* genes (*dum19 dum25*). High transformation efficiency was achieved (100 − 250 transformants/μg DNA), the best results being obtained with linearized plasmid DNA. Whatever the recipient strain or the construct used, the selection process was extremely slow and even after 2 months in the dark, heteroplasmic clones could still be found. To speed up the selection process, several attempts were made to develop a co-transformation strategy allowing a primary selection in the light, by bombarding the recipient strain simultaneously with two plasmids, one carrying a nuclear marker and the other containing the mtDNA of interest. Nuclear transformants were first selected in the light and then tested for their growth in the dark or by TTC staining to detect concomitant mitochondrial transformation events. This type of selection, commonly used for yeast transformation (Bonnefoy et al., 2007), was not successful (Remacle et al., 2006). A curious molecular feature of recombination following mitochondrial transformation in *Chlamydomonas* is that a strain lacking the *cob* gene and the left terminus could be rescued, even if the transforming DNA is nearly completely devoid of the left terminus. This suggests that the right terminus can be copied to reconstruct the left terminus by recombination. In addition, both non-deleterious and deleterious mutations could be introduced. Myxothiazol-resistant transformants have been created by introducing a nucleotide substitution into the *cob* gene. Similarly, an in-frame deletion of 23 codons was created in the *nd4* genes of both deleted and frameshift mutants (see sections III.D and III.F). These results open the way to reverse genetics in *Chlamydomonas* mitochondria and more specifically, to site-directed mutagenesis of mitochondrially encoded subunits of complex I. This is of special interest because the yeast *S. cerevisiae*, whose mitochondrial genome can be manipulated virtually at will, lacks complex I.

ACKNOWLEDGMENTS

This work was supported by grants from the Belgian FRFC-FNRS (2.4587.04 and 2.4582.05 to C.R. and F.4735.06 to P.C). P.C. is post-doctoral researcher of the Belgian FNRS.

REFERENCES

Alexander, N.J., Gillham, N.W., and Boynton, J.E. (1974). The mitochondrial genome of *Chlamydomonas*. Induction of minute colony mutations by acriflavine and their inheritance. *Mol. Gen. Genet.* **130**, 275–290.

Aoyama, H., Hagiwara, Y., Misumi, O., Kuroiwa, T., and Nakamura, S. (2006). Complete elimination of maternal mtDNA during meiosis resulting in the paternal inheritance of the mitochondrial genome in *Chlamydomonas* species. *Protoplasma* **228**, 231–242.

Atkins, J.F., Nichols, B.P., and Thompson, S. (1983). The nucleotide sequence of the first externally suppressible-1 frameshift mutant, and of some nearby leaky frameshift mutants. *EMBO J.* **2**, 1345–1350.

Beckers, M.C., Munaut, C., Minet, A., and Matagne, R.F. (1991). The fate of mtDNAs of mt^+ and mt^- origin in gametes and zygotes of *Chlamydomonas*. *Curr. Genet.* **20**, 239–243.

Bennoun, P. and Delosme, M. (1999). Chloroplast suppressors that act on a mitochondrial mutation in *Chlamydomonas reinhardtii*. *Mol. Gen. Genet.* **262**, 85–89.

Bennoun, P., Delosme, M., and Kück, U. (1991). Mitochondrial genetics of *Chlamydomonas reinhardtii*: Resistance mutations marking the cytochrome *b* gene. *Genetics* **127**, 335–343.

Bennoun, P., Delosme, M., Godehardt, I., and Kück, U. (1992). New tools for mitochondrial genetics of *Chlamydomonas reinhardtii*: Manganese mutagenesis and cytoduction. *Mol. Gen. Genet.* **234**, 147–154.

Boer, P.H. and Gray, M.W. (1986). The URF 5 gene of *Chlamydomonas reinhardtii* mitochondria: DNA sequence and mode of transcription. *EMBO J.* **5**, 21–28.

Boer, P.H. and Gray, M.W. (1988a). Genes encoding a subunit of respiratory NADH dehydrogenase (ND1) and a reverse transcriptase-like protein (RTL) are linked to ribosomal RNA gene pieces in *Chlamydomonas reinhardtii* mtDNA. *EMBO J.* **7**, 3501–3508.

Boer, P.H. and Gray, M.W. (1988b). Scrambled ribosomal RNA gene pieces in *Chlamydomonas reinhardtii* mtDNA. *Cell* **55**, 399–411.

Boer, P.H. and Gray, M.W. (1988c). Transfer RNA genes and the genetic code in *Chlamydomonas reinhardtii* mitochondria. *Curr. Genet.* **14**, 583–590.

Boer, P.H. and Gray, M.W. (1991). Short dispersed repeats localized in spacer regions of *Chlamydomonas reinhardtii* mtDNA. *Curr. Genet.* **19**, 309–312.

Bonnefoy, N., Remacle, C., and Fox, T.D. (2007). Genetic transformation of *Saccharomyces cerevisiae* and *Chlamydomonas* mitochondria. In: *Mitochondria*, Vol. 80 (L.A. Pon and E. Schon, Eds.), pp. 525–548. Academic Press, San Diego.

Boynton, J.E. and Gillham, N.W. (1996). Genetics and transformation of mitochondria in the green alga *Chlamydomonas*. *Methods Enzymol.* **264**, 279–296.

Boynton, J.E., Harris, E.H., Burkhart, B.D., Lamerson, P.M., and Gillham, N.W. (1987). Transmission of mitochondrial and chloroplast genomes in crosses of *Chlamydomonas*. *Proc. Natl. Acad. Sci. U. S. A.* **84**, 2391–2395.

Bullerwell, C.E. and Gray, M.W. (2004). Evolution of the mitochondrial genome: Protist connections to animal, fungi and plants. *Curr. Opin. Microbiol.* **7**, 528–534.

Burger, G., Gray, M.W., and Lang, B.F. (2003). Mitochondrial genomes: Anything goes. *Trends Genet.* **19**, 709–713.

Bussieres, J., Lemieux, C., Lee, R.W., and Turmel, M. (1996). Optional elements in the chloroplast DNAs of *Chlamydomonas eugametos* and *C. moewusii*: Unidirectional gene conversion and co-conversion of adjacent markers in high-viability crosses. *Curr. Genet.* **30**, 356–365.

Cardol, P., Matagne, R.F., and Remacle, C. (2002). Impact of mutations affecting ND mitochondria-encoded subunits on the activity and assembly of complex I in *Chlamydomonas*. Implication for the structural organization of the enzyme. *J. Mol. Biol.* **319**, 1211–1221.

Carroll, J., Fearnley, I.M., Shannon, R.J., Hirst, J., and Walker, J.E. (2003). Analysis of the subunit composition of complex I from bovine heart mitochondria. *Mol. Cell. Proteomics* **2**, 117–126.

Clifton, S.W., Minx, P., Fauron, C.M., Gibson, M., Allen, J.O., Sun, H., Thompson, M., Barbazuk, W.B., Kanuganti, S., Tayloe, C., Meyer, L., Wilson, R.K., and Newton, K.J. (2004). Sequence and comparative analysis of the maize NB mitochondrial genome. *Plant Physiol.* **136**, 3486–3503.

Colin, M., Dorthu, M.P., Duby, F., Remacle, C., Dinant, M., Wolwertz, M.R., Duyckaerts, C., Sluse, F., and Matagne, R.F. (1995). Mutations affecting the mitochondrial genes encoding the cytochrome oxidase subunit I and apocytochrome *b* of *Chlamydomonas reinhardtii*. *Mol. Gen. Genet.* **249**, 179–184.

Colleaux, L., Michel-Wolwertz, M.R., Matagne, R.F., and Dujon, B. (1990). The apocytochrome *b* gene of *Chlamydomonas smithii* contains a mobile intron related to both *Saccharomyces* and *Neurospora* introns. *Mol. Gen. Genet.* **223**, 288–296.

Denovan-Wright, E.M. and Lee, R.W. (1992). Comparative analysis of the mitochondrial genomes of *Chlamydomonas eugametos* and *Chlamydomonas moewusii*. *Curr. Genet.* **21**, 197–202.

Denovan-Wright, E.M. and Lee, R.W. (1995). Evidence that the fragmented ribosomal RNAs of *Chlamydomonas* mitochondria are associated with ribosomes. *FEBS Lett.* **370**, 222–226.

Denovan-Wright, E.M., Nedelcu, A.M., and Lee, R.W. (1998). Complete sequence of the mtDNA of *Chlamydomonas eugametos*. *Plant Mol. Biol.* **36**, 285–295.

Dorthu, M.P., Remy, S., Michel-Wolwertz, M.R., Colleaux, L., Breyer, D., Beckers, M.C., Englebert, S., Duyckaerts, C., Sluse, F.E., and Matagne, R.F. (1992). Biochemical, genetic and molecular characterization of new respiratory-deficient mutants in *Chlamydomonas reinhardtii*. *Plant Mol. Biol.* **18**, 759–772.

Duby, F. and Matagne, R.F. (1999). Alteration of dark respiration and reduction of phototrophic growth in a mtDNA deletion mutant of *Chlamydomonas* lacking *cob*, *nd4*, and the 3′ end of *nd5*. *Plant Cell* **11**, 115–125.

Duby, F., Cardol, P., Matagne, R.F., and Remacle, C. (2001). Structure of the telomeric ends of mt DNA, transcriptional analysis and complex I assembly in the *dum24* mitochondrial mutant of *Chlamydomonas reinhardtii*. *Mol. Genet. Genomics*, **266**, 109–114.

Dujon, B. (1981). Mitochondrial genetics and functions. In: *The Molecular Biology of the Yeast Saccharomyces cerevisiae-Life Cycle and Inheritance* (J.N. Strathern, E.W. Jones, and J.R. Broach, Eds.), pp. 505–635. Cold Spring Harbor Laboratory Press, Cold-Spring Harbor, New York.

Ebersold, W. (1967). *Chlamydomonas reinhardtii*: Heterozygous diploid strains. *Science* **157**, 447–449.

Ehara, T., Osafune, T., and Hase, E. (1995). Behavior of mitochondria in synchronized cells of *Chlamydomonas reinhardtii* (Chlorophyta). *J. Cell Sci.* **108**, 499–507.

Fan, J. and Lee, R.W. (2002). Mitochondrial genome of the colorless green alga *Polytomella parva*: Two linear DNA molecules with homologous inverted repeat termini. *Mol. Biol. Evol.* **19**, 999–1007.

Fan, J., Schnare, M.N., and Lee, R.W. (2003). Characterization of fragmented mitochondrial ribosomal RNAs of the colorless green alga *Polytomella parva*. *Nucleic Acids Res.* **31**, 769–778.

Gabay-Laughnan, S. and Newton, K.J. (2005). Mitochondrial mutations in maize. *Maydica* **50**, 349–359.

Gillham, N.W. (1994). *Organelle Genes and Genomes*. Oxford University Press, New York.

Gillham, N.W., Boynton, J.E., and Harris, E.H. (1987). Specific elimination of mtDNA from *Chlamydomonas* by intercalating dyes. *Curr. Genet.* **12**, 41–47.

Grant, D. and Chiang, K.S. (1980). Physical mapping and characterization of *Chlamydomonas* mtDNA molecules: Their unique ends, sequence homogeneity, and conservation. *Plasmid* **4**, 82–96.

Gray, M.W. and Boer, P.H. (1988). Organization and expression of algal (*Chlamydomonas reinhardtii*) mtDNA. *Philos. Trans. R. Soc. Lond. B Biol. Sci.* **319**, 135–147.

Gutierres, S., Sabar, M., Lelandais, C., Chetrit, P., Diolez, P., Degand, H., Boutry, M., Vedel, F., de Kouchkovsky, Y., and De Paepe, R. (1997). Lack of mitochondrial and nuclear-encoded subunits of complex I and alteration of the respiratory chain in *Nicotiana sylvestris* mitochondrial deletion mutants. *Proc. Natl. Acad. Sci. U. S. A.* **94**, 3436–3441.

Gutierres, S., Combettes, B., De Paepe, R., Mirande, M., Lelandais, C., Vedel, F., and Chetrit, P. (1999). In the *Nicotiana sylvestris* CMSII mutant, a recombination-mediated change 5′ to the first exon of the mitochondrial *nad1* gene is associated with lack of the NADH:ubiquinone oxidoreductase (complex I) NAD1 subunit. *Eur. J. Biochem.* **261**, 361–370.

Hiramatsu, T., Nakamura, S., Misumi, O., and Kuroiwa, T. (2006). Morphological changes in mitochondrial and chloroplast nucleoids and mitochondria during the *Chlamydomonas reinhardtii* (Chlorophyceae) cell cycle. *J. Phycol.* **42**, 1048–1058.

Jacquier, A. and Dujon, B. (1985). An intron-encoded protein is active in a gene conversion process that spreads an intron into a mitochondrial gene. *Cell* **41**, 383–394.

Jansch, L., Kruft, V., Schmitz, U.K., and Braun, H.P. (1996). New insights into the composition, molecular mass and stoichiometry of the protein complexes of plant mitochondria. *Plant J.* **9**, 357–368.

Karpova, O.V. and Newton, K.J. (1999). A partially assembled complex I in ND4-deficient mitochondria of maize. *Plant J.* **17**, 511–521.

Knoop, V. (2004). The mt DNA of land plants: Peculiarities in phylogenetic perspectives. *Curr. Genet.* **46**, 123–139.

Kroymann, J. and Zetsche, K. (1997). The apocytochrome-*b* gene in *Chlorogonium elongatum* (Chlamydomonadaceae): An intronic GIY-YIG ORF in green algal mitochondria. *Curr. Genet.* **31**, 414–418.

Kroymann, J. and Zetsche, K. (1998). The mitochondrial genome of *Chlorogonium elongatum* inferred from the complete sequence. *J. Mol. Evol.* **47**, 431–440.

Kubo, T., Nishizawa, S., Sugawara, A., Itchoda, N., Estiati, A., and Mikami, T. (2000). The complete nucleotide sequence of the mitochondrial genome of sugar beet (*Beta vulgaris* L.) reveals a novel gene for tRNA(Cys)(GCA). *Nucleic Acids Res.* **28**, 2571–2576.

Kück, U. and Neuhaus, H. (1986). Universal genetic code evidenced in mitochondria of *Chlamydomonas reinhardtii*. *Appl. Microbiol. Biotechnol.* **23**, 462–469.

Kurokawa, S., Bessho, Y., Higashijima, K., Shirouzu, M., Yokoyama, S., Watanabe, K.I., and Ohama, T. (2005). Adaptation of intronic homing endonuclease for successful horizontal transmission. *FEBS J.* **272**, 2487–2496.

Kurokawa, S., Yamasaki, T., Komatsu, T., Watanabe, K.I., and Ohama, T. (2006). Degenerated recognition property of a mitochondrial homing enzyme in the unicellular green alga *Chlamydomonas smithii*. *Plant Mol. Biol.* **62**, 141–150.

Lang, B.F., Burger, G., O'Kelly, C.J., Cedergren, R., Golding, G.B., Lemieux, C., Sankoff, D., Turmel, M., and Gray, M.W. (1997). An ancestral mtDNA resembling a eubacterial genome in miniature. *Nature* **387**, 493–497.

Ma, D.P., Yang, Y.W., King, Y.T., and Hasnain, S.E. (1989). Nucleotide sequence of cloned *nad4* (urf4) gene from *Chlamydomonas reinhardtii* mtDNA. *Gene* **85**, 363–370.

Ma, D.P., Yang, Y-W., King, T.Y., and Hasnain, S.E. (1990). The mitochondrial apocytochrome *b* gene from *Chlamydomonas reinhardtii*. *Plant Mol. Biol.* **15**, 357–359.

Ma, D.P., King, Y.T., Kim, Y., and Luckett, W.S., Jr. (1992a). The group I intron of apocytochrome *b* gene from *Chlamydomonas smithii* encodes a site-specific endonuclease. *Plant Mol. Biol.* **18**, 1001–1004.

Ma, D.P., King, Y.T., Kim, Y., Luckett, W.S., Jr., Boyle, J.A., and Chang, Y.F. (1992b). Amplification and characterization of an inverted repeat from the *Chlamydomonas reinhardtii* mitochondrial genome. *Gene* **119**, 253–257.

Martin, A.N. and Li, Y. (2007). RNase MRP RNA and human genetic diseases. *Cell Res.* **173**, 219–226.

Matagne, R.F. and Baurain, D. (2001). A mutation in the GTPase domain of the large subunit rRNA is involved in the suppression of a -1 T frameshift mutation affecting a mitochondrial gene in *Chlamydomonas reinhardtii*. *Mol. Genet. Genomics* **266**, 103–108.

Matagne, R.F. and Remacle, C. (2002). The genetics and molecular biology of mitochondria in *Chlamydomonas*. In: *Recent Research Developments in Plant Biology*, Vol. 2 (S.G. Pandalai, Ed.), pp. 15–32. Research Signpost, Trivandrum, India.

Matagne, R.F., Michel-Wolwertz, M.R., Munaut, C., Duyckaerts, C., and Sluse, F. (1989). Induction and characterization of mtDNA mutants in *Chlamydomonas reinhardtii*. *J. Cell Biol.* **108**, 1221–1226.

Matagne, R.F., Remacle, C., and Dinant, M. (1991). Cytoduction in *Chlamydomonas reinhardtii*. *Proc. Natl. Acad. Sci. U. S. A.* **88**, 7447–7450.

Michaelis, G., Vahrenholz, C., and Pratje, E. (1990). MtDNA of *Chlamydomonas reinhardtii*: The gene for apocytochrome *b* and the complete functional map of the 15.8 kb DNA. *Mol. Gen. Genet.* **223**, 211–216.

Nedelcu, A.M. and Lee, R.W. (1998a). A degenerate group II intron in the intronless mitochondrial genome of *Chlamydomonas reinhardtii*: Evolutionary implications. *Mol. Biol. Evol.* **15**, 918–922.

Nedelcu, A.M. and Lee, R.W. (1998b). Short repetitive sequences in green algal mitochondrial genomes: Potential roles in mitochondrial genome evolution. *Mol. Biol. Evol.* **15**, 690–701.

Newman, S.M., Boynton, J.E., Gillham, N.W., Randolph-Anderson, B.L., Johnson, A.M., and Harris, E.H. (1990). Transformation of chloroplast ribosomal RNA genes in *Chlamydomonas*: Molecular and genetic characterization of integration events. *Genetics* **126**, 875–888.

Nishimura, Y., Higashiyama, T., Suzuki, L., Misumi, O., and Kuroiwa, T. (1998). The biparental transmission of the mitochondrial genome in *Chlamydomonas reinhardtii* visualized in living cells. *Eur. J. Cell Biol.* **77**, 124–133.

Notsu, Y., Masood, S., Nishikawa, T., Kubo, N., Akiduki, G., Nakazono, M., Hirai, A., and Kadowaki, K. (2002). The complete sequence of the rice (*Oryza sativa* L.)

mitochondrial genome: Frequent DNA sequence acquisition and loss during the evolution of flowering plants. *Mol. Genet. Genomics,* **268**, 434–445.

Pla, M., Mathieu, C., De Paepe, R., Chetrit, P., and Vedel, F. (1995). Deletion of the last two exons of the mitochondrial *nad7* gene results in lack of the NAD7 polypeptide in a *Nicotiana sylvestris* CMS mutant. *Mol. Gen. Genet.* **248**, 79–88.

Pineau, B., Mathieu, C., Gerard-Hirne, C., De Paepe, R., and Chetrit, P. (2005). Targeting the NAD7 subunit to mitochondria restores a functional complex I and a wild type phenotype in the *Nicotiana sylvestris* CMS II mutant lacking *nad7*. *J. Biol. Chem.* **280**, 25994–26001.

Popescu, C.E. and Lee, R.W. (2007). Mitochondrial genome sequence evolution in *Chlamydomonas*. *Genetics* **175**, 819–826.

Pratje, E., Schnierer, S., and Dujon, B. (1984). MtDNA of *Chlamydomonas reinhardtii*: The DNA sequence of a region showing homology with mammalian URF2. *Curr. Genet.* **9**, 75–82.

Randolph-Anderson, B.L., Boynton, J.E., Gillham, N.W., Harris, E.H., Johnson, A.M., Dorthu, M.P., and Matagne, R.F. (1993). Further characterization of the respiratory deficient *dum-1* mutation of *Chlamydomonas reinhardtii* and its use as a recipient for mitochondrial transformation. *Mol. Gen. Genet.* **236**, 235–244.

Remacle, C. and Matagne, R.F. (1993). Transmission, recombination and conversion of mitochondrial markers in relation to the mobility of a group I intron in *Chlamydomonas*. *Curr. Genet.* **23**, 518–525.

Remacle, C. and Matagne, R. (1998). Mitochondrial genetics. In: *The Molecular Biology of Chloroplasts and Mitochondria in Chlamydomonas* (J. Rochaix, M. Goldschmidt-Clermont, and S. Merchant, Eds.), pp. 661–674. Kluwer Academic Publishers, Dordrecht.

Remacle, C., Bovie, C., Michel-Wolwertz, M.R., Loppes, R., and Matagne, R.F. (1990). Mitochondrial genome transmission in *Chlamydomonas* diploids obtained by sexual crosses and artificial fusions: Role of the mating type and of a 1 kb intron. *Mol. Gen. Genet.* **223**, 180–184.

Remacle, C., Colin, M., and Matagne, R.F. (1995). Genetic mapping of mitochondrial markers by recombinational analysis in *Chlamydomonas reinhardtii*. *Mol. Gen. Genet.* **249**, 185–190.

Remacle, C., Colin, M., and Matagne, R.F. (1998). Suppression of a +1 T mutation by a nearby substitution in the mitochondrial *cox1* gene of *Chlamydomonas reinhardtii*: A new type of frameshift suppression in an organelle genome. *Mol. Gen. Genet.* **259**, 294–298.

Remacle, C., Baurain, D., Cardol, P., and Matagne, R.F. (2001a). Mutants of *Chlamydomonas reinhardtii* deficient in mitochondrial complex I: Characterization of two mutations affecting the *nd1* coding sequence. *Genetics* **158**, 1051–1060.

Remacle, C., Duby, F., Cardol, P., and Matagne, R.F. (2001b). Mutations inactivating mitochondrial genes in *Chlamydomonas reinhardtii*. *Biochem. Soc. Trans.* **29**, 442–446.

Remacle, C., Gloire, G., Cardol, P., and Matagne, R.F. (2004). Impact of a mutation in the mitochondrial LSU rRNA gene from *Chlamydomonas reinhardtii* on the activity and the assembly of respiratory-chain complexes. *Curr. Genet.* **45**, 323–330.

Remacle, C., Cardol, P., Coosemans, N., Gaisne, M., and Bonnefoy, N. (2006). High-efficiency biolistic transformation of *Chlamydomonas* mitochondria can be used to insert mutations in complex I genes. *Proc. Natl. Acad. Sci. U. S. A.* **103**, 4771–4776.

Rich, P.R., Mischis, L.A., Purton, S., and Wiskich, J.T. (2001). The sites of interaction of triphenyltetrazolium chloride with mitochondrial respiratory chains. *FEMS Microbiol. Lett.* **202**, 181–187.

Ryan, R., Grant, D., Chiang, K.S., and Swift, H. (1978). Isolation and characterization of mtDNA from *Chlamydomonas reinhardtii*. *Proc. Natl. Acad. Sci. U. S. A.* **75**, 3268–3272.

Sazanov, L.A. and Hinchliffe, P. (2006). Structure of the hydrophilic domain of respiratory complex I from *Thermus thermophilus*. *Science* **311**, 1430–1436.

Sazanov, L.A., Peak-Chew, S.Y., Fearnley, I.M., and Walker, J.E. (2000). Resolution of the membrane domain of bovine complex I into subcomplexes: Implications for the structural organization of the enzyme. *Biochemistry* **39**, 7229–7235.

Schagger, H. and von Jagow, G. (1991). Blue native electrophoresis for isolation of membrane protein complexes in enzymatically active form. *Anal. Biochem.* **199**, 223–231.

Schönfeld, C., Wobbe, L., Borgstädt, R., Kienast, A., Nixon, P.J., and Kruse, O. (2004). The nucleus-encoded protein MOC1 is essential for mitochondrial light acclimation in *Chlamydomonas reinhardtii*. *J. Biol. Chem.* **279**, 50366–50374.

Slater, T.F., Sawyer, B., and Sträuby, U. (1963). Studies on succinate-tetrazolium reductase systems. III. Points of coupling of four different tetrazolium salts. *Biochim. Biophys. Acta,* **77**, 383–393.

Sugiyama, Y., Watase, Y., Nagase, M., Makita, N., Yagura, S., Hirai, A., and Sugiura, M. (2005). The complete nucleotide sequence and multipartite organization of the tobacco mitochondrial genome: Comparative analysis of mitochondrial genomes in higher plants. *Mol. Genet. Genomics,* **272**, 603–615.

Tracy, R.L. and Stern, D.B. (1995). Mitochondrial transcription initiation: Promoter structures and RNA polymerases. *Curr. Genet.* **28**, 205–216.

Unseld, M., Marienfeld, J.R., Brandt, P., and Brennicke, A. (1997). The mitochondrial genome of *Arabidopsis thaliana* contains 57 genes in 366,924 nt. *Nat. Genet.* **15**, 57–61.

Vahrenholz, C., Riemen, G., Pratje, E., Dujon, B., and Michaelis, G. (1993). MtDNA of *Chlamydomonas reinhardtii*: The structure of the ends of the linear 15.8-kb genome suggests mechanisms for DNA replication. *Curr. Genet.* **24**, 241–247.

Yamasaki, T., Kurokawa, S., Watanabe, K.I., Ikuta, K., and Ohama, T. (2005). Shared molecular characteristics of successfully transformed mitochondrial genomes in *Chlamydomonas reinhardtii*. *Plant Mol. Biol.* **58**, 515–527.

Yu, F., Fu, A., Aluru, M., Park, S., Xu, Y., Liu, H., Liu, X., Foudree, A., Nambogga, M., and Rodermel, S. (2007). Variegation mutants and mechanisms of chloroplast biogenesis. *Plant Cell Environ.* **30**, 350–365.

CHAPTER 13

Oxidative Phosphorylation: Building Blocks and Related Components

Pierre Cardol,[1,2] Francisco Figueroa,[3] Claire Remacle,[1] Lars-Gunnar Franzén[4] and Diego González-Halphen[3]

[1]Genetics of Microorganisms, [2]Plant Biochemistry and Photobiology, Institute of Plant Biology B22, University of Liège, Belgium,
[3]Departamento de Genética Molecular, Instituto de Fisiología Celular, Universidad Nacional Autónoma de México, México,
[4]School of Business and Engineering, Naturrum, University of Halmstad, Sweden

CHAPTER CONTENTS

I. INTRODUCTION

Oxidative phosphorylation (OXPHOS) is defined as an electron transfer chain driven by substrate oxidation that is coupled to the synthesis of ATP through an electrochemical transmembrane gradient (Figure 13.1). Historically, bovine heart mitochondria have been the system of choice for the structural characterization of eukaryotic OXPHOS complexes (Saraste, 1999), because they can be purified in relatively large quantities. The yeast *Saccharomyces cerevisiae*, which is amenable to a large variety of molecular genetic tools (Bonnefoy and Fox, 2002), has become the model organism to study the biogenesis of mitochondrial complexes (Barrientos et al., 2002; Herrmann and Neupert, 2003) and the effect of mutations on OXPHOS components (Fisher et al., 2004). By contrast, mitochondria of photosynthetic organisms have been poorly characterized from a biochemical point of view, mainly due to the difficulties in obtaining preparations free of chloroplast contaminants. Nevertheless, the characterization of *Arabidopsis* mitochondrial components through proteomic approaches has advanced significantly (Millar et al., 2005).

Chlamydomonas is an attractive model for studying mitochondria in unicellular photosynthetic organisms, because it allows combining classical biochemical approaches with genetics, and because photosynthesis-deficient mutants may be grown heterotrophically, and respiratory-deficient mutants as obligate photoautotrophs. Prior work with *Chlamydomonas* mitochondria involved the sequencing of its 15.8 kb linear genome (Gray and Boer, 1988; Michaelis et al., 1990; see Chapter 12), the biochemical characterization of enriched fractions of several OXPHOS complexes (Atteia, 1994), and the sequencing of several nuclear genes encoding polypeptides that participate in OXPHOS (Amati et al., 1988; Franzén and Falk, 1992; Atteia and Franzén, 1996; Nurani and Franzén, 1996; Pérez-Martínez et al., 2000, 2001; Dinant et al., 2001; Funes et al., 2002a; Atteia et al., 2003). In parallel, several *Chlamydomonas* mutants affected in mitochondrial components were characterized (Matagne et al., 1989; Dorthu et al., 1992; Colin et al., 1995; Duby and Matagne, 1999; Remacle et al., 2001, 2004; Cardol et al., 2002, 2006, 2008; Remacle et al., 2004; see Chapter 12), and procedures to obtain purified mitochondria were developed (Eriksson et al., 1995; Nurani and Franzén, 1996). In particular, wall-deficient mutants allowed the isolation of mitochondrial preparations free of thylakoid components. Thereafter, the major mitochondrial complexes of *Chlamydomonas* were resolved with Blue Native Polyacrylamide Gel Electrophoresis (BN-PAGE) (van Lis et al., 2003; Cardol et al., 2004), allowing N-terminal sequencing or tandem mass spectrometry analysis of individual subunits.

Genome sequencing has allowed the prediction of the *Chlamydomonas* OXPHOS proteome (Cardol et al., 2005), whose major components are shown schematically in Figure 13.1. This chapter summarizes knowledge of the OXPHOS constituents of *Chlamydomonas* and of the components

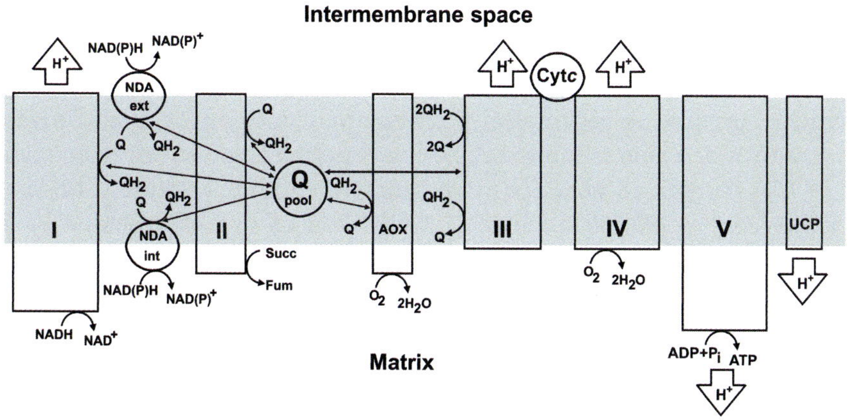

FIGURE 13.1 *The oxidative phosporylation (OXPHOS) system in the inner membrane of plant mitochondria. Electrons are transferred from NADH and NAD(P)H to coenzyme Q (Q) via type-I NADH: ubiquinone oxidoreductase or complex I (I) and type-II alternative NADH dehydrogenases (NDA; located on the internal (int) or external (ext) faces of the inner membrane). Some electrons from organic acid oxidation are also transferred to Q via the succinate dehydrogenase or complex II (II). Electrons are thereafter transferred from reduced coenzyme Q (QH_2) to cytochrome c (cyt c) via the ubiquinol: cytochrome c oxidoreductase or complex III (III) and then to oxygen via the cytochrome c oxidase or complex IV (IV). An alternative oxidase (AOX) can bypass complexes III and IV by transferring electrons directly from QH_2 to oxygen. The oxidoreduction reactions mediated by complexes I, III, and IV are coupled with protons (H^+) transfer from the matrix to the intermembrane space. The energy associated with the proton gradient is used for the production of ATP by complex V (ATP synthase) or dissipated by uncoupling proteins (UCP).*

involved in their biogenesis. We also address alternate dehydrogenases and oxidases which are particular to photosynthetic organisms, and several other mitochondrial components related to OXPHOS. When appropriate, we make reference to the components of *Polytomella* sp., a colorless alga closely related to *Chlamydomonas* (Pröschold et al., 2001; see also Volume 1, Chapter 1).

II. COMPLEXES OF THE MITOCHONDRIAL RESPIRATORY CHAIN

A. Complex I, NADH:ubiquinone oxidoreductase

Complex I (rotenone-sensitive NADH:ubiquinone oxidoreductase; EC 1.6.5.3) is the largest membrane-bound multisubunit complex of the mitochondrial respiratory chain. With an apparent molecular mass between 950 kD and 1000 kD, the *Chlamydomonas* enzyme comprises 42 proteins of molecular masses ranging from 7 kD to 77 kD (van Lis et al., 2003; Cardol et al., 2004). Complex I from several organisms has been characterized, including beef heart (45 subunits; Carroll et al., 2006), *Arabidopsis* and rice (42 subunits; Heazlewood et al., 2003; Meyer et al., 2008), and the fungi *Neurospora crassa* (39 subunits; Videira and Duarte, 2002; Marques et al.,

2005) and *Yarrowia lipolytica* (37 subunits; Abdrakhmanova et al., 2004). Five to nine Complex I subunits (the ND subunits) are usually encoded in the mitochondrial genome, whereas the remaining subunits are nuclear gene products. Because of its complexity, information on Complex I structure and mechanism of action is limited. A simpler enzyme is found in bacteria with only 14 subunits, all of which have counterparts in eukaryotic Complex I (Dupuis et al., 1998; Friedrich et al., 1998). Low resolution structures based on electron microscopy have revealed that mitochondrial NADH:ubiquinone oxidoreductase shares an L-shape with its bacterial counterpart, with one arm embedded in the mitochondrial inner membrane and another arm protruding into the matrix (see e.g. Guenebaut et al., 1998; Sazanov et al., 2003; Radermacher et al., 2006). In bacteria, the membrane arm is constituted by the homologues of the hydrophobic ND1, 2, 3, 4, 4L, 5, and 6 subunits, usually encoded in the mitochondrial genomes of eukaryotes. It also contains proton translocation machinery with at least one ubiquinone binding site sensitive to rotenone. The matrix arm components contribute to the NADH dehydrogenase segment of the electron pathway by binding a flavin mononucleotide (FMN) and nine iron-sulfur clusters (Hinchliffe and Sazanov, 2005).

Including the 14 highly conserved subunits homologous to the prokaryotic type-I NADH dehydrogenase enzyme, all eukaryotic enzymes contain at least 33 subunits in common (Cardol et al., 2005). *Chlamydomonas* Complex I thus contains nine components that are not found in all kingdoms: NUO21, NUOP1, NUOP3–6, and CAG1–3 (Table 13.1). While NUO21 and NUOP1 are found in land plants and fungi but not in mammals, NUOP3 and CAG1–3 seem to be common only to plants, and NUOP4–6 could be specific to the green algal lineage (Cardol et al., 2004, 2005, 2008). A polypeptide that comigrated with Complex I in *Chlamydomonas*, originally named NUOP6, was subsequently suggested to be a Complex V subunit in both *Chlamydomonas* and *Polytomella* sp., and renamed ASA7 (Vázquez-Acevedo et al., 2006).

The additional subunits associated with Complex I, the so-called supernumerary subunits, are considered to participate in assembly, stability, or regulation, rather than in enzyme activity (Friedrich et al., 1998; Heazlewood et al., 2003). CAG1–3 are three structurally related subunits representing gamma-type carbonic anhydrases (see Chapter 8, section III.C). In *Arabidopsis*, five gamma-type carbonic anhydrase subunits have been assigned to the membrane arm of Complex I (Parisi et al., 2004; Sunderhaus et al., 2006). One of them (NP_175159) has also been shown to be important for Complex I assembly and to be possibly involved in mitochondrial one-carbon metabolism (Perales et al., 2005). Single particle electron microscopy of Complex I from *Polytomella* and *Arabidopsis* suggested that these gamma-type carbonic anhydrase subunits could constitute a plant-specific spherical domain which is attached to the central part of the membrane arm of Complex I, and exposed to the matrix (Perales et al., 2005; Sunderhaus et al., 2006).

In plants, L-galactono-1,4-lactone dehydrogenase (GLDH) forms part of Complex I. GLDH catalyses the final oxidation of galactono-gamma-lactone to ascorbic acid (Ostergaard et al., 1997; Siendones et al., 1999); its activity decreases with leaf age and is modulated by the availability of oxidized cytochrome *c* (Bartoli et al., 2000). Ascorbate, the major soluble antioxidant and redox buffer in mammals and plants, plays important roles in plant development, photoprotection, and cell expansion (Smirnoff, 2000; Arrigoni and De Tullio, 2002). Complex I from *Arabidopsis* exhibits two forms of different mobilities in BN-PAGE. The form with higher mobility has associated GLDH, and the activity of this enzyme could be inhibited by rotenone (Heazlewood et al., 2003). These observations pointed towards the existence of a subpopulation of Complex I whose function could include regulating ascorbate synthesis by monitoring the rate of NADH-driven electron flow through Complex I (Millar et al., 2003). There is a GLDH protein homologue encoded in *Chlamydomonas* (EDP08950), whose possible association with Complex I remains to be explored.

Another Complex I-related enzyme is rhodanese (thiosulfate: cyanide sulfur transferase), which catalyses the transfer of a sulphane sulfur atom from thiosulfate to cyanide. Rhodanese is involved in cyanide detoxification, maintenance of the sulfate pool, and selenium metabolism, and participates in the formation of Fe-S clusters. Thus, it may regulate OXPHOS by controlling the status of Fe-S clusters of respiratory chain enzymes (Ogata and Volini, 1990). A 34.6-kD protein exhibiting rhodanese activity has been found associated with Complex I of the yeast *Y. lipolytica*. Deletion of the corresponding gene demonstrated that it is not essential for Complex I biosynthesis (Abdrakhmanova et al., 2005), but it could be important for the repair, regulation, or quality control of the Fe-S proteins of Complex I (Ogata and Volini, 1990). A gene encoding rhodanese is present in *Chlamydomonas* (EDO98946).

Only three chaperones involved in Complex I assembly have been identified to date, CIA30 and CIA84 of *N. crassa* (Kuffner et al., 1998), and a paralogue (B17.2L) of the bovine B17.2 structural subunit (Ogilvie et al., 2005). In *Chlamydomonas*, a gene encoding the CIA30 homologue is present (Table 13.1). Based on subcomplexes identified in various mutants, partial models of Complex I assembly have been proposed for human (Antonicka et al., 2003; Vogel et al., 2007), *Neurospora* (Videira and Duarte, 2002), and *Chlamydomonas* (Cardol et al., 2006, 2008). In *Chlamydomonas*, the analysis of mutants deficient in the synthesis of mitochondrial-encoded hydrophobic components ND1, ND4, ND5, and ND6 (Cardol et al., 2002, 2008; Remacle et al., 2006), or mutants that have lost the expression of nuclear genes encoding ND3 and ND4L (Cardol et al., 2006) led to the identification of two subcomplexes. A 200-kD soluble subcomplex, bearing a rotenone-insensitive NADH dehydrogenase activity, could correspond to a fragment of the matrix-exposed arm. In ND4 and ND5-deficient cells, a

Table 13.1 Complex I constituents and assembly factors

Chlamydomonas			Counterparts in other organisms		
Gene	**Protein product**	**Accession #**	***Homo sapiens/ Bos taurus***	***Neurospora/ Yarrowia***	***Arabidopsis***
		Constituents (Subunits)			
NUO10	Subunit 10	EDO98029	NDUFS7/PSST	19.3/NUKM	NP_196738
NUO8	Subunit 8	EDO96988	NDUFS8/TYKY	21.3c/NUIM	NP_173114 NP_178022
NUO5	24-kD subunit	EDP08001	NDUFV2/24	24/NUHM	NP_567244
NUO9	ND9 subunit	EDP05911	NDUFS3/30	31/NUGM	Q95748
NUO7	49-kD ND7 subunit	EDP00001	NDUFS2/49	49/NUCM	P93306
NUO6	51-kD subunit	EDP06369	NDUFV1/51	51/NUBM	NP_196470
NUOS1	76-kD subunit	EDP03454	NDUFS1/75	78/NUAM	NP_851103 NP_568550
nd1	Subunit 1	AAB93446	ND1	ND1/NU1M	NP_085565
nd2	Subunit 2	AAB93444	ND2	ND2/NU2M	NP_085584
ND3	ND3 subunit	EDP02792	ND3	ND3/NU3M	P92533
nd4	Subunit 4	AAB93441	ND4	ND4/NU4M	NP_085518
ND4L	ND4L subunit	EDP00050	ND4L	ND4L/NULM	NP_085525
nd5	Subunit 5	AAB93442	ND5	ND5/NU5M	NP_085478
nd6	Subunit 6	AAB93445	ND6	ND6/NU6M	NP_085495
NUOA1	8-kD subunit	EDP07892	NDUFA1/MWFE	9.8/NIMM	NP_566330
ACP1	Acyl-carrier protein	EDO98915	NDUFAB1/SDAP	9.6/ACPM	NP_176708
NUOB8	11-kD subunit	EDP01212	NDUFA2/B8	10.5/NI8M	NP_199600
NUOB12	7-kD subunit	EDP07174	NDUFB3/B12	10.6/NB2M	NP_178355
NUOB13	18-kD subunit	EDP08707	NDUFA5/B13	29.9/NUFM	NP_568778
NUOS6	16.3-kD subunit	EDP06101	NDUFS6/13a	18.4/NUMM	NP_566191
NUOB14	B14 subunit	EDP02978	NDUFA6/B14	14.8/NB4M	NP_566416
TIM22-like	B14.7 subunit	EDP09567	NDUFA11/B14.7	21.3b/N.I.	NP_565968
NUO17	17-kD subunit	EDO99854	Q8WZ96/ESSS	11.6/NUWM	NP_567054
NUOS5	10-kD subunit	EDP05506	NDUFS5/PFFD	11.5/NIPM	NP_191838 NP_182291
NUOP2	13-kD subunit	EDP03217	NDUFB4/B15	7/NUVM	NP_565726
NUOB16	B16.6 subunit	EDO97447	NDUFA12/B16.6	14/NB6M	NP_171957 NP_565761
NUO13	18-kD subunit	EDO98585	DAP13/B17.2	13.4/N7BM	NP_566192
NUOB18	10-kD subunit	EDO99667	NDUFB7/B18	11.3/NB8M	NP_565280
NUOS4	18-kD subunit	EDP01388	NDUFS4/AQDQ	21/NUYM	NP_201560

(Continued)

Table 13.1 *Continued*

	Chlamydomonas		Counterparts in other organisms		
Gene	**Protein product**	**Accession #**	***Homo sapiens/ Bos taurus***	***Neurospora/ Yarrowia***	***Arabidopsis***
			Constituents (Subunits)		
NUOS4b	18-kD-like subunit				
NUOA8	12.5-kD subunit	EDO98408	NDUFA8/PGIV	20.8/NUJM	NP_850849
NUOB22	14-kD subunit	EDO99297	NDUFB9/B22	11.7/N.I.	NP_567970
NUOB10	17.8-kD subunit	EDP02977	NDUFB10/PDSW	12.3/NIDM	NP_564540 NP_566608
NUOA9	39-kD subunit	EDP06432	NDUFA9/39	40/NUEM	NP_565469
NUO21	13-kD subunit	EDP08510	N.I.	NUO20.9/NUXM	NP_567500
NUOP1	10-kD subunit	EDP08728	N.I.	NUO10.4/N.I.	NP_567596
NUOP3	22-kD subunit	EDP07513	N.I.	N.I.	NP_566309
CAG1	Gamma carbonic anhydrase	EDO96533	N.I.	N.I.	NP_175159
CAG2	Gamma carbonic anhydrase	EDO97352	N.I.	N.I.	NP_201156
CAG3	Gamma carbonic anhydrase	EDP00854	N.I.	N.I.	NP_569036
CAH9	Carbonic anhydrase 9	EDP07163	N.I.	N.I.	NP_190437
NUOP4	9-kD subunit	EDP02796	N.I.	N.I.	N.I.
NUOP5	16-kD subunit	EDP02802	N.I.	N.I.	N.I.
			Assembly factors		
NUOAF1	Complex I intermediate-associated CIA30 protein	EDP06825	NDUFAF1/CIA30	CIA30	NP_177386

Note: N.I. (not identified) indicates that no sequence of significant similarity was found.

700-kD membrane-associated subcomplex also demonstrates NADH dehydrogenase activity (Cardol et al., 2002; Remacle et al., 2006). This subcomplex is less tightly bound to the membrane than the wild-type enzyme, and is mainly composed of subunits classically belonging to the matrix-exposed arm (Cardol et al., 2008).

B. Complex II, succinate:ubiquinone oxidoreductase

Complex II or succinate:ubiquinone oxidoreductase (EC 1.3.99.1) is usually presented as an enzyme involved both in the Krebs cycle and respiratory chain. Classically, Complex II comprises four distinct nucleus-encoded polypeptides, SDH1 (the flavoprotein subunit), SDH2 (the iron-sulfur subunit),

Table 13.2 Complex II constituents

Chlamydomonas			Counterparts in other organisms		
Gene	Protein product	Accession #	*Homo sapiens/ Bos taurus*	Yeast	*Arabidopsis*
SDH1	Subunit A	EDP09580	SDHA	SDH1	NP_179435 NP_201477
SDH2	Iron-sulfur subunit	EDP08267	SDHB	SDH2	NP_198881 NP_189374
SDH3	Subunit b_{560}, alternative splicing variant 1	EDP09244	SDHC (QPS1)	SDH3	NP_196522
SDH3	Subunit b_{560}, alternative splicing variant 2	EDP09245	SDHC (QPS1)	SDH3	NP_196522
SDH4	Subunit D	EDP09690	SDHD	SDH4	NP_566077

and two hydrophobic membrane anchors, SDH3 (subunit III) and SDH4 (subunit IV) (Cecchini et al., 2002; Lemire and Oyedotun, 2002). The genes encoding the corresponding subunits are all present in *Chlamydomonas* (Cardol et al., 2005) (Table 13.2). A single Complex II chaperone, TCM62, has been identified in yeast mitochondria (Dibrov et al., 1998), but no counterpart gene could be identified in *Chlamydomonas*.

C. Complex III, ubiquinol:cytochrome *c* oxidoreductase

Ubiquinol-cytochrome *c* oxidoreductase (EC 1.10.2.2), bc_1 complex, or mitochondrial Complex III, has been well characterized from a variety of sources including mammals, yeast, and higher plants. Composed of ten well-conserved subunits, the oligomeric membrane protein complex contains three subunits carrying redox groups: cytochrome *b*, cytochrome c_1, and a Rieske iron-sulfur protein (ISP) (Braun and Schmitz, 1995; Iwata et al., 1998; Eubel et al., 2003). The bc_1 complex from *Chlamydomonas* was isolated and resolved into nine subunits with molecular masses ranging from 10 kD to 50 kD (Atteia, 1994; van Lis et al., 2003), and four of these polypeptides were identified: QCR1 or core I (53 kD), whose N-terminal sequence was obtained; cytochrome *b*, cytochrome c_1, and the ISP (Gray and Boer, 1988; Atteia et al., 2003). Six other Complex III components were identified by searching the genome: QCR2 (core II), QCR7 (subunit VI), QCR8 (subunit VII or "ubiquinol-binding" protein), QCR6 (subunit VIII or "hinge protein"), QCR9 (subunit X), and QCR10 (subunit XI) (Cardol et al., 2005; Table 13.3). Thus, on the basis of this putative subunit composition, Complex III of *Chlamydomonas* is very similar to that of other eukaryotes.

Table 13.3 Complex III constituents and assembly factors

Chlamydomonas			Counterparts in other organisms		
Gene	**Protein product**	**Accession #**	***Homo sapiens/ Bos taurus***	**Yeast**	***Arabidopsis***
		Constituents (Subunits)			
QCR1	50-kD core 1 subunit Peptidase beta subunit (beta-MPP)	EDP00713	UQCRC1/ Core1	COR1/ beta-MPP	NP_186858
QCR2	core 2 subunit		UQCRC2/ Core2	QCR2	NP_566548 NP_175610
MPPA1 *MPPA2*	Peptidase alpha subunit	EDP00385 EDP05489			
cob	Cytochrome *b*	AAB93440	CYB/III	COB	P42792
CYC1	Cytochrome *c1*	EDO96204	CYC1/IV	CYT1	NP_189360 NP_198897
RIP1	Rieske iron-sulfur protein	EDP09520	UQCRCFS1/ RISP,IX	RIP1	NP_196848 NP_568288
QCR7	14-kD subunit	EDP08285	UQCRB/VI, 14 kD	QCR7	NP_849484 NP_194973
QCR8	9-kD subunit	EDP00113	UQCRQ/VII	QCR8	NP_187697 NP_196156
QCR6	8-kD subunit	EDO99747	UQCRH/VIII	QCR6	NP_178219 NP_172964
QCR9	7-kD subunit	EDO98612	UQCR10/X	QCR9	NP_190841
QCR10	10-kD subunit	EDP00790	6.4 kD/XI	QCR10	NP_565941
		Assembly factors			
CBP3	Chaperone	EDP09522	CBP3	CBP3	NP_199935
BCS1	Biogenesis factor	EDO97908	BCS1	BCS1	NP_850841
ABC1	Ubiquinone biosynthesis protein	EDP06299	ABC1	ABC1	NP_192075

QCR1 and QCR2 homologues from higher plants possess a proteolytic activity (mitochondrial processing peptidase or MPP) that cleaves the transit peptide of preproteins when they are imported into mitochondria (Glaser and Dessi, 1999). In *Chlamydomonas*, QCR1 exhibits consensus sequences typical of β-MPPs while QCR2 does not share consensus sequences for α-MPP activity (van Lis et al., 2003). Nevertheless, a third component, with the characteristics of an MPP α subunit (MPPA) is also encoded in *Chlamydomonas* (Table 13.3).

Several proteins involved in Complex III biogenesis have been identified in yeast: COQ8 (previously ABC1: Bousquet et al., 1991; Do et al., 2001), the assembly factor BCS1, required for the expression of a functional ISP (Nobrega et al., 1992; Cruciat et al., 1999); the chaperone CBP3 (Shi et al., 2001); CBP1, 2, and 6, CBS1 and 2, required for proper expression, maturation, assembly, and stability of cytochrome *b* (Dieckmann et al., 1982; McGraw and Tzagoloff, 1983; Dieckmann and Tzagoloff, 1985; Rodel, 1986) and the assembly factor CBP4 (Crivellone, 1994). Three homologues of the yeast proteins were found to be encoded in *Chlamydomonas*: COQ8, BCS1, and CBP3. Their participation in Complex III biogenesis remains to be confirmed.

D. Cytochrome *c*

Soluble cytochrome *c* from *Chlamydomonas* is encoded by a single nuclear gene (Table 13.4) very similar to those of higher plants (Amati et al., 1988). The gene has three introns, two of which are located at almost the same positions in the corresponding genes of *Arabidopsis* and rice, suggesting that these two introns predate the divergence of green algae and higher plants. Expression of cytochrome *c* in *Chlamydomonas* is induced by acetate and light. Inhibitors of the mitochondrial respiratory chain do not induce its expression but DCMU, an inhibitor of the photosynthetic electron transport chain, does (Felitti et al., 2000).

Three distinct pathways of cytochrome *c* biogenesis have been described (Kranz et al., 1998). System I occurs in mitochondria from higher plants and protozoa, system II is found in many Gram-positive bacteria, cyanobacteria,

Table 13.4 Cytochrome *c* and assembly factors

Chlamydomonas			Counterparts in other organisms		
Gene	Protein product	Accession #	*Homo sapiens/ Bos taurus*	Yeast	*Arabidopsis*
		Cytochrome *c*			
CYC	Mitochondrial cytochrome *c*	EDP00604	CYC/CytC	CYC1	NP_173697 NP_192742
		Assembly factors			
HCS1	Cytochrome *c* heme lyase	EDP00694	CCHL	CYT2 (CC1HL)	N.I.
HCS2	Cytochrome *c* heme lyase	ED098886	N.I.	CYC3	N.I.
HCS3	Cytochrome *c* heme lyase	EDP00575	N.I.	N.I.	N.I.

Note: N.I. (not identified) indicates that no sequence of significant similarity was found.

and chloroplasts from higher plants and green algae, and system III occurs in mitochondria from yeasts (CYC3 and CYT2) and animals (CCHL). In *Chlamydomonas*, whereas no system I counterparts could be identified, three heme lyase enzymes homologous to system III components (yeast CYC3 and CYT2, and animal CCHL) are encoded (Cardol et al., 2005).

E. Complex IV, cytochrome *c* oxidase

Complex IV (cytochrome *c* oxidase, EC 1.9.3.1) catalyzes electron transfer from soluble cytochrome *c* to oxygen and contributes to the formation of the electrochemical potential. In yeast and mammals, the complex is composed of 12 (Taanman and Capaldi, 1992) and 13 (Yanamura et al., 1988) subunits, respectively. In *Arabidopsis*, the enzyme is composed of 10–12 subunits but only 7 have been identified (Eubel et al., 2003). Three subunits, usually encoded in the mitochondrial genome, comprise the catalytic core: COX1, which binds heme *a*, heme a_3, and Cu_B; COX2, which binds Cu_A; and COX3. The other subunits, all of which are nucleus-encoded, are considered to be regulatory proteins or polypeptides that participate in biogenesis of the complex. Exceptions to mitochondrial coding of the catalytic core exist. In some legumes only subunits COX1 and COX3 are mitochondrially-encoded (Daley et al., 2002), whereas in *Chlamydomonas* (Michaelis et al., 1990) and *Polytomella parva* (Fan and Lee, 2002) only COX1 is encoded in the mitochondrial genome.

The *Chlamydomonas* cytochrome *c* oxidase complex was resolved by BN-PAGE into 10 polypeptides with molecular masses ranging from 8 kD to 40 kD (van Lis et al., 2003). Eight coding sequences for these polypeptides have been identified in *Chlamydomonas*, the orthologues of bovine subunits COX1, 2, 3, 4, 5b, 6a, and 6b (Cardol et al., 2005) and the plant-specific subunit COX5c (Jansch et al., 1996; Hamanaka et al., 1999; Table 13.5). The subunit composition of *Chlamydomonas* Complex IV deduced from genome sequencing is quite similar to the subunit composition of Complex IV from *Arabidopsis*. Nevertheless, important differences do exist. In *Chlamydomonas* and *Polytomella*, COX2 is encoded by two separate nuclear genes (*COX2a* and *COX2b*), whereas in the vast majority of eukaryotes COX2 is encoded by a single mitochondrial gene. The chlorophycean genes encode COX2A, which comprises the N-terminal, hydrophobic half of a conventional COX2 protein; and COX2B, which corresponds to the soluble C-terminal half (Pérez-Martínez et al., 2001). These algal subunits contain unique C- and N-terminal extensions, respectively – not present in conventional COX2 subunits – that may participate in the assembly and stabilization of the split COX2 subunit in the mature complex (Pérez-Martínez et al., 2001). Split *COX2a* and *COX2b* genes have also been identified in other *Chlamydomonas*-related algae and in apicomplexan parasites (Funes et al., 2002b; Gardner et al., 2002).

Table 13.5 Complex IV constituents and assembly factors

Chlamydomonas			Counterparts in other organisms		
Gene	**Protein product**	**Accession #**	***Homo sapiens/ Bos taurus***	**Yeast**	***Arabidopsis***
		Constituents (Subunits)			
COX1	Subunit 1	AAB93443	COI	COI	P60620
COX2a	Subunit II, protein IIa of split subunit	EDP00208	COII	COII	P93285
COX2b	Subunit II, protein IIb of split subunit	EDP09974	COII	COII	P93285
COX3	Subunit 3	EDP04132	COIII	COIII	P92514
COX5b	Subunit 4, 13 kD	EDP08953	COX5B	IV (COX4)	NP_188185 NP_178140
COX5c	11-kD subunit	EDP00106	N.I.	N.I.	NP_850738 NP_182260
COX13	12-kD subunit	EDP03436	COX6A	COXVIa (COX13)	NP_195496
COX12	COX subunit	EDP05132	COX6B	COXVIb (COX12)	NP_173661 NP_194535
COX90	COX subunit	ED098724	N.I.	N.I.	N.I.
		Assembly factors			
SHY1	COX assembly protein	ED097446	Surfeit1	SHY1	NP_566592
SCO1	COX assembly factor	ED097370	SCO1, SCO2	SCO1, SCO2	NP_566339
COX10	COX assembly protein	ED096513	COX10	COX10	NP_566019
COX11	COX assembly protein	ED098350	COX11	COX11	NP_171743
COX15	COX assembly factor	EDP06678	COX15	COX15	NP_200420
COX16	COX assembly factor	EDP06214	AAH01702	COX16	NP_680684
COX17	COX assembly protein	ED097130	COX17	COX17	NP_175711
COX18	COX assembly factor	ED098825	OXA1	COX18, OXA1 (PET1402)	NP_201011
COX19	COX assembly protein	EDP08032	COX19	COX19	NP_849852
COX23	COX assembly protein	ED098664	NP_077276	COX23	NP_171718
PET191	COX assembly protein	ED098359	AAH47722	PET191	BAC43353

Note: N.I. (not identified) indicates that no sequence of significant similarity was found.

Chlamydomonas Complex IV supernumerary subunits show some unusual characteristics. One is that the first 60 residues of the mature subunit COX6b are highly hydrophobic and have counterparts in plant sequences, but not in animals or yeast. Second, COX4 lacks the three conserved cysteines that are known to bind a zinc atom in cytochrome *c* oxidase (Rizzuto et al., 1991). Finally, COX6a possesses an N-terminal region with high similarity to COX5a from mammals, while the C-terminal region is more similar to the COX6a subunit. The functional significance of these differences remains to be explored.

The Complex IV assembly pathway is well-studied in *S. cerevisiae*, with more than 20 chaperones involved in its biogenesis. Several homologues of these proteins have been identified in *Chlamydomonas* (Cardol et al., 2005): those that play a role in copper delivery to cytochrome *c* oxidase such as COX11, COX17, COX19, COX23, SCO1, and SCO2 (Nobrega et al., 2002; Barros et al., 2004; Horng et al., 2004; Leary et al., 2004); those required for heme *a* biosynthesis, such as COX10 and COX15 (Glerum and Tzagoloff, 1994; Barros and Tzagoloff, 2002); SHY1, which is believed to act as a translational activator for the mitochondrially-encoded subunits (Barrientos et al., 2002); those required for the insertion of various hydrophobic proteins in the inner mitochondrial membrane such as COX18 and OXA1 (He and Fox, 1997; Saracco and Fox, 2002; Funes et al., 2004); and two chaperones, COX16 and Pet191 (McEwen et al., 1993; Carlson et al., 2003). Lastly, COX90 is a protein identified in *Chlamydomonas* which has no counterpart in other organisms apart from green algae (Lown et al., 2001).

F. Complex V, ATP synthase

Complex V (F_0F_1-ATP synthase, EC 3.6.1.3) catalyzes the phosphorylation of ADP by inorganic phosphate, using the proton motive force generated by the substrate-driven electron transfer chain. F_0F_1-ATP synthase is considered a molecular motor in which a central rotor-stalk [$\gamma/\delta/\varepsilon/c_{10}$] (subunits of beef mitochondria) rotates around an axis perpendicular to the plane of the membrane, while other polypeptides constitute the fixed elements of the motor: subunit *a*, the catalytic core [α_3/β_3], the peripheral stator-stalk [OSCP/*b*/*d*/F_6], and the so-called minor subunits [A6L/*e*/*f*/*g*] (Golden and Pedersen, 1998; Ko et al., 2000; Walker and Dickson, 2006). Proton flux across subunit *a* causes a rotary movement of the membrane-embedded ring c_{10} and its bound, protruding central rotor-stalk. In a full cycle, three sequential 120° movements of subunit γ induce consecutive conformational changes in the three catalytic β subunits leading to substrate binding (ADP + Pi), ATP synthesis, and ATP release (Itoh et al., 2004). Additional subunits play regulatory roles, such as the inhibitory protein IF_1 (Gledhill and Walker, 2005), or are involved in the formation and stabilization of a dimeric complex, such as subunits *e* and *g* (Brunner et al., 2002; Fronzes et al., 2006).

The eukaryotic Complex V contains components similar to those found in bacteria, but possesses additional subunits. In mammals, ATPase includes 16 subunits (Collinson et al., 1994) whereas the yeast enzyme comprises 20 components, 13 being essential for the structure of the complex (Velours et al., 1998; Velours and Arselin, 2000). Mitochondrial ATP synthase preparations from *Chlamydomonas* exhibited an activity of 2.87 μmoles ATP hydrolyzed/minute/mg protein, and was 40% inhibited by oligomycin (Nurani and Franzén, 1996). The algal mitochondrial ATP synthase contains 16 subunits of 7–60 kD (Franzén and Falk, 1992; Nurani and Franzén, 1996; Funes et al., 2002a; van Lis et al., 2003; Cardol et al., 2005; Vázquez-Acevedo et al., 2006; Table 13.6). Eight polypeptides were identified as subunits α, β, γ, δ, ε, *a* (ATP6), *c* (ATP9), and OSCP, while no homologues of the *b*, *d*, *e*, *f*, *g*, A6L, and F6 subunits were found. Nine novel polypeptides (named ASA for **A**TP **S**ynthase **A**ssociated proteins), ASA1–9, are associated with chlorophycean mitochondrial ATP synthases and exhibit no counterparts in other eukaryotes (Cardol et al., 2005; Vázquez-Acevedo et al., 2006). The ASA1 polypeptide (61 kD), initially named MASAP, MAAP67, or P60, was found in the isolated mitochondrial ATP synthases of both *Chlamydomonas* and *Polytomella* sp. It was hypothesized that ASA1 is responsible for the extraordinary stability of the chlorophycean Complex V dimer (van Lis et al., 2003, 2005). Complex V of chlorophycean algae such as *Chlamydomonas*, *Polytomella* sp., *Scenedesmus obliquus*, *Chlorogonium elongatum*, and *Chlorococcum ellipsoideum* migrates in BN-PAGE as a stable dimer of 1600 kD (van Lis et al., 2003; Dudkina et al., 2005; van Lis et al., 2005; Vázquez-Acevedo et al., 2006). The dimer was further confirmed in *Polytomella* sp. by electron microscopy (Dudkina et al., 2005) and biochemical studies (Vázquez-Acevedo et al., 2006). Models suggesting the topological arrangement of subunits ASA1–9 in Complex V have been proposed (Vázquez-Acevedo et al., 2006; Van Lis et al., 2007). Thus, chlorophycean algae contain novel constituents in their mitochondrial ATP synthase which give rise to a highly stable, dimeric enzyme.

The chlorophycean OXPHOS complex exhibits three additional particular features. First, in contrast to other eukaryotes, it is entirely nucleus-encoded (Vázquez-Acevedo et al., 2006). Second, two isoforms of ATP9 (subunit *c*) could be present in *Chlamydomonas*, since two genes are predicted to encode *c*-like subunits. These genes are adjacent on linkage group XIX, and are divergently transcribed. Nevertheless, the N-terminal sequences of the ATP9 subunits (van Lis et al., 2003) do not allow predictions regarding whether both or only one of the polypeptides are present in the mature complex. Third, the β subunit of *Chlamydomonas* and *Polytomella* Complex V has a C-terminal extension not found in other organisms (Franzén and Falk, 1992; Atteia et al., 1997), which shows sequence similarity to the inhibitor protein of the F_0F_1-ATPase from other eukaryotes, and to the ε subunit of the *E. coli* enzyme. Therefore, the C-terminal extension may play a regulatory role in these algae.

Eight assembly factors for Complex V have been identified in yeast: two chaperones for the F_1 domain, ATP11 and ATP12 (Ackerman and Tzagoloff, 1990a; Wang et al., 2001); FMC1, which is required for the assembly or stability of F_1 in heat stress conditions (Lefebvre-Legendre et al., 2001); ATP10 and ATP22, which are required for the assembly of F_0 (Ackerman

Table 13.6 Complex V constituents and assembly factors

Chlamydomonas			Counterparts in other organisms		
Gene	**Protein product**	**Accession #**	***Homo sapiens/ Bos taurus***	**Yeast**	***Arabidopsis***
		Constituents (Subunits)			
ATP6	Subunit 6	EDP09230	ATP6/a	ATPA (ATP6)	P93298 P92547
ATP9A	Subunit 9, isoform A	EDO97408	ATP5G3/c	ATPC (ATP9)	P60112 NP_178769
ATP9B	Subunit 9, isoform B	EDO97377	ATP5G3/c	ATPC (ATP9)	P60112 NP_178769
ATP1	Alpha subunit	EDP07337	ATP5A1/α	α (ATP1)	P92549 NP_178788
ATP1B	Alpha subunit isoform B	EDP06403	ATP5A1/α	α (ATP1)	P92549 NP_17878
ATP2	Beta subunit	EDP04740	ATP5B/β	β (ATP2)	NP_568203 NP_680155 NP_568204
ATP3	Gamma subunit	EDO97956	ATP5C1/γ	γ (ATP3)	NP_180863
ATP16	Delta subunit	EDO99236	ATP5D/δ	δ (ATP16)	NP_199514
ATP15	Epsilon subunit	EDP06388	ATP5E/ε	ε (ATP15)	NP_175576
ATP5	Subunit 5, OSCP subunit	EDP01322	ATP5O/OSCP	ATP5	NP_196849
ASA1	Associated 60.6-kD protein	EDP03873	N.I.	N.I.	N.I.
ASA2	Associated 45.5-kD protein	EDP00850	N.I.	N.I.	N.I.
ASA3	Associated 36.3-kD protein	EDO98373	N.I.	N.I.	N.I.
ASA4	Associated 31.2-kD protein	EDP08830	N.I.	N.I.	N.I.
ASA5	Associated 14.3-kD protein	EDP00370	N.I.	N.I.	N.I.
ASA6	Associated 13.3-kD protein	EDP06853	N.I.	N.I.	N.I.
ASA7	Associated 19.5-kD protein	EDP00858	N.I.	N.I.	N.I.
ASA8	Associated 10.0-kD protein	EDP01930	N.I.	N.I.	N.I.
ASA9	Associated 12.0-kD protein	EDP02597	N.I.	N.I.	N.I.
		Assembly factors			
ATP11	Assembly factor 1 for F1 component	EDP05655	ATPAF1	ATP11	NP_565778
ATP12a	Assembly factor 2, splice variant a	EDP00509	ATPAF2	ATP12	NP_198882
ATP12b	Assembly factor 2, splice variant b	EDP00510	ATPAF2	ATP12	NP_198882

Note: N.I. (not identified) indicates that no sequence of significant similarity was found.

and Tzagoloff, 1990b; Helfenbein et al., 2003; Tzagoloff et al., 2004); AEP1 and AEP2, which are required for the correct expression of ATP9 (Payne et al., 1991; Finnegan et al., 1995); and AEP3, which stabilizes the mitochondrial dicistronic mRNA encoding subunits 6 and 8 (Ellis et al., 2004). Genes for only two of them, ATP11 and ATP12 (with two splice variants), could be identified in *Chlamydomonas* (Table 13.6). The rest of the genes encoding ATP synthase biogenesis factors may not be present at all, or their sequences may have highly diverged. Also, specific chaperones involved in the biogenesis of the ASA subunits, unique to algal ATP synthase but yet unidentified, may be present in *Chlamydomonas*.

G. Other components related to OXPHOS

1. Alternative NAD(P)H dehydrogenases

In addition to the proton-pumping Complex I, plant and fungal mitochondria contain several type II NAD(P)H dehydrogenases. These enzymes, located at the surface of the mitochondrial inner membrane and facing either the intermembrane space or the matrix, allow electron transfer from NAD(P)H to ubiquinone. They are single, low molecular weight polypeptides whose activity is insensitive to rotenone (Møller et al., 1993). In *S. cerevisiae*, three enzymes have been characterized (Marres et al., 1991; Luttik et al., 1998) whereas in *Arabidopsis*, three gene families (*NDA*, *NDB*, and *NDC*), with different evolutionary origins, encode NADH dehydrogenases that are exclusively targeted to mitochondria (Michalecka et al., 2003; Elhafez et al., 2006).

Chlamydomonas encodes six proteins with amino acid similarity to known type II – NAD(P)H dehydrogenases (NDA1, 2, 3, 5, 6, and 7; Cardol et al., 2005; Table 13.7). In particular, the *NDA6* and *NDA7* genes are adjacent and presumably arose through a duplication event. Biochemical studies suggest that at least one type-II NAD(P)H dehydrogenase could be involved in plastoquinone reduction in *Chlamydomonas* chloroplasts (Peltier and Cournac, 2002; Mus et al., 2005). This situation differs with land plants where a bacterial Complex I-like enzyme, absent in *Chlamydomonas*, acts in the chloroplast (Sazanov et al., 1998).

2. Alternative oxidases

Besides the classical terminal aa_3-type oxidase, mitochondria from plants, several fungi, and several protists also possess an alternative oxidase (AOX) that drives the electrons from the ubiquinol pool directly to molecular oxygen and does not contribute to the formation of an electrochemical gradient. The enzyme is thought to regulate mitochondrial respiratory electron flow and to protect plant cells from oxidative damage (Maxwell et al., 1999). In plants, various developmental and environmental conditions enhance AOX expression (Berthold et al., 2000; Clifton et al., 2006; Elhafez et al., 2006),

Table 13.7 Alternative electron transfer constituents

Chlamydomonas			Counterparts in other organisms		
Gene	Protein product	Accession #	*Homo sapiens/ Bos taurus*	Yeast	*Arabidopsis*
AOX1	Alternative oxidase	EDP02600	N.I.	Q01355 (Nc)	NP_188876 (AOX1a)
AOX2	Alternative oxidase	EDP06011	N.I.	N.I.	NP_188875 (AOX1b)
NDA1	Mitochondrial type-II NADH dehydrogenase	EDO99183	N.I.	ND1	NP_180560 (NDA2)
NDA2	Type-II calcium-dependent NADH dehydrogenase	EDO96450	N.I.	N.I.	NP_567283 (NDB2)
NDA3	Mitochondrial NADH dehydrogenase	EDO97072	N.I.	N.I.	NP_179673 (NDB4)
NDA5	Type-II NADH dehydrogenase	EDP04459	N.I.	N.I.	NP_568205 (NDC1)
NDA6	Type-II NADH dehydrogenase	EDO96700	N.I.	N.I.	NP_190005
NDA7	Type-II NADH dehydrogenase	EDO96701	N.I.	N.I.	NP_974819

Note: N.I. (not identified) indicates that no sequence of significant similarity was found.

whereas its induction in fungi is mainly dependent on cytochrome pathway restriction and triggering by reactive oxygen species (Helmerhorst et al., 2005; Juárez et al., 2006).

In *Chlamydomonas*, a cyanide and carbon monoxide-resistant respiratory pathway was recognized (Hommersand and Thimann, 1965), and later shown to be sensitive to salicyl hydroxamic acid (Weger et al., 1990). Furthermore, the alternate respiratory pathway was present under normal growth conditions but decreased considerably at high CO_2 concentrations (Goyal and Tolbert, 1989). Several *Chlamydomonas* respiratory-deficient mutants with an impaired cyanide-sensitive cytochrome pathway still exhibit a fully functional alternative pathway (Dorthu et al., 1992).

Evidence for the presence of an AOX in *Chlamydomonas* has been provided (Matagne et al., 1989; Eriksson et al., 1995; Derzaph and Weger, 1996) and the expression of two genes, *AOX1* and *AOX2*, has been reported (Table 13.7), *AOX1* being more actively transcribed than *AOX2* (Dinant et al., 2001). By fusing the *AOX1* promoter region to the promoterless arylsulfatase reporter gene, it was shown that this promoter is unresponsive to H_2O_2-induced oxidative stress, antimycin A and other inducers known to act on AOX in higher plants (Baurain et al., 2003). In contrast, the reporter gene was strongly activated by changing the nitrogen source from NH_4 to NO_3, and the level of the AOX protein concomitantly increased. However, RNA gel blot experiments demonstrated that the *AOX1* mRNA level was increased not only by changing the nitrogen source but also by the presence of H_2O_2, cold stress and antimycin A (Molen et al., 2006). This may indicate that key regulatory elements exist downstream of the transcriptional start, which escaped the detection using the reporter gene. Nevertheless, and unlike higher plants, even when *AOX* is transcriptionally activated by antimicyn A, there is no accumulation of AOX protein, suggesting a post-ranscriptional control of AOX levels (Molen et al., 2006). In contrast to *Chlamydomonas*, *Polytomella* seems to lack AOX, which may be related to the loss of photosynthesis in this alga (Reyes-Prieto et al., 2002).

3. Uncoupling proteins

Uncoupling proteins (UCPs) belong to the mitochondrial anion carrier protein family. UCPs are anchored in the inner membrane and compete with the F_0F_1-ATP synthase for the proton electron gradient built up by the activity of respiratory-chain Complexes I, III, and IV. They are present in all eukaryotes investigated except yeast, and their physiological roles are still under investigation and debate. A consensus function for UCPs might be the maintenance of energy balance during phosphorylating conditions, indirectly protecting the mitochondrion against reactive oxygen species by diminishing the reduced state of the respiratory chain (Sluse et al., 2006). To date, three UCP-encoding genes have been identified in Chlamydomonas: *UCP1, UCP2, and UCP3* (Table 13.8).

Table 13.8 Other mitochondrial constituents

Chlamydomonas			Counterparts in other organisms		
Gene	Protein product	Accession #	*Homo sapiens/ Bos taurus*	Yeast	*Arabidopsis*
UCP1	Uncoupling protein	EDP01675	NP_068605(UCP1)	N.I.	NP_190979 (UCP1)
UCP2	Uncoupling protein	EDP08337	NP_003346 (UCP2)	N.I.	NP_974962 (UCP2)
UCP3	Uncoupling protein	EDP04204	095847 (UCP4)	N.I.	NP_172866
ANT1	Adenine nucleotide translocator	EDP01885	ADT1,2,3	ADT1,2,3	NP_194568
ANT2	ADP/ATP carrier protein	EDP05104	ADT1,2,3	ADT1,2,3	NP_196853
PFL1	Pyruvate-formate lyase	EDP09457	N.I.	N.I.	N.I.
PAT1	Phosphotransacetylase	EDP02499	N.I.	N.I.	N.I.
ACK1	Acetate kinase	EDP02500	N.I.	N.I.	N.I.
ACK2	Acetate kinase	EDP04790	N.I.	N.I.	N.I.

Note: N.I. (not identified) indicates that no sequence of significant similarity was found.

4. Pyruvate formate-lyase

When *Chlamydomonas* is subjected to anaerobic conditions in the dark – in artificial systems where oxygen is replaced by inert gases such as argon or nitrogen – fermentative products such as formate, ethanol, acetate, small amounts of lactate and glycerol, and low amounts of CO_2 and H_2 can be detected (Gfeller and Gibbs, 1984; Kreuzberg, 1984). The production of formate in the ratio of 2:1:1 relative to ethanol and acetate was predicted to be carried out by pyruvate formate-lyase (PFL) (Kreuzberg, 1984), the enzyme that cleaves pyruvate into acetyl-CoA and formate. PFL is widespread in bacteria but rarely found in eukaryotes. The genes encoding PFL, phosphotransacetylase, and acetate kinase are present in *Chlamydomonas* (Table 13.8) and biochemical evidence exists for the presence of the corresponding protein products in mitochondria (Atteia et al., 2006). Thus, an alternative route of mitochondrial ATP synthesis is present in *Chlamydomonas* involving pyruvate, acetyl-CoA, acetyl-phosphate, and acetate as end product.

5. Ubiquinone biosynthesis pathway components

Ubiquinone (Coenzyme Q or CoQ) is a lipid composed of a redox-active benzoquinone ring connected to a long isoprenoid side chain (usually 6–10 units). CoQ plays a central role in the mitochondrial respiratory chain since it shuttles electrons from Complex I, Complex II, and type-II NAD(P)H dehydrogenases to Complex III and AOX. In addition, the redox state of CoQ regulates the activity of UCPs in land plants, mammals, and protists (Turunen et al., 2004; Sluse et al., 2006). In yeast, the terminal biosynthesis of CoQ starts from farnesyl pyrophosphate, a common lipid precursor, and requires at least nine enzymes (Coq1–9). However, the exact functions of several *COQ* genes are still unknown and their corresponding products may either catalyze an individual step in CoQ biosynthesis (e.g. *COQ5*, *COQ6*) or have a regulatory role (e.g. *COQ8*, *COQ9*) (Turunen et al., 2004; Johnson et al., 2005). In addition, the yeast Coq10 protein could be a Q-binding protein required for proper localization of the quinone within mitochondria (Barros et al., 2005). While the size of the isoprenoid side chain is not known in *Chlamydomonas*, the CoQ biosynthesis pathway is well conserved based on genome sequence, which includes homologues of *COQ1–6* and *COQ8–10* (Table 13.9).

III. IMPORT OF NUCLEUS-ENCODED OXPHOS PROTEINS

Nucleus-encoded mitochondrial proteins are synthesized as precursors with N-terminal extensions called mitochondrial targeting sequences (MTSs), mitochondrial targeting peptides or presequences. These MTSs are required for import into the organelle, and are usually removed by intramitochondrial proteases during or shortly after import (Neupert, 1997).

Table 13.9 Putative constituents of the ubiquinone biosynthesis pathway

Chlamydomonas			Counterparts in other organisms		
Gene	Protein product	Accession #	*Homo sapiens/ Bos taurus*	Yeast	*Arabidopsis*
COQ1	*trans*-Prenyltransferase	EDP05515	TPT	COQ1	NP_850234
COQ2	Polyprenyl-4-hydroxybenzoatetransferase	EDP01665	COQ2	COQ2	NP_849431
COQ3	2,3-Dihydroxy-5-polyprenylbenzoate-methyltransferase	EDP06526	COQ3	COQ3	NP_180649
COQ4	Unknown function	EDP00126	COQ4	COQ4	NP_178465
COQ5	C-Methyltransferase	EDP07004 EDP01746 EDP06889 EDP09311 ED099272 ED095817	COQ5	COQ5	NP_200540
COQ6	C-Hydroxylase	EDP07603	COQ6	COQ6	NP_189062
COQ8	Unknown function	EDP06299	ABC1	COQ8/ABC1	NP_192075(ABC1)
COQ9	Unknown function	EDP08484	COQ9	COQ9	NP_973861
COQ10	Q-binding protein	EDP04032	COQ10	COQ10	NP_193500

Chlamydomonas MTSs appear to range from relatively short to unusually long. Among the OXPHOS proteins, MTSs of the following lengths have been found: Complex V β subunit, 26 amino acids (Franzén and Falk, 1992); Complex V α subunit, 45 amino acids (Nurani and Franzén, 1996); Rieske ISP, 54 amino acids (Atteia and Franzén, 1996); cytochrome c_1, 70 amino acids (Atteia et al., 2003); ATP6, 107 amino acids (Funes et al., 2002a); COX3, 110 amino acids (Pérez-Martínez et al., 2000); and COX2A, 130 amino acids (Pérez-Martínez et al., 2001). COX2B lacks a cleavable MTS, but the protein contains an N-terminal extension of 43 amino acids that is not present in (mitochondria-encoded) COX2 proteins from other organisms (Pérez-Martínez et al., 2001).

An analysis of these MTSs reveals a number of common features including a low number of acidic residues, and a high proportion of alanines, leucines, serines, glycines, and arginines. They also contain regions with the potential to form positively charged, amphiphilic α-helices, a characteristic feature of MTSs (von Heijne et al., 1989). For the MTS of the F_0F_1-ATPase β subunit (Franzén and Falk, 1992), this analysis is strengthened by NMR and circular dichroism studies showing that a synthetic peptide corresponding to the MTS adopts an α-helical structure in 2,2,2-trifluoroethanol/water, a solvent supposed to mimic biological membranes (Lancelin et al., 1996). A similar analysis of peptides corresponding to chloroplast targeting sequences from *Chlamydomonas* ferredoxin (Lancelin et al., 1994) and Rubisco activase (Krimm et al., 1999) showed that a helix-coil motif or a coil-helix motif is formed in the solvent/water mixture. Although these helices are amphiphilic they have weaker hydrophobic sectors, and therefore they are less amphiphilic than the helices predicted in MTSs. These differences between the mitochondrial and chloroplast targeting sequences are probably important for the sorting of proteins between these organelles.

The shorter *Chlamydomonas* MTSs are those that seem to direct the protein exclusively to the mitochondrial matrix, such as in the case of the F_0F_1-ATP synthase α and β subunits. The rest of the MTSs seem to have bipartite structures, with N-terminal regions containing positively charged, amphiphilic α-helices that may direct the precursor to the mitochondrial matrix (mitochondrial targeting signal), followed by C-terminal regions that may direct the proteins to their final destination in the inner mitochondrial membrane. Bipartite MTSs seem to be unusually large in those preproteins that are relatively more hydrophobic (with two or more putative transmembrane stretches) and where the corresponding genes have been transferred to the nucleus relatively recently in evolution: *ATP6*, *COX2A*, *COX3*, *NAD3*, and *NAD4L*. A long MTS could improve the interaction of the precursor with the mitochondrial import machinery (Claros and Vincens, 1996) or could help the unfolding of the protein to increase its importability (Claros et al., 1995).

After import into the mitochondrion, the MTS is removed by MPP, which was mentioned in section II.C. In most organisms, this protease is a soluble enzyme, but in higher plants it is membrane-bound. It has been found that the QCR1 and QCR2 subunits of the higher plant Complex III are identical to the MPP subunits. Thus in higher plants, Complex III is a bifunctional enzyme, involved in both electron transport and protein processing (Glaser and Dessi, 1999). In most organisms, including animals and *S. cerevisiae*, there is a soluble MPP, the subunits of which show sequence homology to the core proteins of Complex III (Neupert, 1997). As mentioned above, *Chlamydomonas* QCR1 appears to be a typical β-MPP protein while QCR2 is not a typical α-MPP protein (van Lis et al., 2003). The situation in *Chlamydomonas* appears to be similar to the one in *N. crassa*, in which the β-MPP protein has two different roles; together with the α-MPP protein it functions as a soluble MPP; it also functions as the QCR1 protein of Complex III. The QCR2 protein of *N. crassa* is different from the α-MPP (Neupert, 1997; Glaser and Dessi, 1999). This analysis is consistent with the results of Nurani et al. (1997), who reported that the *Chlamydomonas* MPP is a soluble enzyme. Either MPPA1 (EDP00385) or MPPA2 (EDP05489), additional QCR/MPP proteins distinct from QCR1 and QCR2, could play the role of the α subunit of a soluble MPP.

The bacterial equivalent of Complex III lacks core proteins. Following endosymbiosis and the evolution of mitochondria, a soluble protease with two subunits may have been attached to Complex III, giving rise to the situation seen in higher plants. In most organisms, the genes of both protease subunits were duplicated. One set of genes evolved to encode core proteins without protease activity, and the other set to encode a soluble MPP. *Chlamydomonas* and *N. crassa* may represent an intermediate stage in this process. Here, the *QCR2/α*-MPP gene has been duplicated, and the two copies have evolved into genes encoding a core protein and a soluble protease subunit, respectively. The *QCR1/β-MPP* gene is not duplicated, and some copies of the gene product integrate into Complex III together with QCR2, while other copies remain soluble and interact with α-MPP.

IV. CONCLUSIONS AND PERSPECTIVES

Overall, the OXPHOS structural components of *Chlamydomonas* are similar to those of higher plants. The main complexes involved in electron transport seem to share a similar number of subunits, and many of the algal polypeptides have plant homologues. Some differences are apparent, such as the presence of a fragmented COX2 subunit, which seems to be unique to chlorophyte algae. More notable is the presence of 8–9 novel subunits in mitochondrial ATP synthase that have no counterparts in plants or other eukaryotes. These subunits, of an as-yet unknown evolutionary origin, appear to form a scaffold that gives rise to a highly stable dimeric complex of 1600 kD, the largest OXPHOS component described to date. The functional consequences (is it more active?, is it more tightly regulated?) of such an ATP synthase in chlorophyte mitochondria remain to be ascertained.

More elusive are the proteins involved in the biogenesis of OXPHOS components. *Chlamydomonas* does encode homologues of some yeast components possibly involved in the biogenesis of mitochondrial OXPHOS proteins. The functions of these components remain to be proven, and other biogenesis proteins, perhaps unique to chlorophyte algae, may well exist.

As a unicellular organism, *Chlamydomonas* offers the unique opportunity to study organelle–organelle interactions, particularly between mitochondria and chloroplasts. It has become evident that crosstalk between these organelles takes place, mainly through intracellular metabolite pools (Raghavendra and Padmasree, 2003). In particular, mitochondrial respiration plays an important role in optimizing photosynthetic efficiency in the chloroplast. Evidence for mitochondria-chloroplast crosstalk in

Chlamydomonas is starting to accumulate. Schönfeld et al. (2004) identified a novel mitochondrial transcription termination factor, MOC1, which regulates the composition of the mitochondrial respiratory chain in the light so that it can act as an effective sink for reductant produced by the chloroplast. Loss of MOC1 causes a high light-sensitive phenotype and disrupts the transcription and expression profiles of cytochrome *c* oxidase and the rotenone-insensitive external NAD(P)H dehydrogenase. Other experimental approaches have shown that activation of *Chlamydomonas* respiration by acetate in the dark induces the activation of mRNAs of nucleus-encoded photosynthetic genes (Matsuo and Obokata, 2006). In addition, inhibitors of the cytochrome pathway (rotenone, antimycin A, or cyanide) inhibited photosynthetic mRNA induction, but inhibitors of the alternate pathway (salicyl hydroxamate) enhanced it. Therefore, nucleus-encoded photosynthetic gene mRNAs are induced in response to activation of the mitochondrial cytochrome pathway (Matsuo and Obokata, 2006). These glimpses into the fine-tuned regulation of mitochondrial function as related to chloroplast components and vice versa, will be expanded in the years to come and will lead us to a more profound understanding of the metabolic interplay between these two organelles. *Chlamydomonas* will be one of the models of choice for these studies.

ACKNOWLEDGMENTS

Work in our laboratories is supported by grants 56619 from CONACyT (Mexico), IN217108 from DGAPA, UNAM (Mexico) and by grants from the Belgian FRFC (2.4587.04 and 2.4582.05) C.R and from the Belgian FNRS (F.4735.06) to P.C. P.C. is postdoctoral researcher of the Belgian FNRS. The technical expertise of Miriam Vázquez-Acevedo (D.G.-H. laboratory) is gratefully acknowledged.

REFERENCES

Abdrakhmanova, A., Zickermann, V., Bostina, M., Radermacher, M., Schägger, H., Kerscher, S., and Brandt, U. (2004). Subunit composition of mitochondrial complex I from the yeast. *Yarrowia lipolytica. Biochim. Biophys. Acta* **1658**, 148–156.

Abdrakhmanova, A., Dobrynin, K., Zwicker, K., Kerscher, S., and Brandt, U. (2005). Functional sulfurtransferase is associated with mitochondrial complex I from *Yarrowia lipolytica*, but is not required for assembly of its iron-sulfur clusters. *FEBS Lett.* **579**, 6781–6785.

Ackerman, S.H. and Tzagoloff, A. (1990a). Identification of two nuclear genes (*ATP11, ATP12*) required for assembly of the yeast F_1-ATPase. *Proc. Natl. Acad. Sci. U. S. A.* **87**, 4986–4990.

Ackerman, S.H. and Tzagoloff, A. (1990b). *ATP 10*, a yeast nuclear gene required for the assembly of the mitochondrial F_1-F_0 complex. *J. Biol. Chem.* **265**, 9952–9959.

Amati, B.B., Goldschmidt-Clermont, M., Wallace, C.J., and Rochaix, J.D. (1988). cDNA and deduced amino acid sequences of cytochrome c from *Chlamydomonas reinhardtii*: Unexpected functional and phylogenetic implications. *J. Mol. Evol.* **28**, 151–160.

Antonicka, H., Ogilvie, I., Taivassalo, T., Anitori, R.P., Haller, R.G., Vissing, J., Kennaway, N.G., and Shoubridge, E.A. (2003). Identification and characterization of a common set of complex I assembly intermediates in mitochondria from patients with complex I deficiency. *J. Biol. Chem.* **278**, 43081–43088.

Arrigoni, O. and De Tullio, M.C. (2002). Ascorbic acid: Much more than just an antioxidant. *Biochim. Biophys. Acta* **1569**, 1–9.

Atteia, A. (1994). Identification of mitochondrial respiratory proteins from the green alga *Chlamydomonas reinhardtii*. *C.R. Acad. Sci. Paris* **317**, 11–19.

Atteia, A. and Franzén, L.-G. (1996). Identification, cDNA sequence and deduced amino acid sequence of the mitochondrial Rieske iron-sulfur protein from the green alga *Chlamydomonas reinhardtii*. Implications for protein targeting and subunit interaction. *Eur. J. Biochem.* **237**, 792–799.

Atteia, A., Dreyfus, G., and González-Halphen, D. (1997). Characterization of the alpha and beta-subunits of the F_0F_1-ATPase from the alga *Polytomella* spp., a colorless relative of *Chlamydomonas reinhardtii*. *Biochim. Biophys. Acta* **1320**, 275–284.

Atteia, A., van Lis, R., Wetterskog, D., Gutiérrez-Cirlos, E.B., Ongay-Larios, L., Franzén, L.G., and González-Halphen, D. (2003). Structure, organization and expression of the genes encoding mitochondrial cytochrome c_1 and the Rieske iron-sulfur protein in *Chlamydomonas reinhardtii*. *Mol. Genet. Genomics* **268**, 637–644.

Atteia, A., van Lis, R., Gelius-Dietrich, G., Adrait, A., Garin, J., Joyard, J., Rolland, N., and Martin, W. (2006). Pyruvate formate-lyase and a novel route of eukaryotic ATP synthesis in *Chlamydomonas* mitochondria. *J. Biol. Chem.* **281**, 9909–9918.

Barrientos, A., Korr, D., and Tzagoloff, A. (2002). Shy1p is necessary for full expression of mitochondrial COX1 in the yeast model of Leigh's syndrome. *EMBO J.* **21**, 43–52.

Barros, M.H. and Tzagoloff, A. (2002). Regulation of the heme A biosynthetic pathway in *Saccharomyces cerevisiae*. *FEBS Lett.* **516**, 119–123.

Barros, M.H., Johnson, A., and Tzagoloff, A. (2004). COX23, a homologue of COX17, is required for cytochrome oxidase assembly. *J. Biol. Chem.* **279**, 31943–31947.

Barros, M.H., Johnson, A., Gin, P., Marbois, B.N., Clarke, C.F., and Tzagoloff, A. (2005). The *Saccharomyces cerevisiae COQ10* gene encodes a START domain protein required for function of coenzyme Q in respiration. *J. Biol. Chem.* **280**, 42627–42635.

Bartoli, C.G., Pastori, G.M., and Foyer, C.H. (2000). Ascorbate biosynthesis in mitochondria is linked to the electron transport chain between complexes III and IV. *Plant Physiol.* **123**, 335–344.

Baurain, D., Dinant, M., Coosemans, N., and Matagne, R.F. (2003). Regulation of the alternative oxidase *Aox1* gene in *Chlamydomonas reinhardtii*. Role of the nitrogen source on the expression of a reporter gene under the control of the Aox1 promoter. *Plant Physiol.* **131**, 1418–1430.

Berthold, D.A., Andersson, M.E., and Nordlund, P. (2000). New insight into the structure and function of the alternative oxidase. *Biochim. Biophys. Acta* **1460**, 241–254.

Bonnefoy, N. and Fox, T.D. (2002). Genetic transformation of *Saccharomyces cerevisiae* mitochondria. *Methods Enzymol.* **350**, 97–111.

Bousquet, I., Dujardin, G., and Slonimski, P.P. (1991). *ABC1*, a novel yeast nuclear gene has a dual function in mitochondria: It suppresses a cytochrome b mRNA translation defect and is essential for the electron transfer in the bc_1 complex. *EMBO J.* **10**, 2023–2031.

Braun, H.P. and Schmitz, U.K. (1995). The bifunctional cytochrome c reductase/processing peptidase complex from plant mitochondria. *J. Bioenerg. Biomembr.* **27**, 423–436.

Brunner, S., Everard-Gigot, V., and Stuart, R.A. (2002). Su e of the yeast F_1F_0-ATP synthase forms homodimers. *J. Biol. Chem.* **277**, 48484–48489.

Cardol, P., Matagne, R.F., and Remacle, C. (2002). Impact of mutations affecting ND mitochondria-encoded subunits on the activity and assembly of complex I in *Chlamydomonas*. Implication for the structural organization of the enzyme. *J. Mol. Biol.* **319**, 1211–1221.

Cardol, P., Vanrobaeys, F., Devreese, B., Van Beeumen, J., Matagne, R., and Remacle, C. (2004). Higher plant-like subunit composition of the mitochondrial complex I from *Chlamydomonas reinhardtii*: 31 conserved components among eukaryotes. *Biochim. Biophys. Acta* **1658**, 212–224.

Cardol, P., González-Halphen, D., Reyes-Prieto, A., Baurain, D., Matagne, R.F., and Remacle, C. (2005). The mitochondrial oxidative phosphorylation proteome of *Chlamydomonas reinhardtii* deduced from the Genome Sequencing Project. *Plant Physiol.* **137**, 447–459.

Cardol, P., Lapaille, M., Minet, P., Matagne, R.F., Franck, F., and Remacle, C. (2006). ND3 and ND4L subunits of mitochondrial complex I, both nucleus-encoded in *Chlamydomonas reinhardtii,* are required for the activity and assembly of the enzyme. *Eukaryotic Cell* **5**, 1460–1467.

Cardol, P., Boutaffala, L., Memmy, S., Devreese, B., Matagne, R.F., and Remacle, C. (2008). In *Chlamydomonas*, the loss of ND5 subunit prevents the assembly of whole mitochondrial complex I and leads to the formation of a low abundant 700 kD subcomplex. *Biochim. Biophys. Acta*, **1777**, 388–396.

Carlson, C.G., Barrientos, A., Tzagoloff, A., and Glerum, D.M. (2003). *COX16* encodes a novel protein required for the assembly of cytochrome oxidase in *Saccharomyces cerevisiae*. *J. Biol. Chem.* **278**, 3770–3775.

Carroll, J., Fearnley, I.M., Skehel, J.M., Shannon, R.J., Hirst, J., and Walker, J.E. (2006). Bovine complex I is a complex of forty-five different subunits. *J. Biol. Chem.* **281**, 32724–32727.

Cecchini, G., Schroder, I., Gunsalus, R.P., and Maklashina, E. (2002). Succinate dehydrogenase and fumarate reductase from *Escherichia coli*. *Biochim. Biophys. Acta* **1553**, 140–157.

Claros, M.G. and Vincens, P. (1996). Computational method to predict mitochondrially imported proteins and their targeting sequences. *Eur. J. Biochem.* **241**, 779–786.

Claros, M.G., Perea, J., Shu, Y., Samatey, F.A., Popot, J.L., and Jacq, C. (1995). Limitations to *in vivo* import of hydrophobic proteins into yeast mitochondria. The case of a cytoplasmically synthesized apocytochrome b. *Eur. J. Biochem.* **228**, 762–771.

Clifton, R., Millar, A.H., and Whelan, J. (2006). Alternative oxidases in *Arabidopsis*: A comparative analysis of differential expression in the gene family provides new insights into function of non-phosphorylating bypasses. *Biochim. Biophys. Acta* **1757**, 730–741.

Colin, M., Dorthu, M.P., Duby, F., Remacle, C., Dinant, M., Wolwertz, M.R., Duyckaerts, C., Sluse, F., and Matagne, R.F. (1995). Mutations affecting the mitochondrial genes encoding the cytochrome oxidase subunit I and apocytochrome b of *Chlamydomonas reinhardtii*. *Mol. Gen. Genet.* **249**, 179–184.

Collinson, I.R., Runswick, M.J., Buchanan, S.K., Fearnley, I.M., Skehel, JM., van Raaij, M.J., Griffiths, D.E., and Walker, J.E. (1994). F_0 membrane domain of ATP synthase from bovine heart mitochondria: Purification, subunit composition, and reconstitution with F_1-ATPase. *Biochemistry* **33**, 7971–7978.

Crivellone, M.D. (1994). Characterization of *CBP4*, a new gene essential for the expression of ubiquinol-cytochrome c reductase in *Saccharomyces cerevisiae*. *J. Biol. Chem.* **269**, 21284–21292.

Cruciat, C.M., Hell, K., Folsch, H., Neupert, W., and Stuart, R.A. (1999). Bcs1p, an AAA-family member, is a chaperone for the assembly of the cytochrome bc_1 complex. *EMBO J.* **18**, 5226–5233.

Daley, D.O., Clifton, R., and Whelan, J. (2002). Intracellular gene transfer: reduced hydrophobicity facilitates gene transfer for subunit 2 of cytochrome c oxidase. *Proc. Natl. Acad. Sci. U. S. A.* **99**, 10510–10515.

Derzaph, K.L.M. and Weger, H.G. (1996). Immunological identification of the alternative oxidase in *Chlamydomonas reinhardtii* (Chlorophyta). *J. Phycol.* **32**, 621–623.

Dibrov, E., Fu, S., and Lemire, B.D. (1998). The *Saccharomyces cerevisiae TCM62* gene encodes a chaperone necessary for the assembly of the mitochondrial succinate dehydrogenase (complex II). *J. Biol. Chem.* **273**, 32042–32048.

Dieckmann, C.L. and Tzagoloff, A. (1985). Assembly of the mitochondrial membrane system. *CBP6*, a yeast nuclear gene necessary for synthesis of cytochrome b. *J. Biol. Chem.* **260**, 1513–1520.

Dieckmann, C.L., Pape, L.K., and Tzagoloff, A. (1982). Identification and cloning of a yeast nuclear gene (*CBP1*) involved in expression of mitochondrial cytochrome b. *Proc. Natl. Acad. Sci. U. S. A.* **79**, 1805–1809.

Dinant, M., Baurain, D., Coosemans, N., Joris, B., and Matagne, R.F. (2001). Characterization of two genes encoding the mitochondrial alternative oxidase in *Chlamydomonas reinhardtii*. *Curr. Genet.* **39**, 101–108.

Do, T.Q., Hsu, A.Y., Jonassen, T., Lee, P.T., and Clarke, C.F. (2001). A defect in coenzyme Q biosynthesis is responsible for the respiratory deficiency in *Saccharomyces cerevisiae abc1* mutants. *J. Biol. Chem.* **276**, 18161–18168.

Dorthu, M.P., Remy, S., Michel-Wolwertz, M.R., Colleaux, L., Breyer, D., Beckers, M.C., Englebert, S., Duyckaerts, C., Sluse, F.E., and Matagne, R.F. (1992). Biochemical, genetic and molecular characterization of new respiratory-deficient mutants in *Chlamydomonas reinhardtii*. *Plant Mol. Biol.* **18**, 759–772.

Duby, F. and Matagne, R.F. (1999). Alteration of dark respiration and reduction of phototrophic growth in a mitochondrial DNA deletion mutant of *Chlamydomonas* lacking *cob*, *nd4*, and the 3′ end of *nd5*. *Plant Cell* **11**, 115–125.

Dudkina, N.V., Heinemeyer, J., Keegstra, W., Boekema, E.J., and Braun, H.P. (2005). Structure of dimeric ATP synthase from mitochondria: an angular association of monomers induces the strong curvature of the inner membrane. *FEBS Lett.* **579**, 5769–5772.

Dupuis, A., Chevallet, M., Darrouzet, E., Duborjal, H., Lunardi, J., and Issartel, J.P. (1998). The complex I from Rhodobacter capsulatus. *Biochim. Biophys. Acta* **1364**, 147–165.

Elhafez, D., Murcha, M.W., Clifton, R., Soole, K.L., Day, D.A., and Whelan, J. (2006). Characterization of mitochondrial alternative NAD(P)H dehydrogenases in *Arabidopsis*: intraorganelle location and expression. *Plant Cell. Physiol.* **47**, 43–54.

Ellis, T.P., Helfenbein, K.G., Tzagoloff, A., and Dieckmann, C.L. (2004). Aep3p stabilizes the mitochondrial bicistronic mRNA encoding subunits 6 and 8 of the H+-translocating ATP synthase of *Saccharomyces cerevisiae*. *J. Biol. Chem.* **279**, 15728–15733.

Eriksson, M., Gardeström, P., and Samuelsson, G. (1995). Isolation, purification and characterization of mitochondria from *Chlamydomonas reinhardtii*. *Plant Physiol.* **107**, 479–483.

Eubel, H., Jansch, L., and Braun, H.P. (2003). New insights into the respiratory chain of plant mitochondria. Supercomplexes and a unique composition of complex II. *Plant Physiol.* **133**, 274–286.

Fan, J. and Lee, R.W. (2002). Mitochondrial genome of the colorless green alga *Polytomella parva*: Two linear DNA molecules with homologous inverted repeat termini. *Mol. Biol. Evol.* **19**, 999–1007.

Felitti, S.A., Chan, R.L., Sierra, M.G., and Gonzalez, D.H. (2000). The cytochrome c gene from the green alga *Chlamydomonas reinhardtii,* Structure and expression in wild-type cells and in obligate photoautotrophic (*dk*) mutants. *Plant Cell Physiol.* **41**, 1149–1156.

Finnegan, P.M., Ellis, T.P., Nagley, P., and Lukins, H.B. (1995). The mature AEP2 gene product of *Saccharomyces cerevisiae*, required for the expression of subunit 9 of ATP synthase, is a 58 kDa mitochondrial protein. *FEBS Lett.* **368**, 505–508.

Fisher, N., Bourges, I., Hill, P., Brasseur, G., and Meunier, B. (2004). Disruption of the interaction between the Rieske iron-sulfur protein and cytochrome b in the yeast bc_1 complex owing to a human disease-associated mutation within cytochrome b. *Eur. J. Biochem.* **271**, 1292–1298.

Franzén, L.G. and Falk, G. (1992). Nucleotide sequence of cDNA clones encoding the beta subunit of mitochondrial ATP synthase from the green alga *Chlamydomonas reinhardtii*: The precursor protein encoded by the cDNA contains both an N-terminal presequence and a C-terminal extension. *Plant Mol. Biol.* **19**, 771–780.

Friedrich, T., Abelmann, A., Brors, B., Guenebaut, V., Kintscher, L., Leonard, K., Rasmussen, T., Scheide, D., Schlitt, A., Schulte, U., and Weiss, H. (1998). Redox components and structure of the respiratory NADH:ubiquinone oxidoreductase (complex I). *Biochim. Biophys. Acta* **1365**, 215–219.

Fronzes, R., Weimann, T., Vaillier, J., Velours, J., and Brethes, D. (2006). The peripheral stalk participates in the yeast ATP synthase dimerization independently of e and g subunits. *Biochemistry* **45**, 6715–6723.

Funes, S., Davidson, E., Claros, M.G., van Lis, R., Pérez-Martínez, X., Vázquez-Acevedo, M., King, M.P., and González-Halphen, D. (2002a). The typically mitochondrial DNA-encoded ATP6 subunit of the F_1F_0-ATPase is encoded by a nuclear gene in *Chlamydomonas reinhardtii*. *J. Biol. Chem.* **277**, 6051–6058.

Funes, S., Davidson, E., Reyes-Prieto, A., Magallon, S., Herion, P., King, M.P., and González-Halphen, D. (2002b). A green algal apicoplast ancestor. *Science* **298**, 2155.

Funes, S., Nargang, F.E., Neupert, W., and Herrmann, J.M. (2004). The Oxa2 protein of *Neurospora crassa* plays a critical role in the biogenesis of cytochrome oxidase and defines a ubiquitous subbranch of the Oxa1/YidC/Alb3 protein family. *Mol. Biol. Cell* **15**, 1853–1861.

Gardner, M.J., Hall, N., Fung, E., White, O., Berriman, M., Hyman, R.W., Carlton, J.M., Pain, A., Nelson, K.E., Bowman, S. et al. (2002). Genome sequence of the human malaria parasite *Plasmodium falciparum*. *Nature* **419**, 498–511.

Gfeller, R.P. and Gibbs, M. (1984). Fermentative metabolism of *Chlamydomonas reinhardtii*: I. Analysis of fermentative products from starch in dark and light. *Plant Physiol.* **75**, 212–218.

Glaser, E. and Dessi, P. (1999). Integration of the mitochondrial-processing peptidase into the cytochrome bc_1 complex in plants. *J. Bioenerg. Biomembr.* **31**, 259–274.

Gledhill, J.R. and Walker, J.E. (2005). Inhibition sites in F_1-ATPase from bovine heart mitochondria. *Biochem. J.* **386**, 591–598.

Glerum, D.M. and Tzagoloff, A. (1994). Isolation of a human cDNA for heme A: farnesyltransferase by functional complementation of a yeast *cox10* mutant. *Proc. Natl. Acad. Sci. U. S. A.* **91**, 8452–8456.

Golden, T.R. and Pedersen, P.L. (1998). The oligomycin sensitivity conferring protein of rat liver mitochondrial ATP synthase: arginine 94 is important for the binding of OSCP to F_1. *Biochemistry* **37**, 13871–13881.

Goyal, A. and Tolbert, N.E. (1989). Variations in the alternative oxidase in *Chlamydomonas* grown in air or high CO_2. *Plant Physiol.* **89**, 958–962.

Gray, M.W. and Boer, P.H. (1988). Organization and expression of algal (*Chlamydomonas reinhardtii*) mitochondrial DNA. *Phil. Trans. R. Soc. Lond. B* **319**, 135–147.

Guenebaut, V., Schlitt, A., Weiss, H., Leonard, K., and Friedrich, T. (1998). Consistent structure between bacterial and mitochondrial NADH:ubiquinone oxidoreductase (complex I). *J. Mol. Biol.* **276**, 105–112.

Hamanaka, S., Ohtsu, K., Kadowaki, K., Nakazono, M., and Hirai, A. (1999). Identification of cDNA encoding cytochrome c oxidase subunit 5c (COX5c) from rice: Comparison of its expression with nuclear-encoded and mitochondrial-encoded COX genes. *Genes Genet. Syst.* **74**, 71–75.

He, S. and Fox, T.D. (1997). Membrane translocation of mitochondrially coded Cox2p: Distinct requirements for export of N and C termini and dependence on the conserved protein Oxa1p. *Mol. Biol. Cell* **8**, 1449–1460.

Heazlewood, J.L., Howell, K.A., and Millar, A.H. (2003). Mitochondrial complex I from *Arabidopsis* and rice: Orthologs of mammalian and fungal components coupled with plant-specific subunits. *Biochim. Biophys. Acta* **1604**, 159–169.

Helfenbein, K.G., Ellis, T.P., Dieckmann, C.L., and Tzagoloff, A. (2003). *ATP22*, a nuclear gene required for expression of the F_0 sector of mitochondrial ATPase in *Saccharomyces cerevisiae*. *J. Biol. Chem.* **278**, 19751–19756.

Helmerhorst, E.J., Stan, M., Murphy, M.P., Sherman, F., and Oppenheim, F. G. (2005). The concomitant expression and availability of conventional and alternative, cyanide-insensitive, respiratory pathways in *Candida albicans*. *Mitochondrion* **5**, 200–211.

Herrmann, J.M. and Neupert, W. (2003). Protein insertion into the inner membrane of mitochondria. *IUBMB Life* **55**, 219–225.

Hinchliffe, P. and Sazanov, L.A. (2005). Organization of iron-sulfur clusters in respiratory complex I. *Science* **309**, 771–774.

Hommersand, M.H. and Thimann, K.V. (1965). Terminal respiration of vegetative cells and zygospores in *Chlamydomonas reinhardtii*. *Plant Physiol.* **40**, 1220–1227.

Horng, Y.C., Cobine, P.A., Maxfield, A.B., Carr, H.S., and Winge, D.R. (2004). Specific copper transfer from the Cox17 metallochaperone to both Sco1 and Cox11 in the assembly of yeast cytochrome C oxidase. *J. Biol. Chem.* **279**, 35334–35340.

Itoh, H., Takahashi, A., Adachi, K., Noji, H., Yasuda, R., Yoshida, M., and Kinosita, K. (2004). Mechanically driven ATP synthesis by F_1-ATPase. *Nature* **427**, 465–468.

Iwata, S., Lee, J.W., Okada, K., Lee, J.K., Iwata, M., Rasmussen, B., Link, T.A., Ramaswamy, S., and Jap, B.K. (1998). Complete structure of the 11-subunit bovine mitochondrial cytochrome bc_1 complex. *Science* **281**, 64–71.

Jansch, L., Kruft, V., Schmitz, U.K., and Braun, H.P. (1996). New insights into the composition, molecular mass and stoichiometry of the protein complexes of plant mitochondria. *Plant J.* **9**, 357–368.

Johnson, A., Gin, P., Marbois, B.N., Hsieh, E.J., Wu, M., Barros, M.H., Clarke, C. F., and Tzagoloff, A. (2005). *COQ9*, a new gene required for the biosynthesis of coenzyme Q in *Saccharomyces cerevisiae*. *J. Biol. Chem.* **280**, 31397–31404.

Juárez, O., Guerra, G., Velázquez, I., Flores-Herrera, O., Rivera-Pérez, R.E., and Pardo, J.P. (2006). The physiologic role of alternative oxidase in *Ustilago maydis*. *FEBS J.* **273**, 4603–4615.

Ko, Y.H., Hullihen, J., Hong, S., and Pedersen, P.L. (2000). Mitochondrial F_0F_1 ATP synthase. Subunit regions on the F_1 motor shielded by F_0, functional significance, and evidence for an involvement of the unique F_0 subunit F_6. *J. Biol. Chem.* **275**, 32931–32939.

Kranz, R., Lill, R., Goldman, B., Bonnard, G., and Merchant, S. (1998). Molecular mechanisms of cytochrome c biogenesis: Three distinct systems. *Mol. Microbiol.* **29**, 383–396.

Kreuzberg, K. (1984). Starch fermentation via formate producing pathway *Chlamydomonas reinhardii*, *Chlorogonium elongatum*, and *Chlorella fusca*. *Plant Physiol.* **61**, 87–94.

Krimm, I., Gans, P., Hernandez, J.F., Arlaud, G.J., and Lancelin, J.M. (1999). A coil-helix instead of a helix-coil motif can be induced in a chloroplast transit peptide from *Chlamydomonas reinhardtii*. *Eur. J. Biochem.* **265**, 171–180.

Kuffner, R., Rohr, A., Schmiede, A., Krull, C., and Schulte, U. (1998). Involvement of two novel chaperones in the assembly of mitochondrial NADH:ubiquinone oxidoreductase (complex I). *J. Mol. Biol.* **283**, 409–417.

Lancelin, J.M., Bally, I., Arlaud, G.J., Blackledge, M., Gans, P., Stein, M., and Jacquot, J.P. (1994). NMR structures of ferredoxin chloroplastic transit peptide from *Chlamydomonas reinhardtii* promoted by trifluoroethanol in aqueous solution. *FEBS Lett.* **343**, 261–266.

Lancelin, J.M., Gans, P., Bouchayer, E., Bally, I., Arlaud, G.J., and Jacquot, J.P. (1996). NMR structures of a mitochondrial transit peptide from the green alga *Chlamydomonas reinhardtii*. *FEBS Lett.* **391**, 203–208.

Leary, S.C., Kaufman, B.A., Pellecchia, G., Guercin, G.H., Mattman, A., Jaksch, M., and Shoubridge, E.A. (2004). Human SCO1 and SCO2 have independent, cooperative functions in copper delivery to cytochrome c oxidase. *Hum. Mol. Genet.* **13**, 1839–1848.

Lefebvre-Legendre, L., Vaillier, J., Benabdelhak, H., Velours, J., Slonimski, P.P., and di Rago, J.P. (2001). Identification of a nuclear gene (FMC1) required for the assembly/stability of yeast mitochondrial F(1)-ATPase in heat stress conditions. *J. Biol. Chem.* **276**, 6789–6796.

Lemire, B.D. and Oyedotun, K.S. (2002). The *Saccharomyces cerevisiae* mitochondrial succinate:ubiquinone oxidoreductase. *Biochim. Biophys. Acta* **1553**, 102–116.

Lown, F.J., Watson, A.T., and Purton, S. (2001). *Chlamydomonas* nuclear mutants that fail to assemble respiratory or photosynthetic electron transfer complexes. *Biochem. Soc. Trans.* **29**, 452–455.

Luttik, M.A., Overkamp, K.M., Kotter, P., de Vries, S., van Dijken, J.P., and Pronk, J.T. (1998). The *Saccharomyces cerevisiae NDE1* and *NDE2* genes encode separate mitochondrial NADH dehydrogenases catalyzing the oxidation of cytosolic NADH. *J. Biol. Chem.* **273**, 24529–24534.

Marques, I., Duarte, M., Assuncao, J., Ushakova, A.V., and Videira, A. (2005). Composition of complex I from *Neurospora crassa* and disruption of two "accessory" subunits. *Biochim. Biophys. Acta* **1707**, 211–220.

Marres, C.A., de Vries, S., and Grivell, L.A. (1991). Isolation and inactivation of the nuclear gene encoding the rotenone-insensitive internal NADH:ubiquinone oxidoreductase of mitochondria from *Saccharomyces cerevisiae*. *Eur. J. Biochem.* **195**, 857–862.

Matagne, R.F., Michel-Wolwertz, M.R., Munaut, C., Duyckaerts, C., and Sluse, F. (1989). Induction and characterization of mitochondrial DNA mutants in *Chlamydomonas reinhardtii*. *J. Cell. Biol.* **108**, 1221–1226.

Matsuo, M. and Obokata, J. (2006). Remote control of photosynthetic genes by the mitochondrial respiratory chain. *Plant J.* **47**, 873–882.

Maxwell, D.P., Wang, Y., and McIntosh, L. (1999). The alternative oxidase lowers mitochondrial reactive oxygen production in plant cells. *Proc. Natl. Acad. Sci. U. S. A.* **96**, 8271–8276.

McEwen, J.E., Hong, K.H., Park, S., and Preciado, G.T. (1993). Sequence and chromosomal localization of two PET genes required for cytochrome c oxidase assembly in *Saccharomyces cerevisiae*. *Curr. Genet.* **23**, 9–14.

McGraw, P. and Tzagoloff, A. (1983). Assembly of the mitochondrial membrane system. Characterization of a yeast nuclear gene involved in the processing of the cytochrome b pre-mRNA. *J. Biol. Chem.* **258**, 9459–9468.

Meyer, E.H., Taylor, N.L., and Millar, A.H. (2008). Resolving and Identifying protein components of plant mitochondrial respiratory complexes using three dimensions of gel electrophoresis. *J. Proteome Res.* **7**, 786–794.

Michaelis, G., Vahrenholz, C., and Pratje, E. (1990). Mitochondrial DNA of *Chlamydomonas reinhardtii*: The gene for apocytochrome b and the complete functional map of the 15.8 kb DNA. *Mol. Gen. Genet.* **223**, 211–216.

Michalecka, A.M., Svensson, A.S., Johansson, F.I., Agius, S.C., Johanson, U., Brennicke, A., Binder, S., and Rasmusson, A.G. (2003). *Arabidopsis* genes encoding mitochondrial type II NAD(P)H dehydrogenases have different evolutionary origin and show distinct responses to light. *Plant Physiol.* **133**, 642–652.

Millar, A.H., Mittova, V., Kiddle, G., Heazlewood, J.L., Bartoli, C.G., Theodoulou, F.L., and Foyer, C.H. (2003). Control of ascorbate synthesis by respiration and its implications for stress responses. *Plant Physiol.* **133**, 443–447.

Millar, A.H., Heazlewood, J.L., Kristensen, B.K., Braun, H.P., and Møller, I.M. (2005). The plant mitochondrial proteome. *Trends Plant Sci.* **10**, 36–43.

Molen, T.A., Rosso, D., Piercy, S., and Maxwell, D.P. (2006). Characterization of the alternative oxidase of *Chlamydomonas reinhardtii* in response to oxidative stress and a shift in nitrogen source. *Physiol. Plant.* **127**, 174–186.

Moller, I.M., Rasmusson, A.G., and Fredlund, K.M. (1993). NAD(P)H-ubiquinone oxidoreductases in plant mitochondria. *J. Bioenerg. Biomembr.* **25**, 377–384.

Mus, F., Cournac, L., Cardettini, V., Caruana, A., and Peltier, G. (2005). Inhibitor studies on non-photochemical plastoquinone reduction and H_2 photoproduction in *Chlamydomonas reinhardtii*. *Biochim. Biophys. Acta* **1708**, 322–332.

Neupert, W. (1997). Protein import into mitochondria. *Annu. Rev. Biochem.* **66**, 863–917.

Nobrega, F.G., Nobrega, M.P., and Tzagoloff, A. (1992). *BCS1*, a novel gene required for the expression of functional Rieske iron-sulfur protein in *Saccharomyces cerevisiae*. *EMBO J.* **11**, 3821–3829.

Nobrega, M.P., Bandeira, S.C., Beers, J., and Tzagoloff, A. (2002). Characterization of *COX19*, a widely distributed gene required for expression of mitochondrial cytochrome oxidase. *J. Biol. Chem.* **277**, 40206–40211.

Nurani, G. and Franzén, L.-G. (1996). Isolation and characterization of the mitochondrial ATP synthase from *Chlamydomonas reinhardtii*. cDNA sequence and deduced protein sequence of the alpha subunit. *Plant Mol. Biol.* **31**, 1105–1116.

Nurani, G., Eriksson, M., Knorpp, C., Glaser, E., and Franzen, L.-G. (1997). Homologous and heterologous protein import into mitochondria isolated from the green alga *Chlamydomonas reinhardtii*. *Plant Mol. Biol.* **35**, 973–980.

Ogata, K. and Volini, M. (1990). Mitochondrial rhodanese: Membrane-bound and complexed activity. *J. Biol. Chem.* **265**, 8087–8093.

Ogilvie, I., Kennaway, N.G., and Shoubridge, E.A. (2005). A molecular chaperone for mitochondrial complex I assembly is mutated in a progressive encephalopathy. *J. Clin. Invest.* **115**, 2784–2792.

Ostergaard, J., Persiau, G., Davey, M.W., Bauw, G., and Van Montagu, M. (1997). Isolation of a cDNA coding for L-galactono-γ-lactone dehydrogenase, an enzyme involved in the biosynthesis of ascorbic acid in plants. Purification, characterization, cDNA cloning, and expression in yeast. *J. Biol. Chem.* **272**, 30009–30016.

Parisi, G., Perales, M., Silvina Fornasari, M., Colaneri, A., Gonzalez-Schain, D., Gomez-Casati, D., Zimmermann, S., Brennicke, A., Araya, A., Ferry, J.G., Echave, J., and Zabaleta, E. (2004). Gamma carbonic anhydrases in plant mitochondria. *Plant Mol. Biol.* **55**, 193–207.

Payne, M.J., Schweizer, E., and Lukins, H.B. (1991). Properties of two nuclear *pet* mutants affecting expression of the mitochondrial *oli1* gene of *Saccharomyces cerevisiae*. *Curr. Genet.* **19**, 343–351.

Peltier, G. and Cournac, L. (2002). Chlororespiration. *Annu. Rev. Plant. Biol.* **53**, 523–550.

Perales, M., Eubel, H., Heinemeyer, J., Colaneri, A., Zabaleta, E., and Braun, H.P. (2005). Disruption of a nuclear gene encoding a mitochondrial gamma carbonic anhydrase reduces complex I and supercomplex I + III_2 levels and alters mitochondrial physiology in *Arabidopsis*. *J. Mol. Biol.* **350**, 263–277.

Pérez-Martínez, X., Vázquez-Acevedo, M., Tolkunova, E., Funes, S., Claros, M.G., Davidson, E., King, M.P., and González-Halphen, D. (2000). Unusual location of a mitochondrial gene, Subunit III of cytochrome C oxidase is encoded in the nucleus of Chlamydomonad algae. *J. Biol. Chem.* **275**, 30144–30152.

Pérez-Martínez, X., Antaramian, A., Vázquez-Acevedo, M., Funes, S., Tolkunova, E., d'Alayer, J., Claros, M.G., Davidson, E., King, M.P., and González-Halphen, D. (2001). Subunit II of cytochrome c oxidase in Chlamydomonad algae is a heterodimer encoded by two independent nuclear genes. *J. Biol. Chem.* **276**, 11302–11309.

Pröschold, T., Marin, B., Schlösser, U.G., and Melkonian, M. (2001). Molecular phylogeny and taxonomic revision of *Chlamydomonas* (Chlorophyta). I. Emendation of *Chlamydomonas* Ehrenberg and *Chloromonas* Gobi, and description of *Oogamochlamys* gen. nov. and *Lobochlamys* gen. nov. *Protist* **152**, 265–300.

Radermacher, M., Ruiz, T., Clason, T., Benjamin, S., Brandt, U., and Zickermann, V. (2006). The three-dimensional structure of complex I from *Yarrowia lipolytica*: A highly dynamic enzyme. *J. Struct. Biol.* **154**, 269–279.

Raghavendra, A.S. and Padmasree, K. (2003). Beneficial interactions of mitochondrial metabolism with photosynthetic carbon assimilation. *Trends Plant Sci.* **8**, 546–553.

Remacle, C., Baurain, D., Cardol, P., and Matagne, R.F. (2001). Mutants of *Chlamydomonas reinhardtii* deficient in mitochondrial complex I. Characterization of two mutations affecting the *nd1* coding sequence. *Genetics* **158**, 1051–1060.

Remacle, C., Gloire, G., Cardol, P., and Matagne, R. (2004). Impact of a mutation in the mitochondrial LSU rRNA gene from *Chlamydomonas reinhardtii* on the activity and the assembly of respiratory-chain complexes. *Curr. Genet.* **45**, 323–330.

Remacle, C., Cardol, P., Coosemans, N., Gaisne, M., and Bonnefoy, N. (2006). High-efficiency biolistic transformation of *Chlamydomonas* mitochondria can be used to insert mutations in complex I genes. *Proc. Natl. Acad. Sci. U. S. A.* **103**, 4771–4776.

Reyes-Prieto, A., El-Hafidi, M., Moreno-Sánchez, R., and González-Halphen, D. (2002). Characterization of oxidative phosphorylation in the colorless chlorophyte *Polytomella* sp. Its mitochondrial respiratory chain lacks a plant-like alternative oxidase. *Biochim. Biophys. Acta* **1554**, 170–179.

Rizzuto, R., Sandona, D., Brini, M., Capaldi, R.A., and Bisson, R. (1991). The most conserved nuclear-encoded polypeptide of cytochrome c oxidase is the putative zinc-binding subunit: primary structure of subunit V from the slime mold *Dictyostelium discoideum*. *Biochim. Biophys. Acta* **1129**, 100–104.

Rodel, G. (1986). Two yeast nuclear genes, *CBS1* and *CBS2*, are required for translation of mitochondrial transcripts bearing the 5′-untranslated COB leader. *Curr. Genet.* **11**, 41–45.

Saracco, S.A. and Fox, T.D. (2002). Cox18p is required for export of the mitochondrially encoded *Saccharomyces cerevisiae* Cox2p C-tail and interacts with Pnt1p and Mss2p in the inner membrane. *Mol. Biol. Cell* **13**, 1122–1131.

Saraste, M. (1999). Oxidative phosphorylation at the fin de siecle. *Science* **283**, 1488–1493.

Sazanov, L.A., Burrows, P.A., and Nixon, P.J. (1998). The plastid *ndh* genes code for an NADH-specific dehydrogenase: Isolation of a complex I analogue from pea thylakoid membranes. *Proc. Natl. Acad. Sci. U. S. A.* **95**, 1319–1324.

Sazanov, L.A., Carroll, J., Holt, P., Toime, L., and Fearnley, I.M. (2003). A role for native lipids in the stabilization and two-dimensional crystallization of the *Escherichia coli* NADH-ubiquinone oxidoreductase (complex I). *J. Biol. Chem.* **278**, 19483–19491.

Schönfeld, C., Wobbe, L., Borgstadt, R., Kienast, A., Nixon, P.J., and Kruse, O. (2004). The nucleus-encoded protein MOC1 is essential for mitochondrial light acclimation in *Chlamydomonas reinhardtii*. *J. Biol. Chem.* **279**, 50366–50374.

Shi, G., Crivellone, M.D., and Edderkaoui, B. (2001). Identification of functional regions of Cbp3p, an enzyme-specific chaperone required for the assembly of ubiquinol-cytochrome c reductase in yeast mitochondria. *Biochim. Biophys. Acta* **1506**, 103–116.

Shrager, J., Hauser, C., Chang, C.W., Harris, E.H., Davies, J., McDermott, J., Tamse, R., Zhang, Z., and Grossman, A.R. (2003). *Chlamydomonas reinhardtii* genome project. A guide to the generation and use of the cDNA information. *Plant Physiol.* **131**, 401–408.

Siendones, E., González-Reyes, J.A., Santos-Ocana, C., Navas, P., and Córdoba, F. (1999). Biosynthesis of ascorbic acid in kidney bean. L-galactono-γ-lactone dehydrogenase is an intrinsic protein located at the mitochondrial inner membrane. *Plant Physiol.* **120**, 907–912.

Sluse, F.E., Jarmuszkiewicz, W., Navet, R., Douette, P., Mathy, G., and Sluse-Goffart, C.M. (2006). Mitochondrial UCPs: New insights into regulation and impact. *Biochim. Biophys. Acta* **1757**, 480–485.

Smirnoff, N. (2000). Ascorbate biosynthesis and function in photoprotection. *Phil. Trans. R. Soc. Lond. B. Biol. Sci.* **355**, 1455–1464.

Sunderhaus, S., Dudkina, N.V., Jansch, L., Klodmann, J., Heinemeyer, J., Perales, M., Zabaleta, E., Boekema, E.J., and Braun, H.P. (2006). Carbonic anhydrase subunits form a matrix-exposed domain attached to the membrane arm of mitochondrial complex I in plants. *J. Biol. Chem.* **281**, 6482–6488.

Taanman, J.W. and Capaldi, R.A. (1992). Purification of yeast cytochrome c oxidase with a subunit composition resembling the mammalian enzyme. *J. Biol. Chem.* **267**, 22481–22485.

Turunen, M., Olsson, J., and Dallner, G. (2004). Metabolism and function of coenzyme Q. *Biochim. Biophys. Acta* **1660**, 171–199.

Tzagoloff, A., Barrientos, A., Neupert, W., and Herrmann, J.M. (2004). Atp10p assists assembly of Atp6p into the F_0 unit of the yeast mitochondrial ATPase. *J. Biol. Chem.* **279**, 19775–19780.

van Lis, R., Atteia, A., Mendoza-Hernández, G., and González-Halphen, D. (2003). Identification of novel mitochondrial protein components of *Chlamydomonas reinhardtii*. A proteomic approach. *Plant Physiol.* **132**, 318–330.

van Lis, R., González-Halphen, D., and Atteia, A. (2005). Divergence of the mitochondrial electron transport chains from the green alga *Chlamydomonas reinhardtii* and its colorless close relative *Polytomella* sp. *Biochim. Biophys. Acta* **1708**, 23–34.

van Lis, R., Mendoza-Hernández, G., Groth, G., and Atteia, A. (2007). New insights into the unique structure of the F_0F_1-ATP synthase from the chlamydomonad algae *Polytomella* sp. and *Chlamydomonas reinhardtii*. *Plant Physiol.* **144**, 1190–1199.

Vázquez-Acevedo, M., Cardol, P., Cano-Estrada, A., Lapaille, M., Reyes-Prieto, A., Remacle, C., and González-Halphen, D. (2006). The mitochondrial ATP synthase of chlorophycean algae contains eight subunits of unknown origin involved in the formation of an atypical stator-stalk and in the dimerization of the complex. *J. Bioenerg. Biomembr.* **38**, 271–282.

Velours, J. and Arselin, G. (2000). The *Saccharomyces cerevisiae* ATP synthase. *J. Bioenerg. Biomembr.* **32**, 383–390.

Velours, J., Spannagel, C., Chaignepain, S., Vaillier, J., Arselin, G., Graves, P.V., Velours, G., and Camougrand, N. (1998). Topography of the yeast ATP synthase F_0 sector. *Biochimie* **80**, 793–801.

Videira, A. and Duarte, M. (2002). From NADH to ubiquinone in *Neurospora* mitochondria. *Biochim. Biophys. Acta* **1555**, 187–191.

Vogel, R.O., Smeitink, J.A., and Nijtmans, L.G. (2007). Human mitochondrial complex I assembly: A dynamic and versatile process. *Biochim Biophys Acta* **1767**, 1215–1227.

von Heijne, G., Steppuhn, J., and Herrmann, R.G. (1989). Domain structure of mitochondrial and chloroplast targeting peptides. *Eur. J. Biochem.* **180**, 535–545.

Walker, J.E. and Dickson, V.K. (2006). The peripheral stalk of the mitochondrial ATP synthase. *Biochim. Biophys. Acta* **1757**, 286–296.

Wang, Z.G., White, P.S., and Ackerman, S.H. (2001). Atp11p and Atp12p are assembly factors for the F_1-ATPase in human mitochondria. *J. Biol. Chem.* **276**, 30773–30778.

Weger, H.G., Chadderton, A.R., Lin, M., Guy, R.D., and Turpin, D.H. (1990). Cytochrome and alternative pathway respiration during transient ammonium assimilation by N-limited *Chlamydomonas reinhardtii*. *Plant Physiol.* **94**, 1131–1136.

Yanamura, W., Zhang, Y.Z., Takamiya, S., and Capaldi, R.A. (1988). Tissue-specific differences between heart and liver cytochrome c oxidase. *Biochemistry* **27**, 4909–4914.

CHAPTER 14

Light-Harvesting Proteins

Jun Minagawa
Institute of Low Temperature Science,
Hokkaido University, Japan

CHAPTER CONTENTS

I. INTRODUCTION

Photosynthesis is a process that involves the conversion of light energy (harvested by pigments that are organized in a protein complex) into a chemically stable form. Photosystems, where the photochemical reactions take place, are formed by large multisubunit chlorophyll-carotenoid–protein complexes embedded in the thylakoid membranes of chloroplasts. During the course of evolution, both photosystem I (PS I) and photosystem II (PS II) acquired a peripherally located multisubunit light-harvesting antenna apparatus that is responsible for capturing light energy and delivering it to the photochemical reaction centers, where it is trapped. In green plants, the light-harvesting antennae for PS I and PS II are composed of light-harvesting complex I (LHCI) and light-harvesting complex II (LHCII) proteins, respectively, which bind chlorophyll *a*, chlorophyll *b*, and carotenoids. Both LHCI and LHCII proteins share significant homology and form an extended light-harvesting complex (LHC) superfamily together with other LHC-like proteins, suggesting monophyletic origin with a common architecture (Wolf et al., 1994). Although the primary role of the LHC proteins is collecting light energy, they are also known to have another crucial role in photosynthesis, protection of the photosynthetic apparatus (Niyogi, 1999; Melkozernov and Blankenship, 2005). When the excitation rate exceeds the rate of its utilization for photochemistry, light absorption may be harmful to the photosynthetic apparatus (see Chapter 23). To limit the amount of light energy transferred to the reaction center, light-harvesting activity is down-regulated. Moreover, LHCII proteins are postulated to act in redistribution of excitation energy between the two photosystems (state transitions, see Chapter 22) and regulate thylakoid membrane stacking.

Historically, the term CAB had been widely used to describe the chlorophyll *a/b*-binding proteins for light-harvesting. However, CAB proteins are more correctly termed LHC proteins (Jansson et al., 1992), because there are additional structurally related proteins that bind chlorophyll *a* and *c* or only chlorophyll *a* instead of *chlorophyll a* and *b*. Moreover, the chlorophyll-*a/b* binding antenna proteins discovered in some cyanobacteria, including *Prochlorococcus*, are not structurally related to chlorophyll*a/b*-binding LHC proteins in photosynthetic eukaryotes (La Roche et al., 1996). The bacterial chlorophyll *a/b*-binding proteins are thus called prochlorophyte chlorophyll-binding proteins. Therefore, the term LHC is used throughout this chapter to describe the chlorophyll-binding antenna complex consisting of the monophyletic eukaryotic polypeptides carrying three transmembrane helices.

Although the LHC proteins are widely distributed among photosynthetic eukaryotes except glaucophytes, the type of bound chlorophyll varies among species, for example chlorophyll *a* and *b* in green plants (chlorophytes, streptophytes, and prasinophytes), chlorophyll *a* in red algae (rhodophytes), and chlorophyll *a* and *c* in chromophytes, haptophytes, and dinoflagellates (dinophytes) (Durnford et al., 1999). The variety of bound carotenoid species is even more divergent. In addition, each organism has different numbers of *LHC* genes, whose products are presumably organized differently. Although there is an amazingly complex composition of LHC polypeptides, technical advances are making it possible to draw a comprehensive understanding of the evolution and diversity of LHC polypeptides, especially where molecular biology has been integrated with genomics. This chapter provides an overview of light-harvesting proteins including their structure, function, assembly, and regulation, by focusing on those in *Chlamydomonas*. Supplementary material on fluorescence quenching analysis, a versatile technique to diagnose light-harvesting functions, is provided on the companion web site (http://www.elsevierdirect.com/companions/9780123708731).

II. STRUCTURE AND COMPOSITION OF THE LIGHT-HARVESTING APPARATUS

A. Light-harvesting antenna for PS II

1. Structure and composition

The PS II–LHCII supercomplex is a large chlorophyll–protein complex comprising more than 30 proteins in the thylakoid membranes of oxygen-evolving photosynthetic eukaryotes. The supercomplex collects light energy, converts it into electrochemical energy and drives electron transfer from water to plastoquinone. At the center of this supercomplex is a reaction center where the charge separation occurs. The reaction center is composed of a heterodimer of the D1 and D2 proteins (depicted in *yellow* and *orange* in Figure 14.1; see Chapter 16 for more information on these proteins) and a few low molecular weight polypeptides. Surrounding the reaction center are other subunits that form the PS II core complex, which includes the chlorophyll *a*-containing core antenna proteins CP43 and CP47 (depicted in *green* and *red* in Figure 14.1) that harvest or mediate transfer of excitation energy to the reaction center, lumenal extrinsic proteins that protect and facilitate oxygen-evolving reactions, and several other low molecular weight subunits. At the periphery of the PS II core in green plants is the additional light-harvesting antenna composed of LHCII proteins, which are the most abundant membrane proteins on earth, accounting for up to 50% of the total chlorophyll in plant and algal thylakoid membranes. The light-harvesting apparatus for PS II is formed by three intramembrane domains of protein complexes: (1) the core antenna immediately associated with reaction center (chlorophyll *a*-containing CP47 and CP43); (2) the minor "less abundant" monomeric antenna (chlorophyll *a*/*b*-containing CP29 and CP26) bordering the core antenna and major trimeric antenna (Harrer et al., 1998; Yakushevska et al., 2003); and (3) the major "more abundant" trimeric antenna residing at the outer edge of the PS II–LHCII supercomplex (Figure 14.1).

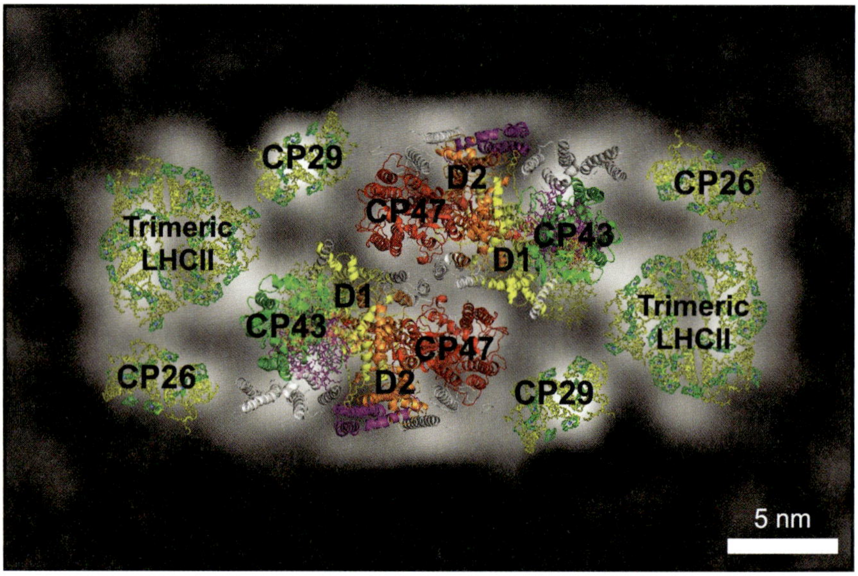

FIGURE 14.1 *Top view of the PS II-LHCII supercomplex. Overlaid X-ray structures of the cyanobacterial PS II core (Ferreira et al., 2004) without extrinsic proteins, and the spinach LHCII complex (Liu et al., 2004), onto the lumenal top view of the three-dimensional structure of the spinach PS II–LHCII supercomplex derived from cryoelectron microscopy and single particle analysis (Nield et al., 2002). A monomer taken from the X-ray structure of the LHCII co-ordinates 1RWT.pdb is used as a model for CP29 and CP26. The transmembrane helices of the D1 (yellow), D2 (orange), CP43 (green), CP47 (red), and α- and β-subunits of cytochrome b_{559} (purple) subunits are displayed as ribbons. Courtesy of J. Nield.*

The three-dimensional structure of the *Chlamydomonas* PS II–LHCII supercomplex has been determined to a resolution of 30Å by averaging negatively-stained single-particle images obtained by transmission electron microscopy (Figures 14.1 and 14.2). It was revealed that the supercomplex comprises two LHCII trimers and four LHCII monomers per PS II core dimer, with dimensions of 350Å × 175Å × 135Å (Nield et al., 2000b). This PS II–LHCII supercomplex corresponds to the C_2S_2-type supercomplex observed in higher plants (Boekema et al., 1995). In higher plants, there are also $C_2S_2M_1$- and $C_2S_2M_2$-type supercomplexes. ("C", "S", and "M" refer to core complex, strongly and moderately associated LHCII trimers, respectively.) As discussed below, higher plants have another monomeric LHCII protein, CP24, which is present only in $C_2S_2M_1$- and $C_2S_2M_2$-type supercomplexes and is not found in green algae. Therefore, CP24 seems to be prerequisite for binding M-trimers to the core in higher plants. The isolated C2S2-type supercomplex in spinach was measured to contain ~100 chlorophylls (Hankamer et al., 1997), but in the membrane, a range of 130–250 chlorophylls per reaction center, which includes all monomeric LHCIIs and 1–3 trimeric LHCII, has been proposed (Jansson et al., 1997).

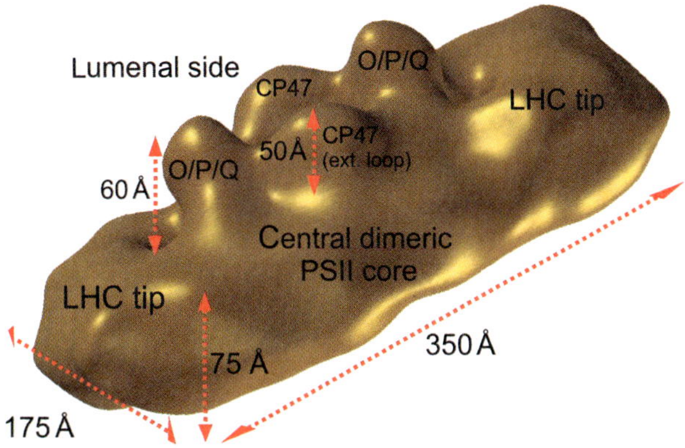

FIGURE 14.2 *Three-dimensional map of the Chlamydomonas PS II-LHCII supercomplex. A three-dimensional structure of the PS II–LHCII supercomplex from Chlamydomonas at 30 Å determined by single-particle analysis (Nield et al., 2000b) is viewed from an oblique angle to visualize the lumenal surface, the extrinsic PsbO, PsbP, and PsbQ proteins (O/P/Q), and the lumenally exposed region of the trimeric LHCII proteins (LHC tip) and CP47 (CP47 ext. loop). Courtesy of J. Nield.*

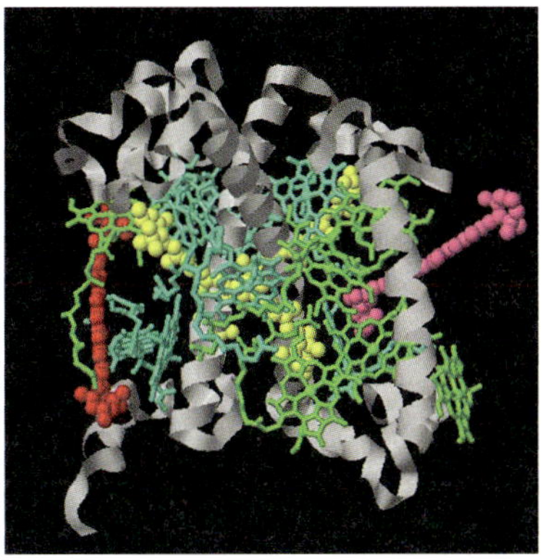

FIGURE 14.3 *Structural model of a plant LHCII protein at 2.5 Å. A side view of a monomer LHCII taken from the LHCII trimer coordinates 2NHW.pdb is represented (Standfuss et al., 2005). The observing point is the center of a trimer. The protein backbone (purple), chlorophyll* a *(blue-green), chlorophyll* b *(yellow-green), luteins (yellow), 9′-cis-neoxanthin (pink), and violaxanthin (red) are shown.*

2. Atomic structure of the major LHCII protein

The near atomic structure of major LHCII trimer was determined for the first time by electron crystallography of two-dimensional crystals from pea (Kühlbrandt et al., 1994) at 3.4 Å, and then by X-ray crystallography of three-dimensional crystals from spinach at 2.7 Å (Liu et al., 2004) and from pea at 2.5 Å (Standfuss et al., 2005; Figure. 14.3). Each LHCII monomer has three transmembrane helices in the order of "stromally exposed N-terminus – transmembrane helix B – amphipathic helix E – transmembrane helix C– transmembrane helix A – amphipathic helix D – lumenally exposed C-terminus," and binds eight chlorophyll *a* (Chl 1–8), six chlorophyll *b* (Chl 9–14), two luteins (L1–2), and one 9′-cis-neoxanthin (N) (Figure 14.4).

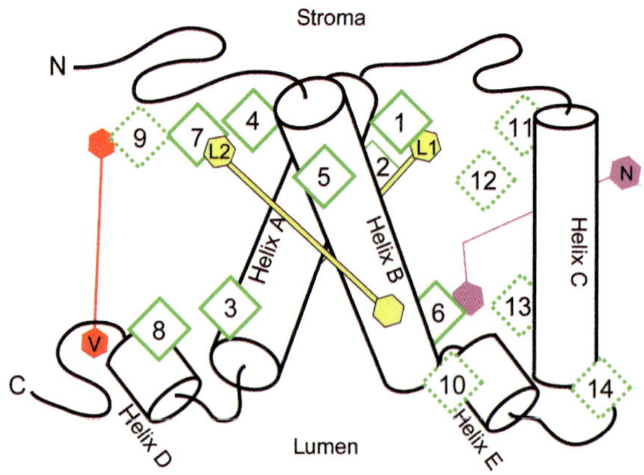

FIGURE 14.4 *Schematic model of the LHCII protein. Transmembrane (A–C) and amphipathic (D–E) helices determined by Liu et al. (2004) are shown as cylinders. Chlorophyll a (Chl 1–8) and chlorophyll b (Chl 9–14) are represented by diamonds and dotted diamonds, respectively. Chl numbers follow the proposed nomenclature of Standfuss et al. (2005). Two luteins (L1 and L2, yellow), one 9′-cis-neoxanthin (N, pink), and one violaxanthin (V, red) are displayed.*

Amino acid side chains or backbone carbonyls to coordinate the central Mg of the 14 chlorophylls have been unequivocally determined (Liu et al., 2004). The monomer–monomer interface accommodates most of the chlorophyll *b* and additional molecules including phosphatidylglycerol and violaxanthin (V). The phosphatidylglycerol may stabilize the trimer structure (Dubertret et al., 1994, 2002) and the violaxanthin may quench excess energy non-photochemically (Wentworth et al., 2001).

Helices A and B have a motif of chlorophyll-binding residues, the so-called "generic LHC-motif" (Jansson, 1999), and a pattern of ionic bridges between the Glu on one helix and the Arg on the other holds the two helices together (Kühlbrandt et al., 1994). This X-shape structure seems to be reinforced by a brace made of two lutein molecules. On the stromal side, helices A and B bind three pairs of symmetrically located chlorophylls (Chl 1/Chl 4, Chl 2/Chl 5, and Chl 7/Chl 12), while on the lumenal side there is one pair of chlorophyll *a* (Chl 3/Chl 6). Although the three-dimensional structure of a monomeric LHC protein has not been determined, it is generally believed that the related monomeric LHC proteins including minor LHCII proteins and LHCI have similar topology. Pigment compositions are, however, significantly different: CP29 binds eight chlorophylls (six chlorophyll *a* and two chlorophyll *b*; Sandona et al., 1998) and CP26 binds nine chlorophylls (six chlorophyll *a* and three chlorophyll *b*; Croce et al., 2002a), while LHCI polypeptides bind eight chlorophyll *a*, two chlorophyll *b*, and two carotenoids (Bassi et al., 1993).

3. Excitation energy transfer

Energy transfer within and between LHCII proteins has been studied extensively in higher plants, where most of the singlet excitation states created either on lutein or chlorophyll *b* are transferred with time constants <100 fs (Croce et al., 2001) or <3 ps (Kleima et al., 1997), respectively, to chlorophyll *a* within the same monomeric subunit. Neoxanthin transfers its energy exclusively toward chlorophyll *b* (Gradinaru et al., 2000; Croce et al., 2001), while the loosely bound violoxanthin, whose role is postulated to be in photoprotection, does not (Caffarri et al., 2001). The intramolecular transfer of excitation then takes place almost exclusively between chlorophyll *a* molecules at hundreds of femtoseconds to over 10 picoseconds (van Amerongen and van Grondelle, 2001). Further spatial redistribution of excitation energy throughout and between LHCII trimers occurs with a time constant of 24–28 ps (Bittner et al., 1994; Barzda et al., 2001). This redistribution is relaxed to a quasiequilibrium state among the antenna chlorophylls and the primary electron donor in PS II (P680), and also between the excited state of P680 and the radical pair; oxidized P680 ($P680^+$) and reduced pheophytin. This is because the equilibration of excitation in the photosynthetic unit is an order of magnitude faster than the process of charge separation occurring at P680 (Holzwarth, 1990). From this "exciton-radical pair equilibrium" model, it can be concluded that the excitation can visit P680 many times prior to the occurrence of photochemistry, and that some of the excited states are rapidly transferred back to the antenna pigments where they are emitted as fluorescence (Schatz et al., 1987).

B. Light-harvesting antenna for PS I

1. Structure and composition

The core of the PS I–LHCI supercomplex is formed by the PsaA/PsaB heterodimer with a few small subunits surrounding it, and this core contains all the components necessary for plastocyanin-ferredoxin oxidoreductase activity. All eukaryotic PS I cores characterized so far are monomers (Amunts et al., 2007), while in cyanobacteria trimeric PS I cores have been observed (Boekema et al., 1987; Jordan et al., 2001). Except for a few chlorophylls required for the photochemistry, all chlorophylls bound in the core subunits function as light-harvesting pigments. Attached to the core are peripheral light-harvesting antennae composed of several LHCI proteins, which account for up to 20% of the total chlorophyll in the thylakoid membrane. Light energy captured by these light-harvesting chlorophylls is delivered to the P700 chlorophylls, where it is trapped as a stable charge separation.

The light-harvesting antenna for PS I in higher plants is divided into two subcomplexes, LHCI-730 and LHCI-680; with characteristic 77K fluorescence peaks at 730 and 680 nm, respectively (Knoetzel et al., 1992). LHCI-730 is known to be formed by a Lhca1/Lhca4 heterodimer, and LHCI-680 by a homo- or heterodimer of Lhca2 and/or Lhca3 (Knoetzel et al., 1992; Ihalainen et al., 2000). On the other hand, the *Chlamydomonas*

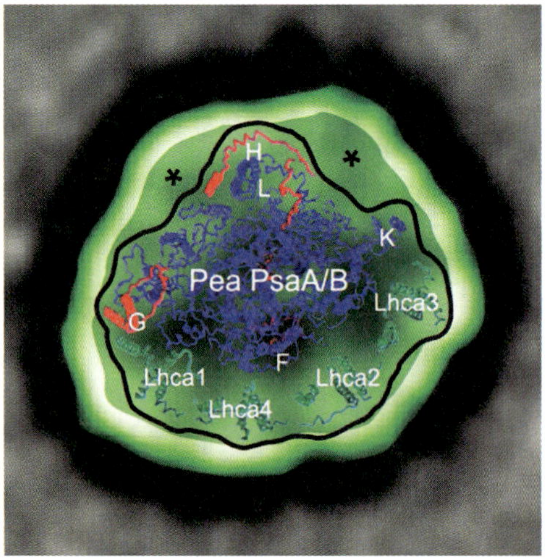

FIGURE 14.5 *Top view of the PS I–LHCI supercomplex. A Chlamydomonas PS I–LHCI supercomplex observed by Kargul et al. (2003), grayscale, overlaid to scale by the pea PS I–LHCI structure at 4.4 Å (Ben-Shem et al., 2003). The pea structure, emphasized by the heavy outline, has been positioned by placing its four copies of LHCI (labeled LHCA1–4; in green) into the four-domain asymmetric crescent of density. The remaining densities visualized in the Chlamydomonas structure are labeled with asterisks. The green outer band represents the putative detergent shell of ~15 Å. From Nield et al. (2004).*

antenna is composed of LHCI-705 and LHCI-680 subcomplexes (Bassi et al., 1992), formed by nearly the same composition of LHCI polypeptides. It has been suggested by single-particle analysis that the LHCI proteins are asymmetrically bound to the PS I core in *Chlamydomonas* (Germano et al., 2002; Kargul et al., 2003) and spinach (Boekema et al., 2001). The crescent-shaped "LHCI belt,", which is bound to the PsaG/F/K side of the PS I core, is formed by four LHCI proteins, sequentially placed in the belt, from the PsaG end, in the order Lhca1, Lhca4, Lhca2, and Lhca3 as found in the 3.4 Å resolution crystal structure from pea (Amunts et al., 2007; Figure 14.5). The crystal structure also indicates clefts between two adjacent Lhca proteins and between the surface of PS I core and each Lhca protein, suggesting relatively weak interactions. It is likely that green algae and higher plants share the same overall architecture of the PS I–LHCI supercomplex; however, there are a few differences between the green algal and higher plant PS I–LHCI supercomplexes. One major difference is the number of bound LHCI polypeptides per PS I core. Biochemical data suggested that 4–8 LHCI polypeptides are bound to the PS I core in higher plants (Jansson et al., 1997; Boekema et al., 2001; Croce et al., 2002b; Klimmek et al., 2005), versus 7–9 in *Chlamydomonas* (Takahashi et al., 2004; Tokutsu et al., 2004). While the crystal structure of the pea PS I–LHCI supercomplex shows only 4 LHCI polypeptides attached to the core (Amunts et al., 2007), single-particle images of the PS I–LHCI supercomplex from *Chlamydomonas*

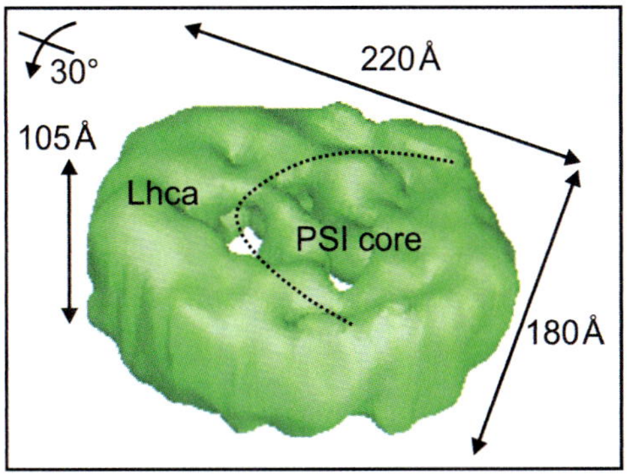

FIGURE 14.6 *Three-dimensional map of the PS I–LHCI supercomplex. Surface representation of the reconstructed single-particle images from cryoelectron microscopy of the Chlamydomonas LHCI–PS I supercomplex at 30 Å resolution, as viewed from an oblique angle. From Kargul et al. (2003).*

suggest 14 (Germano et al., 2002), 11 (Kargul et al., 2003), or 6 LHCI polypeptides (Kargul et al., 2005) bound to the core. The crystal structure of the pea PS I–LHCI supercomplex defined 103 chlorophylls on the core proteins, 14 chlorophylls each on Lhca1–4, and 9 chlorophylls in the cleft between LHCI and the core (Amunts et al., 2007), while Kargul et al. (2003) estimated 95 chlorophyll *a* in the core, and 80 chlorophyll *a*, and 40 chlorophyll *b* in the LHCI proteins in *Chlamydomonas*. The light-harvesting apparatus for PS I is therefore possibly larger in green algae than in higher plants, and may contain a second row of the "LHCI belt" (Kargul et al., 2003; Figure 14.6).

2. Structure of the LHCI monomer

Although the number of pigments and orientations is significantly different, the general structural features of LHCI proteins are shared with LHCII proteins. Besides binding the same number of chlorophylls on the polypeptides, LHCI proteins also bind extra pigments including "gap chlorophylls" located in the cleft between LHCI and the core. These extra pigments are postulated to increase the connectivity of the network formed by LHCI proteins and the core (Melkozernov et al., 2006). A three-dimensional model of the Lhca4 protein from higher plants was constructed on the basis of similarity with LHCII proteins (Melkozernov and Blankenship, 2003). This model features three transmembrane helices and a central domain formed by helices A and B, which binds six chlorophylls and two carotenoids.

3. Red chlorophylls

Green algae, higher plants, and cyanobacteria exhibit fluorescence in the far-red region (>700 nm), which is a fingerprint of PS I *in vivo* (Butler, 1961) and *in vitro* (Lam et al., 1984). The "red chlorophylls" responsible for the far-red fluorescence exist both in the PS I core (~720 nm) and LHCI proteins (~735 nm) in higher plants (van Grondelle et al., 1994), of which 80% is due to LHCI chlorophylls (Croce and Bassi, 1998). *Chlamydomonas* PS I shows a fluorescence peak at 715–717 nm at low temperature, which is a shorter wavelength than observed in higher plants (Wollman and Bennoun, 1982). It has not yet been determined whether the 715–717 nm fluorescence arises from chlorophyll(s) in the PS I core (Ihalainen et al., 2005) or in LHCI protein(s) (Melkozernov et al., 2004). An isolated LHCI complex from *Chlamydomonas* has a low temperature fluorescence peak at 705 nm (Bassi et al., 1992), suggesting that the peripheral LHCI antenna binds significantly fewer red chlorophylls than in higher plants. The origin of far-red fluorescence in LHCI proteins in higher plants was localized to the pigment pair Chl 5/Chl 12, which is located on the periphery of an LHCI protein as well as at the interface between LHCI proteins and the PS I core (Morosinotto et al., 2002, 2003; Schmid et al., 2002; Melkozernov and Blankenship, 2005; Figures 14.4 and 14.5). The function of these red chlorophylls is under intensive debate, but they probably form a series of excitation traps (van Grondelle et al., 1994). While the red chlorophylls in the PS I core act as deep excitation traps at low temperature (van Grondelle et al., 1994), those in LHCI proteins efficiently localize the excitation energy even at physiological temperatures (Croce et al., 2002b) and transfer it towards the PS I core or quench it nonphotochemically.

4. Excitation energy transfer

Extensive spectroscopic studies have shown several energy transfer processes with different lifetimes in PS I (Melkozernov, 2001b; Melkozernov et al., 2006). Light energy captured by peripherally located chlorophyll *b* in LHCI proteins rapidly populates chlorophyll *a* molecules that absorb in the 680–710 nm region within 0.2–0.7 ps (Du et al., 1993; Melkozernov et al., 2000). The intramolecular redistribution of excitation energy that involves different pools of chlorophylls then occurs within 3–4 ps (Klug et al., 1989; Melkozernov et al., 2002) and intersubunit energy equilibration is completed within 30–50 ps (Melkozernov, 2001). The energy transfer from LHCI proteins to the core occurs rather slowly at ~100 ps in *Chlamydomonas* at room temperature, because the LHCI proteins are isoenergetic to the PS I core (Melkozernov et al., 2004). This energy coupling of LHCI proteins and the core is not observed at 77K, indicating that at low temperatures the peripheral antenna is energetically uncoupled from the core (Melkozernov et al., 2005). The photochemical trapping time is reported to be 15–40 ps in higher plants (van Grondelle et al., 1994) and 15–25 ps in *Chlamydomonas* (Melkozernov et al., 2005). Thus, a diffusion-limited excitation dynamics is suggested in PS I (Croce et al., 2000).

III. LIGHT-HARVESTING PROTEINS

A. Overview

The *Chlamydomonas* genome sequence predicts 9 *LHCA* and 12 *LHCB* genes encoding LHC polypeptides associated with PS I and PS II, respectively, and 9 *LHC*-like genes. Previously, various non-consensus and thus confusing designations have been given to these genes; for example, the first such gene cloned in *Chlamydomonas* was named *cabII-1* (now called *LHCBM6*; Imbault et al., 1988), where *cab* stands for chlorophyll *a*/*b*-binding protein. The proliferation of the sequences revealed, however, that there are a number of homologous three-helix proteins in eukaryotic algae, the ones that bind chlorophyll *a*/*b* in Chlorophyta, and ones that bind only chlorophyll *a* in Rhodophyta and chlorophyll *a*/*c* in Chromophyta. Thus, a nomenclature based on light-harvesting complex was proposed in higher plants, Lhca1–6 and Lhcb1–8 (Jansson et al., 1992; Klimmek et al., 2006). The increasing data obtained at the protein level and the diversity among the photosynthetic organisms have made it possible to correlate each gene identified in *Chlamydomonas* to a specific LHC polypeptide within the LHC nomenclature. The rationales behind the designations are described in the section where each gene is mentioned. Correlations with most of the previous designations are given by Nield et al. (2004) for LHCI and Minagawa and Takahashi (2004) for LHCII and LHC-like sequences.

Table 14.1 Genes encoding LHCII proteins in *Chlamydomonas*

Gene	MW[a]	No. of TMH	Related gene(s) in higher plants	Protein	Accession #
LHCBM1	27565.43	3	*Lhcb1–Lhcb3*	LHCII type IV (p16)	BAB64418
LHCBM2	26650.43	3	*Lhcb1–Lhcb3*	LHCII type III (p17)	BAB64417
LHCBM3	27380.11	3	*Lhcb1–Lhcb3*	LHCII type I (p11/p13)	BAB64416
LHCBM4	26948.72	3	*Lhcb1–Lhcb3*	LHCII type I (p11/p13)	AAD03731
LHCBM5	28686.92	3	*Lhcb1–Lhcb3*	LHCII type II	AAD03732
LHCBM6	26905.55	3	*Lhcb1–Lhcb3*	LHCII type I (p11/p13)	AAM18056
LHCBM7	26686.42	3	*Lhcb1–Lhcb3*	LHCII type III (p17)	AAK01125
LHCBM8	27075.78	3	*Lhcb1–Lhcb3*	LHCII type I (p11/p13)	AAL88457
LHCBM9	27104.91	3	*Lhcb1–Lhcb3*	LHCII type I (p11/p13)	EDP01724
LHCB4	29937.91	3	*Lhcb4*	CP29 (p9)	BAB64415
LHCB5	30713.97	3	*Lhcb5*	CP26 (p10)	BAB20613
LHCB7	42467.91	3	*Lhcb7*	LHCB7	EDP07628

[a] *Predicted MW of precursor.*

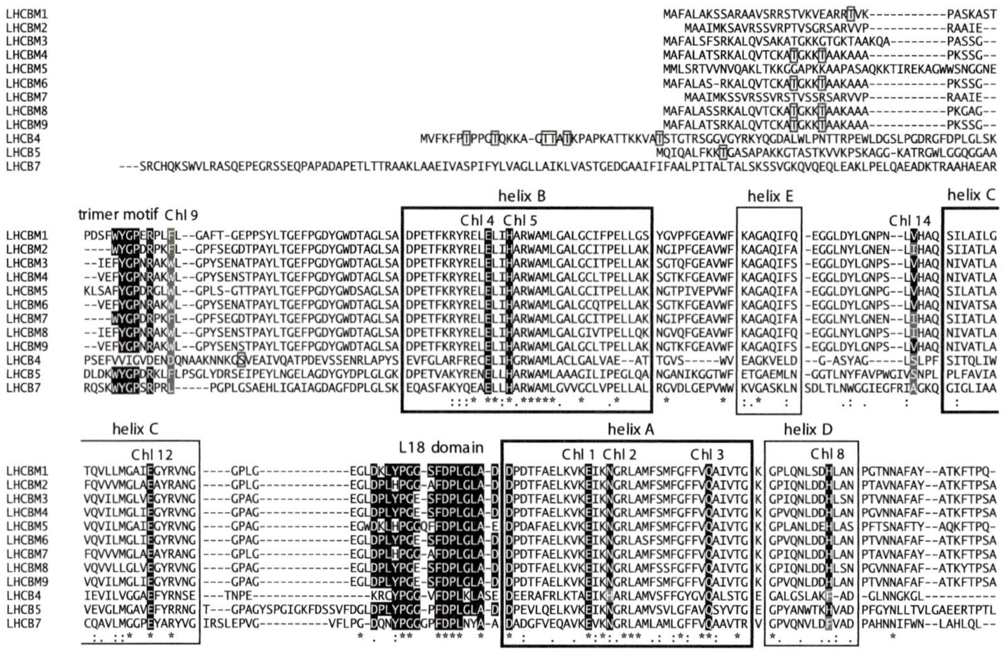

FIGURE 14.7 *Sequence alignment of Chlamydomonas LHCII proteins. Protein sequences were aligned using Clustal X (Thompson et al., 1997), using gaps (–) to maximize homology. Boxed regions correspond to putative transmembrane (A–C, thick line) and amphipathic (D–E, thin line) helices (Liu et al., 2004). Residues contributing their side chains or backbone carbonyls to coordinate the central Mg of chlorophyll in LHCII (Standfuss et al., 2005) are shaded in black or gray, respectively. Chl numbers follow the proposed nomenclature of Standfuss et al. (2005). The trimer motif (WYXXXR; Hobe et al., 1995) and L18 domain (VDPLYPGGSFDPLGLADD; DeLille et al., 2000) are indicated; conserved residues are shaded. Phosphorylatable Thr and Ser residues are boxed (Turkina et al., 2006). Amino acid sequences shown correlate with the accession numbers given in Table 14.1.*

B. Major LHCII proteins (outer trimeric antenna): LHCBM1–LHCBM9

The *Chlamydomonas* nuclear genome encodes nine genes for the major LHCII polypeptides (Minagawa and Takahashi, 2004; Table 14.1). The current nomenclature for these genes, *LHCBM1–9* (*M* = Major LHCII), follows a rough order of their abundance in the *Chlamydomonas* EST database (Asamizu et al., 1999). Figure 14.7 shows an alignment of the 12 predicted LHCII polypeptides and Figure 14.8 shows a neighbor-joining tree (Saitou and Nei, 1987) displaying their phylogenetic relationships. The tree indicates four distinct branches for major LHCII proteins. Type I comprises five duplicated genes including *LHCBM3*, *LHCBM4*, *LHCBM6*, *LHCBM8*, and *LHCBM9* with sequence identities of 89–99%. All type I genes except *LHCBM3* are linked in the genome. *LHCBM4* and *LHCBM8* lie only 3.7 kb apart and share 92% identity, whereas *LHCBM4* and *LHCBM9* are 99% identical. *LHCBM2* and *LHCBM7*, encoding major

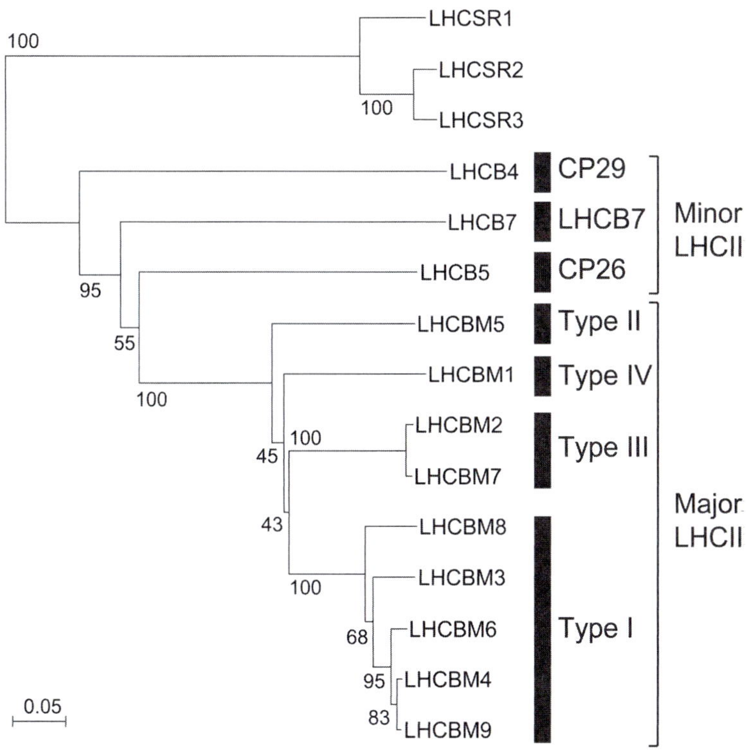

FIGURE 14.8 *Unrooted phylogenetic tree of Chlamydomonas LHCII proteins. Neighbor joining-distance tree of LHCII proteins constructed with bootstrap values (% of 1000 replicates) displayed at the appropriate nodes. Amino acid sequences correlate with accession numbers given in Tables 14.1 and 14.3.*

LHCII type III proteins, are also duplicated (99% identity) and close together in the genome. Major LHCII polypeptides type II and type IV are encoded by single genes, *LHCBM5* and *LHCBM1*, respectively. In the polypeptide sequences of LHCBM1–9, all residues contributing to the coordination of the central Mg of chlorophyll are conserved (Standfuss et al., 2005). The striking structural motif, WYGPDR (or WYXXXR), which is essential for the trimerization of LHCII proteins *in vitro* (Hobe et al., 1995), is fully conserved in major LHCII polypeptides types III and IV, but only partially conserved in the type I and II proteins in *Chlamydomonas* (Figure 14.7), although the entire form is conserved in all the major LHCII polypeptides in higher plants (Hobe et al., 1995).

In contrast to green algal major LHCII polypeptides, the major trimeric LHCII polypeptides in higher plants are classified into three types, type I (Lhcb1.1–1.5 in *Arabidopsis*), type II (Lhcb2.1–2.3), and type III (Lhcb3), which usually occur in a ratio of about 8:3:1 (Jansson, 1994). This ratio is roughly proportional to the number of the duplicated genes. These proteins

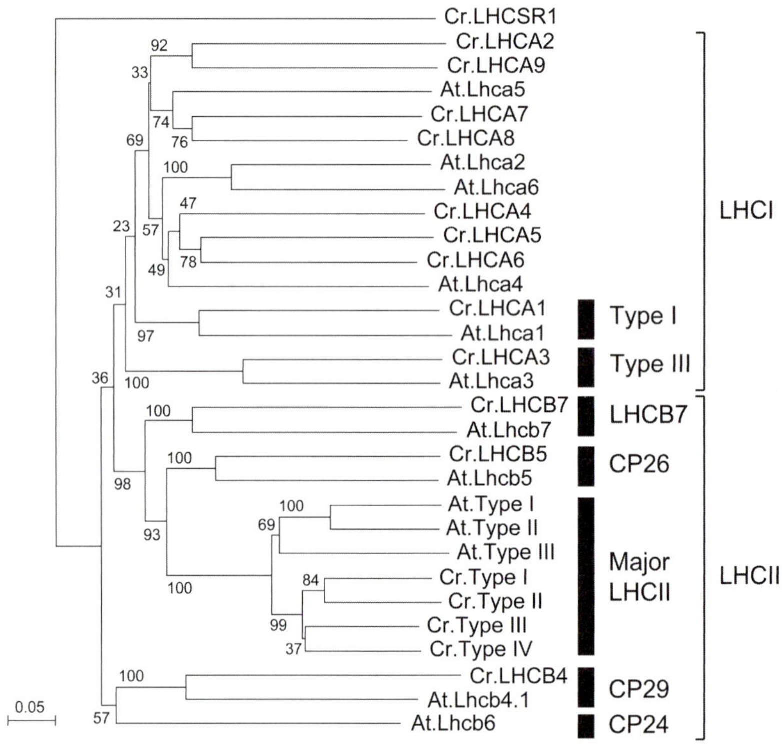

FIGURE 14.9 *Unrooted phylogenetic tree of LHC proteins from Chlamydomonas and Arabidopsis. LHC family proteins from Chlamydomonas (Cr) and Arabidopsis (At) were used to construct a neighbor-joining distance tree, with bootstrap values (% of 1000 replicates) displayed at the appropriate nodes. The protein sequences used are from the following databases: The Arabidopsis Information Resource – At.Lhca1 (At3g54890), At.Lhca2 (At3g61470), At.Lhca3 (At1g61520), At.Lhca4 (At3g47470), At.Lhca5 (At1g45474), At.Lhca6 (At1g19150), At.Type I (At1g29920), At.Type II (At2g05100), At. Type III (At5g54270), At.LHCB4.1 (At5g01530), At.Lhcb5 (At4g10340), At.Lhcb6 (At1g15820), and At. Lhcb7 (At1g76570). Other genes are described in Tables 14.1 and 14.2.*

are organized into homotrimers of Lhcb1, heterotrimers composed of a combination of Lhcb1 and Lhcb2, or Lhcb1 and Lhcb3 (Larsson et al., 1987). In *Arabidopsis*, Lhcb1 and Lhcb2 have only 14 differences in their amino acid sequences, while Lhcb3 is smaller and its amino acid sequence diverges somewhat more. The three types of the major LHCII polypeptides in higher plants do not correspond to the four types in *Chlamydomonas* (Figure 14.9; Teramoto et al., 2001). Thus, the genes encoding major LHCII polypeptides may have diversified into four and three types in Chlorophyta (Chlorophyceae and Ulvophyceae) and Streptophyta (land plants and charophytes), respectively, after the separation of the two phyla.

Bassi and Wollman (1991) identified five LHCII polypeptides in purified LHCII fractions including p10, p11, p13, p16, and p17 (nomenclature of Chua and Bennoun, 1975). One additional LHC protein, p9, copurified with PS II core fractions, bringing the total to six LHC proteins associated with PS II.

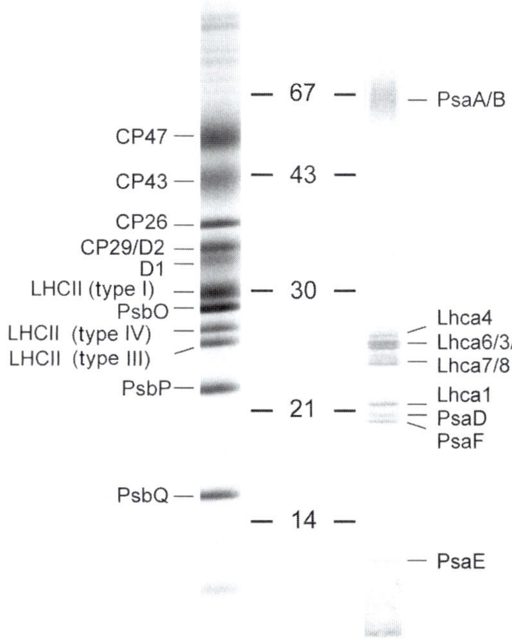

FIGURE 14.10 *Denaturing gel analysis of PS I–LHCI and PS II–LHCII supercomplexes. The polypeptide compositions of the purified PS II–LHCII (left) and PS I–LHCI (right) supercomplexes as resolved by high-resolution SDS-polyacrylamide gel electrophoresis. The gel was stained with Coomassie blue. From Iwai, M. and Minagawa, J., unpublished (left) and Tokutsu et al. (2004) (right).*

Proteins p11, p13, p16, and p17 oligomerize into the high molecular weight form of LHCII, suggesting that these four polypeptides constitute the major trimeric antenna. Protein gel analysis together with microsequencing and/or mass fingerprinting detected major LHCII proteins types I, IV, and III, which correspond to p11/p13, p16, and p17, respectively (Minagawa and Takahashi, 2004; Vink et al., 2004; Figure 14.10). A detailed two-dimensional map of LHCII polypeptides identified six LHCII proteins including type I (LHCBM3, LHCBM4, LHCBM6), type III (LHCBM2, LHCBM7), and type IV (LHCBM1), and two minor LHCII proteins, CP26 and CP29 (Stauber et al., 2003). These results suggest that the predominant components of the *Chlamydomonas* outer trimeric antenna are the major LHCII proteins type I, type III, and type IV. It was not until the discovery of the mobile LHCII polypeptides for state transitions that the LHCII protein type II (LHCBM5) was detected in the light-harvesting complexes in *Chlamydomonas* (Takahashi et al., 2006; see Section V.A.3. and Chapter 22).

C. Minor LHCII proteins (inner monomeric antenna): LHCB4, LHCB5, and LHCB7

Polypeptides p9 and p10 in *Chlamydomonas* exist only in monomeric form, have a higher chlorophyll *a* content, and exhibit red-shifted absorption maxima at 676–677 nm, which led Bassi and Wollman (1991) to conclude that p9 and p10 correspond to CP29 and CP26, respectively. The correspondence of p10 and CP26 was later confirmed by protein sequencing (Minagawa et al., 2001). While the two minor antenna proteins are conserved throughout the Chlorophyta, a homolog (E-value less than 0.1) for another higher plant minor antenna protein, CP24, has not been found in algae (Minagawa and Takahashi, 2004). On the other hand, LHCB7, which has 31 and 46% identity to CP29 and CP26, respectively, is maintained in several green algae (Minagawa and Takahashi, 2004; Klimmek et al., 2006). In *Arabidopsis* and poplar, *LHCB7* expression is not clustered with other highly expressed *LHC* genes, suggesting a specific function during some stages of plastid development or forming different light-harvesting antennae under aberrant conditions (Klimmek et al., 2006). Although the gene product has not been identified in any organism, it is hypothesized to encode an LHCII antenna protein.

CP26 exists predominantly as a monomer, but trimerization was observed in an *Arabidopsis* transformant engineered to be deficient in one of the major LHCII polypeptides (Ruban et al., 2003). This is consistent with the fact that CP26 has the "trimer motif" mentioned above (Hobe et al., 1995; Minagawa et al., 2001). *Chlamydomonas* LHCB7 also carries the entire motif (Figure 14.7) and might trimerize under certain conditions. On the other hand, the motif is completely absent in all CP29 proteins, which might suggest a specific role of this monomeric protein. The sequence alignment shows that all the side-chain chlorophyll ligands determined in the crystal structure of major LHCII are conserved except for His instead of Asn as a ligand for Chl 2 in CP29, and Phe instead of His as a ligand for Chl 8 in CP29 and LHCB7 (Figure 14.7). The Glu-Arg ion bridges within the central portions of helices A and B are well-conserved (Kühlbrandt et al., 1994).

Nucleus-encoded integral thylakoid membrane proteins are imported from the cytosol by virtue of information contained in the transit peptide, which is usually cleaved off after transport (Douwe de Boer and Weisbeek, 1991). The transit peptides of *Chlamydomonas* CP26 and CP29, however, were shown not to be removed upon localization within the thylakoid membranes (Minagawa et al., 2001; Turkina et al., 2006). The N-terminal sequence of CP29 is subject to a series of processing events: excision of Met1, acetylation of Val2, and phosphorylation of Thr7 (Turkina et al., 2004). These alterations could be the reason why the transit peptides are retained (Turkina et al., 2006).

While the genes encoding major LHCII polypeptides diverged into several subtypes after the separation of Chlorophyta and Streptophyta, those encoding minor LHCII polypeptides including *LHCB4*, *LHCB5*, and possibly *LHCB7*, appear to be conserved across the green lineage (Figure 14.9), arguing in favor of specific role(s) for each one. On the other hand, the most divergent LHCII polypeptide CP24 (Huber et al., 2001) only exists in

Streptophyta, suggesting that it arose only after plants were exposed to the environmental challenges of a terrestrial lifestyle.

The two monomeric LHCII proteins (CP26 and CP29), which have been viewed as belonging solely to the PS II complex, and an unreported major LHCII protein (LHCBM5) are proposed to be shuttled between the two photosystems and act as docking sites for the major trimeric LHCII antenna for both photosystems during state transitions (Takahashi et al., 2006; see Section V.A.3. and Chapter 22).

D. LHCI proteins: LHCA1–LHCA9

The light-harvesting antenna for PS I is formed by nine polypeptides including LHCA1 (p22.1), LHCA3 (p14.1), LHCA4 (p14), LHCA5 (p15.1), LHCA6 (p18.1), LHCA7 (p15), and LHCA8 (p18) in stoichiometric amounts, and LHCA2 (p19) and LHCA9 (p22.2) in substoichiometric amounts (Bassi et al., 1992; Hippler et al., 2001; Stauber et al., 2003; Takahashi et al., 2004; Tokutsu et al., 2004; Figures 14.10 and 14.11; Table 14.2). Among them, LHCA1 (type I) and LHCA3 (type III) are maintained, but all the other LHCA proteins do not correlate with each other in green algae and higher plants (Takahashi et al., 2004; Tokutsu et al., 2004; Figure 14.9). LHCA1 and LHCA3 are indeed the most abundant LHCI proteins in

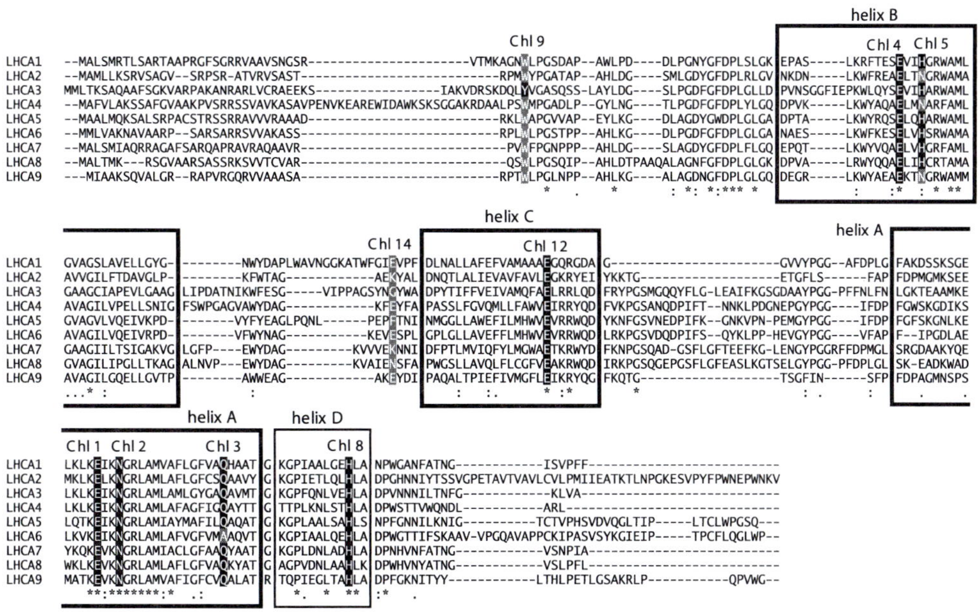

FIGURE 14.11 *Alignment of Chlamydomonas LHCI proteins. Chlamydomonas LHCI protein sequences were aligned as described in the legend to Figure 14.7. Accession numbers are given in Table 14.2.*

Table 14.2 Genes encoding LHCI proteins in *Chlamydomonas*

Gene	MW[a]	No. of TMH	Related gene(s) in higher plants	Protein	Accession #
LHCA1	23900.28	3	*Lhca1*	p22/p22.1	BAD06923
LHCA2	26922.26	3	*Lhca1–Lhca4*	p19	EDP05477
LHCA3	28886.33	3	*Lhca3*	p14.1	BAD06919
LHCA4	28701.92	3	*Lhca1–Lhca4*	p14	BAD06918
LHCA5	28228.62	3	*Lhca1–Lhca4*	p15.1	BAD06920
LHCA6	27781.42	3	*Lhca1–Lhca4*	p18.1	BAD06922
LHCA7	26222.88	3	*Lhca1–Lhca4*	p15	BAD06924
LHCA8	25920.87	3	*Lhca1–Lhca4*	p18	BAD06921
LHCA9	22844.34	3	*Lhca1–Lhca4*	p22.2	EDP04026

[a]*Predicted MW of precursor.*

Chlamydomonas (Hippler et al., 2001) and have specific features present in higher plant homologues, such as a very short sequence between helices C and A (type I), and a six-residue insert at the beginning of helix B (type III) (Figure 14.11). LHCA3 and the two minor LHCI proteins LHCA2 and LHCA9 are suggested to be stabilized by the PS I core when assembled in the light-harvesting antenna, since they are not present in the stable oligomeric structure of the PS I light-harvesting antenna (Takahashi et al., 2004).

The other LHCI polypeptides including seven *Chlamydomonas* LHCI proteins, LHCA2, LHCA4, LHCA5, LHCA6, LHCA7, LHCA8, LHCA9 and four of the *A. thaliana* LHCI proteins, Lhca2, Lhca4, Lhca5, and Lhca6, are only loosely grouped into a single lineage (Figure 14.9). A comparable organization is observed for the proteins in the LHCII complex, where minor monomeric LHCII polypeptides form separate branches from major trimeric LHCII polypeptides. By analogy to the minor LHCII polypeptides, which lie between major LHCII trimers and PS II core groups (Harrer et al., 1998), LHCI type I and type III proteins might function as proximal antennae for PS I, and the five other *Chlamydomonas* LHCI polypeptides might function as distal antennae (Tokutsu et al., 2004).

There are two additional LHCA proteins in higher plants (Jansson, 1999). One, Lhca5, has been detected in substoichiometric amounts in thylakoids of *Arabidopsis* (Ganeteg et al., 2004) and in tomato thylakoids (Ganeteg et al., 2004; Storf et al., 2004). This polypeptide has been suggested to assemble peripherally with PS I–LHCI supercomplexes under high light conditions (Ganeteg et al., 2004; Lucinski et al., 2006). The product of the *Lhca6* gene was identified in tomato (Zolla et al., 2002), contravening an earlier suggestion that it was a pseudogene derived from *Lhca2*

(Jansson, 1999). The substoichiometric proteins, LHCA2 and LHCA9 in *Chlamydomonas* and Lhca5 and Lhca6 in *Arabidopsis*, might form a second row of the LHCI belt that is subject to environmental regulation.

The PS I light-harvesting complex is reorganized in response to environmental factors. The effects of iron deficiency are particularly well documented in *Chlamydomonas*. When placed under iron-deficient conditions, the LHCI complex reduces functional coupling with the PS I core to prevent damage to the iron-deficient PS I core proteins (Moseley et al., 2002). This is caused by remodeling of the LHCI complex in which the N-terminus of LHCA3 is processed, LHCA5 is depleted, and LHCA4 and LHCA9 are upregulated (Naumann et al., 2005).

IV. LHC-LIKE PROTEINS

A. Overview

There are several distant relatives of the LHC superfamily collectively called LIL (light-harvesting-like) proteins (Table 14.3), although their primary roles may not be light-harvesting but rather dissipation of absorbed energy in the form of heat or fluorescence (Jansson, 1999; Adamska, 2001). Unlike LHC proteins, they are not assigned within the crystal structure or single-particle images of PS I and PS II complexes. Therefore, LILs are neither tightly nor constitutively associated with the PS I or PS II particles. Biochemical approaches, however, have suggested that some LILs interact with the

Table 14.3 Genes encoding LHC like proteins in *Chlamydomonas*

Gene	MW[a]	No. of TMH	Related gene(s) in higher plants	Accession #
HLI	11395.34	1	*Ohp1, Ohp2*	BAD67135
LIL3	27492.21	2	*Lil3*	BAD67136
LHCSR1	27556.63	3	–	CAA64632
LHCSR2	28224.19	3	–	EDP01013
LHCSR3	21713.75	3	–	EDP01087
LHL4	30157.61	3	–	BAD67137
ELI1	18340.35	3	*Elip1–Elip3*	EDP01315
ELI2	18746.81	3	*Elip1–Elip3*	BAD67134
ELI3	20707.8	3	*Elip1–Elip3*	EDP02265
ELI4	29808.34	3	*Elip1–Elip3*	EDP02136
PSBS	34809.28	4	*PsbS*	EDP09214

[a] *Predicted MW of precursor.*

photosystems. Therefore, their absence from the published structures might reflect transient or weak interactions.

It is generally accepted that the LHC superfamily is monophyletic (Wolfe et al., 1994), and the discoveries of one-helix HLIPs (high-light inducible proteins) in cyanobacteria, and the four-helix PSBS protein, support this idea (Green, 2003). Within the superfamily, helices A and B are most conserved, with the N-terminal, helix C, and lumenal and stromal loop regions exhibiting more divergence (see Figure 14.4). One scenario for the evolution of the LHC superfamily would thus be as follows: (1) A common ancestor with a single transmembrane domain (HLIPs) appeared, corresponding to helix B. (2) Another sequence encoding a potential transmembrane helix was added to generate SEPs (stress-enhanced proteins) with a B-C structure. (3) A tandem gene duplication-fusion gave rise to a four-helix protein (e.g. PSBS) with the structure B-C-B′-C′. (4) The C′ helix was lost, resulting in three-helix proteins such as LHC and ELIP (early light-inducible protein), which carry B-C-B′ or B-C-A structures. However, the limited appearance of the potential ancestor protein PSBS throughout the algal species and no sequence similarities observed in the "C" part of SEP, LHC, ELIP, and PSBS indicate a more complex evolutional history (Green, 2003).

B. One-helix protein HLIP

HLIPs were discovered in *Synechococcus* PCC 7942 as proteins with a single transmembrane helix with similarity to the LHC motif (Dolganov et al., 1995). Four genes encode HLIPs in *Synechococcus*. These are *hliA*, *hliB*, *hliC*, and *hliD* (Dolganov et al., 1995), also known as *scpC*, *scpD*, *scpB*, and *scpE* in *Synechocystis* (small cab-like protein; Funk and Vermaas, 1999). Since the *hli* genes are transiently expressed under various stress conditions, they were proposed to be involved in high-light acclimation (Dolganov et al., 1995). In fact, more than twice as many *hli* genes are found in the marine cyanobacterium *Prochlorococcus* strain MED4, which is adapted to high-light conditions in surface waters of the open ocean, than in several other cyanobacteria adapted to growth in moderate- or low-light conditions (Bhaya et al., 2002). Studies of deletion mutants indicate that HLIPs are involved in the regulation of tetrapyrrole synthesis in response to pigment availability (Xu et al., 2002, 2004). HLIPs are thought to form dimers in the thylakoid membranes and bind carotenoids (He et al., 2001). HliB has been localized to the vicinity of PsbH and CP47 in the PS II core dimer of *Synechocystis* PCC6803, and has been hypothesized to participate in PS II assembly/repair (Promnares et al., 2006).

HLIP genes are ubiquitous among photosynthetic organisms, being present in cyanelle genomes in glaucophytes (Stirewalt et al., 1995), nucleomorph genomes in cryptophytes (Douglas et al., 2001), chloroplast genomes in rhodophytes (Maid and Zetsche, 1992), and in the nuclear genomes of

Chlamydomonas (Teramoto et al., 2004), and *Arabidopsis* (Jansson et al., 2000; Andersson et al., 2003). The transcript levels of a *Chlamydomonas HLIP* gene reach a maximum within 1 hour after transfer to high light, and rapidly return to the initial level upon return to normal conditions, suggesting a photoprotective role (Teramoto et al., 2004). The detailed function is, however, largely unknown. Toward understanding its function in photosynthetic eukaryotes, it is of note that one of the HLIP proteins in *Arabidopsis*, Ohp2, is related to helix B of Lhca4 and is associated with PS I (Andersson et al., 2003).

C. Two-helix protein Sep

Sep1 and *Sep2* are *Arabidopsis* ESTs which encode proteins that include consensus sequences from ELIPs (Heddad and Adamska, 2000). Hydropathy plots of these Sep proteins predict that they possess two transmembrane helices, with the first helix carrying the LHC motif. The *Sep1* and *Sep2* genes (also known as *Lil4* and *Lil5* in *Arabidopsis*; Klimmek et al., 2006) are up-regulated by high light stress, and were therefore suggested to be involved in photoprotection (Heddad and Adamska, 2000). *Chlamydomonas* does not encode Sep homologs in the available genome sequence.

D. Two-helix protein LIL3

Lil3 was first described in the context of two *Arabidopsis* ESTs (Jansson, 1999). More information was provided by the *Chlamydomonas LIL3* gene sequence (Teramoto et al., 2004). The *Chlamydomonas* LIL3 polypeptide is composed of 249 amino acids and includes two putative transmembrane helices. Although *Arabidopsis* Sep proteins also carry two transmembrane helices, Lil3 and Sep proteins do not show significant similarity. In contrast to other stress-related genes including *HLI* and *ELI*, which are transiently induced upon a shift to high light conditions, the regulation of *LIL3* is rather similar to the *LHC* genes in *Chlamydomonas* (Teramoto et al., 2004). After transfer to high light the mRNA level decreased rapidly to 30–40% of the initial level, and gradually recovered. This might suggest a role in light harvesting rather than stress response. Based on the phylogenetic tree, however, LIL3 clearly clusters with other LHC-like proteins that are thought to be stress-related.

E. Three-helix protein LHCSR

LHCSR genes, distantly related to the LHC superfamily, include the *Chlamydomonas eugametos* LHC-related light-inducible gene *LI818*, whose expression is regulated by a nonphotosynthetic photoreceptor (Gagne and Guertin, 1992). Homologues have been found in various eukaryotic algae including chlorophytes, prasinophytes, and several chromophytes.

C. reinhardtii encodes *LHCSR1* to *LHCSR3*. *LHCSR1* transcription is regulated in a substantially different manner compared with other *LHC* genes in *C. reinhardtii* (Savard et al., 1996). It is strongly induced by exposure to high light by a redox-independent mechanism (Elrad and Grossman, 2004; Teramoto et al., 2004). These limited data are suggestive of a photoprotective function for LHCSRs, but further experimentation is clearly required.

F. Three-helix protein LHL4

A later addition to the LHC-like gene family, *LHL4*, was identified as a *Chlamydomonas* EST encoding a 285-amino acid protein (Teramoto et al., 2004). The predicted protein contains three transmembrane helices, but is not similar to any other proteins in the LHC superfamily except that it retains the LHC motif in helices A and B. Structurally, it is unique due to a long stromal loop between helices C and A. Upon transfer of cells grown under low light to high light conditions, *LHL4* mRNA abundance increases rapidly more than 10-fold, then returns rapidly to a low level (Teramoto et al., 2004). Detailed expression analysis and the action spectra of its light sensitivity revealed that *LHL4* is regulated neither by the redox status of the intersystem electron transport chain nor reactive oxygen species, but instead by a blue light receptor with a flavin-based chromophore (see Volume 1, Chapter 6) with absorption bands at 375 and 450 nm (Teramoto et al., 2006). The unique stromal loop might be related to its regulation by the blue light receptor.

G. Three-helix protein ELIP

The *ELI* genes were among the first nuclear genes found to be transiently expressed during the greening of etiolated plants (Meyer and Kloppstech, 1984; Adamska, 2001). Strong light causing photoinhibition also induces *ELI* transcription and accumulation of their protein products in mature plants (Potter and Kloppstech, 1993). Biochemical studies suggest that ELIPs bind chlorophyll *a* and lutein, but not chlorophyll *b* (Adamska et al., 1999), which could be due to the unusual chlorophyll *b* binding stretches including shorter N-terminus, extra membrane loop, and C-terminal sequences. It has been suggested that ELIPs function in photoprotection, either by scavenging free chlorophylls (or chlorophyll degradation products) or by promoting the proper assembly of pigment–protein complexes during exposure of cells to high light (Adamska, 2001; Hutin et al., 2003).

Chlamydomonas contains four *ELI* genes (Table 14.3). Like other ELIPs, helices A and B have high sequence similarity with LHC polypeptides, whereas helix C shows only a little. As in higher plants, *ELI2* (*Lhl1*) and *ELI4* expression are transiently induced by high light or low CO_2 (Miura et al., 2004; Teramoto et al., 2004). Although the expression of *ELI* genes in higher plants is regulated by the redox state of the intersystem electron transport chain (Montane et al., 1998) and hydrogen peroxide, a well-described sensor of oxidative stress (Kimura et al., 2001), neither of these signals affects the expression of *Chlamydomonas ELI2* (Teramoto et al., 2004).

H. Four-helix protein PSBS

PsbS was described as a 22-kD intrinsic membrane protein that binds chlorophyll and carotenoid, and therefore denoted CP22 (Funk et al., 1994). However, later studies cast doubt on its pigment-binding ability (Aspinall-O'Dea et al., 2002; Dominici et al., 2002). PsbS interacts weakly with PS II (Funk et al., 1995), but is not present in the PS II–LHCII supercomplex and is located in the LHCII-rich regions that interconnect the supercomplexes in the membrane (Nield et al., 2000a). The PsbS sequence predicts four transmembrane helices, but most of the chlorophyll ligands determined from the LHCII crystal structure are not conserved. An *Arabidopsis* mutant (*npq4*) deficient in the ability to induce q_E quenching (see Section V.A.2.) has defects in the *PsbS* gene (Li et al., 2000), indicating that this protein is essential for thermal dissipation of excess energy and is probably a sensor of lumenal pH (Niyogi et al., 2005). The putative PSBS protein in *Chlamydomonas* (Table 14.3) shares 45% identity over 180 amino acids with *Arabidopsis* PsbS (Anwaruzzaman et al., 2004; Minagawa and Takahashi, 2004) including the proposed pH-sensing Glu residues (Li et al., 2004). However, the gene product has not been detected in *Chlamydomonas* thylakoid membrane preparations. Its apparently low expression levels might account for the limited capacity for q_E quenching in *Chlamydomonas* (Finazzi et al., 2006).

V. REGULATION OF THE LIGHT-HARVESTING COMPLEX

A. Photoacclimation

1. Introduction

For photosynthetic organisms, the ability to survive fluctuations in light quality and quantity is crucial. In order to maximize the efficiency of light harvesting while minimizing damage to the photosynthetic apparatus by excess light (photoinhibition) under natural variable light conditions, the light-harvesting apparatus is dynamically regulated by the process of photoacclimation. Three well-known photoacclimation mechanisms are: (1) energy-dependent non-photochemical quenching to downregulate light-harvesting ability within seconds (q_E-quenching); (2) rapid adjustment of the absorption cross-sections of PS I and PS II by redistributing mobile LHCII proteins (state transition); and (3) transcriptional and translational regulation of *LHC* gene expression (antenna size adjustment).

2. q_E-quenching

The rapid downregulation of light harvesting ability, which can be measured as uncoupler-sensitive nonphotochemical quenching (NPQ), is called q_E-quenching (Govindjee et al., 1967; Murata and Sugahara, 1969). Wraight and Crofts (1971) showed that this NPQ is induced by ΔpH across the thylakoid membranes. Its extent and capacity were correlated with the accumulation of the xanthophyll, zeaxanthin (Demmig et al., 1987). Zeaxanthin is formed in thylakoids by reversible de-epoxidation of violaxanthin in a process called the xanthophyll cycle (Yamamoto et al., 1962), which is activated by low lumenal pH (Eskling et al., 1997). Lumenal acidification is not only required for the de-epoxidation reaction, but also for the activation of another key factor for the induction of q_E-quenching, PsbS. The protein(s) with which the quencher is associated has been a subject of intense research and four proposals have been made: the PS II core complex (Finazzi et al., 2004), minor monomeric LHCII proteins (Ahn et al., 2008), major trimeric LHCII proteins (Pascal et al., 2005; Ruban et al., 2008), and PsbS (Niyogi et al., 2005). The minor LHCII proteins have been studied extensively because they contain a high content of xanthophyll cycle pigments (Bassi et al., 1993). In fact, Crimi et al. (2001) demonstrated that binding of zeaxanthin decreases the lifetime of chlorophyll fluorescence in CP29, suggesting xanthophyll-dependent q_E-quenching. An extensive search for NPQ mutants in *Chlamydomonas* and *Arabidopsis* identified several, but none mapped to CP29, CP26, or CP24 (Baroli and Niyogi, 2000). The *Chlamydomonas npq5* mutant is unable to establish rapid and reversible NPQ (Niyogi et al., 1997). Since *npq5* maps to the *LHCBM1* gene encoding the major LHCII type IV polypeptide, this protein could be the quencher in *Chlamydomonas* (Elrad et al., 2002). Additional details regarding thermal dissipation mechanism(s) can be found in Chapter 23.

3. State transition and phosphorylation

The distribution of absorbed light energy between the two photosystems is dynamically balanced, ensuring maximum efficiency for photosynthetic electron transport in changing light environments (Bonaventura and Myers, 1969; Murata, 1969). This balance is regulated by a process termed state transition; state I and state II are induced by preferential excitation of PS I and PS II, respectively. State I is recognized by a higher chlorophyll fluorescence yield at room temperature, a higher ratio of F685 and F695 to F720 at 77K, a higher quantum yield of PS II reactions, and a lower quantum yield of PS I reactions at low light intensities. Conversely, state II is recognized by a lower chlorophyll fluorescence yield at room temperature, a lower ratio of F685 and F695 to F720 at 77K, a lower quantum yield of PS II, and a higher quantum yield of PS I reactions. The molecular mechanisms of these physiological responses are described in detail in Chapter 22. This section focuses specifically on phosophorylation of LHCII proteins.

Reversible phosphorylation of LHCII proteins is significant in the migration of mobile LHCII proteins during state transitions (Andersson et al., 1982; Depège et al., 2003). Both the major and minor LHCII polypeptides in *Chlamydomonas* are subject to phosphorylation/dephosphorylation (Wollman and Delepelaire, 1984; Fleischmann et al., 1999; Stauber et al., 2003; Vink et al., 2004; Takahashi et al., 2006; Turkina et al., 2006). The phosphorylation sites have been precisely determined (Turkina et al., 2006): two Thr, one Thr, seven Thr, and one Ser residues in major LHCII protein type I, type IV, CP29, and CP26, respectively, were shown to be phosphorylated under certain conditions (Figure 14.7; p. 514). Most of these phosphorylatable residues are clustered in the N-terminal portion of the polypeptide. The heavy phosphorylation of CP29 is particularly well-documented under various conditions, suggesting that environmentally induced phosphorylation could control the reorganization of photosynthetic complexes (Turkina et al., 2004, 2006; Kargul et al., 2005).

4. Antenna size adjustment

Transcriptional regulation of *LHC* genes plays a central role in antenna size adjustment. Many factors affect *LHC* transcription in green algae, including circadian rhythms (Kindle, 1987), chlorophyll synthesis (Johanningmeier and Howell, 1984), and intensity of incident light (Elrad et al., 2002; Teramoto et al., 2002; Durnford et al., 2003; McKim and Durnford, 2006). These factors also influence gene expression in higher plants, but in this case the regulation acts through phytochrome, which is not found in *Chlamydomonas.*

Teramoto et al. (2002) showed that the expression of six *LHCB* genes, *LHCBM1, LHCBM2, LHCBM3, LHCBM6, LHCB4,* and *LHCB5,* was reduced dramatically after 6 hours in high light. The light intensity that triggers down-regulation decreases when cells are grown at lower temperatures or lower ambient CO_2 concentrations, suggesting that excitation energy, CO_2 assimilation, and *LHCB* transcription are balanced by a negative feedback mechanism. Since this negative feedback was only marginally affected by electron transport inhibitors or mutations in PS I and/or PS II, expression of these genes is mainly regulated by a redox-independent mechanism (Teramoto et al., 2002). Redox-independent regulation has also been described for other light-responsive genes in *Chlamydomonas* (Im and Grossman, 2002) and *Dunaliella salina* (Masuda et al., 2003). Two possible candidates for the redox-independent signal have been proposed; reactive oxygen species and precursors of chlorophyll synthesis. Numerous chlorophyll molecules bound to the LHCII complex or precursor(s) of chlorophyll synthesis could be a potent source of reactive oxygen species, and levels of precursors of chlorophyll synthesis may also be affected by reactive oxygen species.

Kinetics of *LHCBM* gene expression were studied following a shift to high-light conditions (Durnford et al., 2003; Elrad and Grossman, 2004). The results indicate a transient decline in mRNA abundance during the first 2 hours, followed by recovery from 6 hours to 24 hours. The expression of these genes is also subject to translational regulation, where the transcripts are off-loaded from polysomes within 2 hours, followed by global polysome disassembly after 4 hours (McKim and Durnford, 2006). Therefore, transcript level recovery after 6 hours would not be reflected at the protein level until polysomes reform after 8 hours. The possible cause for polysome disassembly was suggested to be reactive oxygen species. The *LHCB4* and *LHCB5* transcripts are coordinately regulated with transcripts of *LHCBM* genes, although they are less responsive during a light shift (Teramoto et al., 2002), and their accumulation is not dramatically altered upon photoacclimation (Minagawa et al., 2001).

B. Genes involved in biosynthesis of the LHCII complex

LHC proteins are encoded in the nucleus, and therefore imported post-translationally into chloroplasts, where the complex is integrated into thylakoid membranes after pigment binding. A gene that may regulate thylakoid integration of LHCII proteins in *Chlamydomonas* maps to the *ALB3* locus, which encodes a chloroplast homolog of Oxa1 in plant mitochondria and YidC in *E. coli* (Bellafiore et al., 2002). In higher plants, Alb3 is required for the insertion of SRP (signal recognition particle)-dependent proteins such as LHCB4.1 and LHCB5 (Yi and Dalbey, 2005). The *Chlamydomonas alb3* mutant has a 10-fold reduction in the level of LHCII proteins, suggesting that Alb3 and the chloroplast SRP facilitate translocation of LHCII proteins. In fact, the L18 domain, which is required for reconstitution of a SRP/LHC transit complex in pea (DeLille et al., 2000), is conserved in all LHC proteins associated with PS II in *Chlamydomonas* (Figure 14.7).

ACKNOWLEDGMENTS

I thank Dr. Wesley Swingley for his constructive comments and critical reading of the manuscript and Masakazu Iwai for unpublished results. This work was supported in part by Grants-in-Aid for Scientific Research from the Ministry of Education, Culture, Sports, Science and Technology (18GS0318, 17570029).

REFERENCES

Adamska, I. (2001). The Elip family of stress proteins in the thylakoid membranes of pro-and eukaryota. In: *Advances in Photosynthesis and Respiration-Regulation of Photosynthesis*, Vol. 11 (E.M. Aro and B. Andersen, Eds.), pp. 487–505. Kluwer Academic Publishers, Dordrecht.

Adamska, I., Roobol-Boza, M., Lindahl, M., and Andersson, B. (1999). Isolation of pigment-binding early light-inducible proteins from pea. *Eur. J. Biochem.* **260**, 453–460.

Ahn, T.K., Avenson, T.J., Ballottari, M., Cheng, Y.-C., Niyogi, K.K., Bassi, R., and Fleming, G.R. (2008). Architecture of a charge-transfer state regulating light harvesting in a plant antenna protein. *Science* **320**, 794–797.

Amunts, A., Drory, O., and Nelson, N. (2007). The structure of a plant photosystem I supercomplex at 3.4Å resolution. *Nature* **447**, 58–63.

Andersson, B., Åkerlund, H.-E., Jergil, B., and Larsson, C. (1982). Differential phosphorylation of the light-harvesting chlorophyll-protein complex in appressed and non-appressed regions of the thylakoid membrane. *FEBS Lett.* **149**, 181–185.

Andersson, U., Heddad, M., and Adamska, I. (2003). Light stress-induced one-helix protein of the chlorophyll *a/b*-binding family associated with Photosystem I. *Plant Physiol.* **132**, 811–820.

Anwaruzzaman, M., Chin, B.L., Li, X.P., Lohr, M., Martinez, D.A., and Niyogi, K.K. (2004). Genomic analysis of mutants affecting xanthophyll biosynthesis and regulation of photosynthetic light harvesting in *Chlamydomonas reinhardtii*. *Photosynth. Res.* **82**, 265–276.

Asamizu, E., Nakamura, Y., Sato, S., Fukuzawa, H., and Tabata, S. (1999). A large scale structural analysis of cDNAs in a unicellular green alga, *Chlamydomonas reinhardtii*. I. Generation of 3433 non-redundant expressed sequence tags. *DNA Res.* **6**, 369–373.

Aspinall-O'Dea, M., Wentworth, M., Pascal, A., Robert, B., Ruban, A., and Horton, P. (2002). *In vitro* reconstitution of the activated zeaxanthin state associated with energy dissipation in plants. *Proc. Natl. Acad. Sci. U. S. A.* **99**, 16331–16335.

Baroli, I. and Niyogi, K.K. (2000). Molecular genetics of xanthophyll-dependent photoprotection in green algae and plants. *Phil. Trans. R. Soc. Lond. B. Biol. Sci.* **355**, 1385–1394.

Barzda, V., de Grauw, C.J., Vroom, J., Kleima, F.J., van Grondelle, R., van Amerongen, H., and Gerritsen, H.C. (2001). Fluorescence lifetime heterogeneity in aggregates of LHCII revealed by time-resolved microscopy. *Biophys. J.* **81**, 538–546.

Bassi, R. and Wollman, F.-A. (1991). The chlorophyll-*a/b* proteins of photosystem II in *Chlamydomonas reinhardtii*. Isolation, characterization and immunological cross-reactivity of higher-plant polypeptides. *Planta* **183**, 423–433.

Bassi, R., Soen, S.Y., Frank, G., Zuber, H., and Rochaix, J.-D. (1992). Characterization of chlorophyll *a/b* proteins of photosystem I from *Chlamydomonas reinhardtii*. *J. Biol. Chem.* **267**, 25714–25721.

Bassi, R., Pineau, B., Dainese, P., and Marquardt, J. (1993). Carotenoid-binding proteins of photosystem II. *Eur. J. Biochem.* **212**, 297–303.

Bellafiore, S., Ferris, P., Naver, H., Gohre, V., and Rochaix, J.-D. (2002). Loss of Albino3 leads to the specific depletion of the light-harvesting system. *Plant Cell* **14**, 2303–2314.

Ben-Shem, A., Frolow, F., and Nelson, N. (2003). Crystal structure of plant photosystem I. *Nature* **426**, 630–635.

Bhaya, D., Dufresne, A., Vaulot, D., and Grossman, A. (2002). Analysis of the *hli* gene family in marine and freshwater cyanobacteria. *FEMS Microbiol. Lett.* **215**, 209–219.

Bittner, T., Irrgang, K.-D., Renger, G., and Wasielewski, M.R. (1994). Ultrafast excitation energy transfer and exciton-exciton annihilation processes in isolated light harvesting complexes of photosystem II (LHC II) from spinach. *J. Phys. Chem.* **98**, 11821–11826.

Boekema, E.J., Dekker, J.P., van Heel, M.G., Rogner, M., Saenger, W., Witt, I., and Witt, H.T. (1987). Evidence for a trimeric organization of the photosystem I complex from the thermophilic cyanobacterium *Synechococcus* sp. *FEBS Lett.* **217**, 283–286.

Boekema, E.J., Hankamer, B., Bald, D., Kruip, J., Nield, J., Boonstra, A.F., Barber, J., and Rögner, M. (1995). Supramolecular structure of the photosystem II complex from green plants and cyanobacteria. *Proc. Natl. Acad. Sci. U. S. A.* **92**, 175–179.

Boekema, E.J., Jensen, P.E., Schlodder, E., van Breemen, J.F., van Roon, H., Scheller, H.V., and Dekker, J.P. (2001). Green plant photosystem I binds light-harvesting complex I on one side of the complex. *Biochemistry* **40**, 1029–1036.

Bonaventura, C. and Myers, J. (1969). Fluorescence and oxygen evolution from *Chlorella pyrenoidosa*. *Biochim. Biophys. Acta,* **189**, 366–383.

Butler, W.L. (1961). A far-red absorbing form of chlorophyll *in vivo*. *Arch. Biochem. Biophys.* **93**, 413–422.

Caffarri, S., Croce, R., Breton, J., and Bassi, R. (2001). The major antenna complex of photosystem II has a xanthophyll binding site not involved in light harvesting. *J. Biol. Chem.* **276**, 35924–35933.

Chua, N.H. and Bennoun, P. (1975). Thylakoid membrane polypeptides of *Chlamydomonas reinhardtii*: Wild-type and mutant strains deficient in photosystem II reaction center. *Proc. Natl. Acad. Sci. U. S. A.* **72**, 2175–2179.

Crimi, M., Dorra, D., Bosinger, C.S., Giuffra, E., Holzwarth, A.R., and Bassi, R. (2001). Time-resolved fluorescence analysis of the recombinant photosystem II antenna complex CP29. Effects of zeaxanthin, pH and phosphorylation. *Eur. J. Biochem.* **268**, 260–267.

Croce, R. and Bassi, R. (1998). The light-harvesting complex of photosystem I: Pigment composition and stoicheometry. In: *Photosynthesis: Mechanisms and Effects* (G. Garab, Ed.), pp. 503–508. Kluwer Academic Publishers, Dordrecht.

Croce, R., Dorra, D., Holzwarth, A.R., and Jennings, R.C. (2000). Fluorescence decay and spectral evolution in intact photosystem I of higher plants. *Biochemistry* **39**, 6341–6348.

Croce, R., Muller, M.G., Bassi, R., and Holzwarth, A.R. (2001). Carotenoid-to-chlorophyll energy transfer in recombinant major light-harvesting complex (LHCII) of higher plants. I. Femtosecond transient absorption measurements. *Biophys. J.* **80**, 901–915.

Croce, R., Canino, G., Ros, F., and Bassi, R. (2002a). Chromophore organization in the higher-plant photosystem II antenna protein CP26. *Biochemistry* **41**, 7334–7343.

Croce, R., Morosinotto, T., Castelletti, S., Breton, J., and Bassi, R. (2002b). The Lhca antenna complexes of higher plants I. *Biochim. Biophys. Acta* **1556**, 29–40.

DeLille, J., Peterson, E.C., Johnson, T., Moore, M., Kight, A., and Henry, R. (2000). A novel precursor recognition element facilitates posttranslational binding to the signal recognition particle in chloroplasts. *Proc. Natl. Acad. Sci. U. S. A.* **97**, 1926–1931.

Demmig, B., Winter, K., Kruger, A., and Czygan, F.C. (1987). Photoinhibition and zeaxanthin formation in intact leaves: A possible role of the xanthophyll cycle in the dissipation of excess light energy. *Plant Physiol.* **84**, 218–224.

Depège, N., Bellafiore, S., and Rochaix, J.-D. (2003). Role of chloroplast protein kinase Stt7 in LHCII phosphorylation and state transition in *Chlamydomonas*. *Science* **299**, 1572–1575.

Dolganov, N.A., Bhaya, D., and Grossman, A.R. (1995). Cyanobacterial protein with similarity to the chlorophyll *a*/*b* binding proteins of higher plants: Evolution and regulation. *Proc. Natl. Acad. Sci. U. S. A.* **92**, 636–640.

Dominici, P., Caffarri, S., Armenante, F., Ceoldo, S., Crimi, M., and Bassi, R. (2002). Biochemical properties of the PsbS subunit of photosystem II either purified from chloroplast or recombinant. *J. Biol. Chem.* **277**, 22750–22758.

Douglas, S., Zauner, S., Fraunholz, M., Beaton, M., Penny, S., Deng, L.T., Wu, X., Reith, M., Cavalier-Smith, T., and Maier, U.G. (2001). The highly reduced genome of an enslaved algal nucleus. *Nature* **410**, 1091–1096.

Douwe de Boer, A. and Weisbeek, P.J. (1991). Chloroplast protein topogenesis: Import, sorting and assembly. *Biochim. Biophys. Acta.* **1071**, 221–253.

Du, M., Xie, X., Jia, Y., Mets, L., and Fleming, G.R. (1993). Direct observation of ultrafast energy transfer in PSI core antenna. *Chem. Phys. Lett.* **201**, 535–542.

Dubertret, G., Mirshahi, A., Mirshahi, M., Gerard-Hirne, C., and Trémolières, A. (1994). Evidence from *in vivo* manipulations of lipid composition in mutants that the Δ^3-*trans*-hexadecenoic acid-containing phosphatidylglycerol is involved in the biogenesis of the light-harvesting chlorophyll *a*/*b*-protein complex of *Chlamydomonas reinhardtii*. *Eur. J. Biochem.* **226**, 473–482.

Dubertret, G., Gerard-Hirne, C., and Trémolières, A. (2002). Importance of *trans*-Δ^3-hexadecenoic acid containing phosphatidylglycerol in the formation of the trimeric light-harvesting complex in *Chlamydomonas*. *Plant Physiol. Biochem.* **40**, 829–836.

Durnford, D.G., Deane, J.A., Tan, S., McFadden, G.I., Gantt, E., and Green, B.R. (1999). A phylogenetic assessment of the eukaryotic light-harvesting antenna proteins, with implications for plastid evolution. *J. Mol. Evol.* **48**, 59–68.

Durnford, D.G., Price, J.A., McKim, S.M., and Sarchfield, M.L. (2003). Light-harvesting complex gene expression is controlled by both transcriptional and post-transcriptional mechanisms during photoacclimation in *Chlamydomonas reinhardtii*. *Physiol. Plant.* **118**, 193–205.

Elrad, D. and Grossman, A.R. (2004). A genome's-eye view of the light-harvesting polypeptides of *Chlamydomonas reinhardtii*. *Curr. Genet.* **45**, 61–75.

Elrad, D., Niyogi, K.K., and Grossman, A.R. (2002). A major light-harvesting polypeptide of photosystem II functions in thermal dissipation. *Plant Cell* **14**, 1801–1816.

Eskling, M., Arvidsson, P.-O., and Åkerlund, H.-E. (1997). The xanthophyll cycle, its regulation and components. *Physiol. Plant.* **100**, 806–816.

Ferreira, K.N., Iverson, T.M., Maghlaoui, K., Barber, J., and Iwata, S. (2004). Architecture of the photosynthetic oxygen-evolving center. *Science* **303**, 1831–1838.

Finazzi, G., Johnson, G.N., Dall'Osto, L., Joliot, P., Wollman, F.-A., and Bassi, R. (2004). A zeaxanthin-independent nonphotochemical quenching mechanism localized in the photosystem II core complex. *Proc. Natl. Acad. Sci. U. S. A.* **101**, 12375–12380.

Finazzi, G., Johnson, G.N., Dall'Osto, L., Zito, F., Bonente, G., Bassi, R., and Wollman, F.-A. (2006). Nonphotochemical quenching of chlorophyll fluorescence in *Chlamydomonas reinhardtii*. *Biochemistry* **45**, 1490–1498.

Fleischmann, M.M., Ravanel, S., Delosme, R., Olive, J., Zito, F., Wollman, F.-A., and Rochaix, J.-D. (1999). Isolation and characterization of photoautotrophic mutants of *Chlamydomonas reinhardtii* deficient in state transition. *J. Biol. Chem.* **274**, 30987–30994.

Funk, C. and Vermaas, W. (1999). A cyanobacterial gene family coding for single-helix proteins resembling part of the light-harvesting proteins from higher plants. *Biochemistry* **38**, 9397–9404.

Funk, C., Schröder, W.P., Green, B.R., Renger, G., and Andersson, B. (1994). The intrinsic 22 kDa protein is a chlorophyll-binding subunit of photosystem II. *FEBS Lett.* **342**, 261–266.

Funk, C., Schroder, W.P., Napiwotzki, A., Tjus, S.E., Renger, G., and Andersson, B. (1995). The PSII-S protein of higher plants: a new type of pigment-binding protein. *Biochemistry* **34**, 11133–11141.

Gagne, G. and Guertin, M. (1992). The early genetic response to light in the green unicellular alga *Chlamydomonas eugametos* grown under light/dark cycles involves genes that represent direct responses to light and photosynthesis. *Plant Mol. Biol.* **18**, 429–445.

Ganeteg, U., Klimmek, F., and Jansson, S. (2004). Lhca5 – an LHC-type protein associated with Photosystem I. *Plant Mol. Biol.* **54**, 641–651.

Germano, M., Yakushevska, A.E., Keegstra, W., van Gorkom, H.J., Dekker, J.P., and Boekema, E.J. (2002). Supramolecular organization of photosystem I and light-harvesting complex I in *Chlamydomonas reinhardtii*. *FEBS Lett.* **525**, 121–125.

Govindjee, Papageorgiou G. and Rabinowitch, E. (1967). In: *Fluorescence; Theory, Instrumentation and Practice* (G.G. Guilbault, Ed.), pp. 511–564. Marcel Dekker, New York.

Gradinaru, C.C., van Stokkum, I.H.M., Pascal, A.A., van Grondelle, R., and van Amerongen, H. (2000). Identifying the pathways of energy transfer between carotenoids and chlorophylls in LHCII and CP29. A multicolor, femtosecond pump-probe study. *J. Phys. Chem. B* **104**, 9330–9342.

Green, B.R. (2003). The evolution of light-harvesting antennas. In: *Light-harvesting Antennas* (B.R. Green and W.W. Parson, Eds.), pp. 129–168. Kluwer Academic Publishers, Dordrecht.

Hankamer, B., Nield, J., Zheleva, D., Boekema, E., Jansson, S., and Barber, J. (1997). Isolation and biochemical characterisation of monomeric and dimeric photosystem II complexes from spinach and their relevance to the organisation of photosystem II *in vivo*. *Eur. J. Biochem.* **243**, 422–429.

Harrer, R., Bassi, R., Testi, M.G., and Schäfer, C. (1998). Nearest-neighbor analysis of a photosystem II complex from *Marchantia polymorpha* L. (liverwort), which contains reaction center and antenna proteins. *Eur. J. Biochem.* **255**, 196–205.

He, Q., Dolganov, N., Bjorkman, O., and Grossman, A.R. (2001). The high light-inducible polypeptides in *Synechocystis* PCC6803. Expression and function in high light. *J. Biol. Chem.* **276**, 306–314.

Heddad, M. and Adamska, I. (2000). Light stress-regulated two-helix proteins in *Arabidopsis thaliana* related to the chlorophyll *a/b*-binding gene family. *Proc. Natl. Acad. Sci. U. S. A.* **97**, 3741–3746.

Hippler, M., Klein, J., Fink, A., Allinger, T., and Hoerth, P. (2001). Towards functional proteomics of membrane protein complexes: Analysis of thylakoid membranes from *Chlamydomonas reinhardtii*. *Plant J.* **28**, 595–606.

Hobe, S., Förster, R., Klingler, J., and Paulsen, H. (1995). N-proximal sequence motif in light-harvesting chlorophyll *a/b*-binding proteins is essential for the trimerization of light-harvesting chlorophyll *a/b* complex. *Biochemistry* **34**, 10224–10228.

Holzwarth, A.R. (1990). The functional organization of the antenna systems in higher plants and green algae as studied by time-resolved fluorescence techniques. In: *Current Research in Photosynthesis*, Vol. 2 (M. Baltscheffsky, Ed.), pp. 223–230. Kluwer Academic Publishers, Dordrecht.

Huber, C.G., Timperio, A.M., and Zolla, L. (2001). Isoforms of photosystem II antenna proteins in different plant species revealed by liquid chromatography-electrospray ionization mass spectrometry. *J Biol. Chem.* **276**, 45755–45761.

Hutin, C., Nussaume, L., Moise, N., Moya, I., Kloppstech, K., and Havaux, M. (2003). Early light-induced proteins protect *Arabidopsis* from photooxidative stress. *Proc. Natl. Acad. Sci. U. S. A.* **100**, 4921–4926.

Ihalainen, J.A., Gobets, B., Sznee, K., Brazzoli, M., Croce, R., Bassi, R., van Grondelle, R., Korppi-Tommola, J.E., and Dekker, J.P. (2000). Evidence for two spectroscopically different dimers of light-harvesting complex I from green plants. *Biochemistry* **39**, 8625–8631.

Ihalainen, J.A., Klimmek, F., Ganeteg, U., van Stokkum, I.H., van Grondelle, R., Jansson, S., and Dekker, J.P. (2005). Excitation energy trapping in photosystem I complexes depleted in Lhca1 and Lhca4. *FEBS Lett.* **579**, 4787–4791.

Im, C.S. and Grossman, A.R. (2002). Identification and regulation of high light-induced genes in *Chlamydomonas reinhardtii*. *Plant J.* **30**, 301–313.

Imbault, P., Wittemer, C., Johanningmeier, U., Jacobs, J.D., and Howell, S.H. (1988). Structure of the *Chlamydomonas reinhardtii cabII-1* gene encoding a chlorophyll-*a/b*-binding protein. *Gene* **73**, 397–407.

Jansson, S. (1994). The light-harvesting chlorophyll *a/b*-binding proteins. *Biochim. Biophys. Acta* **1184**, 1–19.

Jansson, S. (1999). A guide to the *Lhc* genes and their relatives in *Arabidopsis*. *Trends Plant Sci.* **4**, 236–240.

Jansson, S., Pichersky, E., Bassi, R., Green, B.R., Ikeuchi, M., Melis, A., Simpson, D.J., Spangfort, M., Staehelin, L.A., and Thornber, J.P. (1992). A nomenclature for the genes encoding the chlorophyll *a/b*-binding proteins of higher plants. *Plant Mol. Biol. Rep.* **10**, 242–253.

Jansson, S., Stefánsson, H., Nyström, U., Gustafsson, P., and Albertsson, P.-Å. (1997). Antenna protein composition of PS I and PS II in thylakoid subdomains. *Biochim. Biophys. Acta* **1320**, 297–309.

Jansson, S., Andersson, J., Kim, S.J., and Jackowski, G. (2000). An *Arabidopsis thaliana* protein homologous to cyanobacterial high-light-inducible proteins. *Plant Mol. Biol.* **42**, 345–351.

Johanningmeier, U. and Howell, S.H. (1984). Regulation of light-harvesting chlorophyll-binding protein mRNA accumulation in *Chlamydomonas reinhardi*. Possible involvement of chlorophyll synthesis precursors. *J. Biol. Chem.* **259**, 13541–13549.

Jordan, P., Fromme, P., Witt, H.T., Klukas, O., Saenger, W., and Krauss, N. (2001). Three-dimensional structure of cyanobacterial photosystem I at 2.5Å resolution. *Nature* **411**, 909–917.

Kühlbrandt, W., Wang, D.N., and Fujiyoshi, Y. (1994). Atomic model of plant light-harvesting complex by electron crystallography. *Nature* **367**, 614–621.

Kargul, J., Nield, J., and Barber, J. (2003). Three-dimensional reconstruction of a light-harvesting complex I-photosystem I (LHCI-PSI) supercomplex from the green alga *Chlamydomonas reinhardtii*. Insights into light harvesting for PSI. *J. Biol. Chem.* **278**, 16135–16141.

Kargul, J., Turkina, M.V., Nield, J., Benson, S., Vener, A.V., and Barber, J. (2005). Light-harvesting complex II protein CP29 binds to photosystem I of *Chlamydomonas reinhardtii* under State 2 conditions. *FEBS J.* **272**, 4797–4806.

Kimura, M., Yoshizumi, T., Manabe, K., Yamamoto, Y.Y., and Matsui, M. (2001). *Arabidopsis* transcriptional regulation by light stress via hydrogen peroxide-dependent and -independent pathways. *Genes Cells* **6**, 607–617.

Kindle, K.L. (1987). Expression of a gene for a light-harvesting chlorophyll *a/b*-binding protein in *Chlamydomonas reinhardtii*: Effect of light and acetate. *Plant Mol. Biol.* **9**, 547–563.

Kleima, F.J., Gradinaru, C.C., Calkoen, F., van Stokkum, I.H., van Grondelle, R., and van Amerongen, H. (1997). Energy transfer in LHCII monomers at 77K studied by sub-picosecond transient absorption spectroscopy. *Biochemistry* **36**, 15262–15268.

Klimmek, F., Ganeteg, U., Ihalainen, J.A., van Roon, H., Jensen, P.E., Scheller, H.V., Dekker, J.P., and Jansson, S. (2005). Structure of the higher plant light harvesting complex I: In vivo characterization and structural interdependence of the Lhca proteins. *Biochemistry* **44**, 3065–3073.

Klimmek, F., Sjodin, A., Noutsos, C., Leister, D., and Jansson, S. (2006). Abundantly and rarely expressed Lhc protein genes exhibit distinct regulation patterns in plants. *Plant Physiol.* **140**, 793–804.

Klug, D.R., Giorgi, L.B., Crystall, B., Barber, J., and Porter, G. (1989). Energy transfer to low energy chlorophyll species prior to trapping by P700 and subsequent electron transfer. *Photosynth. Res.* **22**, 277–284.

Knoetzel, J., Svendsen, I., and Simpson, D.J. (1992). Identification of the photosystem I antenna polypeptides in barley. Isolation of three pigment-binding antenna complexes. *Eur. J. Biochem.* **206**, 209–215.

Lam, E., Oritz, W., Mayfield, S., and Malkin, R. (1984). Isolation and characterization of a light-harvesting chlorophyll *a/b* Protein complex associated with Photosystem I. *Plant Physiol.* **74**, 650–655.

La Roche, J., van der Staay, G.W.M., Partensky, F., Ducret, A., Aebersold, R., Li, R., Golden, S.S., Hiller, R.G., Wrench, P.M., Larkum, A.W.D., and Green, B.R. (1996). Independent evolution of the prochlorophyte and green plant chlorophyll *a/b* light-harvesting proteins. *Proc. Natl. Acad. Sci. U. S. A.* **93**, 15244–15248.

Larsson,, U.K., Anderson, J.M., and Andersson, B. (1987). Variations in the relative content of the peripheral and inner light-harvesting chlorophyll *a/b*-protein complex (LHC II) subpopulations during thylakoid light adaptation and development. *Biochim. Biophys. Acta* **894**, 69–75.

Li, X.P., Björkman, O., Shih, C., Grossman, A.R., Rosenquist, M., Jansson, S., and Niyogi, K.K. (2000). A pigment-binding protein essential for regulation of photosynthetic light harvesting. *Nature* **403**, 391–395.

Li, X.P., Gilmore, A.M., Caffarri, S., Bassi, R., Golan, T., Kramer, D., and Niyogi, K.K. (2004). Regulation of photosynthetic light harvesting involves intrathylakoid lumen pH sensing by the PsbS protein. *J. Biol. Chem.* **279**, 22866–22874.

Liu, Z., Yan, H., Wang, K., Kuang, T., Zhang, J., Gui, L., An, X., and Chang, W. (2004). Crystal structure of spinach major light-harvesting complex at 2.72Å resolution. *Nature* **428**, 287–292.

Lucinski, R., Schmid, V.H., Jansson, S., and Klimmek, F. (2006). Lhca5 interaction with plant photosystem I. *FEBS Lett.* **580**, 6485–6488.

Maid, U. and Zetsche, K. (1992). A 16kb small single-copy region separates the plastid DNA inverted repeat of the unicellular red alga *Cyanidium caldarum*:

Physical mapping of the IR-flanking regions and nucleotide sequence of the *psbD-psbC*, rps16, 5S rRNA and *rpl21* genes. *Plant Mol. Biol.* **19**, 1001–1010.

Masuda, T., Tanaka, A., and Melis, A. (2003). Chlorophyll antenna size adjustments by irradiance in *Dunaliella salina* involve coordinate regulation of chlorophyll a oxygenase (CAO) and *Lhcb* gene expression. *Plant Mol. Biol.* **51**, 757–771.

McKim, S.M. and Durnford, D.G. (2006). Translational regulation of light-harvesting complex expression during photoacclimation to high-light in *Chlamydomonas reinhardtii*. *Plant Physiol. Biochem.* **44**, 857–865.

Melkozernov, A.N. (2001). Excitation energy transfer in photosystem I from oxygenic organisms. *Photosynth. Res.* **70**, 129–153.

Melkozernov, A.N. and Blankenship, R.E. (2003). Structural modeling of the Lhca4 subunit of LHCI-730 peripheral antenna in photosystem I based on similarity with LHCII. *J. Biol. Chem.* **278**, 44542–44551.

Melkozernov, A.N. and Blankenship, R.E. (2005). Structural and functional organization of the peripheral light-harvesting system in Photosystem I. *Photosynth. Res.* **85**, 33–50.

Melkozernov, A.N., Lin, S., Schmid, V.H., Paulsen, H., Schmidt, G.W., and Blankenship, R.E. (2000). Ultrafast excitation dynamics of low energy pigments in reconstituted peripheral light-harvesting complexes of Photosystem I. *FEBS Lett.* **471**, 89–92.

Melkozernov, A.N., Schmid, V.H.R., Lin, S., Paulsen, H., and Blankenship, R.E. (2002). Excitation energy transfer in the Lhca1 subunit of LHC I-730 peripheral antenna of I. *J. Phys. Chem. B* **106**, 4313–4317.

Melkozernov, A.N., Kargul, J., Lin, S., Barber, J., and Blankenship, R.E. (2004). Energy coupling in the PSI-LHCI supercomplex from the green alga *Chlamydomonas reinhardtii*. *J. Phys. Chem. B* **108**, 10547–10555.

Melkozernov, A.N., Kargul, J., Lin, S., Barber, J., and Blankenship, R.E. (2005). Spectral and kinetic analysis of the energy coupling in the PS I-LHC I supercomplex from the green alga *Chlamydomonas reinhardtii* at 77 K. *Photosynth. Res.* **86**, 203–215.

Melkozernov, A.N., Barber, J., and Blankenship, R.E. (2006). Light harvesting in photosystem I supercomplexes. *Biochemistry* **45**, 331–345.

Meyer, G. and Kloppstech, K. (1984). A rapidly light-induced chloroplast protein with a high turnover coded for by pea nuclear DNA. *Eur. J. Biochem.* **138**, 201–207.

Minagawa, J. and Takahashi, Y. (2004). Structure, function and assembly of photosystem II and its light-harvesting proteins. *Photosynth. Res.* **82**, 241–263.

Minagawa, J., Han, K.C., Dohmae, N., Takio, K., and Inoue, Y. (2001). Molecular characterization and gene expression of *lhcb5* gene encoding CP26 in the light-harvesting complex II of *Chlamydomonas reinhardtii*. *Plant Mol. Biol.* **46**, 277–287.

Miura, K., Yamano, T., Yoshioka, S., Kohinata, T., Inoue, Y., Taniguchi, F., Asamizu, E., Nakamura, Y., Tabata, S., Yamato, K.T., Ohyama, K., and Fukuzawa, H. (2004). Expression profiling-based identification of CO_2-responsive genes regulated by CCM1 controlling a carbon-concentrating mechanism in *Chlamydomonas reinhardtii*. *Plant Physiol.* **135**, 1595–1607.

Montane, M.H., Tardy, F., Kloppstech, K., and Havaux, M. (1998). Differential control of xanthophylls and light-induced stress proteins, as opposed to light-harvesting chlorophyll *a/b* proteins, during photosynthetic acclimation of barley leaves to light irradiance. *Plant Physiol.* **118**, 227–235.

Morosinotto, T., Castelletti, S., Breton, J., Bassi, R., and Croce, R. (2002). Mutation analysis of Lhca1 antenna complex. Low energy absorption forms originate from pigment-pigment interactions. *J. Biol. Chem.* **277**, 36253–36261.

Morosinotto, T., Breton, J., Bassi, R., and Croce, R. (2003). The nature of a chlorophyll ligand in Lhca proteins determines the far red fluorescence emission typical of I. *J. Biol. Chem.* **278**, 49223–49229.

Moseley, J.L., Allinger, T., Herzog, S., Hoerth, P., Wehinger, E., Merchant, S., and Hippler, M. (2002). Adaptation to Fe-deficiency requires remodeling of the photosynthetic apparatus. *EMBO J.* **21**, 6709–6720.

Murata, N. (1969). Control of excitation transfer in Photosynthesis I. Light-induced change of chlorophyll *a* fluoresence in *Porphyridium cruentum*. *Biochim. Biophys. Acta,* **172**, 242–251.

Murata, N. and Sugahara, K. (1969). Control of excitation transfer in photosynthesis III. Light-induced decrease of chlorophyll *a* fluorescence related to photophosphorylation system in spinach chloroplasts. *Biochim. Biophys. Acta* **189**, 182–192.

Naumann, B., Stauber, E.J., Busch, A., Sommer, F., and Hippler, M. (2005). N-terminal processing of Lhca3 is a key step in remodeling of the photosystem I-light-harvesting complex under iron deficiency in *Chlamydomonas reinhardtii*. *J. Biol. Chem.* **280**, 20431–20441.

Nield, J., Funk, C., and Barber, J. (2000a). Supermolecular structure of photosystem II and location of the PsbS protein. *Philos. Trans. R. Soc. Lond. B Biol. Sci.* **355**, 1337–1344.

Nield, J., Kruse, O., Ruprecht, J., da Fonseca, P., Büchel, C., and Barber, J. (2000b). Three-dimensional structure of *Chlamydomonas reinhardtii* and *Synechococcus elongatus* photosystem II complexes allows for comparison of their oxygen-evolving complex organization. *J Biol. Chem.* **275**, 27940–27946.

Nield, J., Balsera, M., De Las Rivas, J., and Barber, J. (2002). Three-dimensional electron cryo-microscopy study of the extrinsic domains of the oxygenevolving complex of spinach: Assignment of the PsbO protein. *J. Biol. Chem.* **277**, 15006–15012.

Nield, J., Redding, K., and Hippler, M. (2004). Remodeling of light-harvesting protein complexes in *Chlamydomonas* in response to environmental changes. *Eukaryot. Cell* **3**, 1370–1380.

Niyogi, K.K. (1999). Photoprotection revisited: Genetic and molecular approaches. *Annu. Rev. Plant Physiol. Plant Mol. Biol.* **50**, 333–359.

Niyogi, K.K., Björkman, O., and Grossman, A.R. (1997). *Chlamydomonas* xanthophyll cycle mutants identified by video imaging of chlorophyll fluorescence quenching. *Plant Cell* **9**, 1369–1380.

Niyogi, K.K., Li, X.P., Rosenberg, V., and Jung, H.S. (2005). Is PsbS the site of non-photochemical quenching in photosynthesis? *J. Exp. Bot.* **56**, 375–382.

Pascal, A.A., Liu, Z., Broess, K., van Oort, B., van Amerongen, H., Wang, C., Horton, P., Robert, B., Chang, W., and Ruban, A. (2005). Molecular basis of photoprotection and control of photosynthetic light-harvesting. *Nature* **436**, 134–137.

Potter, E. and Kloppstech, K. (1993). Effects of light stress on the expression of early light-inducible proteins in barley. *Eur. J. Biochem.* **214**, 779–786.

Promnares, K., Komenda, J., Bumba, L., Nebesarova, J., Vacha, F., and Tichy, M. (2006). Cyanobacterial small chlorophyll-binding protein ScpD (HliB) is located on the periphery of photosystem II in the vicinity of PsbH and CP47 Subunits. *J. Biol. Chem.* **281**, 32705–32713.

Ruban, A.V., Wentworth, M., Yakushevska, A.E., Andersson, J., Lee, P.J., Keegstra, W., Dekker, J.P., Boekema, E.J., Jansson, S., and Horton, P. (2003). Plants lacking the main light-harvesting complex retain photosystem II macro-organization. *Nature* **421**, 648–652.

Ruban, A.V., Berera, R., Ilioaia, C., van Stokkum, I.H., Kennis, J.T., Pascal, A.A., van Amerongen, H., Robert, B., Horton, P., and van Grondelle, R. (2007). Identification of a mechanism of photoprotective energy dissipation in higher plants. *Nature* **450**, 575–578.

Saitou, N. and Nei, M. (1987). The neighbor-joining method: A new method for reconstructing phylogenetic trees. *Mol. Biol. Evol.* **4**, 406–425.

Sandonà, D., Croce, R., Pagano, A., Crimi, M., and Bassi, R. (1998). Higher plants light harvesting proteins. Structure and function as revealed by mutation analysis of either protein or chromophore moieties. *Biochim. Biophys. Acta* **1365**, 207–214.

Savard, F., Richard, C., and Guertin, M. (1996). The *Chlamydomonas reinhardtii* LI818 gene represents a distant relative of the *cabI/II* genes that is regulated during the cell cycle and in response to illumination. *Plant Mol. Biol.* **32**, 461–473.

Schatz, G.H., Brock, H., and Holzwarth, A.R. (1987). Picosecond kinetics of fluorescence and absorbance changes in photosystem II particles excited at low photon density. *Proc. Natl. Acad. Sci. U. S. A.* **84**, 8414–8418.

Schmid, V.H., Potthast, S., Wiener, M., Bergauer, V., Paulsen, H., and Storf, S. (2002). Pigment binding of photosystem I light-harvesting proteins. *J. Biol. Chem.* **277**, 37307–37314.

Standfuss, J., Terwisscha van Scheltinga, A.C., Lamborghini, M., and Kühlbrandt, W. (2005). Mechanisms of photoprotection and nonphotochemical quenching in pea light-harvesting complex at 2.5 Å resolution. *EMBO J.* **24**, 919–928.

Stauber, E.J., Fink, A., Markert, C., Kruse, O., Johanningmeier, U., and Hippler, M. (2003). Proteomics of *Chlamydomonas reinhardtii* light-harvesting proteins. *Eukaryotic Cell*, **2**, 978–994.

Stirewalt, V.L., Michalowski, C.B., Löffelhardt, W., Bohnert, H.J., and Bryant, D.A. (1995). Nucleotide sequence of the cyanelle genome from *Cyanophora paradoxa*. *Plant Mol. Biol. Rep.* **13**, 327–332.

Storf, S., Stauber, E.J., Hippler, M., and Schmid, V.H. (2004). Proteomic analysis of the photosystem I light-harvesting antenna in tomato (*Lycopersicon esculentum*). *Biochemistry* **43**, 9214–9224.

Takahashi, Y., Yasui, T.A., Stauber, E.J., and Hippler, M. (2004). Comparison of the subunit compositions of the PSI-LHCI supercomplex and the LHCI in the green alga *Chlamydomonas reinhardtii*. *Biochemistry* **43**, 7816–7823.

Takahashi, H., Iwai, M., Takahashi, Y., and Minagawa, J. (2006). Identification of the mobile light-harvesting complex II polypeptides for state transitions in *Chlamydomonas reinhardtii*. *Proc. Natl. Acad. Sci. U. S. A.* **103**, 477–482.

Teramoto, H., Ono, T.-A., and Minagawa, J. (2001). Identification of *Lhcb* gene family encoding the light-harvesting chlorophyll-*a*/*b* proteins of photosystem II in *Chlamydomonas reinhardtii*. *Plant Cell Physiol.* **42**, 849–856.

Teramoto, H., Nakamori, A., Minagawa, J., and Ono, T.-A. (2002). Light-intensity-dependent expression of *Lhc* gene family encoding light-harvesting chlorophyll-*a*/*b* proteins of photosystem II in *Chlamydomonas reinhardtii*. *Plant Physiol.* **130**, 325–333.

Teramoto, H., Itoh, T., and Ono, T.-A. (2004). High-intensity-light-dependent and transient expression of new genes encoding distant relatives of light-harvesting chlorophyll-*a/b* proteins in *Chlamydomonas reinhardtii*. *Plant Cell Physiol.* **45**, 1221–1232.

Teramoto, H., Ishii, A., Kimura, Y., Hasegawa, K., Nakazawa, S., Nakamura, T., Higashi, S.-i., Watanabe, M., and Ono, T.-A. (2006). Action spectrum for expression of the high intensity light-inducible *Lhc*-like gene *Lhl4* in the green alga *Chlamydomonas reinhardtii*. *Plant Cell Physiol.* **47**, 419–425.

Thompson, J.D., Gibson, T.J., Plewniak, F., Jeanmougin, F., and Higgins, D.G. (1997). The CLUSTAL_X windows interface: Flexible strategies for multiple sequence alignment aided by quality analysis tools. *Nucleic Acids Res.* **25**, 4876–4882.

Tokutsu, R., Teramoto, H., Takahashi, Y., Ono, T.-A., and Minagawa, J. (2004). The light-harvesting complex of photosystem I in *Chlamydomonas reinhardtii:* Protein composition, gene structures and phylogenic implications. *Plant Cell Physiol.* **45**, 138–145.

Turkina, M.V., Villarejo, A., and Vener, A.V. (2004). The transit peptide of CP29 thylakoid protein in *Chlamydomonas reinhardtii* is not removed but undergoes acetylation and phosphorylation. *FEBS Lett.* **564**, 104–108.

Turkina, M.V., Kargul, J., Blanco-Rivero, A., Villarejo, A., Barber, J., and Vener, A.V. (2006). Environmentally modulated phosphoproteome of photosynthetic membranes in the green alga *Chlamydomonas reinhardtii*. *Mol. Cell Proteomics* **5**, 1412–1425.

van Amerongen, H. and van Grondelle, R. (2001). Understanding the energy transfer function of LHCII, the major light-harvesting complex of green plants. *J. Phys. Chem. B* **105**, 604–617.

van Grondelle, R., Dekker, J.P., Gillbro, T., and Sundstrom, V. (1994). Energy transfer and trapping in photosynthesis. *Biochim. Biophys. Acta* **1187**, 1–65.

Vink, M., Zer, H., Alumot, N., Gaathon, A., Niyogi, K., Herrmann, R.G., Andersson, B., and Ohad, I. (2004). Light-modulated exposure of the light-harvesting complex II (LHCII) to protein kinase(s) and state transition in *Chlamydomonas reinhardtii* xanthophyll mutants. *Biochemistry* **43**, 7824–7833.

Wentworth, M., Ruban, A.V., and Horton, P. (2001). Kinetic analysis of nonphotochemical quenching of chlorophyll fluorescence. 2. Isolated light-harvesting complexes. *Biochemistry* **40**, 9902–9908.

Wolfe, G.R., Cunningham, F.X., Durnfordt, D., Green, B.R., and Gantt, E. (1994). Evidence for a common origin of chloroplasts with light-harvesting complexes of different pigmentation. *Nature* **367**, 566–568.

Wollman, F.-A. and Bennoun, P. (1982). A new chlorophyll-protein complex related to Photosystem I in *Chlamydomonas reinhardii*. *Biochim. Biophys. Acta* **680**, 352–360.

Wollman, F.-A. and Delepelaire, P. (1984). Correlation between changes in light energy distribution and changes in thylakoid membrane polypeptide phosphorylation in *Chlamydomonas reinhardtii*. *J. Cell Biol.* **98**, 1–7.

Wraight, C.A. and Crofts, A.R. (1971). Delayed fluorescence and the high-energy state of chloroplasts. *Eur. J. Biochem.* **19**, 386–397.

Xu, H., Vavilin, D., Funk, C., and Vermaas, W. (2002). Small Cab-like proteins regulating tetrapyrrole biosynthesis in the cyanobacterium *Synechocystis* sp. PCC 6803. *Plant Mol. Biol.* **49**, 149–160.

Xu, H., Vavilin, D., Funk, C., and Vermaas, W. (2004). Multiple deletions of small Cab-like proteins in the cyanobacterium *Synechocystis* sp. PCC 6803: Consequences for pigment biosynthesis and accumulation. *J. Biol. Chem.* **279**, 27971–27979.

Yakushevska, A.E., Keegstra, W., Boekema, E.J., Dekker, J.P., Andersson, J., Jansson, S., Ruban, A.V., and Horton, P. (2003). The structure of photosystem II in *Arabidopsis*: localization of the CP26 and CP29 antenna complexes. *Biochemistry* **42**, 608–613.

Yamamoto, H.Y., Nakayama, T.O., and Chichester, C.O. (1962). Studies on the light and dark interconversions of leaf xanthophylls. *Arch. Biochem. Biophys.* **97**, 168–173.

Yi, L. and Dalbey, R.E. (2005). Oxa1/Alb3/YidC system for insertion of membrane proteins in mitochondria, chloroplasts and bacteria (review). *Mol. Membr. Biol.* **22**, 101–111.

Zolla, L., Rinalducci, S., Timperio, A.M., and Huber, C.G. (2002). Proteomics of light-harvesting proteins in different plant species. Analysis and comparison by liquid chromatography-electrospray ionization mass spectrometry. Photosystem I. *Plant Physiol.* **130**, 1938–1950.

CHAPTER 15

Photosystem I

Kevin E. Redding
Department of Chemistry & Biochemistry
Arizona State University, Tempe, Arizona, USA

CHAPTER CONTENTS

I. INTRODUCTION

A. Function of photosystem I

Photosystem I (PS I) is a chlorophyll (Chl)–protein complex that functions as a light-driven plastocyanin:ferredoxin oxidoreductase. Electron transfer from plastocyanin ($E_m \approx +370\,mV$) to ferredoxin ($E_m \approx -430\,mV$) would normally be very endergonic ($\Delta G \approx +87$ kJ/mol), but is rendered favorable by coupling to absorption of a photon of visible light. For example, the energy of a red photon ($\lambda = 700\,nm$) is ~170 kJ/mol, which is more than sufficient to drive the otherwise-unfavorable reaction. Thus, PS I can be thought of as a light-driven electron pump, transferring electrons from plastocyanin (or cytochrome c_6) on the lumenal side to ferredoxin on the stromal side, both across the thylakoid membrane and over an energy barrier. PS I can function as part of the linear or cyclic electron transport pathways.

B. Electron donors: plastocyanin and cytochrome c_6

Chlamydomonas can use two different electron donors to PS I: plastocyanin or cytochrome (cyt) c_6. Plastocyanin is a small, acidic, blue copper protein found in most oxygenic phototrophs. The structure of *Chlamydomonas* plastocyanin has been determined by X-ray crystallography to a resolution of 1.5 Å (Redinbo et al., 1993). Like plastocyanin from other species, it is an antiparallel β-barrel possessing a single copper atom, which is coordinated by two imidazole nitrogens (His37 and His87), a thioether sulfur (Met92), and thiolate sulfur (Cys84). The polypeptide structure distorts the coordination environment around the Cu away from ideal tetrahedral geometry, stabilizing the Cu(I) form, which explains the high reduction potential of this protein (~+370 mV). *Chlamydomonas* chloroplast cyt c_6 has a similar midpoint potential and its structure also been determined by X-ray diffraction (Kerfeld et al., 1995). The 1.9-Å structure revealed it to be a typical (class I) cytochrome c_6, with four α-helices connected by short loops with tight turns and a very short (2-strand) β-sheet. This short β-sheet may mediate oligomerization of cyt c_6 *in vivo*, as evidenced by the fact that two crystal forms displayed oligomerization with contacts near the heme crevice.

Although plastocyanin is thought of as the preferred intermediary between cyt b_6f and the reaction center (see Chapter 17), cyt c_6 replaces it under conditions of copper limitation (Merchant et al., 2006; see Chapter 10). Although their structures and electron transfer cofactors are completely different, their reduction potentials are almost the same. Moreover, their overall sizes (~10 kD), shape, charge distribution, and surface polarity patterns are remarkably similar, allowing for PS I to dock both of them (see section IV.B for details). *Chlamydomonas*, along with many other green algae, thus represents an intermediate between photosynthetic bacteria, many of which use a cyt *c* as the electron donor to their photosynthetic reaction center (although most cyanobacteria can also use a plastocyanin) and higher plants, which use plastocyanin exclusively.

C. Electron acceptor: ferredoxin

The electron acceptor of PS I is ferredoxin, a small (94-residue) Fe_2S_2 cluster protein that is used in the chloroplast as an electron donor to $NADP^+$(via ferredoxin-NADP reductase) and thioredoxin (via ferredoxin-dependent thioredoxin reductase, see Chapter 11). Ferredoxin from *Chlamydomonas* has been purified and sequenced (Schmitter et al., 1988), and seems to be typical of other chloroplast ferredoxins. Although it has not yet been crystallized, a structure based on homology modeling has been presented (Garcia-Sanchez et al., 2000). Some species of cyanobacteria can express a flavoprotein called flavodoxin as an alternate electron acceptor during conditions of iron limitation (Sétif, 2006). Although there are potentially four flavodoxin-like genes in the *Chlamydomonas* genome, so far there is no evidence that a flavodoxin serves as acceptor for PS I in *Chlamydomonas* (Meimberg et al., 1999).

D. *In vivo* requirement for PS I

Many of the early mutants lacking PS I were later shown to be deficient in post-transcriptional processes of expression, in particular splicing and translation of chloroplast mRNAs encoding PS I components (Table 15.1; also see Chapters 26 and 28). Chloroplast genome transformation allowed precise deletions to be made in genes encoding the three core subunits (PsaA, B, C) (Takahashi et al., 1991; Fischer et al., 1996), and the resultant mutants were nonphotosynthetic, as expected.

Although the "Z-scheme" of linear photosynthetic electron transfer was established >45 years ago, in the mid-1990s some data from *Chlamydomonas* mutants supposedly lacking PS I gave support to the idea that PS I was not strictly required for the photo-reduction of ferredoxin. The *B4* and *F18* mutants are both deficient in *trans*-splicing of the *psaA* mRNA. The fact that they seemed to be capable of H_2 evolution (Greenbaum et al., 1995), and then later CO_2 fixation (J.W. Lee et al., 1996), led to the conclusion that PS I was not necessary for linear electron transfer from PS II to ferredoxin/$NADP^+$. There were later questions about residual PS I being present in these strains (or in pseudo-revertants derived from them), based upon presence of immunoreactive PsaA polypeptide (Cournac et al., 1997) and the fact that the action spectrum of the H_2 photoevolution activity resembled that of PS I, not PS II (Boichenko, 1998). Deletion of the third exon of *psaA* (*psaA-3*) in these strains resulted in loss of photoautotrophic growth (Cournac et al., 1997), suggesting further that PS I was required for that ability. Finally, the demonstration that derivatives of the *B4* or *F8* mutants – which had been previously capable of H_2 photoevolution, CO_2 uptake, and

Table 15.1 Photosystem I genes and corresponding mutants

Gene[a]	Accession #	Mutants (location)[b]
psaA	NP_958375	*Chloroplast* ***psaA-3*Δ**: deletion of *psaA* exon 3 (Fischer et al., 1996) *Nuclear* ***raa1***: defective in splicing of all 3 exons (Perron et al., 2004; Merendino et al., 2006) ***raa2* (formerly *maa2*)**: defective in splicing of exons 1 and 2 (Perron et al., 1999) ***raa3***: defective in splicing of exons 1 and 2 (Rivier et al., 2001) ***TR72***: defective in splicing of exons 1 and 2 (maturation of *tscA*) (Balczun et al., 2005)
psaB	NP_958404	*Chloroplast* ***psaB*Δ**: deletion of *psaB* gene (Fischer et al., 1996) *Nuclear* ***tab1***: no translation of *psaB* mRNA (Rochaix et al., 2004) ***tab2***: no translation of *psaB* mRNA (Dauvillée et al., 2003)
psaC	NP_958423	*Chloroplast* ***psaC*Δ**: deletion of *psaC* gene (Takahashi et al., 1991; Fischer et al., 1996)
PSAD	EDO99874	*Nuclear* ***cen***: lack of *psaD* mRNA, among others (Hahn et al., 1996)
PSAE	EDP06390	*Nuclear* No mutants identified
PSAF	EDP00906	*Nuclear* ***3bf***: deletion of the gene (Farah et al., 1995)
PSAG	EDO96591	*Nuclear* No mutants identified
PSAH	EDP05888	*Nuclear* No mutants identified
PSAI	EDP06049	*Nuclear* No mutants identified
psaJ	NP_958417	*Chloroplast* ***psaJ*Δ**: deletion of *psaJ* gene (Fischer et al., 1999a)
PSAK	EDP00485	*Nuclear* No mutants identified
PSAL	EDP05530	*Nuclear* No mutants identified
PSAN	EDP06675	*Nuclear* No mutants identified
PSAO	EDO98403	*Nuclear* No mutants identified

[a] *Note that genes names in all capitals denote a nuclear localization of the gene. For PS I, all subunit names are based on the gene name. (i.e. psaB encodes PsaB, PSAE encodes PsaE, etc. These have also been referred to as "PSI-B" and "PSI-E", etc.)*

[b] *Only mutants defective in expression – either chloroplast deletion mutants or nuclear mutants for which the gene/protein have been identified – are described above. Many point mutants are also available, as described in the text.*

photoautotrophic growth – lost all of those abilities upon deletion of *psaA-3* (or *psaC*) and regained them upon reintroduction of the gene (in the case of *psaA-3*), and that these activities all correlated with detectable PsaA polypeptide and P_{700} photochemical activity in whole cells, provided definitive proof that active PS I was required for all of them (Redding et al., 1999).

II. COMPOSITION OF PS I

A. Purification of PS I from *Chlamydomonas*

Traditionally, PS I has been purified by use of sucrose density gradients after detergent solubilization (usually β-dodecylmaltoside). Typical protocols are exemplified by those of Takahashi et al. (1991) and Hippler et al. (1997). Good success has also been obtained using column chromatography, including weak anion exchange (Ramesh et al., 2004a) and molecular sieve (Krabben et al., 2000) matrices, as well as an immobilized-metal affinity matrix for a His-tagged version of PS I (Gulis et al., 2008).

B. Components

1. Polypeptides

In eukaryotes, PS I is composed of 14 polypeptides: PsaA, B, C, D, E, F, G, H, I, J, K, L, N, and O. Of these, four (PsaA, PsaB, PsaC, and PsaJ) are encoded by the chloroplast genome in *Chlamydomonas*; the first three of these form the core of PS I, binding all of the electron transfer cofactors. The PsaA/B heterodimer comprises most of the mass of PS I, with seven small membrane-inserted polypeptides (PsaF, G, H, I, J, K, L) on the periphery (Amunts et al., 2007; Jordan et al., 2001). Note that PSAG, PSAH, and PSAN are specific to eukaryotes and are involved in binding LHC polypeptides (Amunts et al., 2007; see Chapters 14 and 22). The four chloroplast genes have all been studied through deletion and site-directed mutation. The other eight polypeptides are encoded by the nuclear genome. Of these, site-directed mutants have so far been made only in PsaF (see section IV.B).

2. Core cofactors

Absorption of a quantum of electromagnetic energy (i.e. photon) by a Chl within PS I produces an excited state, which can then migrate through the densely packed pigment molecules via a radiationless tunneling process known as Förster energy transfer. Arrival of the excitation energy at the reaction center results in an electron transfer event between the excited Chl and a neighboring Chl, producing a separation of charge (Chl* Chl → Chl^+ Chl^-). This charge-separated state, which would persist only transiently in a solution of Chl molecules, is stabilized by further electron transfer involving another Chl (→ Chl^+ Chl Chl^-). All of the initial energy and electron transfer events occur within the sub-picosecond and picosecond timescales. This process is often referred to as "trapping" of excitation energy. In a few tens of picoseconds, the charge-separated state is stabilized even further by transfer of the electron to a quinone molecule. Note that this sequence of events is essentially the same for all photosynthetic reaction centers. The type 2 reaction centers fully reduce quinone to quinol, which is their final product. The type 1 reaction centers, of which PS I is the best-known member, contain FeS clusters, which serve as the terminal electron acceptors and allow the reduction of low-potential electron acceptors, such as ferredoxin.

The arrangement of the cofactors involved in electron transfer within PS I is shown in Figure 15.1A.There are several features about the environment of the core cofactors that will be referred to later, and most are indicated in Figure 15.1B. P_{700} is the ultimate electron donor within PS I, and is presumed to be the primary electron donor as well (although see section III.C for an alternate hypothesis). The P_{700} Chls, P_A and P_B, are coordinated by the axial ligands, PsaA-His676 and PsaB-His656, respectively. (Note that all residue numbering in this chapter is for the *Chlamydomonas*

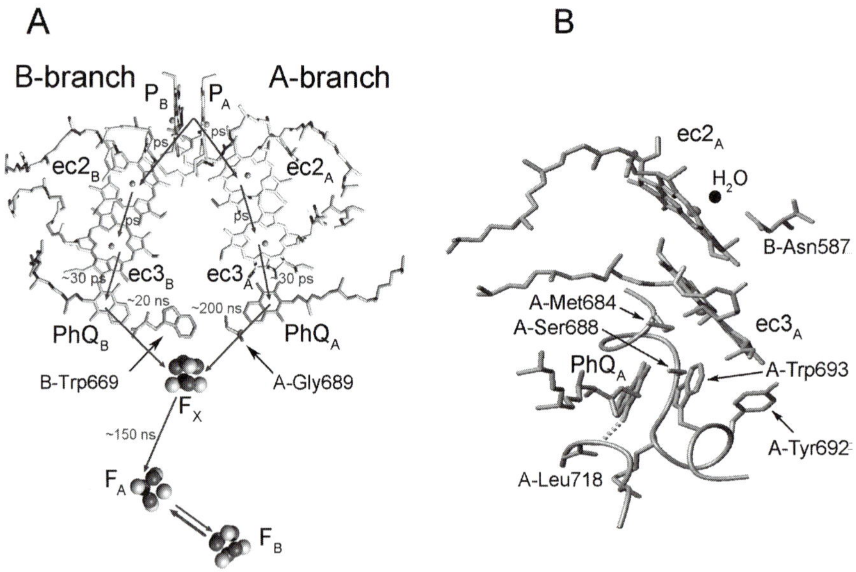

FIGURE 15.1 *Structure of the PS I core from Thermosynechococcus elongatus (Jordan et al., 2001). Panel A: The two branches of electron transfer cofactors in the PS I core. P_{700} is composed of the P_A and P_B chlorophylls. On each branch is a pair of chlorophylls ($ec2_Aec3_A$ or $ec2_Bec3_B$) and a phylloquinone (PhQ_A or PhQ_B). The F_X cluster is shared by PsaA and PsaB, while the two terminal Fe_4S_4 clusters (F_A and F_B) are bound by PsaC. An asymmetry between the phylloquinone sites (PsaA-Gly689/PsaB-Trp669) is also indicated. Panel B: The environments of the ec2, ec3, and phylloquinone cofactors (A-branch). The view is from behind the PhQ_A headgroup in order to see its interactions with Trp693 (π-stack), Ser688 (H-bond to Trp693), and Leu718 (H-bond). Also shown are the axial ligand (Met684) and H-bond donor (Tyr692) to $ec3_A$, as well as the water molecule that serves as axial ligand to $ec2_A$ and the residue (PsaB-Asn587) coordinating it.*

polypeptides.) The former also receives a H-bond from PsaA-Thr739, which is part of a larger H-bonding network near P_A, and introduces asymmetry into the special pair of PS I, as there is nothing similar near P_B (see section III.B). The environments near the other cofactors on either branch are much more similar. On each branch there is a pair of Chls (ec2 and ec3) and a phylloquinone (PhQ). Although it is generally accepted that electron transfer occurs from P_{700} to ec2 to ec3, the exact order of events is not yet clear (see section III.C). It is thought that the ec2 Chls must be involved, because the distance between P_{700} and ec3 is too large for electron transfer between them in picoseconds (Klukas et al., 1999). The ec2 Chls are not H-bonded and both are coordinated by water molecules that are H-bonded to the oxygen of an Asn side chain amide (PsaB-Asn587 for $ec2_A$ and PsaA-Asn601 for $ec2_B$). Methionine side chains (PsaA-Met684 and PsaB-Met664) serve as the axial ligands to $ec3_A$ and $ec3_B$, respectively. It was very surprising that such a soft base (S of a thioether) would be paired with such a hard acid (Mg^{2+}), and it is suspected that this may have something to do with the properties of these Chls. The 13^1-keto oxygens of $ec3_A$ and $ec3_B$ receive H-bonds from the phenolic hydroxyl group of PsaA-Tyr692 and PsaA-Tyr672, respectively. Electron transfer from ec3 to PhQ occurs in tens of picoseconds. The PhQs are also H-bonded to the protein, but only to one carbonyl (the one *meta* from the 2-methyl group) from the backbone N-H of PsaA-Leu718 (PhQ_A) and PsaB-Leu702 (PhQ_B). The benzoquinone moieties of PhQ_A and PhQ_B are π-stacked largely with the 5-membered rings of PsaA-Trp693 and PsaB-Trp673, respectively, with the indole N almost directly above C-1 of PhQ. Both Trp indole nitrogens also appear to donate a H-bond to the hydroxyl O of nearby Ser residues (PsaA-Ser688 and PsaB-Ser668). The Fe_4S_4 cluster, F_X, is bound by 2 Cys residues each in a highly conserved loop between the eighth and ninth transmembrane alpha-helices of PsaA and PsaB (PsaA-Cys575, 584; PsaB-Cys561, 570). Electron transfer from the PhQs to F_X occurs in the nanosecond timescale, and this is likely to be the case as well for the subsequent transfer to the terminal Fe_4S_4 clusters, F_A and F_B. These last two cofactors are bound by the peripheral subunit, PsaC.

C. Structure of PS I–LHCI supercomplex

The structure of PS I from *Chlamydomonas* has not yet been solved by X-ray crystallography, so the cyanobacterial PS I (Jordan et al., 2001) and pea PS I–LHCI (Amunts et al., 2007) structures still serve as useful models for the algal protein. PSI–LHCI supercomplexes have been isolated from *Chlamydomonas* cells in State 1 and State 2. Based on image analysis of electron micrographs of negatively stained particles, it was deduced that the State 1 supercomplex consists of a PS I monomer with 11 LHCA polypeptides arranged on the side where the PsaF and PsaJ polypeptides would be, between the PsaG and PsaK subunits (Kargul et al., 2003). This arrangement of LHCI on one side of PS I is very similar to that of pea

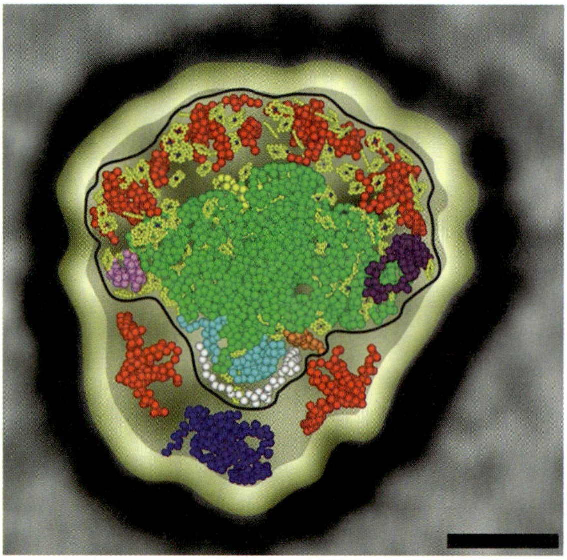

FIGURE 15.2 *Detailed modeling of the projection map for the LHCI–PS I supercomplex isolated from* Chlamydomonas *cells placed in State 2. Modeling is based on higher plant coordinates (1qzv.pdb (Ben-Shem et al., 2003)) with PS I core (green), LHCI antenna (red), PsaJ (yellow), PSAK (magenta), PSAG (purple), PSAI (orange), PSAL (cyan) and PSAH (white). The additional density observed in State 2 LHCI–PS I supercomplex, which is able to accommodate an additional LHC subunit, is colored blue and was suggested to be phospho-CP29 (Kargul et al., 2005). Chlorophylls are shown in yellow, but were excluded from the additional density attributed to the LHCI and CP29 subunits. The detergent shell surrounding the particles in the hydrophobic membrane plane sits within any stain present and this shell is assigned here as a 15-Å wide outer contour (cream). Scale bar is 50 Å. Reprinted from Kargul et al. (2005) with permission.*

PS I (Ben-Shem et al., 2003), but the algal supercomplex seems to have a larger antenna complement. These particles contained ~215 Chl molecules (of which ~40 were Chl b) and their estimated mass was ~700 kD. If one assumes that the PS I core antenna/reaction center consists of ~100 Chl *a*, as in cyanobacterial PS I (Jordan et al., 2001), then the attached LHCI polypeptides bind an additional ~75 Chl *a* and ~40 Chl *b*. The particles isolated from State 2 cells appeared to contain CP29, normally thought to be a component of PS II, and the site for its association was on the opposite side from LHCI, near the eukaryotic-specific PsaG subunit (Figure 15.2) (Kargul et al., 2005).

III. ENERGY TRANSFER AND ELECTRON TRANSFER WITHIN PS I

A. Energy transfer and trapping

Before PS I can perform the electron transfer events at the reaction center, it must absorb the energy of incident photons and delivery this energy to the reaction center. This is the job of the antenna pigments, some of which are in the core antenna of PS I and some of which are in the external antenna bound by LHC proteins. Ultra-fast absorption and fluorescence spectroscopic techniques have been used to probe the dynamics of energy transfer and trapping within PS I and LHCI–PS I. Spectral overlap between the Chl *a* molecules used as antenna and early electron transfer cofactors, and the temporal overlap between energy transfer processes and early electron transfer events, has complicated interpretation of the data (reviewed in Gobets and van Grondelle, 2001). In order to simplify analysis and focus on processes occurring within PS I, mutants with low expression of LHCs were widely used in earlier studies. The fact that *Chlamydomonas* PS I preparations lacking LHCs are devoid of long-wavelength pigments ($\lambda_{max} > 700$ nm) or so-called "red Chls," unlike those from cyanobacteria, has greatly facilitated resolution of core antenna and reaction center processes (Müller et al., 2003). Time-resolved absorption measurements usually detected 1 or 2 fast processes (lifetimes of <1 ps and 2–8 ps) of equilibration among the Chl species in the PS I core antenna, with energy coming to reside largely on longer wavelength Chls (Hastings et al., 1995; Gibasiewicz et al., 2001; Ihalainen et al., 2005), followed by a slower process (lifetime of 20–30 ps) attributed to trapping and primary charge separation (Hastings et al., 1995; Melkozernov et al., 1997; Gibasiewicz et al., 2001). Fluorescence decay kinetics in thylakoid membranes from a mutant lacking PS II and most of the LHCs is dominated by a component with a ~25-ps decay time, consistent with the assignment of this decay to energy trapping (Melkozernov et al., 1998). Mutants in which the axial ligand of P_B (PsaB-His656) had been converted to Asn or Leu slowed this decay ~2-fold or ~3-fold, respectively. The PsaB-H656N mutant was also seen to slow trapping by ~2-fold in the transient absorption data (Melkozernov et al., 1997). This was interpreted in terms of a model in which trapping is limited by the trap itself ("trap-limited") rather than by migration of energy to the trap ("diffusion-limited"); mutation of the P_{700} axial ligand presumably lowered the trapping rate by decreasing the rate of primary charge separation.

At very low temperatures (10 K), the trapping time is only slightly longer (Gibasiewicz et al., 2002), and it becomes possible to examine the excitonic coupling between the reaction center chlorophylls. The close proximity of the molecules allows an interaction in which the individual pigment molecules lose their individuality; this splits the lowest excited state into lower and higher energy states, in effect lowering the energy of the lowest excited state, and thus leading to an overall red-shift of the absorption of the interacting pigment molecules. The initial difference spectrum immediately (~160 fs) after excitation at 700 nm by a 50-fs pulse had four major negative peaks centered at 667, 675, 684, and 695 nm, as well as four positive peaks at higher energy (maxima at 634, 645, 652, and 661 nm). These features were interpreted in terms of excitonic coupling between the six chlorophylls of the reaction center, with the four negative peaks corresponding to the lowest four of the six theoretically calculated 1-exciton states (Gibasiewicz et al., 2002). The positive peaks were assigned to absorption by the lowest excited state. This model was tested by use of mutations that affect the reaction center chlorophylls, as well as the so-called "connecting chlorophylls" located between the reaction center and the ring of antenna pigments. The 684-nm band was largely absent in a double mutant in which the axial ligands for both connecting chlorophylls (PsaA-His730 and PsaB-His714) were simultaneously converted to Leu, indicating that these chlorophylls are also coupled with the reaction center system (Gibasiewicz et al., 2003). Surprisingly, neither chemical oxidation of P_{700} nor mutation of an axial ligand (PsaB-H656S) affected the low-temperature excitonic spectrum, indicating that P_{700} does not significantly contribute to it. Mutation of either the axial ligands (Gibasiewicz et al., 2003) or H-bond donors (Li et al., 2006) to the ec3 Chls provoked several changes in the low-temperature excitonic spectrum, consistent with their coupling to ec2 and connecting Chls. At this point, the theoretical calculations of the band positions do not match experimental values, but these energies can be very sensitive to changes in position, and the calculations are based on the structure of PS I from the thermophilic cyanobacterium *Thermosynechococcus elongatus*, which is likely different from the algal protein.

A new interpretation of the ultra-fast data was put forward by Müller et al. (2003), in which charge separation would occur with a higher rate (time constant of 6–9 ps). In this work, for the first time the observation of three post-trapping states was reported, putatively corresponding to the three radical pair states existing before P_{700}^{+} F_X^{-}. The last one was assigned to P_{700}^{+} PhQ^{-}, and the second one to P_{700}^{+} A_0^{-} (presumably P_{700}^{+} ec3^{-}), in agreement with other interpretations (Gobets and van Grondelle, 2001). It has been appreciated for some time that the distance between P_{700} and ec3 precludes fast electron transfer between (<1 ns), and it has been widely assumed that the ec2 Chl serves as an intermediate (Klukas et al., 1999). In this view, the first radical pair would be P_{700}^{+} ec2^{-}. An alternate interpretation is that charge separation occurs within ec2/ec3 to generate ec2^{+} ec3^{-} as the first radical pair, followed by electron transfer from P_{700} to give P_{700}^{+} ec3^{-}. Analysis of the transient absorption data gave a lifetime of ~20 ps for the first radical pair and ~35 ps for the second. It was also important to include a significant charge recombination rate for the first radical pair back to the excited reaction center (RC*); this was confirmed in subsequent time-resolved fluorescence measurements (Holzwarth et al., 2005), in which estimates of ~440 ns^{-1} and ~50 ns^{-1} were calculated for the effective rates of charge separation and back-reaction, respectively. The PsaB-H656C mutation (P_B axial ligand) did not slow the rate of primary charge separation, as seen by transient absorption methods, but slowed the secondary electron transfer process approximately twofold (Holzwarth et al., 2006). It also slowed the back-reaction rate approximately fourfold. Two other mutations near P_{700} (PsaA-T739V and PsaA-W679A) had weaker effects on the back-reaction rate (~1.5-fold), and even smaller effects on secondary electron transfer, but neither affected primary charge separation. This suggests that the alternate model is correct and that P_A/P_B pair is not the primary electron donor. The possible presence of two traps in PS I ($ec2_A/ec3_A$ versus $ec2_B/ec3_B$) also leads naturally to the question of the origins of directionality of electron transfer within PS I (see section III.C). This analysis has contributed to the debate about whether or not charge separation in PS I is limited by the trap itself or by transfer of energy to the trap. If the alternate model is correct, then it is trap-limited in the PS I core antenna/reaction center complex from *Chlamydomonas* (Müller et al., 2003; Holzwarth et al., 2005). Only time will tell if this model holds up to further scrutiny, but the absence of long-wavelength pigments in PS I from *Chlamydomonas* has truly proven to be a great advantage in unraveling the ultra-fast steps of electron transfer.

The situation changes when one performs ultra-fast measurements with LHCI–PS I supercomplexes, since LHCI from algae and plants contains Chls absorbing at longer wavelengths, which will slow transfer of excitation energy to the trap. Using time-resolved absorption and fluorescence spectroscopy, Melkozernov et al. (2004) resolved two decay components assigned to photochemical trapping: a ~25-ps component corresponding to trapping in the core antenna, and a ~100-ps component corresponding to trapping from the peripheral antenna. Similarly, Ihalainen et al. (2005) reported the presence of a minor component of fluorescence decay with a ~70-ps lifetime, in addition to the major component at 28 ps, which was also assigned to trapping. Interestingly, the slow phase of trapping disappears at 77 K, indicating that the peripheral antenna becomes uncoupled from the core antenna at such low temperatures (Melkozernov et al., 2005). Another low-temperature (4–8 K) fluorescence study of LHCI–PS I supercomplexes came to a similar conclusion with regards to the low-energy pool of chlorophylls: they are almost isoenergetic with the primary donor (Gibasiewicz et al., 2005). Moreover, this pool seems to be relatively small (~5–6 Chls).

B. P_{700}: asymmetric nature and effect of mutations

P_{700} is defined rigorously only as a spectroscopic entity – the pigment that becomes bleached upon photooxidation. With the advent of the crystal structure, we have come to identify it with the pair of chlorophylls (P_A and P_B) in the crystal structure occupying the analogous position in the purple bacterial reaction center. However, mutations in *Chlamydomonas* PS I first made the connection between the spectroscopic entity and the specific region of the protein, before the high-resolution structure was available. Webber and colleagues demonstrated that mutation of PsaB-His656 to

Asn or Ser affected the spectroscopic properties of P_{700}, including shifts of the midpoint potential (+40 mV), appearance of a new bleaching band (at 667 nm) in the P_{700}^{+}–P_{700} optical difference spectrum, and an alteration of the P_{700}^{+} electron-nuclear double resonance (ENDOR) spectrum, consistent with their tentative assignment of this residue as the axial ligand to one of the P_{700} Chls (Webber et al., 1996). It was later shown that mutation of the analogous histidine in PsaA (PsaA-His676) to Gln affected the P_{700}^{+}–P_{700} Fourier-transform infrared (FTIR) difference spectrum (~5 cm^{-1} upshift of a negative peak at ~1700 cm^{-1} assigned to the 13^1-keto group), while mutation to Leu broadened the electron paramagnetic resonance (EPR) spectrum of the P_{700}^{+} radical (Redding et al., 1998). In that study, seven pairs of histidines, which were predicted to be in membrane-integrated alpha-helices and conserved in all species as well as between PsaA and PsaB, were converted to Gln and the double mutants were screened for spectroscopic changes in the P_{700} EPR and FTIR spectra. Only the mutations in the tenth helix (PsaA-His676/PsaB-His656) produced such changes. A later study provided a more in-depth characterization of the mutations in PsaA-His676 and PsaB-His656 (Krabben et al., 2000), which are now known to be the axial ligands to P_A and P_B (Jordan et al., 2001). Pulsed EPR and ENDOR studies of the P_{700}^{+} radical indicate that the unpaired electron spin (and, consequently, the positive charge) is localized to one of the Chls of P_{700} (reviewed by (Webber and Lubitz, 2001). Thus, this Chl should be P_B, since mutation of PsaB-His656 had a much larger effect than mutation of PsaA-His676 upon several parameters of the P_{700}^{+} cation radical, including an increase in the hyperfine coupling assigned to the 12-methyl protons (Webber et al., 1996; Krabben et al., 2000) and a decrease of the g_{zz} principal component of the **g**-tensor of P_{700}^{+} (Petrenko et al., 2004). Although the reasons for these changes are not the same (and beyond the scope of this chapter), the logic of the assignment of P_B as the molecule on which the unpaired electron resides is clear, as EPR is selective for the unpaired spin and thus for the species on which it resides.

Thanks to the X-ray crystal structure, we now know that PsaA-His676 and PsaB-His656 are the axial ligands of P_A and P_B, respectively (Figure 15.3). Although the axial ligands are arranged symmetrically, the two Chls are H-bonded asymmetrically. While P_A receives a H-bond from PsaA-Thr739, which is part of a H-bonding network, there is nothing comparable in the vicinity of P_B. The asymmetry of P_{700} is also manifested chemically and spectroscopically. As discussed above, most of the unpaired electron spin in the P_{700}^{+} radical is localized to P_B. P_{700} is also a chemical heterodimer, consisting of one Chl *a* (P_B) and one Chl *a′* (P_A). While the presence of Chl *a′* in PS I had been detected previously, it was not generally appreciated that it was a component of P_{700} until the cyanobacterial crystal structure confirmed its presence (Jordan et al., 2001). The presence of one Chl *a′* in PS I has now been shown to be a trait in many species, including *Chlamydomonas* (Nakamura et al., 2003, 2002).

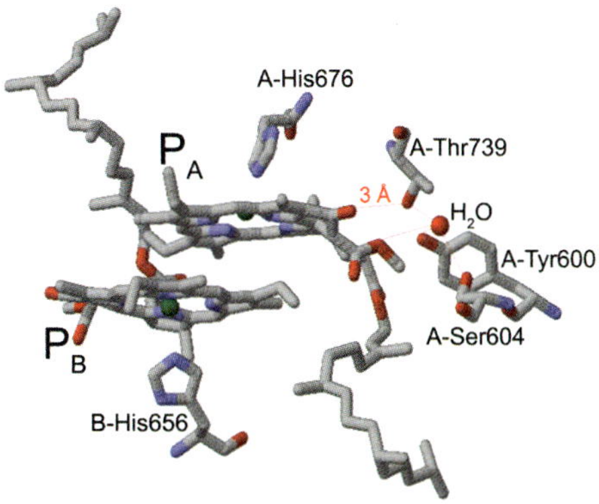

FIGURE 15.3 *Environment of P_{700}, with important residues indicated, using the T. elongatus coordinates (Jordan et al., 2001). P_A is a Chl a′ coordinated by PsaA-His676, and P_B is a Chl a coordinated by PsaB-His656. A H-bonding network next to P_A is indicated, with PsaA-Thr739 playing an important role by donating a H-bond to the 13^1-keto oxygen of P_A, as well as coordinating a bound water molecule, along with PsaA-Tyr600 and PsaA-Ser604, which is H-bonded to the 13^2-ester group.*

It has largely been assumed that the asymmetry in the protein environment could explain the functional asymmetry of P_{700}. The H-bond from PsaA-Thr739 could facilitate enolization of the Chl *a*, allowing for its conversion the Chl *a* isomer, which would be stabilized by increased H-bonding of the 13^2 methyl ester to the H-bonded water (Webber and Lubitz, 2001). Likewise, asymmetric H-bonding to P_{700} could cause asymmetry in the charge and spin distribution within $P_{700}{}^{+}$. This has been tested directly in *Chlamydomonas* by mutation of PsaA-Thr739 to Tyr, His, Val (Witt et al., 2002), and Ala (Li et al., 2004). Unexpectedly, the loss of the H-bond donor did not result in loss of the Chl *a′* in any of the mutants tested, indicating that it is not essential for isomerization of Chl *a*. Since there is no known mechanism for biosynthesis of Chl *a′*, it is still unclear how it is made and specifically inserted into the P_A site. All of the mutations resulted in a decrease of the hyperfine coupling assigned to the 12-methyl protons, with the Ala substitution having the strongest effect, and His and Tyr having the weakest (Witt et al., 2002; Li et al., 2004). This can be explained by a small shift of the unpaired electron spin density to P_A. The changes observed in the $P_{700}{}^{+}$–P_{700} FTIR difference spectra (Witt et al., 2002; Wang et al., 2003; Li et al., 2004) are consistent with both loss of the H-bond and a small shift of the positive charge from P_B to P_A (~16% in the PsaA-T739A mutant), but the interpretation of the FTIR spectra is somewhat controversial (see Breton et al., 1999; Hamacher et al., 1996; Breton, 2001; Pantelidou et al., 2004; Wang et al., 2004) and beyond the scope of this chapter. However, it is clear that the H-bonding network near P_A is largely inadequate to explain the asymmetry between the two sides of P_{700}, contrary to expectations.

There are several spectroscopic differences between P_{700} in *Chlamydomonas* versus other organisms, including the position and width of the main bleaching band (e.g. "P_{700}" does not always peak at 700 nm – it peaks at 696 nm in *Chlamydomonas*). Witt et al. (2003) attempted to understand the structural origin of some of these differences. Inspection of the *T. elongatus* crystal structure reveals that under the P_{700} Chls is a pair of Trp residues. These are conserved in all organisms examined to date except *Chlamydomonas*, where the B-side residue is a Tyr (PsaB-Trp659). Significantly, conversion of this residue to a Trp was not tolerated in *Chlamydomonas* – the mutant protein failed to accumulate. Mutation to Ala or His allowed assembly of PS I and the mutant complexes exhibited a blue shift in the main P_{700} bleaching band, among other rather subtle changes. The bottom line from this study was that one could not attribute species-specific spectroscopic differences to particular amino acid residues, with the further proviso that while the *T. elongatus* structure is likely a good model for PS I from other species, it cannot be used for detailed simulations where sub-Å deviations can have a major impact.

C. Secondary electron transfer events

Once a P_{700}^{+} $ec3^{-}$ charge-separated state is formed, the next step is electron transfer to the nearby PhQ to form P_{700}^{+} PhQ^{-}. This has been measured by ultra-fast pump-probe visible spectroscopy in PS I particles from *Chlamydomonas* and the time constant of this step is on the order of a few tens of picoseconds [e.g. ~20 ps (Hastings et al., 1995), ~35 ps (Müller et al., 2003)]. Because the decay of the P_{700}^{+} $ec3^{-}$ state is about as fast as its formation, it has been difficult to observe it. However, mutations of the axial ligand to ec3 can slow this rate drastically. In both the PsaA-M684H and PsaB-M664H mutants, a bleaching signal peaking at ~682 nm (i.e. similar to "A_0"; Hastings et al., 1995) was still visible 200–300 ps after excitation, by which time this bleaching has long vanished in wild-type PS I (Ramesh et al., 2004b). This effect was interpreted as a slowing or blockage of electron transfer from the affected ec3 Chl to its nearby PhQ, and it was later seen in mutants that replaced either axial ligand with Ser or Leu (Ramesh et al., 2007). The fact that this bleaching signal doubled in size under reducing conditions (which presumably reduces both PhQs, blocking forward electron transfer from both ec3 Chls) provided further evidence that it represented the reduced "A_0" Chl. Later measurements taken on a longer timescale demonstrated that these mutations did not block electron transfer from the ec3 Chls, but rather slowed it down by a factor of ~100, since the bleaching decayed with a time constant of 1–2 ns (Ramesh et al., 2007). Since mutations in both PsaA-Met684 and PsaB-Met664 produce this effect, it would seem that electron transfer can be slowed in either branch.

Transient EPR spectroscopy has also been used to observe intermediate charge-separated states in PS I. The advantage of this technique is that each such state is a radical pair and thus accessible by this form of spectroscopy, which is otherwise blind to the electrochromic band-shifted pigments, unconnected chlorophylls, and other complications of visible spectroscopy. The disadvantage is the temporal resolution, which is on the order of tens of nanoseconds, precluding direct observation of the P_{700}+ ec3$^-$ state. Despite this, however, it is possible to extract information from the subsequent radical pair (P_{700}+ PhQ$^-$), because singlet-triplet mixing in the earlier state will affect the spectrum of the later state (van der Est, 2006). This was examined in the PsaA-M684H and PsaB-M664H mutants, where it was found that the PsaA mutation had a strong effect on the shape of the spin-polarized EPR spectrum of P_{700}^+ PhQ$^-$, while the corresponding B-side mutant had a minimal effect (McConnell et al., 2005), comparable to results obtained with similar mutants in *Synechocystis* (Cohen et al., 2004).

Mutations have also been created to remove the H-bond to the ec3 Chls: PsaA-Y692F and PsaB-Y672F (Li et al., 2006). Neither mutation appeared to produce the types of effects on the shape of the P_{700}^+ PhQ$^-$ spectrum associated with slowing of electron transfer from A_0, but the PsaB-Y672F mutant displayed a ~680-nm bleaching with a 1.5-ns decay time. These conflicting results raise the question of the identity of the ~680-nm band seen in the ultra-fast measurements, and it was suggested that it might result from uncoupled Chls (Li et al., 2006). Ramesh et al. (2007) tried to address this by performing experiments under oxidizing conditions, where charge separation does not occur, and found that the ~680-nm bleaching band did not appear. Also, since the P_{700}^+ PhQ_B^- state should decay much faster than the P_{700}^+ PhQ_A^- state (see below), it is likely that any changes in the EPR spectrum of the former would be difficult, if not impossible, to see by transient EPR. This topic is still controversial, but the combination of *Chlamydomonas* mutants and spectroscopy should eventually contribute to its resolution.

Below a temperature of ~150 K, forward electron transfer from PhQ is blocked in a fraction of PS I reaction centers, resulting in a back reaction from the P_{700}^+ PhQ$^-$ state (Schlodder et al., 1998). In the other fraction of reaction centers, electron transfer proceeds generating P_{700}^+ F_A/F_B^-, which does not decay for hours. The origin of this bifurcation is unknown, but it has been postulated that at low temperature PS I is stuck in one of two conformations: in one, electron transfer uses only the A-branch, while the B-branch is used exclusively in the other. This would explain the low temperature results, with the B-branch population leading to irreversible forward electron transfer, while the A-branch population back-reacts. It would also explain why transient EPR at low temperature using repetitive flashes seems to access only P_{700}^+ PhQ_A^-, since the back-reacting population would form a charge-separated state with each laser flash (Redding and van der Est, 2006). The spin-polarized P_{700}^{+} PhQ^{-} spectrum observed at low temperature decays with a ~20 μs time constant, but upon illumination in reducing conditions (dithionite at pH 8), which should reduce all the FeS clusters, a ~2.5 μs decay component appeared, accounting for up to 70% of the decay (Muhiuddin et al., 2001). The authors assigned this to the P_{700}^{+} PhQ_B^{-} state, which would only be observed in repetitive flashes when forward electron transfer to F_X is blocked.

In order to test this hypothesis, the experiment was repeated in PhQ site mutants. Dithionite reduction in PsaA-W693H/L resulted in a single decay component with a decay time of ~4 μs, while only a 18-μs component was seen in the PsaB-W673L mutant (Fairclough et al., 2003). Presumably, the affected PhQ is doubly reduced under these reducing conditions. The authors assigned these processes to decay of the spin-polarized P_{700}^{+} PhQ_A^{-} (~20 μs) and P_{700}^{+} PhQ_B^{-} (~2.5 μs) radical pairs, due to a combination of back reaction and/or loss of spin polarization. It should be mentioned that the presence of reduced F_X would be expected to accelerate this decay in either pair. In the PsaA-M684H mutant, a spin-polarized signal appeared only after illumination with dithionite, and it decayed with a time of ~2.5 μs. The PsaB-M664H mutant displayed a ~19-μs component before and after illumination. Analysis of the out-of-phase electron spin-echo modulation of these signals indicated that the distance between the radicals in the P_{700}^{+} PhQ^{-} radical pair is different in the two cases: 25.43±0.01 Å in the PsaB-M664H mutant and 24.27±0.02 Å in the PsaA-M684H mutants (Santabarbara et al., 2005). The interpretation of these results was that mutation of the axial ligand to an ec3 Chl blocked electron transfer down that branch, but it is difficult to reconcile this with the observation that these mutants only appear to slow electron transfer to PhQ, not block it (Ramesh et al., 2007).

Measurement of forward electron transfer from the phylloquinones within PS I has been contentious. Although a kinetic component with a decay time of 150–250 ns and positive amplitude in the 380–400 nm region is accepted as representing reoxidation of PhQ^{-}, a faster decay component (~10–25 ns) has been attributed to heterogeneity in PS I or to fast equilibria between PhQ and F_X (reviewed in Rappaport et al., 2006). Observation of biphasic decay in living algal cells (Joliot and Joliot, 1999; Guergova-Kuras et al., 2001) has eliminated heterogeneity induced during biochemical purification as an explanation. In *Chlamydomonas* the two decay components had time constants of ~18 ns and ~190 ns *in vivo* (Guergova-Kuras et al., 2001). The overall shapes of the two decay-associated spectra (DAS) were similar, with positive amplitudes in the 370–410 nm and 450–500 nm regions, respectively assigned primarily to PhQ^{-} absorption and electrochromic bandshifts of nearby pigments. A later study (Byrdin et al., 2006) found that the slower DAS had significant absorption at 340–350 nm that was absent in the faster DAS. The decay rate of the slower component was decreased in the PsaA-W693F mutant (to 700 ns) without effect upon the faster component. In the PsaB-W673F mutant the decay could be fit, albeit poorly, by a single exponential decay constant of ~160 ns.The situation was clarified by examination of the double mutant (PsaA-W693F/PsaB-W673F), in which there were clearly two decay components with time constants of ~100 ns and ~700 ns. Assuming that the A-side mutation specifically slowed the slow component ~3.5-fold and the B-side mutation had a similar effect (~5-fold slowing) upon the fast component, then the phenotype of the PsaB-W673F mutant could be explained by temporal overlap of the two decays. These results allowed assignment of the faster decay to forward electron electron transfer from PhQ_B ($P_{700}{}^+$ $PhQ_B{}^-$ → $P_{700}{}^+$ $F_X{}^-$) and the slower to electron transfer from PhQ_A. Each mutation also provoked changes in the specific DAS: an absorption increase at ~400 nm. Other mutations in the PhQ have since been examined, and the general trend so far holds true: mutations near PhQ_A (e.g. PsaA-S692A, PsaA-G689W) affect the slower decay component, while those near PhQ_B (e.g. PsaB-S672A) affect the faster decay component (Byrdin et al., 2006; Rappaport et al., 2006).

It is important to note that the ratio of the amplitudes of the two decay components (35–40% fast) was not significantly altered in any of the PhQ site mutants. This would be expected if the ratio were determined by the inherent directionality of electron transfer within PS I. Several studies have attempted to make use of this ratio as an indicator of directionality. This was first tried with the PsaA-T739A mutation, which removes the H-bond to P_A, but it had no effect upon the ratio of the two decay components (Li et al., 2004). Mutation of the H-bond donors to the ec3 Chls, however, had a striking effect, as seen by both *in vivo* pump-probe spectroscopy and transient EPR (Li et al., 2006). The PsaA-Y692F mutation caused an increase in the fraction of fast electron transfer (to 65–70%), while the PsaB-Y672F mutation caused a decrease in fast electron transfer (to 17–25%).

If loss of the H-bond destabilized an $ec3^-$ radical anion, then it should also destabilize the specific $P_{700}{}^+ec3^-$ radical pair, thus implicating that radical pair in the primary charge separation mechanism. The facts that the ec3 H-bond mutants affect both directionality and rate of primary charge separation, and that the P_{700} H-bond mutant does neither, provide strong support for the alternate model of charge separation, in which $ec2^+$ $ec3^-$ is the first radical pair. The PsaA-M684H and PsaB-M664H mutations produced analogous effects on the *in vivo* kinetic components: fast electron transfer increased in the former (to 76%) and decreased in the latter (to 17%) (Byrdin et al., 2006). Presumably, mutation of the axial ligand also destabilizes the targeted $ec3^-$ anion, but analysis is complicated by the significant amount of back reaction seen in the PsaA-Met684 mutants, generating the P_{700} triplet state. This was estimated to account for ~25% of charge separations in this mutant (Byrdin et al., 2006). However, no such back reaction was seen in the PsaB-M664H mutant, for reasons that are still unclear.

It is not obvious why the rate of electron transfer from the two quinones should differ by an order of magnitude, since there are few differences in primary structure of the two quinone sites. There is, however, one major asymmetry between the PhQ sites, consisting of the pair PsaA-Gly689/PsaB-Trp669, which face each other in the space just above the F_X cluster (see Figure 15.1A, p. 546) . In fact, this asymmetry is demanded by structural constraints, as there would not be adequate space for two Trp residues. Ivashin and Larsson (2003) calculated electronic coupling factors between cofactors and suggested that PsaB-Trp669 is the key, allowing fast electron transfer reaction through it to F_X from PhQ_B but not PhQ_A. The fact that the PsaA-G689W mutation slowed electron transfer from PhQ_A considerably, as seen by *in vivo* pump-probe visible spectroscopy (Byrdin et al., 2006), does not seem to support this hypothesis, although it is likely that the introduced Trp is not in the same structural context as PsaB-Trp669, due to the fact that the protein must likely rearrange itself to accommodate the introduced indole group. Electron transfer in PsaB-W669G was examined by pulsed EPR spectroscopy at 265 K (Ali et al., 2006). In the wild type, an electron spin echo signal assigned to $P_{700}{}^+$ PhQ^- decays with a time constant of ~280 ns, which is assigned primarily to $P_{700}{}^+$ $PhQ_A{}^-$ → $P_{700}{}^+$ $F_X{}^-$, but this decay became biphasic in PsaB-W669G, with decay times of ~250 and ~890 ns. The interpretation was that electron transfer from PhQ_B had been drastically slowed by the mutation. Thus, this residue may play an important role in determining the electron transfer rate difference between the two PhQ sites, which has been so useful in analysis of directionality.

There are good reasons that bidirectionality of electron transfer within PS I was first discovered in *Chlamydomonas* rather than in cyanobacteria. First, the fraction of B-side electron transfer is higher in PS I from *Chlamydomonas*, and second, the rate of electron transfer from PhQ_B is faster in cyanobacteria, which makes it even more difficult to see (Agalarov and Brettel, 2003; Rappaport et al., 2006). These two facts no doubt go far to explain why this concept was controversial in the field at first, with results from algal and cyanobacterial mutants often seeming to disagree (Agalarov and Brettel, 2003; Rappaport et al., 2006; Redding and van der Est, 2006). However, it must be pointed out that, although there now seems to be general agreement that both branches in PS I can be used, there is no consensus on the factors that control directionality. For example, taken at face value, the ultra-fast (Ramesh et al., 2007) and EPR (Fairclough et al., 2003) results on the ec3 axial ligand mutants would indicate that they do not alter the branching ratio, while *in vivo* analysis of forward electron transfer from the PhQs (Byrdin et al., 2006) would indicate that they do. See Redding and van der Est (2006) for a more detailed discussion of this issue.

IV. ELECTRON TRANSFER TO AND FROM PS I

A. Acceptor side: FeS clusters and the reduction of ferredoxin

It has been difficult to observe electron transfer between the FeS clusters of PS I due to the fact that they all absorb in a similar region of the spectrum, resulting in very low amplitudes for absorbance differences. Another reason is the temporal overlap with the preceding electron transfer events. This problem was solved with the use of mutations that slow electron transfer from PhQ_A by a factor of >4, revealing a new kinetic component with $\tau \approx 180$ ns and a DAS with low positive amplitude in the 340–400 nm region and negative amplitude in the 420–460 nm region (Byrdin et al., 2006). This shape was what would have been predicted for the $P_{700}^+ F_X^- \rightarrow P_{700}^+ (F_A/F_B)^-$ electron transfer step based upon the difference spectra of the $P_{700}^+ F_X^-$ and $P_{700}^+ (F_A/F_B)^-$ states (Lüneberg et al., 1994), and a rate similar to the $P_{700}^+ PhQ_{(A)}^- \rightarrow P_{700}^+ F_X^-$ step that had been delimited by photovoltage studies (Leibl et al., 1995). It is important to note that this should represent electron transfer within PS I reaction centers that had previously used the B-branch to deliver an electron to F_X, since it was only seen in mutants in which the $P_{700}^+ PhQ_A^- \rightarrow P_{700}^+ F_X^-$ step had been slowed to >1 μs. This implies that the ~200-ns kinetic component in wild-type PS I is a convolution of the $PhQ_A \rightarrow F_A$ and $F_X \rightarrow F_A$ steps, which may partially explain the differences in the DAS of the ~20-ns and ~200-ns components. It is widely assumed that once an electron has arrived at the PsaC subunit, there is a rapid equilibrium between reduction of the F_A and F_B clusters, with F_A^- being somewhat more stable (see below).

In wild-type PS I in the absence of an electron acceptor, back-reaction from the $P_{700}^+ (F_A/F_B)^-$ state occurs with a time constant of ~100 ms at room temperature. However, in site-directed mutants that replaced the Trp residues that interact with either PhQ this reaction was accelerated – to 20 ms in PsaA-W693F, 14 ms in PsaB-W673F and 8 ms in the double mutant (Boudreaux et al., 2001). This was explained by a stabilization of PhQ^- in the mutants, and provides evidence that the back reaction can also make use of both cofactor branches.

The polypeptide region responsible for coordination of the F_X cluster was the only one possible to pick out before the crystal structure was available. The sequence FPCDGPGRGGTC between transmembrane helices 8 and 9 is perfectly conserved in all PsaA and PsaB polypeptides, and is the only sequence well conserved in the distantly related reaction center genes of heliobacteria and green sulfur bacteria (Liebl et al., 1993). In the cyanobacterial structure, it was later seen that the two Cys residues of this sequence in PsaA and the two in PsaB together coordinate the Fe_4S_4 cluster (Jordan et al., 2001). Several mutations were made in this sequence to probe structure/function relationships in the cluster. The first site-directed mutations made in *Chlamydomonas* PS I targeted PsaB-Pro560 and Cys561 (Webber et al., 1993). Although conversion of the Pro to Ala or Leu had surprisingly little effect, mutation of the Cys to His resulted in no accumulation of PS I. Mutation of the analogous Cys residue in PsaA (Cys575) to Asp also resulted in no PS I (Hallahan et al., 1995), but the PsaA-C575H mutant accumulated some PS I, although it was not possible to photoreduce the terminal F_A/F_B clusters in this mutant. The PsaA-C575H and PsaA-C575S mutants also displayed a much slower decay of the P_{700}^+ PhQ^- spin-polarized EPR signal at 260 K (~2 μs versus 355 ns in wild type), which was tentatively attributed to back reaction (Evans et al., 1999). Thus, mutation of this Cys seemed to result in either no F_X cluster or one so altered that it could not serve as an electron acceptor from the PhQs.

The PsaA-D576L mutant had similar back-reaction kinetics to the wild type (Evans et al., 1999) but the F_A/F_B^- EPR spectrum was altered, indicating that the nearby F_A and F_B centers were sensitive to the loss of Asp576. Unfortunately, the authors were unable to observe the reduced F_X cluster directly by EPR, so it difficult to know the effect of this mutation upon the cluster. Thus, as one might expect, the PS I reaction center is extremely sensitive to mutation of the cysteines serving as ligands to the irons of the F_X cluster. The conserved dodecamer also serves as part of the binding site for the PsaC subunit to the PsaA/B heterodimer. Although the PsaB-R566E and PsaB-D562N mutants did not accumulate enough PS I to study, substitution of the two Pro residues for Leu allowed assembly of PS I (Rodday et al., 1995). These mutants showed evidence for a weakened association with PsaC in reconstitution experiments with recombinant cyanobacterial PsaC after removal of the endogenous subunit with urea.

The PsaC subunit contains the last two electron transfer cofactors of PS I: the F_A and F_B clusters. Before the PS I structure was available, the similarity to bacterial ferredoxins containing 2 Fe_4S_4 clusters allowed use of a homology-based model to drive mutagenesis experiments. PsaC contains two CxxCxxCxxxCP motifs; the first three Cys residues of one motif along with the last Cys of the other motif (PsaC-Cys 11, 14, 17, 58) coordinate F_B, while the other four (PsaC-Cys 21, 48, 51, 54) form F_A. In *Chlamydomonas* as in plant PS I, photoreduction of the terminal clusters at low temperature favors reduction of F_A (Fischer et al., 1997), despite the fact that F_B is the cluster that is further out and that reduces ferredoxin (Fischer et al., 1999b). The ratio of F_A/F_B seems to be ~1.5 at 15 K in *Chlamydomonas* PS I, as measured by EPR (the clusters can be distinguished by their different *g*-tensors), but it became ~0.5 in the PsaC-K52S/R53A mutant, which was created to make the region near F_A look more similar to the cluster in the green sulfur bacterial reaction center (Fischer et al., 1997). The double mutation did not significantly modify the principal component of the g-tensor of either F_A^- or F_B^-, nor did it alter the kinetics of ferredoxin reduction (see below). However, the double mutant was more light-sensitive than expected in the presence of oxygen, and it was suggested that the positive residues at positions 52–53 raise the reduction potential of F_A in wild-type PS I, thereby pulling electron density away from F_B and reducing the rate of the Mehler reaction (reduction of O_2 to superoxide, O_2^-, at the acceptor side).

Ferredoxin interacts with the acceptor side of PS I, which includes the peripheral PsaC, PsaD, and PsaE subunits. Docking of ferredoxin requires several charged residues on the PsaD subunit (Sétif, 2006). It also makes use of a specific Lys residue (35) on PsaC (Fischer et al., 1998). This residue was converted to Arg, Thr, Glu, and Asp, and all of these mutations blocked crosslinking of PsaC with ferredoxin *in vitro* using amide-forming reagents, indicating that the major crosslink formed between these proteins is between PsaC-Lys35 and specific Asp and/or Glu residues of ferredoxin. This interaction is critical to the association between PS I and ferredoxin, since the Asp and Glu substitutions also blocked crosslinking of ferredoxin with PsaD and PsaE; the PsaC-K35T mutation had a weaker effect upon this interaction. PsaC-K35R had almost no effect upon docking, as expected, since it preserves the electrostatic interaction between PsaC and ferredoxin (although the guanidinium group of Arg is incapable of forming an amide crosslink).

Kinetic analysis with purified PS I and ferredoxin allowed measurement of ferredoxin reduction after a laser flash: two first-order rates ($<1\,\mu s$ and $\sim 5\,\mu s$) correspond to electron transfer to pre-docked ferredoxin, while the second-order rate ($\sim 2.3 \times 10^8$/M/s) reflects electron transfer after docking to unbound ferredoxin (Fischer et al., 1997, 1998). The rates were unaffected in PsaC-K35R, but no first-order reactions were seen with the K35T, K35D, or K35E mutants (Fischer et al., 1998). With the latter, the second-order rate constants were decreased by about 5-, 40-, and 80-fold, respectively. The location of Lys35 next to F_B suggested that this is the cluster that reduces ferredoxin, and this was confirmed in a subsequent study in which multiple replacements were made near both clusters (Fischer et al., 1999b). Mutation of three residues near F_B to make them look like the corresponding residues near F_A (PsaC-I12V/T15K/Q16R) resulted in a ~60-fold increase in affinity for ferredoxin. Crosslinking between ferredoxin and the triple mutant PsaC was greatly increased compared to the wild type. The converse mutation (PsaC-V49I/K52T/R53Q) had no effect upon ferredoxin reduction kinetics, but it did affect the F_A^- EPR spectrum (Fischer et al., 1999b).

B. Donor side: oxidation of plastocyanin and cytochrome c_6

The last event in PS I electron transfer is the re-reduction of P_{700}^+ by the electron donor, which is either cyt c_6 or plastocyanin. In eukaryotes, the reduction of P_{700}^+ by plastocyanin takes place with both fast first-order kinetics reflecting pre-bound plastocyanin and second-order kinetics corresponding to docking and reduction by free plastocyanin. Farah et al. (1995) isolated a strain lacking PsaF by screening transformants and found that the mutant exhibited a much slower oxidation of cyt *f in vivo*, suggesting a limitation in electron transfer between PS I and cyt b_6f. Experiments with PS I purified from this mutant, which appeared otherwise normal except for lack of PsaF, demonstrated that the first-order kinetics (4–5 μs for both donor proteins with wild-type PS I) were absent. Also absent was a crosslinked product between cyt c_6 or plastocyanin and PS I, which was shown to be bridged via PsaF, and which exhibited the same fast first-order kinetics as free PS I with high concentrations of donor, verifying that these complexes represent the physiological docked state (Hippler et al., 1997). A similar active crosslinked complex with cyt c_6 was also demonstrated using a zero-length crosslinker (Sommer et al., 2006). PS I lacking PsaF exhibited slower second-order kinetics with rate constants (in the wild type, 6.5×10^7/M/s for plastocyanin and 3.5×10^7/M/s for cyt c_6) reduced by a factor of ~80 or ~34, respectively.

The possibility of transformation of the PsaF-deficient mutant with mutated versions of the *PsaF* gene has allowed dissection of its function. Eukaryotic PsaF has an N-terminal extension compared to its cyanobacterial counterpart, and this extension has been shown to be necessary and sufficient to confer tight binding of plastocyanin and cyt c_6 in cyanobacteria with chimeric PsaF containing the N-terminus from *Chlamydomonas* (Hippler et al., 1999). In *Chlamydomonas*, this extension has five Lys residues (12, 16, 19, 23, 30), which would be predicted to be on one side of an α-helix, and likely help to dock the acidic donor proteins. These Lys residues were individually mutated and introduced into algae lacking the endogenous *PSAF* gene (Hippler et al., 1998). Although no transformants expressing mutations at Lys19 could be found, mutant PS I at every other site could be purified and studied. The K23Q mutation drastically inhibited crosslinking with plastocyanin and electron transfer from both donor proteins. Although it did not affect the rate of the first-order reaction, the mutation did increase its amplitude by a factor of 2–3 and decreased the second-order rate constants for reaction with plastocyanin or cyt c_6 by a factor of 13 or 7, respectively. The K12P mutant was similar to the wild type in most respects, while the other two mutants (K16Q and K30Q) were intermediate (Hippler et al., 1998). An interesting observation is that plastocyanin seemed more sensitive to changes in PsaF than cyt c_6. The sites of linkage between cyt c_6 and PsaF in the crosslinked product were identified by mass spectrometry to be PsaA-Lys27/cyt c_6-Glu69 and PsaA-Lys23/cyt c_6-Glu69/70 (Sommer et al., 2006).

The other main interaction of the donor proteins is with the lumenal face of the PsaA/B heterodimer, which is rather flat and hydrophobic, matching the character of the face of the donor proteins closest to the electron transfer cofactors. In between P_{700} and the lumenal face are two α-helices that are parallel to the membrane plane, and just over P_{700} are two Trp residues in analogous positions in PsaA and PsaB in these helices: PsaA-Trp651 and PsaB-Trp627. Mutation of either of these to Phe resulted in abolishment of the tight complex with plastocyanin giving fast first-order kinetics, and reduced the second-order rate constants by a factor of 5–15 (Sommer et al., 2002, 2004). The effects on cyt c_6 were somewhat different: first-order kinetics were still observed although they were slowed by a factor of 5–10, and the affinity for cyt c_6 was reduced (~13-fold in PsaA-W651F and ~2-fold in PsaB-W627F). Mutation of PsaA-Trp651 to Ser further decreased affinity for cyt c_6 (compared to PsaA-W651F) but increased interaction with plastocyanin, providing further evidence for differences in the way the two proteins dock with the donor side of PS I (Sommer et al., 2004). It would appear that PsaA-Trp651 is more important for interaction with cyt c_6, while PsaB-Trp627 plays a similar role for plastocyanin.

The function of the donor side relies also upon interactions between subunits of PS I. The subunit closest to PsaF is PsaJ, a small hydrophobic polypeptide with a single transmembrane helix. Deletion of the chloroplast *psaJ* gene had no gross effect upon PS I assembly or function (Fischer et al., 1999a). PS I isolated from the deletion mutant behaved as if it consisted of two populations: 30% of the particles behaved like the wild type in terms of first- and second-order rates of re-reduction by plastocyanin or cyt c_6, while the remaining reaction centers behaved like the *psaF* deletion mutant. Both donor proteins could be crosslinked to PsaF in the particles lacking PsaJ. Thus, the PsaJ subunit seems to play a role positioning the N-terminal domain of PsaF to optimize proper docking of donor proteins with the lumenal face of PS I. On this face is another critical feature of the docking interaction: PsaB-Glu613. Mutation of this residue to Lys resulted in electron transfer rates from both plastocyanin and cyt c_6 that were comparable to those lacking PsaF, although both donor proteins could still be crosslinked to PsaF, albeit at lower efficiency (Sommer et al., 2002). These results indicated that PsaB-Glu613, like PsaJ, plays a role in properly positioning the PsaF N-terminal helix so that it can dock donor proteins. Interestingly, the PsaB-E613N mutant exhibited higher affinity for both electron donors (K_D was reduced ~3-fold for plastocyanin and ~10-fold for cyt c_6), suggesting that the negative charge on the Glu residue might help to unbind the donor protein after electron transfer to P_{700} (Sommer et al., 2002).

In pump-probe experiments in living cells with long durations between actinic flashes (>500 ms), $P_{700}{}^+$ re-reduction is monophasic with a time constant of ~5 μs (Guergova-Kuras et al., 2001; Ramesh et al., 2002). In copper-replete media, the electron donor will be plastocyanin (see Chapter 10), so these kinetics would indicate that the vast majority of PS I reaction centers have a pre-bound plastocyanin when the laser flash occurs. This fact allows one to address the effect of the driving force upon the electron transfer rate in the plastocyanin–PS I complex by looking at $P_{700}{}^+$ re-reduction in mutants with mutations near P_{700} that change its oxidation potential. Ramesh et al. (2002) examined a series of mutants with replacements of the axial ligand for P_B (PsaB-His656), and found that the rate of electron transfer from plastocyanin to $P_{700}{}^+$ increased as the oxidation potential increased, as expected. A similar effect was also seen *in vitro* with artificial electron donors.

V. ASSEMBLY AND DISASSEMBLY OF PS I

Relatively little is known about post-translational processes of PS I biogenesis. Although it is likely that many genes and proteins are involved in this process, so far only two have been identified (reviewed by Rochaix et al., 2004). Both genes (*ycf3* and *ycf4*) are chloroplast-encoded and deletion of either results in loss of PS I (Boudreau et al., 1997). There were no detectable deficiencies in transcription or translation of core subunits (PsaA, B, C) in the deletion mutants. Both gene products are localized to the thylakoid membrane, but neither appears to be a component of PS I. Ycf4 behaves as an integral membrane protein, and a large proportion of the polypeptide from detergent-solubilized membranes fractionated as part of a complex much larger than PS I. In contrast, Ycf3 could be washed from the membrane by carbonate or chaotropes, and is present at only about 5% the level of PS I. Both Ycf3 (172 residues) and Ycf4 (197 residues) are conserved in other oxygenic phototrophs. Ycf3 contains three tetratricopeptide repeats (TPR), which are thought to function as sites for protein–protein interactions. Site-directed mutations in the second and third TPRs led to some growth defects, although the mutants appeared to accumulate fully functional PS I at near-normal levels (Naver et al., 2001). Random mutagenesis of the N-terminus produced a temperature-sensitive mutation, which allowed one to distinguish between instability and inefficient assembly. After a shift of the mutant to the nonpermissive temperature, PS I remained stable, indicating that the effect was upon assembly (Naver et al., 2001).

The process of disassembly and degradation of the PS I complex is also relatively poorly characterized. An *in vitro* assay was created to follow the process (Henderson et al., 2003), where disassembly of the PS I complex (as visualized by green gel electrophoresis and low-temperature fluorescence) occurred after loss of photochemical activity, and before proteolysis of the subunit polypeptides. It is clear that degradation of PS I is not simply due to proteolysis, as addition of trypsin/chymotrypsin did not cause disassembly of PS I, even after all the polypeptides had been clipped. The disassembly process was metal-dependent: it was completely inhibited by EDTA and could be restored by Zn^{2+}, with Co^{2+} and Ca^{2+} being somewhat less efficient. The latter result might indicate a role for a metalloprotease, such as FtsH, a Zn-metalloprotease thought to be involved in degradation of PS II (see Chapter 19).

It is well known that many alterations in the structure of PS I will lead to poor accumulation. For example, deletion of 10 residues from, or addition of 27 residues to, the C-terminus of PsaB resulted in loss of PS I (H. Lee et al., 1996). Selection of phototrophic revertants from the latter resulted in variants with shorter extensions. However, it is possible to select second-site mutants that accumulate more of the mutant PS I. For example, phototrophic revertants of PsaA-D576L, which accumulates lowered amounts of PS I, were due to nuclear suppressor mutations (Evans et al., 1999). It may well be that such mutations affect the components of the system used to degrade these aberrant PS I complexes.

VI. FRONTIERS OF PS I RESEARCH USING THE *CHLAMYDOMONAS* SYSTEM

As can be seen in this chapter, *Chlamydomonas* has been an incredibly powerful and productive platform from which to study PS I. This is due to several factors, including the ability to make mutations in the major subunits with ease, and to purify the protein in large amounts. These have allowed researchers to ask fundamental questions about the operation of PS I and will no doubt continue in the future to contribute to ongoing questions, such as the following: How does energy transfer limit (and perhaps even direct) charge separation? What is the mechanism of primary charge separation, and how many states participate in it? What states are produced by back-reaction from different cofactors within PS I? What is the origin of the "bifurcation" of electron transfer observed at low temperature? What factors optimize docking, electron transfer, and release of the electron transfer donors and acceptors?

The resemblance of PS I from this organism to that of higher plants also allows the study of questions that would not be easy or possible in either plants or cyanobacteria. These include: the mechanisms by which external antenna proteins (e.g. LHCI) attach to the PS I, and how these interactions are regulated; the structural arrangement of these antenna proteins with PS I, and how these relate to pathways of excitation quenching; and the role of specialized membrane domains (e.g. granal and stromal thylakoids) in the function of PS I. Moreover, the fact that algal chloroplasts have much in common with those of plants has opened the door to more wide-ranging studies, such as those having to do with how PS I is expressed and assembled in an efficient fashion, how it participates in both linear and cyclic electron flow pathways, how these are regulated, and how PS I is disassembled and degraded.

ACKNOWLEDGMENTS

I acknowledge financial support from the U.S. National Science Foundation (grant MCB-0347935) and Department of Energy (grant DE-FG02-00ER15097). I thank Joanna Kargul, Jon Nield, and James Barber for providing Figure 15.2, and Art van der Est for assisting with Figure 15.1.

REFERENCES

Agalarov, R. and Brettel, K. (2003). Temperature dependence of biphasic forward electron transfer from the phylloquinone(s) A_1 in photosystem I: Only the slower phase is activated. *Biochim. Biophys. Acta* **1604**, 7–12.

Ali, K., Santabarbara, S., Heathcote, P., Evans, M.C., and Purton, S. (2006). Bidirectional electron transfer in photosystem I: Replacement of the symmetry-breaking tryptophan close to the PsaB-bound phylloquinone (A_{1B}) with a glycine residue alters the redox properties of A_{1B} and blocks forward electron transfer at cryogenic temperatures. *Biochim. Biophys. Acta* **1757**, 1623–1633.

Amunts, A., Drory, O., and Nelson, N. (2007). The structure of a plant photosystem I supercomplex at 3.4 Å resolution. *Nature* **447**, 58–63.

Balczun, C., Bunse, A., Hahn, D., Bennoun, P., Nickelsen, J., and Kück, U. (2005). Two adjacent nuclear genes are required for functional complementation of a chloroplast *trans*-splicing mutant from *Chlamydomonas reinhardtii*. *Plant J.* **43**, 636–648.

Ben-Shem, A., Frolow, F., and Nelson, N. (2003). Crystal structure of plant photosystem I. *Nature* **426**, 630–635.

Boichenko, V.A. (1998). Photosynthetic generation of O_2 and H_2 by photosystem I-deficient *Chlamydomonas* mutants. *Biochemistry (Engl. Transl. Biokhimiya)* **63**, 164–170.

Boudreau, E., Takahashi, Y., Lemieux, C., Turmel, M., and Rochaix, J.D. (1997). The chloroplast *ycf3* and *ycf4* open reading frames of *Chlamydomonas reinhardtii* are required for the accumulation of the photosystem I complex. *EMBO J.* **16**, 6095–6104.

Boudreaux, B., MacMillan, F., Teutloff, C., Agalarov, R., Gu, F., Grimaldi, S., Bittl, R., Brettel, K., and Redding, K. (2001). Mutations in both sides of the photosystem I reaction center identify the phylloquinone observed by electron paramagnetic resonance spectroscopy. *J. Biol. Chem.* **276**, 37299–37306.

Breton, J. (2001). Fourier transform infrared spectroscopy of primary electron donors in type I photosynthetic reaction centers. *Biochim. Biophys. Acta* **1507**, 180–193.

Breton, J., Nabedryk, E., and Leibl, W. (1999). FTIR study of the primary electron donor of photosystem I (P_{700}) revealing delocalization of the charge in $P_{700}{}^{+}$ and localization of the triplet character in $^{3}P_{700}$. *Biochemistry* **38**, 11585–11592.

Byrdin, M., Santabarbara, S., Gu, F., Fairclough, W.V., Heathcote, P., Redding, K., and Rappaport, F. (2006). Assignment of a kinetic component to electron transfer between iron-sulfur clusters F_X and $F_{A/B}$ of Photosystem I. *Biochim. Biophys. Acta* **1757**, 1529–1538.

Cohen, R.O., Shen, G., Golbeck, J.H., Xu, W., Chitnis, P.R., Valieva, A.I., van der Est, A., Pushkar, Y., and Stehlik, D. (2004). Evidence for asymmetric electron transfer in cyanobacterial photosystem I: Analysis of a methionine-to-leucine mutation of the ligand to the primary electron acceptor A_0. *Biochemistry* **43**, 4741–4754.

Cournac, L., Redding, K., Bennoun, P., and Peltier, G. (1997). Limited photosynthetic electron flow but no CO_2 fixation in *Chlamydomonas* mutants lacking Photosystem I. *FEBS Lett.* **416**, 65–68.

Dauvillée, D., Stampacchia, O., Girard-Bascou, J., and Rochaix, J.D. (2003). Tab2 is a novel conserved RNA binding protein required for translation of the chloroplast psaB mRNA. *EMBO J.* **22**, 6378–6388.

Evans, M.C.W., Purton, S., Patel, V., Wright, D., Heathcote, P., and Rigby, S.E.J. (1999). Modification of electron transfer from the quinone electron carrier, A(1), of Photosystem 1 in a site directed mutant D576→L within the Fe-S_x binding site of PsaA and in second site suppressors of the mutation in *Chlamydomonas reinhardtii*. *Photosynth. Res.* **61**, 33–42.

Fairclough, W.V., Forsyth, A., Evans, M.C., Rigby, S.E., Purton, S., and Heathcote, P. (2003). Bidirectional electron transfer in photosystem I: Electron transfer on the PsaA side is not essential for phototrophic growth in *Chlamydomonas*. *Biochim. Biophys. Acta* **1606**, 43–55.

Farah, J., Rappaport, F., Choquet, Y., Joliot, P., and Rochaix, J.D. (1995). Isolation of a *psaF*-deficient mutant of *Chlamydomonas reinhardtii*: Efficient interaction of plastocyanin with the photosystem I reaction center is mediated by the PsaF subunit. *EMBO J.* **14**, 4976–4984.

Fischer, N., Stampacchia, O., Redding, K., and Rochaix, J.D. (1996). Selectable marker recycling in the chloroplast. *Mol. Gen. Genet.* **251**, 373–380.

Fischer, N., Sétif, P., and Rochaix, J.D. (1997). Targeted mutations in the *psaC* gene of *Chlamydomonas reinhardtii*: Preferential reduction of F_B at low temperature is not accompanied by altered electron flow from photosystem I to ferredoxin. *Biochemistry* **36**, 93–102.

Fischer, N., Hippler, M., Sétif, P., Jacquot, J.P., and Rochaix, J.D. (1998). The PsaC subunit of photosystem I provides an essential lysine residue for fast electron transfer to ferredoxin. *EMBO J.* **17**, 849–858.

Fischer, N., Boudreau, E., Hippler, M., Drepper, F., Haehnel, W., and Rochaix, J.D. (1999a). A large fraction of PsaF is nonfunctional in photosystem I complexes lacking the PsaJ subunit. *Biochemistry* **38**, 5546–5552.

Fischer, N., Sétif, P., and Rochaix, J.D. (1999b). Site-directed mutagenesis of the PsaC subunit of photosystem I. F_B is the cluster interacting with soluble ferredoxin. *J. Biol. Chem.* **274**, 23333–23340.

Garcia-Sanchez, M.I., Diaz-Quintana, A., Gotor, C., Jacquot, J.P., De la Rosa, M.A., and Vega, J.M. (2000). Homology predicted structure and functional interaction of ferredoxin from the eukaryotic alga *Chlamydomonas reinhardtii* with nitrite reductase and glutamate synthase. *J. Biol. Inorg. Chem.* **5**, 713–719.

Gibasiewicz, K., Ramesh, V.M., Melkozernov, A.N., Lin, S., Woodbury, N.W., Blankenship, R.E., and Webber, A.N. (2001). Excitation dynamics in the core antenna of PS I from *Chlamydomonas reinhardtii* CC 2696 at room temperature. *J. Phys. Chem. B* **105**, 11498–11506.

Gibasiewicz, K., Ramesh, V.M., Lin, S., Woodbury, N.W., and Webber, A.N. (2002). Excitation dynamics in eukaryotic PS I from *Chlamydomonas reinhardtii* CC 2696 at 10K. Direct detection of the reaction center exciton states. *J. Phys. Chem. B* **106**, 6322–6330.

Gibasiewicz, K., Ramesh, V.M., Lin, S., Redding, K., Woodbury, N.W., and Webber, A.N. (2003). Excitonic interactions in wild-type and mutant PSI reaction centers. *Biophys. J.* **85**, 2547–2559.

Gibasiewicz, K., Szrajner, A., Ihalainen, J.A., Germano, M., Dekker, J.P., and van Grondelle, R. (2005). Characterization of low-energy chlorophylls in the PSI-LHCI supercomplex from *Chlamydomonas reinhardtii*. A site-selective fluorescence study. *J. Phys. Chem. B* **109**, 21180–21186.

Gobets, B. and van Grondelle, R. (2001). Energy transfer and trapping in photosystem I. *Biochim. Biophys. Acta* **1507**, 80–99.

Greenbaum, E., Lee, J.W., Tevault, C.V., Blankinship, S.L., and Mets, L.J. (1995). CO_2 fixation and photoevolution of H_2 and O_2 in a mutant of *Chlamydomonas* lacking Photosystem I. *Nature* **376**, 438–441.

Guergova-Kuras, M., Boudreaux, B., Joliot, A., Joliot, P., and Redding, K. (2001). Evidence for two active branches for electron transfer in photosystem I. *Proc. Natl. Acad. Sci. U. S. A.* **98**, 4437–4442.

Gulis, G., Narasimhulu, K.V., Fox, L.N., and Redding, K.E. (2008). Purification of His6-tagged Photosystem I from *Chlamydomonas reinhardtii*. *Photosynth. Res.* **96**, 51–60.

Hahn, D., Bennoun, P., and Kück, U. (1996). Altered expression of nuclear genes encoding chloroplast polypeptides in non-photosynthetic mutants of *Chlamydomonas reinhardtii*: Evidence for post-transcriptional regulation. *Mol. Gen. Genet.* **252**, 362–370.

Hallahan, B.J., Purton, S., Ivison, A., Wright, D., and Evans, M.C.W. (1995). Analysis of the proposed Fe-S_X binding region of Photosystem 1 by site directed mutation of PsaA in *Chlamydomonas reinhardtii*. *Photosynth. Res.* **46**, 257–264.

Hamacher, E., Kruip, J., Rögner, M., and Mäntele, W. (1996). Characterization of the primary electron donor of photosystem I, P_{700}, by electrochemistry and Fourier transform infrared (FTIR) difference spectroscopy. *Spectrochimica Acta, Part A* **52A**, 107–121.

Hastings, G., Hoshina, S., Webber, A.N., and Blankenship, R.E. (1995). Universality of energy and electron transfer processes in Photosystem I. *Biochemistry* **34**, 15512–15522.

Henderson, J.N., Zhang, J., Evans, B.W., and Redding, K. (2003). Disassembly and degradation of photosystem I in an *in vitro* system are multievent, metal-dependent processes. *J. Biol. Chem.* **278**, 39978–39986.

Hippler, M., Drepper, F., Farah, J., and Rochaix, J.D. (1997). Fast electron transfer from cytochrome c_6 and plastocyanin to photosystem I of *Chlamydomonas reinhardtii* requires PsaF. *Biochemistry* **36**, 6343–6349.

Hippler, M., Drepper, F., Haehnel, W., and Rochaix, J.D. (1998). The N-terminal domain of PsaF: Precise recognition site for binding and fast electron transfer from cytochrome c_6 and plastocyanin to photosystem I of *Chlamydomonas reinhardtii*. *Proc. Natl. Acad. Sci. U. S. A.* **95**, 7339–7344.

Hippler, M., Drepper, F., Rochaix, J.D., and Muhlenhoff, U. (1999). Insertion of the N-terminal part of PsaF from *Chlamydomonas reinhardtii* into photosystem I from *Synechococcus elongatus* enables efficient binding of algal plastocyanin and cytochrome c6. *J. Biol. Chem.* **274**, 4180–4188.

Holzwarth, A.R., Müller, M.G., Niklas, J., and Lubitz, W. (2005). Charge recombination fluorescence in photosystem I reaction centers from *Chlamydomonas reinhardtii*. *J. Phys. Chem. B* **109**, 5903–5911.

Holzwarth, A.R., Muller, M.G., Niklas, J., and Lubitz, W. (2006). Ultrafast transient absorption studies on photosystem I reaction centers from *Chlamydomonas reinhardtii*. 2: Mutations near the P_{700} reaction center chlorophylls provide new insight into the nature of the primary electron donor. *Biophys. J.* **90**, 552–565.

Ihalainen, J.A., van Stokkum, I.H., Gibasiewicz, K., Germano, M., van Grondelle, R., and Dekker, J.P. (2005). Kinetics of excitation trapping in intact Photosystem I of *Chlamydomonas reinhardtii* and *Arabidopsis thaliana*. *Biochim. Biophys. Acta* **1706**, 267–275.

Ivashin, N. and Larsson, S. (2003). Electron transfer pathways in photosystem I reaction centers. *Chem. Phys. Lett.* **375**, 383–387.

Joliot, P. and Joliot, A. (1999). *In vivo* analysis of the electron transfer within photosystem I: Are the two phylloquinones involved? *Biochemistry* **38**, 11130–11136.

Jordan, P., Fromme, P., Witt, H.T., Klukas, O., Saenger, W., and Krauß, N. (2001). Three-dimensional structure of cyanobacterial photosystem I at 2.5 Å resolution. *Nature* **411**, 909–917.

Kargul, J., Nield, J., and Barber, J. (2003). Three-dimensional reconstruction of a light-harvesting complex I-photosystem I (LHCI-PSI) supercomplex from the green alga *Chlamydomonas reinhardtii*. Insights into light harvesting for PSI. *J. Biol. Chem.* **278**, 16135–16141.

Kargul, J., Turkina, M.V., Nield, J., Benson, S., Vener, A.V., and Barber, J. (2005). Light-harvesting complex II protein CP29 binds to photosystem I of *Chlamydomonas reinhardtii* under State 2 conditions. *FEBS J.* **272**, 4797–4806.

Kerfeld, C.A., Anwar, H.P., Interrante, R., Merchant, S., and Yeates, T.O. (1995). The structure of chloroplast cytochrome c6 at 1.9 Å resolution: Evidence for functional oligomerization. *J. Mol. Biol.* **250**, 627–647.

Klukas, O., Schubert, W.D., Jordan, P., Krau, N., Fromme, P., Witt, H.T., and Saenger, W. (1999). Localization of two phylloquinones, QK and QK′, in an improved electron density map of photosystem I at 4-Å resolution. *J. Biol. Chem.* **274**, 7361–7367.

Krabben, L., Schlodder, E., Jordan, R., Carbonera, D., Giacometti, G., Lee, H., Webber, A.N., and Lubitz, W. (2000). Influence of the axial ligands on the spectral properties of P_{700} of photosystem I: a study of site-directed mutants. *Biochemistry* **39**, 13012–13025.

Lee, H., Bingham, S.E., and Webber, A.N. (1996). Site-directed mutagenesis and analysis of second-site revertants indicates a requirement for C-terminal amino acids of PsaB for stable assembly of the photosystem I reaction center complex in *Chlamydomonas reinhardtii*. *Photochem. Photobiol.* **64**, 46–52.

Lee, J.W., Tevault, C.V., Owens, T.G., and Greenbaum, E. (1996). Oxygenic photoautotrophic growth without Photosystem I. *Science* **273**, 364–367.

Leibl, W., Toupance, B., and Breton, J. (1995). Photoelectric characterization of forward electron transfer to iron-sulfur centers in Photosystem I. *Biochemistry* **34**, 10237–10244.

Li, Y., Lucas, M.-G., Konovalova, T., Abbott, B., MacMillan, F., Petrenko, A., Sivakumar, V., Wang, R., Hastings, G., Gu, F., van Tol, J., Brunel, L.-C., Timkovich, R., Rappaport, F., and Redding, K. (2004). Mutation of the putative hydrogen-bond donor to P_{700} of photosystem I. *Biochemistry* **43**, 12634–12647.

Li, Y., van der Est, A., Lucas, M.G., Ramesh, V.M., Gu, F., Petrenko, A., Lin, S., Webber, A.N., Rappaport, F., and Redding, K. (2006). Directing electron transfer within Photosystem I by breaking H-bonds in the cofactor branches. *Proc. Natl. Acad. Sci. U. S. A.* **103**, 2144–2149.

Liebl, U., Mockensturm-Wilson, M., Trost, J.T., Brune, D.C., Blankenship, R.E., and Vermaas, W. (1993). Single core polypeptide in the reaction center of the photosynthetic bacterium *Heliobacillus mobilis*: Structural implications and relations to other photosystems. *Proc. Natl. Acad. Sci. U. S. A.* **90**, 7124–7128.

Lüneberg, J., Fromme, P., Jekow, P., and Schlodder, E. (1994). Spectroscopic characterization of the PS I core complexes from the thermophilic *Synechococcus* sp.: Identical reoxidation kinetics of A_1^- before and after removal of the iron-sulfur clusters F_A and F_B. *FEBS Letters* **338**, 197–202.

McConnell, M.D., Wyndhamn, I., Brown, S.A.E., Ramesh, V.M., van der Est, A., and Webber, A.N. (2005). Directionality of electron transfer in photosystem I of *Chlamydomonas reinhardtii* probed by electron paramagnetic resonance. In: *Photosynthesis: Fundamental Aspects to Global Perspectives* (A. van der Est and D. Bruce, Eds.). Alliance Communications Group, Lawrence, Kansas.

Meimberg, K., Fischer, N., Rochaix, J.D., and Muhlenhoff, U. (1999). Lys35 of PsaC is required for the efficient photoreduction of flavodoxin by photosystem I from *Chlamydomonas reinhardtii*. *Eur. J. Biochem.* **263**, 137–144.

Melkozernov, A.N., Su, H., Lin, S., Bingham, S., Webber, A.N., and Blankenship, R.E. (1997). Specific mutation near the primary donor in photosystem I from

Chlamydomonas reinhardtii alters the trapping time and spectroscopic properties of P_{700}. *Biochemistry* **36**, 2898–2907.

Melkozernov, A.N., Su, H., Webber, A.N., and Blankenship, R.E. (1998). Excitation energy transfer in thylakoid membranes from *Chlamydomonas reinhardtii* lacking chlorophyll *b* and with mutant photosystem I. *Photosynth. Res.* **56**, 197–207.

Melkozernov, A.N., Kargul, J., Lin, S., Barber, J., and Blankenship, R.E. (2004). Energy coupling in the PSI-LHCI supercomplex from the green alga *Chlamydomonas reinhardtii*. *J. Phys. Chem. B* **108**, 10547–10555.

Melkozernov, A.N., Kargul, J., Lin, S., Barber, J., and Blankenship, R.E. (2005). Spectral and kinetic analysis of the energy coupling in the PS I-LHC I supercomplex from the green alga *Chlamydomonas reinhardtii* at 77 K. *Photosynth. Res.* **86**, 203–215.

Merchant, S.S., Allen, M.D., Kropat, J., Moseley, J.L., Long, J.C., Tottey, S., and Terauchi, A.M. (2006). Between a rock and a hard place: Trace element nutrition in *Chlamydomonas*. *Biochim. Biophys. Acta* **1763**, 578–594.

Merendino, L., Perron, K., Rahire, M., Howald, I., Rochaix, J.D., and Goldschmidt-Clermont, M. (2006). A novel multifunctional factor involved in *trans*-splicing of chloroplast introns in *Chlamydomonas*. *Nucleic Acids Res.* **34**, 262–274.

Muhiuddin, I.P., Heathcote, P., Carter, S., Purton, S., Rigby, S.E.J., and Evans, M.C.W. (2001). Evidence from time resolved studies of the $P_{700}{}^{+}/A1^{-}$ radical pair for photosynthetic electron transfer on both the PsaA and PsaB branches of the photosystem I reaction centre. *FEBS Lett.* **503**, 56–60.

Müller, M.G., Niklas, J., Lubitz, W., and Holzwarth, A.R. (2003). Ultrafast transient absoprtion studies on photosystem I reaction centers from *Chlamydomonas reinhardtii* I. A new interpretation of the energy trapping and early electron transfer steps in Photosystem I. *Biophys. J.* **85**, 3899–3922.

Nakamura, A., Yoshida, E., Taki, T., and Watanabe, T. (2002). Universal existence of one chlorophyll *a'* molecule in photosystem I of oxygenic photosynthetic organisms. *J. Photoscience* **9**, 367–369.

Nakamura, A., Akai, M., Yoshida, E., Taki, T., and Watanabe, T. (2003). Reversed-phase HPLC determination of chlorophyll *a'* and phylloquinone in photosystem I of oxygenic photosynthetic organisms. *Eur. J. Biochem.* **270**, 2446–2458.

Naver, H., Boudreau, E., and Rochaix, J.D. (2001). Functional studies of Ycf3: Its role in assembly of photosystem I and interactions with some of its subunits. *Plant Cell* **13**, 2731–2745.

Pantelidou, M., Chitnis, P.R., and Breton, J. (2004). FTIR spectroscopy of *Synechocystis* 6803 mutants affected on the hydrogen bonds to the carbonyl groups of the PsaA chlorophyll of P_{700} supports an extensive delocalization of the charge in $P_{700}{}^{+}$. *Biochemistry* **43**, 8380–8390.

Perron, K., Goldschmidt-Clermont, M., and Rochaix, J.D. (1999). A factor related to pseudouridine synthases is required for chloroplast group II intron trans-splicing in *Chlamydomonas reinhardtii*. *EMBO J.* **18**, 6481–6490.

Perron, K., Goldschmidt-Clermont, M., and Rochaix, J.D. (2004). A multiprotein complex involved in chloroplast group II intron splicing. *RNA* **10**, 704–711.

Petrenko, A., Maniero, A.L., van Tol, J., MacMillan, F., Li, Y., Brunel, L.C., and Redding, K. (2004). A high-field EPR study of $P_{700}{}^{+}$ in wild-type and mutant photosystem I from *Chlamydomonas reinhardtii*. *Biochemistry* **43**, 1781–1786.

Ramesh, V.M., Guergova-Kuras, M., Joliot, P., and Webber, A.N. (2002). Electron transfer from plastocyanin to the photosystem I reaction center in mutants with increased potential of the primary donor in *Chlamydomonas reinhardtii*. *Biochemistry* **41**, 14652–14658.

Ramesh, V.M., Bingham, S.E., and Webber, A.N. (2004a). A simple method for chloroplast transformation in *Chlamydomonas reinhardtii*. *Methods Mol. Biol.* **274**, 301–307.

Ramesh, V.M., Gibasiewicz, K., Lin, S., Bingham, S.E., and Webber, A.N. (2004b). Bidirectional electron transfer in photosystem I: Accumulation of A_0^- in A-Side or B-Side mutants of the axial ligand to chlorophyll A_0. *Biochemistry* **43**, 1369–1375.

Ramesh, V.M., Gibasiewicz, K., Lin, S., Bingham, S.E., and Webber, A.N. (2007). Replacement of the methionine axial ligand to the primary electron acceptor A_0 slows the A_0^- reoxidation dynamics in Photosystem I. *Biochim. Biophys. Acta* **1767**, 151–160.

Rappaport, F., Diner, B.A., and Redding, K. (2006). Optical measurements of secondary electron transfer in Photosystem I. In: *Photosystem I: The Plastocyanin: Ferredoxin Oxidoreductase in Photosynthesis* (J. Golbeck, Ed.). Kluwer Academic Publishers, Dordrecht.

Redding, K. and van der Est, A. (2006). The directionality of electron transfer in photosystem I. In: *Photosystem I: The Plastocyanin:Ferredoxin Oxidoreductase in Photosynthesis* (J. Golbeck, Ed.), pp. 413–437. Kluwer Academic Publishers, Dordrecht.

Redding, K., MacMillan, F., Leibl, W., Brettel, K., Hanley, J., Rutherford, A.W., Breton, J., and Rochaix, J.D. (1998). A systematic survey of conserved histidines in the core subunits of Photosystem I by site-directed mutagenesis reveals the likely axial ligands of P_{700}. *EMBO J.* **17**, 50–60.

Redding, K., Cournac, L., Vassiliev, I.R., Golbeck, J.H., Peltier, G., and Rochaix, J.D. (1999). Photosystem I is indispensable for photoautotrophic growth, CO_2 fixation, and H_2 photoproduction in *Chlamydomonas reinhardtii*. *J. Biol. Chem.* **274**, 10466–10473.

Redinbo, M.R., Cascio, D., Choukair, M.K., Rice, D., Merchant, S., and Yeates, T.O. (1993). The 1.5-Å crystal structure of plastocyanin from the green alga *Chlamydomonas reinhardtii*. *Biochemistry* **32**, 10560–10567.

Rivier, C., Goldschmidt-Clermont, M., and Rochaix, J.D. (2001). Identification of an RNA-protein complex involved in chloroplast group II intron *trans*-splicing in *Chlamydomonas reinhardtii*. *EMBO J.* **20**, 1765–1773.

Rochaix, J.D., Perron, K., Dauvillee, D., Laroche, F., Takahashi, Y., and Goldschmidt-Clermont, M. (2004). Post-transcriptional steps involved in the assembly of photosystem I in *Chlamydomonas*. *Biochem. Soc. Trans.* **32**, 567–570.

Rodday, S.M., Webber, A.N., Bingham, S.E., and Biggins, J. (1995). Evidence that the F_X domain in photosystem I interacts with the subunit PsaC: Site-directed changes in PsaB destabilize the subunit interaction in *Chlamydomonas reinhardtii*. *Biochemistry* **34**, 6328–6334.

Santabarbara, S., Kuprov, I., Fairclough, W.V., Purton, S., Hore, P.J., Heathcote, P., and Evans, M.C. (2005). Bidirectional electron transfer in photosystem I: Determination of two distances between P_{700}^+ and A_1^- in spin-correlated radical pairs. *Biochemistry* **44**, 2119–2128.

Schlodder, E., Falkenberg, K., Gergeleit, M., and Brettel, K. (1998). Temperature dependence of forward and reverse electron transfer from A_1^-, the reduced secondary electron acceptor in Photosystem I. *Biochemistry* **37**, 9466–9476.

Schmitter, J.M., Jacquot, J.P., de Lamotte-Guery, F., Beauvallet, C., Dutka, S., Gadal, P., and Decottignies, P. (1988). Purification, properties and complete amino acid sequence of the ferredoxin from a green alga, *Chlamydomonas reinhardtii*. *Eur. J. Biochem.* **172**, 405–412.

Sétif, P. (2006). Electron transfer from the bound iron-sulfur clusters to ferredoxin/flavodoxin: Kinetic and structural properties of ferredoxin/flavodoxin reduction by Photosystem I. In: *Photosystem I: The Plastocyanin:Ferredoxin Oxidoreductase in Photosynthesis* (J. Golbeck, Ed.), pp. 439–545. Kluwer Academic Publishers, Dordrecht.

Sommer, F., Drepper, F., and Hippler, M. (2002). The luminal helix l of PsaB is essential for recognition of plastocyanin or cytochrome c_6 and fast electron transfer to photosystem I in *Chlamydomonas reinhardtii*. *J. Biol. Chem.* **277**, 6573–6581.

Sommer, F., Drepper, F., Haehnel, W., and Hippler, M. (2004). The hydrophobic recognition site formed by residues PsaA-Trp651 and PsaB-Trp627 of photosystem I in *Chlamydomonas reinhardtii* confers distinct selectivity for binding of plastocyanin and cytochrome c_6. *J. Biol. Chem.* **279**, 20009–20017.

Sommer, F., Drepper, F., Haehnel, W., and Hippler, M. (2006). Identification of precise electrostatic recognition sites between cytochrome c_6 and the photosystem I subunit PsaF using mass spectrometry. *J. Biol. Chem.* **281**, 35097–35103.

Takahashi, Y., Goldschmidt-Clermont, M., Soen, S.Y., Franzen, L.G., and Rochaix, J.D. (1991). Directed chloroplast transformation in *Chlamydomonas reinhardtii*: Insertional inactivation of the *psaC* gene encoding the iron sulfur protein destabilizes Photosystem I. *EMBO J.* **10**, 2033–2040.

van der Est, A. (2006). Electron transfer involving phylloquinone in Photosystem I. In: *Photosystem I: The Plastocyanin:Ferredoxin Oxidoreductase in Photosynthesis* (J. Golbeck, Ed.). Kluwer Academic Publishers, Dordrecht.

Wang, R., Sivakumar, V., Li, Y., Redding, K., and Hastings, G. (2003). Mutation induced modulation of hydrogen bonding to P_{700} studied using FTIR difference spectroscopy. *Biochemistry* **42**, 9889–9897.

Wang, R., Sivakumar, V., Johnson, T.W., and Hastings, G. (2004). FTIR difference spectroscopy in combination with isotope labeling for identification of the carbonyl modes of P_{700} and $P_{700}{}^{+}$ in Photosystem I. *Biophys. J.* **86**, 1061–1073.

Webber, A.N. and Lubitz, W. (2001). P_{700}: The primary electron donor of Photosystem I. *Biochim. Biophys. Acta* **1507**, 61–79.

Webber, A.N., Gibbs, P.B., Ward, J.B., and Bingham, S.E. (1993). Site-directed mutagenesis of the photosystem I reaction center in chloroplasts. The proline-cysteine motif. *J. Biol. Chem.* **268**, 12990–12995.

Webber, A.N., Su, H., Bingham, S.E., Kass, H., Krabben, L., Kuhn, M., Jordan, R., Schlodder, E., and Lubitz, W. (1996). Site-directed mutations affecting the spectroscopic characteristics and midpoint potential of the primary donor in Photosystem I. *Biochemistry* **35**, 12857–12863.

Witt, H., Schlodder, E., Teutloff, C., Niklas, J., Bordignon, E., Carbonera, D., Kohler, S., Labahn, A., and Lubitz, W. (2002). Hydrogen bonding to P_{700}: Site-directed mutagenesis of threonine A739 of photosystem I in *Chlamydomonas reinhardtii*. *Biochemistry* **41**, 8557–8569.

Witt, H., Bordignon, E., Carbonera, D., Dekker, J.P., Karapetyan, N., Teutloff, C., Webber, A., Lubitz, W., and Schlodder, E. (2003). Species-specific differences of the spectroscopic properties of P_{700}: Analysis of the influence of non-conserved amino acid residues by site-directed mutagenesis of photosystem I from *Chlamydomonas reinhardtii*. *J. Biol. Chem.* **278**, 46760–46771.

CHAPTER 16

Photosystem II, a Structural Perspective

Zoee Gokhale[1] and Richard T. Sayre[1,2]

[1]Biophysics Program, Ohio State University, Columbus, Ohio, USA

[2]Donald Danforth Plant Science Center, St. Louis, Missouri, USA

CHAPTER CONTENTS

I. INTRODUCTION

During the last decade substantial progress has been made in the elucidation of the structure and function of the photosystem II (PS II) apparatus. Much of this progress has been achieved using either prokaryotic cyanobacterial

PS II complexes, *Chlamydomonas* chloroplast PS II complexes, or the detergent-fractionated thylakoid membranes of flowering plants such as spinach. Undoubtedly, the 3.5-Å resolution crystal structure from thermophilic cyanobacteria has had the greatest impact on our understanding of structure–function relationships in PS II (Zouni et al., 2001; Kamiya and Shen, 2003; Ferreira et al., 2004). This structure has also served as a model for the chloroplast PS II complex, although there are substantive differences between the two PS II types including the extrinsic subunits of the water-oxidizing complex, and some of the small (10–15 kD) structural membrane protein subunits. Electron diffraction studies of chloroplast PS II complexes indicate, however, that its reaction center core is essentially identical to that of cyanobacteria, with the exception of the extrinsic loops connected to the transmembrane helices (Büchel and Kühlbrandt, 2005). Thus, the cyanobacterial PS II structure can serve as a model for the chloroplast PS II reaction center core, with the exception of the antennae and extrinsic protein complexes.

Several reviews have been published on PS II structure and function, covering the early events in primary charge separation (Diner and Rappaport, 2002), structural studies (Ruffle and Sayre, 1998; Minagawa and Takahashi, 2004), and the mechanism of water oxidation (Nelson and Yocum, 2006). In this chapter, we focus primarily on the use of *Chlamydomonas* PS II complexes as a tool to elucidate PS II structure and function, with a primary focus on the reaction center and associated proximal antenna complex proteins. Less emphasis is placed on the functions of the small structural membrane polypeptides of PS II. Their functions in cyanobacteria and chloroplasts (of *Chlamydomonas*) have been covered in prior reviews (Erickson and Rochaix, 1992; Ruffle and Sayre, 1998; Minagawa and Takahashi, 2004). Models for the PS II complex are presented and discussed in the context of the partial reactions leading to the trapping of excited states, charge separation, and the generation of molecular oxygen and reduced plastoquinone.

II. OVERVIEW OF THE PHOTOSYSTEM II COMPLEX

A. Introduction

PS II is a multi-subunit protein–cofactor complex embedded in the thylakoid membranes, which catalyzes the oxidation of water and the reduction of plastoquinone. The holocomplex consists of no less than 20 protein subunits and has at least 77 cofactors. X-ray diffraction studies have led to the elucidation of cofactor–protein interactions at near-atomic resolution (3.8–3.0 Å; Jordan et al., 2001; Zouni et al., 2001; Kamiya and Shen, 2003; Ferreira et al., 2004; Biesiadka et al., 2004; Loll et al., 2005). Substantial knowledge regarding the identity and function of the redox-active components (chlorophyll, pheophytin, plastoquinone, tyrosine, and manganese) had previously been generated using biophysical approaches (reviewed in Diner and Rappaport, 2002). Significantly, the predictions inferred from biophysical investigations were largely consistent with the structural information obtained from PS II crystals.

The initial PS II-specific events leading to the generation of a charge separation involve excitation energy transfer from the proximal antenna chlorophyll–protein complexes, CP43 and CP47, to the chlorins of the PS II reaction center. Energy transfer to the primary electron donor, P680, generates P680*. Subsequent charge transfer results in the reduction of pheophytin a ($Pheo_{D1}$) in several picoseconds. The reduced primary acceptor then transfers an electron to the first stable (long-lived) electron acceptor, a bound plastoquinone (Q_A) molecule. The rate-limiting step in PS II electron transfer is the transfer of an electron to the plastoquinone bound at the Q_B-binding site. Following two rounds of reduction and subsequent protonation, the plastoquinol molecule at the Q_B site leaves and is replaced by a new plastoquinone molecule.

The electron-deficient primary donor P680$^+$ is subsequently reduced via a four-step (S-state transitions) oxidation of water to molecular oxygen by the water-oxidizing complex (also known as the oxygen-evolving complex, OEC) that includes an Mn_4-Ca cluster (reviewed in Bricker and Ghanotakis, 1996; Nelson and Yocum, 2006). The OEC is oxidized by a redox-active and neutral tyrosine radical (Tyr_Z; Y161) located on the D1 polypeptide of the reaction center (Barry and Babcock, 1987; Debus et al., 1988). Finally P680$^+$ is reduced and becomes available for another round of the light-driven charge separation.

B. Photosystem II particles and crystal structures

The first oxygen-evolving PS II preparations ("BBY" type) were from spinach thylakoid membranes solubilized with the non-ionic detergent Triton X-100 in the presence of 5 mM $MgCl_2$ (Berthold et al., 1981). BBY-type particles were capable of evolving oxygen at rates (500–700 micromoles O_2/mg Chl/hour) three-fold greater than those of thylakoid membranes. These preparations contained the reaction center polypeptides (D1 and D2), the proximal antenna complexes (CP43 and CP47), multiple small structural membrane proteins including cytochrome b_{559}, and the distal light-harvesting chlorophyll proteins (LHC). Subsequent enrichments of PS II particles led to the isolation of core particles that lacked the distal antenna complexes and had slightly higher rates of oxygen evolution (500–1000 micromoles O_2/mg Chl/hour; Ghanotakis et al., 1987).

With *Chlamydomonas* gaining popularity as a model organism for photosynthesis structure–function studies, protocols were developed for the isolation of BBY particles from *Chlamydomonas* (Shim et al., 1990) (see detailed methods descriptions at http://www.elsevierdirect.com/companions/9780123708731). A twofold enrichment in PS II and a ten-fold reduction of PS I was obtained by detergent fractionation of thylakoid membranes. These PS II particles had 250–300 chlorophylls per reaction center.

Subsequently, enriched PS II particles (40–50 Chls per Q_A) were isolated from *Chlamydomonas*, comprising CP43, CP47, D1, and D2 and several low molecular weight polypeptides, but these particles were unable to evolve oxygen (de Vitry et al., 1984, 1991). PS II core particles capable of high rates of oxygen evolution (>1000 micromoles O_2/hour/mg chlorophyll) were first isolated from *Chlamydomonas* by Bumann and Oesterhelt (1994). These particles were prepared by solubilization of thylakoid membranes with dodecyl maltoside in the presence of $CaCl_2$, followed by density gradient centrifugation and anion-exchange chromatography. They contained 40 Chl *a* molecules per reaction center. This PS II isolation protocol is appealing for potential crystallization attempts since only one detergent was used.

The minimal structural complex capable of photosynthetic electron transfer is the photosynthetic reaction center. PS II reaction centers are particularly useful for spectroscopic investigations of the primary photochemical events in photosynthesis, given that they have no antenna chlorophylls and are capable of generating a stable charge-separated state ($P680^+/Pheo^-$). PS II reaction centers lack plastoquinone in the Q_A- and Q_B-binding sites and also lack the Mn of the water-splitting complex. PS II reaction centers were first isolated from spinach thylakoids solubilized in 4% (w/v) Triton X-100 followed by anion exchange chromatography (Nanba and Satoh, 1987). They contained the D1, D2, Cyt b_{559} α and β subunits, and the product of the chloroplast-encoded *psbI* gene (Nanba and Satoh, 1987). PS II reaction centers were also prepared from a *Chlamydomonas* mutant strain (*F54-14*) lacking the PS I and ATP synthase complexes, both of which were often found to contaminate PS II particle preparations (Alizadeh et al., 1995). However a higher Triton X-100 to chlorophyll ratio (50:1) was required for membrane solubilization than was used for spinach thylakoids. Once bound to the anion exchange column, the solubilized mixture was washed with buffer containing dodecyl maltoside and eluted using a salt gradient, yielding PS II reaction centers with Chl *a*, Pheo, Cyt b_{559}, and carotenoid ratio of 6/2/1/1-2. Subsequently, PS II reaction center preparations were developed for wild type thylakoid membranes from *Chlamydomonas* using similar techniques but reduced detergent concentrations (Wang et al., 2002).

The PS II complex was crystallized as a dimer in which each monomer is characterized by a pseudo-twofold symmetry that equates the D1, CP47, and PsbI subunits to the D2, CP43, and PsbX subunits (Figure 16.1; Ferreira et al., 2004). The reaction center polypeptides D1 and D2 form the core of the complex, each consisting of five transmembrane helices that structurally resemble the L and M chains of the bacterial reaction center. On either side of the reaction center core are the proximal antenna subunits CP43 (PsbC) and CP47 (PsbB) comprising six transmembrane helices each (Sayre and Wroebel-Boerner, 1994). The six transmembrane spans of CP43 and CP47 are arranged in a manner similar to the N-terminal six transmembrane spans of the PsaA and PsaB subunits of PS I and have a similar light-harvesting

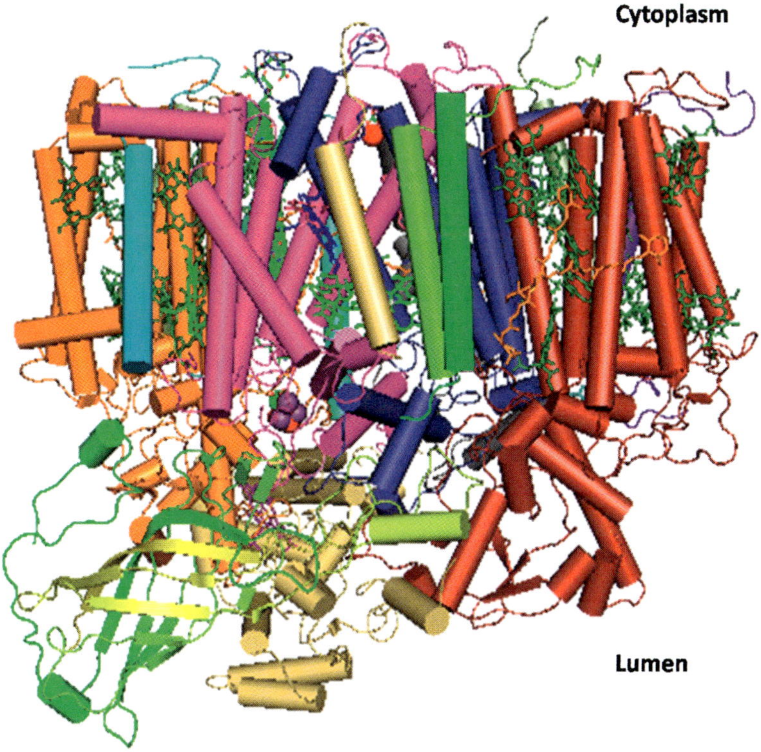

FIGURE 16.1 *View of the PS II complex. Helices are represented as cylinders with D1 in pink; D2 in violet; CP47 in firebrick red; CP43 in orange; Cyt b_{559} in grey, PsbH in purple; PsbI, PsbJ, and PsbK (barely visible) in cyan; PsbL, PsbM, PsbN, and PsbO in lime green; PsbT, PsbU, and PsbV in yellow orange and PsbX and PsbZ are scarcely discernible in the backdrop in green. The reaction center chlorophylls are shown in light green, chlorophylls of the antenna in forest green, pheophytins in blue, quinones in purple, β-carotenes in orange, non-heme iron is shown in red and the hemes in magenta. The oxygen atoms of the oxygen-evolving complex are shown as spheres of light red, the manganese ions in purple, and the calcium ion in lime green. (Figure based on Protein Data Bank coordinates 1S5L, Ferreira et al. (2004).)*

function (see Chapter 15). Surrounding the reaction center core and antenna are several low molecular weight membrane polypeptides that include the α and β subunits of cytochrome b_{559} (Cyt b_{559}) (reviewed in Ruffle and Sayre, 1998). Later crystallographic studies also demonstrated the presence of lipids within the PS II complex. The lipids are concentrated in the region between the reaction center and proximal antenna (Loll et al., 2005). Owing to the composition and location of the lipid molecules, they have been implicated in providing structural flexibility to the complex and aiding in the turnover of protein subunits (D1) during photooxidative stress.

The PS II reaction center core binds six chlorophylls (P_{D1}, P_{D2}, Chl_{D1}, Chl_{D2}, Chl_{zD1}, and Chl_{zD2}), two pheophytins ($Pheo_{D1}$ and $Pheo_{D2}$), two plastoquinones (Q_A and Q_B), two carotenoids and a non-heme iron (Figure 16.2).

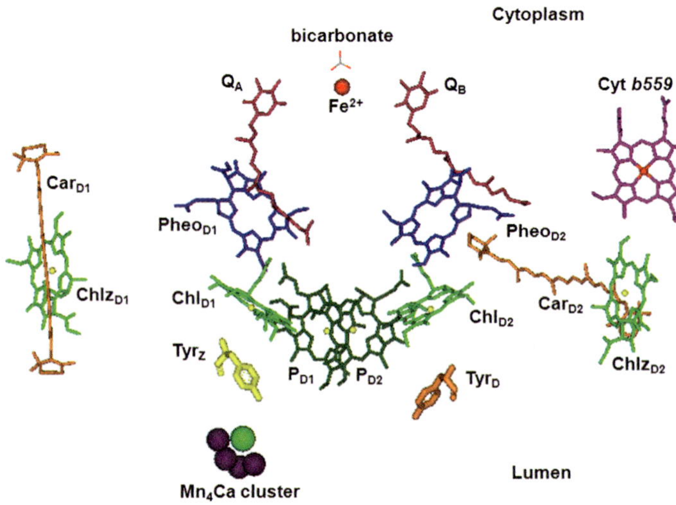

FIGURE 16.2 *The organization of electron transfer cofactors in cyanobacterial PS II reaction center crystals. The D1 subscript refers to the active branch pathway most closely associated with the D1 protein. The D2 subscript refers to the inactive branch pathway most closely associated with the D2 protein. (Figure based on Protein Data Bank coordinates 2AXT, Loll et al. (2005).)*

The cofactors are arranged in parallel C2 symmetry-related branches with nearly identical cofactor orientations and center-to-center distances between cofactors. Interestingly, only one of these branches (D1 side) is active in electron transfer. It is hypothesized that local asymmetries in protein–cofactor interactions account for the unidirectional electron transfer. Similar structural relationships exist in the PS I complex (Chapter 15).

C. D1 and D2 polypeptides

Originally, many polypeptides of the PS II reaction center were identified through comparative analyses of SDS-PAGE profiles from wild-type and PS II mutant thylakoid membranes isolated from *Chlamydomonas* (Chua and Bennoun, 1975). The D1 and D2 reaction center polypeptides were first resolved using a continuous urea gradient which separated the bands migrating at ~30 kD into two distinct species (Satoh et al., 1983). The D1 polypeptide (diffuse band 1), as it came to be known, was demonstrated to include the Q_B herbicide-binding site through radiolabeling studies using an azido derivative of atrazine as a photoaffinity label (Pfister et al., 1981), and was shown to be rapidly synthesized and turned over in illuminated chloroplasts (Grebanier et al., 1978; Satoh et al., 1983). Subsequently, it was shown that the 32-kD D1 protein was encoded by the chloroplast *psbA* gene (Zurawski et al., 1982). The gene encoding the D2 protein, *psbD*, was identified when an open reading frame was discovered in the chloroplast genome that was highly similar to the *psbA* gene of *Chlamydomonas* (Rochaix et al., 1984). The D1 and D2 polypeptides have 27% primary sequence identity and similar hydropathy plot profiles. Many of the conserved residues are clustered near the cofactor-binding sites suggesting that they share similar functions.

When the D1 and D2 amino acid sequences were revealed, their relatedness to the sequences of the L and M subunits of the *Rhodobacter sphaeroides* bacterial reaction center was quickly recognized (Hearst and Sauer, 1984). This observation became especially valuable when the type II bacterial reaction center from *Rhodopseudomonas viridis* was crystallized and its structure determined at 3Å and later 1.6-Å resolution (Deisenhofer et al., 1985; Michel and Deisenhofer, 1988). The quinone-type bacterial reaction center has four subunits; H (heavy), M (medium), L (light), and cytochrome c. The L and M subunits have five transmembrane helices each and form the core of the complex. They coordinate and/or ligate four bacteriochlorophylls, two bacteriopheophytins, two quinones, one non-heme iron, and one carotenoid. The H subunit has one membrane-spanning helix and a globular domain associated with the L/M heterodimer on the cytoplasmic side. The cytochrome subunit covalently binds four heme groups and is attached to the L/M complex on the periplasmic side. Upon photoexcitation, primary charge separation is initiated by the primary electron donor, a bacteriochlorophyll dimer. Electrons travel up the active branch through intermediate cofactors finally reducing the first stable electron acceptor, a quinone. The bacteriochlorophyll dimer is then reduced by cytochrome c.

The sequence similarity between the D1 and D2 subunits and the L and M polypeptides is most conserved in the environment surrounding the reaction center chlorophylls and the non-heme iron. This information in conjunction with hydropathy plots was used to predict the topology of the D1 and D2 polypeptides in PS II (Trebst, 1986a,b). The presence of five transmembrane units per chain was confirmed by mapping D1 and D2 proteolytic fragments identified by antibodies generated against synthetic peptides (Sayre et al., 1986; Trebst, 1986a). The isolation of a PS II reaction center from spinach established that the D1 and D2 proteins bound the chlorins and quinones of PS II (Nanba and Satoh, 1987).

Before the architecture of the PS II reaction center complex was known, a number of homology-based structural models for D1 and D2 had been proposed based on their similarities to the L and M subunits of the bacterial reaction center (Svensson et al., 1990, 1996; Roffey et al., 1991; Ruffle et al., 1992; Xiong et al., 1996). When the cyanobacterial PS II structure was solved, it was unambiguously confirmed that D1 and D2 formed a heterodimer with five transmembrane spans each, that coordinated or bound several electron transfer cofactors arranged in two parallel branches across the membrane.

Unlike the bacterial reaction center, PS II possesses a tetra-Mn cluster capable of catalyzing the oxidation of water, yielding molecular oxygen. In addition, symmetry-related and redox-active Tyr residues have been identified on the D1 (Tyr_Z) and D2 (Tyr_D) polypeptides. Other conspicuous differences between the bacterial reaction center and PS II reaction center polypeptides are the residues that ligate or which are located nearest to the Mg atom of the monomeric chlorophylls (Chl_{D1} and Chl_{D2} in the PS II reaction center), being histidine residues in the bacterial reaction center and D1-T179 and D2-I178 in PS II, and the lack of sequence similarity in the amino terminal regions of the L and M and D1 and D2 proteins (Svensson et al., 1990). Also, D1 and D2 bind two additional chlorophylls (Chl_{ZD1} and Chl_{ZD2}) located on their peripheries, which are coordinated by D1-H118 and D2-H117 (Hutchison et al., 1995).

D. P680, the primary donor

PS II is unique in its capability of catalyzing the oxidation of water to molecular oxygen. This process is driven by the high redox potential (1.2 E_V) of the primary donor of PS II, also known as P680. (The name is based on the primary donor chlorophyll's maximal Qy absorption peak.) The P680/P680$^+$ midpoint potential is approximately 370 mV more positive than the midpoint potential of free chlorophyll a in solution (0.8 E_V for Chl *a* in dimethylformamide). In all other reaction center types the midpoint potential of the primary donor is more negative than free Chl *a* in dimethylformamide. On the other hand, the six chlorins located in the core of the reaction center complex are nearly isoenergetic and are spectrally indistinguishable from each other (Ishikita and Knapp, 2005; Ishikita et al., 2005). It has been proposed that P680 is likely a multimer composed of a set of weakly coupled pigments including P_{D1}, P_{D2}, Chl_{D1}, Chl_{D2}, $Pheo_{D1}$, and $Pheo_{D2}$. In spite of the obvious homology between the two reaction centers, this model for the primary donor in the PS II reaction center differs substantially from that of the bacterial reaction center, in which it has been shown that the primary donor is the bacteriochlorophyll a special pair (Durrant et al., 1995). By and large the organization of P_{D1} and P_{D2} (Chl_{SP}) is comparable to the bacteriochlorophyll special pair of the bacterial reaction center, however the relative orientation and center-to-center distance between their head groups differs slightly from their bacterial counterparts, the distance being longer (8.2 Å) for the chlorophylls of the special pair in PS II (Figure 16.3; Ferreira et al., 2004). In comparison, the center-to center distance between the chlorophylls of the special pair in the bacterial reaction center is 7.5 Å. This difference may account for the weaker coupling between P_{D1} and P_{D2} and the monomeric properties exhibited by P680 (Barter et al., 2003).

The true identity of P680 is still unknown, however experimental data support two different hypotheses. Several lines of evidence suggest that P680 is the Chl_{D1} monomer including: (a) the lack of any significant red-shift in the Qy absorption band as expected for an energetically coupled chlorophyll dimer; (b) results from Stark effect and hole burning experiments that are not expected for a chlorophyll dimer; and (c) electron paramagnetic resonance (EPR) spectroscopic evidence for the localization of a

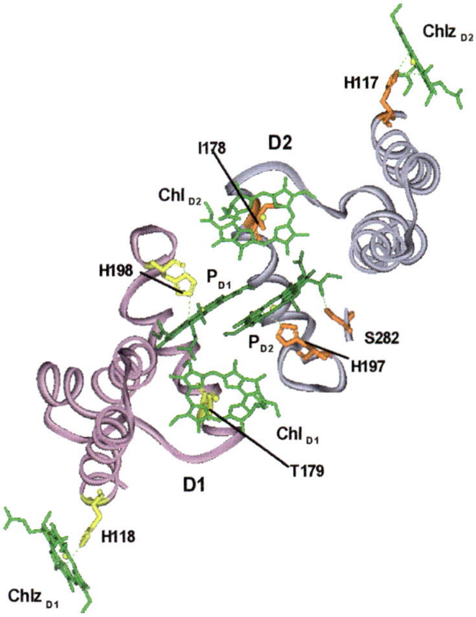

FIGURE 16.3 *The chlorophyll special pair and chlorophyll monomers and associated ligands and hydrogen bonding residues from the cyanobacterial PS II reaction center crystal structure (Ferreira et al., 2004). Note that the chlorophylls of the special pair (P_{D1} and P_{D2}) are both coordinated by histidine residues (D1-H198 and D2-H197) as in the bacterial reaction center. In contrast to the histidine ligands of the BRC, the PS II Chl_{D1} monomer is coordinated by a threonine (D1-T179). The Chl_{D2} monomer has D2-I178 located in the mirror image position relative to the Chl_{D1} ligand, D1-T179. A water molecule is thought to provide the fifth ligand to the Chl_{D2} Mg atom. The $Chlz_{D1\ and\ D2}$ cofactors are not part of the multimer complex.*

chlorophyll triplet on a chlorophyll monomer oriented 30° relative to the membrane plane normal (van Meighem et al., 1991; van Gorkom and Schelvis, 1993). This orientation is not consistent with the orientation of the chlorophylls of the special pair, whose ring axes are nearly parallel to the normal of the membrane plane. These geometries are, however, consistent with the primary donor being a chlorophyll monomer, presumably Chl_{D1} of the active branch pathway. Recent visible-pump, mid-infrared probe spectroscopy experiments on the femtosecond timescale have provided evidence that the reduction of pheophytin (primary electron acceptor) occurs on a 0.6- to 0.8-ps time scale, whereas $P680^+$ is generated after approximately 6 ps (Groot et al., 2005). Hence, the authors concluded that in the PS II reaction center, primary charge separation occurs between the chlorophyll monomer, Chl_{D1}, and the pheophytin on the "active" branch. If Chl_{D1} is the primary donor then Chl_{D1}^+ is presumably very rapidly reduced by P_{D1}. Thus, it is kinetically difficult to discriminate between Chl_{D1}^+ and P_{D1}^+ as the primary donor. Hypotheses suggesting that P680 has a dimeric nature originate from the observations that the histidine ligands for the chlorophyll of the special pair in PS II are analogous to the bacteriochlorophyll special pair ligands, and that selective excitation of the long wavelength pigments shows double the amplitude of the bleaching expected for a Chl monomer (Schelvis et al., 1994).

Mutagenesis that alters the redox potential of the chlorophyll dimer in PS II has been very useful in deciphering the organization of P680. The D1-H198 and D2-H197 residues are ligands to the chlorophylls of the Chl_{SP} in the PS II reaction center and are analogous to L-H173 and M-H200 of the bacterial reaction center that ligate the $BChl_{SP}$. Both these residues in the PS II reaction center were replaced by amino acids that introduced changes in the reduction potential of the special pair chlorophylls as well as a blue shift in their absorption spectrum, explained by the loss of the polarizable His ligand that presumably stabilizes the excited state (Diner et al., 2001). Substitution of the P_{D1} ligand with glutamine induced the largest blue shifts in the Chl Soret band and Qy transitions. The observed kinetics of charge separation between Q_A^- and the primary donor, as well as the kinetics of $P680^+$ reduction by Tyr_Z were altered in this mutant, suggesting modification in the E_M of $P680^+$. The kinetic and spectral effects of mutating the Mg axial ligand to P_{D2} were much less severe than those resulting from mutagenesis of the P_{D1} Mg axial ligand, suggesting that $P680^+$ cation is localized primarily on P_{D1} of the Chl_{SP}. The assignment of P680 to the P_{D1} Chl has also been supported by Fourier transform infrared spectroscopy analyses of the charge-delocalized state on the Chl_{SP} (Noguchi et al., 1993, 2001).

The substitution of the histidine ligands of P_{D1} and P_{D2} to non-coordinating residues like alanine and leucine were surprisingly unremarkable. For example, the replacement of D1-H198 and D1-H197 with Ala led to the formation of reaction centers that functionally resembled those of the wild type (Diner et al., 2001). However, in the case of substitutions with larger non-ligating residues like Leu and Tyr, the mutant reaction centers failed to assemble (Vermaas et al., 1988a; Diner et al., 2001). It was reasoned that a water molecule was providing the Mg axial ligand necessary for P_{D1} coordination in the mutant reaction centers with the substitution to Ala, which was displaced when the ligand was mutated to a larger non-ligating residue. A similar type of coordination was seen by water in the reaction centers of L-H173G and M-H200G mutants of *R. sphaeroides*. Substitution of the P_{D1} and P_{D2} axial ligands with Leu did not, however, corroborate results from analogous replacements in the bacterial reaction center, where the reaction center assembled but with the incorporation of bacteriopheophytin in the place of the bacteriochlorophylls of the special pair.

The first site-directed mutants of the *Chlamydomonas* chloroplast genome were designed to determine the protein–cofactor interactions that affect the high midpoint potential of the $P680^+/P680$ couple (Roffey et al., 1994a; Hutchison et al., 1996a). Mutation of D1-H195 to an asparagine

(D1-H195N) or tyrosine (D1-H195Y) had no effect on oxygen evolution or kinetics of $P680^+$ reduction. However, D1-H195D mutants exhibited 100-fold slower $P680^+$ reduction kinetics, suggesting that the introduction of a negatively charged residue in the vicinity of the Tyr_Z and P680 shifted the redox potential of the $Tyr_Z^{\bullet}/Tyr_Z$ or $P680^+/P680$ couple resulting in slower rates of charge transfer (Roffey et al., 1994a,b).

Computational studies have addressed the same issue in which the E_M^{ox} of the PS II reaction center chlorophylls were calculated using the 3.2- and 3.5-Å crystal structure coordinates via a linearized Poisson-Boltzmann approach (Ishikita and Knapp 2005; Ishikita et al., 2005). These analyses revealed that the calculated midpoint potentials of the four chlorophylls participating in primary charge separation are almost identical, lending support to the multimer model for charge separation. The upshifted redox potentials of the PS II reaction center chlorophylls can by and large be attributed to multiple cofactor–protein and cofactor–cofactor interactions, and it is conceivable that any of these four chlorophylls (P_{D1}, P_{D2}, Chl_{D1} or Chl_{D2}) could initiate charge separation following excitation. Specific amino acid residues, including ligands to P_{D1} and P_{D2} and the D2-R180 residue, have been suggested by Ishikita et al. to have an influence on the redox potential of select cofactors, and through experimentation have been shown to substantially affect the quantum yield and kinetics of charge separation in PS II reaction centers. It should be noted, however, that mutagenesis of other amino acid residues in D1 and D2 can also affect the E_M of P680.

E. Pheophytin

It was originally thought that in quinone-type reaction centers the primary electron acceptor of PS II was a bound plastoquinone. However, in the late 1970s it was discovered that the principal photochemical event is marked by electron transfer from the excited primary donor to $Pheo_{D1}$, located on the active branch (Klimov et al., 1977; Klimov and Krasnovskii, 1981; Klimov, 2003). Electron transfer occurs on a picosecond time scale, however, the rate of primary charge transfer has been the topic of much debate and ongoing research. In isolated bacterial reaction center preparations, primary charge separation occurs in 2.8 ps at room temperature and 0.7–1.2 ps at 10 K (Woodbury et al., 1985; Martin et al., 1986). Previously it was shown using 0.5-ps time resolution that primary charge separation occurs with a time constant of 3.0 ± 0.6 ps at 277 K (Wasielewski et al., 1989). Formation of the charge-separated state ($P680^+$-$Pheo_{D1}^-$) was monitored by transient absorption spectroscopy between wavelengths of 450 and 840 nm following excitation with a 500 fs pulse at 610 nm, and was characterized by an increase in absorption at 820 nm due to generation of $P680^+$ as well as absorbance changes (Q_X band) due to the formation of the pheophytin anion. The value of the time constant for the appearance of the charge-separated state was comparable to the time constant for the decay of the lowest singlet excited state of P680. A different rate constant of 21 ps for primary charge separation was interpreted from similar ultrafast experiments carried out by Durrant et al. (1992, 1993). This rate constant for primary charge separation was based on the observed rate of bleaching of the pheophytin ground state at 545 nm, the appearance of the pheophytin anion at 460 nm, and the loss of stimulated emission from Chl proposed to be caused by the formation of the oxidized primary donor. The kinetics of charge separation at many wavelengths and especially in the 545 nm region attributed to the Q_X band of pheophytin are multiphasic in nature and can be exponentially fit to yield three lifetime components of 3 ps, 20 ps, and between 50ps and 100 ps (Greenfield et al., 1997). Additional slower (50 ps) lifetime components have been assigned to charge separation that is limited by energy transfer processes ("equilibration").

Biochemical analyses as well as evidence from crystal structures reveal that PS II has two pheophytins per reaction center and these can be distinguished on the basis of their optical properties. The active-branch pheophytin has an absorption peak centered at 676 nm, whereas the inactive-branch pheophytin contributes to the peak at 680 nm (Breton, 1990; Tang et al., 1990; van Kan et al., 1990; van Gorkom and Shelvis, 1993). The two pheophytins are related pseudo-symmetrically with subtle differences in their hydrogen bonding interactions between the Pheo head groups and the surrounding protein (Figure 16.4). In higher plants and *Chlamydomonas* there is a strong hydrogen bond between the ring V carbonyl group of the

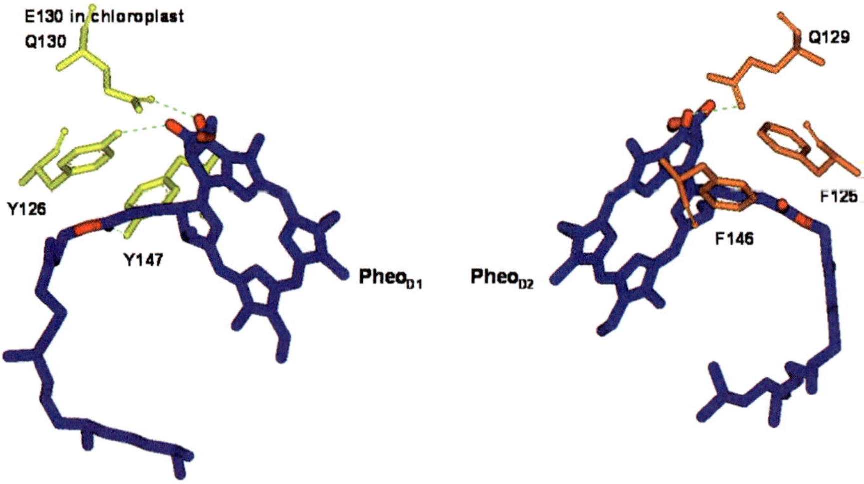

FIGURE 16.4 *Amino acid residues present in the near protein environment (cyanobacterial model) of the primary acceptor $Pheo_{D1}$ of the active branch and the mirror image cofactor, $Pheo_{D2}$ of the inactive branch. Note residue D1-Q130 is a glutamate residue in the chloroplast PS II complex. (Figure based on Protein Data Bank coordinates 1S5L, Ferreira et al. (2004).)*

active-branch $Pheo_{D1}$ and D1-E130. The equivalent residue on the inactive-branch $Pheo_{D2}$ is a glutamine. Two additional putative H-bonds may be provided by D1-Y126 and D1-Y147, one to the $Pheo_{D1}$ head group and the other to the phytyl tail. $Pheo_{D2}$, however, lacks these H-bonds. The analogous residues on the inactive-branch $Pheo_{D2}$ are phenylalanines (Ferreira et al., 2004). The homology between the D1-130 residue of PS II and L-104 residue of the bacterial reaction center as well as the non-covalent interactions of $BPheo_L$ and $Pheo_{D1}$ with the surrounding protein environment is striking and was originally realized from sequence alignments between the D1 and the bacterial reaction center L subunit (Hearst and Sauer, 1984; Deisenhofer and Michel, 1989). Results from several spectroscopic studies also have suggested that in cyanobacterial PS II the photoreducible $Pheo_{D1}$ is hydrogen bonded via its 9-keto group to D1-130, but a glutamine residue in cyanobacterial PS II reaction centers and a glutamic acid residue in chloroplast PS II reaction centers (Moënne-Loccoz et al., 1989). This hydrogen-bonding interaction is similar to the hydrogen bonding between the keto-carbonyl group of ring V of $BPheo_{active}$ and L-E104 of the *R. viridis* bacterial reaction center. Structural studies estimate the distance between the oxygen atom of the 9-keto group and the nitrogen of D1-Q130 (cyanobacterial residue) at 3.2 Å, consistent with a weak hydrogen bond (Ferreira et al., 2004). High field EPR studies of the radical anion of $Pheo_{D1}$ from *Chlamydomonas* D1-E130 mutants (Q, H, and L) showed a strong effect of the hydrogen bond strength on the *g* tensor of the radical (Dorlet et al., 2001). The D1-E130Q mutant showed a shift in the g_x value as compared to wild type, consistent with a weaker hydrogen bond, and the D1-E130L mutant showed a loss of this bond as well as a near-complete loss in the ability to evolve oxygen. Weakening of hydrogen bond strength was implicated in a decrease in the redox potential of $Pheo^-$ and was correlated with a reduced rate of oxygen evolution in these mutants.

Previous experiments using PS II reaction centers from wild-type chloroplasts, and analogous *Synechocystis* sp. PCC 6803 mutants, showed that the quantum yield for radical pair ($P680^+$ $Pheo^-$) formation was highest in PS II complexes having a D1-E130 hydrogen bond to $Pheo_{D1}$, followed by the D1-E130H and D1-Q130L mutants (Giorgi et al., 1996). In cyanobacteria, the D1-Q130E mutation caused a red shift in the Pheo Q_x band from 541.5 nm to 544 nm, while the time constant for P680 to $Pheo_{D1}$ electron transfer was unaffected. An 20–30% increase in quantum yield for the formation of $P680^+$ $Pheo^-$ in the cyanobacterial D1-Q130E mutant mimics a naturally occurring interconversion between the low light and high light versions of the D1 polypeptide in *Synchechococcus*, in which the quantum yield for charge separation is increased by switching expression from a D1 protein containing Q130 (in the low light form) to one containing E130 residue (in the high light form; Clarke et al., 1993).

In PS II it is assumed that localized differences in the protein environment surrounding the cofactors determine the directionality of electron transfer. *Chlamydomonas* PS II reaction centers in which the pheophytin on the inactive branch ($Pheo_{D2}$) had been replaced by a chlorophyll were used to characterize the factors affecting energy distribution among the core pigment molecules (Xiong et al., 2004). A D1 mutation substituting histidine for leucine (D1-L210H) provided a ligand for the Mg ion of chlorophyll. Unexpectedly, a substitution of a chlorophyll for a pheophytin on the inactive branch resulted in mutant PS II complexes deficient in oxygen evolution. In addition, reductions in the variable component of chlorophyll fluorescence decay (following a flash) and an inability to photoaccumulate reduced Q_A indicated that electron transfer in D1-L210H PS II complexes was severely perturbed. The rate constant for primary charge separation ($P680^+/Pheo^-$), however, was not significantly affected although its amplitude was substantially reduced. These results suggested that the energetics of the primary electron donor and acceptor pair were not substantially changed in the mutant but that the quantum yield of charge separation was dramatically reduced.

Circular dichroism analyses of pigment–pigment interactions in D1-L210H demonstrated a loss of the ellipticity associated with P680, and the emergence of new energy coupling bands attributed to interactions between the inactive branch chlorophyll monomer and the chlorophyll occupying the site normally occupied by the inactive branch pheophytin (Xiong et al., 2004). These observations were attributed to a redistribution of the excited state energy among the reaction center chromophores to the strongly H-bonded chlorophyll occupying the inactive branch pheophytin site. The energy trapped by this new chlorophyll was ultimately dissipated via chlorophyll fluorescence. Since the new chlorophyll occupying the $Pheo_{D2}$ site has a strong hydrogen bond (D2-Q129) unlike any other chlorophylls of the reaction center, it is not unexpected that this chlorophyll would serve as an energy trap.

F. Q_A and Q_B, the stable electron acceptor of photosystem II and the herbicide-binding site

During light-driven charge separation a plastoquinone molecule in the Q_A site, which is usually capable of accepting only a single electron, accepts an electron from $Pheo_{D1}$. Due to the higher midpoint potential of plastoquinone in the Q_A site than $Pheo^-$ the charge-separated state is stabilized, reducing the probability of a back reaction between Q_A^- and $P680^+$. Under high light conditions, however, Q_A can accept two electrons and become doubly reduced, prompting the turnover of D1 (Keren et al., 1995). Electron transfer from Q_A to Q_B is considered to be the rate-limiting step with a time constant of 100 microseconds for the initial electron transfer from Q_A^- to Q_B, and a slower rate of 200 microseconds for the transfer from Q_A^- to Q_B^- that is likely due to electrostatic effects. The architecture of the Q_A site is known from the cyanobacterial crystal structure and includes mainly hydrophobic

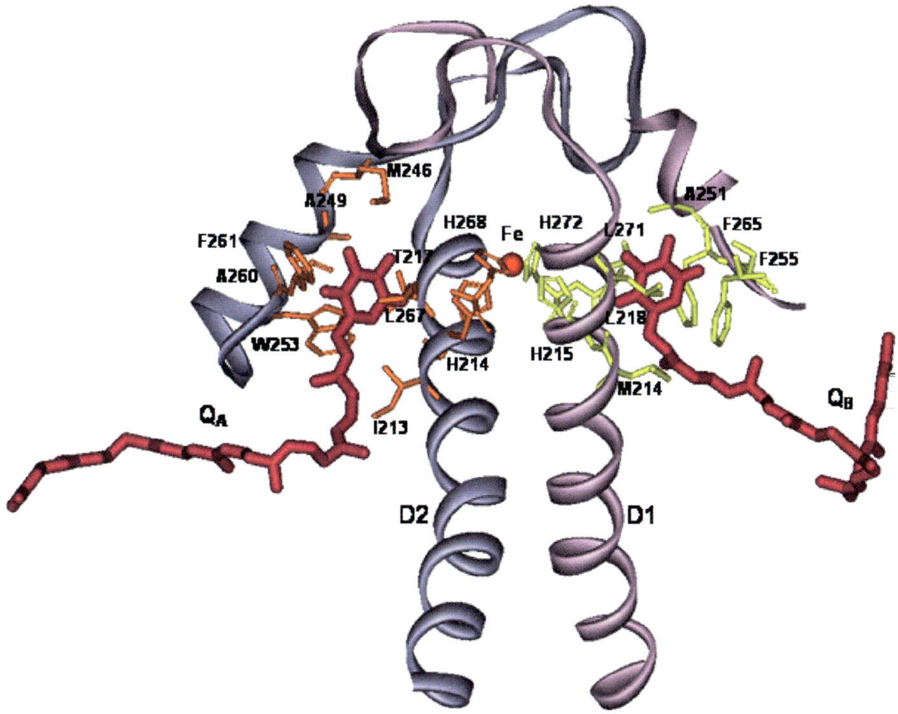

FIGURE 16.5 *View of the non-heme iron and the Q_A- and Q_B-binding pockets. The non-heme iron is shown as a red sphere. The four histidines D1-H215, D1-H272 (yellow) and D2-H214, D2-H268 (orange) are involved in coordinating the non-heme iron. The two quinones Q_A and Q_B are shown in magenta with the hydrophobic residues lining the binding pockets in orange and yellow respectively. (Figure based on Protein Data Bank coordinates 1S5L, Ferreira et al. (2004).)*

residues of the D2 polypeptide (Figure 16.5; Ferreira et al., 2004). Although PS II and the bacterial reaction center have different bound quinone molecules, the binding pockets are structurally similar. Residues that are conserved in the Q_A site in PS II and the bacterial reaction center include the PS II D2-T217, D2-F252, D2-W253, and D2-H214 residues. These residues correspond in the bacterial reaction center to M-T220, M-F249, M-W250, and M-H217. Crystallographic data suggests that in PS II, the distance between Q_A and the non-heme iron is approximately 9 Å (Ferreira et al., 2004; Loll et al., 2005). Earlier studies involving EPR measurements of the Q_A^-/Fe(II) signal indicated that Q_A^- was also in close (7 Å) proximity to the non-heme iron (Nugent et al., 1981). The non-heme iron is coordinated by four histidine residues, two each from the D1 and D2 proteins.

The Q_B site in PS II and bacterial reaction centers is less conserved when compared with the Q_A site of the two reaction center types. Several quinone hydrogen bonding residues are conserved, however. D1-H215 and D1-S264 in the PS II reaction center are hydrogen bonded to Q_B and are conserved in the bacterial reaction center, the analogous residues being

L-H190 and L-S223. A superimposition of the C_α atoms of the residues comprising the Q_B site in the two types of reaction centers shows an average displacement of 3.5Å, attributed to the slightly larger hydrophobic-binding pocket in PS II than in the bacterial reaction center (Ferreira et al., 2004).

In addition to plastoquinone, the PS II Q_B-binding pocket can bind the herbicides atrazine, DCMU, bromacil, and metribuzin. Spontaneous or chemically induced herbicide resistant mutations have been shown to be clustered in the Q_B-binding cavity spanning residues 211–275 of D1 (reviewed in Diner et al., 1991). Herbicide resistance-conferring mutations have an energetic cost, however. The D1-S264A mutant exhibits increased tolerance toward metribuzin and atrazine but $Q_A^- \rightarrow Q_B$ and $Q_A^- \rightarrow Q_B^-$ electron transfer rates are slower than in the wild type. The atrazine resistance-conferring D1-G256D mutant also exhibits slower $Q_A^- \rightarrow Q_B^-$ but not $Q_A^- \rightarrow Q_B$ electron transfer, implying varying effects of the two mutations on Q_B protontation (Crofts et al., 1993). Other site-specific mutations in the Q_B-binding pocket such as D1-A251V produced phenotypes resistant to atrazine but with slower $Q_A^- \rightarrow Q_B$ electron transfer rates, whereas D1-V219I, D1-F255Y, and D1-L275F herbicide tolerant mutants showed only modest changes in $Q_A^- \rightarrow Q_B$ and $Q_A^- \rightarrow Q_B^-$ electron transfer rates (Johanningmeier et al., 1987; Erickson et al., 1989; Govindjee et al., 1991; Strasser et al., 1992).

Mutagenesis of the *Chlamydomonas psbA* gene has also led to the identification of D1 residues involved in herbicide binding. Double (D1-N266T/D1-I259S) and triple (D1-N266T/D1-S264A/D1-I259S) mutants with reduced metribuzin sensitivity have been identified (Przibilla et al., 1991). The triple mutant had lower resistance to metribuzin than D1-S264A but higher tolerance to phemediphan. Replacements of the Q_B-binding site residue, D1-A251, with 12 different amino acids resulted in mutants that were unable to carry out photosynthesis as well some that were resistant to herbicides (Lardans et al., 1997). Five D1-A251 mutants (Arg, Asp, Gln, Glu, and His) were incapable of photosynthesis. Of these, D1-A251R was able to synthesize a less stable but full-length 32-kD D1 protein, whereas the other four mutants synthesized shorter (24–25 kD) and unstable D1 polypeptides. The substitution of D1-A251 (part of the loop region between transmembrane helices IV and V) with valine conferred herbicide resistance in both cyanobacteria and *Chlamydomonas*. In cyanobacteria, however, mutations and/or deletions in this loop region do not lead to the loss of photosynthetic competence in contrast to the corresponding *Chlamydomonas* mutants, indicating that there are intrinsic differences between the cyanobacterial and chloroplast PS II reaction centers.

The non-heme iron is located between Q_A and Q_B and is involved in mediating electron flow. Analogous to the bacterial reaction center it is coordinated by four histidines, D1-H215, D1-H272, D2-H214, and D2-H268 (Figure 16.5). In contrast to the bacterial reaction center, where a glutamate residue provides a fifth Fe ligand, in PS II reaction centers bicarbonate apparently serves the same function (Ferreira et al., 2004). In addition, it has been suggested that bicarbonate ions form hydrogen bonds with neighboring amino acid residues of the D1 and D2 polypeptides at physiological pH, helping to stabilize the complex and aid in electron transfer (reviewed in van Rensen et al., 1999). The displacement of bicarbonate by formate decreases the kinetics of Q_B reduction by 100-fold. The formate/bicarbonate effect has been studied by monitoring EPR spectra of the Q_A-Fe-Q_B complex (Bowden et al., 1991). Bicarbonate has also been implicated in the protonation of Q_B^-, since the kinetics for oxidation of Q_A^- by Q_B is less affected by the absence of bicarbonate than the kinetics of Q_B^- protonation (Xu et al., 1991). Apart from mediating electron transfer from Q_A to Q_B, the non-heme Fe also aids in the stabilization of Q_A^- and the PS II structure (Dutton et al., 1978).

Prior to the localization of the bicarbonate-binding site by crystallography, attempts were made to localize it by site-directed mutagenesis. Positively charged residues located near the non-heme Fe were targeted for mutagenesis. One such residue is D1-R269, which is towards the stromal surface of D1. A non-conservative glycine mutation was engineered into *Chlamydomonas* (Hutchison et al., 1996b). The D1-R269G mutant was unable to evolve oxygen with dichlorobenzoquinone as the external electron acceptor, had an increased susceptibility to photoinhibition or light-dependent turnover of the D1 protein, and lacked manganese associated with the water-splitting apparatus. In dark-grown cells the Tyr_D and Q_A^-/Fe(II) EPR signals were only 30% that of wild-type cells indicating reduced levels of functional PS II complexes in the mutant. In addition, Chl *a* fluorescence decay kinetics revealed a blockage in electron transfer from Q_A to Q_B consistent with the loss of bicarbonate. However, contradictory evidence – that this mutant was capable of binding bicarbonate – was obtained from the formate-enhanced Q_A^-/Fe(II) EPR signal whose amplitude was consistent with the predicted levels of functional PS II complexes in the membrane. It was apparent from these studies, however, that mutations on the stromal side of the PS II complex could affect charge transfer processes on the lumenal side of the membrane, possibly due to a reduced ability to stabilize charge-separated states associated with the assembly of the water-splitting complex.

G. Redox-active tyrosines, the tetra-manganese complex and water oxidation

The catalyst for water oxidation is the charge-accumulating, tetra-Mn complex which is bound to the D1 and CP43 proteins. The complex is stabilized by chloride, calcium and the extrinsic 33-, 23-, and 17-kD OEE proteins (Mayfield et al., 1987a,b, 1989; Bricker and Ghanotakis, 1996). The driving force for water oxidation is $P680^+$ which oxidizes Tyr_Z, which in turn oxidizes the tetra-Mn complex. The oxidation of two molecules of water requires four quantum events which drive the S-state complex through five redox transitions ($S_0 \rightarrow S_4$) beginning with the dark-stable S_1 state (Kok et al., 1970). In addition to Mn, the inorganic cofactors calcium and chloride are required for water oxidation. PS II particles contain both low ($K_D = 2\,mM$) and high (K_D = 100 micromolar) affinity calcium-binding sites. Depletion of calcium from PS II particles blocks oxygen evolution and results in the loss of the normal $g = 2$ multiline EPR signal and its replacement by a unique dark stable multiline spectrum (Dorlet et al., 1999). Calcium and chloride have also been shown to be required for the photoactivation or the light-dependent assembly of the tetra-Mn complex (Homann, 1988). Extended X-ray absorption fine structure and Fourier transform infrared spectroscopy experiments indicated that calcium is in close proximity to manganese (Noguchi et al., 1993), results that were later confirmed by X-ray diffraction. Chloride depletion also inhibits oxygen evolution. Unlike calcium, however, chloride may be replaced by other anions such as bromide, though oxygen evolution becomes less effective (Homann, 1988). A modified S_2 state can be formed in chloride-depleted PS II complexes, but it does not advance without re-addition of chloride (Sandusky and Yocum, 1986; Homann, 1988). The location of chloride in the crystal structure has not been determined.

The tetra-Mn complex requires at least 32 ligands if it is octahedrally coordinated. Mutagenesis and biophysical studies have led to the identification of several Mn ligands. At least 12 residues located on the lumenal side of the D1 protein were implicated as possible Mn ligands prior to the resolution of the crystal structure including D59, D61, E65, D170, E189, H190, H331, E332, H336, D342, and the C-terminal A344 (Figure 16.6;

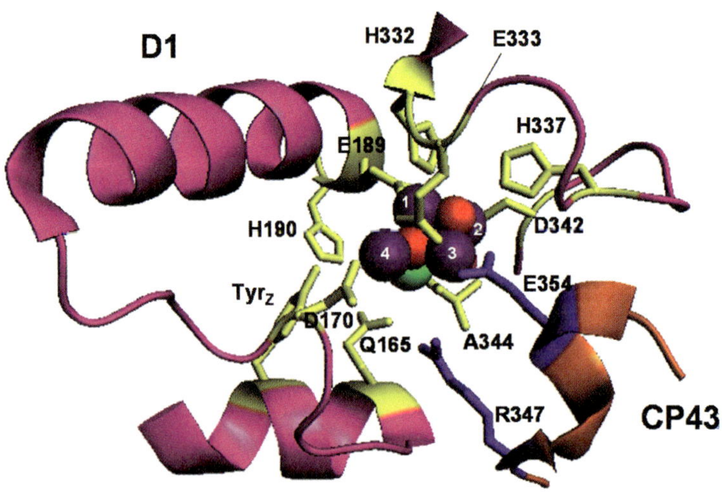

FIGURE 16.6 *The oxygen-evolving center shown with its side-chain ligands. In the Mn_4-Ca metal ion cluster, Mn is shown in purple, Ca in green and the oxygen atoms as red spheres. Side-chain ligands of the D1 polypeptide (light magenta) are shown as yellow sticks and those of the CP43 chain (orange) in deep purple. Figure based on Protein Data Bank coordinates 1S5L, (Ferreira et al., 2004) and numbering of the Mn atoms is based on the crystal structure from Loll et al. (2005).*

Diner et al., 1991; Debus, 1992; Nixon and Diner, 1992; Roffey et al., 1994b; Whitelegge et al., 1995). A D2 residue, E69, had also been implicated as a possible Mn ligand based on mutagenesis studies (Vermaas et al., 1990). With the subsequent resolution of the 3.0-Å PS II crystal structure it became evident that nearly all of the aforementioned site-specific PS II mutations were located near the tetra-Mn-Ca cluster with the exception of D2-E69. In addition to the D1 tetraMn-Ca complex ligands, the CP43-E354 residue is proposed to provide a Mn ligand or function in water oxidation near the Mn4-binding site. Mn4, presumably the first Mn assembled in the cluster and the one most susceptible to X-ray radiation damage, is coordinated by D1-D170 and D1-E333 (Zouni et al., 2001). Mn3 is coordinated by D1-E333 and CP43-E354. Additional D1 residues in the first coordination sphere include D342, E189, and H332, which coordinate Mn1, and D342, A344, and CP43-D354, which coordinate Mn2. Possible calcium ligands include Tyr_Z, A344, and E189 of the D1 protein.

Several models have been proposed for water oxidation by the tetraMn-Ca complex. One of the more attractive models that accommodates both structural and biophysical data is the metallo-radical model proposed by Babcock and Britt (Babcock et al., 1989; Britt, 1996). According to this model, the oxidation of Tyr_Z by $P680^+$ generates a neutral tyrosine radical concomitant with the exchange of a proton between Tyr_Z and D1-H190. The D1-H190 residue functions as a weak base which shuttles a proton from Tyr_Z to the thylakoid lumen (Tang et al., 1994; Britt, 1996; Babcock et al., 1989). The stepwise reduction and protonation of oxidized Tyr_Z generates higher S- (oxidation) states of the tetra-Mn complex leading to the oxidation of water, presumably located near the calcium-binding site and Mn4. Support for this model came from EPR studies that demonstrated that oxidized Tyr_Z was a neutral radical, and from mutagenesis of D1-H190 in *Chlamydomonas* and later cyanobacteria that demonstrated its critical role in water oxidation. *Chlamydomonas* D1-H190 mutants are unable to evolve oxygen and cannot assemble a tetra-Mn cluster (Roffey et al., 1994b). The D1-H190 mutants also were unable to generate the A_T thermoluminescence band attributed to oxidation of Tyr_Z (Kramer et al., 1994). Structural homology models had predicted that D1-H190 was located near Tyr_Z, however D1-H190 mutations had no effect on the $Tyr_Z^{\bullet}$ EPR hyperfine structure. These results were in contrast to those observed with the symmetry-related D2-H189 residue. Non-conservative mutations of D2-H189 altered the EPR-detectable hyperfine structure of Tyr_D associated with the loss of the strong hydrogen bond (Vermaas et al., 1988b; Tommos et al., 1993; Tang et al., 1993, 1996). These results indicated that the H-bond distance between D1-H190 and Tyr_Z is greater than the H-bond distance between D2-H189 and $Tyr_{D.}$ This prediction was later supported by the crystal structure. Subsequently, it was shown that Tyr_Z oxidation and proton release was a concerted reaction in wild-type PS II complexes but not in D1-H190 mutants (Hays et al., 1999).

Later, an alternative metallo-radical water oxidation model was proposed (McEvoy et al., 2005). According to this model D1-H190 does not participate in proton shunting to the lumen but acts as a proton rocking residue, accepting a proton from Tyr_Z with its oxidation and donating it back to $Tyr_Z^{\bullet}$ during its reduction. This proton rocking mechanism allows Tyr_Z to maintain a favorable redox potential for oxidation of the tetra-Mn complex. Substrate water is bound both to Mn4 as well as coordinated by the calcium ion. Beyond the S_2 state, the water coordinated by the Mn is twice oxidized by proton-coupled electron transfer, followed by a nucleophilic attack by the calcium-bound water on the $Mn4^V$ oxo-oxygen producing molecular oxygen. The CP43-R357 residue is proposed to play an analogous role to the Tyr_Z residue, functioning as a deprotonated base catalyst. The argument favoring the CP43-R357 residue as the base catalyst for proton extraction from water is the location of this residue near the Ca-binding site, the apparent proton channel leading to the lumen, and the lack of an apparent proton channel leading from Tyr_Z to the lumen.

Mutagenesis of CP43-R357 to a serine impairs but does not block oxygen evolution, and is associated with an 11-fold larger Kok double-miss parameter and accelerated decay of advanced S-states. Unlike wild-type thylakoids, however, oxygen evolution in CP43-R357S is insensitive to bicarbonate depletion. Based on these observations Ananyev et al. (2005) suggested that CP43-R357 binds bicarbonate, which is required for oxygen evolution. It should be noted, however, that the magnitude of the bicarbonate effect on oxygen evolution in wild-type membranes is less than the magnitude of the effect of the CP43-R357S mutation. Thus, the CP43-R357S mutation may mask the effect of bicarbonate depletion in this mutant. The CP43-R357S mutant phenotype also has implications for the CP43-R357 base catalyst, metallo-radical water oxidation model. The non-conservative CP43-R357S mutant would not be expected to function as a base catalyst for water oxidation. Each of the models for water oxidation will require further investigation and possible modification to account for the structural and functional observations.

H. The Chl_Z cycle

Thompson and Brudvig (1988) first proposed that a chlorophyll monomer known as Chl_Z could reduce $P680^+$ with low quantum yield. The Chl_Z cation is re-reduced by Cyt b_{559}, which in turn is reduced by Q_B. In addition a bound PS II carotenoid participates in the Chl_Z electron transfer cycle around PS II (Vrettos et al., 1999). The Chl_Z cycle has been proposed to reduce photoinhibition or the light-dependent turnover of D1. Consistent with this idea, it has been shown that Chl_Z^+ cation is a potent quencher of chlorophyll fluorescence in PS II (Schweitzer and Brudvig, 1997; Schweitzer et al., 1998). The peripheral chlorophylls coordinated by D1-H118 and D2-H117 were proposed to be the redox-active chlorophylls (Chl_Z) based on the measured distance between Chl_Z^+ and the non-heme Fe (II) in PS II (Koulougliotis et al., 1994; Stewart et al., 1998). Significantly, the proposed chlorophyll ligands (D1-H118 and D2-H117 in *Chlamydomonas* PS II) are not conserved in the analogous L and M subunits of the bacterial reaction center. Thus, this pair of conserved and C_2 symmetry-related histidine residues was proposed to coordinate the extra pair of the peripheral chlorophyll (Schelvis et al., 1994). Later, it was demonstrated by site-directed mutagenesis that D1-H118 and D2-H117 indeed coordinated two additional chlorophylls not present in the bacterial reaction center (Hutchison and Sayre, 1995; Lince and Vermaas, 1998; Stewart et al., 1998). This assignment was confirmed by the high resolution PS II crystal structure.

The location and identity of Chl_Z is of particular interest because of its potential involvement in energy funneling and participation in the cyclic electron transfer pathway around PS II. In principle, both of the symmetry-related peripheral Chls could be photooxidizable. However, EPR studies indicate that only one Chl_Z cation is accumulated per $Y_D^{\bullet}$ (Schweitzer et al., 1998). Therefore, it is likely that only one of the two peripheral accessory chlorophylls participates in the electron transfer cycle around PS II. Stewart et al. (1998) proposed that the chlorophyll coordinated by D1-H118 is the Chl_Z species. This conclusion was based on the observation that the Chl_Z^+ resonance Raman spectrum was altered in D1-H118 mutants but not in D2-H117 mutants. The PS II model structures (Xiong et al., 1996) indicated, however, that the chlorophyll coordinated by the D2-H117 residue was closer (25 Å) to Q_B and Cyt b_{559} and that the D1-carotenoid was most closely associated with the chlorophyll special pair. Since Q_B indirectly reduces Chl_Z, it is to be expected that the Chl_Z peripheral chlorophyll would be nearer to Q_B than to Q_A. Consistent with this speculation, Shigemori et al. (1998) demonstrated that the distance between $Y_D^{\bullet}$ and the Chl_Z cation (29 Å) was consistent with the coordination of Chl_{ZD2} by the D2-H117. Moreover, the efficient oxidation of Cyt b_{559} by Chl_Z suggests that Chl_Z is in close proximity to the heme of Cyt b_{559} (de Paula et al, 1985; Thompson and Brudvig, 1988). Finally, Hutchison and Sayre (1995) and Wang et al. (2002) substituted the symmetrically located D1-H118 and D2-H117 with conservative amino acids (D1-H118Q, D2-H117N, and D2-H117Q in *Chlamydomonas*). They demonstrated that D2-H117 mutants had a reduced capacity to quench chlorophyll fluorescence (F695). This was interpreted as being due to reduced Chl_Z^+ levels relative to wild-type and D1-H118Q mutant PS II complexes. Finally, the D2-H117 mutants had reduced light-harvesting efficiency relative to wild-type and the D1-H118Q mutant (Johnston et al., 2000; Ruffle et al., 2001). As a result, the D2-H117 mutants were less sensitive to photoinhibitory light treatments than wild-type and the D1-H118Q mutant. In summary, the accumulated structural and biophysical studies support the assignment of the Chl_{ZD2} pigment as the chlorophyll monomer involved in the Chl_Z cycle.

III. BEYOND THE PHOTOSYSTEM II CRYSTAL STRUCTURE

While the cyanobacterial PS II crystal structures have provided profound insights, the extrinsic subunits of the water-splitting apparatus and the proximal antenna complexes, in particular, differ substantially from those of chloroplast PS II complexes. So is it important to study chloroplast PS II complexes and if so, why? The thermodynamically feasible increases in photosynthetic efficiency achievable by better regulation of light-harvesting efficiency and associated energy dissipation at high light intensities, reductions in PS II-specific photoinhibition, and the enhanced quantum efficiencies of chloroplast PS II complexes relative to cyanobacteria all have practical implications for improving chloroplastic PS II efficiency and in turn crop and biofuel yields. But there also remain major basic scientific questions regarding PS II structure and function which have not been resolved. How is water split? What is the chemistry of charge accumulation in the tetra-Mn water-splitting apparatus? What is the function of the inactive parallel electron transfer branch in PS II? Does it provide protection against over excited states? What protein–cofactor interactions account for the high midpoint potential of the primary donor, P680, in PS II and how is energy trapped by the primary donor? The ability to engineer efficiently the chloroplast genome of *Chlamydomonas* allows investigators working in the "inner sanctum" of photosynthesis to address the most fundamental questions that make possible the harnessing of solar energy into biologically useful molecules.

REFERENCES

Alizadeh, S., Nixon, P.J., Telfer, A., and Barber, J. (1995). Isolation and characterisation of the photosystem two reaction centre complex from a double mutant of *Chlamydomonas reinhardtii*. *Photosynth. Res.* **43**, 165–171.

Ananyev, G., Nguyen, T., Putnam Evans, C., and Dismukes, G.C. (2005). Mutagenesis of CP43-arginine-357 to serine reveals new evidence for (bi)carbonate functioning in the water oxidizing complex of Photosystem II. *Photochem. Photobiol. Sci.* **4**, 991–998.

Babcock, G.T., Barry, B.A., Debus, R.J., Hoganson, C.W., Atamian, M., McIntosh, L., Sithole, I., and Yocum, C.F. (1989). Water oxidation in photosystem II: From radical chemistry to multi-electron chemistry. *Biochemistry* **28**, 9557–9565.

Barry, B.A. and Babcock, G.T. (1987). Tyrosine radicals are involved in the photosynthetic oxygen-evolving system. *Proc. Natl. Acad. Sci. U. S. A.* **84**, 7099–7103.

Barter, L.M.C., Durrant, J.R., and Klug, D.R. (2003). A quantitative structure-function relationship for the Photosystem II reaction center: Supermolecular behaviour in natural photosynthesis. *Proc. Natl. Acad. Sci. U. S. A.* **100**, 946–951.

Berthold, D.A., Babcock, G.T., and Yocum, C.F. (1981). A highly resolved, oxygen-evolving PS II preparation from spinach thylakoid membranes. *FEBS. Lett.* **134**, 231–234.

Biesiadka, J., Loll, B., Kern, J., Irrgang, K.-D., and Zouni, A. (2004). Crystal structure of cyanobacterial photosystem II at 3.2Å resolution: A closer look at the Mn-cluster. *Phys. Chem. Chem. Phys.* **6**, 4733–4736.

Bowden, S.J., Hallahan, B.J., Ruffle, S.V., Evans, M.C.W., and Nugent, J.H.A. (1991). Preparation and characterization of photosystem II core particles with and without bound bicarbonate. *Biochim. Biophys. Acta,* **1060**, 89–96.

Breton, J. (1990). Orientation of the pheophytin primary electron acceptor and of the cytochrome b-559 in the D1-D2 Photosystem II reaction center. In: *Perspectives in Photosynthesis* (J. Jortner and B. Pullman, Eds.), pp. 23–28. Kluwer Academic Publishers, Dordrecht.

Bricker, T.M. and Ghanotakis, D.F. (1996). Introduction to oxygen evolution. In: *Advances in Photosynthesis: Oxygenic Photosynthesis: The Light Reactions* (D.R. Ort and and C.F. Yocum, Eds.), pp. 113–136. Kluwer Academic Press, Dordrecht.

Britt, R.D. (1996). Oxygen evolution. In: *Advances in Photosynthesis: Oxygenic Photosynthesis: The Light Reactions* (D.R. Ort and C.F. Yocum, Eds.), pp. 137–164. Kluwer Academic Press, Dordrecht.

Büchel, C. and Kühlbrandt, W. (2005). Structural differences in the inner part of Photosystem II between higher plants and cyanobacteria. *Photosynth. Res.* **85**, 3–13.

Bumann, D. and Oesterhelt, D. (1994). Purification and characterization of oxygen-evolving photosystem II core complexes from the green alga *Chlamydomonas reinhardtii*. *Biochemistry* **33**, 10906–10910.

Clarke, A.K., Hurry, V.M., Gustafsson, P., and Oquist, G. (1993). Two functionally distinct forms of the photosystem II reaction-center protein D1 in the cyanobacterium *Synechococcus* sp. PCC 7942. *Proc. Natl. Acad. Sci. U. S. A.* **90**, 11985–11989.

Chua, N.H. and Bennoun, P. (1975). Thylakoid membrane polypeptides of *Chlamydomonas reinhardtii*: Wild-type and mutant strains deficient in Photosystem II reaction center. *Proc. Natl. Acad. Sci. U. S. A.* **72**, 2175–2179.

Debus, R.J. (1992). The manganese and calcium ions of photosynthetic oxygen evolution. *Biochim. Biophys. Acta,* **1102**, 269–352.

Debus, R.J., Barry, B.A., Sithole, I., Babcock, G.T., and McIntosh, L. (1988). Directed mutagenesis indicates that the donor to $P680^+$ in Photosystem II is Tyrosine-161 of the D1 polypeptide. *Biochemistry* **27**, 9071–9074.

Deisenhofer, J. and Michel, H. (1989). The photosynthetic reaction centre from the purple bacterium *Rhodopseudomonas viridis*. *EMBO J.* **8**, 2149–2170.

Deisenhofer, J., Epp, O., Miki, K., Huber, R., and Michel, H. (1985). Structure of the protein subunits in the photosynthetic reaction centre of *Rhodopseudomonas viridis* at 3Å resolution. *Nature* **318**, 618–624.

de Paula, J.C., Innes, J.B., and Brudvig, G.W. (1985). Electron transfer in Photosystem II at cryogenic temperatures. *Biochemistry* **24**, 8114–8120.

de Vitry, C., Wollman, F.-A., and Delepelaire, P. (1984). Function of the polypeptides in the Photosystem II reaction center of *Chlamydomonas reinhardtii*. *Biochim. Biophys. Acta* **767**, 415–422.

de Vitry, C., Diner, B.A., and Popot, J.-L. (1991). Photosystem II particles from *Chlamydomonas reinhardtii*. *J. Biol. Chem.* **266**, 16614–16621.

Diner, B.A. and Rappaport, F. (2002). Structure, dynamics, and energetics of the primary photochemistry of photosystem II of oxygenic photosynthesis. *Ann. Rev. Plant. Biol.* **53**, 551–580.

Diner, B.A., Nixon, P.J., and Farchaus, J.W. (1991). Site-directed mutagenesis of photosynthetic reaction centers. *Curr. Opin. Struct. Biol.* **1**, 546–554.

Diner, B., Schlodder, E., Nixon, P.J., Coleman, W.J., Rappaport, F., Lavergne, J., Vermaas, W.F.J., and Chisholm, D.A. (2001). Site-directed mutations at D1-His198 and D2-His197 of photosystem II in *Synechocystis* PCC 6803: Site of primary charge separation and cation and triplet stabilization. *Biochemistry* **40**, 9265–9281.

Dorlet, P., Boussac, A., Rutherford, A.W., and Un, S. (1999). Multifrequency high-field EPR study of the interaction between the tyrosyl Z radical and the manganese cluster in plant photosystem II. *J. Phys. Chem. B.* **103**, 10945–10954.

Dorlet, P., Xiong, L., Sayre, R.T., and Un, S. (2001). High-field EPR study of the pheophytin anion radical in wild type and D1-130 mutants of Photosystem II in *Chlamydomonas reinhardtii*. *J. Biol. Chem.* **276**, 22313–22316.

Durrant, J.R., Hastings, G., Joseph, D.M., Barber, J., Porter, G., and Klug, D.R. (1992). Subpicosecond equilibration of excitation energy in isolated Photosystem II reaction centers. *Proc. Natl. Acad. Sci. U. S. A.* **89**, 11632–11636.

Durrant, J.R., Hastings, G., Joseph, D.M., Barber, J., Porter, G., and Klug, D. R. (1993). Rate of oxidation of P680 in isolated Photosystem 2 reaction centers monitored by loss of chlorophyll stimulated emission. *Biochemistry* **32**, 8259–8267.

Durrant, J.R., Klug, D.R., Kwa, S.L.S., van Grondelle, R., Porter, G., and Dekker, J.P. (1995). A multimer model for P680, the primary electron donor of photosystem II. *Proc. Natl. Acad. Sci. U. S. A.* **92**, 4798–4802.

Dutton, P.L., Prince, R.C., and Tiede, D.M. (1978). The reaction center of photosynthetic bacteria. *Photochem. Photobiol.* **28**, 939–949.

Erickson, J. and Rochaix, J.-D. (1992). The molecular biology of photosystem II. In: *The Photosystems: Structure, Function and Molecular Biology* (J. Barber, Ed.), pp. 137–164. Elsevier, Amsterdam.

Erickson, J.M., Pfister, K., Rahire, M., Togasaki, R.K., Mets, L., and Rochaix, J.-D. (1989). Molecular and biophysical analysis of herbicide-resistant mutants of *Chlamydomonas reinhardtii*: Structure-function relationship of the photosystem II D1 polypeptide. *Plant Cell* **1**, 361–371.

Ferreira, K.N., Iverson, T.M., Maghlaoui, K., Barber, J., and Iwata, S. (2004). Architecture of the photosynthetic oxygen-evolving center. *Science* **303**, 1831–1838.

Ghanotakis, D.F., Waggoner, C.M., Bowlby, N.R., Demetriou, D.M., Babcock, G.T., and Yocum, C.F. (1987). Comparative structural and catalytic properties of oxygen-evolving photosystem II preparations. *Photosynth. Res.* **14**, 191–199.

Ghanotakis, D., Tsiotisz, S. and Bricker, T.M. (1998). Polypeptides of Photosystem II: Structure and Function. In: *Plant Photobiology: Photosynthesis and Photomorphogenesis* (G.S. Singhal, G. Renger, S.K. Sopory, K.-D. Irrgang, and Govindjee, Eds.) pp. 264–291, Narosa Publishing House, New Delhi.

Giorgi, L.B., Nixon, P.J., Merry, S.A.P., Joseph, D.M., Durrant, J.R., Rivas, J.D.L., Barber, J., Porter, G., and Klug, D.R. (1996). Comparison of primary charge separation in the Photosystem II reaction center complex isolated from wild-type and D1-130 mutants of the cyanobacterium *Synechocystis* PCC 6803. *J. Biol. Chem.* **271**, 2093–2101.

Govindjee, Schwarz, B., Rochaix, J.-D., and Strasser, R.J. (1991). The herbicide-resistant D1 mutant L275F of *Chlamydomonas reinhardtii* fails to show the bicarbonate-reversible formate effect on chlorophyll *a* fluorescence transients. *Photosynth. Res.* **27**, 199–208.

Grebanier, A.E., Coen, D.M., Rich, A., and Bogorad, L. (1978). Membrane proteins synthesized but not processed by isolated maize chloroplasts. *J. Cell Biol.* **78**, 734–746.

Greenfield, S.R., Seibert, M., Govindjee, and Wasielewski, M.R. (1997). Direct measurement of the effective rate constant for primary charge separation in isolated Photosystem II reaction centers. *J. Phys. Chem. B.* **101**, 2251–2255.

Groot, M.L., Pawlowicz, N.P., van Wilderen, L.J.G.W., Breton, J., van Stokkum, I.H.M., and van Grondelle, R. (2005). Initial electron donor and acceptor in isolated Photosystem II reaction centers identified with femtosecond mid-IR spectroscopy. *Proc. Natl. Acad. Sci. U. S. A.* **102**, 13087–13092.

Hays, A.-M., Vassiliev, I., Golbeck, J.H., and Debus, R.J. (1999). Role of D1-His190 in the proton-coupled oxidation of tyrosine YZ in manganese-depleted photosystem II. *Biochemistry* **38**, 11851–11865.

Hearst, J.E. and Sauer, K. (1984). Protein sequence homologies between portions of the L and M subunits of reaction centers of *Rhodopseudomonas capsulata* and the QB-protein of chloroplast thylakoid membranes: A proposed relation to quinone-binding sites. *Z. Naturforsch.* **39C**, 421–424.

Homann, P. (1988). The chloride and calcium requirement of photosynthetic water oxidation. *Biochim. Biophys. Acta,* **809**, 311–319.

Hutchison, R.S. and Sayre, R.T. (1995). Site specific mutagenesis at histidine 118 of the Photosystem II D1 protein of *Chlamydomonas reinhardtii*. In: *Photosynthesis: from Light to Biosphere* (P. Mathis, Ed.), pp. 471–474. Kluwer Academic Publishers, Dordrecht.

Hutchison, R., Roffey, R.A., and Sayre, R.T. (1996a). Chloroplast transformation. In: *Molecular Genetics of Photosynthesis, Frontiers in Molecular Biology* (B. Andersson, A.H. Salter, and J. Barber, Eds.), pp. 180–196. Oxford University Press, Oxford.

Hutchison, R.S., Xiong, J., Sayre, R.T., and Govindjee (1996b). Construction and characterization of a Photosystem II D1 mutant (arginine-269-glycine) of *Chlamydomonas reinhardtii*. *Biochim. Biophys. Acta* **1277**, 83–92.

Ishikita, H. and Knapp, E.-W. (2005). Redox potentials of chlorophylls and β-carotene in the antennae complexes of photosystem II. *J. Am. Chem. Soc.* **127**, 1963–1967.

Ishikita, H., Loll, B., Biesiadka, J., Saenger, W., and Kanpp, E.-W. (2005). Redox potential of chlorophylls in the photosystem II reaction center. *Biochem.* **44**, 4118–4124.

Johanningmeier, U., Bodner, U., and Wildner, G.F. (1987). A new mutation in the gene encoding for the herbicide-binding protein in *Chlamydomonas*. *FEBS Lett.* **211**, 221–224.

Johnston, H.G., Wang, J., Ruffle, S.V., Sayre, R.T., and Gustafson, T.L. (2000). Fluorescence decay kinetics of wild type and D2-H117N mutant photosystem II reaction centers isolated from *Chlamydomonas reinhardtii*. *J. Phys. Chem.* **104**, 4777–4781.

Jordan, P., Fromme, P., Witt, H.T., Klukas, O., Saenger, W., and Krauß, N. (2001). Three-dimensional structure of cyanobacterial photosystem I at 2.5 Å resolution. *Nature* **411**, 909–917.

Kamiya, N. and Shen, J.-R. (2003). Crystal structure of oxygen-evolving photosystem II from *Thermosynechococcus vulcanus* at 3.7-Å resolution. *Proc. Natl. Acad. Sci. U. S. A.* **100**, 98–103.

Keren, N., Gong, H., and Ohad, I. (1995). Oscillations of reaction center II-D1 protein degradation *in vivo* induced by repetitive light flashes. Correlation between the level of RCII-Q_B^- and protein degradation in low light. *J. Biol. Chem.* **270**, 806–814.

Klimov, V.V. (2003). Discovery of pheophytin function in the photosynthetic energy conversion as the primary electron acceptor of Photosystem II. *Photosynth. Res.* **76**, 247–253.

Klimov, V.V. and Krasnovskii, A.A. (1981). Pheophytin as the primary electron acceptor in Photosystem 2 reaction centers. *Photosynthetica* **15**, 592–609.

Klimov, V.V., Klevanik, A.V., and Shuvalov, V.A. (1977). Reduction of pheophytin in the primary light reaction of Photosystem II. *FEBS Lett.* **82**, 183–186.

Kok, B., Forbush, B., and McGloin, M.P. (1970). Cooperation of charges in photoynthesis in photosynthetic O_2 evolution. A linear four step mechanism. *Photochem. Photobiol.* **11**, 457–475.

Koulougliotis, D., Innes, J.B., and Brudvig, G.W. (1994). Location of chlorophyll$_Z$ in photosystem II. *Biochem.* **33**, 11814–11822.

Kramer, D.M., Roffey, R.A., Govindjee, and Sayre, R.T. (1994). The A_T thermoluminescence band from *Chlamydomonas reinhardtii* and the effects of mutagenesis of histidine residues on the donor side of Photosystem II D1 polypeptide. *Biochim. Biophys. Acta* **1185**, 228–237.

Lardans, A., Gillham, N.W., and Boynton, J.E. (1997). Site-directed mutations at residue 251 of the photosystem II D1 protein of *Chlamydomonas* that result in a nonphotosynthetic phenotype and impair D1 synthesis and accumulation. *J. Biol. Chem.* **272**, 210–216.

Lince, M.T. and Vermaas, W. (1998). Association of His117 in the D2 protein of photosystem II with a chlorophyll that affects excitation-energy transfer efficiency to the reaction center. *Eur. J. Biochem.* **256**, 595–602.

Loll, B., Kern, J., Saenger, W., Zouni, A., and Biesiadka, J. (2005). Towards a complete cofactor arrangement in the 3.0 Å resolution structure of photosystem II. *Nature* **438**, 1040–1044.

Martin, J.L., Breton, J., Holt, A.J., Migus, A., and Antonetti, A. (1986). Femtosecond spectroscopy of electron transfer in the reaction center of the photosynthetic bacterium *Rhodopseudomonas sphaeroides* R-26: Direct electron transfer from the dimeric bacteriochlorophyll primary donor to the bacteriopheophytin acceptor with a time constant of 2.8 ± 0.2 psec. *Proc. Natl. Acad. Sci. U. S. A.* **83**, 957–961.

Mayfield, S.P., Bennoun, P., and Rochaix, J.-D. (1987a). Expression of the nuclear encoded OEE1 protein is required for oxygen evolution and stability of Photosystem II particles in *Chlamydomonas reinhardtii*. *EMBO J.* **6**, 313–318.

Mayfield, S.P., Rahire, M., Frank, H., Zuber, H., and Rochaix, J.-D. (1987b). Expression of the nuclear gene encoding oxygen-evolving enhancer protein 2 is required for high levels of photosynthetic oxygen evolution in *Chlamydomonas reinhardtii*. *Proc. Natl. Acad. Sci. U. S. A.* **84**, 749–753.

Mayfield, S.P., Schirmer-Rahire, G., Frank, H., Zuber, H., and Rochaix, J.-D. (1989). Analysis of the genes of the OEE1 and OEE3 proteins of the photosystem II complex in *Chlamydomonas reinhardtii*. *Plant. Mol. Biol.* **12**, 683–693.

McEvoy, J.P., Gascón, J.A., Batista, V.S., and Brudvig, G.W. (2005). The mechanism of photosynthetic water splitting. *Photochem. Photobiol. Sci.* **4**, 940–949.

Michel, H. and Deisenhofer, J. (1988). Relevance of the photosynthetic reaction center from purple bacteria to the structure of Photosystem II. *Biochemistry* **27**, 1–7.

Minagawa, J. and Takahashi, Y. (2004). Structure, function and assembly of Photosystem II and its light-harvesting proteins. *Photosynth. Res.* **82**, 241–263.

Moënne-Loccoz, P., Robert, B., and Lutz, M. (1989). A resonance Raman characterization of the primary electron acceptor in photosystem II. *Biochemistry* **28**, 3641–3645.

Nanba, O. and Satoh, K. (1987). Isolation of a Photosystem II reaction center consisting of D-1 and D-2 polypeptides and cytochrome b-559. *Proc. Natl. Acad. Sci. U. S. A.* **84**, 109–112.

Nelson, N. and Yocum, C.F. (2006). Structure and functions of photosystem I and II. *Annu. Rev. Plant. Biol.* **57**, 521–565.

Nixon, P.J. and Diner, B.A. (1992). Aspartate 170 of Photosystem II reaction center polypeptide D1 is involved in the assembly of the oxygen evolving manganese cluster. *Biochemistry* **31**, 942–948.

Noguchi, T., Inoue, Y., and Satoh, K. (1993). FTIR studies on the triplet state of P680 in the Photosystem II reaction center: Triplet equilibrium within a chlorophyll dimer. *Biochemistry* **32**, 7186–7195.

Noguchi, T., Tomo, T., and Kato, C. (2001). Triplet formation on a monomeric chlorophyll in the photosystem II reaction center as studied by time-resolved infrared spectroscopy. *Biochemistry* **40**, 2176–2185.

Nugent, J.H.A., Diner, B.A., and Evans, M.C.W. (1981). Direct detection of the electron acceptor of Photosystem II complex. Evidence that Q is an iron-quinone complex. *FEBS Lett.* **124**, 241–244.

Pfister, K., Steinback, K.E., Gardner, G., and Arntzen, C.J. (1981). Photoaffinity labeling of an herbicide receptor protein in chloroplast membranes. *Proc. Natl. Acad. Sci. U. S. A.* **78**, 981–985.

Przibilla, E., Heiss, S., Johanningmeier, U., and Trebst, A. (1991). Site-specific mutagenesis of the D1 subunit of photosystem II in wild-type *Chlamydomonas*. *Plant Cell* **3**, 169–174.

Rochaix, J.-D., Dron, M., Rahire, M., and Malnoë, P.M. (1984). Sequence homology between the 32 kDa and D2 chloroplast membrane polypeptides of *Chlamydomonas reinhardtii*. *Plant Mol. Biol.* **3**, 363–370.

Roffey, R.A., Golbeck, J.H., Hille, C.R., and Sayre, R.T. (1991). Photosynthetic electron transport in genetically altered photosystem II reaction centers of chloroplasts. *Proc. Natl. Acad. Sci. U. S. A.* **88**, 9122–9126.

Roffey, R., van Wijk, K., Sayre, R., and Styring, S. (1994a). Spectroscopic characterization of Tyrosine-Z in Histidine 190 mutants of the D1 protein in Photosystem II (PS II) in *Chlamydomonas reinhardtii*. *J. Biol. Chem.* **269**, 5115–5121.

Roffey, R.A., Kramer, D.M., Govindjee, and Sayre, R.T. (1994b). Lumenal side histidine mutations in the D1 protein of Photosystem II affect donor side electron transfer in *Chlamydomonas reinhardtii*. *Biochim. Biophys. Acta* **1185**, 257–270.

Ruffle, S.V. and Sayre, R.T. (1998). Functional analysis of photosystem II. In: *The Molecular Biology of Chloroplasts and Mitochondria in Chlamydomonas* (J.D. Rochaix, M. Goldschmidt-Clermont, and S. Merchant, Eds.), pp. 287–322. Kluwer Academic Publishers, Dordrecht.

Ruffle, S.V., Donnelly, D., Blundell, T.L., and Nugent, J.H.A. (1992). A three-dimensional model of the Photosystem II reaction centre of *Pisum sativum*. *Photosynth. Res.* **34**, 287–300.

Ruffle, S.V., Wang, J., Johnston, H., Gustafson, T.G., Hutchison, R., Minagawa, J., Crofts, A., and Sayre, R.T. (2001). Photosystem II peripheral accessory chlorophyll mutants in *Chlamydomonas reinhardtii*. Biochemical characterization and sensitivity to photo-inhibition. *Plant Physiol.* **127**, 633–644.

Sandusky, P.O. and Yocum, C.F. (1986). The chloride requirement for photosynthetic oxygen evolution: Factors affecting the nucleophilic displacement of chloride from the oxygen evolving complex. *Biochim. Biophys. Acta* **849**, 85–93.

Satoh, K., Nakatani, H.Y., Steinback, K.E., Watson, J., and Arntzen, C.J. (1983). Polypeptide composition of a Photosystem II core complex: Presence of a herbicide-binding protein. *Biochim. Biophys. Acta* **724**, 142–150.

Sayre, R.T. and Wrobel-Boerner, E.A. (1994). Molecular topology of the photosystem II chlorophyll a binding protein, CP43: Topology of a thylakoid membrane protein. *Photosynth. Res.* **40**, 11–19.

Sayre, R.T., Andersson, B., and Bogorad, L. (1986). The topology of a membrane protein: The orientation of the 32 kd Qb-binding chloroplast thylakoid membrane protein. *Cell* **47**, 601–608.

Schelvis, J., van Noort, P., Aartsma, T., and van Gorkom, H. (1994). Energy transfer, charge separation and pigment arrangement in the reaction center of Photosystem II. *Biochim. Biophys. Acta*, **1184**, 242–250.

Schweitzer, R.H. and Brudvig, G.W. (1997). Fluorescence quenching by chlorophyll cations in photosystem II. *Biochemistry* **36**, 11351–11359.

Schweitzer, R.H., Melkozernov, A.N., Blankenship, R.E., and Brudvig, G.W. (1998). Time-resolved fluorescence measurements of Photosystem II: The effect of quenching by oxidized chlorophyll Z. *J. Phys. Chem. B* **102**, 8320–8326.

Shigemori, K., Hara, H., Kawamori, A., and Akabori, K. (1998). Determination of distances from tyrosine D to QA and chlorophyll Z in photosystem II studied by '2 + 1' pulsed EPR. *Biochim. Biophys. Acta* **1363**, 187–198.

Shim, H., Cao, J., Govindjee, and Debrunner, P.G. (1990). Purification of highly active oxygen evolving photosystem II from *Chlamydomonas reinhardtii*. *Photosynth. Res.* **26**, 223–228.

Stewart, D.H., Cua, A., Chisholm, D.A., Diner, B.A., Bocian, D.F., and Brudvig, G. W. (1998). Identification of histidine 118 in the D1 polypeptide of photosystem II as the axial ligand to chlorophyll Z. *Biochemistry* **37**, 10040–10046.

Strasser, R.J., Eggenberg, P., Pfister, K., and Govindjee (1992). An equilibrium model for electron transfer in the Photosystem II acceptor complex: An application to *Chlamydomonas reinhardtii* cells of D1 mutants and those treated with formate. *Arch. Sci. Genève*, **42**, 207–224.

Svensson, B., Vass, I., Cedergren, E., and Styring, S. (1990). Structure of donor side components in photosystem II predicted by computer modelling. *EMBO J.* **9**, 2051–2059.

Svensson, B., Etchebest, C., Tuffery, P., van Kan, P., Smith, J., and Styring, S. (1996). A model for the photosystem II reaction center core including the structure of the primary donor P680. *Biochemistry* **35**, 14486–14502.

Tang, D., Jankowiak, R., Seibert, M., Yocum, C.F., and Small, G.J. (1990). Excited-state structure and energy-transfer dynamics of two different preparations of the reaction center of Photosystem II: A hole-burning study. *J. Phys. Chem.* **94**, 6519–6522.

Tang, X.-S., Chisholm, D.A., Dismukes, G.C., Brudvig, G.W., and Diner, B.A. (1993). Spectroscopic evidence from site-directed mutants of *Synechocystis* PCC6803 in favor of a close interaction between Histidine 189 and redox-active

Tyrosine 160, both of polypeptide D2 of the Photosystem II reaction center. *Biochemistry* **32**, 13742–13748.

Tang, X.-S., Diner, B.A., Larsen, B.S., Gilchrist, M.L., Jr., Lorigan, G.A., and Britt, R.D. (1994). Identification of histidine at the catalytic site of the photosynthetic oxygen-evolving complex. *Proc. Natl. Acad. Sci. U. S. A.* **91**, 704–708.

Tang, X.-S., Zheng, M., Chisholm, D.A., Dismukes, G.C., and Diner, B.A. (1996). Investigation of the differences in the local protein environments surrounding tyrosine radicals, Y_Z and Y_D, in Photosystem II using wild type and the D2-Tyr160Phe mutant of *Synechocystis* 6803. *Biochemistry* **35**, 1475–1484.

Thompson, L.K. and Brudvig, G.W. (1988). Cytochrome b_{559} may function to protect photosystem II from photoinhibition. *Biochemistry* **27**, 6653–6658.

Tommos, C., Davidsson, L., Svensson, B., Madsen, C., Vermaas, W., and Styring, S. (1993). Modified EPR spectra of the $Tyrosine_D$ radical in Photosystem II in site-directed mutants of *Synechocystis* sp. PCC 6803: Identification of side chains in the immediate vicinity of $Tyrosine_D$ on the D2 protein. *Biochemistry* **32**, 5436–5441.

Trebst, A. (1986a). The topology of plastoquinone and herbicide binding polypeptides of photosystem II in the thylakoid membrane. *Z. Naturforsch.* **42c**, 240–245.

Trebst, A. (1986b). The three dimensional structure of the herbicide binding niche on the reaction center polypeptides of photosystem II. *Z. Naturforsch.* **42c**, 742–750.

van Gorkom, H.J. and Schelvis, J. (1993). Kok's oxygen clock: What makes it tick? The structure of P680 and consequences of its oxidizing power. *Photosynth. Res.* **38**, 297–301.

van Kan, P.J.M., Otte, S.C.M., Klienherenbrink, F.A.M., Nieveen, M.C., Aartsma, T.J., and van Gorkom, H.J. (1990). Time resolved spectroscopy at 10K of the Photosystem II reaction center; deconvolution of the red absorption band. *Biochim. Biophys. Acta* **1020**, 146–152.

van Mieghem, F.J.E., Satoh, K., and Rutherford, A.W. (1991). A chlorophyll tilted 30° relative to the membranes in the Photosystem II reaction centre. *Biochim. Biophys. Acta* **1058**, 379–385.

van Rensen, J.J.S., Xu, C., and Govindjee (1999). Role of bicarbonate in photosystem II, the water-plastoquinone oxido-reductase of plant photosynthesis. *Physiol. Plant.* **105**, 585–592.

Vermaas, W.F.J., Ikeuchi, M., and Inoue, Y. (1988a). Protein composition of the photosystem II core complex in genetically engineered mutants of the cyanobacterium *Synechocystis* sp. PCC 6803. *Photosynth. Res.* **17**, 97–113.

Vermaas, W.F.J., Rutherford, A.W., and Hanson, O. (1988b). Site-directed mutagenesis in Photosystem II of the cyanobacterium *Synechocystis* sp. PCC6803: Donor D is a tyrosine residue in the D2 protein. *Proc. Natl. Acad. Sci. U. S. A.* **85**, 8477–8481.

Vermaas, W., Charite, J., and Shen, G.Z. (1990). Glu-69 of the D2 protein in photosystem II is a potential ligand to Mn involved in photosynthetic oxygen evolution. *Biochemistry* **29**, 5325–5332.

Vrettos, J.S., Stewart, D.H., de Paula, J.C., and Brudvig, G.W. (1999). Low-temperature optical and resonance Raman spectra of a carotenoid cation radical in Photosystem II. *J. Phys. Chem. B.* **103**, 6403–6406.

Wang, J., Gosztola, D., Ruffle, S.V., Hemann, C. , Seibert, M., Wasielewski, M.R., Hille, R., Gustafson, T.L., and Sayre, R.T. (2002). Functional asymmetry of

Photosystem II D1 and D2 peripheral chlorophyll mutants of *Chlamydomonas reinhardtii*. *Proc. Natl. Acad. Sci. U. S. A.* **99**, 4091–4096.

Wasielewski, M.R., Johnson, D.G., Seibert, M., and Govindjee (1989). Determination of the primary charge separation rate in isolated Photosystem II reaction centers with 500-fs time resolution. *Proc. Natl. Acad. Sci. U. S. A.* **86**, 524–528.

Whitelegge, J.P., Koo, D., Diner, B.A., Domian, I., and Erickson, J.M. (1995). Assembly of the photosystem II oxygen-evolving complex is inhibited in *psb*A site-directed mutants of *Chlamydomonas reinhardtii*. Aspartate 170 of the D1 polypeptide. *J. Biol. Chem.* **270**, 225–235.

Woodbury, N.W., Becker, M., Middendorf, D., and Parson, W.W. (1985). Picosecond kinetics of the initial photochemical electron-transfer reaction in bacterial photosynthetic reaction centers. *Biochemistry* **24**, 7516–7521.

Xiong, J., Subramaniam, S., and Govindjee (1996). Modeling of the D1/D2 proteins and cofactors of the Photosystem II reaction center: Implications for herbicide and bicarbonate binding. *Protein Sci.* **5**, 2054–2073.

Xiong, L., Johnston, H.G., Seibert, M., Wasielewski, M.R., Gustafson, T.L., and Sayre, R.T. (2004). Substitution of a chlorophyll into the inactive branch pheophytin-binding site impairs charge separation in photosystem II. *J. Phys. Chem. B*, **108**, 16904–16911.

Xu, C., Taoka, S., Crofts, A.R., and Govindjee (1991). Kinetic characteristics of formate/formic acid binding at the plastoquinone reductase site in spinach thylakoids. *Biochim. Biophys. Acta* **1098**, 32–40.

Zouni, A., Witt, H.-T., Kern, J., Fromme, P., Krauss, N., Saenger, W., and Orth, P. (2001). Crystal structure of photosystem II from *Synechococcus elongatus* at 3.8 Å resolution. *Nature* **409**, 739–743.

Zurawski, G., Bohnert, H.J., Whitfeld, P.R., and Bottomley, W. (1982). Nucleotide sequence of the gene for the M_r 32,000 thylakoid membrane protein from *Spinacia oleracea* and *Nicotiana debneyi* predicts a totally conserved primary translation product of M_r 38,950. *Proc. Natl. Acad. Sci. U. S. A.* **79**, 7699–7703.

CHAPTER 17

The Cytochrome b_6f Complex

Catherine de Vitry and Richard Kuras
Centre National de la Recherche Scientifique,
Institut de Biologie Physico-Chimique, Paris, France

CHAPTER CONTENTS

I. INTRODUCTION

A common feature of the electron transfer chain components of oxygenic photosynthesis is that they were first identified as photoinduced absorption changes and characterized on a spectral basis. Over time these components were biochemically identified in terms of subunits, cofactors, and protein complexes. Ultimately, crystal structures showing three-dimensional representations of these components were obtained, which while answering some long-standing questions, also raised new ones. The cytochrome b_6f complex fits this progression, and the ease with which it is possible to generate and screen photosynthesis mutants in *Chlamydomonas* has been crucial for the associated structure-function and biogenesis studies.

Several reviews have used the b_6f complex crystal structures as a basis for discussing its functions (Cramer et al., 2004, 2005, 2006; Crofts, 2004; Smith et al., 2004; Merchant and Sawaya, 2005; Cramer and Zhang, 2006), whereas prior reviews had primarily considered its biogenesis (Merchant and Dreyfuss, 1998; Wollman, 1998; Xie and Merchant, 1998; de Vitry and Vallon, 1999; Wollman et al., 1999; Choquet and Vallon, 2000; Nakamoto et al., 2000). Although quite similar to the bc_1 complex, its homologue found in bacteria and mitochondria, the b_6f complex contains extra cofactors which point to additional functions. This chapter summarizes key features of the structural data, and develops insights into b_6f complex function and biogenesis based on work performed in *Chlamydomonas*.

II. STRUCTURE OF THE b_6f COMPLEX

A. Subunits of the complex

Each monomeric complex is composed of four large subunits [cytochrome b_6, subunit IV, cytochrome *f*, and the Rieske iron sulfur protein (ISP)] and four small subunits (PetG, PetL, PetM, and PetN). Three of the large subunits (cytochrome b_6, cytochrome *f*, and the Rieske ISP) bind redox-active prosthetic groups. In *Chlamydomonas* five of the subunits (cytochrome b_6, subunit IV, cytochrome *f*, PetG, and PetL) are encoded in the chloroplast, whereas PetM, PetN, and ISP are encoded in the nucleus (Table 17.1). The nuclear localization of the *PETN* gene differs from higher plants, where PetN is chloroplast-encoded.

An important milestone occurred in 2003, when b_6f complex crystal structures were obtained from the thermophilic bacterium *Mastigocladus laminosus* (Kurisu et al., 2003) and *Chlamydomonas* (Stroebel et al., 2003). The structures are very similar and feature a dimeric organization of the complex. In both structures, each monomer is composed of 13 transmembrane helices whose overall positions are rather similar (Figure 17.1). Nine of these helices are attributed to the four large subunits, the four remaining

Table 17.1 Cytochrome b_6f subunits in *Chlamydomonas*

Gene name	Gene location	Protein name	Redox cofactor	TMH[a]	M[b] kD	Notes	Protein accession #
petA	Chloroplast	Cytochrome *f*	1 *c*-heme	1	31.2	Sec translocation	CAA51422
petB	Chloroplast	Cytochrome b_6	2 *b*-hemes 1 *c*-heme	4	24.2		CAA51423
PETC	Nucleus	Rieske iron-sulfur protein	1 Fe_2S_2 cluster	1	18.4	Tat translocation	CAA53947
petD	Chloroplast	Subunit IV		3	17.4		CAA51424
petG	Chloroplast	PetG subunit		1	4.0		CAA46979
petL	Chloroplast	PetL subunit		1	3.4		AAB17716
PETM	Nucleus	PetM subunit		1	4.0		Q42496
PETN	Nucleus	PetN subunit		1	3.5	Chloroplast-encoded in most eukaryotes	P0C1D4
PETO	Nucleus	PetO, subunit V		1	15.2	In Volvocales, phosphoprotein	Q9LLC6

[a]*TMH, number of transmembrane α-helices.*

[b]*M, molecular mass of the mature subunit calculated from the sequence without cofactors.*

helices (shown in green in Figure 17.1B and 17.1D) are located away from the center of the dimer and form a loose bundle. Even if their positions are comparable in the two complexes, the protein assignments for three of them differ depending on the organism.

While both structures agree on the position of PetM, the electron densities assigned to PetG, PetL, and PetN in *Chlamydomonas* (Figure 17.1B) were respectively assigned to PetL, PetN, and PetG in *M. laminosus* (Figure 17.1D). In the *Chlamydomonas* structure, PetL and PetM are found at the periphery of the complex, at the opposite side from the central cavity formed by the junction of the two monomers. In the *M. laminosus* structure, PetL is by contrast embedded more deeply in the structure and surrounded by helices A and B from cytochrome b_6, helix G from subunit IV and the PetM subunit. The assignments proposed in *Chlamydomonas* for the small subunits seem the strongest since they are supported by Hg binding to cysteine residues of PetG and PetN (Stroebel et al., 2003), and mutant phenotypes (Table 17.2) induced by the absence of the small subunits. While the mutants can grow photosynthetically in the absence of PetL or PetM (assigned to peripheral locations, see Figure 17.1B), the absence of either PetG or PetN prevents the accumulation of the b_6f complex and leads to a non-photosynthetic growth phenotype. Finally,

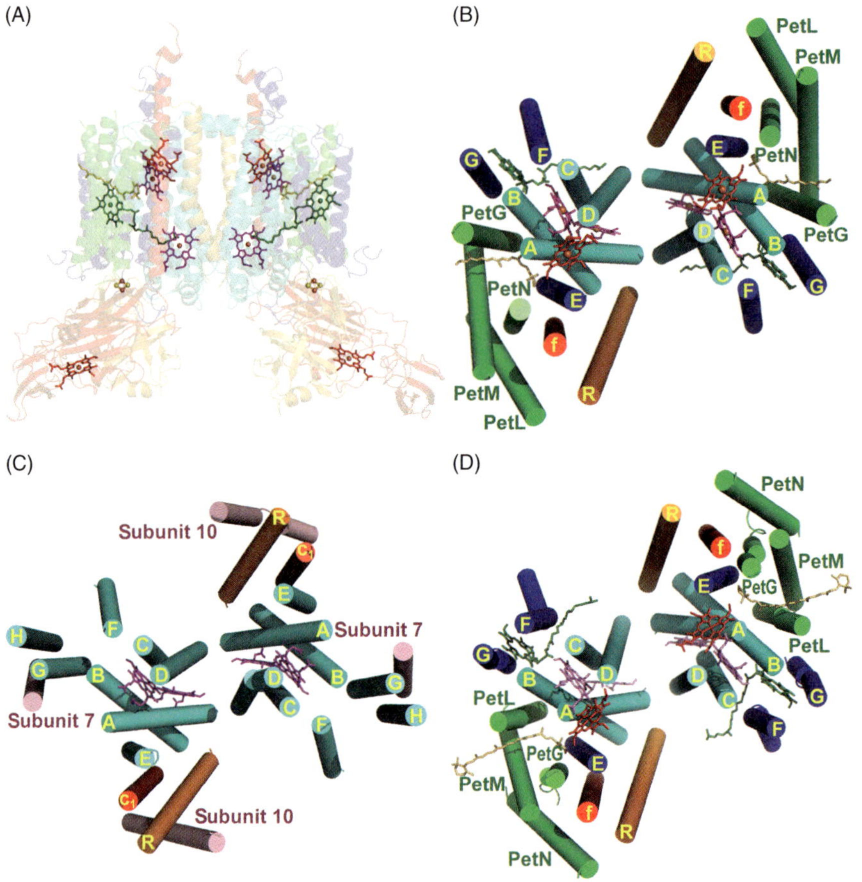

FIGURE 17.1 *Cytochrome b_6f and bc_1 complexes. (A) Side view of the Chlamydomonas cytochrome b_6f complex in the plane of the membrane is represented with the lumenal side at the bottom. (B) Top view from the n-side of the transmembrane region of the Chlamydomonas b_6f complex, the chicken bc_1 complex (c) and the M. laminosus b_6f complex (D), represented by showing only the transmembrane helices as cylinders. Coordinates used to generate the figures of the Chlamydomonas, M. laminosus and chicken complexes are from pdb files 1Q90, 2D2C and 1BCC, respectively. The figure was generated using PyMol. Helices for cytochrome b and cytochrome b_6 are colored in cyan and labeled from A to D for the latter and from A to H for the former. Helices for subunit IV are shown in blue and labeled from E to G, the Rieske ISP is shown in orange and labeled R, cytochrome f and cytochrome c_1, are shown in red and respectively labeled f and c_1, the PetG, PetL, PetM and PetN subunits are shown in green whereas subunit 7 and subunit 10 of the bc_1 complex are shown in light pink. Shown as wireframes are the β-carotenes colored in brown, the chlorophylls colored in green and the hemes colored either in pink (b-type hemes) or in red (c-type hemes). In panel (A), the Fe_2S_2 center is shown as a ball-and-stick model (yellow and orange).*

the *Chlamydomonas* assignments are supported by prediction of lipid-facing surfaces, which combines lipophilicity, conservation of residues in the helix, analysis of packing, and H-bonding interactions (Adamian and Liang, 2006). In summary, since the overall *Chlamydomonas* and *M. laminosus* structures are quite similar, the different positioning of the small subunits

Table 17.2 Function of very small subunits of the cytochrome b_6f complex as deduced from knockout mutants

	Chlamydomonas reinhardtii	*Synechocystis* PCC 6803	*Nicotiana tabacum*
PetG	Essential (Berthold et al., 1995)	Essential (Schneider et al., 2007)	Essential (Schwenkert et al., 2007)
PetL	Not essential (Takahashi et al., 1996)	Not essential (Schneider et al., 2007)	Not essential (Fiebig et al., 2004; Schottler et al., 2007; Schwenkert et al., 2007)
PetM	n.d.	Not essential (Schneider et al., 2001)	n.d.
PetN	n.d.	Essential (Schneider et al., 2007)	Essential (Hager et al., 1999; Schwenkert et al., 2007)

Note: Shown, where known, is if the protein is essential or not for cytochrome b_6f complex biogenesis, stability and/or function. n.d., no data.

is more likely to reflect a misattribution of electron density than a true difference in their arrangement.

B. Prosthetic groups

Seven prosthetic groups are found in the monomeric b_6f structures from both *Chlamydomonas* and *M. laminosus*. Two of them, one chlorophyll *a* molecule and one β-carotene, are pigments whose roles in the b_6f complex remain to be elucidated. The position of the chlorophyll *a* seems to be conserved since in both structures, its chlorin ring is located between the second and third helices of subunit IV. It is oriented perpendicularly to the membrane plane approximately in the middle of the lipid bilayer. The ring is not completely buried inside the protein and in both structures, is partially exposed to the lipid phase. At the other extremity of the chlorophyll molecule, the phytyl tail ends at the entrance of the Q_o site and appears to reduce, albeit to a lower extent in the *M. laminosus* structure, the accessibility of the Q_o site to the quinol molecule. The β-carotene is situated relatively far from the chlorophyll (at least 14 Å in the *Chlamydomonas* structure). The ring of the β-carotene is located in close proximity to both the first helix of subunit IV and the transmembrane helix attributed to the PetN subunit (in the *Chlamydomonas* structure). The β-carotene tail is pointing out of the complex between the PetG and PetM helices.

The five other prosthetic groups present per monomer of b_6f complex are one $[Fe_2S_2]$ cluster, two *b*-hemes and two *c*-hemes, all of which are cofactors with redox properties. The two *b*-hemes are located in the center of the four-helix bundle formed by cytochrome b_6. Each of these hemes is non-covalently bound to the protein by two histidyl residues situated on transmembrane helices B and D. The two hemes are located on either side of the membrane, with heme b_L closer to the electrochemically positive side (p-side) of the membrane, and heme b_H closer to the electrochemically negative side (n-side). The *c*-heme covalently attached to the cytochrome *f* and the $[Fe_2S_2]$ bound to the polar head of the Rieske ISP are located at the periphery of the membrane in the lumenal space. The second *c*-heme is covalently attached to the transmembrane helix A of cytochrome b_6.

C. Comparison with bc_1 complexes

1. Similar dimer organization but different small subunits

Structural data for the mammalian mitochondrial bc_1 complex became available in 1997 (Xia et al., 1997) and based on previous molecular, biochemical and functional analyses, the core of the b_6f complex was expected to share some structural attributes with the bc_1 complex core. Indeed, the core of the bc_1 complex is essentially formed of its cytochrome *b* subunit whose first half is homologous to cytochrome b_6, whereas the second half is homologous to subunit IV. The resolution of the b_6f structure confirmed that transmembrane helices for the main subunits were almost identically organized in both bc_1 and b_6f complexes (Figure 17.1). The structural data also revealed that the respective positions and orientation of the two *b*-hemes is highly conserved between the bc_1 and b_6f complexes. However this conserved organization does not apply to the small subunits, and the membrane space occupied by the four helices of PetG, PetL, PetM, and PetN appears empty in the bc_1 complex (Figure 17.1C).

Concerning the extrinsic part, there are important differences. It was previously known that the extrinsic parts of cytochrome *f* and cytochrome c_1, its functional homologue, do not share sequence similarity and are in fact quite different in structure (Xia et al., 1997; Chi et al., 2000). As a consequence, the position of the *c*-heme bound to cytochrome *f* in *Chlamydomonas* differs from that of the cofactor associated with cytochrome c_1 when the two structures are superimposed. The positions of the Rieske ISPs are not easily comparable, either. Indeed, the catalytic cycle of the bc_1-complexes involves the movement of the polar head of the Rieske ISP, and the position of this protein is known to differ from crystal to crystal (Iwata et al., 1998; Kim et al., 1998; Xia et al., 1998; Zhang et al., 1998; Breyton, 2000; Brugna et al., 2000). However, in the presence of Q_o site inhibitors such as stigmatellin, which locks the head of the Rieske ISP in the proximal position in the bc_1 complex or its chemical derivative tridecylstigmatellin (TDS), which acts similarly in b_6f complexes, the overall positions of the Rieske ISP and the iron-sulfur cluster are comparable in the b_6f and bc_1 complexes.

2. Three additional cofactors: chlorophyll *a*, β-carotene, and heme c_i'

As compared to the bc_1 complex, the b_6f complex contains three additional cofactors which were introduced above. These are chlorophyll *a*, a β-carotene, and a *c*-type heme. While the pigments had been found through biochemical analyses (Pierre et al., 1997; Zhang et al., 1999), the presence of an additional *c*-type heme, bound to cytochrome b_6, was unexpected in the crystal structure. This potentially redox-active molecule is located in the membrane space, close to the n-side of the membrane and is covalently attached to helix A of cytochrome b_6 via a thioether bond between the sulfydryl group of Cys35 (*Chlamydomonas* numbering) and one of the vinyl groups of the heme. It is outside the four-helix bundle formed by cytochrome b_6 but in close proximity to the b_H heme. The two heme planes are oriented perpendicularly to each other.

This additional heme was first identified by *in vivo* spectroscopy as redox center G (Lavergne, 1983) and is usually referred as heme c_i' in *Chlamydomonas*, but can also be found under the names of heme *x* (which was its original designation in the b_6f structure from *M. laminosus*) and heme c_n. The attachment of this heme to the protein moiety presents some unusual features: it is covalently bound only to a single cysteine residue and in addition, this cysteine is not part of any conserved CxxCH motif usually known to act as the covalent ligands for *c*-type hemes. The fact that this heme is covalently attached to cytochrome b_6 is consistent with previous biochemical data suggesting the presence of a covalent heme in cytochrome b_6 and the existence of several biogenesis factors assisting heme binding to cytochrome b_6 (Kuras et al., 1997). The covalent binding of heme c_i' to Cys35 of cytochrome b_6 as revealed by the structures is also consistent with data showing that a heme remained tightly associated with a proteolytic peptide containing Cys35 following the digestion of *Chlamydomonas* cytochrome b_6 with Lys-C protease (de Vitry et al., 2004a). None of the crystal structures shows a residue close enough to the heme c_i' to provide an axial ligand to the heme iron. A possible axial ligand, either a hydroxyl or a water molecule, is found on the side of heme c_i' oriented towards heme b_H and was suggested to bridge the heme iron of heme c_i' to the carboxyl group of one of the propionates of heme b_H (see Figure 17.4). As expected for a pentacoordinated heme, heme c_i' displays high spin spectral properties (de Vitry et al., 2004a; Zhang et al., 2004; Alric et al., 2005; Baymann et al., 2007).

3. Heme c_i' in the Q_i site

Since the polypeptide chains around the Q_i site, where the quinone is reduced (Figure 17.2), display the same topology in bc_1 and b_6f complexes, their Q_i active sites could have been expected to share some similarities. Quite surprisingly, the position filled by heme c_i' in the b_6f structure is in the pocket attributed to the Q_i site in the chicken cytochrome bc_1 structure. Indeed, in the bc_1 structures, this cavity is occupied either by a ubiquinone molecule which is the natural substrate for the bc_1 complex or by antimycin A, an inhibitor which is known to displace the quinone at

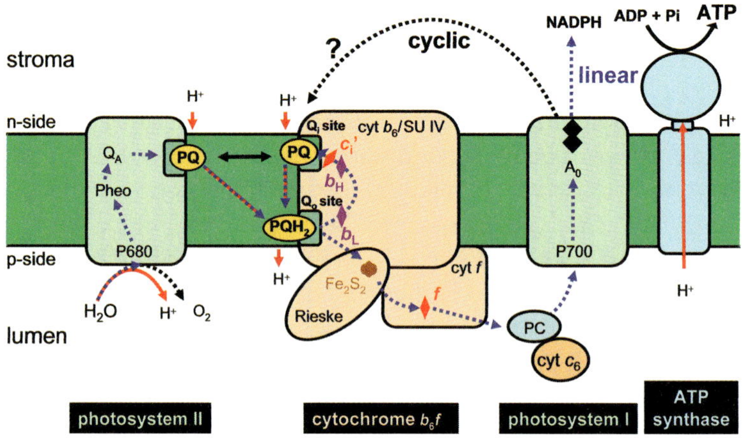

FIGURE 17.2 *Schematic representation of the photosynthetic electron transport chain in the thylakoid membrane. Abbreviations are as follows: Pheo, pheophytin; PQ, plastoquinone; PQH_2, plastoquinol; PC, plastocyanin. Hemes are shown as diamonds. The linear electron transfer pathway from water to NADPH is indicated by dotted blue arrows. The red arrows indicate when proton release or uptake occurs. The re-injection of electron from photosystem 1 into the PQ pool during cyclic electron transfer is shown as a dotted black arrow. The question mark indicates, as described in the text (section III.B.2), that the mechanism is not fully elucitated and that several routes are possible.*

the Q_i site of the bc_1 complexes but not in the b_6f complexes. Therefore in the b_6f complex, heme c_i' is directly between the b_H heme and the quinone exchange cavity located between the two monomers, thus making difficult the direct access from a plastoquinone molecule to heme b_H. In the *M. laminosus* structure the quinone molecule at the Q_i site is not buried inside the Q_i cavity as in the bc_1 complex, but rather is close to heme c_i' at the entrance of the cavity containing hemes b_H and c_i'. A weak electron density close to the face of heme c_i' oriented toward the Q_i entrance was proposed to be accounted for by either plastoquinone or another ligand in the *Chlamydomonas* structure (Stroebel et al., 2003). Quinone analogue inhibitors [2-*n*-nonyl-4-hydroxyquinoline-*N*-oxide (NQNO) and TDS] were recently localized to the Q_i site close to a face of heme c_i' in a new *M. laminosus* structure (Yamashita et al., 2007).

4. A chlorophyll phytyl tail in the Q_o site

The overall shape of the b_6f Q_o site is more similar to its bc_1 equivalent than to the Q_i site. In both the bc_1 and b_6f complexes, the Q_o site corresponds to the volume space where the quinol molecule is oxidized and protons are released and in both cases, this volume presents a bifurcated shape with two lobes, either proximal or distal from heme b_L. In both the *Chlamydomonas* b_6f structure and the chicken bc_1 structure, the head groups of the stigmatellin inhibitor (TDS in *Chlamydomonas*) are located in the distal domain of the Q_o site close to the Rieske ISP docking interface.

It has been proposed that in this conformation the interaction of the head group of the stigmatellin with the mobile domain of the Rieske ISP mimics the reaction complex formed by the quinol substrate with the oxidized Rieske ISP (Crofts et al., 1999). In the *Chlamydomonas* b_6f complex, the position of the TDS tail is slightly shifted as compared to the tail of the stigmatellin in the bc_1 complex because the entrance of the Q_o site is partially obstructed by the phytyl tail of the chlorophyll molecule.

By contrast, the orientation proposed for the TDS in the *M. laminosus* b_6f structure is rather unexpected, that is, it is upside down as compared to the *Chlamydomonas* structure and it is the tail of TDS that extends into the Q_o site, whereas the head group remains close to the Q_o entrance. It was initially proposed that the inhibitory effect on the b_6f complex was due to the tail of TDS preventing quinol access to the Q_o site (Kurisu et al., 2003). This hypothesis was later ruled out when site-directed mutants of the cytochrome b_6 and the Rieske ISP in the cyanobacterium *Synechococcus* sp. PCC 7002 (Yan and Cramer, 2004) clearly showed that TDS inhibition requires the inhibitor to be in the orientation seen in both the bc_1 structures and the *Chlamydomonas* b_6f structure, with its head group buried deep in the Q_o site and close to the Rieske ISP. A subsequent *M. laminosus* structure confirmed the "head-in" orientation of TDS (Yamashita et al., 2007).

III. THE b_6f COMPLEX AT THE CROSSROADS OF ELECTRON TRANSPORT AND SIGNAL TRANSDUCTION

A. Introduction

Although the main structural and functional features of the bc_1 and b_6f complexes are quite similar, several characteristics in their cofactor compositions and structural organizations ensure the specific roles they play in terms of electron transfer mechanism, physiological function and regulation. Here we discuss studies in *Chlamydomonas* that have addressed the specificities of the b_6f complex functions and the intriguing roles of its additional cofactors.

B. Electron transfer mechanisms

1. *Linear electron transfer*

The b_6f complex is a plastoquinol-plastocyanin reductase (EC 1.10.99.1) and a key enzyme of the linear electron transfer chain (Figure 17.2) that participates in oxygenic photosynthesis in cyanobacteria and chloroplasts. The function of the linear electron transfer chain is to oxidize water into molecular oxygen and to reduce NADP+ to NADPH while building up a transmembrane proton gradient that will be converted by ATP synthase into chemical energy (ATP). The b_6f complex catalyses electron transfer between photosystem (PS) II and PS I, the two photosynthetic reaction centers of oxygenic photosynthesis, and participates in the build-up of the transmembrane proton gradient (see Chapters 15, 16, 18 and 22 for discussion of other electron transfer components).

The current mechanistic model for the b_6f complex derives from the Q-cycle model originally proposed by Mitchell (1975) for the cytochrome bc_1 complex, later modified by Crofts et al. (1983). The mechanism involves two distinct catalytic sites and two separate chains of redox carrier molecules. The "high potential chain" is located on the lumenal side close to the thylakoid membrane and includes the Rieske ISP and cytochrome *f*, two proteins with close redox midpoint potentials of 300–350 mV. The "low potential chain" spans the thylakoid membrane and is formed by the b_L-heme (Em,7 = –130 mV) and the b_H-heme (Em,7 = –35 mV). The heme c_i' (Em,7 = 100 mV) probably also participates in the low potential chain, but its role needs to be clarified. The two catalytic sites are located on opposite side of the thylakoid membrane, and both bind the plastoquinone/plastoquinol lipo-soluble substrate. The Q_o site is on the lumenal side of the membrane and has a higher affinity for substrate in its reduced form plastoquinol (PQH_2), whereas the Q_i site located on the stromal side has a higher affinity for plastoquinone (PQ).

Turnover of the b_6f complex is indirectly induced by illumination. Upon oxidation of the PS I reaction center by a flash, the oxidizing equivalent is first transferred to plastocyanin then to the high potential chain of the b_6f complex via the oxidation of cytochrome *f* and re-reduction of plastocyanin. The positive charge is then transferred to the Rieske ISP whose soluble head moves close to the Q_o site. In a bifurcated reaction, one electron of a PQH_2 molecule located in the Q_o site is transferred to the Rieske ISP while the second electron is transferred via the low potential chain to heme b_H. In the process, two protons are released into the lumenal space while the PQ molecule will diffuse out of the Q_o site, to be replaced by PQH_2. Completion of b_6f turnover requires a second turnover of the Q_o site which is made possible by the injection of a second oxidizing equivalent to the high potential chain through plastocyanin. During oxidation of the second PQH_2 in the Q_o site, one electron is transferred to the high potential chain and will ultimately reduce the PS I reaction center, one electron is given to the b_L-heme and two additional protons are released into the lumen. Concomitantly the two electrons stored in the low potential chain are transferred to a quinone molecule bound in the Q_i site. During the reduction of the quinone, the uptake of two protons from the stroma is necessary to form PQH_2. Therefore, the net result of one complete turnover of the b_6f complex is the complete oxidation of one of the PQH_2 molecules coming from PS II upon light induction.

Although this model is widely accepted (Malkin, 1992; Cramer et al., 2006), some authors have reported discrepancies between this theoretical mechanism and some experimental data. Thus it was proposed that under highly reducing conditions when no plastoquinone is available at the Q_i site, the charged semiquinone formed at the Q_o site would be transferred to the Q_i where it would be reduced (Joliot and Joliot, 1994). Evidence for a proton pumping pathway acting in addition to the Q-cycle, was also reported (Joliot and Joliot 1988; Deniau and Rappaport, 2000).

2. Cyclic electron transfer around PS I

The b_6f complex is involved not only in photosynthetic linear electron transfer (water $\rightarrow$ PS II $\rightarrow$ b_6f $\rightarrow$ PS I $\rightarrow$ NADPH) but also in a cyclic pathway around PS I (Figure 17.2). Whereas in linear electron flow, electrons from the donor side of PS I are transferred to NADPH through ferredoxin (Fd) and are used in carbon fixation, in cyclic electron flow the electrons from PS I are transferred back to plastoquinone in the thylakoid membrane and/or to the b_6f complex. The cyclic pathway is not fully elucidated and several routes that involve either Fd or NADPH have been proposed to re-inject electrons from PS I into plastoquinone (Bendall and Manasse, 1995). The most common model is that the electrons are transferred back to the plastoquinone pool through two distinct pathways, one relying on a ferredoxin-quinone oxidoreductase (FQR) and the other on a NAD(P)H dehydrogenase (NDH). The presence of an enzymatically active ferredoxin:$NADP^+$ oxidoreductase (FNR) bound stoichiometrically to the b_6f complex purified from spinach thylakoids was reported by Zhang et al. (2001). In the presence of both ferredoxin and NADPH, the authors were able to obtain *in vitro* a partial reduction of the *b*-heme complement of the b_6f complex, suggesting that FNR bound to the b_6f complex provided the link to the main electron transfer chain for ferredoxin-dependent cyclic electron transfer. However, whether a cofactor in the b_6f complex is directly reduced and/or whether the complex participates indirectly via the plastoquinone pool is a matter of debate (Kramer et al., 2004).

From studies on an *Arabidopsis* mutant, a new small protein named PGR5 was proposed to be involved in the cyclic electron pathway mediated by the ferredoxin-plastoquinone reductase (Munekage et al., 2002). PGR5 appears to be involved somewhere in the electron transfer between ferredoxin and plastoquinone, not as an electron carrier but rather as a factor necessary for either the stabilization or the activation of the complex that catalyzes ferredoxin-dependent plastoquinone reduction. The other proposed route for cyclic electron transport involves the chloroplast NDH. The discovery of genes homologous to those encoding subunits of the mitochondrial complex I (NADH dehydrogenase) in the chloroplast genome (Ohyama et al., 1986; Shinozaki et al., 1986) raised the possibility of an NDH-dependent pathway in chloroplasts, but the cyclic NDH route was first identified in cyanobacteria (Ogawa, 1991). Several years later, the presence of a NADH-specific dehydrogenase in pea thylakoid membranes that contained subunits encoded by chloroplast *ndh* genes was reported (Sazanov et al., 1998). The same year, disruption of chloroplast-encoded *ndh* genes from tobacco led to the conclusion that the encoded proteins were involved in an NDH complex acting as a chloroplast respiratory chain that could also participate in cyclic flow around PS I (Burrows et al., 1998). The presence of the NDH complex was also reported in mesophyll and bundle sheath chloroplasts from *Zea mays* (Darie et al., 2005) and others reported that cylic electron flow through NDH seems to play a central role in driving the CO_2-concentrating mechanism in C_4 photosynthesis (Takabayashi et al., 2005), probably by providing the necessary ATP.

In *Chlamydomonas*, all genes predicted to encode the chloroplast NDH enzyme seem to be absent. The 11 *ndh* genes previously found in plant chloroplast genomes (Ohyama et al., 1986; Shinozaki et al., 1986) are absent from the *Chlamydomonas* chloroplast genome (Maul et al., 2002). The three nuclear genes *NdhM*, *NdhN*, and *NdhO* reported to encode subunits of the plastid NDH complex in both tobacco and *Arabidopsis* (Rumeau et al., 2005), are also missing from *Chlamydomonas*. This suggests that in *Chlamydomonas* the cyclic pathway around PS I does not proceed through a plastidic NDH complex. A partial purification of a putative NADPH-plastoquinone reductase from *Chlamydomonas* thylakoid membranes was reported (Godde, 1982), but it seems likely that the thylakoid preparation was contaminated by mitochondrial membranes. An alternative hypothesis would be that the *Chlamydomonas* NDH complex present in thylakoids is highly divergent from those already characterized and cannot be revealed by sequence similarity searches.

In plants, the cyclic electron pathway is usually considered as a means to deal with stress conditions such as excessive light that over-saturates the photochemical processes leading to CO_2 fixation. Cyclic electron transfer is purported to increase the proton gradient (ΔpH) across the thylakoid membrane using the Q-cycle catalyzed by the b_6f complex. The increased ΔpH results in an acidification of the lumen, which is known to participate in the induction of the non-photochemical quenching of chlorophyll fluorescence, by which the excess energy is released as heat (see Chapter 23). Therefore it seems that the cyclic pathway is necessary to induce the thermal dissipation of excess energy under adverse conditions (Munekage et al., 2002; Miyake et al., 2005a,b; Yamamoto et al., 2006).

The physiological role of the cyclic pathway under normal conditions is a matter of controversy. It was largely accepted that under normal conditions, at least in C_3 plants, linear flow is prevalent (Bendall and Manasse, 1995). Cyclic flow, however, has been reported to be necessary following dark adaptation to provide the ATP needed to activate the Benson-Calvin cycle (Joliot and Joliot, 2002). These authors also suggested that even under continuous light the contribution of the cyclic pathway should be significant, since the ATP produced by linear photosynthesis does not appear sufficient to fulfill the ATP requirements of the stroma. A subsequent proposition was that the cyclic pathway is essential to prevent an over-reduction of the stroma through NADPH accumulation during photosynthesis (Munekage et al., 2004).

The regulation of partitioning between cyclic and linear electron flow is also controversial. Joliot and Joliot suggested that in C_3 plants, the balance between cyclic and linear flow is regulated through the formation of PS I supercomplexes controlled by the ATP concentration (Joliot and Joliot, 2002). A later proposal was that cyclic and linear pathways are in competition for the re-oxidation of the ferredoxin that freely diffuses in the stromal compartment (Joliot and Joliot, 2006), and that the regulation occurs through the redox state of the stroma (Breyton et al., 2006). In green algae, cyclic flow operates under anaerobic conditions (Joliot and Joliot, 1988), and it was reported that in *Chlamydomonas* the switch between linear and cyclic electron transfer is tightly coupled to state transitions (Finazzi et al., 2002), which involve the reversible migration of light-harvesting antennas between PS II and PS I, and are considered a means to redistribute excitation energy between the two photosystems (see Chapter 22). Correlations between state transitions and shifts between cyclic and linear flow were not observed in *Arabidopsis* (Lunde et al., 2003) and it was proposed that in *Chlamydomonas*, the existence of the phenomenon could be related to a shift from an oxygenic type of photosynthesis generating both reducing power and ATP to an anaerobic photosynthesis generating only ATP, which could be required to cope with environmental changes (Finazzi, 2005).

3. Spectral and redox characterization of heme c_i'

The first insight into the presence of an extra heme in the b_6f complex came from *in vivo* spectroscopic evidence of a redox center "G" proposed to be in equilibrium with heme b_H and located near the stromal side of the membrane (Lavergne, 1983). "G" was later characterized as a cytochrome c' (Joliot and Joliot, 1988) but there was still no direct evidence that it was part of the b_6f complex. The discovery of an unusual extra heme bound to the protein backbone of cytochrome b_6 (Kurisu et al., 2003; Stroebel et al., 2003) was a major surprise that not only renewed interest in "G" but also raised new questions about the spectral, redox and functional properties of this unexpected heme (Figures 17.3 and 17.4).

Resonance Raman spectroscopy performed on b_6f complexes from *Chlamydomonas* (de Vitry et al., 2004a) and EPR (electron paramagnetic resonance) spectroscopy on purified b_6f complexes from *M. laminosus* (Zhang et al., 2004) and *Chlamydomonas* (Baymann et al., 2007) identified this extra heme as being a c'-like heme under a five-coordinated high-spin state. The absorption spectra monitored *in vitro* as a function of the redox potential

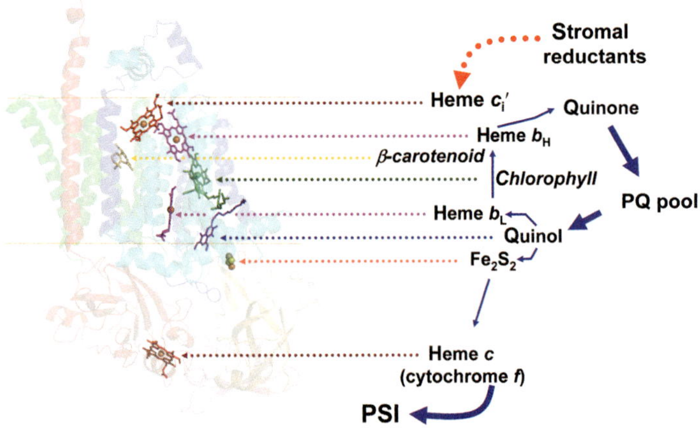

FIGURE 17.3 *Potential pathways of electron transfer in the b_6f complex represented as an overlay on the structural organization of the cofactors identified as being present by the crystal structure. A monomer with the Rieske ISP head of the functional monomer from Chlamydomonas drawn with PyMol is shown in the color code as in Figure 1. The labels of the pigments not involved in electron transfer are italicized.*

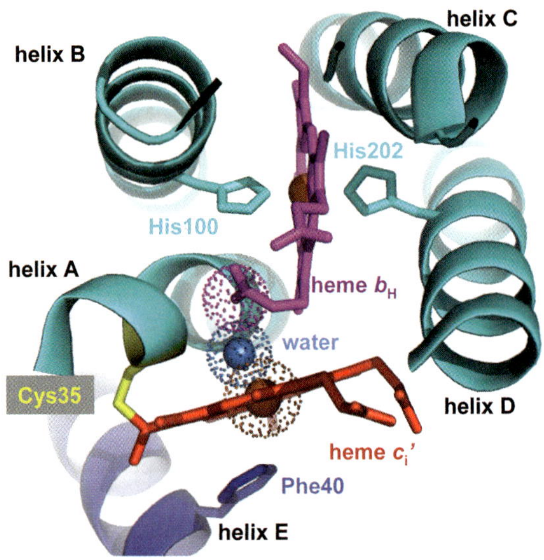

FIGURE 17.4 *Overview of the heme c_i' region in the b_6f structure of Chlamydomonas. Heme c_i' is covalently attached by a unique thioether bond to Cys35 of the cytochrome b_6 subunit and its iron is bridged to a b_H heme propionate through a water molecule. The figure was generated using PyMol and the coloring is as in Figure 1, with Cys35 shown in yellow and the water as a blue sphere. The heme irons are shown as orange spheres. Dots representing electron densities have been drawn around the water molecule, the heme iron of heme c_i' and an oxygen atom of the carboxyl group of one of the propionates of heme b_H, to highlight the water bridge between heme c_i' and heme b_H.*

showed that the differential spectrum of heme c_i' in *Chlamydomonas* consists of a broad absorption increase centered at 425 nm with no obvious change in the green region, as observed *in vivo* for the previously described redox center "G". The midpoint potential of heme c_i' at pH 7.0 was measured at +100 mV with a pH dependence of –60 mV per pH unit in the pH 6–9 range (Alric et al., 2005). The Q_i site inhibitor NQNO, a semiquinone analogue that only partially blocks linear electron flow (Joliot and Joliot, 1986; Jones and Whitmarsh, 1988), lowers heme c_i' potential by about 225 mV which is thermodynamically equivalent to a stronger binding to the oxidized form of the heme (Alric et al., 2005).

Based on this decrease of heme c_i' potential in presence of NQNO, a plus/minus NQNO difference spectrum of heme c_i' has been elegantly detected in spinach b_6f complexes containing FNR, in the presence of reduced ferredoxin (Yamashita et al., 2007). EPR spectra of the c_i'/b_H moiety show a pair of strongly interacting hemes (Zatsman et al., 2006; Baymann et al., 2007). It was proposed that the tightly packed pair of redox centers (heme c_i'/heme b_H) provides the two electrons required to fully reduce plastoquinone, explaining the apparent absence of a semiquinone in the Q_i site of wild-type b_6f complexes. The presence of semiquinone is expected to lead to a photosensitive strain in the presence of oxygen because of the generation of reactive oxygen species, analogous to the superoxide production described in the bc_1 complex (Forquer et al., 2006). A role of heme c_i' in limiting production of reactive oxygen species would be consistent with heme c_i' presence in species that perform oxygenic photosynthesis, and generate oxygen from water oxidation in the membrane where heme c_i' is located. The Q_i site location on the stromal membrane side and the redox properties of heme c_i' are consistent with its involvement in a variant of the turnover of the Q_i site as described in the bc_1 complex or in cyclic electron transfer.

An altered Q_i site was obtained in the cytochrome b_6 *petB-R214H* mutant of *Synechococcus* sp. PCC7002, which has partially blocked electron transfer at the Q_i site (Nelson et al., 2005). In the *Chlamydomonas* structure, the cytochrome b_6 arginine residue R207 (*Synechococcus* R214) lies close to the c_i' heme. Characterization of a diversity of mutants of the Q_i pocket is a promising approach to unravel the role of heme c_i'.

C. Source of regulatory signals

1. Kinase activation and state transitions

The b_6f complex is a key element in state transition signaling. As introduced in section III.B.2, this process controls the allocation of light-harvesting proteins between PS I and PS II, with the Q_o site of cytochrome b_6f acting as the primary sensor triggering activation of a kinase specific for the phosphorylation of light-harvesting proteins. In *Chlamydomonas* this also regulates the switch between cyclic and linear transfer regimes, in response to physiological needs as expressed by the redox state of the plastoquinone pool (Wollman, 2001; Chapter 22).

Chlamydomonas b_6f mutants have greatly contributed to the characterization of state transition regulation. Mutants lacking the b_6f complex established its requirement for kinase activation (Lemaire et al., 1987; Wollman and Lemaire, 1988). The key role of the b_6f complex Q_o site in kinase activation was shown in *petD-pwye* mutants of subunit IV (Zito et al., 1999) and in the Rieske ISP deletion mutant Δ*petC* (de Vitry et al., 1999), and by blocking the Q_o site using inhibitors (Finazzi et al., 2001). The kinase responsible for state transitions in *Chlamydomonas* is a chloroplast thylakoid-associated serine-threonine protein kinase, STT7, which is a membrane protein of 76 kD either containing a transmembrane helix or firmly bound to the stromal side of the membrane (Depège et al., 2003). The lack of kinase activation and the wild-type like PQH_2 oxidase activity in a mutant in which PetL was fused to the C-terminus of subunit IV (Zito et al., 2002) suggested that kinase docking to the b_6f complex occurred in the transmembrane region of subunit IV, where the chlorophyll ring is exposed to the lipids. The phytyl chain of the chlorophyll bound to the b_6f complex was proposed to provide a direct pathway for signaling the binding of the quinol from the Q_o site to a peripheral region of the complex where kinase docking is expected to take place (Finazzi, 2005).

A model for kinase activation implicates plastoquinol binding to the Q_o site and the movement of the Rieske ISP (Wollman, 2001). Deletion of the Rieske ISP flexible domain decreased electron transfer efficiency but did not represent a limiting step for kinase activation in the time scale tested (de Vitry et al., 2004b). Although the high affinity of the Q_o site for PQH_2 (Kramer et al., 1994; Finazzi et al., 1997) should allow kinase activation in aerated *Chlamydomonas* cells when the plastoquinone pool is partly reduced, this was not observed (Zito et al., 1999). Simultaneous plastoquinol binding to the two monomer Q_o sites might be needed for kinase activation. State transitions via the b_6f complex were also reported in cyanobacteria (Mao et al., 2002). However, the mechanisms of these state transitions must be different from that of photosynthetic eukaryotes, since cyanobacteria have different types of antenna complexes and kinases.

2. Signaling chlorophyll biosynthesis

It was reported that defects in the b_6f complex prevented the light-induced expression of nuclear genes involved in chlorophyll biosynthesis (Shao et al., 2006). The signal that affected the expression of these genes was found to depend only on the integrity of the b_6f complex Q_o site, in a signaling pathway that differs from the pathway regulating state transitions. The molecular nature of the signal generated and its subsequent transduction to the cytosol/nucleus is currently unknown.

3. Photoprotection

The function of the β-carotene is still a matter of debate. Since the β-carotene was shown to be able to protect the bound chlorophyll *a* from photobleaching in the presence of oxygen, its role in photoprotection was suggested (Zhang et al., 1999). Chlorophyll-to-carotenoid triplet excitation energy transfer was observed for the b_6f complex from spinach (Soriano et al., 1999) but not from *Synechocystis* sp. PCC 6803 (Peterman et al., 1998). In the b_6f complex from the marine green alga *Bryopsis corticulans*, which contains α-carotene instead of β-carotene, a chlorophyll *a*-to-α-carotene triplet excitation transfer was observed (Li et al., 2006). Since the structural data indicate that the chlorophyll is too far (14 Å) from the β-carotene for direct energy transfer, it has been proposed that residues close to the chlorophyll, especially a nearby tyrosine, rather than the β-carotene, might reduce the yield of the chlorophyll triplet state and singlet oxygen (Dashdorj et al., 2005). Alternatively, a long distance triplet energy transfer between the chlorophyll and β-carotene in the b_6f complex was proposed to occur via an O_2 molecule trapped in an intraprotein channel (Kim et al., 2005). Mutagenesis of residues close to the chlorophyll in the *Chlamydomonas* b_6f complex should elucidate additional aspects of the mechanism of photoprotection in the b_6f complex.

IV. BIOGENESIS OF THE b_6f COMPLEX

A. Introduction

The b_6f complex forms a hetero-oligomeric integral membrane protein dimer, whose biogenesis involves the assembly of at least eight different protein subunits and two *b*-type hemes, two *c*-type hemes, one Fe_2S_2, one chlorophyll *a*, and one β-carotene per monomer, as discussed in section IV. It requires the coordination of the chloroplast and nuclear genomes which both encode subunits (Table 17.1), and factors that coordinate their expression or catalyze the assembly of cofactors (Table 17.3). Heme, carotenoids, and chlorophyll biosynthesis are reviewed in Chapters 20 and 21. Iron-sulfur cluster biosynthesis occurs in the stroma and implicates both the SUF-system and other proteins (Kessler and Papenbrock, 2005).

B. Subunit assembly and degradation

1. Targeting and translocation of b_6f subunits

The b_6f subunits are transmembrane proteins that have to be targeted to their functional site in the thylakoid membrane. Cytochrome b_6f precursor subunits encoded by the nuclear genome are imported from the cytoplasm through the chloroplast envelope TIC/TOC complexes (Gutensohn et al., 2006). Their N-terminal transit peptides are cleaved by the stromal

Table 17.3 Nuclear and chloroplast proteins participating in the expression of chloroplast *pet* genes in *Chlamydomonas*

Gene	Function	Accession #	References	Notes
MCA1	*petA* mRNA stability/ maturation	AAY44840	Raynaud et al. (2007)	Target 5′UTR, PPR repeats
MCD1	*petD* mRNA stability/ maturation	AAK14341	Murakami et al. (2005)	Target 5′UTR, restricted to Volvocales
TCA1	*petA* mRNA translation	ABQ57415	Raynaud et al. (2007)	Target 5′UTR, restricted to Volvocales
ccsA[a]	*c*-Heme attachment to cyt *f* and c_6	AAB03815	Hamel et al. (2003); Xie and Merchant (1996)	Heme delivery, 5 TMH
CCS1	*c*-Heme attachment to cyt *f* and c_6	AAB95196	Dreyfuss et al. (2003); Inoue et al. (1997)	Heme delivery, 3 TMH
CCS2[b]	*c*-Heme attachment to cyt *f* and c_6		Xie et al. (1998)	
CCS3[b]	*c*-Heme attachment to cyt *f* and c_6		Xie et al. (1998)	
CCS4[b]	*c*-Heme attachment to cyt *f* and c_6		Xie et al. (1998)	Redox chemistry, motif RCXXC
CCS5[b]	*c*-Heme attachment to cyt *f* and c_6		Dreyfuss and Merchant (1999)	Redox chemistry
CCS6[b]	*c*-Heme attachment to cyt *f* and c_6		Page et al. (2004)	
CCDA[d]	*c*-Heme attachment to cyt *f* and c_6	AAL84598	Page et al. (2004)	Thiol disulfide transporter, 6 TMH[c]
HCF164[d]	*c*-Heme attachment to cyt *f* and c_6	EDP04704	Lennartz et al. (2001)	Thioredoxin like, 1 TMH
CCB1	c_i'-Heme attachment to cyt b_6	ABP57441	Kuras et al. (2007)	Heme delivery, 3 TMH
CCB2	c_i'-Heme attachment to cyt b_6	ABP57442	Kuras et al. (2007)	Heme delivery, 2 TMH
CCB3	c_i'-Heme attachment to cyt b_6	ABP57443	Kuras et al. (2007)	Heme delivery, 2 TMH
CCB4	c_i'-Heme attachment to cyt b_6	ABP57444	Kuras et al. (2007)	Heme delivery, 2 TMH

[a] *ccsA is chloroplast-encoded, all the others are nucleus-encoded.*

[b] *CCS2–CCS6 have not yet been molecularly characterized.*

[c] *TMH, number of transmembrane* α*-helices.*

[d] *CCDA and HCF164 were characterized in Arabidopsis mutants but not yet in Chlamydomonas mutants and are therefore only putative orthologues.*

processing peptidase, and folding is mediated by chaperones (see Chapter 19). Cytochrome b_6 and subunit IV have stromal N-termini, whereas the lumenal N-termini of PetA and PetO are targeted by a hydrophobic transit peptide with a consensus cleavage site AXA↓ for the lumenal processing peptidase (Kuras et al., 1995; Hamel et al., 2000). The N-termini of the four small hydrophobic subunits (PetG, PetL, PetM, and PetN) are located in the lumen (de Vitry et al., 1996) as predicted by the positive inside rule (Gavel et al., 1991) with no transit peptide in case of PetG, PetL, and PetN and no hydrophobic transit peptide in the case of PetM.

Protein translocation into or across the thylakoid membrane takes place by at least four independent pathways designated Sec, SRP, Tat, and spontaneous (Gutensohn et al., 2006). Cytochrome *f* is translocated by the Sec pathway which requires ATP and imports unfolded proteins (Rohl and van Wijk, 2001). The Rieske ISP has an uncleaved twin arginine (Tat) signal and is translocated by the Tat pathway, which imports folded proteins with their cofactors (Molik et al., 2001). Changes in efficiency of thylakoid targeting were observed *in vivo* after mutating the *Chlamydomonas* Rieske ISP Tat signal (Finazzi et al., 2003). Fe_2S_2 cluster binding is not a prerequisite for Rieske ISP translocation nor essential for its stable assembly into the b_6f complex, in isolated pea chloroplasts (Kapazoglou et al., 2000).

2. Concerted accumulation and steps of assembly

The attachment of cofactors is an early process in the biosynthesis of the b_6f complex since holocytochromes *f* and b_6 are detected after a 5-minute pulse labeling (Kuras et al., 1995). Cofactor attachment can occur independently of complex assembly, since the holo cytochromes are detected in deletion mutants of the other subunits (Kuras et al., 1997). The cofactors contribute to protein stability, given that the corresponding apocytochromes are unstable and poorly accumulated following site-directed mutagenesis of the heme ligands (Kuras et al., 1995, 1997; de Vitry et al., 2004a). The integration of the iron-sulfur cluster into the Rieske ISP occurs after chloroplast import and probably in the stroma, which contains both the SUF system and other proteins implicated in iron-sulfur cluster biosynthesis (Kessler and Papenbrock, 2005). The Rieske holoprotein would then be an appropriate substrate for the Tat translocation pathway.

The major transmembrane subunits encoded in the chloroplast, cytochrome *f*, cytochrome b_6, and subunit IV accumulate in a concerted manner (Kuras and Wollman, 1994). The concerted accumulation results from an assembly-controlled regulation of cytochrome *f* synthesis (see Chapter 29), and rapid degradation of most unassembled subunits (Choquet et al., 1998, 2001, 2003). Many factors that control the expression of chloroplast-encoded b_6f subunits at the level of mRNA stability/maturation, translation or post-translationally have been identified from mutants of *Chlamydomonas* and several corresponding genes have been cloned (Table 17.3). Expression of the *petA* gene depends on two specific nucleus-encoded factors in limiting concentrations: the short-lived pentatricopeptide repeat (PPR; sequence motif involved in protein/RNA interactions) protein MCA1 (Raynaud et al., 2007), required for stable accumulation of *petA* transcripts and the "pioneer" (no known homologue) protein TCA1 (Wostrikoff et al., 2001; Raynaud et al., 2007), required for its translation. The expression of *petD* depends on a specific nuclear factor MCD1 (Drager et al., 1998, 1999; Murakami et al., 2005), required for stable accumulation of the *petD* transcript and for limiting *petD* transcript degradation by 5′-3′-exonuclease activity. A small *cis*-acting sequence in the 5′-untranslated region (UTR) of *petD* mRNA is essential for its stability (Drager et al., 1999) and forms a small stem-loop structure (Higgs et al., 1999). Nuclear suppressor mutations of *mcd1* mutants or of *petD* 5′ UTR mutants define four nuclear loci *MCD2–MCD5* which can promote *petD* transcript stability, although *mcd3–mcd5* also have pleiotropic effects (Esposito et al., 2001; Rymarquis et al., 2006). *MCD2* was shown to encode a nonsense suppressor of an *mcd1* nonsense mutant (Murakami et al., 2005). Cytochrome *c* maturation factors have been characterized in *Chlamydomonas* which mediate covalent heme binding to cytochromes *f* and b_6 (see section IV.C).

Nuclear factors that control the expression of chloroplast-encoded b_6f subunits were also identified in higher plants. The PPR protein CRP1 was described in maize, and is required for *petA* and *petD* translation and processing of the *petD* transcript from a polycistronic precursor (Barkan et al., 1994; Fisk et al., 1999). An analogous *Arabidopsis* PPR protein, HCF152, is involved in the processing of the *psbB-psbT-psbH-petB-petD* transcript (Meierhoff et al., 2003; Nakamura et al., 2003), as are the 15-kD protein HCF153 (Lennartz et al., 2006), the thioredoxin-like protein HCF164 (Lennartz et al., 2001), and the disulfide transporter CCDA (Page et al., 2004). These factors are not ubiquitously found in chloroplasts; some such as TCA1 and MCD1 have so far been found only in Volvocales, whereas others such as HCF153 are found only in higher plants.

Chlamydomonas mutants deficient in the Rieske ISP accumulate a relatively stable dimer subcomplex (de Vitry et al., 1999). Mutants deficient in Rieske ISP have been also reported in *Arabidopsis* (Maiwald et al., 2003) and in cyanobacteria, where PetC1 appears to be the major Rieske ISP (Schneider et al., 2004). A first step of b_6f complex assembly comprises the concerted accumulation of a cytochrome subcomplex, with which the Rieske ISP subsequently assembles. This observation is in line with the rather peripheral position of the transmembrane helix of the Rieske ISP (Figure 17.1), and with fact that the Rieske ISP dissociates easily from purified b_6f complexes, especially in mutants with altered structures (Takahashi et al., 1996; Finazzi et al., 1997). The limited stability of the Rieske ISP-depleted dimer indicates that it probably contributes to dimer stability. Indeed, the transmembrane segment associated with one monomer traverses the membrane obliquely, and is connected by a flexible hinge to the soluble head associated with the other monomer (Kurisu et al., 2003; Stroebel et al., 2003). The Rieske ISP transmembrane helix (de Vitry et al., 2004b) was also reported to play a role in assembly mediated control of cytochrome *f* synthesis (Choquet et al., 2003).

3. Role of the small hydrophobic subunits PetG, PetL, PetM, and PetN

Knockout mutants obtained in *Chlamydomonas, Synechocystis* PCC6803, and *Nicotiana tabacum* (summarized in Table 17.2) indicate that only PetG and PetN are essential for b_6f complex biogenesis, stability and function. This conclusion is in perfect agreement with the positions of the small subunits in the *Chlamydomonas* structure where PetL and PetM are assigned more peripheral locations that PetG and PetN.

4. The loosely bound and reversibly phosphorylated PetO subunit

PetO is considered as a cytochrome b_6f subunit because (i) it copurifies with the other cytochrome b_6f subunits in early purification steps, (ii) it is deficient in mutants accumulating little of the b_6f complex, and (iii) it colocalizes with cytochrome *f*, which migrates between stacked and unstacked membrane regions during state transitions (Hamel et al., 2000). However, PetO is loosely bound and not present in highly purified b_6f complexes (Pierre et al., 1995; Stroebel et al., 2003). PetO is a single transmembrane helix with two large hydrophilic domains extending into the stromal and lumenal sides of the thylakoid membrane. The protein is reversibly phosphorylated upon state transition, which points to a possible role in the redox-controlled adaptation of the photosynthetic apparatus (Hamel et al., 2000). The absence of PetO homologues outside of the Volvocales suggests it that has a specific role in these organisms. A PetO mutant or deficient strain will be necessary to elucidate its function.

A possible distant functional orthologue of PetO is the 7.2-kD protein Ssr2298, reported to be loosely associated with the b_6f complex in *Synechocystis* sp. PCC6803, and to be important for complex function but not its assembly. This protein is restricted to cyanobacteria and was proposed to be involved in the regulation of phycobillisome movement during state transitions (Volkmer et al., 2007). It should be kept in mind, however, that the mechanisms of state transitions are very different in cyanobacteria and in eukaryotes.

5. Degradation of the b_6f complex

There is a rapid proteolytic degradation of unassembled subunit IV and cytochrome b_6 in mutants lacking one of the major transmembrane subunits (Kuras and Wollman, 1994). In wild-type *Chlamydomonas*, b_6f complexes tend to be lost when cultures are in late stationary phase (Finazzi et al., 1997) whereas in mutants with structural modifications in the cd loop of cytochrome b_6 (Finazzi et al., 1997), lacking PetL (Takahashi et al., 1996), or some Rieske ISP mutants (de Vitry et al., 1999), the complexes are lost upon entering stationary phase. A chloroplast-encoded ClpP protease was implicated in the degradation of assembled b_6f complex during nitrogen starvation, a condition which induces gametogenesis in the presence of light. ClpP does not seem to be involved in the degradation of unassembled subunits, but was shown to participate in the degradation of the b_6f complex in Rieske ISP mutants (Majeran et al., 2000). Unassembled Rieske ISP imported *in organello* and associated with the stromal side of the thylakoid membrane appeared to be degraded by the thylakoid membrane-bound protease FtsH (Ostersetzer and Adam, 1997).

C. Discovery of two cytochrome *c* maturation pathways

1. Introduction

Heme (Fe-protoporphyrin IX) is an essential cofactor for many fundamental biological processes. It is a hydrophobic and cytotoxic macrocycle, which therefore requires specific pathways to be safely delivered to its subcellular destinations. The *c*-type cytochromes are characterized by the covalent attachment of heme to the protein most frequently by two thioether bonds formed between the heme vinyl groups and the thiols of two cysteine residues in a conserved CXXCH motif, with histidine as one of the axial ligands to heme iron. Heme *c* binding comprises several critical steps including the reduction of the cysteine residues, the ferrous heme supply, and the heme binding. Factors involved in covalent heme binding have been extensively studied and this maturation usually occurs through one of three multifactorial systems known to operate on the p-side of the membrane, which faces the bacterial periplasm, the mitochondrial intermembrane space, or the plastid lumen (Kranz et al., 1998), each of which has a characteristic heme concentration dependence (Richard-Fogal et al., 2007).

The corresponding n-sides are the bacterial cytoplasm, the mitochondrial matrix, and the plastid stroma. Organisms performing oxygenic photosynthesis require a particular mechanism for heme binding on the n-side of the photosynthetic membrane, which involves two independent *c*-heme maturation pathways in the chloroplast (Table 17.3; Figure 17.5). These are system II (CCS; c-type cytochrome synthesis) for hemes *f*, c_6 and probably c_{6A}, and system IV (CCB; Cofactor assembly on complex C subunit B) for heme c_i'. Cytochrome c_6, widespread in cyanobacteria and algae, is functionally similar to the copper protein plastocyanin and its role is to transfer electrons between the b_6f complex and PS I (Kerfeld and Krogmann, 1998). In organisms encoding both plastocyanin and cytochrome c_6, the latter is

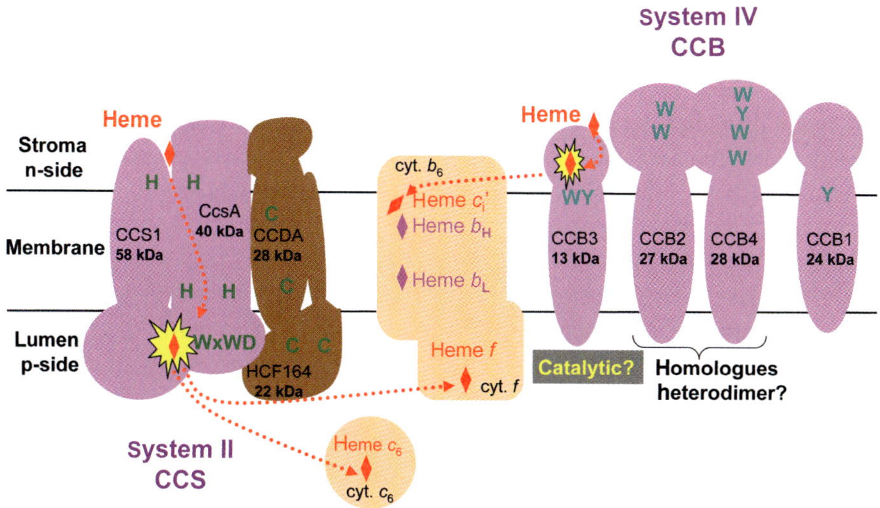

FIGURE 17.5 *Schematic representation of the two multi-factorial systems of c-heme ligation in the chloroplast. The system II (CCS) pathway ligates hemes f and c_6 on the lumenal face (p-side) of the membrane. The system IV (CCB) pathway ligates heme c_i' on the stromal face (n-side) of the membrane. Factors involved in a redox relay for the reduction of the cysteine residues are in brown. Factors involved in ferrous heme supply and heme binding are in purple. Hemes are shown as diamonds. The presumed active sites are shown in yellow with heme inside. The name, molecular mass and important residues of cloned factors are indicated. Important and hypothetically important residues are respectively labeled in green and blue.*

expressed under copper deficient conditions when plastocyanin is no longer available (see Chapter 10). Cytochrome c_{6A} is a dithio-cytochrome of low abundance located in the thylakoid lumen (reviewed in Marcaida et al., 2006) which has been identified both in higher plants (Wastl et al., 2002) and *Chlamydomonas* (Wastl et al., 2004).

The ease with which it is possible to generate and screen photosynthesis mutants in *Chlamydomonas* has been crucial for the discovery of cytochrome *c* maturation pathways. Most of the system II and system IV components were originally uncovered in *Chlamydomonas*: the *ccsA* and *CCS1* genes (Xie and Merchant 1996; Inoue et al., 1997), and the later characterization of four *CCB* genes (Kuras et al., 2007).

2. The system II (CCS) pathway

In *Chlamydomonas*, there are at least seven partners involved in the CCS system. Six of them (CCS1–CCS6) are nucleus-encoded (Inoue et al., 1997; Xie et al., 1998), and the seventh, CcsA, is chloroplast-encoded (Xie and Merchant, 1996). All are required for heme attachment to the apoforms of plastid *c*-type cytochromes (*f* and c_6) in the lumen. In *Arabidopsis*, the thioredoxin-like protein HCF164 (Lennartz et al., 2001) and the homologue of the prokaryotic thiol disulfide transporter CCDA (Page et al., 2004) are required for the maturation of holocytochrome *f*. The *CCS1*, *ccsA*, *CCDA*, and *HCF164* genes have been cloned while *CCS2*, *CCS3*, *CCS4*, *CCS5*, and *CCS6* are still only genetically defined (Table 17.3).

Several studies have addressed the transmembrane topology and the interactions between CCS partners. Functionally important residues such as Trp, His, Cys, and Asp (indicated in system II of Figure 17.5) have been defined by site-directed mutagenesis. Molecular investigations revealed that membrane-bound CCS1 (Dreyfuss et al., 2003) and CcsA (Hamel et al., 2003) are likely to together possess the heme delivery and cytochrome *c* ligation functions. The membrane-anchored HCF164 (Lennartz et al., 2001; Motohashi and Hisabori, 2006) with the transmembrane CCDA (Page et al., 2004) would be part of a redox relay required for the reduction of the two cysteines in the heme binding site of apocytochrome *c* (Figure 17.5). HCF164 is localized in the thylakoid membrane and not in the chloroplast envelope (Lennartz et al., 2001); its accumulation does not depend of the presence of CCDA (Page et al., 2004). Exogenous reduced thiol compounds can rescue *c*-type cytochrome maturation in *ccs4* and *ccs5* mutants (Page et al., 2004) suggesting their participation in this redox relay. In *E. coli*, a minimal recombinant system II comprising a single fused *ccsBA* polypeptide can replace the eight system I genes and perform heme delivery and periplasmic cytochrome *c* heme ligation (Feissner et al., 2006). CCS1 is proposed to interact with CcsA in the context of a 200-kD thylakoid membrane complex (Hamel et al., 2003). The accumulation of this complex was shown to be reduced in *ccs2*, *ccs3*, *ccs4*, and *ccsA* mutants which suggests that the complex includes other CCS factors (Dreyfuss et al., 2003).

3. The system IV (CCB) pathway

Genetic studies led to the identification of four *Chlamydomonas* nuclear *CCB* loci that control heme c_i' maturation Gumpel et al., 1995; Kuras et al., 1997, 2007; de Vitry et al., 2004a). The signature of apocytochrome b_6 impaired in heme c_i' binding was described for the *petB-C35V* mutant (de Vitry et al., 2004a), where the unique covalent cysteine ligand to heme c_i' was substituted. The signature consists of a double band in urea-SDS-polyacrylamide gels, or a single band of lower apparent molecular mass than in the wild-type in SDS-polyacrylamide gels concomitant with the lack of peroxidase activity. The same signature is observed in all *ccb* mutants, indicating that heme c_i' binding is prevented in each case. This entirely new biogenesis pathway comprises four protein factors (Table 17.3), the CCB factors, for which no function had been previously ascribed although they are present in all photosynthetic organisms having b_6f complexes (Kuras et al., 2007). After CCB pathway (system IV) characterization in *Chlamydomonas* (Kuras et al., 2007), a study of an *Arabidopsis ccb2* mutant was reported (Lyska et al., 2007).

CCB1 is ubiquitous among organisms performing oxygenic photosynthesis. CCB3 belongs to the large YGGT protein family (InterPro

IPR003425) found in both plastids and bacteria. This family of integral membrane proteins of unknown function was named after the *E. coli yggT* gene and its members are characterized by the eponymous YGGT motif. Photosynthetic eukaryotes encode up to four YGGT proteins, but for each organism only a single protein clusters with *Chlamydomonas* CCB3, defining a group of putative orthologues. CCB3 is the most conserved CCB factor, while CCB2 and CCB4 are paralogs with low similarity that derive from a unique cyanobacterial ancestor. We suggest that CCB2 and CCB4 probably act as an heterodimer in the chloroplast as in the case of PsaA and PsaB in PS I (Tsukatani et al., 2004).

The membrane localization of the four known CCB factors is consistent with their role in heme binding to cytochrome b_6 on its stromal face (Figure 17.5). In cytochrome *c* maturation systems I and II conserved Trp and Tyr residues have been identified as critical in heme interactions (Kranz et al., 1998; Uchida et al., 2004). The conserved Trp and Tyr residues in CCB1–CCB4 on the stromal face suggest, but remain to be tested by site-directed mutagenesis, a role of CCB1–CCB4 in heme chaperoning and delivery to apocytochrome b_6. We note that there are no conserved cysteine residues in any of the four CCB factors. This is in contrast to previously characterized maturation systems in which some cysteine residues have been identified as being critical for apocytochrome and heme reduction. There are several possible bases for this observation: it may reflect the higher reducing potential of the stroma and cyanobacterial cytoplasm versus that of the lumen or periplasm, the particular heme-binding process by a unique thioether bond, the easier access to ferrous heme produced by the neighboring ferrochelatase, or merely the existence of an additional protein factor that has so far escaped genetic screens. Generation of new *ccb* mutants should either establish that the four CCB factors represent the entire pathway, or identify additional CCB partners.

The presence of CCB homologues is restricted to organisms performing oxygenic photosynthesis. This applies even to *Bacillus subtilis*, whose atypical respiratory *bc* complex was shown to contain a covalent heme bound to its cytochrome b_6 subunit (Yu and Le Brun, 1998) and whose cytochrome *c* maturation system on the p-side (called RES) is similar to chloroplast CCS (Kranz et al., 1998). *B. subtilis* does encode a YGGT family member (YlmG) but it clusters away from the CCB3 branch in the phylogenetic tree (Kuras et al., 2007). The CCB proteins are among the few proteins ubiquitous to all organisms performing oxygenic photosynthesis.

V. CONCLUSION

While most b_6f subunits and cofactor maturation pathways (such as *c*-heme attachment CCB and CCS systems) seem conserved in all organisms performing oxygenic photosynthesis, some factors controlling b_6f complex gene expression or possibly signal transduction seem restricted to particular subclasses of organisms and are probably indicative of evolutionary adaptation to specific environmental factors. Examples are TCA1, MCD1, and PetO in *Chlamydomonas*, which have no homologues except in Volvocales, HCF153 which is specific to higher plants, and Ssr2998 which is restricted to cyanobacteria. Additional studies are required to elucidate the mechanisms of electron transfer in the b_6f complex, the cascade of events leading to kinase activation and state transitions, the molecular nature of the signals generated by the b_6f complex and its subsequent transduction to the cytosol/nucleus, and the biogenesis of the b_6f complex. The available *Chlamydomonas*, higher plant and cyanobacteria genome information and mutants will certainly continue to contribute in a complementary manner to these studies.

ACKNOWLEDGMENTS

Many thanks to Mariana Guergova-Kuras, Daniel Picot, and Francis-André Wollman for critical reading of the manuscript and to our colleagues of UMR 7141 for fruitful discussions. This work was supported by the Unité Mixte de Recherche 7141, Centre National de la Recherche Scientifique – Université Pierre et Marie Curie (Paris 6).

REFERENCES

Adamian, L. and Liang, J. (2006). Prediction of transmembrane helix orientation in polytopic membrane proteins. *BMC Struct. Biol.* **6**, 13.

Alric, J., Pierre, Y., Picot, D., Lavergne, J., and Rappaport, F. (2005). Spectral and redox characterization of the heme c_i of the cytochrome b_6f complex. *Proc. Natl. Acad. Sci. U. S. A.* **102**, 15860–15865.

Barkan, A., Walker, M., Nolasco, M., and Johnson, D. (1994). A nuclear mutation in maize blocks the processing and translation of several chloroplast mRNAs and provides evidence for the differential translation of alternative mRNA forms. *EMBO J.* **13**, 3170–3181.

Baymann, F., Giusti, F., Picot, D., and Nitschke, W. (2007). The c_i/b_H moiety in the b_6f complex studied by EPR: A pair of strongly interacting hemes. *Proc. Natl. Acad. Sci. U. S. A.* **104**, 519–524.

Bendall, D.S. and Manasse, R.S. (1995). Cyclic photophosphorylation and electron-transport. *Biochim. Biophys. Acta* **1229**, 23–38.

Berthold, D.A., Schmidt, C.L., and Malkin, R. (1995). The deletion of petG in *Chlamydomonas reinhardtii* disrupts the cytochrome *bf* complex. *J. Biol. Chem.* **270**, 29293–29298.

Breyton, C. (2000). Conformational changes in the cytochrome b_6f complex induced by inhibitor binding. *J. Biol. Chem.* **275**, 13195–13201.

Breyton, C., Nandha, B., Johnson, G.N., Joliot, P., and Finazzi, G. (2006). Redox modulation of cyclic electron flow around photosystem I in C3 plants. *Biochemistry* **45**, 13465–13475.

Brugna, M., Rodgers, S., Schricker, A., Montoya, G., Kazmeier, M., Nitschke, W., and Sinning, I. (2000). A spectroscopic method for observing the domain movement of the Rieske iron-sulfur protein. *Proc. Natl. Acad. Sci. U. S. A.* **97**, 2069–2074.

Burrows, P.A., Sazanov, L.A., Svab, Z., Maliga, P., and Nixon, P.J. (1998). Identification of a functional respiratory complex in chloroplasts through analysis of tobacco mutants containing disrupted plastid *ndh* genes. *EMBO J.* **17**, 868–876.

Chi, Y.I., Huang, L.S., Zhang, Z., Fernandez-Velasco, J.G., and Berry, E.A. (2000). X-ray structure of a truncated form of cytochrome *f* from *Chlamydomonas reinhardtii*. *Biochemistry* **39**, 7689–7701.

Choquet, Y. and Vallon, O. (2000). Synthesis, assembly and degradation of thylakoid membrane proteins. *Biochimie* **82**, 615–634.

Choquet, Y., Stern, D.B., Wostrikoff, K., Kuras, R., Girard-Bascou, J., and Wollman, F.A. (1998). Translation of cytochrome *f* is autoregulated through the 5′ untranslated region of *petA* mRNA in *Chlamydomonas* chloroplasts. *Proc. Natl. Acad. Sci. U. S. A.* **95**, 4380–4385.

Choquet, Y., Wostrikoff, K., Rimbault, B., Zito, F., Girard-Bascou, J., Drapier, D., and Wollman, F.A. (2001). Assembly-controlled regulation of chloroplast gene translation. *Biochem. Soc. Trans.* **29**, 421–426.

Choquet, Y., Zito, F., Wostrikoff, K., and Wollman, F.A. (2003). Cytochrome *f* translation in *Chlamydomonas* chloroplast is autoregulated by its carboxyl-terminal domain. *Plant Cell* **15**, 1443–1454.

Cramer, W.A. and Zhang, H. (2006). Consequences of the structure of the cytochrome b_6f complex for its charge transfer pathways. *Biochim. Biophys. Acta* **1757**, 339–345.

Cramer, W.A., Zhang, H., Yan, J., Kurisu, G., and Smith, J.L. (2004). Evolution of photosynthesis: Time-independent structure of the cytochrome b_6f complex. *Biochemistry* **43**, 5921–5929.

Cramer, W.A., Yan, J., Zhang, H., Kurisu, G., and Smith, J.L. (2005). Structure of the cytochrome b_6f complex: New prosthetic groups, Q-space, and the "hors d'oeuvres hypothesis" for assembly of the complex. *Photosynth. Res.* **85**, 133–143.

Cramer, W.A., Zhang, H., Yan, J., Kurisu, G., and Smith, J.L. (2006). Transmembrane traffic in the cytochrome b_6f complex. *Annu. Rev. Biochem.* **75**, 769–790.

Crofts, A.R. (2004). The cytochrome bc_1 complex: Function in the context of structure. *Annu. Rev. Physiol.* **66**, 689–733.

Crofts, A.R., Meinhardt, S.W., Jones, K.R., and Snozzi, M. (1983). The role of the quinone pool in the cyclic electron-transfer chain in *Rhodopseudomonas sphaeroides*: A modified Q-cycle mechanism. *Biochim. Biophys. Acta* **723**, 202–218.

Crofts, A.R., Guergova-Kuras, M., Huang, L., Kuras, R., Zhang, Z., and Berry, E.A. (1999). Mechanism of ubiquinol oxidation by the bc_1 complex: role of the iron sulfur protein and its mobility. *Biochemistry* **38**, 15791–15806.

Darie, C.C., Biniossek, M.L., Winter, V., Mutschler, B., and Haehnel, W. (2005). Isolation and structural characterization of the Ndh complex from mesophyll and bundle sheath chloroplasts of *Zea mays*. *FEBS J.* **272**, 2705–2716.

Dashdorj, N., Zhang, H., Kim, H., Yan, J., Cramer, W.A., and Savikhin, S. (2005). The single chlorophyll a molecule in the cytochrome b_6f complex: Unusual optical properties protect the complex against singlet oxygen. *Biophys. J.* **88**, 4178–4187.

de Vitry, C. and Vallon, O. (1999). Mutants of *Chlamydomonas*: Tools to study thylakoid membrane structure, function and biogenesis. *Biochimie* **81**, 631–643.

de Vitry, C., Breyton, C., Pierre, Y., and Popot, J.L. (1996). The 4-kDa nuclear-encoded PetM polypeptide of the chloroplast cytochrome b_6f complex. Nucleic acid and protein sequences, targeting signals, transmembrane topology. *J. Biol. Chem.* **271**, 10667–10671.

de Vitry, C., Finazzi, G., Baymann, F., and Kallas, T. (1999). Analysis of the nucleus-encoded and chloroplast-targeted Rieske protein by classic and site-directed mutagenesis of *Chlamydomonas*. *Plant Cell* **11**, 2031–2044.

de Vitry, C., Desbois, A., Redeker, V., Zito, F., and Wollman, F.A. (2004a). Biochemical and spectroscopic characterization of the covalent binding of heme to cytochrome b_6. *Biochemistry* **43**, 3956–3968.

de Vitry, C., Ouyang, Y., Finazzi, G., Wollman, F.A., and Kallas, T. (2004b). The chloroplast Rieske iron-sulfur protein. At the crossroad of electron transport and signal transduction. *J. Biol. Chem.* **279**, 44621–44627.

Deniau, C. and Rappaport, F. (2000). New insights on the proton pump associated with cytochrome b_6f turnovers from the study of H/D substitution effects on the electrogenicity and electron transfer reactions. *Biochemistry* **39**, 3304–3310.

Depège, N., Bellafiore, S., and Rochaix, J.D. (2003). Role of chloroplast protein kinase Stt7 in LHCII phosphorylation and state transition in *Chlamydomonas*. *Science* **299**, 1572–1575.

Drager, R.G., Girard-Bascou, J., Choquet, Y., Kindle, K.L., and Stern, D.B. (1998). In vivo evidence for $5' \rightarrow 3'$ exoribonuclease degradation of an unstable chloroplast mRNA. *Plant J.* **13**, 85–96.

Drager, R.G., Higgs, D.C., Kindle, K.L., and Stern, D.B. (1999). 5′ to 3′ exoribonucleolytic activity is a normal component of chloroplast mRNA decay pathways. *Plant J.* **19**, 521–531.

Dreyfuss, B.W. and Merchant, S. (1999). *CCS5*, a new locus required for chloroplast *c*-type cytochrome synthesis. In: *Photosynthesis: Mechanisms and Effects*, Vol. IV (G. Garab, Ed.), pp. 3221–3226. Kluwer Academic Publishers, The Netherlands.

Dreyfuss, B.W., Hamel, P.P., Nakamoto, S.S., and Merchant, S. (2003). Functional analysis of a divergent system II protein, Ccs1, involved in *c*-type cytochrome biogenesis. *J. Biol. Chem.* **278**, 2604–2613.

Esposito, D., Higgs, D.C., Drager, R.G., Stern, D.B., and Girard-Bascou, J. (2001). A nucleus-encoded suppressor defines a new factor which can promote *petD* mRNA stability in the chloroplast of *Chlamydomonas reinhardtii*. *Curr. Genet.* **39**, 40–48.

Feissner, R.E., Richard-Fogal, C.L., Frawley, E.R., Loughman, J.A., Earley, K.W., and Kranz, R.G. (2006). Recombinant cytochromes c biogenesis systems I and II and analysis of haem delivery pathways in *Escherichia coli*. *Mol. Microbiol.* **60**, 563–577.

Fiebig, A., Stegemann, S., and Bock, R. (2004). Rapid evolution of RNA editing sites in a small non-essential plastid gene. *Nucleic Acids Res.* **32**, 3615–3622.

Finazzi, G. (2005). The central role of the green alga *Chlamydomonas reinhardtii* in revealing the mechanism of state transitions. *J. Exp. Bot.* **56**, 383–388.

Finazzi, G., Büschlen, S., de Vitry, C., Rappaport, F., Joliot, P., and Wollman, F.A. (1997). Function-directed mutagenesis of the cytochrome b_6f complex in *Chlamydomonas reinhardtii*: Involvement of the cd loop of cytochrome b_6 in quinol binding to the Q_o site. *Biochemistry* **36**, 2867–2874.

Finazzi, G., Zito, F., Barbagallo, R.P., and Wollman, F.A. (2001). Contrasted effects of inhibitors of cytochrome b_6f complex on state transitions in *Chlamydomonas reinhardtii*: The role of Q_o site occupancy in LHCII kinase activation. *J. Biol. Chem.* **276**, 9770–9774.

Finazzi, G., Rappaport, F., Furia, A., Fleischmann, M., Rochaix, J.D., Zito, F., and Forti, G. (2002). Involvement of state transitions in the switch between linear and cyclic electron flow in *Chlamydomonas reinhardtii*. *EMBO Reports* **3**, 280–285.

Finazzi, G., Chasen, C., Wollman, F.A., and de Vitry, C. (2003). Thylakoid targeting of Tat passenger proteins shows no delta pH dependence *in vivo*. *EMBO J.* **22**, 807–815.

Fisk, D.G., Walker, M.B., and Barkan, A. (1999). Molecular cloning of the maize gene *crp1* reveals similarity between regulators of mitochondrial and chloroplast gene expression. *EMBO J.* **18**, 2621–2630.

Forquer, I., Covian, R., Bowman, M.K., Trumpower, B.L., and Kramer, D.M. (2006). Similar transition states mediate the Q-cycle and superoxide production by the cytochrome bc_1 complex. *J. Biol. Chem.* **281**, 38459–38465.

Gavel, Y., Steppuhn, J., Herrmann, R., and von Heijne, G. (1991). The "positive-inside rule" applies to thylakoid membrane proteins. *FEBS Lett.* **282**, 41–46.

Godde, D. (1982). Evidence for a membrane-bound NADH-plastoquinone-oxidoreductase in *Chlamydomonas reinhardii CW-15*. *Arch. Microbiol.* **131**, 197–202.

Gumpel, N.J., Ralley, L., Girard-Bascou, J., Wollman, F.A., Nugent, J.H., and Purton, S. (1995). Nuclear mutants of *Chlamydomonas reinhardtii* defective in the biogenesis of the cytochrome b_6f complex. *Plant Mol. Biol.* **29**, 921–932.

Gutensohn, M., Fan, E., Frielingsdorf, S., Hanner, P., Hou, B., Hust, B., and Klosgen, R.B. (2006). Toc, Tic, Tat et al.: structure and function of protein transport machineries in chloroplasts. *J. Plant Physiol.* **163**, 333–347.

Hager, M., Biehler, K., Illerhaus, J., Ruf, S., and Bock, R. (1999). Targeted inactivation of the smallest plastid genome-encoded open reading frame reveals a novel and essential subunit of the cytochrome b_6f complex. *EMBO J.* **18**, 5834–5842.

Hamel, P., Olive, J., Pierre, Y., Wollman, F.A., and de Vitry, C. (2000). A new subunit of cytochrome b_6f complex undergoes reversible phosphorylation upon state transition. *J. Biol. Chem.* **275**, 17072–17079.

Hamel, P.P., Dreyfuss, B.W., Xie, Z., Gabilly, S.T., and Merchant, S. (2003). Essential histidine and tryptophan residues in CcsA, a system II polytopic cytochrome c biogenesis protein. *J. Biol. Chem.* **278**, 2593–2603.

Higgs, D.C., Shapiro, R.S., Kindle, K.L., and Stern, D.B. (1999). Small *cis*-acting sequences that specify secondary structures in a chloroplast mRNA are essential for RNA stability and translation. *Mol. Cell. Biol.* **19**, 8479–8491.

Inoue, K., Dreyfuss, B.W., Kindle, K.L., Stern, D.B., Merchant, S., and Sodeinde, O.A. (1997). *Ccs1*, a nuclear gene required for the post-translational assembly of chloroplast *c*-type cytochromes. *J. Biol. Chem.* **272**, 31747–31754.

Iwata, S., Lee, J.W., Okada, K., Lee, J.K., Iwata, M., Rasmussen, B., Link, T.A., Ramaswamy, S., and Jap, B.K. (1998). Complete structure of the 11-subunit bovine mitochondrial cytochrome bc_1 complex. *Science* **281**, 64–71.

Joliot, P. and Joliot, A. (1986). Proton pumping and electron transfer in the cytochrome b_6f complex of algae. *Biochim. Biophys. Acta* **849**, 211–222.

Joliot, P. and Joliot, A. (1988). The low-potential electron-transfer chain in the cytochrome b_6f complex. *Biochim. Biophys. Acta* **933**, 319–333.

Joliot, P. and Joliot, A. (1994). Mechanism of electron transfer in the cytochrome b_6f complex of algae: Evidence for a semiquinone cycle. *Proc. Natl. Acad. Sci. U. S. A.* **91**, 1034–1038.

Joliot, P. and Joliot, A. (2002). Cyclic electron transfer in plant leaf. *Proc. Natl. Acad. Sci. U. S. A.* **99**, 10209–10214.

Joliot, P. and Joliot, A. (2006). Cyclic electron flow in C3 plants. *Biochim. Biophys. Acta* **1757**, 362–368.

Jones, R.W. and Whitmarsh, J. (1988). Inhibition of electron transfer and electrogenic reaction in the cytochrome b/f complex by 2-*n*-nonyl-4-hydroxyquinoline *N*-oxide (NQNO) and 2,5-dibromo-3-methyl-6-isopropyl-p-benzoquinone (DBMIB). *Biochim. Biophys. Acta* **933**, 258–268.

Kapazoglou, A., Mould, R.M., and Gray, J.C. (2000). Assembly of the Rieske iron-sulphur protein into the cytochrome *bf* complex in thylakoid membranes of isolated pea chloroplasts. *Eur. J. Biochem.* **267**, 352–360.

Kerfeld, C.A. and Krogmann, D.W. (1998). Photosynthetic cytochromes *c* in cyanobacteria, algae, and plants. *Annu. Rev. Plant Physiol. Plant Mol. Biol.* **49**, 397–425.

Kessler, D. and Papenbrock, J. (2005). Iron-sulfur cluster biosynthesis in photosynthetic organisms. *Photosynth. Res.* **86**, 391–407.

Kim, H., Xia, D., Yu, C.A., Xia, J.Z., Kachurin, A.M., Zhang, L., Yu, L., and Deisenhofer, J. (1998). Inhibitor binding changes domain mobility in the iron-sulfur protein of the mitochondrial bc_1 complex from bovine heart. *Proc. Natl. Acad. Sci. U. S. A.* **95**, 8026–8033.

Kim, H., Dashdorj, N., Zhang, H., Yan, J., Cramer, W.A., and Savikhin, S. (2005). An anomalous distance dependence of intraprotein chlorophyll-carotenoid triplet energy transfer. *Biophys. J.* **89**, L28–L30.

Kramer, D.M., Joliot, A., Joliot, P., and Crofts, A.R. (1994). Competition among plastoquinol and artificial quinone/quinol couples at the quinol oxidizing site of the cytochrome *bf* complex. *Biochim. Biophys. Acta* **1184**, 251–262.

Kramer, D.M., Avenson, T.J., and Edwards, G.E. (2004). Dynamic flexibility in the light reactions of photosynthesis governed by both electron and proton transfer reactions. *Trends Plant Sci.* **9**, 349–357.

Kranz, R., Lill, R., Goldman, B., Bonnard, G., and Merchant, S. (1998). Molecular mechanisms of cytochrome *c* biogenesis: Three distinct systems. *Mol. Microbiol.* **29**, 383–396.

Kuras, R. and Wollman, F.A. (1994). The assembly of cytochrome b_6f complexes: An approach using genetic transformation of the green alga *Chlamydomonas reinhardtii*. *EMBO J.* **13**, 1019–1027.

Kuras, R., Büschlen, S., and Wollman, F.A. (1995). Maturation of pre-apocytochrome *f in vivo*. A site-directed mutagenesis study in *Chlamydomonas reinhardtii*. *J. Biol. Chem.* **270**, 27797–27803.

Kuras, R., de Vitry, C., Choquet, Y., Girard-Bascou, J., Culler, D., Buschlen, S., Merchant, S., and Wollman, F.A. (1997). Molecular genetic identification of a pathway for heme binding to cytochrome b_6. *J. Biol. Chem.* **272**, 32427–32435.

Kuras, R., Saint-Marcoux, D., Wollman, F.A., and de Vitry, C. (2007). A specific c-type cytochrome maturation system is required for oxygenic photosynthesis. *Proc. Natl. Acad. Sci. U. S. A.* **104**, 9906–9910.

Kurisu, G., Zhang, H., Smith, J.L., and Cramer, W.A. (2003). Structure of the cytochrome b_6f complex of oxygenic photosynthesis: Tuning the cavity. *Science* **302**, 1009–1014.

Lavergne, J. (1983). Membrane potential-dependent reduction of cytochrome b_6 in an algal mutant lacking photosystem I centers. *Biochim. Biophys. Acta* **725**, 25–33.

Lemaire, C., Girard-Bascou, J., and Wollman, F.-A. (1987). Characterization of the b_6f complex subunits and studies on the LHC-kinase in *Chlamydomonas reinhardtii* using mutant strains altered in the b_6f complex. In: *Progress in Photosynthesis Research*, (J. Biggins, Ed.), pp. 655–658. Martinus Nijhoff: Dordrecht, The Netherlands.

Lennartz, K., Plucken, H., Seidler, A., Westhoff, P., Bechtold, N., and Meierhoff, K. (2001). HCF164 encodes a thioredoxin-like protein involved in the biogenesis of the cytochrome b_6f complex in *Arabidopsis*. *Plant Cell* **13**, 2539–2551.

Lennartz, K., Bossmann, S., Westhoff, P., Bechtold, N., and Meierhoff, K. (2006). HCF153, a novel nuclear-encoded factor necessary during a post-translational step in biogenesis of the cytochrome *bf* complex. *Plant J.* **45**, 101–112.

Li, B.X., Zuo, P., Chen, X.B., Li, L.B., Zhang, J.P., and Kuang, T.Y. (2006). Study on energy transfer between carotenoid and chlorophyll a in cytochrome b_6f complex from *Bryopsis corticulans*. *Photosynth. Res.* **88**, 43–50.

Lunde, C., Jensen, P.E., Rosgaard, L., Haldrup, A., Gilpin, M.J., and Scheller, H.V. (2003). Plants impaired in state transitions can to a large degree compensate for their defect. *Plant Cell Physiol.* **44**, 44–54.

Lyska, D., Paradies, S., Meierhoff, K., and Westhoff, P. (2007). HCF208: An homologue of *Chlamydomonas* CCB2 is required for accumulation of native cytochrome b_6 in *Arabidopsis thaliana*. *Plant Cell Physiol.* **48**, 1737–1746.

Maiwald, D., Dietzmann, A., Jahns, P., Pesaresi, P., Joliot, P., Joliot, A., Levin, J.Z., Salamini, F., and Leister, D. (2003). Knock-out of the genes coding for the Rieske protein and the ATP-synthase delta-subunit of *Arabidopsis*. Effects on photosynthesis, thylakoid protein composition, and nuclear chloroplast gene expression. *Plant Physiol.* **133**, 191–202.

Majeran, W., Wollman, F.A., and Vallon, O. (2000). Evidence for a role of ClpP in the degradation of the chloroplast cytochrome b_6f complex. *Plant Cell* **12**, 137–150.

Malkin, R. (1992). Cytochrome bc_1 and b_6f complexes of photosynthetic membranes. *Photosynth. Res.* **33**, 121–136.

Mao, H.B., Li, G.F., Ruan, X., Wu, Q.Y., Gong, Y.D., Zhang, X.F., and Zhao, N.M. (2002). The redox state of plastoquinone pool regulates state transitions via cytochrome b_6f complex in *Synechocystis* sp. PCC 6803. *FEBS Lett.* **519**, 82–86.

Marcaida, M.J., Schlarb-Ridley, B.G., Worrall, J.A., Wastl, J., Evans, T.J., Bendall, D.S., Luisi, B.F., and Howe, C.J. (2006). Structure of cytochrome c_{6A}, a novel dithio-cytochrome of *Arabidopsis thaliana*, and its reactivity with plastocyanin: Implications for function. *J. Mol. Biol.* **360**, 968–977.

Maul, J.E., Lilly, J.W., Cui, L., dePamphilis, C.W., Miller, W., Harris, E.H., and Stern, D.B. (2002). The *Chlamydomonas reinhardtii* plastid chromosome: Islands of genes in a sea of repeats. *Plant Cell* **14**, 2659–2679.

Meierhoff, K., Felder, S., Nakamura, T., Bechtold, N., and Schuster, G. (2003). HCF152, an *Arabidopsis* RNA binding pentatricopeptide repeat protein involved in the processing of chloroplast *psbB-psbT-psbH-petB-petD* RNAs. *Plant Cell* **15**, 1480–1495.

Merchant, S. and Dreyfuss, B.W. (1998). Posttranslational assembly of photosynthetic metalloproteins. *Annu. Rev. Plant Physiol. Plant Mol. Biol.* **49**, 25–51.

Merchant, S. and Sawaya, M.R. (2005). The light reactions: A guide to recent acquisitions for the picture gallery. *Plant Cell* **17**, 648–663.

Mitchell, P. (1975). Protonmotive redox mechanism of cytochrome bc_1 complex in the respiratory chain: Protonmotive ubiquinone cycle. *FEBS Lett.* **56**, 1–6.

Miyake, C., Horiguchi, S., Makino, A., Shinzaki, Y., Yamamoto, H., and Tomizawa, K. (2005a). Effects of light intensity on cyclic electron flow around PSI and its relationship to non-photochemical quenching of Chl fluorescence in tobacco leaves. *Plant Cell Physiol.* **46**, 1819–1830.

Miyake, C., Miyata, M., Shinzaki, Y., and Tomizawa, K. (2005b). CO_2 response of cyclic electron flow around PSI (CEF-PSI) in tobacco leaves – relative electron fluxes through PSI and PSII determine the magnitude of non-photochemical quenching (NPQ) of Chl fluorescence. *Plant Cell Physiol.* **46**, 629–637.

Molik, S., Karnauchov, I., Weidlich, C., Herrmann, R.G., and Klosgen, R.B. (2001). The Rieske Fe/S protein of the cytochrome b_6f complex in chloroplasts: Missing link in the evolution of protein transport pathways in chloroplasts? *J. Biol. Chem.* **276**, 42761–42766.

Motohashi, K. and Hisabori, T. (2006). HCF164 receives reducing equivalents from stromal thioredoxin across the thylakoid membrane and mediates reduction of target proteins in the thylakoid lumen. *J. Biol. Chem.* **281**, 35039–35047.

Munekage, Y., Hojo, M., Meurer, J., Endo, T., Tasaka, M., and Shikanai, T. (2002). PGR5 is involved in cyclic electron flow around photosystem I and is essential for photoprotection in *Arabidopsis*. *Cell* **110**, 361–371.

Munekage, Y., Hashimoto, M., Miyake, C., Tomizawa, K., Endo, T., Tasaka, M., and Shikanai, T. (2004). Cyclic electron flow around photosystem I is essential for photosynthesis. *Nature* **429**, 579–582.

Murakami, S., Kuehnle, K., and Stern, D.B. (2005). A spontaneous tRNA suppressor of a mutation in the *Chlamydomonas reinhardtii* nuclear *MCD1* gene required for stability of the chloroplast *petD* mRNA. *Nucleic Acids Res.* **33**, 3372–3380.

Nakamoto, S.S., Hamel, P., and Merchant, S. (2000). Assembly of chloroplast cytochromes *b* and *c*. *Biochimie* **82**, 603–614.

Nakamura, T., Meierhoff, K., Westhoff, P., and Schuster, G. (2003). RNA-binding properties of HCF152, an *Arabidopsis* PPR protein involved in the processing of chloroplast RNA. *Eur. J. Biochem.* **270**, 4070–4081.

Nelson, M.E., Finazzi, G., Wang, Q.J., Middleton-Zarka, K.A., Whitmarsh, J., and Kallas, T. (2005). Cytochrome b_6 arginine 214 of *Synechococcus* sp. PCC 7002, a key residue for quinone-reductase site function and turnover of the cytochrome *bf* complex. *J. Biol. Chem.* **280**, 10395–10402.

Ogawa, T. (1991). A gene homologous to the subunit-2 gene of NADH dehydrogenase is essential to inorganic carbon transport of *Synechocystis* PCC6803. *Proc. Natl. Acad. Sci. U. S. A.* **88**, 4275–4279.

Ohyama, K., Fukuzawa, H., Kohchi, T., Shirai, H., Sano, T., Sano, S., Umesono, K., Shiki, Y., Takeuchi, M., Chang, Z., Aota, S., Inokuchi, H., and Ozeki, H. (1986). Chloroplast gene organization deduced from complete sequence of liverwort *Marchantia polymorpha* chloroplast DNA. *Nature* **322**, 572–574.

Ostersetzer, O. and Adam, Z. (1997). Light-stimulated degradation of an unassembled Rieske FeS protein by a thylakoid-bound protease: The possible role of the FtsH protease. *Plant Cell* **9**, 957–965.

Page, M.L., Hamel, P.P., Gabilly, S.T., Zegzouti, H., Perea, J.V., Alonso, J.M., Ecker, J.R., Theg, S.M., Christensen, S.K., and Merchant, S. (2004). A homolog of prokaryotic thiol disulfide transporter CcdA is required for the assembly of the cytochrome b_6f complex in *Arabidopsis* chloroplasts. *J. Biol. Chem.* **279**, 32474–32482.

Peterman, E.J., Wenk, S.O., Pullerits, T., Palsson, L.O., van Grondelle, R., Dekker, J.P., Rogner, M., and van Amerongen, H. (1998). Fluorescence and absorption spectroscopy of the weakly fluorescent chlorophyll a in cytochrome b_6f of *Synechocystis* PCC6803. *Biophys. J.* **75**, 389–398.

Pierre, Y., Breyton, C., Kramer, D., and Popot, J.L. (1995). Purification and characterization of the cytochrome b_6f complex from *Chlamydomonas reinhardtii*. *J. Biol. Chem.* **270**, 29342–29349.

Pierre, Y., Breyton, C., Lemoine, Y., Robert, B., Vernotte, C., and Popot, J.L. (1997). On the presence and role of a molecule of chlorophyll *a* in the cytochrome b_6f complex. *J. Biol. Chem.* **272**, 21901–21908.

Raynaud, C., Loiselay, C., Wostrikoff, K., Kuras, R., Girard-Bascou, J., Wollman, F.-A., and Choquet, Y. (2007). Evidence for regulatory function of nucleus-encoded factors on mRNA stabilization and translation in the chloroplast. *Proc. Natl. Acad. Sci. U. S. A.* **104**, 9093–9098.

Richard-Fogal, C.L., Frawley, E.R., Feissner, R.E., and Kranz, R.G. (2007). Heme concentration dependence and metalloporphyrin inhibition of the system I and II cytochrome *c* assembly pathways. *J. Bacteriol.* **189**, 455–463.

Rohl, T. and van Wijk, K.J. (2001). *In vitro* reconstitution of insertion and processing of cytochrome *f* in a homologous chloroplast translation system. *J. Biol. Chem.* **276**, 35465–35472.

Rumeau, D., Becuwe-Linka, N., Beyly, A., Louwagie, M., Garin, J., and Peltier, G. (2005). New subunits NDH-M, -N, and -O, encoded by nuclear genes, are essential for plastid Ndh complex functioning in higher plants. *Plant Cell* **17**, 219–232.

Rymarquis, L.A., Higgs, D.C., and Stern, D.B. (2006). Nuclear suppressors define three factors that participate in both 5′ and 3′ end processing of mRNAs in *Chlamydomonas* chloroplasts. *Plant J.* **46**, 448–461.

Sazanov, L.A., Burrows, P.A., and Nixon, P.J. (1998). The plastid *ndh* genes code for an NADH-specific dehydrogenase: Isolation of a complex I analogue from pea thylakoid membranes. *Proc. Natl. Acad. Sci. U. S. A.* **95**, 1319–1324.

Schneider, D., Berry, S., Rich, P., Seidler, A., and Rogner, M. (2001). A regulatory role of the PetM subunit in a cyanobacterial cytochrome b_6f complex. *J. Biol. Chem.* **276**, 16780–16785.

Schneider, D., Berry, S., Volkmer, T., Seidler, A., and Rogner, M. (2004). PetC1 is the major Rieske iron-sulfur protein in the cytochrome b_6f complex of *Synechocystis* sp. PCC 6803. *J. Biol. Chem.* **279**, 39383–39388.

Schneider, D., Volkmer, T., and Rogner, M. (2007). PetG and PetN, but not PetL, are essential subunits of the cytochrome b_6f complex from *Synechocystis* PCC 6803. *Res. Microbiol.* **158**, 45–50.

Schottler, M.A., Flugel, C., Thiele, W., and Bock, R. (2007). Knock-out of the plastid-encoded PetL subunit results in reduced stability and accelerated leaf age-dependent loss of the cytochrome b_6f complex. *J. Biol. Chem.* **282**, 976–985.

Schwenkert, S., Legen, J., Takami, T., Shikanai, T., Herrmann, R.G., and Meurer, J. (2007). Role of the low-molecular-weight subunits PetL, PetG, and PetN in assembly, stability, and dimerization of the cytochrome b_6f complex in tobacco. *Plant Physiol.* **144**, 1924–1935.

Shao, N., Vallon, O., Dent, R., Niyogi, K.K., and Beck, C.F. (2006). Defects in the cytochrome b_6f complex prevent light-induced expression of nuclear genes involved in chlorophyll biosynthesis. *Plant Physiol.* **141**, 1128–1137.

Shinozaki, K., Ohme, M., Tanaka, M., Wakasugi, T., Hayashida, N., Matsubayashi, T., Zaita, N., Chunwongse, J., Obokata, J., Yamaguchi-Shinozaki, K. et al. (1986).

The complete nucleotide sequence of the tobacco chloroplast genome – its gene organization and expression. *EMBO J.* **5**, 2043–2049.

Smith, J.L., Zhang, H., Yan, J., Kurisu, G., and Cramer, W.A. (2004). Cytochrome *bc* complexes: A common core of structure and function surrounded by diversity in the outlying provinces. *Curr. Opin. Struct. Biol.* **14**, 432–439.

Soriano, G.M., Ponamarev, M.V., Carrell, C.J., Xia, D., Smith, J.L., and Cramer, W.A. (1999). Comparison of the cytochrome bc_1 complex with the anticipated structure of the cytochrome b_6f complex: le plus ça change le plus c'est la même chose. *J. Bioenerg. Biomembr.* **31**, 201–213.

Stroebel, D., Choquet, Y., Popot, J.L., and Picot, D. (2003). An atypical haem in the cytochrome b_6f complex. *Nature* **426**, 413–418.

Takabayashi, A., Kishine, M., Asada, K., Endo, T., and Sato, F. (2005). Differential use of two cyclic electron flows around photosystem I for driving CO_2-concentration mechanism in C4 photosynthesis. *Proc. Natl. Acad. Sci. U. S. A.* **102**, 16898–16903.

Takahashi, Y., Rahire, M., Breyton, C., Popot, J.L., Joliot, P., and Rochaix, J.D. (1996). The chloroplast *ycf7* (*petL*) open reading frame of *Chlamydomonas reinhardtii* encodes a small functionally important subunit of the cytochrome b_6f complex. *EMBO J.* **15**, 3498–3506.

Tsukatani, Y., Miyamoto, R., Itoh, S., and Oh-Oka, H. (2004). Function of a PscD subunit in a homodimeric reaction center complex of the photosynthetic green sulfur bacterium *Chlorobium tepidum* studied by insertional gene inactivation. Regulation of energy transfer and ferredoxin-mediated NADP+ reduction on the cytoplasmic side. *J. Biol. Chem.* **279**, 51122–51130.

Uchida, T., Stevens, J.M., Daltrop, O., Harvat, E.M., Hong, L., Ferguson, S.J., and Kitagawa, T. (2004). The interaction of covalently bound heme with the cytochrome c maturation protein CcmE. *J. Biol. Chem.* **279**, 51981–51988.

Volkmer, T., Schneider, D., Bernat, G., Kirchhoff, H., Wenk, S.O., and Rogner, M. (2007). Ssr2998 of *Synechocystis* sp. PCC 6803 is involved in regulation of cyanobacterial electron transport and associated with the cytochrome b_6f complex. *J. Biol. Chem.* **282**, 3730–3737.

Wastl, J., Bendall, D.S., and Howe, C.J. (2002). Higher plants contain a modified cytochrome c_6. *Trends Plant Sci.* **7**, 244–245.

Wastl, J., Purton, S., Bendall, D.S., and Howe, C.J. (2004). Two forms of cytochrome c_6 in a single eukaryote. *Trends Plant Sci.* **9**, 474–476.

Wollman, F.A. (1998). The structure, function and biogenesis of cytochrome b_6f complexes. In: *The Molecular Biology of Chloroplasts and Mitochondria in Chlamydomonas* (J.D. Rochaix, M. Goldschmidt-Clermont, and S. Merchant, Eds.), pp. 459–476. Kluwer Academic Publishers, The Netherlands.

Wollman, F.A. (2001). State transitions reveal the dynamics and flexibility of the photosynthetic apparatus. *EMBO J.* **20**, 3623–3630.

Wollman, F.A. and Lemaire, C. (1988). Studies on kinase-controlled state transitions in Photosystem II and b_6f mutants from *Chlamydomonas reinhardtii* which lack quinone-binding proteins. *Biochim. Biophys. Acta* **933**, 85–94.

Wollman, F.A., Minai, L., and Nechustai, R. (1999). The biogenesis and assembly of photosynthetic proteins in thylakoid membranes. *Biochim. Biophys. Acta* **1411**, 21–85.

Wostrikoff, K., Choquet, Y., Wollman, F.A., and Girard-Bascou, J. (2001). TCA1, a single nuclear-encoded translational activator specific for *petA* mRNA in *Chlamydomonas reinhardtii* chloroplast. *Genetics* **159**, 119–132.

Xia, D., Yu, C.A., Kim, H., Xia, J.Z., Kachurin, A.M., Zhang, L., Yu, L., and Deisenhofer, J. (1997). Crystal structure of the cytochrome bc_1 complex from bovine heart mitochondria. *Science* **277**, 60–66.

Xia, D., Kim, H., Yu, C.A., Yu, L., Kachurin, A., Zhang, L., and Deisenhofer, J. (1998). A novel electron transfer mechanism suggested by crystallographic studies of mitochondrial cytochrome bc_1 complex. *Biochem. Cell Biol.* **76**, 673–679.

Xie, Z. and Merchant, S. (1996). The plastid-encoded *ccsA* gene is required for heme attachment to chloroplast *c*-type cytochromes. *J. Biol. Chem.* **271**, 4632–4639.

Xie, Z. and Merchant, S. (1998). A novel pathway for cytochrome *c* biogenesis in chloroplasts. *Biochim. Biophys. Acta* **1365**, 309–318.

Xie, Z., Culler, D., Dreyfuss, B.W., Kuras, R., Wollman, F.A., Girard-Bascou, J., and Merchant, S. (1998). Genetic analysis of chloroplast *c*-type cytochrome assembly in *Chlamydomonas reinhardtii*: One chloroplast locus and at least four nuclear loci are required for heme attachment. *Genetics* **148**, 681–692.

Yamamoto, H., Kato, H., Shinzaki, Y., Horiguchi, S., Shikanai, T., Hase, T., Endo, T., Nishioka, M., Makino, A., Tomizawa, K., and Miyake, C. (2006). Ferredoxin limits cyclic electron flow around PSI (CEF-PSI) in higher plants stimulation of CEF-PSI enhances non-photochemical quenching of Chl fluorescence in transplastomic tobacco. *Plant Cell Physiol.* **47**, 1355–1371.

Yamashita, E., Zhang, H., and Cramer, W.A. (2007). Structure of the cytochrome b_6f complex: Quinone analogue inhibitors as ligands of heme c_n. *J. Mol. Biol.* **370**, 39–52.

Yan, J. and Cramer, W.A. (2004). Molecular control of a bimodal distribution of quinone-analogue inhibitor binding sites in the cytochrome b_6f complex. *J. Mol. Biol.* **344**, 481–493.

Yu, J. and Le Brun, N.E. (1998). Studies of the cytochrome subunits of menaquinone:cytochrome *c* reductase (*bc* complex) of *Bacillus subtilis*. Evidence for the covalent attachment of heme to the cytochrome *b* subunit. *J. Biol. Chem.* **273**, 8860–8866.

Zatsman, A.I., Zhang, H., Gunderson, W.A., Cramer, W.A., and Hendrich, M.P. (2006). Heme-heme interactions in the cytochrome b_6f complex: EPR spectroscopy and correlation with structure. *J. Am. Chem. Soc.* **128**, 14246–14247.

Zhang, Z., Huang, L., Shulmeister, V.M., Chi, Y.I., Kim, K.K., Hung, L.W., Crofts, A.R., Berry, E.A., and Kim, S.H. (1998). Electron transfer by domain movement in cytochrome bc_1. *Nature* **392**, 677–684.

Zhang, H., Huang, D., and Cramer, W.A. (1999). Stoichiometrically bound beta-carotene in the cytochrome b_6f complex of oxygenic photosynthesis protects against oxygen damage. *J. Biol. Chem.* **274**, 1581–1587.

Zhang, H., Whitelegge, J.P., and Cramer, W.A. (2001). Ferredoxin: NADP+ oxidoreductase is a subunit of the chloroplast cytochrome b_6f complex. *J. Biol. Chem.* **276**, 38159–38165.

Zhang, H., Primak, A., Cape, J., Bowman, M.K., Kramer, D.M., and Cramer, W.A. (2004). Characterization of the high-spin heme x in the cytochrome b_6f complex of oxygenic photosynthesis. *Biochemistry* **43**, 16329–16336.

Zito, F., Finazzi, G., Delosme, R., Nitschke, W., Picot, D., and Wollman, F.A. (1999). The Q_o site of cytochrome b_6f complexes controls the activation of the LHCII kinase. *EMBO J.* **18**, 2961–2969.

Zito, F., Vinh, J., Popot, J.L., and Finazzi, G. (2002). Chimeric fusions of subunit IV and PetL in the b_6f complex of *Chlamydomonas reinhardtii*: Structural implications and consequences on state transitions. *J. Biol. Chem.* **277**, 12446–12455.

CHAPTER 18

The CF_0F_1 ATP Synthase Complex of Photosynthesis

Giovanni Finazzi, Dominique Drapier and Fabrice Rappaport
Centre National de la Recherche Scientifique,
Institut de Biologie Physico-Chimique, Paris, France

CHAPTER CONTENTS

I. INTRODUCTION

Activity of the CF_0F_1 ATP synthase complex (hereafter, "ATP synthase") in thylakoid membranes is an essential feature of photosynthesis. Photosynthetic electron flow, which ultimately generates reducing power, is coupled to the production of a transmembrane electrochemical potential difference. This proton motive force is at the origin of a transmembrane proton flux which is coupled by the ATP synthase to ATP synthesis.

Understanding the activity of this enzyme is therefore essential for the comprehension of overall cellular energetics. However, this can be a challenging task, particularly *in vivo*, due to the stringent requirement for light-driven ATP synthesis in photosynthetic organisms. Clearly, the use of a facultative autotroph such as *Chlamydomonas* confers a great advantage for both structure/function studies and physiological characterization (e.g. Strotmann et al., 1998; Wollman et al., 1999). Indeed, mutants where the ATP synthase has been modified or is completely absent can still grow with respiratory metabolism on medium containing a reduced carbon source. In this chapter, we outline the current understanding of ATP synthase assembly and function.

II. COMPOSITION, STRUCTURE AND SPECIFIC FEATURES OF THE ATP SYNTHASE

A. Subunit composition

In chloroplasts, the ATP synthase comprises nine different subunits arranged in two subcomplexes. The first is CF_0, which is embedded in the thylakoid membranes and responsible for proton translocation across the thylakoid membrane into the lumen. The second is the catalytic moiety CF_1 which is extrinsic, exposed to the stroma and responsible for the production of the ATP or its hydrolysis to ADP and phosphate molecules.

CF_1 consists of five subunits, α, β, γ, δ, and ε, assembled in a 3:3:1:1:1 stoichiometry (Figure 18.1). ATP synthesis requires energy from a transmembrane electrochemical proton gradient. As discussed below, the transmembrane electrochemical potential is converted into mechanical work by the CF_0 subcomplex. Thus CF_1 cannot catalyze ATP synthesis when it is detached from CF_0. It can, however, catalyze ATP hydrolysis in this form. In mitochondria, by contrast, F_1 becomes inactive when detached from F_0. The three catalytic nucleotide-binding sites are borne by the three β subunits, and located at their interfaces with the α subunits. Each of the three α subunits possesses one non-catalytic adenine nucleotide binding site located at the interface with the β subunits.

CF_0 is composed of four subunits designated I, II, III, and IV. Whereas subunits I, II, and IV are present in single copies, subunit III is an

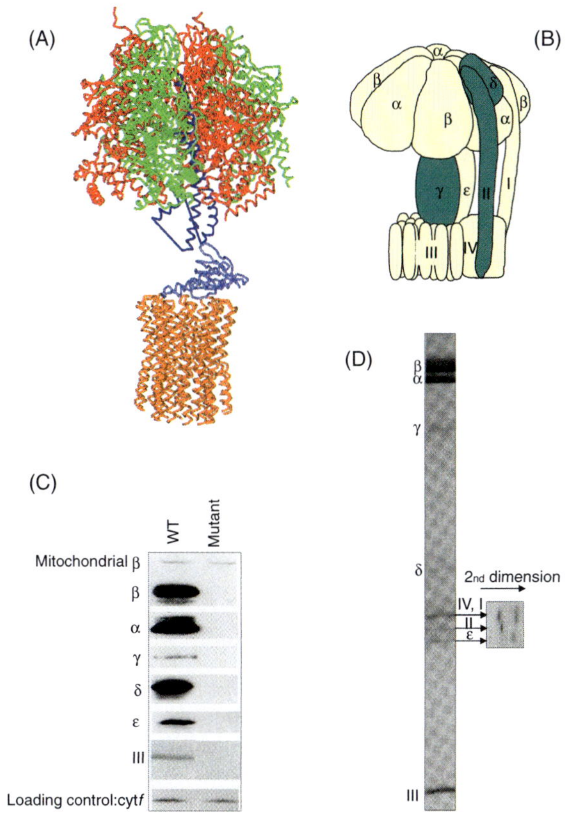

FIGURE 18.1 *Subunit composition and structure of the (C)F_0F_1 complex. (A) Structure of the yeast mitochondrial F_0F_1 ATPase showing the c_{10}-ring of the F_0 complex and the $\alpha\beta_3\gamma\varepsilon$ subunits; redrawn from the pdb file 2HLD (Stock et al., 1999). (B) Illustration of the chloroplast protein complex from Wollman et al. (1999); the nucleus-encoded subunits are in green. (C) Immunoblot detection of different subunits of the complex, performed with samples from whole wild type (WT) and mda1–thm24 mutant cells, analyzed in a 12–18% polyacrylamide gel containing 8 M urea. (D) Biochemical analysis, through a 12–18% polyacrylamide gel containing 8 M urea, followed by a second dimension in a 7.5–15% polyacrylamide gel for the 15–20-kD region, and Coomassie blue staining.*

oligomer arranged in a ring-like structure with organism-dependent stoichiometry, ranging from 10 to 15 monomeric units (Figure 18.1). Subunit III was reported to form a cylinder of 13 monomers in *Chlamydomonas* (Meyer zu Tittingdorf et al., 2004), whereas 14 monomers are present in the CF_0 of spinach (Seelert et al., 2003).

B. Mutants

With the exception of subunit II, at least one mutant for each subunit of the complex has been isolated in *Chlamydomonas* (Strotmann et al., 1998; Harris, 2001; Dent et al., 2005). No additional subunit in the complex has been identified, either by biochemical studies or by genetic screens. One should keep in mind, however, that the screens used to date select only strains with a totally deficient enzyme, thus mutants partially impaired for the ATP synthase function or strains having thylakoid membranes with a reduced density of ATP synthase complexes, would not have been detected. Screening for *Chlamydomonas* ATP synthase mutants is based initially on an acetate requirement, then on a fluorescence test following several minutes exposure to $70 \mu E/m^2/s$ illumination. In dim light, the fluorescence pattern of an ATP synthase mutant is quite similar to that of the wild type, whereas an enhanced fluorescence yield occurs upon exposure to strong light (Bennoun and Béal, 1998; Majeran et al., 2001). Strains accumulating less than half the number of complexes present in the wild type do not show such light-dependent changes in their fluorescence patterns, and retain the ability to grow phototautotrophically (Drapier et al., 2002). This is indicative of the great efficiency of the chloroplast ATP synthase in *Chlamydomonas*, whose thylakoid membranes are less rich in ATP synthase complexes than those of spinach (Strotmann et al., 1998). No changes in the subunit composition of the chloroplast ATP synthase complex have been reported among different photosynthetic eukaryotes, in marked contrast to the mitochondrial complex (see Chapter 13).

C. Genetic origin, organization, and regulation of the *ATP* genes

One characteristic of the chloroplast ATP synthase is the dual genetic origin of both CF_1 and CF_0 subunits. This contrasts with mitochondria, where all F_1 subunits are nucleus-encoded and all F_0 subunits are mitochondrially encoded in most organisms (although not in *Chlamydomonas*; see Chapter 12). The arrangement of the chloroplast-localized ATP synthase genes is variable: two clusters are present in land plants, a large one which includes *atpI* (subunit IV), *atpH* (subunit III), *atpF* (subunit I), *atpA* (subunit α), and a small one containing *atpB* (subunit β) and *atpE* (subunit ε), while independent dispersed genes are present in *Chlamydomonas*, with the exception of *atpA* and *atpH*. The *ATPC* (subunit γ) gene was probably the first to migrate to the nuclear genome following endosymbiosis, followed by *ATPD* (subunit δ) and *ATPG* (subunit II) (Strotmann et al., 1998; Table 18.1). In the *Chlamydomonas* chloroplast genome, *atpA* and *atpH* belong to a tetracistronic transcription unit grouping *atpA*, *psbI*, *cemA*, and *atpH* as first described by Levy et al. (1997); mono-, as well as polycistronic transcripts are used for translation of the α subunit, *atpA* being the first gene of the transcription unit (Drapier et al., 1998). On the other hand, the last gene of the transcription unit, *atpH*, has its own promoter and cannot be translated from the polycistronic mRNAs. This is observed in the nuclear mutant *ac46*, in which the monocistronic *atpH* transcript is destabilized and subunit III is absent, despite the normal accumulation of polycistronic transcripts containing *atpH* (Majeran et al., 2001).

Table 18.1 ATP synthase genes and corresponding mutants

Gene	Subunit	Accession #	mutants	References
atpA	α	NP_958406	*Chloroplast*	
			Δ*atpA*: deletion of the gene	Drapier et al. (2007)
			CW15Δ*atpA*: partial replacement of *atpA* by *aadA* cassette	Fiedler et al. (1997)
			***FUD16*:** two missense mutations in the coding region	Ketchner et al. (1995)
			Nuclear	
			***F54*:** impaired translation of *atpA* mRNA	Drapier et al. (1992)
			***ncc1*:** reduced accumulation of *atpA* mRNA	Drapier et al. (2002)
atpB	β	NP_958414	*Chloroplast*	
			***ac-u-c-2-21*, *FUD50*:** 2.3 kbp deletion in *atpB* gene	Woessner et al. (1984)
			CW15Δ*atpB*: partial replacement of *atpB* by *aadA* cassette	Fiedler et al. (1997)
			***ac-u-c-2-9*:** one point mutation in the coding region	Robertson et al. (1989)
			Nuclear	
			***thm24*:** no accumulation of *atpB* mRNA	Drapier et al. (1992)
ATPC	γ	EDP08312	*Nuclear*	
			***ATPC1*:** DNA insertion into *ATPC*	Smart and Selman (1991); Drapier et al. (2007)
ATPD	δ	ED097493	*Nuclear*	
			CAL007.02.03: DNA insertion into *ATPD*	Dent et al. (2005)
atpE	ε	CAA37928	*Chloroplast*	
			***FUD17*:** frameshift in the coding sequence	Robertson et al. (1990); Johnson and Melis (2004)
atpF	Su I	AAN41265	*Chloroplast*	
			FUD18: mutation not determined	Lemaire and Wollman (1989)
ATPG	Su II	EDP00272	No isolated mutant	
atpH	Su III	CAA62149	No known mutant in chloroplast gene;	
			Nuclear	
			***ac46*:** mutant shows no accumulation of monocistronic *atpH* mRNA	Lemaire and Wollman (1989); Majeran et al. (2001)
atpI	Su IV	AAC16328	*Chloroplast*	
			FUD23: mutation not determined	Lemaire and Wollman (1989)

The phenotypes of nuclear mutants affected in the synthesis of the ATP synthase point to a similar regulation of the chloroplast *atp* genes and of chloroplast genes encoding other photosynthetic complex subunits. For example, specific nuclear factors control stability of the transcripts. This is the case for *MDB1*, the mutation of which, in *thm24*, prevents stabilization of *atpB* mRNA. Similarly, *MDH1*, absent in *ac46*, is required for stabilization of *atpH* monocistronic transcripts (Majeran et al., 2001). Other factors promote translation initiation such as *TDA1*, whose absence in the mutant *F54* prevents *atpA* translation (Drapier et al., 1992).

The role of the 3′ untranslated region (UTR) of the *atpB* gene in the stabilization and the translatability of the mRNA has been extensively studied and is a model for 3′ end maturation: an endonucleolytic cleavage site is located downstream of the mature 3′ end. This 3′ end coincides with a protective stem-loop-forming inverted repeat, and the cleavage is followed by 3′-5′-exonucleolytic resection (Stern and Kindle, 1993; Hicks et al., 2002). The *atpB* transcript was also shown to be transiently polyadenylated: 12 poly(A) addition sites were mapped in the 3′ UTR and one in the stem-loop. Cleavage in the stem-loop would potentiate mRNA degradation (Komine et al., 2000), but this would not be the case for poly(A) tails added downstream of the inverted repeat (Komine et al., 2002). It was reported that normally processed *atpB* mRNA was translatable whereas longer *atpB* transcripts that were rescued in a suitable genetic context were not loaded on polysomes (Rott et al., 1998). A nuclear mutation, *crp3*, has been found to act at the 3′ end of *atpB* mRNA, making *atpB* transcripts lacking a 3′ inverted repeat more stable. However this mutation is not gene-specific since it affects the formation of other transcript 3′ ends, in particular those of *psbI* and *cemA*, which are both in the *atpA* transcription unit (Levy et al., 1999).

A nuclear factor with unusual properties has been revealed by the analysis of the spontaneous *ncc1* mutation. This mutation defines a locus encoding the factor MDA1, whose mutated allele acts in the coding region of *atpA* to destabilize the transcript (Drapier et al., 2002). Although only 1/10 of the wild-type *atpA* monocistronic transcript level is found in the mutant, the high efficiency of translation still allows accumulation of 50% of active ATP synthase complexes in the thylakoid membrane. The presence of *cis*-acting elements in a coding region of a gene (like the one in *atpA*) has been also reported for prokaryotes, animal, and fungal cells to control mRNA half-life (Ross, 1995; Hennigan and Jacobson, 1996; Kulkarni and Golden, 1997; Tierney and Medcalf, 2001).

Another unusual characteristic is that the *ncc1* mutation is semidominant. This allows interpretation of the mutant phenotype as either the loss of protection of an endonucleolytic cleavage site by a MDA1 multimeric protein complex, or if MDA1 were itself an endonuclease, changes in the regulation or target specificity of the mutated protein. The phenotype of *ncc1* illustrates the idiosyncratic behavior of the *atpA* transcript, which has been found to be rather short-lived, with only a small fraction of the mRNA pool engaged in translation (Eberhard et al., 2002).

In *Chlamydomonas*, there are three in-frame ATGs that could serve as an initiation codons for β-subunit synthesis. Despite the high overall conservation of the *atpB* coding region, the N-terminal of the protein is rather divergent, with a shorter N-terminal sequence in other unicellular algae as compared to *Chlamydomonas*. Protein sequencing suggested that β-subunit synthesis initiates at the second methionine codon (Fiedler et al., 1995). This was confirmed in studies using chloroplast transformation, where mutagenesis demonstrated that only the second ATG was able to initiate translation *in vivo* (Rimbault et al., 2000).

D. Complex assembly

Biogenesis of the ATP synthase occurs in distinct contexts in chloroplasts and mitochondria. In mitochondria, the F_0 and F_1 subcomplexes can assemble independently and as mentioned above, all F_1 subunits are synthesized in the cytosol before import, while all F_0 subunits are generally synthesized in the organelle (Ackerman and Tzagoloff, 2005). The situation is more complicated for the chloroplast ATP synthase as each subcomplex requires the two genetic compartments to deliver the constituent subunits in the proper stoichiometry. In addition, neither CF_1 nor CF_0 can accumulate in the absence of the other (Lemaire and Wollman, 1989). For the major chloroplast-encoded subunits of the other photosynthetic complexes, a coupling between the rates of synthesis of several subunits encoded by the chloroplast genome has been documented, and a negative feedback exerted by some unassembled subunits on the initiation of their translation has been described in the cases of the PS I, PS II, and b_6f complexes (see Chapter 29).

The coupling between the rates of α and β synthesis develops in a different context. First, instead of assembling subunits in a 1:1 stoichiometry as in the other photosynthetic complexes, there is a need to produce α and β in threefold excess with respect to their assembly partners. Second, assembly of CF_1 occurs in the stroma whereas the CF_0 is assembled instead within the lipid bilayer. How the α and β subunits assemble in correct stoichiometry with the γ subunit is now understood (see Drapier et al., 2007 for a discussion). The lack of γ subunits results in a significant decrease in the synthesis of β subunits, which in turn represses synthesis of α subunits (Figure 18.2). The biosynthesis of the two main subunits of CF_1, and consequently of the whole complex, as the two subcomplexes cannot accumulate separately, is therefore under the control of the nucleus, the γ subunit being nucleus-encoded. At variance with the negative feedback mechanisms described for the biogenesis of the two photosystems and the b_6f complex, subunit α does

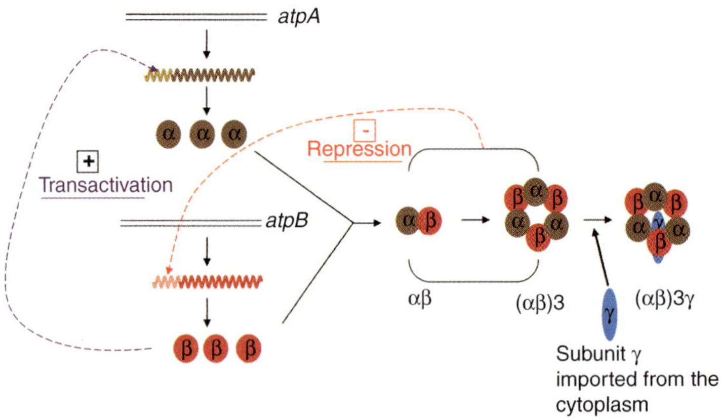

FIGURE 18.2 *Model for the regulation of subunits α and β synthesis, from Drapier et al. (2007). Two intertwined regulation loops participate in the biogenesis of the ATP synthase complex: (A) Subunit β trans-activates translation initiation of atpA mRNA by acting on its 5′ UTR through an unknown mechanism. (B) Oligomers of subunits α and β, when accumulated in the chloroplast in excess over subunit γ, negatively regulate the synthesis of subunit β, at the level of atpB mRNA translation initiation. Consequently the synthesis of subunit α is also decreased.*

not autoregulate; instead its synthesis is *trans*-activated by the β subunit. This mechanism presumably circumvents the propensity of the α subunit to self-aggregate (Ketchner et al., 1995; Lefebvre-Legendre et al., 2005), which would prevent an efficient autoregulatory negative feedback.

Coupling between the synthesis of the chloroplast-encoded subunits III and IV has also been observed, with the absence of subunit III resulting in a decrease of subunit IV synthesis (Lemaire and Wollman, 1989). Again, this coupling involves subunits that assemble in greatly uneven stoichiometries, and further studies should be conducted to elucidate the biogenesis of CF_0 and the assembly of the other subunits δ, ε, I, and II. Subunit γ has emerged as having a major role in the assembly process, and in the coordination of expression from the two genomes. The dramatic drop in β subunit synthesis, followed by that of subunit α, that occurs when *Chlamydomonas* cells are briefly incubated with cycloheximide, an inhibitor of cytosolic translation and thus of γ subunit synthesis (Drapier et al., 2007), illustrates the powerful control of a nucleus-encoded subunit on the synthesis of the chloroplast complex and the possibility of a rapid adaptation of the cells to metabolic change.

The synthesis and assembly of the chloroplast α3/β3 CF_1 hexamer follows a distinct pathway as compared to the mitochondrial hexamer α3/β3, whose assembly in yeast requires the chaperones Atp11p and Atp12p (Ackerman and Tzagoloff, 2005). No such chaperones have been found for the chloroplast counterpart. Other differences probably exist at the level of CF_0 and F_0 assembly. Not only can F_0 accumulate in the absence of F_1, but the ring of subunit c can accumulate in the mitochondrial membranes in the absence of other F_0 subunits (Ackerman and Tzagoloff, 2005).

E. Structure of the CF_0F_1 complex: monomer and dimer

The structure of the (C)F_1 $\alpha3/\beta3/\gamma$ heptamer is well known from its crystallization (Abrahams et al., 1994) and will not be discussed in detail here. Reviews are available for the mitochondrial (Leslie and Walker, 2000) and chloroplast (Groth and Pohl, 2001) complexes. The roles of the other subunits and their interactions in the complex are under intensive study.

Escherichia coli has been a useful model for studies of the δ subunit (Wilkens et al., 2000; Weber et al., 2003) and led to its positioning at the top of the F_1, with the helices 1 and 5 of the 6 α-helices of the N-terminal ensuring the binding with F_1; the two b subunits (the *E. coli* equivalent of chloroplast subunits I and II) of the F_0 are stretched between the membranes to the top of the F_1 to make the contact with the δ subunit. Work with spinach has shown that the N-terminal of the δ subunit interacts with the CF_1 subcomplex, while its C-terminal interacts with CF_0 (Ni et al., 2004).

Subunit ε must be necessary for ATP synthesis, partly because membranes with CF_1 lacking subunit ε are leaky to protons; the β sandwich N-terminal domain of the molecule interacts with the c-ring of F_0 and subunit γ (Cipriano and Dunn, 2006). The C-terminal of the protein (an antiparallel coil of two α helices) is not required for ATP synthesis *per se* but likely plays a regulatory role (Nowak et al., 2002). Using an artificial system containing CF_1-stripped thylakoids, CF_1-ε subunits and recombinant ε subunits expressed in *E. coli*, these authors observed a decreased inhibition of ATPase activity when the ε subunit lacked its last 45 amino acids.

At variance with the bacterial system, ε subunit is required for the association of CF_1 and CF_0 in chloroplasts, as no ATP synthase subunits are assembled in a *Chlamydomonas* mutant impaired in ε subunit synthesis (Lemaire and Wollman, 1989). Characterization of *Chlamydomonas* transformants with ε subunit C-terminal truncations revealed no alterations in the ATP synthase assembly or function with six amino acids removed, but truncations of more than ten amino acids prevented the recovery of phototrophic strains, probably because of a lack of enzyme assembly (Johnson and Melis, 2004).

Subunits I and II are part of the CF_0 subcomplex and the counterparts of subunit b in *E. coli* where two copies of the gene exist; this presence of two different subunits in the photosynthetic organisms should allow a better adaptation to the thylakoid membranes. If subunit I is more divergent, there is a strong homology between subunits II and b. It is the C-terminal of the proteins that interact with the CF_1, while their N-termini are anchored in the membrane by a single α helix. Interactions between coiled-coil regions stabilize the association between subunits I and II. The coiled-coil region predicted for *Chlamydomonas* subunit I comprises 60 residues, as compared to 100 residues in *E. coli* subunit b or 40 residues for spinach or *Arabidopsis* subunit I (Poetsch et al., 2007). The subunit II coiled-coil region is similar to that of subunit b and comprises 100 residues in higher plants (Poetsch et al., 2007). A conserved pattern with two positive charges within a coiled-coil region, first identified at the C-terminal of spinach subunit II, is present in the primary sequences of subunit II of most photosynthetic organisms but not *Volvox* or *Chlamydomonas* (Poetsch et al., 2007). Subunit IV is the subunit of the stator which is embedded in the thylakoid membrane; the structure and function of subunit a (its equivalent in *E. coli*) has been studied by several groups (reviewed in Junge et al., 1997; see also Vik and Ishmukhametov, 2005; Angevine et al., 2007). Subunit a consists of five transmembrane helices and plays a role in proton translocation across the membrane, in interaction with subunits c.

Dimerization of the ATP synthase was first described for the yeast mitochondrial complex (Arnold et al., 1998), and was considered as a unique characteristic of the mitochondrial ATP synthase. Dimerization requires the involvement of two specific subunits (e and g) (Giraud et al., 2002), and plays a role in the stability of the complex (van Lis et al., 2003). It also helps modulate the local membrane structure (Paumard et al., 2002). No relationship between changes in complex dimerization and in the metabolic state of the cell has been reported so far (van Lis et al., 2003). Work with *Chlamydomonas* yielded the first evidence for dimerization in the chloroplast (Rexroth et al., 2004); these authors did not note any novel subunits in the chloroplast dimer, at variance with the mitochondrial counterpart. The dimer dissociates when thylakoids are prepared in presence of vanadate or phosphate (but not molybdate), while the *Chlamydomonas* mitochondrial dimer is very stable in the presence of the same molecules. This sensitivity of the chloroplast dimer to vanadate, a phosphatase inhibitor, led to a model where the dimerization interfaces are located in regions of CF_1 that bear the phosphate and vanadate-binding sites (Rexroth et al., 2004). These authors proposed a role for subunit I in association with the $\alpha3/\beta3$ hexamers, as it shows homology with subunits that are involved in dimerization of the mitochondrial enzyme. For the mitochondrial dimer, the dimerization interface has been located on or close to the peripheral stalk of the complex (Fronzes et al., 2006).

Dimerization could allow a better efficiency of the enzyme in minimizing the loss of energy for ATP synthesis due to the rotation of the stator (Rexroth et al., 2004). Schwaßmann et al. (2007) proposed a relationship between the proportion of dimers in the chloroplast and the physiological state of the cells. The proportion of dimers is higher in *Chlamydomonas* cells grown photoautotrophically than in cells grown in the light on acetate-containing medium, where the ATP is produced both by photosynthesis and metabolism of acetate. Stable isotope labeling further showed that the dimers are formed from pre-existing monomers.

III. MECHANISTIC ASPECTS OF THE SYNTHESIS/ HYDROLYSIS OF ATP

A. ATP synthesis

Along the lines of the basic principle of the chemiosmotic theory established by Mitchell (1961, 1966), the driving force for ATP synthesis is provided by the $\Delta\tilde{\mu}_{H^+}$ generated by the photosynthetic or respiratory electron transport chain. The ATP synthase couples the transmembrane proton flux driven by $\Delta\tilde{\mu}_{H^+}$ to the synthesis of ATP. Accordingly, the coupling between ATP synthesis and the transmembrane proton transport is described by equation (18.1).

$$\mathrm{ADP} + P_i + \mathrm{nH_{in}}^+ \leftrightarrow \mathrm{ATP} + \mathrm{H_2O} + \mathrm{nH_{out}}^+ \qquad (18.1)$$

The Gibbs free energy, ΔG_p, associated with this reaction is given by equation (18.2):

$$\Delta G_p = \Delta G_p^0 + RT \ln \frac{[ATP]}{[ADP].[P_i]} - n.\Delta\tilde{\mu}_{H^+} \qquad (18.2)$$

where $[ATP]$, $[ADP]$, $[P_i]$ are the sums of the molar concentrations of all the ionic species of each compound, ΔG_p^0 is the standard Gibbs free energy of the reaction and n, the number of translocated protons per synthesized ATP. The success of the chemiosmotic theory resides in its ability to formally and simply describe the function of the enzyme without any assumptions as to the chemical nature of the transient states involved, nor on the precise mechanisms underlying the coupling between the transmembrane proton current and the enzymatic reaction. Results from numerous studies aimed at unraveling the structure and function of the ATP synthase now allow one to go far beyond this phenomenological description and to unravel the coupling mechanisms at the molecular level.

An important issue raised by equation (18.2) is that at equilibrium:

$$\Delta G_p^0 + RT \ln \frac{[ATP]}{[ADP].[P_i]} = n.\Delta\tilde{\mu}_{H^+} \qquad (18.3)$$

so that, n, the H^+/ATP ratio, is the key parameter which determines, for a given $[ATP]/[ADP].[P_i]$ ratio, the amplitude of the transmembrane electrochemical potential. In principle, this parameter directly follows from the structure and subunit composition of the enzyme. Indeed, all F-type, H^+-ATPases have similar structures, consisting of two main entities, the membrane-inserted part (F_0), and the hydrophilic one (F_1), which binds the nucleotide and phosphate. As discussed above, the subunit composition of the (C)F_1 subcomplex is $(\alpha\,\beta)_3\gamma\,\delta\,\varepsilon$. The (C)$F_0$ subcomplex is composed of essentially four subunits (I)b, (II)b, an oligomer of subunit (III)c assembled in a ring-like structure and subunit (IV)a. Subunits I and II are anchored in the thylakoid membrane and connect CF_1 and CF_0 at the periphery of the complex, stabilizing the entire edifice (see Poetsch et al., 2007 for discussion). Functional studies demonstrated that the catalytic nucleotide-binding sites are borne by the three β subunits which, during the turnover of the enzyme, sequentially adopt three conformations differing in their substrate binding properties (open, tight or loose; Boyer, 1997). These conformational changes were proposed to result from the rotation of the central stalk, the γ subunit, which would successively interact with the three β subunits in each of the three $\alpha\beta$ pairs. The γ subunit would thus act as an axle-tree and provide the mechanical work which would be converted into chemical work and bond formation by conformation changes of the β subunit.

Numerous studies using a wide range of approaches have supported the idea that ATP hydrolysis induces the rotation of the γ subunit (reviewed in Boyer, 1997; Junge et al., 1997; Capaldi and Aggeler, 2002; Dimroth et al., 2003). These include the sequential formation of a disulfide bridge between engineered γ and β subunits (Duncan et al., 1995; Figure 18.3), the relaxation of the polarization anisotropy of a probe attached to the γ subunit of an immobilized $(\alpha\beta)_3\gamma$ complex (Sabbert et al., 1996), and fluorescence microscopy with a fluorescent actin filament attached to the γ subunit of an immobilized $(\alpha\beta)_3\gamma$ complex (Noji et al., 1997). Furthermore, a fluorescent actin filament linked to the c subunit showed that the rotation of the γ subunit is effectively transmitted to the c-ring (Sambongi et al., 1999). Since ATP hydrolysis induces rotation of the γ subunit which itself drags

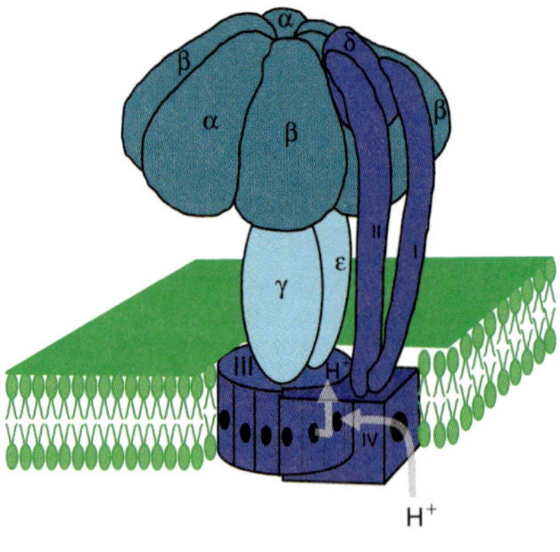

FIGURE 18.3 *H^+ driven ATP synthesis in the CF_0F_1 complex. The CF_0F_1 complex can be seen as a molecular motor, in which rotation of the transmembrane CF_0 subcomplex (subunits III) is transmitted via a rotating stalk (γ and ε) to the extrinsic CF1 portion. There, conformational changes in the $\alpha_3\beta_3$ subcomplex lead to the synthesis of ATP from ADP and Pi. $\alpha_3\beta_3$ rotation is prevented by a stator (subunits δ, I, II and IV). Rotation is due to protonation of the III subunits, which are accessible to H^+ by two independent channels, opened to the lumen and the stroma. In the presence of a $\Delta\tilde{\mu}_{H^+}$, protonation via the lumenal channel (horizontal arrow) induces a rotation of CF_0, which allows deprotonation on the stromal side (vertical arrow).*

along the rotation of the (III)c-ring, obvious reversibility requirements for the function of the enzyme imply that ATP synthesis is driven by the rotation of the (III)c-ring and the γ central stalk. This was experimentally demonstrated by forcing the rotation of the c-ring and ensuing ATP synthesis by magnetic tweezers propelling a magnetic bead linked to one of the subunits of the c-ring (Rondelez et al., 2005). The present view of the coupling between the transmembrane proton current and the synthesis of ATP is thus that the proton flux induces the rotation of the (III)c-ring, which acts as an electric motor and drags along the rotation of the central stalk γ, which itself spins against the F_1 entity and mechanically induces the conformational changes of the β subunits where catalysis occurs.

The (III)c-ring is an oligomer, the stoichiometry of which will be discussed below. Each monomer is folded as a helical hairpin with the loop at the negatively charged side of the membrane and the C- and N-termini at the positively charged side (Hoppe et al., 1984; Girvin et al., 1998). The structure shows a cylindrical arrangement of monomers with a hydrophobic cavity at the center of the ring, which is filled by lipids when embedded in the membrane (Oberfeld et al., 2006). Each monomer bears, at the middle of one of the two branches of the hairpin, a highly conserved acidic residue (Glu or Asp), which is essential for H^+ translocation, acting as the binding site for H^+ (Valiyaveetil and Fillingame, 1998). Subunit IV(a) of the F_0 would allow the access of a proton from the positively charged side of the membrane to this binding site. Once protonated, the site would be able to move out of the (IV)a/(III)c interface into the lipid phase, a move which would induce a stepped rotation of the entire ring. Simultaneously a protonated site would enter the interface, allowing the release of the proton to the negatively charged side of the membrane through an outlet channel borne by subunit (IV)a. The two inlet and outlet channels would be spatially and electrostatically separated by an Arg residue on the (IV)a subunit, a separation which forces the clock-like turn of the (III)c-ring before an H^+ reaches the exit channel. This would be the mechanism that prevents leakage at the (IV)a/(III)c interface (see Junge et al., 1997; Dimroth et al., 2006, for the discussion of a slightly different model for the Na^+ ATPase).

Based on the mechanism discussed above, a 360° rotation of the (III)c-ring would correspond to full rotation of the central stalk and thus to the synthesis of three ATP per (C)F_0F_1 ATP synthase. In this framework, *n*, the H^+/ATP ratio, would thus be entirely determined by the stoichiometry of (III)c-monomers in the (III)c-ring. Recognition of this idea, combined with the threefold symmetry of the (C)F_1 moiety, has led to the lasting assumption that the number of (III)c-monomers in the ring should also be a multiple of three. Yet, structural studies revealed that this is not the case. For example, 10 monomers have been found in yeast mitochondria (Stock et al., 1999), the thermophilic bacterium *Bacillus* PS3 (Mitome et al., 2004) and possibly in *E. coli* (Jiang et al., 2001); 14 monomers were found in the III-ring of spinach chloroplast ATP synthase (Seelert et al., 2000), whereas in *Chlamydomonas*, the molecular mass of the oligomer suggests 13 monomers (Meyer zu Tittingdorf et al., 2004). Fifteen monomers were found in the cyanobacterium *Spirulina platensis* (Pogoryelov et al., 2005), and 11 monomers in the Na^+ ATP synthase from the bacteria *Ilyobacter tartaricus* (Stahlberg et al., 2001) and *Propionigenum modestum* (Meier et al., 2003). Such variability points to a relative plasticity of $(C)F_0F_1$ ATP synthase function, since a variation of the (III)c-ring stoichiometry is expected to lead to different H^+/ATP ratios. A larger number of monomers should result in a larger H^+/ATP ratio, which would be advantageous for ATP synthesis at low electrochemical potential. Conversely, a small number of (III)c-monomers in the ring might prevail in organisms with high ion motive force.

Recognition of this situation has led to the proposal that the stoichiometry of the (III)c-ring could vary depending on physiological conditions. Implicitly, it was suggested that the stoichiometry of the (III)c-ring found by structural analysis could be significantly different from that prevailing *in vivo*, yet single molecule imaging suggests a fixed stoichiometry (Seelert et al., 2000). Moreover, the possibility of loss of (III)c-monomers during purification was rendered unlikely by the finding that in the thermophilic bacterium *Bacillus* PS3, which allows the purification of highly stable oligomers, the stoichiometry of the (III)c-ring was 10, as in *E. coli* (Mitome et al., 2004). Furthermore, although different carbon sources indeed led to different stoichiometries of the (III)c-ring in the case of *E. coli*, this finding cannot be readily extrapolated since in *Chlamydomonas*, (III)c-ring stoichiometry was found to be constant over a wide range of metabolic states (Meyer zu Tittingdorf et al., 2004).

It can thus be concluded that a mismatch between the symmetry of the (III)c-ring and the threefold symmetry of the $(C)F_1$ is probably the general case, and the mechanistic consequences have been hotly debated. As discussed by Stock et al. (1999), such a mismatch might be expected provided the power transmission between $(C)F_0$ and $(C)F_1$ includes elasticity and is, to a certain degree, permissive. In this case, the H^+/ATP ratio would, on the average, be non-integral (e.g. $4.7 = 14/3$ when the (III)c-ring is a 14-mer); but in effect, 4 or 5 five steps in the rotation of the (III)c-ring could be coupled, with different probabilities, to ATP synthesis. This prediction has been experimentally substantiated by the finding that the γ subunit may be elastically twisted or bent and still allow ATP synthesis (Junge et al., 2001; Mitome et al., 2004). Based on the assumption of a strict correlation between monomer stoichiometry in the (III)c-ring and the H^+/ATP ratio, the latter has been estimated at $14/3 = 4.7$ in photosynthetic eukaryotes, and at $13/3 = 4.3$ in *Chlamydomonas*. Combined with the number of three H^+ transferred to the lumen per electron transfer from water to $NADP^+$ (i.e. 12 H^+ per 2 NADPH), this figure yields the number of ATP synthesized per NADPH: $3 \times 6/14 \approx 1.29$, suggesting a shortfall in the amount of ATP produced by linear photosynthetic electron flow with respect to NADPH to meet the requirement of the Calvin-Benson cycle, which is 1.5 ATP per NADPH. In turn, this points to a mandatory involvement of ~20% alternative electron flow processes (PS I-driven cyclic electron flow, Mehler reaction, etc.; see Allen, 2002, for a discussion).

It should be noted, however, that the H^+/ATP ratio may not follow simply from the stoichiometry of (III)c-monomers in (C)F_0. The possibility of mismatch between these two figures was raised by studying a fully characterized chemiosmotic model system consisting of liposomes containing one CF_0F_1 and an H^+/ATP ratio of 3.9–4.0, in agreement with previous reports (Van Walraven et al., 1996), that is, significantly smaller than $4.7 = 14/3$ that would be expected from the number of III-monomers in the ring (Turina et al., 2003). Yet, the current views on the mechanism underlying the stepwise rotation of the III-ring make the energetic penalty resulting from the exposure of a non-protonated III monomer to the lipid bilayer enormous. Thus, the full rotation of the III-ring with two non-protonated III-monomers seems highly unlikely. A situation where the full catalytic cycle of the (C)F_1 could be driven by an incomplete cycle of the III-ring is also extremely difficult to envision since as discussed above, both are coupled by a single central stalk, the γ subunit. One of the remaining possibilities is that on average, two H^+ bound to the acidic side chains of two out of the fourteen III-monomers, would overcome the electrostatic barrier thought to prevent leakage between the inlet and outlet channels and would be driven along more than one cycle by the rotation of the ensemble.

B. ATPase activity and *in vitro* and *in vivo* regulation of the CF_0F_1 complex

Like any enzyme, the CF_0F_1 complex is reversible. It can thus either synthesize ATP at the expense of a $\Delta\tilde{\mu}_{H^+}$, or catalyze the reverse reaction, namely ATP hydrolysis, and build up a $\Delta\tilde{\mu}_{H^+}$ across the thylakoid membranes. Whereas in the respiratory chain the $\Delta\tilde{\mu}_{H^+}$ is constantly built up by the overall electron transfer between NADPH and O_2, so that the mitochondrial enzyme is generally oriented toward ATP synthesis rather than its hydrolysis, in the photosynthetic chain, the driving force at the origin of the $\Delta\tilde{\mu}_{H^+}$, light, naturally fluctuates during the day and is absent at night, and the ensuing catalytic activity is thus prone to fluctuate between synthesis and hydrolysis. Surprisingly, although ATP hydrolysis by the CF_0F_1 complex has been described in isolated thylakoids, no substantial ATP hydrolysis can be detected in isolated chloroplasts in the dark below a critical ADP/ATP ratio (reviewed in Strotmann et al., 1998). This suggests that the chloroplast enzyme is tightly regulated. As will be discussed below, several processes may contribute to this regulation which would ultimately

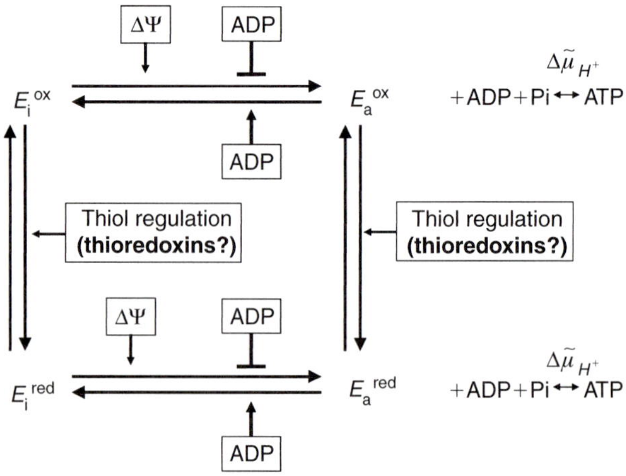

FIGURE 18.4 *Positive and negative regulation of CF_0F_1 complex activity by substrates (ADP and $\Delta\tilde{\mu}_{H^+}$) and redox power ("thiol regulation"). See text for further details. E_i, inactive complex; E_a, active complex; ox, oxidized; red, reduced. The active form of the complex catalyzes either the synthesis of ATP starting from ADP, phosphate and protons provided by the trans-thylakoid electrochemical proton gradient, or ATP hydrolysis. Activation is indicated by an arrow. Deactivation is indicated by a bar. (Redrawn from Junesh and Gräber, 1987).*

enhance ATP synthesis and inhibit the wasteful consumption of ATP in the dark.

A commonly observed regulatory mechanism in enzymology is inactivation or inhibition by the product or substrate, and the CF_0F_1 complex is no exception. It has been shown that ADP deactivates the enzyme by the tight binding to F_1 (Schumann and Strotmann, 1981; Bar-Zvi and Shavit, 1982; see also Figure 18.4). The binding of the di-nucleotide to a single site out of the three is sufficient to attain full deactivation (Zhou et al., 1988). This deactivation process is retarded by the addition of inorganic phosphate (Dunham and Selman, 1981) suggesting that as expected, the balanced availability of the two substrates for ATP synthesis is critical in regulation of the enzyme.

The $\Delta\tilde{\mu}_{H^+}$ also helps to modulate enzyme activity (Figure 18.4). Thermodynamically speaking, the two components ΔpH and ΔΨ are equivalent as both can drive ATP synthesis by promoting the electrogenic translocation of H^+ across the membrane. The first study taken as evidence that ATP synthesis could be driven entirely by a ΔpH, is the acid bath procedure (Jagendorf and Uribe, 1966). In this case, ATP synthesis could be promoted by resuspending thylakoids in an alkaline buffer after having incubated them in an acidic one. Similarly, simply imposing a transmembrane ΔΨ is sufficient to promote ATP synthesis (Witt et al., 1976). Yet, several reports point to the fact that the two components of the $\Delta\tilde{\mu}_{H^+}$ may not fully substitute each other, and that the activation of the enzyme specifically requires

$\Delta\Psi$. Re-investigation showed that the procedure used to generate the ΔpH is not as specific as initially thought and also generates a significant $\Delta\Psi$, leading to the conclusion that ATP synthesis requires a minimum transmembrane voltage (Kaim and Dimroth, 1999). The threshold level for $\Delta\Psi$ was estimated at ~20 mV in the case of chloroplast ATP synthase and 40 mV in the case of *E. coli* ATP synthase or *P. modestum* Na^+ ATP synthase (Kaim and Dimroth, 1999). This finding is in line with previous reports where a suspension of spinach thylakoids was exposed to a series of light flashes to gradually increase the $\Delta\Psi$, revealing that a threshold level must be reached to allow ATP synthesis (Junge, 1970; Junge et al., 1970). Above this threshold, the rate of ATP synthesis increases as a sigmoidal function of $\Delta\Psi$ or ΔpH.

Besides these regulatory features which allow the transition of the enzyme from an idling state to an activated one, and rely on the availability of the "substrate," the ATP synthase from photosynthetic organisms is subject to other regulation mechanisms, which are typical of F-ATPases and rely mainly on conformational changes. It is well known that membrane energization induces conformational changes in the ε subunit of chloroplast and bacterial enzymes (Richter and McCarty, 1987; Aggeler and Capaldi, 1996; Zimmermann et al., 2005). Comparison of the structures of the $\gamma\varepsilon$ complex from *E. coli* (Rodgers and Wilce, 2000), in which ε participates in enzyme regulation, and the bovine mitochondrial F_1 (Gibbons et al., 2000), in which it does not, is insightful since it suggests that the two C-terminal helices of the ε subunit may adopt two conformations, extended and contracted. The extended conformation was shown to inhibit ATPase activity but not ATP synthesis, whereas the contracted one allowed both activities (Tsunoda et al., 2001). This led to the proposal that the extended configuration would act as a ratchet that would differentially regulate ATP hydrolysis and synthesis, and that rotation of the γ subunit would determine whether subunit ε is in the extended or contracted configuration (Feniouk and Junge, 2005). In addition to this regulatory switch which is common to the bacterial and chloroplast enzymes, inactivation of the ATPase activity of the CF_0F_1 complex in the dark has been shown to stem from a specific mechanism called thiol regulation (Figure 18.4). This process describes a change in the redox state of a thiol bridge on the γ subunit of the CF_1 moiety (reviewed in Richter et al., 2005), which decreases the sensitivity of the complex to the regulatory effect exerted by the $\Delta\tilde{\mu}_{H^+}$ (see above). Indeed, illumination of thylakoid membranes in the presence of a reducing agent such as dithiothreitol decreases the $\Delta\tilde{\mu}_{H^+}$ value required for complex activation by more than 0.5 pH units (Junesh and Gräber, 1985, 1987). Consequently, it is conceivable that the concomitant decrease of the $\Delta\tilde{\mu}_{H^+}$ and of the stromal reducing power that follows dark adaptation of thylakoid membranes results in a complete inactivation of the CF_0F_1 complex.

This would inhibit ATP hydrolysis, which would otherwise be substantially enhanced by the increase of the $[ATP]/[ADP].[P_i]$ ratio.

The redox-sensitive thiol bridge on the γ subunit of the CF_1 occurs within a CXXXXXC motif present in plants as well as in *Chlamydomonas* (Cys 198 and 204 in *Chlamydomonas*; Yu and Selman, 1988). The critical role of these cysteines in *Chlamydomonas* thiol regulation was revealed by site-directed mutagenesis, where substitution of either or both cysteines with a serine completely abolished redox regulation (Ross et al., 1995). Consistent with this finding, no thiol regulation is observed in *Synechocystis* sp. 6803, where the putative regulatory S–S bridge is absent (Cozens and Walker, 1987). Furthermore, introduction of a plant-like regulatory motif by mutagenesis of the *atpC* gene results in the acquisition of thiol regulation (Werner-Grune et al., 1994). Besides the essential function of the two cysteines in thiol regulation, a role for the spacer between them has been proposed, also based on site-directed mutagenesis. For example, replacement of *Chlamydomonas* γ subunit Asp 199 with either a basic or a neutral residue, or replacement of Lys 203 with a negatively charged residue, results in the complete loss of thiol regulation (Ross et al., 1996). Thiol regulation *in vivo* is likely modulated by thioredoxins (Mills et al., 1980), which are directly responsible for the reduction of the S–S bridge (Schwarz et al., 1997), as well as for its oxidation (Dann and McCarty, 1992). Apparently, thiol regulation of the CF_0F_1 complex acts downstream of the $\Delta\tilde{\mu}_{H^+}$ control. This is suggested by the finding that the S–S bridge on the γ subunit becomes accessible to reductants only after membrane energization (Schumann et al., 1985). Thioredoxins are discussed in detail in Chapter 11.

Despite the rather complete description of the mechanism leading to both the $\Delta\tilde{\mu}_{H^+}$ and thiol-linked modulation of the ATP synthase complex, its regulation *in vivo* is still not fully understood. For example, incubation of green algae under reducing conditions results in its progressive inactivation (Velthuys, 1978). In anaerobically grown *Chlamydomonas*, this inactivation parallels a substantial increase of stromal reducing power, manifested in part by a twofold increase of the NADPH/NADP ratio (Forti et al., 2003). This enhancement is likely due to the activation of glycolysis, owing to the decrease of the cytoplasmic ATP/ADP ratio (Bulté et al., 1990). Owing to the redox equilibration between stromal electron carriers, reduction of the NADP pool should be mirrored by an increased concentration of reduced thioredoxin, that is, by the activation of the ATPase by thiol regulation. This apparent contradiction suggests that the major regulation of ATP hydrolysis *in vivo* is exerted by the $\Delta\tilde{\mu}_{H^+}$, with thiol regulation only providing a downstream feedback process. A possible interpretation is to assume that changes of the ATP/ADP ratio that mirror variation of the $\Delta\tilde{\mu}_{H^+}$ directly control ATP synthase activity. As discussed above, direct binding of ADP to the complex results in the progressive inhibition of ATP hydrolysis activity that follows the dark adaptation of isolated thylakoids

(Strotmann et al., 1987) and intact chloroplasts (Altavater-Mackensen and Strotmann, 1988).

The relatively minor role exerted by thiol regulation *in vivo* is confirmed by the finding that no major phenotype is associated with *Chlamydomonas* mutants where the cysteines responsible for thiol regulation had been removed (Chen et al., 1995). On the other hand, reduced growth in low light has been reported in *Arabidopsis* mutants where thiol regulation was impaired (Gabrys et al., 1994). This suggests that thiol regulation may have a larger impact on this enzyme in plants than in freshwater algae, perhaps reflecting a different interplay between redox state and ATP metabolism (see Noctor and Foyer, 2000; Peltier and Cournac, 2002 for further discussion). Alternatively, the different sensitivities of plants and algae to thiol regulation may indicate a more rapid deactivation of the CF_0F_1 complex upon dark adaptation, owing to a faster relaxation of the $\Delta\tilde{\mu}_{H^+}$. Consistent with this hypothesis, it has been shown that light-triggered ATP hydrolyzing activity of the complex is faster in *Chlamydomonas* than in plants (Fiedler et al., 1997).

The hypothesis of a rather weak ATP hydrolytic activity of the ATP synthase complex in *Chlamydomonas* due to the fast generation of the $\Delta\tilde{\mu}_{H^+}$ is contradicted by well-established evidence that a permanent $\Delta\tilde{\mu}_{H^+}$ is maintained in dark-adapted algae under both aerobic and anaerobic conditions (Diner and Joliot, 1976; Joliot and Joliot, 1980). This phenomenon has been interpreted as reflecting hydrolysis by the chloroplast ATP synthase of ATP produced either by aerobic respiratory metabolism, or by glycolysis under anaerobic conditions. Thus, the presence of a $\Delta\tilde{\mu}_{H^+}$ provides in principle strong evidence that a non-negligible fraction of the ATP synthase retains an active conformation even after a long dark adaptation in *Chlamydomonas*. However, another process has been proposed to account for the presence of a $\Delta\tilde{\mu}_{H^+}$ in the dark in the absence of active ATP synthase complexes. This process, chlororespiration, refers to a series of oxido-reduction processes that occur in the dark in the thylakoid membranes (Bennoun, 1982). The permanent gradient present in the absence of ATP synthase would result from a coupled electron transfer through a chloroplast respiratory pathway, in which plastoquinone is reduced by a chloroplast NAD(P)H dehydrogenase and reoxidized at the expense of oxygen by a chloroplast oxidase (reviewed in Peltier and Cournac, 2002). This pathway is similar to that prevailing in photosynthetic prokaryotes, which associate photosynthesis and respiration in the same membranes (Kallas, 1994). Consistent with predictions of chlororespiration, a $\Delta\tilde{\mu}_{H^+}$ has been observed in the *FUD50* mutant of *Chlamydomonas*, which is devoid of the ATP synthase complex (Bennoun, 1982, 1994). Yet as discussed below, the relative balance between the $\Delta\Psi$ and ΔpH components of the $\Delta\tilde{\mu}_{H^+}$ is different in this mutant as compared to the wild type. Thus, a contribution of ATP synthase to the establishment of the $\Delta\tilde{\mu}_{H^+}$ in the dark cannot be ruled out.

C. Consequences of ATP hydrolysis: amplitude, nature, and consequences of the dark-generated electrochemical proton gradient

The properties and origin of the $\Delta\tilde{\mu}_{H^+}$ established in the dark were first studied in the green alga *Chlorella sorokiniana* (Joliot and Joliot, 1989). These authors estimated the amplitude of $\Delta\tilde{\mu}_{H^+}$ at 20–50 mV. This estimate was later extended to the ΔpH component, showing that its amplitude (1.5 pH units) was twice as large as that of the ΔΨ (Finazzi and Rappaport, 1998). Thus, the overall amplitude of the $\Delta\tilde{\mu}_{H^+}$ is 110–140 mV, which fits well with the estimate of the ΔG_P in this alga, based on the ATP/ADP ratio measured in cellular extracts under the same conditions. The analysis was later confirmed in *Chlamydomonas* (Finazzi et al., 2003). As stated above, this is consistent with ATP hydrolysis being essentially controlled by the ATP/ADP ratio, *in vivo* as well as in isolated thylakoids (Strotmann et al., 1987) and intact chloroplasts (Altavater-Mackensen and Strotmann, 1988). However, at variance with the *in vitro* data, the *in vivo* results suggest that the size of the $\Delta\tilde{\mu}_{H^+}$ is essentially governed by the thermodynamic equilibrium of the proton concentration across the thylakoid membranes and the cellular ATP content, rather than by kinetic effects due to ADP binding on the complex. Indeed, the amplitude of the permanent $\Delta\tilde{\mu}_{H^+}$ has been shown not to depend on the fraction of active ATP synthase complexes (Rappaport et al., 1999).

What is the rationale for maintaining a $\Delta\tilde{\mu}_{H^+}$ in the dark at the expense of ATP hydrolysis, *in vivo*? Several thermodynamic advantages can be considered. Because of the existence of a non-negligible proton motive force in darkness, the light-induced $\Delta\tilde{\mu}_{H^+}$ increase required to activate ATP synthesis (see above) may be smaller than figures derived from *in vitro* experiments (reviewed in Witt, 1979). This might have physiological consequences for photosynthesis, as a very small threshold for ATP synthesis may improve photosynthetic efficiency in low light, where the size of the light-induced $\Delta\tilde{\mu}_{H^+}$ may be significantly smaller than in normal or high light.

Another advantage of maintaining a $\Delta\tilde{\mu}_{H^+}$ in the dark can be envisaged in the case of photoprotection. Under conditions of excess light, a variety of mechanisms are initiated which protect chloroplasts from damage. These processes, which are associated with PS II, are collectively referred to as non-photochemical quenching (NPQ; see Chapter 23). In plants, it is acknowledged that the main component of NPQ is an increased thermal dissipation within the PS II light-harvesting apparatus (Demmig-Adams et al., 1996; Niyogi, 1999), which is triggered by the generation of a ΔpH (Briantais et al., 1979) and associated with the formation of zeaxanthin via de-epoxidation of violaxanthin (Yamamoto et al., 1962). The ΔpH has two effects on NPQ, as it is required for both the activation of violaxanthin de-epoxidase (Pfundel et al., 1994), the key enzyme that allows synthesis of zeaxanthin, and for the enhancement of the thermal dissipation process itself (Demmig-Adams et al., 1996; Niyogi, 1999). It is therefore conceivable that maintaining a stable $\Delta\tilde{\mu}_{H^+}$ in the dark may enhance the NPQ capacity (Gilmore and Yamamoto, 1992). Indeed a permanent $\Delta\tilde{\mu}_{H^+}$ of similar amplitude as the one characterized in freshwater algae would in general be sufficient to keep violaxanthin de-epoxidase in an active conformation even in the dark. Because of preloading of the antenna complexes with zeaxanthin, even a small $\Delta\tilde{\mu}_{H^+}$ increase due to illumination would lead to a significant and very rapid NPQ response. This is exactly what is required by unicellular photosynthetic organisms, which are exposed to rapidly changing light conditions, and by plants exposed to "sunflecks" (periods of direct sunlight against a background of low light).

Besides the specific possible interplay between the existence of a permanent $\Delta\tilde{\mu}_{H^+}$ in the dark and the onset of NPQ during a dark-light transition, different types of interplay between the turnover of the CF_0F_1 complex and photoprotection have been discussed. Kramer et al. (2003, 2004) proposed that a down-regulation of ATPase activity might account for the maintenance of a sustained ΔpH, and therefore of a high NPQ under conditions where the overall rate of electron flow to CO_2 is restricted. In particular, they have proposed that a reduced proton conductivity may stem from a possible restriction of the stromal $[P_i]$ during CO_2 limitation. This could limit H^+ flux through the ATPase either directly, by affecting ATP synthesis, or allosterically, and therefore enhance the amplitude of the ΔpH to levels where a sustained NPQ response may occur.

D. Origin of the $\Delta\tilde{\mu}_{H^+}$ observed *in vivo*

It is known that the relative size of the electric component of the proton motive force measured *in vivo* is larger than *in vitro,* where the $\Delta\Psi$ is almost negligible (Avenson et al., 2005). The reasons for the different partitioning of the $\Delta\tilde{\mu}_{H^+}$ *in vivo* and *in vitro* are still under discussion. It has been argued that changes in the ionic stromal balance, or in the lumenal buffer capacity, are likely to be the major effectors of this phenomenon (Kramer et al., 2003, 2004). As an alternative explanation, it has been proposed that turnover of the ATP synthase may be responsible for the generation of an electrical gradient across the thylakoid membranes *in vivo,* which is lost *in vitro.* This possibility is consistent with the data demonstrating a $\Delta\tilde{\mu}_{H^+}$ across thylakoid membranes in mutants devoid of the ATP synthase (Bennoun, 1982). As stated above, this finding was first interpreted as an evidence for the involvement of chlororespiration rather than ATP hydrolysis in the generation of the dark $\Delta\tilde{\mu}_{H^+}$. However, later studies performed in both *Chlamydomonas* and *Chlorella* have convincingly shown that the $\Delta\tilde{\mu}_{H^+}$ observed in ATP synthase mutants is not due to chlororespiration, but is still linked to ATP hydrolysis. It has been shown that this $\Delta\tilde{\mu}_{H^+}$ does not comprise a ΔpH, but is entirely due to the electric component ΔΨ (Finazzi and Rappaport, 1998; Rappaport et al., 1999).

This dominance of ΔΨ has been interpreted in terms of the activity of a different ATPase (the "X" pump; Rappaport et al., 1999), the nature of which is still elusive. It is, however, clear that this enzyme (or enzymatic activity) has an ion selectivity distinct from protons (Rappaport et al., 1999). Moreover, it has been shown that its contribution to the generation and dissipation of the $\Delta\tilde{\mu}_{H^+}$ is almost negligible during illumination of wild-type cells. This could imply that the "X pump" has an unfavorable ATP/ion ratio as compared to the CF_0F_1 enzyme (Rappaport et al., 1999). Alternatively, it can be proposed that its activation requires a higher $\Delta\tilde{\mu}_{H^+}$ than the CF_0F_1 (Finazzi et al., 2006). Thus, although the activity of the "X pump" would be negligible under normal physiological conditions, it might contribute significantly to $\Delta\tilde{\mu}_{H^+}$ turnover under conditions of restricted ATP metabolism. In these conditions, sustained generation of the electrochemical proton gradient coupled to reduced ATP consumption might transiently bring the $\Delta\tilde{\mu}_{H^+}$ to the threshold required for activation of the "X pump." This activation should lead to a specific dissipation of ΔΨ. Consistent with this hypothesis, a different partitioning of the $\Delta\tilde{\mu}_{H^+}$ (with a significant reduction of the ΔΨ component) has been observed in tobacco leaves (Avenson et al., 2005) and *Chlamydomonas* cells (Finazzi et al., 2006) under conditions of restricted photosynthetic activity. The former result indicates that the "X pump" may also be present in vascular plants, where it could also be involved in $\Delta\tilde{\mu}_{H^+}$ plasticity.

E. Consequences of the electrochemical proton gradient in the dark: protein targeting to the chloroplast

Another possible consequence of the presence of an electrochemical proton gradient in the dark is the ability to assemble the photosynthetic apparatus in the absence of any photosynthetic light reactions. It is known that some nucleus-encoded subunits of the three major photosynthetic complexes (the PS II and PS I reaction centers and the cytochrome b_6f complex) are translocated into the lumen via the so called "ΔpH-dependent Tat pathway" (reviewed in Cline and Henry, 1996), after entering the stroma via the outer (TOC) and inner (TIC) membrane translocon complexes (reviewed in Reumann and Keegstra, 1999). Along with the Tat pathway, thylakoid or lumenal targeting can occur via two other pathways with bacterial homologues, the Sec and SRP routes (Michl et al., 1994; also see Chapter 19). These three pathways differ not only in their protein specificities but also from an energetic point of view. The Sec and SRP pathways require hydrolysis of nucleotide triphosphates – ATP and GTP respectively – although a proton motive force may also play a role (Ernst et al., 1994). In contrast, protein translocation via the Tat pathway has been reported to be insensitive to

NTP hydrolysis, being strictly dependent on a ΔpH (Mould and Robinson, 1991). Given the above considerations, it is conceivable that the presence of a sustained, dark-stable $\Delta\tilde{\mu}_{H^+}$ (having a significant ΔpH component) may help maintain an efficient housekeeping of their photosynthetic apparatus in the dark.

Although this hypothesis has some attractive features, work performed in *Chlamydomonas* has challenged it, by showing that elimination of the *trans*-thylakoid ΔpH does not impair translocation of Tat passenger proteins *in vivo*. Two possibilities can be proposed for the continued operation of the Tat pathway in thylakoids without a ΔpH. The *in vivo* experiments may contain some factor(s) that are missing in the *in vitro* system that alter the energetic requirements of the transport reaction (Theg et al., 2005). This interpretation invokes an entropic role for the ΔpH, consistent with indications that a ΔpH is required for the assembly of the Tat translocase in isolated thylakoids (Mori and Cline, 2002). The existence of currently unknown factors (likely chaperones) is reasonable given that the Tat thylakoid pathway is adapted for the transport of folded proteins (reviewed in Robinson and Bolhuis, 2001), which are rarely used as *in vitro* substrates. This in turn might account for the very different kinetic performances of the Tat pathway *in vitro/in organello*, where a large fraction of the protein precursor accumulates in the stroma (Berks et al., 2000), and *in vivo*, where no such accumulation is seen (Finazzi et al., 2003). Alternatively the overall $\Delta\tilde{\mu}_{H^+}$, rather than ΔpH alone, may be able to power the Tat pathway. This explanation is supported by experiments in which the ΔΨ was inferred to contribute to powering the Tat pathway *in vitro* under select experimental conditions (Theg et al., 2005).

IV. CONCLUDING REMARKS

The ATPase-ATP synthase enzyme CF_0F_1 is one of the most complicated and fascinating photosynthetic membrane complexes. As discussed above, progress in the structural characterization of the both the respiratory and photosynthetic complexes, combined with the development of highly refined functional approaches, has led to a significant improvement of our understanding of structure/function relationships. However, several major issues remain largely unresolved. These include the mechanism(s) of its assembly and targeting to thylakoid membranes, the stoichiometric relationship between the H^+ translocation through the CF_0 moiety and the synthesis of ATP synthesis in the CF_1, and the physiological roles of both its ATP synthase and ATPase activities in modulating cellular energetic and $\Delta\tilde{\mu}_{H^+}$ homeostasis.

In principle, the basic mechanisms of ATP synthesis are similar in respiratory and photosynthetic systems. Thus, studies in bacteria, yeast, chloroplasts, and mitochondria all contribute to our understanding. However, the photosynthetic complex has specific regulatory mechanisms to which its respiratory counterpart is not subjected. *Chlamydomonas* has been widely and successfully employed to address these issues in the past, and is likely to play a major role in the future.

REFERENCES

Abrahams, J.P., Leslie, A.G., Lutter, R., and Walker, J.E. (1994). Structure at 2.8 Å resolution of F_1-ATPase from bovine heart mitochondria. *Nature* **370**, 621–628.

Ackerman, S.H. and Tzagoloff, A. (2005). Function, structure, and biogenesis of mitochondrial ATP synthase. *Prog. Nucl. Acids Res. Mol. Biol.* **80**, 95–133.

Aggeler, R. and Capaldi, R.A. (1996). Nucleotide-dependent movement of the epsilon subunit between alpha and beta subunits in the *Escherichia coli* F_1F_0-type ATPase. *J. Biol. Chem.* **271**, 13888–13891.

Allen, J.F. (2002). Photosynthesis of ATP-electrons, proton pumps, and poise. *Cell* **110**, 273–276.

Altavater-Mackensen, R. and Strotmann, H. (1988). Contents of endogenous nucleotides and ATPase activity of isolated whole chloroplasts. *Biochim. Biophys. Acta* **934**, 213–219.

Angevine, C.M., Herold, K.A., Vincent, O.D., and Fillingame, R.H. (2007). Aqueous access pathways in ATP synthase subunit a. Reactivity of cysteine substituted into transmembrane helices 1, 3, and 5. *J. Biol. Chem.* **282**, 9001–9007.

Arnold, I., Pfeiffer, K., Neupert, W., Stuart, R.A., and Schagger, H. (1998). Yeast mitochondrial F_1F_0-ATP synthase exists as a dimer: Identification of three dimer-specific subunits. *EMBO J.* **17**, 7170–7178.

Avenson, T.J., Cruz, J.A., Kanazawa, A., and Kramer, D.M. (2005). Regulating the proton budget of higher plant photosynthesis. *Proc. Natl. Acad. Sci. U. S. A.* **102**, 9709–9713.

Bar-Zvi, D. and Shavit, N. (1982). Modulation of the chloroplast ATPase by tight ADP binding. Effect of uncouplers and ATP. *J. Bioenerg. Biomembr.* **14**, 467–478.

Bennoun, P. (1982). Evidence for a respiratory chain in the chloroplast. *Proc. Natl. Acad. Sci. U. S. A.* **79**, 4352–4356.

Bennoun, P. (1994). Chlororespiration revisited – mitochondrial-plastid interactions in *Chlamydomonas*. *Biochim. Biophys. Acta* **1186**, 59–66.

Bennoun, P. and Béal, D. (1998). New digital imaging instrument for measuring fluorescence and delayed luminescence. In: *The Molecular Biology of Chloroplasts and Mitochondria in Chlamydomonas* (J.D. Rochaix, M. Goldschmidt-Clermont, and S. Merchant, Eds.), pp. 451–458. Kluwer Academic Publishers Dordrecht, The Netherlands.

Berks, B.C., Sargent, F., and Palmer, T. (2000). The Tat protein export pathway. *Mol. Microbiol.* **35**, 260–274.

Boyer, P.D. (1997). The ATP synthase: A splendid molecular machine. *Annu. Rev. Biochem.* **66**, 717–749.

Briantais, J.M., Vernotte, C., Picaud, M., and Krause, G.H. (1979). A quantitative study of the slow decline of chlorophyll a fluorescence in isolated chloroplasts. *Biochim. Biophys. Acta* **548**, 128–138.

Bulté, L., Gans, P., Rebeille, F., and Wollman, F.-A. (1990). ATP control on state transitions *in vivo* in *Chlamydomonas reinhardtii*. *Biochim. Biophys. Acta* **1020**, 72–80.

Capaldi, R.A. and Aggeler, R. (2002). Mechanism of the F_1F_0-type ATP synthase, a biological rotary motor. *Trends Biochem. Sci.* **27**, 154–160.

Chen, Z., Govender, D., Gross, R., and Birge, R.R. (1995). Advances in protein-based three-dimensional optical memories. *BioSystems* **35**, 145–151.

Cipriano, D.J. and Dunn, S.D. (2006). The role of the epsilon subunit in the *Escherichia coli* ATP synthase. The C-terminal domain is required for efficient energy coupling. *J. Biol. Chem.* **281**, 501–507.

Cline, K. and Henry, R. (1996). Import and routing of nucleus-encoded chloroplast proteins. *Annu. Rev. Cell Dev. Biol.* **12**, 1–26.

Cozens, A.L. and Walker, J.E. (1987). The organization and sequence of the genes for ATP synthase subunits in the cyanobacterium *Synechococcus* 6301. Support for an endosymbiotic origin of chloroplasts. *J. Mol. Biol.* **194**, 359–383.

Dann, M.S. and McCarty, R.E. (1992). Characterization of the activation of membrane-bound and soluble CF_1 by thioredoxin. *Plant Physiol.* **99**, 153–160.

Demmig-Adams, B., Gilmore, A.M., and Adams, W.W., 3rd (1996). Carotenoids 3: *in vivo* function of carotenoids in higher plants. *FASEB J.* **10**, 403–412.

Dent, R.M., Haglund, C.M., Chin, B.L., Kobayashi, M.C., and Niyogi, K.K. (2005). Functional genomics of eukaryotic photosynthesis using insertional mutagenesis of *Chlamydomonas reinhardtii*. *Plant Physiol.* **137**, 545–556.

Dimroth, P., von Ballmoos, C., Meier, T., and Kaim, G. (2003). Electrical power fuels rotary ATP synthase. *Structure* **11**, 1469–1473.

Dimroth, P., von Ballmoos, C., and Meier, T. (2006). Catalytic and mechanical cycles in F-ATP synthases. Fourth in the Cycles Review Series. *EMBO Rep.* **7**, 276–282.

Diner, B. and Joliot, P. (1976). Effect of the transmembrane electric field on the photochemical and quenching properties of photosystem II *in vivo*. *Biochim. Biophys. Acta* **423**, 479–498.

Drapier, D., Girard-Bascou, J., and Wollman, F.A. (1992). Evidence for nuclear control of the expression of the *atpA* and *atpB* chloroplast genes in *Chlamydomonas*. *Plant Cell* **4**, 283–295.

Drapier, D., Suzuki, H., Levy, H., Rimbault, B., Kindle, K.L., Stern, D.B., and Wollman, F.A. (1998). The chloroplast *atpA* gene cluster in *Chlamydomonas reinhardtii*. Functional analysis of a polycistronic transcription unit. *Plant Physiol.* **117**, 629–641.

Drapier, D., Girard-Bascou, J., Stern, D.B., and Wollman, F.A. (2002). A dominant nuclear mutation in *Chlamydomonas* identifies a factor controlling chloroplast mRNA stability by acting on the coding region of the *atpA* transcript. *Plant J.* **31**, 687–697.

Drapier, D., Rimbault, B., Vallon, O., Wollman, F.A., and Choquet, Y. (2007). Intertwined translational regulations set uneven stoichiometry of chloroplast ATP synthase subunits. *EMBO J.* **26**, 3581–3591.

Duncan, T.M., Bulygin, V.V., Zhou, Y., Hutcheon, M.L., and Cross, R.L. (1995). Rotation of subunits during catalysis by *Escherichia coli* F_1-ATPase. *Proc. Natl. Acad. Sci. U. S. A.* **92**, 10964–10968.

Dunham, K.R. and Selman, B.R. (1981). Interactions of inorganic phosphate with spinach coupling factor 1. Effects on ATPase and ADP binding activities. *J. Biol. Chem.* **256**, 10044–10049.

Eberhard, S., Drapier, D., and Wollman, F.A. (2002). Searching limiting steps in the expression of chloroplast-encoded proteins: Relations between gene copy number, transcription, transcript abundance and translation rate in the chloroplast of *Chlamydomonas reinhardtii*. *Plant J.* **31**, 149–160.

Ernst, F., Hoffschulte, H.K., Thome-Kromer, B., Swidersky, U.E., Werner, P.K., and Muller, M. (1994). Precursor-specific requirements for SecA, SecB, and delta muH+ during protein export of *Escherichia coli*. *J. Biol. Chem.* **269**, 12840–12845.

Feniouk, B.A. and Junge, W. (2005). Regulation of the F_0F_1-ATP synthase: The conformation of subunit epsilon might be determined by directionality of subunit gamma rotation. *FEBS Lett.* **579**, 5114–5118.

Fiedler, H.R., Schmid, R., Leu, S., Shavit, N., and Strotmann, H. (1995). Isolation of CF_0CF_1 from *Chlamydomonas reinhardtii cw15* and the N-terminal amino acid sequences of the CF_0CF_1 subunits. *FEBS Lett.* **377**, 163–166.

Fiedler, H.R., Schlesinger, J., Strotmann, H., Shavit, N., and Leu, S. (1997). Characterisation of *atpA* and *atpB* deletion mutants produced in *Chlamydomonas reinhardtii cw15*: Electron transport and photophosphorylation activities of isolated thylakoids. *Biochim. Biophys. Acta* **1319**, 109–118.

Finazzi, G. and Rappaport, F. (1998). *In vivo* characterization of the electrochemical proton gradient generated in darkness in green algae and its kinetics effects on the cytochrome b_6f turnover. *Biochemistry* **37**, 9999–10005.

Finazzi, G., Chasen, C., Wollman, F.A., and de Vitry, C. (2003). Thylakoid targeting of Tat passenger proteins shows no delta pH dependence *in vivo*. *EMBO J.* **22**, 807–815.

Finazzi, G., Johnson, G.N., Dall'Osto, L., Zito, F., Bonente, G., Bassi, R., and Wollman, F.A. (2006). Nonphotochemical quenching of chlorophyll fluorescence in *Chlamydomonas reinhardtii*. *Biochemistry* **45**, 1490–1498.

Forti, G., Furia, A., Bombelli, P., and Finazzi, G. (2003). *In vivo* changes of the oxidation-reduction state of NADP and of the ATP/ADP cellular ratio linked to the photosynthetic activity in *Chlamydomonas reinhardtii*. *Plant Physiol.* **132**, 1464–1474.

Fronzes, R., Weimann, T., Vaillier, J., Velours, J., and Brethes, D. (2006). The peripheral stalk participates in the yeast ATP synthase dimerization independently of e and g subunits. *Biochemistry* **45**, 6715–6723.

Gabrys, H., Kramer, D.M., Crofts, A.R., and Ort, D.R. (1994). Mutants of chloroplast coupling factor reduction in *Arabidopsis*. *Plant Physiol.* **104**, 769–776.

Gibbons, C., Montgomery, M.G., Leslie, A.G., and Walker, J.E. (2000). The structure of the central stalk in bovine F_1-ATPase at 2.4 Å resolution. *Nat. Struct. Biol.* **7**, 1055–1061.

Gilmore, A.M. and Yamamoto, H.Y. (1992). Dark induction of zeaxanthin-dependent nonphotochemical fluorescence quenching mediated by ATP. *Proc. Natl. Acad. Sci. U. S. A.* **89**, 1899–1903.

Giraud, M.F., Paumard, P., Soubannier, V., Vaillier, J., Arselin, G., Salin, B., Schaeffer, J., Brethes, D., di Rago, J.P., and Velours, J. (2002). Is there a relationship between the supramolecular organization of the mitochondrial ATP synthase and the formation of cristae? *Biochim. Biophys. Acta* **1555**, 174–180.

Girvin, M.E., Rastogi, V.K., Abildgaard, F., Markley, J.L., and Fillingame, R.H. (1998). Solution structure of the transmembrane H^+-transporting subunit c of the F_1F_0 ATP synthase. *Biochemistry* **37**, 8817–8824.

Groth, G. and Pohl, E. (2001). The structure of the chloroplast F_1-ATPase at 3.2 Å resolution. *J. Biol. Chem.* **276**, 1345–1352.

Harris, E.H. (2001). *Chlamydomonas* as a model organism. *Annu. Rev. Plant Physiol. Plant Mol. Biol.* **52**, 363–406.

Hennigan, A.N. and Jacobson, A. (1996). Functional mapping of the translation-dependent instability element of yeast MATα1 mRNA. *Mol. Cell Biol.* **16**, 3833–3843.

Hicks, A., Drager, R.G., Higgs, D.C., and Stern, D.B. (2002). An mRNA 3′ processing site targets downstream sequences for rapid degradation in *Chlamydomonas* chloroplasts. *J. Biol. Chem.* **277**, 3325–3333.

Hoppe, J., Brunner, J., and Jorgensen, B.B. (1984). Structure of the membrane-embedded F_0 part of F_1F_0 ATP synthase from *Escherichia coli* as inferred from labeling with 3-(trifluoromethyl)-3-(*m*-[^{125}I]iodophenyl)diazirine. *Biochemistry* **23**, 5610–5616.

Jagendorf, A.T. and Uribe, E. (1966). ATP formation caused by acid-base transition of spinach chloroplasts. *Proc. Natl. Acad. Sci. U. S. A.* **55**, 170–177.

Jiang, W., Hermolin, J., and Fillingame, R.H. (2001). The preferred stoichiometry of c subunits in the rotary motor sector of *Escherichia coli* ATP synthase is 10. *Proc. Natl. Acad. Sci. U. S. A.* **98**, 4966–4971.

Johnson, E. and Melis, A. (2004). Functional characterization of *Chlamydomonas reinhardtii* with alterations in the *atpE* gene. *Photosynth. Res.* **82**, 131–140.

Joliot, P. and Joliot, A. (1980). Dependence of delayed luminescence upon adenosine triphosphatase activity in *Chlorella*. *Plant Physiol.* **65**, 691–696.

Joliot, P. and Joliot, A. (1989). Characterization of linear and quadratic electrochromic probes in *Chlorella sorokiniana* and *Chlamydomonas reinhardtii*. *Biochim. Biophys. Acta* **975**, 355–360.

Junesh, U. and Gräber, P. (1985). The ATP synthesis as a function of the DpH in normal and dithiothreitol-modified chloroplasts. *Biochim. Biophys. Acta* **809**, 429–434.

Junesh, U. and Gräber, P. (1987). Influence of the redox state and the activation of the chloroplast ATP synthase on proton transport-coupled ATP synthesis/hydrolysis. *Biochim. Biophys. Acta* **893**, 275–288.

Junge, W. (1970). The critical electric potential difference for photophosphorylation. Its relation to the chemiosmotic hypothesis and to the triggering requirements of the ATPase system. *Eur. J. Biochem.* **14**, 582–592.

Junge, W., Rumberg, B., and Schroder, H. (1970). The necessity of an electric potential difference and its use for photophosphorylation in short flash groups. *Eur. J. Biochem.* **14**, 575–581.

Junge, W., Lill, H., and Engelbrecht, S. (1997). ATP synthase: An electrochemical transducer with rotatory mechanics. *Trends Biochem. Sci.* **22**, 420–423.

Junge, W., Panke, O., Cherepanov, D.A., Gumbiowski, K., Muller, M., and Engelbrecht, S. (2001). Inter-subunit rotation and elastic power transmission in F_0F_1-ATPase. *FEBS Lett.* **504**, 152–160.

Kaim, G. and Dimroth, P. (1999). ATP synthesis by F-type ATP synthase is obligatorily dependent on the transmembrane voltage. *EMBO J.* **18**, 4118–4127.

Kallas, T. (1994). The cytochrome b_6f complex. In: *The Molecular Biology of Cyanobacteria* (D.A. Bryant, Ed.), pp. 259–317. Kluwer Academic Publishers, Dordrecht.

Ketchner, S.L., Drapier, D., Olive, J., Gaudriault, S., Girard-Bascou, J., and Wollman, F.A. (1995). Chloroplasts can accommodate inclusion bodies. Evidence from a mutant of *Chlamydomonas reinhardtii* defective in the assembly of the chloroplast ATP synthase. *J. Biol. Chem.* **270**, 15299–16330.

Komine, Y., Kwong, L., Anguera, M.C., Schuster, G., and Stern, D.B. (2000). Polyadenylation of three classes of chloroplast RNA in *Chlamydomonas reinhardtii*. *RNA* **6**, 598–607.

Komine, Y., Kikis, E., Schuster, G., and Stern, D. (2002). Evidence for *in vivo* modulation of chloroplast RNA stability by 3′-UTR homopolymeric tails in *Chlamydomonas reinhardtii*. *Proc. Natl. Acad. Sci. U. S. A.* **99**, 4085–4090.

Kramer, D.M., Cruz, J.A., and Kanazawa, A. (2003). Balancing the central roles of the thylakoid proton gradient. *Trends Plant Sci.* **8**, 27–32.

Kramer, D.M., Avenson, T.J., and Edwards, G.E. (2004). Dynamic flexibility in the light reactions of photosynthesis governed by both electron and proton transfer reactions. *Trends Plant Sci.* **9**, 349–357.

Kulkarni, R.D. and Golden, S.S. (1997). mRNA stability is regulated by a coding-region element and the unique 5′ untranslated leader sequences of the three *Synechococcus psbA* transcripts. *Mol. Microbiol.* **24**, 1131–1142.

Lefebvre-Legendre, L., Salin, B., Schaeffer, J., Brethes, D., Dautant, A., Ackerman, S.H., and di Rago, J.P. (2005). Failure to assemble the $\alpha_3\beta_3$ subcomplex of the ATP synthase leads to accumulation of the α and β subunits within inclusion bodies and the loss of mitochondrial cristae in *Saccharomyces cerevisiae*. *J. Biol. Chem.* **280**, 18386–18392.

Lemaire, C. and Wollman, F.A. (1989). The chloroplast ATP synthase in *Chlamydomonas reinhardtii*. II. Biochemical studies on its biogenesis using mutants defective in photophosphorylation. *J .Biol. Chem.* **264**, 10235–10242.

Leslie, A.G. and Walker, J.E. (2000). Structural model of F_1-ATPase and the implications for rotary catalysis. *Philos. Trans. R. Soc. Lond. B. Biol. Sci.* **355**, 465–471.

Levy, H., Kindle, K.L., and Stern, D.B. (1997). A nuclear mutation that affects the 3′ processing of several mRNAs in *Chlamydomonas* chloroplasts. *Plant Cell* **9**, 825–836.

Levy, H., Kindle, K.L., and Stern, D.B. (1999). Target and specificity of a nuclear gene product that participates in mRNA 3′-end formation in *Chlamydomonas* chloroplasts. *J. Biol. Chem.* **274**, 35955–35962.

Majeran, W., Olive, J., Drapier, D., Vallon, O., and Wollman, F.A. (2001). The light sensitivity of ATP synthase mutants of *Chlamydomonas reinhardtii*. *Plant Physiol.* **126**, 421–433.

Meier, T., Matthey, U., von Ballmoos, C., Vonck, J., Krug von Nidda, T., Kuhlbrandt, W., and Dimroth, P. (2003). Evidence for structural integrity in the undecameric c-rings isolated from sodium ATP synthases. *J. Mol. Biol.* **325**, 389–397.

Meyer zu Tittingdorf, J.M., Rexroth, S., Schafer, E., Schlichting, R., Giersch, C., Dencher, N.A., and Seelert, H. (2004). The stoichiometry of the chloroplast ATP synthase oligomer III in *Chlamydomonas reinhardtii* is not affected by the metabolic state. *Biochim. Biophys. Acta* **1659**, 92–99.

Michl, D., Robinson, C., Shackleton, J.B., Herrmann, R.G., and Klosgen, R.B. (1994). Targeting of proteins to the thylakoids by bipartite presequences: CF_0II is imported by a novel, third pathway. *EMBO J.* **13**, 1310–1317.

Mills, J.D., Mitchell, P., and Schürmann, P. (1980). Modulation of coupling factor ATPase activity in intact chloroplasts: the role of the thioredoxin systems. *FEBS Lett.* **112**, 173–177.

Mitchell, P. (1961). Coupling of phosphorylation to electron and hydrogen transfer by a chemi-osmotic type of mechanism. *Nature* **191**, 144–148.

Mitchell, P. (1966). Chemiosmotic coupling in oxidative and photosynthetic phosphorylation. *Biol. Rev. Camb. Philos. Soc.* **41**, 445–502.

Mitome, N., Suzuki, T., Hayashi, S., and Yoshida, M. (2004). Thermophilic ATP synthase has a decamer c-ring: Indication of noninteger 10:3 H^+/ATP ratio and permissive elastic coupling. *Proc. Natl. Acad. Sci. U. S. A.* **101**, 12159–12164.

Mori, H. and Cline, K. (2002). A twin arginine signal peptide and the pH gradient trigger reversible assembly of the thylakoid ΔpH/Tat translocase. *J. Cell Biol.* **157**, 205–210.

Mould, R.M. and Robinson, C. (1991). A proton gradient is required for the transport of two lumenal oxygen-evolving proteins across the thylakoid membrane. *J. Biol. Chem.* **266**, 12189–12193.

Ni, Z.L., Shi, X.B., and Wei, J.M. (2004). Functional consequences of N- or C-terminal deletions of the delta subunit of chloroplast ATP synthase. *Biochemistry* **43**, 2272–2278.

Niyogi, K.K. (1999). Photoprotection revisited: Genetic and molecular approaches. *Annu. Rev. Plant Physiol. Plant Mol. Biol.* **50**, 333–359.

Noctor, G. and Foyer, C.H. (2000). Homeostasis of adenylate status during photosynthesis in a fluctuating environment. *J. Exp. Bot. 51 Spec No*, 347–356.

Noji, H., Yasuda, R., Yoshida, M., and Kinosita, K., Jr. (1997). Direct observation of the rotation of F_1-ATPase. *Nature* **386**, 299–302.

Nowak, K.F., Tabidze, V., and McCarty, RE. (2002). The C-terminal domain of the epsilon subunit of the chloroplast ATP synthase is not required for ATP synthesis. *Biochemistry* **41**, 15130–15134.

Oberfeld, B., Brunner, J., and Dimroth, P. (2006). Phospholipids occupy the internal lumen of the c ring of the ATP synthase of *Escherichia coli*. *Biochemistry* **45**, 1841–1851.

Paumard, P., Vaillier, J., Coulary, B., Schaeffer, J., Soubannier, V., Mueller, D.M., Brethes, D., di Rago, J.P., and Velours, J. (2002). The ATP synthase is involved in generating mitochondrial cristae morphology. *EMBO J.* **21**, 221–230.

Peltier, G. and Cournac, L. (2002). Chlororespiration. *Annu. Rev. Plant Biol.* **53**, 523–550.

Pfundel, E.E., Renganathan, M., Gilmore, A.M., Yamamoto, H.Y., and Dilley, R.A. (1994). Intrathylakoid pH in isolated pea chloroplasts as probed by violaxanthin deepoxidation. *Plant Physiol.* **106**, 1647–1658.

Poetsch, A., Berzborn, R.J., Heberle, J., Link, T.A., Dencher, N.A., and Seelert, H. (2007). Biophysics and bioinformatics reveal structural differences of the two peripheral stalk subunits in chloroplast ATP synthase. *J. Biochem. (Tokyo)* **141**, 411–420.

Pogoryelov, D., Yu, J., Meier, T., Vonck, J., Dimroth, P., and Muller, D.J. (2005). The c15 ring of the *Spirulina platensis* F-ATP synthase: F_1/F_0 symmetry mismatch is not obligatory. *EMBO Rep.* **6**, 1040–1044.

Rappaport, F., Finazzi, G., Pierre, Y., and Bennoun, P. (1999). A new electrochemical gradient generator in thylakoid membranes of green algae. *Biochemistry* **38**, 2040–2047.

Reumann, S. and Keegstra, K. (1999). The endosymbiotic origin of the protein import machinery of chloroplastic envelope membranes. *Trends Plant Sci.* **4**, 302–307.

Rexroth, S., Meyer zu Tittingdorf, J.M., Schwassmann, H.J., Krause, F., Seelert, H., and Dencher, N.A. (2004). Dimeric H^+-ATP synthase in the chloroplast of *Chlamydomonas reinhardtii*. *Biochim. Biophys. Acta* **1658**, 202–211.

Richter, M.L. and McCarty, R.E. (1987). Energy-dependent changes in the conformation of the epsilon subunit of the chloroplast ATP synthase. *J. Biol. Chem.* **262**, 15037–15040.

Richter, M.L., Samra, H.S., He, F., Giessel, A.J., and Kuczera, K.K. (2005). Coupling proton movement to ATP synthesis in the chloroplast ATP synthase. *J. Bioenerg. Biomembr.* **37**, 467–473.

Rimbault, B., Esposito, D., Drapier, D., Choquet, Y., Stern, D., and Wollman, F.A. (2000). Identification of the initiation codon for the *atpB* gene in *Chlamydomonas* chloroplasts excludes translation of a precursor form of the beta subunit of the ATP synthase. *Mol. Gen. Genet.* **264**, 486–491.

Robertson, D., Woessner, J.P., Gillham, N.W., and Boynton, J.E. (1989). Molecular characterization of two point mutants in the chloroplast *atpB* gene of the green alga *Chlamydomonas reinhardtii* defective in assembly of the ATP synthase complex. *J. Biol. Chem.* **264**, 2331–2337.

Robertson, D., Boynton, J.E., and Gillham, N.W. (1990). Cotranscription of the wild-type chloroplast *atpE* gene encoding the CF_1/CF_0 epsilon subunit with the 3′ half of the *rps7* gene in *Chlamydomonas reinhardtii* and characterization of frameshift mutations in *atpE*. *Mol. Gen. Genet.* **221**, 155–163.

Robinson, C. and Bolhuis, A. (2001). Protein targeting by the twin-arginine translocation pathway. *Nat. Rev. Mol. Cell Biol.* **2**, 350–356.

Rodgers, A.J. and Wilce, M.C. (2000). Structure of the gamma-epsilon complex of ATP synthase. *Nat. Struct. Biol.* **7**, 1051–1054.

Rondelez, Y., Tresset, G., Nakashima, T., Kato-Yamada, Y., Fujita, H., Takeuchi, S., and Noji, H. (2005). Highly coupled ATP synthesis by F_1-ATPase single molecules. *Nature* **433**, 773–777.

Ross, J. (1995). mRNA stability in mammalian cells. *Microbiol. Rev.* **59**, 423–450.

Ross, S.A., Zhang, M.X., and Selman, B.R. (1995). Role of the *Chlamydomonas reinhardtii* coupling factor 1 gamma-subunit cysteine bridge in the regulation of ATP synthase. *J. Biol. Chem.* **270**, 9813–9818.

Ross, S.A., Zhang, M.X., and Selman, B.R. (1996). A role for the disulfide bond spacer region of the *Chlamydomonas reinhardtii* coupling factor 1 gamma-subunit in redox regulation of ATP synthase. *J. Bioenerg. Biomembr.* **28**, 49–57.

Rott, R., Levy, H., Drager, R.G., Stern, D.B., and Schuster, G. (1998). 3′-Processed mRNA is preferentially translated in *Chlamydomonas reinhardtii* chloroplasts. *Mol. Cell. Biol.* **18**, 4605–4611.

Sabbert, D., Engelbrecht, S., and Junge, W. (1996). Intersubunit rotation in active F-ATPase. *Nature* **381**, 623–625.

Sambongi, Y., Iko, Y., Tanabe, M., Omote, H., Iwamoto-Kihara, A., Ueda, I., Yanagida, T., Wada, Y., and Futai, M. (1999). Mechanical rotation of the c subunit oligomer in ATP synthase (F_0F_1): Direct observation. *Science* **286**, 1722–1724.

Schumann, J. and Strotmann, H. (1981). The mechanism of induction and deactivation of light-triggered ATPase. In: *Photosynthesis II: Photosynthetic Electron Transport and Photophosphorylation* (G. Akoyunoglou, Ed.), pp. 881–892. Balaban, Philadelphia.

Schumann, J., Richter, M.L., and McCarty, R.E. (1985). Partial proteolysis as a probe of the conformation of the gamma subunit in activated soluble and membrane-bound chloroplast coupling factor 1. *J. Biol. Chem.* **260**, 11817–11823.

Schwarz, O., Schurmann, P., and Strotmann, H. (1997). Kinetics and thioredoxin specificity of thiol modulation of the chloroplast H^+-ATPase. *J. Biol. Chem.* **272**, 16924–16927.

Schwaßmann, H.J., Rexroth, S., Seelert, H., and Dencher, N.A. (2007). Metabolism controls dimerization of the chloroplast F_0F_1 ATP synthase in *Chlamydomonas reinhardtii*. *FEBS Lett.* **581**, 1391–1396.

Seelert, H., Poetsch, A., Dencher, N.A., Engel, A., Stahlberg, H., and Muller, D.J. (2000). Structural biology. Proton-powered turbine of a plant motor. *Nature* **405**, 418–419.

Seelert, H., Dencher, N.A., and Muller, D.J. (2003). Fourteen protomers compose the oligomer III of the proton-rotor in spinach chloroplast ATP synthase. *J. Mol. Biol.* **333**, 337–344.

Smart, E.J. and Selman, B.R. (1991). Isolation and characterization of a *Chlamydomonas reinhardtii* mutant lacking the gamma-subunit of chloroplast coupling factor 1 (CF1). *Mol. Cell. Biol.* **11**, 5053–5058.

Stahlberg, H., Muller, D.J., Suda, K., Fotiadis, D., Engel, A., Meier, T., Matthey, U., and Dimroth, P. (2001). Bacterial Na^+-ATP synthase has an undecameric rotor. *EMBO Rep.* **2**, 229–233.

Stern, D.B. and Kindle, K.L. (1993). 3′ end maturation of the *Chlamydomonas reinhardtii* chloroplast *atpB* mRNA is a two-step process. *Mol. Cell. Biol.* **13**, 2277–2285.

Stock, D., Leslie, A.G., and Walker, J.E. (1999). Molecular architecture of the rotary motor in ATP synthase. *Science* **286**, 1700–1705.

Strotmann, H., Kleefeld, S., and Lohse, D. (1987). Control of ATP hydrolysis in chloroplasts. *FEBS Lett.* **221**, 265–269.

Strotmann, H., Shavit, N., and Leu, S. (1998). Assembly and function of the chloroplast ATP synthase. In: *The Molecular Biology of Chloroplasts and Mitochondria in Chlamydomonas* (J.D. Rochaix, M. Goldschmidt-Clermont, and S. Merchant, Eds.), pp. 477–500. Kluwer Academic Publishers, Dordrecht, The Netherlands.

Theg, S.M., Cline, K., Finazzi, G., and Wollman, F.A. (2005). The energetics of the chloroplast Tat protein transport pathway revisited. *Trends Plant Sci.* **10**, 153–154.

Tierney, M.J. and Medcalf, R.L. (2001). Plasminogen activator inhibitor type 2 contains mRNA instability elements within exon 4 of the coding region. Sequence homology to coding region instability determinants in other mRNAs. *J. Biol. Chem.* **276**, 13675–13684.

Tsunoda, S.P., Rodgers, A.J., Aggeler, R., Wilce, M.C., Yoshida, M., and Capaldi, R.A. (2001). Large conformational changes of the epsilon subunit in the bacterial F_1F_0 ATP synthase provide a ratchet action to regulate this rotary motor enzyme. *Proc. Natl. Acad. Sci. U. S. A.* **98**, 6560–6564.

Turina, P., Samoray, D., and Graber, P. (2003). H^+/ATP ratio of proton transport-coupled ATP synthesis and hydrolysis catalysed by CF_0F_1-liposomes. *EMBO J.* **22**, 418–426.

Valiyaveetil, F.I. and Fillingame, R.H. (1998). Transmembrane topography of subunit a in the *Escherichia coli* F_1F_0 ATP synthase. *J. Biol. Chem.* **273**, 16241–16247.

van Lis, R., Atteia, A., Mendoza-Hernandez, G., and González-Halphen, D. (2003). Identification of novel mitochondrial protein components of *Chlamydomonas reinhardtii*. A proteomic approach. *Plant Physiol.* **132**, 318–330.

Van Walraven, H.S., Strotmann, H., Schwarz, O., and Rumberg, B. (1996). The H^+/ATP coupling ratio of the ATP synthase from thiol-modulated chloroplasts and two cyanobacterial strains is four. *FEBS Lett.* **379**, 309–313.

Velthuys, B.R. (1978). A third site of proton translocation in green plant photosynthetic electron transport. *Proc. Natl. Acad. Sci. U. S. A.* **75**, 6031–6034.

Vik, S.B. and Ishmukhametov, R.R. (2005). Structure and function of subunit a of the ATP synthase of *Escherichia coli*. *J. Bioenerg. Biomembr.* **37**, 445–449.

Weber, J., Wilke-Mounts, S., and Senior, A.E. (2003). Identification of the F_1-binding surface on the delta-subunit of ATP synthase. *J. Biol. Chem.* **278**, 13409–13416.

Werner-Grune, S., Gunkel, D., Schumann, J., and Strotmann, H. (1994). Insertion of a "chloroplast-like" regulatory segment responsible for thiol modulation into gamma-subunit of F_0F_1-ATPase of the cyanobacterium *Synechocystis* 6803 by mutagenesis of *atpC*. *Mol. Gen. Genet.* **244**, 144–150.

Wilkens, S., Zhou, J., Nakayama, R., Dunn, S.D., and Capaldi, R.A. (2000). Localization of the delta subunit in the *Escherichia coli* F_1F_0-ATP synthase by immuno electron microscopy: The delta subunit binds on top of the F_1. *J. Mol. Biol.* **295**, 387–391.

Witt, H.T. (1979). Energy conversion in the functional membrane of photosynthesis. Analysis by light pulse and electric pulse methods. *Biochim. Biophys. Acta* **505**, 355–427.

Witt, H.T., Schlodder, E., and Gräber, P. (1976). Membrane-bound ATP synthesis generated by an external electrical field. *FEBS Lett.* **69**, 272–276.

Woessner, J.P., Masson, A., Harris, E.H., Bennoun, P., Gillham, N.W., and Boynton, J.E. (1984). Molecular and genetic analysis of the chloroplast ATPase of *Chlamydomonas*. *Plant Mol. Biol.* **3**, 177–190.

Wollman, F.A., Minai, L., and Nechushtai, R. (1999). The biogenesis and assembly of photosynthetic proteins in thylakoid membranes. *Biochim. Biophys. Acta* **1411**, 21–85.

Yamamoto, H.Y., Nakayama, T.O., and Chichester, C.O. (1962). Studies on the light and dark interconversions of leaf xanthophylls. *Arch. Biochem. Biophys.* **97**, 168–173.

Yu, L.M. and Selman, B.R. (1988). cDNA sequence and predicted primary structure of the gamma subunit from the ATP synthase from *Chlamydomonas reinhardtii*. *J. Biol. Chem.* **263**, 19342–19345.

Zhou, J.M., Xue, Z.X., Du, Z.Y., Melese, T., and Boyer, P.D. (1988). Relationship of tightly bound ADP and ATP to control and catalysis by chloroplast ATP synthase. *Biochemistry* **27**, 5129–5135.

Zimmermann, B., Diez, M., Zarrabi, N., Graber, P., and Borsch, M. (2005). Movements of the epsilon-subunit during catalysis and activation in single membrane-bound H(+)-ATP synthase. *EMBO J.* **24**, 2053–2063.

CHAPTER 19

Chaperones and Proteases

Michael Schroda[1] *and Olivier Vallon*[2]

[1]Institute of Biology II/Plant Biochemistry, University of Freiburg, Germany

[2]Institut de Biologie Physico-Chimique, Centre National de la Recherche Scientifique/Université Pierre et Marie Curie, Paris, France

CHAPTER CONTENTS

I. INTRODUCTION

Protein folding is a spontaneous process, but many proteins need assistance to prevent aggregation at early or late stages of their life. Chaperones bind proteins and maintain them in a state where folding paths can be explored. They also resolve protein aggregates formed under stress, or help to form or disrupt specific interactions between proteins. Most cellular chaperones use ATP to induce conformational changes in the substrate, with multiple rounds of hydrolysis usually necessary to lead it to a stable "native" conformation (Bukau et al., 2006). Sometimes chaperones are intimately associated with proteolytic systems that function in quality control (the "refold or destroy" paradigm). In these compartmentalized proteases, the peptidase active sites are shielded within a proteolytic chamber. A distinct domain or an associated protein with chaperone activity is responsible for recognizing the substrate and feeding it, in an unfolded form, into the proteolytic chamber. In all compartments of the cell, a wide variety of proteases participate in various steps in protein biogenesis and degradation, from maturation of precursors, to activation of signaling components, to degradation of damaged proteins.

This chapter summarizes current knowledge of chaperones and proteases found in *Chlamydomonas*, building on an earlier study based on version 1.0 of the genome sequence (Schroda, 2004). We also make use of the available genome data from *Volvox* and two *Ostreococcus* species (Palenik et al., 2007). Comparison with the gene repertoire of higher plants will be used to predict function and to compare the complexity of the plant and algal systems. We will emphasize chaperones and proteases that are targeted to the organelles. Note that available prediction programs do not reliably distinguish between chloroplast- and mitochondrially-targeted proteins, so we will often limit ourselves to assuming an organellar localization unless experimental data are available or clear orthology exists with a higher plant protein of known localization. Some important proteins involved in protein folding will not be reviewed here, such as the protein-disulfide and peptidyl-prolyl *cis–trans* isomerases (Vallon, 2005; also see Chapter 11). Similarly, we will not describe the cytosolic proteasome and the ubiquitination pathway, nor the many carboxy- and amino-peptidases, caspases, subtilisin-like peptidases, etc. encoded in the genome. Gene Ontology lists 377 *Chlamydomonas* genes as having peptidase activity, and this figure is probably an underestimation of the repertoire of proteases in *Chlamydomonas*.

II. REGULATION OF THE HEAT SHOCK RESPONSE

Chlamydomonas is usually grown at approximately 25°C, but the optimum for growth in TAP medium probably lies between 30°C and 32°C. Abrupt transfer from 25°C to temperatures above 39°C results in heat shock, and adaptive responses are triggered that allow resumption of growth after a lag period, provided the new temperature is compatible with cellular mechanisms. A common feature of heat shock and other stress responses is the increased synthesis of molecular chaperones (also referred to as heat shock proteins, HSPs). No microarray study of the heat shock response has been performed yet with *Chlamydomonas*, but *in vivo* pulse labeling experiments show extensive changes in the pattern of cytosolic protein translation (Schulz-Raffelt et al., 2007).

In eukaryotes, the expression of genes encoding HSPs is mediated by a specific class of transcription factors, termed heat shock factors (HSFs; reviewed in Morimoto, 1998). In both *Chlamydomonas* and *Volvox* two genes encoding HSFs are present; these were termed *HSF1* and *HSF2*. Although both genes are expressed at the mRNA level, only the *HSF1* cDNA has been cloned to date (Schulz-Raffelt et al., 2007). *Chlamydomonas* HSFs appear to be situated evolutionarily between yeast and higher plants: in contrast to higher plants, which contain at least 21 genes encoding HSFs (Nover et al., 2001), yeast and *Chlamydomonas* have only one and two, respectively. Also in contrast to higher plant HSFs, which need to undergo trimerization for gaining transcriptional activation competence (Lee et al., 1995), yeast HSF and *Chlamydomonas* HSF1 form trimers constitutively (Sorger and Nelson, 1989; Schulz-Raffelt et al., 2007). However, like higher plant HSFs (Scharf et al., 1990; Hübel and Schöffl, 1994) and in contrast to yeast HSF (Wiederrecht et al., 1988), *Chlamydomonas* HSFs are heat-shock inducible; and whereas the amino acid sequence of yeast HSF is distinct from that of higher plants, the amino acid sequence of *Chlamydomonas* HSF1 is closely related to that of its higher plant counterparts (Schulz-Raffelt et al., 2007).

The accumulation of *HSP* transcripts in *Chlamydomonas* during heat shock correlates with HSF1 hyperphosphorylation (Schulz-Raffelt et al., 2007). Hyperphosphorylation is essential for the activation of HSFs in all organisms studied to date (Holmberg et al., 2002). Hence, phosphorylation appears to be an ancient requirement for HSF activation, whereas trimerization might represent an additional level of regulation characteristic of multicellular organisms (Morimoto, 1998). *Chlamydomonas* HSF1 constitutively interacts with cytosolic HSP70A (Schulz-Raffelt et al., 2007; see section VI.B.3).

The analysis of strains underexpressing *HSF1* by RNAi allowed the conclusion that HSF1 is the key regulator of the heat stress response in *Chlamydomonas* (Schulz-Raffelt et al., 2007). Cells in which HSF1 was strongly downregulated were highly sensitive to heat stress. This correlated with a complete lack in these cells of heat shock gene induction and heat shock-induced HSP accumulation.

III. SMALL HSPs

Small heat shock proteins (sHSPs) contain as a signature motif the α-crystallin domain of ~90 amino acids at their C-termini (reviewed in Haslbeck et al., 2005; Sun and MacRae, 2005). sHSPs form dimers, which under non-stress conditions assemble into homo-oligomers of 230 kD and more (Chen et al., 1994). sHSPs are strongly induced by heat shock upon which their oligomeric storage form disassembles into dimers. These dimers expose at their N-termini hydrophobic surfaces for the interaction with denatured proteins, with which sHSPs then assemble into large complexes to prevent their irreversible aggregation (Lee et al., 1997; Haslbeck et al., 2004). In addition, sHSPs in such complexes facilitate the access of ATP-dependent chaperones (ClpB and HSP70; see section IX), which catalyze the refolding of the complexed, denatured proteins back to the native state (Mogk et al., 2003).

Plants contain a large number of different *sHSP* genes (e.g. 19 in *Arabidopsis*; Scharf et al., 2001). These are thought to originate from duplications and subsequent diversification of a nuclear *sHSP* gene in a process that is thought to have occurred only in the lineage leading to higher plants (Waters et al., 1996). The sHSPs in *Chlamydomonas* are more homologous to one another than to any of the sHSPs in *Arabidopsis*, suggesting independent diversification in the two lineages after they diverged (Schroda, 2004; see also Waters et al., 1996; Waters and Vierling, 1999; Waters and Rioflorido, 2007). This contrasts with the other chaperone families HSP60,

HSP70, HSP90, and HSP100, where the *Chlamydomonas* proteins are more closely related to the *Arabidopsis* homologue of same intracellular localization than to other members of that family found in other compartments (Schroda, 2004).

In addition to the six *sHSP* genes present in version 1.0 of the *Chlamydomonas* genome sequence (Schroda, 2004; now termed *HSP22A–HSP22E* and *HSP22G*), two more genes termed *HSP22F* and *HSP22H* were found in version 3.0 (Table 19.1). The *HSP22H* gene model is incomplete and found only 25 kb from *HSP22A* in a region containing many sequence gaps. The *HSP22A* and *HSP22B* genes, whose products are 70% identical (Schroda, 2004), are arranged head-to-head. This feature is shared by the *HSP22E* and *HSP22F* genes which within their coding regions share 98% identity at the nucleotide level and are linked to *HSP22D*. These arrangements are indicative of recent gene duplication events on the way to the evolution of a higher diversity of *sHSP* genes in *Chlamydomonas*. As judged from predictions by the TargetP program (Emanuelsson et al., 2000), HSP22C–HSP22G appear to be organelle-targeted whereas HSP22A, HSP22B, and HSP22H appear to be cytosolic (Table 19.1).

The *Chlamydomonas* HSP22A protein was initially reported to be located in thylakoid membranes and to play a role in the protection of photosystem II (PS II) against photoinhibition (Kloppstech et al., 1985; Schuster et al., 1988). Its import into the chloroplast was believed to take place without a

Table 19.1 Small heat shock proteins

#	Name	Introns[a]	cDNA support	Localization[b]	ORF (aa)	kD (pre)	pI[c] (pre)	Accession #	Reference
1	HSP22A	2	Complete	cyt (4)	157	16.7	6.08	P12811	Grimm et al. (1989)
2	HSP22B	1	Complete	cyt (5)	183	20.0	8.14	EDP07084	
3	HSP22C	1	Complete	cp (5)	291	30.0	6.85	EDP01504	
4	HSP22D	0	Complete	cp (2)	178	18.6	5.29	EDP09693	
5	HSP22E	1	Complete	cp (3)	241	25.2	9.09	EDP09559	
6	HSP22F	1	Complete	cp (3)	241	25.2	9.09	EDP10019	
7	HSP22G	3	None	mt (4)	362	36.7	8.6	EDP08996	
8	HSP22H	≥2	None	cyt (3)	180*	?	?	EDP07081	

aa, amino acids; pre, precursor; pI, isoelectric point; cyt, nucleo-cytosol; cp, chloroplast; mt, mitochondria.

[a] *Introns within coding region.*

[b] *According to TargetP prediction (Emanuelsson et al., 2000); reliability class is given in parentheses, the lower the value, the higher the reliability of the prediction.*

[c] *Determined with ExPASy get pI/Mw.*

* *Incomplete because of gap in sequence encoding C-terminus.*

processing step (Grimm et al., 1989). The protective role of sHSPs was initially supported by the report that a tomato chloroplast-targeted sHSP protected PS II electron transport during heat stress (Heckathorn et al., 1998). However, as the protective role of sHSPs could not be reproduced using recombinant chloroplast-targeted pea sHSP, these findings were questioned (Härndahl and Sundby, 2001). Furthermore, thermoprotection of PS II oxygen evolving capacity in *Chlamydomonas* was demonstrated to be mediated by proteins induced at a growth temperature of 35°C, which is below that required to induce HSP22A (≥39°C; Tanaka et al., 2000). And finally, when the thylakoid targeting of HSP22A in *Chlamydomonas* was reinvestigated, it was suggested to be instead a cytosolic protein that becomes associated with the chloroplast outer envelope during heat shock, and dissociates during recovery (Eisenberg-Domovich et al., 1994). Similarly, a sHSP of *Synechocystis* has been shown to specifically associate with thylakoid membranes upon heat shock, leading to a rigidification of the membranes (Tsvetkova et al., 2002), which potentially might reduce thermodamage of PS II. Hence, association of *Chlamydomonas* HSP22A with the chloroplast outer envelope might at least improve the thermotolerance of processes occurring at this membrane, such as protein import.

Microarray analysis performed with *Chlamydomonas* revealed that genes encoding the putatively chloroplast-targeted HSP22E and HSP22F are induced by oxidative stress (Fischer et al., 2005), sulfur starvation (Zhang et al., 2004), and phosphorus starvation (Moseley et al., 2006). In the latter two studies, induction occurred after prolonged starvation and was stronger and earlier in mutants deficient in the starvation response (*sac1* and *psr1*, respectively). This suggests that HSP22 proteins participate in the response to severe stress, when normal regulatory mechanisms are no longer sufficient to maintain cellular homeostasis. The other *HSP22* gene represented on the microarray, *HSP22C*, showed no significant induction in any of these studies.

IV. HSP33: A REDOX-REGULATED CHAPERONE

HSP33 is a redox-regulated molecular chaperone (reviewed in Ilbert et al., 2006). Under reducing conditions, HSP33 is an inactive monomer. Oxidative stress leads to the oxidation of four highly reactive cysteines, two of which form intramolecular disulfides, and the other two of which form intermolecular disulfide bridges with another HSP33 monomer, leading to homodimerization (Graumann et al., 2001). Under oxidative stress and elevated temperatures, conditions under which bacterial HSP70 apparently is inactive, HSP33 homodimers are active and act ATP-independently as "holdases"; that is, they prevent unfolding proteins from aggregation, but cannot actively refold them. Under reducing conditions and lower temperatures,

HSP70 appears to take over substrates from the now-reduced HSP33 to catalyze refolding in an ATP-dependent process (Hoffmann et al., 2004).

HSP33s are widespread in prokaryotes, whereas in eukaryotes genes encoding HSP33 homologues have so far been reported only in the genome sequences of *Dictyostelium discoideum*, three different species of the *Trypanosomatidae*, and of *Chlamydomonas* (Ilbert et al., 2006). All these eukaryotic HSP33 proteins contain N-terminal extensions, which based on PSORT predictions are targeted to mitochondria (Ilbert et al., 2006). Genes encoding HSP33 proteins are also present in *Volvox* and *Ostreococcus*, and organellar targeting of HSP33 is predicted in *Chlamydomonas* and *Volvox* by TargetP and ChloroP.

V. HSP60s

Type I chaperonins, which are found in eubacteria, mitochondria, and chloroplasts, assemble into two stacked rings, each containing seven subunits of about 60 kD (reviewed in Bukau and Horwich, 1998; Frydman, 2001). In bacteria, the chaperonin is called GroEL, in plastids and mitochondria CPN60 (or HSP60 to account for its heat shock inducibility; Viitanen et al., 1995). The tetradecameric double ring cooperates with a co-chaperonin, which forms one heptameric ring consisting of 10-kD subunits termed GroES in bacteria, and CPN10 or HSP10 in mitochondria and plastids.

Compared to those of mitochondria and bacteria, the plastid chaperonins have evolved some peculiarities: The ~60-kD subunits of the tetradecameric double ring may be either of the α- or of the β-type, which share only about 50% identity. In contrast to the β subunits, which may autoassemble into homotetradecamers, the α-subunits will oligomerize only when β subunits are present (Dickson et al., 2000). Another peculiarity of the plastid chaperonins concerns the co-chaperonin, which consists of a head-to-tail fusion of two CPN10 domains giving a subunit of ~21 kD (also termed CPN20; Bertsch et al., 1992). In addition to CPN20, conventional chloroplast CPN10 co-chaperonins have been described in *Arabidopsis*, tomato, and soybean (Koumoto et al., 2001).

Protein folding occurs within the confined cavity formed by the chaperonin and co-chaperonin rings, where the chaperonin subunits cycle ATP-dependently between binding-active and folding-active states (Bukau and Horwich, 1998). The most prominent chloroplast protein that requires folding by chaperonins to assume the native, oligomerization-competent state is the Rubisco large subunit (see Boston et al., 1996 and references therein). However, several other plastid substrates have been identified that require stromal chaperonins and HSP70 to assume the active state (see section VI.C.2.a).

As judged from homology data and TargetP prediction, the HSP60 family in *Chlamydomonas* consists of three chloroplast-targeted members now

Table 19.2 Chaperonins and co-chaperonins

#	Name	Introns[a]	cDNA support	Localization[b]	ORF (aa)	Mat (aa)	kD (mat)	pI[c]	Accession #	Reference
1	CPN60A	17	Complete	cp	580	547	58.1	5.04	Q42694	Thompson et al. (1995)
2	CPN60B1	16	Complete	cp	580	550	58.9	5.35	EDO97815	Thompson et al. (1995)
3	CPN60B2	15	Complete	cp	577	546	58.5	5.11	EDP03982	Thompson et al. (1995)
4	CPN60C	≥10	None	mt	571*	542	?	?	EDP05086	
1	CPN23	4	Complete	cp	238	212	23	5.14	EDP05269	
2	CPN20	5	Complete	cp	216	194	20.5	6.26	EDO96350	
3	CPN11	4	Complete	cp	128	103	10.9	5.4	EDP01237	
4	CPN10	2	Complete	mt	99	93	10	4.92	EDP05941	

ORF, open reading frame; mat, mature protein; aa, amino acids; pI, isoelectric point; cp, chloroplast; mt, mitochondrial

[a] *Introns within coding region.*

[b] *According to localization of* ***Arabidopsis*** *orthologs.*

[c] *Determined with ExPASy get pI/Mw.*

* *Incomplete because of gap in internal sequence encoding 58 aa.*

termed CPN60A, CPN60B1, and CPN60B2, and one member termed CPN60C that functions in the mitochondrion (van Lis et al., 2003); moreover, three co-chaperones termed CPN11, CPN20, and CPN23 appear to be chloroplast-targeted and one, termed CPN10, appears to be targeted to mitochondria (Table 19.2; Schroda, 2004). All chloroplast CPN60s are induced by heat shock, both at the mRNA and protein levels (Thompson et al., 1995; Tanaka et al., 2000). CPN60A was found to specifically interact with group II intron RNA, and suggested to play a role as a general organelle splicing factor (Balczun et al., 2006). The interaction with RNA appeared to be mediated via the two ATPase domains of the chaperonin.

VI. HSP70s

A. Mechanism of HSP70 function and classification

Except for some Archaea, chaperones of the HSP70 family are found in all organisms and are present in every protein-containing compartment of the eukaryotic cell. HSP70 chaperones consist of an N-terminal ATPase domain which is functionally coupled to a C-terminal substrate-binding domain (reviewed in Wegele et al., 2004; Mayer and Bukau, 2005). Proteins recognized as substrates by HSP70 expose stretches of four to five hydrophobic amino acids that are flanked by basic residues (Rüdiger et al., 1997). HSP70s bind substrates with low affinity in the ATP state, and with high affinity in the ADP state. Upon binding of HSP70, exposed hydrophobic surfaces on the substrate are shielded from the environment, thus protecting substrates from aggregation. Moreover, by the power stroke exerted by HSP70s following ATP hydrolysis (Pierpaoli et al., 1997), and/or by their intrinsic secondary amide peptide bond *cis–trans* isomerase activity (Schiene-Fischer et al., 2002), the conformation of bound substrates may be altered, which in turn may allow non-native proteins to convert to the native state. Thus, HSP70s play a major role in the folding of nascent chains and in the renaturation of non-native proteins that have accumulated during stress situations such as heat shock (Frydman, 2001).

HSP70s also have highly specialized functions such as the regulation of the heat stress response (Tomoyasu et al., 1998), the uncoating of clathrin-coated vesicles (Ungewickell et al., 1995), and membrane translocation of proteins (Kang et al., 1990). Specificity of HSP70 function is mediated largely by its co-chaperones, of which the J-domain co-chaperones represent an important class (see section VI.H). HSP70s can be divided into DnaK-type and HSP110-type HSP70s. DnaK-type HSP70s again can be subdivided into those which are regulated by GrpE-type nucleotide exchange factors (the main bacterial and organellar HSP70s), those regulated by BAG-1 (HSP70s in the eukaryotic cytosol), and those that do not need any nucleotide exchange factor (specialized bacterial and organellar HSP70s) (Brehmer et al., 2001). Nine genes encode HSP70-like proteins in *Chlamydomonas* (see also Schroda, 2004). These have been termed HSP70A–HSP70G, BIP1, and BIP2 (Table 19.3). Information on the genes and their products is summarized in the following section.

B. HSP70A

1. *HSP70A: the only cytosolic DnaK-type HSP70 in Chlamydomonas*

The *HSP70A* gene was originally identified as one of three *HSP70* genes inducible by heat shock and light (Von Gromoff et al., 1989). HSP70A appears to be the only cytosolic DnaK-type HSP70 in *Chlamydomonas*, as judged by its high similarity with cytosolic HSP70s from other organisms (Schroda, 2004), and from the presence of characteristic signature sequences like the C-terminal EEVD motif (Müller et al., 1992). Two J-domain proteins, GLSA and RSP16, are the only co-chaperones of HSP70A that have been characterized to date (see sections VI.H.2 and VI.H.3).

2. *The many facets of the HSP70A promoter*

The promoter of the *HSP70A* gene certainly is one of the best-studied in *Chlamydomonas*. What makes it so fascinating on the one hand is its inducibility by various cues including stress, light, hydrogen peroxide, singlet oxygen, deflagellation and in *Volvox*, the stage of development (Von Gromoff et al., 1989; Kropat et al., 1995; Leisinger et al., 2001; Cheng et al.,

Table 19.3 HSP70s and GrpE/HEP-type co-factors

#	Name	Introns[a]	cDNA support	Localization[b]	ORF (aa)	Mat (aa)	kD (mat)	pI[c] (mat)	Accession #	Reference
1	HSP70A	6	Complete	cyt	649	649	70.9	5.16	EDO97621	Müller et al. (1992)
2	HSP70B	5	Complete	cp(e)	679	639	67.9	5.05	AA65356	Drzymalla et al. (1996)
3	HSP70C	16	Only 3′ half	mt(e)	*633	626	67.2	5.52	EDP02463	von Gromoff et al. (1989)
4	BIP1	7	Complete	ER	656	632	70	4.96	EDP06660	
5	BIP2	9	Only 3′ end	ER	662	638	70.3	5.02	EDP06859	
6	HSP70D	13	None	cp/mt	659	632	67.5	5.36	EDP03456	
7	HSP70E	19	Complete	cyt	803	803	87.1	5.21	EDP01195	
8	HSP70F	?	None	cp/mt	?	?	?	?	EDP00821	
9	HSP70G	17	Partial	ER	1109	1089	117.7	5.22	EDP05802	
1	CGE1	7	Complete	cp (e)	258 (1a) 60 (1b)	219 (1a) 221 (1b)	23.8 24.0	4.71 4.71	Q945T2, Q945T1	Schroda et al. (2001b)
2	CGE2	≥2	None	cp	*132	?	?	?	EDP03672	
3	MGE1	5	Almost complete	mt	264	220	23.5	5.06	EDO97605	
1	HEP1	4	Almost complete	mt	166	92	10.4	9.17	EDO98451	Willmund et al. (2008)
2	HEP2	3	Complete	cp(e)	205	135	14.9	5.16	ABS84949	Willmund et al. (2008)

ORF, open reading frame; mat, mature protein; aa, amino acids; pI, isoelectric point; cyt, nucleo-cytoplasm; cp, chloroplast, mt, mitochondrial.

[a] *Introns within coding region.*

[b] *According to localization of orthologs; (e) determined experimentally.*

[c] *Determined with ExPASy get pI/Mw.*

* *Incomplete: prediction of extreme N-terminus impossible due to lacking EST support.*

2005, 2006; Shapiro et al., 2005; Shao et al., 2007). On the other hand, the *HSP70A* promoter has the amazing potential to activate other promoters in its vicinity (Schroda et al., 1999, 2000, 2002; Lodha et al., 2008).

In fact, the *Chlamydomonas HSP70A* gene is regulated by two independent promoters, P_{A1} and P_{A2}, with distinct transcription initiation sites (von Gromoff et al., 2006). Stress inducibility is mediated by the upstream P_{A1} promoter via its four heat shock elements (HSEs) (Kropat et al., 1995; von Gromoff et al., 2006; Lodha et al., 2008); light inducibility is mediated by the downstream P_{A2} promoter via two plastid response elements (PREs), which share the consensus sequence 5′-GCGACNAN$_{15}$TA-3′ (von Gromoff et al., 2006). As PREs are employed for induction of *HSP70A* by both MgProto and light, it appears that light induction of this gene is mediated via MgProto (Kropat et al., 1997). And as light inducibility was conferred by the PREs in an orientation- and distance-independent manner to non-light inducible promoters (i.e. promoters *RBCS2*, *TUB2*, and *CYC6*), the PREs qualify as enhancers (Schroda et al., 2000; von Gromoff et al., 2006). In addition to HSEs and PREs, as-yet unidentified sequence motifs within the *HSP70A* promoter mediate its inducibility by hydrogen peroxide and singlet oxygen, respectively (Shao et al., 2007).

Upon attempting to express the *Chlamydomonas HSP70B* gene ectopically in *Chlamydomonas*, expression was observed very rarely and then at low levels (Schroda et al., 2000). This was the case when the *HSP70B* gene was driven by its own promoter, or by the *RBCS2* or *TUB2* promoters. However, when the *HSP70A* promoter was used, high level expression (after heat shock) was observed at high frequency. Moreover, fusion of the *HSP70A* promoter upstream of the otherwise non-active *HSP70B*, *RBCS2*, and *TUB2* promoters activated them efficiently (Schroda et al., 1999, 2000).

To understand the mechanism underlying the activating effect of the *HSP70A* promoter, the bacterial resistance gene *ble* (Lumbreras et al., 1998) was used to assess promoter efficiencies by measuring transformation rates. It turned out that the *ble* gene driven by the *HSP70A-RBCS2* promoter fusion yielded about twice as many transformants as the *ble* gene driven by the *RBCS2* promoter alone (Schroda et al., 2002). Surprisingly, however, average *ble* transcript levels in transformants generated with either construct were the same. This apparent contradiction was resolved by determining the percentage of transformants that actually expressed the integrated transgene: the *RBCS2* promoter-driven *ble* construct was expressed in only 20% of the transformants, whereas the *HSP70A-RBCS2* promoter-driven gene was expressed in more than 60% (Schroda et al., 2002). These data suggested that the stimulatory effect of the *HSP70A* promoter on other promoters was due to the *HSP70A* promoter's ability to counteract transcriptional silencing. Two elements within the *HSP70A* promoter were mapped that independently counteracted transgene silencing: a proximal element bounded by nt −23 to −285 relative to the translational start codon, and a distal element located upstream of position −285 (Schroda et al., 2002). Whereas the proximal element exhibited a strong spacing dependence with respect to the downstream *RBCS2* promoter, the distal element appeared to act in a spacing-independent manner.

To identify *cis*-acting elements that might mediate the stimulatory effect through binding of specific protein factors, DNase I hypersensitivity assays were carried out on the native *HSP70A* gene locus, revealing up to four hypersensitive sites (Lodha and Schroda, 2005). The two strongest ones were constitutively detected in dark-grown cells, after a dark-to-light shift, in light, and after heat shock. The strongest hypersensitive site mapped close to HSE1 and the TATA-box, while the second-strongest mapped close to HSE4. As HSE1 and HSE4 lie within the proximal and distal elements, respectively, these data suggested that the stimulatory effect of the *HSP70A* promoter might be exerted via heat shock factors constitutively binding to the HSEs. Mutation and deletion analyses indeed demonstrated that HSE4 and HSE2, aided by HSE1, HSE3 and the TATA-box, were the major *cis*-regulatory elements involved (Lodha et al., 2008).

The working hypothesis emerging from these data is as follows: transgene promoters in *Chlamydomonas* are frequently silenced by the formation of repressive chromatin structures that occlude *cis*-regulatory sequences normally accessed by *trans*-activation factors (Schroda et al., 2002). This may, for example, be caused by tightly bound, unfavorably positioned nucleosomes (Richmond and Davey, 2003; Li et al., 2006). Heat shock factors constitutively binding to the HSEs in the *HSP70A* promoter recruit chromatin remodeling activities that open up the chromatin structure at the promoter itself and in surrounding sequences. This may relieve repressive chromatin structures downstream, thereby rendering the *cis*-regulatory sequences of a downstream promoter accessible to transcription factors.

A constitutively open chromatin structure has been observed for heat shock gene promoters in yeast (Gross et al., 1993), *Drosophila* (Lis and Wu, 1993), *Xenopus* (Landsberger and Wolffe, 1995), and humans (Xing et al., 2005). In yeast and in humans, the open chromatin structure is mediated by heat shock factors constitutively binding to HSEs within the promoters. Maintaining the chromatin structure of heat shock gene promoters in an open conformation may keep them readily accessible for transcription factors to allow for a rapid stress response. As an important side effect of the *Chlamydomonas* work described above, *HSP70A-RBCS2* promoter fusions emerged as a powerful tool for transgene expression in both *Chlamydomonas* and *Volvox* (e.g. Sizova et al., 2001; Zhao et al., 2001; Lechtreck et al., 2002; Koblenz et al., 2003; Pfannenschmid et al., 2003; Reisdorph and Small, 2004; Dent et al., 2005; Murakami et al., 2005; Schoppmeier et al., 2005; Glanz et al., 2006; Hallmann and Wodniok, 2006).

3. Functions of HSP70A

Despite the fact that HSP70A is apparently the only DnaK-like HSP70 in the nucleus and cytosol of *Chlamydomonas* and thus presumably broadly involved in the cellular machinery, only a few roles have been characterized to date. The first was a possible function in folding of flagellar proteins, as HSP70A was localized to flagella (Bloch and Johnson, 1995; Pazour et al., 2005) and found to be induced by deflagellation (Shapiro et al., 2005). Specifically, HSP70A exhibited a discontinuous, punctuate pattern along the length of the flagella, with a concentration in the flagellar tip (Shapiro et al., 2005). HSP70A may interact with dimeric RSP16, a flagellar J-domain protein that is part of the radial spoke (Yang et al., 2005, 2006; see section VI.H.2; see also Volume 3, Chapter 7).

Another specialized function of HSP70A has been reported in *Volvox*, where it cooperates with a specialized J-domain protein termed GlsA in mediating asymmetric cell divisions in embryonic cells (Cheng et al., 2005;

see section VI.H.3). Finally, HSP70A in *Chlamydomonas* was found to interact with HSF1 (Schulz-Raffelt et al., 2007; see section II). This finding is in line with observations in animals and higher plants, where HSFs have been found to interact with members of the HSP70 and HSP90 families (Abravaya et al., 1992; Baler et al., 1992; Ali et al., 1998; Zou et al., 1998; Kim and Schöffl, 2002). Apparently, the interaction of HSF with HSP90 prevents its activation (Duncan, 2005), whereas the interaction with HSP70 abolishes its transactivation competence (Shi et al., 1998). Inactivation by HSP90 prior to heat shock, and by HSP70 following heat shock, were suggested to be mediated by preventing HSF trimerization and by catalyzing HSF monomerization, respectively (Mosser et al., 1993; Lee and Schöffl, 1996; Voellmy, 2004). However, since *Chlamydomonas* HSF1 appears to be constitutively trimeric (Schulz-Raffelt et al., 2007), the interaction of HSF1 with HSP70A most likely does not affect the oligomerization state, but may serve a different purpose. For example, HSP70A may only facilitate folding of HSF1 to the native state and the regulation of the stress response via chaperones may occur at the level of stress kinases mediating HSF1 hyperphosphorylation (see section II). Clearly, more experimental data are required to clarify this issue.

C. HSP70B

1. Expression and localization of HSP70B

HSP70B and *HSP70A* were found concomitantly as genes inducible by heat shock and light (von Gromoff et al., 1989). The *HSP70B* gene was also found to be induced by Mg-Proto (Kropat et al., 1997; Vasileuskaya et al., 2004), oxidative stress (Fischer et al., 2005), and high light intensity (Yokthongwattana et al., 2001), and to be under circadian control (Jacobshagen et al., 2001). The HSP70B protein is targeted to the chloroplast, where it accumulates mainly in the stroma, but is also found associated with low-density membranes (containing inner envelopes, Zerges and Rochaix, 1998), and thylakoids (Drzymalla et al., 1996; Schroda et al., 2001b; Liu et al., 2005; Willmund and Schroda, 2005). As judged from an estimated contribution to total cellular protein of ~0.15%, HSP70B is relatively abundant (Willmund and Schroda, 2005; Liu et al., 2007). Four (co)chaperones have so far been reported to interact with HSP70B: the nucleotide exchange factor CGE1, the escorting protein HEP2, the J-domain protein CDJ2, and HSP90C (see sections VI.H.4–VI.H.6 and VII.D; Figure 19.1).

2. Functions of HSP70B

a. Folding of chloroplast proteins

The following circumstantial evidence suggests a role for HSP70B in the refolding of stress-denatured chloroplast proteins. First, HSP70B is induced

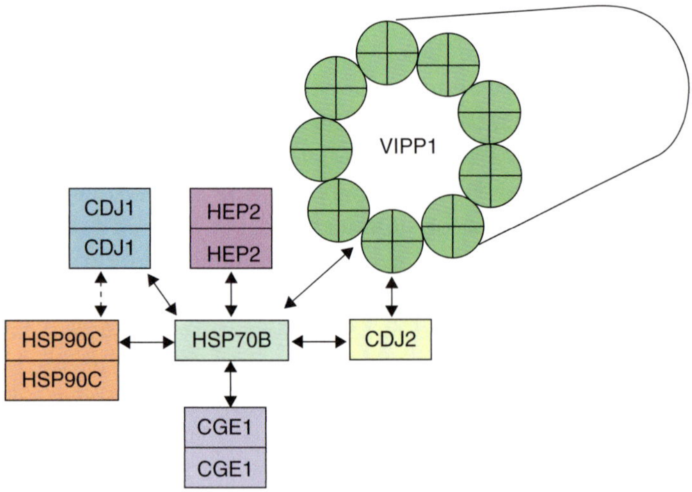

FIGURE 19.1 *Interaction of chloroplast HSP70B with its co-chaperones (CGE1, CDJ1, CDJ2, HEP2), substrates (VIPP1), and with other chaperone systems (HSP90C). The oligomeric state of the proteins is indicated. VIPP1 presumably forms rings consisting of 9 × 4 VIPP1 subunits, which can assemble into rod-like structures.*

about threefold at the protein level after heat shock (Drzymalla et al., 1996). Second, in ATP-depleted cell extracts much higher concentrations of HSP70B were in complex with its co-chaperone CGE1 when cells had experienced heat shock, as compared to non-stressed cells (Schroda et al., 2001b). This suggests that under heat shock conditions much more HSP70B is in a folding-active state than under non-stress conditions.

Although HSP70 in the matrix is essential for protein import into mitochondria (Kang et al., 1990), evidence for a analogous function of stromal HSP70 is lacking. In contrast, a role for stromal HSP70 in protein maturation has been documented for the subunits of the coupling factor CF_1 (Chen and Jagendorf, 1994), ferredoxin-$NADP^+$ reductase (Tsugeki and Nishimura, 1993), the Rieske protein (Madueño et al., 1993), and phytoene desaturase (Bonk et al., 1996). In all these cases, the chloroplast HSP70 and HSP60 systems appear to act in a coordinate manner.

b. Protection and repair of photosystem II

The pronounced light-inducibility of *Chlamydomonas* HSP70B (about two-fold at the protein level, Drzymalla et al., 1996) suggested that it might have a protective role under conditions where light leads to protein damage. For example, this is the case for PS II when cells are exposed to excess light in a process called photoinhibition. Indeed, PS II photodamage after high light treatment was less pronounced and PS II activity recovered faster in cells overexpressing HSP70B than in the wild-type, whereas the opposite was observed in cells underexpressing HSP70B (Schroda et al., 1999). Presumably, HSP70B overexpression allowed for faster repair of photodamaged PS II by facilitating a coordinated exchange of damaged D1 proteins for *de novo*-synthesized D1 (Schroda et al., 2001a). In support of this view, HSP70B in the green alga *Dunaliella salina* was found to be part of a ~320-kD complex containing photodamaged D1, D2, and CP47 proteins (Yokthongwattana et al., 2001). Moreover, *Chlamydomonas* strains underexpressing the Alb3.2 protein, which appears to be involved in the assembly/maintenance of PS I and PS II, contained elevated levels of HSP70B (Göhre et al., 2006).

c. Assembly/disassembly of VIPP1 oligomers

In *Chlamydomonas*, HSP70B was found to interact with the vesicle-inducing protein in plastids 1 (VIPP1), to which it appears to be targeted via its J-domain co-chaperone CDJ2 (Liu et al., 2005; see section VI.H.6; Figure 19.1). VIPP1 got its name from the phenotype of underexpressing *Arabidopsis* mutants (Kroll et al., 2001). Mutant plants could not grow photoautotrophically because of a distorted thylakoid structure and were deficient in the formation of vesicles. In wild-type plants, vesicles bud off from the inner envelope and are believed to transport galactolipids to the thylakoids (for reviews see Vothknecht and Soll, 2005; Benning et al., 2006). VIPP1 is conserved from cyanobacteria to higher plants and since reducing VIPP1 expression in *Synechocystis* resulted in distorted thylakoids as well, it appears to be functionally conserved (Westphal et al., 2001). Like its ancestor, the phage shock protein A (PspA), VIPP1 was shown to assemble into rotationally symmetric rings of >1 Md (Aseeva et al., 2004; Hankamer et al., 2004).

In ATP-depleted cell extracts, CDJ2 formed complexes with VIPP1 of >>670 kD and <230 kD, whereas in cell extracts supplemented with ATP CDJ2/VIPP1-containing complexes were ~670 kD; moreover, HSP70B coprecipitated with CDJ2 and with VIPP1 as complexes >230 kD, but only in the presence of ATP (Liu et al., 2007). This behavior suggested a role for the chaperones in the assembly and disassembly of VIPP1 rings, in which CDJ2 binds to VIPP1 rings or VIPP1 monomers/dimers and delivers them to HSP70B in the ATP state. HSP70B would then convert rings (>>670 kD) via intermediate assembly states (~670 kD) to monomers/dimers (<230 kD), and monomers/dimers via intermediate assembly states into rings, using ATP hydrolysis as an energy source. Under limiting ATP, CDJ2 bound to VIPP1 rings (>>670 kD) and VIPP1 monomers/dimers (<230 kD) would accumulate, since HSP70B/CGE1 could not process them. The suggested role of the HSP70B/CDJ2/CGE1 chaperones in VIPP1 assembly/disassembly was confirmed by the finding that the purified chaperones converted purified VIPP1 monomers/dimers into oligomers, and purified VIPP1 oligomers to monomers/dimers, in an ATP-dependent manner (Liu et al., 2007).

Electron microscopy studies using purified VIPP1 revealed that in contrast to PspA, VIPP1 rings assemble into huge, rod-shaped supercomplexes (Liu et al., 2007). VIPP1 rods, too, were disassembled by purified chaperones in an ATP-dependent process. In ATP-depleted *Chlamydomonas* whole cell extracts, VIPP1 also was found to form oligomers that migrated similarly to the rods formed by purified VIPP1 in sucrose gradient centrifugation. The VIPP1 oligomers in cell extracts were partially disassembled by the addition of ATP and essentially completely disassembled by the addition of ATP and purified chaperones. These observations suggested that VIPP1 also may form rod-shaped supercomplexes *in vivo* that are disassembled by the native HSP70B-CDJ2-CGE1 chaperones in an ATP-dependent process.

Rod-shaped supercomplexes referred to as microtubule-like structures (MTLs) were observed earlier by electron microscopy in plastids from algae and various plant tissues (e.g. Schnepf, 1961; Newcomb, 1967; Pickett-Heaps, 1968; Sprey, 1968; Lunney et al., 1975; Rivera and Arnott, 1982; Whatley et al., 1982; Lawrence and Possingham, 1984). Their morphological parameters are very similar to those of VIPP1 rods: both have outer diameters of 150–400 Å, a hole in the middle of ~100 Å, lengths ranging from 0.05 to several μm, and rough margins and transverse striations originating from the fusion of particles (rings?). Many MTLs appear to originate at the inner chloroplast membrane, from which they run perpendicularly into the stroma and eventually may contact thylakoids. Consistently, VIPP1 in *Chlamydomonas* chloroplasts was detected in the stroma, low density membranes, and thylakoids (Liu et al., 2005). Although direct proof is lacking, the similarity of morphological properties of VIPP1 rods and MTLs raises the attractive possibility that the two structures might be identical. The functions attributed to MTLs are diverse: it was suggested that they might be involved in plastid division (Sprey, 1968; Rivera and Arnott, 1982), in the formation of the prolamellar body (Newcomb, 1967), in the formation of thylakoids during greening (Sprey, 1968), or that they might serve a transport function between the plastid envelope and internal lamellae and vesicles (Lunney et al., 1975; Carde et al., 1982; Whatley et al., 1982).

At a first glance, a role for VIPP1 rods as tracks for the transport of proteins and/or lipids would be consistent with the thylakoid biogenesis defects in *Arabidopsis vipp1* mutants (Kroll et al., 2001; Westphal et al., 2001). However, several lines of evidence suggest that VIPP1 may have different or additional functions: (i) PspA significantly improved the performance of the bacterial Sec and of bacterial and thylakoidal TAT protein export pathways (DeLisa et al., 2004; Vrancken et al., 2007). VIPP1 also improved export via bacterial and thylakoidal TAT pathways (DeLisa et al., 2004). As defects in components of Sec and TAT pathways led to a similar thylakoid structure as observed in the *Arabidopsis vipp1* mutant (Settles et al., 1997), the mutant phenotype might be due to inefficient functioning of those protein translocation pathways rather than to vesicle formation. (ii) Some membrane proteins of the bacterial plasma membrane and the thylakoid membrane require YidC or Alb3, respectively, for correct membrane insertion

(Kuhn et al., 2003). *E. coli* mutants lacking YidC induced the expression of the *psp* operon (Van der Laan et al., 2003) and *Chlamydomonas* RNAi strains in which the Alb3.2 protein was downregulated had elevated levels of VIPP1, HSP70B, and CDJ2 (Göhre et al., 2006). These results rather suggest an involvement of PspA and of VIPP1/HSP70B/CDJ2 in processes related to membrane protein sorting.

d. Cooperation with plastidic HSP90C

HSP70B was found to interact with the chloroplast HSP90 homologue, HSP90C (Willmund and Schroda, 2005). As outlined in section VII.D, HSP70B might mediate the early folding steps of a subset of chloroplast proteins, which are folded to their functional states by HSP90C.

D. HSP70C

As judged from its similarity with mitochondrial HSP70s in *Arabidopsis*, HSP70C (formerly hsp68) appears to be targeted to mitochondria (Schroda, 2004). The *HSP70C* gene is inducible by heat shock and light (von Gromoff et al., 1989; Vasileuskaya et al., 2004).

E. BIP1 and BIP2

BIP (Immunoglobulin heavy chain binding protein) is a HSP70 protein found in the ER lumen. *Chlamydomonas* encodes *BIP1* and *BIP2*, whose products are similar to the three BIP proteins from *Arabidopsis*, including predicted N-terminal signal peptides, and canonical C-terminal HDEL ER retention signals (Schroda, 2004). *BIP1* is strongly induced by light (Vasileuskaya et al., 2004). The BIP1 and BIP2 amino acid sequences are 92% identical and 95% similar, suggesting a rather recent gene duplication. Accordingly, both genes are located in a cluster which also includes the gene encoding HSP90B (see section VII.C).

F. HSP70D and HSP70F

HSP70D and *HSP70F* are the only *HSP70* genes identified in the *Chlamydomonas* genome that are not supported by ESTs (Table 19.3). The deduced HSP70D amino acid sequence is most similar to plant stromal and cyanobacterial HSP70s (around 42% identity and 60% similarity) and contains an N-terminal extension that may be an organellar transit peptide. However, it completely lacks the C-terminal peptide-binding domain, which is replaced by an unrelated highly charged domain of 113 amino acids. This organization is also found in *Ostreococcus* and *Volvox* homologues, but no such genes can be found in other taxa and nothing is known about the role of this class of HSP70 proteins. The *HSP70F* sequence contains large gaps;

the available deduced amino acid sequence also is most similar to plant stromal and cyanobacterial HSP70s (around 35% identity and 52% similarity). HSP70F also appears to contain an N-terminal extension for organelle targeting.

G. HSP70E and HSP70G

HSP70E and HSP70G appear to be members of the HSP110/SSE family of HSP70-like proteins. HSP110/SSE proteins are larger and more divergent in sequence than DnaK-type HSP70s and are abundant in eukaryotes (Easton et al., 2000). They differ from other HSP70 proteins in the C-terminal peptide-binding domain, and contain an additional C-terminal domain of unknown function. They do not display folding activity by themselves, but rather interact with cytosolic DnaK-type HSP70s, and by acting as nucleotide exchange factors improve the folding efficiency of these HSP70s (Raviol et al., 2006). The *Chlamydomonas HSP70E* and *HSP70G* genes encode proteins of 87 and 119.6 kD, respectively, that share only 28% identical and 48% similar amino acids. HSP70G contains multiple repeats of a KKEAP motif near the C-terminus. TargetP predicts HSP70G to be targeted to the ER, which is supported by its high similarity to ER-resident GRP170/ORP150/Hyou1 proteins and the potential ER retention motif KGEEL at its C-terminus. Many ESTs are available for *HSP70E*, but few for *HSP70G*. The *HSP70E* gene was shown to be induced by Mg-Proto (Vasileuskaya et al., 2004).

H. HSP70 co-chaperones

1. Functions and categories of HSP70 co-chaperones

HSP70 co-chaperones are usually involved in one or a combination of the following processes: they regulate the ATPase activity of their HSP70 partner, supply it with substrates, and/or connect it with other factors involved in protein folding or degradation. The J-domain proteins constitute an important class of co-chaperones (reviewed in Miernyk, 2001; Hennessy et al., 2005). These interact via their J-domain with their HSP70 partner, stimulate its ATPase activity and supply it with specific substrates (Liberek et al., 1991; Han and Christen, 2003). In doing so, J-domain proteins mediate substrate specificity and thereby determine HSP70 function. At least 63 *Chlamydomonas* genes potentially encode proteins harboring a J domain. Of these, only three also contain a Zn finger domain, whereas in *Arabidopsis* eight proteins with J and Zn finger domains are present (Miernyk, 2001). Of the 63 J-domain proteins found, at least five are predicted to be targeted to the chloroplast and therefore were termed CDJ1–CDJ5 (chloroplast DnaJ proteins).

Similarly important are co-chaperones that regulate the affinity for ATP of their HSP70 partner. For HSP70s in the eukaryotic cytosol these are BAG1 and HIP1, which stimulate nucleotide exchange (Gässler et al., 2001) or stabilize the ADP-bound state (Höhfeld et al., 1995), respectively. Most prokaryotic HSP70s require a GrpE-type nucleotide exchange factor, which is termed MGE in mitochondria (Laloraya et al., 1994) and CGE in chloroplasts (Schroda et al., 2001b). Three genes encoding GrpE-type nucleotide exchange factors were found in the *Chlamydomonas* genome sequence (Table 19.3). One of them, MGE1, contains signature sequences characteristic of mitochondrial GrpEs; the others, CGE1 and CGE2, contain signatures characteristic of chloroplast GrpE homologues (Schroda et al., 2001b; Schroda, 2004).

Other co-chaperones exhibit a variety of functions. CHIP1 inhibits HSP70's ATPase activity and mediates ubiquitinylation of kinetically trapped substrates for degradation by the proteasome (Ballinger et al., 1999; Höhfeld et al., 2001). Co-chaperones like HOP (Smith et al., 1993) or the J-domain protein TPR2 (Brychzy et al., 2003) mediate the interaction of the HSP70 chaperone system with HSP90. A homologue of HOP is encoded in the genome. In *Chlamydomonas* and *Volvox* the most intensely studied HSP70 co-chaperones are the J-domain proteins RSP16, CDJ2, and GLSA, and the nucleotide exchange factor CGE1.

2. J-domain protein RSP16 is a constitutive component of radial spokes

The radial spoke is a T-shaped, structurally conserved macromolecular complex in "9 + 2" axonemes required to control flagellar movement (see also Volume 3, Chapter 7). The radial spoke sediments as a 20S particle and contains at least 23 distinct polypeptides (Yang et al., 2001, 2006), one of which is the ~40-kD J-domain protein RSP16 (Yang et al., 2005). In addition to its N-terminal J domain, RSP16 contains the DnaJ C-terminal domain, but lacks the Cys-rich, Zn-finger forming domain characteristic of *bona fide* DnaJ/HSP40 proteins. RSP16 was found to form dimers and to be a constitutive component of radial spokes, but it was not found in 12S spoke precursor complexes containing all other known spoke proteins. Hence, it was suggested that RSP16 may be involved in converting spoke precursors into mature spokes (Yang et al., 2005, 2006). This process might take place at the flagellar tip, where the putative HSP70 chaperone partner of RSP16, HSP70A, was found to be enriched (Bloch and Johnson, 1995; Shapiro et al., 2005).

3. GlsA mediates unequal divisions of embryonic cells in *Volvox*

In *Volvox carteri*, asymmetric cell divisions determine the fate of embryonic cells in development. Cells larger than 8 μm develop into immortal gonidia, whereas cells smaller than 8 μm differentiate into mortal somatic cells (Kirk et al., 1993). In the *glsA* mutant (*gonidialess*), asymmetric cell division is abolished and all embryonic cells differentiate into somatic cells. The gene affected encodes the 82-kD GlsA protein, which contains a J domain at its N-terminus and two so-called SANT domains and an M domain in its C-terminal part (Miller and Kirk, 1999; Cheng et al., 2005). GlsA expression begins in mature gonidia just as they prepare to initiate cell division, and continues through most of the cleavage period (Miller and Kirk, 1999). A *Chlamydomonas* gene encoding a homologue of GlsA, *GAR1*, complemented the *gonidialess* phenotype in the *Volvox* mutant (Cheng et al., 2003). GAR1 shares 72% identical and 80% similar residues with GlsA. The GlsA J domain is essential, as mutations in the highly conserved HPD motif of the J domain abolished the protein's ability to complement the *gonidialess* phenotype (Miller and Kirk, 1999). This implied that GlsA, as expected for a J-domain protein, acts in concert with its HSP70 partner. Accordingly, GlsA was found to interact with HSP70A (Cheng et al., 2005; see section VI.B.3). GlsA is equally distributed in all blastomeres until the 16-cell embryo, as is the case for HSP70A. In 32-, 64-, and 128-cell stages, however, HSP70A is more abundant in the anterior region of the embryo where asymmetrically dividing cells are located. Both proteins co-localized with histones during interphase, which was consistent with findings that SANT domains participate in regulating chromatin structure. It was suggested that the level of HSP70A-GlsA complexes needs to exceed a certain threshold to modify chromatin in such a way that a gene product required to bring about a shift in the division plane is produced (Cheng et al., 2005).

4. CGE1 is the nucleotide exchange factor of HSP70B

A chloroplast homologue of the bacterial GrpE nucleotide exchange factor, CGE1, was identified as a ~28-kD protein co-precipitating with HSP70B from ATP-depleted *Chlamydomonas* extracts (Schroda et al., 2001b). Like HSP70B, most CGE1 was stromal, and like the *HSP70B* gene, *CGE1* was inducible by heat shock and light (Schroda et al., 2001b; Vasileuskaya et al., 2004). Native PAGE and pulse labeling revealed that CGE1 monomers assemble rapidly into dimers and presumably also into tetramers; moreover, CGE1 formed ATP-sensitive complexes of ~120 kD with HSP70B, suggesting an interaction of dimeric CGE1 with monomeric HSP70B (Figure 19.1). As mentioned in section VI.C.2.a, the ~120-kD complexes accumulated dramatically after heat shock (Schroda et al., 2001b).

Mass spectrometry separated a and b isoforms of CGE1, where CGE1b possesses additional valine and glutamine residues inserted between amino acids 3 and 4 of mature CGE1a. This difference is due to temperature-dependent alternative splicing, with *CGE1b* transcript levels increasing upon heat shock (Schroda et al., 2001b). The same mechanism also leads to temperature-dependent protein accumulation: CGE1a is dominant at low temperatures, whereas CGE1b becomes equally abundant above 30°C (Willmund et al., 2007). The two isoforms have different affinities for

HSP70B, with that of CGE1b being about 25% higher. The biological significance of this, if any, remains enigmatic.

Chloroplast, mitochondrial, and bacterial GrpE homologues have similar structures (Harrison et al., 1997; Padidam et al., 1999; Borges et al., 2003; Willmund et al., 2007), assemble into stable dimers and tetramers which form ATP-sensitive complexes with HSP70 (Osipiuk et al., 1993; Deloche and Georgopoulos, 1996; Wu et al., 1996; Azem et al., 1997; Schroda et al., 2001b; Willmund et al., 2007), and are functionally interchangeable (Deloche et al., 1997; Schroda et al., 2001b; Willmund et al., 2007). Chloroplast CGE1 (Schroda et al., 2001b; Willmund et al., 2007) is additionally light-inducible, exists as two isoforms, and forms dimers via an N-terminal coiled-coil, whereas *E. coli* GrpE dimerizes via a 4-helix-bundle.

5. HEP2 is required for activating HSP70B

The characterization of HSP70B was severely hampered by the apparent non-functionality of the *E. coli*-expressed protein which unlike HSP70B purified from *Chlamydomonas*, could not form complexes with CGE1 (Willmund et al., 2007). Sichting et al. (2005) reported that a specialized protein containing a Zn finger motif and termed HSP70 escort protein 1 (HEP1) was essential for maintaining mitochondrial HSP70s in a functional state in yeast, by preventing their aggregation. Thus, it appeared possible that a chloroplast HEP1 homologue was analogously required. *Chlamydomonas* and higher plants encode two and three HEP1 homologues, respectively (Willmund et al., 2008). TargetP analysis of the two *Chlamydomonas* homologues revealed that one of them (now termed HEP2) was predicted to be chloroplast-targeted, the other to mitochondria (Table 19.3). Moreover, homologues predicted to be mitochondrially versus chloroplast-targeted formed separate clades.

Chlamydomonas HEP2 can be detected in stromal extracts, and cross-linking and coimmunoprecipitation studies suggested that it forms dimers and indeed interacts with HSP70B *in vitro* and in *Chlamydomonas* extracts (Willmund et al., 2008; Figure 19.1). Like the interaction of mitochondrial HSP70 with HEP1 (Sichting et al., 2005), the HSP70B–HEP2 interaction was ATP-dependent. Although HEP2 interacted with non-functional HSP70B, the latter could not be converted to the active state. However, when HEP2 was coexpressed with HSP70B in *E. coli*, functional HSP70B could be recovered quantitatively. Two observations suggested that HEP2 was involved in *de novo* folding of HSP70B rather than in maintaining it in an active state. First, functional HSP70B remained functional in the absence of HEP2; and second, HSP70B expressed with an N-terminal fusion protein normally was completely non-functional after fusion protein cleavage. However, if HEP2 was present during fusion protein cleavage, functional HSP70B was recovered. As HEP homologues do not exist in (cyano)bacteria, they appear to represent a eukaryotic invention, presumably to cope with folding problems of HSP70s that are targeted to organelles (Willmund et al., 2008).

6. CDJ2 mediates the interaction of HSP70B with VIPP1

CDJ2 contains an N-terminal J domain, but lacks the Gly/Phe-rich, the Cys-rich, and the C-terminal DnaJ domains typical of Hsp40-like proteins (Liu et al., 2005). Its C-terminal region is predicted to mediate coiled-coil interactions (Lupas, 1996). Homologues of CDJ2 exist in higher plants but not in (cyano)bacteria, suggesting that CDJ2 (like HEP2) evolved following endosymbiosis. *Chlamydomonas HSP70B*, *CDJ1*, and *CGE1* were shown to be inducible by light and heat shock, but neither condition induced *CDJ2*. Like CGE1, CDJ2 is mainly in the stroma. A prominent protein interacting with CDJ2 was identified as VIPP1, and both CDJ2 and VIPP1 were found to interact with HSP70B and CGE1, which catalyze the assembly and disassembly of VIPP1 oligomers (Liu et al., 2005, 2007; see section VI.C.2.c; Figure 19.1).

VII. HSP90s

A. Mechanism of HSP90 function and classification

HSP90 family members are highly conserved (reviewed in Young et al., 2001; Wegele et al., 2004). They contain N-terminal ATP-binding domains, where ATP is bound in a unique, kinked conformation that is mimicked by the anti-tumor drugs geldanamycin and radicicol (Prodromou et al., 1997; Roe et al., 1999). The active form of HSP90 is a dimer, and dimerization is mediated by the C-terminal ~190 amino acids (Minami et al., 1994). The HSP90 dimer acts as a "molecular clamp" driven by ATP binding and hydrolysis, which opens and closes via transient interactions of the two N-termini (Chadli et al., 2000; Prodromou et al., 2000). Client proteins recognized by HSP90 are thought to be in a near-native state and thus represent late folding intermediates (Jakob et al., 1995). Cytosolic HSP90 interacts with a large set of cohort proteins, of which HSP70, HSP40, HOP, p23, CHIP, CDC37, and several immunophilins are well-studied examples (Pratt and Toft, 2003; Wegele et al., 2004). In contrast, chloroplast HSP70 is yet the only partner identified for chloroplast HSP90s.

A major function attributed to cytosolic HSP90 is a role in the maturation of signal transduction proteins such as hormone receptors and kinases (Richter and Buchner, 2001; Pratt and Toft, 2003; Wegele et al., 2004). It also participates in the regulation of the stress response (Ali et al., 1998) and has been implicated in the general refolding of denatured proteins (Jakob et al., 1995). *Chlamydomonas* contains three genes encoding members of the HSP90 family, termed *HSP90A–HSP90C* (Table 19.4; Schroda, 2004; Willmund and Schroda, 2005).

Table 19.4 HSP90s and co-chaperones

#	Name	Introns[a]	cDNA support	Localization[b]	ORF (aa)	Mat (aa)	kD (mat)	pI[c]	Accession #	Reference
1	HSP90A	7	Almost complete	Cyt	705	705	80.7	4.99	EDP01972	Schulz-Raffelt et al. (2007)
2	HSP90B	10	Almost complete	ER	792	768	87.4	4.8	EDP06860	
3	HSP90C	9	Complete	cp (e)	810	739	82.4	5.05	ED096722	Willmund and Schroda (2005)
4	HOP1	13	Complete	cyt	567	567	62.9	5.61	EDP04359	

ORF, open reading frame; mat, mature protein; aa, amino acids; pI, isoelectric point, cyt, nucleo-cytoplasm; cp, chloroplast.

[a]*Introns within coding region.*

[b]*According to localization of orthologs; (e) determined experimentally.*

[c]*Determined with ExPASy get pI/Mw.*

B. HSP90A is likely localized to cytosol and flagella

Chlamydomonas HSP90A lacks an N-terminal extension, contains the C-terminal EEVD motif characteristic of cytosolic HSP90s, and exhibits about 80% identity to *Arabidopsis* cytosolic HSP90s. Hence, HSP90A appears to be localized to the cytosol (Schroda, 2004; Willmund and Schroda, 2005). HSP90A was also found in the flagellar proteome (Pazour et al., 2005). The *HSP90A* gene is strongly inducible by heat shock (Schulz-Raffelt et al., 2007) and oxidative stress (Fischer et al., 2005). A homologue of the cytosolic co-chaperone HOP/TPR2/pSti1 is found in the genome, with ESTs mostly from the deflagellation and stress II libraries.

C. HSP90B is likely to be targeted to the ER

As mentioned in section VI.E, the *Chlamydomonas HSP90B* gene is flanked by genes encoding the ER-targeted HSP70 homologues BIP1 and BIP2 (Schroda, 2004). Moreover, HSP90B exhibits high similarity to the ER-localized HSP90 of *Arabidopsis*, its N-terminal part contains an extension predicted with high probability to be an ER signal peptide, and the C-terminal part ends with the KDEL retention motif typical of ER proteins. Hence, HSP90B appears to be targeted to the ER (Schroda, 2004; Willmund and Schroda, 2005).

D. Chloroplast HSP90C interacts with chloroplast HSP70B

Fractionation studies revealed that *Chlamydomonas* HSP90C is targeted to the chloroplast (Willmund and Schroda, 2005), with significant amounts in stromal, low-density membrane and thylakoid fractions. The *HSP90C* gene was induced by heat shock and light; heat shock increased HSP90C levels 3.9-fold, slightly more than chloroplast HSP70B (3-fold). HSP90C formed homodimers in *Chlamydomonas* extracts, which also was the case for recombinant HSP90C purified from *E. coli* (Willmund and Schroda, 2005). The latter exhibited a weak ATPase activity with a K_m of ~48 μM and a k_{cat} of ~0.71/minute at 30°C. These values were similar to those measured for other members of the HSP90 family (Panaretou et al., 1998; Felts et al., 2000). HSP90C's ATPase activity was inhibited by radicicol, an HSP90-specific drug.

As mentioned in section VI.C.2.d, HSP90C in *Chlamydomonas* cell extracts was found to interact with HSP70B (Willmund and Schroda, 2005). It is not yet clear how the interaction between chloroplast HSP70B and HSP90C is mediated. The two proteins appear not to interact *in vitro* (Willmund and Schroda, 2005), suggesting the need for an "organizing protein" like HOP (Smith et al., 1993) or TPR2 (Brychzy et al., 2003). Since genes encoding chloroplast homologues of HOP/TPR2 have not been found in *Chlamydomonas*, the existence of (a) specialized chloroplast protein(s) is likely.

Phylogenetic analyses suggest that chloroplast-targeted HSP90 is not derived from cyanobacterial HSP90 (HtpG), but rather from an ER-targeted HSP90 gene that had duplicated and later acquired a chloroplast transit peptide (Emelyanov, 2002; Stechmann and Cavalier-Smith, 2004; Chen et al., 2006). Since chloroplast HSP70 is derived from the cyanobacterial endosymbiont (Schroda, 2004), this means that in the chloroplast we observe the interaction of two chaperones of distinct evolutionary origins.

A role for the chloroplast HSP70/HSP90 system in stress and light signaling is suggested by phenotypes of the *Arabidopsis cr88* (chlorate resistant) mutant, which carries a point mutation in the C-terminal dimerization domain of a chloroplast-targeted HSP90 (Cao et al., 2003). The mutant exhibits reduced light-inducible expression of the nuclear *NR2*, *CAB*, and *RBCS* genes, and of the plastid *rbcL* gene. Furthermore, *cr88* shows retarded de-etiolation in red light and reduced nitrate reductase activity (Lin and Cheng, 1997; Cao et al., 2000). The authors concluded that chloroplast HSP90 may be involved in the transduction of a light signal responsible for the regulation of a subset of photosynthesis-related genes. In *Chlamydomonas*, chlorophyll biogenesis intermediates have been shown to serve as chloroplast signaling molecules for the induction of nuclear genes by light (Kropat et al., 1997; Beck, 2005; Shao et al., 2007). As cyanobacterial HSP90 has been shown to control chlorophyll biogenesis by regulating the activity of the HemE protein, this might also be the mechanism by which light induction of nuclear genes is influenced by chloroplast HSP90 (Watanabe et al., 2007).

VIII. ELEVATED LEVELS OF sHSPs, HSP60, AND HSP70 IN *CHLAMYDOMONAS ACIDOPHILA*

Chlamydomonas acidophila is a dominant phytoplankton species in acidic mining and volcanic lakes, which are characterized by very low pH and high concentrations of dissolved heavy metals and sulfate (see references in Gerloff-Elias et al., 2006). Levels of sHSPs, HSP60, and HSP70 in *C. acidophila* were tested over a pH range from 1.5 to 7; under all pH conditions, HSP levels were generally higher than in *C. reinhardtii* and accumulated to even higher levels at "extreme" pH values of 1.5 and 7 (Gerloff-Elias et al., 2006). Unfortunately, the antibodies used against HSP70 and HSP60 could not distinguish between individual family members. It could be determined that the sHSP expressed constitutively in *C. acidophila* had a molecular weight of ~25 kD and therefore might be distinct from the sHSP of ~20 kD accumulating in *C. reinhardtii* after heat shock. However, homologous proteins in the two species might be of different molecular weights. The elevated HSP levels were interpreted as an adaptation of *C. acidophila* to its extreme environment (Gerloff-Elias et al., 2006).

IX. HSP100 PROTEINS FUNCTION AS CHAPERONES AND/OR COMPONENTS OF ATP-DEPENDENT PROTEASES

A. Structure and classification of HSP100 chaperones

Many cytosolic and organellar chaperones belong to the large AAA+ family of ATPases associated with various cellular activities (reviewed in Schirmer et al., 1996; Iyer et al., 2004; Sauer et al., 2004; Bukau et al., 2006). The AAA+ module consists of an $\alpha\beta$-nucleotide–binding domain with the Walker A and B consensus sequences, followed by a purely α-helical domain containing the sensor-2 sequence. HSP100/Clp proteins contain either one (ClpX, ClpY = HslU), or two AAA+ domains in tandem (ClpA–ClpE). They assemble into hexamers and use the energy released by ATP hydrolysis to unfold substrate proteins and thread them through the central pore of the hexameric ring. Ten genes encoding HSP100/Clp proteins were found in *Chlamydomonas*, three of them of the ClpX/Y-type, five of the ClpB, and one each of the ClpC- and ClpD-types (Table 19.5).

B. ClpB

ClpB proteins are characterized by a large coiled-coiled M domain inserted near the end of the first AAA+ domain (Weibezahn et al., 2005). They bind to protein aggregates and extract polypeptide chains (Glover and Lindquist, 1998) that may then be refolded by the HSP70 chaperone system. The ability

Table 19.5 ORFs encoding Clp/Hsp 100 proteins

#	Name	Introns[a]	cDNA support	Localization[b]	ORF (aa)	Mat (aa)	kD (mat)[c]	Accession #
1	CLPX	15	Almost complete	mt	585	?	?	EDO99845
2	HSLU1	12	5′ and 3′	mt	559	?	?	EDP05503
3	HSLU2	11	None	mt	516	?	?	EDP02176
4	CLPB1/ HSP101	4	Complete	cyt	925	925	100.8	EDO99306
5	CLPB2	?	None	?	?	?	?	EDP06950
6	CLPB3	11	Almost complete	cp/mt	1040	925	100.7	EDP06752
7	CLPB4	≥18	Only 3′ end	cp/mt	?	?	?	EDO97520
8	CLPB5	?	None	?	?	?	?	EDP03578
9	CLPC	?	Only 5′ end	cp	?	?	?	
10	CLPD	18	Only 5′ end and 3′ UTR	cp	997	954	102.4	EDP08126

ORF, open reading frame; mat, mature protein; aa, amino acids; pI, isoelectric point; cyt, nucleo-cytoplasm; cp, chloroplast; mt, mitochondria.

[a]Introns within coding region.

[b]According to localization of orthologs.

[c]Determined with ExPASy get pI/Mw.

of ClpB to disassemble protein aggregates is fundamental for induced thermotolerance mediated by cytosolic ClpBs (also termed HSP101 and HSP104) (Sanchez and Lindquist, 1990). Chloroplast ClpB may also be involved in thermotolerance (Yang et al., 2006; see however Lee et al., 2007). Also in *Chlamydomonas* the lack of synthesis of protein(s) of ~100 kD (presumably ClpB-like proteins) in *HSF1*-RNAi strains correlated with the inability of these strains to survive heat stress (Schulz-Raffelt et al., 2007). At least five *Chlamydomonas* genes encode ClpB-type proteins. CLPB1 is orthologous to the cytosolic HSP101 isoform, while CLPB3 and CLPB4 are predicted to be located in organelles, based on targeting of their closest *Arabidopsis* orthologs ClpB-p and ClpB-m (Lee et al., 2007). The gene products of two other incomplete ClpB-type sequences with no EST support, called CLPB2 and CLPB5, are of uncertain intracellular localization. CLPB2 lacks a conserved DE sequence in the Walker B motif of the second AAA domain, and might thus be inactive. Its gene is linked to that encoding CLPB3 on linkage group II.

C. ClpC, ClpD, and the ClpCP protease

Chlamydomonas has two HSP100 proteins of the ClpA type (no M domain), both presumably located in the chloroplast, termed CLPC and CLPD. The *Chlamydomonas CLPC* gene contains many sequence gaps, including in version 4. *Volvox* only has a single *CLPC* gene. By comparison *Arabidopsis* has three genes of that type, the closely related *ClpC1* and *ClpC2*, and *ClpD* (*ERD1*). In higher plants, *ClpD* is induced by drought and senescence (Kiyosue et al., 1993; Weaver et al., 1999), and a ClpC protein has been shown to be essential for chloroplast protein import (Nielsen et al., 1997). Cyanobacterial ClpC can prevent protein aggregation and perform refolding of denatured proteins independently of the HSP70 system (Andersson et al., 2006). In addition, these chaperones have been proposed to interact with the ClpP peptidase to form an ATP-dependent protease, like their *E. coli* homologue ClpA. The three *Arabidopsis* proteins ClpC1, ClpC2, and ClpD contain in their second AAA+ domain the (I/L)GF motif that has been implicated in ClpP binding (Kim et al., 2001), and evidence has been presented for an interaction of cyanobacterial ClpC with a ClpP3/ClpR peptidase (Stanne et al., 2007). We note, however, that the green algal ClpD proteins (in *Chlamydomonas*, *Volvox* and *Ostreococcus*) all lack this IGF motif, while the unique ClpC protein has it. In green algae, ClpC is thus the best candidate for the chaperone partner in the Clp protease, in addition to other roles it may play in chloroplast biogenesis.

The *Chlamydomonas* genome also encodes two proteins homologous to *E. coli* ClpS, which has been found to modulate interaction between the ClpA chaperone and its protein substrates (Dougan, 2002). The two *Chlamydomonas* homologues are predicted to be directed to an organelle. Most cyanobacterial and plant genomes contain two or more *CLPS* genes, except *Arabidopsis* where *ClpS2* appears to be missing.

The chloroplast ClpP complex, which carries the peptidase activity of the Clp protease, is essential for cell survival in *Chlamydomonas* (Huang et al., 1994) and in plants (Shikanai et al., 2001, Rudella et al., 2006; Zheng et al., 2006; but see Cahoon et al., 2003). At variance with the *E. coli* protein which is composed of 14 identical subunits, the chloroplast complex is a hetero-oligomer (Peltier et al., 2001, 2004; Majeran et al., 2005). Except for the chloroplast-encoded ClpP1, all other subunits are nuclear-encoded and imported into the organelle. Based on the presence or absence of the catalytic Ser-His-Asp triad, the subunits can be classified as ClpP (all three residues present) or ClpR (at least one of them mutated). The *Chlamydomonas* enzyme is composed of three active (ClpP1, CLPP4, and CLPP5) and five inactive (CLPR1, CLPR2, CLPR3, CLPR4, and CLPR6) subunits (Majeran et al., 2005). These subunits are believed to form a barrel-shaped tetradecamer, with the various subunits distributed between the two heptameric rings. CLPR1, CLPR3, and CLPR4 are evolutionarily related to the single ClpR protein of cyanobacteria, while ClpP1 and CLPR2 derive from the cyanobacterial ClpP3 (in the case of CLPR2 via loss of the catalytic triad). The phylogenetic origin of the other proteins (CLPP4, CLPP5, and CLPR6) is less clear, but probably they arose early in chloroplast evolution as they have relatives in the rhodophytes. Duplication of *clpP* genes and loss of catalytic activity have occurred multiple times in photosynthetic eukaryotes (Majeran et al., 2005). For example, *Chlamydomonas* and *Ostreococcus* CLPR6 are phylogenetically related to higher plant CLPP6, but have suffered a secondary loss of catalytic activity. No such phenomenon is observed for the mitochondrial ClpP complex, which appears to be a homo-oligomer of CLPP2 (see below).

A unique structure of the chloroplast *clpP1* gene was observed in *Chlamydomonas reinhardtii* and *C. eugametos* (Huang et al., 1994). A 35 kD insertion sequence, IS1, occurs just before the second β-strand. IS1 is also present in *Volvox*, but not in *Ostreococcus* (Majeran et al., 2005). In addition to IS1, the ClpP1 protein of *C. eugametos* contains an intein (protein intron) called IS2. After modification of a key residue, this intein has been shown to excise in *E. coli* (Wang and Liu, 1997). In contrast, IS1 in both species lacks the catalytic residues necessary for intein excision. In *C. reinhardtii*, immunoblotting identifies a low MW form of ClpP1, in addition to the expected 59-kD translation product, and both the high and low MW forms are present in the ClpP complex in a ratio that is stable over time when translation is blocked (Majeran et al., 2000). The IS1 domain is essential and cannot be deleted from the gene, which again distinguishes it from inteins. Because it is inserted on the apical surface of the complex (Majeran et al., 2005), it should prevent interaction between ClpP and the chaperone. Yet, the ClpCP protease could form if the IS1-containing ClpP1 were absent from one of the heptamers, or processed in a way that would allow interaction with the chaperone. While the *Arabidopsis* ClpP1 protein lacks IS1, the ClpP complex includes two tightly-bound subunits, ClpT1 and ClpT2 (initially called ClpS1 and ClpS2, Peltier et al., 2004 but now renamed to avoid confusion). They resemble the N-terminal domain of ClpC and have been proposed to regulate chaperone binding. Distant homologues of these proteins, called CLPT3 and CLPT4, are encoded in the *Chlamydomonas* genome and may also be associated with the ClpP complex.

D. The ClpXP and HslUV proteases

Chlamydomonas contains a single *CLPX* gene (Table 19.5) which based on its higher plant homologue (Halperin et al., 2001), is directed to the mitochondrial matrix where it forms an ATP-dependent protease with CLPP2. Both *CLPX* and *CLPP2* originated from the proteobacterial ancestor of mitochondria. *Chlamydomonas*, *Volvox*, and *Ostreococcus* CLPX lack the N-terminal Zn finger domain involved in substrate recognition. Instead, the *Chlamydomonas* and *Volvox* proteins contain a large insertion between the first helix of the AAA+ domain and the "Walker A" consensus sequence, not found in *Arabidopsis* or *Ostreococcus*. An insertion at this position is also seen in the red alga *Cyanidioschyzon* and animal ClpX, but its role has not been investigated.

HslU (also called ClpY) is another AAA+ chaperone associated in bacteria with a peptidase, HslV (ClpQ). HslV is not related to ClpP-type serine proteases, but belongs to the same NTN hydrolase family as the peptidase subunits of the proteasome. While HslU and HslV are both absent from higher plants and animals, they can be found in *Chlamydomonas* and a variety of unicellular eukaryotes including *Plasmodium*, *Dictyostelium*, *Cyanidioschyzon*, and *Ostreococcus*, as well as in the moss *Physcomitrella patens*. Two *HSLU* and one *HSLV* genes are found in *Chlamydomonas* and *Volvox*, and TargetP analysis of their non-conserved N-terminal extensions suggests that the products are targeted to mitochondria or chloroplasts.

X. FTSH PROTEASES

FtsH is a membrane-bound chaperone and protease. Its ATPase domain also belongs to the AAA+ family, is closely related to the N-terminal AAA+ domain of ClpC (Iyer et al., 2004) and forms a hexameric ring (Niwa et al., 2002). In addition, FtsH contains a C-terminal metalloprotease domain with the typical Zn-binding HEXXH motif. The many FtsH isoforms of *Arabidopsis* distribute between chloroplast and mitochondrion (Adam et al., 2005; Janska, 2005). The thylakoid enzyme exposes its catalytic site on the stromal surface and is formed of a hetero-oligomer of two type A (FtsH1 or FtsH5) and four type B (FtsH2 or FtsH8) subunits. In *Chlamydomonas*, these two types are represented by *FTSH1* and *FTSH2*, respectively, whose products were identified in a proteomic study of a thylakoid-enriched fraction (Allmer et al., 2006).

The mitochondrion typically harbors two types of FtsH proteases with respect to topology and function. One exposes its catalytic sites to the mitochondrial matrix ("m-AAA" protease) and is hetero-oligomeric in higher plants and yeast (FtsH3/FtsH10 and Yta10/Yta12, respectively). In *Chlamydomonas* and *Ostreococcus*, this enzyme appears to be formed of a single FTSH3 protein, which has been identified in the proteomics survey of the mitochondrion. In contrast, the i-AAA protease has its catalytic site in the intermembrane space, and is probably a hetero-oligomer in *Arabidopsis* as well as in *Chlamydomonas*, formed by the products of the *FTSH4* and *FTSH11* genes (*FTHS11* is homologous to yeast Yta11 = Yme1). The *Chlamydomonas FTSH11* sequence is partial, but comparison with the *Volvox* ortholog supports its identification. *Chlamydomonas* encodes an additional FtsH-like protein which lacks the HEXXH motif, and has homologues in *Volvox* and *Ostreococcus*. It shows a complete AAA+ domain and may therefore retain a chaperone function, but is presumably inactive as a protease. It has been annotated in the *Chlamydomonas* genome as FTSHi1, but is not an ortholog of the *Arabidopsis* FtsHi proteins (Sokolenko et al., 2002) that have independently evolved towards loss of peptidase activity.

XI. Lon

The Lon protein combines an AAA+ domain (related to the C-terminal domain in ClpA) with a serine protease domain (S16 clan) that uses a Ser-Lys dyad, instead of the classical triad of serine proteases (Rotanova et al., 2006). An additional N-terminal domain has been implicated in substrate recognition. Lon is found in eubacteria, archaea, and eukaryotic organelles and has been implicated in the degradation of short-lived and abnormal proteins in bacteria and mitochondria. The protein forms a homo-oligomer, with four to eight (most probably six) subunits associated in a ring. In spite of the fact that TargetP predicts that three of the four *Arabidopsis* homologues and all those in rice are located in the chloroplast (Tripathi and Sowdhamini, 2006), they probably distribute between the mitochondrion, the chloroplast and the peroxisome (Sarria et al., 1998; Heazlewood et al., 2004; Janska, 2005). The *Chlamydomonas* genome contains a single gene encoding a Lon protease, *LON1*, whose product probably is located in the mitochondrion. It is most closely related to the mitochondrial higher plant Lon proteins, and possesses a putative transit peptide. It may contribute to the degradation of misfolded polypeptides or of specific substrates with a regulatory role, as found for the bean mitochondrial cytoplasmic male sterility protein orf239 (Sarria et al., 1998).

XII. DEGP AND C-TERMINAL PROCESSING PEPTIDASES

DegP proteins were initially described in the periplasm of eubacteria but are also found in eukaryotes where they localize to the organelles and possibly to the cytosol (Huesgen et al., 2005). Studies of the *E. coli* enzyme suggest that it functions as a hexamer, carrying out a chaperone-like activity at low temperatures, and switching to a proteolytic function at higher temperatures when its catalytic sites become exposed. In addition to the S1 serine protease domain, bacterial and eukaryotic DegP proteins possess C-terminal PDZ domains, which form the sides of the hexamer and are believed to be important in the functional switch. No ATPase domain is found in these enzymes, which do not hydrolyze ATP when they interact with their substrates.

Ten genes for DegP proteases are present in *Chlamydomonas*, named *DEGn* based on numbering of their closest relative in *Arabidopsis* (Huesgen et al., 2005). Figure 19.2 shows that *Chlamydomonas* DEG10 belongs to a large clade comprising most of the mitochondrial higher plant DegPs (DegP3, DegP4, DegP10, DegP11, DegP12). It is probably the only mitochondrial

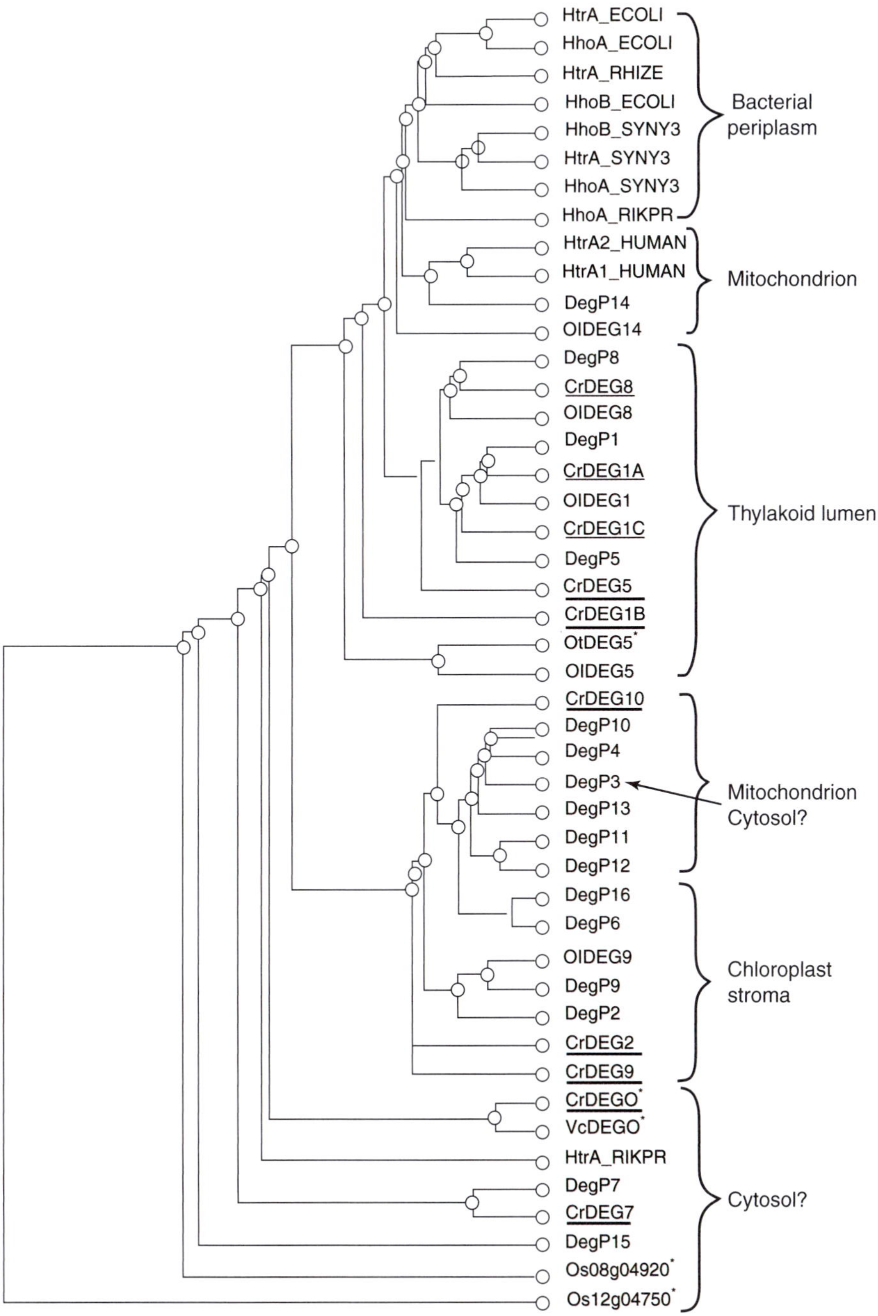

FIGURE 19.2 *DegP phylogenetic tree, constructed at http://bioinfo.hku.hk/services/analyseq/cgi-bin/phylomat_in.pl with the Kitsch method, using a molecular clock and the Dayhoff matrix. Sequences are from Chlamydomonas (Cr), Volvox carteri (Vc, partial), Ostreococcus lucimarinus (Ol), Ostreococcus tauri (Ot), Oryza sativa (Os), Arabidopsis thaliana (no prefix), Homo sapiens (HUMAN), E. coli (ECOLI), Synechocystis sp. PCC 6803 (SYNY3), Rickettsia prowazekii (RIKPR) and Rhizobium etli (RHIZE). Chlamydomonas sequences are underlined, inactive isoforms are labelled*, putative intracellular localization is indicated.*

DegP in *Chlamydomonas*, as no ortholog exists for the other type of mitochondrial DegP, Deg14 in *Arabidopsis* and HTRA1/HTRA2 in humans. *Ostreococcus* has a *DEG14*-type gene, but lacks *DEG10*.

The *Chlamydomonas DEG2* and *DEG9* genes also belong to this clade but they are more related to *Arabidopsis DEGP2/9*, whose products are associated with the stromal side of the thylakoids (Haussuhl et al., 2001; Huesgen et al., 2005). The naming of the *Chlamydomonas DEG2* and *DEG9* is meant to convey their overall similarity to *Arabidopsis DEGP2* and *DEGP9* but does not imply orthology *sensu stricto* as gene diversification in this clade occurred independently in the chlorophytes and angiosperms (*Ostreococcus* has a single gene, *DEG9*). In higher plants, DegP2 has been implicated in the initial cleavage leading to D1 turnover (Haussuhl et al., 2001), but this assumption has been challenged (Barker et al., 2006) and this role attributed to DegP1 (Kapri-Pardes et al., 2007). Another pair of stromal DegPs has evolved from this clade in *Arabidopsis* (Deg6/Deg16), but they have no orthologs in green algae.

The third main branch of the DegP phylogenetic tree comprises isoforms directed to the thylakoid lumen, DegP1, DegP5, and DegP8. *Chlamydomonas* has three genes of the *DEG1* type, called *DEG1A, DEG1B* and *DEG1C*, along with single *DEG8* and *DEG5* genes. Note that divergence is high among DEG5/DegP5 sequences, and *Ostreococcus* DEG5 even lacks the catalytic His and Asp residues. Their main common characteristic is the absence of a PDZ domain. Based on their N-terminal sequences and the localization of their *Arabidopsis* orthologues, *Chlamydomonas* DEG1, DEG5, and DEG8 are probably targeted to the thylakoid lumen. The rather divergent *DEG1B* has no EST support and gene structure is uncertain. In microarray experiments, the *DEG1C* mRNA was found to increase during treatment with the photosensitizer Neutral Red, but not with Rose Bengal (Fischer et al., 2005). Modest expression increases have also been found for this gene during S and P starvation (Zhang et al., 2004; Moseley et al., 2006). None of the other *DEG* genes present on the microarray showed significant changes during these treatments. In *Arabidopsis*, several *DegP* genes have been shown to be induced by heat stress at the mRNA level, but discordant behavior of the mRNA and protein levels has been observed under some conditions (Huesgen et al., 2005).

Another type of DegP is DEG7, with no organelle targeting sequence, three PDZ domains, and no known function. *Chlamydomonas* and *Volvox* have one *DEG7* gene, like *Arabidopsis*, but *DEG7* genes are absent from *Ostreococcus*. *Chlamydomonas* encodes yet another type of DegP protein, represented by two paralogs which we propose to call DEGO1 and DEGO2 rather than DEGP because they lack at least part of the catalytic triad. The residues His, Asp, and Ser are replaced respectively by Ala/Gly, Glu/Asp, and Leu/Trp, which suggests that the proteins are catalytically inactive as peptidases, but act as chaperones. The connection between the peptidase and PDZ domains is expanded, with a Gly-rich region that is also extremely rich in Arg residues in DEGO1, and in Glu/Asp residues in DEGO2, suggesting a possible interaction between the two. Orthologs are encoded in *Volvox*, but none can be found in other organisms. DEGOs clearly have arisen independently from another type of DegP-like sequence that has been described in rice as lacking the catalytic residues (Tripathi and Sowdhamini, 2006).

The C-terminal processing peptidase is involved in the C-terminal processing of the D1 protein between Ala-344 and Ser-345, allowing liganding of the Mn cluster (Hatano-Iwasaki et al., 2000; Takahashi et al., 1996). It also associates a serine peptidase domain of the S41 family with a PDZ domain which in the *Scenedesmus* enzyme appears to bind the D1 C-terminus (Liao et al., 2000). Three genes for C-terminal processing peptidases can be identified in *Chlamydomonas*, called *TSP1*, *TSP2*, and *TSP3* after the *E. coli* homologue Tail Specific Protease *TspA*. All three proteins are predicted to be located in the thylakoid lumen, and TSP1 has been found associated with the thylakoid membrane (Allmer et al., 2006). Three TSP proteins are also encoded in higher plants (Tripathi and Sowdhamini, 2006), *Ostreococcus* and *Synechocystis*, but it is not clear whether gene diversification has occurred before or after the lineages diverged. In *Chlamydomonas*, it is unclear which of the TSP isozymes carries out D1 maturation.

XIII. N-TERMINAL MATURATION OF ORGANELLE-ENCODED PROTEINS

Polypeptides translated in the chloroplast or the mitochondrion carry an N-terminal formylmethionine residue contributed by the initiator fMet-tRNA. In most cases N-terminal maturation follows, starting with the removal of the formyl group by the metalloprotease-like enzyme peptide deformylase (PDF). Subsequent cleavage of the Met residue is carried out by methionine aminopeptidases (MAP), which are also present in the cytosol. The N-terminal maturation pathway of plastids and mitochondria is homologous to that of eubacteria and inherited from the endosymbionts (Giglione et al., 2000). *Chlamydomonas* harbors two genes for type I PDF, *PDF1A* and *PDF1B*. By analogy to their *Arabidopsis* homologues, PDF1A is predicted mitochondrial and PDF1B dually targeted to organelles. Similarly, the *MAP1D* gene probably encodes a methionine aminopeptidase with a dual chloroplast/mitochondrial location. Two other *MAP* genes, *MAP1A* and *MAP2*, are predicted to encode cytosolic products. In *Arabidopsis*, there are two cytosolic MAP2 and four MAP1 proteins, routed to the cytosol (MAP1A), chloroplast (MAP1B), or both mitochondrion and chloroplast (MAP1C, MAP1D).

The importance of N-terminal maturation in the chloroplast has been studied in *Chlamydomonas* using actinonine, an inhibitor of PDF (Giglione et al., 2003). It was found that unlike in *Arabidopsis*, PDF inhibition did not prevent assembly of the photosynthetic apparatus, although it did cause progressive loss of PS II and retarded recovery after photoinhibition. Pulse-chase labeling of chloroplast-encoded proteins revealed a retarded biogenesis rate of PS II complexes and decreased stability of PS II proteins, in particular D2. The destabilizing effect of actinonine was probably due to a secondary impairment of methionine removal rather than to the continued presence of the formyl group: D2 mutants carrying residues at position 2 that are known to prevent MAP action without affecting deformylation (T2D and T2E) showed high instability independent of actinonine addition. In contrast, a control mutation (T2V) that allows methionine removal behaved similarly to the wild-type. Note that D2 phosphorylation, which occurs on Thr2, is impaired in the three mutants, which indicates that phosphorylation is not a major determinant of D2 stability (Fleischmann and Rochaix, 1999).

Actinonine also affected N-terminal maturation of RbcL, as evidenced by a reduced electrophoretic mobility in gels, but this time without affecting the stability or activity of the protein (Giglione et al., 2003). Here again, a cascade effect is likely: the drug, by blocking removal of the formyl group, prevents excision of the methionine. This in turn will affect further processing of RbcL which is normally cleaved after the second residue, leaving an N-terminal proline that becomes acetylated. The identities of the enzymes carrying out the final cleavage and the acetylation are unknown.

XIV. PEPTIDASES IN ORGANELLAR IMPORT AND MEMBRANE TRANSLOCATION

A key step in protein import into organelles is the removal of the targeting peptide, cTP for plastids and mTP for mitochondria. This occurs in the organelles and is performed by large metalloproteases of the M16 family (Gakh et al., 2002; Richter and Lamppa, 2003). *Chlamydomonas* contains a single chloroplast processing enzyme gene (*CPE1*) whose product is also called "stromal processing peptidase." *Chlamydomonas* CPE1 shows an extended cTP, as found in higher plants and *Ostreococcus*. The finding of a unique *CPE1* gene is at odds with an earlier report describing at least four stromal processing peptidase activities in *Chlamydomonas* (Rüfenacht and Boschetti, 2000). The mitochondrial processing peptidase (MPP) is a more complex enzyme, as it associates homologous α- and β-subunits. Because the former lacks the HXXEH Zn-binding motif, the β-subunit is thought to be solely responsible for catalysis. In *Chlamydomonas* as in *Arabidopsis*, a single *MPPB* gene encodes the β-subunit, while *MPPA1* and *MPPA2* genes encode the α-subunits. The two MPPA sequences are more similar to each other within than across species, indicating independent duplications subsequent to higher plant/chlorophyte divergence, in line with the presence of a single *MPPA* gene in *Ostreococcus*. In plant and mammalian mitochondria, the MPP enzyme forms the matrix-exposed part of the mitochondrial complex III (ubiquinol-cytochrome c oxidoreductase), of which they are known as the "core proteins." This is true in *Chlamydomonas* as well: the products of the *MPPA2* and *MPPB* genes have been identified by N-terminal sequencing as core proteins II and I (QCR2 and QCR1), respectively (van Lis et al., 2003). The core II band apparently contained more than one protein, so it is possible that the *Chlamydomonas* MPP is heterogeneous, associating MPPB with either MPPA1 or MPPA2. Note, however, that MPPA1 is represented by fewer ESTs than MPPA2. *Chlamydomonas* has at least four additional genes coding for M16 metalloproteases, but they are more related to plant insulinase-like proteins than to MPP subunits.

Once cleaved from the precursor, the targeting peptide is degraded by another M16 protease, called "presequence protease" (PreP; Stahl et al., 2005). There are two very similar enzymes in *Arabidopsis*, PreP1 and PreP2, both dually targeted to chloroplasts and mitochondria (Bhushan et al., 2005). The crystal structure of *Arabidopsis* PreP1 revealed a large proteolytic chamber whose access is regulated by substrate binding (Johnson et al., 2006), and the term peptidasome was coined for the enzyme. It represents a paradigm for the entire M16 class and can be considered a compartmentalized protease even though it is not ATP-dependent. *Chlamydomonas* and *Volvox* contain a single PreP gene, *PREP1*, which is consistent with dual targeting, although no experimental data are available.

Some proteins imported into the mitochondrial matrix require further processing of an N-terminal octapeptide. This step is carried out by mitochondrial intermediate peptidase (MIP), a member of the M3 family of thiol-dependent metalloproteases which includes thimet oligopeptidase (Isaya et al., 1992). The *Chlamydomonas MIP* gene appears to be expressed although it shows evidence for an active retrotransposon in its third intron. Three other thimet-like peptidases are encoded in the genome, called THO1, THO2, and THO3. The first two are closely related to bacterial oligopeptidases, while THO3 is most similar to higher plant enzymes. THO2 is probably located in the cytosol, but an organellar location is likely for the other two. Thimet oligopeptidase is believed to play an important role in the final step of intracellular protein degradation, in particular in the degradation of peptides produced by the large self-compartmentalized proteases.

Once in the chloroplast stroma or mitochondrial matrix, proteins destined to cross the thylakoid or inner mitochondrial membrane need to interact with one of several translocation systems via their hydrophobic signal peptides. On the lumenal side of the membrane, a thylakoid processing peptidase (TPP) cleaves the newly imported protein at the C-terminus of the signal peptide, immediately after the AXA consensus sequence, to generate the mature N-terminus. These enzymes belong to the S26A family of serine proteases, typified by the bacterial Type I signal peptidase (LepB in *E. coli*) and employ a Ser-Lys catalytic dyad. TPP is encoded by the single *TPP1* gene in *Chlamydomonas*. In contrast, *Arabidopsis* has three such genes, of which two have been characterized respectively as TPP and Plsp1, the latter being found in both thylakoid and envelope membranes (Inoue et al., 2005). The mitochondrial inner membrane peptidase (IMP) also belongs to the S26A family. *Chlamydomonas* has two genes of this type (*IMP1* and *IMP2*). Another group of related enzymes forms the S26B family of type II signal peptidases, which employ a Ser-His dyad. In eukaryotes, they are located in the ER where they constitute the catalytic subunit of a complex that comprises the transmembrane proteins SPC12, SPC22, and SPC25 (Kalies and Hartmann, 1996) and process secreted proteins. *Chlamydomonas* has one gene for each of the subunits of the complex.

XV. INTRAMEMBRANE PROTEASES: SITE2 PEPTIDASE, PARL/RHOMBOID, SPP, SPPA

Because a water molecule is necessary for hydrolysis of a peptide bond, it was a surprise to discover several classes of peptidases that act in the transmembrane region of their substrates (Wolfe and Kopan, 2004). Integral membrane-spanning proteins can be cleaved within their transmembrane regions by serine, aspartyl, and metallo-proteases. Because cleavage often is regulated and generates signaling molecules of major importance, this process has been designated regulated intramembrane proteolysis. A similar mechanism leads to the clearance of hydrophobic signal peptides after they are cleaved from preproteins which they direct to the secretory pathway. Below we review four known classes of such enzymes, with emphasis on those that may be present in the organelles.

Sterol-regulatory element binding protein (SREBP) site-2 protease (S2P) is a hydrophobic metalloprotease (M50 family) involved in regulation of lipidemia in mammals. S2P has a clear ortholog in *Arabidopsis* but not in *Chlamydomonas*, *Volvox*, or *Ostreococcus*. This, in line with the absence of a recognizable SREBP homologue, suggests that green algae do not use this pathway for regulation of their lipid metabolism. However, two other families of M50 metalloproteases can be found in green algae and higher plants, exemplified by EGY1 and RSEP. In *Arabidopsis* EGY1 is located in the thylakoid membrane and its mutation leads to defects in chloroplast development and gravitropism (Chen et al., 2005). A *Chlamydomonas* EGY1 ortholog was found in a thylakoid proteomic study (Allmer et al., 2006).

The other type of M50 metalloprotease found in plants and algae is also present in bacteria, where the best-studied examples are RseP (YaeL) in

E. coli and SpoIVFB in *B. subtilis* (Makinoshima and Glickman, 2006). They also have signaling roles, as they cleave the membrane-attached anti-sigma factor E and pro-sigmaK, respectively. In both cases, intramembrane cleavage requires a prior proteolytic event, as is the case for site-2 protease. Three proteins related to RseP are encoded in *Chlamydomonas*. RSEP1 is probably directed to the mitochondrion, and has no ortholog in *Ostreococcus* or *Arabidopsis*. RSEP2 is found in *Chlamydomonas* and *Ostreococcus* where it shows clear organelle-targeting sequences, like the two related proteins in *Arabidopsis* (AraSP and At2g32480). AraSP was shown to be in the chloroplast inner envelope membrane and necessary for normal chloroplast development (Bolter et al., 2006). A surprising feature of plant *RSEP2* genes is the high degree of conservation of the N-terminal sequences between the two *Arabidopsis* paralogs and between the *Volvox* and *Chlamydomonas* orthologs, which is unusual for a cleaved cTP. This suggests that these extensions are maintained in the mature protein and are important for the enzyme's anchoring or function. In addition, the *Chlamydomonas* genome encodes a third protein in this family, *RSEP3* (duplicated as *RSEP3* and *RSEP4* in *Ostreococcus*). The RSEP3 protein has no organelle-targeting sequence and does not appear to have an ortholog in *Arabidopsis*.

Rhomboid is the prototype of the second class of intramembrane peptidases, which use serine as a nucleophile. In *Drosophila*, rhomboid acts on a membrane-tethered EGF to liberate the free ligand for signaling (Urban, 2006). It belongs to a ubiquitous family of serine proteases found in eubacteria, Archaea, and eukaryotes, with evidence for multiple horizontal gene transfer events (Koonin et al., 2003). One of the best-characterized members of the family is PARL, a mitochondrial inner membrane protease that participates in maintenance of mitochondrial structure by regulated intramembrane proteolysis of a dynamin-type GTPase (Mgm1/OPA1). Ten genes encoding Rhomboid-like proteins are present in *Chlamydomonas*, and are named *RBL1–RBL10*. None of the predicted RBL proteins are likely orthologs of PARL, and no *MGM1*-like gene is present. *Chlamydomonas* RBL1, RBL4, RBL5, RBL7, RBL9, and RBL10 are predicted to be organelle-targeted. Several lack at least one of four conserved residues (the Ser-His catalytic dyad and important His and Asn residues), and are thus probably inactive (RBL1, RBL3, RBL8 and possibly RBL5). This has been found for several human and *Arabidopsis* homologues as well (Tripathi and Sowdhamini, 2006; Urban, 2006).

The eukaryotic-type signal peptide peptidase (SPP) resides in the ER and is responsible for degradation of signal peptides after they are cleaved off their cargo proteins. SPP belongs to the A22 family of aspartyl proteases which also contains presenilin, whose action on β-amyloid precursor protein causes Alzheimer's disease (Verdile et al., 2007). While SPP and SPP-like enzymes cleave type II substrates (where the transmembrane helix has its N-terminal end facing the cytosol) (Friedmann et al., 2004), presenilins cleave type I substrate only (C-terminus towards the cytosol) owing to their reverse orientation. *Chlamydomonas* contains a single presenilin gene *PSN1*, compared to two in *Arabidopsis* and one in *Ostreococcus*. The highly variable region between helices VI and VII is probably the site of an activating cleavage event, as in humans. Presenilin belongs to the gamma-secretase complex. Genes encoding its other components, nicastrin, APH1 and PEN2, can be found in *Chlamydomonas*, suggesting conservation of this ancient eukaryotic system. Yet, no algal homologue can be found for the best-characterized substrates of the enzyme, APP and Notch, and the role of gamma-secretase in plants remains unknown. An insertional mutant in *Chlamydomonas PSN1* shows slight Rose Bengal and metronidazole sensitivity (Dent et al., 2005).

In *Chlamydomonas*, three genes can be found for the SPP-type of A22 enzyme, called *PSL1, PSL2*, and *PSL3*. The first two are on linkage group VII. Like mammalian SPP, PSL1 has no signal peptide, while PSL2 has a predicted cleavable signal sequence and extracellular N-terminal domain, similar to human SPPL2. PSL3 is unusual in having a long insertion after the GxGD helix, and the *Chlamydomonas* or *Ostreococcus* proteins are not orthologs of human SPPL3. They are predicted by TargetP and Predotar to be organelle-targeted, which suggests that PSL3 participates in signal peptide clearance in the thylakoid or mitochondrial membrane. No *PSL3*-like gene is present in *Arabidopsis*.

The proteins encoded by the bacterial and plant *SppA* genes are also called signal peptide peptidase (Sokolenko, 2005) but are not related to the SPP mentioned above. SppA belongs to the S49 family of serine proteases, which are distantly related to ClpP. In *E. coli*, proteases IV (SppA1) and protease V (SppA2 or SohB) have been found as high MW membrane-bound oligomers, and have been implicated in the processing of signal peptides. While SppA1 has two peptidase domains in tandem, SppA2 has only one. The single *Arabidopsis* SppA1 is associated with the thylakoid membrane and only its second peptidase domain appears catalytically active (Lensch et al., 2001). Based on the location of its active site residues in a putative intramembrane anchor, it has been proposed that SppA1 cleaves substrates inside the membrane. In *Synechocystis*, SppA1 has been implicated in the degradation of linker proteins of the phycobilisome antenna (Sokolenko, 2005). In contrast, the single SPP2 protein in *Ostreococcus* is of the SppA2 type.

Chlamydomonas contains three genes encoding *SppA* proteins, designated *SPP1A, SPP1B*, and *SPP1C* to stress their belonging to the SppA1 type. It is unusual that *Chlamydomonas* has more members in a family of proteases or chaperones than *Arabidopsis*. Note that there are orthologs of *SPP1A* and *SPP1B* in *Volvox*, but not of *SPP1C*. *Chlamydomonas* SPP1A is very similar to the higher plant enzyme and thus probably anchored in the thylakoid membrane, but SPP1B and SPP1C are divergent from other

SPPA1 sequences. The genes are close together on linkage group VI and probably arose through a local duplication. *SPP1B/C* are closely linked to the mating type locus, and the *SPP1C* cDNA was sequenced as "probe 6" by Ferris et al. (2002). While the protein deduced from the mt^--linked cDNA appears complete, the *SPP1C* allele found in the mt^+ locus encodes a frame-shifted and truncated protein, and thus is probably a pseudogene. However, even the protein generated from the mt^- allele of *SPP1C* is probably catalytically inactive, as the essential Ser of the active site in the second S49 domain is replaced by Gly. Based on TargetP analysis and comparison of N-terminal sequences, both SPP1B and SPP1C appear to be targeted to an organelle, probably the mitochondrion. SPP1B indeed has been identified in a survey of mitochondrial membranes. While *SPP1C* mRNA is undetectable in vegetative cells, the gene is expressed in gametes of both mating types, with a higher expression level in mt^- gametes (Ferris et al., 2002). The expression is even stronger at an early stage of zygote formation. These authors studied a rare recombinant that carries the truncated mt^+ allele of *SPP1C* in its mt^- locus, and found that the inheritance of chloroplast mutations was normal. Based on the putative mitochondrial localization of SPP1C, it might be worthwhile examining the inheritance of a mitochondrial marker in such crosses.

XVI. PROTEOLYTIC PROCESSES IN THE ORGANELLES OF *CHLAMYDOMONAS*: WHODUNNIT?

The preceding text describes a wide spectrum of proteases, and what little we know of their functions is mostly inferred from studies in other organisms. In fact, little direct evidence is available regarding the relationship between specific degradation processes in *Chlamydomonas* and known classes of proteases.

In most of the photosynthetic mutants that have been characterized in *Chlamydomonas*, a specific defect in the biogenesis of a single subunit of a complex leads to a more or less complete loss of all the other subunits. In some cases, a subunit's translation rate will decrease, owing to a negative feedback of the unassembled polypeptide on translation initiation. This mechanism is generally known as control by epistasy of synthesis (CES) and operates in a variety of photosynthetic complexes in *Chlamydomonas* and possibly in higher plants (see Chapter 29). However, the predominant mechanism explaining the concerted accumulation of photosynthetic proteins is the rapid degradation of non-assembled subunits (Jensen et al., 1986; de Vitry et al., 1989; Kuras and Wollman, 1994). Presumably, the absence of its assembly partners affects the conformation and internal motion of the polypeptide, making it a substrate for a protease that normally does not recognize it when assembled. The identity of the protease(s) involved will vary depending on the location of the substrate (compartment), on its physical state (membrane-bound or soluble, interacting with other subunits or not) and possibly on its sequence (specific cleavage site for endopeptidases, N- or C-terminal binding site for processive proteases).

One way to address these questions is to alter the abundance of a protease and look for effects on the degradation rates of specific substrates. This approach has been used with some success on the ClpCP protease. The chloroplast *clpP1* gene is essential, but attenuation of its expression was achieved by mutating its AUG initiation codon to AUU (Majeran et al., 2000). In the *clpP1-AUU* strain, accumulation of the ClpP complex is reduced to 40% of the wild-type level. In this context, the unassembled mutant Rieske protein produced in the *PETC-ac21* mutant accumulated to some extent, at least during exponential growth. In addition, the partially assembled cytochrome b_6f complex lacking Rieske protein was more stable. In a mutant lacking ATPase CF_0, the level of CF_1 complex accumulating in the stroma was not affected by the *clpP1-AUU* mutation (Majeran et al., 2001). When wild-type cells are starved for nitrogen, the cytochrome b_6f complex is rapidly degraded, but in the *clpP1-AUU* mutant proteolysis was slowed (Majeran et al., 2000). However, ClpP attenuation had no effect in mutants that lack cytochrome b_6f accumulation because of impaired heme attachment to cytochrome f or to cytochrome b_6, nor in PS II mutants lacking CP47 or D1 (Majeran et al., 2000).

Taken together, these data suggest that ClpP can participate in the turnover of integral membrane proteins, either fully assembled or destabilized due to a missing subunit. Accordingly, a fraction of *Chlamydomonas* ClpP is always found associated with the thylakoid membrane. The role of ClpP has been examined in other stress situations where a photosynthetic complex is known to be actively degraded. In ATPase mutants, moderate light causes PS II inactivation followed by its degradation, a process which is ClpP-dependent (Majeran et al., 2001). The mechanism of PS II degradation under these conditions is probably different from that prevailing during photoinhibition, where the *clpP1-AUU* mutation has no effect (Majeran et al., 2001). In higher plants and cyanobacteria, FtsH has been implicated as a major player in photoinhibition-induced PS II degradation (Adam et al., 2005), and DegP could also play a role in an initial cleavage event (Huesgen et al., 2005).

Other cases of stress-induced proteolysis in the *Chlamydomonas* literature remain to be associated with specific protease(s). In Cu-deficient cells, for example, plastocyanin is synthesized normally but does not accumulate, due to induction of a proteolytic system (Li and Merchant, 1995). Sulfur starvation leads to a decrease in accumulation of several chloroplast mRNAs. This effect depends on signaling by the SAC3 kinase and appears to involve proteolytic inactivation of the chloroplast sigma factor Sig1

(Irihimovitch and Stern, 2006). Again, the culprit protease is unknown, but microarray analysis (Zhang et al., 2004) identified two chloroplast-directed proteases whose mRNAs are induced during S starvation, DEG5A and the aspartyl aminopeptidase AAP1. The aspartyl protease ASP1 and the cysteine endopeptidase CEP1 are also induced, but they are directed to the secretory pathway.

Rubisco is also a target of active degradation, in particular during oxidative stress (Houtz and Portis, 2003). Direct cleavage has been observed around Gly-329 both *in vitro* and *in vivo* (Ishida et al., 1998; Nakano et al., 2006), presumably resulting from the generation of hydroxide radicals at the catalytic site via a Fenton-type reaction. An alternative proposal is that oxidative damage causes a structural change in the enzyme that makes it susceptible to proteolytic attack (Desimone et al., 1996), and mutational studies in *Chlamydomonas* have identified residues contributing to stability *in vitro* and *in vivo* (Esquivel et al., 2006; Marin-Navarro and Moreno, 2006). As a major nitrogen store, Rubisco is expected to be degraded during N starvation, a situation where intracellular protein catabolism is enhanced in *Chlamydomonas* (Cullimore and Sims, 1980). In tobacco, the pepsin-type aspartyl protease CND41 has been implicated in Rubisco degradation (Murakami et al., 2000; Kato et al., 2004). However, no convincing evidence exists for the chloroplast localization of this protease, which in fact is preceded by a signal peptide that should direct it to the secretory pathway. CND41 may rather participate in lysosomal degradation of Rubisco during leaf senescence (Chiba et al., 2003).

In *Chlamydomonas*, three genes encode homologous pepsin-type aspartyl proteases with similarity to CND41; these are ASP1 ("Chlapsin"), ASP2 and ASP3. The ASP proteins are probably also directed to the secretory pathway, as are several acidic cysteine endoproteases (CPE1–CPE5, CPR1, and the vacuolar processing enzyme, VPE). Lysosomal degradation thus probably exists in *Chlamydomonas*, even though no intracellular compartment has been identified as the equivalent of the plant or yeast vacuole. One possible route for Rubisco degradation would be the inclusion of fragments of the chloroplast into autophagy vesicles that would then fuse with a degradation vacuole. A (near)-complete complement of autophagy genes (Reggiori and Klionsky, 2005) is found in *Chlamydomonas*, including genes for the APG1 kinase and the APG5 and APG12 proteins involved in formation of the autophagosome, albeit no clear orthologue is found for the WD repeat-containing APG16 that forms a complex with APG5 and APG12. The genome also encodes the APG8 ubiquitin-like molecule that unconventionally ligates to phosphatidylethanolamine, as well as the APG4 cysteine protease (necessary for activation of APG8) and the associated E1- and E2-type ligases APG7 and APG3. There is also a gene for APG6/VPS30/beclin, which together with class III phosphoinositide 3-kinases may mediate the amino acid-dependent regulation of autophagy. The mTOR (target of rapamycin) kinase may play a role, since its inactivation by the rapamycin/FKB12 complex triggers autophagic-like processes in *Chlamydomonas* (Crespo et al., 2005). It will be of interest to determine whether the degradation of cytosolic and chloroplast ribosomes during N starvation (Martin et al., 1976) involves a form of specialized autophagy.

In vitro assays have also been used to characterize the degradation of *Chlamydomonas* chloroplast proteins. PS I was degraded in isolated membranes incubated at 30°C (Henderson et al., 2003). Proteolytic activity was EDTA-sensitive and could be restored by Zn^{2+}, suggesting the intervention of a metalloprotease. ATP, however, was not required, ruling out AAA+ proteases as the effectors. In contrast, experiments using a truncated D1 protein (Preiss et al., 2001) showed ATP dependence of its degradation *in vitro*, as well as EDTA sensitivity. *In vivo*, truncated D1 was degraded much more rapidly than the full-length protein, and degradation was slowed by treatments aimed at reducing the stromal ATP level (dark incubation, uncoupler addition).

XVII. CONCLUSION

The wide variety of chaperone and proteolytic systems found in *Chlamydomonas* makes it an excellent system to study basic processes of the eukaryotic cell. Their multiple origins (from the ancestral archaea-like host and the cyanobacterial and proteobacterial endosymbionts) is largely reflected in their known or predicted present-day subcellular localizations (cytosol, organelles). Amplification of protease gene families has often been accompanied by the appearance of apparently inactive isoforms, which may either play a structural role in a complex, or be specialized in another, non-hydrolytic, function. As a general trend, chaperone and protease systems in *Chlamydomonas* are closely related to those in higher plants, but with a smaller number of genes. This relative simplicity, together with the ease of genetic approaches in *Chlamydomonas*, should bolster research in these areas in the coming years.

REFERENCES

Abravaya, K.A., Myers, M., Murphy, S.P., and Morimoto, R.I. (1992). The human heat shock protein 70 interacts with HSF, the transcription factor that regulates heat shock gene expression. *Genes Dev.* **6**, 1153–1164.

Adam, Z., Zaltsman, A., Sinvany-Villalobo, G., and Sakamoto, W. (2005). FtsH proteases in chloroplasts and cyanobacteria. *Physiol. Plant.* **123**, 386–390.

Ali, A., Bharadwaj, S., O'Carroll, R., and Ovsenek, N. (1998). Hsp90 interacts with and regulates the activity of heat shock factor 1 in Xenopus oocytes. *Mol. Cell. Biol.* **18**, 4949–4960.

Allmer, J., Naumann, B., Markert, C., Zhang, M., and Hippler, M. (2006). Mass spectrometric genomic data mining: Novel insights into bioenergetic pathways in *Chlamydomonas*. *Proteomics* **6**, 6207–6220.

Andersson, F.I., Blakytny, R., Kirstein, J., Turgay, K., Bukau, B., Mogk, A., and Clarke, A.K. (2006). Cyanobacterial ClpC/HSP100 protein displays intrinsic chaperone activity. *J. Biol. Chem.* **281**, 5468–5475.

Aseeva, E., Ossenbühl, F., Eichacker, L.A., Wanner, G., Soll, J., and Vothknecht, U.C. (2004). Complex formation of Vipp1 depends on its α-helical PspA-like domain. *J. Biol. Chem.* **279**, 35535–35541.

Azem, A., Oppliger, W., Lustig, A., Jenö, P., Feifel, B., Schatz, G., and Horst, M. (1997). The mitochondrial hsp70 chaperone system. Effect of adenine nucleotides, peptide substrate, and mGrpE on the oligomeric state of mhsp70. *J. Biol. Chem.* **272**, 20901–20906.

Balczun, C., Bunse, A., Schwarz, C., Piotrowski, M., and Kück, U. (2006). Chloroplast heat shock protein Cpn60 from *Chlamydomonas* exhibits a novel function as a group II intron-specific RNA-binding protein. *FEBS Lett.* **580**, 4527–4532.

Baler, R., Welch, W.J., and Voellmy, R. (1992). Heat shock gene regulation by nascent polypeptides and denatured proteins: hsp70 as a potential autoregulatory factor. *J. Cell Biol.* **117**, 1151–1159.

Ballinger, C.A., Connell, P., Wu, Y., Hu, Z., Thompson, L.J., Yin, L.-Y., and Patterson, C. (1999). Identification of CHIP, a novel tetratricopeptide repeat-containing protein that interacts with heat shock proteins and negatively regulates chaperone functions. *Mol. Cell. Biol.* **19**, 4535–4545.

Barker, M., de Vries, R., Nield, J., Komenda, J., and Nixon, P.J. (2006). The deg proteases protect *Synechocystis* sp. PCC 6803 during heat and light stresses but are not essential for removal of damaged D1 protein during the photosystem two repair cycle. *J. Biol. Chem.* **281**, 30347–30355.

Beck, C.F. (2005). Signaling pathways from the chloroplast to the nucleus. *Planta* **222**, 743–756.

Benning, C., Xu, C., and Awai, K. (2006). Non-vesicular and vesicular trafficking involving plastids. *Curr. Opin. Plant Biol.* **9**, 241–247.

Bertsch, U., Soll, J., Seetharam, R., and Viitanen, P.V. (1992). Identification, characterization, and DNA sequence of a functional double groES-like chaperonin from chloroplasts of higher plants. *Proc. Natl. Acad. Sci. U.S.A.* **89**, 8696–8700.

Bhushan, S., Stahl, A., Nilsson, S., Lefebvre, B., Seki, M., Roth, C., McWilliam, D., Wright, S.J., Liberles, D.A., Shinozaki, K., Bruce, B.D., Boutry, M., and Glaser, E. (2005). Catalysis, subcellular localization, expression and evolution of the targeting peptides degrading protease, AtPreP2. *Plant Cell Physiol.* **46**, 985–996.

Bloch, M.A. and Johnson, K.A. (1995). Identification of a molecular chaperone in the eukaryotic flagellum and its localization to the site of microtubule assembly. *J. Cell Sci.* **108**, 3541–3545.

Bolter, B., Nada, A., Fulgosi, H., and Soll, J. (2006). A chloroplastic inner envelope membrane protease is essential for plant development. *FEBS Lett.* **580**, 789–794.

Bonk, M., Tadros, M., Vandekerckhove, J., Al-Babili, S., and Beyer, P. (1996). Purification and characterization of chaperonin 60 and heat-shock protein 70 from chromoplasts of *Narcissus pseudonarcissus*. *Plant Physiol.* **111**, 931–939.

Borges, J.C., Fischer, H., Craievich, A.F., Hansen, L.D., and Ramos, C.H. (2003). Free human mitochondrial GrpE is a symmetric dimer in solution. *J. Biol. Chem.* **278**, 35337–35344.

Boston, R.S., Viitanen, P.V., and Vierling, E. (1996). Molecular chaperones and protein folding in plants. *Plant Mol. Biol.* **32**, 191–222.

Brehmer, D., Rüdiger, S., Gässler, C.S., Klostermeier, D., Packschies, L., Reinstein, J., and Bukau, B. (2001). Tuning of chaperone activity of Hsp70 proteins by modulation of nucleotide exchange. *Nat. Struct. Biol.* **8**, 427–432.

Brychzy, A., Rein, T., Winklhofer, K.F., Hartl, F.U., Young, J.C., and Obermann, W.M.J. (2003). Cofactor Tpr2 combines two TPR domains and a J domain to regulate the Hsp70/Hsp90 chaperone system. *EMBO J.* **22**, 3613–3623.

Bukau, B. and Horwich, A.L. (1998). The Hsp70 and Hsp60 chaperone machines. *Cell* **92**, 351–366.

Bukau, B., Weissman, J., and Horwich, A. (2006). Molecular chaperones and protein quality control. *Cell* **125**, 443–451.

Cahoon, A.B., Cunningham, K.A., and Stern, D.B. (2003). The plastid clpP gene may not be essential for plant cell viability. *Plant Cell Physiol.* **44**, 93–95.

Cao, D., Lin, Y., and Cheng, C.-L. (2000). Genetic interactions between the chlorate-resistant mutant cr88 and the photomorphogenic mutants cop1 and hy5. *Plant Cell* **12**, 199–210.

Cao, D., Froehlich, J.E., Zhang, H., and Cheng, C.-L. (2003). The chlorate-resistant and photomorphogenesis-defective mutant cr88 encodes a chloroplast-targeted HSP90. *Plant J.* **33**, 107–118.

Carde, J.-P., Joyard, J., and Douce, R. (1982). Electron microscopic studies of envelope membranes from spinach plastids. *Biol. Cell* **44**, 315–324.

Chadli, A., Bouhouche, I., Sullivan, W., Stensgard, B., McMahon, N., Catelli, M.G., and Toft, D.O. (2000). Dimerization and N-terminal domain proximity underlie the function of the molecular chaperone heat shock protein 90. *Proc. Natl. Acad. Sci. U.S.A.* **97**, 12524–12529.

Chen, B., Zhong, D., and Monteiro, A. (2006). Comparative genomics and evolution of the HSP90 family of genes across all kingdoms of organisms. *BMC Genomics* **7**, 156.

Chen, G.G. and Jagendorf, A.T. (1994). Chloroplast molecular chaperone-assisted refolding and reconstitution of an active multisubunit coupling factor CF1 core. *Proc. Natl. Acad. Sci. U.S.A.* **91**, 11497–11501.

Chen, G., Bi, Y.R., and Li, N. (2005). EGY1 encodes a membrane-associated and ATP-independent metalloprotease that is required for chloroplast development. *Plant J.* **41**, 364–375.

Chen, Q., Osteryoung, K., and Vierling, E. (1994). A 21-kDa chloroplast heat shock protein assembles into high molecular weight complexes *in vivo* and in organelle. *J. Biol. Chem.* **269**, 13216–13223.

Cheng, Q., Fowler, R., Tam, L., Edwards, L., and Miller, S.M. (2003). The role of GlsA in the evolution of asymmetric cell division in the green alga *Volvox carteri*. *Dev. Genes Evol.* **213**, 328–335.

Cheng, Q., Pappas, V., Hallmann, A., and Miller, S.M. (2005). Hsp70A and GlsA interact as partner chaperones to regulate asymmetric division in *Volvox*. *Dev. Biol.* **286**, 537–548.

Cheng, Q., Hallmann, A., Edwards, L., and Miller, S.M. (2006). Characterization of a heat-shock-inducible hsp70 gene of the green alga *Volvox carteri*. *Gene* **371**, 112–120.

Chiba, A., Ishida, H., Nishizawa, N.K., Makino, A., and Mae, T. (2003). Exclusion of ribulose-1,5-bisphosphate carboxylase/oxygenase from chloroplasts by specific bodies in naturally senescing leaves of wheat. *Plant Cell Physiol.* **44**, 914–921.

Choquet, Y., Stern, D.B., Wostrikoff, K., Kuras, R., Girard-Bascou, J., and Wollman, F.A. (1998). Translation of cytochrome *f* is autoregulated through the 5′ untranslated region of *petA* mRNA in *Chlamydomonas* chloroplasts. *Proc. Natl. Acad. Sci. U. S. A.* **95**, 4380–4385.

Crespo, J.L., Diaz-Troya, S., and Florencio, F.J. (2005). Inhibition of target of rapamycin signaling by rapamycin in the unicellular green alga *Chlamydomonas*. *Plant Physiol.* **139**, 1736–1749.

Cullimore, J.V. and Sims, A.P. (1980). An association between photorespiration and protein catabolism: Studies with *Chlamydomonas*. *Planta* **150**, 392–396.

de Vitry, C., Olive, J., Drapier, D., Recouvreur, M., and Wollman, F.-A. (1989). Posttranslational events leading to the assembly of photosystem II protein complex: A study using photosynthesis mutants from *Chlamydomonas*. *J. Cell Biol.* **109**, 991–1006.

DeLisa, M.P., Lee, P., Palmer, T., and Georgiou, G. (2004). Phage shock protein PspA of *Escherichia coli* relieves saturation of protein export via the Tat pathway. *J. Bacteriol.* **186**, 366–373.

Deloche, O. and Georgopoulos, C. (1996). Purification and biochemical properties of *Saccharomyces cerevisiae*'s Mge1p, the mitochondrial co-chaperone of Ssc1p. *J. Biol. Chem.* **271**, 23960–23966.

Deloche, O., Kelley, W.L., and Georgopoulos, C. (1997). Structure-function analyses of the Ssc1p, Mdj1p, and Mge1p *Saccharomyces cerevisiae* mitochondrial proteins in *Escherichia coli*. *J. Bacteriol.* **179**, 6066–6075.

Dent, R.M., Haglund, C.M., Chin, B.L., Kobayashi, M.C., and Niyogi, K.K. (2005). Functional genomics of eukaryotic photosynthesis using insertional mutagenesis of *Chlamydomonas*. *Plant Physiol.* **137**, 545–556.

Desimone, M., Henke, A., and Wagner, E. (1996). Oxidative stress induces partial degradation of the large subunit of ribulose-1,5-bisphosphate carboxylase/oxygenase in isolated chloroplasts of barley. *Plant Physiol.* **111**, 789–796.

Dickson, R., Weiss, C., Howard, R.J., Alldrick, S.P., Ellis, R.J., Lorimer, G., Azem, A., and Viitanen, P.V. (2000). Reconstitution of higher plant chloroplast chaperonin 60 tetradecamers active in protein folding. *J. Biol. Chem.* **275**, 11829–11835.

Dougan, D.A., Reid, B.G., Horwich, A.L., and Bukau, B. (2002). ClpS, a substrate modulator of the ClpAP machine. *Mol. Cell* **9**, 673–683.

Drzymalla, C., Schroda, M., and Beck, C.F. (1996). Light inducible gene HSP70B encodes a chloroplast-localized heat shock protein in *Chlamydomonas*. *Plant Mol. Biol.* **31**, 1185–1194.

Duncan, R.F. (2005). Inhibition of Hsp90 function delays and impairs recovery from heat shock. *FEBS J.* **272**, 5244–5256.

Easton, D.P., Kaneko, Y., and Subjeck, J.R. (2000). The Hsp110 and Grp170 stress proteins: newly recognized relatives of the Hsp70s. *Cell Stress Chaperones* **5**, 276–290.

Eisenberg-Domovich, Y., Kloppstech, K., and Ohad, I. (1994). Reversible membrane association of heat-shock protein 22 in *Chlamydomonas* during heat shock and recovery. *Eur. J. Biochem.* **222**, 1041–1046.

Emanuelsson, O., Nielsen, H., Brunak, S., and von Heijne, G. (2000). Predicting subcellular localization of proteins based on their N-terminal amino acid sequence. *J. Mol. Biol.* **300**, 1005–1016.

Emelyanov, V.V. (2002). Phylogenetic relationship of organellar Hsp90 homologs reveal fundamental differences to organellar Hsp70 and Hsp60 evolution. *Gene* **299**, 125–133.

Esquivel, M.G., Pinto, T.S., Marin-Navarro, J., and Moreno, J. (2006). Substitution of tyrosine residues at the aromatic cluster around the betaA-betaB loop of rubisco small subunit affects the structural stability of the enzyme and the *in vivo* degradation under stress conditions. *Biochemistry* **45**, 5745–5753.

Felts, S.J., Owen, B.A.L., Nguyen, P., Trepel, J., Donner, D.B., and Toft, D.O. (2000). The hsp90-related protein TRAP1 is a mitochondrial protein with distinct functional properties. *J. Biol. Chem.* **275**, 3305–3312.

Ferris, P.J., Armbrust, E.V., and Goodenough, U.W. (2002). Genetic structure of the mating-type locus of *Chlamydomonas*. *Genetics* **160**, 181–200.

Fischer, B.B., Krieger-Liszkay, A., and Eggen, R.I. (2005). Oxidative stress induced by the photosensitizers neutral red (type I) or rose bengal (type II) in the light causes different molecular responses in *Chlamydomonas*. *Plant Sci.* **168**, 747–759.

Fleischmann, M.M. and Rochaix, J.D. (1999). Characterization of mutants with alterations of the phosphorylation site in the D2 photosystem II polypeptide of *Chlamydomonas*. *Plant Physiol.* **119**, 1557–1566.

Friedmann, E., Lemberg, M.K., Weihofen, A., Dev, K.K., Dengler, U., Rovelli, G., and Martoglio, B. (2004). Consensus analysis of signal peptide peptidase and homologous human aspartic proteases reveals opposite topology of catalytic domains compared with presenilins. *J. Biol. Chem.* **279**, 50790–50798.

Frydman, J. (2001). Folding of newly translated proteins *in vivo*: The role of molecular chaperones. *Annu. Rev. Biochem.* **70**, 603–649.

Gakh, O., Cavadini, P., and Isaya, G. (2002). Mitochondrial processing peptidases. *Biochim. Biophys. Acta* **1592**, 63–77.

Gässler, C.S., Wiederkehr, T., Brehmer, D., Bukau, B., and Mayer, M.P. (2001). Bag-1 M accelerates nucleotide release for human Hsc70 and Hsp70 and can act concentration dependent as positive and negative cofactor. *J. Biol. Chem.* **276**, 32538–32544.

Gerloff-Elias, A., Barua, D., Molich, A., and Spijkerman, E. (2006). Temperature- and pH-dependent accumulation of heat-shock proteins in the acidophilic green alga *Chlamydomonas acidophila*. *FEMS Microbiol. Ecol.* **56**, 345–354.

Giglione, C., Serero, A., Pierre, M., Boisson, B., and Meinnel, T. (2000). Identification of eukaryotic peptide deformylases reveals universality of N-terminal protein processing mechanisms. *EMBO J.* **19**, 5916–5929.

Giglione, C., Vallon, O., and Meinnel, T. (2003). Control of protein life-span by N-terminal methionine excision. *EMBO J.* **22**, 13–23.

Glanz, S., Bunse, A., Wimbert, A., Balczun, C., and Kück, U. (2006). A nucleosome assembly protein-like polypeptide binds to chloroplast group II intron RNA in *Chlamydomonas*. *Nucleic Acids Res.* **34**, 5337–5351.

Glover, J.R. and Lindquist, S. (1998). Hsp104, Hsp70 and Hsp40: A novel chaperone system that rescues previously aggregated proteins. *Cell* **94**, 73–82.

Göhre, V., Ossenbühl, F., Crevecoeur, M., Eichacker, L.A., and Rochaix, J.D. (2006). One of two Alb3 proteins is essential for the assembly of the photosystems and for cell survival in *Chlamydomonas*. *Plant Cell* **18**, 1454–1466.

Graumann, J., Lilie, H., Tang, X., Tucker, K.A., Hoffmann, J.H., Vijayalakshmi, J., Saper, M., Bardwell, J.C., and Jakob, U. (2001). Activation of the redox-regulated molecular chaperone Hsp33-a two-step mechanism. *Structure* **9**, 377–387.

Grimm, B., Ish-Shalom, D., Even, D., Glaczinski, H., Ottersbach, P., Ohad, I., and Kloppstech, K. (1989). The nuclear-coded chloroplast 22 kDa heat shock protein of *Chlamydomonas*. Evidence for translocation into the organelle without a processing step. *Eur. J. Biochem.* **182**, 539–546.

Gross, D.S., Adams, C.C., Lee, S., and Stentz, B. (1993). A critical role for heat shock transcription factor in establishing a nucleosome-free region over the TATA-initiation site of the yeast HSP82 heat shock gene. *EMBO J.* **12**, 3931–3945.

Hallmann, A. and Wodniok, S. (2006). Swapped green algal promoters: aphVIII-based gene constructs with *Chlamydomonas* flanking sequences work as dominant selectable markers in *Volvox* and vice versa. *Plant Cell Rep.* **25**, 582–591.

Halperin, T., Zheng, B., Itzhaki, H., Clarke, A.K., and Adam, Z. (2001). Plant mitochondria contain proteolytic and regulatory subunits of the ATP-dependent Clp protease. *Plant Mol. Biol.* **45**, 461–468.

Han, W. and Christen, P. (2003). Mechanism of the targeting action of DnaJ in the DnaK molecular chaperone system. *J. Biol. Chem.* **278**, 19038–19043.

Hankamer, B.D., Elderkin, S., Buck, M., and Nield, J. (2004). Organization of the AAA+ adaptor protein PspA is an oligomeric ring. *J. Biol. Chem.* **279**, 8862–8866.

Härndahl, U. and Sundby, C. (2001). Does the chloroplast small heat shock protein protect photosystem II during heat stress *in vitro*? *Physiol. Plant.* **111**, 273–275.

Harrison, C.J., Hayer-Hartl, M., Di Liberto, M., Hartl, F.-U., and Kuriyan, J. (1997). Crystal structure of the nucleotide exchange factor GrpE bound to the ATPase domain of the molecular chaperone DnaK. *Science* **276**, 431–435.

Haslbeck, M., Ignatiou, A., Saibil, H., Helmich, S., Frenzl, E., Stromer, T., and Buchner, J. (2004). A domain in the N-terminal part of Hsp26 is essential for chaperone function and oligomerization. *J. Mol. Biol.* **343**, 445–455.

Haslbeck, M., Franzmann, T., Weinfurtner, D., and Buchner, J. (2005). Some like it hot: The structure and function of small heat-shock proteins. *Nat. Struct. Mol. Biol.* **12**, 842–846.

Hatano-Iwasaki, A., Minagawa, J., Inoue, Y., and Takahashi, Y. (2000). Characterization of chloroplast *psbA* transformants of *Chlamydomonas* with impaired processing of a precursor of a photosystem II reaction center protein, D1. *Plant Mol. Biol.* **42**, 353–363.

Haussuhl, K., Andersson, B., and Adamska, I. (2001). A chloroplast DegP2 protease performs the primary cleavage of the photodamaged D1 protein in plant photosystem II. *EMBO J.* **20**, 713–722.

Heazlewood, J.L., Tonti-Filippini, J.S., Gout, A.M., Day, D.A., Whelan, J., and Millar, A.H. (2004). Experimental analysis of the *Arabidopsis* mitochondrial proteome highlights signaling and regulatory components, provides assessment of targeting prediction programs, and indicates plant-specific mitochondrial proteins. *Plant Cell* **16**, 241–256.

Heckathorn, S.A., Downs, C.A., Sharkey, T.D., and Coleman, J.S. (1998). The small, methionine-rich chloroplast heat-shock protein protects photosystem II electron transport during heat stress. *Plant Physiol.* **116**, 439–444.

Henderson, J.N., Zhang, J., Evans, B.W., and Redding, K. (2003). Disassembly and degradation of photosystem I in an *in vitro* system are multievent, metal-dependent processes. *J. Biol. Chem.* **278**, 39978–39986.

Hennessy, F., Nicoll, W.S., Zimmermann, R., Cheetham, M.E., and Blatch, G.L. (2005). Not all J domains are created equal: Implications for the specificity of Hsp40–Hsp70 interactions. *Protein Sci.* **14**, 1697–1709.

Hoffmann, J.H., Linke, K., Graf, P.C., Lilie, H., and Jakob, U. (2004). Identification of a redox-regulated chaperone network. *EMBO J.* **23**, 160–168.

Höhfeld, J., Minami, Y., and Hartl, F.-U. (1995). Hip, a novel co-chaperone involved in the eukaryotic Hsc70/Hsp40 reaction cycle. *Cell* **83**, 589–598.

Höhfeld, J., Cyr, D.M., and Patterson, C. (2001). From the cradle to the grave: Molecular chaperones that may chose between folding and degradation. *EMBO Rep.* **2**, 885–890.

Holmberg, C.I., Tran, S.E., Eriksson, J.E., and Sistonen, L. (2002). Multisite phosphorylation provides sophisticated regulation of transcription factors. *Trends Biochem. Sci.* **27**, 619–627.

Houtz, R.L. and Portis, A.R. (2003). The life of ribulose 1,5-bisphosphate carboxylase/oxygenase – posttranslational facts and mysteries. *Arch. Biochem. Biophys.* **414**, 150–158.

Huang, C., Wang, S., Chen, L., Lemieux, C., Otis, C., Turmel, M., and Liu, X.Q. (1994). The *Chlamydomonas* chloroplast *clpP* gene contains translated large insertion sequences and is essential for cell growth. *Mol. Gen. Genet.* **244**, 151–159.

Hübel, A. and Schöffl, F. (1994). *Arabidopsis* heat shock factor: Isolation and characterization of the gene and the recombinant protein. *Plant Mol. Biol.* **26**, 353–362.

Huesgen, P.F., Schuhmann, H., and Adamska, I. (2005). The family of Deg proteases in cyanobacteria and chloroplasts of higher plants. *Physiol. Plant.* **123**, 413–420.

Ilbert, M., Graf, P.C., and Jakob, U. (2006). Zinc center as redox switch – new function for an old motif. *Antioxid. Redox Signal.* **8**, 835–846.

Inoue, K., Baldwin, A.J., Shipman, R.L., Matsui, K., Theg, S.M., and Ohme-Takagi, M. (2005). Complete maturation of the plastid protein translocation channel requires a type I signal peptidase. *J. Cell Biol.* **171**, 425–430.

Irihimovitch, V. and Stern, D.B. (2006). The sulfur acclimation SAC3 kinase is required for chloroplast transcriptional repression under sulfur limitation in *Chlamydomonas*. *Proc. Natl. Acad. Sci. U.S.A.* **103**, 7911–7916.

Isaya, G., Kalousek, F., and Rosenberg, L.E. (1992). Amino-terminal octapeptides function as recognition signals for the mitochondrial intermediate peptidase. *J. Biol. Chem.* **267**, 7904–7910.

Ishida, H., Shimizu, S., Makino, A., and Mae, T. (1998). Light-dependent fragmentation of the large subunit of ribulose-1,5-bisphosphate carboxylase/oxygenase in chloroplasts isolated from wheat leaves. *Planta* **204**, 305–309.

Iyer, L.M., Leipe, D.D., Koonin, E.V., and Aravind, L. (2004). Evolutionary history and higher order classification of AAA+ ATPases. *J. Struct. Biol.* **146**, 11–31.

Jacobshagen, S., Whetstine, J.R., and Boling, J.M. (2001). Many but not all genes in *Chlamydomonas* are regulated by the circadian clock. *Plant Biol.* **3**, 592–597.

Jakob, U., Lilie, H., Meyer, I., and Buchner, J. (1995). Transient interactions of Hsp90 with early unfolding intermediates of citrate synthase. Implications for heat shock *in vivo*. *J. Biol. Chem.* **270**, 7288–7294.

Janska, H. (2005). ATP-dependent proteases in plant mitochondria: What do we know about them today? *Physiol. Plant.* **123**, 399–405.

Jensen, K.H., Herrin, D.L., Plumley, F.G., and Schmidt, G.W. (1986). Biogenesis of photosystem II complexes: Transcriptional, translational, and posttranslational regulation. *J. Cell Biol.* **103**, 1315–1325.

Johnson, K.A., Bhushan, S., Stahl, A., Hallberg, B.M., Frohn, A., Glaser, E., and Eneqvist, T. (2006). The closed structure of presequence protease PreP forms a unique 10,000 Å chamber for proteolysis. *EMBO J.* **25**, 1977–1986.

Kang, P.J., Ostermann, J., Shilling, J., Neupert, W., Craig, E.A., and Pfanner, N. (1990). Requirement for hsp70 in the mitochondrial matrix for translocation and folding of precursor proteins. *Nature* **348**, 137–143.

Kalies, K.-U. and Hartmann, E. (1996). Membrane topology of the 12- and the 25-kDa subunits of the mammalian signal peptidase complex. *J. Biol. Chem.* **271**, 3925–3929.

Kapri-Pardes, E., Naveh, L., and Adam, Z. (2007). The thylakoid lumen protease Deg1 is involved in the repair of photosystem II from photoinhibition in *Arabidopsis*. *Plant Cell* **19**, 1039–1047.

Kato, Y., Murakami, S., Yamamoto, Y., Chatani, H., Kondo, Y., Nakano, T., Yokota, A., and Sato, F. (2004). The DNA-binding protease, CND41, and the degradation of ribulose-1,5-bisphosphate carboxylase/oxygenase in senescent leaves of tobacco. *Planta* **220**, 97–104.

Kim, B.H. and Schöffl, F. (2002). Interaction between Arabidopsis heat shock transcription factor 1 and 70 kDa heat shock proteins. *J. Exp. Bot.* **53**, 371–375.

Kim, Y.I., Levchenko, I., Fraczkowska, K., Woodruff, R.V., Sauer, R.T., and Baker, T.A. (2001). Molecular determinants of complex formation between Clp/Hsp100 ATPases and the ClpP peptidase. *Nat. Struct. Biol.* **8**, 230–233.

Kirk, M.M., Ransick, A., McRae, S.E., and Kirk, D.L. (1993). The relationship between cell size and cell fate in *Volvox carteri*. *J. Cell Biol.* **123**, 191–208.

Kiyosue, T., Yamaguchi-Shinozaki, K., and Shinozaki, K. (1993). Characterization of cDNA for a dehydration-inducible gene that encodes a CLP A, B-like protein in *Arabidopsis thaliana* L. *Biochem. Biophys. Res. Commun.* **196**, 1214–1220.

Kloppstech, K., Meyer, G., Schuster, G., and Ohad, I. (1985). Synthesis, transport and localization of a nuclear coded 22-kd heat-shock protein in the chloroplast membranes of peas and *Chlamydomonas*. *EMBO J.* **4**, 1901–1909.

Koblenz, B., Schoppmeier, J., Grunow, A., and Lechtreck, K.F. (2003). Centrin deficiency in *Chlamydomonas* causes defects in basal body replication, segregation and maturation. *J. Cell Sci.* **116**, 2635–2646.

Koonin, E.V., Makarova, K.S., Rogozin, I.B., Davidovic, L., Letellier, M.C., and Pellegrini, L. (2003). The rhomboids: A nearly ubiquitous family of intramembrane serine proteases that probably evolved by multiple ancient horizontal gene transfers. *Genome Biol.* **4**, R19.

Koumoto, Y., Shimada, T., Kondo, M., Hara-Nishimura, I., and Nishimura, M. (2001). Chloroplasts have a novel Cpn10 in addition to Cpn20 as co-chaperonins in *Arabidopsis thaliana*. *J. Biol. Chem.* **276**, 29688–29694.

Kroll, D., Meierhoff, K., Bechtold, N., Kinoshita, M., Westphal, S., Vothknecht, U.C., Soll, J., and Westhoff, P. (2001). VIPP1, a nuclear gene of *Arabidopsis thaliana* essential for thylakoid membrane formation. *Proc. Natl. Acad. Sci. U.S.A.* **98**, 4238–4242.

Kropat, J., von Gromoff, E.D., Müller, F.W., and Beck, C.F. (1995). Heat shock and light activation of a *Chlamydomonas HSP70* gene are mediated by independent regulatory pathways. *Mol. Gen. Genet.* **248**, 727–734.

Kropat, J., Oster, U., Rüdiger, W., and Beck, C.F. (1997). Chlorophyll precursors are signals of chloroplast origin involved in light induction of nuclear heat-shock genes. *Proc. Natl. Acad. Sci. U.S.A.* **94**, 14168–14172.

Kuhn, A., Stuart, R., Henry, R., and Dalbey, R.E. (2003). The Alb3/Oxa1/YidC protein family: Membrane-localized chaperones facilitating membrane protein insertion? *Trends Cell Biol.* **13**, 510–516.

Kuras, R. and Wollman, F.A. (1994). The assembly of cytochrome b_6f complexes: An approach using genetic transformation of the green alga *Chlamydomonas*. *EMBO J.* **13**, 1019–1027.

Laloraya, S., Gambill, B.D., and Craig, E.A. (1994). A role for a eukaryotic GrpE-related protein, Mge1p, in protein translocation. *Proc. Natl. Acad. Sci. U.S.A.* **91**, 6481–6485.

Landsberger, N. and Wolffe, A.P. (1995). Role of chromatin and *Xenopus laevis* heat shock transcription factor in regulation of transcription from the *X. laevis* hsp70 promoter *in vivo*. *Mol. Cell. Biol.* **15**, 6013–6024.

Lawrence, M.E. and Possingham, J.V. (1984). Observations of microtubule-like structures within spinach plastids. *Biol. Cell* **52**, 77–82.

Lechtreck, K.F., Rostmann, J., and Grunow, A. (2002). Analysis of *Chlamydomonas* SF-assemblin by GFP tagging and expression of antisense constructs. *J. Cell Sci.* **115**, 1511–1522.

Lee, G.J., Roseman, A.M., Saibil, H.R., and Vierling, E. (1997). A small heat shock protein stably binds heat-denatured model substrates and can maintain a substrate in a folding-competent state. *EMBO J.* **16**, 659–671.

Lee, J.H. and Schöffl, F. (1996). An Hsp70 antisense gene affects the expression of HSP70/HSC70, the regulation of HSF, and the acquisition of thermotolerance in transgenic *Arabidopsis thaliana*. *Mol. Gen. Genet.* **252**, 11–19.

Lee, J.H., Hübel, A., and Schöffl, F. (1995). Derepression of the activity of genetically engineered heat shock factor causes constitutive synthesis of heat shock proteins and increased thermotolerance in transgenic *Arabidopsis*. *Plant J.* **8**, 603–612.

Lee, U., Rioflorido, I., Hong, S.W., Larkindale, J., Waters, E.R., and Vierling, E. (2007). The *Arabidopsis* ClpB/Hsp100 family of proteins: Chaperones for stress and chloroplast development. *Plant J.* **49**, 115–127.

Leisinger, U., Rufenacht, K., Fischer, B., Pesaro, M., Spengler, A., Zehnder, A.J., and Eggen, R.I. (2001). The glutathione peroxidase homologous gene from *Chlamydomonas* is transcriptionally up-regulated by singlet oxygen. *Plant Mol. Biol.* **46**, 395–408.

Lensch, M., Herrmann, R.G., and Sokolenko, A. (2001). Identification and characterization of SppA, a novel light-inducible chloroplast protease complex associated with thylakoid membranes. *J. Biol. Chem.* **276**, 33645–33651.

Li, H.H. and Merchant, S. (1995). Degradation of plastocyanin in copper-deficient *Chlamydomonas*. Evidence for a protease-susceptible conformation of the apoprotein and regulated proteolysis. *J. Biol. Chem.* **270**, 23504–23510.

Li, J., Längst, G., and Grummt, I. (2006). NoRC-dependent nucleosome positioning silences rRNA genes. *EMBO J.* **25**, 5735–5741.

Liao, D.I., Qian, J., Chisholm, D.A., Jordan, D.B., and Diner, B.A. (2000). Crystal structures of the photosystem II D1 C-terminal processing protease. *Nat. Struct. Biol.* **7**, 749–753.

Liberek, K., Marszalek, J., Ang, D., Georgopoulos, C., and Zylicz, M. (1991). *Escherichia coli* DnaJ and GrpE heat shock proteins jointly stimulate ATPase activity of DnaK. *Proc. Natl. Acad. Sci. U.S.A.* **88**, 2874–2878.

Lin, Y. and Cheng, C.L. (1997). A chlorate-resistant mutant defective in the regulation of nitrate reductase gene expression in *Arabidopsis* defines a new HY locus. *Plant Cell* **9**, 21–35.

Lis, J. and Wu, C. (1993). Protein traffic on the heat shock promoter: Parking, stalling and trucking along. *Cell* **74**, 1–4.

Liu, C., Willmund, F., Whitelegge, J.P., Hawat, S., Knapp, B., Lodha, M., and Schroda, M. (2005). J-domain protein CDJ2 and HSP70B are a plastidic chaperone pair that interacts with vesicle inducing protein in plastids 1 (VIPP1). *Mol. Biol. Cell* **16**, 1165–1177.

Liu, C., Willmund, F., Golecki, J.R., Cacace, S., Heß, B., Markert, C., and Schroda, M. (2007). The chloroplast HSP70B-CDJ2-CGE1 chaperones catalyze assembly and disassembly of VIPP1 oligomers in *Chlamydomonas*. *Plant J.* **50**, 265–277.

Lodha, M. and Schroda, M. (2005). Analysis of chromatin structure in the control regions of the *Chlamydomonas HSP70A* and *RBCS2* genes. *Plant Mol. Biol.* **59**, 501–513.

Lodha, M., Schulz-Raffelt, M., and Schroda, M. (2008). A new assay for promoter analysis in *Chlamydomonas* reveals a role for heat shock elements and TATA-box in HSP70A promoter mediated activation of transgene expression. *Eukaryotic Cell* **7**, 172–176.

Lumbreras, V., Stevens, D.R., and Purton, S. (1998). Efficient foreign gene expression in *Chlamydomonas* mediated by an endogenous intron. *Plant J.* **14**, 441–447.

Lunney, C.A., Davis, G.J., and Jones, M.N. (1975). Unusual structures associated with peripheral reticulum in chloroplasts of *Myriophyllum spicatum* L. *J. Ultrastruct. Res.* **50**, 293–296.

Lupas, A. (1996). Coiled coils: New structures and new functions. *Trends Biochem. Sci.* **21**, 375–382.

Madueño, F., Napier, J.A., and Gray, J.C. (1993). Newly imported iron-sulfur protein associates with both Cpn60 and Hsp70 in the chloroplast stroma. *Plant Cell* **5**, 1865–1876.

Majeran, W., Wollman, F.-A., and Vallon, O. (2000). Evidence for a role of ClpP in the degradation of the chloroplast cytochrome b_6f complex. *Plant Cell* **12**, 137–150.

Majeran, W., Olive, J., Drapier, D., Vallon, O., and Wollman, F.A. (2001). The light sensitivity of ATP synthase mutants of *Chlamydomonas*. *Plant Physiol.* **126**, 421–433.

Majeran, W., Friso, G., van Wijk, K.J., and Vallon, O. (2005). The chloroplast ClpP complex in *Chlamydomonas* contains an unusual high molecular mass subunit with a large apical domain. *FEBS J.* **272**, 5558–5571.

Makinoshima, H. and Glickman, M.S. (2006). Site-2 proteases in prokaryotes: Regulated intramembrane proteolysis expands to microbial pathogenesis. *Microbes Infect.* **8**, 1882–1888.

Marin-Navarro, J. and Moreno, J. (2006). Cysteines 449 and 459 modulate the reduction-oxidation conformational changes of ribulose 1.5-bisphosphate carboxylase/oxygenase and the translocation of the enzyme to membranes during stress. *Plant Cell Environ.* **29**, 898–908.

Martin, N.C., Chiang, K.S., and Goodenough, U.W. (1976). Turnover of chloroplast and cytoplasmic ribosomes during gametogenesis in *Chlamydomonas*. *Dev. Biol.* **51**, 190–201.

Mayer, M.P. and Bukau, B. (2005). Hsp70 chaperones: Cellular functions and molecular mechanism. *Cell. Mol. Life Sci.* **62**, 670–684.

Miernyk, J.A. (2001). The J-domain proteins of *Arabidopsis thaliana*: An unexpectedly large and diverse family of chaperones. *Cell Stress Chaperones* **6**, 209–218.

Miller, S.M. and Kirk, D.L. (1999). *glsA*, a *Volvox* gene required for asymmetric division and germ cell specification, encodes a chaperone-like protein. *Development* **126**, 649–658.

Minai, L., Wostrikoff, K., Wollman, F.A., and Choquet, Y. (2006). Chloroplast biogenesis of photosystem II cores involves a series of assembly-controlled steps that regulate translation. *Plant Cell* **18**, 159–175.

Minami, Y., Kimura, Y., Kawasaki, H., Suzuki, K., and Yahara, I. (1994). The carboxy-terminal region of mammalian HSP90 is required for its dimerization and function *in vivo*. *Mol. Cell. Biol.* **14**, 1459–1464.

Mogk, A., Schlieker, C., Friedrich, K.L., Schönfeld, H.-J., Vierling, E., and Bukau, B. (2003). Refolding of substrates bound to small Hsps relies on a disaggregation reaction mediated most effeciently by ClpB/DnaK. *J. Biol. Chem.* **278**, 31033–31042.

Morimoto, R.I. (1998). Regulation of the heat shock transcriptional response: Cross talk between a family of heat shock factors, molecular chaperones, and negative regulators. *Genes Dev.* **12**, 3788–3796.

Moseley, J.L., Chang, C.W., and Grossman, A.R. (2006). Genome-based approaches to understanding phosphorus deprivation responses and PSR1 control in *Chlamydomonas*. *Eukaryotic Cell* **5**, 26–44.

Mosser, D.D., Duchaine, J., and Massie, B. (1993). The DNA-binding activity of the human heat shock transcription factor is regulated *in vivo* by hsp70. *Mol. Cell. Biol.* **13**, 5427–5438.

Müller, F.W., Igloi, G.L., and Beck, C.F. (1992). Structure of a gene encoding heat-shock protein HSP70 from the unicellular alga *Chlamydomonas*. *Gene* **111**, 165–173.

Murakami, S., Kondo, Y., Nakano, T., and Sato, F. (2000). Protease activity of CND41, a chloroplast nucleoid DNA-binding protein, isolated from cultured tobacco cells. *FEBS Lett.* **468**, 15–18.

Murakami, S., Kuehne, K., and Stern, D.B. (2005). A spontaneous tRNA suppressor of a mutation in the *Chlamydomonas* nuclear *MCD1* gene required for stability of the chloroplast *petD* mRNA. *Nucleic Acids Res.* **33**, 3372–3380.

Nakano, R., Ishida, H., Makino, A., and Mae, T. (2006). *In vivo* fragmentation of the large subunit of ribulose-1,5-bisphosphate carboxylase by reactive oxygen species in an intact leaf of cucumber under chilling-light conditions. *Plant Cell Physiol.* **47**, 270–276.

Newcomb, E.H. (1967). Fine structure of protein-storing plastids in bean root tips. *J. Cell Biol.* **33**, 143–163.

Nielsen, E., Akita, M., Davila-Aponte, J., and Keegstra, K. (1997). Stable association of chloroplastic precursors with protein translocation complexes that contain proteins from both envelope membranes and a stromal Hsp100 molecular chaperone. *EMBO J.* **16**, 935–946.

Niwa, H., Tsuchiya, D., Makyio, H., Yoshida, M., and Morikawa, K. (2002). Hexameric ring structure of the ATPase domain of the membrane-integrated metalloprotease FtsH from *Thermus thermophilus* HB8. *Structure* **10**, 1415–1423.

Nover, L., Bharti, K., Döring, P., Mishra, S.K., Ganguli, A., and Scharf, K.D. (2001). *Arabidopsis* and the heat stress transcription factor world: How many heat stress transcription factors do we need? *Cell Stress Chaperones* **6**, 177–189.

Osipiuk, J., Georgopoulos, C., and Zylicz, M. (1993). Initiation of λ DNA replication. The *Escherichia coli* small heat shock proteins, DnaJ and GrpE, increase DnaK's affinity for the λP protein. *J. Biol. Chem.* **268**, 4821–4827.

Padidam, M., Reddy, V.S., Beachy, R.N., and Fauquet, C.M. (1999). Molecular characterization of a plant mitochondrial chaperone GrpE. *Plant Mol. Biol.* **39**, 871–881.

Palenik, B., Grimwood, J., Aerts, A., Rouzé, P., Salamov, A., Putnam, N., Dupont, C., Jorgensen, R., Derelle, E., Rombauts, S. et al. (2007). The tiny eukaryote *Ostreococcus* provides genomic insights into the paradox of plankton speciation. *Proc. Natl. Acad. Sci. U.S.A.* **104**, 7705–7710.

Panaretou, B., Prodromou, C., Roe, S.M., O´Brien, R., Ladbury, J.E., Piper, P.W., and Pearl, L.H. (1998). ATP binding and hydrolysis are essential to the function of the Hsp90 molecular chaperone *in vivo*. *EMBO J.* **17**, 4829–4836.

Pazour, G.J., Agrin, N., Leszyk, J., and Witman, G.B. (2005). Proteomic analysis of a eukaryotic cilium. *J. Cell Biol.* **170**, 103–113.

Peltier, J.B., Ytterberg, J., Liberles, D.A., Roepstorff, P., and van Wijk, K.J. (2001). Identification of a 350 kDa ClpP protease complex with 10 different Clp isoforms in chloroplasts of *Arabidopsis thaliana*. *J. Biol. Chem.* **276**, 16318–16327.

Peltier, J.B., Ripoll, D.R., Friso, G., Rudella, A., Cai, Y., Ytterberg, J., Giacomelli, L., Pillardy, J., and van Wijk, K.J. (2004). Clp protease complexes from photosynthetic and non-photosynthetic plastids and mitochondria of plants, their predicted three-dimensional structures, and functional implications. *J. Biol. Chem.* **279**, 4768–4781.

Pfannenschmid, F., Wimmer, V.C., Rios, R.M., Geimer, S., Kröckel, U., Leiherer, A., Haller, K., Nemcova, Y., and Mages, W. (2003). *Chlamydomonas* DIP13 and human NA14: A new class of proteins associated with microtubule structures is involved in cell division. *J. Cell Sci.* **116**, 1449–1462.

Pickett-Heaps, J.D. (1968). Microtubule-like structures in the growing plastids of two algae. *Planta* **81**, 193–200.

Pierpaoli, E.V., Sandmeier, E., Baici, A., Schonfeld, H.J., Gisler, S., and Christen, P. (1997). The power stroke of the DnaK/DnaJ/GrpE molecular chaperone system. *J. Mol. Biol.* **269**, 757–768.

Pratt, W.B. and Toft, D.O. (2003). Regulation of signaling protein function and trafficking by the hsp90/hsp70-based chaperone machinery. *Exp. Biol. Med.* **228**, 111–133.

Preiss, S., Schrader, S., and Johanningmeier, U. (2001). Rapid, ATP-dependent degradation of a truncated D1 protein in the chloroplast. *Eur. J. Biochem.* **268**, 4562–4569.

Prodromou, C., Roe, S.M., O'Brien, O., Ladbury, J.E., Piper, P.W., and Pearl, L.H. (1997). Identification and structural charactarization of the ATP/ADP-binding site in the Hsp90 molecular chaperone. *Cell* **90**, 65–75.

Prodromou, C., Panaretou, B., Chohan, S., Siligardi, G., O´Brian, R., Ladburry, J.E., Roe, S.M., Piper, P.W., and Pearl, L.H. (2000). The ATPase cycle of Hsp90 drives a molecular clamp via transient dimerisation of the N-terminal domains. *EMBO J.* **19**, 4383–4392.

Raviol, H., Sadlish, H., Rodriguez, F., Mayer, M.P., and Bukau, B. (2006). Chaperone network in the yeast cytosol: Hsp110 is revealed as an Hsp70 nucleotide exchange factor. *EMBO J.* **25**, 2510–2518.

Reggiori, F. and Klionsky, D.J. (2005). Autophagosomes: Biogenesis from scratch? *Curr. Opin. Cell Biol.* **17**, 415–422.

Reisdorph, N.A. and Small, G.D. (2004). The *CPH1* gene of *Chlamydomonas* encodes two forms of cryptochrome whose levels are controlled by light-induced proteolysis. *Plant Physiol.* **134**, 1546–1554.

Richmond, T.J. and Davey, C.A. (2003). The structure of DNA in the nucleosome core. *Nature* **423**, 145–150.

Richter, K. and Buchner, J. (2001). Hsp90: Chaperoning signal transduction. *J. Cell. Phys.* **188**, 281–290.

Richter, S. and Lamppa, G.K. (2003). Structural properties of the chloroplast stromal processing peptidase required for its function in transit peptide removal. *J. Biol. Chem.* **278**, 39497–39502.

Rivera, E.R. and Arnott, H.J. (1982). Tubular structures in the plastids of *Echinomastus intertextus* Brit. & Rose (Cactaceae). *New Phytol.* **90**, 551–561.

Roe, S.M., Prodromou, C., O'Brien, R., Ladbury, J.E., Piper, P.W., and Pearl, L.H. (1999). Structural basis for inhibition of the Hsp90 molecular chaperone by the antitumor antibiotics radicicol and geldanamycin. *J. Med. Chem.* **42**, 260–266.

Rotanova, T.V., Botos, I., Melnikov, E.E., Rasulova, F., Gustchina, A., Maurizi, M.R., and Wlodawer, A. (2006). Slicing a protease: Structural features of the ATP-dependent Lon proteases gleaned from investigations of isolated domains. *Protein Sci.* **15**, 1815–1828.

Rudella, A., Friso, G., Alonso, J.M., Ecker, J.R., and van Wijk, K.J. (2006). Down-regulation of ClpR2 leads to reduced accumulation of the ClpPRS protease complex and defects in chloroplast biogenesis in *Arabidopsis*. *Plant Cell* **18**, 1704–1721.

Rüdiger, S., Germeroth, L., Schneider-Mergener, J., and Bukau, B. (1997). Substrate specificity of the DnaK chaperone determined by screening of cellulose-bound peptide libraries. *EMBO J.* **16**, 1501–1507.

Rüfenacht, A. and Boschetti, A. (2000). Chloroplasts of the green alga *Chlamydomonas* possess at least four distinct stromal processing proteases. *Photosynth. Res.* **63**, 249–258.

Sanchez, Y. and Lindquist, S.L. (1990). HSP104 required for induced thermotolerance. *Science* **248**, 1112–1115.

Sarria, R., Lyznik, A., Vallejos, C.E., and Mackenzie, S.A. (1998). A cytoplasmic male sterility-associated mitochondrial peptide in common bean is post-translationally regulated. *Plant Cell* **10**, 1217–1228.

Sauer, R.T., Bolon, D.N., Burton, B.M., Burton, R.E., Flynn, J.M., Grant, R.A., Hersch, G.L., Joshi, S.A., Kenniston, J.A., Levchenko, I., Neher, S.B., Oakes, E.S., Siddiqui, S.M., Wah, D.A., and Baker, T.A. (2004). Sculpting the proteome with AAA(+) proteases and disassembly machineries. *Cell* **119**, 9–18.

Scharf, K.-D., Rose, S., Zott, W., Schöffl, F., and Nover, L. (1990). Three tomato genes code for heat stress transcription factors with a region of remarkable homology to the DNA-binding domain of the yeast HSF. *EMBO J.* **9**, 4495–4501.

Scharf, K.-D., Siddique, M., and Vierling, E. (2001). The expanding family of *Arabidopsis thaliana* small heat stress proteins and a new family of proteins containing α-crystallin domains (Acd proteins). *Cell Stress Chaperones* **6**, 225–237.

Schiene-Fischer, C., Habazettl, J., Schmid, F.X., and Fischer, G. (2002). The hsp70 chaperone DnaK is a secondary amide peptide bond *cis-trans* isomerase. *Nat. Struct. Biol.* **9**, 419–424.

Schirmer, E.C., Glover, J.R., Singer, M.A., and Lindquist, S. (1996). HSP100/Clp proteins: A common mechanism explains diverse functions. *Trends Biochem. Sci.* **21**, 289–296.

Schnepf, E. (1961). Plastidenstrukturen bei Passiflora. *Protoplasma* **54**, 310–313.

Schoppmeier, J., Mages, W., and Lechtreck, K.F. (2005). GFP as a tool for the analysis of proteins in the flagellar basal apparatus of *Chlamydomonas*. *Cell. Motil. Cytoskeleton* **61**, 189–200.

Schroda, M. (2004). The *Chlamydomonas* genome reveals its secrets: Chaperone genes and the potential roles of their gene products in the chloroplast. *Photosynth. Res.* **82**, 221–240.

Schroda, M., Vallon, O., Wollman, F.-A., and Beck, C.F. (1999). A chloroplast-targeted heat shock protein 70 (HSP70) contributes to the photoprotection and repair of photosystem II during and after photoinhibition. *Plant Cell* **11**, 1165–1178.

Schroda, M., Blöcker, D., and Beck, C.F. (2000). The HSP70A promoter as a tool for the improved expression of transgenes in *Chlamydomonas*. *Plant J.* **21**, 121–131.

Schroda, M., Kropat, J., Oster, U., Rüdiger, W., Vallon, O., Wollman, F.-A., and Beck, C.F. (2001a). A role for molecular chaperones in assembly and repair of photosystem II? *Biochem. Soc. Trans.* **29**, 413–418.

Schroda, M., Vallon, O., Whitelegge, J.P., Beck, C.F., and Wollman, F.-A. (2001b). The chloroplastic GrpE homolog of *Chlamydomonas*: Two isoforms generated by differential splicing. *Plant Cell* **13**, 2823–2839.

Schroda, M., Beck, C.F., and Vallon, O. (2002). Sequence elements within an HSP70 promoter counteract transcriptional transgene silencing in *Chlamydomonas*. *Plant J.* **31**, 445–455.

Schulz-Raffelt, M., Lodha, M., and Schroda, M. (2007). Heat shock factor 1 is a key regulator of the stress response in *Chlamydomonas*. *Plant J.* **52**, 286–295.

Schuster, G., Even, D., Kloppstech, K., and Ohad, I. (1988). Evidence for protection by heat-shock proteins against photoinhibition during heat-shock. *EMBO J.* **7**, 1–6.

Settles, A.M., Yonetani, A., Baron, A., Bush, D.R., Cline, K., and Martienssen, R. (1997). Sec-independent protein translocation by the maize Hcf106 protein. *Science* **278**, 1467–1469.

Shao, N., Krieger-Liszkay, A., Schroda, M., and Beck, C.F. (2007). A reporter system for the individual detection of hydrogen peroxide and singlet oxygen: Its use for the assay of reactive oxygen species produced *in vivo*. *Plant J.* **50**, 475–487.

Shapiro, J., Ingram, J., and Johnson, K.A. (2005). Characterization of a molecular chaperone present in the eukaryotic flagellum. *Eukaryotic Cell* **4**, 1591–1594.

Shi, Y., Mosser, D.D., and Morimoto, R.I. (1998). Molecular chaperones as HSF1-specific transcriptional repressors. *Genes Dev.* **12**, 654–666.

Shikanai, T., Shimizu, K., Ueda, K., Nishimura, Y., Kuroiwa, T., and Hashimoto, T. (2001). The chloroplast *clpP* gene, encoding a proteolytic subunit of ATP-dependent protease, is indispensable for chloroplast development in tobacco. *Plant Cell Physiol.* **42**, 264–273.

Sichting, M., Mokranjac, D., Azem, A., Neupert, W., and Hell, K. (2005). Maintenance of structure and function of mitochondrial Hsp70 chaperones requires the chaperone Hep1. *EMBO J.* **24**, 1046–1056.

Sizova, I., Fuhrmann, M., and Hegemann, P. (2001). A *Streptomyces rimosus* aphVIII gene coding for a new type phosphotransferase provides stable antibiotic resistance to *Chlamydomonas*. *Gene* **277**, 221–229.

Smith, D.F., Sullivan, W.P., Marion, T.N., Zaitsu, K., Madden, B., McCormick, D.J., and Toft, D.O. (1993). Identification of a 60-kilodalton stress-related protein, p60, which interacts with hsp90 and hsp70. *Mol. Cell. Biol.* **13**, 869–876.

Sokolenko, A. (2005). SppA peptidases: Family diversity from heterotrophic bacteria to photoautotrophic eukaryotes. *Physiol. Plant.* **123**, 391–398.

Sokolenko, A., Pojidaeva, E., Zinchenko, V., Panichkin, V., Glaser, V.M., Herrmann, R.G., and Shestakov, S.V. (2002). The gene complement for proteolysis in the cyanobacterium *Synechocystis* sp. PCC 6803 and *Arabidopsis thaliana* chloroplasts. *Curr. Genet.* **41**, 291–310.

Sorger, P.K. and Nelson, H.C. (1989). Trimerization of a yeast transcriptional activator via a coiled-coil motif. *Cell* **59**, 807–813.

Sprey, B. (1968). Zur Feinstruktur des Plastidenstromas von *Hordeum vulgare* L. *Protoplasma* **66**, 469–479.

Stahl, A., Nilsson, S., Lundberg, P., Bhushan, S., Biverstahl, H., Moberg, P., Morisset, M., Vener, A., Maler, L., Langel, U., and Glaser, E. (2005). Two novel targeting peptide degrading proteases, PrePs, in mitochondria and chloroplasts, so similar and still different. *J. Mol. Biol.* **349**, 847–860.

Stanne, T.M., Pojidaeva, E., Andersson, F.I., and Clarke, A.K. (2007). Distinctive types of ATP-dependent Clp proteases in cyanobacteria. *J. Biol. Chem.* **282**, 14394–14402.

Stechmann, A. and Cavalier-Smith, T. (2004). Evolutionary origins of Hsp90 chaperones and a deep paralogy in their bacterial ancestors. *J. Euk. Microbiol.* **51**, 364–373.

Sun, Y. and MacRae, T.H. (2005). Small heat shock proteins: Molecular structure and chaperone function. *Cell. Mol. Life Sci.* **62**, 2460–2476.

Takahashi, Y., Utsumi, K., Yamamoto, Y., Hatano, A., and Satoh, K. (1996). Genetic engineering of the processing site of D1 precursor protein of photosystem II reaction center in *Chlamydomonas*. *Plant Cell Physiol.* **37**, 161–168.

Tanaka, Y., Nishiyama, Y., and Murata, N. (2000). Acclimation of the photosynthetic machinery to high temperature in *Chlamydomonas* requires synthesis de novo of proteins encoded by the nuclear and chloroplast genomes. *Plant Physiol.* **124**, 441–449.

Thompson, M.D., Paavola, C.D., Lenvik, T.R., and Gantt, J.S. (1995). *Chlamydomonas* transcripts encoding three divergent plastid chaperonins are heat-inducible. *Plant Mol. Biol.* **27**, 1031–1035.

Tomoyasu, T., Ogura, T., Tatsuta, T., and Bukau, B. (1998). Levels of DnaK and DnaJ provide tight control of heat shock gene expression and protein repair in *Escherichia coli*. *Mol. Microbiol.* **30**, 567–581.

Tripathi, L.P. and Sowdhamini, R. (2006). Cross genome comparisons of serine proteases in *Arabidopsis* and rice. *BMC Genomics* **7**, 200.

Tsugeki, R. and Nishimura, M. (1993). Interaction of homologues of Hsp70 and Cpn60 with ferredoxin-NADP+ reductase upon its import into chloroplasts. *FEBS Lett.* **320**, 198–202.

Tsvetkova, N.M., Horváth, I., Török, Z., Wolkers, W.F., Balogi, Z., Shigapova, N., Crowe, L.M., Tablin, F., Vierling, E., Crowe, J.H., and Vigh, L. (2002). Small heat-shock proteins regulate membrane lipid polymorphism. *Proc. Natl. Acad. Sci. U.S.A.* **99**, 13504–13509.

Ungewickell, E., Ungewickell, H., Holstein, S.E.H., Lindner, R., Prasad, K., Barouch, W., Martin, B., Greene, L.E., and Eisenberg, E. (1995). Role of auxilin in uncoating clathrin-coated vesicles. *Nature* **378**, 632–635.

Urban, S. (2006). Rhomboid proteins: Conserved membrane proteases with divergent biological functions. *Genes Dev.* **20**, 3054–3068.

Vallon, O. (2005). *Chlamydomonas* immunophilins and parvulins: Survey and critical assessment of gene models. *Eukaryotic Cell* **4**, 230–241.

Van der Laan, M., Urbanus, M.L., Ten Hagen-Jongman, C.M., Nouwen, B., Oudega, B., Harms, N., Driessen, A.J., and Luirink, J. (2003). A conserved function of YidC in the biogenesis of respiratory chain complexes. *Proc. Natl. Acad. Sci. U.S.A.* **100**, 5801–5806.

van Lis, R., Atteia, A., Mendoza-Hernandez, G., and González-Halphen, D. (2003). Identification of novel mitochondrial protein components of *Chlamydomonas*. A proteomic approach. *Plant Physiol.* **132**, 318–330.

Vasileuskaya, Z., Oster, U., and Beck, C.F. (2004). Involvement of tetrapyrroles in inter-organellar signaling in plants and algae. *Photosynth. Res.* **82**, 289–299.

Verdile, G., Gandy, S.E., and Martins, R.N. (2007). The role of presenilin and its interacting proteins in the biogenesis of Alzheimer's beta amyloid. *Neurochem. Res.* **32**, 609–623.

Viitanen, P.V., Schmidt, M., Buchner, J., Suzuki, T., Vierling, E., Dickson, R., Lorimer, G., Gatenby, A., and Soll, J. (1995). Functional characterization of the higher plant chloroplast chaperonins. *J. Biol. Chem.* **270**, 18158–18164.

Voellmy, R. (2004). On mechanisms that control heat shock transcription factor activity in metazoan cells. *Cell Stress Chaperones* **9**, 122–133.

von Gromoff, E.D., Treier, U., and Beck, C.F. (1989). Three light-inducible heat shock genes of *Chlamydomonas*. *Mol. Cell. Biol.* **9**, 3911–3918.

von Gromoff, E.D., Schroda, M., Oster, U., and Beck, C.F. (2006). Identification of a plastid response element that acts as an enhancer within the *Chlamydomonas* HSP70A promoter. *Nucleic Acids Res.* **34**, 4767–4779.

Vothknecht, U.C. and Soll, J. (2005). Chloroplast membrane transport: Interplay of prokaryotic and eukaryotic traits. *Gene* **354**, 99–109.

Vrancken, K., De Keersmaeker, S., Geukens, N., Lammertyn, E., Anne, J., and Van Mellaert, L. (2007). PspA overexpression in *Streptomyces lividans* improves both Sec- and Tat-dependent protein secretion. *Appl. Microbiol. Biotechnol.* **73**, 1150–1157.

Wang, S. and Liu, X.Q. (1997). Identification of an unusual intein in chloroplast ClpP protease of *Chlamydomonas eugametos*. *J. Biol. Chem.* **272**, 11869–11873.

Watanabe, S., Kobayashi, T., Saito, M., Sato, M., Nimura-Matsune, K., Chibazakura, T., Taketani, S., Nakamoto, H., and Yoshikawa, H. (2007). Studies on the role of HtpG in the tetrapyrrole biosynthesis pathway of the cyanobacterium *Synechococcus elongatus* PCC 7942. *Biochem. Biophys. Res. Commun.* **352**, 36–41.

Waters, E.R. and Rioflorido, I. (2007). Evolutionary analysis of the small heat shock proteins in five complete algal genomes. *J. Mol. Evol.* **65**, 162–174.

Waters, E. and Vierling, E. (1999). Chloroplast small heat shock proteins: Evidence for atypical evolution of an organelle-localized protein. *Proc. Natl. Acad. Sci. U.S.A.* **96**, 14394–14399.

Waters, E.R., Lee, G.J., and Vierling, E. (1996). Evolution, structure and function of the small heat shock proteins in plants. *J. Exp. Bot.* **296**, 325–338.

Weaver, L.M., Froehlich, J.E., and Amasino, R.M. (1999). Chloroplast-targeted ERD1 protein declines but its mRNA increases during senescence in *Arabidopsis*. *Plant Physiol.* **119**, 1209–1216.

Wegele, H., Müller, L., and Buchner, J. (2004). Hsp70 and Hsp90-a relay team for protein folding. *Rev. Physiol. Biochem. Pharmacol.* **151**, 1–44.

Weibezahn, J., Schlieker, C., Tessarz, P., Mogk, A., and Bukau, B. (2005). Novel insights into the mechanism of chaperone-assisted protein disaggregation. *Biol. Chem.* **386**, 739–744.

Westphal, S., Heins, L., Soll, J., and Vothknecht, U.C. (2001). Vipp1 deletion mutant of *Synechocystis*: A connection between bacterial phage shock and thylakoid biogenesis? *Proc. Natl. Acad. Sci. U.S.A.* **98**, 4243–4248.

Whatley, J.M., Hawes, C.R., Horne, J.C., and Kerr, J.D.A. (1982). The establishment of the plastid thylakoid system. *New Phytol.* **90**, 619–629.

Wiederrecht, G., Seto, D., and Parker, C.S. (1988). Isolation of the gene encoding the *S. cerevisiae* heat shock transcription factor. *Cell* **54**, 841–853.

Willmund, F. and Schroda, M. (2005). HSP90C is a bona-fide Hsp90 that interacts with plastidic HSP70B in *Chlamydomonas*. *Plant Physiol.* **138**, 2310–2322.

Willmund, F., Mühlhaus, T., Wojciechowska, M., and Schroda, M. (2007). The NH2-terminal domain of the chloroplast GrpE homolog CGE1 is required for dimerization and co-chaperone function *in vivo*. *J. Biol. Chem.* **282**, 11317–11328.

Willmund, F., Hinnenberger, M., Nick, S., Schulz-Raffelt, M., Mühlhaus, T., and Schroda, M. (2008). Assistance for a chaperone: HEP2 activates plastidic HSP70B for cochaperone binding in *Chlamydomonas*. *J. Biol. Chem.* **283**, 16363–16373

Wolfe, M.S. and Kopan, R. (2004). Intramembrane proteolysis: Theme and variations. *Science* **305**, 1119–1123.

Wu, B., Wawrzynow, A., Zylicz, M., and Georgopoulos, C. (1996). Structure-function analysis of the *Escherichia coli* GrpE heat shock protein. *EMBO J.* **15**, 4806–4816.

Xing, H., Wilkerson, D.C., Mayhew, C.N., Lubert, E.J., Skaggs, H.S., Goodson, M.L., Hong, Y., Park-Sarge, O.K., and Sarge, K.D. (2005). Mechanism of hsp70i gene bookmarking. *Science* **307**, 421–423.

Yang, C., Compton, M.M., and Yang, P. (2005). Dimeric novel HSP40 is incorporated into the radial spoke complex during the assembly process in flagella. *Mol. Biol. Cell* **16**, 637–648.

Yang, J.Y., Sun, Y., Sun, A.Q., Yi, S.Y., Qin, J., Li, M.H., and Liu, J. (2006). The involvement of chloroplast HSP100/ClpB in the acquired thermotolerance in tomato. *Plant Mol. Biol.* **62**, 385–395.

Yang, P., Diener, D.R., Rosenbaum, J.L., and Sale, W.S. (2001). Localization of calmodulin and dynein light chain LC8 in flagellar radial spokes. *J. Cell Biol.* **153**, 1315–1326.

Yang, P., Diener, D.R., Yang, C., Kohno, T., Pazour, G.J., Dienes, J.M., Agrin, N.S., King, S.M., Sale, W.S., Kamiya, R., Rosenbaum, J.L., and Witman, G.B. (2006). Radial spoke proteins of *Chlamydomonas* flagella. *J. Cell Sci.* **119**, 1165–1174.

Yokthongwattana, K., Chrost, B., Behrman, S., Casper-Lindley, C., and Melis, A. (2001). Photosystem II damage and repair cycle in the green alga *Dunaliella salina*: Involvement of a chloroplast-localized HSP70. *Plant Cell Physiol.* **42**, 1389–1397.

Young, J.C., Moarefi, I., and Hartl, F.U. (2001). Hsp90: A specialized but essential protein-folding tool. *J. Cell Biol.* **154**, 267–273.

Zerges, W. and Rochaix, J.-D. (1998). Low density membranes are associated with RNA-binding proteins and thylakoids in the chloroplast of *Chlamydomonas*. *J. Cell Biol.* **140**, 101–110.

Zhang, Z., Shrager, J., Jain, M., Chang, C.W., Vallon, O., and Grossman, A. (2004). Insights into the survival of *Chlamydomonas* during sulfur starvation based on microarray analysis of gene expression. *Eukaryotic Cell* **3**, 1331–1348.

Zhao, H., Lu, M., Singh, R., and Snell, W.J. (2001). Ectopic expression of a *Chlamydomonas* mt+-specific homeodomain protein in mt- gametes initiates zygote development without gamete fusion. *Genes Dev.* **15**, 2767–2777.

Zheng, B., MacDonald, T.M., Sutinen, S., Hurry, V., and Clarke, A.K. (2006). A nuclear-encoded ClpP subunit of the chloroplast ATP-dependent Clp protease is essential for early development in *Arabidopsis thaliana*. *Planta* **224**, 1103–1115.

Zou, J., Guo, Y., Guettouche, T., Smith, D.F., and Voellmy, R. (1998). Repression of heat shock transcription factor Hsf1 activation by Hsp90 (Hsp90 complex) that forms a stress-sensitive complex with HSF1. *Cell* **94**, 471–480.

CHAPTER 20

Biosynthesis of Chlorophylls and Hemes

Samuel I. Beale
Division of Biology and Medicine,
Brown University, Providence, Rhode Island, USA

CHAPTER CONTENTS

I. INTRODUCTION

Chlamydomonas, like all chloroplast-containing photosynthetic eukaryotes, synthesizes chlorophyll pigments. In higher plants and green algae, including *Chlamydomonas*, the predominant chlorophylls are chlorophyll *a* and chlorophyll *b*. These chlorophylls consist of two components, a tetrapyrrole pigment and the isoprenoid phytol, which are brought together and covalently linked at a late stage of chlorophyll biosynthesis. All of the carbon atoms of the tetrapyrroles are derived from glutamate in a multi-step biosynthetic pathway that occurs entirely within the chloroplast. Most, but not all, of the enzymes and other macromolecules that participate in chlorophyll biosynthesis are encoded by the nuclear genome. In addition to producing chlorophylls, the tetrapyrrole biosynthetic pathway has two branch points that lead to non-chlorophyll end products such as hemes. Chlorophyll biosynthesis is tightly regulated to be capable of providing the appropriate quantities of chlorophylls to the cells under varying demands for the pigments, while avoiding the accumulation of potentially harmful intermediates. The regulatory mechanisms are only poorly understood at present. This chapter is focused on tetrapyrrole metabolism as a general process that occurs in all chloroplast-containing organisms, but stresses results that have been obtained with *Chlamydomonas* as the experimental system, and results that may be unique to *Chlamydomonas*.

II. OUTLINE OF TETRAPYRROLE BIOSYNTHESIS

An outline of chlorophyll *a* biosynthesis is shown in Figure 20.1. Conceptually the pathway can be divided into several sections, each leading to a key intermediate or branch point: the initial steps (Figure 20.1, steps 1–6) that divert general metabolic intermediates into the formation of the first cyclic tetrapyrrole, uroporphyrinogen III; transformation of uroporphyrinogen III to protoporphyrin IX along the oxidative branch (steps 7–9); insertion of Mg^{2+} into protoporphyrin IX to begin the branch leading to chlorophylls (step 10); formation of the isocyclic "fifth" ring that is present in all chlorophylls (steps 11 and 12); reduction of a peripheral vinyl group to an ethyl group (step 14), followed or preceded by reduction of the macrocyclic ring system to form a chlorin, the defining oxidation state of true chlorophylls (step 13); and addition of the polyisoprene alcohol phytol to the tetrapyrrole to complete the structure of chlorophyll *a* (step 15). Chlorophyll *b* is derived from a chlorophyll *a* precursor, and minor chlorophylls present in photosynthetic reaction centers of plants and green algae are probably derived from chlorophyll *a*. An early branch point occurs at uroporphyrinogen III and leads to siroheme, the prosthetic group of assimilatory nitrite and sulfite reductases. Other products of this "reduced tetrapyrrole" pathway branch are heme d_1 and Factor F_{430} (which are found only in prokaryotes), as well as corrinoids such as cofactor B_{12} (which is synthesized only in prokaryotes, although vitamin B_{12} can be taken up by *Chlamydomonas* and used as a cofactor of methionine synthase; see Croft et al., 2005). A later branch point occurs at protoporphyrin IX and leads to hemes, which have many roles in the cells, including respiratory and photosynthetic electron transport. Catabolism of heme produces another class of end products, bilins. Although *Chlamydomonas* might be able to catabolize heme, the cells apparently do not accumulate or make use of any end product bilins, which in plants and most other algae are the prosthetic groups of phycobiliproteins and phytochromes.

III. ALA FORMATION FROM GLUTAMATE

A. Overview

5-Aminolevulinic acid (ALA), the first universal tetrapyrrole precursor, can be formed by two different pathways. Members of the α-proteobacterial group (which includes photosynthetic bacteria of the *Rhodobacter* and *Rhodopseudomonas*, and *Rhodospirillum* genera as well as the nonphotosynthetic genera *Agrobacterium*, *Rhizobium*, and *Bradyrhizobium*), and all eukaryotic organisms that do not contain chloroplasts (animals, yeasts, fungi), form ALA by condensation of succinyl-coenzyme A with glycine in a reaction catalyzed by the pyridoxal-P-containing enzyme ALA synthase

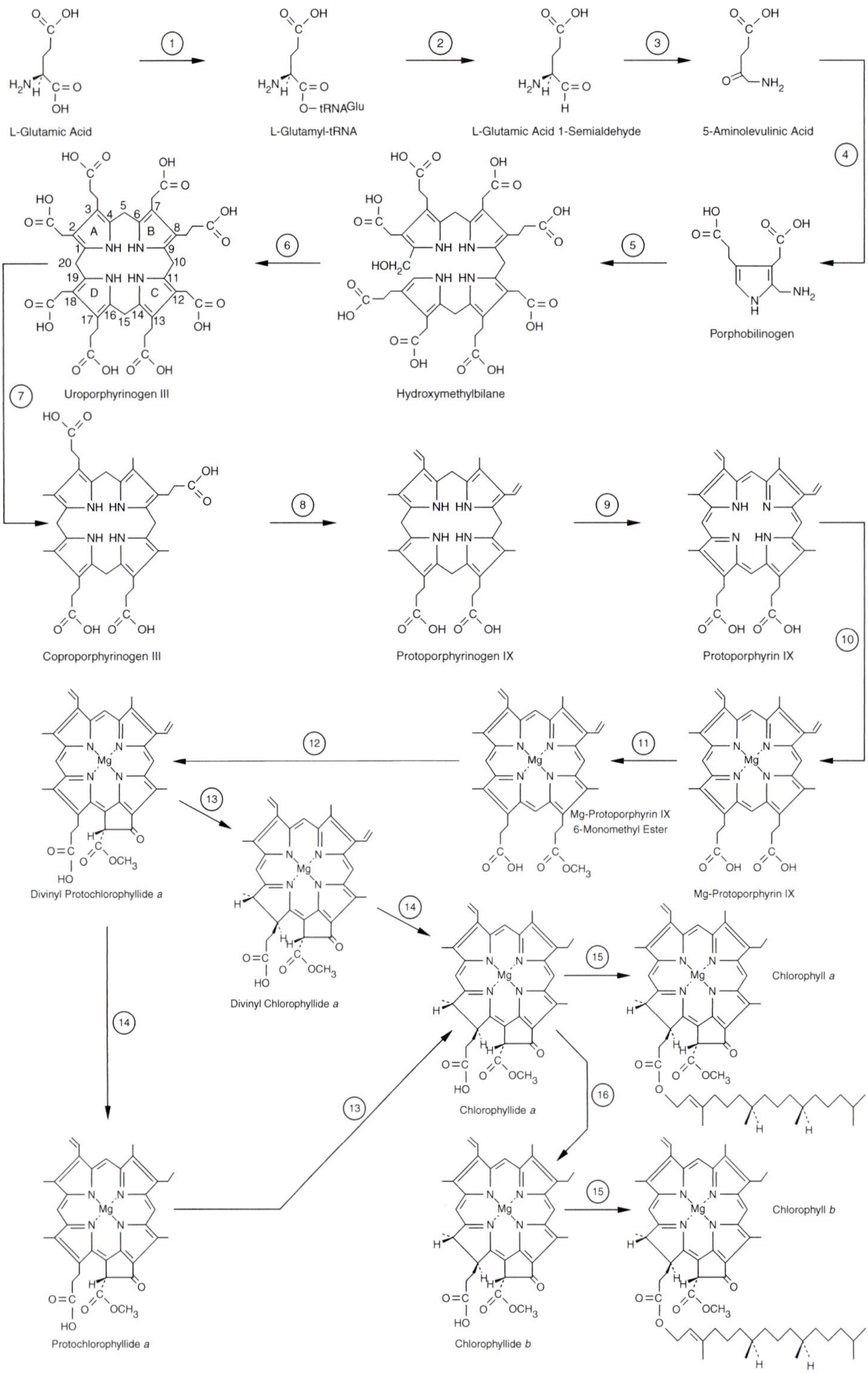

FIGURE 20.1 *Outline of chlorophyll biosynthesis from glutamate. The numbers assigned to enzymes that catalyze the individual reactions are named in Table 20.1. The customary letter designations of the pyrrole rings and the IUPAC number designations of tetrapyrrole carbon atoms are illustrated for uroporphyrinogen III.*

(EC 2.3.1.37) (Gibson et al., 1958; Kikuchi et al., 1958). In contrast, all plants and algae, and all bacteria that are not in the α-proteobacterial group, including cyanobacteria, many photosynthetic bacteria, and archaea, form ALA by a different route that begins with the five-carbon precursor, the amino acid Glu (Beale and Castelfranco, 1974; Beale et al., 1975; Meller et al., 1975). The three-step process begins with activation of Glu by ligation to $tRNA^{Glu}$ (Kannangara et al., 1984), followed by reduction of the α-carboxyl group of the activated Glu to form Glu 1-semialdehyde (GSA) (Pontoppidan and Kannangara, 1994), and then transamination of GSA to form ALA (Kannangara and Gough, 1978).

B. Activation of glutamate

Glu-tRNA synthetase (GTS; Figure 20.1, step 1) has been extensively studied in connection with its role in protein synthesis. Like other aminoacyl-tRNA synthetases, GTS requires the cognate amino acid and tRNA as substrates, and the reaction requires the energy of ATP hydrolysis. The single GTS in barley chloroplasts has a subunit molecular mass of 54 kD (Bruyant and Kannangara, 1987). Similar enzymes have been obtained from other chloroplast-containing organisms and also from cyanobacteria (Weinstein et al., 1987; Chang et al., 1990; Rieble and Beale, 1991). Two conflicting reports have appeared on the structure of the *Chlamydomonas* chloroplast GTS: one describes a unimolecular GTS with a molecular mass of 62 kD (Chen et al., 1990a), and the other describes a homodimeric GTS having native and subunit molecular masses of 60 kD and 32 kD, respectively (Chang et al., 1990). A homodimeric GTS would be consistent with other characterized plastid and prokaryotic GTSs, whereas a unimolecular GTS would not. The *Chlamydomonas* genome contains two genes for GTS (Table 20.1). Neither gene has been unequivocally identified as the one that specifically encodes chloroplast GTS. However, *GTS1* seems to be the more favorable candidate, because its product more closely resembles bifunctional GTSs of plant chloroplasts and some bacteria. These bifunctional enzymes are able to charge two different tRNAs with Glu, $tRNA^{Glu}$, and $tRNA^{Gln}$. The Glu that is bound to $tRNA^{Gln}$ is subsequently converted to Gln (Schön and Söll, 1988; Schön et al., 1988).

In all species examined, the tRNA required for ALA formation contains the UUC Glu anticodon. This was determined for *Synechocystis* sp. PCC 6803 by affinity purification using a ligand directed against the UUC Glu anticodon (Schneegurt and Beale, 1988). Only one $tRNA^{Glu}$ gene is present in the barley plastid genome, and this gene hybridizes with the RNA that supports ALA formation (Berry-Lowe, 1987). In the barley chloroplast system, the required RNA was purified, sequenced, and characterized as $tRNA^{Glu}$ bearing the UUC Glu anticodon (Schön et al., 1986). The *Chlamydomonas* plastid genome contains two genes (NCBI accession numbers X54406 and

Table 20.1 Enzymes involved in tetrapyrrole metabolism as it occurs in *Chlamydomonas*. References to reactions shown in Figure 20.1 are given in the first column for enzymes catalyzing steps of chlorophyll biosynthesis

Step (Figure 20.1)	Reaction	EC number	Abbreviated protein name(s)	Accession #
	Chlorophyll synthesis:			
1	Glutamyl-tRNA synthetase	6.1.1.17	GTS1 GTS2	EDO97532 EDP07219
2	Glutamyl-tRNA reductase	1.2.1.70	GTR (HEMA)	AAG41962, EDP03885
3	Glutamate 1-semialdehyde aminotransferase (Glutamate 1-semialdehyde aminomutase)	5.4.3.8	GSAT (GSA)	AAA18861, EDP00181
4	Porphobilinogen synthase (ALA dehydratase)	4.2.1.24	PBGS (ALAD)	AAA79515, EDP06754
5	Hydroxymethylbilane synthase (PBG deaminase)	4.3.1.8	HMBS (PBGD)	EDO97482
6	Uroporphyrinogen III synthase	4.2.1.75	UROS	EDP00789
7	Uroporphyrinogen III decarboxylase	4.1.1.37	UROD1 UROD2 UROD3	EDO98596 EDP06613 EDP06796
8	Coproporphyrinogen III oxidative decarboxylase	1.3.3.3	CPX1 CPX2	AAD28475, EDP06704 AAY42146, EDP06766
9	Protoporphyrinogen IX oxidase	1.3.3.4	PPX (PPO)	AAC79685, EDP02493
10	Mg chelatase	6.1.1.1	CHLD	EDP07156
			CHLH1	CAC69552, EDP07149
			CHLH2	EDP00299
			CHLI1	AAK69657, EDP04965
			CHLI2	EDP05319
	Mg chelatase activating protein		GUN4	EDP07013
11	Mg-protoporphyrin IX methyltransferase	2.1.1.11	PPMT	EDO97000
12	Mg-protoporphyrin IX monomethylester cyclase	1.14.13.81	CHL27A (CRD1)	AAF65221, EDP04035
			CHL27B (CTH1)	AAK32150, EDP05493

(*Continued*)

Table 20.1 *Continued*

Step (Figure 20.1)	Reaction	EC number	Abbreviated protein name(s)	Accession #
13	Light-dependent protochlorophyllide oxidoreductase	1.3.1.33	POR	AAB04951, EDP09202
13	Light-independent protochlorophyllide reductase	1.18.-.–	CHLB	AAB06265 (plastid genome)
			CHLL	CAA43020 (plastid genome)
			CHLN	AAN41268 (plastid genome)
14	Divinyl (proto)chlorophyllide vinyl reductase	1.3.1.75	DVR	EDP09906
15	Chlorophyll synthase	2.5.1.62	CHS (CHLG)	EDO97346
16	Chlorophyllide *a* oxidase		CAO	BAA33964, EDP09913
	Geranylgeranyl reductase		GGR	EDP09986
	Chlorophyll catabolism:			
	Chlorophyllase	3.1.1.14		EDP01364
	Pheophorbide demethoxycarbonylase		PDC	
	Pyropheophorbide oxygenase			
	Siroheme synthesis:			
	Precorrin-2 synthase	2.1.1.107	UPM	EDP02286
	Precorrin-2 dehydrogenase	1.3.1.76		
	Sirohydrochlorin ferrochelatase	4.99.1.4	SIRB	EDO99391
	Protoporphyrin IX-based heme synthesis:			
	Ferrochelatase	4.99.1.1	FeC	AAK16728, EDP03868
	Heme *c* ligase	4.4.1.17		
	Protoheme farnesyltransferase (Heme *o* synthase)	2.5.1.-		EDO96513
	Heme *a* synthase	1.-.– –		
	Heme catabolism:			
	Heme oxygenase	1.14.99.3	HMOX1	EDP06362
			HMOX2	EDO98578

X54407) that encode identical $tRNA^{Glu}$ molecules. Post-transcriptional processing of the tRNAs results in UUC anticodon triplets with an unspecified modification to the first U and an unmodified second U (O'Neill et al., 1990).

C. Reduction of the glutamate α-carboxyl group

Glu-tRNA reductase (GTR) converts the tRNA-ligated α-carboxyl group of Glu to the aldehyde, forming GSA, using electrons derived from NADPH (Figure 20.1, step 2). The GTR enzyme is able to recognize the tRNA portion of the Glu-tRNA substrate, and not all Glu-tRNAs are able to serve as substrates. GTRs differ in their spectrum of acceptable $tRNA^{Glu}$. For example, *E. coli* $tRNA^{Glu}$ functions *in vitro* with GTRs from *Chlamydomonas* and *Chlorobium vibrioforme*, but not with those from barley, *Chlorella vulgaris*, *Synechocystis* sp. PCC 6803, and *Euglena gracilis*. *E. gracilis* plastid $tRNA^{Glu}$, but not cytosolic $tRNA^{Glu}$, functions with *E. gracilis* GTR. The most striking evidence for tRNA recognition is the finding that mutation of a single base of *E. gracilis* plastid $tRNA^{Glu}$ renders it inactive as a GTR substrate *in vivo* even though it still functions in protein synthesis (Stange-Thomann et al., 1994). Several specific nucleotides of $tRNA^{Glu}$ have been identified that are required for activity with barley GTR (Willows et al., 1995). A detailed study of $tRNA^{Glu}$ bases required for recognition by *E. coli* GTR concluded that the entire anticodon stem-loop is not required (Randau et al., 2004). As noted above, the *Chlamydomonas* plastid genome contains two genes that encode identical $tRNA^{Glu}$ molecules.

GTR requires a divalent metal ion for activity. Mg^{2+}, Mn^{2+}, and Ca^{2+} supported *in vitro* activity of *C. vulgaris* GTR, but micromolar concentrations of Zn^{2+} strongly inhibited the reaction (Mayer et al., 1994). Barley GTR was also inhibited by Zn^{2+} (Pontoppidan and Kannangara, 1994). Purified recombinant barley GTR contains a tightly bound heme molecule that could be reduced by NADPH and oxidized by air (Vothknecht et al., 1996). The role of this heme has not been determined. However, heme is an important feedback regulator of GTR activity in many organisms (see section XIII.A.1). In a well-studied example, recombinant *C. vibrioforme* GTR expressed in *E. coli* contains one tightly but noncovalently bound heme molecule per enzyme subunit (Srivastava and Beale, 2005). The heme-complexed enzyme has GTR activity. However, expression of the protein in *E. coli* cells in which heme synthesis was inhibited produced an enzyme that was largely devoid of heme and had higher specific activity than the heme-complexed enzyme (Srivastava and Beale, 2005).

GTR from the hyperthermophilic archaeon *Methanopyrus kandleri* was crystallized and its structure was determined to 1.9-Å resolution (Moser et al., 2001). The homodimeric enzyme has an extended V-shape, with the C-termini of the subunits tightly joined. The structure suggests a reaction mechanism wherein the tRNA-activated Glu is transferred to an active-site Cys residue, and then the thioester-linked α-carbon of Glu is reduced by NADPH. GTR-encoding genes have been cloned from several plants (Ilag et al., 1994; Bougri and Grimm, 1996; Kumar et al., 1996; Vothknecht et al., 1996; Tanaka et al., 1996, 1997) and *Chlamydomonas* (Srivastava et al., 2005). There are at least two differentially expressed genes for GTR in all plants examined to date, but only one was detected in *Chlamydomonas*. The plant and *Chlamydomonas* genes all encode proteins that have N-terminal chloroplast transit peptides.

All predicted GTRs contain a conserved Cys residue that for the *E. coli* enzyme, forms a thioester with the Glu moiety that is transferred from the Glu-tRNA substrate (Schauer et al., 2002). In the solved structure of the *M. kandleri* GTR, the conserved Cys is too far away from the proposed NADPH binding site to allow direct reduction of the Glu moiety (Moser et al., 2001). It was proposed that during catalysis, the protein undergoes a conformational change which brings the NADPH binding site closer to the Glu-Cys (see section III.D).

Reported native molecular masses of GTR from plant, algal, and bacterial sources vary over a wide range and are not consistent with peptide molecular masses predicted from gene sequences, which are in the range of 45–60 kD (Drolet et al., 1989; Elliott, 1989; Li et al., 1989; Verkamp and Chelm, 1989; Petricek et al., 1990; Majumdar et al., 1991; Ilag et al., 1994). Barley GTR was purified to apparent homogeneity, and has a native molecular mass of 270 kD and is composed of identical 54-kD subunits (Pontoppidan and Kannangara, 1994). In contrast, the *Chlamydomonas* GTR was initially reported to be a 130-kD monomer (Chen et al., 1990b), but subsequent characterization revealed that the native enzyme is a homodimer of 52.5-kD subunits (Srivastava et al., 2005).

D. The transamination step

GSA aminotransferase (GSAT) catalyzes the rearrangement of the atoms of GSA to form the isomer, ALA (Figure 20.1, step 3). GSAT contains a pyridoxal-P or pyridoxamine-P cofactor. In the reaction, GSA is the sole substrate and ALA the sole product. It is possible to hypothesize a reaction in which the amino and oxo moieties of a single GSA molecule are interchanged during the course of the reaction to form ALA. However, labeling experiments with ^{13}C- and ^{15}N-labeled Glu have revealed that this does not happen. Instead, the amino group and carbon skeleton of each ALA molecule are derived from different GSA molecules (Mau and Wang, 1988; Mayer et al., 1993). This result can be explained by a mechanism in which the enzyme catalyzes two successive transamination reactions. In the first reaction, an amino group is transferred from a pyridoxamine-P cofactor of the enzyme to C-1 of GSA to produce a diamino intermediate,

4,5-diaminovaleric acid, and pyridoxal-P. Then, the amino group that was initially present on the C-2 of GSA is transferred to the pyridoxal-P, to form ALA and regenerate pyridoxamine-P. A detailed kinetic analysis based on time-resolved spectrophotometric measurements supports this model (Smith et al., 1998). Friedmann et al. (1992) synthesized both (*S*)- and (*R*)-4,5-diaminovaleric acid and showed that the (*S*)-isomer is the preferred substrate for *Synechococcus* sp. PCC 6301 GSAT. The (*S*)-isomer corresponds to the product derived from L-Glu. Recombinant *Synechococcus* sp. PCC 6301 GSAT showed a nearly complete preference for L-GSA as a substrate, although the D-isomer was able to bind at the GSAT active site and elicit spectral changes (Smith et al., 1992).

The crystal structure of *Synechococcus* sp. PCC 6301 GSAT reveals that the enzyme has a homodimeric structure and the two active sites are situated at the interfaces of the subunits (Hennig et al., 1994, 1997). There is an interesting asymmetry and the active sites are slightly different. It has been proposed that the active sites interact conformationally, and that when one is in an open conformation that can accept substrate or release product, the other is in a closed conformation that firmly binds the diamino intermediate during the double transamination reaction and prevents its release. GSAT is a member of the aspartate aminotransferase enzyme family (Elliott et al., 1990; Mehta and Christen, 1994).

All of the GSATs that have been examined are inhibited by very low concentrations of the mechanism-based suicide inhibitor gabaculine (3-amino-2,3-dihydrobenzoic acid), both *in vitro* and *in vivo*, as indicated by inhibition of chlorophyll formation in plants to which gabaculine is administered. Gabaculine forms a Schiff base with pyridoxal-P that then irreversibly isomerizes to a secondary amine. Gabaculine-inactivated GSAT from *C. vulgaris* can be reactivated by removal of the inactivated cofactor and addition of pyridoxal-P (Avissar and Beale, 1989a). Gabaculine-resistant mutants of *Chlamydomonas* contain elevated levels of GSAT activity that is still sensitive to gabaculine *in vitro* (Kahn and Kannangara, 1987). In contrast, gabaculine-resistant mutants of *Synechococcus* sp. PCC 6301 contain a variant GSAT that is relatively insensitive to gabaculine (Bull et al., 1989).

E. Protein interactions and substrate channeling

Although each of the three reactions in the conversion of Glu to ALA can occur independently, there is evidence that the enzymes channel the intermediates rather than release them into the medium. *Chlamydomonas* GTR forms a complex with GTS in the presence of Glu-tRNA (Chen et al., 1990b), and the complex migrates as a single entity during glycerol gradient centrifugation (Jahn, 1992). A complex between GTS and GTR might facilitate the channeling of Glu-tRNA toward ALA biosynthesis and regulate competition for Glu-tRNA with the protein synthesizing apparatus.

The structure of the GSA has not been unequivocally determined. Although most workers have assumed that the product is free GSA or its hydration product (hemiacetal), Jordan et al. (1993) proposed that the product is the cyclic ester formed from the γ-carboxyl group and the hydrated aldehyde group. The cyclic structure does not contain free aldehyde or carboxylic acid functions, and is more compatible with some previously reported properties of the chemically synthesized product at low pH (stability in aqueous solution, heat stability) than the free or hydrated α-aminoaldehyde. It seems probable that in solution GSA occurs as an equilibrium mixture of the free aldehyde, hydrated form, and cyclic compound, analogously to aldose sugars. However, free GSA might never occur normally *in vivo*. Using the solved structures for *M. kandleri* GTR and *Synechococcus* sp. PCC 6301 GSAT, a complex between the two homodimeric proteins was modeled (Moser et al., 2001). In the model, their active sites align such that GSA could be transferred from GTR to GSAT. Formation of a complex between *E. coli* GTR and GSAT has been described (Lüer et al., 2005).

Physical and kinetic evidence obtained with purified recombinant *Chlamydomonas* GTR and GSAT indicates that the two enzymes form a stable dimeric complex and that in the complex, GSA is transferred between the two enzymes without being released into the medium (Nogaj and Beale, 2005). In addition to being able to channel the intermediate between GTR and GSAT, the two-enzyme complex of *Chlamydomonas* has several fold more GTR activity than GTR alone (Nogaj and Beale, 2005). Even inactive mutant GSAT stimulated GTR activity (as measured independently of GSAT activity by Glu-tRNA-dependent NADPH oxidation) when added to GTR assays. It was concluded that GTR and GSAT should be considered as a single catalytic entity and that GSA is a normally protein-bound intermediate.

There is evidence that GTR interacts with at least two other proteins in addition to GTS and GSAT. One of these is the thylakoid membrane-bound inhibitory protein FLP (see section XIII.A.1). Another protein or proteins that appears to interact with GTR (or with GTS or GSAT) is an unidentified component in gel-filtered *Chlamydomonas* cell extract supernatant, that must be added to the incubation mixture in order for heme to inhibit ALA synthesis in incubations containing recombinant *Chlamydomonas* GTR (Srivastava et al., 2005; see section XIII.A.1).

In summary, there is evidence suggesting that *Chlamydomonas* GTR interacts *in vivo* with at least four other proteins and that these interactions are important for protecting the product GSA, enhancing the enzymatic efficiency of GTR, regulating GTR activity, and possibly directing precursor Glu-tRNA to GTR.

IV. STEPS COMMON TO THE BIOSYNTHESIS OF ALL TETRAPYRROLES

A. Overview

Three enzymes working sequentially convert eight molecules of ALA to one molecule of the first macrocyclic tetrapyrrole, uroporphyrinogen III. The enzymatic steps can be categorized as a dehydrating dimerization to form porphobilinogen (PBG), then a deaminating oligomerization to form the open-chain tetrapyrrole hydroxymethylbilane (HMB, also known as preuroporphyrinogen), and finally a complex isomerization and dehydration reaction to form the uroporphyrinogen III macrocycle.

B. Porphobilinogen formation

PBG synthase (PBGS; also known as ALA dehydratase) catalyzes the asymmetric condensation of two ALA molecules to form PBG, with the release of two molecules of H_2O (Figure 20.1, step 4). In single turnover experiments, Jordan and Seehra (1980) showed that in the reaction catalyzed by *Rhodobacter sphaeroides* PBGS, the first ALA molecule to be bound is the one that contributes the propionic acid side chain of the product. During the formation of PBG, removal of hydrogen to form the aromatic pyrrole ring must occur on the enzyme, as is indicated by the stereospecific retention of the hydrogen atom that is derived from the *pro-S* C_5 hydrogen of ALA (Abboud and Akhtar, 1976).

Earlier work had indicated that native PBGS is an oligomer of 6 to 8 identical subunits. The X-ray crystallographic structure of *E. coli* PBGS revealed it to be a homooctamer structurally related to aldolases (Erskine et al., 1997). The native PBGS contains eight active sites, all situated at the surface and at the interfaces of two subunits. Each substrate molecule is bound to the enzyme via a Schiff base link to one of two Lys residues at each active site. The enzyme contains no organic cofactor, and the reaction requires no external energy source.

Despite an overall high sequence similarity of all PBGSs, there are significant species-specific differences with regard to pH optimum (which ranges from 6.5 to 8.5) and the requirement for or enhancement of activity by divalent metals (Jaffe, 1993). It is hypothesized that all PBGSs contain tightly bound Zn^{2+} atoms that are not released under normal conditions of protein purification and exposure to chelators such as EDTA (Jaffe, 1993, 1995). In addition, PBGSs from all mammals and birds examined, yeast, and some bacteria, including cyanobacteria (Jones et al., 1994), have an additional requirement for Zn^{2+} in the micromolar concentration range, which is presumably bound at a second site. In contrast, PBGSs from all plants and algae, and some bacteria, have no demonstrable requirement for Zn^{2+} in the incubation medium but instead require micromolar concentrations of

Mg^{2+} for activity. Both classes of PBGS that require divalent metal ions in the incubation medium are inhibited by EDTA, which presumably removes the comparatively loosely bound metal atoms at the second site. Among the PBGSs that require Zn^{2+} in the incubation medium, some (e.g. that from *E. coli*), but not all, are stimulated approximately twofold by Mg^{2+}, which is proposed to bind at a third site on the protein. In the case of human PBGS, binding of this Mg^{2+} and activation of the enzyme are accompanied by a quaternary change in the enzyme structure from hexameric to octameric (Breinig et al., 2003). Finally, a fourth class of PBGS, represented by the enzyme from *Rhodobacter capsulatus*, neither requires nor is stimulated by micromolar concentrations of either Zn^{2+} or Mg^{2+} (Nandi and Shemin, 1973). *Chlamydomonas* contains a single gene for PBGS (Matters and Beale, 1995). Although the properties of this enzyme have not been examined directly, analysis of the predicted amino acid sequence indicates that, like PBGSs from plants, it is an Mg^{2+}-requiring enzyme (Matters and Beale, 1995).

C. Linear tetrapyrrole formation

HMB synthase (HMBS, also known as PBG deaminase and preuroporphyrinogen synthase) condenses four PBG molecules to form the first tetrapyrrole, HMB (Figure 20.1, step 5). Although this enzyme is active in isolation, the released product, free HMB, rapidly cyclizes spontaneously and irreversibly to form the nonphysiological product uroporphyrinogen I (Battersby et al., 1979a). Biosynthesis of the biologically relevant isomer, uroporphyrinogen III, requires the presence of a second enzyme, uroporphyrinogen III synthase (see below) during or immediately after release of the initial tetrapyrrole product of HMBS.

HMBS was purified 200-fold from *E. gracilis* (Williams et al., 1981). The native enzyme, a soluble 41-kD monomer, does not contain a dissociable prosthetic group or require metal ions for activity. HMBS from wheat germ and spinach leaves is about 40 kD (Higuchi and Bogorad, 1975), whereas the *Chlorella regularis* enzyme is 35 kD (Shioi et al., 1980).

HMBS from *Rhodobacter sphaeroides* and *E. gracilis* was used to establish that the order of assembly of the four porphobilinogen units is ABCD, as they appear in uroporphyrinogen (Figure 20.1; Jordan and Seehra, 1979; Battersby et al., 1979b). Nascent monopyrrole- through tetrapyrrole-enzyme complexes exist (Scott et al., 1980; Jordan and Berry, 1981; Battersby et al., 1983a) that are linked via a dipyrromethane cofactor (Hart et al., 1987; Jordan and Warren, 1987; Warren and Jordan, 1988). In HMBS from *E. coli*, the dipyrromethane cofactor is post-translationally attached to a Cys residue (C_{242} of the *E. coli* enzyme) of the apoprotein, and it remains permanently attached to the enzyme. After the hexapyrrole stage is reached, binding of the next PBG by the enzyme triggers the release of HMB and prepares the enzyme to accept new substrate molecules to be added to the free end of the dipyrromethane (Battersby et al., 1983b). The hydroxymethyl group of the released HMB is formed from the aminomethyl group of PBG with overall retention of the configuration about the carbon atom (Schauder et al., 1987). The dipyrromethane cofactor was reported to be derived from HMB (Shoolingin-Jordan et al., 1996). Although this seems to present a "chicken and egg" dilemma, since cofactor-containing HMBS is needed to form HMB, the apoenzyme can also construct the cofactor from PBG, although at a much lower rate.

The *E. coli* HMBS crystal structure revealed a large active site cleft that can contain the growing polypyrrole chain, while the dipyrromethane cofactor remains covalently attached to the enzyme (Louie et al., 1992, 1996). Plant and algal genes for HMBS encode products that are targeted to the chloroplast. *Chlamydomonas* contains a single candidate gene for HMBS (Table 20.1); however, there is no experimental information available on the gene or enzyme.

D. Uroporphyrinogen III formation

Uroporphyrinogen III synthase (UROS, also known by the earlier name uroporphyrinogen III cosynthase) catalyzes the cyclization of HMB to form the first macrocyclic tetrapyrrole, uroporphyrinogen III (Figure 20.1, step 6). As discussed earlier, free HMB rapidly cyclizes spontaneously and irreversibly to form the nonphysiological product uroporphyrinogen I unless UROS is present to direct its conversion to the correct isomer.

The reaction catalyzed by UROS is fascinating because the ends of the linear substrate molecule, HMB, are not directly ligated together; instead, the D pyrrole ring first undergoes a rearrangement that effectively "flips" it within the plane as it is drawn in Figure 20.1. The rearrangement is not simply an exchange of the acetate and propionate residues at the D-pyrrole β-positions, but also involves the interchange of the carbon atoms at the α-positions. The mechanism of ring inversion has been the subject of intensive investigation and is now generally believed to involve a *spiro* intermediate (Crockett et al., 1991; Stark et al., 1993; Petersen et al., 1998).

The crystal structure of human UROS indicates that the active site is situated in a flexible region between two α/β domains that could facilitate the large movements of the substrate during the reaction (Mathews et al., 2001). Site-specific mutation of acidic and basic amino acids near the active site did not inactivate the enzyme, suggesting that the reaction does not involve acid/base catalysis. Contrary to earlier reports (Kohashi et al., 1984), UROS contains no reversibly bound cofactor or metal ions.

HMBS and UROS appear to form a complex that facilitates transfer of unstable HMB between the two enzymes. The sedimentation velocity of wheat germ HMBS was influenced by the presence of wheat germ UROS

(Higuchi and Bogorad, 1975). Also, the presence of *E. gracilis* UROS influenced the affinity of *E. gracilis* HMBS for PBG (Battersby et al., 1979b). The presence of *R. sphaeroides* UROS was reported to facilitate release of HMB from HMBS (Rosé et al., 1988). *Chlamydomonas* contains a single candidate gene for UROS (Table 20.1), but there are no reports characterizing *Chlamydomonas* UROS either from the expressed enzyme or in cell extracts, nor of interactions with HMBS.

V. STEPS FROM UROPORPHYRINOGEN III TO PROTOPORPHYRIN IX

A. Overview

The three enzymes that sequentially convert uroporphyrinogen III to protoporphyrin IX comprise the "oxidative" branch of the pathway and are responsible for transforming the compound from a hydrophilic, poorly metal binding, photochemically unreactive species into one that is hydrophobic, binds metals tightly, and is photochemically reactive, all of which traits are needed to perform the light-harvesting and photochemical functions of chlorophyll. The use of porphyrinogens rather than porphyrins as biosynthetic intermediates in this part of the pathway is advantageous to the organism because the porphyrinogens are much less photoreactive than porphyrins. Aberrant conditions that lead to the accumulation of porphyrins cause phototoxicity in plants (see below) and also in animals, where the result is severe porphyric skin damage. Only at the last step, where the superior metal chelating properties of the porphyrin are needed, is the porphyrinogen converted to a porphyrin.

B. Uroporphyrinogen III decarboxylation

Uroporphyrinogen III decarboxylase (UROD) catalyzes the stepwise decarboxylation of the four acetate residues of uroporphyrinogen III to form coproporphyrinogen III (Figure 20.1, step 7). At physiological substrate concentrations, the decarboxylations occur sequentially, in the order of the residues on rings D, A, B, and C (Luo and Lim, 1993). At higher substrate concentrations, the decarboxylations occur in random order. Each decarboxylation occurs with retention of the configuration about the α carbon atom (Barnard and Akhtar, 1975).

UROD activity was measured in soluble leaf extracts of several plants and purified 72-fold from tobacco leaves (Chen and Miller, 1974). No metal requirements were observed, and EDTA or other metal chelating agents enhanced activity. The 43-kD tobacco UROD is 33% similar to other URODs, and has an N-terminal chloroplast transit sequence

(Mock et al., 1995). The protein was imported into chloroplasts and processed to a 39-kD product that was located in the stroma. The crystal structures of tobacco (Martins et al., 2001) and human (Phillips et al., 2003) URODs indicate that the substrate and partially decarboxylated intermediates lie in a hydrophobic cleft and are H-bonded to the enzyme by an Asp carboxyl oxygen linked to pyrrole nitrogens. This bonding arrangement would provide approximately equal affinity for the substrate and the partially decarboxylated intermediates. The active site(s) for decarboxylation has not been identified.

Chlamydomonas contains three candidate genes for UROD (Table 20.1), and sequences corresponding to all of them are present in the EST database. It is not apparent why there would be a need for three URODs in *Chlamydomonas*. However, it should be noted that there have been no reports characterizing UROD activity from expression of any of these genes, nor on the subcellular targeting of their products, and it is possible that one or two of the gene products catalyze other reactions.

C. Oxidative decarboxylation of coproporphyrinogen III

Coproporphyrinogen III oxidative decarboxylase (CPX) catalyzes O_2-dependent decarboxylation of two of the four propionate residues on coproporphyrinogen III, those on rings A and B, to yield the divinyl compound protoporphyrinogen IX (Figure 20.1, step 8). In the reaction, O_2 is reduced to H_2O_2 (Breckau et al., 2003). Oxidative decarboxylation of the two propionate groups requires the removal of four electrons, and therefore the reaction requires the reduction of two O_2 to two H_2O_2, but the actual stoichiometry has not been reported. *E. gracilis* CPX accepts ring A monovinyl (3-vinyl) porphyrinogen more readily than ring B monovinyl (8-vinyl) porphyrinogen as a substrate, suggesting that oxidative decarboxylation of the ring A propionate of coproporphyrinogen occurs before that of ring B (Cavaleiro et al., 1974).

Seehra et al. (1983) proposed a CPX reaction mechanism involving pyrrolic N-assisted removal of single protons as hydride ions from the β-carbons of the propionate groups. The α-protons of the propionate groups do not appear to be involved: in the reaction catalyzed by an avian blood extract, both of the α-protons of both propionate groups were retained on the terminal carbon atoms of the protoporphyrinogen vinyl groups (Zaman and Akhtar, 1976). Although O_2 is the obligatory electron acceptor in aerobic CPX, extracts of anaerobic *R. sphaeroides* and yeast can carry out the reaction anaerobically in the presence of ATP, oxidized pyridine nucleotide and methionine (Tait, 1972; Poulson and Polglase, 1974). In facultative anaerobes such as *E. coli* and *Salmonella typhimurium*, aerobic and anaerobic protoporphyrinogen IX synthesis is catalyzed by different enzymes, products of the *hemF* and *hemN* genes, respectively (Xu and Elliott, 1993, 1994).

Genes for CPX have been cloned and sequenced from plants (Madsen et al., 1993; Kruse et al., 1995) and *Chlamydomonas* (Hill and Merchant, 1995; Quinn et al., 1999). The encoded 43–45-kD peptides contain chloroplast transit sequences, and the predicted proteins more closely resemble CPX than the anaerobic bacterial protoporphyrinogen IX-producing enzymes. Tobacco CPX was imported into pea chloroplasts and processed to a 39 kD final size. The mature chloroplast CPX is located exclusively in the stroma (Kruse et al., 1995). The yeast CPX crystal structure (Phillips et al., 2004) reveals that after binding the substrate in a deep catalytic cleft, the enzyme assumes a completely closed configuration. The active enzyme is probably dimeric (Phillips et al., 2004), and the *E. coli* CPX is also dimeric (Breckau et al., 2003).

Chlamydomonas contains two genes for CPX (Table 20.1). One of these, *CPX1*, has been expressed and its product was shown to be localized to the chloroplast (Hill and Merchant, 1995; Quinn et al., 1999). *CPX1* expression increases under growth at subnormal concentrations of Cu^{2+} or O_2 (see section XIII.C). The role of *CPX2* is unknown, and neither ESTs nor expression analysis have been reported. When the expressed *E. coli* CPX-encoding *hemF* gene product was exposed to O_2 and then purified anaerobically, subsequent anaerobic incubation with coproporphyrinogen III resulted in a single turnover, that is, one molecule of protoporphyrinogen IX was formed (Breckau et al., 2003). This result shows that after exposure to O_2, the anaerobically purified enzyme contains either bound O_2 or an oxidized cofactor, such as a metal. Earlier, mammalian CPX was reported to contain one essential Cu^{2+} atom per polypeptide (Kohno et al., 1996), but this finding was challenged (Medlock and Dailey, 1996). It seems unlikely that the *Chlamydomonas* CPX1 requires Cu^{2+}, because both the abundance and the activity of this enzyme are elevated in Cu-deficient cells (see section XIII.C). *E. coli* HemF was reported to require Mn for activity (Breckau et al., 2003). However, no metal was detected in the yeast CPX crystal structure (Phillips et al., 2004). Because of these conflicting results, the universality of a metal requirement for CPX activity is currently unresolved.

D. Protoporphyrinogen IX oxidation

Protoporphyrinogen IX oxidase (PPX) catalyzes the removal of six electrons from protoporphyrinogen IX to form protoporphyrin IX (Figure 20.1, step 9). Molecular oxygen is the obligatory electron acceptor in the reaction catalyzed by plant PPX. However, the respiratory chain can function as the electron acceptor for protoporphyrin IX formation in anaerobically grown *E. coli*, and presumably also in anaerobic photosynthetic bacteria. The anaerobic reaction appears to be catalyzed by a different enzyme than the aerobic O_2-dependent PPX (Klemm and Barton, 1987). PPX from all sources contains noncovalently bound flavin cofactor (Siepker et al., 1987; Camadro et al.,

1994). In the six-electron oxidation of protoporphyrinogen IX, three molecules of O_2 are consumed and H_2O_2 is produced (Ferreira and Dailey, 1988). The hydrogen atom that is removed from the 10-meso position of protoporphyrinogen IX is removed from the opposite face of the molecule than the ones that are removed from the 5-, 15-, and 20-meso positions (Jones et al., 1984). It was hypothesized that three of the meso hydrogens are lost by an oxidation process occurring on one face of the molecule, and that the fourth proton is lost from the other face by a tautomerization reaction.

Plant PPX is of agricultural interest because it is the target of some commercial herbicides, including diphenylethers, that inhibit the enzyme and cause a massive buildup of cellular protoporphyrinogen IX that diffuses to the cell membrane where it is oxidized, probably by a nonspecific peroxidase (Yamato et al., 1995), to protoporphyrin IX, which is a photodynamic sensitizer that causes light-induced damage to the membrane (Becerril and Duke, 1989; Matringe et al., 1989). The observation that an inhibitor of PPX raises the concentration of protoporphyrin IX, the product of the enzyme reaction, appears paradoxical. However, is explained by the fact that when PPX is inhibited, protoporphyrinogen IX diffuses away from the normal site of its formation, and in other cellular regions it is nonspecifically oxidized to protoporphyrin IX but this product is inaccessible to the normal biosynthetic enzymes. The unavailability of protoporphyrin IX at the sites of heme and chlorophyll biosynthesis creates a deficiency of these products, which can cause a compensatory release of feedback control over ALA synthesis and thereby lead to a massive overproduction of protoporphyrinogen IX. A gene for an herbicide-resistant mutant PPX has been isolated from *Chlamydomonas* (Randolph-Anderson et al., 1998). The PPX crystal structures from tobacco (Koch et al., 2004) and the bacterium *Myxococcus xanthus* (Corradi et al., 2006) are largely conserved, although their membrane anchoring sites might be different. Because the substrate cannot move within the confining active site, it was proposed that all three of the oxidation steps occur at a single site on the substrate (Koch et al., 2004). PPX is discussed further in the context of heme biosynthesis (see section XI).

VI. STEPS SPECIFIC TO CHLOROPHYLL FORMATION

A. Mg^{2+} chelation

The insertion of Mg^{2+} into the protoporphyrin IX macrocycle, the first committed step of chlorophyll synthesis, is catalyzed by Mg chelatase (Figure 20.1, step 10). Mg chelation is probably the most complex step of chlorophyll synthesis, and it requires the products of at least three genes, as well as ATP to drive the reaction. Chemically, insertion of Mg^{2+} into the porphyrin nucleus is very difficult, in part because of the difficulty of removing H_2O molecules coordinated to the Mg^{2+} (Fleischer et al., 1964; Hambright,

1975). One role of ATP might be to provide the energy needed to remove this H_2O, possibly by consuming the H_2O for the hydrolysis of ATP.

ATP-dependent Mg chelatase activity was first detected in intact chloroplasts (Castelfranco et al., 1979). In addition to protoporphyrin IX, deuteroporphyrin IX (vinyl groups replaced by protons) and mesoporphyrin IX (vinyl groups replaced by ethyl), and porphyrins in which either one of the two vinyl groups is replaced by ethyl (Walker and Weinstein, 1991a) were effective substrates. The enzyme uses only Mg^{2+} as the metal substrate (Fuesler et al., 1981). In chloroplast lysates, Mg chelatase is associated loosely with membranes. In the absence of Mg^{2+}, Mg chelatase readily dissociates from membranes and it remains soluble when Mg^{2+} is added for the assay (Walker and Weinstein, 1995; Gibson et al., 1996; Kannangara et al., 1997; Guo et al., 1998). The enzyme was localized to the envelope membrane at Mg^{2+} concentrations at or above 5 mM (Nakayama et al., 1998). The Mg^{2+}-dependent membrane association of Mg chelatase might have a physiological significance, and can explain an earlier observation that activity extracted from cucumber cotyledon chloroplasts at high Mg^{2+} concentration depended only on the membrane fraction (Lee et al., 1992).

Mg chelation was determined to require at least three separable protein components in chloroplast lysates (Walker and Weinstein, 1991b). That these corresponded to three gene products was first elucidated in studies of the "photosynthesis gene cluster" of the photosynthetic bacteria *R. capsulatus* and *R. sphaeroides*, which contains many genes for structural components and biosynthetic enzymes of the photosynthetic apparatus (Alberti et al., 1995). Insertional mutagenesis revealed that disruption of *bchD*, *bchH*, or *bchI* caused the cells to accumulate protoporphyrin IX and rendered them unable to synthesize Mg-containing bacteriochlorophyll precursors (Bollivar et al., 1994a,b). The combined recombinant *R. sphaeroides* *bchD*, *bchH*, and *bchI* gene products were subsequently reconstituted as Mg chelatase (Gibson et al., 1995; Willows et al., 1996). Essentially similar results were obtained with the homologous gene products from *Synechocystis* sp. PCC 6803 (Jensen et al., 1996). Because the end product in cyanobacteria is chlorophyll rather than bacteriochlorophyll, the genes were designated *chlD*, *chlH*, and *chlI*.

The three protein components of Mg chelatase have predicted molecular masses of 60–87 kD, 120–155 kD, and 36–46 kD for CHLD, CHLH, and CHLI, respectively. The degree of sequence identity among homologous proteins from different sources is 28–58%, 38–86%, and 50–90% for CHLD, CHLH, and CHLI, respectively. The N-terminal portion of CHLD has significant sequence similarity to CHLI (see below). The Mg chelatase proteins have significant similarity to predicted proteins involved in the biosynthesis of the Co-containing corrinoids and Ni-containing Factor F_{430}, metalloporphyrins which are derived from the branch of the tetrapyrrole pathway that diverges at uroporphyrinogen III (see section X): CHLH is similar to a protein of aerobic Co chelatase (Schubert et al., 1999), and all three Mg chelatase proteins are similar to proteins of Ni chelatase (Walker and Willows, 1997).

CHLH binds protoporphyrin IX in a 1:1 molar ratio (Willows et al., 1996), and expression of recombinant CHLH in *E. coli* causes the cells to become reddish colored due to the accumulation of the porphyrin-protein complex (Gibson et al., 1995; Willows et al., 1996). From these results, it can be concluded that CHLH binds the porphyrin substrate and therefore probably contains the catalytic site (or at least part of it).

A fluorescence-based continuous assay for Mg chelatase detected a lag phase of several minutes before activity commenced (Walker et al., 1992). The lag was eliminated by pre-incubating the enzyme with ATP. ATP-γ-*S* could be substituted for ATP in the pre-incubation to eliminate the lag, but not for the actual catalysis of Mg chelation, indicating that the lag elimination does not require ATP hydrolysis (Walker and Weinstein, 1994). With separated protein components, it was determined that only CHLD and CHLI needed to be pre-incubated with ATP to eliminate the lag upon subsequent addition of protoporphyrin IX and CHLH (Willows et al., 1996; Guo et al., 1998; Jensen et al., 1998). CHLI contains recognizable "Walker A" and "Walker B" Mg-ATP-binding motifs (Hansson et al., 2002) similar to those of DNA-dependent ATPases (Koonin, 1993, 1997), and purified CHLI has ATPase activity *in vitro* (Hansson and Kannangara, 1997; Jensen et al., 1999; Petersen et al., 1999; Sirijovski et al., 2006). Purified CHLH also was earlier reported to have weak ATPase activity (Hansson and Kannangara, 1997; Jensen et al., 1999; Petersen et al., 1999), but this was later determined to be artifactual (Sirijovski et al., 2006). Although the N-terminal portion of CHLD resembles CHLI, CHLD does not contain Walker Mg-ATP-binding motifs and the protein does not have ATPase activity.

Results with the *Synechocystis* sp. PCC 6803 Mg chelatase indicated that the optimal ratio for the CHLI, CHLD, and CHLH subunits is 2:1:4 (Jensen et al., 1998). Electron micrographic and X-ray crystallographic analysis of *R. capsulatus* BCHI and *Synechocystis* sp. PCC 6803 CHLI revealed that they assemble into hexameric and heptameric rings, respectively, resembling a family of proteins known as AAA$^+$ proteins, which are thought to function in the remodeling of other proteins (Fodje et al., 2001; Ried et al., 2003). In the presence of Mg^{2+} plus ATP or ADP, CHLI and CHLD are proposed to form a stable layered ring structure (Walker and Willows, 1997; Fodje et al., 2001). In the absence of CHLH, the ATPase activity of CHLI in the CHLI-CHLD complex is inhibited, and adenylate hydrolysis is not required for maintenance of this structure. However, with the addition of protoporphyrin IX and CHLH, ATPase is activated and ATP hydrolysis is needed for Mg chelation. Similar results have been obtained with the Mg chelatases of *R. sphaeroides* and *Synechocystis* sp. PCC 6803 (Gibson et al., 1999; Jensen et al., 1999). That *R. capsulatus* BCHI is multimeric in its functional state was shown by the fact that incubation mixtures containing both wild-type and ATPase-deficient mutant BCHI subunits plus all other components required for Mg chelatase activity are inactive in Mg chelation (Hansson et al., 2002). The ability of the mutant BCHI subunits to inhibit Mg chelatase indicates that they form mixed BCHI multimers with wild-type BCHI and that the mixed multimers are catalytically defective. The results also explain the semidominant Mg chelatase-deficient phenotype of heterozygous barley plants that contain one wild-type and one mutant allele of the CHLI-encoding gene, *xantha-h*. Preincubation of *Synechocystis* sp. PCC 6803 CHLH with protoporphyrin IX in the presence of ATP and Mg^{2+} significantly enhanced Mg chelatase activity after reconstitution of CHLI and CHLD that had been preincubated with ATP and Mg^{2+} (Jensen et al., 1998). The mechanism for this enhancement is unknown.

The *Chlamydomonas* nuclear genome contains one gene for CHLH and two genes each for CHLD and CHLI (Table 20.1). Some algae contain *CHLI*-like genes in their plastid genome, but *Chlamydomonas* does not (Grossman et al., 2004). The rationale for the apparently redundant *CHLD* and *CHLI* genes in *Chlamydomonas* is not known. However, CHLI might have functions additional to its role in Mg chelation, as is suggested by the expression of a CHLI-like protein in the nonphotosynthetic alga *Prototheca wickerhamii*, which contains a remnant nongreen plastid (Borza et al., 2005). ChlD also might have additional functions, suggested by its reported association with ribosomes (Kannangara et al., 1997). Finally, CHLH might also have additional functions. *Arabidopsis* CHLH specifically binds the phytohormone abscisic acid, and was reported to be an ABA receptor (Shen et al., 2006). *Chlamydomonas* is known to respond to abscisic acid, which seems to ameliorate effects of stress (Yoshida et al., 2003, 2004). CHLH-deficient *Chlamydomonas* mutants are viable in the dark but are sensitive to light due to overaccumulated protoporphyrin IX (Chekounova et al., 2001). The responses of these mutant cells to abscisic acid have not been reported.

Although CHLD, CHLH, and CHLI are sufficient to catalyze Mg chelation *in vitro*, another protein, GUN4, greatly increased the Mg chelation rate when added to incubations containing these essential components (Larkin et al., 2003). GUN4 was originally detected in a screen for *Arabidopsis* mutants that have defective coordination of the expression of nuclear and chloroplast genes (Susek et al., 1993). All of the *GUN* mutants have reduced chlorophyll contents. The *GUN4* product is targeted to the chloroplasts, and *GUN4* genes are found in plants, cyanobacterial and red algae, but not in photosynthetic bacteria. *Synechocystis* sp. PCC 6803 *gun4* mutants have low chlorophyll content and accumulate protoporphyrin IX (Wilde et al., 2004). GUN4 is required for Mg chelation at physiological Mg^{2+} concentrations *in vitro*, and it was concluded that GUN4 is required for effective Mg chelation *in vivo* (Davison et al., 2005). The GUN4 crystal structure from *Thermosynechococcus elongatus* reveals the presence of a likely binding site for protoporphyrin IX and/or Mg-protoporphyrin IX and suggests that GUN4 might function to direct the substrate to the enzyme or aid in removing the product (Davison et al., 2005). Taken together, it seems reasonable to consider GUN4 to be a component of Mg chelatase. *Chlamydomonas* contains one *GUN4* gene (Table 20.1), and corresponding sequences are present in the EST database.

B. Methylation of the C-13^3 propionate

S-Adenosyl-L-methionine:Mg-protoporphyrin IX methyltransferase (PPMT) catalyzes the methyl esterification of the 13^3-propionate carboxyl group of Mg protoporphyrin IX (Figure 20.1, step 11). The methylation is thought to be necessary to prevent spontaneous decarboxylation of the propionate at a later step when a β-keto group is formed on the propionate. PPMT has been isolated as a soluble enzyme from several plants and algae (Radmer and Bogorad, 1967; Ebbon and Tait, 1969; Ellsworth and St. Pierre (1976); Shieh et al., 1978; Hinchigeri et al., 1981) and photosynthetic bacteria (Gorchein, 1972). The *bchM* gene of *R. capsulatus* was identified as encoding PPMT by expression in *E. coli* (Bollivar et al., 1994a; Gibson and Hunter, 1994). PPMT has amino acid similarity to other methyltransferases. *Chlamydomonas* contains one candidate PPMT gene (Table 20.1), but there have been no reports on the enzyme activity from the expressed gene or from *Chlamydomonas* cells.

In addition to its role as a methylating enzyme, PPMT has important interactions with Mg chelatase. *R. capsulatus* PPMT stabilized BCHH *in vitro* and activated Mg chelatase several fold (Hinchigeri et al., 1997). Tobacco PPMT significantly stimulated Mg chelatase when added to *in vitro* assays (Alawady et al., 2005). Conversely, *Synechocystis* sp. PCC 6803 CHLH stimulated PPMT *in vitro* (Shepherd et al., 2005). *R. capsulatus bchH* knockout mutants did not have detectable PPMT activity, indicating that BCHH stabilizes PPMT *in vivo* (Hinchigeri et al., 1997). Direct interaction of tobacco PPMT with CHLH was detected by a yeast two-hybrid assay (Alawady et al., 2005). Together, these observations suggest that PPMT associates with CHLH *in vivo* and that each enzyme is required for efficient function of the other. The observations also suggests that Mg chelatase and PPMT might act to channel the intermediate Mg-protoporphyrin IX.

C. Formation of the isocyclic ring

Formation of the isocyclic ring of chlorophylls from the 13-methyl propionate in most aerobic photosynthetic organisms is catalyzed by Mg-protoporphyrin IX monomethyl ester oxidative cyclase. The cyclase step (Figure 20.1, step 12) is among the least-well characterized components of chlorophyll biosynthesis. By analogy to fatty acid β-oxidation, Granick (1950) proposed that isocyclic ring formation occurs via β-oxidation of the 13-methyl propionate group to a 13^1-keto group. The 13^2-methylene group would then attach to the 15-meso bridge of the porphyrin in an oxidative cyclization. Cyclization introduces chirality at C-13^2 and the absolute configuration is *R* (Fleming, 1967). In fatty acid β-oxidation, the β-keto group is introduced by desaturation of the α-carbon-β-carbon bond, followed by hydration of the double bond to form a β-hydroxy group. The β-hydroxy is then oxidized to form the β-keto group. Although early accounts reported the accumulation of intermediates corresponding to the acrylate, hydroxy, and keto derivatives of Mg-protoporphyrin IX monomethyl ester in a series of chlorophyll-deficient *C. vulgaris* mutants (Ellsworth and Aronoff, 1968, 1969), later experiments with chloroplast extracts have not supported the participation of the acrylate (Wong et al., 1985).

Formation of the isocyclic ring entails removal of six electrons from Mg-protoporphyrin IX monomethyl ester. An O_2 requirement for the cyclase step was demonstrated in whole plants (Spiller et al., 1982) and *in vitro* (Chereskin and Castelfranco, 1982; Chereskin et al., 1982, 1983; Bollivar and Beale, 1995). Atmospheric oxygen was incorporated into the isocyclic ring keto oxygen atom of divinyl protochlorophyllide *a* synthesized by intact cucumber cotyledons (Walker et al., 1989). and into chlorophylls synthesized by whole plants and green algae (Schneegurt and Beale, 1992; Porra et al., 1993). Accumulation of Mg-protoporphyrin IX monomethyl ester in plant tissues treated with Fe^{2+} chelators, and inhibition of *in vitro* cyclase by Fe^{2+} chelators, indicates the participation of labile Fe^{2+} in the reaction (Chereskin and Castelfranco, 1982; Chereskin et al., 1982, 1983; Walker et al., 1991; Whyte et al., 1992; Bollivar and Beale, 1995). In most *in vitro* assays, activity requires both a soluble component and a membrane-associated component (Wong and Castelfranco, 1984; Walker et al., 1991; Whyte et al., 1992). However, in *Chlamydomonas* cell extracts, all activity was membrane-associated (Bollivar and Beale, 1996). Washing the membranes with buffer or salt solution increased their cyclase activity, and adding back the chloroplast supernatant fraction inhibited the enzyme. Activity of the *Chlamydomonas* cyclase was inhibited by EDTA and other metal chelators, and the inhibition could be reversed by the addition of Fe^{2+} to the sample.

A putative gene for cyclase (or a component of cyclase) was identified in a chlorophyll-deficient *Arabidopsis* mutant, *CHL27* (Tottey et al., 2003). Similar genes were identified in barley (Rzeznicka et al., 2005) and *Chlamydomonas* (*CHL27A*, also known as *CRD1*; Moseley et al., 2000). *CHL27* mutants, and plants with antisense-induced lowered expression of *CHL27*, are chlorophyll-deficient and accumulate Mg-protoporphyrin IX monomethyl ester (Tottey et al., 2003; Rzeznicka et al., 2005). An extract of the *CHL27* barley mutant, *xantha-l*, had no detectable cyclase activity (Rzeznicka et al., 2005). *Chlamydomonas* has a second *CHL27* gene, *CHL27B* (also known as *CTH1*; Moseley et al., 2002). Expression of *CHL27A* and *CHL27B* is reciprocally responsive to Cu nutrition and O_2 status. In normally grown cells *CHL27B* expression is predominant, but in Cu-deficient or hypoxic cells, *CHL27A* expression is predominant (Moseley et al., 2002; Merchant et al., 2006). In Cu-deprived cells, CHL27A-deficient mutants have decreased levels of PS I and LHCI, for unexplained reasons (Moseley et al., 2000).

CHL27 genes encode putative membrane-bound proteins that contain a di-iron active site (Moseley et al., 2000). The relatively labile Fe in di-iron complexes correlates with the *in vivo* and *in vitro* inhibition by Fe^{2+} chelators. To date, there are no reports of *in vitro* cyclase activity by expressed CHL27 proteins. Separation of plant extracts into soluble and membrane fractions that are both required for *in vitro* cyclase activity suggests that the cyclase system requires one or more gene products, in addition to CHL27. A second locus in barley, *viridis-k*, has been described that has a cyclase-deficient phenotype (Gough, 1972). It is possible that the products of both *xantha-l* and *viridis-k* comprise the plant cyclase enzyme. As discussed above, *Chlamydomonas* extracts could not be separated into two required fractions. Thus far, a second genetic locus for Mg-protoporphyrin IX monomethyl ester accumulation, indicating a second required component for cyclase activity, has not been identified in *Chlamydomonas*. Together, these observations leave open the possibility that the *Chlamydomonas* cyclase requires only one gene product. Anaerobic (bacterio)chlorophyll-synthesizing organisms do not contain *CHL27* homologues and O_2-independent cyclase is presumably an entirely different reaction which is associated with the *bchE* gene in *R. capsulatus* (Bollivar et al., 1994).

D. Reduction of the 8-vinyl group

Reduction of the 8-vinyl substituent of the tetrapyrrole to an ethyl group is catalyzed by divinyl(proto)chlorophyllide 8-vinyl reductase (DVR). The vinyl reduction can occur either before or after reduction of the D pyrrole ring (Figure 20.1, step 14). DVR activity was detected in isolated plastid membranes, but not stroma, from cucumber cotyledons, maize, and barley (Parham and Rebeiz, 1992, 1995), and specifically required NADPH as reductant. Endogenous (already present on the isolated membranes) and exogenous divinyl chlorophyllide *a* were both effectively converted to monovinyl chlorophyllide *a*, but no reactivity was detected toward divinyl protochlorophyllide *a*. However, intact chloroplasts from barley and cucumber cotyledons could reduce the vinyl group of endogenous divinyl protochlorophyllide (Tripathy and Rebeiz, 1988).

There is a necrotic but photosynthetically active maize mutant that contains only divinyl chlorophylls *a* and *b* (Bazzaz, 1981; Bazzaz et al., 1982; Brereton et al., 1983). The pigment composition of the maize mutant suggests that a single gene product is responsible for 8-vinyl reduction of both divinyl protochlorophyllide *a* and divinyl chlorophyllide *a*. An *Arabidopsis* gene, *DVR*, was identified that is required for vinyl reduction, and the recombinant gene product is active *in vitro*, converting divinyl chlorophyllide *a* to chlorophyllide *a* in the presence of NADPH (Nagata et al., 2005). The ability of DVR to also reduce the 8-vinyl group of divinyl protochlorophyllide *a* was not reported, but it was assumed that the enzyme could reduce both protochlorophyllide *a* and chlorophyllide *a*, and DVR was named divinyl protochlorophyllide reductase (Nagata et al., 2005). A different vinyl reductase, having no resemblance to *Arabidopsis* DVR, was identified in *Synechocystis* sp. PCC 6083 (Ito et al., 2008). Previously, a gene was identified in *R. capsulatus*, *bchJ*, that is required for the formation of monovinyl pigments (Suzuki and Bauer, 1995). However, the predicted *bchJ* product bears no resemblance to DVR, and its function is unknown. A gene similar to *DVR* is present in the *Chlamydomonas* genome (Table 20.1).

It should be noted that although monovinyl chlorophylls are nearly ubiquitous in oxygenic photosynthetic organisms, certain marine *Synechococcus* species accumulate divinyl chlorophylls *a* and *b* instead of the monovinyl pigments (Goericke and Repeta, 1992). These organisms do not contain a *DVR* gene (Nagata et al., 2005).

E. Reduction of pyrrole ring D

Reduction of pyrrole ring D converts the porphyrin molecule protochlorophyllide *a* into the chlorin chlorophyllide *a* (Figure 20.1, step 13). The reduction can occur by two entirely different processes, one that requires light and is catalyzed by a single-polypeptide enzyme, and another that is independent of light but requires at least three polypeptides. All chlorophyll-containing organisms appear to possess the light-dependent pathway, and all except the angiosperms and some algae also have the light-independent route. Bacteriochlorophyll-containing photosynthetic bacteria appear to contain only the light-independent pathway (Suzuki and Bauer, 1995). The products of both reduction processes have the same stereoconfiguration, with the three chiral centers being 7*S*, 8*S*, and 10*R* (Fleming, 1967).

1. Light-dependent pathway

NADPH-protochlorophyllide oxidoreductase (POR) is a plastid (pro)thylakoid membrane-associated protein that can be isolated from etiolated tissues, where it comprises a major portion of total etioplast membrane proteins (Dehesh and Ryberg, 1985). In early work, a membrane fragment fraction containing POR was referred to as "protochlorophyllide holochrome." As isolated, POR contains bound NADPH and protochlorophyllide *a* in a ternary complex (Griffiths, 1978; Apel et al., 1980; Oliver and Griffiths, 1981). Exposure of the complex to light that can be absorbed by the bound protochlorophyllide (the red absorption maximum is at 650 nm) causes a rapid conversion of the pigment to chlorophyllide. Active, detergent-solubilized POR has a molecular mass of approximately 35–38 kD on glycerol gradients

(Ikeuchi and Murakami, 1982) or gel filtration, a value that corresponds to the predicted product of POR-encoding genes (Apel et al., 1980; Beer and Griffiths, 1981). Therefore, the monomeric enzyme is active. However, aggregated states of the enzyme exist which can be distinguished from the monomer by spectral shifts, and the state of aggregation might be influenced by protein phosphorylation (Wiktorsson et al., 1996). On the basis of structural similarities deduced from the gene sequences, POR has been categorized as a member of the secondary alcohol reductase family (Baker, 1993).

The deduced structure of POR also suggests that it is a peripheral, rather than integral membrane protein (Spano et al., 1992a). Although POR is normally found in association with membranes, when recombinant POR is expressed in bacteria it does not require membrane association for activity (Schulz et al., 1989; Knaust et al., 1993; Martin et al., 1997).

NADPH is the proximal electron donor in the POR-catalyzed photoreduction of protochlorophyllide, and the *pro-S* hydride of NADPH is transferred to C-17 of the pigment (Valera et al., 1987; Begley and Young, 1989). The proton that is introduced into C-18 of the product is derived from H_2O. A single photon is required for the reaction (Griffiths et al., 1996). A reaction intermediate that absorbs maximally at 695 nm has been detected spectrally, but its structure has not been determined (Franck and Mathis, 1980). After the reduction is completed, the $NADP^+$ product is exchanged for a NADPH molecule and then the chlorophyllide is released from the enzyme (Oliver and Griffiths, 1982). Contrary to earlier suggestions that POR is a flavoprotein (Walker and Griffiths, 1988), active expressed pea and cyanobacterial PORs do not contain a flavin cofactor, even though the enzyme is inhibited by the flavin antagonist quinacrine (Martin et al., 1997; Townley et al., 1998). Substrate specificity experiments showed that the substrate pigment must have a free C-17 carboxylic acid (Griffiths, 1974, 1980). Both mono- and di-vinyl protochlorophyllides were effective substrates (Griffiths and Jones, 1975), but protochlorophyllide *a′*, which has an inverted configuration at the C-13^2 – derived chiral carbon atom in the isocyclic ring, was not (Helfrich et al., 1996). The metal-free derivative of protochlorophyllide is not a substrate, and of the several metal-substituted pigments tested, only Zn-protochlorophyllide was active as a substrate (Griffiths, 1980). Zn-protochlorophyllide *b* was also a substrate, even though Mg-protochlorophyllide *b* was not (Schoch et al., 1995).

Most or all angiosperm plants have at least two *POR* genes, and the products of these genes have different roles in the greening of etiolated tissues and the maintenance of lower levels of chlorophyll synthesis at later developmental stages. Multiple *POR* genes are also present in some gymnosperms that can form chlorophyll in the dark (Spano et al., 1992b), but different roles for the products of these genes have not been reported. Only a single POR-encoding gene is present in other organisms that can form chlorophyll in the dark, including *Synechocystis* sp. PCC 6803 (He et al., 1998) and *Chlamydomonas* (Li and Timko, 1996).

2. Light-independent pathway

All organisms that can reduce ring D of protochlorophyllide in the absence of light require three genes first identified by insertional mutagenesis of *R. capsulatus* and named *bchB*, *bchL*, and *bchN* (Zsebo and Hearst, 1984; Burke et al., 1993a). In cyanobacteria and eukaryotic organisms, the homologous genes are named *chlB*, *chlL*, and *chlN*, and in eukaryotes they are chloroplast-encoded (Lidholm and Gustafsson, 1991; Choquet et al., 1992; Huang and Liu, 1992; Suzuki and Bauer, 1992; Li et al., 1993). These genes are not present in angiosperms. Targeted disruption of the genes has demonstrated their essentiality for light-independent protochlorophyllide reduction (Li et al., 1993; Liu et al., 1993; Fujita et al., 1996).

It is notable that all three predicted gene products have similarity to components of the nitrogenase enzyme system (Peters et al., 1995). ChlB and ChlN resemble each other and also NifD and NifK, which are subunits of the nitrogenase enzyme, a native a_2b_2 complex that binds N_2 and reduces it to NH_3. ChlL resembles NifH, nitrogenase reductase, a native homodimer that acts to convey reducing equivalents from reduced ferredoxin to nitrogenase with energy derived from ATP hydrolysis. By analogy with the nitrogen fixation system, it can be hypothesized that the light-independent protochlorophyllide reduction system consists of a $ChlB_2ChlN_2$ component that binds protochlorophyllide, and a $ChlL_2$ component that acts to convey reducing equivalents (presumably from reduced ferredoxin) to the $ChlB_2ChlN_2$-protochlorophyllide complex with concomitant ATP hydrolysis. In support of this model, ChlL in *Chlamydomonas* can be functionally replaced *in vivo* by *NifH* from *Klebsiella pneumoniae* (Cheng et al., 2005). Also, a mutant strain of *Synechocystis* sp. PCC 6803 that is deficient in both POR and ChlL can make a small amount of chlorophyll, indicating that the ChlL function can be replaced to a limited extent by another protein (He et al., 1998).

Wild-type *Chlamydomonas* cells grown in the light contain ChlB and ChlN, but not ChlL (Cahoon and Timko, 2000). ChlL was present only in cells grown in the dark or at low light intensities. Cells grown in the light or dark had comparable transcript levels for all three proteins. It was proposed that translation of ChlL was blocked by the energy state or redox potential in the chloroplasts of light-grown cells. However, it is also possible that photosynthetically produced O_2 has a destabilizing effect on ChlL (see below). Light-independent protochlorophyllide reduction in the cyanobacterium *Leptolyngbya boryana* is sensitive to light (Yamazaki et al., 2006). POR mutant cells were unable to grow at high light intensity unless they were grown anaerobically or at very low O_2 concentration. POR-deficient *Chlamydomonas* mutant cells accumulated about half as much chlorophyll as wild-type cells in the light, which suggests that the *Chlamydomonas* light-independent system is more resistant to O_2 than that of *L. boryana* (Ford et al., 1983).

In vitro light-independent protochlorophyllide reduction was obtained in incubations containing BchB, BchN, and BchL, all of which were purified from *R. capsulatus* cells overexpressing the corresponding affinity-tagged genes (Fujita and Bauer, 2000). Activity required all three proteins, plus ATP, and dithionite. In later experiments, reduced ferredoxin could be substituted for dithionite (Nomata et al., 2005). Co-expressed BchB and BchN co-purified, indicating that they form a tight complex (Fujita and Bauer, 2000). *In vitro* light-independent protochlorophyllide reduction was also obtained with the ChlB, ChlN, and ChlL proteins from *Plectonema boryanum* and *L. boryana* (Yamazaki et al., 2006), but, so far, there are no reports of *in vitro* activity from *Chlamydomonas* or its expressed genes.

In addition to chloroplast-encoded ChlB, ChlN, and ChlL, light-independent protochlorophyllide reduction in *Chlamydomonas* requires nucleus-encoded components. Mutations at any of at least seven nuclear loci, *Y1* and *Y5–Y10*, result in the inability of the cells to reduce protochlorophyllide in the dark (Ford and Wang, 1980; Cahoon and Timko, 2000). All of the *y* mutants contain wild-type levels of ChlB and ChlN, but all except *y7* lack ChlL (Cahoon and Timko, 2000). It was proposed that the products of the *Y* genes are needed for translation of the chloroplast-encoded ChlL. It is likely that the products of some of these genes are involved in the synthesis of complex metal clusters, which are likely to be present in the ChlB-ChlN complex and ChlL, by analogy to the nitrogenase system. *R. capsulatus* BchL, like NifH, was reported to contain an Fe-S cluster and was inactivated by O_2 (Nomata et al., 2006). However, the *Chlamydomonas* ChlL must differ in some respects because its light-independent protochlorophyllide reductase is active in the presence of O_2.

In addition to circumventing the O_2-sensitivity of the light-independent protochlorophyllide reducing system, there are other reasons why organisms that are able to form chlorophyll without the need for light might benefit from having POR. One possibility is that POR is energetically more efficient than the light-independent system. Another possibility is that POR has an additional role. Starting from the observation that all organisms that accumulate chlorophyll have POR and all organisms that accumulate bacteriochlorophyll lack POR, it seems reasonable to propose that in organisms other than angiosperms (which rely on POR for all protochlorophyllide reduction), POR primarily has a signaling function rather than a biosynthetic one. Since protochlorophyllide has an absorption spectrum that is similar to that of chlorophyll (but very different from that of bacteriochlorophyll), light activation of POR in chlorophyll-containing species could provide a signal to the cells that photosynthetically useful light is available and induce the cells to initiate the developmental program needed to exploit that light.

F. Phytylation

The last step of chlorophyll *a* formation is esterification of the 17-propionic acid with the C_{20} polyisoprene alcohol phytol (Figure 20.1, step 15). The initial C_{20} product of the polyisoprene biosynthetic pathway is Δ2,6,10,14-*all trans*-geranylgeranyl-PP. Phytol differs from geranylgeraniol in that the distal three of the four double bonds in geranylgeraniol are saturated in phytol. The initial chlorophyll that is formed upon exposure of etiolated angiosperm tissues to light contains geranylgeranyl as the isoprene component (Schoch et al., 1977; Schoch, 1978). Shortly after the esterification occurs, the pigment-bound geranylgeranyl is reduced sequentially at the 6, 10, and 14 positions to yield (2*E*, 7*R*, 11*R*)-phytyl, in a process that requires NADPH and ATP (Schoch and Schäfer, 1978; Benz et al., 1980). A similar post-esterification reduction of geranylgeraniol bound to chlorophyllides *a* and *b* after the initial exposure to light occurs in dark-grown mutant *Chlamydomonas* cells that require light for greening (Maloney et al., 1989). At later stages of greening chloroplasts contain phytyl-PP, and the chlorophyll that is formed at this time is directly esterified with phytyl (Rüdiger, 1987). Reduction of geranylgeranyl-PP occurs in the envelope membranes of spinach chloroplasts, whereas reduction of geranylgeranyl-chlorophyllide occurs in the thylakoid membranes (Soll et al., 1983). Both activities require NADPH as the electron donor. An expressed *Arabidopsis* geranylgeranyl reductase catalyzed the reduction of geranylgeranyl-PP as well as geranylgeranyl-chlorophyll *a* (Keller et al., 1998).

A single thylakoid-localized esterifying enzyme, chlorophyll synthase (CHS), is able to use either geranylgeranyl-PP or phytol-PP as a substrate (Rüdiger et al., 1980; Soll et al., 1983; Oster and Rüdiger, 1997; Oster et al., 1997). The *Chlamydomonas* genome and EST databases encode a similar sequence (Table 20.1). Expressed CHS from *Synechocystis* sp. PCC 6803 and *Arabidopsis* can use geranylgeranyl-PP as the isoprene substrate, but the enzyme prefers phytol-PP. CHS can use both chlorophyllide *a* and chlorophyllide *b* as the pigment substrate, but not bacteriochlorophyllide (Benz and Rüdiger, 1981; Oster et al., 1997; Oster and Rüdiger, 1997). *In vivo* results suggest that the CHS of cucumber cotyledon chloroplasts can use both the divinyl- and monovinyl forms of chlorophyllide *a* as substrate (Adra and Rebeiz, 1998). CHS cannot use metal-free chlorophyllide *a* derivative (pheophorbide *a*) or chlorophyllide *a′* (the C-13^2 epimer) (Helfrich et al., 1994). Of the metal complexes tested, only the Mg- and Zn-containing pigments were suitable substrates for CHS (Helfrich and Rüdiger, 1992). Sequence alignments indicate that CHS is related to other polyprenyltransferases involved in the biosynthesis of menaquinone and ubiquinone (Lopez et al., 1996).

G. Can *Chlamydomonas* synthesize chlorophyll anaerobically?

Three steps in the general chlorophyll biosynthetic process described here use O_2 as a substrate. These are the steps catalyzed by CPX, PPX, and CHL27 (Figure 20.1, steps 8, 9, and 12; see sections V.B, V.C, and VI.C). Organisms such as photosynthetic bacteria that can grow anaerobically use different O_2-independent reactions, catalyzed by different enzymes; however, *Chlamydomonas* does not encode any of them. Formation of chlorophyll *b* from chlorophyll *a* also requires O_2, and chlorophyll *b* appears to be restricted to aerobic organisms (see section VII). Although *Chlamydomonas* cells are remarkably tolerant of anaerobiosis, there are no reports of actual anaerobic growth of *Chlamydomonas*. From the above, it can be concluded that *Chlamydomonas* cannot synthesize chlorophyll anaerobically.

VII. BIOSYNTHESIS OF CHLOROPHYLL *b*

All plants and green algae contain chlorophyll *b* (7-desmethyl-7-formyl chlorophyll *a*) as approximately 15–25% of their total chlorophyll content. The existence of viable chlorophyll *b*-deficient mutants indicates that chlorophyll *b* is not essential for photosynthesis, but the pigment undoubtedly contributes importantly to the success of terrestrial and shallow-water plants by absorbing light in the 425–475 nm region, where light is plentiful and chlorophyll *a* absorbs poorly, and transferring the excitation energy to chlorophyll *a* with high efficiency owing to the high spectral overlap in the red region.

In vivo studies with $^{18}O_2$ in *C. vulgaris* and *Zea mays* indicate that the formyl oxygen atom of chlorophyll *b* is derived from O_2 (Schneegurt and Beale, 1992; Porra et al., 1993). These results are consistent with the observation that all chlorophyll *b*-containing organisms are aerobic. Experiments with chlorophylls made from chirally labeled ALA showed that the H^{Si} proton of the methyl group is stereospecifically removed in the conversion of the hydroxymethyl intermediate to the formyl group (Shoolingin-Jordan and Williams, 2003). A *Chlamydomonas* gene, *CAO*, that is essential for chlorophyll *b* accumulation was identified by insertional mutagenesis and rescue by transformation and its product was initially named chlorophyll *a* reductase (Tanaka et al., 1998). The predicted CAO is a 51-kD protein that is related to methyl monooxygenases and contains both a [2Fe-2S] center and mononuclear Fe. A similar gene was also identified in *Arabidopsis* (Espineda et al., 1999).

In vitro CAO activity was obtained in anaerobic conditions with unfractionated *E. coli* cells expressing recombinant *Arabidopsis* CAO in the presence of NADPH and a reduced ferredoxin-generating system (Oster et al., 2000). The substrate was determined to be chlorophyllide *a* rather than the phytylated chlorophyll *a*. Therefore, CAO was renamed chlorophyllide *a* oxidase. Protochlorophyllide *a* was not a substrate. The reaction produced some 7-desmethyl-7-hydroxymethyl chlorophyllide *a*, and this compound could serve as a substrate, suggesting that it is an intermediate in the chlorophyll *a*-to-*b* conversion.

VIII. MINOR CHLOROPHYLL *a* DERIVATIVES IN REACTION CENTERS

In addition to the predominant pigments, chlorophylls *a* and *b*, plants and green algae contain small amounts of two other chlorophyll-derived pigments. Each PS II reaction center contains two molecules of pheophytin *a*, the Mg-free analog of chlorophyll *a* (Omata et al., 1984), and each PS I reaction center contains two molecules of chlorophyll *a′*, the C-13^2 epimer of chlorophyll *a* (Maeda et al., 1992). There is no reliable evidence for specific enzymes that catalyze the demetalation or C-13^2 epimerization of chlorophyll *a* or any of its precursors. Moreover, enzymes that catalyze the last several steps of chlorophyll *a* synthesis do not accept the demetalated or C-13^2 epimerized derivatives as substrates (Helfrich et al., 1994; Rüdiger, 2006). However, both pheophytin *a* and chlorophyll *a′* are known to be formed from chlorophyll *a* spontaneously *in vitro*, particularly under mild acidic conditions. Therefore, it seems probable that the pheophytin *a* and chlorophyll *a′* in reaction centers are formed from chlorophyll *a* spontaneously *in vivo* upon associating with their respective binding sites within the reaction centers.

IX. CHLOROPHYLL DEGRADATION AND INTERCONVERSION OF CHLOROPHYLLS *a* AND *b*

In senescing higher plant leaves, chlorophylls are degraded by a specific pathway involving enzymes that are elevated in senescing leaves (Hörtensteiner, 2006). Obviously, unicellular algae like *Chlamydomonas* do not undergo organ senescence. Nevertheless, there are circumstances where chlorophyll degradation would be advantageous by preventing the accumulation of photosensitizing free chlorophyll. These circumstances include periods of nutritional deficiency, when nonessential parts of the photosynthetic apparatus (e.g., LHCs) are selectively degraded so that the protein amino acids can be redeployed for more pressing needs. Also, during normal photosynthesis, labile chlorophyll protein complexes, such as the D1 subunit of PS II, are turned over.

The first step of chlorophyll *a* is catabolism in plants is hydrolysis of the phytyl ester bond by chlorophyllase (Table 20.1) to form chlorophyllide *a* and phytol. Although it had been thought that chlorophyll degradation is irreversible once it had been dephytylated, later evidence indicated that in *Synechocystis* sp. PCC 6803 a portion of the hydrolysis products can be recycled to form new chlorophyll *a* molecules (Vavilin and Vermaas, 2007).

In the second step of chlorophyll degradation, chlorophyllide *a* is demetalized to form pheophorbide *a*. An enzyme that catalyzes the demetalation has not been identified, and the reaction might occur nonenzymatically. Next, the macrocycle ring of pheophorbide *a* is opened at the 5 position by oxygen-dependent pheophorbide *a* monooxygenase (Hörtensteiner et al., 1998). The open-chain product (a bilin) is then modified in several ways and ultimately transported to the vacuole (Hinder et al., 1996). Chlorophyll *b* catabolites are not detected in senescing leaves (Kräutler and Matile, 1999), and pheophorbide *a* monooxygenase does not accept pheophorbide *b* as a substrate (Hörtensteiner et al., 1995). These observations suggest that catabolism of chlorophyll *b* must begin with its conversion to chlorophyll *a*.

In vitro conversion of chlorophyll *b* to chlorophyll *a* was detected in preparations from a mutant barley strain that is unable to form chlorophyll *b*, indicating that conversion of chlorophyll *b* to chlorophyll *a* must occur by a different process than reversal of the reaction that synthesizes chlorophyllide *b* from chlorophyllide *a* (Rudoi and Shcherbakov, 1998). Conversion of chlorophyll *b* to chlorophyll *a* in extracts of cucumber cotyledon etioplasts required both the membrane and stromal fractions as well as ATP (Ito et al., 1994). The ATP was probably required to form reduced ferredoxin (see below). A likely intermediate in the reduction of chlorophyll *b* is 7-hydroxymethyl chlorophyll *a*. NADPH, but not ATP, was required for reduction of chlorophyll *b* to 7-hydroxymethyl chlorophyll *a* by intact barley etioplasts and etioplast membranes (Ito et al., 1996; Scheumann et al., 1996). In lysed etioplasts, reduced ferredoxin was required for the reduction of 7-hydroxymethyl chlorophyll *a* to chlorophyll *a* (Scheumann et al., 1999). The reverse reaction, conversion of chlorophyll *a* to chlorophyll *b*, was not detected in these systems. Chlorophyllide *b* was also a substrate in the lysed barley etioplast system, being converted to chlorophyllide *a* (Scheumann et al., 1999).

In the green alga *Chlorella protothecoides*, degradation products of both chlorophyll *a* and chlorophyll *b* are excreted into the medium, which indicates that in contrast to senescing plant leaves, this alga can degrade chlorophyll *b* without its having to be converted to chlorophyll *a* (Gossauer and Engel, 1996). Chlorophyll degradation products that resemble bilins were excreted into the medium from a chlorophyll *b*-less *Chlamydomonas* mutant (Doi et al., 1997). The structures of these products have not been completely identified. When held under anaerobic conditions, *Chlamydomonas* cells accumulate the 13^2-demethoxycarbonylation product of pheophorbide *a*, pyropheophorbide *a* (Doi et al., 2001). The pyropheophorbide *a* accumulates presumably because its conversion to bilins requires oxygen, as does the formation of bilins in the catabolism of chlorophylls in plants (Hörtensteiner, 2006).

A *Chlamydomonas* enzyme has been characterized that removes the 13^2-methoxycarbonyl group of pheophorbide *a* to form pyropheophorbide *a* (Suzuki et al., 2002). This enzyme, named pheophorbide demethoxycarbonylase, has a somewhat different reaction mechanism than the pheophorbidase enzymes that have been described in a few plants (Suzuki et al., 2002), and it appears to be restricted to green algae. The role of pheophorbidases in plants is not known, but the pheophorbide demethoxycarbonylase in *Chlamydomonas* is probably involved in chlorophyll catabolism. Pheophorbide demethoxycarbonylase is present in only trace amounts in light-grown cells, but it is induced when cells are kept in the dark for 3 days (Doi et al. 1997). The enzyme has been extracted from cells and characterized as a 170-kD homodimer. Its ability to use pheophorbide *b*, in addition to pheophorbide *a*, as a substrate has not been reported. A *Chlamydomonas* gene for this enzyme has not been identified, nor has a gene for a ring-opening pheophorbide oxygenase. However, an expressed putative chlorophyllase gene has been identified (Table 20.1).

The ability of plants to interconvert chlorophylls *b* and *a* suggests the possible existence of a "chlorophyll cycle," which might function to adjust the chlorophyll *a*/*b* ratio in response to changing external conditions (Ito et al., 1996; Ohtsuka et al., 1997). At this time, there is no evidence to suggest that such a cycle operates in *Chlamydomonas*.

X. THE BRANCH FROM UROPORPHYRINOGEN III TO SIROHEME

Organisms that are able to assimilate oxidized forms of sulfur and nitrogen must be able to convert them to their reduced forms sulfide and ammonium, respectively. The pathways for reducing sulfur and nitrogen have a similar last step, a 6-electron reduction, which in plants and algae is catalyzed by the reduced ferredoxin-dependent chloroplast enzymes sulfite reductase (EC 1.8.7.1) and nitrite reductase (EC 1.7.7.1), respectively. Assimilatory sulfite and nitrite reductases have been described in higher plants and a *Chlamydomonas* nitrite reductase has been characterized (Romero et al., 1987; see also Chapter 3), but there are no descriptions of sulfite reductase in *Chlamydomonas*, although three putative sulfite reductase-encoding genes have been reported (Zhang et al., 2004). Some nitrite reductases can also reduce sulfite (Janick et al., 1983), and the same might be true for the *Chlamydomonas* enzyme. Assimilatory sulfite and nitrate reductases all contain the prosthetic group siroheme (Figure 20.2). Unlike the more familiar protoporphyrin IX-derived hemes which are prosthetic groups of proteins that participate in respiratory and photosynthetic electron transport as well as other cellular redox and O_2-binding proteins, siroheme is not derived from protoporphyrin IX, but instead is formed from uroporphyrinogen III by addition of *S*-adenosylmethionine-derived methyl groups at positions 2 and 7 on the macrocycle, removal of two electrons to

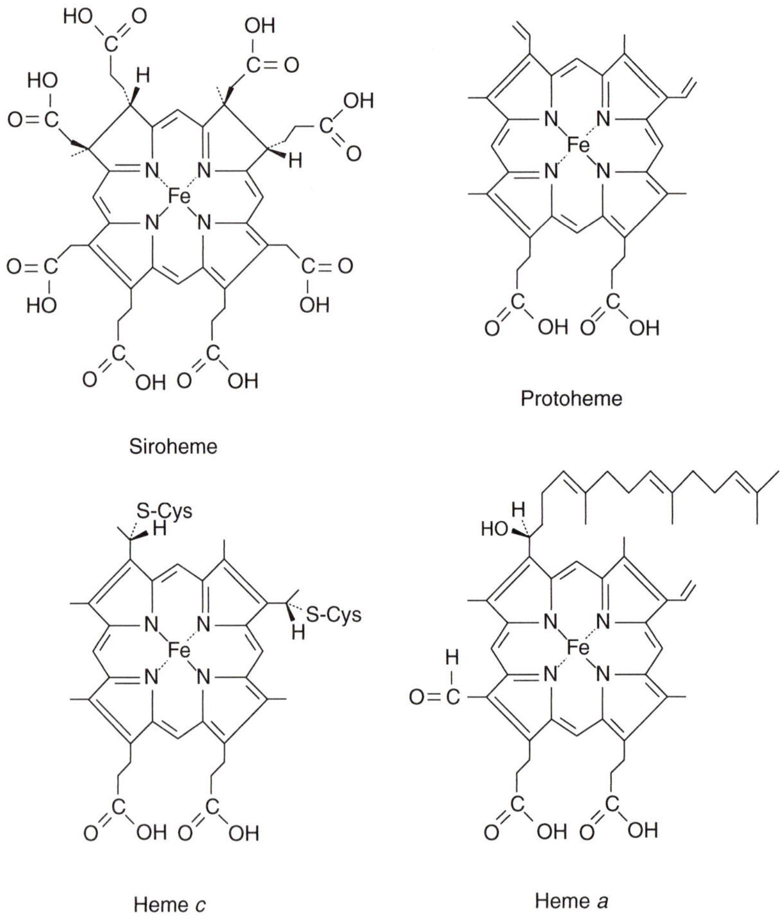

FIGURE 20.2 *Hemes of plants and green algae.*

yield the tetrahydroporphyrin (dihydrochlorin) macrocycle, and insertion of Fe^{2+} into the tetrapyrrole.

Chemical arguments suggest that the order of the steps of siroheme formation from uroporphyrinogen III is probably that given above: first methylation, then dehydrogenation, and finally Fe^{2+} chelation. Methylation of uroporphyrinogen III first at position 2 to form precorrin-1 and then at position 7 yields precorrin-2 (which is also a precursor of heme d_1, cofactor B_{12} and Factor F_{430} in prokaryotes) and would effectively limit subsequent dehydrogenation beyond the tetrahydroporphyrin state. All four asymmetric carbons of precorrin-2 have *S* configurations (Scott et al., 1978). Dehydrogenation of precorrin-2 to sirohydrochlorin produces a compound that has the aromaticity and metal-binding properties necessary for efficient chelation of Fe^{2+}. Some organisms including *Bacillus megaterium* use three separate enzymes to convert uroporphyrinogen III to siroheme: an *S*-adenosylmethionine-uroporphyrinogen III 2,7-methyltransferase (SirA; referred to here as precorrin-2 synthase); an $NADP^+$-precorrin-2 oxidoreductase (SirC; referred to here as precorrin-2 dehydrogenase); and a sirohydrochlorin ferrochelatase (SirB; Raux et al., 2003). In other organisms (possibly including *Synechocystis* sp. PCC 6803; Panek and O'Brian, 2002), the methyltransferase step is catalyzed by a bifunctional enzyme that also has uroporphyrinogen III synthase activity (Fujino et al., 1995; Anderson et al., 2001). In *Saccharomyces cerevisiae* the final two steps, oxidation of precorrin-2 and Fe-chelation of sirohydrochlorin, are catalyzed by a single enzyme (Raux et al., 1999) and in *Clostridium josui*, these last two steps are catalyzed by a trifunctional enzyme that also contains GTR (Fujino et al., 1995). Finally, in *E. coli* and *S. typhimurium*, all three of the reactions leading from uroporphyrinogen III to siroheme are catalyzed by a single trifunctional enzyme, CysG (Woodcock et al., 1998; Stroupe et al., 2003).

The *CysG* gene of *E. coli* and *S. typhimurium* encodes a homodimeric enzyme with 50-kD subunits. The C-terminal domain, named $CysG^A$, has precorrin-2 synthase activity, and the N-terminal domain, $CysG^B$, catalyzes the other two steps of siroheme synthesis. During the methylation reactions, *S*-adenosylmethionine apparently becomes covalently bound to CysG (Woodcock and Warren, 1996). The 2.2-Å crystal structure of *S. typhimurium* CysG showed that each monomer of the homodimeric protein consists of two independent modules (Stroupe et al., 2003). The C-terminal module that catalyzes the methylations is a deep groove with a hydrophobic patch surrounded by H-bonding residues, and the active sites are highly charged. The N-terminal dehydrogenase-ferrochelatase module is at the dimer interface and contains a NAD^+-binding Rossman fold. The crystal structure also revealed that CysG is a phosphoprotein. From analysis of mutants in which the phosphorylated Ser was mutated, it was suggested that phosphorylation inhibits dehydrogenase activity and can control the partitioning of precorrin-2 between the siroheme and cobalamin pathways (Stroupe et al., 2003).

Genes homologous to the portion of *CysG* that encodes precorrin-2 synthase are present in *Arabidopsis* (Leustek et al., 1997), maize (Fan et al., 2006), and *Chlamydomonas* (Table 20.1). A sirohydrochlorin ferrochelatase gene is also present in *Arabidopsis* (Raux-Deery et al., 2005) and *Chlamydomonas* (protein accession EDP03868), but a gene for precorrin-2 dehydrogenase has not been identified in *Chlamydomonas*.

XI. THE BRANCH FROM PROTOPORPHYRIN IX TO HEMES

A. Fe^{2+} chelation by protoporphyrin IX

Synthesis of protoheme (Figure 20.2) from protoporphyrin IX by the insertion of Fe^{2+} into the macrocycle is catalyzed by the enzyme protoporphyrin

IX ferrochelatase (FeC). In contrast to Mg chelatase, FeC is composed of a single protein type and Fe chelation does not require any energy input. FeCs from various sources vary widely in their primary structures and physical properties (Dailey et al., 2000; Dailey and Dailey, 2002). Most FeCs are dimeric integral membrane proteins, but *Bacillus subtilis* FeC is a soluble monomer. Many FeCs have C-terminal extensions that bear a [2Fe-2S] cluster, although some have the extension but not the cluster. Crystal structures of *B. subtilis* (Al-Karadaghi et al., 1997) and human (Wu et al., 2001) FeCs have been obtained. Even though the two proteins have less than 10% sequence identity, their structures are similar. The enzyme holds the substrate protoporphyrin IX in a distorted conformation, which was proposed to facilitate the abstraction of two protons prior to the entry of Fe^{2+} into the macrocycle (Wu et al., 2001).

Most plants contain at least two FeC genes, whose functions are discussed in section XI. In contrast to plants, *Chlamydomonas* contains a single FeC gene (Table 20.1; van Lis et al., 2005). The predicted protein most closely resembles FeCs that are expressed in photosynthetic tissues of plants. The C-terminal extension does not contain a [2Fe-2S] cluster binding motif, but it does contain an "LHC" motif, a conserved hydrophobic stretch that corresponds to the third transmembrane segment found in the light-harvesting complex family of proteins and which likely plays a role in anchoring the protein in membranes (Jansson, 1999). An antibody raised against recombinant *Chlamydomonas* FeC detected the protein only in chloroplasts, and not in the mitochondria (van Lis et al., 2005).

B. Chloroplast heme *c* synthesis

In addition to protoheme, chloroplasts contain heme *c* (Figure 20.2). In contrast to protoheme, heme *c* is covalently bound to the apocytochromes via one or (more usually) two thioether links between protein-Cys residues and the α-carbon atoms of the ethyl groups at positions 3 and 8 (the positions of the vinyl groups in the precursor protoheme). The introduced chirality at both C-3^1 and C-8^1 is *S* in all *c*-type hemes for which the configuration was reported (Timkovich et al., 1998). *Chlamydomonas* chloroplasts contain three known *c*-type hemes: one in cytochrome c_6, one located on the cytochrome *f* moiety of the b_6f complex, and a third, named heme c_i (or alternatively, heme x), that is also located in the b_6f complex near the high-potential heme *b* on the stromal side of the membrane (Stroebel et al., 2003). Heme c_i is unusual in that it is bound to the apoprotein by a single thioether link, the one at the 3 position. Also, examination of the electron density map of the *Chlamydomonas* b_6f complex in the heme c_i region suggests that the introduced chirality at C-3^1 is *R*, rather than the more usual *S* configuration in other c-type hemes. In *Chlamydomonas*, both cytochrome *f* and cytochrome c_6 are assembled in the thylakoid lumen (Howe and Merchant, 1992, 1994; Howe et al., 1995).

Three different cytochrome *c* heme ligating systems have been described (Xie et al., 1998; Nakamoto et al., 2000; Hamel et al., 2003). Type I occurs in α- and γ-proteobacteria, Type III occurs in fungal and animal mitochondria, and Type II occurs in chloroplasts and some bacteria including cyanobacteria. Type II ligating systems consist of a membrane protein with essential Cys residues, that is thought to maintain the ligating Cys residues of apocytochrome *c* in the reduced state that is necessary for ligation; another membrane protein with a characteristic Trp-rich motif that is thought to convey heme to the apocytochrome *c* and catalyze the ligation; and other proteins with unknown functions (Thöny-Meyer, 2000). Mutant screens in *Chlamydomonas* have identified at least five nuclear genes and one chloroplast gene that are involved in chloroplast heme *c* synthesis (Howe and Merchant, 1992; Nakamoto et al., 2000). All of the mutants are deficient in both cytochrome b_6f and cytochrome c_6, indicating that the same factors are required for synthesizing heme *c* in both of these cytochromes, although a gene coding specifically for heme *c* lyase has not been identified. All of these mutants still contain mitochondrial *c*-type cytochromes, which indicates that the two organelles have different heme *c* ligation factors. None of the *Chlamydomonas* genes required for heme *c* ligation to cytochromes *f* and c_6 resemble the Type III yeast lyase genes or candidate Type I bacterial lyase genes (Xie et al., 1998).

Before the discovery of the heme c_i in cytochrome b_6f, four nuclear genes were found that are required for cytochrome b_6 assembly in *Chlamydomonas*, and it was suggested that cytochrome b_6 contains a covalently bound heme (Kuras et al., 1997). At least some of these genes, named *CCB1–CCB4*, are probably involved in heme c_i ligation. For a fuller discussion of cytochrome b_6f assembly and function, see Chapter 17.

C. Mitochondrial protoheme synthesis from chloroplast-derived precursors

In plants and algae, heme is required in chloroplasts mainly for cofactors of PS II and the cytochrome b_6f complex. Heme is also needed for components of the mitochondrial electron transport chain and as a cofactor of various non-chloroplast hemoproteins (e.g. catalase, peroxidases). In animal and fungal cells, the last two or three steps of heme synthesis occur in mitochondria. Although all of the steps of heme synthesis leading to protoporphyrinogen IX are generally believed to occur exclusively in the chloroplasts in plants, the site(s) of the last two steps of heme synthesis in plants have not been well determined. Many plants contain two PPX-encoding genes. In tobacco, the products of two genes, *PPX*-I and *PPX*-II, are specifically targeted to chloroplasts and mitochondria, respectively (Lermontova et al., 1997). Spinach also contains two PPX-encoding genes, and PPX-I is targeted to the chloroplast thylakoids, but PPX-II is dually targeted to the chloroplast envelope and mitochondria by alternative translation initiation at two in-frame start codons (Watanabe et al., 2001). For FeC in plants, reports of mitochondrial targeting and localization have been disputed, and the situation remains unresolved (Chow et al., 1998; Lister et al., 2001; Cornah et al., 2002; Masuda et al., 2003). It is difficult to understand a biosynthetic role for mitochondrial PPX if FeC is absent from mitochondria. In contrast to plants, *Chlamydomonas* contains only one gene for PPX and one for FeC (van Lis et al., 2005). Moreover, PPX and FeC were detected only in chloroplasts of *Chlamydomonas* cells. These results imply that the chloroplast must be capable of supplying all cellular hemes in this eukaryotic photosynthetic organism.

If all cellular hemes are synthesized exclusively in chloroplasts, export of heme from plastids to mitochondria and the cytosol must occur. Heme export from chloroplasts is poorly understood, but *in vitro* export of heme from isolated pea chloroplasts has been described (Thomas and Weinstein, 1990). Two pathways of heme transport from plastids to mitochondria can be envisioned. One is active transport via the cytosol. Another is through a channel formed at the junctions between mitochondria and chloroplasts. In *Chlamydomonas*, mitochondria are mostly closely associated to the chloroplast, which occupies 60% of the cell volume. This close association could facilitate direct transport between the organelles. Active transport could involve a putative heme chaperone, which has been described in *Arabidopsis* and proposed to function in mitochondrial heme import (Spielewoy et al., 2001). Other candidate plant mitochondrial heme transport genes have been identified in mitochondrial genomes (Schuster, 1994; Jekabsons and Schuster, 1995). A *Chlamydomonas* nuclear gene that encodes an expressed putative mitochondrial ABC transporter (protein accession EDP03867) immediately precedes the FeC gene. This transporter might be a partner in the transport of heme into the mitochondrion. Although it is not common for functionally related nuclear genes to occur in clusters in *Chlamydomonas* there is precedent, and examples include genes involved in uniparental inheritance (Armbrust et al., 1993), genes for histones (Walther and Hall, 1995), hydrogenase cofactor assembly (Posewitz et al., 2004), and components of the nitrate assimilation system (Quesada et al., 1993).

D. Synthesis of other mitochondrial hemes

Regardless of whether mitochondrial protoheme is synthesized within the mitochondria or imported from the chloroplast, mitochondria must be able to synthesize two other hemes: heme *c*, which is covalently attached to cytochromes *c* and c_1; and heme *a*, which exists only in the mitochondria-specific cytochrome oxidase.

Chlamydomonas mitochondria contain cytochrome *c* and cytochrome c_1 (Atteia, 1994). The best-characterized mitochondrial *c* heme ligation systems are the Type III systems of yeast, for which different lyases are responsible for heme ligation to mitochondrial cytochrome *c* and cytochrome c_1 (Dumont et al., 1987; Zollner et al., 1992) (see discussion in section XI.B). Candidate genes for plant mitochondrial heme *c* lyases resemble candidate genes for Type I bacterial heme *c* lyases, and are unlike the Type III system of yeast or the Type II system of chloroplasts (Oda et al., 1992; Gonzalez et al., 1993; Schuster et al. 1993; Hamel et al., 2003). The candidate plant mitochondrial heme *c* lyase genes are encoded in the mitochondrial genome, but genes for *Chlamydomonas* mitochondrial heme *c* lyase(s) have not been described. It is very likely that they are not the same ones that are involved in chloroplast heme *c* ligations, because several nuclear gene mutations that disrupt the latter do not affect mitochondrial cytochrome *c* synthesis (Howe and Merchant, 1992; Nakamoto et al., 2000).

Heme *a*, the heme cofactor of mitochondrial terminal oxidase, is synthesized from protoheme in two steps (Mogi et al., 1994). First, farnesyl alcohol is attached at the C-3^2 position of the 3-vinyl group to form 3-desvinyl-3-hydroxyfarnesylethyl protoheme, which is also known as heme *o*. This condensation reaction is catalyzed by heme *o* synthase and uses farnesyl-PP as the source of the farnesyl group (Table 20.1). From the crystal structure of bovine heart cytochrome oxidase, it was discerned that the introduced chirality at C-3^1 is *S* (Yamashita et al., 2005). Some organisms use heme *o* rather than heme *a* in cytochrome oxidase, and *E. coli* cytochrome *bo* contains heme *o*. Heme *a* is formed from heme *o* by oxidation of the 18-methyl group to a formyl group (Figure 20.2), catalyzed by the heme-containing enzyme heme *a* synthase (Table 20.1). Results with expressed *B. subtilis* heme *a* synthase show that the methyl oxidation requires O_2 and proceeds via the partially oxidized hydroxymethyl intermediate. However, unlike the analogous monooxygenase reactions that occur in chlorophyll *b* synthesis, these oxidations apparently do not result in O_2 incorporation into the formyl group, and therefore heme *a* synthase is not a monooxygenase (Brown et al., 2004). Genes for enzymes that catalyze each of these steps, named protoheme farnesyltransferase and heme *a* synthase, respectively, have been identified in bacteria, fungi, and animal cells (Mogi et al., 1994). *Chlamydomonas* contains a candidate protoheme farnesyltransferase gene (Table 20.1) but so far, a heme *a* synthase gene has not been identified.

XII. HEME CATABOLISM

A major role of heme catabolism in plants, cyanobacteria, and red algae is to provide the bilin chromophores for phytochromes and phycobiliproteins. Even though *Chlamydomonas* apparently contains no phytochromes or other bilin-containing proteins, there might still be a need for heme catabolism. Heme turnover is an essential requirement in models for feedback regulation of early steps of tetrapyrrole synthesis by heme (see section XIII. A.2). *Chlamydomonas* possesses candidate genes for two heme oxygenases, *HMOX1* and *HMOX2* (Table 20.1). Sequences corresponding to *HMOX1*, but not *HMOX2*, are in the EST database. This apparent expression pattern is reminiscent of the two *HMOX* genes in *Synechocystis* sp. PCC 6803, only one of which is expressed under normal growth conditions (Cornejo et al., 1998).

XIII. REGULATION

A. Overview

In discussions about regulation of tetrapyrrole biosynthesis, it is useful to distinguish between on the one hand, regulation of specific biosynthetic steps that cause changes in the amounts of end products produced and on the other hand, changes in the amounts of enzymes and the transcripts of their encoding genes that have little or no effect on the amounts of end products formed. This section will focus on the former and largely exclude the latter. Also, because regulatory responses and mechanisms are not as highly conserved between *Chlamydomonas* and higher plants as are the biosynthetic steps themselves, this section will concentrate on *Chlamydomonas*.

B. Enzyme-level regulation of chlorophyll formation

1. Regulation of ALA formation

The most consistently observed regulatory effect at the enzyme level in a wide range of organisms including *Chlamydomonas* is feedback inhibition of GTR by heme (Gough and Kannangara, 1979; Wang et al., 1984; Weinstein and Beale, 1985; Rieble and Beale, 1988; Rieble et al., 1989). Micromolar or submicromolar heme concentrations inhibit ALA formation from Glu in most of the unfractionated enzyme systems studied. Other tetrapyrroles such as protoporphyrin, Mg-protoporphyrin, protochlorophyllide, and chlorophyllide, are much less effective or completely ineffective at inhibiting GTR (Weinstein and Beale, 1985; Huang and Wang, 1986). There are some exceptions to this general pattern. For example, the *E. gracilis* system is relatively unaffected by heme below 25 μM, a concentration considered to be too high to be physiologically relevant (Mayer et al., 1987). On the other hand, the sensitivity of the *C. vulgaris* system toward heme was increased several fold by physiological concentrations of glutathione in the incubation mixture (Weinstein et al., 1993). Other thiol-containing compounds had no effect. The N-terminal 30 amino acid residues of the barley GTR protein have been shown to be involved in heme regulation but not in catalysis (Vothknecht et al., 1998). The key role of heme in regulating ALA formation *in vivo* was demonstrated in *Synechocystis* sp. PCC 6803. Mutant cells containing a truncated ferrochelatase with greatly reduced activity had highly elevated levels of ALA synthesis and the cells accumulated large amounts of protoporphyrin IX (Sobotka et al., 2008). The levels of chlorophyll were also increased, suggesting that abnormally high levels of protoporphyrin IX can partially overpower the regulatory mechanisms that control its entry into the chlorophyll branch.

In section III.D, FLP was described as a membrane-bound *Chlamydomonas* protein that inhibits GTR. FLP (FLU-Like Protein) is similar to the *Arabidopsis* protein FLU (Falciatore et al., 2005) and a barley protein coded by the *tigrina-d* gene (Lee et al., 2003). The *FLU* gene was originally identified in an *Arabidopsis* genetic screen for photosensitive mutants (Meskauskiene et al., 2001). Such mutants overaccumulate protochlorophyllide *a* in the dark and thereby become photosensitive when exposed to light. FLU was shown to interact with GTR but not GSAT by yeast two-hybrid experiments (Meskauskiene and Apel, 2002). FLU appears to regulate ALA synthesis independently of heme (Meskauskiene et al., 2001). The fact that *Arabidopsis flu* mutants overaccumulate protochlorophyllide *a* but not heme was interpreted to indicate that FLU affects only the Mg^{2+} branch of tetrapyrrole biosynthesis (Meskauskiene et al., 2001). However, because normal heme levels might be maintained in the *flu* mutant by increases in both the synthesis and degradation of heme, this inference might not be valid. In an *Arabidopsis hy1* background that is deficient in heme catabolism, protochlorophyllide *a* accumulation in the *flu* mutant was suppressed (Goslings et al., 2004). The fact that yeast two-hybrid experiments showed that FLU interacts with GTR but not GSAT suggests that FLU might act by disrupting the GTR-GSAT interaction (as described in section III.D for *Chlamydomonas* GTR-GSAT) and thereby lower GTR activity. In higher plants, FLU clearly plays an important role in the protective mechanism that prevents chlorophyll precursor accumulation in the dark.

Because *Chlamydomonas* normally synthesizes chlorophyll in the dark as well as light, it was surprising to find a FLU-like protein in the alga. The *Chlamydomonas FLP* gene produces two transcripts by alternative splicing that are translated into short and long forms of FLP, which differ in length by 12 amino acids (Falciatore et al., 2005). In GST pulldown assays, the long form of FLP binds to GTR more strongly than the short form. FLP expression was increased in mutants that are defective in later steps of chlorophyll synthesis and accumulate chlorophyll precursors. Antisense suppression of FLP levels caused overaccumulation of chlorophyll precursors and the cells became photosensitive, akin to *Arabidopsis flu* mutants. Together, these results indicate that FLP interacts with GTR and inhibits its activity in a physiologically meaningful way in *Chlamydomonas*, similar to the role of FLU in plants (Falciatore et al., 2005). To date, there are no reports that FLU or FLP affects GTR activity *in vitro*.

As described above, observations of the GTR that was extracted from *Chlamydomonas* cells showed that like others, its activity was inhibited by heme (Wang et al., 1984). Unexpectedly, recombinant *Chlamydomonas* GTR was not inhibited by heme (Srivastava et al., 2005). However, heme inhibition was restored upon addition of gel-filtered *Chlamydomonas* cell extract supernatant to the incubation mixture. In *Chlamydomonas*, heme was reported to inhibit GTS (Chang et al., 1990). The physiological significance of the effect of heme on GTS has not been further investigated.

2. Downstream consequences of heme-regulated ALA formation

In a model that was first proposed for *R. sphaeroides*, which synthesizes ALA by a different route than plants and algae, heme could function as an effective early-stage feedback regulator of tetrapyrrole biosynthesis even though the pathway produces bacteriochlorophyll in addition to heme (Lascelles and Hatch, 1969). For physiologically relevant regulation under varying demands for heme and bacteriochlorophyll, the model requires that heme be turned over and that Mg chelatase have a higher affinity than ferrochelatase for protoporphyrin IX. The steady-state level of "free" heme would be influenced by its turnover and incorporation into hemoproteins, and also by diversion of its precursor, protoporphyrin IX, for bacteriochlorophyll synthesis. Any tendency for the free heme level to increase would be countered by increased inhibition of ALA synthesis by heme. Conversely, any tendency for a decrease in the free heme level would decrease heme inhibition of ALA synthesis and thereby increase the rate of heme synthesis.

In accord with this model, Mg chelatase has a much higher affinity for protoporphyrin IX than does ferrochelatase (Guo et al., 1998). The large affinity difference implies that when Mg chelatase is active, protoporphyrin IX is preferentially directed toward chlorophyll formation and away from chloroplast heme synthesis. Heme is known to be degraded in intact leaf tissue (Castelfranco and Jones, 1975) and in isolated chloroplasts (Thomas and Weinstein, 1990), and of course it is also consumed in the formation of hemoproteins. The possession of heme-catabolizing heme oxygenase genes by *Chlamydomonas* even though the cells apparently contain no phytochromes or other bilin-containing proteins (see section XII) is also in accord with the model.

In light of the newer observations on the effects of FLU-like proteins on ALA synthesis *in vivo*, this model is now known to be incomplete, although still partially applicable, for plants and algae. The model remains completely applicable for photosynthetic bacteria, which do not contain *FLU*-like genes.

C. Regulation of the activity of other enzymes

The largest gap in our understanding of the regulation of chlorophyll synthesis is the absence of information about how Mg chelation is regulated.

In general, Mg chelatase is not inhibited *in vitro* by protochlorophyllide, chlorophyllide, Mg-protoporphyrin IX, or heme (e.g. Fuesler et al., 1981). Barley Mg chelatase, however, was inhibited by the chlorophyll catabolite pheophorbide (Pöpperl et al., 1997). Although this observation is difficult to understand in terms of providing end-product feedback modulation of chlorophyll synthesis, it might indicate a mechanism for inhibiting chlorophyll formation under conditions when chlorophyll is being degraded. The *Arabidopsis* Mg chelatase CHLL1 protein has been reported to undergo thioredoxin-mediated reversible oxidation and reduction of a specific pair of Cys residues (Ikegami et al. 2007). The ATPase activity of isolated CHLL1 was completely inactivated upon oxidation and was fully recovered upon thioredoxin-mediated reduction. Moreover, Mg chelatase activity of isolated chloroplasts was significantly stimulated by reducing conditions. Light-dependent changes in the *in vivo* oxidation-reduction state of CHLL1 were reported (Ikegami et al., 2007). These observations suggest that chlorophyll synthesis can be regulated at the Mg chelatase step in response to light-mediated changes in chloroplast redox status. Discoveries showing that GUN4 and PPMT activate Mg chelatase have increased our understanding of the requirements for efficient Mg chelation (see sections VI.A and VI.B). It remains to be determined whether PPMT, and especially GUN4, undergo changes in their concentrations in response to environmental and developmental variables that affect chlorophyll content, and thereby control the rate of chlorophyll synthesis. It is clear that progress in understanding how chlorophyll synthesis is regulated will depend on increased knowledge about how Mg chelatase, the first committed step of the chlorophyll pathway, is regulated.

Other tetrapyrrole biosynthetic enzymes have not been found to be feedback-inhibited by physiologically relevant concentrations of intermediates or end products.

D. Regulation of enzyme abundance

Higher plant genomes generally have two or more copies of most genes for chlorophyll biosynthetic steps, and these genes are differentially expressed at the RNA and protein level depending on tissue and developmental stage of photosynthetic competence. The importance of gene expression in regulating chlorophyll synthesis in *Chlamydomonas* is much less well understood. A number of studies on the expression of tetrapyrrole biosynthetic genes have revealed significant changes in the mRNA levels for many of the genes in response to changes in light, nutritional status, and other environmental variables (Matters and Beale, 1994, 1995a,b; Lohr et al., 2005; Nogaj et al., 2005; van Lis et al., 2005). Generally, the changes in the levels of enzymes encoded by these genes are smaller than the changes in the mRNA levels, and changes in enzyme levels cannot account for the varying rates of chlorophyll accumulation. For example, in cells that are synchronized to a 12-hour light:12-hour dark schedule, virtually no chlorophyll is synthesized during the dark phase even though there are adequate levels of the several enzymes examined, including GTR and GSAT, to support a significant rate of chlorophyll synthesis (Nogaj et al., 2005; van Lis et al., 2005).

As noted in section V.B, *Chlamydomonas* has two genes for CPX. *CPX1* mRNA, CPX protein level and CPX activity all increase several fold as part of the cellular response to Cu deficiency (Hill and Merchant, 1995). Cu deficiency also induces the expression of cytochrome c_6, a chloroplast lumenal protein that functionally replaces the Cu-containing electron carrier, plastocyanin, in Cu-deficient cells. It is clear that the cells would need to synthesize some additional heme under Cu deficiency to supply cytochrome c_6, but it is not readily apparent why this small increase in heme synthesis would require such a great increase in CPX activity. Although it has been noted that after ALA synthesis, CPX is the next rate-limiting step of heme synthesis (Merchant et al., 2006), the amount of heme synthesis required for cytochrome c_6 is very small compared to the total demand for tetrapyrroles for other hemes and chlorophylls.

CPX1 is also upregulated in cells growing at low O_2 concentration (Merchant et al., 2006). This response can be understood as being compensatory, because CPX requires O_2 as a reactant, and under low O_2 the CPX reaction rate might decrease sufficiently to become rate-limiting for tetrapyrrole synthesis, unless the amount of CPX enzyme is increased to compensate for the lower activity.

The reciprocal effects of Cu deprivation and hypoxia on the expression of the two Mg-protoporphyrin IX monomethylester oxidative cyclase enzymes, CHL27A and CHL27B, was described in section VI.C. Although conditions of Cu deprivation or hypoxia versus normal growth cause the relative expression levels of CHL27A and CHL27B to vary greatly, the cells accumulate similar amounts of chlorophylls (Moseley et al., 2002; Merchant et al., 2006). Therefore in wild-type cells, the relative levels of CHL27A and CHL27B cannot be considered to regulate chlorophyll synthesis.

An interesting observation was made in *Arabidopsis* transformants overexpressing CAO. Fully green leaves from these plants had significantly more chlorophyll *b* and light-harvesting antenna pigment-proteins (e.g. LHCII) than control leaves (Tanaka et al., 2001). This observation suggests that plants might be able to adjust their antenna size through modulation of chlorophyll *b* synthesis by changing the expression level of *CAO*. To date, a similar effect has not been reported in *Chlamydomonas*.

It can be concluded that regulation of the levels of biosynthetic enzymes plays only a minor role, if any, in overall regulation of chlorophyll synthesis in *Chlamydomonas*. However, certain adjustments in chloroplast composition, such as the amounts of antenna complexes, might be regulated at the level of the abundance of certain enzymes such as CAO.

REFERENCES

Abboud, M.M. and Akhtar, M. (1976). Stereochemistry of hydrogen elimination in the enzymatic formation of the C-2–C-3 double bond of porphobilinogen. *J. Chem. Soc. Chem. Commun.* **1976**, 1007–1008.

Adra, A.N. and Rebeiz, C.A. (1998). Chloroplast biogenesis 81: Transient formation of divinyl chlorophyll *a* following a 2.5 ms light flash treatment of etiolated cucumber cotyledons. *Photochem. Photobiol.* **68**, 852–856.

Alberti, M., Burke, D.H., and Hearst, J.E. (1995). Structure and sequence of the photosynthesis gene cluster. In: *Anoxygenic Photosynthetic Bacteria* (R.E. Blankenship, M.T. Madigan, and C.E. Bauer, Eds.), pp. 1083–1106. Kluwer Academic Publishers, Dordrecht, The Netherlands.

Alawady, A., Reski, R., Yaronskaya, E., and Grimm, B. (2005). Cloning and expression of the tobacco CHLM sequence encoding Mg protoporphyrin IX methyltransferase and its interaction with Mg chelatase. *Plant. Mol. Biol.* **57**, 679–691.

Al-Karadaghi, S., Hansson, M., Nikonov, S., Jonsson, B., and Hederstedt, L. (1997). Crystal structure of ferrochelatase: The terminal enzyme in heme biosynthesis. *Structure* **51**, 1501–1510.

Anderson, P.J., Entsch, B., and McKay, D.B. (2001). A gene, *cobA* + *hemD*, from *Selenomonas ruminantium* encodes a bifunctional enzyme involved in the synthesis of vitamin B_{12}. *Gene* **281**, 63–70.

Apel, K., Santel, H.J., Redlinger, T.E., and Falk, K. (1980). The protochlorophyllide holochrome of barley (*Hordeum vulgare* L.). Isolation and characterization of the NADPH:protochlorophyllide oxidoreductase. *Eur. J. Biochem.* **111**, 251–258.

Armbrust, E.V., Ferris, P.J., and Goodenough, U.W. (1993). A mating-type linked gene-cluster expressed in *Chlamydomonas* zygotes participates in the uniparental inheritance of the chloroplast genome. *Cell* **74**, 801–811.

Atteia, A. (1994). Identification of mitochondrial respiratory proteins from the green alga *Chlamydomonas reinhardtii*. *C. R. Acad. Sci.* **317**, 11–19.

Avissar, Y.J. and Beale, S.I. (1989a). Biosynthesis of tetrapyrrole pigment precursors: Pyridoxal requirement of the aminotransferase step in the formation of δ-aminolevulinate from glutamate in extracts of *Chlorella vulgaris*. *Plant Physiol.* **89**, 852–859.

Baker, M.E. (1993). Protochlorophyllide reductase is homologous to human carbonyl reductase and pig 20β-hydroxysteroid dehydrogenase. *Biochem. J.* **300**, 605–607.

Barnard, G.F. and Akhtar, M. (1975). Stereochemistry of porphyrinogen carboxylyase reaction in haem biosynthesis. *J. Chem. Soc. Chem. Commun.* **1975**, 494–496.

Battersby, A.R., Fookes, C.J.R., Gustafson-Potter, K.E., Matcham, G.W.J., and McDonald, E. (1979a). Proof by synthesis that unrearranged hydroxymethylbilane is the product from deaminase and the substrate for cosynthetase in the biosynthesis of Uro'gen-III. *J. Chem. Soc. Chem. Commun.* **1979**, 1155–1158.

Battersby, A.R., Fookes, C.J.R., Matcham, G.W.J., and McDonald, E. (1979b). Order of assembly of the four pyrrole rings during biosynthesis of the natural porphyrins. *J. Chem. Soc. Chem. Commun.* **1979**, 539–541.

Battersby, A.R., Fookes, C.J.R., Matcham, G.W.J., McDonald, E., and Hollenstein, R. (1983a). Biosynthesis of porphyrins and related molecules. Part 20. Purification of deaminase and studies on its mode of action. *J. Chem. Soc. Perkin. Trans. I*, **1983**, 3031–3040.

Battersby, A.R., Fookes, C.J.R., Hart, G., Matcham, G.W.J., and Pandey, P.S. (1983b). Biosynthesis of porphyrins and related macrocycles. Part 21. The interaction of deaminase and its product (hydroxymethylbilane) and the relationship between deaminase and cosynthetase. *J. Chem. Soc. Perkin. Trans. I,* **1983**, 3041–3047.

Bazzaz, M.B. (1981). New chlorophyll chromophores isolated from a chlorophyll-deficient mutant of maize. *Photobiochem. Photobiophys.* **2**, 199–207.

Bazzaz, M.B., Bradley, C.V., and Brereton, R.G. (1982). 4-Vinyl-4-desethyl chlorophyll *a*: Characterization of a new naturally occurring chlorophyll using fast atom bombardment, field desorption and "in beam" electron impact mass spectroscopy. *Tetrahedron Lett.* **23**, 1211–1214.

Beale, S.I. and Castelfranco, P.A. (1974). The biosynthesis of δ-aminolevulinic acid in higher plants. II. Formation of ^{14}C-δ-aminolevulinic acid from labeled precursors in greening plant tissues. *Plant Physiol.* **53**, 297–303.

Beale, S.I., Gough, S.P., and Granick, S. (1975). The biosynthesis of δ-aminolevulinic acid from the intact carbon skeleton of glutamic acid in greening barley. *Proc. Natl. Acad. Sci. U.S.A.* **72**, 2719–2723.

Becerril, J.M. and Duke, S.O. (1989). Protoporphyrin IX content correlates with activity of photobleaching herbicides. *Plant Physiol.* **90**, 1175–1181.

Beer, N.S. and Griffiths, W.T. (1981). Purification of the enzyme NADPH:protochlorophyllide oxidoreductase. *Biochem. J.* **195**, 83–92.

Begley, T.P. and Young, H. (1989). Protochlorophyllide reductase. 1. Determination of the regiochemistry and the stereochemistry of the reduction of protochlorophyllide to chlorophyllide. *J. Am. Chem. Soc.* **111**, 3095–3096.

Benz, J. and Rüdiger, W. (1981). Chlorophyll biosynthesis: Various chlorophyllides as exogenous substrates for chlorophyll synthase. *Z. Naturforsch.* **36c**, 51–57.

Benz, J., Wolf, C., and Rüdiger, W. (1980). Chlorophyll biosynthesis: Hydrogenation of geranylgeraniol. *Plant Sci. Lett.* **19**, 225–230.

Berry-Lowe, S. (1987). The chloroplast glutamate tRNA gene required for δ-aminolevulinate synthesis. *Carlsberg Res. Commun.* **52**, 197–210.

Bollivar, D.W. and Beale, S.I. (1995). Formation of the isocyclic ring of chlorophyll by isolated *Chlamydomonas reinhardtii* chloroplasts. *Photosynth. Res.* **43**, 113–124.

Bollivar, D.W. and Beale, S.I. (1996). The chlorophyll biosynthetic enzyme Mg-protoporphyrin IX monomethyl ester (oxidative) cyclase: Characterization and partial purification from *Chlamydomonas reinhardtii* and *Synechocystis* sp. PCC 6803. *Plant Physiol.* **112**, 105–114.

Bollivar, D.W., Jiang, Z.Y., Bauer, C.E., and Beale, S.I. (1994a). Heterologous expression of the *bchM* gene product from *Rhodobacter capsulatus* and demonstration that it encodes *S*-adenosyl-L-methionine:Mg-protoporphyrin IX methyltransferase. *J. Bacteriol.* **176**, 5096–5290.

Bollivar, D.W., Suzuki, J.Y., Beatty, J.T., Dobrowolski, J.M., and Bauer, C.E. (1994b). Directed mutational analysis of bacteriochlorophyll *a* biosynthesis in *Rhodobacter capsulatus*. *J. Mol. Biol.* **237**, 622–640.

Borza, T., Popescu, C.E., and Lee, R.W. (2005). Multiple metabolic roles for the nonphotosynthetic plastid of the green alga *Prototheca wickerhamii*. *Eukaryotic Cell* **4**, 253–261.

Bougri, O. and Grimm, B. (1996). Members of a low-copy number gene family encoding glutamyl-tRNA reductase are differentially expressed in barley. *Plant J.* **9**, 867–878.

Breckau, D., Mahlitz, E., Sauerwald, A., Layer, G., and Jahn, D. (2003). Oxygen-dependent coproporphyrinogen III oxidase (HemF) from *Escherichia coli* is stimulated by manganese. *J. Biol. Chem.* **47**, 46625–46631.

Breinig, S., Kervinen, J., Stith, L., Wasson, A.S., Fairman, R., Wlodawer, A., Zdanov, A., and Jaffe, E.K. (2003). Control of tetrapyrrole biosynthesis by alternate quaternary forms of porphobilinogen synthase. *Nature Struct. Biol.* **10**, 757–763.

Brereton, R.G., Bazzaz, M.B., Santikarn, S., and Williams, D.H. (1983). Positive and negative ion fast atom bombardment mass spectrometric studies on chlorophylls: Structure of 4-vinyl-4-desethyl chlorophyll *b*. *Tetrahedron Lett.* **24**, 5775–5778.

Brown, K.R., Brown, B.M., Hoagland, E., Mayne, C.L., and Hegg, E.L. (2004). Heme A synthase does not incorporate molecular oxygen into the formyl group of heme A. *Biochemistry* **43**, 8616–8624.

Bruyant, P. and Kannangara, C.G. (1987). Biosynthesis of δ-aminolevulinate in greening barley leaves, VIII: Purification and characterization of the glutamate-tRNA ligase. *Carlsberg Res. Commun.* **52**, 99–109.

Bull, A.D., Pakes, J.F., Hoult, R.C., Rogers, L.J., and Smith, A.J. (1989). Tetrapyrrole biosynthesis in a gabaculin-tolerant mutant of *Synechococcus* 6301. *Biochem. Soc. Trans.* **17**, 911–912.

Burke, D.H., Alberti, M., and Hearst, J.E. (1993a). *bchFNBH* bacteriochlorophyll synthesis genes of *Rhodobacter capsulatus* and identification of the third subunit of light-independent protochlorophyllide reductase in bacteria and plants. *J. Bacteriol.* **175**, 2414–2422.

Cahoon, A.B. and Timko, M.P. (2000). yellow-in-the-dark mutants of *Chlamydomonas* lack the CHLL subunit of light-independent protochlorophyllide reductase. *Plant Cell* **12**, 559–568.

Camadro, J.-M., Thome, F., Brouillet, N., and Labbe, P. (1994). Purification and properties of protoporphyrinogen oxidase from the yeast *Saccharomyces cerevisiae*. Mitochondrial location and evidence for a precursor form of the protein. *J. Biol. Chem.* **269**, 32085–32091.

Castelfranco, P.A. and Jones, O.T.G. (1975). Protoheme turnover and chlorophyll synthesis in greening barley tissue. *Plant Physiol.* **55**, 485–490.

Castelfranco, P.A., Weinstein, J.D., Schwarcz, S., Pardo, A.D., and Wezelman, B.E. (1979). The Mg insertion step in chlorophyll biosynthesis. *Arch. Biochem. Biophys.* **192**, 592–598.

Cavaleiro, J.A.S., Kenner, G.W., and Smith, K.M. (1974). Pyrroles and related compounds. Part XXXII. Biosynthesis of protoporphyrin-IX from coproporphyrinogen-III. *J. Chem. Soc. Perkin Trans. I,* **1974**, 1188–1194.

Chang, T.-E., Wegmann, B., and Wang, W.-Y. (1990). Purification and characterization of glutamyl-tRNA synthetase: An enzyme involved in chlorophyll biosynthesis. *Plant Physiol.* **93**, 1641–1649.

Chekounova, E., Voronetskaya, V., Papenbrock, J., Grimm, B., and Beck, C.F. (2001). Characterization of *Chlamydomonas* mutants defective in the H subunit of Mg-chelatase. *Mol. Genet. Genomics,* **266**, 363–373.

Chen, T.C. and Miller, G.W. (1974). Purification and characterization of uroporphyrinogen decarboxylase from tobacco leaves. *Plant Cell. Physiol.* **15**, 993–1005.

Chen, M.-W., Jahn, D., Schön, A., O'Neill, G.P., and Söll, D. (1990a). Purification and characterization of *Chlamydomonas reinhardtii* chloroplast glutamyl-tRNA synthetase, a natural misacylating enzyme. *J. Biol. Chem.* **265**, 4054–4057.

Chen, M.-W., Jahn, D., O'Neill, G.P., and Söll, D. (1990b). Purification of the glutamyl-tRNA reductase from *Chlamydomonas reinhardtii* involved in δ-aminolevulinic acid formation during chlorophyll biosynthesis. *J. Biol. Chem.* **265**, 4058–4063.

Cheng, Q., Day, A., Dowson-Day, M., Shen, G.F., and Dixon, R. (2005). The *Klebsiella pneumoniae* nitrogenase Fe protein gene (*nifH*) functionally substitutes for the *chlL* gene in *Chlamydomonas reinhardtii*. *Biochem. Biophys. Res. Commun.* **329**, 966–975.

Chereskin, B.A. and Castelfranco, P.A. (1982). Effects of iron and oxygen on chlorophyll biosynthesis. II. Observations on the biosynthetic pathway in isolated etiochloroplasts. *Plant Physiol.* **69**, 112–116.

Chereskin, B.A., Wong, Y.-S., and Castelfranco, P.A. (1982). *In vitro* synthesis of the chlorophyll isocyclic ring: Transformation of magnesium-protoporphyrin IX and magnesium-protoporphyrin IX monomethyl ester into magnesium-2,4-divinyl pheoporphyrin a_5. *Plant Physiol.* **70**, 987–993.

Chereskin, B.A., Castelfranco, P.A., Dallas, J.L., and Straub, K.M. (1983). Mg-2,4-divinyl pheoporphyrin a_5: The product of a reaction catalyzed *in vitro* by developing chloroplasts. *Arch. Biochem. Biophys.* **226**, 10–18.

Choquet, Y., Rahire, M., Girard-Bascou, J., Erikson, J., and Rochaix, J.-D. (1992). A chloroplast gene is required for the light-independent accumulation of chlorophyll in *Chlamydomonas reinhardtii*. *EMBO J.* **11**, 1697–1704.

Chow, K.S., Singh, D.P., Walker, A.R., and Smith, A.G. (1998). Two different genes encode ferrochelatase in *Arabidopsis*: Mapping, expression and subcellular targeting of the precursor proteins. *Plant J.* **15**, 531–541.

Cornah, J.E., Roper, J.M., Singh, D.P., and Smith, A.G. (2002). Measurement of ferrochelatase activity using a novel assay suggests that plastids are the major site of haem biosynthesis in both photosynthetic and nonphotosynthetic cells of pea (*Pisum sativum* L.). *Biochem. J.* **362**, 423–432.

Cornejo, J., Willows, R.D., and Beale, S.I. (1998). Phytobilin biosynthesis: cloning and expression of a gene encoding soluble ferredoxin-dependent heme oxygenase from *Synechocystis* sp. PCC 6803. *Plant J.* **15**, 99–107.

Corradi, H.R., Corrigall, A.V., Boix, E., Mohan, C.G., Sturrock, E.D., Meissner, P.N., and Acharya, K.R. (2006). Crystal structure of protoporphyrinogen oxidase from *Myxococcus xanthus* and its complex with the inhibitor acifluorfen. *J. Biol. Chem.* **281**, 38625–38633.

Crockett, N., Alefounder, P.R., Battersby, A.R., and Abell, C. (1991). Uroporphyrinogen III synthase: Studies on its mechanism of action, molecular biology and biochemistry. *Tetrahedron* **47**, 6003–6014.

Croft, M.T., Lawrence, A.D., Raux-Deery, E., Warren, M.J., and Smith, A.G. (2005). Algae acquire vitamin B_{12} through a symbiotic relationship with bacteria. *Nature* **438**, 90–93.

Dailey, H.A. (1990). Conversion of coproporphyrinogen to protoheme in higher eukaryotes and bacteria: Terminal three enzymes. In: *Biosynthesis of Heme and Chlorophylls* (H.A. Dailey, Ed.), pp. 123–161. McGraw-Hill, New York.

Dailey, T.A. and Dailey, H.A. (2002). Identification of [2Fe-2S] clusters in microbial ferrochelatases. *J. Bacteriol.* **184**, 2460–2464.

Davison, P.A., Schubert, H.L., Reid, J.D., Iorg, C.D., Heroux, A., Holl, C.P., and Hunter, C.N. (2005). Structural and biochemical characterization of Gun4 suggests a mechanism for its role in chlorophyll biosynthesis. *Biochemistry* **44**, 7603–7612.

Dehesh, K. and Ryberg, M. (1985). The NADPH-protochlorophyllide oxidoreductase is the major constituent of prolamellar bodies in wheat (*Triticum aestivum* L.). *Planta* **164**, 396–399.

Doi, M., Shima, S., Egashira, T., Nakamura, K., and Okayama, S. (1997). New bile pigment excreted by a *Chlamydomonas reinhardtii* mutant: A possible breakdown catabolite of chlorophyll *a*. *J. Plant Physiol.* **150**, 504–508.

Doi, M., Inage, T., and Shioi, Y. (2001). Chlorophyll degradation in a *Chlamydomonas reinhardtii* mutant: An accumulation of pyropheophorbide *a* by anaerobiosis. *Plant Cell Physiol.* **42**, 469–474.

Drolet, M., Péloquin, L., Echelard, Y., Cousineau, L., and Sasarman, A. (1989). Isolation and nucleotide sequence of the *hemA* gene of *Escherichia coli* K12. *Mol. Gen. Genet.* **216**, 347–352.

Dumont, M.E., Ernst, J.F., Hampsey, D.M., and Sherman, F. (1987). Identification and sequence of the gene encoding cytochrome-*c* heme lyase in the yeast *Saccharomyces cerevisiae*. *EMBO J.* **6**, 235–241.

Ebbon, J.G. and Tait, G.H. (1969). Studies on *S*-adenosylmethionine-magnesium protoporphyrin methyltransferase in *Euglena gracilis* strain *Z*. *Biochem. J.* **111**, 573–582.

Elliott, T. (1989). Cloning, genetic characterization, and nucleotide sequence of the *hemA-prfA* operon of *Salmonella typhimurium*. *J. Bacteriol.* **171**, 3948–3960.

Elliott, T., Avissar, Y.J., Rhie, G., and Beale, S.I. (1990). Cloning and sequence of the *Salmonella typhimurium hemL* gene and identification of the missing enzyme in *hemL* mutants as glutamate-1-semialdehyde aminotransferase. *J. Bacteriol.* **172**, 7071–7084.

Ellsworth, R.K. and Aronoff, S. (1968). Investigation on the biogenesis of chlorophyll. III. Biosynthesis of Mg-vinyl-phaeoporphine a_5 methylester from Mg-protoporphine IX monomethyl ester as observed in *Chlorella* mutants. *Arch. Biochem. Biophys.* **125**, 269–277.

Ellsworth, R.K. and Aronoff, S. (1969). Investigation on the biogenesis of chlorophyll *a*. IV. Isolation and partial characterization of some biosynthetic intermediates between Mg-protoporphine IX monomethyl ester and Mg-vinyl-phaeoporphine a_5 methylester, obtained from *Chlorella* mutants. *Arch. Biochem. Biophys.* **130**, 374–383.

Ellsworth, R.K. and St. Pierre, M.E. (1976). Biosynthesis and inhibition of (–)-*S*-adenosyl-L-methionine:magnesium-protoporphyrin methyltransferase of wheat. *Photosynthetica* **10**, 291–301.

Erskine, P.T., Senior, N., Awan, S., Lambert, R., Lewis, G., Tickle, I.J., Sarwar, M., Spencer, P., Thomas, P., Warren, M.J., Shoolingin-Jordan, P.M., Wood, S.P., and Cooper, J.B. (1997). X-ray structure of 5-aminolaevulinate dehydratase, a hybrid aldolase. *Nature Struct. Biol.* **4**, 1025–1031.

Espineda, C.E., Linford, A.S., Devine, D., and Brusslan, J. (1999). The *AtCAO* gene, encoding chlorophyll *a* oxygenase, is required for chlorophyll *b* synthesis in *Arabidopsis thaliana*. *Proc. Natl. Acad. Sci. U. S. A.* **96**, 10507–10511.

Falciatore, A., Merendino, L., Barneche, F., Ceol, M., Meskauskiene, R., Apel, K., and Rochaix, J.-D. (2005). The FLP proteins act as regulators of chlorophyll synthesis in response to light and plastid signals in *Chlamydomonas*. *Genes Devel.* **19**, 176–187.

Fan, J., Wang, D., Liang, Z., Guo, M., Teng, M., and Niu, L. (2006). Maize uroporphyrinogen III methyltransferase: Overexpression of the functional gene fragments in *Escherichia coli* and one-step purification. *Prot. Expr. Purif.* **46**, 40–46.

Ferreira, G.C. and Dailey, H.A. (1988). Mouse protoporphyrinogen oxidase. Kinetic parameters and demonstration of inhibition by bilirubin. *Biochem. J.* **250**, 597–603.

Fleischer, E.B., Choi, E.I., Hambright, P., and Stone, A. (1964). Porphyrin studies: Kinetics of metalloporphyrin formation. *Inorg. Chem.* **3**, 1284–1287.

Fleming, I. (1967). Absolute configuration and the structure of chlorophyll. *Nature* **216**, 151–152.

Fodje, M.N., Hansson, A., Hansson, M., Olsen, J.G., Gough, S., Willows, R.D., and Al-Karadaghi, S. (2001). Interplay between an AAA module and an integrin I domain may regulate the function of magnesium chelatase. *J. Mol. Biol.* **311**, 111–122.

Ford, C. and Wang, W.-Y. (1980). Temperature-sensitive *yellow* mutants of *Chlamydomonas reinhardtii*. *Mol. Gen. Genet.* **180**, 5–10.

Ford, C., Mitchell, S., and Wang, W.-Y. (1983). Characterization of NADPH: protochlorophyllide oxidoreductase in the *y-7* and *pc-1 y-7* mutants of *Chlamydomonas reinhardtii*. *Mol. Gen. Genet.* **192**, 290–292.

Franck, F. and Mathis, P. (1980). A short-lived intermediate in the photo-enzymatic reduction of protochlorophyll(ide) into chlorophyll(ide) at a physiological temperature. *Photochem. Photobiol.* **32**, 799–803.

Friedmann, H.C., Duban, M.E., Valasinas, A., and Frydman, B. (1992). The enantioselective participation of (*S*)- and (*R*)-diaminovaleric acids in the formation of δ-aminolevulinic acid in cyanobacteria. *Biochem. Biophys. Res. Commun.* **185**, 60–68.

Fuesler, T.P., Wright, L.A., Jr, and Castelfranco, P.A. (1981). Properties of magnesium chelatase in greening etioplasts. Metal ion specificity and effect of substrate concentration. *Plant Physiol.* **67**, 246–249.

Fujino, E., Fujino, T., Karita, S., Sakka, K., and Ohmiya, K. (1995). Cloning and sequencing of some genes responsible for porphyrin biosynthesis from the anaerobic bacterium *Clostridium josui*. *J. Bacteriol.* **177**, 5169–5175.

Fujita, Y. and Bauer, C.E. (2000). Reconstitution of light-independent protochlorophyllide reductase from purified BchL and BchN-BchB subunits. *In vitro* confirmation of nitrogenase-like features of a bacteriochlorophyll biosynthesis enzyme. *J. Biol. Chem.* **275**, 23583–23588.

Fujita, Y., Takagi, H., and Hase, T. (1996). Identification of the *chlB* gene product essential for light-independent chlorophyll biosynthesis in the cyanobacterium *Plectonema boryanum*. *Plant Cell. Physiol.* **37**, 313–323.

Gibson, L.C. and Hunter, C.N. (1994). The bacteriochlorophyll biosynthesis gene, *bchM*, of *Rhodobacter sphaeroides* encodes *S*-adenosyl-L-methionine: Mg-protoporphyrin IX methyltransferase. *FEBS Lett.* **352**, 127–130.

Gibson, K.D., Laver, W.G., and Neuberger, A. (1958). Formation of δ-aminolaevulic acid *in vitro* from succinyl-coenzyme a and glycine. *Biochem. J.* **70**, 71–81.

Gibson, L.C., Willows, R.D., Kannangara, C.G., von Wettstein, D., and Hunter, C.N. (1995). Magnesium-protoporphyrin chelatase of *Rhodobacter sphaeroides*: Reconstitution of activity by combining the products of the *bchH*, *-I*, and *-D* genes expressed in *Escherichia coli*. *Proc. Natl. Acad. Sci. U. S. A.* **92**, 1941–1944.

Gibson, L.C., Marrison, J.L., Leech, R.M., Jensen, P.E., Bassham, D.C., Gibson, M., and Hunter, C.N. (1996). A putative Mg chelatase subunit from *Arabidopsis thaliana* cv C24. Sequence and transcript analysis of the gene, import of the protein into chloroplasts, and *in situ* localization of the transcript and protein. *Plant Physiol.* **111**, 61–71.

Gibson, L.C.D., Jensen, P.E., and Hunter, C.N. (1999). Magnesium chelatase from *Rhodobacter sphaeroides*: Initial characterization of the enzyme using purified subunits and evidence for a BchI-BchD complex. *Biochem. J.* **337**, 243–251.

Goericke, R. and Repeta, D.J. (1992). The pigments of *Prochlorococcus marinus*: The presence of divinylchlorophyll *a* and *b* in a marine procaryote. *Limnol. Oceanogr.* **37**, 425–433.

Gonzalez, D.H., Bonnard, G., and Grienenberger, J.-M. (1993). A gene involved in the biogenesis of cytochromes is co-transcribed with a ribosomal protein gene in wheat mitochondria. *Curr. Genet.* **24**, 248–255.

Gorchein, A. (1972). Magnesium protoporphyrin chelatase activity in *Rhodopseudomonas spheroides*: Studies with whole cells. *Biochem. J.* **127**, 97–106.

Goslings, D., Meskauskiene, R., Kim, C., Lee, K.P., Nater, M., and Apel, K. (2004). Concurrent interactions of heme and FLU with Glu tRNA reductase (HEMA1), the target of metabolic feedback inhibition of tetrapyrrole biosynthesis, in dark- and light-grown *Arabidopsis* plants. *Plant J.* **40**, 957–967.

Gossauer, A. and Engel, N. (1996). Chlorophyll catabolism: Structures, mechanisms, conversions. *J. Photochem. Photobiol. B,* **32**, 141–151.

Gough, S. (1972). Defective synthesis of porphyrins in barley plastids caused by mutation in nuclear genes. *Biochim. Biophys. Acta,* **286**, 36–54.

Gough, S.P. and Kannangara, C.G. (1979). Biosynthesis of δ-aminolevulinate in greening barley leaves. III. The formation of δ-aminolevulinate in *tigrina* mutants of barley. *Carlsberg Res. Commun.* **44**, 403–416.

Granick, S. (1950). The structural and functional relationships between heme and chlorophyll. *Harvey Lect.* **44**, 220–245.

Griffiths, W.T. (1974). Protochlorophyll and protochlorophyllide as precursors for chlorophyll synthesis *in vitro*. *FEBS Lett.* **49**, 196–200.

Griffiths, W.T. (1978). Reconstitution of chlorophyll formation by isolated etioplast membranes. *Biochem. J.* **174**, 681–692.

Griffiths, W.T. (1980). Substrate-specificity studies on protochlorophyllide reductase in barley (*Hordeum vulgare*) etioplast membranes. *Biochem. J.* **186**, 267–278.

Griffiths, W.T. and Jones, O.T.G. (1975). Magnesium 2,4-divinyl phaeoporphyrin a_5 as a substrate for chlorophyll biosynthesis *in vitro*. *FEBS Lett.* **50**, 355–358.

Griffiths, W.T., McHugh, T., and Blankenship, R.E. (1996). The light intensity dependence of protochlorophyllide photoconversion and its significance to the catalytic mechanism of protochlorophyllide reductase. *FEBS Lett.* **398**, 235–238.

Grossman, A.R., Lohr, M., and Im, C.-S. (2004). *Chlamydomonas reinhardtii* in the landscape of pigments. *Annu. Rev. Genet.* **38**, 119–173.

Guo, R., Luo, M., and Weinstein, J.D. (1998). Magnesium-chelatase from developing pea leaves. Characterization of a soluble extract from chloroplasts and resolution into three required protein fractions. *Plant Physiol.* **116**, 605–615.

Hambright, P. (1975). Dynamic coordination chemistry of metalloporphyrins. In: *Porphyrins and Metalloporphyrins* (K.M. Smith, Ed.), pp. 233–278. Elsevier Scientific Publishing Co, Amsterdam.

Hamel, P.P., Dreyfuss, B.W., Xie, Z., Gabilly, S.T., and Merchant, S. (2003). Essential histidine and tryptophan residues in CcsA, a System II polytopic cytochrome *c* biogenesis protein. *J. Biol. Chem.* **278**, 2593–2603.

Hansson, M. and Kannangara, C.G. (1997). ATPases and phosphate exchange activities in magnesium chelatase subunits of *Rhodobacter sphaeroides*. *Proc. Natl. Acad. Sci. U. S. A.* **94**, 13351–13356.

Hansson, A., Willows, R.D., Roberts, T.H., and Hansson, M. (2002). Three semi-dominant barley mutants with single amino acid substitutions in the smallest magnesium chelatase subunit form defective AAA(+) hexamers. *Proc. Natl. Acad. Sci. U. S. A.* **99**, 13944–13949.

Hart, G.J., Miller, A.D., Leeper, F.J., and Battersby, A.R. (1987). Biosynthesis of natural porphyrins: Proof that hydroxymethylbilane synthase (porphobilinogen deaminase) uses a novel binding group in its catalytic action. *J. Chem. Soc. Chem. Commun.* **1987**, 1762–1765.

He, Q., Brune, D., Nieman, R., and Vermaas, W. (1998). Chlorophyll *a* synthesis upon interruption and deletion of *por* coding for the light-dependent NADPH: protochlorophyllide oxidoreductase in a photosystem-I-less/*chlL*$^-$ strain of *Synechocystis* sp. *PCC 6803. Eur. J. Biochem.* **253**, 161–172.

Helfrich, M. and Rüdiger, W. (1992). Various metallopheophorbides as substrates for chlorophyll synthetase. *Z. Naturforsch.* **47c**, 231–238.

Helfrich, M., Schoch, S., Lempert, U., Cmiel, E., and Rüdiger, W. (1994). Chlorophyll synthetase cannot synthesize chlorophyll *a'*. *Eur. J. Biochem.* **219**, 267–275.

Helfrich, M., Schoch, S., Schäfer, W., Ryberg, M., and Rüdiger, W. (1996). Absolute configuration of protochlorophyllide *a* and substrate specificity of NADPH-protochlorophyllide oxidoreductase. *J. Am. Chem. Soc.* **118**, 2606–2611.

Hennig, M., Grimm, B., Jenny, M., Müller, R., and Jansonius, J.N. (1994). Crystallization and preliminary X-ray analysis of wild-type and K272A mutant glutamate 1-semialdehyde aminotransferase from *Synechococcus*. *J. Mol. Biol.* **242**, 591–594.

Hennig, M., Grimm, B., Contestabile, R., John, R.A., and Jansonius, J.N. (1997). Crystal structure of glutamate-1semialdehyde aminomutase: An a_2-dimeric vitamin-B_6-dependent enzyme with asymmetry in structure and active site reactivity. *Proc. Natl. Acad. Sci. U. S. A.* **94**, 4866–4871.

Higuchi, M. and Bogorad, L. (1975). The purification and properties of uroporphyrinogen I synthases and uroporphyrinogen III cosynthase. Interactions between the enzymes. *Ann. NY Acad. Sci.* **244**, 401–418.

Hill, K.L. and Merchant, S. (1995). Coordinate expression of coproporphyrinogen oxidase and cytochrome c_6 in the green alga *Chlamydomonas reinhardtii* in response to changes in copper availability. *EMBO J.* **14**, 857–865.

Hinchigeri, S.B., Chan, J.C.-S., and Richards, W.R. (1981). Purification of *S*-adenosyl-L-methionine:magnesium protoporphyrin methyltransferase by affinity chromatography. *Photosynthetica* **15**, 351–359.

Hinchigeri, S.B., Hundle, B., and Richards, W.R. (1997). Demonstration that the BchH protein of *Rhodobacter capsulatus* activates *S*-adenosyl-L-methionine: magnesium protoporphyrin IX methyltransferase. *FEBS Lett.* **407**, 337–342.

Hinder, B., Schellenberg, M., Rodoni, S., Ginsburg, S., Vogt, E., Martinoia, E., Matile, P., and Hörtensteiner, S. (1996). How plants dispose of chlorophyll catabolites. Directly energized uptake of tetrapyrrolic breakdown products into isolated vacuoles. *J. Biol. Chem.* **271**, 27233–27236.

Hörtensteiner, S. (2006). Chlorophyll degradation during senescence. *Annu. Rev. Plant Biol.* **57**, 55–77.

Hörtensteiner, S., Vicentini, F., and Matile, P. (1995). Chlorophyll breakdown in senescent cotyledons of rape, *Brassica napus* L.: Enzymatic cleavage of phaeophorbide *a in vitro*. *New Phytol.* **129**, 237–246.

Hörtensteiner, S., Wüthrich, K.L., Matile, P., Ongania, K.H., and Kräutler, B. (1998). The key step in chlorophyll breakdown in higher plants; Cleavage of pheophorbide *a* macrocycle by a monooxygenase. *J. Biol. Chem.* **273**, 15335–15339.

Howe, G. and Merchant, S. (1992). The biosynthesis of membrane and soluble plastidic *c*-type cytochromes of *Chlamydomonas reinhardtii* is dependent on multiple common gene products. *EMBO J.* **11**, 2789–2801.

Howe, G. and Merchant, S. (1994). Role of heme in the biosynthesis of cytochrome c_6. *J. Biol. Chem.* **269**, 5824–5832.

Howe, G., Mets, L., and Merchant, S. (1995). Biosynthesis of cytochrome *f* in *Chlamydomonas reinhardtii*: Analysis of the pathway in gabaculine-treated cells and in the heme attachment mutant B6. *Mol. Gen. Genet.* **246**, 156–165.

Huang, C. and Liu, X.-Q. (1992). Nucleotide sequence of the *frxC*, *petB* and *trnL* genes in the chloroplast genome of *Chlamydomonas reinhardtii*. *Plant Mol. Biol.* **18**, 985–988.

Huang, D.-D. and Wang, W.-Y. (1986). Genetic control of chlorophyll biosynthesis: Regulation of δ-aminolevulinate synthesis in *Chlamydomonas*. *Mol. Gen. Genet.* **205**, 217–220.

Ikegami, A., Yoshimura, N., Motohashi, K., Takahashi, S., Romano, P.G.N., Hisabori, T., Takamiya, K., and Masuda, T. (2007). The CHLI1 subunit of *Arabidopsis thaliana* magnesium chelatase is a target protein of the chloroplast thioredoxin. *J. Biol. Chem.* **282**, 19282–19291.

Ikeuchi, M. and Murakami, S. (1982). Measurement and identification of NADPH: protochlorophyllide oxidoreductase solubilized with Triton X-100 from etioplast membranes of squash cotyledons. *Plant Cell. Physiol.* **23**, 1089–1099.

Ilag, L.L., Kumar, A.M., and Söll, D. (1994). Light regulation of chlorophyll biosynthesis at the level of 5-aminolevulinate formation in *Arabidopsis*. *Plant Cell* **6**, 265–275.

Ito, H., Takaichi, S., Tsuji, H., and Tanaka, A. (1994). Properties of synthesis of chlorophyll *a* from chlorophyll *b* in cucumber etioplasts. *J. Biol. Chem.* **269**, 22034–22038.

Ito, H., Ohtsuka, T., and Tanaka, A. (1996). Conversion of chlorophyll *b* to chlorophyll *a* via 7-hydroxymethyl chlorophyll. *J. Biol. Chem.* **271**, 1475–1479.

Ito, H., Yokono, M., Tanaka, R., and Tanaka, A. (2008). Identification of a novel vinyl reductase gene essential for the biosynthesis of monovinyl chlorophyll in *Synechocystis* sp. PCC 6803. *J. Biol. Chem.* **283**, 9002–9011.

Jaffe, E.K. (1993). Predicting the Zn(II) ligands in metalloproteins: Case study, porphobilinogen synthase. *Comments Inorg. Chem.* **15**, 67–92.

Jaffe, E.K. (1995). Porphobilinogen synthase, the first source of heme's asymmetry. *J. Bioenerg. Biomembr.* **27**, 169–179.

Jahn, D. (1992). Complex formation between glutamyl-tRNA synthetase and glutamyl-tRNA reductase during tRNA-dependent synthesis of 5-aminolevulinic acid in *Chlamydomonas*. *FEBS Lett.* **314**, 77–80.

Janick, P.A., Rueger, D.C., Krueger, R.J., Barber, M.J., and Siegel, L.M. (1983). Characterization of complexes between *Escherichia coli* sulfite reductase hemoprotein subunit and its substrates sulfite and nitrite. *Biochemistry* **22**, 396–408.

Jansson, S. (1999). A guide to the *Lhc* genes and their relatives in *Arabidopsis*. *Trends Plant Sci.* **4**, 236–240.

Jekabsons, W. and Schuster, W. (1995). *orf250* encodes a second subunit of an ABC-type heme transporter in *Oenothera* mitochondria. *Mol. Gen. Genet.* **246**, 166–173.

Jensen, P.E., Gibson, L.C.D., Henningsen, K.W., and Hunter, C.N. (1996). Expression of the *chlI*, *chlD*, and *chlH* genes from the cyanobacterium *Synechocystis* PCC6803 in *Escherichia coli* and demonstration that the three cognate proteins are required for magnesium-protoporphyrin chelatase activity. *J. Biol. Chem.* **271**, 16662–16667.

Jensen, P.E., Gibson, L.C.D., and Hunter, C.N. (1998). Determinants of catalytic activity with the use of purified I, D, and H subunits of the magnesium protoporphyrin IX chelatase from *Synechocystis* PCC6803. *Biochem. J.* **334**, 335–344.

Jensen, P.E., Gibson, L.C.D., and Hunter, C.N. (1999). ATPase activity associated with the magnesium-protoporphyrin IX chelatase enzyme of *Synechocystis* PCC6803: Evidence for ATP hydrolysis during Mg^{2+} insertion, and the Mg-ATP-dependent interaction of the ChlI and ChlD subunits. *Biochem. J.* **339**, 127–134.

Jones, M.C., Jenkins, J.M., Smith, A.G., and Howe, C.J. (1994). Cloning and characterisation of genes for tetrapyrrole biosynthesis from the cyanobacterium *Anacystis nidulans* R2. *Plant Mol. Biol.* **24**, 435–448.

Jordan, P.M. and Berry, A. (1981). Mechanism of action of porphobilinogen deaminase. The participation of stable enzyme substrate covalent intermediates between porphobilinogen and the porphobilinogen deaminase from *Rhodopseudomonas spheroides*. *Biochem. J.* **195**, 177–181.

Jordan, P.M. and Seehra, J.S. (1979). The biosynthesis of uroporphyrinogen III: Order of assembly of the four porphobilinogen molecules in the formation of the tetrapyrrole ring. *FEBS Lett.* **104**, 364–366.

Jordan, P.M. and Seehra, J.S. (1980). Mechanism of action of 5-aminolevulinic acid dehydratase: Stepwise order of addition of the two molecules of 5-aminolevulinic acid in the enzymic synthesis of porphobilinogen. *J. Chem. Soc. Chem. Commun.* **1980**, 240–242.

Jordan, P.M., and Warren, M.J. (1987). Evidence for a dipyrromethane cofactor at the catalytic site of *Escherichia* coli porphobilinogen deaminase. *FEBS Lett.* **225**, 87–92.

Jordan, P.M., Cheung, K.-M., Sharma, R.P., and Warren, M.J. (1993). 5-Amino-6-hydroxy-3,4,5,6-tetrahydropyran-2-one (HAT). A stable, cyclic form of glutamate 1-semialdehyde, the natural precursor for tetrapyrroles. *Tetrahedron Lett.* **34**, 1177–1180.

Kahn, A. and Kannangara, C.G. (1987). Gabaculine-resistant mutants of *Chlamydomonas reinhardtii* with elevated glutamate 1-semialdehyde aminotransferase activity. *Carlsberg Res. Commun.* **52**, 73–81.

Kannangara, C.G. and Gough, S.P. (1978). Biosynthesis of δ-aminolevulinate in greening barley leaves: Glutamate 1-semialdehyde aminotransferase. *Carlsberg Res. Commun.* **43**, 185–194.

Kannangara, C.G., Gough, S.P., Oliver, R.P., and Rasmussen, S.K. (1984). Biosynthesis of δ-aminolevulinate in greening barley leaves. VI. Activation of glutamate by ligation to RNA. *Carlsberg Res. Commun.* **49**, 417–437.

Kannangara, C.G., Vothknecht, U.C., Hansson, M., and von Wettstein, D. (1997). Magnesium chelatase: Association with ribosomes and mutant complementation studies identify barley subunit Xantha-G as a functional counterpart of *Rhodobacter* subunit BchD. *Mol. Gen. Genet.* **254**, 85–92.

Keller, Y., Bouvier, F., D'Harlingue, A., and Camara, B. (1998). Metabolic compartmentation of plastid prenyllipid biosynthesis. Evidence for the involvement of a multifunctional geranylgeranyl reductase. *Eur. J. Biochem.* **251**, 413–417.

Kikuchi, G., Kumar, A., Talmage, P., and Shemin, D. (1958). The enzymatic synthesis of δ-aminolevulinic acid. *J. Biol. Chem.* **233**, 1214–1219.

Klemm, D.J. and Barton, L.L. (1987). Purification and properties of protoporphyrinogen oxidase from an anaerobic bacterium, *Desulfovibrio gigas*. *J. Bacteriol.* **169**, 5209–5215.

Knaust, R., Seyfried, B., Schmidt, L., Schulz, R., and Senger, H. (1993). Phototransformation of monovinyl and divinyl protochlorophyllide by NADPH: protochlorophyllide oxidoreductase of barley expressed in *Escherichia coli*. *J. Photochem. Photobiol. B.* **20**, 161–166.

Koch, M., Breithaupt, C., Kiefersauer, R., Freigang, J., Huber, R., and Messerschmidt, A. (2004). Crystal structure of protoporphyrinogen IX oxidase: A key enzyme in haem and chlorophyll biosynthesis. *EMBO J.* **23**, 1720–1728.

Kohashi, M., Clement, R.P., Tse, J., and Piper, W.N. (1984). Rat hepatic uroporphyrinogen III co-synthase. Purification and evidence for a bound folate coenzyme participating in the biosynthesis of uroporphyrinogen III. *Biochem. J.* **220**, 755–765.

Kohno, H., Furukawa, T., Tokunaga, R., Taketani, S., and Yoshinaga, T. (1996). Mouse coproporphyrinogen oxidase is a copper-containing enzyme: Expression in *Escherichia coli* and site-directed mutagenesis. *Biochim. Biophys. Acta,* **1292**, 156–162.

Koonin, E.V. (1993). A common set of conserved motifs in a vast variety of putative nucleic acid-dependent ATPases including MCM proteins involved in the initiation of eukaryotic DNA replication. *Nucleic Acids Res.* **21**, 2541–2547.

Koonin, E.V. (1997). Evidence for a family of archaeal ATPases. *Science* **275**, 1489–1490.

Kräutler, B. and Matile, P. (1999). Solving the riddle of chlorophyll breakdown. *Acc. Chem. Res.* **32**, 35–43.

Kruse, E., Mock, H.-P., and Grimm, B. (1995). Coproporphyrinogen III oxidase from barley and tobacco – sequence analysis and initial expression studies. *Planta* **196**, 796–803.

Kumar, A.M., Csankovszki, G., and Söll, D. (1996). A second and differentially expressed glutamyl-tRNA reductase gene from *Arabidopsis thaliana*. *Plant Mol. Biol.* **30**, 419–426.

Kuras, R., de Vitry, C., Choquet, Y., Girard-Bascou, J., Culler, D., Büschlen, S., Merchant, S., and Wollman, F.-A. (1997). Molecular genetic identification of a pathway for heme binding to cytochrome b_6. *J. Biol. Chem.* **272**, 32427–32435.

Larkin, R.M., Alonso, J.M., Ecker, J.R., and Chory, J. (2003). *GUN4*, a regulator of chlorophyll synthesis and intracellular signaling. *Science* **299**, 902–906.

Lascelles, J. and Hatch, T.P. (1969). Bacteriochlorophyll and heme synthesis in *Rhodopseudomonas spheroides*: Possible role of heme in regulation of the branched biosynthetic pathway. *J. Bacteriol.* **98**, 712–720.

Lee, H.J., Ball, M.D., Parham, R., and Rebeiz, C.A. (1992). Chloroplast biogenesis 65. Enzymic conversion of protoporphyrin IX to Mg-protoporphyrin IX in a subplastidic membrane fraction of cucumber etiochloroplasts. *Plant Physiol.* **99**, 1134–1140.

Lee, K.P., Kim, C., Lee, D.W., and Apel, K. (2003). *TIGRINA d*, required for regulating the biosynthesis of tetrapyrroles in barley, is an ortholog of the *FLU* gene of *Arabidopsis thaliana*. *FEBS Lett.* **553**, 119–124.

Lermontova, I., Kruse, E., Mock, H.-P., and Grimm, B. (1997). Cloning and characterization of a plastidal and a mitochondrial isoform of tobacco protoporphyrinogen IX oxidase. *Proc. Natl. Acad. Sci. U. S. A.* **94**, 8895–8900.

Leustek, T., Smith, M., Murillo, M., Singh, D.P., Smith, A.G., Woodcock, S.C., Awan, S.J., and Warren, M.J. (1997). Siroheme biosynthesis in higher plants.

Analysis of an *S*-adenosyl-L-methionine-dependent uroporphyrinogen III methyltransferase from *Arabidopsis thaliana*. *J. Biol. Chem.* **272**, 2744–2752.

Li, J. and Timko, M.P. (1996). The *pc-1* phenotype of *Chlamydomonas reinhardtii* results from a deletion mutation in the nuclear gene for NADPH:protochlorophyllide oxidoreductase. *Plant Mol. Biol.* **30**, 15–37.

Li, J.-M., Russell, C.S., and Cosloy, S.D. (1989). Cloning and structure of the *hemA* gene of *Escherichia coli* K-12. *Gene* **82**, 209–217.

Li, J., Goldschmidt-Clermont, M., and Timko, M.P. (1993). Chloroplast-encoded *chlB* is required for light-independent protochlorophyllide reductase activity in *Chlamydomonas reinhardtii*. *Plant Cell* **5**, 1817–1829.

Lidholm, J. and Gustafsson, P. (1991). Homologues of the green algal *gidA* gene and the liverwort *frxC* gene are present on the chloroplast genomes of conifers. *Plant Mol. Biol.* **17**, 787–798.

Liu, X.-Q., Xu, H., and Huang, C. (1993). Chloroplast *chlB* is required for light-independent chlorophyll accumulation in *Chlamydomonas reinhardtii*. *Plant Mol. Biol.* **23**, 297–308.

Lister, R., Chew, O., Rudhe, C., Lee, M.-N., and Whelan, J. (2001). *Arabidopsis thaliana* ferrochelatase-I and -II are not imported into *Arabidopsis* mitochondria. *FEBS Lett.* **506**, 291–295.

Lohr, M., Im, C.-S., and Grossman, A.R. (2005). Genome-based examination of chlorophyll and carotenoid biosynthesis in *Chlamydomonas reinhardtii*. *Plant Physiol.* **138**, 490–515.

Lopez, J.C., Ryan, S., and Blankenship, R.E. (1996). Sequence of the *bchG* gene from *Chloroflexus aurantiacus*: Relationship between chlorophyll synthase and other polyprenyltransferases. *J. Bacteriol.* **178**, 3369–3373.

Louie, G.V., Brownlie, P.D., Lambert, R., Cooper, J.B., Blundell, T.L., Wood, S.P., Warren, M.J., Woodcock, S.C., and Jordan, P.M. (1992). Structure of porphobilinogen deaminase reveals a flexible multidomain polymerase with a single catalytic site. *Nature* **359**, 33–39.

Louie, G.V., Brownlie, P.D., Lambert, R., Cooper, J.B., Blundell, T.L., Wood, S.P., Malashkevich, V.N., Hadener, A., Warren, M.J., and Shoolingin-Jordan, P.M. (1996). The three-dimensional structure of *Escherichia coli* porphobilinogen deaminase at 1.76-Å resolution. *Proteins* **25**, 48–78.

Lüer, C., Schauer, S., Möbius, K., Schulze, J., Schubert, W.-D., Heinz, D.W., Jahn, D., and Moser, J. (2005). Complex formation between glutamyl-tRNA reductase and glutamate-1-semialdehyde 2,1-aminomutase in *Escherichia coli* during initial reactions of porphyrin biosynthesis. *J. Biol. Chem.* **280**, 18568–18572.

Luo, J. and Lim, C.K. (1993). Order of uroporphyrinogen III decarboxylation on incubation of porphobilinogen and uroporphyrinogen III with erythrocyte uroporphyrinogen decarboxylase. *Biochem. J.* **289**, 529–532.

Madsen, O., Sandal, L., Sandal, N.N., and Marcker, K.A. (1993). A soybean coproporphyrinogen oxidase gene is highly expressed in root nodules. *Plant Mol. Biol.* **23**, 35–43.

Maeda, H., Watanabe, T., Kobayashi, M., and Ikegami, I. (1992). Presence of two chlorophyll *a′* molecules at the core of I. *Biochim. Biophys. Acta*, **1099**, 74–80.

Majumdar, D., Avissar, Y.J., Wyche, J.H., and Beale, S.I. (1991). Structure and expression of the *Chlorobium vibrioforme hemA* gene. *Arch. Microbiol.* **156**, 281–289.

Maloney, M.A., Hoober, J.K., and Marks, D.B. (1989). Kinetics of chlorophyll accumulation and formation of chlorophyll–protein complexes during greening of *Chlamydomonas reinhardtii y-1* at 38°C. *Plant Physiol.* **91**, 1100–1106.

Martin, G.E., Timko, M.P., and Wilks, H.M. (1997). Purification and kinetic analysis of pea (*Pisum sativum* L.) NADPH:protochlorophyllide oxidoreductase expressed as a fusion with maltose-binding protein in *Escherichia coli*. *Biochem. J.* **325**, 139–145.

Martins, B.M., Grimm, B., Mock, H.P., Huber, R., and Messerschmidt, A. (2001). Crystal structure and substrate binding modeling of the uroporphyrinogen-III decarboxylase from *Nicotiana tabacum*. Implications for the catalytic mechanism. *J. Biol. Chem.* **276**, 44108–44116.

Masuda, T., Suzuki, T., Shimada, H., Ohta, H., and Takamiya, K. (2003). Subcellular localization of two types of ferrochelatase in cucumber. *Planta* **217**, 602–609.

Mathews, M.A.A., Schubert, H.L., Whitby, F.G., Alexander, K.J., Schadick, K., Bergonia, H.A., Phillips, J.D., and Hill, C.P. (2001). Crystal structure of human uroporphyrinogen III synthase. *EMBO J.* **20**, 5832–5839.

Matringe, M., Camadro, J.-M., Labbe, P., and Scalla, R. (1989). Protoporphyrinogen oxidase as a molecular target for diphenyl ether herbicides. *Biochem. J.* **260**, 231–235.

Matters, G.L. and Beale, S.I. (1994). Structure and light-regulated expression of the *gsa* gene encoding the chlorophyll biosynthetic enzyme, glutamate-1-semialdehyde aminotransferase, in *Chlamydomonas reinhardtii*. *Plant Mol. Biol.* **24**, 617–629.

Matters, G.L. and Beale, S.I. (1995a). Structure and expression of the *Chlamydomonas reinhardtii alad* gene encoding the chlorophyll biosynthetic enzyme, δ-aminolevulinic acid dehydratase (porphobilinogen synthase). *Plant Mol. Biol.* **27**, 607–617.

Matters, G.L. and Beale, S.I. (1995b). Blue-light-regulated expression of genes for two early steps of chlorophyll biosynthesis in *Chlamydomonas reinhardtii*. *Plant Physiol.* **109**, 471–479.

Mau, Y.-H.L. and Wang, W.-Y. (1988). Biosynthesis of δ-aminolevulinic acid in *Chlamydomonas reinhardtii*. Study of the transamination mechanism using specifically labeled glutamate. *Plant Physiol.* **86**, 793–797.

Mayer, S.M., Weinstein, J.D., and Beale, S.I. (1987). Enzymatic conversion of glutamate to δ-aminolevulinate in soluble extracts of *Euglena gracilis*. *J. Biol. Chem.* **262**, 12541–12549.

Mayer, S.M., Gawlita, E., Avissar, Y.J., Anderson, V.E., and Beale, S.I. (1993). Intermolecular nitrogen transfer in the enzymatic conversion of glutamate to δ-aminolevulinic acid by extracts of *Chlorella vulgaris*. *Plant Physiol.* **101**, 1029–1038.

Mayer, S.M., Rieble, S., and Beale, S.I. (1994). Metal requirements of the enzymes catalyzing conversion of glutamate to δ-aminolevulinic acid in extracts of *Chlorella vulgaris* and *Synechocystis* sp. PCC 6803. *Arch. Biochem. Biophys.* **312**, 203–209.

Medlock, A.E. and Dailey, H.A. (1996). Human coproporphyrinogen oxidase is not a metalloprotein. *J. Biol. Chem.* **271**, 32507–32510.

Mehta, P.K. and Christen, P. (1994). Homology of 1-aminocyclopropane-1-carboxylate synthase, 8-amino-7-oxononanoate synthase, 2-amino-6-caprolactam racemase, 2,2-dialkylglycine decarboxylase, glutamate-1-semialdehyde 2,1-aminomutase and isopenicillin-*N*-epimerase with aminotransferases. *Biochem. Biophys. Res. Commun.* **198**, 138–143.

Meller, E., Belkin, S., and Harel, E. (1975). The biosynthesis of δ-aminolevulinic acid in greening maize leaves. *Phytochemistry* **14**, 2399–2402.

Merchant, S.S., Allen, M.D., Kropat, J., Moseley, J.L., Long, J.C., Tottey, S., and Terauchi, A.M. (2006). Between a rock and a hard place: Trace element nutrition in *Chlamydomonas*. *Biochim. Biophys. Acta,* **1763**, 578–594.

Meskauskiene, R. and Apel, K. (2002). Interaction of FLU, a negative regulator of tetrapyrrole biosynthesis, with the glutamyl-tRNA reductase requires the tetratricopeptide repeat domain of FLU. *FEBS Lett.* **532**, 27–30.

Meskauskiene, R., Nater, M., Goslings, D., Kessler, F., op den Camp, R., and Apel, K. (2001). FLU: A negative regulator of chlorophyll biosynthesis in *Arabidopsis thaliana*. *Proc. Natl. Acad. Sci. U. S. A.* **98**, 12826–12831.

Mock, H.-P., Trainotti, L., Kruse, E., and Grimm, B. (1995). Isolation, sequencing and expression of cDNA sequences encoding uroporphyrinogen decarboxylase from tobacco and barley. *Plant Mol. Biol.* **28**, 245–256.

Mogi, T., Saiki, K., and Anraku, Y. (1994). Biosynthesis and functional role of heme *o* and heme *a*. *Mol. Microbiol.* **14**, 391–398.

Moseley, J., Quinn, J., Eriksson, M., and Merchant, S. (2000). The *Crd1* gene encodes a putative di-iron enzyme for photosystem I accumulation in copper deficiency and hypoxia in *Chlamydomonas reinhardtii*. *EMBO J.* **19**, 2139–2151.

Moseley, J.L., Page, M.D., Adler, N.P., Eriksson, M., Quinn, J., Soto, F., Theg, S.M., Hippler, M., and Merchant, S. (2002). Reciprocal expression of two candidate di-iron enzymes affecting photosystem I and light-harvesting complex accumulation. *Plant Cell* **14**, 673–688.

Moser, J., Schubert, W.D., Beier, V., Bringemeier, I., Jahn, D., and Heinz, D.W. (2001). V-shaped structure of glutamyl-tRNA reductase, and first enzyme of tRNA-dependent tetrapyrrole biosynthesis. *EMBO J.* **20**, 6583–6590.

Nagata, N., Tanaka, R., Satoh, S., and Tanaka, A. (2005). Identification of a vinyl reductase gene for chlorophyll synthesis in *Arabidopsis thaliana* and implications for the evolution of *Prochlorococcus* species. *Plant Cell* **17**, 233–240.

Nakamoto, S.S., Hamel, P., and Merchant, S. (2000). Assembly of chloroplast cytochromes *b* and *c*. *Biochimie* **82**, 603–614.

Nakayama, M., Masuda, T., Bando, T., Yamagata, H., Ohta, H., and Takamiya, K. (1998). Cloning and expression of the soybean *chlH* gene encoding a subunit of Mg-chelatase and localization of the Mg^{2+} concentration-dependent ChlH protein within the chloroplast. *Plant Cell Physiol.* **39**, 275–284.

Nandi, D.L. and Shemin, D. (1973). δ-Aminolevulinic acid dehydratase of *Rhodopseudomonas capsulata*. *Arch. Biochem. Biophys.* **158**, 305–311.

Nogaj, L.A. and Beale, S.I. (2005). Physical and kinetic interactions between glutamyl-tRNA reductase and glutamate-1-semialdehyde aminotransferase of *Chlamydomonas reinhardtii*. *J. Biol. Chem.* **280**, 24301–24307.

Nogaj, L.A., Srivastava, A., van Lis, R., and Beale, S.I. (2005). Cellular levels of glutamyl-tRNA reductase and glutamate-1-semialdehyde aminotransferase do not control chlorophyll synthesis in *Chlamydomonas reinhardtii*. *Plant Physiol.* **139**, 389–396.

Nomata, J., Swem, L.R., Bauer, C.E., and Fujita, Y. (2005). Overexpression and characterization of dark-operative protochlorophyllide reductase from *Rhodobacter capsulatus*. *Biochim. Biophys. Acta,* **1708**, 229–237.

Nomata, J., Kitashima, M., Inoue, K., and Fujita, Y. (2006). Nitrogenase Fe protein-like FeS cluster is conserved in L-protein (BchL) of dark-operative protochlorophyllide reductase from *Rhodobacter capsulatus*. *FEBS Lett.* **580**, 6151–6154.

Oda, K., Yamato, K., Ohta, E., Nakamura, Y., Takemura, M., Nozato, N., Akashi, K., Kanegae, T., Ogura, Y., Kohchi, Y., and Ohyama, K. (1992). Gene organization deduced from the complete sequence of liverwort *Marchantia polymorpha* mitochondrial DNA. A primitive form of plant mitochondrial genome. *J. Mol. Biol.* **223**, 1–7.

Ohtsuka, T., Ito, H., and Tanaka, A. (1997). Conversion of chlorophyll *b* to chlorophyll *a* and the assembly of chlorophyll with apoproteins by isolated chloroplasts. *Plant Physiol.* **113**, 137–147.

Oliver, R.P. and Griffiths, W.T. (1981). Covalent labelling of the NADPH:protochlorophyllide oxidoreductase from etioplast membranes with [3H]*N*-phenylmaleimide. *Biochem. J.* **195**, 93–101.

Oliver, R.P. and Griffiths, W.T. (1982). Pigment-protein complexes of illuminated etiolated leaves. *Plant Physiol.* **70**, 1019–1025.

Omata, T., Murata, N., and Satoh, K. (1984). Quinone and pheophytin in the photosynthetic reaction center II from spinach chloroplasts. *Biochim. Biophys. Acta,* **765**, 403–405.

O'Neill, G.P., Schön, A., Chow, H., Chen, M.-W., Kim, Y.-C., and Söll, D. (1990). Sequence of $tRNA^{Glu}$ and its genes from the chloroplast genome of *Chlamydomonas reinhardtii*. *Nucleic Acids Res.* **18**, 5893.

Oster, U. and Rüdiger, W. (1997). The *G4* gene of *Arabidopsis thaliana* encodes a chlorophyll synthase of etiolated plants. *Bot. Acta,* **110**, 420–423.

Oster, U., Bauer, C.E., and Rüdiger, W. (1997). Characterization of chlorophyll *a* and bacteriochlorophyll *a* synthases by heterologous expression in *Escherichia coli*. *J. Biol. Chem.* **272**, 9671–9676.

Oster, U., Tanaka, R., Tanaka, A., and Rüdiger, W. (2000). Cloning and functional expression of the gene encoding the key enzyme for chlorophyll *b* biosynthesis (CAO) from *Arabidopsis thaliana*. *Plant J.* **21**, 305–310.

Panek, H. and O'Brian, M.R. (2002). A whole genome view of prokaryotic haem biosynthesis. *Microbiology* **148**, 2273–2282.

Parham, R. and Rebeiz, C.A. (1992). Chloroplast biogenesis: [4-Vinyl] chlorophyllide *a* reductase is a divinyl chlorophyllide *a*-specific, NADPH-dependent enzyme. *Biochemistry* **31**, 8460–8464.

Parham, R. and Rebeiz, C.A. (1995). Chloroplast biogenesis 72: [4-Vinyl] chlorophyllide *a* reductase assay using divinyl chlorophyllide *a* as an exogenous substrate. *Anal. Biochem.* **231**, 164–169.

Peters, J.W., Fisher, K., and Dean, D.R. (1995). Nitrogenase structure and function: A biochemical-genetic perspective. *Annu. Rev. Microbiol.* **49**, 335–366.

Petersen, P.M., Hawker, C.J., Stamford, N.P.J., Leeper, F.J., and Battersby, A.R. (1998). Biosynthesis of porphyrins and related macrocycles. Part 50. Synthesis of the *N*-formyl-dihydro analogue of the spiro-intermediate and its interaction with uroporphyrinogen III synthase. *J. Chem. Soc. Perkin. Trans. I,* **1998**, 1531–1539.

Petersen, B.L., Kannangara, C.G., and Henningsen, K.W. (1999). Distribution of ATPase and ATP-to-ADP phosphate exchange activities in magnesium chelatase subunits of *Chlorobium vibrioforme* and *Synechocystis* PCC6803. *Arch. Microbiol.* **171**, 146–150.

Petricek, M., Rutberg, L., Schröder, I., and Hederstedt, L. (1990). Cloning and characterization of the *hemA* region of the *Bacillus subtilis* chromosome. *J. Bacteriol.* **172**, 2250–2258.

Phillips, J.D., Whitby, F.G., Kushner, J.P., and Hill, C.P. (2003). Structural basis for tetrapyrrole coordination by uroporphyrinogen decarboxylase. *EMBO J.* **22**, 6225–6233.

Phillips, J.D., Whitby, F.G., Warby, C.A., Labbe, P., Yan, C., Pflugrath, J.W., Ferrara, J.D., Robinson, H., Kushner, J.P., and Hill, C.P. (2004). Crystal structure of the oxygen-dependent coproporphyrinogen oxidase (Hem13p) of *Saccharomyces cerevisiae*. *J. Biol. Chem.* **279**, 38960–38968.

Pontoppidan, B. and Kannangara, C.G. (1994). Purification and partial characterization of barley glutamyl-tRNA^{Glu} reductase, the enzyme that directs glutamate to chlorophyll biosynthesis. *Eur. J. Biochem.* **225**, 529–537.

Pöpperl, G., Oster, U., Blos, I., and Rüdiger, W. (1997). Magnesium chelatase of *Hordeum vulgare* L. is not activated by light but inhibited by pheophorbide. *Z. Naturforsch.* **52c**, 144–152.

Porra, R.J., Schäfer, W., Cmiel, E., Katheder, I., and Scheer, H. (1993). Derivation of the formyl-group oxygen of chlorophyll *b* from molecular oxygen in greening leaves of a higher plant (*Zea mays*). *FEBS Lett.* **323**, 31–34.

Posewitz, M.C., King, P.W., Smolinski, S.L., Zhang, L., Seibert, M., and Ghirardi, M.L. (2004). Discovery of two novel radical *S*-adenosylmethionine proteins required for the assembly of an active [Fe] hydrogenase. *J. Biol. Chem.* **279**, 25711–25720.

Poulson, R. and Polglase, W.J. (1974). Aerobic and anaerobic coproporphyrinogenase activities in extracts from *Saccharomyces cerevisiae*. *J. Biol. Chem.* **249**, 6367–6371.

Quesada, A., Galván, A., Schnell, R.A., Lefebvre, P.A., and Fernández, E. (1993). Five nitrate assimilation-related loci are clustered in *Chlamydomonas reinhardtii*. *Mol. Gen. Genet.* **240**, 387–394.

Quinn, J.M., Nakamoto, S.S., and Merchant, S. (1999). Induction of coproporphyrinogen oxidase in *Chlamydomonas* chloroplasts occurs via transcriptional regulation of *Cpx1* mediated by copper response elements and increased translation from a copper deficiency-specific form of the transcript. *J. Biol. Chem.* **274**, 14444–14454.

Radmer, R.J. and Bogorad, L. (1967). (–)*S*-adenosyl-L-methionine-magnesium protoporphyrin methyltransferase, an enzyme in the biosynthetic pathway of chlorophyll in *Zea mays*. *Plant Physiol.* **42**, 463–465.

Randau, L., Schauer, S., Ambrogelly, A., Salazar, J.C., Moser, J., Sekine, S., Yokoyama, S., Söll, D., and Jahn, D. (2004). tRNA recognition by glutamyl-tRNA reductase. *J. Biol. Chem.* **279**, 34931–34937.

Randolph-Anderson, B.L., Sato, R., Johnson, A.M., Harris, E.H., Hauser, C.R., Oeda, K., Ishige, F., Nishio, S., Gillham, N.W., and Boynton, J.E. (1998). Isolation and characterization of a mutant protoporphyrinogen oxidase gene from *Chlamydomonas reinhardtii* conferring resistance to porphyric herbicides. *Plant Mol. Biol.* **38**, 839–859.

Raux, E., McVeigh, T., Peters, S.E., Leustek, T., and Warren, M.J. (1999). The role of *Saccharomyces cerevisiae* Met1p and Met8p in sirohaem and cobalamin biosynthesis. *Biochem. J.* **338**, 701–708.

Raux, E., Leech, H.K., Beck, R., Schubert, H.L., Santander, P.J., Roessner, C.A., Scott, A.I., Martens, J.H., Jahn, D., Thermes, C., Rambach, A., and Warren, M.J. (2003). Identification and functional analysis of enzymes required for precorrin-2 dehydrogenation and metal ion insertion in the biosynthesis of sirohaem and cobalamin in *Bacillus megaterium*. *Biochem. J.* **370**, 505–516.

Raux-Deery, E., Leech, H.K., Nakrieko, K.-A., Mclean, K.J., Munro, A.W., Heathcote, P., Rigby, S.E.J., Smith, A.G., and Warren, M.J. (2005). Identification and characterisation of the terminal enzyme of siroheme biosynthesis from *Arabidopsis thaliana*; a plastid-located sirohydrochlorin ferrochelatase containing a 2Fe-2S centre. *J. Biol. Chem.* **280**, 4713–4721.

Rieble, S. and Beale, S.I. (1988). Enzymatic transformation of glutamate to δ-aminolevulinic acid by soluble extracts of *Synechocystis* sp. 6803 and other oxygenic prokaryotes. *J. Biol. Chem.* **263**, 8864–8871.

Rieble, S. and Beale, S.I. (1991). Separation and partial characterization of enzymes catalyzing δ-aminolevulinic acid formation in *Synechocystis* sp. PCC 6803. *Arch. Biochem. Biophys.* **289**, 289–297.

Reid, J.D., Siebert, C.A., Bullough, P.A., and Hunter, C.N. (2003). The ATPase activity of the ChlI subunit of magnesium chelatase and formation of a heptameric AAA(+) ring. *Biochemistry* **42**, 6912–6920.

Rieble, S., Ormerod, J.G., and Beale, S.I. (1989). Transformation of glutamate to δ-aminolevulinic acid by soluble extracts of *Chlorobium vibrioforme*. *J. Bacteriol.* **171**, 3782–3787.

Romero, L.C., Galván, F., and Vega, J.M. (1987). Purification and properties of the siroheme-containing ferredoxin-nitrite reductase from *Chlamydomonas reinhardtii*. *Biochim. Biophys. Acta,* **914**, 55–63.

Rosé, S., Frydman, R.B., de los Santos, C., Sburlati, A., Valasinas, A., and Frydman, B. (1988). Spectroscopic evidence for a porphobilinogen deaminase-tetrapyrrole complex that is an intermediate in the biosynthesis of uroporphyrinogen III. *Biochemistry (USA)* **27**, 4871–4879.

Rüdiger, W. (1987). Chlorophyll synthetase and its implication for regulation of chlorophyll biosynthesis. In: *Progress in Photosynthesis Research*, Vol. IV (J. Biggins, Ed.), pp. 461–467. Martinus Nijhoff, Boston.

Rüdiger, W. (2006). Biosynthesis of chlorophylls *a* and *b*: The last steps. In: *Chlorophylls and Bacteriochlorophylls* (B. Grimm, R.J. Porra, W. Rüdiger, and H. Scheer, Eds.), pp. 189–200. Springer, Dordrecht.

Rüdiger, W., Benz, J., and Guthoff, C. (1980). Detection and partial characterization of activity of chlorophyll synthetase in etioplast membranes. *Eur. J. Biochem.* **109**, 193–200.

Rudoi, A.B. and Shcherbakov, R.A. (1998). Analysis of the chlorophyll biosynthetic system in a chlorophyll *b*-less barley mutant. *Photosynth. Res.* **58**, 71–80.

Rzeznicka, K., Walker, C.J., Westergren, T., Kannangara, C.G., von Wettstein, D., Merchant, S., Gough, S.P., and Hansson, M. (2005). *Xantha-l* encodes a membrane subunit of the aerobic Mg-protoporphyrin IX monomethyl ester cyclase involved in chlorophyll biosynthesis. *Proc. Natl. Acad. Sci. U. S. A.* **102**, 5886–5891.

Schauder, J.-R., Jendrezejewski, S., Abell, A., Hart, G.J., and Battersby, A.R. (1987). Stereochemistry of formation of the hydroxymethyl group of hydroxymethylbilane, the precursor of Uro'gen-III. *J. Chem. Soc. Chem. Commun.* **1987**, 436–438.

Schauer, S., Chaturvedi, S., Randau, L., Moser, J., Kitabatake, M., Lorenz, S., Verkamp, E., Schubert, W.-D., Nakayashiki, T., Murai, M., Wall, K., Thomann, H.-U., Hieinz, D.W., Inokuchi, H., Söll, D., and Jahn, D. (2002). *Escherichia coli* glutamyl-tRNA reductase. Trapping the thioester intermediate. *J. Biol. Chem.* **277**, 48658–48663.

Scheumann, V., Ito, H., Tanaka, A., Schoch, S., and Rüdiger, W. (1996). Substrate specificity of chlorophyll(ide) *b* reductase in etioplasts of barley (*Hordeum vulgare* L.). *Eur. J. Biochem.* **242**, 163–170.

Scheumann, V., Schoch, S., and Rüdiger, W. (1999). Chlorophyll *b* reduction during senescence of barley seedlings. *Planta* **209**, 364–370.

Schneegurt, M.A. and Beale, S.I. (1988). Characterization of the RNA required for biosynthesis of δ-aminolevulinic acid from glutamate. Purification by anticodon-based affinity chromatography and determination that the UUC glutamate anticodon is a general requirement for function in ALA biosynthesis. *Plant Physiol.* **86**, 497–504.

Schneegurt, M.A. and Beale, S.I. (1992). Origin of the chlorophyll *b* formyl oxygen atom in *Chlorella vulgaris*. *Biochemistry* **31**, 11677–11683.

Schoch, S. (1978). The esterification of chlorophyllide *a* in greening bean leaves. *Z. Naturforsch.* **33c**, 712–714.

Schoch, S. and Schäfer, W. (1978). Tetrahydrogeranylgeraniol, a precursor of phytol in the biosynthesis of chlorophyll *a* – localization of the double bonds. *Z. Naturforsch.* **33c**, 408–412.

Schoch, S., Lempert, U., and Rüdiger, W. (1977). Über die letzten Stufen der Chlorophyll-biosynthese zwischenprodkte zwischen Chlorophyllid und phytolhaltigem Chlorophyll. *Z. Pflanzenphysiol.* **83**, 427–436.

Schoch, S., Helfrich, M., Wiktorsson, B., Sundqvist, C., Rüdiger, W., and Ryberg, M. (1995). Photoreduction of zinc protopheophorbide *b* with NADPH-protochlorophyllide oxidoreductase from etiolated wheat (*Triticum aestivum* L.). *Eur. J. Biochem.* **242**, 674–681.

Schön, A. and Söll, D. (1988). tRNA specificity of a mischarging aminoacyl-tRNA synthetase: Glutamyl-tRNA synthetase from barley chloroplasts. *FEBS Lett.* **228**, 241–244.

Schön, A., Krupp, G., Gough, S., Berry-Lowe, S., Kannangara, C.G., and Söll, D. (1986). The RNA required in the first step of chlorophyll biosynthesis is a chloroplast glutamate tRNA. *Nature* **322**, 281–284.

Schön, A., Kannangara, C.G., Gough, S., and Söll, D. (1988). Protein biosynthesis in organelles requires misaminoacylation of tRNA. *Nature* **331**, 187–190.

Schubert, H.L., Raux, E., Wilson, K.S., and Warren, M.J. (1999). Common chelatase design in the branched tetrapyrrole pathways of heme and anaerobic cobalamin synthesis. *Biochemistry* **38**, 10660–10669.

Schulz, R., Steinmüller, K., Klaas, M., Forreiter, C., Rasmussen, S., Hiller, C., and Apel, K. (1989). Nucleotide sequence of a cDNA coding for the NADPH-protochlorophyllide oxidoreductase (PCR) of barley (*Hordeum vulgare* L.) and its expression in *Escherichia coli*. *Mol. Gen. Genet.* **217**, 355–361.

Schuster, W. (1994). The highly edited *orf206* in *Oenothera* mitochondria may encode a component of a heme transporter involved in cytochrome *c* biogenesis. *Plant Mol. Biol.* **25**, 33–42.

Schuster, W., Combettes, B., Flieger, K., and Brennicke, A. (1993). A plant mitochondrial gene encodes a protein involved in cytochrome *c* biogenesis. *Mol. Gen. Genet.* **239**, 49–57.

Scott, A.I., Irwin, A.J., Siegel, L.M., and Shoolery, J.N. (1978). Sirohydrochlorin. Prosthetic group of sulfite and nitrite reductases and its role in the biosynthesis of vitamin B_{12}. *J. Am. Chem. Soc.* **100**, 7987–7994.

Scott, A.I., Burton, G., Jordan, P.M., Matsumoto, H., Fagerness, P.E., and Pryde, L. M. (1980). N.M.R. spectroscopy as a probe for the study of enzyme-catalysed reactions. Further observations of preuroporphyrinogen, a substrate for uroporphyrinogen III cosynthetase. *J. Chem. Soc. Chem. Commun.* **1980**, 384–387.

Seehra, J.S., Jordan, PM, and Akhtar, M. (1983). Anaerobic and aerobic coproporphyrinogen-III oxidases of *Rhodopseudomonas spheroides*. Mechanism, and stereochemistry of vinyl group formation. *Biochem. J.* **209**, 709–718.

Shen, Y.-Y., Wang, X.-F., Wu, F.-Q., Du, S.-Y., Cao, Z., Shang, Y., Wang, X.-L., Peng, C.-C., Yu, X.-C., Zhu, S.-Y., Fan, R.-C., Xu, Y.-H., and Zhang, D.-P. (2006). The Mg-chelatase H subunit is an abscisic acid receptor. *Nature* **443**, 823–826.

Shepherd, M., McLean, S., and Hunter, C.N. (2005). Kinetic basis for linking the first two enzymes of chlorophyll biosynthesis. *FEBS J.* **272**, 4532–4539.

Shieh, J., Miller, G.W., and Psenak, M. (1978). Properties of S-adenosyl-L-methionine-magnesium protoporphyrin IX methyltransferase from barley. *Plant Cell Physiol.* **19**, 1051–1059.

Shioi, Y., Nagamine, M., Kuroki, M., and Sasa, T. (1980). Purification by affinity chromatography and properties of uroporphyrinogen-I synthase from *Chlorella regularis*. *Biochim. Biophys. Acta* **616**, 300–309.

Shoolingin-Jordan, P.M. and Williams, S.C. (2003). Stereochemical course of the formation of the C(7)-formyl group from a chiral methyl group during the transformation of chlorophyllide *a* into chlorophyllide *b*. *Helv. Chim. Acta,* **86**, 4185–4194.

Shoolingin-Jordan, P.M., Warren, M.J., and Awan, S.J. (1996). Discovery that the assembly of the dipyrromethane cofactor of porphobilinogen deaminase holoenzyme proceeds initially by the reaction of preuroporphyrinogen with the apoenzyme. *Biochem. J.* **316**, 373–376.

Siepker, L.J., Ford, M., de Kock, R., and Kramer, S. (1987). Purification of bovine protoporphyrinogen oxidase: Immunological cross-reactivity and structural relationship to ferrochelatase. *Biochim. Biophys. Acta,* **913**, 349–358.

Sirijovski, N., Olsson, U., Lundqvist, J., Al-Karadaghi, S., Willows, R.D., and Hansson, M. (2006). ATPase activity associated with the magnesium chelatase H-subunit of the chlorophyll biosynthetic pathway is an artefact. *Biochem. J.* **400**, 477–484.

Smith, M.A., Kannangara, C.G., and Grimm, B. (1992). Glutamate 1-semialdehyde aminotransferase: Anomalous enantiomeric reaction and enzyme mechanism. *Biochemistry* **31**, 11249–11254.

Smith, M.A., King, P.J., and Grimm, B. (1998). Transient-state kinetic analysis of *Synechococcus* glutamate 1-semialdehyde aminotransferase. *Biochemistry* **37**, 319–329.

Sobotka, R., McLean, S., Zuberova, M., Hunter, C.N., and Tichy, M. (2008). The C-terminal extension of ferrochelatase is critical for enzyme activity and for functioning of the tetrapyrrole pathway in *Synechocystis* strain PCC 6803. *J. Bacteriol.* **190**, 2086–2095.

Soll, J., Schultz, G., Rüdiger, W., and Benz, J. (1983). Hydrogenation of geranylgeraniol. Two pathways exist in spinach chloroplasts. *Plant Physiol.* **71**, 849–854.

Spano, A.J., He, Z., Michel, H., Hunt, D.F., and Timko, M.P. (1992a). Molecular cloning, nuclear gene structure, and developmental expression of NADPH:protochlorophyllide oxidoreductase in pea (*Pisum sativum* L.). *Plant Mol. Biol.* **18**, 967–972.

Spano, A.J., He, Z., and Timko, M.P. (1992b). NADPH:protochlorophyllide oxidoreductases in white pine (*Pinus strobus*) and loblolly pine (*P. teada*). *Mol. Gen. Genet.* **236**, 86–95.

Spielewoy, N., Schulz, H., Grienenberger, J.M., Thöny-Meyer, L., and Bonnard, G. (2001). CCME, a nuclear-encoded heme-binding protein involved in cytochrome *c* maturation in plant mitochondria. *J. Biol. Chem.* **276**, 5491–5497.

Spiller, S.C., Castelfranco, A.M., and Castelfranco, P.A. (1982). Effects of iron and oxygen of chlorophyll biosynthesis. I. *In vivo* observations on iron and oxygen-deficient plants. *Plant Physiol.* **69**, 107–111.

Srivastava, A. and Beale, S.I. (2005). Glutamyl-tRNA reductase of *Chlorobium vibrioforme* is a dissociable homodimer that contains one tightly bound heme per subunit. *J. Bacteriol.* **187**, 4444–4450.

Srivastava, A., Lake, V., Nogaj, L.A. , Mayer, S.M., Willows, R.D., and Beale, S.I. (2005). The *Chlamydomonas reinhardtii gtr* gene encoding the tetrapyrrole biosynthetic enzyme glutamyl-tRNA reductase: Structure of the gene and properties of the expressed enzyme. *Plant Mol. Biol.* **58**, 643–658.

Stange-Thomann, N., Thomann, H.-U., Lloyd, A.J., Lyman, H., and Söll, D. (1994). A point mutation in *Euglena gracilis* chloroplast $tRNA^{Glu}$ uncouples protein and chlorophyll biosynthesis. *Proc. Natl. Acad. Sci. U. S. A.* **91**, 7947–7951.

Stark, W.M., Hawker, C.J., Hart, G.J., Philippides, A., Petersen, P.M., Lewis, J.D., Leeper, F.J., and Battersby, A.R. (1993). Biosynthesis of porphyrins and related macrocycles. Part 40. Synthesis of a spiro-lactam related to the proposed spiro-intermediate for porphyrin biosynthesis: Inhibition of cosynthetase. *J. Chem. Soc. Perkin Trans. I,* **1993**, 2875–2891.

Stroebel, D., Choquet, Y., Popot, J.-L., and Picot, D. (2003). An atypical haem in the cytochrome b_6f complex. *Nature* **426**, 413–418.

Stroupe, M.E., Leech, H.K., Daniels, D.S., Warren, M.J., and Getzoff, E.D. (2003). CysG structure reveals tetrapyrrole-binding features and novel regulation of siroheme biosynthesis. *Nature Struct. Biol.* **10**, 1064–1073.

Susek, R.E., Ausubel, F.M., and Chory, J. (1993). Signal transduction mutants of *Arabidopsis* uncouple nuclear CAB and RBCS gene expression from chloroplast development. *Cell* **74**, 787–799.

Suzuki, J.Y. and Bauer, C.A. (1992). Light-independent chlorophyll synthesis: Involvement of the chloroplast gene *chlL* (*frxC*). *Plant Cell* **4**, 929–940.

Suzuki, J.Y. and Bauer, C.A. (1995). Altered monovinyl and divinyl protochlorophyllide pools in *bchJ* mutants of *Rhodobacter capsulatus*. Possible monovinyl substrate discrimination of light-independent protochlorophyllide reductase. *J. Biol. Chem.* **270**, 3732–3740.

Suzuki, Y., Doi, M., and Shioi, Y. (2002). Two enzymatic reaction pathways in the formation of pyropheophorbide *a*. *Photosynth. Res.* **74**, 225–233.

Tait, G.H. (1972). Coproporphyrinogenase activities in extracts of *Rhodopseudomonas spheroides* and *Chromatium* strain D. *Biochem. J.* **128**, 1159–1169.

Tanaka, R., Yoshida, K., Nakayashiki, T., Masuda, T., Tsuji, H., Inokuchi, H., and Tanaka, A. (1996). Differential expression of two *hemA* mRNAs encoding glutamyl-tRNA reductase proteins in greening cucumber seedlings. *Plant Physiol.* **110**, 1223–1230.

Tanaka, R., Yoshida, K., Nakayashiki, T., Tsuji, H., Inokuchi, H., Okada, K., and Tanaka, A. (1997). The third member of the *hemA* family encoding glutamyl-tRNA reductase is primarily expressed in roots in *Hordeum vulgare*. *Photosynth. Res.* **53**, 161–171.

Tanaka, A., Ito, H., Tanaka, R., Tanaka, N.K., Yoshida, K., and Okada, K. (1998). Chlorophyll *a* oxygenase (CAO) is involved in chlorophyll *b* formation from chlorophyll *a*. *Proc. Natl. Acad. Sci. U. S. A.* **95**, 12719–12723.

Tanaka, R., Koshino, Y., Sawa, S., Ishiguro, S., Okada, K., and Tanaka, A. (2001). Overexpression of chlorophyllide *a* oxygenase (CAO) enlarges the antenna size of photosystem II in *Arabidopsis thaliana*. *Plant J.* **26**, 365–373.

Thomas, J. and Weinstein, J.D. (1990). Measurement of heme efflux and heme content in isolated developing chloroplasts. *Plant Physiol.* **94**, 1414–1423.

Timkovich, R., Bergmann, D., Arciero, D.M., and Hooper, A.B. (1998). Primary sequence and solution conformation of ferrocytochrome *c*-552 from *Nitrosomonas europaea*. *Biophys. J.* **75**, 1964–1972.

Thöny-Meyer, L. (2000). Haem-polypeptide interactions during cytochrome *c* maturation. *Biochim. Biophys. Acta,* **1459**, 316–324.

Tottey, S., Block, M.A., Allen, A., Westergren, T., Albrieux, C., Scheller, H.V., Merchant, S., and Jensen, P.E. (2003). *Arabidopsis* CHL27, located in both envelope and thylakoid membranes, is required for the synthesis of protochlorophyllide. *Proc. Natl. Acad. Sci. U. S. A.* **100**, 16119–16124.

Townley, H.E., Griffiths, W.T., and Nugent, J.P. (1998). A reappraisal of the mechanism of the photoenzyme protochlorophyllide reductase based on studies with the heterologously expressed protein. *FEBS Lett.* **422**, 19–22.

Tripathy, B.C. and Rebeiz, C.A. (1988). Chloroplast biogenesis 60. Conversion of divinyl protochlorophyllide to monovinyl protochlorophyllide in green(ing) barley, a dark monovinyl/light divinyl plant species. *Plant Physiol.* **87**, 89–94.

Valera, V., Fung, M., Wessler, A.N., and Richards, W.R. (1987). Synthesis of 4*R*- and 4*S*-tritium labeled NADPH for the determination of the coenzyme stereospecificity of NADPH:protochlorophyllide oxidoreductase. *Biochem. Biophys. Res. Commun.* **148**, 515–520.

van Lis, R., Atteia, A., Nogaj, L.A., and Beale, S.I. (2005). Subcellular localization and light-regulated expression of protoporphyrinogen IX oxidase and ferrochelatase in *Chlamydomonas reinhardtii*. *Plant. Physiol.* **139**, 1946–1958.

Vavilin, D. and Vermaas, W. (2007). Continuous chlorophyll degradation accompanied by chlorophyllide and phytol reutilization for chlorophyll synthesis in *Synechocystis* sp. PCC 6803. *Biochim. Biophys. Acta,* **1767**, 920–929.

Verkamp, E. and Chelm, B.K. (1989). Isolation, nucleotide sequence, and preliminary characterization of the *Escherichia coli* K-12 *hemA* gene. *J. Bacteriol.* **171**, 4728–4735.

Vothknecht, U.C., Kannangara, C.G., and von Wettstein, D. (1996). Expression of catalytically active barley glutamyl $tRNA^{Glu}$ reductase in *Escherichia coli* as a fusion protein with glutathione *S*-transferase. *Proc. Natl. Acad. Sci. U. S. A.* **93**, 9287–9291.

Vothknecht, U.C., Kannangara, C.G., and von Wettstein, D. (1998). Barley glutamyl $tRNA^{Glu}$ reductase: Mutations affecting haem inhibition and enzyme activity. *Phytochemistry* **47**, 513–519.

Walker, C.J. and Griffiths, W.T. (1988). Protochlorophyllide reductase: A flavoprotein? *FEBS Lett.* **239**, 259–262.

Walker, C.J. and Weinstein, J.D. (1991a). Further characterization of the magnesium chelatase in isolated developing cucumber chloroplasts. Substrate specificity, regulation, intactness, and ATP requirements. *Plant Physiol.* **95**, 1189–1196.

Walker, C.J. and Weinstein, J.D. (1991b). *In vitro* assay of the chlorophyll biosynthetic enzyme Mg-chelatase: Resolution of the activity into soluble and membrane-bound fractions. *Proc. Natl. Acad. Sci. U. S. A.* **88**, 5789–5793.

Walker, C.J. and Weinstein, J.D. (1994). The magnesium insertion step of chlorophyll biosynthesis is a two-stage reaction. *Biochem. J.* **299**, 277–284.

Walker, C.J. and Weinstein, J.D. (1995). Re-examination of the localization of Mg-chelatase within the chloroplast. *Physiol. Plant.* **94**, 419–424.

Walker, C.J. and Willows, R.D. (1997). Mechanism and regulation of Mg-chelatase. *Biochem. J.* **327**, 321–333.

Walker, C.J., Mansfield, K.E., Smith, K.M., and Castelfranco, P.A. (1989). Incorporation of atmospheric oxygen into the carbonyl functionality of the protochlorophyllide isocyclic ring. *Biochem. J.* **257**, 599–602.

Walker, C.J., Castelfranco, P.A., and Whyte, B.J. (1991). Synthesis of divinyl protochlorophyllide. Enzymological properties of the Mg-protoporphyrin IX monomethyl ester oxidative cyclase system. *Biochem. J.* **276**, 691–697.

Walker, C.J., Hupp, L.R., and Weinstein, J.D. (1992). Activation and stabilization of Mg-chelatase activity by ATP as revealed by a novel *in vitro* continuous assay. *Plant Physiol. Biochem.* **30**, 263–269.

Walther, Z. and Hall, J.L. (1995). The uni chromosome of *Chlamydomonas*: Histone genes and nucleosome structure. *Nucleic Acids Res.* **23**, 3756–3763.

Wang, W.-Y., Huang, D.-D., Stachon, D., Gough, S.P., and Kannangara, C.G. (1984). Purification, characterization, and fractionation of the δ-aminolevulinic acid synthesizing enzymes from light-grown *Chlamydomonas reinhardtii* cells. *Plant Physiol.* **74**, 569–575.

Warren, M.J. and Jordan, P.M. (1988). Further evidence for the involvement of a dipyrromethane cofactor at the active site of porphobilinogen deaminase. *Biochem. Soc. Trans.* **16**, 963–965.

Watanabe, N., Che, F.-S., Iwano, M., Takayama, S., Yoshida, S., and Isogai, A. (2001). Dual targeting of spinach protoporphyrinogen oxidase II to mitochondria and chloroplasts by alternative use of two in-frame initiation codons. *J. Biol. Chem.* **276**, 20474–20481.

Weinstein, J.D. and Beale, S.I. (1985). Enzymatic conversion of glutamate to δ-aminolevulinate in soluble extracts of the unicellular green alga, *Chlorella vulgaris*. *Arch. Biochem. Biophys.* **237**, 454–464.

Weinstein, J.D., Mayer, S.M., and Beale, S.I. (1987). Formation of δ-aminolevulinic acid from glutamic acid in algal extracts. Separation into an RNA and three required enzyme components by serial affinity chromatography. *Plant Physiol.* **84**, 244–250.

Weinstein, J.D., Howell, R.W., Leverette, R.D., Grooms, S.Y., Brignola, P.S., Mayer, S.M., and Beale, S.I. (1993). Heme inhibition of δ-aminolevulinic acid synthesis is enhanced by glutathione in cell-free extracts of *Chlorella*. *Plant Physiol.* **101**, 657–665.

Whyte, B.J., Vijayan, P., and Castelfranco, P.A. (1992). *In vitro* synthesis of protochlorophyllide: Effects of Mg^{2+} and other cations in the reconstituted (oxidative) cyclase. *Plant Physiol. Biochem.* **30**, 279–284.

Wiktorsson, B., Ryberg, M., and Sundqvist, C. (1996). Aggregation of NADPH-protochlorophyllide oxidoreductase-pigment complexes is favoured by protein phosphorylation. *Plant Physiol. Biochem.* **34**, 23–34.

Wilde, A., Mikolajczyk, S., Alawady, A., Lokstein, H., and Grimm, B. (2004). The *gun4* gene is essential for cyanobacterial porphyrin metabolism. *FEBS Lett.* **571**, 119–123.

Williams, D.C., Morgan, G.S., McDonald, E., and Battersby, A.R. (1981). Purification of porphobilinogen deaminase from *Euglena gracilis* and studies of its kinetics. *Biochem. J.* **193**, 301–310.

Willows, R.D., Kannangara, C.G., and Pontoppidan, B. (1995). Nucleotides of tRNA (Glu) involved in recognition by barley chloroplast glutamyl-tRNA synthetase and glutamyl-tRNA reductase. *Biochim. Biophys. Acta,* **1263**, 228–234.

Willows, R.D., Gibson, L.C.D., Kannangara, C.G., Hunter, C.N., and von Wettstein, D. (1996). Three separate proteins constitute the magnesium chelatase of *Rhodobacter sphaeroides*. *Eur. J. Biochem.* **235**, 438–443.

Wong, Y.-S. and Castelfranco, P.A. (1984). Resolution and reconstitution of the Mg-protoporphyrin IX monomethyl ester (oxidative) cyclase, the enzyme system responsible for the formation of the chlorophyll isocyclic ring. *Plant Physiol.* **75**, 658–661.

Wong, Y.S., Castelfranco, P.A., Goff, D.A., and Smith, K.M. (1985). Intermediates in the formation of the chlorophyll isocyclic ring. *Plant Physiol.* **79**, 725–729.

Woodcock, S.C. and Warren, M.J. (1996). Evidence for a covalent intermediate in the *S*-adenosyl-L-methionine-dependent transmethylation reaction catalysed by sirohaem synthase. *Biochem. J.* **313**, 415–421.

Woodcock, S.C., Raux, E., Levillayer, F., Thermes, C., Rambach, A., and Warren, M.J. (1998). Effect of mutations in the transmethylase and dehydrogenase/chelatase domains of sirohaem synthase (CysG) on sirohaem and cobalamin biosynthesis. *Biochem. J.* **330**, 121–129.

Wu, C.-K., Dailey, H.A., Rose, J.P., Burden, A., Sellers, V.M., and Wang, B.-C. (2001). The 2-Å structure of human ferrochelatase, the terminal enzyme of heme biosynthesis. *Nature Struct. Biol.* **8**, 156–160.

Xie, Z., Culler, D., Dreyfuss, B.W., Kuras, R., Wollman, F.-A., Girard-Bascou, J., and Merchant, S. (1998). Genetic analysis of chloroplast *c*-type cytochrome assembly in *Chlamydomonas reinhardtii*: One chloroplast locus and at least four nuclear loci are required for heme attachment. *Genetics* **148**, 681–692.

Xu, K. and Elliott, T. (1993). An oxygen-dependent coproporphyrinogen oxidase encoded by the *hemF* gene of *Salmonella typhimurium*. *J. Bacteriol.* **175**, 4990–4999.

Xu, K. and Elliott, T. (1994). Cloning, DNA sequence, and complementation analysis of the *Salmonella typhimurium hemN* gene encoding a putative oxygen-independent coproporphyrinogen III oxidase. *J. Bacteriol.* **176**, 3196–3203.

Yamashita, E., Aoyama, H., Yao, M., Muramoto, K., Shinzawa-Itoh, K., Yoshikawa, S., and Tsukihara, T. (2005). Absolute configuration of the hydroxyfarnesylethyl group of haem A, determined by X-ray structural analysis of bovine heart cytochrome *c* oxidase using methods applicable at 2.8 Å resolution. *Acta Crystal.* **D61**, 1373–1377.

Yamato, S., Ida, T., Katagiri, M., and Ohkawa, H. (1995). A tobacco soluble protoporphyrinogen-oxidizing enzyme similar to plant peroxidases in their amino acid sequences and immunochemical reactivity. *Biosci. Biotechnol. Biochem.* **59**, 558–559.

Yamazaki, S., Nomata, J., and Fujita, Y. (2006). Differential operation of dual protochlorophyllide reductases for chlorophyll biosynthesis in response to environmental oxygen levels in the cyanobacterium *Leptolyngbya boryana*. *Plant Physiol.* **142**, 911–922.

Yoshida, K., Igarashi, E., Mukai, M., Hirata, K., and Miyamoto, K. (2003). Induction of tolerance to oxidative stress in the green alga, *Chlamydomonas reinhardtii*, by abscisic acid. *Plant Cell Environ.* **26**, 451–457.

Yoshida, K., Igarashi, E., Wakatsuki, E., Miyamoto, K., and Hirata, K. (2004). Mitigation of osmotic and salt stresses by abscisic acid through reduction of stress-derived oxidative damage in *Chlamydomonas reinhardtii*. *Plant Sci.* **167**, 1335–1341.

Zaman, Z. and Akhtar, M. (1976). Mechanism and stereochemistry of vinyl-group formation in haem biosynthesis. *Eur. J. Biochem.* **61**, 215–223.

Zhang, Z., Shrager, J., Jain, M., Chang, C.-W., Vallon, O., and Grossman, A.R. (2004). Insights into the survival of *Chlamydomonas reinhardtii* during sulfur starvation based on microarray analysis of gene expression. *Eukaryotic Cell* **3**, 1331–1348.

Zollner, A., Rodel, G., and Haid, A. (1992). Molecular cloning and characterization of the *Saccharomyces cerevisiae CYT2* gene encoding cytochrome c_1-heme lyase. *Eur. J. Biochem.* **207**, 1093–1100.

Zsebo, K.M. and Hearst, J.E. (1984). Genetic-physical mapping of a photosynthetic gene cluster from *R. capsulata*. *Cell,* **37**, 937–947.

CHAPTER 21

Carotenoids

Martin Lohr
Institut für Allgemeine Botanik,
Johannes Gutenberg-Universität Mainz, Germany

CHAPTER CONTENTS

I. INTRODUCTION

Carotenoids are an essential component of all photosynthetic organisms due to their eminent photoprotective and antioxidant properties. However, their occurrence is not restricted to plants, algae, and cyanobacteria, as some fungi and non-photosynthetic bacteria can synthesize carotenoids as well, and many animals rely on food-borne carotenoids as visual pigments, antioxidants, or colorants. The carotenoids belong to the isoprenoids and their basic structure is made up of eight isoprene units, resulting in a C_{40} backbone. Formally, two types of carotenoids can be discerned: Carotenes are pure hydrocarbons while xanthophylls are derivatives that contain one or more oxygen functions. Although carotenoid-deficient mutants of *Chlamydomonas* were among the earliest carotenoid-deficient plants and algae under investigation, current knowledge on the biosynthesis of these compounds has been gained mostly from studies of carotenogenic bacteria and vascular plants. Still, work on *Chlamydomonas* and other closely related green algae added significantly to the present picture of algal carotenoids, and the purpose of this chapter is to give an overview of what is known about carotenoid metabolism in *Chlamydomonas* with reference to other green algae and vascular plants.

II. OCCURRENCE AND ANALYSIS

In vegetative cells of *Chlamydomonas* growing under low to moderate light conditions, the major carotenoids are β-carotene, violaxanthin, neoxanthin, lutein, and loroxanthin (19-hydroxy-lutein). Earlier results that were obtained by thin-layer chromatographic (TLC) methods (Sager and Zalokar, 1958; Krinsky and Levine, 1964; Stolbova, 1971; Sirevåg and Levine, 1973; Francis et al., 1973; Nitzsche, 1974) were subsequently confirmed by high-performance liquid chromatography (HPLC) (Eichenberger et al., 1986; Chunaev et al., 1991). In nutrient-replete cultures grown under moderate light intensities, lutein and β-carotene each account for about 25% of total carotenoids, while the share of the other three xanthophylls is between 10% and 20% each (Eichenberger et al., 1986; Niyogi et al., 1997b). Neoxanthin is present exclusively as the 9′-*cis*-isomer (Takaichi and Mimuro, 1998), whereas the conformation of the other xanthophylls is all-*trans* (see Figure 21.1 for chemical structures); a minor fraction of β-carotene is present as *cis*-isomers (see below). Under high light or other stress conditions which lead to an increased photosynthetic excitation pressure, part of the violaxanthin pool of *Chlamydomonas* can be de-epoxidized to zeaxanthin via the intermediate antheraxanthin due to operation of the xanthophyll cycle (Niyogi et al., 1997b; see Chapter 23). The reaction sequence is reversed in subsequent low light. Thus, the general pattern of photosynthetic pigments in *Chlamydomonas* is the same as in vascular plants with the exception of loroxanthin. The occurrence of loroxanthin is restricted to certain families of green algae (Goodwin, 1980; Fawley and Buchheim, 1995; Schagerl and Angeler, 1998; Schagerl et al., 2003).

While the carotenoids in *Chlamydomonas* display only stoichiometric changes throughout the vegetative cell cycle (Francis et al., 1975) or during senescence under nutrient limitation (Czygan, 1968; Lohr et al., 2005), preliminary analyses of mature zygospores in our laboratory have indicated a significantly different pigment pattern including a number of additional carotenoids which share the presence of one or two ketogroups in the molecule. Among them, we tentatively identified canthaxanthin, astaxanthin, and fritschiellaxanthin (4-keto-lutein), as well as numerous putative acyl esters of the latter two pigments. The ability to synthesize ketocarotenoids

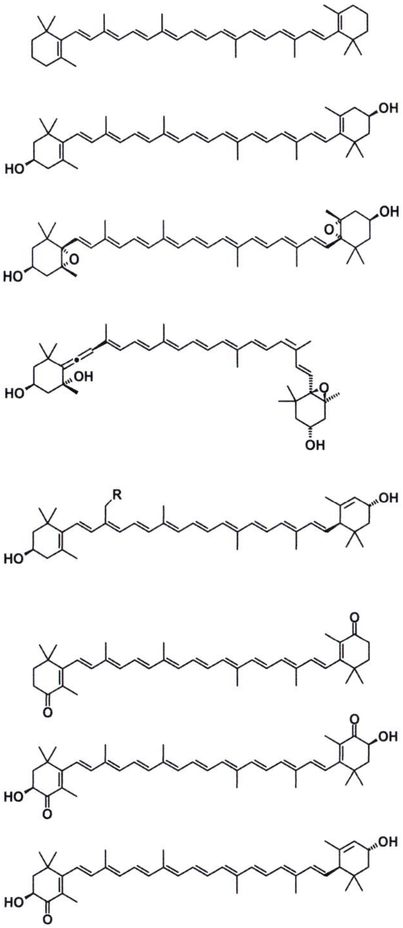

FIGURE 21.1 *Chemical structures of the major carotenoids that are present in vegetative cells and of the ketocarotenoids tentatively identified in the zygospore resting stage of Chlamydomonas.*

has also been demonstrated for various other green algae, for example *Chlamydomonas nivalis*, and members of the genera *Haematococcus* and *Scenedesmus* (Goodwin, 1980; Kessler et al., 1997; Remias et al., 2005; Jin et al., 2006).

For analysis of the carotenoids that are encountered in *Chlamydomonas* numerous protocols are available in the literature. Among them are methods for separation of the isomers lutein and zeaxanthin by HPLC (e.g. Thayer and Björkman, 1990; Gilmore and Yamamoto, 1991; Kraay et al., 1992; Fraser et al., 2000) or TLC (Hager and Meyer-Bertenrath, 1966), the separation of neoxanthin and loroxanthin by HPLC (e.g. Eichenberger et al., 1986; Plumley and Schmidt, 1995; Kraay et al., 1992) or TLC (Janero and

Table 21.1 Summary of the putative carotenogenic enzymes

Step	Enzyme	Full name	Synonyms	Accession #[a]	Reference	*Chlamydomonas* Mutants
1	DXS	1-deoxy-D-xylulose-5-phosphate synthase	CLA1, DEF, DXPS	CAA07554	Schwender et al. (1999)	
2	DXR	1-deoxy-D-xylulose-5-phosphate reductoisomerase	IspC, YaeM	EDP02894		
3	CMS	4-diphosphocytidyl-2-C-methyl-D-erythritol synthase	IspD, YgbP	ED099224		
4	CMK	4-diphosphocytidyl-2-C-methyl-D-erythritol kinase	IspE, YchB, CDPMEK	EDP02028		
5	MCS	2-C-methyl-D-erythritol 2,4-cyclodiphosphate synthase	IspF, YgbB	EDP05431		
6	HDS	4-hydroxy-3-methylbut-2-en-1-yl diphosphate synthase	IspG, GcpE	AAX54905	Lohr et al. (2005)	
7	IDS	Isopentenyl-/dimethylallyl-diphosphate synthase	IspH, LytB	ED097597		
8	IDI	Isopentenyl-diphosphate delta-isomerase	Ipi	AAC32601	Sun et al. (1998)	
9	GGPPS	Geranylgeranyl-diphosphate synthase	GGPS, CrtE	ED096545		
10	PSY	Phytoene synthase	CrtB, Pys	AAT38474	McCarthy et al. (2004)	*lts1*
11	PDS	Phytoene desaturase	CrtP	AAT38476	McCarthy et al. (2004)	
12	Z-ISO	ζ-Carotene isomerase		Not identified	Nikulina et al. (1999)	*ac5* (?)
13	ZDS	ζ-Carotene desaturase	CrtQ, CrtQb	EDP07040		
14	CRTISO	Carotenoid isomerase	CrtH	ED099516		
15	LCYB	Lycopene β-cyclase	LBC, CrtL-b	AAX54906	Lohr et al. (2005)	
16	LCYE	Lycopene ε-cyclase	LEC, CrtL-e	AAT46065	Anwaruzzaman et al. (2004)	*lor1*
17	CHYB	Carotene β-hydoxylase	CrtZ, CrtR-b	AAX54907	Lohr et al. (2005)	
18	CYP97A5	Carotene β-hydoxylase	LUT5	ED098047		

(Continued)

Table 21.1 *Continued*

Step	Enzyme	Full name	Synonyms	Accession #[a]	Reference	*Chlamydomonas* Mutants
19	CYP97C3	Carotene ε-hydoxylase	LUT1	ABQ59243		
20	ZEP	Zeaxanthin epoxidase	ABA1	AAO34404	Baroli et al. (2003)	*npq2*
21	VDE	Violaxanthin de-epoxidase	AVDE1	no orthologue		*npq1*
22	NSY	Neoxanthin synthase	NXS	no orthologue		
23	LSY	Loroxanthin synthase		not identified		
24	BKT	Carotene β-ketolase	CrtW	AAX54908	Lohr et al. (2005)	

[a]*no orthologue = no orthologue of the corresponding genes from vascular plants in available Chlamydomonas sequence data. See Figure 21.2 and text for enzyme functions.*

Barrnett, 1981), and the analysis of ketocarotenoids and their acyl esters by HPLC (Yuan and Chen, 1998; Fraser et al., 2000) or TLC (Weber, 1975).

III. BIOSYNTHESIS

As mentioned earlier, our current knowledge on the biosynthesis of carotenoids has been gained mainly from studies of carotenogenic bacteria (Armstrong and Hearst, 1996) and vascular plants (Cunningham and Gantt, 1998; Hirschberg, 2001; Bouvier et al., 2005). For most of the genes from vascular plants that are known to be involved in carotenogenesis, putative or known orthologs have been identified in *Chlamydomonas* (Grossman et al., 2004; Table 21.1), and a detailed comparison of the respective genes with those from the cyanobacterium *Synechocystis* sp. PCC 6803 and from *Arabidopsis* has been performed (Lohr et al., 2005). Figure 21.2 depicts the putative pathway of carotenoid biosynthesis in *Chlamydomonas*. As indicated already by experimental data of Sirevåg and Levine (1973) and as is the case for vascular plants, the enzymes involved are all nucleus-encoded and consequently are synthesized in the cytosol. An analysis of the deduced open reading frames showed that they are predicted to encode precursor proteins containing cleavable N-terminal extensions which target them to the plastid (Lohr et al., 2005).

Briefly, carotenogenesis can be subdivided into five stages: (i) formation of the building block which is central to all isoprenoids, the active isoprene (Figure 21.2, steps 1–7), (ii) stepwise condensation of eight isoprene units leading to phytoene, the first carotenoid (steps 9–10), (iii) extension of the conjugated π-electron system by four desaturation steps and interposed isomerization reactions to yield all-*trans* lycopene (steps 11–14), (iv) cyclization of the ends of lycopene leading to the formation of carotenes (steps 15–16), and (v) oxygenation reactions generating various xanthophylls (steps 17–24). Here, only a brief description of the biosynthetic events will be presented. Details of the individual reactions have been addressed in several reviews (Cunningham and Gantt, 1998; Hirschberg, 2001; Grossman et al., 2004; Bouvier et al., 2005).

As has been indicated by the particular labeling pattern of carotenoids (Disch et al., 1998; Schwender et al., 2001), in green algae the active isoprene is synthesized exclusively in the plastid from the substrates pyruvate and glyceraldehyde-3-phosphate via the methyl-erythritol-phosphate (MEP) pathway, which comprises seven steps (Lichtenthaler, 1999; Rodríguez-Concepción and Boronat, 2002; Rohmer, 2003). The respective genes are represented by homologues in *Chlamydomonas* (see Table 21.1, steps 1–7), whereas homologues of genes encoding the enzymes of the mevalonate pathway could not be detected (Grossman et al., 2004; Lohr et al., 2005).

The active isoprene comprises two isomers, isopentenyl diphosphate (IPP) and dimethylallyl diphosphate (DMAPP), and the enzyme isopentenyl diphosphate isomerase catalyzes the rapid equilibration between the two isoforms (step 8). Two molecules of DMAPP and six molecules of IPP are utilized in the subsequent condensation reactions (steps 9–10) leading to formation of the first carotenoid of the pathway, phytoene. Next, the system of conjugated double bonds is extended by the consecutive action of two desaturases (steps 11 and 13) followed by a *cis–trans* isomerization that is catalyzed by a carotenoid isomerase (step 14), resulting in the linear polyene lycopene. Observations on a maize pale-yellow endosperm mutant (*y9*, see below) suggest that an additional isomerase is interposed between the two desaturases, catalyzing the *cis–trans* isomerization of ζ-carotene (step 12), but the enzyme/gene has not yet been identified (Li et al., 2007).

A screen for *Chlamydomonas* mutants unable to perform the synthesis of colored carotenoids (McCarthy et al., 2004) led to the identification of several clones that were defective in phytoene synthase (PSY, step 10); all of the these mutants proved to be allelic to a series of light-sensitive carotenoid-deficient mutants that had been characterized before with the defective locus being designated as *lts1* (Stolbova, 1972; Chemerilova, 1978). Surprisingly, no mutants defective in phytoene desaturase (PDS, step 11) were picked up, suggesting that the screen was not saturating or that in contrast to vascular plants, either some mechanism of lethality might have prevented their isolation, or the PDS may be redundant with some other enzyme (McCarthy et al., 2004). Another mutant which was isolated based on its acetate requirement for growth and therefore was termed *ac5*, has been shown to

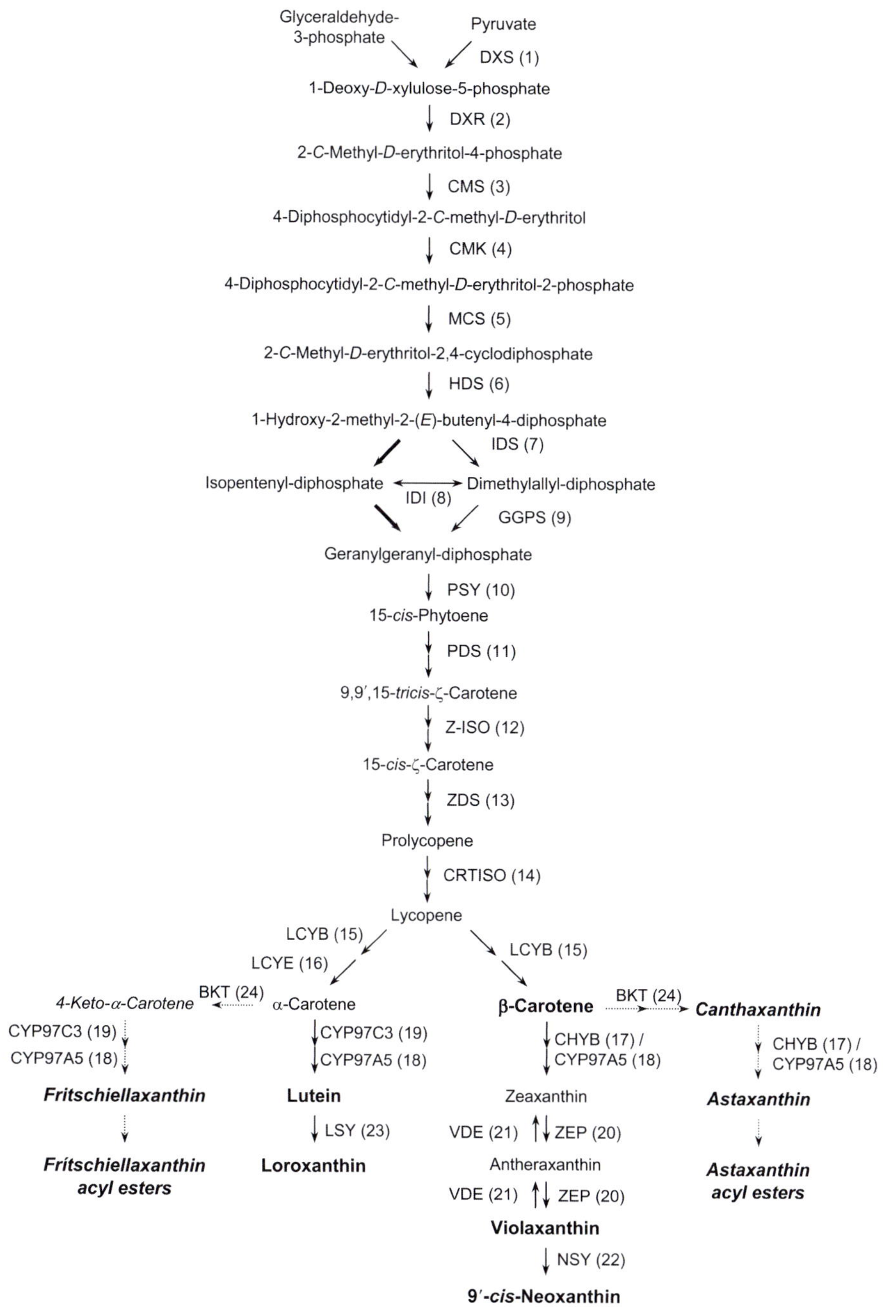

FIGURE 21.2 *Putative pathways of carotenoid biosynthesis in Chlamydomonas (based on Lohr et al., 2005). The names of major carotenoids that are present in vegetative cells are in bold. Names of the tentatively identified secondary carotenoids which accumulate only in zygospores are in bold italics; hypothetical zygospore-specific pathways are indicated by dotted arrows. For the enzymes of the pathways only abbreviations are given (see Table 21.1 for full names).*

be affected in the desaturation of ζ-carotene to lycopene (Nikulina et al., 1999), which is catalyzed by ζ-carotene desaturase (ZDS, step 13). However, the mutant accumulated ζ-carotene only in the dark, suggesting that the gene encoding ZDS is not the primary target of the mutation. A similar behavior has been reported for the maize pale-yellow endosperm mutant *y9* (Robertson, 1975), which accumulates various isomers of ζ-carotene specifically in dark-grown tissues such as roots and etiolated leaves (Li et al., 2007), and although the identity of the *y9* locus in maize is yet unknown, it was proposed to encode a ζ-carotene isomerase (Z-ISO, step 12). Since the *ac5* mutant was also shown to accumulate several ζ-carotene isomers in the dark (Nikulina et al., 1999), the corresponding locus in *Chlamydomonas* might be homologous to the *y9* locus in maize.

At the level of lycopene, the pathway splits into two branches: (i) cyclization of lycopene by lycopene-β-cyclase (LCYB, step 15) results in the formation of β-carotene (β,β-carotene), which is the precursor of the xanthophylls violaxanthin and neoxanthin, and (ii) the concerted action of LCYB and lycopene-ε-cyclase (LCYE, step 16) leads to the formation of α-carotene (β,ε-carotene), the precursor of lutein and loroxanthin. The carotenes are hydroxylated at the C3-position (steps 17–19) by two types of evolutionary unrelated hydroxylases, the non-heme di-iron monoxygenases that catalyze only the hydroxylation of β-ionon rings (carotene-β-hydroxylases, CHYB; Sun et al., 1996), and the heme-containing cytochrome P450 monoxygenases (P450) comprising both an ε-ring hydroxylase (CYP97C1 in *Arabidopsis*; Tian et al., 2004) and a β-ring hydroxylase (CYP97A3 in *Arabidopsis*; Kim and DellaPenna, 2006). In *Arabidopsis*, CHYB appears to be mainly responsible for hydroxylation of β-carotene to yield zeaxanthin, while the two P450 enzymes preferentially catalyze the hydroxylation of α-carotene to lutein. *Chlamydomonas* contains candidate genes for CHYB as well as for the two P450 enzymes (Table 21.1); however, no functional analyses of the respective gene products have been performed.

While for *Chlamydomonas* the gene encoding zeaxanthin epoxidase (ZEP, step 20) has been cloned (Baroli et al., 2003), in version 3.0 of the genome there are no genes encoding putative orthologues of violaxanthin de-epoxidase (VDE, step 21) and neoxanthin synthase (NSY, step 22) from vascular plants (Anwaruzzaman et al., 2004; Lohr et al., 2005). Furthermore, the enzyme that is responsible for the formation of loroxanthin by hydroxylation of lutein (LSY, step 23) is not known.

For the late steps of carotenogenesis in vegetative cells, a number of mutants are available: on the α-branch, the *Chlamydomonas* mutant *lor1* (formerly designated as *pg-101*) lacks lutein and loroxanthin (Chunaev et al., 1991), and the *lor1* phenotype has been shown to be caused by a defective LCYE (Anwaruzzaman et al., 2004); the corresponding LCYE-deficient mutant of *Arabidopsis* is known as *lut2* (Pogson and Rissler, 2000). On the β-branch, mutants defective in either the ZEP or the VDE reaction have been isolated by a video-imaging screen allowing detection of aberrant non-photochemical quenching (NPQ) of chlorophyll fluorescence (Niyogi et al., 1997a). Accordingly, these mutants were termed *npq1* (VDE) and *npq2* (ZEP); the latter corresponds to the ZEP-deficient *aba1* mutant of *Arabidopsis* (Pogson and Rissler, 2000).

The formation of the secondary ketocarotenoids in zygospores of *Chlamydomonas* is likely to be catalyzed by the product of the *BKT* gene (step 24) that was discovered in the course of the genome project (Lohr et al., 2005), but is awaiting experimental confirmation. The last steps of ketocarotenoid biosynthesis may take place in the cytosol, as has been shown for the green microalga *Haematococcus pluvialis* (syn. *H. lacustris*; Grünewald et al., 2001). Available experimental data suggest that in *H. pluvialis* the CHYB (Linden, 1999) and P450 enzymes (Schoefs et al., 2001) may both contribute to the hydroxylation reactions during astaxanthin formation. No information is available on the enzyme(s) involved in the subsequent acylation of the ketocarotenoids. The functional significance of acylation might be to trigger the sequestration of carotenoids to lipid globules and to avoid crystallization of the planar chromophores, thus allowing the storage of large amounts of pigments.

Concerning the regulatory aspects of carotenogenesis in *Chlamydomonas* vegetative cells, available experimental data demonstrate control at the transcriptional as well as the post-translational levels. The mRNAs encoding PSY and PDS were shown to each have a half-life of about 2–3 hours both in the dark and under high-light conditions (Bohne and Linden, 2002), making them likely targets for transcriptional regulation. In fact, an increase of transcript levels for both PSY and PDS has been observed during exposure of cells pre-grown either in darkness or under low light (30 micromoles photons/m^2/s) to high-intensity white light (200 micromoles photons/m^2/s) and during incubation of dark-grown cells under actinic (80 micromoles photons/m^2/s) blue light, but not red light (Bohne and Linden, 2002). Furthermore, induction is not sensitive to inhibitors of photosynthetic electron transport such as 3-(3,4-dichlorophenyl)-1,1-dimethylurea (DCMU), indicating that it is not governed by photosynthetic redox control (Bohne and Linden, 2002; Im et al., 2006). In this respect, *Chlamydomonas* differs from *H. pluvialis*, for which a transcriptional regulation of PSY, PDS, LCYB, and CHYB by photosynthetic redox control has been shown (Steinbrenner and Linden, 2001, 2003). A detailed analysis for *Chlamydomonas* using different light intensities and light qualities demonstrated that non-actinic very-low-fluence (0.01 micromoles photons/m^2/s) white light and low-fluence (0.2 micromoles photons/m^2/s) blue light are similarly efficient in increasing the transcript levels for HDS, PSY, PDS, and ZDS (Im et al., 2006). Moreover, a comparison of the response of a wild-type *Chlamydomonas* strain with that of three

RNAi strains with decreased levels of the photoreceptor phototropin (PHOT) indicated that PHOT is involved in blue light regulation of carotenoid biosynthesis (Im et al., 2006).

At the post-translational level, work on chlorophyll *b*-deficient mutants of *Chlamydomonas* suggests that synthesis of neoxanthin from violaxanthin and the accumulation of loroxanthin are linked to the assembly of the particular light-harvesting complexes to which the xanthophylls are bound, presumably by interaction of the antenna proteins with the enzyme catalyzing the formation of neoxanthin and by a reduced turnover of protein-bound loroxanthin, respectively (Plumley and Schmidt, 1995).

In *Chlamydomonas*, regulation of ketocarotenoid formation during zygospore maturation has not yet been investigated. For *H. pluvialis*, abiotic stress factors have been demonstrated to promote the induction of cyst formation and the concomitant accumulation of ketocarotenoids. Biosynthesis of secondary carotenoids has been shown to be inducible by high light, salt stress, nitrogen deficiency, phosphate or sulfate starvation (Boussiba, 2000; Jin et al., 2006). It appears reasonable to assume that one or more of these factors similarly affect the formation of ketocarotenoids in zygospores of *Chlamydomonas*.

IV. LOCALIZATION AND FUNCTIONAL SIGNIFICANCE

With respect to subcellular distribution, the carotenoids in vegetative cells localize to the chloroplast where they either serve as photosynthetic pigments bound to the protein complexes of the two photosystems (see Chapter 14), or as components of the eyespot apparatus (see Volume 3, Chapter 13). In the thylakoid membranes, the bulk of β-carotene is located in the core complexes of the two photosystems (Sugiura et al., 1998; Pineau et al., 2001), but it is also present in the light-harvesting antenna of photosystem I (PS I) (Pineau et al., 2001) and in the cytochrome b_6f complex where it probably adopts a 9-*cis* conformation (Stroebel et al., 2003). Some of the carotenes in the core complex of PS I and the two carotenes of the PS II reaction center also contain one or two *cis*-bonds at different positions (Koyama and Fujii, 1999; Jordan et al., 2001). In the core antennae, the primary function of β-carotene appears to be quenching of triplet chlorophyll (Groot et al., 1995, 1999; Jordan et al., 2001), while the two carotenes in the PS II reaction center act as quenchers of singlet oxygen (Telfer, 2002) and are involved in cyclic electron transfer in PS II (Tracewell and Brudvig, 2003; Ishikita et al., 2007). Experiments with *Chlamydomonas* exposed to high light indicate that the biosynthetic supply of β-carotene is essential for the assembly of the D1 protein during the PS II repair cycle (Trebst and Depka, 1997; Depka et al., 1998). The role of the single molecule of β-carotene in the cytochrome b_6f complex is still under investigation (Kim et al., 2005; Shao et al., 2006).

The xanthophylls lutein and violaxanthin are bound to the light-harvesting antennae of both photosystems, while loroxanthin and neoxanthin appear to be confined to the light-harvesting complexes of PS II (Pineau et al., 2001). The particular significance of loroxanthin for the photosynthetic apparatus of *Chlamydomonas* and other green algae is not known. On a timescale of femto- to picoseconds, excited xanthophylls in the light-harvesting complexes transfer energy to chlorophylls, probably from both the S_2- and the S_1-excited states (Walla et al., 2002), but they also protect the photosynthetic apparatus by quenching triplet chlorophyll in the range of nanoseconds (Peterman et al., 1995, 1997).

As in vascular plants, in high light-stressed cells of *Chlamydomonas* the formation of zeaxanthin due to increased VDE activity correlates with an enhanced thermal dissipation of singlet chlorophyll in the light-harvesting antennae (Niyogi et al., 1997b), and zeaxanthin has also been implicated in the scavenging of reactive oxygen species (Baroli et al., 2003). Contrary to the situation in vascular plants, however, the carotenoid lutein (and possibly loroxanthin) appears to contribute significantly to NPQ (Niyogi et al., 1997b; for further details, see Chapter 23).

The eyespot, which appears to be attached to the inner plastid envelope membrane, consists of β-carotene–containing lipid globules (Ohad et al., 1969; Schmidt et al., 2006) that are organized in two layers. The heavily pigmented and highly refractive globules modulate the light intensity impinging on the photosensory rhodopsins in the adjacent plasma membrane by reflection or absorbance of light depending on its angle of incidence (Kateriya et al., 2004, see also see Volume 3, Chapter 13).

In *Chlamydomonas* zygospores, the chloroplasts are reduced in size and the cytosol is orange-colored (Cavalier-Smith, 1976). This resembles observations on vegetative resting stages of *H. pluvialis*, in which the ketocarotenoids are located in cytoplasmic lipid globules (Santos and Mesquita, 1984), suggesting that the ketocarotenoids in *Chlamydomonas* zygospores are also cytoplasmic, and that secondary carotenoids in both organisms may fulfill similar functions.

The role of astaxanthin and its esters in the aplanospores of *H. pluvialis* is still under debate. The major functions that have been suggested are that it may operate as a sunscreen to shield the photosynthetic pigments from excessive light (Yong and Lee, 1991; Hagen et al., 1994; Wang et al., 2003) or that it might prevent oxidative damage caused by reactive oxygen species. The latter suggestion was inferred from both *in vitro* and *in vivo* studies of the antioxidant activity of astaxanthin and its esters (Kobayashi et al., 1997a; Kobayashi and Sakamoto, 1999). The significance of ketocarotenoid accumulation in *Chlamydomonas* zygospores has not been examined.

V. CAROTENOIDS AS BIOSYNTHETIC PRECURSORS

Carotenoids can serve as precursors of a number of other molecules with important physiological functions in *Chlamydomonas*; the respective cleavage products are referred to as apocarotenoids. Retinal, the chromophore that is covalently linked to the rhodopsin-type photoreceptors of the eyespot apparatus (Foster et al., 1984; Derguini et al., 1991; Kateriya et al., 2004), likely results from symmetric cleavage of β-carotene by a β-carotene-15,15′-oxygenase (BCO). BCOs from several animals have been characterized (Giuliano et al., 2003), and candidate genes related to the animal enzyme have been identified in *Chlamydomonas* (Nambara and Marion-Poll, 2005). Moreover, in proteomic analyses of the isolated eyespot apparatus, peptide fragments of a protein with similarity to a retinal-synthesizing apocarotenoid cleavage oxygenase (ACO) from *Synechocystis* sp. PCC 6803 (Ruch et al., 2005) were identified (Schmidt et al., 2006). BCOs and ACOs are members of an extended family of carotenoid oxygenases that also includes the 9-*cis*-epoxycarotenoid dioxygenases (NECDs). NECDs catalyze the asymmetric cleavage of 9-*cis*-neoxanthin or 9-*cis*-violaxanthin which is the committed step in the biosynthesis of abscisic acid (ABA), and *Chlamydomonas* contains predicted gene models with significant similarity to known NECDs (Grossman et al., 2004).

ABA is a phytohormone that is involved in the adaptation of plants to abiotic stresses (Nambara and Marion-Poll, 2005), and it has been detected in a number of green algae (Hirsch et al., 1989; Tietz et al., 1989; Cowan and Rose, 1991; Pardha-Saradhi et al., 2000). In *Chlamydomonas*, the cellular concentrations of ABA are correlated with faster recovery of photodamaged PS II complexes (Pardha-Saradhi et al., 2000) and an increased tolerance to oxidative stress due to a stimulated activity of antioxidant enzymes (Yoshida et al., 2003, 2004). In drought-stressed *H. pluvialis*, ABA has been reported to be involved in triggering cyst formation (Kobayashi et al., 1997b), and it might play a similar role in *Chlamydomonas* during maturation of the zygospores.

VI. OUTLOOK

While *Chlamydomonas* has been an important model for studying chlorophyll biosynthesis, its research potential has not yet been fully explored with respect to carotenoid metabolism. Based on the publicly available genome sequence and an expanding molecular genetic toolbox, future work on *Chlamydomonas* may contribute important new findings regarding the regulation of carotenogenesis and in particular the coordinated biosynthesis of carotenoids and tetrapyrroles in plants and algae. Other promising research objectives are related specifically to *Chlamydomonas* and closely related green algae. One goal would be to identify the yet missing loroxanthin synthase that catalyzes the hydroxylation of lutein at C19, and to learn more about the functional significance of this pigment in the photosynthetic complexes as compared to lutein. Other examples of important yet unexplored issues would be the mechanism of carotenoid sequestration to the eyespot, the role(s) and signaling pathways of ABA in unicellular photosynthetic organisms, and the significance of secondary carotenoids for the survival of the zygospores as well as their possible modes of action.

ACKNOWLEDGMENTS

I thank Sonja Werner, Matthias Bauch, and Harald Paulsen for critical reading of the manuscript. Part of this work was supported by grants from the German Science Foundation (LO840/1-1) and the University Mainz (Forschungsfonds 2006 and 2008).

REFERENCES

Anwaruzzaman, M., Chin, B.L., Li, X.-P., Lohr, M., Martinez, D.A., and Niyogi, K.K. (2004). Genomic analysis of mutants affecting xanthophyll biosynthesis and regulation of photosynthetic light harvesting in *Chlamydomonas reinhardtii*. *Photosynth. Res.* **82**, 265–276.

Armstrong, G.A. and Hearst, J.E. (1996). Carotenoids. 2. Genetics and molecular biology of carotenoid pigment biosynthesis. *FASEB J.* **10**, 228–237.

Baroli, I., Do, A.D., Yamane, T., and Niyogi, K.K. (2003). Zeaxanthin accumulation in the absence of a functional xanthophyll cycle protects *Chlamydomonas reinhardtii* from photooxidative stress. *Plant Cell* **15**, 992–1008.

Bohne, F. and Linden, H. (2002). Regulation of carotenoid biosynthesis genes in response to light in *Chlamydomonas reinhardtii*. *Biochim. Biophys. Acta,* **1579**, 26–34.

Boussiba, S. (2000). Carotenogenesis in the green alga *Haematococcus pluvialis*: Cellular physiology and stress response. *Physiol. Plant.* **108**, 111–117.

Bouvier, F., Rahier, A., and Camara, B. (2005). Biogenesis, molecular regulation and function of plant isoprenoids. *Prog. Lipid Res.* **44**, 357–429.

Cavalier-Smith, T. (1976). Electron microscopy of zygospore formation in *Chlamydomonas reinhardii*. *Protoplasma* **87**, 297–315.

Chemerilova, V.I. (1978). Study of pigmentation-modifying mutations in *Chlamydomonas reinhardi* strains of different ploidy. II. Compounds for *lts1* mutations and their use for obtaining triploid cultures. *Soviet Genetics* **14**, 110–115, [Russian original *Genetika* **14(1)**, 154–162].

Chunaev, A.S., Mirnaya, O.N., Maslov, V.G., and Boschetti, A. (1991). Chlorophyll *b*- and loroxanthin-deficient mutants of *Chlamydomonas reinhardtii*. *Photosynthetica* **25**, 291–301.

Cowan, A.K. and Rose, P.D. (1991). Abscisic acid metabolism in salt-stressed cells of *Dunaliella salina*. *Plant Physiol.* **97**, 798–803.

Cunningham, F.X., Jr and Gantt, E. (1998). Genes and enzymes of carotenoid biosynthesis in plants. *Annu. Rev. Plant Physiol. Plant Mol. Biol.* **49**, 557–583.

Czygan, F. (1968). Secondary carotenoids in green algae. I. Chemistry, occurrence, and factors controlling the biogenesis of these polyenes. *Arch. Mikrobiol.* **61**, 81–102.

Depka, B., Jahns, P., and Trebst, A. (1998). β-Carotene to zeaxanthin conversion in the rapid turnover of the D1 protein of photosystem II. *FEBS Lett.* **424**, 267–270.

Derguini, F., Mazur, P., Nakanishi, K., Starace, D.M., Saranak, J., and Foster, K.W. (1991). All-*trans*-retinal is the chromophore bound to the photoreceptor of the alga *Chlamydomonas reinhardtii*. *Photochem. Photobiol.* **54**, 1017–1021.

Disch, A., Schwender, J., Müller, C., Lichtenthaler, H.K., and Rohmer, M. (1998). Distribution of the mevalonate and glyceraldehyde phosphate/pyruvate pathways for isoprenoid biosynthesis in unicellular algae and the cyanobacterium *Synechocystis* PCC 6714. *Biochem. J.* **333**, 381–388.

Eichenberger, W., Boschetti, A., and Michel, H.P. (1986). Lipid and pigment composition of a chlorophyll *b*-deficient mutant of *Chlamydomonas reinhardtii*. *Physiol. Plant.* **66**, 589–594.

Fawley, M.W. and Buchheim, M.A. (1995). Loroxanthin, a phylogenetically useful character in *Chlamydomonas* and other chlorophycean flagellates. *J. Phycol.* **31**, 664–667.

Foster, K.W., Saranak, J., Patel, N., Zarilli, G., Okabe, M., Kline, T., and Nakashini, K. (1984). A rhodopsin is the functional photoreceptor for phototaxis in the unicellular eukaryote *Chlamydomonas*. *Nature* **311**, 756–759.

Francis, G.W., Knutsen, G., and Lien, T. (1973). Loroxanthin from *Chlamydomonas reinhardtii*. *Acta Chem. Scand.* **27**, 3599–3601.

Francis, G.W., Strand, L.P., Lien, T., and Knutsen, G. (1975). Variations in the carotenoid content of *Chlamydomonas reinhardtii* throughout the cell cycle. *Arch. Microbiol.* **104**, 249–254.

Fraser, P.D., Pinto, M.E.S., Holloway, D.E., and Bramley, P.M. (2000). Application of high-performance liquid chromatography with photodiode array detection to the metabolic profiling of plant isoprenoids. *Plant J.* **24**, 551–558.

Gilmore, A.M. and Yamamoto, H.Y. (1991). Resolution of lutein and zeaxanthin using a non-endcapped, lightly carbon-loaded C_{18} high-performance liquid chromatographic column. *J. Chromatogr. A* **543**, 137–145.

Giuliano, G., Al-Babili, S., and von Lintig, J. (2003). Carotenoid oxygenases: Cleave it or leave it. *Trends Plant Sci.* **8**, 145–149.

Goodwin, T.W. (1980). *The Biochemistry of the Carotenoids. Vol I: Plants* – 2nd edition. Chapman & Hall, London/New York.

Grossman, A.R., Lohr, M., and Im, C.S. (2004). *Chlamydomonas reinhardtii* in the landscape of pigments. *Annu. Rev. Genet.* **38**, 119–173.

Groot, M.L., Peterman, E.J.G., van Stokkum, I.H.M., Dekker, J.P., and van Grondelle, R. (1995). Triplet and fluorescing states of the CP47 antenna complex of photosystem II studied as a function of temperature. *Biophys. J.* **68**, 281–290.

Groot, M.L., Frese, R.N., de Weerd, F.L., Bromek, K., Pettersson, Å., Peterman, E.J.G., van Stokkum, I.H.M., van Grondelle, R., and Dekker, J.P. (1999). Spectroscopic properties of the CP43 core antenna protein of photosystem II. *Biophys. J.* **77**, 3328–3340.

Grünewald, K., Hirschberg, J., and Hagen, C. (2001). Ketocarotenoid biosynthesis outside of plastids in the unicellular green alga *Haematococcus pluvialis*. *J. Biol. Chem.* **276**, 6023–6029.

Hagen, C., Braune, W., and Björn, L.O. (1994). Functional aspects of secondary carotenoids in *Haematococcus lacustris* (Volvocales). III. Action as a "sun-shade". *J. Phycol.* **30**, 241–248.

Hager, A. and Meyer-Bertenrath, T. (1966). Extraction and quantitative determination of carotenoids and chlorophylls of leaves, algae and isolated chloroplasts with the aid of thin-layer chromatography. *Planta* **69**, 198–217.

Hirsch, R., Hartung, W., and Gimmler, H. (1989). Abscisic acid content of algae under stress. *Bot. Acta.* **102**, 326–334.

Hirschberg, J. (2001). Carotenoid biosynthesis in flowering plants. *Curr. Opin. Plant Biol.* **4**, 210–218.

Im, C.S., Eberhard, S., Huang, K., Beck, C.F., and Grossman, A.R. (2006). Phototropin involvement in the expression of genes encoding chlorophyll and carotenoid biosynthesis enzymes and LHC apoproteins in *Chlamydomonas reinhardtii*. *Plant J.* **48**, 1–16.

Ishikita, H., Loll, B., Biesiadka, J., Kern, J., Irrgang, K.-D., Zouni, A., Saenger, W., and Knapp, E.-W. (2007). Function of two β-carotenes near the D1 and D2 proteins in photosystem II dimers. *Biochim. Biophys. Acta.* **1767**, 79–87.

Janero, D.R. and Barrnett, R. (1981). Analytical prenyl pigment separation from a total green-plant lipid extract. *Anal. Biochem.* **111**, 283–290.

Jin, E., Lee, C.-G., and Polle, J.E.W. (2006). Secondary carotenoid accumulation in *Haematococcus* (Chlorophyceae): Biosynthesis, regulation, and biotechnology. *J. Microbiol. Biotechnol.* **16**, 821–831.

Jordan, P., Fromme, P., Witt, H.T., Klukas, O., Saenger, W., and Krauß, N. (2001). Three-dimensional structure of cyanobacterial photosystem I at 2.5 Å resolution. *Nature* **411**, 909–917.

Kateriya, S., Nagel, G., Bamberg, E., and Hegemann, P. (2004). "Vision" in single-celled algae. *News Physiol. Sci.* **19**, 133–137.

Kessler, E., Schäfer, M., Hummer, C., Kloboucek, A., and Huss, V.A.R. (1997). Physiological, biochemical, and molecular characters for the taxonomy of the subgenera of *Scenedesmus* (Chlorococcales, Chlorophyta). *Bot. Acta.* **110**, 244–250.

Kim, H., Dashdorj, N., Zhang, H., Yan, J., Cramer, W.A., and Savikhin, S. (2005). An anomalous distance dependence of intraprotein chlorophyll-carotenoid triplet energy transfer. *Biophys. J.* **89(4)**, L28–L30.

Kim, J. and DellaPenna, D. (2006). Defining the primary route for lutein synthesis in plants: The role of *Arabidopsis* carotenoid β-ring hydroxylase CYP97A3. *Proc. Natl. Acad. Sci. U. S. A.* **103**, 3474–3479.

Kobayashi, M. and Sakamoto, Y. (1999). Singlet oxygen quenching ability of astaxanthin esters from the green alga *Haematococcus pluvialis*. *Biotechnol. Lett.* **21**, 265–269.

Kobayashi, M., Kakizono, T., Nishio, N., Nagai, S., Kurimura, Y., and Tsuji, Y. (1997a). Antioxidant role of astaxanthin in the green alga *Haematococcus pluvialis*. *Appl. Microbiol. Biotechnol.* **48**, 351–356.

Kobayashi, M., Hirai, N., Kurimura, Y., Ohigashi, H., and Tsuji, Y. (1997b). Abscisic-acid dependent algal morphogenesis in the unicellular green alga *Haematococcus pluvialis*. *Plant Growth Regulation* **22**, 79–85.

Koyama, Y. and Fujii, R. (1999). *Cis-trans* carotenoids in photosynthesis: Configurations, excited-state properties and physiological functions. In: *The Photochemistry of Carotenoids. Advances in Photosynthesis, Volume 8* (H.A. Frank, A.J. Young, G. Britton, and R.J. Cogdell Eds.), pp. 161–188. Kluwer Academic Publishers, Dordrecht.

Kraay, G.W., Zapata, M., and Veldhuis, M.J.W. (1992). Separation of chlorophylls c_1, c_2, and c_3 of marine phytoplankton by reversed-phase-C_{18}-high-performance liquid chromatography. *J. Phycol.* **28**, 708–712.

Krinsky, N.I. and Levine, R.P. (1964). Carotenoids of wild type and mutant strains of the green alga, *Chlamydomonas reinhardtii*. *Plant Physiol.* **39**, 680–687.

Li, F., Murillo, C., and Wurtzel, E. (2007). Maize *Y9* encodes a product essential for 15-cis-ζ-carotene isomerization. *Plant Physiol.* **144**, 1181–1189.

Lichtenthaler, H.K. (1999). The 1-deoxy-d-xylulose-5-phosphate pathway of isoprenoid biosynthesis in plants. *Annu. Rev. Plant Physiol. Plant Mol. Biol.* **50**, 47–65.

Linden, H. (1999). Carotenoid hydroxylase from *Haematococcus pluvialis*: cDNA sequence, regulation and functional complementation. *Biochim. Biophys. Acta.* **1446**, 203–212.

Lohr, M., Im, C.S., and Grossman, A.R. (2005). Genome-based examination of chlorophyll and carotenoid biosynthesis in *Chlamydomonas reinhardtii*. *Plant Physiol.* **138**, 490–515.

McCarthy, S.S., Kobayashi, M.C., and Niyogi, K.K. (2004). White mutants of *Chlamydomonas reinhardtii* are defective in phytoene synthase. *Genetics* **168**, 1249–1257.

Nambara, E. and Marion-Poll, A. (2005). Abscisic acid biosynthesis and catabolism. *Annu. Rev. Plant Biol.* **56**, 165–185.

Nikulina, K., Chunaev, A.S., and Boschetti, A. (1999). Accumulation of ζ-carotene in *Chlamydomonas reinhardtii* under control of the *ac5* nuclear gene. *Plant Cell Rep.* **19**, 37–42.

Nitzsche, H. (1974). The identity of loroxanthin with pyrenoxanthin, trollein and trihydroxy-α-carotene. *Arch. Microbiol.* **95**, 79–90.

Niyogi, K.K., Björkman, O., and Grossman, A.R. (1997a). *Chlamydomonas* xanthophyll cycle mutants identified by video imaging of chlorophyll fluorescence quenching. *Plant Cell* **9**, 1369–1380.

Niyogi, K.K., Björkman, O., and Grossman, A.R. (1997b). The roles of specific xanthophylls in photoprotection. *Proc. Natl. Acad. Sci. U. S. A.* **94**, 14162–14167.

Ohad, I., Goldberg, I., Broza, R., Schuldiner, S., and Gan-Zvi, E. (1969). Changes in lipid and pigment composition and photosynthetic activity during formation of chloroplast lamellae in a mutant of *Chlamydomonas reinhardtii* y-1. In: *Progress in Photosynthesis Research*, Vol. 1 (H. Metzner, Ed.), pp. 284–295. International Union of Biological Sciences, Tübingen.

Pardha-Saradhi, S.P., Suzuki, I., Katoh, A., Sakamoto, A., Sharmila, P., Shi, D.-J., and Murata, N. (2000). Protection against the photo-induced inactivation of the photosystem II complex by abscisic acid. *Plant Cell Environ.* **23**, 711–718.

Peterman, E.J.G., Dukker, F.M., van Grondelle, R., and van Amerongen, H. (1995). Chlorophyll *a* and carotenoid triplet states in light-harvesting complex II of higher plants. *Biophys. J.* **69**, 2670–2678.

Peterman, E.J.G., Gradinaru, C.C., Calkoen, F., Borst, J.C., van Grondelle, R., and van Amerongen, H. (1997). Xanthophylls in light-harvesting complex II of higher plants: Light harvesting and triplet quenching. *Biochemistry* **36**, 12208–12215.

Pineau, B., Gérard-Hirne, C., and Selve, C. (2001). Carotenoid binding to photosystems I and II of *Chlamydomonas reinhardtii* cells grown under weak light or exposed to intense light. *Plant Physiol. Biochem.* **39**, 73–85.

Plumley, F.G. and Schmidt, G.W. (1995). Light-harvesting chlorophyll *a/b* complexes: Interdependent pigment synthesis and protein assembly. *Plant Cell* **7**, 689–704.

Pogson, B.J. and Rissler, H.M. (2000). Genetic manipulation of carotenoid biosynthesis and photoprotection. *Phil. Trans. R. Soc. Lond. B.* **355**, 1395–1403.

Remias, D., Lütz-Meindl, U., and Lütz, C. (2005). Photosynthesis, pigments and ultrastructure of the alpine snow alga *Chlamydomonas nivalis*. *Eur. J. Phycol.* **40**, 259–268.

Robertson, D.S. (1975). Survey of the albino and white-endosperm mutants of maize. Their phenotypes and gene symbols. *J. Hered.* **66**, 127–130.

Rodríguez-Concepción, M. and Boronat, A. (2002). Elucidation of the methylerythritol phosphate pathway for isoprenoid biosynthesis in bacteria and plastids. A metabolic milestone achieved through genomics. *Plant Physiol.* **130**, 1079–1089.

Rohmer, M. (2003). Mevalonate-independent methylerythritol phosphate pathway for isoprenoid biosynthesis. Elucidation and distribution. *Pure Appl. Chem.* **75**, 375–387.

Ruch, S., Beyer, P., Ernst, H., and Al-Babili, S. (2005). Retinal biosynthesis in Eubacteria: *in vitro* characterization of a novel carotenoid oxygenase from *Synechocystis* sp. *PCC 6803*. *Mol. Microbiol.* **55**, 1015–1024.

Sager, R. and Zalokar, M. (1958). Pigments and photosynthesis in a carotenoid-deficient mutant of *Chlamydomonas*. *Nature* **182**, 98–100.

Santos, M.F. and Mesquita, J.F. (1984). Ultrastructural study of *Haematococcus lacustris* (Girod.) Rostafinksi (Volvocales). I. Some aspects of carotenogenesis. *Cytologia* **49**, 215–228.

Schagerl, M. and Angeler, D.G. (1998). The distribution of the xanthophyll loroxanthin and its systematic significance in the colonial Volvocales (Chlorophyta). *Phycologia* **37**, 79–83.

Schagerl, M., Pichler, C., and Donabaum, K. (2003). Patterns of major photosynthetic pigments in freshwater algae. 2. Dinophyta, Euglenophyta, Chlorophyceae and Charales. *Ann. Limnol. – Int. J. Lim.* **39**, 49–62.

Schmidt, M., Geßner, G., Luff, M., Heiland, I., Wagner, V., Kaminski, M., Geimer, S., Eitzinger, N., Reißenweber, T., Voytsekh, O., Fiedler, M., Mittag, M., and Kreimer, G. (2006). Proteomic analysis of the eyespot of *Chlamydomonas reinhardtii* provides novel insights into its components and tactic movements. *Plant Cell* **18**, 1908–1930.

Schoefs, B., Rmiki, N., Rachadi, J., and Lemoine, Y. (2001). Astaxanthin accumulation in *Haematococcus* requires a cytochrome P450 hydroxylase and an active synthesis of fatty acids. *FEBS Lett.* **500**, 125–128.

Schwender, J., Müller, C., Zeidler, J., and Lichtenthaler, H.K. (1999). Cloning and heterologous expression of a cDNA encoding 1-deoxy-D-xylulose-5-phosphate reductoisomerase of *Arabidopsis thaliana*. *FEBS Lett.* **455**, 140–144.

Schwender, J., Gemünden, C., and Lichtenthaler, H.K. (2001). Chlorophyta exclusively use the 1-deoxyxylulose 5-phosphate/2-*C*-methylerythritol 4-phosphate pathway for the biosynthesis of isoprenoids. *Planta* **212**, 416–423.

Shao, N., Vallon, O., Dent, R., Niyogi, K.K., and Beck, C.F. (2006). Defects in the cytochrome b_6/f complex prevent light-induced expression of nuclear genes involved in chlorophyll biosynthesis. *Plant Physiol.* **141**, 1128–1137.

Sirevåg, R. and Levine, R.P. (1973). Transcription and translation for carotenoid synthesis in *Chlamydomonas reinhardtii*. *Planta* **111**, 73–84.

Steinbrenner, J. and Linden, H. (2001). Regulation of two carotenoid biosynthesis genes coding for phytoenen synthase and carotenoid hydroxylase during stress-induced astaxanthin formation in the green alga *Haematococcus pluvialis*. *Plant Physiol.* **125**, 810–817.

Steinbrenner, J. and Linden, H. (2003). Light induction of carotenoid biosynthesis genes in the green alga *Haematococcus pluvialis*: Regulation by photosynthetic redox control. *Plant Mol. Biol.* **52**, 343–356.

Stolbova, A.V. (1971). Genetic analysis of pigment mutations of *Chlamydomonas reinhardi*. Identification of the basic pigments and description of a collection of pigmented forms. *Soviet Genetics* **7**, 1171–1175, [Russian original *Genetika* **7(9)**, 90–94].

Stolbova, A.V. (1972). Genetic analysis of pigment mutations of *Chlamydomonas reinhardi*. Communication III. Analysis of mutations of chlorophyll deficiency and light sensitivity in di- and trihybrid segregations. *Soviet Genetics* **8**, 513–517, [Russian original *Genetika* **8(4)**, 124–129].

Stroebel, D., Choquet, Y., Popot, J.L., and Picot, D. (2003). An atypical haem in the cytochrome b_6f complex. *Nature* **426**, 413–418.

Sugiura, M., Inoue, Y., and Minagawa, J. (1998). Rapid and discrete isolation of oxygen-evolving His-tagged photosystem II core complex from *Chlamydomonas reinhardtii* by Ni^{2+} affinity column chromatography. *FEBS Lett.* **426**, 140–144.

Sun, Z., Gantt, E., and Cunningham, F.X., Jr (1996). Cloning and functional analysis of the β-carotene hydroxylase of *Arabidopsis thaliana*. *J. Biol. Chem.* **271**, 24349–24353.

Sun, Z., Cunningham, F.X., Jr, and Gantt, E. (1998). Differential expression of two isopentenyl pyrophosphate isomerases and enhanced carotenoid accumulation in a unicellular chlorophyte. *Proc. Natl. Acad. Sci. U. S. A.* **95**, 11482–11488.

Takaichi, S. and Mimuro, M. (1998). Distribution and geometric isomerism of neoxanthin in oxygenic phototrophs: 9′-*cis*, a sole molecular form. *Plant Cell Physiol.* **39**, 968–977.

Telfer, A. (2002). What is β-carotene doing in the photosystem II reaction centre? *Phil. Trans. R. Soc. Lond. B.* **357**, 1431–1440.

Thayer, S.S. and Björkman, O. (1990). Leaf xanthophyll content and composition in sun and shade determined by HPLC. *Photosynth. Res.* **23**, 331–343.

Tian, L., Musetti, V., Kim, J., Magallanes-Lundback, M., and DellaPenna, D. (2004). The *Arabidopsis LUT1* locus encodes a member of the cytochrome P450 family that is required for carotenoid ε-ring hydroxylation activity. *Proc. Natl. Acad. Sci. U. S. A.* **101**, 402–407.

Tietz, A., Köhler, R., Ruttkowski, U., and Kasprik, W. (1989). Further investigations on the occurrence and the effects of abscisic acid in algae. *Biochem. Physiol. Pflanzen* **184**, 259–266.

Tracewell, C.A. and Brudvig, G.W. (2003). Two redox-active β-carotene molecules in photosystem II. *Biochemistry* **42**, 9127–9136.

Trebst, A. and Depka, B. (1997). Role of carotene in the rapid turnover and assembly of photosystem II in *Chlamydomonas reinhardtii*. *FEBS Lett.* **400**, 359–362.

Walla, P.J., Linden, P.A., Ohta, K., and Fleming, G.R. (2002). Excited-state kinetics of the carotenoid S_1 state in LHC II and two-photon excitation spectra of lutein and β-carotene in solution: Efficient Car $S_1 \rightarrow$Chl electronic energy transfer via hot S_1 states? *J. Phys. Chem. A* **106**, 1909–1916.

Wang, B., Zarka, A., Trebst, A., and Boussiba, S. (2003). Astaxanthin accumulation in *Haematococcus pluvialis* (Chlorophyceae) as an active photoprotective process under high irradiance. *J. Phycol.* **39**, 1116–1124.

Weber, A. (1975). Chlorophylls and carotenoids of the Chaetophorineae (Chlorophyceae; Ulotrichales). 2. The effect of different nitrogen concentrations on pigment composition and morphogenesis of the green alga *Fritschiella tuberosa* Iyengar. *Arch. Microbiol.* **102**, 45–52.

Yong, Y.Y.R. and Lee, Y.K. (1991). Do carotenoids play a photoprotective role in the cytoplasm of *Haematococcus lacustris* (Chlorophyta)? *Phycologia* **30**, 257–261.

Yoshida, K., Igarashi, E., Mukai, M., Hirata, K., and Miyamoto, K. (2003). Induction of tolerance to oxidative stress in the green alga, *Chlamydomonas reinhardtii*, by abscisic acid. *Plant Cell Environ.* **26**, 451–457.

Yoshida, K., Igarashi, E., Wakatsuki, E., Miyamoto, K., and Hirata, K. (2004). Mitigation of osmotic and salt stresses by abscisic acid through reduction of stress-derived oxidative damage in *Chlamydomonas reinhardtii*. *Plant Sci.* **167**, 1335–1341.

Yuan, J.-P. and Chen, F. (1998). Chromatographic separation and purification of *trans*-astaxanthin from the extracts of *Haematococcus pluvialis*. *J. Agric. Food Chem.* **46**, 3371–3375.

CHAPTER 22

State Transitions

Jean-David Rochaix
Departments of Molecular Biology and Plant Biology,
University of Geneva, Geneva, Switzerland

CHAPTER CONTENTS

I. INTRODUCTION

Plants and algae are constantly subjected to changes in light intensity and quality. In order to survive and to optimize their growth, these photosynthetic organisms have developed both short-term and long-term responses which allow them to adapt to changing environmental conditions. Although they need light for photosynthesis, too much light can be harmful and cause damage to the cells through over-reduction of the electron carriers and production of reactive oxygen species. When the absorbed light energy exceeds the capacity of the photosynthetic system, the excess energy is dissipated as heat. This short-term response, called feedback de-excitation, occurs within minutes and involves the activation of the xanthophyll cycle, in which violaxanthin is converted to zeaxanthin, and the PsbS protein (see Chapter 23; reviewed in Holt et al., 2004). Short-term adaptation also occurs under changing low light conditions and allows plants and algae to configure their photosynthetic system to maintain an optimal photosynthetic yield. This process, called state transition, allows for a balancing of the excitation energy between the light-harvesting systems of photosystem II (PS II) and photosystem I (PS I) and thereby ensures a high photosynthetic performance. In contrast, long-term adaptations of the photosynthetic machinery to changing light conditions occur over several hours or days. They mostly involve an adjustment of the PS II and PS I levels and changes in the ratio between light-harvesting complex II and PS II (Chow et al., 1990; Allen and Pfannschmidt, 2000).

During the past years considerable progress has been achieved in understanding the molecular basis of these acclimation processes using combined physiological, molecular genetic, biochemical and biophysical approaches. This chapter will deal mostly with state transitions in a comparative analysis between *Chlamydomonas* and *Arabidopsis*. Several earlier reviews have appeared on state transitions (Allen, 1992; Wollman, 2001; Zer and Ohad, 2003).

II. WHAT ARE STATE TRANSITIONS?

The primary reactions of photosynthesis occur in the thylakoid membranes and are mediated by PS II, the cytochrome b_6f complex and PS I which are linked in series through the photosynthetic electron transport chain. At one end of this chain the oxidizing power of PS II is used to split water giving rise to electrons, protons, and molecular oxygen. Electrons are shuttled from PS II via the plastoquinone pool to the cytochrome b_6f complex and transferred further to PS I by the soluble electron carrier plastocyanin. To ensure an optimal photosynthetic yield under limiting light conditions, it is essential that the two photosystems operate at the same rate. Because their respective antenna systems have different light absorption properties, changes in light conditions can lead to unequal excitation of the two photosystems and thus to a decreased photosynthetic yield. Plants and algae are able to balance the light excitation energy between the antennae systems of PS II and PS I in the short term through state transitions, a process originally discovered independently in a green alga by Bonaventura and Myers (1969), and in a red alga by Murata (1969). It was shown that upon illumination with 650 nm light, which is preferentially absorbed by PS II, PS I becomes sensitized so that it can absorb more light excitation energy. Reciprocally, illumination with 700 nm light, which is absorbed preferentially by PS I, sensitizes PS II.

The current model for state transitions is outlined in Figure 22.1 (reviewed in Allen and Staehelin, 1992; Wollman, 2001). Under illumination conditions that lead to excess excitation of PS II compared to PS I, the redox state of the plastoquinone pool is changed to a more reduced state. This favors the docking of plastoquinol (PQH_2) to the Q_0 site of the cytochrome b_6f complex, an event that leads to the activation of a thylakoid protein kinase which phosphorylates specific LHCII proteins (Vener et al., 1997; Zito et al., 1999). A fraction of LHCII is then displaced from PS II to PS I, thus re-equilibrating the light excitation energy between the two photosystems. When the mobile

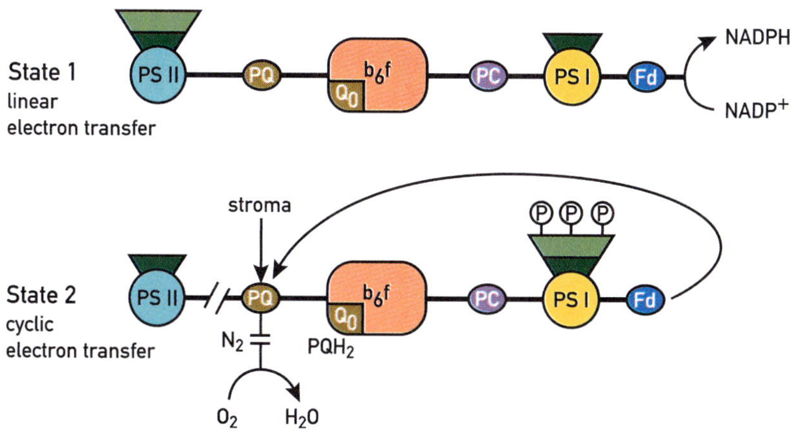

FIGURE 22.1 *State transitions in Chlamydomonas. In state 1 plastoquinone (PQ) is oxidized as a result of the preferential light excitation of PS I relative to PS II. The mobile part of LHCII (light green area) is associated with PS II. In state 1 the photosynthetic electron transfer chain operates in a linear mode ultimately resulting in the reduction of $NADP^+$ to NADPH. Upon preferential excitation of PS II relative to PS I, the plastoquinone pool is reduced and occupancy of the Q_0 site of the cytochrome b_6f complex by plastoquinol (PQH_2) leads to the activation of the LHCII kinase, phosphorylation of LHCII and displacement of the mobile part of LHCII from PS II to PS I. State 2 can also be induced in the dark through the chlororespiratory chain under anaerobic conditions (N_2) to prevent the oxidation of the plastoquinone pool by a plastid oxidase. In state 2 most of PS II is disconnected from the electron transport chain which operates in a cyclic mode as an ATP generator.*

LHCII antenna is associated with PS I, this is referred to as state 2. State transition is a reversible process as over-excitation of PS I leads to dephosphorylation of LHCII and to the return of the mobile LHCII to PS II. This is referred to as state 1.

The PS II and PS I complexes are not equally distributed over the thylakoid membranes. PS II is mostly associated with the grana membranes whereas PS I is found mainly in the thylakoid stromal lamellae (Andersson and Andersson, 1980). Thus the light excitation energy needs to be redistributed over several hundreds of nanometers on the thylakoid membrane during state transitions.

State transitions are not limited to algae but also occur in land plants, and in cyanobacteria which possess a different light-harvesting apparatus. *Chlamydomonas* has emerged as an excellent model system for studying state transitions. One major advantage of this alga is that during state transitions the mobile part of the LHCII antenna is rather large and constitutes 80% of the total antenna (Delosme et al., 1996). This is in contrast to land plants in which the mobile LHCII represents only 15–20% of the PS II antenna (Allen, 1992). Because room temperature fluorescence is mostly associated with PS II, large fluorescence changes occur during state transitions in *Chlamydomonas*. This property makes *Chlamydomonas* particularly attractive for genetic studies and has been used for the screening of mutants deficient in state transitions (see below).

III. PHOSPHORYLATION AND MOBILITY OF LHCII POLYPEPTIDES

A. Phosphorylation of LHCII

A key role in state transitions is played by the redox state of the plastoquinone pool, which is influenced by the relative activities of PS II and PS I. In the early eighties it was shown that reduction of the plastoquinone pool through preferential excitation of PS II relative to PS I, leads to the activation of a protein kinase which phosphorylates LHCII proteins (Horton and Black, 1980; Allen et al., 1981). The light-harvesting system of plants and algae comprises the antennae of PS II (LHCII) and of PS I (LHCI). The composition of these antenna systems in *Chlamydomonas* is shown in Table 22.1.

LHCII and LHCI serve to collect and to transfer the light excitation energy to the reaction centers of PS II and PS I. These systems are involved in the acclimation of cells to changes in light conditions, and participate both in state transitions and thermal dissipation of excess light energy. In *Chlamydomonas*, the major trimeric light-harvesting system of PS II (LHCIIb) comprises polypeptides encoded by a small multigene family whose members share high sequence similarity (Elrad et al., 2002; Minagawa and Takahashi, 2004) (Table 22.1; also see Chapter 14).

Table 22.1 LHCII and LHCI polypeptides

LHCII		LHCI
Major complexes		
LHCBM3	Type I	LHCA1
LHCBM4		LHCA2
LHCBM6		LHCA3
LHCBM8		LHCA4
LHCBM7		LHCA5
LHCBM2	Type III	LHCA6
		LHCA7
LHCBM1	Type IV	LHCA8
LHCBM5	Type II	LHCA9
Minor complexes		
LHCB4	CP29	
LHCB5	CP26	

LHCII, light-harvesting system of PS II; LHCI, light-harvesting system of PS I. See Chapter 14 for additional information on these proteins.

Each of these proteins binds eight chlorophyll *a*, six chlorophyll *b*, and several carotenoid and xanthophyll molecules. The atomic structure of the LHC proteins has revealed that they contain three *trans*-membrane helices and a short amphipathic helix that is exposed to the lumenal side of the thylakoid membrane (Kuhlbrandt et al., 1994; Liu et al., 2004). Moreover in land plants there are three monomeric light-harvesting polypeptides of PS II, CP29, CP26, and CP24, which are encoded by the *LHCB4*, *LHCB5*, and *LHCB6* genes respectively, and which provide a connection between the major light-harvesting system and the reaction center of PS II. While CP29 and CP26 are present in *Chlamydomonas*, no homologue of CP24 could be found (Teramoto et al., 2001; Elrad and Grossman, 2004). LHCI of *Chlamydomonas* comprises polypeptides encoded by nine genes (Table 22.1). Seven distinct LHCI proteins could be resolved (Tokutsu et al., 2004). The LHC family of *Chlamydomonas* also includes four genes encoding early light-induced proteins (ELIPs) and three genes encoding LI818 proteins (see Chapter 14). The latter are strongly induced following sulfur starvation (Elrad and Grossman, 2004).

Several thylakoid membrane proteins from the PS II complex and LHCII undergo light- and redox-dependent phosphorylation (Bennett, 1977, 1979; Allen et al., 1981). Because many of these LHCII proteins have similar sequences and sizes, it is difficult to separate them and to identify those that are phosphorylated during a state 1 to state 2 transition by classical means such as ^{32}P-phosphate labeling, detection of shifts in electrophoretic mobility of individual proteins, or immunoblotting with anti-phosphothreonine antibodies. The ability to map phosphorylation sites of thylakoid proteins of *Chlamydomonas* was rather limited until mass spectrometry provided new powerful tools for identifying not only the proteins that are phosphorylated, but also the phosphorylated residues within the proteins. An elegant method consists in treating isolated thylakoid membranes with trypsin, thus releasing phosphorylated peptides from the stroma-exposed N-terminus of LHC proteins (Turkina and Vener, 2006; Vener, 2006). The negatively charged phosphorylated peptides are enriched on a positively charged metal affinity column and subjected to matrix-assisted laser desorption/ionization and electrospray ionization mass spectrometry.

Using this approach a thorough mapping study was performed on thylakoid membranes from *Chlamydomonas* cells in state 1 and state 2 or exposed to high light. This analysis revealed a total of 19 *in vivo* phosphorylation sites corresponding to 15 genes, mostly from the Lhcb protein family (Turkina et al., 2006). It showed in particular that the PS II core proteins D1 and CP43 are phosphorylated at their N-terminal threonines. The LHCII proteins LHCBM4, LHCBM6 and LHCBM8 are all phosphorylated at Thr3. A surprising result from this analysis was that transition from state 1 to state 2 promotes phosphorylation of D2, whereas earlier studies indicated increased phosphorylation of D2 under oxidizing conditions (Delepelaire and Wollman, 1985). It remains to be seen whether different phosphorylation sites of D2 are involved in these responses. Removal of the Thr2 phosphorylation site by site-directed mutagenesis did not reveal any apparent phenotype (Andronis et al., 1998; Fleischmann et al., 1999). Another protein which is phosphorylated under state 2 conditions is PSBR from the PS II core, where a Ser residue is phosphorylated (Turkina et al., 2006). PSBR appears to contain a single transmembrane domain and to be exposed to the thylakoid lumen where it is required for stabilization of the oxygen evolving complex of PS II (Suorsa et al., 2006). Transition from state 1 to state 2 also leads to quadruple phosphorylation of CP29, and phosphorylation of LHCBM1 which is a component of the major LHCII complex (Turkina et al., 2006) (Table 22.2). An interesting finding is that the principal changes in phosphorylation observed during state transitions are clustered at the interface between the PS II core and the associated light-harvesting antenna proteins which include LHCII, CP26, and CP29. Taken together these results suggest that the differential phosphorylation of CP29 could play a key role in state transitions and that its phosphorylation status is particularly important for this process.

B. Mobility of LHCII

The model of state transitions postulates that upon phosphorylation, a mobile pool of LHCII complexes moves laterally from PS II in the grana to

Table 22.2 Thylakoid protein phosphorylation

Protein	Residue	State 1	State 2	ML	HL
D1	T2			+	+
D2	T2	–	+		
CP43	T2			+	+
PSBR[a]	S42	–	+		
CP26	T10			–	+
CP29[b]	T7	+	+	–	+
	T11			–	+
	T17	–	+		
	T18			–	+
	T20			–	+
	T33	+	+		
	S103	–	+		
LHCBM1[c]	T27	–	+		
LHCBM4	T19			–	+
	T23			–	+
LHCBM6	T18			–	+
	T22			–	+
LHCBM8	T19			–	+
	T23			–	+
LHCBM9	T19			–	+
	T23			–	+
LCI5	S137	–	+		
	S197	–	+		
Unknown A		+	+		
Unknown B		–	+		

Note: Residues which are highlighted are phosphorylated under state 2 conditions. ML, moderate light (120 micromoles/m^2/s); HL, high light (1200 micromoles/m^2/s) (Turkina et al., 2006).

[a]PSBR, PS II protein exposed to the thyakoid lumen where it stabilizes the oxygen evolving complex of PS II.

[b]CP29 linker protein between PS II core and its associated major LHCII antenna, essential for PS II–LHCII supercomplex formation.

[c]LHCBM1 is involved in non-photochemical quenching (Elrad et al., 2001).

PS I in the stromal region of the thylakoid membrane. Such a lateral redistribution of LHCII upon transition from state 1 to state 2 was indeed observed using immuno-electron microscopy in *Chlamydomonas* (Vallon et al., 1991). These studies also revealed that under the same conditions a fraction of the cytochrome b_6f complex migrates from the grana to the stromal lamellae. A further important step was the demonstration of a reversible association of

LHCII with PS I through photoacoustic measurements of the quantum yield of PS I and PS II charge separations in state 1 and state 2 in *Chlamydomonas* (Delosme et al., 1996). This study clearly showed a connection of LHCII to PS I in state 2 which depended on a functional cytochrome b_6f complex. A 50–60% increase in the quantum yield of PS I occurs upon transition from state 1 to state 2, with a clear contribution of LHCII.

It was later that biochemical evidence for an association between LHCII polypeptides and PS I in state 2 was provided. Such an association was first obtained through the isolation of a PSI–LHCII complex from *Arabidopsis* (Zhang et al., 1998). Two to threefold more LHCII is associated with PS I in state 2 than in state 1. Crosslinking studies indicate that the PSAH/I/O region of PS I acts as a docking site for the mobile LHCII antenna under state 2 conditions. These results are compatible with the finding that *Arabidopsis* mutants lacking PSAH are deficient in state transitions (Lunde et al., 2000). It is interesting that the 4.4-Å crystal structure of PS I from land plants localizes PSAH in an exposed hydrophobic region of PS I, and reveals that PSAH binds a chlorophyll molecule which may be involved in the transfer of energy from LHCII to the PS I core (Ben-Shem et al., 2003).

Further support for an association of LHCII with PS I comes from studies with *Chlamydomonas* in which the mobile part of LHCII is significantly larger than in land plants (Kargul et al., 2005; Takahashi et al., 2006). Thylakoid membranes from *Chlamydomonas* cells in state 1 and state 2 were solubilized with a mild detergent and fractionated by sucrose density centrifugation. In state 1 three bands are separated corresponding with increasing density of the gradient to the major LHCII polypeptides, the PS II complex and the PSI–LHCI complex. Thylakoids from cells in state 2 give rise to an additional band which was shown to correspond to a PSI–LHCI supercomplex which is distinctly larger than in state 1. In addition to LHCI polypeptides, three LHCII proteins corresponding to CP29, CP26, and LHCBM5 were only detected in the PSI–LHCI supercomplex from cells in state 2 (Kargul et al., 2005; Takahashi et al., 2006). It is noteworthy that amongst the LHCII proteins associated with PS II these three proteins are the most similar to those of LHCI. LHCBM5 is a LHCII type II protein which appears to accumulate to a similar level as CP29 and CP26 but at a much lower level than the other major LHCII proteins (Takahashi et al., 2006). Moreover CP29 and LHCBM5 associated with the PS I supercomplex were phosphorylated. Based on these results Takahashi et al. proposed that LHCBM5 could occupy a similar position as CP24 in land plants, that is LHCBM5, CP29, and CP26 would act as linker between the PS II core dimer and the trimeric LHCII in state 1. In state 2, upon phosphorylation, these three proteins would move to PS I on the PSAH side and allow for the docking of the LHCII trimer to the PS I complex (Figure 22.2).

Of particular importance for these studies has been the X-ray structure of the land plant LHCI–PSI supercomplex, which revealed that four Lhca

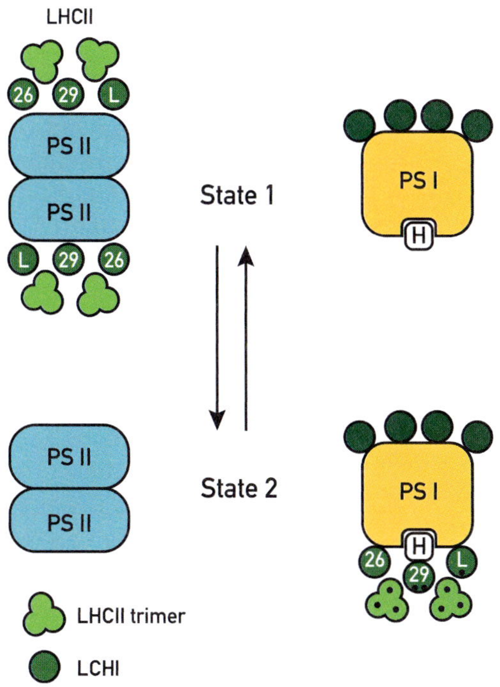

FIGURE 22.2 *Model for state transitions in Chlamydomonas. In state 1 the dimeric PS II complex is connected to the LHCII trimers presumably through the minor monomeric LHCII CP26 (26), CP29 (29), and LHCBM5 (L). During transition from state 1 to state 2 the phosphorylation of several LHCII proteins leads to the detachment of the mobile part of LHCII from PS II in the grana regions, their migration to the stromal lamellae and binding to PS I near PSAH on the opposite side of the LHCI belt of the complex. Phosphorylation sites are indicated by dots. Note that during a state 1 to state 2 transition two additional Thr residues of CP29 are phosphorylated (adapted from Takahashi et al., 2006; Turkina et al., 2006).*

subunits form a light-harvesting belt asymmetrically located on the PSAG/J/K side of the PS I core domain (Ben-Shem et al., 2003). Analysis of PS I–LHCI/LHCII supercomplexes from *Arabidopsis* by electron microscopy revealed in one case a large density on the PSAH/L/A/K side which could correspond to LHCII trimers (Kouril et al., 2005). In another study with *Chlamydomonas*, a smaller density near PSAH was assigned to CP29 (Kargul et al., 2005). It therefore appears that CP29, CP26, and LHCBM5, which were thought to belong exclusively to the PS II complex, can shuttle between PS II and PS I during state transitions and provide docking sites for the trimeric LHCII proteins in both PS I and PS II (Takahashi et al., 2006). CP29 is phosphorylated at Thr6, Thr16, Thr32, and Ser102 in state 2, but only at Thr6 and Thr32 in state 1 (Turkina et al., 2006). Quadruple phosphorylation of CP29 upon a state 1 to state 2 transition would cause the dissociation of CP29-LHCII from PS II and association to PS I. Under high light three additional sites of CP29 are phosphorylated, which could induce dissociation of LHCII from the photosystems and favor thermal energy dissipation within the LHCII trimers (Elrad et al., 2002; Pascal et al., 2005). These data suggest a mechanism of regulation of the algal light-harvesting complex through differential phosphorylation of the PSII-LHCII linker protein CP29. Additionally, the redox-dependent phosphorylations of CP29 may induce conformational changes that alter the relative affinity of CP29 for PS II and PS I. Taken together these newer data reveal that state transitions in *Chlamydomonas* are regulated by reversible multiple protein phosphorylations at the interface between the PS II core and LHCII, and in particular by phosphorylation of the linker protein CP29 and probably also LHCBM5.

Mutants of *Chlamydomonas* deficient in PS I are blocked in state 1 even though they are still able to phosphorylate LHCII under state 2 conditions. However in this case LHCII remains functionally connected to PS II (Delosme et al., 1996). Similarly, in *Arabidopsis* lines lacking PSAH, which are also blocked in state 1, LHCII remains connected to PS II although it is phosphorylated under state 2 conditions (Lunde et al., 2000). These results indicate that the association of the phosphorylated LHCII with PS I and PS II is a competitive process which is determined by the relative binding affinities of phosphorylated LHCII to the two photosystems.

IV. REGULATION OF PHOSPHORYLATION

A. Redox control of phosphorylation

Under increasing light intensities *in vivo,* the LHCII polypeptides and the PS II core proteins have opposite phosphorylation responses (Rintamaki et al., 1997). Under low light, phosphorylation of the LHCII proteins is maximal whereas that of the PS II core proteins is low. Phosphorylation of the PS II proteins reaches its maximum at high light intensities where phosphorylation of LHCII is considerably reduced. Thus at least two kinases for the phosphorylation of LHCII and of the PS II core are present in the thylakoid membranes, and are subjected to different types of regulation.

The redox state of the plastoquinone pool is important for the activation of the LHCII kinase. A key event is the occupancy of the Q_0 site of the cytochrome b_6f complex by plastoquinol (Vener et al., 1997; Zito et al., 1999; also see Chapter 17). The cytochrome b_6f complex is therefore essential for the phosphorylation of LHCII. Indeed this phosphorylation does not occur in mutants deficient in the cytochrome b_6f complex (Wollman and Lemaire, 1988).

Because the Q_0 site is on the lumenal side of the thylakoid membrane and the catalytic domain of the LHCII kinase, together with its target sites, are on the stromal side, the signal for kinase activation needs to be transduced across the thylakoid membrane. Insights into this process came from the crystal structure of the mitochondrial bc_1 complex (Iwata et al., 1998; Zhang et al., 1998) which revealed that the position of the Rieske protein, a key subunit of the complex, changes depending on whether the Q_0 site is occupied by stigmatellin, a Q_0 site inhibitor. This movement appears to be required for shuttling electrons from the membrane soluble quinol (proximal site) to the extra-membrane heme of cytochrome c_1 (distal site). The structure of the closely related chloroplast cytochrome b_6f complex has also been determined (Kurisu et al., 2003; Stroebel et al., 2003). Moreover, analysis of crystals of the cytochrome b_6f complex with and without stigmatellin confirmed the changing position of the Rieske protein and showed that it involves a reorganization of the membrane-spanning portion of the complex (Breyton, 2000).

Another player in *Chlamydomonas* signal transduction could be subunit V of the cytochrome b_6f complex, also called PETO. This protein, which is loosely associated with the cytochrome b_6f complex, contains a single transmembrane domain with two large hydrophilic regions on the stromal and lumenal sides of the thylakoid membrane. PETO undergoes reversible phosphorylation during state transitions, a unique feature among the subunits of the cytochrome b_6f complex (Hamel et al., 2000). Thus it is possible that the lumenal domain of PETO senses the structural changes of the Rieske protein and transmits them through its transmembrane region to its stromal domain, which may interact with the LHCII kinase.

Further insights in the activation of the LHCII kinase came from the careful analysis of the effects of specific inhibitors of the cytochrome b_6f complex on state transitions and LHCII phosphorylation (Finazzi et al., 2001b). By binding tightly to the Q_0 site, stigmatellin locks the Rieske protein in its proximal position and prevents transition from state 1 to state 2 and LHCII phosphorylation. However, PETO is still phosphorylated in the presence of stigmatellin. This is in contrast to plastoquinol (PQH_2), which is released rapidly from the Q_0 site, allowing the Rieske protein to move to its distal position. These results suggest a two-step process for the activation of the LHCII kinase (Finazzi et al., 2001b). The first step would involve the movement of the Rieske protein from the distal to the proximal position upon PQH_2 binding to Q_0, with a concomitant conformational change of the cytochrome b_6f complex leading to the activation of the LHCII kinase. This event could be marked by the phosphorylation of PETO. In the second step, relaxation of the Rieske protein to the distal position would release the activated kinase from the cytochrome b_6f complex and allow it to interact with LHCII. Such a movement would be triggered by the turnover of PQH_2 at the Q_0 site but not by stigmatellin, which blocks the Rieske protein in one conformation. This dynamic model is also compatible with the observation that the LHCII kinase is a low abundance protein compared to the cytochrome b_6f complex (Gal et al., 1997). It is thus necessary that the active kinase phosphorylate several PETO and LHCII substrates before it returns to its inactive state.

The cytochrome b_6f complex contains at least three small subunits, PetG, PetL, and PETM, with still-undefined functions. Cytochrome b_6f complexes containing chimeric proteins in which the C-terminus of subunit IV is fused to the N-terminal end of PetL are still assembled and transfer electrons efficiently (Zito et al., 2002). This indicates that quinol binding to the cytochrome b_6f complex is not affected. However, in this mutant the ability to perform state transitions is lost, suggesting that the transduction of the activating signal from the Q_0 site to the kinase is affected perhaps by impeding the interaction of the kinase with the complex.

The high light-induced inactivation of the LHCII kinase occurs *in vivo* but not with isolated thylakoid membranes. This suggests that stromal components are involved in this inactivation process. When thylakoid membranes isolated from high light-treated plants were exposed to oxidized DTT, LHCII phosphorylation was restored, but was totally prevented when the thylakoids were treated with reduced DTT (Rintamaki et al., 1997). Thus the thiol-redox state is involved in the high light-induced inactivation of LHCII phosphorylation. These data suggest that the ferredoxin/thioredoxin system participates in the regulation of LHCII phosphorylation, especially because it is known that reduction of thioredoxin is generally saturated at light intensities similar to those responsible for the inactivation of LHCII phosphorylation (Rintamaki et al., 1997).

B. Light-induced conformational changes of LHCII

Besides its role in the regulation of protein phosphorylation through the signal transduction chain that connects redox reactions to the protein kinase activation or inactivation, light acts also directly on the chlorophyll protein substrate. By using a partially purified thylakoid protein kinase and isolated LHCII complex, it was shown that illumination of the chlorophyll-protein complex exposes the phosphorylation site to the kinase (Zer et al., 1999, 2003). This process appears to be due to a reversible light-induced conformational change of the stromally exposed N-termini of the LHCII proteins, as evidenced by their increased susceptibility to trypsin cleavage after light exposure.

V. THYLAKOID PROTEIN KINASES

A. Biochemical hunt for the LHCII kinase

A LHCII kinase activity was initially discovered by Bennett (1977). This was followed by a long biochemical hunt for this particular kinase. Several thylakoid-associated kinases were identified, with apparent molecular weights of 25, 38, 58 and 64 kD (Lin et al., 1982; Coughlan and Hind, 1986, 1987). Among these candidates, the 64-kD protein was itself phosphorylated and cofractionated with the cytochrome b_6f complex (Coughlan and Hind, 1986, 1987). However, sequencing revealed that this protein is a polyphenol oxidase

(Sokolenko et al., 1995). The 58-kD protein copurified with the PS II complex and was reported to phosphorylate LHCII (Race and Hind, 1996). In spite of several attempts, the isolation of the LHCII kinase through biochemical means was unsuccessful, although several kinase activities associated with the thylakoid membrane were identified.

B. TAK protein kinases

Therefore, alternative approaches were chosen. One approach used an elaborate screen for proteins that interact with the N-terminal portion of the light-harvesting proteins containing the Thr that is phosphorylated during a state 1 to state 2 transition. This led to identification of a kinase associated with the thylakoid membrane of *Arabidopsis* called thylakoid associated kinase (TAK), which is part of a three-member family (Snyders and Kohorn, 1999). These kinases contain a repeated Gly-Ser motif which is also found in the human transforming growth factor β1 receptor. The three TAK kinases are associated with PS II and phosphorylate LHCII *in vitro*. Both LHCII phosphorylation and autophosphorylation are increased by reductant, indicating a redox control of these kinases (Snyders and Kohorn, 2001). TAK antisense lines were sensitive to high light intensities. They showed decreased phosphorylation of LHCII and were deficient in state transitions (Snyders and Kohorn, 2001). Although the TAK antiserum reacts with a *Chlamydomonas* protein of similar size as in *Arabidopsis*, a gene similar to *TAK* is not present in the *Chlamydomonas* nuclear genome or in the EST database. It is therefore not yet clear whether the TAK kinases are also present in algae.

C. STT7 protein kinase

Because biochemical methods for identifying the LHCII kinase were not successful, new approaches based on a genetic screen for mutants deficient in state transitions were initiated with *Chlamydomonas* with the ultimate aim of identifying not only the LHCII kinase, but also other factors involved in the signal transduction chain of state transitions (Fleischmann et al., 1999; Kruse et al., 1999). *Chlamydomonas* is particularly well suited for a screen of this type because as mentioned above, state transitions are associated with large fluorescence changes. Because the room temperature fluorescence arises mostly from the PS II antenna, cells of *Chlamydomonas* yield a considerably higher fluorescence in state 1 than in state 2. Thus the difference in fluorescence between state 1 and state 2 can be measured with a fluorescence video imaging system and used for screening mutants deficient in state transitions (see Fluorescence Quenching Analysis on the companion web site http://www.elsevierdirect.com/companions/9780123708731). To maximize the signal in this screen, it is important to compare cells which are totally in state 1 with those that are fully in state 2. Addition of DCMU (3-(3,4-dichlorophenyl)1,1-dimethylurea)) in the presence of light leads to the

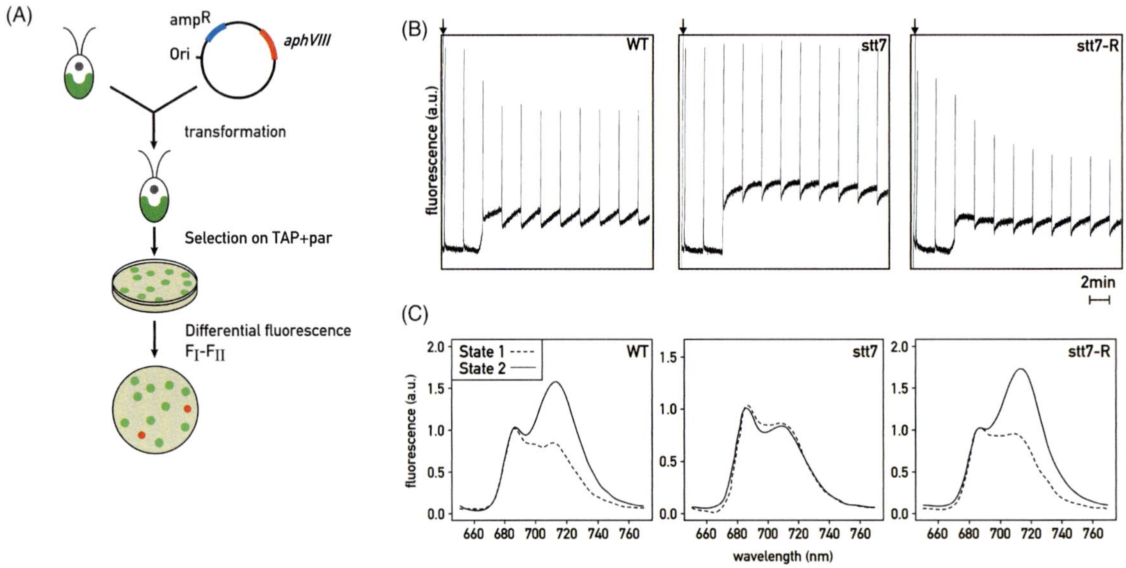

FIGURE 22.3 *(A) Screen for Chlamydomonas state transition mutants. Insertional mutagenesis is achieved by transforming Chlamydomonas cells with a plasmid containing a selectable marker, for example aphVIII conferring resistance to paromomycin (par). Transformants are subjected to state 1 and state 2 conditions. In each case the fluorescence is measured with a video imaging system. The difference between the fluorescence signals in state 1 (FI) and state 2 (FII) is computed and displayed. Wild-type cells yield a strong signal whereas mutant cells do not. (B) Fluorescence changes associated with a state 1 to state 2 transition. Cells in state 1 are incubated in the dark in a closed vessel. At different times a saturating light pulse is given to measure the maximum fluorescence F_{max}. Because of continued respiration the environment becomes anaerobic and transition to state 2 is induced with a marked decrease in F_{max}. No decrease in F_{max} occurs in the stt7 mutant blocked in state 1 under the same conditions. In the stt7 mutant rescued with the wild-type STT7 gene (stt7-R), transition to state 2 is restored accompanied by a decrease of fluorescence as in the wild type (Depège et al., 2003, with permission). (C) Low temperature fluorescence emission spectra in state 1 and state 2. The fluorescence peaks at 685 and 715 nm correspond to PS II and PS I, respectively. Transition from state 1 to state 2 leads to an increase in PS I fluorescence in the wild-type and the rescued strains. In contrast the fluorescence spectra of stt7 are nearly identical in state 1 and state 2 (Depège et al., 2003, with permission).*

complete oxidation of the plastoquinone pool and thus to state 1. To obtain state 2, one can take advantage of the chlororespiratory chain which feeds electrons from the stroma into the plastoquinone pool (Bennoun, 1982). The plastoquinone pool can be fully reduced in the dark in the presence of nitrogen, which prevents its reoxidation by a plastid oxidase. Alternatively, states 1 and 2 can be induced by illuminating algae with light preferentially absorbed by PS I and PS II. Using these screens several mutants deficient in state transitions were obtained using insertional mutagenesis (Figure 22.3A) (Sizova et al., 2001). Several of these mutants have been characterized (Table 22.3) (Fleischmann et al., 1999; Kruse et al., 1999). While some mutations appear to act directly on state transitions, others act in an indirect way.

One of the first state transition mutants isolated, *stt7*, is blocked in state 1 (Fleischmann et al., 1999). The block in state 1 was confirmed by measuring the time course of fluorescence during a transition from state 1 to state 2 at

Table 22.3 *Chlamydomonas* mutants deficient in state transitions

Mutant	Gene	Protein/function	Reference
stt7	*STT7*	STT7 protein kinase	Depège et al. (2002)
stt2	*MEND1*	Phylloquinone synthesis	Lefebvre-Legendre et al. (2007)
stm3	*NAB1*	Repressor of LHCII translation	Mussgnug et al. (2005)
stm6	*MOC1*	Mitochondrial termination factor	Schonfeld et al. (2004)

room temperature and the fluorescence emission spectrum at −77°K (Figure 22.3B), and further verified by performing photoacoustic measurements (Fleischmann et al., 1999). Moreover, the *stt7* mutant fails to phosphorylate LHCII proteins under state 2 conditions. Immunogold labeling of sections of cells with LHCII antisera revealed that the redistribution of LHCII, and to a lesser extent of the cytochrome b_6f complex, from the grana to the stroma regions of the thylakoids during a state 1 to state 2 transition no longer occurs in *stt7* (Figure 22.4) (Fleischmann et al., 1999).

The nuclear gene that is inactivated in *stt7* encodes a thylakoid-associated Ser-Thr protein kinase (Depège et al., 2003). Transformation of the *stt7* mutant with the wild-type *STT7* gene restores state transition and LHCII phosphorylation under state 2 conditions. Because of the presence of a potential transmembrane domain in STT7, one possible orientation is that its N-terminal end is on the lumen side and the kinase domain on the stromal side (Figure 22.5). However an alternative configuration with the kinase associated firmly on the stromal side of the thylakoid membrane cannot be excluded. An STT7 homologue of unknown function, STL1, is also encoded in *Chlamydomonas*. Both STT7 and STL1 are conserved in land plants and in *Arabidopsis* the most similar proteins are STN7 and STN8, respectively. The phylogeny of these kinases is shown in Figure 22.5. The STT7 kinase is also conserved in other plants such as rice, and in marine algae.

Because of the availability of *Arabidopsis* T-DNA insertion lines with disruptions in the *STN7* and *STN8* genes, it was possible to show that the STN7 kinase is required for state transitions in *Arabidopsis* (Bellafiore et al., 2005). Loss of STN7 leads to the specific lack of phosphorylation of several LHCII proteins under state 2 conditions since the phosphorylation of other major thylakoid proteins, CP43, D1, D2, is not affected. Moreover, site-directed mutagenesis of a Lys residue of the catalytic site of the kinase does not affect its accumulation, but abolishes transition from state 1 to state 2, thus demonstrating that the catalytic activity of STN7 is indeed necessary for state transitions. This had been assumed but never proven before. Thus STT7 and STN7 are not only structurally but also functionally related. It is

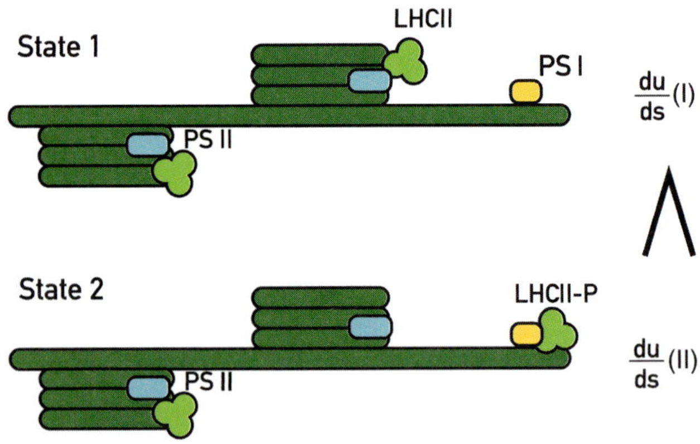

Strain	du/ds	
	α-LHCII	α-cytf+Rieske
WT state 1	0.25±0.07	0.44±0.06
WT state 2	0.81±0.17	0.82±0.17
Stt7 state 1	0.27±0.06	0.54±0.24
Stt7 state 2	0.32±0.06	0.61±0.05

FIGURE 22.4 *Movement of LHCII during state transitions. Displacement of LHCII and the cytochrome b_6f complex from the grana to the stromal thylakoid region during a transition from state 1 to state 2 occurs in the wild-type but is strongly diminished in the stt7 mutant. Wild-type and mutant cells in state 1 and state 2 were examined by immuno-electron microscopy and the distribution of LHCII and the cytochrome b_6f complex was examined using specific antisera and immunogold labeling. The ratio of grains over the lamellar regions (du) and grana regions (ds) was estimated. This ratio du/ds is higher in state 2 than in state 1 in wild-type cells but greatly diminished in stt7 (Fleischmann et al., 1999a).*

not yet known whether these kinases act directly on the light-harvesting proteins or whether they are part of a signaling kinase cascade. The fact that the STT7 kinase was isolated twice independently in a mutant screen suggests that the number of kinases in this potential cascade is limited. In contrast, state transitions still occur in *Arabidopsis* plants lacking STN8. Loss of this kinase leads to a specific decrease in phosphorylation of D1, D2, and CP43. It will be interesting to elucidate the function of STL1 in *Chlamydomonas*, which may fulfill a similar function as STN8 in *Arabidopsis* based on the sequence similarity.

At high light intensities the LHCII kinase is inactivated. This appears to be mediated through the ferredoxin-thioredoxin system (Rintamaki et al., 2000). In this respect it is interesting to note that among a total of seven Cys residues, two residues near the N-terminal are conserved in STT7 and STN7 but not in STL1 and STN8 and might be the targets of thioredoxin (Figure 22.5C). Although it is clear that the STT7/STN7 kinase is required for phosphorylation of the LHCII proteins and for state transitions, the

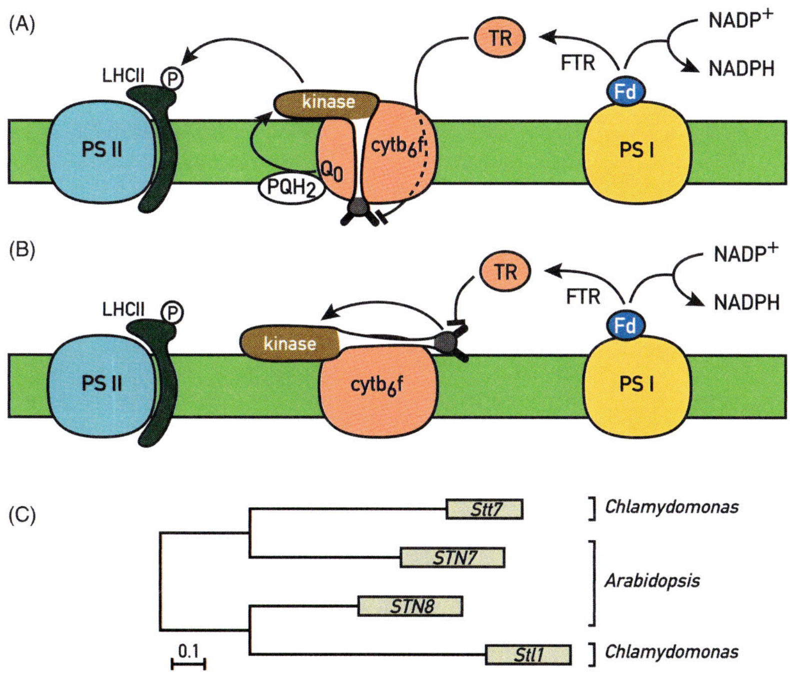

FIGURE 22.5 *Two possible models for the STT7/STN7 kinase in the thylakoid membrane. The STT7 kinase could either (A) contain a transmembrane domain with its N-terminal in the thylakoid lumen and the kinase domain with the C-terminal end on the stromal side, or the (B) kinase could be firmly bound to the stromal side of the membrane. Two conserved Cys residues near the N-terminal end (indicated by bars) are essential for the kinase activity and their redox state could be controlled through the ferredoxin-thioredoxin system. In the case of the transmembrane model one needs to postulate redox signaling across the thylakoid membrane (broken line). (C) Amino acid sequences of the Chlamydomonas STT7 and STL1 and Arabidopsis STN7 and STN8 protein kinases were aligned with ClustalW (Thompson et al., 1994) and the tree was constructed using protein maximum likelihood with the Molphy program (Adachi and Hasegawa, 1996). The scale bar indicates 0.1 substitution per site.*

question whether the LHCII proteins are the direct substrates of STT7/STN7 is not yet solved. In particular, the existence of another kinase acting downstream of STT7/STN7 cannot be excluded.

D. Thylakoid phosphatases

Whereas thylakoid protein kinases have been studied intensively, less is known about protein phosphatases which act on thylakoid proteins. Although several soluble chloroplast protein phosphatases have been characterized which are capable of dephosphorylating the light-harvesting complex (Kieleczawa et al., 1992; Hammer et al., 1997), it is not clear whether the same process is also mediated by thylakoid phosphatases. A physical and functional association has been reported for a thylakoid membrane

PP2A-like phosphatase and a cyclophilin-like lumenal protein, TLP40, providing another example of transmembrane regulation as discussed above for the activation of the LHCII kinase (Fulgosi et al., 1998; Vener et al., 1999). TLP40 has been proposed to coordinate protein folding inside the thylakoid lumen with protein dephosphorylation at the outer thylakoid surface, and thereby control turnover of photodamaged PS II (Vener et al., 1999).

VI. MAINTENANCE OF ATP LEVELS THROUGH STATE TRANSITIONS

The cytochrome b_6f complex functions either in linear electron flow from PS II to PS I or in cyclic electron flow around PS I. During transition from state 1 to state 2, a portion of the cytochrome b_6f complex, together with the mobile LHCII complex, is displaced from the grana to the stromal thylakoid lamellae both *in vivo* and *in vitro* (Vallon et al., 1991). This lateral redistribution of cytochrome b_6f complexes implies a reorganization of the photosynthetic electron transfer chain. Indeed it has been shown that the PS II inhibitor DCMU inhibits electron flow through the cytochrome b_6f complex in state 1 but not in state 2 in *Chlamydomonas* (Finazzi et al., 2001a). This suggests that in this alga most of the excitation energy is used by PS I in state 2, and that transition from state 1 to state 2 induces a switch from linear to cyclic electron flow (see Figure 22.1). This hypothesis was further confirmed by showing that in the *stt7* mutant, inhibition of electron flow through the cytochrome b_6f complex by DCMU still occurs under state 2 conditions (Finazzi et al., 2002).

A marked difference in state transitions between *Chlamydomonas* and land plants is the size of the mobile part of LHCII which is 80 and 15%, respectively. While a displacement of 15–20% of LHCII is compatible with an optimal adjustment of the antenna size for balancing the excitation energy between the two photosystems under changing light conditions, a migration of 80% of LHCII is more difficult to rationalize in these terms. There is evidence, however, that state transitions fulfill another role in *Chlamydomonas*, namely maintenance of intracellular ATP levels (Bulté et al., 1990). An increase in the intracellular demand for ATP leads to state 2 whereas once normal amounts of ATP have been reached, state 1 is restored. This was further shown by the analysis of mitochondrial mutants of *Chlamydomonas* affected in different complexes of the respiratory chain (Cardol et al., 2003). In all of these mutants the cells were in state 2, thereby favoring cyclic over linear electron transport in the chloroplast and compensating for the defect in mitochondrial ATP production. Thus state transitions have a dual role in *Chlamydomonas*, re-balancing of the excitation energy between the two photosystems and meeting the cellular ATP requirements. It is likely that the latter represents the main function of state transitions in this alga. In this sense state transitions would act as a switch between linear electron transfer in state 1 and cyclic electron flow in state 2. The former favors linear photophosphorylation and production of NADPH for carbon fixation and the latter mainly cyclic photophosphorylation. The picture which emerges from these studies is that of a highly dynamic photosynthetic system which can rapidly adjust to meet the intracellular demand for reducing power and ATP. The displacement of a large part of the LHCII antenna from PS II to PS I would greatly enhance cyclic electron flow in state 2. In contrast, in *Arabidopsis*, the role of state transitions appears to be restricted to the balancing of the light excitation energy between the two photosystems, a process which requires only 15–20% of mobile LHCII.

VII. GENETIC APPROACHES

Genetic analysis of state transitions in *Chlamydomonas* led to the identification of key components such as the STT7 kinase, and confirmed the importance of the cytochrome b_6f complex for this process. However, state transitions depend to a large extent on the redox state of the plastoquinone pool, which is influenced by many different factors and pathways in the cell. In particular, redox conditions in the chloroplast are controlled by the movement of metabolites between mitochondria and chloroplasts. It is therefore expected that a genetic analysis may also identify factors that act in an indirect way on state transitions or influence state transitions only under a limited set of conditions. Because one of the screens for state transitions takes advantage of the chlororespiratory chain for inducing state 2, any mutation affecting this pathway would affect the transition from state 1 to state 2. Indeed, several of the mutants identified through the state transition screen are deficient in factors that act only indirectly on state transitions. Some of these mutants are of special interest for understanding the redox control of photosynthesis.

The *menD1* mutant was originally identified as deficient in state transitions and photosynthetically active but light-sensitive. However closer analysis revealed that transition from state 1 to state 2 still occurs after treatment of the cells with the uncoupler FCCP (carbonyl cyanide-*p*-trifluoromethoxyphenylhydrazone), which leads to the depletion of ATP and provides a more stringent conditions for inducing state 2 (Lefebvre-Legendre et al., 2007). Cloning of the gene deficient in *menD1* indicated that it is *MEND*, which encodes 2-succinyl-6-hydroxy-2,4-cyclohexadiene-1-carboxylate (SHCHC) synthase, an enzyme which catalyzes the first committed step of the phylloquinone biosynthetic pathway. The light sensitivity of the *menD1* mutant could be overcome by addition of phylloquinone to the growth medium (Lefebvre-Legendre et al., 2007). Phylloquinone functions as the electron transfer cofactor at the A1 site of PS I. Analysis of pigments by mass spectrometry revealed that phylloquinone is absent in *menD1* whereas total plastoquinone levels were not affected. Analysis of electron transfer by absorption spectroscopy indicated that plastoquinone replaces phylloquinone in PS I and that electron transfer from A1 to the iron–sulfur centers is slowed at least 40-fold (Lefebvre-Legendre et al., 2007). Consistent with a replacement of phylloquinone by plastoquinone, the size of the free plastoquinone pool of *menD1* was reduced by 30%. In contrast to cyanobacterial MenD-deficient mutants, PS I accumulates normally in *menD1* whereas the level of PS II decreases. One possibility is that this decrease is due to the reduced size of the free plastoquinone pool. Another mutant isolated through the same screen appears to be deficient in MENC, which is also involved in the phylloquinone biosynthetic pathway.

Because state transitions involve modulation of the size and composition of the light-harvesting systems, one would expect that some of the mutants deficient in state transitions are affected at the level of LHCII. One mutant of this sort, *stm3*, was indeed identified (Mussgnug et al., 2005). It accumulates increased levels of LHCBM subunits and is affected in the *NAB1* gene. *NAB1* encodes a cytoplasmic protein which binds to *LHCBM* mRNA and thereby represses its translation. The RNA-binding motif of NAB1 which determines the binding affinity and specificity is similar to that of the translation repressor FRG2 from *Xenopus laevis*, a finding which implies similar RNA-masking systems in animals and algae.

The state transition mutant *stm6* is deficient in *MOC1*, a gene whose product is related to the transcription termination factor mTERF from human mitochondria (Schonfeld et al., 2004). In this mutant, which is light-sensitive, expression of the mitochondrial respiratory complexes in the light is markedly perturbed. In particular, the level of the rotenone-insensitive NADPH dehydrogenase is strongly reduced, suggesting that the mitochondria are less able to function as an electron sink for chloroplast metabolites produced by photosynthesis. One possibility is that state transitions are impaired in this mutant because of over-reduction of the stroma, which would inactivate the LHCII kinase (Schonfeld et al., 2004). The *stm6* mutant has proven to be highly promising for photobiological hydrogen production in *Chlamydomonas* (Kruse et al., 2005), which is known to redirect the protons and electrons formed during photosynthesis to produce hydrogen through the chloroplast hydrogenases HYDA1 and HYDA2, which are highly sensitive to oxygen (see Chapter 7). Because *stm3* is blocked in state1, cyclic electron flow is abolished and the possible competition for electrons between this pathway and hydrogenase is eliminated. As a result of its modified respiratory metabolism, the *stm6* mutant accumulates large starch reserves and has a low dissolved O_2 concentration which leads to reduced inhibition of hydrogenase and to significant increase in hydrogen production, approximately five times the wild-type level (Kruse et al., 2005).

VIII. ROLE OF STATE TRANSITIONS

It is becoming clear that state transitions in *Chlamydomonas* have a dual role for balancing the excitation energy between the two photosystems under changing light conditions and for adjusting the cellular ATP level. As mentioned above not only is a large portion of the LHCII antenna displaced from PS II to PS I during a transition from state 1 to state 2, but there is a switch from linear to cyclic electron transfer. This implies that the level of ATP increases massively during such a transition. It is likely that the major function of state transitions in *Chlamydomonas* is the control of ATP/ADP ratio rather than maintaining a balance of the excitation energy between the antennae.

Mutant *stn7 Arabidopsis* plants deficient in state transition do not display a phenotype when grown under standard laboratory conditions. However when plants were subjected to periodic short-term changes with alternating 1-hour light cycles favoring PS II or PS I, growth of *stn7* was impaired as compared to the wild type (Bellafiore et al., 2005). To test whether state transitions are important for plants exposed to outdoor conditions, field tests were performed with *stn7* plants, *stn8* plants, which are impaired in PS II protein phosphorylation, and with the *stn7 stn8* double mutant (Frenkel et al., 2007). Plant fitness estimated by seed production revealed that seed sets of *stn7* mutant plants were 19% smaller than those of wild-type plants, whereas *stn8* had normal seed production. In contrast the double mutant had a significantly lower seed set than *stn7* alone, indicating a synergistic effect between the two thylakoid protein kinases.

IX. CONCLUSIONS AND PROSPECTS

The process of state transitions was discovered years ago using spectroscopic methods (Bonaventura and Myers, 1969; Murata, 1969) This process is widespread among photosynthetic organisms, occurring in cyanobacteria, algae, and land plants. *Chlamydomonas* has been a powerful system for unraveling some of the molecular mechanisms which underlie state transitions. The availability of many mutant strains deficient in photosynthetic activity together with the ease of performing large scale genetic screens have been particularly useful. In this way it was recognized early that the cytochrome b_6f complex plays a key role in state transitions (Wollman and Lemaire, 1988). Moreover the large fluorescence changes associated with state transitions in *Chlamydomonas* have opened the door for a thorough genetic dissection of this process, which has already provided important new insights with the identification of STT7/STN7 protein kinase.

The establishment of nuclear genome sequences from *Chlamydomonas* and several land plants has greatly accelerated the pace of research. Also, the availability of T-DNA insertion lines in *Arabidopsis* allows one to easily compare the phenotype of mutants affected in homologous genes in these two organisms. This is especially important for state transitions, as it allows one to compare related light adaptation processes in a motile unicellular green alga and a sessile land plant. This analysis has revealed that the protein kinases involved in LHCII and PS II core phosphorylation are highly conserved. This was not obvious because algal cells are motile and can adjust to changes in light conditions by moving, which is not possible for land plants. It remains to be seen to what extent the interactants of these kinases and the signaling chain are conserved between algae and plants. A comparative analysis is likely to provide new insights into these processes of adaptation to the changing light environment.

ACKNOWLEDGMENTS

I thank M. Goldschmidt-Clermont for critical reading of the manuscript and N. Roggli for preparing the figures. The work in the author's laboratory was supported by grant 3100-0667763.02 from the Swiss National Fund for Scientific Research.

REFERENCES

Adachi, J. and Hasegawa, M. (1996). Model of amino acid substitution in proteins encoded by mitochondrial DNA. *J. Mol. Evol.* **42**, 459–468.

Allen, J.F. (1992). Protein phosphorylation in regulation of photosynthesis. *Biochim. Biophys. Acta* **1098**, 275–335.

Allen, J.F. and Pfannschmidt, T. (2000). Balancing the two photosystems: Photosynthetic electron transfer governs transcription of reaction centre genes in chloroplasts. *Philos. Trans. R. Soc. Lond. B. Biol. Sci.* **355**, 1351–1359.

Allen, J.F. and Staehelin, L. (1992). Biochemical characterization of photosystem II antenna polypeptides in grana and stroma membranes of spinach. *Plant Physiol.* **100**, 1517–1526.

Allen, J.F., Bennett, J., Steinback, K.E., and Arntzen, C.J. (1981). Chloroplast protein phosphorylation couples plastoquinone redox state to distribution of excitation energy between photosystems. *Nature* **291**, 25–29.

Andersson, B. and Andersson, J. (1980). Lateral heterogeneity in the distribution of chlorophyll-protein complexes of the thylakoid membranes of spinach chloroplasts. *Biochem. Biophys. Acta* **59**, 427–440.

Andronis, C., Kruse, O., Deak, Z., Vass, I., Diner, B.A., and Nixon, P.J. (1998). Mutation of residue threonine-2 of the D2 polypeptide and its effect on photosystem II function in *Chlamydomonas reinhardtii*. *Plant Physiol.* **117**, 515–524.

Bellafiore, S., Barneche, F., Peltier, G., and Rochaix, J.D. (2005). State transitions and light adaptation require chloroplast thylakoid protein kinase STN7. *Nature* **433**, 892–895.

Bennett, J. (1977). Phosphorylation of chloroplast membrane proteins. *Nature* **269**, 344–346.

Bennett, J. (1979). Chloroplast phosphoproteins. Phosphorylation of polypeptides of the light-harvesting chlorophyll protein complex. *Eur. J. Biochem.* **99**, 133–137.

Bennoun, P. (1982). Evidence for a respiratory chain in the chloroplast. *Proc. Natl. Acad. Sci. U. S. A.* **79**, 4352–4356.

Ben-Shem, A., Frolow, F., and Nelson, N. (2003). Crystal structure of plant photosystem I. *Nature* **426**, 630–635.

Bonaventura, C. and Myers, J. (1969). Fluorescence and oxygen evolution from *Chlorella pyrenoidosa*. *Biochim. Biophys. Acta* **189**, 366–383.

Breyton, C. (2000). Conformational changes in the cytochrome b_6f complex induced by inhibitor binding. *J. Biol. Chem.* **275**, 13195–13201.

Bulté, L., Gans, P., Rebeille, F., and Wollman, F.A. (1990). ATP control on state transitions in *Chlamydomonas*. *Biochim. Biophys. Acta* **1020**, 72–80.

Cardol, P., Gloire, G., Havaux, M., Remacle, C., Matagne, R., and Franck, F. (2003). Photosynthesis and state transitions in mitochondrial mutants of *Chlamydomonas reinhardtii* affected in respiration. *Plant Physiol.* **133**, 2010–2020.

Chow, W.S., Melis, A., and Anderson, J.M. (1990). Adjustments of photosystem stoichiometry in chloroplasts improve the quantum efficiency of photosynthesis. *Proc. Natl. Acad. Sci. U. S. A.* **87**, 7502–7506.

Coughlan, S.J. and Hind, G. (1986). Purification and characterization of a membrane-bound protein kinase from spinach thylakoids. *J. Biol. Chem.* **261**, 11378–11385.

Coughlan, S. and Hind, G. (1987). Phosphorylation of thylakoid proteins by a purified kinase. *J. Biol. Chem.* **262**, 8402–8408.

Delepelaire, P. and Wollman, F.A. (1985). Correlations between fluorescence and phosphorylation changes in thylakoid membranes of *Chlamydomonas reinhardtii in vivo*: A kinetic analysis. *Biochim. Biophys. Acta.* **809**, 277–283.

Delosme, R., Olive, J., and Wollman, F.A. (1996). Changes in light energy distribution upon state transitions: An *in vivo* photoacoustic study of the wild type and photosynthesis mutants from *Chlamydomonas reinhardtii*. *Biochim. Biophys. Acta* **1273**, 150–158.

Depège, N., Bellafiore, S., and Rochaix, J.D. (2003). Role of chloroplast protein kinase Stt7 in LHCII phosphorylation and state transition in *Chlamydomonas*. *Science* **299**, 1572–1575.

Elrad, D. and Grossman, A.R. (2004). A genome's-eye view of the light-harvesting polypeptides of *Chlamydomonas reinhardtii*. *Curr. Genet.* **45**, 61–75.

Elrad, D., Niyogi, K.K., and Grossman, A.R. (2002). A major light-harvesting polypeptide of photosystem II functions in thermal dissipation. *Plant Cell* **14**, 1801–1816.

Finazzi, G., Barbagallo, R.P., Bergo, E., Barbato, R., and Forti, G. (2001a). Photoinhibition of *Chlamydomonas reinhardtii* in State 1 and State 2: Damages to the photosynthetic apparatus under linear and cyclic electron flow. *J. Biol. Chem.* **276**, 22251–22257.

Finazzi, G., Zito, F., Barbagallo, R.P., and Wollman, F.A. (2001b). Contrasted effects of inhibitors of cytochrome b_6f complex on state transitions in *Chlamydomonas reinhardtii*: The role of Q_0 site occupancy in LHCII kinase activation. *J. Biol. Chem.* **276**, 9770–9774.

Finazzi, G., Rappaport, F., Furia, A., Fleischmann, M., Rochaix, J.D., Zito, F., and Forti, G. (2002). Involvement of state transitions in the switch between linear and cyclic electron flow in *Chlamydomonas reinhardtii*. *EMBO Rep.* **3**, 280–285.

Fleischmann, M.M., Ravanel, S., Delosme, R., Olive, J., Zito, F., Wollman, F.A., and Rochaix, J.D. (1999). Isolation and characterization of photoautotrophic mutants of *Chlamydomonas reinhardtii* deficient in state transition. *J. Biol. Chem.* **274**, 30987–30994.

Frenkel, M., Bellafiore, S., Rochaix, J.D., and Jansson, S. (2007). Hierarchy amongst photosynthetic acclimation responses for plant fitness. *Physiol. Plant.* **129**, 455–459.

Fulgosi, H., Vener, A.V., Altschmied, L., Herrmann, R.G., and Andersson, B. (1998). A novel multi-functional chloroplast protein: Identification of a 40 kDa immunophilin-like protein located in the thylakoid lumen. *EMBO J* **17**, 1577–1587.

Gal, A., Zer, H., and Ohad, I. (1997). Redox-cntrolled thylakoid protein phosphorylation. News and views. *Plant Physiol.* **100**, 869–885.

Hamel, P., Olive, J., Pierre, Y., Wollman, F.A., and de Vitry, C. (2000). A new subunit of cytochrome b_6f complex undergoes reversible phosphorylation upon state transition. *J. Biol. Chem.* **275**, 17072–17079.

Hammer, M.F., Markwell, J., and Sarath, G. (1997). Purification of a protein phosphatase from chloroplast stroma capable of dephosphorylating the light-harvesting complex-II. *Plant Physiol.* **113**, 227–233.

Holt, N.E., Fleming, G.R., and Niyogi, K.K. (2004). Toward an understanding of the mechanism of nonphotochemical quenching in green plants. *Biochemistry* **43**, 8281–8289.

Horton, P. and Black, M.T. (1980). Activation of adenosine 5 triphosphate-induced quenching of chlorophyll fluorescence by reduced plastoquinone. *FEBS Lett.* **119**, 141–145.

Iwata, S., Lee, J.W., Okada, K., Lee, J.K., Iwata, M., Rasmussen, B., Link, T.A., Ramaswamy, S., and Jap, B.K. (1998). Complete structure of the 11-subunit bovine mitochondrial cytochrome bc_1 complex. *Science* **281**, 64–71.

Kargul, J., Turkina, M.V., Nield, J., Benson, S., Vener, A.V., and Barber, J. (2005). Light-harvesting complex II protein CP29 binds to photosystem I of *Chlamydomonas reinhardtii* under State 2 conditions. *FEBS J.* **272**, 4797–4806.

Kieleczawa, J., Coughlan, S.J., and Hind, G. (1992). Isolation and characterization of an alkaline phosphatase from pea thylakoids. *Plant Physiol.* **99**, 1029–1036.

Kouril, R., Zygadlo, A., Arteni, A.A., de Wit, C.D., Dekker, J.P., Jensen, P.E., Scheller, H.V., and Boekema, E.J. (2005). Structural characterization of a complex of photosystem I and light-harvesting complex II of *Arabidopsis thaliana*. *Biochemistry* **44**, 10935–10940.

Kruse, O., Nixon, P.J., Schmid, G.H., and Mullineaux, C.W. (1999). Isolation of state transition mutants of *Chlamydomonas reinhardtii* by fluorescence video imaging. *Photosynth. Res.* **61**, 43–51.

Kruse, O., Rupprecht, J., Bader, K.P., Thomas-Hall, S., Schenk, PM., Finazzi, G., and Hankamer, B. (2005). Improved photobiological H_2 production in engineered green algal cells. *J. Biol. Chem.* **280**, 34170–34177.

Kuhlbrandt, W., Wang, D.N., and Fujiyoshi, Y. (1994). Atomic model of plant light-harvesting complex by electron crystallography. *Nature* **367**, 614–621.

Kurisu, G., Zhang, H., Smith, J.L., and Cramer, W.A. (2003). Structure of the cytochrome b_6f complex of oxygenic photosynthesis: Tuning the cavity. *Science* **302**, 1009–1014.

Lefebvre-Legendre, L., Rappaport, F., Finazzi, G., Ceol, M., Grivet, C., Hopfgartner, G., and Rochaix, J.D. (2007). Loss of phylloquinone in *Chlamydomonas* affects plastoquinone pool size and PSII synthesis. *J. Biol. Chem.* **28**, 13250–13263.

Lin, Z.F., Lucero, H.A., and Racker, E. (1982). Protein kinases from spinach chloroplasts. I. Purification and identification of two distinct protein kinases. *J. Biol. Chem.* **257**, 12153–12156.

Liu, Z., Yan, H., Wang, K., Kuang, T., Zhang, J., Gui, L., An, X., and Chang, W. (2004). Crystal structure of spinach major light-harvesting complex at 2.72 Å resolution. *Nature* **428**, 287–292.

Lunde, C., Jensen, P.E., Haldrup, A., Knoetzel, J., and Scheller, H.V. (2000). The PS I-H subunit of photosystem I is essential for state transitions in plant photosynthesis. *Nature* **408**, 613–615.

Minagawa, J. and Takahashi, Y. (2004). Structure, function and assembly of photosystem II and its light-harvesting proteins. *Photosynth. Res.* **82**, 241–263.

Murata, N. (1969). Control of excitation transfer in photosynthesis. I. Light-induced change of chlorophyll a fluorescence in *Porphyridium cruentum*. *Biochim. Biophys. Acta* **172**, 242–251.

Mussgnug, J.H., Wobbe, L., Elles, I., Claus, C., Hamilton, M., Fink, A., Kahmann, U., Kapazoglou, A., Mullineaux, C.W., Hippler, M., Nickelsen, J., Nixon, P.J., Kruse, O., Schonfeld, C., Borgstadt, R., and Kienast, A. (2005). NAB1 is an RNA binding protein involved in the light-regulated differential expression of the light-harvesting antenna of *Chlamydomonas reinhardtii*. *Plant Cell* **17**, 3409–3421.

Pascal, A.A., Liu, Z., Broess, K., van Oort, B., van Amerongen, H., Wang, C., Horton, P., Robert, B., Chang, W., and Ruban, A. (2005). Molecular basis of photoprotection and control of photosynthetic light-harvesting. *Nature* **436**, 134–137.

Race, H.L. and Hind, G. (1996). A protein kinase in the core of photosystem II. *Biochemistry* **35**, 13006–13010.

Rintamaki, E., Salonen, M., Suoranta, U.M., Carlberg, I., Andersson, B., and Aro, E.M. (1997). Phosphorylation of light-harvesting complex II and photosystem II core proteins shows different irradiance-dependent regulation *in vivo*. Application of phosphothreonine antibodies to analysis of thylakoid phosphoproteins. *J. Biol. Chem.* **272**, 30476–30482.

Rintamaki, E., Martinsuo, P., Pursiheimo, S., and Aro, E.M. (2000). Cooperative regulation of light-harvesting complex II phosphorylation via the plastoquinol and ferredoxin-thioredoxin system in chloroplasts. *Proc. Natl. Acad. Sci. U. S. A.* **97**, 11644–11649.

Schonfeld, C., Wobbe, L., Borgstadt, R., Kienast, A., Nixon, P.J., and Kruse, O. (2004). The nucleus-encoded protein MOC1 is essential for mitochondrial light acclimation in *Chlamydomonas reinhardtii*. *J. Biol. Chem.* **279**, 50366–50374.

Sizova, I., Fuhrmann, M., and Hegemann, P. (2001). A *Streptomyces rimosus aph-VIII* gene coding for a new type phosphotransferase provides stable antibiotic resistance to *Chlamydomonas reinhardtii*. *Gene* **277**, 221–229.

Snyders, S. and Kohorn, B.D. (1999). TAKs, thylakoid membrane protein kinases associated with energy transduction. *J. Biol. Chem.* **274**, 9137–9140.

Snyders, S. and Kohorn, B.D. (2001). Disruption of thylakoid-associated kinase 1 leads to alteration of light harvesting in *Arabidopsis*. *J. Biol. Chem.* **276**, 32169–32176.

Sokolenko, A., Fulgosi, H., Gal, A., Altschmied, L., Ohad, I., and Herrmann, R.G. (1995). The 64 kDa polypeptide of spinach may not be the LHCII kinase, but a lumen-located polyphenol oxidase. *FEBS Lett.* **371**, 176–180.

Stroebel, D., Choquet, Y., Popot, J.L., and Picot, D. (2003). An atypical haem in the cytochrome b_6f complex. *Nature* **426**, 413–418.

Suorsa, M., Sirpio, S., Allahverdiyeva, Y., Paakkarinen, V., Mamedov, F., Styring, S., and Aro, E.M. (2006). PsbR, a missing link in the assembly of the oxygen-evolving complex of plant photosystem II. *J. Biol. Chem.* **281**, 145–150.

Thompson, J.D., Higgins, D.G., and Gibson, T.J. (1994). CLUSTAL W: Improving the sensitivity of progressive multiple sequence alignment through sequence weighting, position-specific gap penalties and weight matrix choice. *Nucleic Acids Res.* **22**, 4673–4680.

Takahashi, H., Iwai, M., Takahashi, Y., and Minagawa, J. (2006). Identification of the mobile light-harvesting complex II polypeptides for state transitions in *Chlamydomonas reinhardtii*. *Proc. Natl. Acad. Sci. U. S. A.* **103**, 477–482.

Teramoto, H., Ono, T., and Minagawa, J. (2001). Identification of Lhcb gene family encoding the light-harvesting chlorophyll-a/b proteins of photosystem II in *Chlamydomonas reinhardtii*. *Plant Cell Physiol.* **42**, 849–856.

Tokutsu, R., Teramoto, H., Takahashi, Y., Ono, T.A., and Minagawa, J. (2004). The light-harvesting complex of photosystem I in *Chlamydomonas reinhardtii*: Protein composition, gene structures and phylogenic implications. *Plant Cell Physiol.* **45**, 138–145.

Turkina, M.V. and Vener, A.V. (2006). Identification of phosphorylated proteins. In: *Plant Proteomics: Methods and Protocols*, Vol. 335 (H. Thiellement, M. Zivy, C. Damerval, and V. Méchin, Eds.), pp. 305–316. Humana Press Inc, Totowa, NJ.

Turkina, M.V., Kargul, J., Blanco-Rivero, A., Villarejo, A., Barber, J., and Vener, A.V. (2006). Environmentally modulated phosphoproteome of photosynthetic membranes in the green alga *Chlamydomonas reinhardtii*. *Mol. Cell Proteomics* **5**, 1412–1425.

Vallon, O., Bulté, L., Dainese, P., Olive, J., Bassi, R., and Wollman, F.A. (1991). Lateral redistribution of cytochrome b_6/f complexes along thylakoid membranes upon state transitions. *Proc. Natl. Acad. Sci. USA.* **88**, 8262–8266.

Vener, A.V. (2006). Phosphorylation of thylakoid proteins. In: *Photoprotection, Photoinhibition, Gene Regulation and Environment* (B. Demmig-Adams, W. Adams, and A.K. Mattoo, Eds.), pp. 107–126. Springer, Dordrecht, The Netherlands.

Vener, A.V., van Kan, P.J., Rich, P.R., Ohad, I.I., and Andersson, B. (1997). Plastoquinol at the quinol oxidation site of reduced cytochrome *bf* mediates signal transduction between light and protein phosphorylation: Thylakoid protein kinase deactivation by a single-turnover flash. *Proc. Natl. Acad. Sci. USA.* **94**, 1585–1590.

Vener, A.V., Rokka, A., Fulgosi, H., Andersson, B., and Herrmann, R.G. (1999). A cyclophilin-regulated PP2A-like protein phosphatase in thylakoid membranes of plant chloroplasts [erratum appears in *Biochemistry* (2000). **39**, 2130]. *Biochemistry* **38**, 14955–14965.

Wollman, F.A. (2001). State transitions reveal the dynamics and flexibility of the photosynthetic apparatus. *EMBO J.* **20**, 3623–3630.

Wollman, F.A. and Lemaire, C. (1988). Studies on kinase-controlled state transitions in photosystem II and b_6f mutants from *Chlamydomonas reinhardtii* which lack quinone–binding proteins. *Biochim. Biophys. Acta* **933**, 85–94.

Zer, H. and Ohad, I. (2003). Light, redox state, thylakoid-protein phosphorylation and signaling gene expression. *Trends Biochem. Sci.* **28**, 467–470.

Zer, H., Vink, M., Keren, N., Dilly-Hartwig, H., Paulsen, H., Herrmann, R., Andersson, B., and Ohad, I. (1999). Regulation of thylakoid protein phosphorylation at the substrate level: Reversible light-induced conformational changes expose the phosphorylation site of the light-harvesting complex II. *Proc. Natl. Acad. Sci. U. S. A.* **96**, 8277–8282.

Zer, H., Vink, M., Shochat, S., Herrmann, R.G., Andersson, B., and Ohad, I. (2003). Light affects the accessibility of the thylakoid light harvesting complex II (LHCII) phosphorylation site to the membrane protein kinase(s). *Biochemistry* **42**, 728–738.

Zhang, Z., Huang, L., Shulmeister, V.M., Chi, Y.I., Kim, K.K., Hung, L.W., Crofts, A.R., Berry, E.A., and Kim, S.H. (1998). Electron transfer by domain movement in cytochrome bc1. *Nature* **392**, 677–684.

Zito, F., Finazzi, G., Delosme, R., Nitschke, W., Picot, D., and Wollman, F.A. (1999). The Qo site of cytochrome b6f complexes controls the activation of the LHCII kinase. *EMBO J.* **18**, 2961–2969.

Zito, F., Vinh, J., Popot, J.L., and Finazzi, G. (2002). Chimeric fusions of subunit IV and PetL in the b_6f complex of *Chlamydomonas reinhardtii*: Structural implications and consequences on state transitions. *J. Biol. Chem.* **277**, 12446–12455.

CHAPTER 23

Photoprotection and High Light Responses

Krishna K. Niyogi
Department of Plant and Microbial Biology,
University of California, Berkeley, California, USA

CHAPTER CONTENTS

I. INTRODUCTION

Although *Chlamydomonas* cells need light for photosynthesis, absorption of light that exceeds photosynthetic capacity often occurs in nature. In limiting

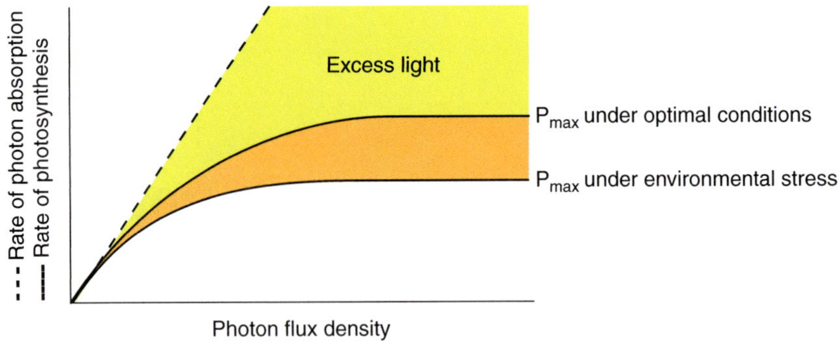

FIGURE 23.1 *Rates of photosynthesis and light absorption as a function of photon flux density.*

light, the rate of photosynthesis increases linearly with photon flux density (PFD), and the initial slope of this light response represents the maximum quantum efficiency of photosynthesis (Figure 23.1). At higher PFDs, photosynthesis saturates at a maximum rate, termed P_{max}. Exposure to even higher PFDs can sometimes cause inhibition of photosynthesis (photoinhibition). In contrast to the rate of photosynthesis, the rate of photon absorption does not saturate at higher PFDs. This results in the absorption of excess light, not only in high light, but also whenever there is an imbalance in the absorption and utilization of light energy. For example, absorption of excess light can occur following dark-to-light transitions and during rapid fluctuations in incident PFD, such as sunflecks. Environmental conditions that constrain P_{max}, such as CO_2 limitation or other nutrient stresses, can make even relatively low PFDs excessive (Figure 23.1).

Excess light is a problem for photosynthetic organisms like *Chlamydomonas* because of the increased potential for photo-oxidative damage (Niyogi, 1999). Oxidizing intermediates and byproducts of photosynthesis such as reactive oxygen species (ROS) can react with nearby proteins, pigments, lipids, and nucleic acids, resulting in decreases in the rate and efficiency of photosynthesis or even cell death (Long et al., 1994). Thus, absorption of excess light necessitates photoprotective mechanisms that prevent, minimize, or repair photo-oxidative damage and allow acclimation to occur.

Most laboratory strains of *Chlamydomonas* can acclimate and grow at PFDs up to ~1000 micromoles photons/m^2/s, but they die at higher PFDs approaching that of full sunlight (~1500–2000 micromoles photons/m^2/s). Different laboratories use different growth conditions, but generally low light (LL) conditions for *Chlamydomonas* are in the range of 50–100 micromoles photons/m^2/s, whereas PFDs of 350–1200 micromoles photons/m^2/s are considered as high light (HL). Following a transition from LL to HL, there is a temporal sequence of changes in post-translational regulation, metabolism, and gene expression that is involved in photoprotection and

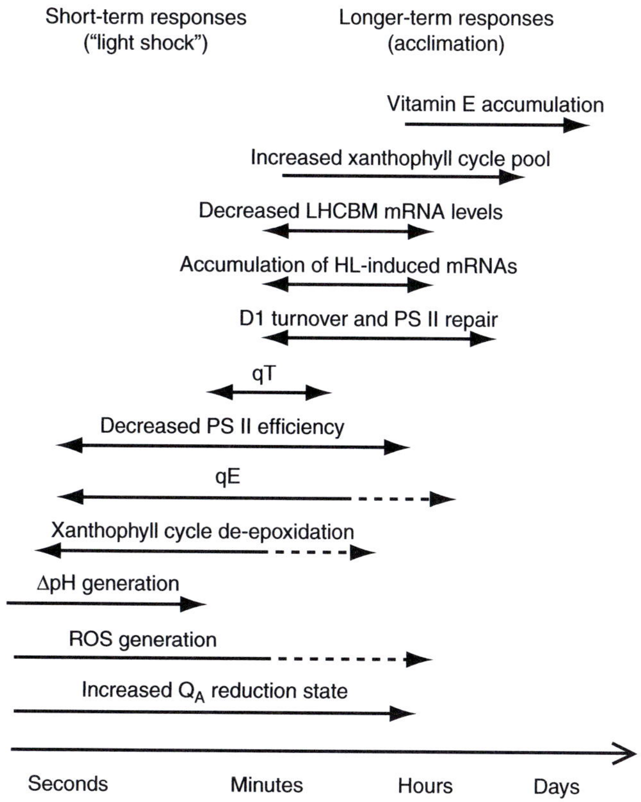

FIGURE 23.2 *Schematic representation of the timescales of specific HL responses in Chlamydomonas.*

acclimation to a higher PFD (Figure 23.2). This chapter focuses on several major photoprotective responses to excess light that have been studied in *Chlamydomonas*, including repair of photo-oxidative damage, thermal dissipation of excess absorbed light energy, and changes in gene expression and metabolites that are involved in photosynthetic acclimation and induction of antioxidant defenses. For information on negative phototaxis and the photophobic motility response in *Chlamydomonas*, please refer to Chapter 13 in Volume 3.

II. HIGH LIGHT-INDUCED DAMAGE AND REPAIR

A. Generation of oxidizing molecules

All organisms must cope with highly reactive compounds that are products of their own metabolism in order to minimize damage to critical cellular constituents. For example, several types of ROS are generated in aerobic environments through electron transfer reactions or excitation energy transfer. Sequential one-electron transfers to molecular oxygen result in formation of superoxide (O_2^-), hydrogen peroxide (H_2O_2), and hydroxyl radical (•OH), whereas singlet oxygen ($^1O_2^*$) is generated by energy transfer in photosensitization reactions. In the chloroplast, potentially damaging intermediates and ROS are generated in the light-harvesting antenna and reaction center of photosystem II (PS II) and the acceptor side of photosystem I (PS I) (Niyogi, 1999; Asada, 2006).

Chlorophyll molecules (and related intermediates in tetrapyrrole biosynthesis) can act as potent endogenous photosensitizers of $^1O_2^*$ formation. In excess light, an increased lifetime of singlet excited chlorophyll (^{1}Chl*) in the PS II antenna will increase the yield of triplet chlorophyll (^{3}Chl*) through intersystem crossing. Relatively long-lived ^{3}Chl* can transfer excitation energy to the triplet ground state of oxygen, resulting in $^1O_2^*$ generation (Foote, 1968). Charge recombination between the $P680^+$/Pheophytin$^-$ radical pair and other backreactions in the PS II reaction center can also generate ^{3}Chl* (^{3}P680*) and then $^1O_2^*$ (Krieger-Liszkay, 2005), which can damage the D1 protein (Vass et al., 1992; Aro et al., 1993; Keren et al., 1997; Keren and Ohad, 1998). Charge recombination can be enhanced by overreduction of the plastoquinone pool in excess light or in mutants that limit downstream photosynthetic electron transport (Hippler et al., 2000). In addition, $P680^+$ itself is a highly oxidizing molecule that has been implicated in donor-side mechanisms of damage to the D1 protein (Anderson et al., 1998).

When light absorption exceeds photosynthetic capacity, a lack of oxidized $NADP^+$ as an acceptor of electrons from PS I favors direct reduction of oxygen by PS I (Mehler, 1951; Asada, 1999) to produce reactive O_2^-. Oxygen reduction is especially prominent in green algae during induction of photosynthesis after dark to light transitions (Radmer and Kok, 1976). O_2^- can be converted enzymatically into H_2O_2 by superoxide dismutase, and a Haber-Weiss or Fenton reaction can lead to metal-catalyzed production of •OH, a type of ROS that is considered to be extremely toxic (Asada, 2006).

B. D1 photodamage and repair

PS II activity is especially susceptible to photoinhibition in excess light (Kyle et al., 1984), and the requirement for synthesis of the chloroplast-encoded D1 protein for recovery of PS II activity was first demonstrated in *Chlamydomonas* (Ohad et al., 1984). Photoinhibition is accompanied by oxidative damage to D1 that necessitates repair of PS II. Rather than degrading and resynthesizing the entire PS II reaction center after photoinactivation, oxygenic photosynthetic organisms have evolved a remarkable repair mechanism for PS II that selectively replaces the damaged D1 protein within a photosynthetic complex that consists of dozens of proteins, pigments, and cofactors. Repair of PS II involves migration of the damaged reaction center from grana to stromal lamellae, degradation of damaged D1, replacement with a newly synthesized D1 protein, religation of pigments and cofactors, and reactivation of the oxygen-evolving complex and electron transport (reviewed in Aro et al., 1993; Keren and Ohad, 1998; Melis, 1991, 1999). Photoinhibition occurs when the rate of PS II inactivation and photodamage exceeds the rate of PS II repair (Ohad et al., 1984), and it is enhanced in excess light and during other abiotic stresses such as salinity stress (Neale and Melis, 1989).

The rate of PS II photodamage increases linearly with PFD (Baroli and Melis, 1996; Tyystjärvi and Aro, 1996), suggesting that there is a probability of damage occurring with each absorbed photon (Park et al., 1995). The probability of D1 photodamage is thought to depend on the redox state of Q_A (Baroli and Melis, 1998; Melis, 1999), which influences charge recombination that can lead to $^1O_2{}^*$ generation (Krieger-Liszkay, 2005). Based on measurement of the action spectrum for photoinactivation, absorption of light by Mn in the oxygen-evolving complex has been proposed as an initial step in photoinactivation of PS II (Hakala et al., 2005; Ohnishi et al., 2005). Inactivation of water splitting might then increase the likelihood of oxidative damage to D1 by $P680^+$ or by $^1O_2{}^*$ generated upon charge recombination (Ohnishi et al., 2005).

A key step in the repair of PS II is the removal and proteolytic degradation of the damaged D1 protein. Degradation of D1 is affected by occupancy of the Q_B site by plastoquinone (Gong and Ohad, 1991; Zer et al., 1994). Experiments with cyanobacteria and plants have led to the identification of FtsH as the primary protease for D1 turnover, with DegP proteases possibly having a role as well (reviewed in Nixon et al., 2005). Homologues of FtsH and DegP are encoded in the *Chlamydomonas* genome (Minagawa and Takahashi, 2004).

The rate of PS II repair depends critically on the synthesis rate of new D1 protein, which is inserted cotranslationally into PS II (Adir et al., 1990). Blocking chloroplast protein synthesis with inhibitors such as chloramphenicol or lincomycin (Ohad et al., 1984) or with mutations that limit chloroplast translation (Heifetz et al., 1997) increases the rate of photoinhibition and results in net loss of D1 protein. Studies with *Chlamydomonas*, *Synechocystis*, and *Arabidopsis* have shown that many situations that enhance photoinhibition, such as blocking the Calvin-Benson cycle (Takahashi and Murata, 2005), lead to inhibition of PS II repair rather than promoting D1 photodamage. Indeed, the primary effect of ROS in photoinhibition has been proposed to be inhibition of D1 synthesis during PS II repair (Nishiyama et al., 2001, 2004, 2006; Takahashi et al., 2007).

During repair of PS II, β-carotene seems to be released from the reaction center and converted into zeaxanthin (Depka et al., 1998), so new synthesis of β-carotene is needed (Trebst and Depka, 1997). Blocking tocopherol (vitamin E) biosynthesis by herbicide treatment of *Chlamydomonas* cells leads to depletion of the tocopherol pool in HL and rapid photoinactivation of PS II and net loss of D1 protein (Trebst et al., 2002), which can be prevented by simultaneous feeding of $^1O_2{}^*$ scavengers or short-chain tocopherol analogs (Trebst et al., 2004; Kruk et al., 2005). Based on these results,

Table 23.1 Selected nuclear genes involved in high light responses

Protein	Gene	Accession #	Mutant	Reference
Glutathione peroxidase homologue	*GPXH (GPX5)*	AAB66330	None	Leisinger et al. (1999)
HSP70B	*HSP70B*	CAA65356	*hsp70B-UE* (antisense)	Schroda et al. (1999)
LHCBM1	*LHCBM1*	AAM18057	*npq5*	Elrad et al. (2002)
Lycopene ε-cyclase	*LCYE*	AAT46065	*lor1*	Eichenberger et al. (1986); Niyogi et al. (1997b); Anwaruzzaman et al. (2004)
PSBS1	*PSBS1*	EDP09214	None	Anwaruzzaman et al. (2004)
PSBS2	*PSBS2*	EDP09661	None	
REP27	*REP27*	ABL73934	*rep27*	Park et al. (2007)
Violaxanthin de-epoxidase	?	–	*npq1*	Niyogi et al. (1997a); Anwaruzzaman et al. (2004)
Zeaxanthin epoxidase	*ZEP1*	AA048940	*npq2*	Niyogi et al. (1997a); Baroli et al. (2003)

a specific role for tocopherol in scavenging $^1O_2{}^*$ and protecting PS II from photoinhibition has been proposed (Krieger-Liszkay and Trebst, 2006).

Several mutants of *Chlamydomonas* that affect PS II photodamage and repair have been identified by screening for enhanced susceptibility to photoinhibition at moderate PFDs (e.g. 150 micromoles photons/m^2/s) (Zhang et al., 1997). One such mutant, *rep27* (Table 23.1), is defective in D1 protein synthesis during PS II repair, and identification of the *REP27* gene revealed that it encodes a conserved chloroplast-targeted protein with two tetratricopeptide repeats (Park et al., 2007). Among PS II subunits, the chloroplast-encoded PsbT protein is important for PS II repair (Ohnishi and Takahashi, 2001), specifically the recovery of Q_A photoreduction activity after photoinhibition (Ohnishi et al., 2007). The HSP70B chaperone (Table 23.1) located in the chloroplast stroma has been shown to function both in protection of PS II from photoinactivation and in PS II repair, perhaps by facilitating D1 synthesis and/or assembly (Schroda et al., 1999; also see Chapter 19).

C. Oxidation of lipids

Lipids can be major targets of photo-oxidative damage in the chloroplast. The lipids of thylakoid membranes in *Chlamydomonas* are enriched in polyunsaturated acyl chains, such as C18:3 and an unusual C16:4 chain

(Trémolières, 1998), which are especially susceptible to damage by $^1O_2^*$. Reaction between $^1O_2^*$ and these lipids produces hydroperoxides (Girotti, 2001) and can lead to initiation of lipid peroxyl radical chain reactions in the thylakoid membrane. Lipid peroxidation products have been detected as thiobarbituric acid-reactive substances or by thermoluminescence after transfer of cells from LL to HL (Baroli et al., 2003, 2004; Fischer et al., 2007). The product of the *GPXH* gene (Table 23.1), which is induced by $^1O_2^*$ and HL (see section IV.B), is potentially involved in repair of peroxidized lipids (Leisinger et al., 1999).

III. REGULATION OF PHOTOSYNTHETIC LIGHT HARVESTING BY NONPHOTOCHEMICAL QUENCHING

A. Overview of NPQ

An increase in light intensity induces short-term responses in algae and plants that provide rapid protection against the harmful effects of excess light, as well as longer-term responses that involve acclimation of the photosynthetic apparatus and chloroplast metabolism to deal with an abundance of light energy. Short-term responses that operate on timescales of seconds to minutes include regulated thermal dissipation of excess absorbed light energy and state transition (Figure 23.2), both of which can be measured as components of nonphotochemical quenching of chlorophyll fluorescence (NPQ). A description of chlorophyll fluorescence quenching analysis can be found on the companion web site (http://www.elsevierdirect.com/companions/9780123708731).

Thermal dissipation is a rapidly reversible component of NPQ, referred to as qE (energy-dependent quenching or feedback de-excitation), that depends on a low thylakoid lumen pH (high ΔpH), specific xanthophylls, and specific members of the LHC protein superfamily. qE involves de-excitation of singlet excited chlorophyll molecules in the PS II antenna, which helps to maintain the balance between dissipation and utilization of absorbed light energy to minimize generation of oxidizing molecules in the photosynthetic apparatus (reviewed in Horton et al., 1996; Müller et al., 2001).

The state transition component of NPQ is called qT, and it is involved in regulating the relative excitation of PS II and PS I (Wollman, 2001) and the ratio of linear to cyclic electron transport (Finazzi et al., 2002). The redox state of plastoquinone, the lipid-soluble electron carrier between PS II and PS I, regulates the activity of a protein kinase(s) that is associated with the cytochrome b_6f complex and involved in phosphorylation of LHCII and other PS II proteins (reviewed in Rochaix, 2007 and discussed in Chapter 22). The specific role of state transitions in response to HL is unclear, although cyclic electron transport might be involved in generating a ΔpH to induce qE (Heber and Walker, 1992; Munekage et al., 2002) and to maintain ATP synthesis during photoinhibition of PS II (Finazzi et al., 2001). The role of state transitions and redox-regulated kinases in acclimation to HL has been questioned, because the *stt7* mutant lacking a thylakoid protein kinase that is necessary for state transitions is not especially sensitive to HL (Fleischmann et al., 1999).

In plants, qE is the major component of NPQ and qT is a relatively minor component, under most experimental conditions. However, the occurrence of qE in *Chlamydomonas* is variable (Niyogi et al., 1997a; Förster et al., 2001; Elrad et al., 2002; Finazzi et al., 2006), being very dependent on growth conditions, and qT can often be the major component of total NPQ. Conditions that favor qE include photoautotrophic growth of cells with air rather than high CO_2 (Niyogi et al., 1997a; Förster et al., 2001) and growth in HL (Iwai et al., 2007), whereas qT predominates when cells are grown in the presence of acetate (Endo and Asada, 1996; Finazzi et al., 2006) or when cells are grown in high CO_2 but measured under conditions of CO_2 limitation (Palmqvist et al., 1990; Falk and Palmqvist, 1992; Iwai et al., 2007). The remainder of this section focuses on qE.

B. The roles of ΔpH and xanthophylls in qE

qE is induced by a low pH in the thylakoid lumen that occurs during illumination with excess light (Horton et al., 1996; Müller et al., 2001). The low pH has two roles: protonation of PS II proteins that are involved in qE and activation of the enzyme violaxanthin de-epoxidase (VDE) (Table 23.1), which converts violaxanthin into antheraxanthin and then zeaxanthin as part of a xanthophyll cycle (Yamamoto et al., 1962, 1999; Eskling et al., 1997). Induction of qE by HL in dark-acclimated *Chlamydomonas* cells, which initially have undetectable or very low levels of zeaxanthin and antheraxanthin, exhibits biphasic kinetics (Niyogi et al., 1997a). The initial, very rapid phase of qE is induced within a few seconds and depends on the buildup of the ΔpH and the presence of lutein (see below) and possibly low levels of zeaxanthin and antheraxanthin. A slower second phase of qE occurs over the next several minutes and correlates with accumulation of additional zeaxanthin and antheraxanthin via VDE activity (Niyogi et al., 1997a). Both phases of qE are inhibited by the uncoupler nigericin, which dissipates the ΔpH. The xanthophyll dependence of qE in *Chlamydomonas* differs somewhat from that in plants (Demmig-Adams, 1990; Niyogi et al., 1998), because the role of zeaxanthin in *Chlamydomonas* is not as prominent.

Chlamydomonas mutants that are defective in qE have been isolated in a forward genetic screen based on video imaging of chlorophyll fluorescence (Niyogi et al., 1997a). Two mutants that are defective in the xanthophyll cycle, *npq1* and *npq2*, were identified by focusing on the second phase of fluorescence quenching occurring between 1 and 10 minutes of HL illumination. The *npq1* mutant (Table 23.1) is blocked in VDE activity, and it fails to convert violaxanthin to antheraxanthin and zeaxanthin in HL. The first phase of qE is still present in *npq1*, whereas the second slower phase is specifically impaired, amounting to an inhibition of only ~25% of the qE. The *npq1* mutation is hypothesized to affect the structural gene encoding VDE, but the molecular genetic basis for the mutation has not yet been determined. Interestingly, *Chlamydomonas* appears to lack an orthologue of the plant-type *VDE* gene (Anwaruzzaman et al., 2004) found in *Arabidopsis* (Bugos and Yamamoto, 1996), *Ostreococcus* (Derelle et al., 2006), and even diatoms (Armbrust et al., 2004), suggesting that the *Chlamydomonas* VDE differs substantially from its counterpart in other algae and plants. Although the *npq1* mutation was generated by insertional mutagenesis, the mutant is not tagged (Niyogi et al., 1997a). A map-based cloning approach is being used to identify the *NPQ1* gene, which has been initially mapped to linkage group IV (Anwaruzzaman et al., 2004).

The *npq2* mutant (Table 23.1) is defective in the *ZEP1* gene (Baroli et al., 2003), which encodes the zeaxanthin epoxidase enzyme that converts zeaxanthin and antheraxanthin into violaxanthin in limiting light. As a result, *npq2* cells contain high levels of zeaxanthin and lack antheraxanthin, violaxanthin, and neoxanthin (Niyogi et al., 1997a). Because of the constitutive accumulation of zeaxanthin, the *npq2* mutant exhibits a very rapid induction of qE that does not increase further upon HL illumination (Niyogi et al., 1997a). A sustained form of NPQ also occurs in *npq2*, even in the presence of the PS II inhibitor 3-(3,4-dichlorophenyl)-1,1-dimethylurea (DCMU), as demonstrated by fluorescence lifetime imaging microscopy (Holub et al., 2000, 2007; Govindjee and Seufferheld, 2002).

A role for lutein in qE was uncovered by analysis of the *lor1* mutant (Niyogi et al., 1997b; Table 23.1), which is defective in the lycopene ε-cyclase (*LCYE*) gene (Anwaruzzaman et al., 2004). The *lor1* mutant (Eichenberger et al., 1986), lacks lutein and loroxanthin and accumulates higher levels of β-xanthophylls, including the xanthophyll cycle carotenoids (Niyogi et al., 1997b). Induction of qE was impaired by ~50% in *lor1* and by ~90% in an *npq1 lor1* double mutant (Niyogi et al., 1997b). These results suggest that zeaxanthin, antheraxanthin, and lutein contribute to qE in *Chlamydomonas* and that essentially all of the qE is xanthophyll-dependent.

The biophysical mechanism of qE and the role of the xanthophylls have been investigated by ultrafast transient absorption spectroscopy in plant thylakoids. The quenching species involved in de-excitation of singlet excited chlorophyll during qE has been proposed to be a chlorophyll-xanthophyll heterodimer, which becomes excited (Ma et al., 2003) and undergoes charge transfer to generate a xanthophyll cation radical and a chlorophyll anion radical (Holt et al., 2005). Charge recombination between the radicals would result in thermal dissipation of excitation energy. By using *Arabidopsis* mutants, the xanthophyll molecule was assigned as zeaxanthin (Holt et al., 2005). Whether a similar mechanism occurs in *Chlamydomonas* remains to be determined.

C. The involvement of antenna proteins in qE

In addition to the xanthophyll mutants described above, several *npq* mutants of *Chlamydomonas* were isolated that are defective in qE but have normal xanthophyll composition and conversion in HL. Such mutants can help to identify protein components of PS II that are necessary for qE. For example, the *npq5* mutant (Table 23.1) exhibits <10% of the wild-type level of qE, but it has a normal xanthophyll cycle and state transition (Elrad et al., 2002). The *npq5* mutant was tagged by insertional mutagenesis, and recovery of flanking DNA sequence revealed an insertion in the *LHCBM1* gene, which encodes one of the 12 members of the light-harvesting complex (LHC) protein family in *Chlamydomonas* that are associated with PS II (Minagawa and Takahashi, 2004). The LHCBM1 protein is a constituent of the trimeric LHCII antenna, and the *npq5* mutant has ~30% fewer trimers than the wild type (Elrad et al., 2002).

The finding of a major component of LHCII trimers that is necessary for qE in *Chlamydomonas* differs from results with *Arabidopsis*, in which no single LHC protein appears to be essential for qE (Andersson et al., 2001, 2003; Kovacs et al., 2006). Instead, mutant analysis in *Arabidopsis* has identified an LHC cousin, the PS II subunit PSBS, as a critical component of the qE mechanism (Li et al., 2000; Peterson and Havir, 2001; Graßes et al., 2002). Protonation of two symmetrical, lumen-facing glutamate residues in PSBS (Li et al., 2004) and possibly other PS II proteins (Horton et al., 1996) is thought to cause a conformational change that is necessary for qE (Müller et al., 2001). *Chlamydomonas* has two closely linked *PSBS* genes (Table 23.1), which encode proteins that are identical except for one amino acid residue in the putative chloroplast transit peptide. Many of the amino acid residues in the four transmembrane domains and two lumenal loops are highly conserved, including the two glutamates that are necessary for function of the protein in *Arabidopsis*. There are only two ESTs corresponding to the *PSBS1* gene (Anwaruzzaman et al., 2004) and none for *PSBS2*, suggesting that the genes are not highly expressed at the RNA level, and attempts to detect the proteins by immunoblotting were not successful (Finazzi et al., 2006). So far, there is no experimental evidence for their function in qE in *Chlamydomonas*. It is possible that the function of PSBS has been somehow replaced by LHCBM1 or another protein in *Chlamydomonas*.

IV. REGULATION OF GENE EXPRESSION AND ACCLIMATION TO HIGH LIGHT

A. Changes in RNA, protein, and metabolite levels during acclimation to high light

During a transition from LL to HL, *Chlamydomonas* cells experience an initial period of "light shock" during which short-term photoprotective responses and regulatory mechanisms such as qE are important for minimizing photodamage. More gradual, acclimatory changes in gene expression and metabolism occur over a time period of hours to days (Figure 23.2). These longer term responses include changes in light-harvesting antenna size, photosynthetic capacity, and antioxidant levels, and they generally require changes in nuclear and chloroplast gene expression to achieve a state of acclimation to HL.

HL-acclimated *Chlamydomonas* cells are characterized by a ~50% lower content of chlorophyll per cell and a slight increase in the chlorophyll *a/b* ratio (Neale and Melis, 1986; Niyogi et al., 1997b; Baroli et al., 2003), which indicate alterations of the light-harvesting antenna systems and/or reaction centers. The chlorophyll antenna size of PS II is lower by ~25% compared to LL cells, and the antenna size of PS I is decreased by >50% (Neale and Melis, 1986). The amount of Q_A per cell is ~15% lower in HL, whereas P700 per cell decreases by ~55%, resulting in a doubling of the PS II:PS I ratio in HL (Neale and Melis, 1986). Thus, most of the decrease in chlorophyll per cell is due to a decrease in PS I content and LHCI antenna size. This contrasts with the situation in other green algae such as *Dunaliella* and *Chlorella*, in which the PS II antenna size decreases and the chlorophyll *a/b* ratio increases dramatically in HL (LaRoche et al., 1991; Escoubas et al., 1995; Maxwell et al., 1995). HL-acclimated *Chlamydomonas* cells have a higher capacity for photosynthetic oxygen evolution and higher rates of respiration (Neale and Melis, 1986). The xanthophyll cycle pool size is ~2x higher in HL on a per chlorophyll basis, and the steady-state de-epoxidation state ([A + Z]/[V + A + Z]) is much higher (~0.6) than in LL-acclimated cells (Niyogi et al., 1997b; Förster et al., 2001; Baroli et al., 2003). The level of lutein is also ~2x higher in HL, whereas loroxanthin decreases (Niyogi et al., 1997b).

The adjustment of PS II chlorophyll antenna size that occurs in *Chlamydomonas* cells during acclimation to HL involves changes in nuclear *LHCBM* gene expression. Compared to *Dunaliella tertiolecta*, in which the response is mainly transcriptional (Escoubas et al., 1995), *Chlamydomonas* exhibits a complicated transcriptional and post-transcription regulation of *LHCBM* genes but does not change its PS II antenna size nearly as much. *Chlamydomonas* has nine genes (*LHCBM1–LHCBM9*) that appear to encode proteins of the major LHCII and three genes that encode components of the inner LHC (LHCB4, LHCB5, and LHCB7) (Elrad and Grossman, 2004; Minagawa and Takahashi, 2004; also see Chapter 14). Expression of these genes in *Chlamydomonas* is coordinately and transiently repressed in HL, with mRNA levels reaching a minimum ~1–2 hours after transfer from LL to HL (Teramoto et al., 2002; Durnford et al., 2003; Elrad and Grossman, 2004). In contrast, some members of the LHC gene family, such as *ELI2* (*LHL1*), *HLI* (*LHL2*), and *LHL4* are transiently induced at the mRNA level (Teramoto et al., 2004). After 6–8 hours in HL, *LHCBM* and *LHCB* mRNA levels recover nearly to LL levels. Experiments with fusions of an *LHCBM* promoter to a reporter gene (*ARS*) have shown that the transient decrease in mRNA levels is due to transcriptional regulation (Durnford et al., 2003). Superimposed on this pattern of transcription, a repression of translation involving dissociation of *LHCBM* mRNAs from polysomes occurs after ~2 hours in HL, followed by a more global translational repression after ~4 hours in HL that can be mimicked by H_2O_2 treatment but not $^1O_2{}^*$ (McKim and Durnford, 2006). Ultimately, the regulation of transcription and translation results in a decrease in PS II antenna size of ~25–40% that is characteristic of the HL-acclimated state (Neale and Melis, 1986; Durnford et al., 2003).

Microarray and differential display analyses of RNA levels have been performed to provide a more global picture of the changes in nuclear gene expression that are involved in the light shock and acclimation responses to HL. Several HL-activated (*HLA*) genes, identified by restriction fragment differential display, encode components of the carbon-concentrating mechanism (CCM; see Chapter 8) that are under control of the regulator CIA5, demonstrating that transfer of *Chlamydomonas* cells from LL to HL in air is accompanied by CO_2 limitation (Im and Grossman, 2001). Induction of these genes also depends on photosynthetic electron transport, whereas induction of other *HLA* genes is independent of the CCM and electron transport (Im and Grossman, 2001). Both cDNA (Im et al., 2003) and oligonucleotide (Fischer et al., 2006b) microarrays have been used to identify dozens of differentially expressed genes following transfer from LL to HL.

Expression of chloroplast genes is also regulated during acclimation to HL, mainly at the level of translation. For example, the synthesis rate of the D1 protein increases substantially in HL (Shapira et al., 1997), whereas RbcL synthesis is inhibited. The inhibition of RbcL translation occurs within 15 minutes and recovers after ~6 hours in HL. This transient repression is associated with the generation of ROS and oxidation of the intracellular glutathione pool (Irihimovitch and Shapira, 2000), resulting in oxidation of thiols in both RbcL and RBCS proteins (Cohen et al., 2005). A conserved RNA recognition motif is exposed upon oxidation of thiols in RbcL, resulting in arrest of *rbcL* mRNA translation (Cohen et al., 2006; see Chapter 9).

Proteomic analysis of *Chlamydomonas* has revealed changes in thylakoid protein phosphorylation (Turkina et al., 2006) in response to HL. After 2 hours in HL, phosphorylation of D1 and CP43 increases, and specific phosphorylation of CP29 occurs at two sites in addition to those found under other conditions. CP26 and several LHCBM proteins are also phosphorylated in HL. These phosphorylation events seem to be localized at the interface between the PS II core and the peripheral antenna system, suggesting that detachment of the antenna might be involved in thermal dissipation and/or PS II repair (Turkina et al., 2006).

During a transition of *Chlamydomonas* cells from LL to HL, the levels of chlorophylls, carotenoids, and tocopherols (vitamin E) change substantially. Carotenoid and tocopherols are major lipid-soluble antioxidant molecules in the chloroplast with complementary photoprotective functions. Within 6 hours in HL, chlorophylls and β-carotene decrease by ~20% and ~40% on a per cell basis, respectively, likely reflecting a decrease in the number of reaction centers (Baroli et al., 2004; Ledford et al., 2004). In addition to the change in xanthophyll cycle de-epoxidation state that accompanies qE, the xanthophyll cycle pool size increases during the first 12 hours in HL (Ledford et al., 2004). After a lag of 6 hours, the α-tocopherol level increases strongly (by threefold) during the first 48 hours of HL (Ledford et al., 2004). The lag might be caused by increased turnover of α-tocopherol (Trebst et al., 2002) during the initial period of light shock upon transfer of cells to HL. The changes in chlorophyll, carotenoid, and tocopherol biosynthetic gene expression, if any, that underlie these responses to HL have not been investigated systematically.

Although *npq1* and *lor1* mutants lacking zeaxanthin and lutein, respectively, are able to acclimate to HL, the *npq1 lor1* double mutant is HL-sensitive and undergoes photo-oxidative bleaching, presumably due to increased production and/or decreased quenching of ROS (Niyogi et al., 1997b), particularly singlet oxygen (Baroli et al., 2004). Following transfer from LL to HL, *npq1 lor1* cells exhibit an initial, transient light shock response in terms of gene expression that is very similar to that seen in the wild type (Ledford et al., 2004). After ~12 hours in HL, however, *npq1 lor1* cells experience a second wave of photo-oxidative stress that appears to overwhelm the limited antioxidant capacity of the mutant. The HL sensitivity of the *npq1 lor1* mutant can be suppressed by *npq2* mutations that confer constitutive accumulation of zeaxanthin (Baroli et al., 2003), which is able to protect against photoinhibition of PS II (Jahns et al., 2000). The inability of the *npq1 lor1* double mutant to survive in high light underscores the multiple roles of xanthophylls as antioxidants (Baroli and Niyogi, 2000).

B. Signaling pathways involved in acclimation to high light

Because perception of excess light occurs in the chloroplast, there must be at least one, and likely more than one, signal transduction pathway from the chloroplast to the nucleus that affects nuclear photosynthetic gene expression (reviewed in Beck, 2005; Nott et al., 2006). These "retrograde" signals could include redox signals from the plastoquinone pool or thioredoxin, chlorophyll biosynthesis intermediates, and ROS.

Evidence for involvement of plastoquinone redox state in regulating *LHCB* transcription in *Dunaliella* was obtained by using the electron transport inhibitors DCMU and 2,5-dibromo-3-methyl-6-isopropyl-*p*-benzoquinone (DBMIB) to oxidize and reduce the plastoquinone pool, respectively (Escoubas et al., 1995). These experiments led to the proposal that a high plastoquinone redox state acts via a chloroplast kinase and a cytosolic factor to repress *LHCB* transcription (Escoubas et al., 1995; Durnford and Falkowski, 1997). Further evidence for this hypothesis was obtained in *Dunaliella salina* and *Chlorella vulgaris* cells in which the plastoquinone redox state was manipulated using a combination of light and temperature changes (Maxwell et al., 1995; Wilson and Huner, 2000). These findings led to the concept of PS II excitation pressure as a signal for *LHCB* gene regulation (Huner et al., 1996, 1998).

However, it appears that the regulation of *LHCBM* expression differs in *Chlamydomonas*. Repression of *LHCBM* and *LHCB* mRNAs by HL still occurs in the presence of DCMU, and it occurs even in the complete absence of both PS II and PS I (Teramoto et al., 2002). Similarly, induction of *LHL4* mRNA is not affected by DCMU in HL or by DBMIB in LL (Teramoto et al., 2004). These results suggest that plastoquinone redox regulation is not the major pathway regulating the levels of *LHCBM*, *LHCB*, and *LHL4* mRNAs in HL, and there must be a redox-independent mechanism possibly involving ROS, a photoreceptor, or chlorophyll biosynthesis intermediates. In the case of *LHL4*, mRNA induction by HL requires cytoplasmic protein synthesis, but a major role of H_2O_2 or O_2^- has been ruled out (Teramoto et al., 2004). Determination of the action spectrum for *LHL4* induction suggests the involvement of a blue/UV-A light photoreceptor in the HL response (Teramoto et al., 2006).

Several lines of evidence have pointed to intermediates in tetrapyrrole biosynthesis as possible plastid-derived signals that might regulate nuclear genes during photoacclimation (reviewed in Vasileuskaya et al., 2004; Beck, 2005; Nott et al., 2006). Early experiments suggested that Mg-protoporphyrin methyl ester and other chlorophyll precursors could inhibit light-induced *LHC* gene expression in *Chlamydomonas* (Johanningmeier and Howell, 1984). The same chlorophyll biosynthesis intermediates can also substitute for light in the induction of an *HSP70* gene in *Chlamydomonas* (Kropat et al., 1997, 2000). Further support for a role of tetrapyrroles as plastid signals has come from the characterization of *gun* mutants in *Arabidopsis* (reviewed in Nott et al., 2006). The *gun* mutants were isolated in a screen for mutants that express the *LHCB1* gene even in the presence of photo-oxidized chloroplasts. The genes affected in several *gun* mutants have been identified, and four of them (*GUN2–GUN5*) affect tetrapyrrole biosynthesis, whereas *GUN1* encodes a chloroplast-localized pentatricopeptide-repeat protein that has been proposed to integrate multiple plastid signals in *Arabidopsis* (Koussevitzky et al., 2007).

A role for $^1O_2^*$ in inducing a subset of nuclear genes involved in acclimation to HL has also been suggested by studies of $^1O_2^*$ signaling in *Chlamydomonas* and other photosynthetic organisms (reviewed in Ledford and Niyogi, 2005). Expression of the nuclear *GPXH* gene is induced by $^1O_2{}^*$ produced either exogenously (Leisinger et al., 2001) or by charge recombination in PS II (Fischer et al., 2006a), and increases in *GPXH* mRNA levels in HL indicate that $^1O_2{}^*$ is generated transiently following a shift from LL to HL (Ledford et al., 2004; Fischer et al., 2006a,b). A region of the *GPXH* promoter that mediates transcriptional upregulation in response to $^1O_2{}^*$ has been identified (Leisinger et al., 2001), and the increase in *GPXH* mRNA level in HL has been attributed to both transcriptional induction and mRNA stabilization (Fischer et al., 2007). A specific region of the *HSP70A* promoter also mediates a transcriptional response to $^1O_2{}^*$, and a reporter construct containing this *cis*-element is induced by HL (Shao et al., 2007). *Chlamydomonas* cells exhibit an acclimation response to $^1O_2{}^*$ stress generated by a photosensitizing dye, Rose Bengal (Fischer et al., 2007; Ledford et al., 2007). This acclimation response involves induction of *GPXH* and other genes by pretreatment of cells with a sublethal concentration of rose bengal in the light (Ledford et al., 2007). Acclimation to $^1O_2{}^*$ can be induced by HL, suggesting that the $^1O_2{}^*$ response might be a subset of the HL response (Ledford et al., 2007). Indeed, there is some overlap between the gene expression responses to HL and $^1O_2{}^*$ (Fischer et al., 2005, 2006b; Ledford et al., 2007).

Mutants that appear to affect regulation of HL responses in *Chlamydomonas* have been identified by selecting for resistance to very high light (VHL; ~2000 micromoles photons/m^2/s). Several nonallelic *VHL^R* mutants in either a wild-type or *psbA-A251L** mutant background are able to grow photoautotrophically at high growth rates in VHL, whereas the parental strains are unable to survive (Förster et al., 1999). These single gene nuclear mutants appear to affect multiple aspects of photosynthesis and photoprotection, including qE and zeaxanthin content, chlorophyll accumulation, PS II function and connectivity, D1 degradation and synthesis, and resistance to ROS (Förster et al., 1999, 2001, 2005). Many changes in soluble protein levels have been detected by comparative proteomic analysis of the wild type and *VHL^R* mutants (Förster et al., 2006). Identification of the genes responsible for the VHL^R phenotype will likely provide new insights into how *Chlamydomonas* acclimates to HL.

ACKNOWLEDGMENTS

This work was supported by the Director, Office of Science, Office of Basic Energy Sciences, Chemical Science Division, of the U.S. Department of Energy under Contract No. DE-AC03-765F00098, by the National Institutes of Health (GM058799 and GM071908), and by the National Science Foundation (MCB-0235878).

REFERENCES

Adir, N., Shochat, S., and Ohad, I. (1990). Light-dependent D1 protein synthesis and translocation is regulated by reaction center II. Reaction center II serves as an acceptor for the D1 precursor. *J. Biol. Chem.* **265**, 12563–12568.

Anderson, J.M., Park, Y.-I., and Chow, W.S. (1998). Unifying model for the photoinactivation of Photosystem II *in vivo* under steady-state photosynthesis. *Photosynth. Res.* **56**, 1–13.

Andersson, J., Walters, R.G., Horton, P., and Jansson, S. (2001). Antisense inhibition of the photosynthetic antenna proteins CP29 and CP26: Implications for the mechanism of protective energy dissipation. *Plant Cell* **13**, 1193–1204.

Andersson, J., Wentworth, M., Walters, R.G., Howard, C.A., Ruban, A.V., Horton, P., and Jansson, S. (2003). Absence of the Lhcb1 and Lhcb2 proteins of the light-harvesting complex of photosystem II: effects on photosynthesis, grana stacking and fitness. *Plant J.* **35**, 350–361.

Anwaruzzaman, M., Chin, B.L., Li, X.-P., Lohr, M., Martinez, D.A., and Niyogi, K.K. (2004). Genomic analysis of mutants affecting xanthophyll biosynthesis and regulation of photosynthetic light harvesting in *Chlamydomonas reinhardtii*. *Photosynth. Res.* **82**, 265–276.

Armbrust, E.V., Berges, J.A., Bowler, C., Green, B.R., Martinez, D., Putnam, N.H., Zhou, S., Allen, A.E., Apt, K.E., Bechner, M. et al. (2004). The genome of the diatom *Thalassiosira pseudonana*: ecology, evolution, and metabolism. *Science* **306**, 79–86.

Aro, E.-M., Virgin, I., and Andersson, B. (1993). Photoinhibition of photosystem II. Inactivation, protein damage and turnover. *Biochim. Biophys. Acta* **1143**, 113–134.

Asada, K. (1999). The water-water cycle in chloroplasts: Scavenging of active oxygens and dissipation of excess photons. *Annu. Rev. Plant Physiol. Plant Mol. Biol.* **50**, 601–639.

Asada, K. (2006). Production and scavenging of reactive oxygen species in chloroplasts and their functions. *Plant Physiol.* **141**, 391–396.

Baroli, I. and Melis, A. (1996). Photoinhibition and repair in *Dunaliella salina* acclimated to different growth irradiances. *Planta* **198**, 640–646.

Baroli, I. and Melis, A. (1998). Photoinhibitory damage is modulated by the rate of photosynthesis and by the photosystem II light-harvesting chlorophyll antenna size. *Planta* **205**, 288–296.

Baroli, I. and Niyogi, K.K. (2000). Molecular genetics of xanthophyll-dependent photoprotection in green algae and plants. *Phil. Trans. R. Soc. Lond. B. Biol. Sci.* **355**, 1385–1394.

Baroli, I., Do, A.D., Yamane, T., and Niyogi, K.K. (2003). Zeaxanthin accumulation in the absence of a functional xanthophyll cycle protects *Chlamydomonas reinhardtii* from photo-oxidative stress. *Plant Cell* **15**, 992–1008.

Baroli, I., Gutman, B.L., Ledford, H.K., Shin, J.W., Chin, B.L., Havaux, M., and Niyogi, K.K. (2004). Photo-oxidative stress in a xanthophyll-deficient mutant of *Chlamydomonas*. *J. Biol. Chem.* **279**, 6337–6344.

Beck, C.F. (2005). Signaling pathways from the chloroplast to the nucleus. *Planta* **222**, 743–756.

Bugos, R.C. and Yamamoto, H.Y. (1996). Molecular cloning of violaxanthin de-epoxidase from romaine lettuce and expression in *Escherichia coli*. *Proc. Natl. Acad. Sci. U. S. A.* **93**, 6320–6325.

Cohen, I., Knopf, J.A., Irihimovitch, V., and Shapira, M. (2005). A proposed mechanism for the inhibitory effects of oxidative stress on Rubisco assembly and its subunit expression. *Plant Physiol.* **137**, 738–746.

Cohen, I., Sapir, Y., and Shapira, M. (2006). A conserved mechanism controls translation of Rubisco large subunit in different photosynthetic organisms. *Plant Physiol.* **141**, 1089–1097.

Demmig-Adams, B. (1990). Carotenoids and photoprotection in plants: A role for the xanthophyll zeaxanthin. *Biochim. Biophys. Acta,* **1020**, 1–24.

Depka, B., Jahns, P., and Trebst, A. (1998). carotene to zeaxanthin conversion in the rapid turnover of the D1 protein of photosystem II. *FEBS Lett.* **424**, 267–270.

Derelle, E., Ferraz, C., Rombauts, S., Rouzé, P., Worden, A.Z., Robbens, S., Partensky, F., Degroeve, S., Echeynié, S., Cooke, R., Saeys, Y., Wuyts, J., Jabbari, K., Bowler, C., Panaud, O., Piégu, B., Ball, S.G., Ral, J.-P., Bouget, F.-Y., Piganeau, G., De Baets, B., Picard, A., Delseny, M., Demaille, J., Van de Peer, Y., and Moreau, H. (2006). Genome analysis of the smallest free-living eukaryote *Ostreococcus tauri* unveils many unique features. *Proc. Natl. Acad. Sci. U. S. A.* **103**, 11647–11652.

Durnford, D.G. and Falkowski, P.G. (1997). Chloroplast redox regulation of nuclear gene transcription during photoacclimation. *Photosynth. Res.* **53**, 229–241.

Durnford, D.G., Price, J.A., McKim, S.M., and Sarchfield, M.L. (2003). Light-harvesting complex gene expression is controlled by both transcriptional and post-transcriptional mechanisms during photoacclimation in *Chlamydomonas reinhardtii*. *Physiol Plant.* **118**, 193–205.

Eichenberger, W., Boschetti, A., and Michel, H.P. (1986). Lipid and pigment composition of a chlorophyll *b*-deficient mutant of *Chlamydomonas reinhardtii*. *Physiol. Plant.* **66**, 589–594.

Elrad, D. and Grossman, A.R. (2004). A genome's-eye view of the light-harvesting polypeptides of *Chlamydomonas reinhardtii*. *Curr. Genet.* **45**, 61–75.

Elrad, D., Niyogi, K.K., and Grossman, A.R. (2002). A major light-harvesting polypeptide of photosystem II functions in thermal dissipation. *Plant Cell* **14**, 1801–1816.

Endo, T. and Asada, K. (1996). Dark induction of the non-photochemical quenching of chlorophyll fluorescence by acetate in *Chlamydomonas reinhardtii*. *Plant Cell Physiol.* **37**, 551–555.

Escoubas, J.-M., Lomas, M., LaRoche, J., and Falkowski, P.G. (1995). Light intensity regulation of *cab* gene transcription is signaled by the redox state of the plastoquinone pool. *Proc. Natl. Acad. Sci. U. S. A.* **92**, 10237–10241.

Eskling, M., Arvidsson, P.-O., and Åkerlund, H.-E. (1997). The xanthophyll cycle, its regulation and components. *Physiol. Plant.* **100**, 806–816.

Falk, S. and Palmqvist, K. (1992). Photosynthetic light utilization efficiency, photosystem II heterogeneity, and fluorescence quenching in *Chlamydomonas reinhardtii* during the induction of the CO_2-concentrating mechanism. *Plant Physiol.* **100**, 685–691.

Finazzi, G., Barbagallo, R.P., Bergo, E., Barbato, R., and Forti, G. (2001). Photoinhibition of *Chlamydomonas reinhardtii* in State 1 and State 2: Damages to the photosynthetic apparatus under linear and cyclic electron flow. *J. Biol. Chem.* **276**, 22251–22257.

Finazzi, G., Rappaport, F., Furia, A., Fleischmann, M., Rochaix, J.-D., Zito, F., and Forti, G. (2002). Involvement of state transitions in the switch between linear and cyclic electron flow in *Chlamydomonas reinhardtii*. *EMBO Rep.* **3**, 280–285.

Finazzi, G., Johnson, G.N., Dall'Osto, L., Zito, F., Bonente, G., Bassi, R., and Wollman, F.-A. (2006). Nonphotochemical quenching of chlorophyll fluorescence in *Chlamydomonas reinhardtii*. *Biochemistry* **45**, 1490–1498.

Fischer, B.B., Krieger-Liszkay, A., and Eggen, R.I.L. (2005). Oxidative stress induced by the photosensitizers neutral red (type I) or rose bengal (type II) in the light causes different molecular responses in *Chlamydomonas reinhardtii*. *Plant Sci.* **168**, 747–759.

Fischer, B.B., Eggen, R.I.L., Trebst, A., and Krieger-Liszkay, A. (2006a). The glutathione peroxidase homologous gene *Gpxh* in *Chlamydomonas reinhardtii* is upregulated by singlet oxygen produced in photosystem II. *Planta* **223**, 583–590.

Fischer, B.B., Wiesendanger, M., and Eggen, R.I.L. (2006b). Growth condition-dependent sensitivity, photodamage and stress response of *Chlamydomonas reinhardtii* exposed to high light conditions. *Plant Cell Physiol.* **47**, 1135–1145.

Fischer, B.B., Dayer, R., Wiesendanger, M., and Eggen, R.I.L. (2007). Independent regulation of the *GPXH* gene expression by primary and secondary effects of high light stress in *Chlamydomonas reinhardtii*. *Physiol Plant.* **130**, 195–206.

Fleischmann, M.M., Ravanel, S., Delosme, R., Olive, J., Zito, F., Wollman, F.-A., and Rochaix, J.-D. (1999). Isolation and characterization of photoautotrophic mutants of *Chlamydomonas reinhardtii* deficient in state transition. *J. Biol. Chem.* **274**, 30987–30994.

Foote, C.S. (1968). Mechanisms of photosensitized oxidation. *Science* **162**, 963–970.

Förster, B., Osmond, C.B., Boynton, J.E., and Gillham, N.W. (1999). Mutants of *Chlamydomonas reinhardtii* resistant to very high light. *J. Photochem. Photobiol. B*, **48**, 127–135.

Förster, B., Osmond, C.B., and Boynton, J.E. (2001). Very high light resistant mutants of *Chlamydomonas reinhardtii*: Responses of Photosystem II, nonphotochemical quenching and xanthophyll pigments to light and CO_2. *Photosynth. Res.* **67**, 5–15.

Förster, B., Osmond, C.B., and Pogson, B.J. (2005). Improved survival of very high light and oxidative stress is conferred by spontaneous gain-of-function mutations in *Chlamydomonas*. *Biochim. Biophys. Acta* **1709**, 45–57.

Förster, B., Mathesius, U., and Pogson, B.J. (2006). Comparative proteomics of high light stress in the model alga *Chlamydomonas reinhardtii*. *Proteomics* **6**, 4309–4320.

Girotti, A.W. (2001). Photosensitized oxidation of membrane lipids: Reaction pathways, cytotoxic effects, and cytoprotective mechanisms. *J. Photochem. Photobiol. B Biol.* **63**, 103–113.

Gong, H. and Ohad, I. (1991). The PQ/PQH_2 ratio and occupancy of photosystem II-Q_B site by plastoquinone control the degradation of D1 protein during photoinhibition *in vivo*. *J. Biol. Chem.* **266**, 21293–21299.

Govindjee, and Seufferheld, M.J. (2002). Non-photochemical quenching of chlorophyll *a* fluorescence: Early history and characterization of two xanthophyll-cycle mutants of *Chlamydomonas reinhardtii*. *Funct. Plant Biol.* **29**, 1141–1155.

Graßes, T., Pesaresi, P., Schiavon, F., Varotto, C., Salamini, F., Jahns, P., and Leister, D. (2002). The role of pH-dependent dissipation of excitation energy in protecting photosystem II against light-induced damage in *Arabidopsis thaliana*. *Plant Physiol. Biochem.* **40**, 41–49.

Hakala, M., Tuominen, I., Keränen, M., Tyystjärvi, T., and Tyystjärvi, E. (2005). Evidence for a role of the oxygen-evolving manganese complex in photoinhibition of photosystem II. *Biochim. Biophys. Acta* **1706**, 68–80.

Heber, U. and Walker, D. (1992). Concerning a dual function of coupled cyclic electron transport in leaves. *Plant Physiol.* **100**, 1621–1626.

Heifetz, P.B., Lers, A., Turpin, D.H., Gillham, N.W., Boynton, J.E., and Osmond, C.B. (1997). *dr* and *spr/sr* mutations of *Chlamydomonas reinhardtii* affecting D1 protein function and synthesis define two independent steps leading to chronic photoinhibition and confer differential fitness. *Plant Cell Environ.* **20**, 1145–1157.

Hippler, M., Biehler, K., Krieger-Liszkay, A., van Dillewjin, J., and Rochaix, J.-D. (2000). Limitation in electron transfer in photosystem I donor side mutants of *Chlamydomonas reinhardtii*. Lethal photo-oxidative damage in high light is overcome in a suppressor strain deficient in the assembly of the light harvesting complex. *J. Biol. Chem.* **275**, 5852–5859.

Holt, N.E., Zigmantas, D., Valkunas, L., Li, X.-P., Niyogi, K.K., and Fleming, G.R. (2005). Carotenoid cation formation and the regulation of photosynthetic light harvesting. *Science* **307**, 433–436.

Holub, O., Seufferheld, M.J., Gohlke, C., Govindjee, and Clegg, R.M. (2000). Fluorescence lifetime imaging (FLI) in real-time: A new technique in photosynthesis research. *Photosynthetica* **38**, 581–599.

Holub, O., Seufferheld, M.J., Gohlke, C., Govindjee, Heiss, G.J., and Clegg, R.M. (2007). Fluorescence lifetime imaging microscopy of *Chlamydomonas reinhardtii*: Non-photochemical quenching mutants and the effect of photosynthetic inhibitors on the slow chlorophyll fluorescence transient. *J. Microscopy* **226**, 90–120.

Horton, P., Ruban, A.V., and Walters, R.G. (1996). Regulation of light harvesting in green plants. *Annu. Rev. Plant Physiol. Plant Mol. Biol.* **47**, 655–684.

Huner, N.P.A., Maxwell, D.P., Gray, G.R., Savitch, L.V., Krol, M., Ivanov, A.G., and Falk, S. (1996). Sensing environmental temperature change through imbalances between energy supply and energy consumption: Redox state of photosystem II. *Physiol. Plant.* **98**, 358–364.

Huner, N.P.A., Öquist, G., and Sarhan, F. (1998). Energy balance and acclimation to light and cold. *Trends Plant Sci.* **3**, 224–230.

Im, C.S. and Grossman, A.R. (2001). Identification and regulation of high light-induced genes in *Chlamydomonas reinhardtii*. *Plant J.* **30**, 301–313.

Im, C.-S., Zhang, Z., Shrager, J., Chang, C.-W., and Grossman, A.R. (2003). Analysis of light and CO_2 regulation in *Chlamydomonas reinhardtii* using genome-wide approaches. *Photosynth. Res.* **75**, 111–125.

Irihimovitch, V. and Shapira, M. (2000). Glutathione redox potential modulated by reactive oxygen species regulates translation of Rubisco large subunit in the chloroplast. *J. Biol. Chem.* **275**, 16289–16295.

Iwai, M., Kato, N., and Minagawa, J. (2007). Distinct physiological responses to a high light and low CO_2 environment revealed by fluorescence quenching in photoautotrophically grown *Chlamydomonas reinhardtii*. *Photosynth. Res.* **94**, 307–314.

Jahns, P., Depka, B., and Trebst, A. (2000). Xanthophyll cycle mutants from *Chlamydomonas reinhardtii* indicate a role for zeaxanthin in the D1 protein turnover. *Plant Physiol. Biochem.* **38**, 371–376.

Johanningmeier, U. and Howell, S.H. (1984). Regulation of light-harvesting chlorophyll-binding protein mRNA accumulation in *Chlamydomonas reinhardi*. Possible involvement of chlorophyll synthesis precursors. *J. Biol. Chem.* **259**, 13541–13549.

Keren, N. and Ohad, I. (1998). State transition and photoinhibition. In: *The Molecular Biology of Chloroplasts and Mitochondria in Chlamydomonas* (J.-D. Rochaix, M. Goldschmidt-Clermont, and S. Merchant, Eds.), pp. 569–596. Kluwer Academic Publishers, Dordrecht.

Keren, N., Berg, A., van Kan, P.J.M., Levanon, H., and Ohad, I. (1997). Mechanism of photosystem II photoinactivation and D1 protein degradation at low light: The role of back electron flow. *Proc. Natl. Acad. Sci. U. S. A.* **94**, 1579–1584.

Koussevitzky, S., Nott, A., Mockler, T.C., Hong, F., Sachetto-Martins, G., Surpin, M., Lim, J., Mittler, R., and Chory, J. (2007). Signals from chloroplasts converge to regulate nuclear gene expression. *Science* **316**, 715–719.

Kovacs, L., Damkjaer, J., Kereiche, S., Ilioaia, C., Ruban, A.V., Boekema, E.J., Jansson, S., and Horton, P. (2006). Lack of the light-harvesting complex CP24 affects the structure and function of the grana membranes of higher plant chloroplasts. *Plant Cell* **18**, 3106–3120.

Krieger-Liszkay, A. (2005). Singlet oxygen production in photosynthesis. *J. Exp. Bot.* **56**, 337–346.

Krieger-Liszkay, A. and Trebst, A. (2006). Tocopherol is the scavenger of singlet oxygen produced by the triplet states of chlorophyll in the PSII reaction centre. *J. Exp. Bot.* **57**, 1677–1684.

Kropat, J., Oster, U., Rüdiger, W., and Beck, C.F. (1997). Chlorophyll precursors are signals of chloroplast origin involved in light induction of nuclear heat-shock genes. *Proc. Natl. Acad. Sci. U. S. A.* **94**, 14168–14172.

Kropat, J., Oster, U., Rüdiger, W., and Beck, C.F. (2000). Chloroplast signalling in the light induction of nuclear *HSP70* genes requires the accumulation of chlorophyll precursors and their accessibility to cytoplasm/nucleus. *Plant J.* **24**, 523–531.

Kruk, J., Holländer-Czytko, H., Oettmeier, W., and Trebst, A. (2005). Tocopherol as singlet oxygen scavenger in photosystem II. *J. Plant Physiol.* **162**, 749–757.

Kyle, D.J., Ohad, I., and Arntzen, C.J. (1984). Membrane protein damage and repair: Selective loss of a quinone-protein function in chloroplast membranes. *Proc. Natl. Acad. Sci. U. S. A.* **81**, 4070–4074.

LaRoche, J., Mortain-Bertrand, A., and Falkowski, P.G. (1991). Light intensity-induced changes in *cab* mRNA and light harvesting complex II apoprotein levels in the unicellular chlorophyte *Dunaliella tertiolecta*. *Plant Physiol.* **97**, 147–153.

Ledford, H.K. and Niyogi, K.K. (2005). Singlet oxygen and photo-oxidative stress management in plants and algae. *Plant Cell Environ.* **28**, 1037–1045.

Ledford, H.K., Baroli, I., Shin, J.W., Fischer, B.B., Eggen, R.I.L., and Niyogi, K.K. (2004). Comparative profiling of lipid-soluble antioxidants and transcripts reveals two phases of photo-oxidative stress in a xanthophyll-deficient mutant of *Chlamydomonas reinhardtii*. *Mol. Gen. Genomics* **272**, 470–479.

Ledford, H.K., Chin, B.L., and Niyogi, K.K. (2007). Acclimation to singlet oxygen stress in *Chlamydomonas reinhardtii*. *Eukaryotic Cell* **6**, 919–930.

Leisinger, U., Rüfenacht, K., Zehnder, A.J.B., and Eggen, R.I.L. (1999). Structure of a glutathione peroxidase homologous gene involved in the oxidative stress response in *Chlamydomonas reinhardtii*. *Plant Sci.* **149**, 139–149.

Leisinger, U., Rüfenacht, K., Fischer, B., Pesaro, M., Spengler, A., Zehnder, A.J.B., and Eggen, R.I.L. (2001). The glutathione peroxidase homologous gene from *Chlamydomonas reinhardtii* is transcriptionally up-regulated by singlet oxygen. *Plant Mol. Biol.* **46**, 395–408.

Li, X.-P., Björkman, O., Shih, C., Grossman, A.R., Rosenquist, M., Jansson, S., and Niyogi, K.K. (2000). A pigment-binding protein essential for regulation of photosynthetic light harvesting. *Nature* **403**, 391–395.

Li, X.-P., Gilmore, A.M., Caffarri, S., Bassi, R., Golan, T., Kramer, D., and Niyogi, K.K. (2004). Regulation of photosynthetic light harvesting involves intrathylakoid lumen pH sensing by the PsbS protein. *J. Biol. Chem.* **279**, 22866–22874.

Long, S.P., Humphries, S., and Falkowski, P.G. (1994). Photoinhibition of photosynthesis in nature. *Annu. Rev. Plant Physiol. Plant Mol. Biol.* **45**, 633–662.

Ma, Y.-Z., Holt, N.E., Li, X.-P., Niyogi, K.K., and Fleming, G.R. (2003). Evidence for direct carotenoid involvement in the regulation of photosynthetic light harvesting. *Proc. Natl. Acad. Sci. U. S. A.* **100**, 4377–4382.

Maxwell, D.P., Laudenbach, D.E., and Huner, N.P.A. (1995). Redox regulation of light-harvesting complex II and *cab* mRNA abundance in *Dunaliella salina*. *Plant Physiol.* **109**, 787–795.

McKim, S.M. and Durnford, D.G. (2006). Translational regulation of light-harvesting complex expression during photoacclimation to high-light in *Chlamydomonas reinhardtii*. *Plant Physiol. Biochem.* **44**, 857–865.

Mehler, A.H. (1951). Studies on reactions of illuminated chloroplasts. I. Mechanism of the reduction of oxygen and other Hill reagents. *Arch. Biochem. Biophys.* **33**, 65–77.

Melis, A. (1991). Dynamics of photosynthetic membrane composition and function. *Biochim. Biophys. Acta* **1058**, 87–106.

Melis, A. (1999). Photosystem-II damage and repair cycle in chloroplasts: What modulates the rate of photodamage *in vivo*? *Trends Plant Sci.* **4**, 130–135.

Minagawa, J. and Takahashi, Y. (2004). Structure, function and assembly of Photosystem II and its light-harvesting proteins. *Photosynth. Res.* **82**, 241–263.

Müller, P., Li, X.-P., and Niyogi, K.K. (2001). Non-photochemical quenching. A response to excess light energy. *Plant Physiol.* **125**, 1558–1566.

Munekage, Y., Hojo, M., Meurer, J., Endo, T., Tasaka, M., and Shikanai, T. (2002). *PGR5* is involved in cyclic electron flow around photosystem I and is essential for photoprotection in *Arabidopsis*. *Cell* **110**, 361–371.

Neale, P.J. and Melis, A. (1986). Algal photosynthetic membrane complexes and the photosynthesis-irradiance curve: A comparison of light-adaptation responses in *Chlamydomonas reinhardtii* (Chlorophyta). *J. Phycol.* **22**, 531–538.

Neale, P.J. and Melis, A. (1989). Salinity-stress enhances photoinhibition of photosystem II in *Chlamydomonas reinhardtii*. *J. Plant Physiol.* **134**, 619–622.

Nishiyama, Y., Yamamoto, H., Allakhverdiev, S.I., Inaba, M., Yokota, A., and Murata, N. (2001). Oxidative stress inhibits the repair of photodamage to the photosynthetic machinery. *EMBO J.* **20**, 5587–5594.

Nishiyama, Y., Allakhverdiev, S.I., Yamamoto, H., Hayashi, H., and Murata, N. (2004). Singlet oxygen inhibits the repair of photosystem II by suppressing the translation elongation of the D1 protein in *Synechocystis* sp. PCC 6803. *Biochemistry* **43**, 11321–11330.

Nishiyama, Y., Allakhverdiev, S.I., and Murata, N. (2006). A new paradigm for the action of reactive oxygen species in the photoinhibition of photosystem II. *Biochim. Biophys. Acta* **1757**, 742–749.

Nixon, P., Barker, M., Boehm, M., de Vries, R., and Komenda, J. (2005). FtsH-mediated repair of the photosystem II complex in response to light stress. *J. Exp. Bot.* **56**, 357–363.

Niyogi, K.K. (1999). Photoprotection revisited: Genetic and molecular approaches. *Annu. Rev. Plant Physiol. Plant Mol. Biol.* **50**, 333–359.

Niyogi, K.K., Björkman, O., and Grossman, A.R. (1997a). *Chlamydomonas* xanthophyll cycle mutants identified by video imaging of chlorophyll fluorescence quenching. *Plant Cell* **9**, 1369–1380.

Niyogi, K.K., Björkman, O., and Grossman, A.R. (1997b). The roles of specific xanthophylls in photoprotection. *Proc. Natl. Acad. Sci. U. S. A.* **94**, 14162–14167.

Niyogi, K.K., Grossman, A.R., and Björkman, O. (1998). *Arabidopsis* mutants define a central role for the xanthophyll cycle in the regulation of photosynthetic energy conversion. *Plant Cell* **10**, 1121–1134.

Nott, A., Jung, H.-S., Koussevitzky, S., and Chory, J. (2006). Plastid-to-nucleus retrograde signaling. *Annu. Rev. Plant Biol.* **57**, 739–759.

Ohad, I., Kyle, D.J., and Arntzen, C.J. (1984). Membrane protein damage and repair: Removal and replacement of inactivated 32-kilodalton polypeptides in chloroplast membranes. *J. Cell Biol.* **99**, 481–485.

Ohnishi, N. and Takahashi, Y. (2001). PsbT polypeptide is required for efficient repair of photodamaged photosystem II reaction center. *J. Biol. Chem.* **276**, 33798–33804.

Ohnishi, N., Allakhverdiev, S.I., Takahashi, S., Higashi, S., Watanabe, M., Nishiyama, Y., and Murata, N. (2005). Two-step mechanism of photodamage to photosystem II: Step 1 occurs at the oxygen-evolving complex and step 2 occurs at the photochemical reaction center. *Biochemistry* **44**, 8494–8499.

Ohnishi, N., Kashino, Y., Satoh, K., Ozawa, S., and Takahashi, Y. (2007). Chloroplast-encoded polypeptide PsbT is involved in the repair of primary electron acceptor Q_A of photosystem II during photoinhibition in *Chlamydomonas reinhardtii*. *J. Biol. Chem.* **282**, 7107–7115.

Palmqvist, K., Sundblad, L.G., Wingsle, G., and Samuelsson, G. (1990). Acclimation of photosynthetic light reactions during induction of inorganic carbon accumulation in the green alga *Chlamydomonas reinhardtii*. *Plant Physiol.* **94**, 357–366.

Park, S., Khamai, P., Garcia-Cerdan, J.G., and Melis, A. (2007). REP27, a tetratricopeptide repeat nuclear-encoded and chloroplast-localized protein, functions in D1/32-kD reaction center protein turnover and photosystem II repair from photodamage. *Plant Physiol.* **143**, 1547–1560.

Park, Y.-I., Chow, W.S., and Anderson, J.M. (1995). Light inactivation of functional photosystem II in leaves of peas grown in moderate light depends on photon exposure. *Planta* **196**, 401–411.

Peterson, R.B. and Havir, E.A. (2001). Photosynthetic properties of an *Arabidopsis thaliana* mutant possessing a defective PsbS gene. *Planta* **214**, 142–152.

Radmer, R.J. and Kok, B. (1976). Photoreduction of O_2 primes and replaces CO_2 assimilation. *Plant Physiol.* **58**, 336–340.

Rochaix, J.-D. (2007). Role of thylakoid protein kinases in photosynthetic acclimation. *FEBS Lett.* **581**, 2768–2775.

Schroda, M., Vallon, O., Wollman, F.-A., and Beck, C.F. (1999). A chloroplast-targeted heat shock protein 70 (HSP70) contributes to the photoprotection and repair of photosystem II during and after photoinhibition. *Plant Cell* **11**, 1165–1178.

Shao, N., Krieger-Liszkay, A., Schroda, M., and Beck, C.F. (2007). A reporter system for the individual detection of hydrogen peroxide and singlet oxygen: Its use for the assay of reactive oxygen species produced *in vivo*. *Plant J.* **50**, 475–487.

Shapira, M., Lers, A., Heifetz, P.B., Irihimovitz, V., Osmond, C.B., Gillham, N.W., and Boynton, J.E. (1997). Differential regulation of chloroplast gene expression in *Chlamydomonas reinhardtii* during photoacclimation: Light stress transiently suppresses synthesis of the Rubisco LSU protein while enhancing synthesis of the PS II D1 protein. *Plant Mol. Biol.* **33**, 1001–1011.

Takahashi, S. and Murata, N. (2005). Interruption of the Calvin cycle inhibits the repair of photosystem II from photodamage. *Biochim. Biophys. Acta* **1708**, 352–361.

Takahashi, S., Bauwe, H., and Badger, M. (2007). Impairment of the photorespiratory pathway accelerates photoinhibition of photosystem II by suppression of repair but not acceleration of damage processes in *Arabidopsis*. *Plant Physiol.* **144**, 487–494.

Teramoto, H., Nakamori, A., Minagawa, J., and Ono, T. (2002). Light-intensity-dependent expression of the *Lhc* gene family encoding light-harvesting chlorophyll-*a/b* proteins of photosystem II in *Chlamydomonas reinhardtii*. *Plant Physiol.* **130**, 325–333.

Teramoto, H., Itoh, T., and Ono, T. (2004). High-intensity-light-dependent and transient expression of new genes encoding distant relatives of light-harvesting chlorophyll-*a/b* proteins in *Chlamydomonas reinhardtii*. *Plant Cell Physiol.* **45**, 1221–1232.

Teramoto, H., Ishii, A., Kimura, Y., Hasegawa, K., Nakazawa, S., Nakamura, T., Higashi, S., Watanabe, M., and Ono, T. (2006). Action spectrum for expression of the high intensity light-inducible *Lhc*-like gene *Lhl4* in the green alga *Chlamydomonas reinhardtii*. *Plant Cell Physiol.* **47**, 419–425.

Trebst, A. and Depka, B. (1997). Role of carotene in the rapid turnover and assembly of photosystem II in *Chlamydomonas reinhardtii*. *FEBS Lett.* **400**, 359–362.

Trebst, A., Depka, B., and Holländer-Czytko, H. (2002). A specific role for tocopherol and of chemical singlet oxygen quenchers in the maintenance of photosystem II structure and function in *Chlamydomonas reinhardtii*. *FEBS Lett.* **516**, 156–160.

Trebst, A., Depka, B., Jäger, J., and Oettmeier, W. (2004). Reversal of the inhibition of photosynthesis by herbicides affecting hydroxyphenylpyruvate dioxygenase by plastoquinone and tocopheryl derivatives in *Chlamydomonas reinhardtii*. *Pest Manag. Sci.* **60**, 669–674.

Trémolières, A. (1998). Glycerolipids: Composition, biosynthesis and function in *Chlamydomonas*. In: *The Molecular Biology of Chloroplasts and Mitochondria in Chlamydomonas* (J.-D. Rochaix, M. Goldschmidt-Clermont, and S. Merchant, Eds.), pp. 415–431. Kluwer Academic Publishers, Dordrecht.

Turkina, M.V., Kargul, J., Blanco-Rivero, A., Villarejo, A., Barber, J., and Vener, A.V. (2006). Environmentally modulated phosphoproteome of photosynthetic membranes in the green alga *Chlamydomonas reinhardtii*. *Mol. Cell. Proteomics* **5**, 1412–1425.

Tyystjärvi, E. and Aro, E.-M. (1996). The rate constant of photoinhibition, measured in lincomycin-treated leaves, is directly proportional to light intensity. *Proc. Natl. Acad. Sci. U. S. A.* **93**, 2213–2218.

Vasileuskaya, Z., Oster, U., and Beck, C.F. (2004). Involvement of tetrapyrroles in inter-organellar signaling in plants and algae. *Photosynth. Res.* **82**, 289–299.

Vass, I., Styring, S., Hundal, T., Koivuniemi, A., Aro, E.-M., and Andersson, B. (1992). Reversible and irreversible intermediates during photoinhibition of photosystem II: Stable reduced Q_A species promote chlorophyll triplet formation. *Proc. Natl. Acad. Sci. U. S. A.* **89**, 1408–1412.

Wilson, K.E. and Huner, N.P.A. (2000). The role of growth rate, redox-state of the plastoquinone pool and the trans-thylakoid ΔpH in photoacclimation of *Chlorella vulgaris* to growth irradiance and temperature. *Planta* **212**, 93–102.

Wollman, F.-A. (2001). State transitions reveal the dynamics and flexibility of the photosynthetic apparatus. *EMBO J.* **20**, 3623–3630.

Yamamoto, H.Y., Nakayama, T.O.M., and Chichester, C.O. (1962). Studies on the light and dark interconversions of leaf xanthophylls. *Arch. Biochem. Biophys.* **97**, 168–173.

Yamamoto, H.Y., Bugos, R.C., and Hieber, A.D. (1999). Biochemistry and molecular biology of the xanthophyll cycle. In: *The Photochemistry of Carotenoids* (H.A. Frank, A.J. Young, G. Britton, and R.J. Cogdell, Eds.), pp. 293–303. Kluwer Academic Publishers, Dordrecht.

Zer, H., Prasil, O., and Ohad, I. (1994). Role of plastoquinol oxidoreduction in regulation of photochemical reaction center II D1 protein turnover *in vivo*. *J. Biol. Chem.* **269**, 17670–17676.

Zhang, L., Niyogi, K.K., Baroli, I., Nemson, J.A., Grossman, A.R., and Melis, A. (1997). DNA insertional mutagenesis for the elucidation of a Photosystem II repair process in the green alga *Chlamydomonas reinhardtii*. *Photosynth. Res.* **53**, 173–184.

CHAPTER 24

The Chloroplast Genome

David C. Higgs
Department of Biological Sciences,
University of Wisconsin – Parkside,
Kenosha, Wisconsin, USA

CHAPTER CONTENTS

I. INTRODUCTION

The chloroplast genomes from *Chlamydomonas* species have provided important insights into the function and evolution of this essential organelle. The field dates to 1962, when Ris and Plaut presented microscopic evidence of chloroplast DNA (cpDNA) in *C. moewusii*. Soon after, Sager and Ishida (1963) characterized cpDNA in purified *C. reinhardtii* chloroplasts and reported its base composition. The focus of this chapter is on *Chlamydomonas* cpDNA structure, sequence, gene content, replication, and evolution, and it is not intended to be comprehensive for all *Chlamydomonas* species. In-depth comparative analyses of cpDNA in algae and vascular plants have been published elsewhere (Nedelcu and Lee, 1998; Simpson and Stern, 2002; Pombert et al., 2005; de Cambiaire et al., 2007). Chloroplast inheritance is covered in Volume 1, Chapter 7. Recombination and replication of cpDNA were reviewed by Sears (1998). Chloroplast gene expression is covered at length in Chapters 25–29 of the present volume.

The utility of *Chlamydomonas* as a model organism for chloroplast biology was dramatically augmented in 1988 when the *C. reinhardtii* chloroplast was successfully transformed (Boynton et al., 1988; reviewed by Goldschmidt-Clermont, 1998). The many subsequent studies involving cpDNA transformation and gene mutation have contributed broadly to our understanding of chloroplast biology and gene function; much of this is relevant to land plants (Swaitek et al., 2001; Merchant et al., 2007).

For this chapter, the complete sequence of *C. reinhardtii* cpDNA (Maul et al., 2002) will be the reference genome. Extensive chloroplast sequences have been obtained from other *Chlamydomonas* species, with approximately half of the *C. moewusii* and *C. eugametos* cpDNAs being available (Boudreau et al., 1994). More limited amounts of cpDNA sequence have been deposited in GenBank from 41 other *Chlamydomonas* species. The cumulative data have and will continue to contribute to our understanding of chloroplast evolution and gene function through comparative genomics and gene manipulation (Turmel et al., 1993; Huang et al., 1994; Kramzar et al., 2006; Pombert et al., 2006).

II. GENOME STRUCTURE AND SEQUENCE

The complete *C. reinhardtii* chloroplast genome has 203,828 bp, 99 genes of known or presumed function (not double-counting genes in the inverted repeats), and eight expressed open reading frames (ORFs), as presented in Figure 24.1 and Table 24.1. The G+C content is 35%, which is similar to the original prediction of 39% by Sager and Ishida (1963) based on buoyant density. The typical structure of chloroplast genomes in plants and algae, including *Chlamydomonas*, is quadripartite with two regions of single-copy DNA and two large inverted repeats (IRs) (Palmer, 1991; Simpson and Stern, 2002; Pombert et al., 2006). Outside of *Chlamydomonas*, exceptions exist including some legumes (Guo et al., 2007; and references within) and the green alga *Chlorella vulgaris* (Wakasugi et al., 1997). For *C. reinhardtii*, the large single-copy region (LSC) and small single-copy regions (SSC) are similar in size, 81,307 bp and 78,088 bp, respectively. The inverted repeats of *C. reinhardtii* are identical and each contains 22,211 bp, including a copy of *psbA*, the five rRNA genes, and two tRNA genes (Maul et al., 2002). In other *Chlamydomonas* species the single-copy regions have greater differences in length (B. Lemieux et al., 1985). The inverted repeats vary in length and

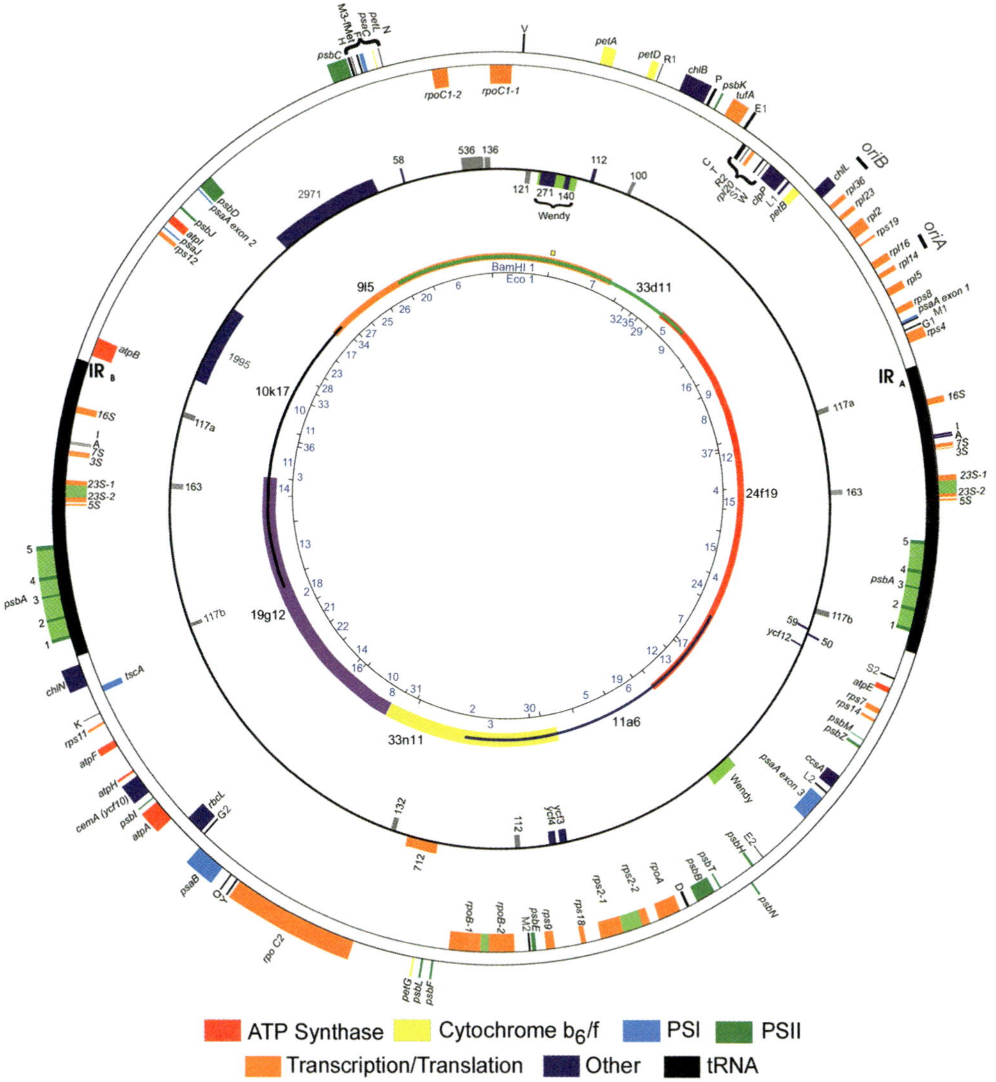

FIGURE 24.1 *The chloroplast chromosome of C. reinhardtii (GenBank accession number NC_005353), adapted from Maul et al. (2002). There are four concentric circles and the 0 position shown with the small orange square is at "12 o'clock" above the inner circle. The inner circle shows BamHI and EcoRI DNA restriction fragments. The second circle represents overlapping BAC clones, and the third circle shows genes and ORFs of unknown function, including ORFs, URFs, and the two Wendy elements. The outer circle shows expressed genes color coded by known or predicted function, see bottom. Genes shown on the same side of a given circle are transcribed from the same DNA strand.*

gene content between *Chlamydomonas* species. For example, *C. moewusii* inverted repeats include *rbcL* while the *C. gelatinosa* inverted repeats include *atpB* (Boudreau et al., 1994; Boudreau and Turmel, 1996). *C. eugametos* and *C. moewusii* both have large non-coding insertions, making their inverted repeats and chloroplast genomes significantly larger (Boudreau and Turmel,

Table 24.1 *Chlamydomonas reinhardtii* chloroplast gene list[a]

Protein-Coding Genes:

Photosystem I	Photosystem II	Cytochrome b_6f	ATP Synthase	Transcription	Translation	Others
psaA[c]	*psbA*[b, c]	*petA*	*atpA*	*rpoA*	*rpl2*	*ccsA*
psaB	*psbB*	*petB*	*atpB*	*rpoB*[c]	*rpl5*	*cemA*
psaC	*psbC*	*petD*	*atpE*	*rpoC1a*	*rpl14*	*chlB*
psaJ	*psbD*	*petG*	*atpF*	*rpoC1b*	*rpl16*	*chlL*
	psbE	*petL*	*atpH*	*rpoC2*	*rpl20*	*chlN*
	psbF		*atpI*		*rpl23*	*clpP*
	psbH				*rpl36*	*rbcL*
	psbI				*rps2*[c]	*ycf3*
	psbJ				*rps3*	*ycf4*
	psbK				*rps4*	*ycf12*
	psbL				*rps7*	
	psbM				*rps8*	
	psbN				*rps9*	
	psbT				*rps11*	
	psbZ				*rps12*	
					rps14	
					rps18	
					rps19	
					tufA	

Functional RNA genes and predicted open reading frames (ORFs):

tRNA		rRNA	Splicing	ORFs[d]
trnA(UGC)	*trnM1(CAU)*	*23S rDNA*[b,c]	*tscA*	*ORF50*
trnA(UGC)	*trnM2(CAU)*	*16S rDNA*[b]		*ORF58*
trnC(GCA)	*trnM3(CAU)-fMet*	*7S rDNA*[b]		*ORF59*
trnD(GUC)	*trnN(GUU)*	*5S rDNA*[b]		*ORF112*
trnE1(UUC)	*trnP(UGG)*	*3S rDNA*[b]		*ORF140*
trnE2(UUC)	*trnQ(UUG)*			*ORF271*
trnF(GAA)	*trnR1(ACG)*			*ORF1995*
trnG1(GCC)	*trnR2(UCU)*			*ORF2971*
trnG2(UCC)	*trnS1(UGA)*			
trnH(GUG)	*trnS2(GCU)*			
trnI(GAU)	*trnT(UGU)*			
trnI(GAU)	*trnV(UAC)*			
trnK(UUU)	*trnW(CCA)*			
trnL1(UAG)	*trnY(GUA)*			
trnL2(UAA)				

[a] *Data are from Maul et al. (2002) Table 1. GenBank accession number for the complete cpDNA sequence is NC_005353.*

[b] *Gene duplicated in large inverted repeats.*

[c] *Intron-containing gene.*

[d] *ORF functions are not known, and numbers correspond to amino acid residues in predicted protein. More ORFs (termed unidentified reading frames, URFs) have been predicted (Maul et al., 2002). However, these were not listed here because they do not appear to be expressed.*

1996). At 243 kb and 285 kb, respectively, the cpDNAs from *C. eugametos* and *C. gelatinosa* both exceed *C. reinhardtii* and at 292 kb, *C. moewusii* has one of the largest chloroplast genomes (Boudreau and Turmel, 1996). On the other hand *C. pitschmannii* cpDNA, at 187 kb, is smaller than that of *C. reinhardtii*.

The majority of protein-coding and tRNA genes are found in the single-copy regions (Figure 24.1). For *C. reinhardtii*, genes are distributed throughout the genome with an average intergenic length of 0.7 kb (Maul et al., 2002). The entire genome can be split into 12 "blocks" in which a block is composed of adjacent genes all transcribed from the same DNA strand (de Cambiaire et al., 2007). When comparing the orientations of adjacent chloroplast genes, *C. reinhardtii* has a high (0.87) sidedness index (C_s) (Cui et al., 2006), like other species in the Chlorophyceae class of green algae. This means that genes next to each other tend (87% of the time) to be transcribed in the same direction. In contrast, the green alga *Chlorella vulgaris* has a C_s value (0.67) that is significantly lower than *C. reinhardtii* (de Cambiaire et al., 2007).

When correlating gene function with order, there is a trend for clustered genes to have related functions. This can be represented with the functional cluster index (C_f), and *C. reinhardtii* has a calculated C_f of 0.03397 (Cui et al., 2006). This is significantly higher than the calculated C_f of 0.01674 for the inferred most recent common ancestor between *C. reinhardtii* and *Chlorella*. To put this in perspective, the C_f for *C. reinhardtii* falls outside of the 95% confidence interval for the inferred most recent common ancestor even when allowing for typical rates of inversion and transposition events. Some of the most striking gene blocks encoding a single functional category include ribosomal proteins (*rpl23-rpl12-rps19*, *rpl16-rpl14-rps8*), Photosystem II (*psbL-psbF*, *psbB-psbT-psbN-psbH*), and translation components (*rrn16-trnI-trnA-rrn23-rrn5*). Cui et al. concluded that the high C_s and C_f support a directionally selective process for *Chlamydomonas* chloroplasts that implicates adaptive evolution. By tracking gene location, the extent of chloroplast rearrangements in *Chlamydomonas* can be compared to other green algae and land plants. Figure 24.2 shows the pronounced number of predicted rearrangements for *C. reinhardtii* and *Chlorella vulgaris*, while far fewer have occurred in land plant cpDNAs. From these examples, it is clear that *C. reinhardtii* cpDNA is a highly rearranged genome, which is apparently tending towards clustering genes with related functions.

Another hallmark of *Chlamydomonas* cpDNAs is the prevalence of short dispersed repeats (SDRs) scattered throughout the genome and mostly within intergenic regions. These were first reported by Gelvin and Howell (1979), but their extent and sequence diversity was not fully appreciated until the complete cpDNA sequence was obtained. The number and distribution of SDRs in *C. reinhardtii* was determined using a self-comparing genome-wide dot plot analysis that predicts 19,500 SDRs (Figure 24.3, blue

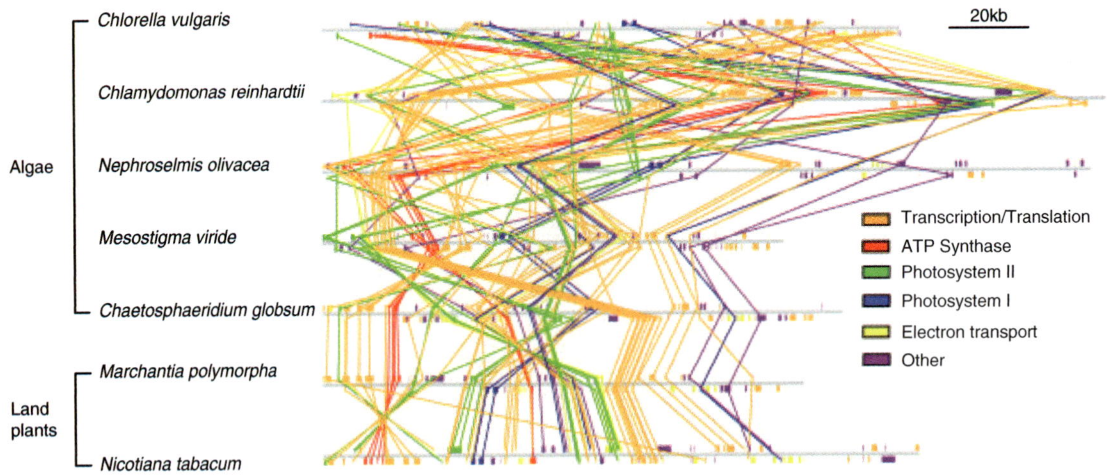

FIGURE 24.2 *Chloroplast rearrangements in selected land plants and algae, including C. reinhardtii and Chlorella vulgaris, adapted from Cui et al. (2006). Order of organisms is based on phylogenetic relationships, and genomes are shown as grey lines drawn to scale (scale bar equals 20 kb). Genes are color coded based on function; those encoded on the positive strand are shown above the line.*

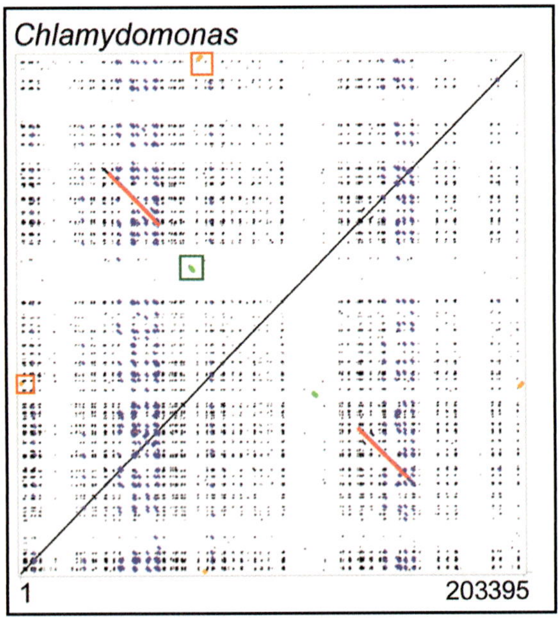

FIGURE 24.3 *Dot-plot analysis of the entire C. reinhardtii cpDNA sequence, adapted from Maul et al. (2002). Key structural features appear as off the diagonal line and are as follows: long red lines, large inverted repeats; blue spots, short dispersed repeats (SDRs); green squares, homologous psaA and psaB genes; and orange squares, Wendy elements. The highly abundant SDRs are found throughout the genome within intergenic DNA, and "clear areas" are the locations of functional genes.*

spots) grouped into ~1000 classes, ultimately account for up to 20% of the chloroplast genome. On average, an individual SDR is 30 bp in length. The most abundant *C. reinhardtii* SDRs are listed in Table 24.2. SDRs are so ubiquitous in intergenic regions, that areas lacking SDRs essentially reveal the locations of genes. Analysis could be refined with the web-based Repeatfinder tool (http://cbsuapps.tc.cornell.edu/repeatfinder.aspx), which was developed to investigate chloroplast SDRs, but is generally applicable to any relatively short input sequence.

The origins of SDRs and their functions in gene expression, if any, are not known. However, from an evolutionary perspective SDRs may have an important role. While very common in *C. reinhardtii* and *C. gelatinosa* cpDNAs, the more distantly related *C. moewusii* and *C. pitschmannii* do not have chloroplast SDR-like sequences (Boudreau and Turmel, 1995, 1996; Maul et al., 2002). They are also abundant in *Chlorella* and the chlorophyte *Pseudendoclonium* (Wakasugi et al., 1997; Pombert et al., 2006). By comparing genomes with a high number of SDRs to those with few SDRs, such as the chlorophyte *Nephroselmis*, the Charophycean green alga *Mesostigma viride*, and vascular plants, a positive correlation between SDRs and genome rearrangement is evident (Pombert et al., 2005; Lemieux et al., 2007). Taken together, SDRs appear to have had, and will continue to have, a significant impact on cpDNA evolution, probably via homologous recombination. Indeed, experiments to study homing of Group I introns demonstrated that SDRs can play a role in double-strand break repair via their recombinogenic properties (Odom et al., 2008).

Table 24.2 Nine of the most abundant chloroplast short dispersed repeats (SDRs) in *C. reinhardtii*[a]

SDR Name	Bases	Sequence (5′–3′)[b]	Similarity[c]	Copies[d]
SDR1	26	CTGCCTCCTTCCCCTTCCCATTCGGG	90	48
SDR2	27	ATATAAATATTGGGCAAGTAAACTTAG	90	28
SDR3	33	ATAACTTTAGTTGCCCGAAGGGGTTTACATAC	90	56
SDR4	25	AGGACAAATTTATTTATTGTGGTAC	90	19
SDR6.1	31	GGACGTCAGTGGCAGTGGTACCGCCACTGCC	90	37
SDR6.2	23	GTGGCAGTGGTACCGCCACTGCC	90	68
SDR7	33	TCCACTAAAATTTATTTGCCGAAGGGGACGTCC	90	51
SDR8	32	TAGGCAGTTGGCAGGCAACTGCACTGACGTCC	90	34
SDR9.1[e]	24	CTGCCAACTGCCGATATTTATATA	100	36

[a] *Data are from Maul et al. (2002) Table 2.*

[b] *Underlined bases are AatII (GACGTC) and KpnI (GGTACC) restriction sites.*

[c] *Percent divergence among those sequences that share similarity based on BLAST analyses.*

[d] *Number of copies in the full chloroplast genome.*

[e] *Decreasing stringency to 90%, allowing three mismatches, detects 15 additional copies of SDR9.1-like repeats.*

A unique feature of *C. reinhardtii* cpDNA is the presence of two copies of "Wendy," a 2.4-kb putative transposable element (Fan et al., 1995). Wendy I and Wendy II exist in the single-copy regions (Figure 24.1), and both have 14-bp terminal inverted repeats that are similar in structure to known transposons. Wendy I contains two putative coding regions (ORF 271 and ORF 140), both of which are expressed at the RNA level. ORF 271 is predicted to encode a protein with similarity to transposase and integrase proteins from known mobile elements (Fan et al., 1995). There is, however, no evidence that either of the Wendy elements is active in transposition. From an evolutionary perspective, it is interesting that no Wendy-like sequence was found in *C. moewusii*, *C. eugametos*, or any other cpDNAs from algae or land plants (Fan et al., 1995), even with the large extant cpDNA sequence database. It seems likely that Wendy invaded *C. reinhardtii* after this species evolved from other Chlamydomonads. Determining whether Wendy elements exist in species closely related to *C. reinhardtii*, such as *C. incerta* and *C. gelatinosa*, could help elucidate the evolution of this putative transposon.

III. GENE CONTENT

C. reinhardtii has a relatively large chloroplast genome, but a smaller than average gene number at 99. (As a point of detail, since the release of the complete cpDNA sequence, the full *rps11* gene was identified and an extra *trnE* gene was found to be have been incorrectly included. These changes keep the gene number at 99.) In comparison, *Synechocystis* sp. PCC6803, a free-living bacterial descendant of the modern chloroplast ancestor, has 3168 predicted genes (Kaneko and Tabata, 1997), revealing the extent of evolutionary gene loss or transfer. *C. reinhardtii* also contrasts with the red alga *Porphyra purpurea* (251 genes), and land plants where the chloroplast gene number ranges from 106 in *Pinus* to 122 in *Marchantia* (Ohyama et al., 1986; Wakasugi et al., 1994; Reith and Munholland, 1995; Maul et al., 2002). Compared to the green algae *Chlorella*, *Mesostigma*, *Nephroselmis*, and *Pseudendoclonium*, the cpDNA from *Chlamydomonas* has the smallest gene complement, lacking genes from different functional categories, including all of the *ndh* genes, two photosystem I genes (*psaI* and *psaM*), eight small and large ribosomal protein genes (*rps* and *rpl*), the *infA* translation initiation factor, the *chlI* chlorophyll biosynthesis gene, and others (Pombert et al., 2005).

The *C. reinhardtii* chloroplast contains 64 protein-coding genes, 29 tRNA genes, 6 rRNA genes, 1 RNA splicing sequence, and 8 conserved ORFs. In addition, 18 URFs were identified, but none appear to be expressed (Maul et al., 2002). For this reason the URFs were not included in Table 24.1. The protein-coding genes can be categorized by function: 37 for photosynthesis and ATP synthesis; 24 for gene expression (transcription, translation, and protein processing); and 3 *ycf* genes, 2 of which (*ycf3* and *ycf4*) are known to function in photosynthesis. Based on partial genome sequences and DNA hybridization data, other *Chlamydomonas* species appear to have similar chloroplast gene content (Nedelcu and Lee, 1998). Details of gene function are discussed in other chapters, and reviewed in Wollman et al. (1999).

The majority of *Chlamydomonas* chloroplast genes have a typical prokaryotic structure, some contain group I or group II introns, and a few have unusual organization and expression mechanisms. Genes with introns are indicated in Table 24.1, and both *cis-* and *trans*-splicing are discussed in Chapter 26. Specifically, the two *psbA* genes contain four *cis*-spliced group I introns. In *C. reinhardtii*, the *23S* rRNA genes (also termed *rrnL*) possess group I introns as well. In other *Chlamydomonas* species the number and location of group I introns in the *rrnL* gene varies, and this variation correlates with the proposal that the *Chlamydomonas* genus is polyphyletic (see Volume 1, Chapter 1). Group II introns exist in the *psaA* gene, which has three exons scattered throughout the genome, and these exons are *trans*-spliced to form the mature *psaA* mRNA. Although the chromosomal locations of the scattered *psaA* exons differ between *C. reinhardtii* and *C. moewusii*, the dependence on *trans*-splicing is conserved. In contrast, land plants and the majority of other algae have *psaA* genes that lack introns (Sugiura et al., 1998; Pombert et al., 2005).

The genes with unusual structure include those encoding RNA polymerase subunits (*rpoB*, *rpoC1*, and *rpoC2*), *rps2*, and *clpP*. The RNA polymerase genes are either split into distinctly expressed genes or gene segments, or they are single genes split by non-consensus intron sequences. In the case of *rpoB*, the gene contains what appears to be two expressed ORFs split by an intervening sequence (Nedelcu and Lee, 1998). This intervening sequence lacks recognizable intron structure, leaving unanswered the question of how, or if, the *rpoB* ORFs are "spliced" to generate a full-length mRNA (Fong and Surzycki, 1992). The *rpoC1* genes of *C. reinhardtii* and *C. moewusii* are more clearly split into *rpoC1a* and *rpoC1b*, which are both expressed (Boudreau et al., 1994; Maul et al., 2002). *rpoC2* is also split into two ORFs, and to determine if this split is the result of an intron or expression from two distinct genes RNA gel blot and RT-PCR analyses were used (Maul et al., 2002). The detected 5.2-kb *rpoC2* transcript is consistent with the predicted mature mRNA size after the putative 3-kb intron is spliced, and RT-PCR confirmed that the two ORFs accumulate. Although these data are consistent with *rpoC2* being split by a spliced intron, they cannot rule out the possibility that the ORFs are expressed on two separate 5.2-kb transcripts. One technical aspect that has complicated analysis of this and other *rpo* genes is the fact that in *Chlamydomonas* these transcripts are difficult to detect using RNA gel blots.

The *rps2* gene is also split into two ORFs, *rps2-1* (previously ORF570) and *rps2-2* (previously ORF208). Pombert et al. (2005) reported that the intervening *rps2* sequence of *C. reinhardtii* does not contain a distinguishable intron. In an effort to determine if both ORFs are expressed within a single transcript, RNA gel blots were hybridized with both *rps2-1* and *rps2-2* specific probes (Maul et al., 2002). In both cases, a 2.5-kb RNA was detected, a length consistent with the expected size of the mature mRNA with the intervening sequence spliced out. These data suggest that *rps2* is split by an intron.

IV. DNA REPLICATION

Shortly after the first reports of *Chlamydomonas* cpDNA, Chiang and Sueoka (1967) provided evidence for semi-conservative replication in *C. reinhardtii*. A characteristic of this replication is sensitivity to ethidium bromide, as first demonstrated by Flechtner and Sager (1973). Using gradient density centrifugation, *Chlamydomonas* cpDNA can be separated from the G+C-rich nuclear DNA (Sager and Ishida, 1963). Furthermore, cpDNA can be specifically labeled with exogenous thymidine that becomes incorporated as a result of the chloroplast-specific thymidine kinase (Swinton and Hanawalt, 1972). Turmel et al. (1981) used $^{15}N/^{14}N$ labeling and gradient density centrifugation in a further study of cpDNA replication, but were not able to identify the discrete heavy DNA strand that was expected for semi-conservative replication. Labeled nucleotides were instead incorporated into both strands. It was concluded that a high frequency of recombination occurs shortly after the cpDNA completes semi-conservative replication. Indeed, recombination is commonly associated with replication in *Chlamydomonas* chloroplasts.

The first mapped cpDNA origins of replication in C. *reinhardtii* were identified by looking for autonomous replication activity in yeast cells (Vallet et al., 1984). Rochaix et al. (1984) went on to characterize fragments that replicated autonomously in *Chlamydomonas* itself. This led to more precise mapping and characterization of the two main cpDNA origins in *C. reinhardtii*, *oriA*, which overlaps the *rpl16* gene, and *oriB*, between the *petB* and *rpl2* genes (Figure 24.1; Rochaix et al., 1984, 1985; Waddell et al., 1984). Further support for these origins came from *C. eugametos* and *C. moewusii* in which homologous sequences were characterized (Wang et al., 1984). Waddell et al. (1984) used electron microscopy to map two displacement loop (D-loop) replicating sequences that corresponded to *oriA* and *oriB*. They showed that both origins are bi-directional and for an individual molecule, replication initiates at only one origin.

An alternative method for identifying replication initiation sequences relies on *in vitro* replication with partially purified *C. reinhardtii* chloroplast proteins (Wu et al., 1986). This helped determine that *Chlamydomonas* chloroplasts, like those of corn and pea, initiate replication at A-T rich sites (Vallet et al., 1984; Wu et al., 1986). This approach also led to the discovery of two additional origins that are insensitive to the replication inhibitor novobiocin (Woelfle et al., 1993). These origins mapped between *psbA* and *rrnL*, and were previously characterized as recombination hotspots. The novobiocin-insensitive origins are unidirectional and dependent on recombination, further distinguishing them from *oriA* and *oriB*.

Chloroplast DNA polymerase activities and replication-related proteins have been identified. Keller and Ho (1981) used purified chloroplasts from the cell wall-deficient strain *cw15* to identify a 180-kD protein complex with DNA polymerase activity. This complex contains proteins of 75, 40, and 15 kD, with the 40-kD protein apparently being the most abundant. In other experiments, partially purified proteins that bound *oriA* revealed a different DNA polymerase activity with a 116-kD protein, an 80-kD protein homologous to mitochondrial c-type DNA polymerase in animals (Wu et al., 1989; Wang et al., 1991), and a DNA primase homologue (Nie and Wu, 1999). It is not known how these two cpDNA polymerase activities relate to each other, nor is it known how each might uniquely regulate replication (reviewed in Sears, 1998). Considering the regulation of cpDNA replication in vegetative cells as compared to zygotic cells, it is possible that the two polymerase activities are active at different times or life cycle stages (Umen and Goodenough, 2001). Nishimura et al. (2002) proposed that regulated cpDNA replication may operate cooperatively with an active digestion mechanism to cause uniparental inheritance in *Chlamydomonas*.

An important function of chloroplast replication is to control the number of genome copies, which in *Chlamydomonas* is approximately 80 copies per chloroplast. This conclusion comes from comparing the absolute amount of cpDNA per chloroplast versus the estimated single genome size (Bastia et al., 1971; Wells and Sager, 1971; Behn and Herrmann, 1977). The copy number can be reduced by treatment with the thymidine analog 5-fluorodeoxyuridine (FdUrd), which inhibits cpDNA replication by binding and inhibiting the chloroplast thymidylate synthetase (Wurtz et al., 1977), but does not interfere with nuclear DNA replication or cell division. Within one chloroplast, the cpDNAs typically cluster into eight to ten nucleoids that can be stained with the fluorochrome 4′,6-diamidino-2-phenylindole (DAPI) (Coleman, 1978). Using a video-intensified microscope photon counting system with wild-type cells and *monokaryotic chloroplast* (*moc*) mutants, cpDNA nucleoids have been observed during the *Chlamydomonas* cell cycle and sexual reproduction (Misumi et al., 1999; Nishimura et al., 2002; Hiramatsu et al., 2006). For *mocG33*, there is no difference in growth, cell division or cell diameter as compared to the wild type, despite the fact that chloroplasts in these cells have only one large nucleoid (Hiramatsu et al., 2006). Furthermore, the *mocA84* mutant, which is similar to *mocG33* in having a single nucleoid but otherwise normal chloroplast structure and growth, has a genome copy number (65–75 per chloroplast) that is similar to wild type (80–90 per chloroplast) (Misumi et al., 1999). Thus, while the *moc* mutants have proven useful in visually monitoring and characterizing the nucleoids and chloroplast uniparental inheritance, the function and importance of these structures remains unclear.

In some *Chlamydomonas* strains, very high copy numbers (thousands per chloroplast) have been observed. Kindle et al. (1994) reported a *C. reinhardtii* strain in which the *atpB* gene was maintained on a small extrachromosomal element at ~2000 copies per chloroplast. This element appears to have been maintained by a replication that initiates in the main cpDNA molecule, not the episome (Suzuki et al., 1997). A model was proposed that relies on recombination-dependent replication for the episomal DNA. Additional studies on copy number have focused on conditions that regulate replication as it relates to transcription and RNA processing (Chang and Wu, 2000), redox modulation (Lau et al., 2000), and methylation-controlled selective replication in germinating zygotes (Umen and Goodenough, 2001). For additional information on this topic, as well as on recombination and repair, see the review by Sears (1998).

V. GENOME EVOLUTION

A. Changes in gene order

Chlamydomonas cpDNA is marked by gene loss, numerous rearrangements, and in some species a dramatic increase in SDRs, with respect to those of land plants (Figure 24.3). The dynamic evolution of *Chlamydomonas* cpDNAs was first demonstrated in a series of reports by Claude Lemieux and colleagues (Lemieux and Lemieux, 1985; B. Lemieux et al., 1985; C. Lemieux et al., 1980, 1985). Although chloroplast gene content is conserved among *Chlamydomonas* species, gene order is not, as shown with DNA gel blot experiments in *C. reinhardtii*, *C. eugametos*, and *C. moewusii* (B. Lemieux et al., 1985; Boudreau et al., 1994). Changes in gene order appear to have resulted mostly from inversions and a limited number of deletions. Chromosome reconstruction models predict stepwise rearrangements that account for the differences in cpDNAs between *Chlamydomonas* species. For example, a three-step model with one deletion and two inversions can account for the differences between *C. moewusii* and *C. pitschmannii* (Boudreau and Turmel, 1995), while a ten-step model with one deletion and nine inversions can account for differences between *C. gelatinosa* and *C. reinhardtii* (Boudreau and Turmel, 1996). Between *C. reinhardtii* and *Chlorella*, a minimum of 72 inversions can explain the differences in gene order (de Cambiaire et al., 2007). Genomic rearrangements correlate with the abundance and locations of repeat sequences, and species with large numbers of SDRs tend to have more rearrangements (Maul et al., 2002; Pombert et al., 2005; Cui et al., 2006).

B. Intragenomic recombination

A well-characterized example of *Chlamydomonas* genome variation within one chloroplast is the "flip-flop" phenomenon, which results from recombination between identical inverted repeat sequences, and inverts the orientation of the single-copy DNA regions (Aldrich et al., 1985; Palmer et al., 1985) and Volume 1, Chapter 7. This inversion does not alter gene sequence. In *C. reinhardtii*, gel blot analysis showed a roughly equal mixture of both "flip-flop" isomers. This presumably continual recombination allows the two inverted repeats to maintain sequence identity, which is also apparent from the fact that a transgene inserted into one inverted repeat is copy-corrected to the other (Blowers et al., 1989).

C. Induced chloroplast rearrangements

Recombination events that cause genomic inversions are not limited to the inverted repeats. Reversion analysis of the non-photosynthetic *FUD6* mutant, which contains a small *petD* deletion that prevents mRNA accumulation, led to isolation of strains in which recombination between homologous SDRs within the two single-copy regions had produced a stable chloroplast genome with a 60-kb inversion (Higgs et al., 1998). At one end of this inversion a novel chimeric *psbZ::petD* gene was formed, which expressed subunit IV from the remnant *petD* region. The inversion also switched the orientation of the inverted repeats, forming direct repeats, yet the genome was genetically stable.

The suppressors of polyadenylation (*spa*) strains are another example of a genome rearrangement that is stable and impacts gene expression (Nishimura et al., 2004). These strains resulted from an original chloroplast transformant (Δ26pAtE) that was manipulated so that the *atpB* mRNA expressed from the Δ26pAtE transgene had a transcribed 3′ poly(A) tail. The 3′ poly(A) tail rendered the *atpB* mRNA unstable and caused these cells to be non-photosynthetic. From this Δ26pAtE strain, photosynthetic suppressors occurred at a high rate, and two unusual strains were found to be heteroplasmic with the original "PS–" Δ26pAtE genome and a suppressing genome termed "PS+". The "PS+" cpDNA contained a 145-kb deletion caused by recombination between the *trnE1* portion of the Δ26pAtE transgene and the endogenous *trnE1*, and was four times more abundant than the "PS–" genome. Ultimately, this large deletion resulted in the synthesis of a unique antisense RNA complementary to the 3′ tail of Δ26pAtE. The antisense RNA was shown to pair with Δ26pAtE mRNA and form a double-stranded

RNA that in turn stabilized the mRNA and promoted photosynthetic growth (Nishimura et al., 2004). These strains remained heteroplasmic under non-selective conditions, showing that they are vegetatively stable. Together, these *spa* strains and the *FUD6* revertants, mentioned earlier, demonstrate that genomic deletions or inversions can (1) impart selective advantage, (2) be stable, and (3) become the dominant form within a population. Furthermore, these provide insight into key processes that have shaped the evolution of chloroplast genomes, as visualized in Figure 24.2.

D. Natural variation

Single-nucleotide polymorphisms (SNPs) and small insertions or deletions (indels) are also observed in *Chlamydomonas* cpDNAs. A limited number of studies have reported cpDNA SNP frequencies within and between *Chlamydomonas* species. The *rbcL* coding region of 932 bp and three intergenic regions (*petD-chlB*, *chlL-rpl23*, and *psbZ-ccsA*) totaling 2180 bp were analyzed among three *C. reinhardtii* strains that diverged from a common ancestor (137c) sometime between 1960 and 1970 (Maul et al., 2002). Over these ~30 years, SNPs accumulated and were reported as a percentage of polymorphic sites. For the intergenic regions, SNP frequencies ranged from 0.4% to 0.9%. In contrast, there were no SNPs in the *rbcL* coding region. A comparison of *C. reinhardtii* 137c to the independently isolated *C. reinhardtii* S1 D2 (see Volume 1, Chapter 1) revealed SNP frequencies of 0.2–0.4% for the intergenic regions and a surprisingly higher 0.7% for the *rbcL* coding region. In a different study, sequences for the upstream intergenic and 5′ untranslated region (UTR) of *petD* were analyzed from four independent *C. reinhardtii* strains (Kramzar et al., 2006). The 362-bp 5′ UTR and partial coding regions were identical in all four strains while the upstream *petA–petD* intergenic region contained both SNPs and indels. To assess interspecies variation, the *petD* 5′ UTRs from three additional *Chlamydomonas* species (*C. incerta*, *C. eugametos*, and *C. moewusii*) were compared to *C. reinhardtii* and extensive sequence variation was found. For example, when comparing *C. reinhardtii* to *C. eugametos*, 37 indels and 41 SNPs were found (Kramzar et al., 2006). This high sequence variation may be useful in revealing conserved sequences that have functional importance for gene regulation.

Nedelcu and Lee (1998) summarized *Chlamydomonas* cpDNA evolution and rates of change in genome size, organization, gene structure, and nucleotide substitutions and compared *Chlamydomonas* to land plants. The rate of nucleotide substitutions was summarized as being "significant," with nearly twice the sequence variation of land plant cpDNAs. This conclusion was based mainly on 23S rRNA sequences from 17 *Chlamydomonas* species (Turmel et al., 1993). The 23S rRNA contains a mix of "variable" and "partly conserved" regions. Of the 2792 aligned bases within the partly conserved regions, 658 bases (23%) were polymorphic, including 282 (10%) binary transitions, 153 (5%) binary transversions, 178 (6%) multistate sites, and there were 45 indels observed.

Taken together, data on protein coding regions, UTRs, functional RNAs, and intergenic regions demonstrate a surprisingly high level of sequence diversity within the *Chlamydomonas* genus. To put this in perspective, the *23S* gene alone has more sequence diversity between *C. reinhardtii* and *C. eugametos* than between the bryophyte *Marchantia* and the evolutionarily distant monocot rice (Turmel et al., 1993).

VI. CONCLUSIONS AND PERSPECTIVES

The complete *Chlamydomonas* chloroplast genome sequence is a resource that complements the many available genetic techniques. In addition to helping answer fundamental questions of function and evolution, the cpDNA sequence is raising new questions, such as the impact of SDRs. Yet, we still do not understand many aspects of chloroplast gene regulation, something which is particularly amenable to testing in *Chlamydomonas* because of the ease of chloroplast transformation. Examples of genes where analysis is incomplete are the unusual split genes *rpoB*, *rpoC1*, *rpoC2*, *rps2*, and *clpP*, whose structures in some cases are apparently unique to *Chlamydomonas*. Other areas ripe for investigation are the chloroplast transcriptional, translational, and DNA replication machinery.

Chloroplast gene order and sidedness are significantly different in *Chlamydomonas*, as compared to other algae and land plants. Cui et al. (2006) concluded that the gene order in *C. reinhardtii* evolved in a non-random way. This implies an adaptive evolution process while also raising the question of how gene order and sidedness provide an evolutionary advantage. It has also been observed that *Chlamydomonas* chloroplast mRNAs are mostly monocistronic while land plants have far more polycistronic mRNAs (Eberhard et al., 2002). Since polycistronic mRNAs are products of gene clusters, it is possible that there is a causative relationship between monocistronic mRNAs and the changes in gene order and sidedness.

The field of chloroplast genomics is growing as a result of the many newly sequenced genomes from many different clades. As of early 2008, NCBI listed 122 complete plastid genome sequences, 83 of which were deposited in the past 3 years alone. A useful resource for comparative chloroplast genomic data is Chloroplast DB (http://chloroplast.cbio.psu.edu). The fact that DNA sequencing is ever faster and less expensive will doubtless accelerate the trend of accumulating chloroplast genome data, eventually including additional *Chlamydomonas* species. The use of pyrosequencing for a completed chloroplast genome has already been reported for land plants (Moore et al., 2006). Having a robust cpDNA sequence database including diverse representative species from within the genus *Chlamydomonas*, would help address unresolved issues of chloroplast evolution, function, and gene expression.

There is, however, a concern of over interpreting phylogenetic trees from taxa-poor datasets that only rely on completed chloroplast genomes. Lemieux et al. (2007) discussed the weakness of using only complete genome sequence from a few taxa versus the more traditional method of using more taxa but with fewer select gene/protein sequences. This can lead to well-supported phylogenetic trees that do not match true evolutionary relationships. This problem was intentionally addressed in a study looking at angiosperm chloroplast phylogeny in order to determine what happens when too few representative taxa are used (Soltis et al., 2004). In the end, the best solution will be to have robust cpDNA datasets that are both taxa-rich and populated with many representative complete cpDNA sequences.

REFERENCES

Aldrich, J., Cherney, B., Merlin, E., Williams, C., and Mets, L. (1985). Recombination within the inverted sequences of the *Chlamydomonas reinhardtii* chloroplast genome produces two orientation isomers. *Curr. Genet.* **9**, 233–238.

Bastia, D., Chiang, K.-S., Swift, H., and Siersma, P. (1971). Heterogeneity, complexity, and repetition of the chloroplast DNA of *Chlamydomonas reinhardtii*. *Proc. Natl. Acad. Sci. U.S.A.* **68**, 1157–1161.

Behn, W. and Herrmann, R.G. (1977). Circular molecules in the b-satellite DNA of *Chlamydomonas reinhardtii*. *Mol. Gen. Genet.* **157**, 25–30.

Blowers, A.D., Bogorad, L., Shark, K.B., and Sanford, J.C. (1989). Studies on *Chlamydomonas* chloroplast transformation: Foreign DNA can be stably maintained in the chromosome. *Plant Cell* **1**, 123–132.

Boudreau, E. and Turmel, M. (1995). Gene rearrangements in *Chlamydomonas* chloroplast DNAs are accounted for by inversions and by the expansion/contraction of the inverted repeat. *Plant Mol. Biol.* **27**, 351–364.

Boudreau, E. and Turmel, M. (1996). Extensive gene rearrangements in the chloroplast DNAs of *Chlamydomonas* species featuring multiple dispersed repeats. *Mol. Biol. Evol.* **13**, 233–243.

Boudreau, E., Otis, C., and Turmel, M. (1994). Conserved gene clusters in the highly rearranged chloroplast genomes of *Chlamydomonas moewusii* and *Chlamydomonas reinhardtii*. *Plant Mol. Biol.* **24**, 585–602.

Boynton, J.E., Gillham, N.W., Harris, E.H., Hosler, J.P., Johnson, A.M., Jones, A.R., Randolph-Anderson, B.L., Robertson, D., Klein, T.M., Shark, K.B., and Sanford, J.C. (1988). Chloroplast transformation in *Chlamydomonas* with high velocity microprojectiles. *Science* **240**, 1534–1538.

Chang, C.H. and Wu, M. (2000). The effects of transcription and RNA processing on the initiation of chloroplast DNA replication in *Chlamydomonas reinhardtii*. *Mol. Gen. Genet.* **263**, 320–327.

Chiang, K.-S. and Sueoka, N. (1967). Replication of chloroplast DNA in *Chlamydomonas reinhardtii* during vegetative cell cycle: Its mode and regulation. *Proc. Natl. Acad. Sci. U.S.A.* **57**, 1506–1513.

Coleman, A.W. (1978). Visualization of chloroplast DNA with two fluorochromes. *Exp. Cell Res.* **114**, 95–100.

Cui, L., Leebans-Mack, J., Wang, L.-S., Tang, J., Rymarquis, L., Stern, D.B., and dePamphilis, C.W. (2006). Adaptive evolution of chloroplast genome structure inferred using a parametric bootstrap approach. *BMC Evol. Biol.* **6**, 13.

de Cambiaire, J.-C., Otis, C., Turmel, M., and Lemieux, C. (2007). The chloroplast genome sequence of the green alga *Leptosira terrestris*: Multiple losses of the inverted repeat and extensive genome rearrangements within the Trebouxiophyceae. *BMC Genomics* **8**, 213.

Eberhard, S., Drapier, D., and Wollman, F.-A. (2002). Searching limiting steps in the expression of chloroplast-encoded proteins: Relations between gene copy number, transcription, transcript abundance and translation rate in the chloroplast of *Chlamydomonas reinhardtii*. *Plant J.* **31**, 149–160.

Fan, W.H., Woelfle, M.A., and Mosig, G. (1995). Two copies of a DNA element, "Wendy", in the chloroplast chromosome of *Chlamydomonas reinhardtii* between rearranged gene clusters. *Plant Mol. Biol.* **29**, 63–80.

Flechtner, V.R. and Sager, R. (1973). Ethidium bromide induced selective and reversible loss of chloroplast DNA. *Nature New Biol.* **241**, 277–279.

Fong, S.E. and Surzycki, S.J. (1992). Organization and structure of plastome *psbF, psbL, petG* and *ORF712* genes in *Chlamydomonas reinhardtii*. *Curr. Genet.* **21**, 527–530.

Gelvin, S.B. and Howell, S.H. (1979). Small repeat sequences in the chloroplast genome of *Chlamydomonas reinhardtii*. *Mol. Gen. Genet.* **173**, 315–322.

Goldschmidt-Clermont, M. (1998). Chloroplast transformation and reverse genetics. In: *The Molecular Biology of Chloroplasts and Mitochondria in Chlamydomonas* (J.D. Rochaix, M. Goldschmidt-Clermont, and S. Merchant, Eds.), pp. 139–149. Kluwer Academic Publishers, Dordrecht, The Netherlands.

Guo, X., Castillo-Ramírez, S., González, V., Bustos, P., Fernández-Vázquez, J.L., Santamaría, R.I., Arellano, J., Cevallos, M.A., and Dávila, G. (2007). Rapid evolutionary change of common bean (*Phaseolus vulgaris* L.) plastome, and the genomic diversification of legume chloroplasts. *BMC Genomics* **8**, 228.

Higgs, D.C., Kuras, R., Kindle, K.L., Wollman, F.-A., and Stern, D.B. (1998). Inversions in the *Chlamydomonas* chloroplast genome suppress a *petD* 5′ untranslated region deletion by creating functional chimeric mRNAs. *Plant J.* **14**, 663–671.

Hiramatsu, T., Nakamura, S., Misumi, O., Kuroiwa, T., and Nakamura, S. (2006). Morphological changes in mitochondrial and chloroplast nucleoids and mitochondria during the *Chlamydomonas reinhardtii* (Chlorophyceae) cell cycle. *J. Phycol.* **42**, 1048–1058.

Huang, C., Wang, S., Ling, C., Lemieux, C., Otis, C., Turmel, M., and Liu, X.-Q. (1994). The *Chlamydomonas* chloroplast *clpP* gene contains translated large insertion sequences and is essential for cell growth. *Mol. Gen. Genet.* **244**, 151–159.

Kaneko, T. and Tabata, S. (1997). Complete genome structure of the unicellular cyanobacterium *Synechocystis* sp. PCC6803. *Plant Cell Physiol.* **38**, 1171–1176.

Keller, S.J. and Ho, C. (1981). Chloroplast DNA replication in *Chlamydomonas reinhardtii*. *Int. Rev. Cytol.* **69**, 157–190.

Kindle, K.L., Suzuki, H., and Stern, D.B. (1994). Gene amplification can correct a photosynthetic growth defect caused by mRNA instability in *Chlamydomonas* chloroplasts. *Plant Cell* **6**, 187–200.

Kramzar, L.M., Mueller, T., Erickson, B., and Higgs, D.C. (2006). Regulatory sequences of orthologous *petD* chloroplast mRNAs are highly specific among *Chlamydomonas* species. *Plant Mol. Biol.* **60**, 405–422.

Lau, K.W., Ren, J., and Wu, M. (2000). Redox modulation of chloroplast DNA replication in *Chlamydomonas reinhardtii*. *Antioxid. Redox Signal.* **2**, 529–535.

Lemieux, B. and Lemieux, C. (1985). Extensive sequence rearrangements in the chloroplast genomes of the green algae *Chlamydomonas eugametos* and *Chlamydomonas reinhardtii*. *Curr. Genet.* **10**, 213–219.

Lemieux, C., Turmel, M., and Lee, R.W. (1980). Characterization of chloroplast DNA in *Chlamydomonas eugametos* and *C. moewusii* and its inheritance in hybrid progeny. *Curr. Genet.* **3**, 139–147.

Lemieux, B., Turmel, M., and Lemieux, C. (1985). Chloroplast DNA variation in *Chlamydomonas* and its potential application to the systematics of this genus. *Biosystems* **18**, 293–298.

Lemieux, C., Turmel, M., Lee, R.W., and Bellemare, G. (1985). A 21 kilobase-pair deletion/addition difference in the inverted repeat sequence of chloroplast DNA from *Chlamydomonas eugametos* and *C. moewusii*. *Plant Mol. Biol.* **5**, 77–84.

Lemieux, C., Otis, C., and Turmel, M. (2007). A clade uniting the green algae *Mesostigma viride* and *Chlorokybus atmophyticus* represents the deepest branch of the Streptophyta in chloroplast genome-based phylogenies. *BMC Biology* **5**, 2.

Martin, W. and Herrmann, R.G. (1998). Gene transfer from organelles to the nucleus: How much, what happens, and why? *Plant Physiol.* **118**, 9–17.

Maul, J.E., Lilly, J.W., Cui, L., dePamphilis, C.W., Miller, W., Harris, E.H., and Stern, D.B. (2002). The *Chlamydomonas reinhardtii* plastid chromosome: Islands of genes in a sea of repeats. *Plant Cell* **14**, 2659–2679.

Merchant, S.S., Prochnik, S.E., Vallon, O., Harris, E.H., Karpowicz, S.J., Witman, G.B., Terry, A., Salamov, A., Fritz-Laylin, L.K., Maréchal-Drouard, L. et al. (2007). The *Chlamydomonas* genome reveals the evolution of key animal and plant functions. *Science* **318**, 245–250.

Misumi, O., Suzuki, L., Nishimura, Y., Sakai, A., Kawano, S., Kuroiwa, H., and Kuroiwa, T. (1999). Isolation and phenotypic characterization of *Chlamydomonas reinhardtii* mutants defective in chloroplast DNA segregation. *Protoplasma* **209**, 273–282.

Moore, M.J., Dhingra, A., Soltis, P.S., Shaw, R., Farmerie, W.G., Folta, K.M., and Soltis, D.E. (2006). Rapid and accurate pyrosequencing of angiosperm plastid genomes. *BMC Plant Biology* **6**, 17.

Nedelcu, A.M. and Lee, R.W. (1998). Modes and tempos of mitochondrial and chloroplast genome evolution in *Chlamydomonas*: A comparative analysis. In: *The Molecular Biology of Chloroplasts and Mitochondria in Chlamydomonas* (J.D. Rochaix, M. Goldschmidt-Clermont, and S. Merchant, Eds.), pp. 63–91. Kluwer Academic Publishers, Dordrecht, The Netherlands.

Nie, Z. and Wu, M. (1999). The functional role of a DNA primase in chloroplast DNA replication in *Chlamydomonas reinhardtii*. *Arch. Biochem. Biophys.* **369**, 174–180.

Nishimura, Y., Misumi, O., Kato, K., Inada, N., Higashiyama, T., Momoyama, Y., and Kuroiwa, T. (2002). An *mt*+ gamete-specific nuclease that targets *mt*− chloroplasts during sexual reproduction in *C. reinhardtii*. *Genes Dev.* **16**, 1116–1128.

Nishimura, Y., Kikis, E.A., Zimmer, S.L., Komine, Y., and Stern, D.B. (2004). Antisense transcript and RNA processing alterations suppress instability of polyadenylated mRNA in *Chlamydomonas* chloroplasts. *Plant Cell* **16**, 2849–2869.

Odom, O.W., Baek, K.-H., Dani, R.N., and Herrin, D.L. (2008). *Chlamydomonas* chloroplasts can use short dispersed repeats and multiple pathways to repair a double-strand break in the genome. *Plant J.* **53**, 842–853.

Ohyama, K., Fukuzawa, H., Kohchi, T., Shirai, H., Sano, T., Sano, S., Umesono, K., Shiki, Y., Takeuchi, M., Chang, Z., Aota, S., Inokuchi, H., and Ozeki, H. (1986). Chloroplast gene sequence deduced from the complete sequence of liverwort *Marchantia polymorpha* chloroplast DNA. *Nature* **322**, 572–574.

Palmer, J.D. (1991). Plastid chromosomes: Structure and evolution. In: *Molecular Biology of Plastids* (L. Bogorad and I.K. Vasil, Eds.), pp. 5–53. Academic Press, San Diego.

Palmer, J.D., Boynton, J.E., Gillham, N.W., and Harris, E.H. (1985). Evolution and recombination of the large inverted repeat in *Chlamydomonas* chloroplast DNA. In: *Molecular Biology of the Photosynthetic Apparatus* (K.E. Steinback, S. Bonitz, C.J. Arntzen, and L. Bogorad, Eds.), pp. 269–278. Cold Spring Harbor Laboratory Press, Cold Spring Harbor NY.

Pombert, J.-F., Otis, C., Lemieux, C., and Turmel, M. (2005). The chloroplast genome sequence of the green alga *Pseudendoclonium akinetum* (Ulvophyceae) reveals unusual structural features and new insights into the branching order of Chlorophyte lineages. *Mol. Biol. Evol.* **22**, 1903–1918.

Pombert, J.-F., Lemieux, C., and Turmel, M. (2006). The complete chloroplast DNA sequence of the green alga *Oltmannsiellopsis viridis* reveals a distinctive quadripartite architecture in the chloroplast genome of early diverging ulvophytes. *BMC Biology* **4**, 3.

Reith, M. and Munholland, J. (1995). Complete nucleotide sequence of the *Porphyra purpurea* chloroplast genome. *Plant Mol. Biol. Rep.* **13**, 333–335.

Ris, H. and Plaut, W. (1962). Ultrastructure of DNA-containing areas in the chloroplast of *Chlamydomonas*. *J. Cell Biol.* **13**, 383–391.

Rochaix, J.D., van Dillewijn, J., and Rahire, M. (1984). Construction and characterization of autonomously replicating plasmids in the green unicellular alga *Chlamydomonas reinhardtii*. *Cell* **36**, 925–931.

Rochaix, J.D., Rahire, M., and Michel, F. (1985). The chloroplast ribosomal intron of *Chlamydomonas reinhardtii* codes for a polypeptide related to mitochondrial maturases. *Nucleic Acids Res.* **13**, 975–984.

Sager, R. and Ishida, M.R. (1963). Chloroplast DNA in *Chlamydomonas*. *Proc. Natl. Acad. Sci. U. S. A.* **50**, 725–730.

Sears, B.B. (1998). Replication, recombination, and repair in the chloroplast genetic system of *Chlamydomonas*. In: *The Molecular Biology of Chloroplasts and Mitochondria in Chlamydomonas* (J.D. Rochaix, M. Goldschmidt-Clermont, and S. Merchant, Eds.), pp. 115–138. Kluwer Academic Publishers, Dordrecht, The Netherlands.

Simpson, C.L. and Stern, D.B. (2002). The treasure trove of algal chloroplast genomes. Surprises in architecture and gene content, and their functional implications. *Plant Physiol.* **129**, 957–966.

Soltis, D.E., Albert, V.A., Savolainen, V., Hilu, K., Qiu, Y.L., Chase, M.W., Farris, J.S., Stefanovic, S., Rice, D.W., Palmer, J.D., and Soltis, P.S. (2004). Genome-scale data, angiosperm relationships, and "ending incongruence": a cautionary tale in phylogenetics. *Trends Plant Sci.* **9**, 477–483.

Sugiura, M., Hirosa, T., and Sugita, M. (1998). Evolution and mechanisms of translation in chloroplasts. *Annu. Rev. Genet.* **32**, 437–459.

Suzuki, H., Ingersoll, J., Stern, D.B., and Kindle, K.L. (1997). Generation and maintenance of tandemly repeated extrachromosomal plasmid DNA in *Chlamydomonas* chloroplasts. *Plant J.* **11**, 635–648.

Swaitek, M., Kuras, R., Sokolenko, A., Higgs, D., Olive, J., Cinque, G., Müller, B., Eichacker, L.A., Stern, D.B., Bassi, R., Herrmann, R.G., and Wollman, F.-A. (2001). The chloroplast gene *ycf9* encodes a PSII core subunit, PsbZ, which participates in PSII supramolecular architecture. *Plant Cell* **13**, 1347–1368.

Swinton, D.C. and Hanawalt, P.C. (1972). *In vivo* specific labeling of *Chlamydomonas* chloroplast DNA. *J. Cell Biol.* **54**, 592–596.

Turmel, M., Lemieux, C., and Lee, R.W. (1981). Dispersive labeling of *Chlamydomonas* chloroplast DNA in ^{15}N-^{14}N density transfer experiments. *Curr. Genet.* **4**, 91–97.

Turmel, M., Gutell, R.R., Mercier, J.-P., Otis, C., and Lemieux, C. (1993). Analysis of the chloroplast large subunit ribosomal RNA gene from 17 *Chlamydomonas* taxa. Three internal transcribed spacers and 12 group I intron insertion sites. *J. Mol. Biol.* **232**, 446–467.

Umen, J.G. and Goodenough, U.W. (2001). Chloroplast DNA methylation and inheritance in *Chlamydomonas*. *Genes Dev.* **15**, 2585–2597.

Vallet, J.-M., Rahire, M., and Rochaix, J.-D. (1984). Localization and sequence analysis of chloroplast DNA sequences of *Chlamydomonas reinhardtii* that promote autonomous replication in yeast. *EMBO J.* **3**, 415–421.

Waddell, J., Wang, X.M., and Wu, M. (1984). Electron microscopic localization of the chloroplast DNA replicative origins in *Chlamydomonas reinhardtii*. *Nucleic Acids Res.* **12**, 3843–3856.

Wakasugi, T., Tsudzuki, J., Ito, S., Nakashima, K., Tsudzuki, T., and Sugiura, M. (1994). Loss of all *ndh* genes as determined by sequencing the entire chloroplast genome of the black pine *Pinus thunbergii*. *Proc. Natl. Acad. Sci. U. S. A.* **91**, 9794–9798.

Wakasugi, T., Nagi, T., Kapoor, M., Sugita, M., Ito, M., Ito, S., Tsudzuki, J., Nakashima, K., Tsudzuki, T., Suzuki, Y., Hamada, A., Ohta, T., Inumura, A., Yoshinga, K., and Sugiura, M. (1997). Complete nucleotide sequence of the chloroplast genome from the green alga *Chlorella vulgaris*: The existence of genes possibly involved in chloroplast division. *Proc. Natl. Acad. Sci. U. S. A.* **94**, 5967–5972.

Wang, X.M., Chang, C.H., Waddell, J., and Wu, M. (1984). Cloning and delimiting one chloroplast DNA replicative origin of *Chlamydomonas*. *Nucleic Acids Res.* **12**, 3857–3872.

Wang, Z.F., Yang, J., Nie, Z.Q., and Wu, M. (1991). Purification and characterization of a gamma-like DNA polymerase from *Chlamydomonas reinhardtii*. *Biochemistry* **30**, 1127–1131.

Wells, R. and Sager, R. (1971). Denaturation and the renaturation kinetics of chloroplast DNA from *Chlamydomonas reinhardtii*. *J. Mol. Biol.* **58**, 611–622.

Woelfle, M.A., Thompson, R.J., and Mosig, G. (1993). Roles of novobiocin-sensitive topoisomerases in chloroplast DNA replication in *Chlamydomonas reinhardtii*. *Nucleic Acids Res.* **21**, 4231–4238.

Wollman, F.-A., Minai, L., and Nechushtai, R. (1999). The biogenesis and assembly of photosynthetic proteins in thylakoid membranes. *Biochim. Biophys. Acta* **1411**, 21–85.

Wu, M., Lou, J.K., Chang, D.Y., Chang, C.H., and Nie, Z.Q. (1986). Structure and function of a chloroplast DNA replication origin of *Chlamydomonas reinhardtii*. *Proc. Natl. Acad. Sci. U. S. A.* **83**, 6761–6765.

Wu, M., Nie, Z.Q., and Yang, J. (1989). The 18-kD protein that binds to the chloroplast DNA replicative origin is an iron-sulfur protein related to a subunit of NADH dehydrogenase. *Plant Cell* **1**, 551–557.

Wurtz, E.A., Boynton, J.E., and Gillham, N.W. (1977). Perturbation of chloroplast DNA amounts and chloroplast gene transmission in *Chlamydomonas reinhardtii* by 5-dluorodeoxyuridine. *Proc. Natl. Acad. Sci. U. S. A.* **74**, 4552–4556.

CHAPTER 25

Chloroplast Transcription

Uwe Klein
Department of Molecular Biosciences,
University of Oslo, Oslo, Norway

CHAPTER CONTENTS

I. INTRODUCTION

The *Chlamydomonas* chloroplast genome contains about 100 genes that are organized and transcribed either singly or as part of multi-gene transcription units. *In vitro* and *in vivo* studies have identified sequence elements and proteins involved in transcription of several of these genes. Not surprisingly, *Chlamydomonas* chloroplasts harbor a prokaryote-like transcription machinery consisting of eubacterial-type promoter sequences and a DNA-dependent

RNA polymerase composed of eubacterial-type subunits. The transcription machinery in *Chlamydomonas* shares many features with that of flowering plant chloroplasts, but also has a number of notable differences. Because chloroplast-specific genes, particularly those encoding thylakoid membrane proteins, have been of special interest, their transcription has been predominantly analyzed. For other groups of genes, for example those encoding ribosomal proteins or transfer RNAs, almost no information on transcription or associated *cis*- and *trans*-acting elements is available to date.

II. TRANSCRIPTION UNITS: SINGLE GENES AND GENE CLUSTERS

Many *Chlamydomonas* chloroplast genes were found to be transcribed into monocistronic transcripts, but complex patterns of transcript accumulation do not seem to be uncommon (e.g. Drapier et al., 1998). Because of inefficient termination of transcription (Rott et al., 1996), genes that are transcribed from their own promoter as monocistronic transcripts may also be transcribed as part of complex primary transcripts that originate from one or more upstream promoters. Processing of primary transcripts is therefore likely to occur frequently in the *Chlamydomonas* chloroplast. Genes that to date were found to be co-transcribed include *psbD* and *psaA-2*; *psaC* and *petL*; *rps7* and *atpE*; *tscA* and *chlN*; *atpA*, *psbI*, *cemA* and *atpH*; *psbZ* and *psbM*; and *psbB*, *psbT* and *psbH* (Table 25.1). Clusters of ribosomal protein and tRNA genes that appear to lack their own promoter sequences might also be transcribed as polycistronic units, but no experimental evidence to support this notion has been reported. Identification of co-transcribed genes has also been shown to be masked by rapid processing of transcripts and might be further complicated by the use of alternative promoters (Sturm et al., 1994).

Table 25.1 Co-transcribed chloroplast genes in *Chlamydomonas*

Gene	Co-transcribed genes	References
psbD	*psaA-2*	Choquet et al. (1988)
psaC	*petL*	Takahashi et al. (1991)
rps7	*atpE*	Robertson et al. (1990)
tscA	*chlN*	Rochaix (1996)
atpA	*psbI*, *cemA*, *atpH*	Drapier et al. (1998)
psbB	*psbT*, *psbH*	Monod et al. (1992); Johnson and Schmidt (1993); Summer et al. (1997); Vaistij et al. (2000)
psbZ	*psbM*	Higgs et al. (1998); Swiatek et al. (2001)

Even though complex transcription units are not unusual in the *Chlamydomonas* chloroplast genome and more complex transcript processing may still be detected, many genes seem to be transcribed from their own promoters. This contrasts with the organization of chloroplast genes in higher plants where operons seem to be predominant (Sugiura, 1992). One reason for the evolution of a high number of single transcription units in *Chlamydomonas* could be the relatively long intergenic sequences (Maul et al., 2002) that would seem to be nonfunctional.

III. THE TRANSCRIPTION MACHINERY

A. Chloroplast promoters

To date three types of chloroplast promoters have been identified in *Chlamydomonas* by deletion, mutational, and functional analyses: A typical sigma^{70}-type promoter, consisting of −35 and −10 sequence elements, has been delimited upstream of the 16S rRNA gene (Klein et al., 1992). A second type of promoter that lacks the canonical −35 sequence of typical sigma^{70}-type promoters, but seems to comprise an extended −10 element with the sequence TATAATAT, has been shown to direct transcription of the *atpB* (Klein et al., 1992), *rbcL* (Klein et al., 1994), and *psbD* genes (Klinkert et al., 2005) *in vivo*. The third type of promoter, which has only been delineated *in vitro*, appears to be an internal sequence of a sub-population of tRNA genes (Jahn, 1992) but has not been characterized further.

A putative fourth type of promoter, termed P_A, is contained in the sequence 5′ - GTTGCCTGCCAACAAATTTATTTATTGTATTAAAAT - 3′, originally identified between the *petA* and *petD* genes (Thompson and Mosig, 1987). The apparently complex P_A promoter has been found to be regulated by changes in DNA superhelicity (Thompson and Mosig, 1987; see section V.B). The promoter also functions in *E. coli* but has not been further dissected, and it is not clear which genes in *Chlamydomonas* are transcribed from it. It is noteworthy that the 36 nt sequence shown above is present in five additional locations: between *psbM* and *rps14*, *rps4* and *URF162*, *atpB* and *URF111*, *ycf12* and *URF123*, and *psaA2* and *psbJ*. There is no evidence, though, for functioning of the sequence as a promoter in these locations.

The sigma^{70}-type promoter preceding the *rrn* gene has been shown to function in a heterologous *E. coli* core RNA polymerase/*Chlamydomonas* RPOD *in vitro* transcription assay (Bohne et al., 2006), indicating that the single nucleus-encoded sigma^{70}-like factor (RPOD) in *Chlamydomonas* (Carter et al., 2004; Bohne et al., 2006; see section III.B.3) is probably involved in transcription of the chloroplast *rrn* gene *in vivo*.

The TATAATAT sequence that drives transcription of the *atpB*, *rbcL*, and *psbD* genes does not seem to require any other element to function as promoter. Sequences in 5′ untranslated regions previously suspected

Table 25.2 Number and distribution of putative promoter sequences

Sequence	Total number in genome	Occurrence in genes (%)	Occurrence in intergenic regions (%)
A and T nucleotides	133,599	65.5	65.8
Presumed promoter TATAATAT	88	16	84
Control sequence TAATTAAT	56	48	52
Presumed promoter element TTGACA	74	36	64
Control sequence TTCAGA	95	60	40

Note: Numbers are based on the chloroplast genome accession number NC_005353.

to represent promoter elements (Klein et al., 1994) have been found to be dispensable for transcription (Salvador et al., 2004). Strong circumstantial evidence for a role of the extended (−10) TATAATAT element in transcription of *Chlamydomonas* chloroplast genes comes from an analysis of the genome sequence (Table 25.2). The sequence TATAATAT occurs in 88 positions, and its distribution is biased towards intergenic regions (84%) even though adenine and thymine nucleotides are equally abundant (~66%) in genes and intergenic regions. A control sequence (TAATTAAT) that has the same number of adenine and thymine nucleotides as the putative promoter sequence is almost equally distributed between genes and intergenic regions (Table 25.2). The uneven distribution of the TATAATAT sequence, together with its proven function as the promoter of the *atpB*, *rbcL*, and *psbD* genes, suggests that other instances of this sequence are also likely to function as promoters. A similar analysis of the distribution of the (–35) TTGACA element of the *rrn* promoter also reveals an uneven distribution between genes and intergenic regions, although not as pronounced as for TATAATAT (Table 25.2). Considering that the (–35) TTGACA sequence represents only one half of the *rrn* gene promoter, this is an expected result.

Additional support for involvement of the TATAATAT sequence in *Chlamydomonas* chloroplast gene expression is provided by analyzing its distribution on the two DNA strands (Figure 25.1). The chloroplast genome contains eight gene clusters that alternate in their strand distribution (Maul et al., 2002). If the TATAATAT sequence is involved in transcription, it should be collocated with the gene clusters on either DNA strand and oriented in the direction of transcription of most genes in the clusters. Because the palindromic TATAATAT sequence is not an inverted repeat, its distribution and orientation on the two DNA strands with respect to direction of transcription of genes can be easily distinguished. These analyses reveal that the TATAATAT sequence is not randomly distributed on the two DNA

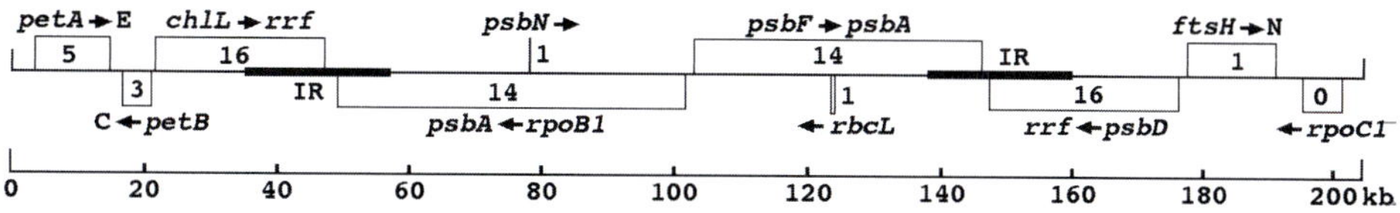

FIGURE 25.1 *Occurrence and orientation of the putative TATAATAT promoter sequence in the Chlamydomonas chloroplast genome. The genome is shown as a linear segment starting in the Wendy transposon upstream of the petA gene. Clusters of genes are drawn as boxes with the first and last gene of each cluster given above or underneath the boxes. The arrows indicate direction of transcription of most of the genes in the clusters. Numbers in the boxes denote the number of TATAATAT sequences in the clusters oriented in the general direction of transcription. The total number of TATAATAT elements on the forward (upper) and complementary (lower) strands is 47 and 41, respectively. Single letters denote tRNA genes. IR, inverted repeat.*

strands but is predominantly (≈80%) collocated with one of the eight gene clusters and in an orientation consistent with the direction of transcription of most of the genes (Figure 25.1).

There are two clusters of the TATAATAT motif consisting each of six repeats separated by short spacer sequences. One sequence with six repeats (5′-taaaaaaaagtaata(**TATAATAT**atttaaaaaaagtaata)$_5$**TATAATAT**atttaaa-3′) is located in the inverted repeat regions (thus, it occurs twice in the genome) 95 bp upstream of two unidentified reading frames (URF162 and URF111) that precede the *rrn* gene. The second sequence contains a cluster of partially overlapping TATAATAT sequences (5′-**TATAATATATAATAT**aatc**TATAATAT**aatc**TATAATAT**aatc**TATAATAT**aatc**TATAATATAATAT**-3′) and is located 908 bp upstream of the *psaB* coding region. The functions of these clusters have not been analyzed, for example by deletion, but they might have a role as redundant promoters or as sites that sequester RNA polymerase holoenzymes.

In most cases a TATAATAT-type or *rrn*-type promoter sequence is located upstream of annotated genes but in some cases no genes appear to be located downstream of putative promoters, or rarely, putative promoter sequences are located within coding regions. These "orphan" promoter sequences may be non-functional in their specific DNA context, they may transcribe genes that have not yet been identified (e.g. RNA genes), or they may have a role as redundant promoters.

There are a number of *Chlamydomonas* chloroplast genes that do not seem to be transcribed from any of the promoter sequences identified to date. These include most tRNA genes and most of the genes encoding ribosomal proteins. Transcription of these genes has not been studied *in vivo* to date and may initiate at a yet-undescribed promoter type. Alternatively, these genes may be co-transcribed as part of unidentified transcription units.

B. Chloroplast RNA polymerases

1. Overview

Plastids of flowering plants harbor two types of RNA polymerases, a plastid-encoded and a nucleus-encoded enzyme (abbreviated PEP and NEP, respectively). The plastid-encoded proteins are homologous to typical eubacterial DNA-dependent RNA polymerases, multisubunit enzymes whose core complex is composed of two α subunits and one β, β′, σ and ω subunit (α_2, β, β′, σ, ω) that add up to a combined molecular weight of at least 390 kD, excluding any factors that modulate promoter specificity and polymerase activity. The NEP is homologous to phage-type DNA-dependent RNA polymerases consisting of a single polypeptide with a molecular weight of about 100 kD (Hess and Börner, 1999; Kühn et al., 2007). Promoters for both types of polymerase are present upstream of some genes in tobacco plastids, while others are transcribed specifically by the bacterial or the phage-type enzyme (Hajdukiewicz et al., 1997).

All functional phage-type RNA polymerases in eukaryotes have been found to be nucleus-encoded. These enzymes have been known to transcribe the mitochondrial genome and so far, NEPs targeted to plastids have been detected only in flowering plants. Lower plants and algae do not seem to import nucleus-encoded RNA polymerases into plastids (Smith and Purton, 2002). This phylogenetic distinction suggests a relatively recent evolutionary path that led to targeting of an isozyme of the mitochondrial RNA polymerase to the plastid compartment in higher plants (Kabeya et al., 2002).

PEP appears to be the only RNA polymerase involved in *Chlamydomonas* chloroplast transcription. Mutants with disrupted *rpo* genes are not viable (Smith and Purton, 2002) indicating that genes essential for chloroplast maintenance and survival of the alga are exclusively transcribed by PEP. It should be noted, however, that the selection of *Chlamydomonas* transformants with disrupted *rpo* genes depends on expression of a cassette harboring the *E. coli aadA* gene (conferring spectinomycin/streptomycin resistance) under control of the *Chlamydomonas atpA* gene promoter (Goldschmidt-Clermont, 1991). The *atpA* promoter is of the extended (−10)-type, most likely requiring PEP for transcription, and can therefore not be used to drive transformants to homoplasmicity that lack this enzyme.

Nevertheless, the notion of only one type of RNA polymerase in the *Chlamydomonas* chloroplast is also supported by analysis of chloroplast transcription in the presence of rifampicin, a compound that inhibits transcription by binding specifically to the β subunit of eubacterial RNA polymerases. Rifampicin completely blocked transcription of the chloroplast *atpA*, *psbA*, *rbcL*, and *petA* genes, showing that these genes are exclusively transcribed by the plastid-encoded enzyme. Inhibition of the synthesis of chloroplast ribosomal RNAs by rifampicin was also shown some time ago (Surzycki, 1969). The drug also reduced considerably the accumulation of transcripts of the genes *psbD*, *petD*, *psaA*, *psaB*, and *atpB* suggesting that these genes are transcribed to a significant extent by the chloroplast-encoded RNA polymerase (Eberhard et al., 2002).

2. RNA polymerase genes

The genes *rpoB* and *rpoC* encode the β and β′ core subunits of eubacterial-type RNA polymerases, and have been found in all plastid genomes sequenced to date. The *rpoA* gene, which encodes the α subunit, is plastid-localized in the majority of organisms, but occasionally found in the nucleus (e.g. in *Euglena*; Hallick et al., 1993). Although organization of the *rpo* genes on plastid chromosomes varies among different groups of organisms[1] *rpoC* is normally, but not always (Ohta et al., 2003), split into *rpoC1* and *rpoC2* (encoding the β′ and β″ subunits, respectively), and usually located adjacent to *rpoB*.

In *Chlamydomonas* the *rpoB* and *rpoC1* genes are further segmented and separated from each other on the chromosome (Fong and Surzycki, 1992; Maul et al., 2002). The *rpoB* gene is split into two sequences, *rpoB1* (2469 bp) and *rpoB2* (1881 bp), separated by a spacer of 617 bp that does not seem to be part of a primary *rpoB1–rpoB2* transcript (Maul et al., 2002). The *rpoC1* gene is divided into two discontinuous segments, *rpoC1a* (1812 bp) and *rpoC1b* (1470 bp), that are separated by a spacer of 2457 bp. Located in the spacer sequence is an open reading frame (536 amino acids) encoding a putative protein of unknown function. *rpoC1a* and *rpoC1b* may be transcribed separately because long (>6 kb) transcripts containing both sequences could not be detected by RT-PCR (Maul et al., 2002). The *rpoC2* gene is a continuous sequence of 9363 bp that potentially encodes a protein of 3120 amino acids with a molecular weight of ≈358 kD. Transcripts of this gene were found to have a size of ≈5.2 kb, suggesting that the gene contains ≈4 kb internal intron sequences in frame with the *rpoC2* coding sequences (Maul et al., 2002).

No genes with homology to *rpoD* and *rpoZ* encoding the σ (sigma) and ω (omega) subunits, respectively, of typical eubacterial RNA polymerases are present in the *Chlamydomonas* chloroplast genome, but a single *RPOD* gene (accession number AJ627180) has been detected in the nuclear genome (see section III.B.4). The organization of *Chlamydomonas* chloroplast *rpo* genes differs from most other plastid chromosomes sequenced to date, but similar arrangements have been found in other chlorophycean algae (Bélanger et al., 2006; de Cambiaire et al., 2006), suggesting that this type of organization may not be unusual in green algae.

3. RNA polymerase core subunits

The polypeptides predicted from the *Chlamydomonas rpoA*, and the split *rpoB* and *rpoC* genes are to a large extent homologous to the eubacterial α, β, and β′ subunits. The primary functions of the subunits, that is assembling the multisubunit complex (α), polymerase activity (β), and unspecific DNA binding (β′), are associated with protein domains that are also present in the *Chlamydomonas* subunits.[2] The predicted *Chlamydomonas* subunits

[1]A complete list of sequenced plastid genomes can be found at http://www.ncbi.nlm.nih.gov/genomes/ORGANELLES/plastids_tax.html.

are larger in size both than their bacterial counterparts and also those predicted from sequences of many other PEP subunit-encoding genes. Most of the expansions in the *Chlamydomonas* subunits are at the N-terminus and/or in the middle. An example of an insertion is given in Figure 25.2 in which the amino acid sequence of the *Chlamydomonas* α subunit is aligned with another green algal sequence (*Scenedesmus obliquus*), and with sequences from a cyanobacterium (*Synechococcus*) and a higher plant (*Arabidopsis*). This insertion increases its predicted molecular weight by 30 kD compared to the other α subunits; however this has not been verified by biochemical means. Sequences of the *Chlamydomonas* β, β' and β'' subunits are also internally expanded compared to the sequences from *Synechococcus* and *Arabidopsis*, while in *Scenedesmus* the β' subunit is predicted to be even larger than in *Chlamydomonas*.

4. Sigma subunits

The core complex of eubacterial DNA-dependent RNA polymerases is able to bind to DNA unspecifically but lacks protein domains that interact with promoter sequences. Promoter selectivity is conferred to the core complex by exchangeable sigma subunits whose synthesis in bacteria is often regulated by environmental signals. Binding of a sigma subunit converts the core complex into a functional promoter-specific RNA polymerase holoenzyme. *E. coli* produces at least seven different sigma subunits, the most prominent being sigma^{70} which recognizes the consensus -35 (TTGACA) and -10 (TATAAT) promoter sequences of housekeeping genes (Gruber and Gross, 2003).

Numerous homologues of bacterial sigma subunits have been identified in plants and algae (Lysenko, 2007). All are nucleus-encoded, in most cases have been shown to be imported into plastids, and all contain protein domains that are homologous to regions 2, 3, and 4 of sigma^{70} (regions 2 and 4 interact with the -10 and -35 elements, respectively, of typical sigma^{70}-type promoters) suggesting that these plant and algal sigma subunits direct the plastid-encoded RNA polymerase to promoters of the sigma^{70} type (Allison, 2000).

Early attempts to isolate sigma subunits from *Chlamydomonas* chloroplasts (Surzycki and Shellenbarger, 1976) resulted in purification of ~51-kD and ~39-kD proteins that stimulated transcription in heterologous and homologous *in vitro* assays with core polymerases from *E. coli* and *Chlamydomonas*. Immunochemical cross-reaction of a partially purified RNA polymerase

[2]Domains found in the subunits of the *Chlamydomonas* plastid-encoded RNA polymerase include the insert and C-terminal domains (α subunit), the lobe, fork, and clamp domains (β subunit), the clamp and pore domain (β' subunit), and the funnel domain (β'' subunit). Domains and their functions can be searched for at CDART (http://www.ncbi.nlm.nih.gov/Structure/lexington/lexington.cgi).

```
C. reinhardtii       MTIYPNLKKIMTKTQSTDFFIACKESRIENNTNFYGCFYLGPFDESLSQTLANDLRRTLLSELTGLAITSIEIEGVLHKFSTLTGMKEPVLDLICNLQNI 100
S. obliquus          -----------MNKKIQDFFLSCKECILENPRSFYGSFSLGPFKNSQSLTVANALRRTLLAELSNIAITHLEIEGVTHEYSTLVGVRESVLDLLLNFKGI
Synechococcus sp.    ---------------MLQYQIDRIEHQITDDRSQTGVFLIGPLERGQATTLGNSLRRMLMGNLAGTAVTAVRIAGVNHEYATIPGVREDVLDILLNCKQL
A. thaliana          ------MVREKVKVSTRTLQWKCVESKRDSKRLYYGRFILSPLMKGQADTIGIAMRRALLGEIEGTCITRAKSENIPHDYSNIAGIQESVHEILMNLNEI

C. reinhardtii       VLRKETISSTNMNYRATKKTYIGFLSVNGPRVIKAADLKLPAGLQCVDPNQYIATLAEDGFLNMKFNINEGKNYIKQKPYNLDVTTLKKRNI-------- 200
S. obliquus          ALKNTSP---------VTKPLFGYLQVRGPGVVRASDLKFPPMIQCVDPDQYIATLNENGKLILKFRISDFKNSQENLNFEVSKTFLNIPNFGHFDTSSF
Synechococcus sp.    TVTSR-----------TDELEIGRLIVSGPAVVKAKDLQFSSQVQVVDGERPIATVSEGHSLELEVHVERGVGYRPVDRHNEDTSAID------------
A. thaliana          VLRSN-----------LYGTRNALICVQGPGYITARDIILPPAVEIIDNTQHIATLTEPIDLCIELKIERNRGYSLKMSNNFEDRSYP------------

C. reinhardtii       -----------------------LLQNFKNKIGLTALKNKQLMSTTMEGKPLLSNSDNLRFKKSFKRMLTSPNQSLKNKTSLGHDTVSNPIPLDAVFMPV 300
S. obliquus          QNFENSNFFSSSFFASQKMIKSKRKTKFSQLKVDYLSSFGFSTQHKRSGLNQTQKFAMQHQNPFRIEKNQQDFNKNKNGNFDSQKNKTNSLWVDPLFNPI
Synechococcus sp.    -----------------------------------------------------------------------------------------LLQIDAVFMPV
A. thaliana          --------------------------------------------------------------------------------------------IDAVFMPV

C. reinhardtii       TKINCIIEENNVYSDFSTDPSLEFSTHLVPSLTTQTNLRQESTHKFVNKANSLLQENNLFRSEKVYLPNIYIPEGEGDALSLKGVSPYSDFKTFLSTLNY 400
S. obliquus          LKVNYIIETIEPMQKNIP----------------------------------------------------------------------------------
Synechococcus sp.    HRVNFTTDETAVAEGGS-----------------------------------------------------------------------------------
A. thaliana          ENANHSIHSYGNGNE-------------------------------------------------------------------------------------

C. reinhardtii       NSLYQTSLFLNQSLGQNKLLPWQANTLFFDVTNFADSQSNDVNMNMDLAGDKTSVQKIHSTGISNTDAQLNKLSLTKIKSFLGSSDKLQNKTYQSFLQPK 500
S. obliquus          ----------------------------------------------------------------------------------------------------
Synechococcus sp.    ----------------------------------------------------------------------------------------------------
A. thaliana          ----------------------------------------------------------------------------------------------------

C. reinhardtii       YASSNLRTLLTQRSLQKNQSVFMHSFLDDQAKHKELTSNTLRANKFQVMKTVVSIPPKTFKTNPINTFHNQSLTFEYAKNFKLKPLRKKSQLILEIWTNG 600
S. obliquus          ----------------------------------------------------------------------------------------NEIVFIELWTNG
Synechococcus sp.    ---------------------------------------------------------------------------------------ARERLRMEVVTDG
A. thaliana          ---------------------------------------------------------------------------------------KQEILFLEIWTNG

C. reinhardtii       SIHPRKALYQALIFLSNNFLKLQTVKMLGSMFKSDLAYGNLKHSITNNYTYYQTNNLSQQSKKTLFKKAEAAASKQALNFLKPSILNTNNLPGGTMDLYL 700
S. obliquus          SIHPRKAFYTALLYLKTMFDKLDCMRLLNYEFSNTMLESEKTSTKFFKTFEYDFRFYNSREDK-----TMKSFTPEKFFLPEEIEDVEKDYLLNLKNQAD
Synechococcus sp.    SMTPDDAIAQAANQLIELFQPLATVTMVE-------------------------------------------------------------EPG--IEPEP
A. thaliana          SLTPKEALHQASRNLINLFIPFLHVEEETFYLENNQHQVTLPFFPFHN---------------------------------------RLVNLRKKKKTKE

C. reinhardtii       QSSLNAPIGILKISLRSYTALKKAGITTLKDLIKYSKKDLFAIKNLGQKSVFEIEQNLALLGLTLVTD----- 773
S. obliquus          NTWNDLPLTNLNLPYRITKILAKNNFFVVGDLLKISPNELKKLSGIGNYCVFILQKRFEKLGLKLGSRNEKNL
Synechococcus sp.    SAEAQIPLEELNLSVRAYNCLKRAQVNSVSDLMGFSYEDLLEIKNFGSKSADEVIEALERIGISIPQSRTSV-
A. thaliana          LAFQYIFIDQLELPPRIYNCLKKSNIHTLLDLLNNSQEDLIKIEHFHVEDVKKILDILEKK------------
```

FIGURE 25.2 *Alignment of PEP α-subunit amino acid sequences from Chlamydomonas, Scenedesmus, Synechococcus, and Arabidopsis. Residues that are identical in all four sequences are shaded. Note the insertion in the middle of the Chlamydomonas α-subunit.*

preparation from *Chlamydomonas* with an antibody against the sigma subunit of the cyanobacterium *Anabaena* sp. provided additional evidence for the presence of at least one sigma-like subunit in *Chlamydomonas* (Troxler et al., 1994). Searches in EST and genome databases revealed a single nuclear gene (*RPOD*, accession EDP03301) which encodes a protein with homology to sigma^{70} (Carter et al., 2004; Bohne et al., 2006). The *RPOD*-encoded protein has an unusually long N-terminal extension compared to other nucleus-encoded sigma proteins, for example those of *Arabidopsis* and the red alga *C. caldarium* (Figure 25.3). Besides an amino-terminal transit peptide, the extension of the *Chlamydomonas* sigma subunit harbors a putative leucine zipper motif and two PEST motifs, the latter suggesting interactions of this region of the protein with other polypeptides (Carter et al., 2004). Immunoblots showed that *RPOD* is expressed at the protein level and when combined with *E. coli* core RNA polymerase, the recombinant RPOD protein was found to bind to typical sigma^{70}-type (−35/ − 10) promoters and to enhance transcription from these promoters *in vitro* (Bohne et al., 2006).

Few *Chlamydomonas* chloroplast genes are transcribed from typical sigma^{70}-type promoters (see section III.A) suggesting either the presence of other types of sigma factors in *Chlamydomonas*, flexibility with regard to promoter recognition by RPOD, or switching of promoter specificity of the single *Chlamydomonas* sigma subunit, possibly induced by interactions with other proteins, as for example proposed for sigma-binding proteins in *Arabidopsis* (Morikawa et al., 2002). To date homology searches did not identify genes encoding sigma subunits other than the sigma^{70} type in plants and algae, but the presence of additional sigma factors in plastids cannot be excluded. For example, the *rpoN*-encoded enhancer-dependent sigma^{54} subunit of *E. coli* lacks all the conserved regions diagnostic of typical sigma^{70} proteins (Gruber and Gross, 2003), and cannot be detected in database searches using sigma^{70} consensus sequences as queries.

IV. INITIATION AND TERMINATION OF TRANSCRIPTION

A. Initiation of transcription

1. Overview

The frequency of transcription initiation at promoter sequences is dependent on the basal promoter sequence and on *cis*- and *trans*-acting elements that facilitate binding and/or promoter clearance of the RNA polymerase. Described below are examples for both promoter sequence-dependent and regulative control of transcription initiation found in *Chlamydomonas*.

2. Transcription start sites

At sigma^{70}-like promoters transcription initiates a few base pairs downstream of the −10 element. For the *Chlamydomonas* chloroplast genes *atpB*

```
                         Transit peptide                                  Leucine zipper motif
C. reinhardtii     MALPGSTMNLTTRCSTTPRSAVVARAVAAPTRPTTKSAVPELLDSRPGERNLNFMEYAQATQMLDRLKGQASDLELLLDQLNALEASLDESVLAPPTVDD 100
Synechococcus sp.  ----------------------------------------------------------------------------------------------------
A. thaliana        -----------------------------------------------------------------------------------MSSCLLPQFKCPPDSFS
C. caldarium       ----------------------------------------------------------------------------------------------------

C. reinhardtii     PKERAARQARRAAKRAERRAQATSATVAAAAGPAMSAVVSHSTPTKAAAAPATSTASSSSSDSGLLDLVSFVGGFDTRPIPATTSAPPAGASSSDVQHLE 200
Synechococcus sp.  ----------------------------------------------------------------------------------------------------
A. thaliana        IHFRTSFCAPKHNKGSVFFQPQCAVSTSPALLTSMLDVAKLRLPSFDTDSDSLISDRQWT-----------------------YTRPDGPSTEAKYLEAL
C. caldarium       ----------------------------------------------------------------------------------------------------

                                                                                                     PEST 1
C. reinhardtii     DLFKLSVGEPDIPRASASAAPAVLRPRKLTPKKPSAAPSAAVTAAPSPAPTLPSTPSTSARIAPAPGSLADELERLLGPTTSREAAESEDEDSFAGPSED 300
Synechococcus sp.  ----------------------------------------------------------------------------------------------------
A. thaliana        ASETLLTSDEAVVVAAAAEAVALARAAVKVAKDATLFKNSNNTNLLTSSTADKRSKWDQFTEKERAGILGHLAVSDNGIVSDKITASASNKESIGDLESE
C. caldarium       -----------------------------------------------------------------------MALTSLGPLPDSFSS--------------

                    PEST 1              PEST 2
C. reinhardtii     DLLALEQEVSRKSSRLPVLDEEDEEDEQQQLEDNEEDAVAGPGSLEASAMATRTSSQLSIMQTGPSLLSLVPASAAPGRSAKARASRRAARNGHASGRLG 400
Synechococcus sp.  ----------------------------------------MPKVNLQQAPNNG-----------------------------------------------
A. thaliana        KQEEVELLEEQPSVSLAVRSTRQTERKARRAKGLEKTASGIPSVKTGSSPKKKR----------------------------------------------
C. caldarium       ----VSSLASAPRVPLFKDELRPFIEDDANDD--DEEGGLVAQEQATMKPRRR-----------------------------------------------

C. reinhardtii     GATANAAGRGKVGSKDGTMNFLGKVESLSTLDVEKEREVTAVCRDFLFLEKVKRQCEKTLHRPATSEEIAAAVAMDVESLKLRYDAGLKAKELLLKSNYK 500
Synechococcus sp.  -----------KSSADMVRTYLREIGRVPLLTHEEEIVYGKQVQRLMALQEKRQTLEESLDRERTREEWAEAAQLTPKQLQRDLQIGERAKRKMIEANLR
A. thaliana        ------LVAQEVDHNDPLRYLRMTTSSSKLLTVREEHELSAGIQDLLKLERLQTELTERSGRQPTFAQWASAAGVDQKSLRQRIHHGTLCKDKMIKSNIR
C. caldarium       ---------AKTSTTCAATAYVREICRLDRLTDEEVASLSMATQDYCALMKQREELCAKLGREPTDTEWSTTVGVSEEEVRKRIERGRRARERLVASNLR

C. reinhardtii     LVMTVCKSFVG-----------------------------KGPHIQDLVSEGVKGLLKGVEKYDATKGFRFGTYAHWWIRQAVSRSLAETGRAVRLPMHM 600
Synechococcus sp.  LVVAIARNIKK-----------------------------RNLEFLDLIQEGSLGLERGVEKFDPTKGYKFSTYAYWWIRQAITRAIAQQGRTIRLPIHI
A. thaliana        LVISIAKNYQG-----------------------------AGMNLQDLVQEGCRGLVRGAEKFDATKGFKFSTYAHWWIKQAVRKSLSDQSRMIRLPFHM
C. caldarium       LVLAIANRYKLRAVVCSAWKSRSKGGMDMPSHEKVRLQRQHNIFLGDLIQEGALGLIRAAEKFDPSKGFKFSTYATWWIRQAIQRCVSEHSRTIRLPNHV

C. reinhardtii     IEQLTRLKNLSAKLQTQLAREPTLPELAKAAGLPVTRVQMLMETARSAASLDTPIGGNELGPTVKDS------------VEDEREAADEEFGSDSLRNDM 700
Synechococcus sp.  TEKLNKIKKTQRQLAQDLGRAATIAEIAEALELEPAQIREYFRVSRHPISLDVRVGDNQDTELMD------------LLED-SGESPDNNVTYELLKQDL
A. thaliana        VEATYRVKEARKQLYSETGKHPKNEEIAEATGLSMKRLMAVLLSPKPPRSLDQKIGMNQNLKPSE------------VIADPEAVTSEDILIKEFMRQDL
C. caldarium       TESLRQIYRMRARLQHVLKRPPTDEEVAEACNMTLKRLKKITQHAWETQTLSLEGPPNSEAHSLYKSQQGEAPLSIGSIIEATEETPEEALERNCMREDV

C. reinhardtii     EAMLLELPEREARVVRLRFGLDDG-KEWTLEEIGEALNVTRERIRQIEAKALRKLRVKTIDVSGKLMEYGENLEMLMDGSREMAARTSSGTRKT 794
Synechococcus sp.  RDLIADLTPQQQEVIILRFGLNDG-KELSLAKIGKRMNLSRERVRQLERQALNHLR--RRRSQIREYLVAS-----------------------
A. thaliana        DKVLDSLGTREKQVIRWRFGMEDG-RMKTLQEIGEMMGVSRERVRQIESSAFRKLKNKKRNNHLQQYLVAQS----------------------
C. caldarium       ECVLRTLPFQEREILRLRYGISDDERPHSTEELSARFQLPVHTIRLIESRALRKLRHPSLAAMLREYVNIL-----------------------
```

FIGURE 25.3 Sequence alignment of sigma^{70}-like subunits of *Chlamydomonas*, *Synechococcus*, *Arabidopsis*, and the red alga *Cyanidium*. Residues identical in all sequences are shaded. Note the N-terminal extension in the *Chlamydomonas* sigma factor which harbors the transit peptide and a leucine zipper motif. In addition, two adjacent PEST sequences were identified further downstream (Carter et al., 1994). These elements are marked by lines above the sequence.

and *rbcL*, the start site of transcription has been determined unambiguously by primer extension of reporter gene transcripts with sigma^{70}-like promoter elements (Blowers et al., 1990; Anthonisen et al., 2001). Transcription of these constructs begins about 5 bp downstream of the −10 element. The 5′ ends of several endogenous *Chlamydomonas* chloroplast transcripts have also been found to lie a few bp downstream of typical −10 sigma^{70} promoter elements (e.g. Matsumoto et al., 1991; Sakamoto et al., 1993; Klinkert et al., 2005). For a number of chloroplast transcripts multiple 5′ ends were found by primer extension (e.g. Nickelsen et al., 1994; Vaistij et al., 2000). These could be formed by processing of primary transcripts or by transcription from more than one promoter.

3. Chloroplast promoter strength

Promoter strength can be estimated by measuring rates of transcription, either in run-on transcription assays with permeabilized cells (Guertin and Bellemare, 1979; Hwang et al., 1996) or *in vivo* (Baker et al., 1984; Blowers et al., 1990) using the rate of incorporation of radiolabeled UTP or phosphate into chloroplast transcripts, respectively. With these approaches relative transcription rates have been determined for *rbcL*, *psbA*, *psbK*, *psaA*, *psaB*, *atpA*, *atpB*, *tufA*, *rpl16*, and *rrn* (Blowers et al., 1990; Salvador et al., 1993; Silk and Wu, 1993; Hwang et al., 1996). According to these analyses the sigma^{70}-type promoter of the *rrn* gene is the strongest, followed by the extended –10 type of promoter of the *rbcL* and *psbA* genes (Blowers et al., 1990). The other genes listed above were transcribed at rates about 20 times lower than the *rrn* gene and at 3–10 times lower than *rbcL* and *psbA*, suggesting that their promoters are considerably weaker. It has been found, however, that initiation from the *rbcL* promoter is augmented approximately 10-fold by an enhancer-like sequence located about 140 bp downstream of the −10 element (see section IV.A.4). Without the enhancing element the *rbcL* promoter is about as robust as, for example, the *atpB* promoter.

4. cis-Elements involved in transcription initiation

Enhancer-like sequences that augment rates of transcription but are dispensable for basic promoter function may not be common in chloroplast genomes. To date only a few have been identified overall (Allison and Maliga, 1995), and only one in *Chlamydomonas* (Klein et al., 1994). The *Chlamydomonas* element is contained in a 35 bp sequence in the *rbcL* coding region, ~140 bp downstream of the transcription start site. It enhances transcription from the *rbcL* promoter approximately 10-fold and also functions with other promoters of the TATAATAT-type, but not with the *rrn* (−35/ −10-type) promoter (Anthonisen et al., 2002). Because the sequence comprising the enhancer has not yet been delimited in detail, it is not possible to do a genome-wide search for putative enhancer sequences.

Sequences containing poly(A) tracts have been reported to function as nonspecific transcription enhancers in *Chlamydomonas* (Lisitsky et al., 2001). These sequences probably cause a regional change in DNA conformation, thereby facilitating initiation at nearby promoters (see section V.B).

B. Transcription termination

1. Overview

Transcription is terminated when RNA polymerase pauses and the newly synthesized transcript is released from the ternary RNA polymerase/DNA/RNA complex. Release of the transcript induces conversion of the open polymerase/DNA complex back to a closed complex, resulting in nonspecific polymerase/DNA contacts and eventual dissociation of the RNA polymerase from the DNA. These processes have been studied in detail in bacteria (Nudler and Gottesman, 2002) and can be assumed to be similar in chloroplasts. However, neither sites of efficient transcription termination nor the events that lead to termination of transcription at these sites are well defined in *Chlamydomonas*.

2. Termination signals

Typical eubacterial transcription terminators consist of an inverted repeat sequence that can fold into an RNA stem-loop structure. Formation of the stem-loop causes RNA polymerase to pause and promotes release of the transcript which, depending on the type of terminator, may be aided (or prevented) by proteins that associate with the ternary transcription complex (Platt, 1986).

Inverted repeats that are present in the 3′ regions of most *Chlamydomonas* chloroplast genes are predicted to form stem-loops within nascent transcripts and might function as transcription terminators. Examples of particularly perfect predicted stem-loop structures in *Chlamydomonas* transcripts are shown in Figure 25.4. Besides the examples shown in this figure, many small perfect and/or imperfect (containing one or more unpaired nucleotide) stem-loop structures can be predicted in 3′ untranslated regions.

The 3′ ends of a number of monocistronic *Chlamydomonas* chloroplast transcripts have been determined to lie a few nucleotides downstream of predicted stem-loops, suggesting termination of transcription and/or processing at these sites (e.g. Stern and Kindle, 1993). However, because *Chlamydomonas* chloroplast transcription has been found to continue beyond predicted RNA stem-loop structures into intergenic regions, the stem-loop structures *per se* are likely not efficient terminators of transcription but function primarily as RNA processing sites, and additional factors are required to create the mature 3′ ends (Rott et al., 1996). A particularly striking example of inefficient termination at a perfect stem-loop structure is seen at the 3′ end of *psbT* (Figure 25.4), which may be the largest found

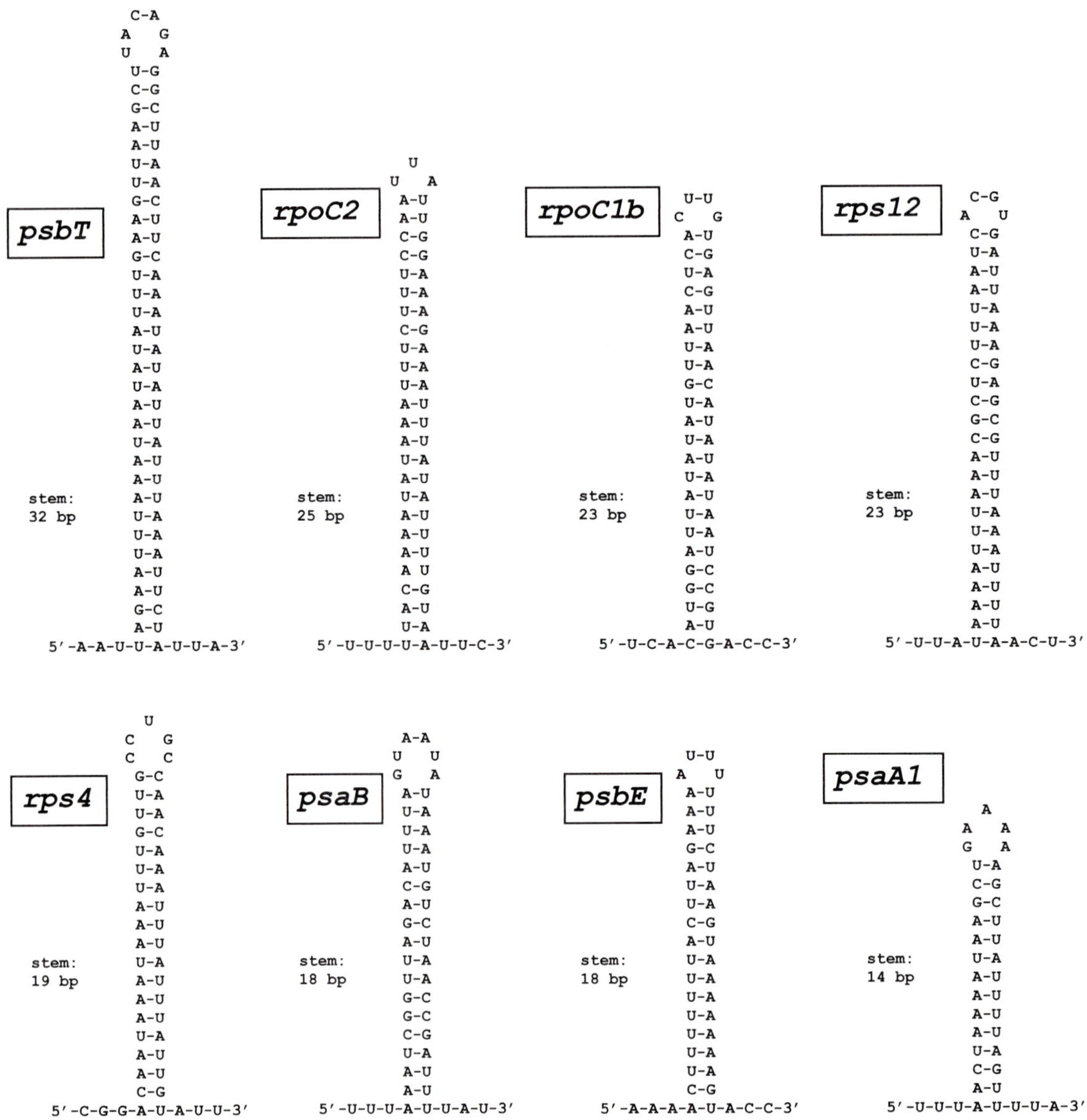

FIGURE 25.4 *Perfect RNA stem-loop structures predicted from the sequences at the ends of Chlamydomonas chloroplast genes. The genes preceding the sequences are indicated to the left of the stem-loops. Whether these stem-loops have a function in transcription termination and/or transcript processing has not yet been determined. Structures shown were predicted by the fast folding DINAMelt Server at the Rensselaer Polytechnic Institute (http://www.bioinfo.rpi.edu/applications/hybrid/quikfold.php).*

at the end of any *Chlamydomonas* chloroplast transcript. Despite its stability, the stem-loop does not seem to terminate transcription very efficiently because the *psbH* gene located 2.7 kb downstream of *psbT* has been found to be co-transcribed with *psbT* as part of a *psbB* - *psbT* - *psbH* transcription unit (Vaistij et al., 2000).

The above example illustrates that stem-loops alone, independent of size, are not sufficient to terminate transcription in the *Chlamydomonas* chloroplast. This is reminiscent of the situation in bacteria where functioning of intrinsic transcription terminators depends not only on structure but also on specific sequences (Cheng et al., 1991). The importance of sequence for functioning of a stem-loop as 3′ RNA processing site in *Chlamydomonas* has been demonstrated with gene constructs that have at their 3′ ends a native inverted repeat sequence inserted in alternate orientations (Blowers et al., 1993; Rott et al., 1998). In opposite orientation the stem portion of the stem-loop will be identical (except where there were G–U base pairs), but the adjacent sequences and the loop will differ. Transformants harboring constructs with "antisense" stem-loops at their 3′ ends did not accumulate transcripts with a defined 3′-terminus, showing that the sequence of the original stem-loop is crucial for its function (Blowers et al., 1993; Rott et al., 1998).

Given the relative inefficiency of transcription termination at predicted RNA stem-loop structures it is not clear at which sites transcription termination occurs in *Chlamydomonas* chloroplasts. Short dispersed repeats that are abundant in intergenic regions of the *Chlamydomonas* chloroplast genome may act as termination sites (Jiao et al., 2004). It is also possible that termination efficiency differs between stem-loops depending on sequence and/or context, and that further analyses will identify stem-loops that act as efficient terminators. In this context it is worth noting that transcription termination has been observed at tRNA sequences in an *in vitro* transcription system from spinach (Stern and Gruissem, 1987).

3. 3′ end maturation

Even though stem-loop structures are not efficient terminators, they define the 3′ ends of transcripts in all cases analyzed to date. Maturation of the 3′ end of chloroplast transcripts occurs at or near stem-loop structures by sequence-dependent processes, as discussed in Chapter 27 .

V. REGULATION OF CHLOROPLAST TRANSCRIPTION

A. Light/dark and circadian regulation

When *Chlamydomonas* cells are grown in a 12-hour light/12-hour dark regime levels of several chloroplast transcript populations fluctuate significantly (Leu et al., 1990; Salvador et al., 1993; Hwang et al., 1996). In most cases transcript levels peak early in the light period and are lowest at the beginning of the dark period. These fluctuations have been found to be caused by changes in rates of both transcription and RNA degradation (Hwang et al., 1996). In several cases fluctuations persisted for at least one 24 hour cycle after moving the cells from the light/dark regime into continuous light or darkness (Hwang et al., 1996), indicating that they are under endogenous control.

Further analyses of these fluctuations concluded that transcription is in most cases under circadian control, whereas RNA degradation seems to be primarily influenced by light conditions (Hwang et al., 1996). The latter would be consistent with the finding that photosynthetic electron transport promotes RNA degradation via a redox-regulated pathway (Salvador and Klein, 1999). Furthermore, circadian-regulated transcription has been found to depend on translation of nucleus-encoded proteins (Kawazoe et al., 2000), suggesting that the nucleus encodes one or more chloroplast transcriptional regulators.

An obvious candidate for such a protein is the unique sigma factor, discussed above. Its abundance has been found to vary, e.g. in response to availability of sulfur, and to correlate with transcriptional activity in the chloroplast (Irihimovitch and Stern, 2006). *RPOD* mRNA levels vary about threefold in light/dark growing cells (Carter et al., 2004) but correlative changes in levels of sigma protein, that would argue for a role of the sigma factor in circadian fluctuations of *Chlamydomonas* chloroplast transcripts, have not been reported. RPOD, however, does appear to play a role in chloroplast transcriptional regulation during sulfur deprivation (see Chapter 5).

B. Regulation of transcription by changes in DNA topology

Regions of negatively supercoiled DNA are much more efficiently transcribed than relaxed DNA, most likely because torsional stress helps to separate DNA strands and facilitates open complex formation within promoter/RNA polymerase complexes (McClure, 1985). Transcription from the *Chlamydomonas* chloroplast P_A promoter (see section III.A) has been found to be affected by changes in DNA superhelicity (Thompson and Mosig, 1987) and light has been shown to alter the topology of *Chlamydomonas* chloroplast DNA (Thompson and Mosig, 1990). Superhelicity of chloroplast DNA fluctuates in cells growing in light/dark cycles concurrent with fluctuations in transcription (Salvador et al., 1998). These fluctuations persisted in cells kept in continuous light, suggesting an endogenously controlled mechanism regulating DNA topology.

Candidates for mediating changes in chloroplast DNA topology are DNA topoisomerases (Thompson and Mosig, 1985). Several type I and type II topoisomerases are predicted to be encoded in the *Chlamydomonas* nuclear genome. Based on sequence homology, a bacteria-like type II topoisomerase, consisting of two subunits (subunit A: EDP04229, subunit B: EDP05794) is a strong candidate for a chloroplast-located enzyme involved in controlling DNA superhelicity. Inhibition of the type II topoisomerase in the *Chlamydomonas* chloroplast by novobiocin, which relaxes supercoiled DNA, affected accumulation of several chloroplast transcripts *in vivo*

(Thompson and Mosig, 1985) suggesting that this enzyme could be involved in endogenous control of transcription.

VI. PERSPECTIVES

Since publication of the first edition of the *Chlamydomonas* Sourcebook considerable progress has been made in elucidating the molecular elements involved in chloroplast transcription. Many of the advances in this field stem from the ease with which the *Chlamydomonas* chloroplast genome can be transformed and manipulated. Progress made in *Chlamydomonas* greatly influenced and aided parallel efforts to analyze the transcriptional machinery in chloroplasts of higher plants, leading to important advances. Most likely, future research will still revolve around identifying *cis*- and *trans*-acting elements that control chloroplast transcription but will increasingly utilize the resources made available through the *Chlamydomonas* genome project. In particular, identification and functional analysis of nucleus-encoded factors that play a role in chloroplast transcription will likely become ever more important. Eventually, understanding transcription and associated molecular processes might contribute to controlled expression of foreign genes in chloroplasts, and to informed and thoughtful manipulations of the chloroplast genome in biotechnological endeavors.

ACKNOWLEDGMENTS

I thank Monique Turmel (Université Laval, Quebec, Canada) for kindly providing the sequence of the *C. reinhardtii rpoA* gene.

REFERENCES

Allison, L.A. (2000). The role of sigma factors in plastid transcription. *Biochimie* **82**, 537–548.

Allison, L.A. and Maliga, P. (1995). Light-responsive and transcription-enhancing elements regulate the plastid *psbD* core promoter. *EMBO J.* **14**, 3721–3730.

Anthonisen, I.L., Salvador, M.L., and Klein, U. (2001). Specific sequence elements in the 5′ untranslated regions of *rbcL* and *atpB* gene mRNAs stabilize transcripts in the chloroplast of *Chlamydomonas reinhardtii*. *RNA* **7**, 1024–1033.

Anthonisen, I.L., Kasai, S., Kato, K., Salvador, M.L., and Klein, U. (2002). Structural and functional characterization of a transcription-enhancing sequence element in the *rbcL* gene of the *Chlamydomonas* chloroplast genome. *Curr. Genet.* **41**, 349–356.

Baker, E.J., Schloss, J.A., and Rosenbaum, J.L. (1984). Rapid changes in tubulin RNA synthesis and stability induced by deflagellation in *Chlamydomonas*. *J. Cell Biol.* **99**, 2074–2081.

Bélanger, A.-S., Brouard, J.-S., Charlebois, P., Otis, C., Lemieux, C., and Turmel, M. (2006). Distinctive architecture of the chloroplast genome in the chlorophycean green alga *Stigeoclonium helveticum*. *Mol. Gen. Genomics* **276**, 464–477.

Blowers, A.D., Ellmore, G.S., Klein, U., and Bogorad, L. (1990). Transcriptional analysis of endogenous and foreign genes in chloroplast transformants of *Chlamydomonas*. *Plant Cell* **2**, 1059–1070.

Blowers, A.D., Klein, U., Ellmore, G.S., and Bogorad, L. (1993). Functional *in vivo* analyses of the 3′ flanking sequences of the *Chlamydomonas* chloroplast *rbcL* and *psaB* genes. *Mol. Gen. Genet.* **238**, 339–349.

Bohne, A.-V., Irihimovitch, V., Weihe, A., and Stern, D.B. (2006). *Chlamydomonas reinhardtii* encodes a single sigma^{70}-like factor which likely functions in chloroplast transcription. *Curr. Genet.* **49**, 333–340.

Carter, M.L., Smith, A.C., Kobayashi, H., Purton, S., and Herrin, D.L. (2004). Structure, circadian regulation and bioinformatic analysis of the unique sigma factor gene in *Chlamydomonas reinhardtii*. *Photosynth. Res.* **82**, 339–349.

Cheng, S.-W.C., Lynch, E.C., Leason, K.R., Court, D.L., Shapiro, B.A., and Friedman, D.I. (1991). Functional importance of sequence in the stem-loop of a transcription terminator. *Science* **254**, 1205–1207.

Choquet, Y., Goldschmidt-Clermont, M., Girard-Bascou, J., Kück, U., Bennoun, P., and Rochaix, J.-D. (1988). Mutant phenotypes support a *trans*-splicing mechanism for the expression of the tripartite *psaA* gene in the *C. reinhardtii* chloroplast. *Cell* **52**, 903–913.

de Cambiaire, J.-C., Otis, C., Lemieux, C., and Turmel, M. (2006). The complete chloroplast sequence of the chlorophycean green alga *Scenedesmus obliquus* reveals a compact gene organization and a biased distribution of genes on the two DNA strands. *BMC Evol. Biol.* **6**, 37.

Drapier, D., Suzuki, H., Levy, H., Rimbault, B., Kindle, K.L., Stern, D.B., and Wollman, F.-A. (1998). The chloroplast *atpA* gene cluster in *Chlamydomonas reinhardtii*. *Plant Physiol.* **117**, 629–641.

Eberhard, S., Drapier, D., and Wollman, F.-A. (2002). Searching limiting steps in the expression of chloroplast-encoded proteins: Relations between gene copy number, transcription, transcript abundance and translation rate in the chloroplast of *Chlamydomonas reinhardtii*. *Plant J.* **31**, 149–160.

Fong, E.F. and Surzycki, S.J. (1992). Chloroplast RNA polymerase genes of *Chlamydomonas reinhardtii* exhibit an unusual structure and arrangement. *Curr. Genet.* **21**, 485–497.

Goldschmidt-Clermont, M. (1991). Transgenic expression of aminoglycoside adenine transferase in the chloroplast: A selectable marker for site-directed transformation of *Chlamydomonas*. *Nucleic Acids Res.* **19**, 4083–4089.

Gruber, T.M. and Gross, C.A. (2003). Multiple sigma subunits and the partitioning of bacterial transcription space. *Annu. Rev. Microbiol.* **57**, 441–466.

Guertin, M. and Bellemare, G. (1979). Synthesis of chloroplast ribonucleic acid in *Chlamydomonas reinhardtii* toluene-treated cells. *Eur. J. Biochem.* **96**, 125–129.

Hallick, R.B., Hong, L., Drager, R.G., Favreau, M.R., Monfort, A., Orsat, B., Spielmann, A., and Stutz, E. (1993). Complete sequence of *Euglena gracilis* chloroplast DNA. *Nucleic Acids Res.* **21**, 3537–3544.

Hajdukiewicz, P.T.J., Allison, L.A., and Maliga, P. (1997). The two RNA polymerases encoded by the nuclear and plastid compartments transcribe distinct groups of genes in tobacco plastids. *EMBO J.* **16**, 4041–4048.

Hess, W.R. and Börner, T. (1999). Organellar RNA polymerases of higher plants. *Int. Rev. Cytol.* **190**, 1–59.

Higgs, D.C., Kuras, R., Kindle, K.L., Wollman, F.-A., and Stern, D. (1998). Inversions in the *Chlamydomonas* chloroplast genome suppress a *petD* 5′ untranslated region deletion by creating functional chimeric mRNAs. *Plant J.* **14**, 663–671.

Hwang, S., Kawazoe, R., and Herrin, D.L. (1996). Transcription of *tufA* and other chloroplast-encoded genes is regulated by a circadian clock in *Chlamydomonas*. *Proc. Natl. Acad. Sci. U. S. A.* **93**, 996–1000.

Irihimovitch, V. and Stern, D.B. (2006) The sulfur acclimation SAC3 kinase is required for chloroplast transcriptional repression under sulfur limitation in *Chlamydomonas reinhardtii*. *Proc. Natl. Acad. Sci. U. S. A.* **103**, 7911–7916.

Jahn, D. (1992). Expression of the *Chlamydomonas reinhardtii* chloroplast tRNA(Glu) gene in a homologous *in vitro* transcription system is independent of upstream promoter elements. *Arch. Biochem. Biophys.* **298**, 505–513.

Jiao, H.S., Hicks, A., Simpson, C., and Stern, D.B. (2004). Short dispersed repeats in the *Chlamydomonas* chloroplast genome are collocated with sites for mRNA 3′ end formation. *Curr. Genet.* **45**, 311–322.

Johnson, C.H. and Schmidt, G.W. (1993). The *psbB* gene cluster of the *Chlamydomonas reinhardtii* chloroplast: Sequence and transcriptional analyses of *psbN* and *psbH*. *Plant Mol. Biol.* **22**, 645–658.

Kabeya, Y., Hashimoto, K., and Sato, N. (2002). Identification and characterization of two phage-type RNA polymerase cDNAs in the moss *Physcomitrella patens*: Implication of recent evolution of nuclear-encoded RNA polymerase of plastids in plants. *Plant Cell Physiol.* **43**, 245–255.

Kawazoe, R., Hwang, S., and Herrin, D.L. (2000). Requirement for cytoplasmic protein synthesis during circadian peaks of transcription of chloroplast-encoded genes in *Chlamydomonas*. *Plant Mol. Biol.* **44**, 699–709.

Klein, U., De Camp, J.D., and Bogorad, L. (1992). Two types of chloroplast gene promoters in *Chlamydomonas reinhardtii*. *Proc. Natl. Acad. Sci. U. S. A.* **89**, 3453–3457.

Klein, U., Salvador, M.L., and Bogorad, L. (1994). Activity of the *Chlamydomonas* chloroplast *rbcL* gene promoter is enhanced by a remote sequence element. *Proc. Natl. Acad. Sci. U. S. A.* **91**, 10819–10823.

Klinkert, B., Schwarz, C., Pohlmann, S., Pierre, Y., Girard-Bascou, J., and Nickelsen, J. (2005). Relationship between mRNA levels and protein accumulation in a chloroplast promoter-mutant of *Chlamydomonas reinhardtii*. *Mol. Gen. Genomics* **274**, 637–643.

Kühn, K., Bohne, A.V., Liere, K., Weihe, A., and Börner, T. (2007). *Arabidopsis* phage-type RNA polymerases: Accurate *in vitro* transcription of organellar genes. *Plant Cell* **19**, 959–971.

Leu, S., White, D., and Michaels, A. (1990). Cell cycle-dependent transcriptional and post-transcriptional regulation of chloroplast gene expression in *Chlamydomonas reinhardtii*. *Biochim. Biophys. Acta* **1049**, 311–317.

Lisitsky, I., Rott, R., and Schuster, G. (2001). Insertion of polyoxyadenosine-rich sequences into an intergenic region increases transcription in *Chlamydomonas reinhardtii*. *Planta* **212**, 851–857.

Lysenko, E.A. (2007). Plant sigma factors and their role in plastid transcription. *Plant Cell Rep.* **26**, 845–859.

Matsumoto, T., Matsuo, M., and Matsuda, Y. (1991). Structural analysis and expression during dark-light transitions of a gene for cytochrome f in *Chlamydomonas reinhardtii*. *Plant Cell Physiol.* **32**, 863–872.

Maul, J.E., Lilly, J.W., Cui, L., dePamphilis, C.W., Miller, W., Harris, E.H., and Stern, D.B. (2002). The *Chlamydomonas reinhardtii* plastid chromosome: Islands of genes in a sea of repeats. *Plant Cell* **14**, 2659–2679.

McClure, W.R. (1985). Mechanism and control of transcription initiation in prokaryotes. *Annu. Rev. Biochem.* **54**, 171–204.

Monod, C., Goldschmidt-Clermont, M., and Rochaix, J.D. (1992). Accumulation of chloroplast *psbB* RNA requires a nuclear factor in *Chlamydomonas reinhardtii*. *Mol. Gen. Genet.* **231**, 449–459.

Morikawa, K., Shiina, T., Murakami, S., and Toyoshima, Y. (2002). Novel nuclear-encoded proteins interacting with a plastid sigma factor, Sig1, in *Arabidopsis thaliana*. *FEBS Lett.* **514**, 300–304.

Nickelsen, J., van Dillewijn, J., Rahire, M., and Rochaix, J.-D. (1994). Determinants for stability of the chloroplast *psbD* RNA are located within its short leader region in *Chlamydomonas reinhardtii*. *EMBO J.* **13**, 3182–3191.

Nudler, E. and Gottesman, M.E. (2002). Transcription termination and antitermination in *E. coli*. *Genes to Cells* **7**, 755–768.

Ohta, N., Matsuzaki, M., Misumi, O., Miyagishima, S., Nozaki, H., Tanaka, K., Shin-I, T., Kohara, Y., and Kuroiwa, T. (2003). Complete sequence and analysis of the plastid genome of the unicellular red alga *Cyanidioschyzon merolae*. *DNA Res.* **10**, 67–77.

Platt, T. (1986). Transcription termination and the regulation of gene expression. *Annu. Rev. Biochem.* **55**, 339–372.

Robertson, D., Boynton, J.E., and Gillham, N.W. (1990). Cotranscription of the wild-type chloroplast *atpE* gene encoding the CF1/CF0 epsilon subunit with the 3′ half of the *rps7* gene in *Chlamydomonas reinhardtii* and characterization of frameshift mutations in *atpE*. *Mol. Gen. Genet.* **221**, 155–163.

Rochaix, J.-D. (1996). Post-transcriptional regulation of chloroplast gene expression in *Chlamydomonas reinhardtii*. *Plant Mol. Biol.* **32**, 327–341.

Rott, R., Drager, R.G., Stern, D.B., and Schuster, G. (1996). The 3′ untranslated regions of chloroplast genes in *Chlamydomonas reinhardtii* do not serve as efficient transcriptional terminators. *Mol. Gen. Genet.* **252**, 676–683.

Rott, R., Liveanu, V., Drager, R.G., Stern, D.B., and Schuster, G. (1998). The sequence and structure of the 3′-untranslated regions of chloroplast transcripts are important determinants of mRNA accumulation and stability. *Plant Mol. Biol.* **36**, 307–314.

Sakamoto, W., Kindle, K., and Stern, D.B. (1993). *In vivo* analysis of *Chlamydomonas* chloroplast *petD* gene expression using stable transformation of β-glucuronidase translational fusions. *Proc. Natl. Acad. Sci. U. S. A.* **90**, 497–501.

Salvador, M.L. and Klein, U. (1999). The redox state regulates RNA degradation in the chloroplast of *Chlamydomonas reinhardtii*. *Plant Physiol.* **121**, 1367–1374.

Salvador, M.L., Klein, U., and Bogorad, L. (1993). Light-regulated and endogenous fluctuations of chloroplast transcript levels in *Chlamydomonas*. Regulation by transcription and RNA degradation. *Plant J.* **3**, 213–219.

Salvador, M.L., Klein, U., and Bogorad, L. (1998). Endogenous fluctuations of DNA topology in the chloroplast of *Chlamydomonas reinhardtii*. *Mol. Cell. Biol.* **18**, 7235–7242.

Salvador, M.L., Suay, L., Anthonisen, I.L., and Klein, U. (2004). Changes in the 5′ untranslated region of the *rbcL* gene accelerate transcript degradation more than 50-fold in the chloroplast of *Chlamydomonas reinhardtii*. *Curr. Genet.* **45**, 176–182.

Silk, G.W. and Wu, M. (1993). Posttranscriptional accumulation of chloroplast *tufA* (elongation factor gene) mRNA during chloroplast development in *Chlamydomonas reinhardtii*. *Plant Mol. Biol.* **23**, 87–96.

Smith, A.C. and Purton, S. (2002). The transcriptional apparatus of algal plastids. *Eur. J. Phycol.* **37**, 301–311.

Stern, D.B. and Gruissem, W. (1987). Control of plastid gene expression: 3′ inverted repeats act as mRNA processing and stabilizing elements, but do not terminate transcription. *Cell* **51**, 1145–1157.

Stern, D.B. and Kindle, K.L. (1993). 3′ end maturation of the *Chlamydomonas reinhardtii* chloroplast *atpB* mRNA is a two-step process. *Mol. Cell. Biol.* **13**, 2277–2285.

Sturm, N.R., Kuras, R., Büschlen, S., Sakamoto, W., Kindle, K., Stern, D.B., and Wollman, F.-A. (1994). The *petD* gene is transcribed by functionally redundant promoters in *Chlamydomonas reinhardtii* chloroplasts. *Mol. Cell. Biol.* **14**, 6171–6179.

Sugiura, M. (1992). The chloroplast genome. *Plant Mol. Biol.* **19**, 149–168.

Summer, E.J., Schmid, V.H.R., Bruns, B.U., and Schmidt, G.W. (1997). Requirement for the H phosphoprotein in photosystem II of *Chlamydomonas reinhardtii*. *Plant Physiol.* **113**, 1359–1368.

Surzycki, S.J. (1969). Genetic functions of the chloroplast of *Chlamydomonas reinhardi*: Effect of rifampin on chloroplast DNA-dependent RNA polymerase. *Proc. Natl. Acad. Sci. U. S. A.* **63**, 1327–1334.

Surzycki, S.J. and Shellenbarger, D.L. (1976). Purification and characterization of a putative sigma factor from *Chlamydomonas reinhardi*. *Proc. Natl. Acad. Sci. U. S. A.* **73**, 3961–3965.

Swiatek, M., Kuras, R., Sokolenko, A., Higgs, D., Olive, J., Cinque, G., Müller, B., Eichacker, L.A., Stern, D.B., Bassi, R., Herrmann, R.G., and Wollman, F.-A. (2001). The chloroplast gene *ycf9* encodes a photosystem II (PSII) core subunit, PsbZ, that participates in PSII supramolecular architecture. *Plant Cell* **13**, 1347–1367.

Takahashi, Y., Goldschmidt-Clermont, M., Soen, S.-Y., Franzén, L.G., and Rochaix, J.-D. (1991). Directed chloroplast transformation in *Chlamydomonas reinhardtii*: Insertional inactivation of the *psaC* gene encoding the iron sulfur protein destabilizes photosystem I. *EMBO J.* **10**, 2033–2040.

Thompson, R.J. and Mosig, G. (1985). An ATP-dependent supercoiling topoisomerase of *Chlamydomonas reinhardtii* affects accumulation of specific chloroplast transcripts. *Nucleic Acids Res.* **13**, 873–891.

Thompson, R.J. and Mosig, G. (1987). Stimulation of a *Chlamydomonas* chloroplast promoter by novobiocin in situ and in *E. coli* implies regulation by torsional stress in the chloroplast DNA. *Cell* **48**, 281–287.

Thompson, R.J. and Mosig, G. (1990). Light affects the structure of *Chlamydomonas* chloroplast chromosomes. *Nucleic Acids Res.* **18**, 2625–2631.

Troxler, R.F., Zhang, F., Hu, J., and Bogorad, L. (1994). Evidence that σ factors are components of chloroplast RNA polymerase. *Plant Physiol.* **104**, 753–759.

Vaistij, F.E., Goldschmidt-Clermont, M., Wostrikoff, K., and Rochaix, J.-D. (2000). Stability determinants in the chloroplast *psbB/T/H* mRNAs of *Chlamydomonas reinhardtii*. *Plant J.* **21**, 469–482.

CHAPTER 26

Chloroplast RNA Splicing

Michel Goldschmidt-Clermont
Departments of Molecular Biology and of Plant Biology,
University of Geneva, Geneva, Switzerland

CHAPTER CONTENTS

I. INTRODUCTION

The introns in the *Chlamydomonas* chloroplast genome belong to group I and group II. Introns of these two groups are also found in other organellar genomes, and in particular in plastids and mitochondria from higher plants. Many group I introns, and some group II introns from bacteria and from fungal mitochondria, are ribozymes that have the remarkable ability to catalyze their own excision from precursor transcripts *in vitro*. Some introns in both groups are mobile genetic elements that can insert into intron-less alleles or into novel sites. However the introns in the two groups differ in their architecture, their mechanisms of splicing and their modes of transposition.

Some introns in group I encode a polypeptide which can act as a maturase to facilitate intron splicing and as an endonuclease to promote intron movement. Likewise some members of group II encode a protein which promotes splicing, and also has endonuclease and reverse transcriptase domains required for intron insertion. However many introns in both groups do not contain such an open reading frame (ORF), or encode a protein that has retained only some of these functions. The structures and functions of group I and group II introns have been described in comprehensive reviews (Bonen and Vogel, 2001; Herrin and Nickelsen, 2004; Lambowitz and Zimmerly, 2004; Haugen et al., 2005; Robart and Zimmerly, 2005; Pyle and Lambowitz, 2006; Houghland et al., 2006). This chapter will focus on aspects of group I and group II splicing that are particular to *C. reinhardtii* and other *Chlamydomonas* spp. (Herrin et al., 1998).

II. GROUP I INTRONS

A. Distribution in *Chlamydomonas* species

In the chloroplast genome of *C. reinhardtii*, there are five group I introns. The 23S ribosomal RNA gene (*rrnL*) is interrupted by an intron named *Cr.LSU* (Rochaix and Malnoë, 1978; Kuo et al., 2006). The *psbA* gene also contains four group I introns, named *Cr.psbA1–Cr.psbA4* (Erickson et al., 1984; Holloway et al., 1999). This is in contrast to vascular plants, where the orthologous *rrnL* and *psbA* genes do not contain introns. In the *C. reinhardtii* chloroplast genome, there are two copies of the *psbA* gene which are located in the inverted repeats, like the *rrnL* genes. The number and position of group I introns differ widely in the chloroplast genomes of other species within the genus *Chlamydomonas* (Turmel et al., 1993a,b; Spiegel et al., 2006) and the same applies to group I introns in the mitochondrial genomes (Nedelcu and Lee, 1998). This suggests that the group I introns have extensively migrated during evolution. The web-based database, FUGOID (Functional Genomics of Organellar Introns Database) includes information on the introns of *Chlamydomonas* spp. (Li and Herrin, 2002; http://web.austin.utexas.edu/fugoid/introndata/main.htm).

B. Group I intron mobility

Many group I introns are mobile: the intron from a gene can insert into an intron-free allele, a process which is known as intron homing. In crosses between the interfertile species *C. moewusii* and *C. eugametos*, which differ in the distribution of group I introns in the chloroplast genome, intron homing and associated gene conversion are observed (Bussieres et al., 1996; Lemieux and Lee, 1987; Lemieux et al., 1988). Efficient homing of a mitochondrial intron is also observed in crosses between *C. smithii* and *C. reinhardtii* (Boynton et al., 1987; Kurokawa et al., 2006; see Chapter 12). The *Cr.LSU* intron is capable of efficiently inserting in an intronless fragment of *rrnL* that has been introduced at an ectopic site in the chloroplast genome by transformation (Dürrenberger and Rochaix, 1991). The polypeptide encoded by the intron is a double-strand DNA endonuclease, I-CreI, which makes a staggered cut in the target sequence near the intron insertion site (Dürrenberger and Rochaix, 1991, 1993; Thompson et al., 1992). The double-stranded break triggers recombination and intron homing *in vivo* (Dürrenberger et al., 1996). The structure and mechanism of this enzyme have been studied in detail (Chevalier et al., 2004). At least two of the introns in *psbA*, *Cr.psbA2* and *Cr.psbA4*, are also mobile and can insert into an intronless copy of the gene (Odom et al., 2001). The DNA endonuclease I-CreII is encoded in *Cr.psbA4*, where it is expressed from its own promoter. The purified enzyme generates staggered double-stranded breaks 4 nt from the intron insertion site (Kim et al., 2005). The endonucleases encoded by several other group I introns from *Chlamydomonas* spp. have also been extensively characterized (Herrin et al., 1998; Spiegel et al., 2006 and references therein, see also the FUGOID database).

These endonucleases can promote efficient intron homing at the DNA level by inducing recombination and gene conversion. However, during evolution an alternative mechanism of intron migration may also operate at the RNA level, involving a reversal of splicing followed by reverse transcription and integration (Haugen et al., 2005). In support of this hypothesis, introns in the *rrnL* genes of *Chlamydomonas* spp. are often inserted at sites that correspond to accessible parts of the 23S rRNA in the ribosome particle, and may thus have been preferred targets for reverse splicing (Turmel et al., 1993a). The *Cr.LSU* intron can react *in vitro* with the cytoplasmic 5.8S rRNA, but only the product of the first step of reverse splicing is observed (Thompson and Herrin, 1994). Thus the full integration of an intron by the RNA-mediated pathway may be a rare event, but nevertheless significant on an evolutionarily timescale.

C. Group I intron splicing

In vivo, the four introns of *psbA* can be spliced in any order, since all combinations of spliced and unspliced exons are found in the precursors that accumulate when cells are grown in the dark (Deshpande et al., 1997). *In vitro*, all five of the *C. reinhardtii* chloroplast group I introns are autocatalytic (Dürrenberger and Rochaix, 1991; Herrin et al., 1990, 1991). Most group I introns that were investigated in other *Chlamydomonas* spp. were also found to be autocatalytic (Cote and Turmel, 1995). The general mechanism of group I splicing is depicted in Figure 26.1A. Efficient self-splicing requires a guanosine co-factor and magnesium ions (Herrin et al., 1998). A detailed analysis of *Cr.LSU* splicing *in vitro* has revealed its unusual metal specificity and structural characteristics (Kuo et al., 2006).

Because chloroplast transformation is relatively easy in *C. reinhardtii*, the same intron mutations can be analyzed for their consequences both *in vitro* and *in vivo*. The effects of many site-directed mutations in *Cr.LSU* that affect self-splicing *in vitro* can be partially compensated by elevating the Mg^{2+} concentration (Li et al., 2002). When the same mutations are introduced *in vivo* by biolistic transformation, most of the strains remain heteroplasmic for mutant and wild-typc copies of the intron, suggesting that the homoplasmic mutations would be lethal. Site-directed mutations in *Cr.psbA4* have less pronounced effects on self-splicing *in vitro* than the corresponding mutations in *Cr.LSU* (Lee and Herrin, 2003). This may be related to the observation that the structure of *Cr.LSU* is comparatively less stable (Kuo et al., 2006). The effects of *Cr.psbA4* mutations on splicing *in vivo* parallel those *in vitro*, but are in general less severe. As discussed in the next section, these differences may be due to the intervention of protein factors that assist splicing *in vivo*.

D. Group I splicing factors

Under physiological conditions self-splicing *in vitro* of some group I introns is very inefficient, and splicing *in vivo* is thought to require the assistance of protein factors. In *C. reinhardtii*, four of the introns (*Cr.psbA2–Cr.psbA4* and *Cr.LSU*) contain large ORFs (Rochaix et al., 1985; Holloway et al., 1999). In *Cr.psbA3*, the ORF is in frame with the preceding exon, and could thus be expressed as a fusion protein. Deletion of the ORF in *Cr.LSU* does not prevent splicing of this intron *in vivo*, and likewise the ORFs in the *psbA* introns were reported to be dispensable for splicing (Lee and Herrin, 2003; Herrin and Nickelsen, 2004;), indicating that none of them is essential for splicing of its host intron. Whether a putative maturase from one intron could facilitate splicing of another intron in a redundant manner may not have been ruled out in these experiments. However as discussed above, in *Chlamydomonas* spp., many of the intron-encoded polypeptides have activities as homing endonucleases.

Considering the numerous nucleus-encoded factors that are required for chloroplast RNA processing (Barkan and Goldschmidt-Clermont, 2000), and in particular for splicing of group II introns (see below), it might be surprising that no mutants specifically defective in group I intron splicing have been described. A possible exception might be the acetate-requiring mutant *ac20*, which is partially deficient in chloroplast ribosomes and

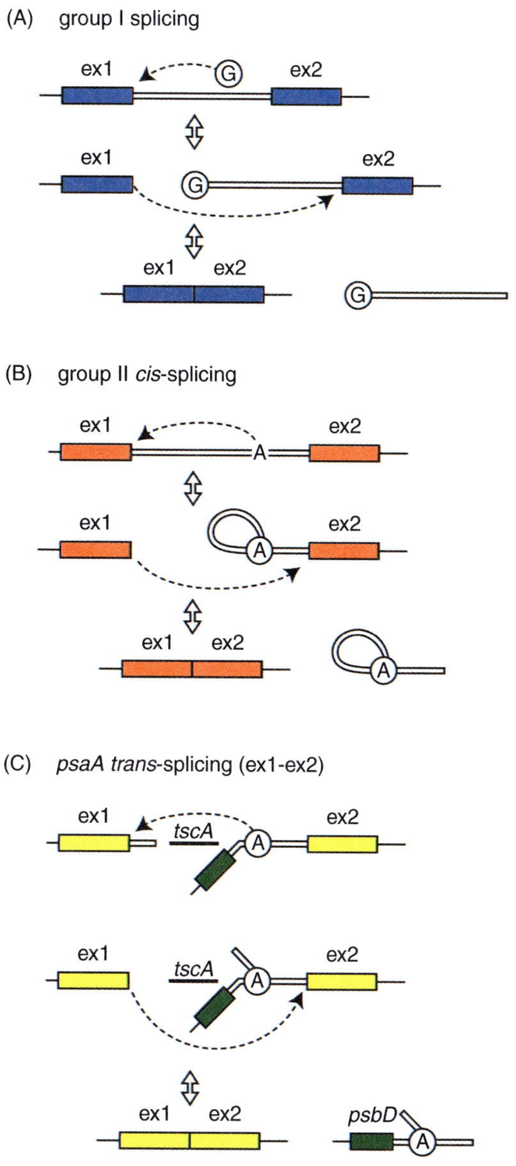

FIGURE 26.1 *Mechanisms of intron splicing. (A) Group I. Splicing is initiated when a guanosine co-factor bound to the intron makes a nucleophilic attack on the 5′ splice site and is attached to the 5′ end of the intron-exon 2 intermediate. In the second step, the 3′ hydroxyl of exon 1 in turn attacks the 3′ intron junction, forming the ligated exons and releasing the linear intron. (B) Group II cis-splicing. In the first step of splicing, the 2′ hydroxyl of an A residue at the branch site attacks the 5′ splice site, resulting in cleavage and release of an intron lariat-exon 2 intermediate. In the second step, a second trans-esterification leads to the junction of the two exons and the release of the intron lariat. In an alternative pathway, the first step proceeds by hydrolysis rather than trans-esterification and the intron remains linear (not shown). (C) psaA trans-splicing (exon 1–exon 2). In C. reinhardtii the first intron of psaA is split into three parts: the precursor of intron 1, the tscA RNA and the precursor of exon 2, which also contains the coding sequence of psbD. Because the intron is split, the two trans-esterification reactions give rise to a Y-shaped branched intron rather than a lariat (compare to B).*

accumulates precursors of the 23S rRNA that still retain the *Cr.LSU* intron (Herrin et al., 1990). However the mutant phenotype is still observed when this intron is deleted from *ac20*, indicating that the primary defect may be in processing of an internal transcribed spacer (ITS1) (Holloway and Herrin, 1998). A simple reason that splicing-defective mutants were not obtained might be that the splicing factors could have redundant functions so that inactivation of a single gene would not significantly affect splicing. However another possibility stems from the observation that site-directed mutants of the *Cr.LSU* intron remain persistently heteroplasmic when splicing is severely affected (Li et al., 2002). This indicates that the homoplasmic mutation would be lethal, most probably because chloroplast translation is essential for cell survival in *Chlamydomonas*. Thus mutants with defects in protein factors required for splicing of *Cr.LSU* would not be recovered because of their lethality. However the argument does not apply to the *psbA* introns, since this gene is involved in photosynthesis and is dispensable for growth on acetate-containing medium (Lee and Herrin, 2003). One possibility is that the putative splicing factors for the *psbA* introns also have a role in *Cr.LSU* splicing. Another likely hypothesis is that the splicing factors are also required for other essential aspects of chloroplast function. The classical examples of such factors with dual functions are the tRNA synthetases that also play a role in splicing of group I introns in fungal mitochondria (Herbert et al., 1988; Lambowitz and Perlman, 1990; Dujardin and Herbert, 1997; Paukstelis et al., 2005).

To circumvent these questions, an alternative genetic approach was used to identify factors involved in splicing of chloroplast group I introns (Li et al., 2002). Using chloroplast transformation, a mutant homoplasmic for a nucleotide substitution in the *Cr.LSU* intron ($P4^{125}A$) was obtained which showed light sensitivity and slow growth. Selection for restored growth in high light allowed the isolation of suppressor mutants in two distinct genes, *css1* and *css2* (chloroplast-splicing suppressor). Significantly, the mutant alleles also suppress a two-nucleotide mutation in loop P4 of the *Cr.psbA4* intron. The *css1* suppressor is dominant, and does not have an apparent phenotype in the presence of the wild-type intron under normal growth conditions.

E. Light-regulated *psbA* splicing

The expression of the D1 subunit of photosystem II, encoded by *psbA*, strongly increases in the light. This is in part due to regulation of translation in response to changes in redox status and in ATP/ADP levels (Bruick and Mayfield, 1999; Trebitsh and Danon, 2001). Splicing of the group I introns in *psbA* is also regulated by light and offers an additional tier of regulation (Deshpande et al., 1997). In the dark, unspliced precursors accumulate, but their levels decrease rapidly when cells are exposed to the light and there is a transient increase in the amount of excised introns. The stimulation of splicing in the light is prevented by inhibitors of photosynthetic electron transport or in a PS I-deficient mutant. The impact of this regulation on the total pool of mature mRNA is limited because the amounts of *psbA* mRNA are already high in the dark. If it is to play a role in the regulation of D1 expression, splicing of *psbA* must under certain circumstances be a limiting step. There is evidence that chloroplast mRNA levels are not limiting for translation and that D1 expression is mostly regulated at the level of translation (Eberhard et al., 2002). However it was reported that for a site-directed mutant of *Cr.psbA4* (P4′-3,4), splicing *in vivo* is impaired and there is a 45% reduction in *psbA* mRNA levels, a 25% decrease in the amounts of D1 polypeptide, and an 18% increase in the doubling time (Lee and Herrin, 2003). The regulation of *psbA* splicing by light could be physiologically important under adverse stress conditions such as excess light. Light-regulated splicing might also be significant if only a separate sub-pool of *psbA* mRNA is active in translation.

III. GROUP II INTRON SPLICING

A. *Trans*-splicing of *psaA*

When Kück and coworkers mapped and sequenced the *psaA* gene of *Chlamydomonas*, they made the surprising discovery that the gene is split into three parts, scattered around the chloroplast genome (Figure 26.2; Kück et al., 1987). The sequences flanking the coding regions have some of the conserved features of group II introns, a first indication that the *psaA* mRNA is assembled by *trans*-splicing of two split introns. This arrangement may be prevalent in most *Chlamydomonas* spp., since the *psaA* gene is also split into three separate exons in *C. eugametos*, which is relatively distantly related to *C. reinhardtii* (Turmel et al., 1995). In plants, the *psaA* gene is not interrupted, but the *rps12* gene contains a split group II intron, while *rps12* is continuous in *C. reinhardtii* (Bonen and Vogel, 2001). Several genes in plant mitochondria also contain *trans*-spliced group II introns.

A large fraction of PS I-deficient mutants lack mature *psaA* mRNA, and accumulate precursors and intermediates of the *trans*-splicing pathway. The analysis of these mutants showed that the three exons of *psaA* are transcribed separately (Figure 26.2). Exon 2 is clearly part of a polycistronic transcription unit, since in some *trans*-splicing mutants large precursor transcripts are observed (Choquet et al., 1988; Herrin and Schmidt, 1988) which contain *psbD*, exon 2, and further sequences downstream. *Trans*-splicing must be rapid and efficient because in the *nac2-26* mutant where *psbD* is unstable, exon 2 is still *trans*-spliced and normal levels of *psaA* mRNA accumulate (Kuchka et al., 1989). This speaks against the hypothesis that co-transcription of the mRNAs for subunits of PS II (*psbD*) and of PS I

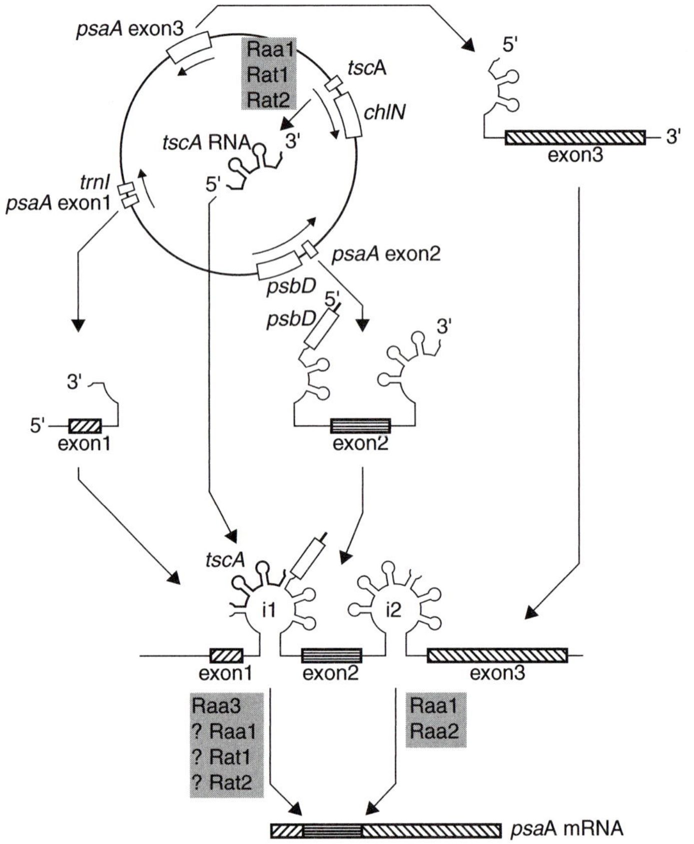

FIGURE 26.2 *psaA trans-splicing. The three exons of psaA are encoded by separate loci in the C. reinhardtii chloroplast genome. Exon 1 is probably co-transcribed with trnI and exon 2 with psbD. The non-coding tscA RNA is transcribed from a separate locus together with chlN. The transcripts assemble to form two split group II introns which are spliced. Processing of tscA requires the nucleus-encoded factors RAA1, RAT1, and RAT2. Trans-splicing of the second intron requires RAA1 and RAA2. Trans-splicing of the first intron involves RAA3, and possibly also RAA1, RAT1, and RAT2, but the requirement for the latter may be indirect, since they participate in tscA maturation. Other nuclear mutations define factors (not shown) required for splicing of the first intron (class C), the second intron (class A) or both (class B). Not drawn to scale (modified from Herrin et al., 1998, with kind permission of Springer Science and Business Media).*

(*psaA*) could be of significance for coordinate regulation. However in double mutants (*trans*-splicing and *nac2*) the entire precursor is degraded. Exon 1 is probably co-transcribed with the downstream *trnI* gene since the 3′ end of the exon 1 precursor maps to the 5′ end of *trnI* (Turmel et al., 1995). Processing at this site must be rapid because the dicistronic precursor is not observed, even in *trans*-splicing mutants.

The precursor transcripts are thought to assemble and form the typical structure of group II introns, with a central core and six helical domains.

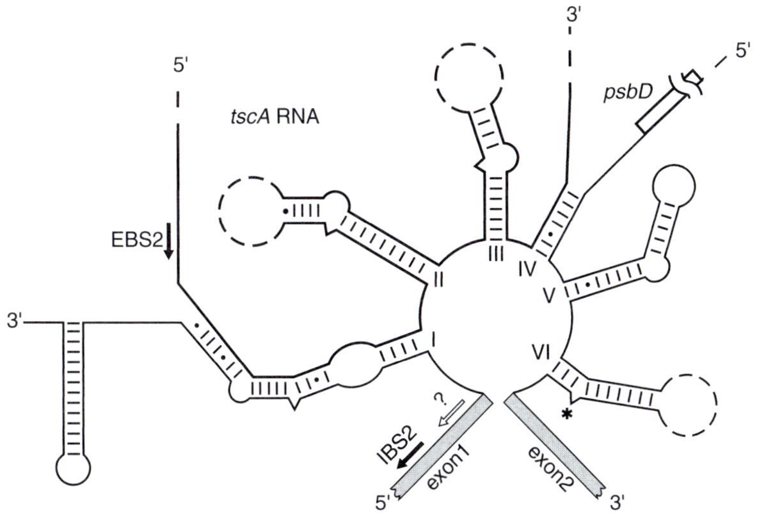

FIGURE 26.3 *Three-part structural model of psaA intron 1. The contribution of three parts of psaA intron 1 to the six domains (roman numerals I–VI) that form the secondary structure of the intron are depicted (precursor of exon 1, thin line; tscA RNA, thick line; precursor of exon 2, thin line). The coding sequence of psbD is in the peripheral part of domain IV, and the branch point in domain VI is indicated with an asterisk. The tertiary interaction between IBS2 and EBS2 is shown by black arrows, but EBS1 which would interact with IBS1 (?, open arrow) has not been identified. The peripheral domains are indicated by dotted lines. Not drawn to scale (modified from Herrin et al., 1998, with kind permission of Springer Science and Business Media).*

However to complete the first intron, a small non-coding transcript is also required, *tscA* (Goldschmidt-Clermont et al., 1991). The *tscA* RNA is co-transcribed from a separate locus together with the *chlN* gene (Figure 26.2) and is indispensable for *trans*-splicing of intron 1 (Roitgrund and Mets, 1990; Hahn et al., 1998). The tripartite intron 1 is formed by the precursor of exon 1, the *tscA* RNA and the precursor of exon 2 (Figure 26.3). This arrangement is interesting from an evolutionary perspective, because the group II introns are thought to be ancestral to the spliceosomal introns in the nucleus. The *tscA* RNA may thus prefigure the snRNAs that are involved *in trans* for splicing of introns in the nucleus (Goldschmidt-Clermont et al., 1991; Sharp, 1991). The proposed structure has some of the conserved features of introns in sub-group II B1, but domain I is atypical and lacks recognizable EBS1 (exon binding sequence) (Figure 26.3). This has led to the suggestion that yet another RNA component could contribute these missing elements (Turmel et al., 1995), but there is to date no experimental evidence to support this hypothesis. Because the intron is split, splicing generates a branched Y-shaped molecule instead of the conventional lariat (Figure 26.1C). This branch can be detected on the 3′ UTR of at least part of the *psbD* RNAs, as expected because *psbD* and exon 2 are co-transcribed so that the *psbD* coding sequence is in the peripheral part of domain IV (Goldschmidt-Clermont et al., 1991).

The predicted structure of the bipartite intron 2 also conforms with the consensus and is composed of the precursors of exons 2 and 3. The conserved A residue in the bulge of the domain VI helix, which is usually involved in the first *trans*-esterification reaction of splicing, is not conserved in intron 2 of *psaA* which has an unpaired bulge with a C and a U in this position. Whether branching occurs on one of these nucleotides, or whether splicing proceeds by hydrolysis rather than by *trans*-esterification has not been resolved (Podar et al., 1998; Vogel and Börner, 2002).

Some group II introns have the ability to self-splice *in vitro*. However, attempts to detect autocatalytic splicing *in vitro* of the introns of *psaA* have not been successful. In contrast, the group II intron in the *psbA* gene from the chloroplast of a psychrophilic *Chlamydomonas* sp. from Antarctica (CCMP-1619) is capable of self-splicing *in vitro*. This group II B2 intron encodes a maturase, which is essential for splicing when the intron is expressed in *E. coli*. Its similarity to an intron inserted in the same position in the *psbA* gene of *Euglena myxocylindracea* suggests that it was acquired by horizontal transfer (Odom et al., 2004).

B. Genetics of *psaA trans*-splicing factors

Video imaging allows screening for mutants that have high fluorescence yield and the fluorescence transients characteristic of defects in PS I or the b_6f complex. Among these, a majority have defects in *psaA trans*-splicing. They fail to accumulate mature *psaA* mRNA, and can be grouped into three phenotypic classes depending on the pattern of *psaA* transcripts that can be observed (Choquet et al., 1988; Herrin and Schmidt, 1988). Mutants in class C are defective in *trans*-splicing of intron 1 specifically; they accumulate the precursor of exon 1 and an intermediate with exon 2 accurately *trans*-spliced to exon 3. Conversely mutants in class A are defective in splicing of intron 2, and exhibit increased levels of the precursor of exon 3, an intermediate with exon 1 accurately spliced to exon 2, and elevated amounts of the precursor of exon 1. Finally, mutants in class B fail to *trans*-splice both introns, and accumulate the three precursors. Complementation analysis between the mutants in each class revealed five nuclear loci for class A, two nuclear loci for class B, and seven nuclear loci for class C, indicating that at least 14 nuclear loci are involved in *psaA trans*-splicing (Goldschmidt-Clermont et al., 1990). In class C there is also one chloroplast locus, which harbors the gene for the small *tscA* RNA which is required for splicing of intron 1 (Roitgrund and Mets, 1990; Goldschmidt-Clermont et al., 1991). Some nuclear genes (*RAA1*, *RAT1*, and *RAT2*) are required for processing of the *tscA* RNA from a polycistronic precursor that includes *chlN* (Figure 26.2; Hahn et al., 1998; Balczun et al., 2005; Merendino et al., 2006). One possibility is that proper processing of *tscA* is a requisite for its function, and hence that the role of these factors in splicing of intron 1 is indirect. However it cannot be ruled out that their requirement for splicing of intron 1 might be direct, and that they would have a second function in *tscA* maturation.

Five nuclear genes involved in *psaA trans*-splicing have been cloned to date: *RAA1*, *RAA2*, *RAA3*, *RAT1*, and *RAT2* (Table 26.1; Perron et al., 1999; Rivier et al., 2001; Balczun et al., 2005; Merendino et al., 2006). *RAA1* encodes a large polypeptide with a set of five repeats of a degenerate 38 amino acid sequence, which is related to the repeats in the translation factor TBC2 (Auchincloss et al., 2002). Although they have a different sequence, these repeats are reminiscent of the PPR (pentatricopeptide) repeats which are found in a large family of plant proteins involved in RNA metabolism, and notably in PPR4 which is required for *trans*-splicing of *rps12* mRNA in maize chloroplasts (Lurin et al., 2004; Schmitz-Linneweber et al., 2006). *RAA1* is required for both *trans*-splicing steps (class B) although one allele, which still expresses part of the gene (*raa1-314B*), is only defective in *trans*-splicing of the second intron (Merendino et al., 2006). Reverse genetics has shown that two partly overlapping domains of RAA1 are each involved in *trans*-splicing of one or the other of the *psaA* introns.

RAA2 is related to pseudouridine synthases, a family of enzymes that modify specific uridines in tRNA, rRNA, and snRNA (Perron et al., 1999).

Table 26.1 Cloned nuclear genes encoding *psaA* trans-splicing factors

	Mutant alleles			Predicted polypeptide			
Gene	**Name**	**Class**	**Functions**	**Size**	**Protein domains**	**References**	**Accession #**
RAA1	*L137H* *HN31* *314B*	B B A	ex1–ex2 ex2–ex3 *tscA* maturation	2103 aa	38 aa repeats similar to TBC2	Hahn et al. (1998); Balczun et al. (2005); Merendino et al. (2006)	CAE53330
RAA2 (formerly named MAA2)	*A18* *L136F* *HN22*	A A A	ex2–ex3	410 aa	Pseudouridine synthase	Perron et al. (1999, 2004)	CAB62387
RAA3	*M18*	C	ex1–ex2	1784 aa	Pyridoxamine 5′ oxidase	Rivier et al. (2001)	AAG40000
RAT1	*TR72* *L135F*	C C	*tscA* maturation	245 aa	Poly(ADP-ribose) polymerase (NAD+ binding domain)	Balczun et al. (2005)	ABK15559
RAT2	*TR72* *L135F*	C C	*tscA* maturation		None	Balczun et al. (2005)	EDP02426

However mutagenesis of residues that are conserved and essential for catalysis in *E. coli* do not affect the competence of RAA2 for *trans*-splicing. Thus the activity of RAA2 as a pseudouridine synthase is not required for its function in *trans*-splicing. During evolution, the protein may have been recruited as a *trans*-splicing factor because of its ability to interact with RNA. Similar evolutionary recruitment of splicing factors may also have prevailed for chloroplast group II introns in higher plants: CRS2 is related to peptidyl-tRNA hydrolases, and CRS1, CAF1, and CAF2 contain multiple copies of a conserved domain (CRM) which is also found in bacterial and archaeal proteins involved in ribosome maturation (Jenkins et al., 1997; Asakura and Barkan, 2006; Barkan et al., 2007). For splicing of group I introns, tRNA synthetases may also have been co-opted during evolution (see section I). RAA3 is a large polypeptide with only a restricted region of sequence similarity to pyridoxamine 5′-phosphate oxidases, but the functional significance of this observation is unclear (Rivier et al., 2001).

RAT1 and *RAT2* form a divergent gene cluster and both are deleted in the mutants *TR72* and *L135F* (Balczun et al., 2005). Complementation by transformation showed that both of these genes are required for *tscA* processing. RAT1 displays sequence similarity to the NAD$^+$-binding domain of poly(ADP-ribose) polymerases, but mutations in this domain do not alter the efficiency of *RAT1* in the complementation assay. Again this suggests that RAT1 may have been recruited during evolution to assist in processing of *tscA* and, directly or indirectly, in *psaA trans*-splicing. RAT2 does not share sequence similarity with known proteins and may be expressed only at very low levels, because the *RAT2* transcript could not be detected in wild-type cells.

Cell fractionation studies have shown that RAA1, RAA2, RAA3, and RAT1 are targeted to the chloroplast. RAA3 is in the stroma, where it forms a large complex which also contains the precursor of exon 1 and the *tscA* RNA. RAA1 and RAA2 are associated in a large complex which is bound to membranes, but can be released with moderate concentrations of salt. RAT1 is also found in a membrane fraction, and in the yeast three-hybrid assay, it shows binding to *tscA RNA*. The picture that emerges from these biochemical studies is that the factors required for *trans*-splicing and for *tscA* maturation are part of large ribonucleoprotein assemblies that may be, for group II splicing in the chloroplast, the counterparts of the spliceosome in the nucleus. The association of some of these complexes with membranes in the chloroplast suggests that splicing may be compartmentalized, perhaps to the light membrane fraction that also contains factors involved in translation (Zerges and Rochaix, 1998; Perron et al., 2004).

To date, the factors shown to be required for group II splicing in *Chlamydomonas* are different from their counterparts in maize and *Arabidopsis* plastids (Asakura and Barkan, 2006; Schmitz-Linneweber et al., 2006). CRS1 is specific to splicing of the group II intron in *atpF* and like

CAF1 and CAF2, it contains CRM domains (Till et al., 2001; Ostheimer et al., 2003). CRS2, which is related to tRNA hydrolases, binds to either CAF1 or CAF2 in complexes that are specific to subsets of group II introns (Jenkins and Barkan, 2001). However some parallels can be drawn with *Chlamydomonas*, since these proteins also engage in large complexes that include their substrate RNAs, and show specificity for subsets of plastid introns. In the plastid genomes of land plants the group II intron in the *trnK* gene contains an ORF called *matK* whose product has sequence similarity to group II intron-encoded polypeptides in the maturase domain, but has a degenerate reverse transcriptase domain and lacks the endonuclease domain. This protein is probably involved in splicing of several plastid group II introns (reviewed in Hausner et al., 2006). However the *Chlamydomonas* chloroplast genome does not contain a *matK* gene.

C. Intron-binding proteins

While the factors described above were revealed using a genetic approach, other candidates have been identified from their intron-binding properties. Using the yeast three-hybrid system with *tscA* RNA as a bait, a chloroplast protein related to the nuclear nucleosome assembly protein (NAP) was identified, called cNAPL (Glanz et al., 2006). Recombinant cNAPL binds to the *tscA* RNA *in vitro*, and with lower affinity to other U-rich RNAs.

Splicing has also been studied in *Chlamydomonas* by inserting heterologous group II introns into the chloroplast genome by transformation. A group II intron from the spinach *atpF* gene was not spliced in the transgenic context (Deshpande et al., 1995). In contrast, the intron *rI1* from *Scenedesmus obliquus* mitochondria was spliced accurately in the *Chlamydomonas* chloroplast (Herdenberger et al., 1994; Holländer and Kück, 1999). It should be noted that the *rI1* intron can self-splice *in vitro* and belongs to group IIB, like the two *psaA* split introns, while the *atpF* intron belongs to group IIA and does not self-splice *in vitro*. The majority of the genetically-identified *psaA* splicing factors are very specific, since most are required for *trans*-splicing of only one of the two introns (classes A and C), and thus would not be expected to recognize the heterologous intron. The two factors that are required for both introns (class B) could be postulated to have a more general activity in splicing. However RAA1 actually has two domains, each specialized for splicing of one of the introns (see above). The other factor in class B (L118B) has not been characterized yet, but it might be a general factor that can also facilitate splicing of the heterologous *rI1* intron. Another approach to identify putative splicing factors is to find proteins that bind to the precursor RNAs *in vitro* (Bunse et al., 2001). In this way, two proteins were identified which bind to domain IV of *rI1*. One of these proteins was identified as CPN60 which is better known as a protein chaperone, but was also reported to participate in RNA metabolism (Balczun et al., 2006). Recombinant CPN60 also binds *in vitro* to *psaA* intron 1, but in this case to domains V–VI.

As discussed above, splicing of the *psbA* introns is regulated by light. Such an effect has not been reported for *psaA trans*-splicing. When cells are treated at 40°C for 80 minutes, the appearance of small amounts of splicing precursors and intermediates indicates that *psaA trans*-splicing is impaired (Choquet et al., 1988). Prolonged treatment causes the disappearance of these transcripts and of the mRNA. If CPN60, which was shown to bind to the first split intron (see above) is required for *trans*-splicing, it might be possible that during temperature treatment this chaperone is sequestered by unfolded proteins and becomes less available for splicing.

In summary, the genetic approach has led to the identification of proteins that are required for group II splicing and are part of large ribonucleoprotein complexes, and are thus directly or indirectly involved in *trans*-splicing of *psaA*. On the other hand proteins that bind to *psaA* intron RNA have been identified, such as CPN60 and cNAPL, but their functions in splicing have not yet been demonstrated.

IV. ARE SPLICING AND ITS REGULATION ESSENTIAL?

The organellar introns can be regarded as genomic parasites that are mobile and can be transmitted horizontally (Haugen et al., 2005; Robart and Zimmerly, 2005). Intron splicing relieves any major negative impact of the insertion on the expression of its host gene, but intron transcription and the involvement of many splicing factors supplied by the host must have some metabolic cost. However in some cases the introns also have been co-opted to participate in gene regulation, as illustrated by the light regulation of the *psbA* introns, and may thus be beneficial to the host (Deshpande et al., 1997). This has gone further on an evolutionary scale, since group II introns are thought to be ancestral to the nuclear introns which play important roles in regulation of gene expression and in allowing protein diversification through alternative splicing. Introns also contribute to genetic diversity as transposable elements and as a source of exon shuffling. Thus the question arises whether the group I and group II introns in the *Chlamydomonas* chloroplast genome have any important physiological function.

Saccharomyces cerevisiae and *Schizosaccharomyces pombe* yeast strains where all introns have been eliminated from the mitochondrial genome retain respiratory competence, showing that mitochondrial splicing is dispensable, although the *S. pombe* derivative grows more slowly than the wild type (Seraphin et al., 1987; Schafer et al., 1991). Chloroplast transformation allows the construction of *Chlamydomonas* strains in which some of the chloroplast introns have been removed. Deletion of the *Cr.Lsu* intron from the two copies of the *rrnL* gene has no apparent effect on cell growth or morphology (Holloway and Herrin, 1998). *Chlamydomonas* strains with an intron-free version of the two *psbA* genes can grow photosynthetically (Johanningmeier and Heiss, 1993; Minagawa and Crofts, 1994). This shows that splicing of *Cr.Lsu* or *Cr.psbA1–Cr.psbA4* can be bypassed and that splicing in itself is not essential for cell viability under the conditions tested.

Likewise, when an intron-free version of the *psaA* gene is introduced in the *Chlamydomonas* chloroplast genome, the cells can grow photosynthetically even in the presence of nuclear mutations that prevent *trans*-splicing (Rivier, 2000). This indicates that *psaA* splicing can also be bypassed under normal laboratory growth conditions, and one might ask how the intriguing current situation evolved, with four separate chloroplast transcripts and at least 14 nuclear genes involved in producing mature *psaA* mRNA. The recruitment of splicing factors may have been facilitated if they were derived from proteins that were involved in other aspects of RNA metabolism (see section III.B). The mobility of introns may have contributed to genome instability, and in particular to the evolutionary appearance of the disrupted introns that are spliced in *trans*. A more extreme example may be the highly fragmented ribosomal RNA genes in the mitochondrial genome of *Chlamydomonas*, which are not spliced but contain a spacer that has retained sequence similarity to group II introns and their reverse transcriptases (Nedelcu and Lee, 1998). This suggests that an ancestral group II intron may have been implicated in the evolution of the mitochondrial genome and its peculiar rRNA gene organization.

V. CONCLUSIONS

Selfish genomic parasites or regulators of gene expression? As described in this chapter, the chloroplast group I introns of *Chlamydomonas* have features of both. Several of the group I introns are capable of homing into intron-less alleles, and such invasions will be nearly silent genetically because RNA splicing is efficient, properties that fit parasites well. On the other hand, the effect of light on *psbA* pre-mRNA levels indicates that splicing is a regulated step of gene expression. Unlike the group I introns, the split group II introns of *psaA* are unlikely to have retained any mobility. The removal of these introns is extensively dependent on nucleus-encoded factors, an indication that *trans*-splicing is intimately integrated with chloroplast gene expression and is under tight nuclear control. Under the fairly benign conditions that have been tested in the laboratory, chloroplast intron splicing can be bypassed and may not play an essential role in the physiology of *Chlamydomonas*. Nevertheless chloroplast RNA splicing may contribute to the fitness of the alga in more challenging environmental conditions or on an evolutionary timescale.

ACKNOWLEDGMENTS

Our work on chloroplast splicing in *Chlamydomonas* is supported by the Swiss National Fund for Scientific Research (3100-0667763.02). I thank J.-D. Rochaix, D. Herrin, and U. Kück for thoughtful comments on the manuscript, and Nicolas Roggli for preparing the figures.

REFERENCES

Asacra, Y. and Barkan, A. (2006). *Arabidopsis* orthologs of maize chloroplast splicing factors promote splicing of orthologous and species-specific group II introns. *Plant Physiol.* **142**, 1656–1663.

Auchincloss, A.H., Zerges, W., Perron, K., Girard-Bascou, J., and Rochaix, J.D. (2002). Characterization of Tbc2, a nucleus-encoded factor specifically required for translation of the chloroplast *psbC* mRNA in *Chlamydomonas reinhardtii*. *J. Cell Biol.* **157**, 953–962.

Balczun, C., Bunse, A., Hahn, D., Bennoun, P., Nickelsen, J., and Kück, U. (2005). Two adjacent nuclear genes are required for functional complementation of a chloroplast trans-splicing mutant from *Chlamydomonas reinhardtii*. *Plant J.* **43**, 636–648.

Balczun, C., Bunse, A., Schwarz, C., Piotrowski, M., and Kück, U. (2006). Chloroplast heat shock protein Cpn60 from *Chlamydomonas reinhardtii* exhibits a novel function as a group II intron-specific RNA-binding protein. *FEBS Lett.* **580**, 4527–4532.

Barkan, A. and Goldschmidt-Clermont, M. (2000). Participation of nuclear genes in chloroplast gene expression. *Biochimie* **82**, 559–572.

Barkan, A., Klipcan, L., Ostersetzer, O., Kawamura, T., Asakura, Y., and Watkins, K.P. (2007). The CRM domain: An RNA binding module derived from an ancient ribosome-associated protein. *RNA* **13**, 55–64.

Bonen, L. and Vogel, J. (2001). The ins and outs of group II introns. *Trends Genet.* **17**, 322–331.

Boynton, J.E., Harris, E.H., Burkhart, B.D., Lamerson, P.M., and Gillham, N.W. (1987). Transmission of mitochondrial and chloroplast genomes in crosses of *Chlamydomonas*. *Proc. Natl. Acad. Sci. U. S. A.* **84**, 2391–2395.

Bruick, R.K. and Mayfield, S.P. (1999). Light-activated translation of chloroplast mRNAs. *Trends Plant Sci.* **4**, 190–195.

Bunse, A.A., Nickelsen, J., and Kück, U. (2001). Intron-specific RNA binding proteins in the chloroplast of the green alga *Chlamydomonas reinhardtii*. *Biochim. Biophys. Acta.* **1519**, 46–54.

Bussieres, J., Lemieux, C., Lee, R.W., and Turmel, M. (1996). Optional elements in the chloroplast DNAs of *Chlamydomonas eugametos* and *C. moewusii*: Unidirectional gene conversion and co-conversion of adjacent markers in high-viability crossses. *Curr. Genet.* **30**, 356–365.

Chevalier, B., Sussman, D., Otis, C., Noel, A.J., Turmel, M., Lemieux, C., Stephens, K., Monnat, R.J., Jr, and Stoddard, B.L. (2004). Metal-dependent DNA cleavage mechanism of the I-CreI LAGLIDADG homing endonuclease. *Biochemistry* **43**, 14015–14026.

Choquet, Y., Goldschmidt-Clermont, M., Girard-Bascou, J., Kück, U., Bennoun, P., and Rochaix, J.D. (1988). Mutant phenotypes support a *trans*-splicing mechanism for the expression of the tripartite *psaA* gene in the *C. reinhardtii* chloroplast. *Cell* **52**, 903–913.

Cote, M.J. and Turmel, M. (1995). *In vitro* self-splicing reactions of chloroplast and mitochondrial group-I introns in *Chlamydomonas eugametos* and *Chlamydomonas moewusii*. *Curr. Genet.* **27**, 177–183.

Deshpande, N.N., Hollingsworth, M., and Herrin, D.L. (1995). The *atpF* group-II intron-containing gene from spinach chloroplasts is not spliced in transgenic *Chlamydomonas* chloroplasts. *Curr. Genet.* **28**, 122–127.

Deshpande, N.N., Bao, Y., and Herrin, D.L. (1997). Evidence for light/redox-regulated splicing of *psbA* pre-RNAs in *Chlamydomonas* chloroplasts. *RNA* **3**, 37–48.

Dujardin, G. and Herbert, C. (1997). Aminoacyl tRNA synthetases involved in group I intron splicing. In: *Ribosomal RNA and Group I Introns* (R. Green and R. Schroeder, Eds.), pp. 179–198. R. G. Landes Company, Austin, TX.

Dürrenberger, F. and Rochaix, J.D. (1991). Chloroplast ribosomal intron of *Chlamydomonas reinhardtii*: *in vitro* self-splicing, DNA endonuclease activity and *in vivo* mobility. *EMBO J.* **10**, 3495–3501.

Dürrenberger, F. and Rochaix, J.D. (1993). Characterization of the cleavage site and the recognition sequence of the I-CreI DNA endonuclease encoded by the chloroplast ribosomal intron of *Chlamydomonas reinhardtii*. *Mol. Gen. Genet.* **236**, 409–414.

Dürrenberger, F., Thompson, A.J., Herrin, D.L., and Rochaix, J.D. (1996). Double strand break-induced recombination in *Chlamydomonas reinhardtii* chloroplasts. *Nucleic Acids Res.* **24**, 3323–3331.

Eberhard, S., Drapier, D., and Wollman, F.A. (2002). Searching limiting steps in the expression of chloroplast-encoded proteins: Relations between gene copy number, transcription, transcript abundance and translation rate in the chloroplast of *Chlamydomonas reinhardtii*. *Plant J.* **31**, 149–160.

Erickson, J.M., Rahire, M., and Rochaix, J.D. (1984). *Chlamydomonas reinhardtii* gene for the 32 000 mol.wt. protein of photosystem II contains four large introns and is located entirely within the chloroplast inverted repeat. *EMBO J.* **3**, 2753–2762.

Glanz, S., Bunse, A., Wimbert, A., Balczun, C., and Kück, U. (2006). A nucleosome assembly protein-like polypeptide binds to chloroplast group II intron RNA in *Chlamydomonas reinhardtii*. *Nucleic Acids Res.* **34**, 5337–5351.

Goldschmidt-Clermont, M., Girard-Bascou, J., Choquet, Y., and Rochaix, J.D. (1990). Trans-splicing mutants of *Chlamydomonas reinhardtii*. *Mol. Gen. Genet.* **223**, 417–425.

Goldschmidt-Clermont, M., Choquet, Y., Girard-Bascou, J., Michel, F., Schirmer-Rahire, M., and Rochaix, J.D. (1991). A small chloroplast RNA may be required for *trans*-splicing in *Chlamydomonas reinhardtii*. *Cell* **65**, 135–143.

Hahn, D., Nickelsen, J., Hackert, A., and Kück, U. (1998). A single nuclear locus is involved in both chloroplast RNA trans-splicing and 3′ end processing. *Plant J.* **15**, 575–581.

Haugen, P., Simon, D.M., and Bhattacharya, D. (2005). The natural history of group I introns. *Trends Genet.* **21**, 111–119.

Hausner, G., Olson, R., Simon, D., Johnson, I., Sanders, E.R., Karol, K.G., McCourt, R.M., and Zimmerly, S. (2006). Origin and evolution of the chloroplast *trnK* (*matK*) intron: A model for evolution of group II intron RNA structures. *Mol. Biol. Evol.* **23**, 380–391.

Herbert, C.J., Labouesse, M., Dujardin, G., and Slonimski, P.P. (1988). The NAM2 proteins from *S. cerevisiae* and *S. douglasii* are mitochondrial leucyl-tRNA synthetases, and are involved in mRNA splicing. *EMBO J.* 7, 473–483.

Herdenberger, F., Holländer, V., and Kück, U. (1994). Correct *in vivo* RNA splicing of a mitochondrial intron in algal chloroplasts. *Nucleic Acids Res.* **22**, 2869–2875.

Herrin, D.L. and Nickelsen, J. (2004). Chloroplast RNA processing and stability. *Photosynth. Res.* **82**, 301–314.

Herrin, D.L. and Schmidt, G.W. (1988). *Trans*-splicing of transcripts for the chloroplast *psaA1* gene. *In vivo* requirement for nuclear gene products. *J. Biol. Chem.* **263**, 14601–14604.

Herrin, D.L., Chen, Y.F., and Schmidt, G.W. (1990). RNA splicing in *Chlamydomonas* chloroplasts. Self-splicing of 23 S preRNA. *J. Biol. Chem.* **265**, 21134–21140.

Herrin, D.L., Bao, Y., Thompson, A.J., and Chen, Y.F. (1991). Self-splicing of the *Chlamydomonas* chloroplast *psbA* introns. *Plant Cell* **3**, 1095–1107.

Herrin, D., Kuo, T.C., and Goldschmidt-Clermont, M. (1998). RNA splicing in the chloroplast. In: *The Molecular Biology of Chloroplasts and Mitochondria in Chlamydomonas* (J.-D. Rochaix, M. Goldschmidt-Clermont, and S. Merchant, Eds.), pp. 183–195. Kluwer Academic Publishers, Dordrecht, The Netherlands.

Holländer, V. and Kück, U. (1999). Group II intron splicing in chloroplasts: Identification of mutations determining intron stability and fate of exon RNA. *Nucleic Acids Res.* **27**, 2345–2353.

Holloway, S.P. and Herrin, D.L. (1998). Processing of a composite large subunit rRNA. Studies with *Chlamydomonas* mutants deficient in maturation of the 23S-like rRNA. *Plant Cell* **10**, 1193–1206.

Holloway, S.P., Deshpande, N.N., and Herrin, D.L. (1999). The catalytic group-I introns of the *psbA* gene of *Chlamydomonas reinhardtii*: Core structures, ORFs and evolutionary implications. *Curr. Genet.* **36**, 69–78.

Houghland, J.L., Piccirilli, J.A., Forconi, M., Jihee, L., and Herschlag, D. (2006). How the group I intron works: A case study of RNA structure and function. In: *The RNA World* (R.F. Gesteland, T.R. Cech, and J.F. Atkins, Eds.), 3rd ed., pp. 133–205. Laboratory Press, Cold Spring Harbor.

Jenkins, B.D. and Barkan, A. (2001). Recruitment of a peptidyl-tRNA hydrolase as a facilitator of group II intron splicing in chloroplasts. *EMBO J.* **20**, 872–879.

Jenkins, B.D., Kulhanek, D.J., and Barkan, A. (1997). Nuclear mutations that block group II RNA splicing in maize chloroplasts reveal several intron classes with distinct requirements for splicing factors. *Plant Cell* **9**, 283–296.

Johanningmeier, U. and Heiss, S. (1993). Construction of a *Chlamydomonas reinhardtii* mutant with an intronless *psbA* gene. *Plant Mol. Biol.* **22**, 91–99.

Kim, H.H., Corina, L.E., Suh, J.K., and Herrin, D.L. (2005). Expression, purification, and biochemical characterization of the intron-encoded endonuclease, I-CreII. *Protein Expr. Purif.* **44**, 162–172.

Kuchka, M.R., Goldschmidt-Clermont, M., van Dillewijn, J., and Rochaix, J.D. (1989). Mutation at the *Chlamydomonas* nuclear *NAC2* locus specifically affects stability of the chloroplast *psbD* transcript encoding polypeptide D2 of PS II. *Cell* **58**, 869–876.

Kück, U., Choquet, Y., Schneider, M., Dron, M., and Bennoun, P. (1987). Structural and transcription analysis of the two homolgous genes for the P700 chlorophyll a-apoprotein in *Chlamydomonas reinhardtii*: Evidence for *in vivo trans*-splicing. *EMBO J.* **6**, 2185–2195.

Kuo, T.C., Odom, O.W., and Herrin, D.L. (2006). Unusual metal specificity and structure of the group I ribozyme from *Chlamydomonas reinhardtii* 23S rRNA. *FEBS J.* **273**, 2631–2634.

Kurokawa, S., Yamasaki, T., Komatsu, T., Watanabe, K.I., and Ohama, T. (2006). Degenerated recognition property of a mitochondrial homing enzyme in the unicellular green alga *Chlamydomonas smithii*. *Plant Mol. Biol.* **62**, 141–150.

Lambowitz, A.M. and Perlman, P.S. (1990). Involvement of aminoacyl-tRNA synthetases and other proteins in group I and group II intron splicing. *Trends Biochem. Sci.* **15**, 440–444.

Lambowitz, A.M. and Zimmerly, S. (2004). Mobile group II introns. *Annu. Rev. Genet.* **38**, 1–35.

Lee, J. and Herrin, D.L. (2003). Mutagenesis of a light-regulated *psbA* intron reveals the importance of efficient splicing for photosynthetic growth. *Nucleic Acids Res.* **31**, 4361–4372.

Lemieux, C. and Lee, R.W. (1987). Nonreciprocal recombination between alleles of the chloroplast 23S rRNA gene in interspecific *Chlamydomonas* crosses. *Proc. Natl. Acad. Sci. U. S. A.* **84**, 4166–4170.

Lemieux, B., Turmel, M., and Lemieux, C. (1988). Unidirectional gene conversions in the chloroplast of *Chlamydomonas* interspecific hybrids. *Mol. Gen. Genet.* **212**, 48–55.

Li, F. and Herrin, D.L. (2002). FUGOID: Functional genomics of organellar introns database. *Nucleic Acids Res.* **30**, 385–386.

Li, F., Holloway, S.P., Lee, J., and Herrin, D.L. (2002). Nuclear genes that promote splicing of group I introns in the chloroplast 23S rRNA and *psbA* genes in *Chlamydomonas reinhardtii*. *Plant J.* **32**, 467–480.

Lurin, C., Andres, C., Aubourg, S., Bellaoui, M., Bitton, F., Bruyere, C., Caboche, M., Debast, C., Gualberto, J., Hoffmann, B., Lecharny, A., Le Ret, M., Martin-Magniette, M.L., Mireau, H., Peeters, N., Renou, J.P., Szurek, B., Taconnat, L., and Small, I. (2004). Genome-wide analysis of *Arabidopsis* pentatricopeptide repeat proteins reveals their essential role in organelle biogenesis. *Plant Cell* **16**, 2089–2103.

Merendino, L., Perron, K., Rahire, M., Howald, I., Rochaix, J.D., and Goldschmidt-Clermont, M. (2006). A novel multifunctional factor involved in *trans*-splicing of chloroplast introns in *Chlamydomonas*. *Nucleic Acids Res.* **34**, 262–274.

Minagawa, J. and Crofts, A. (1994). A robust protocol for site-directed mutagenesis of the D1 protein in *Chlamydomonas reinhardtii*. *Photosynth. Res.* **42**, 121–131.

Nedelcu, A.M. and Lee, R.W. (1998). A degenerate group II intron in the intronless mitochondrial genome of *Chlamydomonas reinhardtii*: Evolutionary implications. *Mol. Biol. Evol.* **15**, 918–922.

Odom, O.W., Holloway, S.P., Deshpande, N.N., Lee, J., and Herrin, D.L. (2001). Mobile self-splicing group I introns from the psbA gene of *Chlamydomonas reinhardtii*: Highly efficient homing of an exogenous intron containing its own promoter. *Mol. Cell. Biol.* **21**, 3472–3481.

Odom, O.W., Shenkenberg, D.L., Garcia, J.A., and Herrin, D.L. (2004). A horizontally acquired group II intron in the chloroplast *psbA* gene of a psychrophilic *Chlamydomonas*: *In vitro* self-splicing and genetic evidence for maturase activity. *RNA* **10**, 1097–1107.

Ostheimer, G.J., Williams-Carrier, R., Belcher, S., Osborne, E., Gierke, J., and Barkan, A. (2003). Group II intron splicing factors derived by diversification of an ancient RNA-binding domain. *EMBO J.* **22**, 3919–3929.

Paukstelis, P.J., Coon, R., Madabusi, L., Nowakowski, J., Monzingo, A., Robertus, J., and Lambowitz, A.M. (2005). A tyrosyl-tRNA synthetase adapted to function in group I intron splicing by acquiring a new RNA binding surface. *Mol. Cell* **17**, 417–428.

Perron, K., Goldschmidt-Clermont, M., and Rochaix, J.D. (1999). A factor related to pseudouridine synthases is required for chloroplast group II intron *trans*-splicing in *Chlamydomonas reinhardtii*. *EMBO J.* **18**, 6481–6490.

Perron, K., Goldschmidt-Clermont, M., and Rochaix, J.D. (2004). A multiprotein complex involved in chloroplast group II intron splicing. *RNA* **10**, 704–710.

Podar, M., Chu, V.T., Pyle, A.M., and Perlman, P.S. (1998). Group II intron splicing *in vivo* by first-step hydrolysis. *Nature* **391**, 891–915.

Pyle, A.M. and Lambowitz, A.M. (2006). Group II introns: Ribozymes that splice RNA and invade DNA. In: *The RNA World* (R.F. Gesteland, T.R. Cech, and J.F. Atkins, Eds.), 3rd ed., pp. 469–505. Cold Spring Harbor Laboratory Press, Cold Spring Harbor, New York.

Rivier, C. (2000). *Characterisation of a factor involved in the trans-splicing of a chloroplast tri-partite group II intron from Chlamydomonas reinhardtii.* Ph.D. Dissertation. University of Geneva, Geneva, Switzerland.

Rivier, C., Goldschmidt-Clermont, M., and Rochaix, J.D. (2001). Identification of an RNA-protein complex involved in chloroplast group II intron *trans*-splicing in *Chlamydomonas reinhardtii*. *EMBO J.* **20**, 1765–1773.

Robart, A.R. and Zimmerly, S. (2005). Group II intron retroelements: Function and diversity. *Cytogenet. Genome Res.* **110**, 589–597.

Rochaix, J.D. and Malnoë, P. (1978). Anatomy of the chloroplast ribosomal DNA of *Chlamydomonas reinhardii*. *Cell* **15**, 661–670.

Rochaix, J.D., Rahire, M., and Michel, F. (1985). The chloroplast ribosomal intron of *Chlamydomonas reinhardii* codes for a polypeptide related to mitochondrial maturases. *Nucleic Acids Res.* **13**, 975–984.

Roitgrund, C. and Mets, L.J. (1990). Localization of two novel chloroplast genome functions: *trans*-splicing of RNA and protochlorophyllide reduction. *Curr. Genet.* **17**, 147–153.

Schafer, B., Merlos-Lange, A.M., Anderl, C., Welser, F., Zimmer, M., and Wolf, K. (1991). The mitochondrial genome of fission yeast: Inability of all introns to splice autocatalytically, and construction and characterization of an intronless genome. *Mol. Gen. Genet.* **225**, 158–167.

Schmitz-Linneweber, C., Williams-Carrier, R.E., Williams-Voelker, P.M., Kroeger, T.S., Vichas, A., and Barkan, A. (2006). A pentatricopeptide repeat protein facilitates the *trans*-splicing of the maize chloroplast *rps12* pre-mRNA. *Plant Cell* **18**, 2650–2663.

Seraphin, B., Boulet, A., Simon, M., and Faye, G. (1987). Construction of a yeast strain devoid of mitochondrial introns and its use to screen nuclear genes involved in mitochondrial splicing. *Proc. Natl. Acad. Sci. U. S. A.* **84**, 6810–6814.

Sharp, P.A. (1991). Five easy pieces. *Science* **254**, 663.

Spiegel, P.C., Chevalier, B., Sussman, D., Turmel, M., Lemieux, C., and Stoddard, B.L. (2006). The structure of I-CeuI homing endonuclease: Evolving asymmetric DNA recognition from a symmetric protein scaffold. *Structure* **14**, 869–880.

Thompson, A.J. and Herrin, D.L. (1994). A chloroplast group I intron undergoes the first step of reverse splicing into host cytoplasmic 5.8 S rRNA. Implications for intron-mediated RNA recombination, intron transposition and 5.8 S rRNA structure. *J. Mol. Biol.* **236**, 455–468.

Thompson, A.J., Yuan, X., Kudlicki, W., and Herrin, D.L. (1992). Cleavage and recognition pattern of a double-strand-specific endonuclease (I-creI) encoded by the chloroplast 23S rRNA intron of *Chlamydomonas reinhardtii*. *Gene* **119**, 247–251.

Till, B., Schmitz-Linneweber, C., Williams-Carrier, R., and Barkan, A. (2001). CRS1 is a novel group II intron splicing factor that was derived from a domain of ancient origin. *RNA* **7**, 1227–1238.

Trebitsh, T. and Danon, A. (2001). Translation of chloroplast *psbA* mRNA is regulated by signals initiated by both photosystems II and I. *Proc. Natl. Acad. Sci. U. S. A.* **98**, 12289–12894.

Turmel, M., Gutell, R.R., Mercier, J.P., Otis, C., and Lemieux, C. (1993a). Analysis of the chloroplast large subunit ribosomal RNA gene from 17 *Chlamydomonas* taxa. Three internal transcribed spacers and 12 group I intron insertion sites. *J. Mol. Biol.* **232**, 446–467.

Turmel, M., Mercier, J.P., and Cote, M.J. (1993b). Group I introns interrupt the chloroplast *psaB* and *psbC* and the mitochondrial *rrnL* gene in *Chlamydomonas*. *Nucleic Acids Res.* **21**, 5242–5250.

Turmel, M., Choquet, Y., Goldschmidt-Clermont, M., Rochaix, J.D., Otis, C., and Lemieux, C. (1995). The *trans*-spliced intron 1 in the *psaA* gene of the *Chlamydomonas* chloroplast: A comparative analysis. *Curr. Genet.* **27**, 270–279.

Vogel, J. and Börner, T. (2002). Lariat formation and a hydrolytic pathway in plant chloroplast group II intron splicing. *EMBO J.* **21**, 3794–3803.

Zerges, W. and Rochaix, J.D. (1998). Low density membranes are associated with RNA-binding proteins and thylakoids in the chloroplast of *Chlamydomonas reinhardtii*. *J. Cell Biol.* **140**, 101–110.

CHAPTER 27

Chloroplast RNA Processing and Stability

David L. Herrin
Section of Molecular Cell and Developmental Biology,
Institute for Cellular and Molecular Biology,
University of Texas, Austin, Texas, USA

CHAPTER CONTENTS

I. CHLOROPLAST RNA PROCESSING

A. Overview

The chloroplast genome of *Chlamydomonas* encodes rRNAs, tRNAs, mRNAs, and a small RNA (*tscA*) that forms part of a *trans*-spliced group II intron. The gene for the 23S RNA of the large ribosomal subunit and two mRNA genes (*psbA* and *psaA*) contain introns whose splicing is discussed in Chapter 26. The focus of this chapter is the processing, degradation, and stability of chloroplast mRNAs, but first there is a discussion of the processing of the uniquely discontinuous 23S rRNA.

Chlamydomonas is an excellent organism for studying chloroplast post-transcriptional processes. Besides the obvious advantages of working with a fast-growing, unicellular organism that allows genetic approaches, there is only one chloroplast per cell in *Chlamydomonas reinhardtii*, and its polyploid genome can be readily manipulated by transformation (Bateman and Purton, 2000; also see Volume 1, Chapter 8). Also, the expression of chloroplast genes in this organism is robust, which facilitates the analysis of specific RNAs, including precursors or intermediates in processing or degradation.

There is little known about the processing of *Chlamydomonas* chloroplast tRNAs, other than the fact that there must be an RNase P activity (which processes the 5′ end of pre-tRNAs), and that the enzyme likely lacks the RNA (ribozyme) subunit found in many bacterial enzymes (based on the apparent absence of a gene encoding the RNA subunit in the *Chlamydomonas* chloroplast genome; de la Cruz and Vioque, 2003). With regard to other types of RNA modifications, such as methylation, there is little information available, and RNA editing has not been reported for *Chlamydomonas* chloroplast RNAs.

Most chloroplast mRNAs in *Chlamydomonas* are subjected to nucleolytic processing at both ends and internally in the case of polycistronic transcripts. Chloroplast mRNAs do not possess a 5′ trimethylguanosine cap like cytoplasmic mRNAs, although they do become polyadenylated at the 3′ terminus as part of a degradation pathway (Komine et al., 2000, 2002). Unlike mRNAs of prokaryotes, which share a common ancestor with green plant chloroplasts, the mature mRNAs of *Chlamydomonas* chloroplasts are relatively long-lived, with half-lives ranging from 20 minutes to 30 minutes for *tufA* to many hours for the *psbA* transcript (Salvador et al., 1993b; Hwang and Herrin, 1994). The stable nature of chloroplast mRNAs signifies that transcription and translation in the organelle are not tightly coupled as in prokaryotes, and there is extensive post-transcriptional control of gene expression. For example, there is evidence for control of chloroplast mRNA half-life by light, translation, nutrient availability, and plastid development, although the mechanisms are not well understood. Much current research is focused on identifying the nuclear genes whose protein products carry out and regulate chloroplast mRNA processing and stability. Notable reviews on these and related topics include those of Rochaix (1996), Goldschmidt-Clermont (1998), Nickelsen (1998, 2003), Stern and Drager (1998), Herrin and Nickelsen (2004), and Bollenbach et al. (2008).

B. Processing of 23S rRNA

The 23S rRNA of green plant chloroplasts is usually fragmented into at least three RNA species, which are generated by the removal of internal transcribed spacer (ITS) sequences without concomitant ligation of the fragments (Turmel et al., 1993; reviewed in Nedelcu and Lee, 1998). Unlike the 23S rRNA itself, the ITSs evolve rapidly and their sequence is not conserved, but they can usually be folded into a structure that has a basal paired region flanked by the cleavage sites (Turmel et al., 1993). In *Chlamydomonas*, the 23S rRNA is processed into four RNAs, called 7S (290 nt), 3S (47 nt), 23Sγ (821 nt) and 23Sδ (1696 nt), which are separated by three ITSs; there is also an 888 nt group I intron in the 23Sδ portion (*Cr.LSU*). The separation of the exons by the ITSs, and therefore the fragmentation of mature 23S rRNA, is conserved in many *Chlamydomonas* species (Turmel et al., 1993), although the functional role of this organization is not clear.

The 23S *rrn* gene is believed to be co-transcribed with the 16S *rrn* and two *trn* genes (Ile and Ala), and it is followed in the gene cluster by 5S *rrn*. The initial processing of the precursor into 16S, tRNAs, 23S and 5S RNAs must be rapid as intermediates containing multiple or larger species were not detected on RNA gel blots (Holloway and Herrin, 1998). However, intermediates of 23S processing could be detected in a wild-type strain, and their levels were increased in the *ac20* mutant, which facilitated their characterization (Holloway and Herrin, 1998). *ac20* is a nuclear mutant of *Chlamydomonas* known to be deficient in chloroplast protein synthesis, and specifically in the assembly of the large (50S) ribosomal subunit (Harris et al., 1974). Holloway and Herrin (1998) presented evidence that processing of the ITSs, particularly ITS-1 (which separates the 7S and 3S rRNAs) is deficient in *ac20*. Splicing of the *Cr.LSU* intron is also reduced in *ac20*, but that character is not responsible for the slow-growth phenotype (as shown by deleting the intron from *ac20* using chloroplast transformation). The data thus suggest that 23S ITS processing is closely linked to subunit assembly. It is noteworthy that a *cis*-acting mutation in the *Cr.LSU* intron that reduced splicing also partially reduced ITS processing (Holloway and Herrin, 1998), suggesting that subunit assembly could influence RNA processing. The authors were able to deduce a preferred pathway for processing of 23S pre-RNA, which is shown in Figure 27.1. This pathway seems to be mainly kinetic, that is, it is not an absolutely required order, but serves to promote efficient processing *in vivo*.

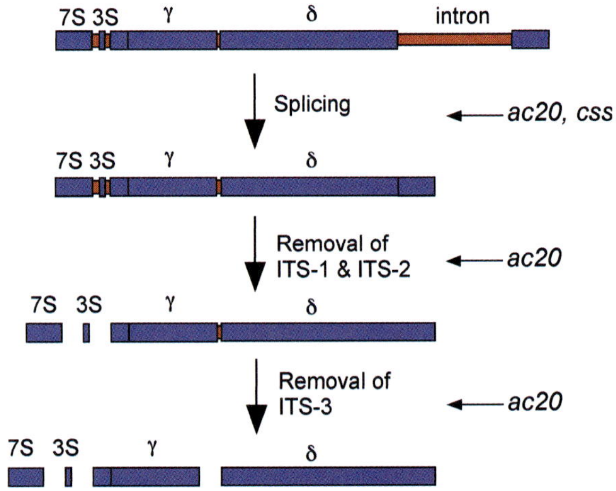

FIGURE 27.1 *Pathway for processing of 23S pre-RNA in Chlamydomonas. The 23S rrn gene is split by three internal transcribed spacers (ITSs) and one group I intron; the four final products (7S, 3S, γ, δ) are indicated. Nuclear genes involved in the processing steps are indicated to the right: ac20 is an acetate-requiring mutant, and the css mutants, of which there are three, are suppressors of a splicing mutation in the intron (adapted from Holloway and Herrin, 1998).*

C. 5′ pre-mRNA processing

Most chloroplast mRNAs in *Chlamydomonas* are probably processed at their 5′ ends, based on the facts that there are usually at least two 5′ ends detected by primer extension, and that successful capping of chloroplast mRNAs with guanylyl transferase, which requires a 5′ di- or triphosphate, has not been reported. Mapping of the 5′ termini of a number of mRNAs, including the *psbA*, *psbB*, *psbD*, *rbcL*, and *atpB* mRNAs, revealed a minor long form, and a more abundant shorter form. It is not yet clear for most mRNAs whether the species with the shorter 5′ untranslated region (UTR) is created by endonuclease or exonuclease digestion. Accurate 5′ processing by a chloroplast extract has not been reported. *Trans*-acting mutants deficient in 5′ processing (and still containing the transcript) have been described by Rymarquis et al. (2006). The cloning of these nuclear genes, which are also implicated in 3′ processing, could shed light on this mechanism.

Evidence has been presented that cleavage of the 5′ end of the long, or precursor, form of *psbA* to the mature mRNA (which is 54 nt shorter) is coupled to association of the pre-mRNA with ribosomes (Bruick and Mayfield, 1998). This conclusion was based on the fact that the mature form is specifically lacking in mutants, both *trans* (nuclear) and *cis* (5′ UTR), that are strongly deficient in translation of *psbA* mRNA. It should be said, however, that binding of the pre-mRNA to ribosomes was not directly demonstrated; this is an important point because translation of the pre-mRNA is apparently poor. Nonetheless, the proposed coupling of 5′ *psbA* processing to translation initiation could indicate that the processing nuclease is associated with ribosomes (see section II.C). Alternatively, ribosome binding to the pre-mRNA could alter its structure such that it becomes more susceptible to nucleolytic cleavage.

The sequence requirements for 5′ processing are not well established, but replacement of four uridines at the presumptive cleavage site with adenines had no effect on the processing of the *psbA* pre-mRNA (Bruick and Mayfield, 1998). Changing 7 nt in the 5′ UTR of the short form of the *psbD* mRNA appeared to inhibit processing of its pre-mRNA *in vivo*; however, the long form did not accumulate to high levels. This can be explained if we assume the precursor is inherently less stable than the mature form. As with *psbA*, mutagenesis of the apparent processing site in *psbD* pre-mRNA had no effect. Also, alterations in the 3′ half of the segment that is removed during maturation were without effect on processing (Nickelsen et al., 1999).

Figure 27.2 summarizes pathways for maturation and degradation of chloroplast mRNAs. An analysis of the role of specific regions in the 5′ UTR on *petD* processing is discussed in the section on polycistronic mRNA processing (see section I.E), and of *psbB* in the section on *cis*-acting elements in mRNA stability (see section II.D). The latter assignment is because stability and 5′ processing appear to be closely linked for this mRNA.

The function of 5′ processing for most chloroplast mRNAs is not clear, but it has been suggested that one function is to produce a better RNA for translation, and there is evidence to support that suggestion (Bruick and Mayfield, 1998). Another role could be to generate a more stable mRNA (Nickelsen et al., 1999). It has also been suggested that 5′ processing of the mRNAs for the photosystem II genes *psbA*, *psbB*, and *psbD* could be a mechanism to promote the coordinated translation of these integral membrane proteins on thylakoid-bound ribosomes. In this scenario, the pre-mRNAs are first bound by a ribosomal subunit (and other factors), and then these complexes bind to the membrane. The pre-mRNAs are then processed and translated on the membrane (Rochaix, 1996).

A number of nuclear mutants have been isolated that have a destabilized chloroplast mRNA. The target of these nuclear genes is usually the 5′ UTR, and some of these nuclear gene products could also play a role in 5′ processing. The mutants are given in Table 27.1, discussed in sections I.F and II.E.

D. 3′ pre-mRNA processing

The inverted repeats (IRs) that occur just downstream of chloroplast genes are inefficient terminators of transcription, but they play important roles in the nucleolytic processing and concomitant stabilization associated with 3′ processing (reviewed in Stern and Drager, 1998). Mapping of the *rbcL*, *psbA*, *psbB*, *petD*, and especially *atpB* RNAs indicated that the 3′ terminus

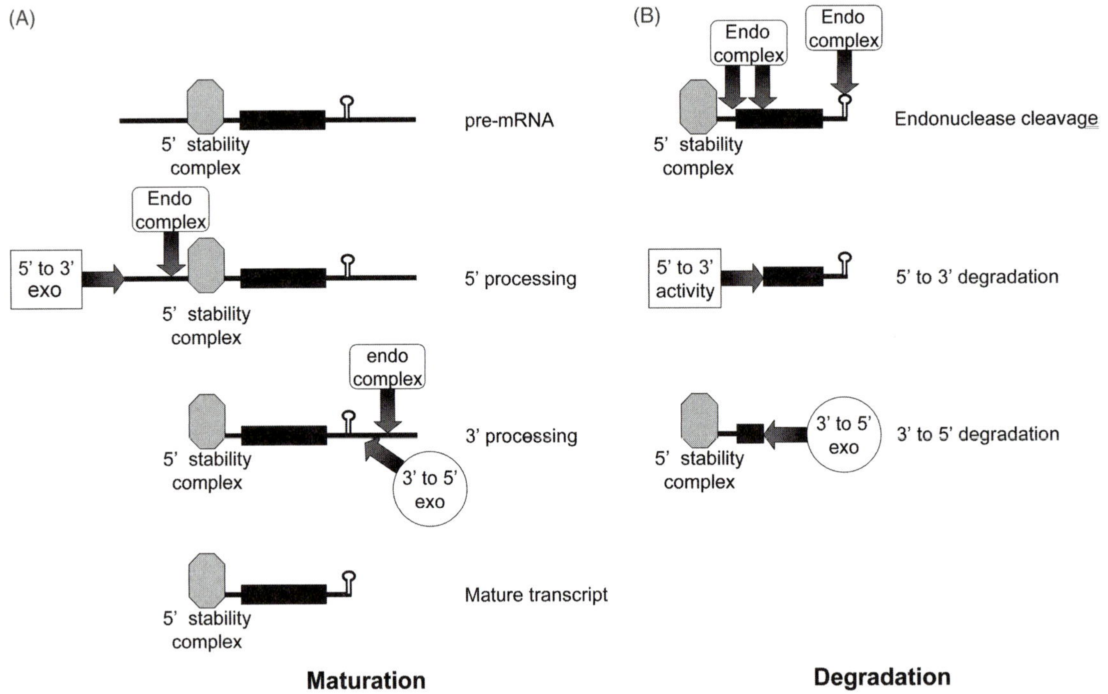

FIGURE 27.2 *Pathways for nucleolytic processing and degradation of mRNAs in Chlamydomonas chloroplasts. Coding regions are shown as filled boxes, and inverted repeats (IRs) as stem-loop structures. (A) Steps involved in maturation of a generic pre-mRNA are shown, and no order is implied. 5′ processing could be performed by an endonuclease complex (endo complex), or a processive 5′→3′ exonuclease, that is ultimately blocked by the stability complex on the 5′ UTR. 3′ processing is performed by the action of an endonuclease complex, followed by a 3′→5′ exonuclease (3′–5′ exo). (B) Degradation could be initiated by an endonuclease complex (endo complex) in the 5′ UTR or other regions of the mRNA. The fragments are then degraded by a 5′→3′ activity (either an exonuclease or a processive endonuclease), or if they lack the 3′ IR, they could be rapidly degraded by a 3′→5′ exonuclease activity (adapted from Rymarquis et al., 2006).*

Table 27.1 Nuclear genes involved in the processing, stability, and degradation of chloroplast RNAs

Mutant or Gene[a]	Accession number of wild-type protein (if known)	Chloroplast gene(s) affected	Defect	Target	References
ac20	–	*23S rrn*	Processing	ITS-1 and ITS-2	Holloway and Herrin (1998)
GE2.10[b]	–	*psbB*	Stability	?	Sieburth et al. (1991)
HN31	–	*tscA*	Processing	3′, bicistronic	Hahn et al. (1998)
mbb1 (222E)[b]	CAC19558	*psbB (& psbH)*	Stability	5′ UTR	Vaistij et al. (2000a)

(Continued)

Table 27.1 *Continued*

Mutant or Gene[a]	Accession number of wild-type protein (if known)	Chloroplast gene(s) affected	Defect	Target	References
6.2z5	–	*psbC*	Stability	?	Sieburth et al. (1991)
mbd1 (nac2–nac26)	CAB96529	*psbD*	Processing and stability	5′ UTR	Boudreau et al. (2000)
suPRB2A-1, suPRB2A-2, suPRB2A-3	–	*psbD*	Processing and stability	5′ UTR	Nickelsen (2000)
mca1 (Mφ11)	AAK14341	*petA*	Stability	5′ UTR	Gumpel et al. (1995); Raynaud et al. (2007)
Mφ37	–	*petB*	Stability	?	Gumpel et al. (1995)
mcd1 (F16)	AAY44840	*petD*	Processing and stability	5′ UTR	Drager et al. (1998)
mcd3, mcd4	–	many	Processing	3′ and 5′ UTR	Rymarquis et al. (2006)
mcd5	–	*petD, psbB, psbT*	Processing	Polycistronic	Rymarquis et al. (2006)
mda1-ncc1	–	*atpA*	Stability	Coding region	Drapier et al. (2002)
thm24	–	*atpB*	Processing and stability	?	Drapier et al. (1992)
rat1,2 (TR72)	ABK15559 (RAT1)	*tscA*	Processing	3′, bicistronic	Balczun et al. (2005)
crp3	–	Several	Processing	3′ UTR	Levy et al. (1999)
tas1, tas2	–	*rbcL*	Processing	3′ UTR	Goldschmidt-Clermont et al. (2008)
			Chloroplast RNases		
Polynucleotide phosphorylase *(PNP1)*	ABF57675	–	–	–	Nishimura et al. (2004)
RNase II homologue *(RNB2)*	AB020871	–	–	–	Yehudai-Resheff et al. (2007)
			Candidate genes for chloroplast RNases		
RAP38	EDP05666	–	–	–	Yamaguchi et al. (2003)
RAP41	EDP05800	–	–	–	Yamaguchi et al. (2003)
XRN1	EDP01687	–	–	–	
RNJ1	ACD36970	–	–	–	

[a]*The original name of the mutant is in parentheses.*

[b]*It is not known whether GE2.10 and mbb1 are allelic.*

usually coincides with, or is just downstream of, a stable stem-loop structure downstream of the coding region. A possible exception is *tufA*, whose 3′ UTR may have two stem-loops followed by a single-stranded extension (Silk and Wu, 1993; Zicker et al., 2007).

Early studies with the *Chlamydomonas atpB* gene involved adding synthetic RNAs to chloroplast extracts that were competent for 3′ processing. These experiments indicated that processing of the *atpB* pre-mRNA occurs in two steps: the first is endonucleolytic cleavage 10 nt downstream of the IR, and the second is 3′→5′ exonuclease digestion back to the IR terminus (Figure 27.2A; Stern and Kindle, 1993). Changing the endonuclease cleavage site (ECS) from an AT-rich sequence to a GC-rich sequence did not abolish processing *in vitro*, but it did shift the position of the 3′ end of the product (Rott et al., 1999). However this change did not affect 3′ processing *in vivo*. On the other hand, deletions that included the IR blocked processing *in vitro* (Stern and Kindle, 1993), and resulted in a massive loss of the mRNA *in vivo* (Stern et al., 1991).

In vivo studies using a GUS reporter gene (*uidA*) and 3′ regions from the *psbA* and *rbcL* genes suggested that the sequence of the IR or flanking regions are important for 3′ processing and RNA accumulation (Blowers et al., 1993). A similar conclusion was reached by Rott et al. (1998b), who examined the *petA* and *petD* 3′ regions fused to the *atpB* coding region as a reporter. As with the earlier *atpB* studies, a complication of the results is that unprocessed RNAs did not readily accumulate. Deletions downstream of *petA* including the IR, however, resulted in significant accumulation of alternatively processed RNAs that terminated at IRs in the intergenic region between *petA* and *petD* (Jiao et al., 2004). Many of these IRs are also short dispersed repeat sequences (SDRs), a series of repeated families found throughout the intergenic regions (see Chapter 24). In addition, the last 67 nt of the *tufA* 3′ UTR is a relatively long SDR (Zicker et al., 2006); thus supporting the suggestion that some chloroplast genes may have evolved to utilize SDRs as 3′ processing signals (Jiao et al., 2004).

The initial step in *atpB* 3′ processing – and possibly that of other mRNAs – generates two products; the upstream product, which becomes the mRNA, and the downstream product, which is degraded. Hicks et al. (2002) inserted the 3′ processing sequence of the *atpB* gene into the 5′ UTR of reporter genes to examine the fate of downstream sequences as well as processing. Both the IR and the ECS of *atpB* were necessary to effectively degrade the reporter RNA, although either one alone permitted some accumulation of the reporter (and therefore gave some processing). Insertion of a poly(G) sequence between the *atpB* processing site and the reporter sequence did not block the latter's degradation, suggesting that it is not degraded by a 5′→3′ exonuclease activity. A poly(G) sequence can inhibit the progress of 5′→3′ exonucleases (Decker and Parker, 1993), and this approach has been successfully used in *Chlamydomonas* chloroplasts (see below). The authors suggested that degradation of RNA downstream of the ECS is catalyzed by a processive endonuclease.

Rymarquis et al. (2007) also used the reporter gene approach to carry out further analyses of the sequences of *atpB* that determine 3′ processing. They found that in addition to the IR, 15–37 nt downstream of the ECS stimulate processing, but the sequence of the ECS is not important. They also showed that the nuclear gene *MCD4* acts on the 3′ processing determinant of *atpB*, because there was increased accumulation of the reporter RNA, and decreased processing when a *mcd4* mutation was crossed into these strains.

Using a new screen for mutants defective in 3′ processing of the *rbcL* mRNA, Goldschmidt-Clermont et al. (2008) discovered that this gene has two IRs downstream of the coding region that can mediate 3′ processing and stabilization of *rbcL* transcripts. *In vivo*, however, the more distal repeat, IR2, is the principal one that is used. Mutations in IR2 that increased the size of the loop at the expense of the base-paired (or stem) region lead to utilization of IR1 for establishing the 3′ end of the *rbcL* transcript. The results are consistent with the two-step model for 3′ processing of the *atpB* mRNA (Stern and Kindle, 1993), which would predict endonucleolytic cleavage downstream of IR2, followed by 3′ exonuclease trimming back to the IR. The authors also identified two nuclear genes (*TAS1* and *TAS2*) that likely play a role in this processing.

In summary, the chloroplast diverges from bacteria, and resembles the nucleus in that the 3′ end of mRNAs is determined post-transcriptionally by processing. One function of 3′ processing is to provide a terminus that is resistant to 3′–5′ exonuclease activity (Drager et al., 1996). There is also evidence that chloroplast mRNAs with the correct 3′ ends are preferentially translated (Rott et al., 1998a).

E. Internal nucleolytic processing of polycistronic mRNAs

There are a number of chloroplast protein-coding genes that are co-transcribed and then processed internally to produce mono- or dicistronic mRNAs (reviewed in Rochaix, 1996). Polycistronic and intermediately processed (i.e. mono- or dicistronic, etc.) mRNAs can accumulate from the same transcription unit. Also, additional promoters internal to the gene cluster can be present, such as with the *atpA* cluster (Drapier et al., 1998), further increasing the complexity of mRNAs produced. Two of the more thoroughly characterized polycistronic gene clusters are the *atpA* and *rpl36* gene clusters. The former is a mixed-function cluster with as many as five genes (Drapier et al., 1998; Rymarquis et al., 2006), whereas the latter is a cluster of four ribosomal protein genes (Rymarquis et al., 2006).

Compared to 5′ and 3′ processing of monocistronic mRNAs, there has been less study of intergenic processing of polycistronic transcripts.

However, the termini of many mRNAs processed from polycistronic RNAs coincide with the termini of mRNAs produced from alternative internal promoters, suggesting that the same machinery is involved in processing both types of transcripts. Processing of *petD* mRNA from a *petA-petD* dicistronic RNA produces the same 5′ end as processing of the precursor synthesized from the *petD* promoter; this was shown by comparing *petD* RNA 5′ ends from a mutant that lacked the *petD* promoter to the *petD* 5′ ends in the wild type (Sakamoto et al., 1994).

Deletion mutagenesis of the 362-nt *petD* 5′ UTR identified three *cis*-acting elements (I–III) required for expression (Sakamoto et al., 1993), and moreover indicated that nt 1–25 (element I) mediated processing (Sakamoto et al., 1994). However, linker-scanning mutagenesis of element I did not yield a mutant that accumulated unprocessed *petD* RNA (Higgs et al., 1999). Instead, mutation of nt 2–8 drastically destabilized the RNA, suggesting that 5′ processing might be required for mature mRNA accumulation. Higgs et al. (1999) also used dimethylsulfate to probe the structure of the *petD* 5′ UTR *in vitro* and *in vivo*, and determined that element I is part of a stem-loop structure. This element is also the target of the *MCD1* gene product, which is needed to stabilize the RNA against 5′ → 3′ exonuclease digestion. This latter point was demonstrated by inserting a poly(G) sequence into the 5′ UTR of *petD* and finding that it impeded degradation in the *mcd1* mutant background (Drager et al., 1998).

Subsequently, the *petD* 5′ UTRs from three other *Chlamydomonas* species (*C. incerta*, *C. eugametos*, and *C. moewusii*) were characterized, revealing sequence conservation of element I (Kramzar et al., 2006). Also, functional conservation was demonstrated for the *C. incerta* 5′ UTR, which was able to mediate processing and stability of a GUS reporter RNA expressed in transformed *C. reinhardtii* chloroplasts. The *C. incerta petD* 5′ UTR also had the most similarity to the *C. reinhardtii* UTR in sequence and spacing of elements I–III.

F. *Trans*-acting factors for pre-mRNA processing

Nuclear genes play important roles in chloroplast gene expression, including mRNA processing. Many of the nuclear genes involved in mRNA processing were identified by using suppressor genetics. In this approach, chloroplast transformation is used to create a strain that has a mutated chloroplast gene whose RNA is poorly processed or unstable, and provides a mutant phenotype (such as light sensitivity or acetate requirement). To obtain suppressors, this primary mutant is mutagenized, either by classical methods (e.g. ultraviolet light or chemicals) or with transformed DNA (see Chapter 10, Volume 1), and then strains with restored processing or stabilization are selected. This approach was first used successfully with Δ26*atpB* as the primary mutant

(Levy et al., 1997); this mutation results in very little *atpB* mRNA because of a deletion of the 3′ IR (Stern et al., 1991). A suppressor strain, Δ26*S*, was isolated that accumulated an *atpB* transcript. This mutant was subsequently shown to be partially defective in 3′ processing of a number of chloroplast transcripts (Levy et al., 1997, 1999). The authors also demonstrated that this mutation is recessive, which is uncommon for a nuclear suppressor. The precise function of the Δ26*S* suppressor mutation (*crp3*) is not yet known, but possible functions include a ribonuclease or an RNA-binding protein needed for 3′ processing (Levy et al., 1999).

Two nuclear genes involved in 3′ processing were discovered during an investigation of mutants defective in *trans*-splicing of the first intron of the *psaA* gene. Part of this tripartite group II intron is a small RNA (*tscA*) that is transcribed as a dicistronic transcript (*tscA-chlN*) that must be processed. As indicated by Hahn et al. (1998), the nuclear *TR72* mutant cannot splice this intron, because *tscA* is not processed at the 3′ end, which liberates it from *chlN*. The *HN31* mutant is similarly deficient in *tscA* processing, but it is also defective in *trans*-splicing of the second *psaA* intron (Hahn et al., 1998).

The *TR72* mutant was generated by insertional mutagenesis with a plasmid (Hahn et al., 1998), so it was relatively straightforward to recover the corresponding DNA from wild-type cells, once the DNA flanking the plasmid had been identified (Balczun et al., 2005). Surprisingly, the locus contained two genes, *RAT1* and *RAT2*, both of which were required for the non-photosynthetic *TR72* to recover a wild-type phenotype. The authors also showed that *RAT1* is expressed, and that the protein localizes to chloroplast membranes. It also has sequence similarity to the NAD^+-binding domain of poly(ATP-ribose) polymerases, but mutagenesis of conserved residues in this region did not affect *RAT1*'s ability to complement *TR72*. The yeast three-hybrid system, which can detect RNA–protein interactions *in vivo*, was employed to show that RAT1 binds *tscA* RNA. Perhaps RAT1 evolved from an enzyme that once modified RNA, but now functions as an RNA-binding protein. The evolution of an RNA-modifying enzyme into a protein with a new RNA-binding function has happened several times during the adaptive evolution of organellar splicing factors, including at least one other example from *Chlamydomonas* chloroplasts (see Chapter 26).

Mutants deficient in the cytochrome b_6f complex (*F16* and *670*) were used to define a nuclear gene, *MCD1*, that is required for 5′ processing and stabilization of the monocistronic *petD* mRNA, which is completely absent in these mutants (Drager et al., 1998). Chimeric genes with the *petD* 5′ UTR fused to a reporter (*uidA* [GUS] and *aadA*) were introduced into the chloroplast genome of *mcd1* strains to identify the target. Analysis of these and other mutant strains showed that the target of *MCD1* is the 5′ end of the 5′′ UTR. Insertion of a poly(G) sequence into the 5′ UTR allowed the accumulation of a *petD* transcript in the *mcd1* background; the

5′ end of this RNA coincided with the poly(G) sequence, suggesting it impeded a 5′→3′ exonuclease (see also below). Esposito et al. (2001) isolated and characterized a suppressor of the *mcd1-2* allele that proved to be a second nuclear gene, *MCD2*. The mutant allele *mcd2-1* increased the accumulation of *petD* mRNA in the *mcd1-2* background, and appeared to stimulate its translation.

The *MCD1* gene was cloned by insertional mutagenesis of an arginine auxotrophic strain with the *ARG7*-containing plasmid (Murakami et al., 2005). DNA flanking the tandem plasmid insert in the *mcd1-3* mutant was recovered, and found to contain part of the *PF28* gene. Hence, the wild-type *MCD1* gene could be identified on cosmid clones containing *PF28*. Characterization of the *MCD1* mRNA by RT-PCR, together with the genomic sequence, indicatcd that the MCD1 protein is very large (>1500 amino acids) and highly basic, and it did not contain recognizable motifs. Sequence analysis of the *mcd1-2* allele, which was suppressed by the *mcd2-1* mutation, revealed an amber mutation of a lysine codon (AAG → TAG) that would generate a highly truncated protein of 112 amino acids. This result also suggested that *mcd2-1* could be an amber suppressor tRNA. Murakami et al. (2005) identified the suppressor tRNA by its tight linkage to the *ARG7* locus (Esposito et al., 2001), and by consulting the *Chlamydomonas* genome sequence (version 2.0); the suppressor is a serine tRNA whose anticodon changed from CGA to CUA.

Two non-photosynthetic strains with mutations in element I of the *petD* 5′ UTR (*LS2* and *LS6*) were used to isolate suppressor strains that were capable of photosynthesis (Rymarquis et al., 2006). The *mcd3* and *mcd4* suppressors were isolated as spontaneous revertants of the *LS2* mutant, and *mcd5* was isolated after UV mutagenesis of *LS6*. Levels of *petD* mRNA were partially restored in all three of these suppressors, and genetic analysis indicated they were distinct nuclear genes. Genetic analysis also showed that the *mcd4* suppressor was recessive, and that the wild-type product, MCD4, might interact with the MCD1 protein. Analysis of other chloroplast mRNAs revealed that many genes had altered 3′ processing in *mcd3* and *mcd4* (less in *mcd5*), and a few genes were affected in 5′ processing. The authors suggested that these genes could encode subunits of large complexes that process RNA ends and internal polycistronic sites (Figure 27.2).

A site-directed mutation in the 5′ UTR of the *psbD* gene that inhibits 5′ processing and reduces stability of the RNA (*PRB2*) was used to isolate three suppressor strains (*SuPRB2A-1*, *SuPRB2A-2*, and *SuPRB2A-3*) with significantly restored *psbD* RNA levels (Nickelsen, 2000). Genetic analysis demonstrated that these suppressors represented three different nuclear genes. It is not known if any of these genes are allelic to the suppressors isolated using a 5′-destabilized *petD* transcript (*mcd3* and *mcd4*), but it seems like a reasonable possibility.

II. CHLOROPLAST RNA STABILITY AND DEGRADATION

A. Overview

All RNAs are subject to degradation, and this can be a major step in gene regulation. In general, mRNAs have shorter half-lives than rRNAs or tRNAs, and degradation provides half the means to regulate accumulation, the other component being the rate of synthesis and processing. Chloroplast mRNA half-lives vary at least 20-fold, from >8 hours for *psbA* to <0.5 hour for *tufA*, and are affected by light, nutrient stress, and translation. RNA degradation is also needed to remove damaged RNAs, recycle nucleotides, and dispose of processing byproducts, such as the RNA downstream of 3′ processing sites. The degradation of some of these RNAs, excised introns for example, may be necessary to release bound proteins so that they can recycle. An advantage to studying RNA turnover in *Chlamydomonas* is that mRNA half-lives can be determined *in vivo* using an inhibitor-chase assay (Hwang and Herrin, 1994; Hwang et al., 1996) or by pulse-chase labeling with ^{32}P-orthophosphate (Salvador et al., 1993b). Also, as it turns out, a number of *Chlamydomonas* mutants affected in photosynthesis and/or chloroplast gene expression have post-transcriptional defects that seem to affect a single mRNA (monocistronic or polycistronic). These mutants are suggestive of a great deal of potential for gene-specific regulation by the nucleus.

B. Degradation pathways

Degradation of mRNAs is generally efficient in chloroplasts and intermediates do not usually accumulate to high levels; however, small amounts (relative to the intact mRNA) can often be detected. It takes some care, however, to distinguish true intermediates from degradation products generated during RNA isolation, although mutants or other conditions that increase intermediates can be helpful in this regard. On the upside, low levels of degradation intermediates suggest that the initial step in RNA degradation is probably rate-limiting, and likely to be the most regulated step. The first step in degrading chloroplast mRNAs is unknown for most genes; however, there is *in vivo* evidence (discussed below) that suggests the degradation of the *petD* and *tufA* mRNAs begins at their 5′ ends.

Following the successful employment of a poly(G) insertion to define the target of the *mcd1* mutant (Drager et al., 1998), Drager et al. (1999) presented evidence that the *petD* mRNA is also degraded by a 5′→3′ exonuclease activity in wild-type cells. This was accomplished by inserting the poly(G) sequence into the *petD* 5′ UTR at nt 25 and 165 in a wild-type strain. Mapping of *petD* 5′ ends in these transformants by RNase protection revealed new transcripts whose 5′ ends terminated at the poly(G) sequence, consistent with degradation by a 5′→3′ exonuclease. Bacteria do not use a 5′→3′ exonuclease to degrade RNA, raising the possibility that the chloroplast co-opted a cytoplasmic enzyme during its evolution. So far, this activity has not been reported in other plants.

Evidence was published indicating that degradation of *tufA* mRNA *in vivo* begins at the 5′ end within the UTR (Zicker et al., 2007). The *tufA* gene is unusual because it is chloroplast-encoded in *Chlamydomonas*, but nucleus-encoded in angiosperms (e.g. Baldauf and Palmer, 1990). The *Chlamydomonas* gene is expressed mainly as a 1.7-kb monocistronic mRNA that has a relatively short half-life in the light (20–30 minutes). Zicker et al. (2007) showed that chloramphenicol, which freezes translating chloroplast ribosomes, causes the nearly quantitative accumulation of a 1.5-kb degradation intermediate that is missing 180 nt from the 5′ end. The intermediate's 5′ end lies within a region of the UTR that is predicted to be highly structured and to encode a small ORF. The intermediate was detected in older cultures without chloramphenicol treatment, indicating it is not an artifact of drug treatment. Also, it did not over-accumulate in a chloroplast ribosome-deficient mutant, indicating its stabilization was specific to elongation-arrested ribosomes. It is not known whether this early degradation intermediate is created by an endonuclease, an exonuclease, or both, but it is likely destroyed by a 5′→3′ exonuclease activity.

As mentioned in the 3′ processing section (I.D), deletion of the 3′ IR destabilized the *atpB* mRNA (Stern et al., 1991). Drager et al. (1996) showed that replacing the IR with a poly(G) sequence of 18 residues caused the accumulation of a discrete *atpB* transcript that terminated with the poly(G) sequence. A similar result was obtained *in vitro* with chloroplast extracts, and both findings argue for degradation of the defective mRNA by a 3′→5′ exonuclease activity (Figure 27.2).

Processing of the *atpB* pre-mRNA at the 3′ end also creates a downstream product that does not accumulate to detectable levels *in vivo* or *in vitro* in a chloroplast extract. Hicks et al. (2002) inserted the 3′ processing signals (IR and ECS) of the *atpB* gene into the 5′ UTR of two different reporter genes (*aadA* and *uidA*) to examine the degradation of the downstream product. The reporter mRNAs did not accumulate when both elements (IR and ECS) were used, but they accumulated partially when only one element (either IR or ECS) was present. Insertion of a poly(G) sequence just downstream of the inserted IR+ECS region did not allow accumulation of the reporter RNA, indicating that degradation of the downstream product is not by an exonuclease. The authors suggested that degradation of RNA downstream of 3′ processing signals is performed by a processive endonuclease, possibly similar to RNase E in *E. coli*. A homologue of RNase E has not been reported for *Chlamydomonas*, however, an RNase J homologue, which could degrade RNA 5′→3′, appears to be present (see section II.C).

An important insight into chloroplast RNA degradation concerned the role of polyadenylation at the 3′ end. Following studies with *E. coli* and spinach, polyadenylated versions of several chloroplast transcripts were detected in *Chlamydomonas* by RT-PCR using an oligo(dT)-adapter primer and nested gene-specific (upstream) primers (Komine et al., 2000). This approach was highly specific, and enabled detection of the very low abundance poly(A) RNAs. Sequencing of the PCR products revealed poly(A) forms for both of the mRNAs that were examined, *atpB* and *petD*. The data also showed that polyadenylation had occurred internally, at the mature 3′ end, and downstream of the mature 3′ end in precursor RNAs. Unlike spinach, the poly(A) tails on *Chlamydomonas* mRNAs were nearly homopolymeric (>98% adenine); they were also detected on tRNAs and 5S rRNA. These investigators also showed that poly(A) RNAs were more prone to degradation in chloroplast extracts than non-poly(A) RNAs, but the differential varied with the transcript (it was large for *atpB* and small for *trnE*).

Subsequently, Komine et al. (2002) demonstrated that a poly(A) tail induces degradation of a chloroplast mRNA *in vivo*, but not if the tail is downstream of a normal 3′ UTR. These experiments involved manipulating the organelle so as to create an mRNA with a defined 3′ end *in vivo*. This was achieved by inserting a $tRNA^{Glu}$ precursor gene downstream of either *atpB* or chimeric *petD-gfp-rbcL* (gfp = green fluorescent protein) genes with terminal poly(A) or poly(U+A) sequences. Cleavage of the tRNA precursor by RNase P thus left the mRNAs with the defined 3′ end. The results showed that having a poly(A) or a poly(U)+AAA at the 3′ end destabilized the RNA, but having poly[A+U] did not, suggesting that the degradation machinery recognizes poly(U) as well as poly(A). The authors also isolated suppressors of the *Δ26pAtE* mutant, which had an *atpB* RNA that was unstable because it lacked the IR and was terminated with a poly(A) tail of 28 residues. These *spa* suppressors showed increased levels of discrete *atpB* mRNAs that were similar in size, or larger than wild-type *atpB* mRNA. Apparently, the restored transcripts had not been cleaved at the RNase P site, and therefore did not have an exposed poly(A) tail, which must be critical for degradation.

Nishimura et al. (2004) characterized a number of *spa* mutants, determining they all had chloroplast-localized mutations, and that two of them, *spa19* and *spa23*, were heteroplasmic with two chloroplast genomes. One of the genomes in these strains had undergone a rearrangement that complemented the other genome, allowing the suppressor strain to be photoautotrophic. The complementation involved the production of an antisense transcript for *atpB* by the rearranged genome. The proposed mechanism was that the antisense RNA annealed to the sense *atpB* RNA, stabilizing it against exonucleases. The *spa19* chloroplast genome was introduced into a nuclear background that had reduced synthesis of the chloroplast polynucleotide phosphorylase (PNPase), because of an antisense gene for *PNP*. The progeny of this cross no longer required the antisense *atpB* genome, suggesting that PNP is largely responsible for the degradation of unstable, poly(A) *atpB* mRNA.

C. Chloroplast RNases

Early work on the function of the 3′ IR led to the discovery that mRNAs can be degraded by a 3′→5′ exonuclease activity if they are not stabilized; a similar activity was observed *in vitro* with chloroplast extracts (Drager et al., 1996). Endonuclease activities that are probably involved in mRNA degradation (Nickelsen et al., 1994) and/or 3′ processing (Stern and Kindle, 1993) have also been demonstrated *in vitro* with chloroplast extracts. A processive 5′→3′ endonuclease activity that degrades the distal cleavage product of 3′ processing has also been proposed (Hicks et al., 2002). Evidence of a 5′→3′ exonuclease was obtained *in vivo*, though this activity has not been demonstrated with chloroplast extracts *in vitro* (Bollenbach et al., 2004). These processes are also indicated in Figure 27.2.

Table 27.1 gives the known and candidate genes for chloroplast RNases. The first chloroplast RNase identified in *Chlamydomonas* was a PNPase, PNP1 (GenBank accession ABF57675), whose depletion in antisense and RNAi strains affected the stability and degradation of poly(A) transcripts, and the ability of *Chlamydomonas* to adapt to phosphate starvation (Nishimura et al., 2004; Yehudai-Resheff et al., 2007). PNPase is well known from *E. coli* and spinach chloroplasts (reviewed in Bollenbach et al., 2004), where it catalyzes the reversible reactions of RNA degradation and polymerization. RNA degradation by PNPase resembles an exonuclease in that it proceeds from 3′ to 5′; however, the reaction is phosphorylytic, and the enzyme requires phosphate. It also needs a poly(A) tail to bind efficiently. RNA polymerization by PNPase produces a long heterogeneous tail on non-structured RNAs. The overall direction of the reaction *in vitro* is affected by the Mg^{2+} concentration and the Pi:NDP ratio; these parameters could also influence PNPase activity *in vivo*. PNP1 in *Chlamydomonas* could thus be involved in both the adenylation and degradation of poly(A) RNA.

Yehudai-Resheff et al. (2007) also presented evidence for a RNase II homologue in *Chlamydomonas* chloroplasts, RNB2 (accession ABO20871), that could be a 3′→5′ hydrolytic exonuclease. The N-terminal peptide of RNB2 directed a GFP fusion protein to chloroplasts of tomato, and expression of the *RNB2* gene was increased in a PNPase-depleted strain, where it may have compensated for the low PNP1 levels. *Chlamydomonas* appears to encode proteins similar to CSP41a and CSP41b of land plants. These proteins co-purified with *Chlamydomonas* chloroplast ribosomes, and were named RAP41 (EDP05800) and RAP38 (EDP05666), respectively, by Yamaguchi et al. (2003). CSP41a from spinach is a chloroplast endoribonuclease that binds and cleaves stem-loops in 3′ UTRs *in vitro* (reviewed in Bollenbach et al., 2004). It may initiate turnover of *rbcL* and other chloroplast mRNAs *in vivo*. CSP41b may have a function similar to CSP41a. It is not yet certain if the *Chlamydomonas* proteins are endonucleases.

The *Chlamydomonas* genome also encodes homologues of bacterial (RNase J1) and yeast (Xrn1p) proteins that catalyze 5′→3′ exonuclease degradation (Parker and Song, 2004; Mathy et al., 2007). The RNase J1 homologue (alternatively spliced variants are RNJ1 (accession EU518648) and RNJ2 (accession EU518649) may be chloroplast targeted, and one of the Xrn1p homologues (XRN1, accession EDP01687) has a predicted organelle targeting sequence.

D. *Cis*-acting stability elements

The analysis of *Chlamydomonas* mutants deficient in the accumulation of chloroplast mRNAs has revealed that degradation of these RNAs can be very rapid if they are not stabilized, and that stabilization is often linked to processing. Most studies of *cis*-acting elements that promote mRNA stability have focused on the 5′ and/or 3′ UTRs, but for the *atpA* gene (and possibly others), the coding region has a stability determinant that is the target of the nuclear *mda1* gene. RNA gel blots showed that the *mda1* (previously *ncc1*) mutant was strongly deficient in intact *atpA* mRNA, although it was photoautotrophic (Drapier et al., 1992). However, when S1 nuclease mapping with a 5′ probe was used to examine *atpA* RNA in the mutant, the levels were much closer to the wild type (Drapier et al., 1992). This result suggested that the *mda1* mutation acts on the *atpA* 3′ UTR or coding region. By using chimeric genes inserted into the chloroplast genome of a *mda1* mutant, Drapier et al. (2002) showed that it is the coding region of *atpA*, rather than the UTRs, that is the target of this dominant mutation.

As discussed in the section on 3′ processing above, deleting the IR at the end of the *atpB* gene resulted in a dramatic reduction in RNA levels. Also, extending the coding region into the IR destabilized the *psaB* mRNA (Lee et al., 1996), presumably because translating ribosomes disrupt the stem-loop. Clearly the 3′ UTR is an important stability element for chloroplast mRNAs (Stern et al., 1991; Blowers et al., 1993). In order to compare the abilities of different 3′ UTRs, Rott et al. (1998b) replaced the 3′ processing region (including the UTR) of *atpB* with the corresponding region from several other chloroplast genes (*petA*, *petD*, *petD-trnR*, *rbcL*) and with an *E. coli* terminator sequence (*thrA*). This was done by adding the various 3′ sequences, in both orientations, to the Δ26 deletion mutant of *atpB*. All of the 3′ regions increased the level of the *atpB* RNA, and the sense strand constructs outperformed the antisense constructs in all cases. The data suggested that both the nucleotide sequence and stem-loop structure are important for RNA stabilization.

A number of studies have underscored the importance of the 5′ UTR in chloroplast gene expression (e.g. Salvador et al., 1993a; Barnes et al., 2005), but the 3′ UTR is also a determinant for the *tufA* gene. The *tufA* mRNA has a relatively short half-life, and its degradation *in vivo* begins at the 5′ end of the RNA within the UTR (Zicker et al., 2007). To determine if the 5′ UTR of *tufA* was destabilizing, the corresponding region of a chimeric *atpA-aadA-rbcL* gene was replaced with the *tufA* sequence, but the level of the mRNA was unchanged. The same result was obtained using the *Renilla luciferase* (*Rluc*) as the reporter. However, replacing the *rbcL* 3′ region with the *tufA* 3′ sequence lowered the mRNA level significantly. Constructs with both UTRs replaced with *tufA* sequences were similar to the 3′ *tufA*-substituted genes, indicating that the 3′ UTR is dominant over the 5′ UTR in controlling *tufA* mRNA levels. It is not clear, however, whether the *tufA* 3′ UTR is also inefficient at promoting 3′ processing, or if it works mainly by promoting turnover of the mature mRNA.

The *rbcL* mRNA has a long half-life (several hours) that is shorter in the light compared to darkness. Site-directed mutagenesis delineated nt 1–63 of the 5′ UTR as necessary to stabilize a GUS reporter mRNA (with a *psaB* 3′ UTR), and confer strong light-dark regulation on the transcript (Anthonisen et al., 2001; Singh et al., 2001). This region was predicted to adopt a secondary structure of two stem-loop domains (nt 1–41 and 49–63, respectively) separated by a short unpaired region. Dimethylsulfate protection mapping of this region was performed *in vitro* using total cellular RNA, and the results were consistent with the predicted structural model (Anthonisen et al., 2001). 3′ deletions into this region indicated that the first stem-loop (nt 1–41) was essential, since RNAs lacking this structure were 50-fold less stable (Salvador et al., 2004). An extensive series of structural alterations revealed that the sequence and structure of the base of the stem (i.e. the bottom 4 nt pairs), plus the short unpaired sequence, were the key regions for RNA stability. The results also suggested that the stem-loop does not protect the RNA by inhibiting a processive exonuclease, but is more likely to function by binding a specific protein (Suay et al., 2005). One possibility not mentioned by the authors is that the two stem-loop domains might stack on top of one another (base to base), leaving the unpaired region as a CCG bulge. If the first C, and the G of this bulge base paired, or interacted in some other way, that could help stabilize the stacked structure; a protein could recognize this juxtaposition of domains and further stabilize the structure. Anthonisen et al. (2001) also mapped a stabilizing element within the first 63 nt of the *atpB* 5′ UTR, specifically to nt 31–42. However, this RNA had two prominent 5′ ends, and they appeared to adopt different structures.

Fargo et al. (2000) performed an extensive genetic analysis of second-site mutations in the *rps7* 5′ UTR that suppressed an unstable *rps7-aadA* mRNA in *E. coli*; the primary mutations were also in the *rps7* 5′ UTR (Fargo et al., 1999). Neither the primary mutants nor the suppressed RNAs showed evidence of a disrupted structure based on their UV melting profiles. However, there were differences in the proteins that bound tightly to these UTRs in a UV cross-linking assay. The primary mutants did not cross-link to 47- and 32-kD proteins like the wild-type *rps7* UTR, but the suppressed UTRs cross-linked to the 32-kD protein. This result suggested that binding of the 32-kD protein is responsible for the stabilization of the mRNAs in the UTR suppressor mutants.

The *psbB* gene is the first in a cluster with two other photosystem II genes, *psbT* and *psbH*. Mapping of *psbB* 5′ RNA ends revealed a minor long form (145-nt UTR), and a much more abundant short form (35-nt UTR) in wild-type cells, but only the minor long form in mutant *222E* (Vaistij et al., 2000b). The long form has two large stem-loops in the 5′ UTR, whereas the 5′ end of the short form maps to the loop of the second stem-loop. Vaistij et al. (2000b) deleted the first stem-loop without much effect on the *psbB* and *psbB/T* mRNAs; however, deletion of the 5′ half of the second stem-loop decreased accumulation of these RNAs, and combining these two deletions abolished them. The authors also inserted a poly(G) sequence in the 5′ UTR of a *psbB-aadA* chimeric gene that was transformed into the chloroplast. The poly(G) caused the accumulation of a transcript that terminated at the G-tract and decreased the level of the short-UTR form, suggesting strongly that the short form is created by 5′–3′ exonuclease digestion of the precursor. The *psbB-aadA* reporter, and other chimeric genes were also used to show that the target of the 222E (*Mbb1*) gene is the 5′ UTR of *psbB*, and that *psbH* also contains a target.

Mutagenesis of the *psbD* 5′ leader revealed two regions required for accumulation of the mature or short form (47-nt UTR): one is at the 5′ end of the precursor (nt 1–12 of the leader) and the other is located at –30 (relative to the start codon). The location of the first element could indicate that recognition and processing of the precursor is necessary to accumulate the mature mRNA. The mutants with changes in the second element had the less abundant precursor, but lacked the mature mRNA. Insertion of a poly(G) sequence upstream of the mature 5′ end caused the accumulation of a new RNA beginning with that sequence, and decreased accumulation of the mature RNA. That result suggested the mature RNA end is generated by 5′→3′ exonuclease digestion of the precursor (Nickelsen et al., 1999). The poly(G) sequence also enabled the accumulation of a *psbD* RNA in the *nac2-26* mutant, which is a nuclear mutant that specifically lacks both forms of the *psbD* mRNA due to instability (Kuchka et al., 1989; Nickelsen et al., 1994). Nickelsen et al. (1994) also showed that chloroplast extracts from *nac2-26* degraded *in vitro*-synthesized *psbD* leader RNA more readily than wild-type extracts, but not when the two were mixed together, suggesting that *nac2-26* lacks a stabilizing factor. UV cross-linking with such extracts identified a 47-kD protein that bound to the *psbD* leader RNA with wild-type, but not *nac2-26* extracts. Surprisingly, the 47-kD protein, which is associated with membranes, cross-linked to the long form of the leader, but not to the short form.

E. *Trans*-acting factors for RNA stability

A number of nuclear mutants of *Chlamydomonas* have been isolated that appear to be affected in the expression of a single chloroplast gene or gene cluster. The defects in these mutants are almost always post-transcriptional, involving RNA translation, splicing, processing and/or stability. Mutants affected in RNA processing (nucleolytic) and/or stability, as well as second-site suppressors, are indicated in Table 27.1. The apparent specificity of most of these nuclear genes for a single chloroplast gene or gene cluster is a remarkable feature of this intergenomic dependency. Presumably, it provides a regulatory advantage to have such potential nuclear control over these organellar genes. The specificity of the mutants is based on the characterization of any growth and/or photosynthetic defects using a variety of physiological and molecular techniques, but it should be noted that chloroplast gene microarrays have been used on only a couple of these mutants to verify the specificity (Erickson et al., 2005). The specific target of these nuclear genes with respect to the part of the RNA that they act on has been determined for several of them, and a few of the genes have been cloned and characterized.

The first *trans*-acting factor for RNA stability that was cloned was *nac2* (*mbd1*), from the *nac2-26* mutant, which lacks mature *psbD* mRNA. It was cloned by complementation of the mutant with a cosmid library of wild-type DNA. The wild-type gene, *MBD1*, encodes a large (1385 amino acids) soluble protein with nine tetratricopeptide repeats (TPRs) in its C-terminal half (Boudreau et al., 2000). TPR, and the related pentatricopeptide repeat (PPR) proteins, are particularly numerous in plants, where they play roles in RNA processing, stability, and translation in organelles (Saha et al., 2007). The repeat structure contains alpha helices that can form a groove or surface for binding RNA or other proteins. Mutation of a conserved alanine in the 4th TPR motif of *MBD1* inhibited folding of the protein, and consequently, *psbD* mRNA accumulation. The MBD1 protein may not bind RNA directly, however (Nickelsen et al., 1999), and instead may act as a scaffold for assembly of a complex that binds the 5′ UTR of *psbD*. Schwarz et al. (2007) described a 550-kD complex that contains MBD1 and the RNA-binding protein, RB38, which does bind to *psbD* mRNA; the binding of RB38 to the mRNA is MBD1-dependent. Schwarz et al. (2007) also suggested that the complex is displaced from *psbD* mRNA before it finds its way into polysomes.

The nuclear gene affected in the *222E* mutant, which lacks the *psbB* and *psbH* mRNAs, was also cloned by complementation, but with an indexed cosmid library (Vaistij et al., 2000a). The wild-type gene, *MBB1*, encodes a protein of 662 amino acids that contains 10 TPR repeats. Most of the protein was found in the soluble phase of fractionated chloroplasts, as part of a high molecular weight complex that contained RNA. Like the

MBD1 protein, MBB1 may not bind RNA with great affinity directly, but instead promote assembly of a complex that has RNA-binding subunits.

The cloning of the *MCD1* gene, which is required for *petD* mRNA processing and stability, was described in section I.F. This gene also encodes a very large protein (like MBD1), although MCD1 does not have recognizable motifs (Murakami et al., 2005). Finally, the protein required for accumulation of the *petA* transcript, MCA1, was cloned using the *mca1-Mφ11* mutant (Gumpel et al., 1995; referenced in Raynaud et al., 2007). This protein is short-lived, and can therefore regulate levels of *petA* mRNA.

F. Regulation of chloroplast RNA stability

The stability of chloroplast mRNAs has been shown to be regulated by endogenous (transcription, translation, plastid development), and exogenous (light, phosphate limitation) factors. Salvador et al. (1993b) reported that 16S rRNA and several mRNAs (*rbcL, atpB, tufA, psaB, psbA*) were considerably (3–4-fold) less stable in the light than in the dark. The authors used *in vivo* pulse-chase labeling with $^{32}PO_4$ and hybridization of labeled RNA to DNA dot-blots to estimate RNA half-lives. This technique avoids potential side effects of rifampicin, which is the drug of choice for inhibitor-chase estimates of chloroplast RNA half-lives; however, the RNA is not directly visualized as it is with the inhibitor-chase protocol. It is also possible that newly synthesized RNA might behave somewhat differently than the bulk pool of RNA. As mentioned above, similar *tufA* mRNA half-lives (20–90 minutes depending on illumination) were obtained with both techniques, but the half-lives determined by pulse-chase for the stable *psbA* and 16S RNAs in the light (1.5 and 7.5 hours, respectively) seem extraordinarily short. For example, a half-life of 7.5 hours for 16S rRNA would mean that most of the RNA present at the beginning of a 12-hour light period would decay by the end of the period. The atypical growth conditions used for the ^{32}P-phosphate labeling (low phosphate medium at 32°C) may have had some effect, or perhaps the extrapolation from a 2-hour chase could have been a factor. But notwithstanding the absolute mRNA half-lives, it seems clear from the data of Salvador et al. (1993b) and others (Hwang et al., 1996; Deshpande et al., 1997) that chloroplast RNA degradation increases in the light.

The mechanism of light stimulation of RNA degradation was studied *in vivo* from the viewpoint that it might be redox controlled (Salvador and Klein, 1999). Inhibition of photosynthetic electron transport with (3-(3,4-dichlorophenyl)-1,1-dimethylurea) DCMU blocked the light-induced decrease in a *rbcL-GUS-psaB* mRNA, whereas the reductant, dithiothreitol, partially overcame the DCMU inhibition. Also, the addition of dithiothreitol in the dark increased degradation of the transgenic mRNA, and slightly increased degradation of three endogenous chloroplast mRNAs (*psaB, atpB,* and *rbcL*). The data thus support the hypothesis that the redox status of the organelle plays a role in regulating mRNA stability, and moreover, could be the mediator of the light effect. The proteins that must be involved in the light-dark regulation of chloroplast RNA stability have not been identified, but direct evidence for involvement of the 5′ UTR was obtained for the *rbcL* mRNA (Salvador et al., 1993a; Singh et al., 2001).

In contrast to light, phosphate starvation increased the levels and stability of the *tufA* and *atpB* mRNAs (Yehudai-Resheff et al., 2007). This response required several hours to manifest, and was due in part to a decrease in the chloroplast PNPase. The PNPase level is regulated in turn by the Phosphorus Starvation Response gene (PSR1), which triggers a decrease in PNPase expression when phosphate is limiting. There is also genetic evidence suggesting that decreased transcription of a chloroplast gene can result in a partial compensation of the mRNA levels by increased mRNA stability. The mutant, *G64*, has a substitution in the *psbD* promoter that caused a 90% decline in *psbD* transcription, but it resulted in only a 65% decrease in the *psbD* mRNA level (Klinkert et al., 2005). The *psbD* protein level correlated closely with the mRNA level in this mutant.

The levels and stability of several chloroplast mRNAs, including *psbD*, *psaB*, and *tufA* increase when their translation is reduced (Herrin and Schmidt, 1987; Kuchka et al., 1988; Silk and Wu, 1988; Xu et al., 1993). Antibiotics that inhibit chloroplast translation, especially chloramphenicol, greatly increased *tufA* mRNA levels (Silk and Wu, 1988). Also, *tufA* mRNA half-life correlates inversely with translation in light-dark cycles, and in cells transferred to continuous light (Breidenbach et al., 1990; Hwang et al., 1996; Lee and Herrin, 2002). It was proposed that translation of an upstream ORF in the 5′ UTR of *tufA* might displace proteins that stabilize the mRNA (Zicker et al., 2007).

tufA mRNA, whose level fluctuates more dramatically than nearly all other chloroplast mRNAs, also provides an example of the role plastid development can play in mRNA stability. *Chlamydomonas* mutants whose chlorophyll synthesis and thylakoid membrane development are light-dependent are well known (see Chapter 8, Volume 1). Light administration to such a strain that had been pre-grown in darkness increased *tufA* mRNA levels by increasing the stability of the mRNA (Silk and Wu, 1993). This light-effect on *tufA* mRNA levels is the opposite of what happens in cells with a fully developed chloroplast.

III. CONCLUSIONS AND PERSPECTIVES

Considerable progress has been made in understanding chloroplast RNA stability in *Chlamydomonas*, especially when compared to other plants, but less is known about RNA processing. Although essentially all chloroplast

RNAs are processed, and such processing is usually critical, the enzymes that process mRNA and rRNA precursors are basically unknown. Accessory proteins like RAT1 might target or orchestrate the action of the processing enzymes. Cloning of other mutant genes, for example *crp3,* should provide additional access to some of these proteins. Also, the development of chloroplast extracts that can process 5′ or ITS regions of pre-RNAs *in vitro* would be a useful tool to complement the genetic approaches.

Some of the enzymes that degrade chloroplast mRNAs (i.e., exonucleases), and some of the proteins that protect the RNAs from degradation have been discovered, and characterized at least partially. The protective proteins are often in large complexes that include PPR/TPR proteins. It also appears that these RNA-binding complexes may contain common as well as unique proteins. A more complete catalogue of the complexes, their subunits and dynamic interactions with the mRNA targets will be necessary to fully understand this important aspect of RNA stabilization.

The mechanism that determines the distinct half-lives of different chloroplast mRNAs remains to be elucidated, but one question that come to mind is as follows. Does mRNA stability reflect the basic affinity of stabilizing proteins for the UTRs? Or does it also involve active displacement of protective proteins by translation or other processes? Another consideration is that endonucleases might catalyze the initial cleavage of many chloroplast mRNAs, and these enzymes have not yet been identified. It should also be noted that most of the aforementioned studies focused on mRNAs coding for photosynthetic proteins; it remains to be seen if these principles will apply to genes for translation and other aspects of chloroplast biology. Finally, a challenging, but important goal for the future will be to understand how different signals are integrated to regulate mRNA stability in the chloroplast.

ACKNOWLEDGMENTS

The author gratefully acknowledges the editors for constructive comments. Also, thanks are due to D. Stern and M. Goldschmidt-Clermont for providing manuscripts prior to publication. The author was supported by grants from the Dept. of Energy (DE-FG03-02ER15352) and the Robert A. Welch Foundation (F-1164).

REFERENCES

Anthonisen, I.L., Salvador, M.L., and Klein, U. (2001). Specific sequence elements in the 5′ untranslated regions of *rbcL* and *atpB* gene mRNAs stabilize transcripts in the chloroplast of *Chlamydomonas reinhardtii*. *RNA* 7, 1024–1033.

Balczun, C., Bunse, A., Hahn, D., Bennoun, P., Nickelsen, J., and Kück, U. (2005). Two adjacent nuclear genes are required for functional complementation of a chloroplast *trans*-splicing mutant from *Chlamydomonas reinhardtii*. *Plant J.* **43**, 636–648.

Baldauf, S.L. and Palmer, J.D. (1990). Evolutionary transfer of the chloroplast *tufA* gene to the nucleus. *Nature* **344**, 262–265.

Barnes, D., Franklin, S., Schultz, J., Henry, R., Brown, E., Coragliotti, A., and Mayfield, S.P. (2005). Contribution of 5′- and 3′-untranslated regions of plastid mRNAs to the expression of *Chlamydomonas reinhardtii* chloroplast genes. *Mol. Genet. Genomics* **274**, 625–636.

Bateman, J.M. and Purton, S. (2000). Tools for chloroplast transformation in *Chlamydomonas*: Expression vectors and a new dominant selectable marker. *Mol. Gen. Genet.* **263**, 404–410.

Blowers, A.D., Klein, U., Ellmore, G.S., and Bogorad, L. (1993). Functional *in vivo* analyses of the 3′ flanking sequences of the *Chlamydomonas* chloroplast *rbcL* and *psaB* genes. *Mol. Gen. Genet.* **238**, 339–349.

Bollenbach, T.J., Schuster, G., and Stern, D.B. (2004). Cooperation of endo- and exoribonucleases in chloroplast mRNA turnover. *Prog. Nucleic Acids Res. Mol. Biol.* **78**, 305–337.

Bollenbach, T., Schuster, G., Portnoy, V., and Stern, D.B. (2008). Processing, degradation and polyadenylation of chloroplast transcripts. *Topics Curr. Genet.* **19**, 175–211.

Boudreau, E., Nickelsen, J., Lemaire, S.D., Ossenbuhl, F., and Rochaix, J.-D. (2000). The *Nac2* gene of *Chlamydomonas* encodes a chloroplast TPR-like protein involved in *psbD* mRNA stability. *EMBO J.* **19**, 3366–3376.

Breidenbach, E., Leu, S., Michaels, A., and Boschetti, A. (1990). Synthesis of EF-Tu and distribution of its mRNA between stroma and thylakoids during the cell cycle of *Chlamydomonas reinhardtii*. *Biochim. Biophys. Acta* **1048**, 209–216.

Bruick, R.K. and Mayfield, S.P. (1998). Processing of the *psbA* 5′ untranslated region in *Chlamydomonas reinhardtii* depends upon factors mediating ribosome association. *J. Cell Biol.* **143**, 1145–1153.

Decker, C.J. and Parker, R. (1993). A turnover pathway for both stable and unstable mRNAs in yeast: Evidence for a requirement for deadenylation. *Genes Dev.* **7**, 1632–1643.

De la Cruz, J. and Vioque, A. (2003). A structural and functional study of plastid RNAs homologous to catalytic bacterial Rnase P RNA. *Gene* **321**, 47–56.

Deshpande, N.N., Bao, Y., and Herrin, D.L. (1997). Evidence for light/redox-regulated splicing of *psbA* pre-RNAs in *Chlamydomonas* chloroplasts. *RNA* **3**, 37–48.

Drager, R.G., Zeidler, M., Simpson, C.L., and Stern, D.B. (1996). A chloroplast transcript lacking the 3′ inverted repeat is degraded by 3′–5′ exoribonuclease activity. *RNA* **2**, 652–663.

Drager, R.G., Girard-Bascou, J., Choquet, Y., Kindle, K.L., and Stern, D.B. (1998). *In vivo* evidence for 5′ → 3′ exoribonuclease degradation of an unstable chloroplast mRNA. *Plant J.* **13**, 85–96.

Drager, R.G., Higgs, D.C., Kindle, K.L., and Stern, D.B. (1999). 5′ to 3′ exoribonucleolytic activity is a normal component of chloroplast mRNA decay pathways. *Plant J.* **19**, 521–531.

Drapier, D., Girard-Bascou, J., and Wollman, F.A. (1992). Evidence for nuclear control of the expression of the *atpA* and *atpB* chloroplast genes in *Chlamydomonas*. *Plant Cell* **4**, 283–295.

Drapier, D., Girard-Bascou, J., Stern, D.B., and Wollman, F.A. (2002). A dominant nuclear mutation in *Chlamydomonas* identifies a factor controlling chloroplast mRNA stability by acting on the coding region of the *atpA* transcript. *Plant J.* **31**, 687–697.

Drapier, D., Suzuki, H., Levy, H., Rimbault, B., Kindle, K.L., Stern, D.B., and Wollman, F.A. (1998). The chloroplast *atpA* gene cluster in *Chlamydomonas reinhardtii*. *Plant Physiol.* **117**, 629–641.

Erickson, B., Stern, D.B., and Higgs, D.C. (2005). Microarray analysis confirms the specificity of a *Chlamydomonas reinhardtii* chloroplast RNA stability mutant. *Plant Physiol.* **137**, 534–544.

Esposito, D., Higgs, D.C., Drager, R.G., Stern, D.B., and Girard-Bascou, J. (2001). A nucleus-encoded suppressor defines a new factor which can promote *petD* mRNA stability in the chloroplast of *Chlamydomonas reinhardtii*. *Curr. Genet.* **39**, 40–48.

Fargo, D.C., Boynton, J.E., and Gillham, N.W. (1999). Mutations altering the predicted secondary structure of a chloroplast 5′ untranslated region affect its physical and biochemical properties as well as its ability to promote translation of reporter mRNAs both in the *Chlamydomonas reinhardtii* chloroplast and in *Escherichia coli*. *Mol. Cell. Biol.* **19**, 6980–6990.

Fargo, D., Hu, E., Boynton, J.E., and Gillham, N.W. (2000). Mutations that alter the higher-order structure of its 5′ untranslated region affect the stability of chloroplast *rps7* mRNA. *Mol. Gen. Genet.* **264**, 291–299.

Goldschmidt-Clermont, M. (1998). Coordination of nuclear and chloroplast gene expression in plant cells. *Int. Rev. Cytol.* **177**, 115–180.

Goldschmidt-Clermont, M., Rahire, M., and Rochaix, J-D. (2008). Redundant *cis*-acting determinants of 3′ processing and RNA stability in the chloroplast *rbcL* mRNA of *Chlamydomonas*. *Plant J.*, **53**, 566–577.

Gumpel, N.J., Ralley, L., Girard-Bascou, J., Wollman, F.-A., Nugent, J.H.A., and Purton, S. (1995). Nuclear mutants of *Chlamydomonas reinhardtii* defective in the biogenesis of the cytochrome b_6f complex. *Plant Mol. Biol.* **29**, 921–932.

Hahn, D., Nickelsen, J., Hackert, A., and Kück, U. (1998). A single nuclear locus is involved in both chloroplast RNA *trans*-splicing and 3′ end processing. *Plant J.* **15**, 575–581.

Harris, E.H., Boynton, J.E., and Gillham, N.W. (1974). Chloroplast ribosome biogenesis in *Chlamydomonas*: Selection and characterization of mutants blocked in ribosome formation. *J. Cell Biol.* **63**, 160–179.

Herrin, D.L. and Nickelsen, J. (2004). Chloroplast RNA processing and stability. *Photosynth. Res.* **82**, 301–314.

Herrin, D.L. and Schmidt, G.W. (1987). Chloroplast gene expression in chloroplast ribosome-deficient mutants of *Chlamydomonas reinhardtii*. In: *Progress in Photosynthesis Research* (J. Biggins, Ed.), pp. 645–648. Martinus Nijhoff, Dordrecht.

Hicks, A., Drager, R.G., Higgs, D.C., and Stern, D.B. (2002). An mRNA 3′ processing site targets downstream sequences for rapid degradation in *Chlamydomonas* chloroplasts. *J. Biol. Chem.* **277**, 3325–3333.

Higgs, D.C., Shapiro, R.S., Kindle, K.L., and Stern, D.B. (1999). Small *cis*-acting sequences that specify secondary structures in a chloroplast mRNA are essential for RNA stability and translation. *Mol. Cell. Biol.* **19**, 8479–8491.

Holloway, S.P. and Herrin, D.L. (1998). Processing of a composite large subunit rRNA: studies with *Chlamydomonas* mutants deficient in maturation of the 23S-like rRNA. *Plant Cell* **10**, 1193–1206.

Hwang, S. and Herrin, D.L. (1994). Circadian regulation of lhc gene transcription in *Chlamydomonas reinhardtii*. *Plant Mol. Biol.* **26**, 557–569.

Hwang, S., Kawazoe, R., and Herrin, D.L. (1996). Transcription of *tufA* and other chloroplast-encoded genes is controlled by a circadian clock in *Chlamydomonas*. *Proc. Natl. Acad. Sci. U. S. A.* **93**, 996–1000.

Jiao, H.S., Hicks, A., Simpson, C., and Stern, D.B. (2004). Short dispersed repeats in the *Chlamydomonas* chloroplast genome are collocated with sites for mRNA 3′ end formation. *Curr. Genet.* **45**, 311–322.

Klinkert, B., Schwarz, C., Pohlman, S., Pierre, Y., Girard-Bascou, J., and Nickelsen, J. (2005). Relationship between mRNA levels and protein accumulation in a chloroplast promoter-mutant of *Chlamydomonas reinhardtii*. *Mol. Genet. Genomics* **274**, 637–643.

Komine, Y., Kikis, E., Schuster, G., and Stern, D.B. (2002). Evidence for *in vivo* modulation of chloroplast RNA stability by 3′-UTR homopolymeric tails in *Chlamydomonas reinhardtii*. *Proc. Natl. Acad. Sci. U. S. A.* **99**, 4085–4090.

Komine, Y., Kwong, L., Anguera, M.C., Schuster, G., and Stern, D.B. (2000). Polyadenylation of three classes of chloroplast RNA in *Chlamydomonas reinhardtii*. *RNA* **6**, 598–607.

Kramzar, L.M., Mueller, T., Erickson, B., and Higgs, D.C. (2006). Regulatory sequences of orthologous *petD* chloroplast mRNAs are highly specific among *Chlamydomonas* species. *Plant Mol. Biol.* **60**, 405–422.

Kuchka, M.R., Mayfield, S.P., and Rochaix, J.-D. (1988). Nuclear mutations specifically affect the synthesis and/or degradation of the chloroplast-encoded D2 polypeptide of photosystem II in *Chlamydomonas reinhardtii*. *EMBO J.* **7**, 319–324.

Kuchka, M. R., Goldschmidt-Clermont, M., van Dillewijn, J., and Rochaix, J.-D. (1989). Mutation at the nuclear *NAC2* locus of *C. reinhardtii* affects the stability of the chloroplast *psbD* transcript encoding polypeptide D2 of photosystem II. *Cell* **58**, 869–876.

Lee, H., Bingham, S.E., and Webber, A.N. (1996). Function of 3′ non-coding sequences and stop codon usage in expression of the chloroplast *psaB* gene in *Chlamydomonas reinhardtii*. *Plant Mol. Biol.* **31**, 337–354.

Lee, J. and Herrin, D.L. (2002). Assessing the relative importance of light and the circadian clock in controlling chloroplast translation in *Chlamydomonas reinhardtii*. *Photosynth. Res.* **72**, 295–306.

Levy, H., Kindle, K.L., and Stern, D.B. (1997). A nuclear mutation that affects the 3′ processing of several mRNAs in *Chlamydomonas* chloroplasts. *Plant Cell* **9**, 825–836.

Levy, H., Kindle, K.L., and Stern, D.B. (1999). Target and specificity of a nuclear gene product that participates in mRNA 3′-end formation in *Chlamydomonas* chloroplasts. *J. Biol. Chem.* **274**, 35955–35962.

Mathy, N., Benard, L., Pellegrini, O., Daou, R., Wen, T., and Condon, C. (2007) 5′-to-3′ exoribonuclease activity in bacteria: Role of RNase J1 in rRNA maturation and 5′ stability of mRNA. *Cell* **129**, 681–692.

Murakami, S., Kuehnle, K., and Stern, D.B. (2005). A spontaneous tRNA suppressor of a mutation in the *Chlamydomonas reinhardtii* nuclear *MCD1* gene required for stability of the chloroplast *petD* mRNA. *Nucleic Acids Res.* **33**, 3372–3380.

Nedelcu, A.M. and Lee, Robert W. (1998). Modes and tempos of mitochondrial and chloroplast genome evolution in *Chlamydomonas*: a comparative analysis.

In: *The Molecular Biology of Chloroplasts and Mitochondria in Chlamydomonas* (J-D. Rochaix, M. Goldschmidt-Clermont, and S. Merchant, Eds.), pp. 63–91. Kluwer Academic Publishers, Dordrecht.

Nickelsen, J. (1998). Chloroplast RNA stability. In: *The Molecular Biology of Chloroplasts and Mitochondria in Chlamydomonas* (J-D. Rochaix, M. Goldschmidt-Clermont, and S. Merchant, Eds.), pp. 151–163. Kluwer Academic Publishers, Dordrecht.

Nickelsen, J. (2000). Mutations at three different nuclear loci of *Chlamydomonas* suppress a defect in chloroplast *psbD* mRNA accumulation. *Curr. Genet.* **37**, 136–142.

Nickelsen, J. (2003). Chloroplast RNA-binding proteins. *Curr. Genet.* **43**, 392–399.

Nickelsen, J., van Dillewijn, J., Rahire, M., and Rochaix, J-D. (1994). Determinants for stability of the chloroplast *psbD* RNA are located within its short leader region in *Chlamydomonas reinhardtii*. *EMBO J.* **13**, 3182–3191.

Nickelsen, J., Fleischmann, M., Boudreau, E., Rahire, M., and Rochaix, J-D. (1999). Identification of *cis*-acting RNA leader elements required for chloroplast *psbD* gene expression in *Chlamydomonas*. *Plant Cell* **11**, 957–970.

Nishimura, Y., Kikis, E.A., Zimmer, S.L., Komine, Y., and Stern, D.B. (2004). Antisense transcript and RNA processing alterations suppress instability of polyadenylated mRNA in *Chlamydomonas* chloroplasts. *Plant Cell* **16**, 2849–2869.

Parker, R. and Song, H. (2004). The enzymes and control of eukaryotic mRNA turnover. *Nat. Struct. Mol. Biol.* **11**, 121–127.

Raynaud, C., Loiselay, C., Wostrikoff, K., Kuras, R., Girard-Bascou, J., Wollman, F-A., and Choquet, Y. (2007). Evidence for regulatory function of nucleus-encoded factors on mRNA stabilization and translation in the chloroplast. *Proc. Natl. Acad. Sci. U. S. A.* **104**, 9093–9098.

Rochaix, J.-D. (1996). Post-transcriptional regulation of chloroplast gene expression in *Chlamydomonas reinhardtii*. *Plant Mol. Biol.* **32**, 327–341.

Rott, R., Levy, H., Drager, R.G., Stern, D.B., and Schuster, G. (1998a). 3′-Processed mRNA is preferentially translated in *Chlamydomonas reinhardtii* chloroplasts. *Mol. Cell. Biol.* **18**, 4605–4611.

Rott, R., Liveanu, V., Drager, R.G., Stern, D.B., and Schuster, G. (1998b). The sequence and structure of the 3′-untranslated regions of chloroplast transcripts are important determinants of mRNA accumulation and stability. *Plant Mol. Biol.* **36**, 307–314.

Rott, R., Livenau, V., Drager, R.G., Higgs, D., Stern, D.B., and Schuster, G. (1999). Altering the 3′ UTR endonucleolytic cleavage site of a *Chlamydomonas* chloroplast mRNA affects 3′-end maturation *in vitro* but not *in vivo*. *Plant Mol. Biol.* **40**, 679–686.

Rymarquis, L.A., Higgs, D.C., and Stern, D.B. (2006). Nuclear suppressors define three factors that participate in both 5′ and 3′ end processing of mRNAs in *Chlamydomonas* chloroplasts. *Plant J.* **46**, 448–461.

Rymarquis, L.A., Webster, B.R., and Stern, D.B. (2007). The nucleus-encoded factor MCD4 participates in degradation of nonfunctional 3′ UTR sequences generated by cleavage of pre-mRNA in *Chlamydomonas* chloroplasts. *Mol. Genet. Genomics* **277**, 329–340.

Saha, D., Prasad, A.M., and Srinivasan, R. (2007). Pentatricopeptide repeat proteins and their emerging roles in plants. *Plant Physiol. Biochem.* **45**, 521–534.

Sakamoto, W., Kindle, K.L., and Stern, D.B. (1993). *In vivo* analysis of *Chlamydomonas* chloroplast *petD* gene expression using stable transformation of beta-glucuronidase translational fusions. *Proc. Natl. Acad. Sci. U. S. A.* **90**, 497–501.

Sakamoto, W., Sturm, N.R., Kindle, K.L., and Stern, D.B. (1994). *petD* mRNA maturation in *Chlamydomonas reinhardtii* chloroplasts: role of 5′ endonucleolytic processing. *Mol. Cell. Biol.* **14**, 6180–6186.

Salvador, M.L. and Klein, U. (1999). The redox state regulates RNA degradation in the chloroplast of *Chlamydomonas reinhardtii*. *Plant Physiol.* **121**, 1367–1374.

Salvador, M.L., Klein, U., and Bogorad, L. (1993a). 5′ sequences are important positive and negative determinants of the longevity of *Chlamydomonas* chloroplast gene transcripts. *Proc. Natl. Acad. Sci. U. S. A.* **90**, 1556–1560.

Salvador, M.L., Klein, U., and Bogorad, L. (1993b). Light-regulated and endogenous fluctuations of chloroplast transcript levels in *Chlamydomonas*. Regulation by transcription and RNA degradation. *Plant J.* **3**, 213–219.

Salvador, M.L., Suay, L., Anthonisen, I.L., and Klein, U. (2004). Changes in the 5′-untranslated region of the *rbcL* gene accelerate transcript degradation more than 50-fold in the chloroplast of *Chlamydomonas reinhardtii*. *Curr. Genet.* **45**, 176–182.

Schwarz, C., Elles, I., Kortmann, J., Piotrowski, M., and Nickelsen, J. (2007). Synthesis of the D2 protein of photosystem II in *Chlamydomonas* is controlled by a high molecular mass complex containing the RNA stabilization factor Nac2 and the translational activator RBP40. *Plant Cell* **19**, 3627–3639.

Sieburth, L.E., Berry-Lowe, S., and Schmidt, G.W. (1991). Chloroplast RNA stability in *Chlamydomonas*: Rapid degradation of *psbB* and *psbC* transcripts in two nuclear mutants. *Plant Cell* **3**, 175–189.

Silk, G.W. and Wu, M. (1988). Darkness and antibiotics increase the steady-state transcripts of the elongation factor gene (*tufA*) in *Chlamydomonas reinhardtii*. *Curr. Genet.* **14**, 119–126.

Silk, G.W. and Wu, M. (1993). Post-transcriptional accumulation of chloroplast *tufA* (elongation factor gene) mRNA during chloroplast development in *Chlamydomonas reinhardtii*. *Plant Mol. Biol.* **33**, 87–96.

Singh, M., Boutanaev, A., Zucchi, P., and Bogorad, L. (2001). Gene elements that affect the longevity of *rbcL* sequence-containing transcripts in *Chlamydomonas reinhardtii* chloroplasts. *Proc. Natl. Acad. Sci. U. S. A.* **98**, 2289–2294.

Stern, D.B. and Drager, R.G. (1998). Chloroplast RNA synthesis and processing. In: *The Molecular Biology of Chloroplasts and Mitochondria in Chlamydomonas* (J-D. Rochaix , M. Goldschmidt-Clermont, and S. Merchant, Eds.), pp. 164–182. Kluwer Academic Publishers, Dordrecht.

Stern, D.B. and Kindle, K.L. (1993). 3′ end maturation of the *Chlamydomonas reinhardtii* chloroplast *atpB* mRNA is a two-step process. *Mol. Cell. Biol.* **13**, 2277–2285.

Stern, D.B., Radwanski, E.R., and Kindle, K. (1991). A 3′ stem/loop structure of the *Chlamydomonas* chloroplast *atpB* gene regulates mRNA accumulation *in vivo*. *Plant Cell* **3**, 285–297.

Suay, L., Salvador, M., Abesha, E., and Klein, U. (2005). Specific roles of 5′ secondary structures in stabilizing transcripts in chloroplasts. *Nucleic Acids Res.* **33**, 4754–4761.

Turmel, M., Gutell, R.R., Mercier, J.P., Otis, C., and Lemieux, C. (1993). Analysis of the chloroplast large subunit ribosomal RNA gene from 17 *Chlamydomonas* taxa: Three internal transcribed spacers and 12 group I intron insertion sites. *J. Mol. Biol.* **232**, 446–467.

Vaistij, F.E., Boudreau, E., Lemaire, S.D., Goldschmidt-Clermont, M., and Rochaix, J.-D. (2000a). Characterization of *Mbb1*, a nucleus-encoded tetratricopeptide-like repeat protein required for expression of the chloroplast *psbB/psbT/psbH* gene cluster in *Chlamydomonas reinhardtii*. *Proc. Natl. Acad. Sci. U. S. A.* **97**, 14813–14818.

Vaistij, F.E., Goldschmidt-Clermont, M., Wostrikoff, K., and Rochaix, J.-D. (2000b). Stability determinants in the chloroplast *psbB/T/H* mRNAs of *Chlamydomonas reinhardtii*. *Plant J.* **21**, 469–482.

Xu, R., Bingham, S.E., and Webber, A.N. (1993). Increased mRNA accumulation in a *psaB* frame-shift mutant of *Chlamydomonas reinhardtii* suggests a role for translation in *psaB* mRNA stability. *Plant Mol. Biol.* **22**, 465–474.

Yamaguchi, K., Beligni, M.V., Prieto, S., Haynes, P.A., McDonald, W.H., Yates, J.R., and Mayfield, S.P. (2003). Proteomic characterization of the *Chlamydomonas reinhardtii* chloroplast ribosome. Identification of proteins unique to the 70S ribosome. *J. Biol. Chem.* **278**, 33774–33785.

Yehudai-Resheff, S., Zimmer, S.L., Komine, Y., and Stern, D.B. (2007). Integration of chloroplast nucleic acid metabolism into the phosphate deprivation response in *Chlamydomonas reinhardtii*. *Plant Cell* **19**, 1023–1038.

Zicker, A.A., Kadakia, C., and Herrin, D.L. (2007). Distinct roles for the 5′ and 3′ untranslated regions in the degradation and accumulation of chloroplast *tufA* mRNA: Identification of an early intermediate in the *in vivo* degradation pathway. *Plant Mol. Biol.* **63**, 689–702.

CHAPTER 28

Protein Synthesis in the Chloroplast

William Zerges[1] and Charles Hauser[2]
[1]Biology Department, Concordia University, Montreal, Quebec, Canada
[2]Bioinformatics Program, St. Edward's University, Austin, Texas, USA

CHAPTER CONTENTS

I. INTRODUCTION

Translation is of fundamental importance as a phase of gene expression and an energetically expensive process whose regulation is critical for cellular energy economy. Translational control functions in diverse biological processes including developmental pattern formation, intracellular protein targeting, the control of homeostasis, and a variety of stress responses. Since the discovery of genomes in plastids, *Chlamydomonas* research has played a central role in the development of our understanding of the plastid translation machinery and the mechanisms that regulate translation in chloroplasts, and this chapter reviews relevant findings and models. The early history of this research is covered in Volume 1, Chapter 7. The regulation of translation for the assembly of photosynthesis complexes is presented in Chapter 29. Reviews have covered other aspects (Harris et al., 1994; Bruick and Mayfield, 1999; Zerges, 2000; Barnes and Mayfield, 2003; Manuell et al., 2004; Alergand et al., 2006; Wobbe et al., 2008). Expression of foreign proteins in transgenic chloroplasts of *Chlamydomonas*, a technology with great potential for agriculture and biopharmaceutical production, is discussed in Volume 1, Chapter 8 (see also Mayfield et al., 2007).

II. METHODOLOGIES USED IN STUDIES OF CHLOROPLAST TRANSLATION

A. Systems reconstituted *in vitro*

Much of our understanding of translation in bacteria and the cytoplasm of eukaryotic cells has relied on *in vitro*-reconstituted systems. These systems have allowed detailed dissections of the initiation, elongation, and termination steps and their regulation. Although no such system has been developed for *Chlamydomonas*, two tobacco chloroplast translation systems could potentially serve as a guide to *Chlamydomonas* researchers (Hirose and Sugiura, 1996; Yukawa et al., 2007).

B. Radioisotope pulse labeling of newly synthesized proteins

1. Overview

Translational regulation has been revealed by changes in the synthesis rates of photosynthesis complex subunits that occur in the absence of corresponding alterations of mRNA abundance. This is advantageous because the abundance of a subunit is generally a poor indicator of its rate of synthesis; most are very stable once assembled into a photosynthesis complex. Moreover, subunits that are synthesized at normal levels, but fail to assemble (i.e. in certain mutants), are degraded and do not accumulate (de Vitry et al., 1989). Thus, protein synthesis rates in the *Chlamydomonas* chloroplast are routinely measured using pulse labeling of intact cells with $^{35}SO_4$ or [^{14}C]acetate, or isolated chloroplasts with [^{35}S]methionine. Inhibition of 80S cytoplasmic ribosomes by cycloheximide allows proteins synthesized by 70S chloroplast ribosomes to be detected by autoradiography of polyacrylamide gels that resolve total cellular proteins.

2. Considerations regarding pulse-labeling

Three caveats should be considered. First, the extremely rapid degradation of a protein that is synthesized at a normal rate could lead to the false conclusion that the mRNA encoding it is not translated. Second, many studies using $^{35}SO_4$ precede labeling with several hours of SO_4 starvation to enhance uptake and incorporation of ^{35}S. Because sulfur starvation drastically alters cell physiology, possible effects on protein synthesis should be considered (Davies et al., 1996; Wykoff et al., 1998). For pulse labeling of PS II subunits (i.e. D1, D2, CP47, and CP43) and the large subunit of Rubisco, cells can be transferred to medium with a low sulfate concentration immediately prior to pulse labeling with $^{35}SO_4$ (Uniacke and Zerges, 2007). The third caveat concerns the possibility that uptake of radiolabel could vary with culture conditions or between wild-type and mutant strains, affecting protein labeling independently of changes in their actual rates of synthesis. Indeed, one study found that a shift from the dark to moderate light or high light enhanced the rate of sulfate uptake by 2- and 2.5-fold, respectively (McKim and Durnford, 2006). Ideally, radioisotope uptake should be allowed to occur under the identical conditions in the experimental and control samples, and then quenched by excess non-radioisotopic compound during the differential conditions under comparison.

C. Analyses of chloroplast polyribosome profiles

The translation of a particular mRNA also can be monitored by its polysomal distribution following sucrose density gradient centrifugation, or simply by partitioning cellular components into bulk polysomal versus nonpolysomal fractions (Yohn et al., 1996, 1998a; Cahoon and Timko, 2000; Trebitsh et al., 2000; Minai et al., 2006). Presence of an mRNA primarily in the nonpolysomal fractions indicates a block at the initiation or pre-initiation step. When the mRNA is present in the polysome fraction, but known to be untranslated by another method (such as pulse labeling), a block at the elongation or termination phases is supported.

D. The use of chloroplast reporter genes

The ability to transform the chloroplast genome allows the use of chimeric reporter genes with the 5′ untranslated region (UTR) or 3′ UTR under study fused to the coding region. Examples of heterologous reporter genes are *aadA* (aminoglycoside adenine transferase), *uidA* (beta-glucuronidase) GFP, and *lucCt* (luciferase) (Goldschmidt-Clermont, 1991; Sakamoto et al., 1993; Franklin et al., 2002; Mayfield and Schultz, 2004; Matsuo et al., 2006). The coding regions of the endogenous *petA* and *atpA* genes have also been used as reporters (Choquet et al., 2003; Wostrikoff et al., 2004; Minai et al., 2006).

E. Genetic approaches

Genetics has been used extensively for the identification and characterization of *trans*-acting factors and *cis*-acting sequences that function in the translation of chloroplast mRNAs. This is covered in sections VI and VII.

III. CHLOROPLAST RIBOSOMES

A. Overview

The conservation of structure, composition, and function of the ribosome is reflected from organisms as phylogenetically distinct as archaea and bacteria. During the ca. 3.5 billion years since divergence of the last common ancestor of prokaryotes and eukaryotes, new ribosomal proteins were incorporated into the eukaryotic apparatus, and numerous substitutions have arisen in the sequences of the rRNAs and ribosomal proteins. Despite these changes, the overall higher order structures of the ribosomal components and their interfaces have been retained.

The crystal and cryoEM structures of the ribosome have revolutionized the field of protein synthesis by providing a three-dimensional reference platform for interpreting biochemical and genetic data. Ribosomes have two subunits that are distinguishable based on sedimentation coefficients (S), ribosomal RNAs (rRNA) and protein complement. The small subunit (SSU), 30S in bacteria and 40S in the eukaryotic cytoplasm, contains a single rRNA and 21–33 proteins. The large subunit (LSU), 50S for prokaryotes and 60S for the eukaryotic ribosome, typically contains two or three rRNAs and from 33 to 47 proteins. Each subunit contains three binding sites for tRNAs: the A (aminoacyl) site, where the incoming aminoacylated tRNA docks; the P (peptidyl) site, which associates the charged tRNA with nascent peptide chain; and the E (exit) site, which contains the deacylated tRNA prior to its exiting the ribosome. The 30S subunit binds mRNA and the tRNA anticodon stem-loop and contributes to the fidelity of translation by monitoring codon-anticodon base pairing, whereas the 50S subunit binds the acceptor arms of tRNA and catalyzes peptide formation between the amino acid on the A-site tRNA and the nascent peptide chain on the P-site tRNA. Both subunits play a role in translocation during which the mRNA and tRNAs are racheted through the ribosome in one-codon increments.

Chloroplasts and mitochondria contain prokaryotic-like translational machinery, consistent with their endosymbiotic origins (Gray, 1992; Gray et al., 1999; Bogorad, 2008). While chloroplast rRNA sequences and the sensitivity of chloroplast ribosomes to bacterial antibiotics reflect this origin, chloroplast translation is in many respects quite distinct from the bacterial process (Zerges, 2000; Manuell et al., 2004). The chloroplast system can also be distinguished from the cytosolic machinery, by number and size of the rRNAs, protein composition, stability in low concentrations of magnesium and sensitivity to antibiotics which inhibit protein synthesis (Table 28.1). Technical advances paired with cryoelectron microscopy have

Table 28.1 Comparison of cytosolic and chloroplast ribosomes

Property	Cytosolic ribosomes		Chloroplast ribosomes	
Optimum [Mg^{2+}]	10–13 mM		25 mM	
Antibiotic sensitivities	Cycloheximide, anisomycin, puromycin		Streptomycin, neamine, kanamycin, spectinomycin, erythromycin, lincomycin, chloramphenicol, puromycin	
Subunit	Small subunit	Large subunit	Small subunit	Large subunit
Sedimentation coefficient	40S	60S	30S	50S
rRNAs	18S (M32703)	25S (AF183463) 5.8S (X02708) 5S (X02706, X02707)	16S (X03269)	23S (J01398) 7S (X16686) 5S (X03271) 3S (X16686)
Proteins	33[a]	47[b]	25[c]	31[d]
Plastid encoded	–	–	11	7
Nucleus encoded	33	47	14	24

Data adapted from Beligni et al. (2004b), Manuell et al. (2005).
[a]*Includes RACK1.*
[b]*Includes acidic stalk proteins (P0, P1, P2).*
[c]*Includes 4 plastid-specific proteins (PSRP-1, PSRP-3, PSRP-4, PSRP-7).*
[d]*Includes 1 plastid-specific protein (PSRP-6).*

produced a detailed accounting of the protein composition of the chloroplast and cytoplasmic ribosomes (Table 28.2; Yamaguchi and Subramanian, 2000; Yamaguchi et al., 2000; Selmer et al., 2006; Manuell et al., 2007).

B. Ribosomal proteins

1. Characterization and nomenclature

While the RNA components of the 70S ribosome are encoded exclusively by the plastid genome, only a limited number of the ribosomal proteins are plastid-encoded. Initial observations suggested a conserved set of 21 plastid-encoded ribosomal proteins that with a few exceptions are found in all other plastid-containing taxa (reviewed in Mache, 1990; Harris et al., 1994; Sugiura et al., 1998). Considerable fluidity in the ribosomal protein gene organization has been revealed as a result of the increasing numbers of complete chloroplast genome sequences available (see de Cambiaire et al., 2007).

The proteins of the small and large *Chlamydomonas* chloroplast ribosomal subunits were first characterized using two-dimensional polyacrylamide gel electrophoresis (Hanson et al., 1974; Brügger and Boschetti, 1975;

Schmidt et al., 1983; Randolph-Anderson et al., 1989) and were numbered based on estimated size and charge. Pulse labeling in the presence of cytoplasmic or chloroplast translation inhibitors was used to determine which proteins are synthesized on chloroplast ribosomes and which in the cytoplasm (Schmidt et al., 1984). Many *Chlamydomonas* chloroplast ribosomal proteins cross-react with antisera directed against proteins from *E. coli*, cyanobacteria and spinach chloroplasts (Schmidt et al., 1984; Randolph-Anderson et al., 1989). In general, the highest apparent similarity was found between *Synechococcus* and *Chlamydomonas*, particularly for chloroplast-synthesized LSU proteins.

The advent of high-throughput proteomic methods later facilitated a more complete accounting of chloroplast ribosomal proteins from *Chlamydomonas* (Yamaguchi and Subramanian, 2003) and spinach (Yamaguchi and Subramanian, 2000; Yamaguchi et al., 2000). Yamaguchi et al. (2002, 2003) identified a minimum of 56 *Chlamydomonas* plastid ribosomal proteins, 38 nucleus-encoded and 18 plastid-encoded (Table 28.2). For consistency with other chloroplast literature, the *Chlamydomonas* full ribosome, LSU and SSU are designated as 70S, 50S, and 30S, respectively, although their sedimentation coefficients have been reported to be somewhat higher (Bourque et al., 1971; Margulies and Tiffany, 1979). The proteomic analysis was accompanied by a revised plastid ribosomal protein (PRP) nomenclature, based on sequence similarity to bacterial ribosomal proteins (Yamaguchi and Subramanian, 2000; Yamaguchi et al., 2000). According to this scheme, chloroplast homologues of *E. coli* L1–L36 are designated PRP L1–L36. For example, the gene for nucleus-encoded PRP L1 in this nomenclature would be *PRPL1*, whereas the gene for plastid-encoded PRP L2 would be *prpL2*. The reader of the older literature should be aware that *Chlamydomonas* ribosomal proteins were previously numbered based on their migration in one- and two-dimensional gels, and that these numbers do not correspond to their equivalents in bacteria (see Harris et al., 1994).

2. Proteins of the 50S ribosomal subunit

Twenty-eight proteins of the 50S subunit have *E. coli* counterparts, except the homologue of spinach PSRP-6. Homologues of bacterial L33, L34, and L36 were not identified by proteomic means but are represented in both EST (Asamizu et al., 1999; Shrager et al., 2003) and genomic sequences, indicating their likely presence in the 50S subunit. Their absence from the proteomic analysis can be attributed to their properties of being small (4.3–8.6 kD) and highly basic (pI 10.2–11.9). Homologues of bacterial L29, L25, L30, and spinach PSRP-5 were not identified either as proteins or genes/transcripts. In addition, genes encoding homologues of L25 and L30 are absent from the *Arabidopsis* genome and the proteome of spinach, indicating their likely absence from chloroplast ribosomes (Yamaguchi and Subramanian, 2000; Yamaguchi et al., 2003).

Table 28.2 Characteristics of *Chlamydomonas* plastid ribosomal proteins

				Mass (kD)		Similarity, %		
Protein	Genome	Accession #	pI[a]	Sequence	SDS-PAGE	*A.t.*	*Syn*	*E. coli*
S1	Nucleus	EDP02480	4.6	32.0	49	64	63	50
S2	Plastid	AAN17820	11.0	63.2	57	52	58	58
S3	Plastid	NP_958402	10.7	81.7	76	63	74	63
S4	Plastid	AAA81363	11.3	29.9	24–26	62	68	55
S5	Nucleus	ED099741	4.0	66.7	84	73	68	63
S6	Nucleus	EDP00617	7.0	13.3	10–11	50	56	48
S7	Plastid	NP_958380	10.4	19.0	14–16	66	67	61
S8	Plastid	NP_958374	10.1	17.1		69	69	69
S9	Plastid	CAA74006	9.8	20.9	18–19	46	46	44
S10	Nucleus	EDP03308	5.8	14.7	14–16	61	72	69
S11	Plastid	NP_958411						
S12	Plastid	AAC_16329	11.5	14.5	11–13	89	90	83
S13	Nucleus	ED098007	10.5	15.4	10–11	66	71	69
S14	Plastid	NP_958381	11.3	11.6		67	81	60
S15	Nucleus	EDP08265	9.9	12.4	10	64	61	61
S16	Nucleus	ED097973	9.9	11.5	10	66	69	58
S17	Nucleus	ED096943	10.1	9.0	10	59	62	56
S18	Plastid	CAA74009	11.9	16.3	14–16	68	75	59
S19	Plastid	NP_958370	10.8	10.3	10	83	87	80
S20	Nucleus	ED098001	9.9	14.2	11–13	62	62	50
S21	Nucleus	EDP09218	4.7	17.7		49	60	59
PSRP-1	Nucleus	ED099886	6.1	25.4		58	55	ND
PSRP-3	Nucleus	EDP06886	4.1	29.3	34	67	74	ND
PSRP-4	Nucleus	EDP02105	10.5	13.5		54	ND	ND
PSRP-7	Nucleus	EDP00635	?	38.0	66	ND	ND	ND
L1	Nucleus	EDP06736	9.4	28.5	26–29	70	63	63
L2	Plastid	NP_958369	11.1	30.8	26–29	71	77	70
L3	Nucleus	ED097839	10.0	25.4	23–24	71	70	63
L4	Nucleus	EDP00320	9.8	23.2	21–22	64	64	54
L5	Plastid	NP_958373	9.9	23.1	17–19	66	73	70
L6	Nucleus	EDP00937	9.8	21.0	17–19	69	66	62
L9	Nucleus	EDP03110	9.6	20.2	14–16	56	62	52

(Continued)

Table 28.2 *Continued*

Protein	Genome	Accession #	pI[a]	Mass (kD) Sequence	SDS-PAGE	Similarity, % A.t.	Syn	E. coli
L10	Nucleus	EDP08451	9.5	>22.2	21–23	52	50	44
L11	Nucleus	EDP06359	9.8	14.7	11–13	76	82	73
L12	Nucleus	EDP08831	4.9	13.4	11–13	58	69	59
L13	Nucleus	EDP08538	9.9	21.3	19–21	77	70	66
L14	Plastid	NP_958372	10.3	13.4	10–11	86	87	75
L15	Nucleus	EDO99032	10.2	23.3	22–23	66	60	53
L16	Plastid	P05726	11.5	15.5	ND	85	87	73
L17	Nucleus	EDP07414	10.5	14.2	11–13	74	74	57
L18	Nucleus	EDP09994	9.7	12.9	10–11	66	74	62
L19	Nucleus	EDP03631	11.2	13.2	11–13	60	68	57
L20	Plastid	NP_958363	11.8	13.4	ND	71	82	75
L21	Nucleus	EDP05027	9.7	14.2	ND	48	67	58
L22	Nucleus	NI	9.6	14.5	11–13	71	75	61
L23	Plastid	Q8HTL3	10.0	11.2	ND	58	65	54
L24	Nucleus	EDO99434	10.0	14.9	11–13	61	74	56
L27	Nucleus	EDP05166	10.4	14.1	13–14	61	75	69
L28	Nucleus	EDP08523	9.8	18.4	16–17	65	75	57
L31	Nucleus	EDP01069	9.2	11.6	<10	62	69	55
L32	Nucleus	EDP04096	9.7	6.9	10–11	76	62	39
L33	Nucleus	EDP08102	10.2	8.6	NI	64	64	60
L34	Nucleus	EDP09319	11.9	6.5	NI	61	86	80
L35	Nucleus	EDO97643	10.5	7.9	<10	47	56	47
L36	Plastid	NP_958367	10.8	4.3	NI	93	91	81
PSRP-6	Nucleus	EDP05091	10.3	5.9	<10	50	ND	ND
RAP38	Nucleus	EDP05666	9.2	40.2	38	84	67	ND
RAP41	Nucleus	EDP05800	9.3	44.3	41	62	ND	ND

Data adapted from Beligni et al. (2004b), Yamaguchi et al. (2002, 2003). ND, not determined; NI, not identified; A.t., Arabidopsis thaliana; Syn, Synechococcus

[a]From predicted mature protein sequence.

3. Proteins of the 30S ribosomal subunit

Proteomic analyses identified 24 SSU proteins, including 20 *E. coli* homologues, homologues of plant PSRP-1, PSRP-3, and PSRP-4, and a novel S1 domain-containing protein, PSRP-7 (Table 28.2). PSRP-7 is produced from the EFT polyprotein, which encodes both the S1 homologue and EF-Ts (see section IV). Comparisons to *E. coli*, *Synechosystis*, and spinach identified four *Chlamydomonas* SSU proteins (S2, S3, S5, and PSRP-3) that are more than twice as large as their homologues due to either N-terminal extensions (NTE), insertions or C-terminal extensions (CTE). An NTE on the plastid-encoded S2 (ORF50) results in a doubling of the mass of the protein as compared to spinach (Yamaguchi et al., 2002). The plastid genome ORFs 208 and 570 both encode products with homology to S2 (Maul et al., 2002). While Leu (1998) reported that ORF570 encoded an S2-like protein with an NTE, proteomic analysis only identified ORF570-derived peptides, indicating that ORF208 is a likely pseudogene (Yamaguchi et al., 2003). S3 contains a 472 amino acid insertion and is encoded by ORF712 of the chloroplast genome. S5 contains an acidic 452 amino acid NTE (pI 3.7) that is proposed to account for its anomalous migration in SDS-polyacrylamide gels (84 kD predicted vs. 67 kD observed). The chloroplast-unique domains from S2, S3, and S5 are modeled to form a structural feature on the ribosome located near S1 and the mRNA entrance channel on the SSU (Wimberly et al., 2000; Manuell et al., 2007). It has been proposed that the additional domain on S2 and the chloroplast-specific PSRP-7, both of which contain RNA-binding domains, augment the mRNA-binding function of S1, which in *Chlamydomonas* contains three RNA-binding motifs as compared to six in *E. coli* (Yamaguchi et al., 2002). *Chlamydomonas* PSRP-3 is approximately twice as large as its spinach counterpart due to the presence of an NTE. In addition, L28 contains a large CTE. These extensions may be relevant to the unique aspects of chloroplast translational regulation.

4. Plastid-specific ribosomal protein functions

Five PSRPs have been identified in *Chlamydomonas*, four associated with the SSU (PSRP-1, PSRP-3, PSRP-4, and PSRP-7) and one with the LSU (PSRP-6). All of them are nucleus-encoded and predicted by ChloroP to contain transit peptides (Emanuelsson et al., 1999; Yamaguchi et al., 2003). Recruitment of these large proteins to the ribosome is a distinct departure from the eubacterial ribosome, suggesting they may have chloroplast-specific functions.

PSRP-1 is partitioned between ribosomal and ribosome-free stromal fractions, at least in spinach; it shows significant sequence similarity to the translation product of the light-repressed transcript (*lrtA*) of *Synechocystis*, and the bacterial transcription specificity factor σ^{54}; and it co-purifies with chloroplast RNase P, suggesting a possible role in tRNA processing (Zhou and Mache, 1989; Tan et al., 1994; Yamaguchi et al., 2003). The function and localization of PSRP-3 and PSRP-4 on the 30S subunit are unresolved.

5. Docking of the SSU on the mRNA for initiation

In *E. coli*, docking of the SSU on the mRNA for initiation involves, in part, RNA binding by S1 (reviewed in Voorma, 1996). In the *Chlamydomonas* chloroplast, SSU docking may also involve S1 as well as S2, a complex with PSRP-7, and S5, because these proteins could be UV-crosslinked to a ^{32}P-labeled *psbA* 5′ UTR probe after it was incubated with ribosomes (Manuell et al., 2007). In addition, recombinant S1 was shown to bind AU-rich sequences in chloroplast mRNAs (Merendino et al., 2003). Although mRNA binding by S7 was not detected in the aforementioned UV-crosslinking assay, as a recombinant protein it was shown to bind the 5′ UTRs of several mRNAs *in vitro*, including the cognizant mRNA *rps7* (Fargo et al., 2001). S7 appears to recognize a secondary structure of the *rps7* 5′ UTR rather than a primary sequence element (Fargo et al., 1999, 2001), and a role for this interaction in *rps7* translation was supported by mutations in its 5′ UTR that abolished both translation and S7 binding, and the reversal of these effects by second-site suppressor mutations (Fargo et al., 2001). It was not determined whether the S7 that bound the 5′ UTR was free or assembled in the small ribosomal subunit. In *E. coli*, S7 has an extra-ribosomal function as a translational feedback repressor (Saito et al., 1994). This possibility should be explored in *Chlamydomonas*, although S7 binding is not specific for the *rps7* 5′ UTR *in vitro* (Fargo et al., 2001).

6. Ribosome-associated proteins

Two additional proteins, RAP38 and RAP41, co-sediment with 70S particles, but are not present in 50S, 30S, or 80S preparations. RAP38 and RAP41 are 35% identical (52% similar) to one another (Yamaguchi et al., 2003) and show significant homology with spinach chloroplast RNA-binding proteins CSP41a and CSP41b, respectively (Bollenbach et al., 2003). While all four proteins contain epimerase domains (Baker et al., 1998) it has not been demonstrated that this is the function provided by the proteins *in vivo*. In addition, an examination of the evolutionary relationships between RAP38, RAP41, CSP41 and members of bacterial epimerase/dehydratase family by neighbor joining revealed three distinct clades: RAP41 homologues, RAP38 homologues, and the epimerase/dehydratase family (Yamaguchi et al., 2003). In spinach, CSP41, CSP29, and CSP55 form a stem-loop RNA-protein complex *in vitro* (Monde et al., 2000), and *in vivo* data suggest a role in RNA degradation (Bollenbach et al., 2003). Thus, RAP38 and RAP41 may be involved in mRNA processing, either directly or indirectly with other factors and/or ribosomal proteins. In addition, RAP41 was the only ribosome-associated protein identified in an initial characterization of the *Chlamydomonas* phosphoproteome (Wagner et al., 2006).

C. Chloroplast rRNAs

1. Organization of ribosomal RNA genes in the chloroplast genome

As in eubacteria, the *Chlamydomonas* chloroplast rRNA genes are organized in an operon and cotranscribed in the order 16S-$tRNA^{Ile}$-$tRNA^{Ala}$-23S-5S, followed by processing to yield mature rRNA molecules (reviewed in Harris et al., 1994). The region between the 3′ end of the 16S rRNA and *trnI* is characterized by a number of short dispersed repeats.

The 23S rRNA is processed into four RNAs: 7S, 3S, 23Sγ, and 23S*, which are separated by three internal transcribed spacers (ITS) (Turmel et al., 1993). The 7S and 3S rRNAs are homologous to the 5′ end of *E. coli* 23S rRNA (Harris et al., 1994). In addition to the ITSs, the 23S rRNA gene is interrupted by a group I intron (*Cr.LSU*) within the 23S* portion, which is capable of self-splicing *in vitro* and intron homing (see Chapter 27). While the functional role of 23S fragmentation remains unclear, Turmel et al. (1993) have shown its conservation in many *Chlamydomonas* species.

Characterization of *rrn* operon processing has been facilitated by the nuclear mutant *ac20* (Holloway and Herrin, 1998), which is deficient in chloroplast protein synthesis and LSU assembly (Harris et al., 1974). While 23S processing intermediates could be detected in a wild-type strain, their levels were increased in *ac20*, particularly ITS-1 (which separates 7S and 3S rRNAs), suggesting that 23S ITS processing is closely linked to subunit assembly. While splicing of the *Cr.LSU* intron is also reduced in *ac20*, this does not appear to be responsible for the observed slow-growth phenotype.

2. 16S rRNA

It is beyond the scope of this chapter to review the structural interactions and requirements for translation that involve the 16S rRNA because these have been better-studied in other systems. Interested readers are referred to reviews on this topic (Caetano-Anolles, 2002; Polacek and Mankin, 2005; Korostelev et al., 2006; Wilson and Nierhaus, 2006; Rodnina et al., 2007), and the cryoEM structure of a translation initiation complex in *E. coli* (Allen et al., 2005).

The secondary structure of the *Chlamydomonas* 16S rRNA is based on comparative sequence analyses with *E. coli* (Figure 28.1). The four secondary structure domains (5′ domain, central domain, 3′-major and 3′-minor domains) largely determine the shape of the SSU (Wimberly et al., 2000; Cannone et al., 2002). A complete description of the domains and location of proteins and their interactions with RNA has been published (Brodersen et al., 2002; Cannone et al., 2002). A few differences exist in domain structures of chloroplast and *E. coli* 16S rRNAs, which are limited to the 5′ domain (Harris et al., 1994).

Translational fidelity is controlled on the small subunit by monitoring the base pairing between the codon and anticodon during the decoding process. The decoding center, which is part of the A-site, is constructed entirely of RNA. It contains the upper section of helix 44 and the 3′ and 5′ ends of 16S rRNA, and is located at the upper part of the body and lower part of the head (Schluenzen et al., 2000). Not surprisingly, several aminoglycoside antibiotics bind helix 44 and perturb decoding (see section III.5).

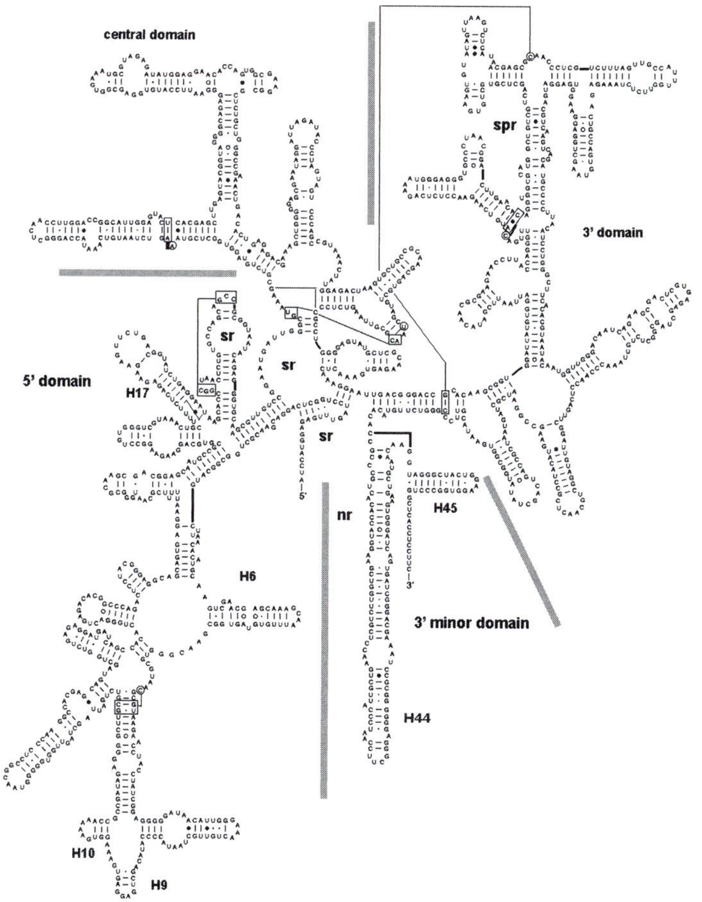

FIGURE 28.1 *Secondary structure of C. reinhardtii 16S rRNA illustrating major functional domains and sites of antibiotic resistance mutations (Table 28.3; sr, streptomycin; spr, spectinomycin; nr, neamine/kanamycin). Sequence from the comparative RNA web site (Cannone et al., 2002).*

A critical function of the small subunit P site is to bind initiator tRNA selectively at the mRNA start codon. In bacteria, initiation factor IF-3 induces tRNA selection through its interaction with the 30S subunit, which promotes binding of initiator tRNA^{fMet} (Hartz et al., 1989; Laursen et al., 2005), probably through 16S rRNA residues G1338 and A1339 (Lancaster and Noller, 2005), both of which are present in *Chlamydomonas* 16S rRNA. While the anti-Shine-Dalgarno (ASD) sequence (CCUCCU) at the 3′ end of the *Chlamydomonas* rRNA likely plays an important role in the initiation of translation of a subset of chloroplast-encoded mRNAs by base pairing with the Shine-Dalgarno (SD) sequence in the mRNA 5′ UTR (see section VII), this is almost certainly less universal in the chloroplast as compared to *E. coli*.

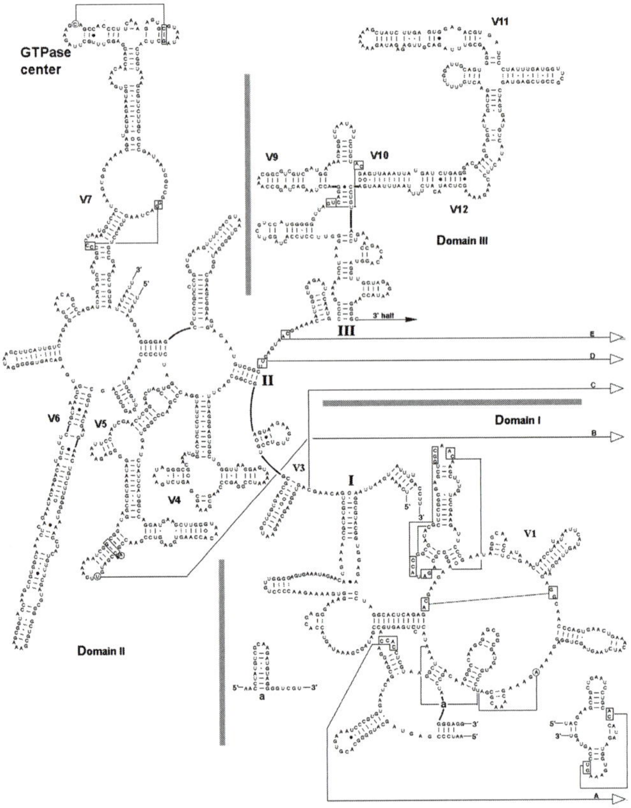

FIGURE 28.2 *Secondary structure of C. reinhardtii 23S rRNA illustrating major functional domains and sites of antibiotic resistance (Table 28.3; cr, chloramphenicol; er, erythromycin). Variable regions are numbered according to Raue et al. (1988). Structure from the Comparative RNA web site (Cannone et al., 2002).*

3. 23S rRNA

The 5S and 23S rRNAs constitute the RNA component of the large ribosomal subunit (Figure 28.2). 23S rRNA is composed of six secondary structure domains (I–VI) (Noller et al., 1981; Gutell and Fox, 1988), with the 5S rRNA being regarded as the seventh domain (Steitz and Moore, 2003). The central loop of Domain V constitutes part of the peptidyl transferase center (PTC), mutations within which confer resistance to the antibiotics erythromycin, lincomycin, and chloramphenicol (see section III.5). A tunnel starts at the PTC where formation of the peptide bond occurs. The narrowest part of this tunnel has been shown to act as a sensor and control point at which the synthesis of certain peptides pauses in response to metabolic status (Gong and Yanofsky, 2002; Tenson and Ehrenberg, 2002; Tenson et al., 2003). This is also the site of action of several classes of antibiotics.

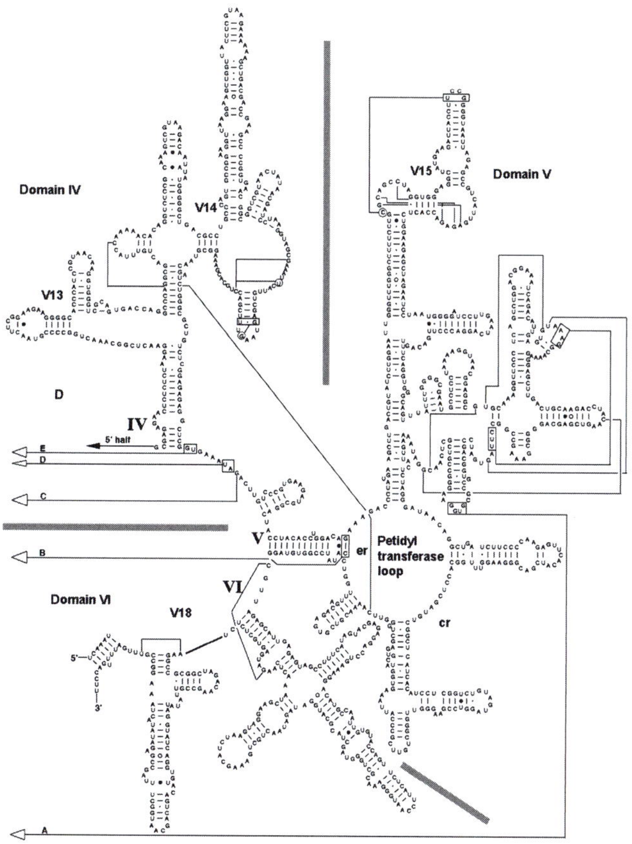

FIGURE 28.2 *(Continued)*

Because the subunit is completely devoid of proteins in the vicinity of the PTC, the ribosome has been designated a ribozyme (Nissen et al., 2000).

Based on comparisons of eubacterial, organelle, archaeal, and eukaryotic LSU RNAs, 18 variable regions in 23S rRNA were identified (Raue et al., 1988). For five regions, chloroplast sequences differ significantly from those of *E. coli*, while for the others the chloroplast sequences are similar to their eubacterial counterparts. The differences listed by 23S domains are:

Domain I: Cyanobacterial and chloroplast 23S rRNAs lack helix 8 (V1) of *E. coli*, and contain an insertion between helices 13 and 14 (V2), which itself can be folded into a helix. **Domain II:** Variable regions V4 and V7 are conserved among eubacteria and chloroplasts, while 3 nt in *E. coli* V6 are replaced by a loop of 5–20 nt in chloroplasts. **Domain III:** Variable region V11 is the most diverse in chloroplast 23S RNAs. Some chloroplast rRNAs have lost part of helix 54. Helix 55 in *Chlorella ellipsoidea* and *Z. mays* contains an insertion compared to *E. coli*. Comparison of eubacterial and plastid rRNAs identifies strong conservation in Domains IV and V.

4. 5S rRNA

5S rRNA is an integral large subunit component of all cytoplasmic and most organellar ribosomes. Regardless of origin, 5S rRNA is ca. 120 nt and can be folded into a common secondary structure consisting of five helices (I–V), two hairpin loops (C and E), two internal loops (B and D) and a hinge region (A), organized in a three-helix junction (comparative RNA web site, http://www.rna.ccbb.utexas.edu). The structure of 5S rRNA is better preserved in eukaryota and archaea than in eubacteria, where much more sequence variability is observed (Szymanski et al., 2003). Examination of the *Chlamydomonas* 5S structure places it within the eukarya (Cannone et al., 2002).

5S rRNA together with proteins L5, L18, and L25 forms the central protuberance of the large ribosomal subunit in bacteria. In *Chlamydomonas*, the absence of an L25 homologue has the potential to expose 5S rRNA to the solvent surface (Manuell et al., 2007).

5. Antibiotic resistance mutations in chloroplast rRNA genes

Antibiotic sensitivity and resistance have been particularly useful in structure-function studies of the ribosome. In addition, 16S rRNA antibiotic resistance mutations have been employed as selectable markers for transformation (Newman et al., 1990; Boynton and Gillham 1993). Typically, antibiotic-binding sites coincide with functional domains such as the decoding center, tRNA-binding sites in the SSU and the PTC, nascent peptide exit tunnel and/or GTPase-activating region of the LSU (reviewed in Poehlsgaard and Douthwaite, 2005), and some have been mapped at atomic resolution (Yonath, 2005). A consistent feature of ribosome structure models is the scarcity of proteins at the 30S–50S interface. In fact, most ribosome inhibitors target this rRNA-rich surface and bind directly to the 16S or 23S rRNAs. Single base mutations and base modifications that alter antibiotic contact surfaces with the ribosome typically confer resistance.

Antibiotics that bind to the *Chlamydomonas* 30S ribosomal subunit perturb its principal function of mRNA decoding. The decoding site is part of the A-site and is located at the end of helix 44 in *E. coli*. Although tRNA or aminoglycosides bound to the A-site protect the same nucleotides in the decoding region (i.e. A1408, A1492, A1493, and G1494; *E. coli* numbering for all mutations) aminoglycosides do not act by sterically hindering tRNA binding to the ribosome (Hobbie et al., 2006). Instead, they induce misincorporation of amino acids into a nascent peptide by scrambling the communication between the decoding and GTPase-associated centers of the ribosome (Carter et al., 2000; Ogle et al., 2002).

Resistance to aminoglycosides can result from alteration of their molecular target, enzymatic inactivation (Walsh, 2000), and mutations in ribosomal

Table 28.3 Sites of antibiotic resistance mutations in the chloroplast ribosome

Class	Antibiotic	Molecule	Position	Reference
30S subunit				
Aminoglycosides	Streptomycin	16S rRNA	U13G; A474C; C856U; A859C; A860G	Harris et al. (1989)
		S12 (sr)	K42T	Liu et al. (1989)
		S12 (sd)	P90L	Liu et al. (1989)
	Neamine Kanamycin	16S rRNA	A1340G; C1341U	Harris et al. (1989)
Other drugs	Spectinomycin	16S rRNA	A1123G; A1123C; G1125A; G1125C; G1125T	Harris et al. (1989); GuhaMajumdar and Sears (2005)
50S subunit				
Macrolides	Erythromycin	23S rRNA	G2014A; A2015G; C2569U; C2569G	Harris et al. (1989)
Lincosamides	Lincomycin	23S rRNA	G2014A; A2015G	Harris et al. (1989)
Nucleoside	Chloramphenicol	23S rRNA	U2514G	Gillham et al. (1991)

proteins, particularly S12 (Wright, 2003; Maisnier-Patin et al., 2007). In *E. coli* methylation of A1408 within 16S rRNA results in resistance to neamine (Beauclerk and Cundliffe, 1987), and mutations in nucleotides equivalent to *E. coli* residues A1408 and A1409 have similarly been shown to confer resistance to neamine and kanamycin in *Chlamydomonas* (Harris et al., 1989). *Chlamydomonas* resistance to another aminoglycoside, streptomycin, is attributed to mutations at positions equivalent to *E. coli* residues 13, 523–525 and 912–915 (Harris et al., 1989), as well as S12 mutations (Liu et al., 1989). Although these sites are widely separated in the primary sequence, they interact with the same ribosomal proteins and are believed to be in close proximity in the 30S subunit (Bonny et al., 1991). Mutations in bases equivalent to *E. coli* residues 1191, 1193, 1064 (Harris et al., 1989; GuhaMajumdar and Sears, 2005) confer resistance to spectinomycin.

Antibiotics targeting the 50S subunit bind within three regions and interfere with its ability to control GTP hydrolysis, formation of peptide bonds and channeling the peptide through the subunit tunnel (reviewed in Poehlsgaard and Douthwaite, 2005; Tenson and Mankin, 2006). The binding of macrolides in the exit tunnel close to the PTC inhibits translation by blocking progression of the nascent peptide through the exit tunnel (Hansen et al., 2002).

Several mutations mapping to the PTC in the *Chlamydomonas* 23S rRNA have been isolated (Harris et al., 1989). Erythromycin-resistant mutants *er-u-AW-17* and *er-u-1* contain a 23S rRNA A2058 → G mutation, and are cross-resistant to lincomycin and clindamycin. An adjacent G2057 → A mutation (*er-u-37*) results in erythromycin resistance, but no cross-resistance. The *er-u-1a* and *er-u-11* mutants that have C2611 → T changes are only moderately resistant to lincomycin and clindamycin. In addition, a mutation in *Chlamydomonas* 23S rRNA, equivalent to *E. coli* nt 2504, confers resistance to the nucleoside-based antibiotic chloramphenicol (Gillham et al., 1991).

Ribosome biogenesis involves numerous processing and modification steps. One conserved modification is the dimethylation of two adjacent adenosine bases in the loop of helix 45 (A1518, A1519) near the 3′ end of the 16S rRNA, producing N^6,N^6-dimethyladenosine. The enzyme responsible for this modification, first identified in bacteria as KsgA, is conserved across all kingdoms (O'Farrell et al., 2004). In *Chlamydomonas*, four gene models encode KsgA-related proteins (see below), of which one (EDP09441) has been annotated as KsgA, suggesting a conservation of 16S methylation in this alga.

In addition to 16S rRNA modifications a diverse range of drug-producing and pathogenic bacteria encode erythromycin resistance determinant genes (*erm*). Members of the Erm gene family encode methyltransferases that specifically methylate the N^6 position of nucleotide A2058 in 23S rRNA (Maravic, 2004), the same residue identified in *er-u-AW-17* and *er-u-1*. Structural analysis of KsgA revealed striking similarities to ErmC and the mitochondrial protein mtTFB, which is both a methylase and a transcription factor (O'Farrell et al., 2004). ClustalW alignments of the three *Chlamydomonas* proteins related to KsgA, and representatives of the KsgA, Erm, and mtTFB families suggest two primary clades, one containing yeast Dim1 and *Chlamydomonas* accession EDP09441, and the other *E. coli* KsgA, *Chlamydomonas* accession EDP04943, and outgroups of mtTFB and ErmC. Dim1 is a KsgA orthologue, since it has been shown to methylate A1518 and A1519 *in vivo*, lending support to the designation of EDP09441 as a KsgA homologue.

IV. GENERAL CHLOROPLAST TRANSLATION FACTORS

A. Overview

Translation requires general translation factors (GTF), many of which are GTPases activated by the ribosome. Despite many years of work, primarily in *E. coli*, the order in which the GTFs bind and are released *in vivo*, and their roles in altering ribosome structure have not been definitively elucidated (Ramakrishnan, 2002). In this section the identities and properties of the *Chlamydomonas* GTFs will be covered.

Translation in eubacteria requires a minimum of 11 GTFs: three initiation factors (IF1, IF2, and IF3) (Gualerzi and Pon, 1990), four elongation factors (EF-Tu, EF-Ts, EF-G, EF-P), three ribosome release factors (RF1, RF2, and RF3) and a ribosome recycling factor (RRF or RF4) (Aoki et al., 1997; Ramakrishnan, 2002). Chloroplast homologues of many eubacterial GTFs have been identified in nuclear and chloroplast genomes of plants and algae (Vallon et al., 1993; Sugiura et al., 1998; Beligni et al., 2004b). *Chlamydomonas* encodes homologues of all 11 GTFs (Table 28.4; Beligni et al., 2004b), but to date only EF-Tu has been characterized (Silk and Wu, 1993; Hwang et al., 1996; Lee and Herrin, 2002; Zicker et al., 2006).

B. Initiation

IF1 and IF3 are important in assembling the preinitiation complex, prevention of initiation of protein synthesis with elongator tRNAs, and the formation of 70S complexes lacking tRNA in the P-site (Laursen et al., 2005).

Table 28.4 *Chlamydomonas* orthologues of *Synechocystis* PCC6803 translation factors

Protein	Genome	MW (kD)[a]	*Chlamydomonas* accession number	*Synechocystis* orthologue	
				Accession	% Similarity[b]
IF-1	Nucleus	10.1	EDP04205	P73301	81
IF-2	Nucleus	24.7	EDP03886	P72689	68
IF-3	Nucleus	N.A.	EDO98179	P72874	63
EF-Tu	Plastid	45.6	CAA36499	P74227	82
EF-Ts	Nucleus	103.4	EDP00633	P74070	73
EF-G	Nucleus	80.0	EDP06820	P74228	73
EF-P	Nucleus	21.1	EDO98769	Q55119	74
RF-1	Nucleus	36.7	EDP01126	P74707	79
RF-2	Nucleus	5.4	EDP09153	P74476	89
RF-3	Nucleus	33.9	EDO97207	P73473	84
RRF	Nucleus	25.8	EDP00115	P74456	72

[a] *Predicted MW.*

[b] *Adapted from Beligni et al. (2004b).*

IF1 also appears to harbor an RNA chaperone activity that may contribute to RNA rearrangements during the early phase of translation initiation (Croitoru et al., 2004, 2006). The role of the G protein IF2 is to promote association of the ribosomal subunits in the presence of an initiator tRNA (Antoun et al., 2006). In addition to these GTFs, *Chlamydomonas* may employ additional factors to regulate initiation as suggested by a mutation in the nuclear *SIM30* gene, that allows more efficient translation of a *petD* mRNA bearing a weakened initiation codon (Chen et al., 1997; see section VI.A).

In *Euglena gracilis* chloroplasts, the homologue of IF3 contains N- and C-terminal extensions relative to *E. coli* IF3 (Yu and Spremulli, 1998). Intramolecular association of these extensions has been proposed to inhibit IF3 activity. While alignment of the *Chlamydomonas* IF3 gene model (EDO98179) with its *Euglena* and *E. coli* homologues suggests that *Chlamydomonas* IF3 lacks these extensions, it is likely that the model is incomplete, given that a consensus splice acceptor site precedes the first codon, hinting at additional 5′ exon(s).

C. Elongation

The initiation steps in eubacteria leave an aminoacylated initiator tRNA in the P site and an empty A site, which potentiates the elongation cycle. Correct codon–anticodon interactions result in conformational changes that stabilize tRNA binding and GTP hydrolysis by EF-Tu. EF-Tu is plastid-encoded in *Chlamydomonas* by the *tufA* gene (Maul et al., 2002). Translation elongation factor P (EF-P) was identified as a protein that stimulates the peptidyltransferase activity of the 70S ribosome in *E. coli* (Glick and Ganoza, 1975). Translocation of the deacylated tRNA in the P site, peptidyl tRNA in the A site and mRNA is facilitated by EF-G (Breitenberger and Spremulli, 1980; Frank and Agrawal, 2000). This prepares the ribosome for a subsequent round of elongation.

During elongation, EF-Tu cycles between the active GTP form in which it interacts with aminoacyl tRNA and the ribosome, and the inactive GDP form. Recycling of EF-Tu·GDP to EF-Tu·GTP is catalyzed by a guanine-nucleotide exchange factor, EF-Ts (Wieden et al., 2002). In *Chlamydomonas* EF-Ts and PSRP-7 are encoded by the nuclear *EFT1* gene, which was originally called *PETs*, but changed during annotation of the genome to avoid confusion with chloroplast *pet* genes. The *EFT1* precursor transcript is alternatively spliced to yield three mRNA variants: a 3 kb *EFTa* transcript encoding a 110-kD polyprotein which is processed into a 55-kD protein (EF-Ts) and a 65-kD protein (PSRP-7); a 1.8 kb *EFTb* transcript encoding a 75-kD protein; and a 1.68 kb *EFTc* transcript encoding

PSRP-7 (Beligni et al., 2004a). EF-Ts contains two GDP exchange domains, in contrast to EF-Ts in bacteria and its eukaryotic analog EF-1b, which are each approximately 30 kD and contain only one such domain. *EFT1* homologues in *Arabidopsis* and rice led these authors to suggest that EF-Ts and PSRP-7 function together in chloroplast translation and that the polyprotein is functional.

D. Termination

Since *Chlamydomonas* encodes close homologues of the eubacterial termination factors, translation termination in chloroplasts is likely to be mechanistically similar (reviewed in Ramakrishnan, 2002). Translation termination begins when a stop codon is encountered in the A-site of the ribosome. Two "class I" release factors, RF1 (which recognize UAA and UAG) and RF2 (which recognizes UAA) together with the "class II" release factor RF3, release the completed polypeptide. Binding of RF1/2 to a ribosome triggers the hydrolysis and release of the peptide chain from the tRNA in the P site. RF3 promotes rapid dissociation of RF1 and RF2. After release of the peptide chain, the ribosome is left with mRNA and a deacylated tRNA in the P site. Ribosome recycling factor (RRF) and EF-G are required to disassemble this complex (Table 28.4; Rolland et al., 1999).

In *Arabidopsis*, screens of albino and high chlorophyll fluorescence (*hcf*) mutants uncovered genes for both RRF1 and RF2 (Meurer et al., 2002; Motohashi et al., 2007). RRF1 was uncovered through a screen of albino or pale-green (*apg*) mutants, where *apg3-1* carried a mutation in the chloroplast homologue AtcpRF1 (Motohashi et al., 2007). Complementation analysis using the *E. coli rf1* mutant revealed that APG3 does function as an RF1 in *E. coli*. Since the chloroplasts of *apg3-1* plants contained few internal thylakoid membranes, and chloroplast proteins related to photosynthesis were not detected by immunoblot analysis, AtcpRF1 is essential for chloroplast development.

The *hcf109* mutation identified a chloroplast-targeted RF2-like protein (AtpfrB), whose absence causes decreased stability of UGA-containing mRNAs (Meurer et al., 2002). This implicates AtpfrB in the regulation of both mRNA stability and protein synthesis. *Chlamydomonas* may have lost this "auxiliary" function of RF2 as only two ORFs in the plastome encode UGA termination codons (Maul et al., 2002). If this is the case, reporter genes containing TGA stop codons would be predicted to not function because of a failure to terminate translation. There are two reports that have shown that UGA stop codons can be functional in *Chlamydomonas* chloroplasts (Lee et al., 1996; Sizova et al., 1996). However, in both of these reports, the possibility of frameshifting and readthrough was not investigated.

V. THE REGULATION OF TRANSLATION BY LIGHT

A. Introduction

Chlamydomonas responds to light in a variety of ways, including changes in chloroplast gene expression at the translational level. This section reviews distinct translational regulatory responses during dark-to-light transitions and under high light stress. Although wild-type *Chlamydomonas* can synthesize chlorophyll in the dark and has a fully developed chloroplast, one study cited here used the *y1* mutant, which does undergo light-dependent chloroplast differentiation (see section V.F).

B. Light perception and signal transduction

It has been proposed that translational control is mediated by the effects of light through photosynthesis on the energy status of the chloroplast, rather than by the photoreceptors and associated signal transduction pathways that mediate most responses to light in animals and the nuclear-cytosolic compartments of plant cells. In this model, the activities of translational regulatory factors are modulated by the redox states of specific components of the photosynthetic electron transport chain, the electrochemical proton gradient across the thylakoid membrane, or the stromal ADP/ATP ratio. These light-modulated components of photosynthesis have been termed "sensors" (Allen, 1993; see also Danon, 1997; Bruick and Mayfield, 1999; Pfannschmidt, 2003; Zerges, 2004).

Much evidence supports this model. Chloroplasts house the molecules involved in light perception because isolated chloroplasts can activate the synthesis of D1 and D2, and other proteins in response to light (Trebitsh and Danon, 2001). Although *Chlamydomonas* has a circadian clock, the diurnal oscillations in protein synthesis are directly cued to light (Lee and Herrin, 2002; Mittag and Wagner, 2003). Similarly, by EM, a 10% increase in ribosome association with thylakoid membranes was observed when cells in the dark phase were illuminated for only 10 minutes, and the opposite effect was observed 10 minutes following a light to dark shift (Chua et al., 1976). Inhibitors of photosynthetic electron transport block the effects of light on the induction of the synthesis of D1, D2, many other proteins, and on the repression of *chlL* translation (Cahoon and Timko, 2000; Trebitsh and Danon, 2001). Lastly, chloroplast protein synthesis was induced in cells in the dark phase of a diurnal regime by the addition of acetate to the medium, which is more likely to affect the sensors mentioned above than a pathway involving a light receptor (Michaels and Herrin, 1990). As the presence of acetate in the medium changes the observed timing of various processes, it would appear that light regulation is best studied with photoautotrophically grown cells (see Heifetz et al., 2000).

C. Examples of translational regulation by light

1. Light induces the translation of mRNAs encoding subunits of photosynthesis complexes

Several approaches have revealed that light activates the translation of chloroplast mRNAs encoding D1 and D2, the large subunit of Rubisco, and other photosynthesis complex proteins. Two studies monitored protein synthesis rates by pulse labeling during a diurnal dark-light regime and found that D1, D2, and the Rubisco large subunit were labeled at much higher rates in the light phase than in the dark phase, supporting the induction of their synthesis by light (Herrin et al., 1986; Lee and Herrin, 2002). In isolated chloroplasts, pulse labeling experiments detected no proteins synthesized in the dark and high levels of synthesis in the light (Trebitsh et al., 2000; Trebitsh and Danon, 2001). Another study observed higher levels of synthesis of D1 and D2 in cells cultured in the light than in the dark (Malnoë et al., 1988). The absence of corresponding changes in the mRNAs encoding these proteins supports translation as the regulated phase of gene expression, with the caveat that uptake of radiolabel may be seriously impeded in the dark. Finally, the mRNAs of 18 genes encoding chloroplast ribosomal proteins accumulate early in the light phase, presumably to populate new ribosomes for the higher levels of protein synthesis. This regulation is truly circadian (Kucho et al., 2005).

2. Light repression of tufA *and* chlL *translation*

During a diurnal regime, ^{35}S-pulse labeling experiments revealed that the synthesis of EF-Tu declined late in the light phase, while the *tufA* mRNA encoding it accumulated (Lee and Herrin, 2002). The preferential loading of the *chlL* mRNA on polysomes in the dark revealed that synthesis of the ChlL subunit of the light-independent protochlorophylide oxidoreductase occurs when this enzyme is required for chlorophyll synthesis (Cahoon and Timko, 2000).

3. psbA *translation is activated for PS II repair during high light stress*

Photochemical reactions within PS II damage its D1 subunit (Aro et al., 1993). Under high intensity light (above 700 μE), a stress condition called photoinhibition results from the dramatic loss of PS II activity and the production of reactive oxygen species (reviewed in Murata et al., 2006; Nishiyama et al., 2006; also see Chapter 23). The acclimation responses to high light stress include the activation of *psbA* translation to provide newly synthesized D1 for PS II repair, a process which probably also occurs at lower levels under moderate light conditions (Keren et al., 1995). However, preferential induction of D1 synthesis in *Chlamydomonas* has not been detected below 700 μE because *de novo* synthesis for PS II assembly predominates. Moreover, the specific stimulation of D1 synthesis by high light is distinct from translation activation by moderate light, which activates the synthesis of several PS II subunits and the large subunit of Rubisco (van Wijk et al., 1994; Minai et al., 2006; Uniacke and Zerges, 2007). High light-stimulated synthesis is also distinct from autoregulatory (CES) regulation of *psbA* translation by unassembled D1, because the translation of a chimeric mRNA with the 5′ and 3′ UTRs of *psbA* flanking the *petA* coding region was not induced during high light stress, while it was induced over 10-fold by the absence of D1, the signal that activates *psbA* translation in the CES hierarchy for PS II assembly (Minai et al., 2006). The above results suggest that the specific activation of D1 synthesis during high light stress occurs after initiation and thus, elongation may be activated during recovery from photoinhibition. In this respect, comparisons of *psbA* mRNA polysome association in cells exposed to moderate and high intensity light might be helpful.

4. rbcL *translational repression during high light stress*

Translation of the *rbcL* mRNA, which encodes the Rubisco large subunit, is repressed during high light stress. This has been linked to an oxidative stress response, because it was induced under low light when ROS production was photosensitized by methyl viologen, and it was prevented by the antioxidant ascorbate (Shapira et al., 1997; Irihimovitch and Shapira, 2000). Glutathione appears to be the redox sensor that mediates this response. Repression of *rbcL* translation is exerted, at least in part, during the entry to, or early in the elongation phase because the mRNA shifts from large to small polysomes and monosomes during high light stress (Cohen et al., 2005).

D. Factors that may mediate translational regulation by light

1. The Rubisco large subunit

The translation factor that mediates *rbcL* translational repression during high light stress may be the Rubisco large subunit itself because it tightly bound RNA following its oxidation *in vitro* (Yosef et al., 2003). RNA binding by the oxidized Rubisco large subunit showed no sequence specificity, although it does have high affinity (Yosef et al., 2003; Cohen et al., 2006). This translational regulation and the RNA-binding activity of Rubisco large subunits are conserved in at least two vascular plants and purple bacteria (Cohen et al., 2006). A model has been advanced in which RNA-binding activity by the large subunit is mediated by an RNA recognition motif in its N-terminal region that becomes exposed during oxidizing conditions due to partial unfolding, or when chaperones fail to bind this region on the nascent protein, for example when they are sequestered by the misfolded proteins that arise during oxidative stress (Yosef et al., 2003; Cohen et al., 2006). A mutually compatible model proposes that the unassembled Rubisco large subunit represses *rbcL* translation to coordinate equimolar synthesis of the large and small subunits (Cohen et al., 2006). Evidence for this regulation was provided by the absence of large subunit synthesis in a *Chlamydomonas* mutant deficient for the Rubisco small subunit (Khrebtukova and Spreitzer, 1996) and similar effects were observed in tobacco (Wostrikoff and Stern, 2007). Both studies indicated that Rubisco large subunit expression follows the CES rules, whereby assembly of the subunit into its complex plays a key role in the control of its synthesis (see Chapter 29).

2. The RB complex

a. Isolation and characterization

The RB complex was proposed to specifically bind the 5′ UTR of the *psbA* mRNA to activate D1 synthesis in response to light (see section V.D.3). This model is based on the detection of RB complex binding activity only when cells were cultured in the light, and on results of competition experiments that led the authors to conclude specificity for the *psbA* 5′ UTR (Danon and Mayfield, 1991, 1994a,b). Genes ascribed to the RB complex proteins RB47 and RB60 were shown to encode homologues of protein disulfide isomerase and poly(A)-binding protein, respectively (Kim and Mayfield, 1997; Yohn et al., 1998a; Somanchi et al., 2005). Genetic evidence supports a specific role of RB47 in D1 translation (Yohn et al., 1996, 1998a, 1998b; Somanchi et al., 2005). Similar analysis was carried out on a 38 kDa protein, called RB38, which was purified on the basis of its binding to the *psbA* 5′ UTR during affinity chromatography. RB38 binds U-rich stretches in RNA and it does not have sequence similarity to any known protein (Barnes et al., 2004). It remains to be demonstrated that RB38, RB47, and RB60 are in a common complex. The proposed role of RB38 is being re-examined in light of the finding that it is one and the same as RBP40, a protein that binds the *psbD* mRNA *in vitro* and *in vivo* and is specifically required for *psbD* translation (Schwarz et al., 2007).

b. Light regulation of the RB complex

The correlation of the level of RNA binding activity of the RB complex detected in extracts and the rate of *psbA* translation *in vivo*, supports translational activation rather than repression (Danon and Mayfield, 1991;

Trebitsh et al., 2000; Trebitsh and Danon, 2001). A model for this regulation is based on redox control (see section V.B), because DTT or reduced thioredoxin activated the complex *in vitro* and oxidizing conditions inactivated it both *in vitro* and *in organello* (Danon and Mayfield, 1994a; Trebitsh et al., 2000). Thus, the light sensor is proposed to be thioredoxin, a protein whose redox state is modulated by the electron flow from the photosynthetic electron transport chain (see Chapter 11). The modulation of RB60 activity by thioredoxin is consistent with their interaction during affinity chromatography (Lemaire et al., 2004). Redox coupling of RB60 and RB47 is proposed to result in the reduction of a disulfide bond between cysteine residues in neighboring RNA-binding domains of RB47, thereby relieving its steric inhibition of RNA binding (Fong et al., 2000; Alergand et al., 2006). Modulation of the stromal pH by the light reactions may further refine this redox control (Alergand et al., 2006). A distinct mode of regulation is proposed to inactivate the complex in the dark, based on the inability of reducing agents to activate the complex following its isolation from dark-grown cells (Trebitsh et al., 2000). This primary repression could be signaled by an oxidized plastoquinone pool, elevated stromal ADP levels (e.g. via ADP-dependent phosphorylation of RB60; Danon and Mayfield, 1994b), or by both sensors (Trebitsh et al., 2000). In the light, photosynthesis is thought to reverse these effects and relieve inhibition of RB60, a step termed "priming" (Trebitsh et al., 2000). According to the model, priming allows RB60 to respond to the redox state of thioredoxin and modulate the RNA-binding activity of RB47 in response to incident light (Danon, 1997; Bruick and Mayfield, 1999).

c. The physiological role of RNA-binding by the RB complex

Although the RB complex has been proposed to specifically activate *psbA* translation for the replacement of the D1 subunit in photodamaged PS II complexes, certain results are more consistent with a role in the translational activation of several mRNAs for *de novo* assembly of PS II and other photosynthesis complexes. First, the activation of the RB complex by light correlated with induced synthesis of several proteins in the chloroplast, including D1, D2, ATPase (CF_1) subunits, and the large subunit of Rubisco (Trebitsh et al., 2000; Trebitsh and Danon, 2001). Second, the binding specificity for the *psbA* mRNA has been reconsidered due to the discovery that RB47 is a member of the PABP family and the results of *in vitro* competition experiments that demonstrate that RB47 binds poly(A) (Yohn et al., 1998a; Barnes et al., 2004). Most chloroplast mRNAs have stretches of A residues, in keeping with the AT-richness of the chloroplast genome (Maul et al., 2002). Third, RB38/RPB40 specifically binds the *psbD* 5′ UTR and is required for its translation (Schwarz et al., 2007). Light regulation of D2 synthesis would concomitantly regulate D1 and CP47 synthesis via the

CES pathway, which is discussed in Chapter 29. The major protein synthesis responses activated by moderate light would be accounted for if RB38/RB40 also regulates the translation of the *psbC* and *rbcL* mRNAs, which is currently unknown. Finally, the high light stress conditions associated with preferential D1 translation for PS II repair are associated with the loss of PS I and PS II activities and oxidizing conditions within the chloroplast, which are predicted by the model to inactivate the RB complex and thereby diminish D1 translation, rather than activate it as has been observed (Trebitsh and Danon, 2001; Haldrup et al., 2003; Förster et al., 2006; Murata et al., 2006; Nishiyama et al., 2006).

3. Other factors

Other chloroplast RBPs could be involved in the light regulation of translation, but they have not been characterized at the molecular level (reviewed in Nickelsen, 2003). One study observed variations in the levels of RNA-binding activities of 15, 38, 47, 56, 62, and 81-kD proteins across conditions that were mixotrophic, heterotrophic, or photoautotrophic and in mutants partially deficient for chloroplast ribosomes (Hauser et al., 1996). Another found that the RNA-binding activities of four RBPs (of 46, 47, 60, and 80 kD) are dramatically stimulated within 5–10 minutes following a shift from dark to light conditions (Zerges et al., 2002). This phenomenon may be mediated by the effect of photosynthesis on the stromal ADP concentration, because it was prevented by two inhibitors that have the common effect of blocking ATP synthesis, and ADP inhibited the RNA binding activities *in vitro* with a K_i of 50 nM. An intermediate in chlorophyll biosynthesis and known light signaling molecule, Mg-protoporphryin IX, might also be involved because it inhibits the RNA-binding activities of the 46- and 47-kD RBPs *in vitro*. Certain factors identified with genetic approaches could potentially mediate the light regulation of translation (see section VI.C.6).

E. Analyses of the light-regulated phase of translation

In the chloroplasts of vascular plants, translational regulation by light can be imposed during the initiation or elongation phases, or at the transition between them (reviewed in Zerges, 2004). In *Chlamydomonas*, however, most evidence supports pre-elongation light regulation. An observed increase in thylakoid-associated polyribosomes during the light phase suggests regulation at the initiation step (Chua et al., 1976). Polysome analyses revealed that the elevated level of *chlL* translation in the dark is imposed by regulation of steps prior to, or during, initiation (Cahoon and Timko, 2000). The 5′ UTRs of *psbA* and *psbD* conferred light regulation to the expression of a reporter gene, thereby demonstrating that regulation occurs prior to the elongation phase (Franklin et al., 2002; Manuell et al., 2004). This could occur at the levels of the processing of the 5′-terminus (see section VI.E and Chapter 27), localization of the mRNA, recruitment of the ribosomal SSU, any step during the assembly of a translation-competent ribosome at the initiation codon, and/or entry of the assembled ribosome into the elongation phase. Moreover, *psbA* mRNA isolated from chloroplasts incubated in the light was detected in a fraction that was presumed to contain polysomes, but this was not the case when chloroplasts were incubated in the dark (Trebitsh et al., 2000). It will be important to compare the complete sucrose density gradient profiles of chloroplast mRNAs encoding thylakoid membrane proteins across dark and light conditions, and to include controls that demonstrate that high sedimentation velocity is due to polysome association rather than to association with another type of RNP particle (see Cahoon and Timko, 2000). It should also be noted that published results do not exclude the regulation of translation elongation, for which there is evidence in vascular plant chloroplasts (Edhofer et al., 1998; Muhlbauer and Eichacker, 1998).

F. Translational regulation during chloroplast differentiation

A third type of translational control occurs in angiosperms during chloroplast differentiation. This has been studied extensively during the greening of etiolated seedlings, where light is required for chlorophyll synthesis, and stimulates synthesis of the subunits of the thylakoid membrane complexes (reviewed in Eckhardt et al., 2004; Zerges, 2004). Wild-type *Chlamydomonas* accumulates chlorophyll in the dark or light and thus has a fully differentiated chloroplast during vegetative growth. Although chloroplast differentiation occurs during the germination of zygospores, translational regulation has not been studied during this process and it would be difficult to generate sufficient numbers of zygospores to do so (Cavalier-Smith, 1976). However, one study characterized protein synthesis and mRNA levels in maturing chloroplasts of the *y1* mutant, which has an undifferentiated chloroplast in the dark because it lacks the light-independent protochlorophyllide oxidoreductase. Comparisons of *y1* cells cultured in darkness or under moderate light revealed substantially higher levels of D1 and D2 synthesis but similar levels of the *psbA* and *psbD* mRNAs, suggesting translation is activated (Malnoë et al., 1988). This regulation has not been revisited. In barley chloroplast differentiation, where this regulation has been most extensively studied, the synthesis and assembly of chloroplast-encoded subunits of PS I and PS II results either from derepression of translation elongation, the short-term stabilization of the apoproteins by newly synthesized chlorophyll, or both processes (Klein et al., 1988a; Mullet et al., 1990).

VI. GENETIC APPROACHES THAT HAVE IDENTIFIED CHLOROPLAST TRANSLATION FACTORS

A. General components of the chloroplast translation machinery

GTFs are recalcitrant to identification by genetic approaches because chloroplast translation is essential for viability, albeit for unknown reasons. This is demonstrated by the toxicity of chloroplast translation inhibitors and by the fact that of the many deletions affecting the chloroplast genome, none removes a gene encoding a ribosomal protein, tRNA, or both copies of the rRNA loci (Gillham et al., 1970; Myers et al., 1982; Barbrook et al., 2006). Hence, only a few mutants with a partial-loss-of-function of GTFs have been isolated. One example is a double mutant for *ac20* and *cr1*, which accumulates approximately 30% of the wild-type level of chloroplast ribosomes, due in part to a defect in 16S rRNA processing (Holloway and Herrin, 1998; Chapter 27). Another example is the nuclear *SIM30* locus, which may encode a GTF because a mutant allele, *sim30-1d*, suppresses the translation defect caused by a mutant AUU *petD* initiation codon (Chen et al., 1997). This suppression phenotype is weakened by antibiotic resistance mutations in the 16S and 23S rRNAs, suggesting that the *SIM30* function involves chloroplast ribosomes.

B. The function of an RNA-binding protein determined by RNA interference

A role of the RNA-binding protein RBP40 (RB38) in *psbD* translation was revealed when RNAi was used to create strains deficient for this protein (Schwarz et al., 2007). RBP40 knockdown strains failed to load the *psbD* mRNA on polysomes, while the mRNAs of *psbA* and *atpB* were unaffected. Moreover, across several RBP40 knockdown strains, the rate of D2 synthesis correlated with RBP40 accumulation. It will be important to use reverse genetic approaches to address the functions of other translation factors originally identified through biochemical approaches.

C. Translation factors with target mRNA specificity identified from mutant screens

1. Identification

The nuclear genome encodes factors that are targeted to the chloroplast to control the translation of a particular mRNA (Table 28.5). These were identified from the analyses of mutants identified in large screens either by an *hcf* phenotype that is characteristic of deficiencies for PS II, PS I,

Table 28.5 Nuclear genes that encode proteins involved in chloroplast translation

Complex	mRNA	Nuclear locus	Protein	Defect occurs	Reference
Cyt. b_6f	*petA*	*TCA1*	Novel protein	Pre-elongation	Raynaud et al. (2007)
	petD	*MCD1*	Novel protein	Pre-elongation (and mRNA stabilization)	Murakami et al. (2005)
PS I	*psaB*	*TAB1 (F15)*	n.d.	Pre-elongation	Stampacchia et al. (1997)
		TAB2 (F14)	Conserved RBP	n.d.	Barneche et al. (2006)
PS II	*psbA*	*TBA1 (hf261)*	Conserved putative oxidoreductase	Pre-elongation	Somanchi et al. (2005)
		TBA2 (F35)	n.d.	Pre-elongation	Yohn et al. (1996)
		hf149	n.d.	Pre-elongation	Yohn et al. (1998b)
		(hf233, hf859, hf1085)	n.d.	Post-initiation	Yohn et al. (1998b)
	psbB	*MBB1*	Novel protein with TPR repeats	Pre-elongation (and mRNA stabilization)	Vaistij et al. (2000)
	psbC	*TBC1 (F34)*	n.d.	Pre-elongation	Zerges et al. (2003)
		TBC2 (F64, G314, B23)	Novel protein	Pre-elongation	Zerges et al. (2003)
		TBC3	n.d.	Pre-elongation	Zerges et al. (2003)
	psbD	*NAC1*	n.d.	Post-initiation	Wu and Kuchka (1995); Cohen et al. (2001)
		AC115	Novel protein	Post-initiation	Rattanachaikunsopon et al. (1999)
		(sup4b)	n.d.	n.d.	Wu and Kuchka (1995)
		NAC2	Novel protein with TPR repeats	Pre-elongation (and mRNA stabilization)	Nickelsen et al. (1999)
ATP synthase	*atpA*	*(F54)*	n.d.	n.d.	Drapier et al. (1992)
LIPOR	*chlL*	*(y1)*	n.d.	Pre-elongation	Cahoon and Timko (2000)
		(y5, y6, y7, y8, y10)	n.d.	Elongation/post-translational	Cahoon and Timko (2000)

Mutant names are given in parentheses. n.d., not determined

or the cytochrome b_6f complex, or as an acetate requirement (the "*ac*" phenotype), which would also capture ATP synthase mutants. In a subset of these mutants, pulse labeling revealed that a particular chloroplast genome-encoded polypeptide is not synthesized, even though the cognizant mRNA accumulates. In another case, the failure of the nuclear *y1* mutant to accumulate ChlL appears to reflect a defect in *chlL* translation (Cahoon and Timko, 2000). The nuclear loci *Y5*, *Y6*, *Y7*, *Y8*, and *Y10* function in translation elongation of the *chlL* mRNA or at post-translational step(s) required for ChlL accumulation (Cahoon and Timko, 2000). mRNA-specific roles for *TBC1*, *TAB1*, and *TCA1* are supported by the isolation of suppressor mutations that either replace or alter the 5′ UTR of the target chloroplast mRNA and restore photosynthesis (Rochaix et al., 1989; Stampacchia et al., 1997; Wostrikoff et al., 2001). Specificity of the *Y1* requirement for *chlL* translation is suggested by wild-type levels of the ChlB and ChlN subunits and the β subunit of the chloroplast ATP synthase in the *y1* mutant (Cahoon and Timko, 2000).

2. Are these mRNA-specific translation factors?

While current thinking envisages a battery of mRNA-specific translation factors, it should be noted that an mRNA-specific translation defect could result from a partial loss-of-function of a GTF for which that mRNA has a stringent requirement. This is not the case, however, for *AC115*, *TAB2*, and, most probably *TBC2*, because null mutations of these genes specifically impair translation of their target mRNAs. If any of these genes encoded a GTF, the the null phenotype would be lethality (see section VI.A). This test should be applied to other factors by analyses of null mutants or strains in which the factor is severely depleted. The degree of specificity in each case could also be resolved by the identification of the chloroplast mRNA(s) that coimmunoprecipitate with each factor by microarray analysis, as has been carried out for Crp1, an analogous translation factor in maize chloroplasts (Schmitz-Linneweber et al., 2005).

3. These translation factors function in different steps

Although a functional requirement for translation does not demonstrate a physical interaction between the nuclear gene product and its target mRNA, evidence of this has been reported in two cases. TAB2 and the *psaB* mRNA were found in a large complex (of 2000 kD) and recombinant TAB2 bound the *psaB* 5′ UTR in an *in vitro* assay (Dauvillée et al., 2003). NAC2 is in a complex with RBP40, and the latter immunoprecipitated with the target *psbD* mRNA (Schwarz et al., 2007). Pre-elongation requirements for seven gene-specific translation factors were revealed by the sufficiency of the 5′ UTR of the target mRNA to confer the requirement for the wild-type function for the expression of another coding region (Table 28.5; Zerges and Rochaix, 1994; Stampacchia et al., 1997; Drager et al., 1998; Nickelsen et al., 1999; Vaistij et al., 2000; Wostrikoff et al., 2001; Dauvillée et al., 2003). Two of the *psbA* translation mutants accumulate only *psbA* mRNAs with a 5′ terminal extension, supporting roles of the wild-type factors in the 5′ terminal processing, which has been proposed to activate translation (see Chapter 27). In three nuclear mutants (*y1*, *tba1*, and *hf149*), the target mRNA is not polysome-associated, pointing to a role in translation initiation or a prerequisite step. Four nuclear gene products appear to function after initiation because their target mRNAs remain polysome-bound in the mutants *nac1*, *hf233*, *hf859*, and *hf1085* (Yohn et al., 1996; Cohen et al., 2001). Roles in stabilization of the protein product of the target mRNA are unlikely in most cases because the products were not labeled in 10 minute pulses, an approach that detected the short-lived D1 or D2 in mutants that cannot assemble them into PS II core complexes (de Vitry et al., 1989; Girard-Bascou et al., 1992). Disassembly of polysomes from the *nac1* mutant (by chelation of Mg^{2+} ions) abolished the high sedimentation coefficient of the *psbD* mRNA, further supporting a post-initiation role for *NAC1* (Cohen et al., 2001), although chelation of Mg^{2+} could also disrupt other RNA-protein interactions. This experiment was not been performed for the *psbD* mRNA in *ac115* or the *psbA* mRNA in *hf233*, *hf859*, *and hf1085*.

4. Localization of mRNA-specific translation factors

Consistent with their proposed roles in translation, TBA1, TCA1, TAB2, and TBC2 cofractionated with chloroplasts (Auchincloss et al., 2002; Dauvillée et al., 2003; Somanchi et al., 2005; Raynaud et al., 2007). Although the intracellular location of AC115 has not been addressed, this protein has a predicted transmembrane domain (Rattanachaikunsopon et al., 1999).

5. Conservation of mRNA-specific translation factors

Is the genetic approach revealing conserved mechanisms of translation control? The most encouraging results have been provided by molecular characterizations of TAB2, a new type of RNA-binding protein. TAB2 homologues with 31–46% amino acid identity were identified in diverse photosynthetic organisms, including a land plant (*Arabidopsis*), a cryptophyte, and three cyanobacterial species (Dauvillée et al., 2003). Conservation at the biochemical level between *Chlamydomonas* and *Arabidopsis* was demonstrated by the ability of *Chlamydomonas TAB2* to partially complement the mutant phenotypes of the *Arabidopsis atab2* mutant (Barneche et al., 2006). Yet, the *Arabidopsis* gene, *ATAB2*, is required for translation of multiple chloroplast mRNAs encoding subunits of PS I and PS II. These broader requirements for *ATAB2* could reflect mutant evolutionary divergence of functional roles of a conserved biochemical activity. In another case, dispersed repeated elements of amino acid similarity in TBC2 were also found in RAA1, a chloroplast *trans*-splicing factor in *Chlamydomonas* and Crp1, a maize protein required for the processing and translation of the chloroplast *petA* and *petD* mRNAs, respectively (Fisk et al., 1999; Auchincloss et al., 2002; Merendino et al., 2006). Finally, AC115 and TCA1 appear to lack homologues, suggesting they may have non-conserved functions (Rattanachaikunsopon et al., 1999; Raynaud et al., 2007).

6. Possible physiological roles of mRNA-specific translational control

A longstanding hypothesis proposes functions of mRNA-specific translation factors in the control of subunit stoichiometry. Given the prevalence of CES control for each chloroplast photosynthetic complex (see Chapter 29), a role for gene-specific activators should be considered in this context. Indeed, it has been proposed that the mRNA-specific factors mediate this translational control (Choquet and Wollman, 2002). Genetic evidence for such a role for TCA1 in the feedback repression of *petA* translation by unassembled cytochrome *f* has been reported (Wostrikoff et al., 2001). Moreover, analogous classes of translation factors exist in *Saccharomyces cerevisiae* for the biogenesis of respiratory chain complexes and in maize for photosynthesis complexes (reviewed in Fox, 1996; Stern et al., 2004).

Another hypothesis proposes that factors localize their target mRNAs within the chloroplast as a prerequisite for translation (see section IX; Zerges and Rochaix, 1998; Zerges, 2000; Dauvillée et al., 2003; Uniacke and Zerges, 2007). In this scenario, rather than being translational regulators *per se*, some are mRNA localizing factors similar to those described in other organisms (reviewed in St. Johnston, 2005).

Roles of two mRNA-specific translation factors in response to physiological conditions have been proposed. TCA1 is a limiting factor for *petA* translation and the levels of this factor and the *petA* product, cytochrome *f*, were found to vary coordinately with the availability of nitrogen (Raynaud et al., 2007). This regulation is superimposed on the regulation of *petA* mRNA abundance by MCA1. The authors noted that these effects may reflect a role of TCA1 in feedback repression of *petA* translation via CES, or distinct physiological regulation. As mentioned above, three of the nuclear genes that are required for *psbA* translation (*TBA1*, *TBA2*, and the mutant gene in *hf149*) are also required for the RNA binding activity or the accumulation of RB47, supporting a role for it in D1 translational activation in response to light (see section IV; Yohn et al., 1996, 1998b; Somanchi et al., 2005). Finally, it should be pointed out that some of these factors may simply partake in the translation of their target mRNAs without a regulatory function. For example, some aspect of the recruitment of the ribosome to the initiation codon of a given mRNA may have evolved independently and require a specific protein. This could be a consequence of only sporadic conservation of SD sequences and poor conservation of 5′ UTR sequences overall.

D. Chloroplast translation functions of nuclear loci identified by suppressor genetics

Nuclear loci identified by suppressor mutations include *SUB4b* in which the *sub4b* mutation suppresses the *psbD* translation defects produced by all *nac1* and *ac115* mutant alleles, but not the *psbC* translation defect produced by a *tbc2* allele (Wu and Kuchka, 1995). Thus, the wild-type function of SUB4b may be specifically related to the roles of NAC1 and AC115 in *psbD* translation. A screen for phenotypic revertants of the *petA* translation defect produced by a *tca1* mutant allele identified only intragenic mutations in *TCA1* and a rearrangement of the chloroplast genome that replaced the *petA* 5′ UTR (Wostrikoff et al., 2001). A dominant mutation at the *TBC3* locus, *tbc3-rb1*, partially suppresses the *psbC* translation defect produced by the *tbc1-F34* allele, and also mutations affecting certain regions of the *psbC* 5′ UTR (see section VII.C.5; Zerges et al., 1997, 2003). Independent of these suppression phenotypes, *tbc3-rb1* decreases by two-fold the synthesis and accumulation of the *psbC* product, reduces the PS II level, and has slight effects on the synthesis of other proteins in the chloroplast (e.g. the Rubisco large subunit), suggesting that it has general functions in chloroplast translation (Zerges et al., 1997).

E. Three mRNA stability factors are also required for translation

Among the photosynthesis-deficient nuclear mutants isolated in the aforementioned screens, several fail to accumulate a specific chloroplast mRNA. While roles of these factors are being explored in the stabilization of their target mRNAs, evidence is emerging that they also function in translation. When these target mRNAs were artificially stabilized in the nuclear mutant backgrounds by the insertion of poly(G) into the 5′ UTR, translation was not restored (Drager et al., 1998; Nickelsen et al., 1999; Vaistij et al., 2000), although the insertions did not abolish translation in wild-type nuclear backgrounds. Further support was provided by the identification of a complex containing one of these factors, NAC2, and the *psbD* translation factor RBP40 (RB38) (Schwarz et al., 2007). These translational roles are being explored in the context of a model which invokes 5′ terminal processing of certain chloroplast mRNAs as a prerequisite for their stabilization and translation.

VII. *CIS*-ACTING TRANSLATION ELEMENTS IN CHLOROPLAST 5′ UTRs

A. Translation initiation regions

1. *Initiation codons*

In *E. coli*, the majority of initiation codons are AUG (90%) and GUG (9%), although AUU has also been identified (*infC* mRNA) (Butler et al., 1987; Hershey, 1987). In addition, AUC, which is not a naturally occurring initiation codon in *E. coli*, was found to be 5% as efficient as AUG for initiating translation (Romero and Garcia, 1991). The cyanobacterium *Synechocystis* uses AUG, GUG, and UUG, and possibly AUU (Sazuka and Ohara, 1996). In the *Chlamydomonas* chloroplast, AUG is the predominant initiation codon, although for *psbC* it is GUG (Rochaix et al., 1989). Mutational studies have revealed that any requirement for AUG as the initiation codon is gene/mRNA specific in *Chlamydomonas*. Site-directed mutagenesis of the initiation codon of *petA* (from AUG to AUU, ACG, ACC, and ACU), or of *petD* (from AUG to AUU or AUC), resulted in inefficient translation (10–20% of wild type) indicating that the second and third positions are not required to specify the site of translation initiation (Chen et al., 1993, 1995). In contrast, changing the *psbD* initiation codon to AUA or AUU abolished its translation, indicating that the third position is critical in this case (Nickelsen et al., 1999).

2. Extended codon–anticodon interactions

Where AUG is not required as the initiation codon, interactions other than codon–anticodon hybridization may assist in recruiting the small ribosomal subunit for initiation complex assembly. Examination of all *Chlamydomonas* chloroplast coding sequences revealed a conserved uridine (75%) at position −1 (Maul et al., 2002; Esposito et al., 2003). Mutation of the $_{-1}$U residue in combination with an AUU initiation codon severely reduced translation of *petA* and *atpB* (Esposito et al., 2001). In *E. coli*, an extended codon–anticodon interaction had been proposed in which a $_{-1}$U forms a fourth base pair with A37 of $tRNA^{fmet}$ (Ganoza et al., 1985). In *Chlamydomonas,* it was shown that cyt *f* was undetectable in strains expressing *petA* mRNAs containing AAUU at residues −1 to +3. However the expression of an ectopic mutant A37U $tRNA^{fmet}$ restored cyt *f* synthesis to nearly wild-type levels, supporting a 5′ extended codon–anticodon interaction model (Esposito et al., 2003).

3. Shine-Dalgarno sequences

a. Conservation

In *E. coli* and related eubacteria, translation initiation requires a ribosome binding site (RBS), whose best-characterized feature is the SD element, a GGAGG sequence centered 7–2 nt upstream of the initiation codon, which interacts with the anti-SD (ASD) sequence at the 3′ end of the 16S rRNA. A second mRNA element defined in *E. coli* is the downstream box (DB), a 15-nt sequence located downstream of the initiation codon and complementary to the anti-DB (ADB) (nt 1469–1483) (Sprengart et al., 1996). While the DB was regarded as a translational enhancer (McCarthy and Brimacombe, 1994), other data suggest that the DB can function efficiently in the absence of a functional SD interaction (Sprengart et al., 1996). In *Synechocystis*, no experimental evidence exists for functional SD sequences. However, the core portions (GAGG, GGAG, and AGGA) of the

SD sequence were identified in the appropriate position preceding 35 of 72 initiation sites evaluated (Sazuka and Ohara, 1996; Osada et al., 1999). ASD sequences are highly conserved between cyanobacteria, green algal chloroplasts, vascular plants, and *E. coli* (Dron et al., 1982; Steege et al., 1982; Maidak et al., 1994). In contrast, SD-like sequences in the leaders of chloroplast mRNAs vary considerably in location, size, base composition, or may be absent altogether (Ruf and Kössel, 1988; Bonham-Smith and Bourque, 1989; Stern et al., 1997; Fargo et al., 1998; Sugiura et al., 1998). The SD sequence variability in systems with conserved ASD sequences in the 16S raises the question of whether chloroplast translation initiation utilizes canonical SD–ASD interactions.

b. Functionality of SD-like sequences

Analyses of chloroplast transformants expressing gene constructs with mutated SD-like sequences in their 5′ UTRs suggest that there is not a universal requirement for an SD-like element in translation. For example, site-directed mutations in the SD-like sequences of several mRNAs (*petD*, *atpB*, *atpE*, *rps4*, and *rps7*) had little or no effect on their translation *in vivo* (Sakamoto et al., 1994; Fargo et al., 1998). Deletion of a SD-like sequence in the *psbA* 5′ UTR abolished translation, but also severely reduced the level of the mRNA (Mayfield et al., 1994). For both *psbD* and *psbC*, mutagenesis of a SD-like sequence reduced expression to approximately 25% of the wild-type level, without affecting mRNA accumulation (Nickelsen et al., 1999; Zerges et al., 2003). Evidence also exists for functionality of SD sequences in tobacco chloroplasts, both as activators and in one case, as a repressor (Hirose et al., 1998; Plader and Sugiura, 2003; Kuroda et al., 2007).

c. Remote SD sequences may exist in chloroplast 5′ UTRs

Roles of remote SD sequences should be considered, based on the finding in barley chloroplasts that initiation complexes form at a *psbA* SD-like sequence with a non-initiator AUG codon located 5′ to the *bona fide* initiation codon (Kim and Mullet, 1994). These authors proposed that the SSU first binds to SD-like sequence and then scans in a 3′ direction to the initiator AUG where it completes the initiation phase.

B. Modes of translational regulation mediated by *cis* elements

1. *Translational repression*

In eubacteria, translation is most frequently regulated by repressive *cis*-elements which form secondary structures or bind proteins that block access of the SSU (Gold, 1988; Kozak, 1999). Similarly, *cis*-elements of three chloroplast mRNAs (*psbA*, *psbD*, and *psaB*) have been proposed to form secondary structures that occlude either the SD sequence or the initiation codon

(Mayfield et al., 1994; Stampacchia et al., 1997; Klinkert et al., 2006). The repression of *rbcL* translation that is associated with oxidative stress may be mediated by the Rubisco large subunit (see section V; Cohen et al., 2005). Autoregulatory (CES) repression also controls the translation of the mRNAs encoding subunits of the four major photosynthetic complexes (see Chapter 29).

2. Translational activation

In many cases 5′ UTR sequences and their cognizant *trans*-acting factors are required for translation (see sections VI and VII.C), giving them *de facto* roles in activation rather than repression. Although the mechanisms are still largely unresolved, these elements could (1) promote the assembly of a ribosome at the initiation codon, (2) relieve repression by other elements and bound factor(s), or (3) have distinct conformations that alternately activate and repress translation. Many chloroplast *cis*-acting translation elements are quite distant from the initiation codon. In this respect, they may serve as ribosomal "landing pads" in some ways analogous to the internal ribosome entry sequences in the 5′ UTRs of picornaviral mRNAs and a few cytoplasmic mRNAs in eukaryotes (reviewed in Komar and Hatzoglou, 2005). An alternate hypothesis for translational activation proposes melting of a repressive RNA structure near the initiation codon, potentiated by the binding of *trans*-acting factors, as reviewed below for the *psbA*, *psbD*, and *psaB* mRNAs.

C. *Cis*-elements in specific chloroplast 5′ UTRs

1. General comparisons

The 5′ UTRs of the *psbC* and *petD* mRNAs have similar organizations; both are relatively long (547 nt and 362 nt, respectively) (Rochaix et al., 1989; Sakamoto et al., 1994) and each contains three distinct elements. The 5′ UTRs of the *psbA* and *psbD* mRNAs are shorter, and each has a single element that is required for translation, in addition to their SD-like sequence (Mayfield et al., 1994; Nickelsen et al., 1994).

2. psbA

A sequence in the *psbA* 5′ UTR has been proposed to repress translation by folding into a secondary structure which prevents access of the small ribosomal subunit to the neighboring SD sequence (Mayfield et al., 1994). Binding of the RB complex in response to light was proposed to activate *psbA* translation by melting this structure (see section V.D.2; Bruick and Mayfield, 1998, 1999). An A-tract near the AUG binds a thylakoid membrane protein called RBP63, possibly to localize D1 synthesis for PS II repair (see section IX; Ossenbühl et al., 2002).

3. psbD

psbD translation requires a stretch of 11 U residues (with a single A residue) that is located 14–25 nt 5′ to the AUG initiation codon (Nickelsen et al., 1999). This U-rich element binds the *psbD* translational activator RBP40 (RB38) both *in vitro* and *in vivo* (Ossenbühl and Nickelsen, 2000; Schwarz et al., 2007). RBP40 appears to function early in polysome assembly because it did not cofractionate with polysomes, yet an RBP40 antiserum immunoprecipitated RNAs of both ribosomal subunits (Schwarz et al., 2007). Its binding to both rRNAs and *psbD* mRNA suggest that RBP40 recruits the ribosomal subunits to the *psbD* initiation region. In addition, genetic and *in vitro* experiments identified a stem-loop structure which occludes the translation initiation codon and may sterically hinder the binding of the small ribosomal subunit (Klinkert et al., 2006). Genetic evidence also supports a model in which RBP40 binding to the U-rich sequence melts or alters this repressive stem-loop structure and exposes the AUG for ribosome assembly and translation initiation (Klinkert et al., 2006).

4. petD

petD 5′ UTR *cis*-elements were defined using site-directed and linker-scanning mutations (Sakamoto et al., 1994; Higgs et al., 1999). Element I is located within the 8 nt at the 5′ end of the mRNA and required for both stability and translation of the mRNA. This element has been proposed to bind a protein factor, possibly an *MCD1*-dependent one (Table 28.5), which would prevent further progression of the 5′–3′ exoribonucleolytic activity that may form the 5′ terminus of the mature mRNA (see section VI.E). Elements II and III are required for translation but not mRNA stability. Element II is located within a 16-nt region in the middle of the UTR, is unstructured, and binds protein(s), based on *in vivo* protection patterns following RNA modification by dimethyl sulfate. Element III spans 14 nt close to the initiation codon. This sequence forms a stem-loop *in vivo*, based on the pattern of dimethyl sulfate modification and the sequence changes caused by intragenic suppressors of Element III mutations. An RNA tertiary structure involving Elements II and III is supported by effects of mutations in Element II on the base-pairing status of nucleotides near Element III (Higgs et al., 1999). The comparison of the *petD* 5′ UTRs of four *Chlamydomonas* species and five independent isolates of *C. reinhardtii* was reported (Kramzar et al., 2006). Although low sequence conservation was observed, and all but one of the heterologous 5′ UTRs were nonfunctional in the *C. reinhardtii* chloroplast, in each *petD* 5′ UTR the three translational control elements reviewed above could be identified by their conserved relative positions and partially conserved sequences. These results indirectly substantiate the roles of these elements in *petD* translation and mRNA stability.

5. psbC

The *psbC* 5′ UTR has an mRNA stability element within the first 92 nt and at least two translation elements, located in the center and immediately 5′ to the initiation region (Zerges et al., 1997, 2003). The central translation element was predicted to form a stable stem-loop secondary structure (Rochaix et al., 1989). Two bulges in one strand of the predicted stem are critical because a mutation that eliminates them (*psbC-Fud34*) completely abolishes translation (Rochaix et al., 1989; Zerges et al., 1997). The other element is located within the 29 nt 5′ to the GUG initiation codon, and includes a SD-like sequence.

Translation of the *psbC* mRNA requires interactions between the 5′ UTR and the products of the nuclear loci *TBC1*, *TBC2*, and *TBC3*. The functions of *TBC1* and *TBC3* appear to be related and distinct from the role of *TBC2* in part because a suppressor point mutation that slightly destabilizes the central stem-loop structure (*psbC-F34suI*) partially reverses the translation block caused by the *tbc1-F34* allele, but not that caused by the *tbc2-F64* allele (Rochaix et al., 1989; Zerges and Rochaix, 1994). Deletion of 50 nt downstream of the central stem-loop also specifically relieves *TBC1* dependence (Zerges et al., 2003). Translational activation involving *TBC1* somehow involves the *psbC* translation initiation region, but for *TBC2* it does not (Zerges et al., 2003). A dominant suppressor mutation at the *TBC3* locus, *tbc3-rb1*, partially reverses the translational blocks caused by the *tbc1-F34* allele and mutations that affect the central translation element (Zerges et al., 1997, 2003). In contrast, sequences upstream of the central translation element and adjacent to the initiation region are required for suppression by *tbc3-rb1*. If *tbc3-rb1* is a gain-of-function for a *psbC* translation factor, as suggested by the dominant suppression phenotype, then these results suggest that *TBC3* acts via these sequences in the wild type.

6. rps7 *mRNA*

A genetic screen was used to isolate mutations in the *rps7* 5′ UTR that reduce its ability to drive expression of a reporter gene in *E. coli* (Fargo et al., 1999). Mutations at seven positions in the 3′ half of the 266-nt 5′ UTR revealed that these residues are required for translation in both *E. coli* and the *Chlamydomonas* chloroplast. A predicted secondary structure in the 5′ UTR affected by these mutations appears to be required for translation, because second site suppressor mutations restored both the structure and translation.

VIII. mRNA TERMINI MAY INTERACT DURING TRANSLATION

The extremities of cytoplasmic mRNAs in eukaryotes are brought together by interactions between PABP bound to the 3′ poly(A) tail and eIF4G bound to the 5′ cap (reviewed in Komarova et al., 2006). Interactions of mRNA termini may facilitate the recycling of terminating ribosomes back to the initiation codon for additional rounds of protein synthesis without disassembly and reassembly. They also might function to prevent translation initiation on truncated or unprocessed mRNAs. Several results raise the intriguing possibility that analogous interactions occur between the ends of chloroplast mRNAs to promote their translation. Circular polysomes were observed by electron microscopy (Chua et al., 1976). Although chloroplast mRNAs do not have 3′ poly(A) tails, unless they are being targeted for degradation by a bacterial-like pathway (Komine et al., 2000), the processing events that form the 3′ ends of mature *atpB* and *rbcL* mRNAs probably stimulate their translation because processed mRNAs were found to be enriched in polyribosome fractions (Rott et al., 1998). A PABP homologue in the chloroplast, RB47, has been proposed to function in translation initiation (see section V.D.2). The binding affinity of the RB complex (which contains RB47) for the *psbA* 5′ UTR is enhanced by the *psbA* 3′ UTR, suggesting that interactions between the termini stabilize this binding interaction (Katz and Danon, 2002).

IX. THE LOCALIZATION OF PROTEIN SYNTHESIS

A. Evidence for translation at the thylakoid membranes

The chloroplast-encoded thylakoid membrane proteins are believed to be co-translationally inserted in *Chlamydomonas*, with some of the earliest evidence being co-fractionation of their polysomes and thylakoids on sucrose density gradients (Chua et al., 1976; Margulies, 1983; Herrin and Michaels, 1985; Breidenbach et al., 1988). Similar results were obtained with chloroplasts of vascular plants (Yamamoto et al., 1981; Klein et al., 1988b; Muhlbauer and Eichacker, 1999; Zhang et al., 1999). At least one chloroplast mRNA encoding a non-membrane protein, the *rbcL* mRNA, also is translated by thylakoid membrane-bound ribosomes, as was also observed in higher plants (Hattori and Margulies, 1986; Breidenbach et al., 1988; Muhlbauer and Eichacker, 1999). This may reflect the localization of translation near sensors that respond to the *trans*-thylakoid membrane electrochemical proton gradient (Muhlbauer and Eichacker, 1999).

Of the thylakoid-associated polysomes in *Chlamydomonas*, approximately 50% were tethered by a nascent polypeptide because this fraction could only be released by the dissociation of ribosomes with puromycin, which induces premature dissociation (Chua et al., 1973; Margulies and Michaels, 1974). The other ribosomes (possibly those translating the *rbcL* mRNA) were bound by electrostatic interactions because they could be dissociated by a high ionic strength (0.5 M KCl) washing step. The latter associations may be physiologically relevant (and not due to non-specific associations formed *in vitro*) because they could not be re-established when the ionic strength was restored (Chua et al., 1973; Margulies and Michaels, 1974).

Translation is believed to occur in association with the non-appressed (stromal) thylakoid membranes. First, ribosomes and polysomes were observed on or near them in EM images of cells (Bourque et al., 1971) and isolated thylakoid membranes (Margulies and Michaels, 1974; Chua et al., 1976). Second, when thylakoid subfractions enriched in either granal or stromal thylakoids were prepared (by differential fragmentation by digitonin), only the stromal thylakoid fractions had the D1 precursor, which is converted to mature D1 by C-terminal proteolytic processing soon after its synthesis (Adir et al., 1990).

B. Other membranes may play a role in translation

Some evidence suggests that the inner membrane of the chloroplast envelope, or a low-density membrane with a similar pigment composition, has a role in protein synthesis (reviewed in Zerges, 2000). The low-density membranes are associated with RNA-binding proteins, including RB47 (Zerges and Rochaix, 1998) and a splicing factor (Perron et al., 1999). An earlier study showed that ca. 50% of chloroplast ribosomes cofractionate with membranes of a similar low buoyant density and chlorophyll level (Margulies and Weistrop, 1980). The low-density membranes may be part of an intra-chloroplast compartment and/or they may also have a role in thylakoid biogenesis, given their association with the homologue of VIPP1, a protein involved in thylakoid biogenesis in *Arabidopsis* (Liu et al., 2005).

C. Results of *in situ* approaches

The localization of translation for PS II *de novo* assembly and repair has been addressed by confocal microscopy of the distributions of specific mRNAs and proteins, which were labeled by fluorescent *in situ* hybridization and indirect immunofluorescence (Uniacke and Zerges, 2007). Locations of translation *in situ* were defined as having colocalized chloroplast mRNAs encoding PS II subunits and proteins of the chloroplast translation machinery, specifically under conditions of PS II subunit synthesis for either *de novo* assembly or the repair of photodamaged D1 subunits. PS II subunit synthesis for *de novo* assembly was detected in discrete regions near the pyrenoid, termed T-zones. Further supporting the localization of PS II assembly near T-zones, two PS II assembly mutants were observed to have unassembled D1 subunits or incompletely assembled PS II complexes near the pyrenoid, where the authors proposed they mark an intermediate compartment of PS II assembly. When cells under high light stress conditions were examined, the *psbA* mRNA colocalized with ribosomal proteins and RB38 (RBP40) throughout much of the chloroplast, consistent with the current model that repair of photodamaged PS II complexes occurs on non-appressed thylakoids (Adir et al., 1990). Using electron microscopy, an earlier study revealed that during the recovery from photoinhibition, clusters of immunogold-labeled D1 appear at transition sites between granal and stromal thylakoid membranes, supporting more highly localized PS II repair sites (Schuster et al., 1988).

X. TRANSLATION AND PROTEIN TARGETING

While much is known about the routing of proteins encoded by the nuclear genome to thylakoids (Kessler and Schnell, 2006), the pathways that target chloroplast genome-encoded proteins are only beginning to be elucidated (Eichacker and Henry, 2001; Vothknecht and Soll, 2005). In vascular plants, D1 is thought to be targeted in a translation-dependent pathway because an orthologue of a highly conserved signal sequence binding protein, cpSRP54, binds nascent D1 (High et al., 1997; Rohl and van Wijk, 2001; Zhang and Aro, 2002). By analogy to characterized pathways involving the signal recognition particle in other organisms, cpSRP54 would direct the ribosome-mRNA-nascent polypeptide complex to the translocon in the thylakoid membrane where the polypeptide is cotranslationally inserted. Indeed, *Chlamydomonas* encodes an apparent orthologue of cpSRP54 (accession number AAK12834). Subsequent steps of cotranslational insertion of D1 into damaged PS II complexes could involve a protein with tetratricopeptide repeats, called Rep27 (Park et al., 2007).

XI. FUTURE PERSPECTIVES

There are many challenges to achieving a mechanistic and regulatory understanding of chloroplast translation. One tool not yet available is a homologous *in vitro* translation system, which some results suggest might require specific chloroplast membrane fractions. Moreover, a close spatiotemporal association of the synthesis and assembly of photosynthesis complex subunits is an emerging theme. Thus, an understanding of how translational regulation relates to targeting and assembly is crucial for an understanding of the biogenesis of chloroplast integral membrane complexes. This relationship, and other fundamental chloroplast processes, could be elucidated by investigations of the molecular and physiological functions of the gene-specific translational regulators. *Chlamydomonas* also would be an ideal system for investigations of the function of cpSRP54 in protein targeting to thylakoid membranes and the functions of the plastid-specific ribosomal proteins. Finally, it will be important to determine whether untranslated mRNAs, which constitute a major fraction of chloroplast mRNAs, are sequestered in RNP complexes, compartmentalized within the chloroplast, or both.

ACKNOWLEDGMENTS

We thank Britta Förster, David Herrin, and Michael Kuchka for helpful comments and Michel Goldschmidt-Clermont, Jörg Nickelsen, Ursula Oberholzer, Jean-David Rochaix, Michal Shapira, and James Uniacke for critical review of the manuscript.

REFERENCES

Adir, N., Shochat, S., and Ohad, I. (1990). Light-dependent D1 protein synthesis and translocation is regulated by reaction center II. Reaction center II serves as an acceptor for the D1 precursor. *J. Biol. Chem.* **265**, 12563–12568.

Alergand, T., Peled-Zehavi, H., Katz, Y., and Danon, A. (2006). The chloroplast protein disulfide isomerase RB60 reacts with a regulatory disulfide of the RNA-binding protein RB47. *Plant Cell Physiol.* **47**, 540–548.

Allen, G.S., Zavialov, A., Gursky, R., Ehrenberg, M., and Frank, J. (2005). The cryo-EM structure of a translation initiation complex from *Escherichia coli*. *Cell* **121**, 703–712.

Allen, J.F. (1993). Redox control of transcription: Sensors, response regulators, activators and repressors. *FEBS Lett.* **332**, 203–207.

Antoun, A., Pavlov, M., Lovmar, M., and Ehrenberg, M. (2006). How initiation factors tune the rate of initiation of protein synthesis in bacteria. *EMBO J.* **25**, 2539–2550.

Aoki, H., Dekany, K., Adams, S.-L., and Ganoza, M.C. (1997). The gene encoding the elongation factor P protein is essential for viability and is required for protein synthesis. *J. Biol. Chem.* **272**, 32254–32259.

Aro, E.M., Virgin, I., and Andersson, B. (1993). Photoinhibition of photosystem II. Inactivation, protein damage and turnover. *Biochim. Biophys. Acta* **1143**, 113–134.

Asamizu, E., Nakamura, Y., Sato, S., Fukuzawa, H., and Tabata, S. (1999). A large scale structural analysis of cDNAs in a unicellular green alga, *Chlamydomonas reinhardtii*. I. Generation of 3433 non-redundant expressed sequence tags. *DNA Res.* **6**, 369–373.

Auchincloss, A.H., Zerges, W., Perron, K., Girard-Bascou, J., and Rochaix, J.-D. (2002). Characterization of Tbc2, a nucleus-encoded factor specifically required for translation of the chloroplast *psbC* mRNA in *Chlamydomonas reinhardtii*. *J. Cell Biol.* **157**, 953–962.

Baker, M.E., Grundy, W.N., and Elkan, C.P. (1998). Spinach CSP41, an mRNA-binding protein and ribonuclease, is homologous to nucleotide-sugar epimerases and hydroxysteroid dehydrogenases. *Biochem. Biophys. Res. Commun.* **248**, 250–254.

Barbrook, A.C., Howe, C.J., and Purton, S. (2006). Why are plastid genomes retained in non-photosynthetic organisms? *Trends Plant Sci.* **11**, 101–108.

Barneche, F., Winter, V., Crevecoeur, M., and Rochaix, J.D. (2006). ATAB2 is a novel factor in the signalling pathway of light-controlled synthesis of photosystem proteins. *EMBO J.* **25**, 5907–5918.

Barnes, D. and Mayfield, S.P. (2003). Redox control of post-transcriptional processes in the chloroplast. *Antioxid. Redox Signal.* **5**, 89–94.

Barnes, D., Cohen, A., Bruick, R.K., Kantardjieff, K., Fowler, S., Efuet, E., and Mayfield, S.P. (2004). Identification and characterization of a novel RNA binding protein that associates with the 5′-untranslated region of the chloroplast *psbA* mRNA. *Biochemistry* **43**, 8541–8550.

Beauclerk, A.A.D. and Cundliffe, E. (1987). Sites of action of two ribosomal RNA methylases responsible for resistance to aminoglycosides. *J. Mol. Biol.* **193**, 661–671.

Beligni, M.V., Yamaguchi, K., and Mayfield, S.P. (2004a). Chloroplast elongation factor ts pro-protein is an evolutionarily conserved fusion with the s1 domain-containing plastid-specific ribosomal protein-7. *Plant Cell* **16**, 3357–3369.

Beligni, M.V., Yamaguchi, K., and Mayfield, S.P. (2004b). The translational apparatus of *Chlamydomonas reinhardtii* chloroplast. *Photosynth. Res.* **82**, 315–325.

Benichou, M., Li, Z., Tournier, B., Chaves, A., Zegzouti, H., Jauneau, A., Delalande, C., Latché, A., Bouzayen, M., Spremulli, L.L., and Pech, J.-C. (2003). Tomato EF-Tsmt, a functional mitochondrial translation elongation factor from higher plants. *Plant Mol. Biol.* **53**, 411–422.

Bogorad, L. (2008). Evolution of early eukaryotic cells: Genomes, proteomes, and compartments. *Photosynth. Res.* **95**, 11–21.

Bollenbach, T.J., Tatman, D.A., and Stern, D.B. (2003). CSP41a, a multifunctional RNA-binding protein, initiates mRNA turnover in tobacco chloroplasts. *Plant J.* **36**, 842–852.

Bonham-Smith, P.C. and Bourque, D.P. (1989). Translation of chloroplast-encoded mRNA: Potential initiation and termination signals. *Nucleic Acids Res.* **17**, 2057–2080.

Bonny, C., Montandon, P.-E., Marc-Martin, S., and Stutz, E. (1991). Analysis of streptomycin-resistance of *Escherichia coli* mutants. *Biochim. Biophys. Acta* **1089**, 213–219.

Bourque, D.P., Boynton, J.E., and Gillham, N.W. (1971). Studies on the structure and cellular location of various ribosome and ribosomal RNA species in the green alga *Chlamydomonas reinhardi*. *J. Cell Sci.* **8**, 153–183.

Boynton, J. and Gillham, N. (1993). Chloroplast transformation in *Chlamydomonas*. *Methods in Enzymology* **217**, 510–536.

Breidenbach, E., Jenni, E., and Boschetti, A. (1988). Synthesis of two proteins in chloroplasts and mRNA distribution between thylakoids and stroma during the cell cycle of *Chlamydomonas reinhardii*. *Eur. J. Biochem.* **177**, 225–232.

Breitenberger, C.A. and Spremulli, L.L. (1980). Purification of *Euglena gracilis* chloroplast elongation factor G and comparison with other prokaryotic and eukaryotic translocases. *J. Biol. Chem.* **255**, 9814–9820.

Brodersen, D.E., Clemons, J.W.M., Carter, A.P., Wimberly, B.T., and Ramakrishnan, V. (2002). Crystal structure of the 30 s ribosomal subunit from *Thermus thermophilus*: Structure of the proteins and their interactions with 16 s RNA. *J. Mol. Biol.* **316**, 725–768.

Brügger, M. and Boschetti, A. (1975). Two-dimensional gel electrophoresis of ribosomal proteins from streptomycin-sensitive and streptomycin-resistant mutants of *Chlamydomonas reinhardi*. *Eur. J. Biochem.* **58**, 603–610.

Bruick, R.K. and Mayfield, S.P. (1998). Processing of the psbA 5′ untranslated region in *Chlamydomonas reinhardtii* depends upon factors mediating ribosome association. *J. Cell Biol.* **143**, 1145–1153.

Bruick, R.K. and Mayfield, S.P. (1999). Light-activated translation of chloroplast mRNAs. *Trends Plant Sci.* **4**, 190–195.

Butler, J.S., Springer, M., and Grunberg-Manago, M. (1987). AUU-To-AUG mutation in the initiator codon of the translation initiation factor IF3 abolishes translational autocontrol of its own gene (infC) *in vivo*. *Proc. Natl. Acad. Sci. U. S. A.* **84**, 4022–4025.

Caetano-Anolles, G. (2002). Tracing the evolution of RNA structure in ribosomes. *Nucleic Acids Res.* **30**, 2575–2587.

Cahoon, A.B. and Timko, M.P. (2000). Yellow-in-the-dark mutants of *Chlamydomonas* lack the ChlL subunit of light-independent protochlorophyllide reductase. *Plant Cell* **12**, 559–568.

Cannone, J., Subramanian, S., Schnare, M., Collett, J., D'Souza, L., Du, Y., Feng, B., Lin, N., Madabusi, L., Muller, K., Pande, N., Shang, Z., Yu, N., and Gutell, R. (2002). The Comparative RNA Web (CRW) Site: An online database of comparative sequence and structure information for ribosomal, intron, and other RNAs. *BMC Bioinformatics* **3**, 2.

Carter, A.P., Clemons, W.M., Brodersen, D.E., Morgan-Warren, R.J., Wimberly, B.T., and Ramakrishnan, V. (2000). Functional insights from the structure of the 30S ribosomal subunit and its interactions with antibiotics. *Nature* **407**, 340–348.

Cavalier-Smith, T. (1976). Electron microscopy of zygospore formation in *Chlamydomonas reinhardii*. *Protoplasma* **87**, 297–315.

Chen, X., Kindle, K., and Stern, D. (1993). Initiation codon mutations in the *Chlamydomonas* chloroplast *petD* gene result in temperature-sensitive photosynthetic growth. *EMBO J.* **12**, 3627–3635.

Chen, X., Kindle, K.L., and Stern, D.B. (1995). The initiation codon determines the efficiency but not the site of translation initiation in *Chlamydomonas* chloroplasts. *Plant Cell* **7**, 1295–1305.

Chen, X., Simpson, C.L., Kindle, K.L., and Stern, D.B. (1997). A dominant mutation in the *Chlamydomonas reinhardtii* nuclear gene *SIM30* suppresses translational defects caused by initiation codon mutations in chloroplast genes. *Genetics* **145**, 935–943.

Choquet, Y. and Wollman, F.-A. (2002). Translational regulations as specific traits of chloroplast gene expression. *FEBS Lett.* **529**, 39–42.

Choquet, Y., Zito, F., Wostrikoff, K., and Wollman, F.A. (2003). Cytochrome *f* translation in *Chlamydomonas* chloroplast is autoregulated by its carboxyl-terminal domain. *Plant Cell* **15**, 1443–1454.

Chua, N.-H., Blobel, G., Siekevitz, P., and Palade, G.E. (1973). Attachment of chloroplast polysomes to thylakoid membranes in *Chlamydomonas reinhardtii*. *Proc. Natl. Acad. Sci. U. S. A.* **70**, 1554–1558.

Chua, N.H., Blobel, G., Siekevitz, P., and Palade, G.E. (1976). Periodic variations in the ratio of free to thylakoid-bound chloroplast ribosomes during the cell cycle of *Chlamydomonas reinhardtii*. *J. Cell Biol.* **71**, 497–514.

Cohen, A., Yohn, C.B., and Mayfield, S. (2001). Translation of the chloroplast-encoded *psbD* is arrested post-initiation in a nuclear mutant of *Chlamydomonas reinhardtii*. *J. Plant Physiol.* **158**, 1069–1075.

Cohen, I., Knopf, J.A., Irihimovitch, V., and Shapira, M. (2005). A proposed mechanism for the inhibitory effects of oxidative stress on Rubisco assembly and its subunit expression. *Plant Physiol.* **137**, 738–746.

Cohen, I., Sapir, Y., and Shapira, M. (2006). A conserved mechanism controls translation of Rubisco large subunit in different photosynthetic organisms. *Plant Physiol.* **141**, 1089–1097.

Croitoru, V., Bucheli-Witschel, M., Hagg, P., Abdulkarim, F., and Isaksson, L.A. (2004). Generation and characterization of functional mutants in the translation initiation factor IF1 of *Escherichia coli*. *Eur. J. Biochem.* **271**, 534–544.

Croitoru, V., Semrad, K., Prenninger, S., Rajkowitsch, L., Vejen, M., Laursen, B.S., Sperling-Petersen, H.U., and Isaksson, L.A. (2006). RNA chaperone activity of translation initiation factor IF1. *Biochimie* **88**, 1875–1882.

Danon, A. (1997). Translational regulation in the chloroplast. *Plant Physiol.* **115**, 1293–1298.

Danon, A. and Mayfield, S.P. (1991). Light regulated translational activators: Identification of chloroplast gene specific mRNA binding proteins. *EMBO J.* **10**, 3993–4001.

Danon, A. and Mayfield, S.P. (1994a). ADP-dependent phosphorylation regulates RNA-binding *in vitro*: Implications in light-modulated translation. *EMBO J.* **13**, 2227–2235.

Danon, A. and Mayfield, S.P. (1994b). Light-regulated translation of chloroplast messenger RNAs through redox potential. *Science* **266**, 1717–1719.

Dauvillée, D., Stampacchia, O., Girard-Bascou, J., and Rochaix, J.D. (2003). Tab2 is a novel conserved RNA binding protein required for translation of the chloroplast *psaB* mRNA. *EMBO J.* **22**, 6378–6388.

Davies, J.P., Yildiz, F.H., and Grossman, A. (1996). Sac1, a putative regulator that is critical for survival of *Chlamydomonas reinhardtii* during sulfur deprivation. *EMBO J.* **15**, 2150–2159.

de Cambiaire, J.-C., Otis, C., Turmel, M., and Lemieux, C. (2007). The chloroplast genome sequence of the green alga *Leptosira terrestris*: Multiple losses of the inverted repeat and extensive genome rearrangements within the Trebouxiophyceae. *BMC Genomics* **8**, 213.

de Vitry, C., Olive, J., Drapier, D., Recouvreur, M., and Wollman, F.A. (1989). Post-translational events leading to the assembly of photosystem II protein complex: A study using photosynthesis mutants from *Chlamydomonas reinhardtii*. *J. Cell Biol.* **109**, 991–1006.

Drager, R.G., Girard-Bascou, J., Choquet, Y., Kindle, K.L., and Stern, D.B. (1998). *In vivo* evidence for 5′→3′ exoribonuclease degradion of an unstable chloroplast mRNA. *Plant. J.* **13**, 85–96.

Drapier, D., Girard-Bascou, J., and Wollman, F.A. (1992). Evidence for nuclear control of the expression of the *atpA* and *atpB* chloroplast genes in *Chlamydomonas*. *Plant Cell* **4**, 283–295.

Dron, M., Rahire, M., and Rochaix, J.D. (1982). Sequence of the chloroplast 16S rRNA gene and its surrounding regions of *Chlamydomonas reinhardii*. *Nucleic Acids Res.* **10**, 7609–7620.

Eberhard, S., Drapier, D., and Wollman, F.A. (2002). Searching limiting steps in the expression of chloroplast-encoded proteins: Relations between gene copy number, transcription, transcript abundance and translation rate in the chloroplast of *Chlamydomonas reinhardtii*. *Plant J.* **31**, 149–160.

Eckhardt, U., Grimm, B., and Hortensteiner, S. (2004). Recent advances in chlorophyll biosynthesis and breakdown in higher plants. *Plant Mol. Biol.* **56**, 1–14.

Edhofer, I., Muhlbauer, S.K., and Eichacker, L.A. (1998). Light regulates the rate of translation elongation of chloroplast reaction center protein D1. *Eur. J. Biochem.* **257**, 78–84.

Eichacker, L.A. and Henry, R. (2001). Function of a chloroplast SRP in thylakoid protein export. *Biochim. Biophys. Acta* **1541**, 120–134.

Emanuelsson, O., Nielsen, H., and von Heijne, G. (1999). ChloroP, a neural network-based method for predicting chloroplast transit peptides and their cleavage sites. *Protein Sci.* **8**, 978–984.

Erickson, J.M., Rahire, M., and Rochaix, J.D. (1984). *Chlamydomonas reinhardii* gene for the 32 000 mol. wt. protein of photosystem II contains four large introns and is located entirely within the chloroplast inverted repeat. *EMBO J.* **3**, 2753–2762.

Esposito, D., Hicks, A.J., and Stern, D.B. (2001). A role for initiation codon context in chloroplast translation. *Plant Cell* **13**, 2373–2384.

Esposito, D., Fey, J.P., Eberhard, S., Hicks, A.J., and Stern, D.B. (2003). *In vivo* evidence for the prokaryotic model of extended codon–anticodon interaction in translation initiation. *EMBO J.* **22**, 651–656.

Fargo, D., Zhang, M., Gillham, N., and Boynton, J. (1998). Shine-Dalgarno-like sequences are not required for translation of chloroplast mRNAs in *Chlamydomonas reinhardtii* chloroplasts or in *Escherichia coli*. *Mol. Gen. Genet.* **257**, 271–282.

Fargo, D.C., Boynton, J.E., and Gillham, N.W. (1999). Mutations altering the predicted secondary structure of a chloroplast 5′ untranslated region affect its physical and biochemical properties as well as its ability to promote translation of reporter mRNAs both in the *Chlamydomonas reinhardtii* chloroplast and in *Escherichia coli*. *Mol. Cell Biol.* **19**, 6980–6990.

Fargo, D., Boynton, J., and Gillham, N. (2001). Chloroplast ribosomal protein S7 of *Chlamydomonas* binds to chloroplast mRNA leader sequences and may be involved in translation initiation. *Plant Cell* **13**, 207–218.

Fisk, D.G., Walker, M.B., and Barkan, A. (1999). Molecular cloning of the maize gene *crp1* reveals similarity between regulators of mitochondrial and chloroplast gene expression. *EMBO J.* **18**, 2621–2630.

Fong, C.L., Lentz, A., and Mayfield, S.P. (2000). Disulfide bond formation between RNA binding domains is used to regulate mRNA binding activity of the chloroplast poly(A)-binding protein. *J. Biol. Chem.* **275**, 8275–8278.

Förster, B., Mathesius, U., and Pogson, B.J. (2006). Comparative proteomics of high light stress in the model alga *Chlamydomonas reinhardtii*. *Proteomics* **6**, 4309–4320.

Fox, T.D. (1996). Genetics of mitochondrial translation. In: *Translational Regulation* (J.W.B. Hershey, M.B. Mathews, and N. Sonenberg, Eds.), pp. 733–758. Cold Spring Harbor Press, Cold Spring Harbor, NY.

Frank, J. and Agrawal, R.K. (2000). A ratchet-like inter-subunit reorganization of the ribosome during translocation. *Nature* **406**, 318–322.

Franklin, S., Ngo, B., Efuet, E., and Mayfield, S. (2002). Development of a GFP reporter gene for *Chlamydomonas reinhardtii* chloroplast. *Plant J.* **30**, 733–744.

Ganoza, M.C., Marliere, P., Kofoid, E.C., and Louis, B.G. (1985). Initiator tRNA may recognize more than the initiation codon in mRNA: A model for translational initiation. *Proc. Natl. Acad. Sci. U. S. A.* **82**, 4587–4591.

Gillham, N.W., Boynton, J.E., and Burkholder, B. (1970). Mutations altering chloroplast ribosome phenotype in *Chlamydomonas*. I. Non-Mendelian mutations. *Proc. Natl. Acad. Sci. U. S. A.* **67**, 1026–1033.

Gillham, N.W., Harris, E.H., Randolph-Anderson, B.L., Boynton, J.E., Hauser, C.R., McElwain, K.B., and Newman, S.M. (1991). Molecular genetics of chloroplast ribosomes in *Chlamydomonas*. In: *The Translational Apparatus of Photosynthetic Organelles*, Vol. 55 (R. Mache, A.R. Subramanian, and E. Shutz, Eds.), pp. 127–144. NATO ASI Series.

Girard-Bascou, J., Pierre, Y., and Drapier, D. (1992). A nuclear mutation affects the synthesis of the chloroplast *psbA* gene production *Chlamydomonas reinhardtii*. *Curr. Genet.* **22**, 47–52.

Glick, B.R. and Ganoza, M.C. (1975). Identification of a soluble protein that stimulates peptide bond synthesis. *Proc. Natl. Acad. Sci. U. S. A.* **72**, 4257–4260.

Gold, L. (1988). Posttranscriptional regulatory mechanisms in *Escherichia coli*. *Annu. Rev. Biochem.* **57**, 199–233.

Goldschmidt-Clermont, M. (1991). Transgenic expression of aminoglycoside adenine transferase in the chloroplast: A selectable marker of site-directed transformation of *Chlamydomonas*. *Nucleic Acids Res.* **19**, 4083–4089.

Gong, F. and Yanofsky, C. (2002). Instruction of translating ribosome by nascent peptide. *Science* **297**, 1864–1867.

Gray, M. (1992). The endosymbiont hypothesis revisited. *Int. Rev. Cytol.* **141**, 233–357.

Gray, M.W., Burger, G., and Lang, B.F. (1999). Mitochondrial evolution. *Science* **283**, 1476–1481.

Gualerzi, C. and Pon, C. (1990). Initiation of mRNA translation in prokaryotes. *Biochemistry* **29**, 5881–5889.

GuhaMajumdar, M. and Sears, B.B. (2005). Chloroplast DNA base substitutions: An experimental assessment. *Mol. Genet. Genomics*, **273**, 177–183.

Gutell, R.R. and Fox, G.E. (1988). A compilation of large subunit RNA sequences presented in a structural format. *Nucleic Acids Res.* **16**, r175–r269.

Haldrup, A., Lunde, C., and Scheller, H.V. (2003). *Arabidopsis thaliana* plants lacking the PSI-D subunit of photosystem I suffer severe photoinhibition, have unstable photosystem I complexes, and altered redox homeostasis in the chloroplast stroma. *J. Biol. Chem.* **278**, 33276–33283.

Hansen, J.L., Ippolito, J.A., Ban, N., Nissen, P., Moore, P.B., and Steitz, T.A. (2002). The structures of four macrolide antibiotics bound to the large ribosomal subunit. *Mol. Cell* **10**, 117–128.

Hanson, M.R., Davidson, J.N., Mets, L.J., and Bogorad, L. (1974). Characterization of chloroplast and cytoplasmic ribosomal proteins of *Chlamydomonas reinhardi* by two-dimensional gel electrophoresis. *Mol. Gen. Genet.* **132**, 105–118.

Harris, E.H., Boynton, J.E., and Gillham, N.W. (1974). Chloroplast ribosome biogenesis in *Chlamydomonas*: Selection and characterization of mutants blocked in ribosome formation. *J. Cell Biol.* **63**, 160–179.

Harris, E.H., Burkhart, B.D., Gillham, N.W., and Boynton, J.E. (1989). Antibiotic resistance mutations in the chloroplast 16S and 23S rRNA genes of *Chlamydomonas reinhardtii*: Correlation of genetic and physical maps of the chloroplast genome. *Genetics* **123**, 281–292.

Harris, E.H., Boynton, J.E., and Gillham, N.W. (1994). Chloroplast ribosomes and protein synthesis. *Microbiol. Rev.* **58**, 700–754.

Hartz, D., McPheeters, D., and Gold, L. (1989). Selection of the initiator tRNA by *Escherichia coli* initiation factors. *Genes Devel.* **12A**, 1899–1912.

Hattori, T. and Margulies, M.M. (1986). Synthesis of large subunit of ribulosebisphosphate carboxylase by thylakoid-bound polyribosomes from spinach chloroplasts. *Arch. Biochem. Biophys.* **244**, 630–640.

Hauser, C.R., Gillham, N.W., and Boynton, J.E. (1996). Translational regulation of chloroplast genes. Proteins binding to the 5′-untranslated regions of chloroplast mRNAs in *Chlamydomonas reinhardtii*. *J. Biol. Chem.* **271**, 1486–1497.

Heifetz, P.B., Forster, B., Osmond, C.B., Giles, L.J., and Boynton, J.E. (2000). Effects of acetate on facultative autotrophy in *Chlamydomonas reinhardtii* assessed by photosynthetic measurements and stable isotope analyses. *Plant Physiol.* **122**, 1439–1445.

Herrin, D.L. and Michaels, A. (1985). *In vitro* synthesis and assembly of the peripheral subunits of coupling factor CF1 (alpha and beta) by thylakoid-bound ribosomes. *Arch. Biochem. Biophys.* **237**, 224–236.

Herrin, D.L., Michaels, A.S., and Paul, A.L. (1986). Regulation of genes encoding the large subunit of ribulose-1,5- bisphosphate carboxylase and the photosystem II polypeptides D-1 and D-2 during the cell cycle of *Chlamydomonas reinhardtii*. *J. Cell Biol.* **103**, 1837–1845.

Hershey, J.W.B. (1987). Protein synthesis. In: *Escherichia coli and Salmonella typhimurium* (F.C. Neidhardt, Ed.), pp. 613–647. American Society for Microbiology, Washington, DC.

Higgs, D.C., Shapiro, R.S., Kindle, K.L., and Stern, D.B. (1999). Small *cis*-acting sequences that specify secondary structures in a chloroplast mRNA are essential for RNA stability and translation. *Mol. Cell Biol.* **19**, 8479–8491.

High, S., Henry, R., Mould, R.M., Valent, Q., Meacock, S., Cline, K., Gray, J.C., and Luirink, J. (1997). Chloroplast SRP54 interacts with a specific subset of thylakoid precursor proteins. *J. Biol. Chem.* **272**, 11622–11628.

Hirose, T. and Sugiura, M. (1996). *Cis*-acting elements and trans-acting factors for accurate translation of chloroplast *psbA* mRNAs: Development of an *in vitro* translation system from tobacco chloroplasts. *EMBO J.* **15**, 1687–1695.

Hirose, T., Kusumegi, T., and Sugiura, M. (1998). Translation of tobacco chloroplast *rps14* mRNA depends on a Shine-Dalgarno-like sequence in the 5′-untranslated region but not on internal RNA editing in the coding region. *FEBS Lett.* **430**, 257–260.

Hobbie, S.N., Pfister, P., Bruell, C., Sander, P., Francois, B., Westhof, E., and Bottger, E.C. (2006). Binding of neomycin-class aminoglycoside antibiotics to mutant ribosomes with alterations in the A site of 16S rRNA. *Antimicrob. Agents Chemother.* **50**, 1489–1496.

Holloway, S.P. and Herrin, D.L. (1998). Processing of a composite large subunit rRNA. Studies with *Chlamydomonas* mutants deficient in maturation of the 23s-like rRNA. *Plant Cell* **10**, 1193–1206.

Hwang, S., Kawazoe, R., and Herrin, D.L. (1996). Transcription of *tufA* and other chloroplast-encoded genes is controlled by a circadian clock in *Chlamydomonas*. *Proc. Natl. Acad. Sci. U. S. A.* **93**, 996–1000.

Irihimovitch, V. and Shapira, M. (2000). Glutathione redox potential modulated by reactive oxygen species regulates translation of Rubisco large subunit in the chloroplast. *J. Biol. Chem.* **275**, 16289–16295.

Jiang, Y., Nock, S., Nesper, M., Sprinzl, M., and Sigler, P.B. (1996). Structure and importance of the dimerization domain in elongation factor Ts from *Thermus thermophilus*. *Biochemistry* **35**, 10269–10278.

Katz, Y.S. and Danon, A. (2002). The 3′-untranslated region of chloroplast *psbA* mRNA stabilizes binding of regulatory proteins to the leader of the message. *J. Biol. Chem.* **277**, 18665–18669.

Keren, N., Gong, H., and Ohad, I. (1995). Oscillations of reaction center II-D1 protein degradation *in vivo* induced by repetitive light flashes. Correlation between the level of RCII-QB- and protein degradation in low light. *J. Biol. Chem.* **270**, 806–814.

Kessler, F. and Schnell, D.J. (2006). The function and diversity of plastid protein import pathways: A multilane GTPase highway into plastids. *Traffic* **7**, 248–257.

Khrebtukova, I. and Spreitzer, R.J. (1996). Elimination of the *Chlamydomonas* gene family that encodes the small subunit of ribulose-1,5-bisphosphate carboxylase/oxygenase. *Proc. Natl. Acad. Sci. U. S. A.* **93**, 13689–13693.

Kim, J. and Mayfield, S.P. (1997). Protein disulfide isomerase as a regulator of chloroplast translational activation. *Science* **278**, 1954–1957.

Kim, J. and Mullet, J. (1994). Ribosome-binding sites on chloroplast *rbcL* and *psbA* mRNAs and light-induced initiation of D1 translation. *Plant Mol. Biol.* **25**, 437–448.

Klein, R.R., Gamble, P.E., and Mullet, J.E. (1988a). Light-dependent accumulation of radiolabeled plastid-encoded chlorophyll a-apoproteins required chlorophyll a. *Plant Physiol.* **88**, 1246–1256.

Klein, R.R., Mason, H.S., and Mullet, J.E. (1988b). Light-regulated translation of chloroplast proteins. I. Transcripts of *psaA-psaB*, *psbA*, and *rbcL* are associated with polysomes in dark-grown and illuminated barley seedlings. *J. Cell Biol.* **106**, 289–301.

Klinkert, B., Elles, I., and Nickelsen, J. (2006). Translation of chloroplast *psbD* mRNA in *Chlamydomonas* is controlled by a secondary RNA structure blocking the AUG start codon. *Nucleic Acids Res.* **34**, 386–394.

Komar, A.A. and Hatzoglou, M. (2005). Internal ribosome entry sites in cellular mRNAs: mystery of their existence. *J. Biol. Chem.* **280**, 23425–23428.

Komarova, A.V., Brocard, M., and Kean, K.M. (2006). The case for mRNA 5′ and 3′ end cross talk during translation in a eukaryotic cell. *Prog. Nucleic Acid Res. Mol. Biol.* **81**, 331–367.

Komine, Y., Kwong, L., Anguera, M.C., Schuster, G., and Stern, D.B. (2000). Polyadenylation of three classes of chloroplast RNA in *Chlamydomonas reinhardtii*. *RNA* **6**, 598–607.

Korostelev, A., Trakhanov, S., Laurberg, M., and Noller, H.F. (2006). Crystal structure of a 70S ribosome-tRNA complex reveals functional interactions and rearrangements. *Cell* **126**, 1065–1077.

Kozak, M. (1999). Initiation of translation in prokaryotes and eukaryotes. *Gene* **234**, 187–208.

Kramzar, L.M., Mueller, T., Erickson, B., and Higgs, D.C. (2006). Regulatory sequences of orthologous *petD* chloroplast mRNAs are highly specific among *Chlamydomonas* species. *Plant Mol. Biol.* **60**, 405–422.

Kucho, K., Okamoto, K., Tabata, S., Fukuzawa, H., and Ishiura, M. (2005). Identification of novel clock-controlled genes by cDNA macroarray analysis in *Chlamydomonas reinhardtii*. *Plant Mol. Biol.* **57**, 889–906.

Kuroda, H., Suzuki, H., Kusumegi, T., Hirose, T., Yukawa, Y., and Sugiura, M. (2007). Translation of psbC mRNAs starts from the downstream GUG, not the upstream AUG, and requires the extended Shine-Dalgarno sequence in tobacco chloroplasts. *Plant Cell Physiol* **48**, 1374–1378.

Lancaster, L. and Noller, H.F. (2005). Involvement of 16S rRNA nucleotides G1338 and A1339 in discrimination of initiator tRNA. *Mol. Cell* **20**, 623–632.

Laursen, B.S., Sorensen, H.P., Mortensen, K.K., and Sperling-Petersen, H.U. (2005). Initiation of protein synthesis in bacteria. *Microbiol. Mol. Biol. Rev.* **69**, 101–123.

Lee, J. and Herrin, D. (2002). Assessing the relative importance of light and the circadian clock in controlling chloroplast translation in *Chlamydomonas reinhardtii*. *Photosynth. Res.* **72**, 295–306.

Lee, H., Bingham, S.E., and Webber, A.N. (1996). Function of 3′ non-coding sequences and stop codon usage in expression of the chloroplast *psaB* gene in *Chlamydomonas reinhardtii*. *Plant Mol. Biol.* **31**, 337–354.

Lemaire, S.D., Guillon, B., Le Marechal, P., Keryer, E., Miginiac-Maslow, M., and Decottignies, P. (2004). New thioredoxin targets in the unicellular photosynthetic eukaryote *Chlamydomonas reinhardtii*. *Proc. Natl. Acad. Sci. U. S. A.* **101**, 7475–7480.

Leu, S. (1998). Extraordinary features in the *Chlamydomonas reinhardtii* chloroplast genome: (1) rps2 as part of a large open reading frame; (2) a *C. reinhardtii* specific repeat sequence. *Biochim. Biophys. Acta* **1365**, 541–544.

Liu, X.Q., Gillham, N.W., and Boynton, J.E. (1989). Chloroplast ribosomal protein gene *rps12* of *Chlamydomonas reinhardtii*. Wild-type sequence, mutation to streptomycin resistance and dependence, and function in *Escherichia coli*. *J. Biol. Chem.* **264**, 16100–16108.

Liu, C., Willmund, F., Whitelegge, J.P., Hawat, S., Knapp, B., Lodha, M., and Schroda, M. (2005). J-domain protein CDJ2 and HSP70B are a plastidic chaperone pair that interacts with vesicle-inducing protein in plastids 1. *Mol. Biol. Cell* **16**, 1165–1177.

Mache, R. (1990). Chloroplast ribosomal proteins and their genes. *Plant Sci.* **72**, 1–12.

Maidak, B.L., Larsen, N., McCaughey, M.J., Overbeek, R., Olsen, G.J., Fogel, K., Blandy, J., and Woese, C.R. (1994). The Ribosomal Database Project. *Nucleic Acids Res.* **22**, 3485–3487.

Maisnier-Patin, S., Paulander, W., Pennhag, A., and Andersson, D.I. (2007). Compensatory evolution reveals functional interactions between ribosomal proteins S12, L14, and L19. *J. Mol. Biol.* **366**, 207–215.

Malnoë, P., Mayfield, S.P., and Rochaix, J.D. (1988). Comparative analysis of the biogenesis of photosystem II in the wild-type and *y-1* mutant of *Chlamydomonas reinhardtii*. *J. Cell Biol.* **106**, 609–616.

Manuell, A., Beligni, M.V., Yamaguchi, K., and Mayfield, S.P. (2004). Regulation of chloroplast translation: Interactions of RNA elements, RNA-binding proteins and the plastid ribosome. *Biochem. Soc. Trans.* **32**, 601–605.

Manuell, A.L., Yamaguchi, K., Haynes, P.A., Milligan, R.A., and Mayfield, S.P. (2005). Composition and structure of the 80S ribosome from the green alga *Chlamydomonas reinhardtii*: 80S ribosomes are conserved in plants and animals. *J. Mol. Biol.* **351**, 266–279.

Manuell, A.L., Quispe, J., and Mayfield, S.P. (2007). Structure of the chloroplast ribosome: Novel domains for translation regulation. *PLoS Biol.* **5**, e209.

Maravic, G. (2004). Macrolide resistance based on the Erm-mediated rRNA methylation. *Curr. Drug Targets – Infectious Disorders* **4**, 193–202.

Margulies, M.M. (1983). Synthesis of photosynthetic membrane proteins directed by RNA from rough thylakoids of *Chlamydomonas reinhardtii*. *Eur. J. Biochem.* **137**, 241–248.

Margulies, M.M. and Michaels, A. (1974). Ribosomes bound to chloroplast membranes in *Chlamydomonas reinhardtii*. *J. Cell Biol.* **60**, 65–77.

Margulies, M.M. and Tiffany, H.L. (1979). Sedimentation behavior of chloroplast ribosomes from *Chlamydomonas reinhardtii*. *Biochim. Biophys. Acta* **563**, 171–180.

Margulies, M.M. and Weistrop, J.S. (1980). Sub-thylakoid fractions containing ribosomes. *Biochim. Biophys. Acta* **606**, 20–33.

Matsuo, T., Onai, K., Okamoto, K., Minagawa, J., and Ishiura, M. (2006). Real-time monitoring of chloroplast gene expression by a luciferase reporter: Evidence for nuclear regulation of chloroplast circadian period. *Mol. Cell Biol.* **26**, 863–870.

Maul, J.E., Lilly, J.W., Cui, L., dePamphilis, C.W., Miller, W., Harris, E.H., and Stern, D.B. (2002). The *Chlamydomonas reinhardtii* plastid chromosome: Islands of genes in a sea of repeats. *Plant Cell* **14**, 2659–2679.

Mayfield, S.P. and Schultz, J. (2004). Development of a luciferase reporter gene, *luxCt*, for *Chlamydomonas reinhardtii* chloroplast. *Plant J.* **37**, 449–458.

Mayfield, S.P., Cohen, A., Danon, A., and Yohn, C.B. (1994). Translation of the *psbA* mRNA of *Chlamydomonas reinhardtii* requires a structured RNA element contained within the 5′ untranslated region. *J. Cell Biol.* **127**, 1537–1545.

Mayfield, S.P., Manuell, A.L., Chen, S., Wu, J., Tran, M., Siefker, D., Muto, M., and Marin-Navarro, J. (2007). *Chlamydomonas reinhardtii* chloroplasts as protein factories. *Curr. Opin. Biotechnol.* **18**, 126–133.

McCarthy, J. and Brimacombe, R. (1994). Prokaryotic translation: The interactive pathway leading to initiation. *Trends Genet.* **10**, 402–407.

McKim, S.M. and Durnford, D.G. (2006). Translational regulation of light-harvesting complex expression during photoacclimation to high-light in *Chlamydomonas reinhardtii*. *Plant Physiol. Biochem.* **44**, 857–865.

Merendino, L., Falciatore, A., and Rochaix, J.D. (2003). Expression and RNA binding properties of the chloroplast ribosomal protein S1 from *Chlamydomonas reinhardtii*. *Plant Mol. Biol.* **53**, 371–382.

Merendino, L., Perron, K., Rahire, M., Howald, I., Rochaix, J.D., and Goldschmidt-Clermont, M. (2006). A novel multifunctional factor involved in *trans*-splicing of chloroplast introns in *Chlamydomonas*. *Nucleic Acids Res.* **34**, 262–274.

Meurer, J., Lezhneva, L., Amann, K., Godel, M., Bezhani, S., Sherameti, I., and Oelmuller, R. (2002). A peptide chain release factor 2 affects the stability of UGA-containing transcripts in *Arabidopsis* chloroplasts. *Plant Cell* **14**, 3255–3269.

Michaels, A. and Herrin, D.L. (1990). Translational regulation of chloroplast gene expression during the light-dark cell cycle of *Chlamydomonas*: Evidence for control by ATP/energy supply. *Biochem. Biophys. Res. Commun.* **170**, 1082–1088.

Minai, L., Wostrikoff, K., Wollman, F.-A., and Choquet, Y. (2006). Chloroplast biogenesis of photosystem II cores involves a series of assembly-controlled steps that regulate translation. *Plant Cell* **18**, 159–175.

Mittag, M. and Wagner, V. (2003). The circadian clock of the unicellular eukaryotic model organism *Chlamydomonas reinhardtii*. *Biol. Chem.* **384**, 689–695.

Monde, R., Schuster, G., and Stern, D. (2000). Processing and degradation of chloroplast mRNA. *Biochimie* **82**, 573–582.

Motohashi, R., Yamazaki, T., Myouga, F., Ito, T., Ito, K., Satou, M., Kobayashi, M., Nagata, N., Yoshida, S., Nagashima, A., Tanaka, K., Takahashi, S., and Shinozaki, K. (2007). Chloroplast ribosome release factor 1 (AtcpRF1) is essential for chloroplast development. *Plant Mol. Biol.* **64**, 481–497.

Muhlbauer, S.K. and Eichacker, L.A. (1998). Light-dependent formation of the photosynthetic proton gradient regulates translation elongation in chloroplasts. *J. Biol. Chem.* **273**, 20935–20940.

Muhlbauer, S.K. and Eichacker, L.A. (1999). The stromal protein large subunit of ribulose-1,5-bisphosphate carboxylase is translated by membrane-bound ribosomes. *Eur. J. Biochem.* **261**, 784–788.

Mullet, J.E., Klein, P.G., and Klein, R.R. (1990). Chlorophyll regulates accumulation of the plastid-encoded chlorophyll apoproteins CP43 and D1 by increasing apoprotein stability. *Proc. Natl. Acad. Sci. U. S. A.* **87**, 4038–4042.

Murakami, S., Kuehnle, K., and Stern, D.B. (2005). A spontaneous tRNA suppressor of a mutation in the *Chlamydomonas reinhardtii* nuclear *MCD1* gene required for stability of the chloroplast *petD* mRNA. *Nucleic Acids Res.* **33**, 3372–3380.

Murata, N., Takahashi, S., Nishiyama, Y., and Allakhverdiev, S.I. (2006). Photoinhibition of photosystem II under environmental stress. *Biochim. Biophys. Acta.* **1767**, 414–421.

Myers, A.M., Grant, D.M., Rabert, D.K., Harris, E.H., Boynton, J.E., and Gillham, N.W. (1982). Mutants of *Chlamydomonas reinhardtii* with physical alterations in their chloroplast DNA. *Plasmid* **7**, 133–151.

Newman, S.M., Boynton, J.E., Gillham, N.W., Randolph-Anderson, B.L., Johnson, A.M., and Harris, E.H. (1990). Transformation of chloroplast ribosomal RNA genes in *Chlamydomonas*: Molecular and genetic characterization of integration events. *Genetics* **126**, 875–888.

Nickelsen, J. (2003). Chloroplast RNA-binding proteins. *Curr. Genet.* **43**, 392–399.

Nickelsen, J., van Dillewijn, J., Rahire, M., and Rochaix, J.D. (1994). Determinants for stability of the chloroplast *psbD* RNA are located within its short leader region in *Chlamydomonas reinhardtii*. *EMBO J.* **13**, 3182–3191.

Nickelsen, J., Fleischmann, M., Boudreau, E., Rahire, M., and Rochaix, J.-D. (1999). Identification of *cis*-acting RNA leader elements required for chloroplast *psbD* gene expression in *Chlamydomonas*. *Plant Cell* **11**, 957–970.

Nishiyama, Y., Allakhverdiev, S.I., and Murata, N. (2006). A new paradigm for the action of reactive oxygen species in the photoinhibition of photosystem II. *Biochim. Biophys. Acta* **1757**, 742–749.

Nissen, P., Hansen, J., Ban, N., Moore, P.B., and Steitz, T.A. (2000). The structural basis of ribosome activity in peptide bond synthesis. *Science* **289**, 920–930.

Noller, H.F., Kop, J., Wheaton, V., Brosius, J., Gutell, R.R., Kopylov, A.M., Dohme, F., Herr, W., Stahl, D.A., Gupta, R., and Woese, C.R. (1981). Secondary structure model for 23S ribosomal RNA. *Nucleic Acids Res.* **9**, 6167–6189.

O'Farrell, H.C., Scarsdale, J.N., and Rife, J.P. (2004). Crystal structure of KsgA, a universally conserved rRNA adenine dimethyltransferase in *Escherichia coli*. *J. Mol. Biol.* **339**, 337–353.

Ogle, J.M., Murphy, F.V., Tarry, M.J., and Ramakrishnan, V. (2002). Selection of tRNA by the ribosome requires a transition from an open to a closed form. *Cell* **111**, 721–732.

Osada, Y., Saito, R., and Tomita, M. (1999). Analysis of base-pairing potentials between 16S rRNA and 5′ UTR for translation initiation in various prokaryotes. *Bioinformatics* **15**, 578–581.

Ossenbühl, F. and Nickelsen, J. (2000). *cis*- and *trans*-acting determinants for translation of *psbD* mRNA in *Chlamydomonas reinhardtii*. *Mol. Cell Biol.* **20**, 8134–8142.

Ossenbühl, F., Hartmann, K., and Nickelsen, J. (2002). A chloroplast RNA binding protein from stromal thylakoid membranes specifically binds to the 5′ untranslated region of the *psbA* mRNA. *Eur. J. Biochem.* **269**, 3912–3919.

Park, S., Khamai, P., Garcia-Cerdan, J.G., and Melis, A. (2007). REP27, a tetratricopeptide repeat nuclear-encoded and chloroplast-localized protein, functions in D1/32-kD reaction center protein turnover and photosystem II repair from photodamage. *Plant Physiol* **143**, 1547–1560.

Perron, K., Goldschmidt-Clermont, M., and Rochaix, J.D. (1999). A factor related to pseudouridine synthases is required for chloroplast group II intron *trans*-splicing in *Chlamydomonas reinhardtii*. *EMBO J.* **18**, 6481–6490.

Pfannschmidt, T. (2003). Chloroplast redox signals: How photosynthesis controls its own genes. *Trends Plant Sci.* **8**, 33–41.

Poehlsgaard, J. and Douthwaite, S. (2005). The bacterial ribosome as a target for antibiotics. *Nat. Rev. Microbiol.* **3**, 870–881.

Polacek, N. and Mankin, A.S. (2005). The ribosomal peptidyl transferase Center: structure, function, evolution, inhibition. *Crit. Rev. Biochem. Mol. Biol.* **40**, 285–311.

Plader, W. and Sugiura, M. (2003). The Shine-Dalgarno-like sequence is a negative regulatory element for translation of tobacco chloroplast rps2 mRNA: An additional mechanism for translational control in chloroplasts. *Plant J* **34**, 377–382.

Ramakrishnan, V. (2002). Ribosome structure and the mechanism of translation. *Cell* **108**, 557–572.

Randolph-Anderson, B.L., Boynton, J.E., and Gillham, N.W. (1989). Electrophoretic and immunological comparisons of chloroplast and prokaryotic ribosomal proteins reveal that certain families of large subunit proteins are evolutionarily conserved. *J. Mol. Evol.* **29**, 68–88.

Rattanachaikunsopon, P., Rosch, C., and Kuchka, M.R. (1999). Cloning and characterization of the nuclear *AC115* gene of *Chlamydomonas reinhardtii*. *Plant Mol. Biol.* **39**, 1–10.

Raue, H.A., Klootwijk, J., and Musters, W. (1988). Evolutionary conservation of structure and function of high molecular weight ribosomal RNA. *Prog. Biophys. Mol. Biol.* **51**, 77–129.

Raynaud, C., Loiselay, C., Wostrikoff, K., Kuras, R., Girard-Bascou, J., Wollman, F.A., and Choquet, Y. (2007). Evidence for regulatory function of nucleus-encoded factors on mRNA stabilization and translation in the chloroplast. *Proc. Natl. Acad. Sci. U. S. A.* **104**, 9093–9098.

Rochaix, J.D., Kuchka, M., Mayfield, S., Schirmer-Rahire, M., Girard-Bascou, J., and Bennoun, P. (1989). Nuclear and chloroplast mutations affect the synthesis or stability of the chloroplast *psbC* gene product in *Chlamydomonas reinhardtii*. *EMBO J.* **8**, 1013–1021.

Rodnina, M.V., Beringer, M., and Wintermeyer, W. (2007). How ribosomes make peptide bonds. *Trends Biochem Sci.* **32**, 20–26.

Rohl, T. and van Wijk, K.J. (2001). *In vitro* reconstitution of insertion and processing of cytochrome *f* in a homologous chloroplast translation system. *J. Biol. Chem.* **276**, 35465–35472.

Rolland, N., Janosi, L., Block, M.A., Shuda, M., Teyssier, E., Miege, C., Cheniclet, C., Carde, J.-P., Kaji, A., and Joyard, J. (1999). Plant ribosome recycling factor homologue is a chloroplastic protein and is bactericidal in *Escherichia coli* carrying temperature-sensitive ribosome recycling factor. *Proc. Natl. Acad. Sci. U. S. A.* **96**, 5464–5469.

Rott, R., Levy, H., Drager, R.G., Stern, D.B., and Schuster, G. (1998). 3′-Processed mRNA is preferentially translated in *Chlamydomonas reinhardtii* chloroplasts. *Mol. Cell Biol.* **18**, 4605–4611.

Ruf, M. and Kössel, K. (1988). Occurrence and spacing of ribosome recognition sites in mRNAs of chloroplasts from higher plants. *FEBS Lett.* **240**, 41–44.

Saito, K., Mattheakis, L.C., and Nomura, M. (1994). Post-transcriptional regulation of the str operon in *Escherichia coli*. Ribosomal protein S7 inhibits coupled translation of S7 but not its independent translation. *J. Mol. Biol.* **235**, 111–124.

Sakamoto, W., Kindle, K.L., and Stern, D.B. (1993). *In vivo* analysis of *Chlamydomonas* chloroplast *petD* gene expression using stable transformation of beta-glucuronidase translational fusions. *Proc. Natl. Acad. Sci. U. S. A.* **90**, 497–501.

Sakamoto, W., Chen, X., Kindle, K.L., and Stern, D.B. (1994). Function of the *Chlamydomonas reinhardtii petD* 5′ untranslated region in regulating the accumulation of subunit IV of the cytochrome b_6/f complex. *Plant J.* **6**, 503–512.

Sazuka, T. and Ohara, O. (1996). Sequence features surrounding the translation initiation sites assigned on the genome sequence of *Synechocystis sp.* strain PCC6803 by amino-terminal protein sequencing. *DNA Res.* **3**, 225–232.

Schluenzen, F., Tocilj, A., Zarivach, R., Harms, J., Gluehmann, M., Janell, D., Bashan, A., Bartels, H., Agmon, I., Franceschi, F., and Yonath, A. (2000). Structure of functionally activated small ribosomal subunit at 3.3 angstroms resolution. *Cell* **102**, 615–623.

Schmidt, R.J., Richardson, C.B., Gillham, N.W., and Boynton, J.E. (1983). Sites of synthesis of chloroplast ribosomal proteins in *Chlamydomonas*. *J. Cell Biol.* **96**, 1451–1463.

Schmidt, R.J., Myers, A.M., Gillham, N.W., and Boynton, J.E. (1984). Immunological similarities between specific chloroplast ribosomal proteins from *Chlamydomonas reinhardtii* and ribosomal proteins from *Escherichia coli*. *Mol. Biol. Evol.* **1**, 317–334.

Schmitz-Linneweber, C., Williams-Carrier, R., and Barkan, A. (2005). RNA immunoprecipitation and microarray analysis show a chloroplast pentatricopeptide repeat protein to be associated with the 5′ region of mRNAs whose translation it activates. *Plant Cell* **17**, 2791–2804.

Schuster, G., Timberg, R., and Ohad, I. (1988). Turnover of thylakoid photosystem II proteins during photoinhibition of *Chlamydomonas reinhardtii*. *Eur. J. Biochem.* **177**, 403–410.

Schwarz, C., Elles, I., Kortmann, J., Piotrowski, M., and Nickelsen, J. (2007). Synthesis of the D2 protein of photosystem II in *Chlamydomonas* is controlled by a high molecular mass complex containing the RNA stabilization factor Nac2 and the translational activator RBP40. *Plant Cell* **19**, 3627–3639.

Selmer, M., Dunham, C.M., Murphy, F.V.I.V., Weixlbaumer, A., Petry, S., Kelley, A.C., Weir, J.R., and Ramakrishnan, V. (2006). Structure of the 70S ribosome complexed with mRNA and tRNA. *Science* **313**, 1935–1942.

Shapira, M., Lers, A., Heifetz, P.B., Irihimovitz, V., Osmond, C.B., Gillham, N.W., and Boynton, J.E. (1997). Differential regulation of chloroplast gene expression in *Chlamydomonas reinhardtii* during photoacclimation: Light stress transiently suppresses synthesis of the Rubisco LSU protein while enhancing synthesis of the PS II D1 protein. *Plant Mol. Biol.* **33**, 1001–1011.

Shrager, J., Hauser, C., Chang, C.-W., Harris, E.H., Davies, J., McDermott, J., Tamse, R., Zhang, Z., and Grossman, A.R. (2003). *Chlamydomonas reinhardtii* genome project. A guide to the generation and use of the cDNA information. *Plant Physiol.* **131**, 401–408.

Silk, G. and Wu, M. (1993). Post-transcriptional accumulation of chloroplast *tufA* (elongation factor gene) mRNA during chloroplast development in *Chlamydomonas reinhardtii*. *Plant Mol. Biol.* **23**, 87–96.

Sizova, I.A., Lapina, T.V., Frolova, O.N., Alexandrova, N.N., Akopiants, K.E., and Danilenko, V.N. (1996). Stable nuclear transformation of *Chlamydomonas reinhardtii* with a *Streptomyces rimosus* gene as the selective marker. *Gene* **181**, 13–18.

Somanchi, A., Barnes, D., and Mayfield, S.P. (2005). A nuclear gene of *Chlamydomonas reinhardtii*, *Tba1*, encodes a putative oxidoreductase required for translation of the chloroplast *psbA* mRNA. *Plant J.* **42**, 341–352.

Sprengart, M.L., Fuchs, E., and Porter, A.G. (1996). The downstream box: An efficient and independent translation initiation signal in *Escherichia coli*. *EMBO J.* **15**, 665–674.

St. Johnston, D. (2005). Moving messages: The intracellular localization of mRNAs. *Nat. Rev. Mol. Cell Biol.* **6**, 363–375.

Stampacchia, O., Girard-Bascou, J., Zanasco, J.L., Zerges, W., Bennoun, P., and Rochaix, J.D. (1997). A nuclear-encoded function essential for translation of the chloroplast *psaB* mRNA in *Chlamydomonas*. *Plant Cell* **9**, 773–782.

Steege, D.A., Graves, M.C., and Spremulli, L.L. (1982). *Euglena gracilis* chloroplast small subunit rRNA. Sequence and base pairing potential of the 3′ terminus, cleavage by colicin E3. *J. Biol. Chem.* **257**, 10430–10439.

Steitz, T.A. and Moore, P.B. (2003). RNA, the first macromolecular catalyst: The ribosome is a ribozyme. *Trends Biochem. Sci.* **28**, 411–418.

Stern, D.B., Higgs, D.C., and Yang, J. (1997). Transcription and translation in chloroplasts. *Trends Plant Sci.* **2**, 308–315.

Stern, D.B., Hanson, M.R., and Barkan, A. (2004). Genetics and genomics of chloroplast biogenesis: Maize as a model system. *Trends Plant Sci.* **9**, 293–301.

Sugiura, M., Hirose, T., and Sugita, M. (1998). Evolution and mechanism of translation in chloroplasts. *Annu. Rev. Genet.* **32**, 437–459.

Szymanski, M., Barciszewska, M., Erdmann, V., and Barciszewski, J. (2003). 5S rRNA: Structure and interactions. *Biochem. J.* **371**, 641–651.

Tan, X., Varughese, M., and Widger, W.R. (1994). A light-repressed transcript found in *Synechococcus* PCC 7002 is similar to a chloroplast-specific small subunit ribosomal protein and to a transcription modulator protein associated with sigma 54. *J. Biol. Chem.* **269**, 20905–20912.

Tenson, T. and Ehrenberg, M. (2002). Regulatory nascent peptides in the ribosomal tunnel. *Cell* **108**, 591–594.

Tenson, T. and Mankin, A. (2006). Antibiotics and the ribosome. *Mol. Microbiol.* **59**, 1664–1677.

Tenson, T., Lovmar, M., and Ehrenberg, M. (2003). The mechanism of action of macrolides, lincosamides and streptogramin B reveals the nascent peptide exit path in the ribosome. *J. Mol. Biol.* **330**, 1005–1014.

Trebitsh, T. and Danon, A. (2001). Translation of chloroplast *psbA* mRNA is regulated by signals initiated by both photosystems II and I. *Proc. Natl. Acad. Sci. U. S. A.* **98**, 12289–12294.

Trebitsh, T., Levitan, A., Sofer, A., and Danon, A. (2000). Translation of chloroplast *psbA* mRNA is modulated in the light by counteracting oxidizing and reducing activities. *Mol. Cell Biol.* **20**, 1116–1123.

Turmel, M., Gutell, R.R., Mercier, J.-P., Otis, C., and Lemieux, C. (1993). Analysis of the chloroplast large subunit ribosomal RNA gene from 17 *Chlamydomonas* taxa: Three internal transcribed spacers and 12 group I intron insertion sites. *J. Mol. Biol.* **232**, 446–467.

Uniacke, J. and Zerges, W. (2007). Photosystem II assembly and repair are differentially localized in *Chlamydomonas*. *Plant Cell* **19**, 3640–3654.

Vaistij, F.E., Goldschmidt-Clermont, M., Wostrikoff, K., and Rochaix, J.D. (2000). Stability determinants in the chloroplast *psbB/T/H* mRNAs of *Chlamydomonas reinhardtii*. *Plant J.* **21**, 469–482.

Vallon, O., Bulte, L., Kuras, R., Olive, J., and Wollman, F.A. (1993). Extensive accumulation of an extracellular L-amino-acid oxidase during gametogenesis of *Chlamydomonas reinhardtii*. *Eur. J. Biochem.* **215**, 351–360.

van Wijk, K.J., Nilsson, L.O., and Styring, S. (1994). Synthesis of reaction center proteins and reactivation of redox components during repair of photosystem II after light-induced inactivation. *J. Biol. Chem.* **269**, 28382–28392.

Voorma, H.O. (1996). Control of translation in prokaryotes. In: *Translational Control* (J.B.W. Hershey, M.B. Matthews, and N. Sonenberg, Eds.), pp. 759–778. Cold Spring Harbor Press, Plainview.

Vothknecht, U.C. and Soll, J. (2005). Chloroplast membrane transport: Interplay of prokaryotic and eukaryotic traits. *Gene* **354**, 99–109.

Wagner, V., Gessner, G., Heiland, I., Kaminski, M., Hawat, S., Scheffler, K., and Mittag, M. (2006). Analysis of the phosphoproteome of *Chlamydomonas reinhardtii* provides new insights into various cellular pathways. *Eukaryotic Cell* **5**, 457–468.

Walsh, C. (2000). Molecular mechanisms that confer antibacterial drug resistance. *Nature* **406**, 775–781.

Wieden, H.-J., Gromadski, K., Rodnin, D., and Rodnina, M.V. (2002). Mechanism of elongation factor (EF)-Ts-catalyzed nucleotide exchange in EF-Tu. Contribution of contacts at the guanine base. *J. Biol. Chem.* **277**, 6032–6036.

Wilson, D.N. and Nierhaus, K.H. (2006). The E-site story: The importance of maintaining two tRNAs on the ribosome during protein synthesis. *Cell. Mol. Life Sci.* **63**, 2725–2737.

Wimberly, B.T., Brodersen, D.E., Clemons, W.M., Morgan-Warren, R.J., Carter, A.P., Vonrhein, C., Hartsch, T., and Ramakrishnan, V. (2000). Structure of the 30S ribosomal subunit. *Nature* **407**, 327–339.

Wobbe, L., Schwarz, C., Nickelsen, J., and Kruse, O. (2008). Translational control of photosynthetic gene expression in phototrophic eukaryotes. *Physiol. Plant.* **113**, 507–515.

Wostrikoff, K. and Stern, D. (2007). Rubisco large-subunit translation is autoregulated in response to its assembly state in tobacco chloroplasts. *Proc. Natl. Acad. Sci. U. S. A.* **104**, 6466–6471.

Wostrikoff, K., Choquet, Y., Wollman, F.A., and Girard-Bascou, J. (2001). TCA1, a single nuclear-encoded translational activator specific for *petA* mRNA in *Chlamydomonas reinhardtii* chloroplast. *Genetics* **159**, 119–132.

Wostrikoff, K., Girard-Bascou, J., Wollman, F.A., and Choquet, Y. (2004). Biogenesis of PSI involves a cascade of translational autoregulation in the chloroplast of *Chlamydomonas*. *EMBO J.* **23**, 2696–2705.

Wright, G.D. (2003). Mechanisms of resistance to antibiotics. *Curr. Opin. Chem. Biol.* **7**, 563–569.

Wu, H.Y. and Kuchka, M.R. (1995). A nuclear suppressor overcomes defects in the synthesis of the chloroplast *psbD* gene product caused by mutations in two distinct nuclear genes of *Chlamydomonas*. *Curr. Genet.* **27**, 263–269.

Wykoff, D.D., Davies, J.P., Melis, A., and Grossman, A.R. (1998). The regulation of photosynthetic electron transport during nutrient deprivation in *Chlamydomonas reinhardtii*. *Plant Physiol.* **117**, 129–139.

Yamaguchi, K. and Subramanian, A.R. (2000). The plastid ribosomal proteins. Identification of all the proteins in the 50 S subunit of an organelle ribosome (chloroplast). *J. Biol. Chem.* **275**, 28466–28482.

Yamaguchi, K. and Subramanian, A.R. (2003). Proteomic identification of all plastid-specific ribosomal proteins in higher plant chloroplast 30S ribosomal subunit. *Eur. J. Biochem.* **270**, 190–205.

Yamaguchi, K., von Knoblauch, K., and Subramanian, A.R. (2000). The plastid ribosomal proteins. Identification of all the proteins in the 30S subunit of an organelle ribosome (chloroplast). *J. Biol. Chem.* **275**, 28455–28465.

Yamaguchi, K., Prieto, S., Beligni, M.V., Haynes, P.A., McDonald, W.H., Yates, J.R., 3rd, and Mayfield, S.P. (2002). Proteomic characterization of the small subunit of *Chlamydomonas reinhardtii* chloroplast ribosome: Identification of a novel S1 domain-containing protein and unusually large orthologs of bacterial S2, S3, and S5. *Plant Cell* **14**, 2957–2974.

Yamaguchi, K., Beligni, M.V., Prieto, S., Haynes, P.A., McDonald, W.H., Yates, J.R., 3rd, and Mayfield, S.P. (2003). Proteomic characterization of the *Chlamydomonas reinhardtii* chloroplast ribosome. Identification of proteins unique to the 70S ribosome. *J. Biol. Chem.* **278**, 33774–33785.

Yamamoto, T., Burke, J., Autz, G., and Jagendorf, A.T. (1981). Bound ribosomes of pea chloroplast thylakoid membranes: Location and release *in vitro* by high salt, puromycin, and RNAse. *Plant Physiol.* **67**, 940–949.

Yohn, C.B., Cohen, A., Danon, A., and Mayfield, S.P. (1996). Altered mRNA binding activity and decreased translational initiation in a nuclear mutant lacking translation of the chloroplast *psbA* mRNA. *Mol. Cell Biol.* **16**, 3560–3566.

Yohn, C.B., Cohen, A., Danon, A., and Mayfield, S.P. (1998a). A poly(A) binding protein functions in the chloroplast as a message-specific translation factor. *Proc. Natl. Acad. Sci. U. S. A.* **95**, 2238–2243.

Yohn, C.B., Cohen, A., Rosch, C., Kuchka, M.R., and Mayfield, S.P. (1998b). Translation of the chloroplast *psbA* mRNA requires the nuclear-encoded poly(A)-binding protein, RB47. *J. Cell Biol.* **142**, 435–442.

Yonath, A. (2005). Antibiotics targeting ribosomes: Resistance, selectivity, synergism and cellular regulation. *Annu. Rev. Biochem.* **74**, 649–679.

Yosef, I., Irihimovitch, V., Knopf, J.A., Cohen, I., Dahan, I., Nahum, E., Keasar, C., and Shapira, M. (2003). RNA binding activity of ribulose-1,5-bisphosphate carboxylase/oxygenase large subunit from *Chlamydomonas reinhardtii*. *J. Biol. Chem.* **279**, 10148–10156.

Yu, N.-J. and Spremulli, L.L. (1998). Regulation of the activity of chloroplast translational initiation factor 3 by NH_2- and COOH-terminal extensions. *J. Biol. Chem.* **273**, 3871–3877.

Yukawa, M., Kuroda, H., and Sugiura, M. (2007). A new *in vitro* translation system for non-radioactive assay from tobacco chloroplasts: Effect of pre-mRNA processing on translation *in vitro*. *Plant J.* **49**, 367–376.

Zerges, W. (2000). Translation in chloroplasts. *Biochimie* **82**, 583–601.

Zerges, W. (2004). Regulation of translation in chloroplasts. In: *Molecular Biology and Biotechnology of Plant Organelles. Chloroplasts and Mitochondria* (H. Daniell and C. Chase, Eds.), pp. 347–384. Kluwer Academic Publishers, Dordrecht.

Zerges, W. and Rochaix, J.D. (1994). The 5′ leader of a chloroplast mRNA mediates the translational requirements for two nucleus-encoded functions in *Chlamydomonas reinhardtii*. *Mol. Cell Biol.* **14**, 5268–5277.

Zerges, W. and Rochaix, J.-D. (1998). Low density membranes are associated with RNA-binding proteins and thylakoids in the chloroplast of *Chlamydomonas reinhardtii*. *J. Cell Biol.* **140**, 101–110.

Zerges, W., Girard-Bascou, J., and Rochaix, J.D. (1997). Translation of the chloroplast *psbC* mRNA is controlled by interactions between its 5′ leader and the nuclear loci *TBC1* and *TBC3* in *Chlamydomonas reinhardtii*. *Mol. Cell Biol.* **17**, 3440–3448.

Zerges, W., Wang, S., and Rochaix, J.D. (2002). Light activates binding of membrane proteins to chloroplast RNAs in *Chlamydomonas reinhardtii*. *Plant Mol. Biol.* **50**, 573–585.

Zerges, W., Auchincloss, A.H., and Rochaix, J.-D. (2003). Multiple translational control sequences in the 5′ leader of the chloroplast *psbC* mRNA interact with nuclear gene products in *Chlamydomonas reinhardtii*. *Genetics* **163**, 895–904.

Zhang, L. and Aro, E.-M. (2002). Synthesis, membrane insertion and assembly of the chloroplast-encoded D1 protein into photosystem II. *FEBS Lett.* **512**, 13–18.

Zhang, L., Paakkarinen, V., van Wijk, K.J., and Aro, E.M. (1999). Co-translational assembly of the D1 protein into photosystem II. *J. Biol. Chem.* **274**, 16062–16067.

Zhou, D. and Mache, R. (1989). Presence in the stroma of chloroplasts of a large pool of a ribosomal protein not structurally related to any *Escherichia coli* ribosomal protein. *Mol. Gen. Genet.* **219**, 204–208.

Zicker, A., Kadakia, C., and Herrin, D. (2006). Distinct roles for the 5′ and 3′ untranslated regions in the degradation and accumulation of chloroplast tufA mRNA: Identification of an early intermediate in the *in vivo* degradation pathway. *Plant Mol. Biol.* **63**, 689–702.

CHAPTER 29

The CES Process

Yves Choquet and Francis-André Wollman
Université Pierre et Marie Curie, Centre National de la Recherche Scientifique
Institut de Biologie Physico-Chimique, Paris, France

CHAPTER CONTENTS

I. INTRODUCTION

Energy transduction in chloroplasts is performed by membrane-associated oligomeric protein complexes that comprise subunits of dual genetic origin, some of them being encoded by the organelle genome, whereas the others are nucleus-encoded. Expression of these two genetic compartments should therefore be coordinated to produce the various subunits of a given protein complex in the stoichiometry required for their functional assembly. The concerted accumulation of the constituent subunits – they are all pleiotropically lost when their assembly is impaired – results from two major regulatory processes. First, proteolysis rapidly removes some unassembled subunits. In this chapter we describe the second regulatory process, the CES process (for control by epistasy of synthesis), which plays a major role in protein assembly and consists in an assembly-dependent regulation of translation. In the absence of certain core subunits, the rate of synthesis of another chloroplast-encoded subunit from the same protein complex, which is then called a CES subunit, is drastically reduced.

We describe present knowledge of the molecular mechanisms responsible for the translational down-regulation of a CES subunit, when its assembly is compromised. CES is a versatile regulatory process, the mode of which may differ from one CES subunit to another, depending on the specific features of the biogenesis of the photosynthetic complex they belong to.

CES was first discovered in the chloroplast of *Chlamydomonas*, where it is well documented. It probably arose with endosymbiosis since cyanobacteria do not exhibit CES processes. We present evidence for its occurence in higher plant chloroplasts and in the mitochondria of the yeast *Saccharomyces cerevisiae*.

II. ASSEMBLY OF MULTISUBUNIT PHOTOSYNTHETIC PROTEIN COMPLEXES

A. Overview

Bioenergetic membranes are the site of ATP synthesis at the expense of a transmembrane proton-motive force that originates from serial electron transfer reactions, coupled to proton translocation across the membrane. Most of these processes develop within, as well as between, membrane-embedded hetero-oligomeric protein complexes. The biogenesis of these molecular machines is particularly complex as they most often comprise more than 10 distinct subunits and accommodate a variety of cofactors: pigments, hemes, iron-sulfur clusters and metal ions, bound either covalently or non-covalently to the protein moiety. For example, photosystem II (PS II), the protein complex that drives the light-induced oxidation of water and initiates photosynthetic electron flow, comprises close to 30 different subunits, most of which are integral membrane polypeptides, and binds about 40 chlorophylls, seven carotenoids, one pheophytin, two quinone molecules, an iron and several manganese atoms (Zouni et al., 2001; Ferreira et al., 2004, see also Chapter 16).

The assembly of the various constituents of such oligomeric complexes requires a temporal and spatial coordination within the cell to deliver the various components in a stoichiometry suitable for their functional assembly. Protein production should be under control to avoid wasteful accumulation of unassembled subunits that might disrupt the integrity of the membrane upon aggregation, or be a source of toxic radical species due to their redox cofactor or pigment complements.

In prokaryotes, the operon organization of many genes is well-suited for fine-tuning of the coupled transcription/translation rates of the various subunits comprising a given protein complex. In contrast, the biogenesis of oligomeric proteins for photosynthesis or respiration in organelles involves an additional level of complexity as their constituent subunits have a dual genetic origin. Subunits encoded by the organellar genome and translated on prokaryotic-like 70S ribosomes are associated in a defined stoichiometry with subunits encoded by nuclear genes that are translated on cytosolic 80S ribosomes and imported into the organelle. Thus, specific mechanisms should operate to allow the stoichiometric accumulation of subunits encoded in two genetic compartments through widely different biosynthetic routes. This issue is particularly critical given the imbalance in gene copy numbers of organellar- versus nucleus-encoded subunits, which can differ by as much as four orders of magnitude – two orders of magnitude in the specific case of *Chlamydomonas*. Furthermore, in chloroplasts, most polycistronic transcription units comprise genes encoding subunits belonging to different protein complexes (Choquet et al., 1988; Robertson et al., 1990; Turmel and Otis, 1994; Drapier et al., 1998) and polycistronic transcripts are often trimmed to monocistronic mRNAs (Sturm et al., 1994; Rymarquis et al., 2006). Consequently, the level of expression of each organelle-encoded subunit is under some specific control. Thus, organelle-based regulatory processes as well as crosstalk between the nucleo-cytosolic and organelle compartments are required to regulate the expression levels of the various subunits of oligomeric proteins.

B. Concerted accumulation of subunits of photosynthetic complexes

Because of its unique potential for genetic studies, *Chlamydomonas* has long been a model for studying the biogenesis of photosynthetic proteins within the chloroplast. Photosynthesis mutants altered in any of the major complexes – PS I and PS II, the cytochrome b_6f complex, the ATP synthase or ribulose bisphosphate carboxylase-oxygenase (Rubisco) – are viable when grown in the presence of acetate as a carbon source. The use of such mutants in studies during the decade beginning in the mid-1970s led to the discovery of a most striking characteristic of the assembly of photosynthesis proteins: most of their subunits displayed a concerted accumulation. This conclusion arose from the exquisite complementarities between the polypeptide profiles of purified photosynthetic protein complexes and the patterns of polypeptide deficiencies observed in various independent mutant strains lacking activity for PS II (Diner and Wollman, 1980; Wollman et al., 1980), PS I (Bennoun et al., 1977; Girard et al., 1980), cytochrome b_6f (Lemaire et al., 1986; Kuras and Wollman, 1994) or ATP synthase (Lemaire and Wollman, 1989a,b).

When these mutants were subsequently characterized at the molecular level, they turned out in most cases to be defective for the expression of an integral core subunit of the photosynthetic complex under study. Further studies with mutant strains defective in small subunits – most of which span the membrane only once and are peripherally located within the protein complex – provided a contrasting picture: some small subunits were strictly required for the concerted accumulation of all subunits from the complex, whereas the absence of others still allowed partial accumulation of a subcomplex. For instance, deletion of genes encoding any one of the three chloroplast-encoded core subunits of the cytochrome b_6f complex, *petA*, *petB*, *petD*, led to the loss of the whole complex (Kuras and Wollman, 1994) as does deletion of the *petG* gene, which encodes a small subunit (Berthold et al., 1995). In contrast, deletion of the *petL* gene still allowed partial accumulation of a functional protein complex (Takahashi et al., 1996). Tobacco or *Arabidopsis* mutants behave as their *Chlamydomonas* equivalents: mutant lines lacking expression of any of the major cytochrome b_6f subunits, but also of PetG and PetN, show impaired accumulation of the whole complex (Hager et al., 1999; Monde et al., 2000; Maiwald et al., 2003; Schwenkert et al., 2007). By contrast, a PetL-less cytochrome

b_6f complex accumulates to almost wild-type level in young leaves from tobacco Δ*petL* lines but this partial complex, like its *Chlamydomonas* counterpart, is less stable than the wild-type one and disappears upon aging (Takahashi et al., 1996; Schottler et al., 2007; Schwenkert et al., 2007).

As a rule, studies of higher plant mutants led to conclusions very similar to those drawn from the study of *Chlamydomonas* mutants. For instance, PS I mutants from *Arabidopsis* lose all PS I subunits in the absence of the reaction center subunits PsaA and PsaB (Lezhneva and Meurer, 2004), as had been shown earlier with similar mutants from *Chlamydomonas* (Girard et al., 1980; Girard-Bascou et al., 1987; Goldschmidt-Clermont et al., 1990). Absence of the extrinsic subunit PSAD also prevents accumulation of PS I subunits in *Arabidopsis* (Ihnatowicz et al., 2004), while partial assembly of functional PS I is retained in higher plant mutants lacking the small subunits PsaE, PsaG, PsaH, PsaJ, and PsaK (Varotto et al., 2002; Hansson et al., 2007; Ihnatowicz et al., 2007).

C. Concerted accumulation of subunits arose with chloroplast compartmentalization

In cyanobacteria, in most cases where gene inactivation targeted a major subunit of a photosynthetic protein there was still a significant – albeit reduced – accumulation of other subunits of the same protein complex. Although cyanobacterial mutants deficient for PsaA or PsaB lack the whole set of PS I subunits as do their chloroplast counterparts (Smart et al., 1991; Nyhus et al., 1993; Smart and McIntosh, 1993), *psaC* deletion strains from *Anabena* or *Synechocystis* still accumulate PsaA and PsaB assembled into chlorophyll–protein complex I (CPI) able to drive P700 light-induced charge separation (Mannan et al., 1991; Yu et al., 1995), in contrast to the *psaC* deletion strain from *Chlamydomonas* which lacks the whole PS I complex because PsaA and PsaB are rapidly degraded (Takahashi et al., 1996). Also *psaD* deletion strains from cyanobacteria can still assemble PS I since they grow photoautotrophically (Chitnis et al., 1989; Xu et al., 1994), at variance with the corresponding mutants from higher plants (Ihnatowicz et al., 2004).

In the case of PS II from *Chlamydomonas*, the absence of subunits D2 or D1 from the reaction center (RC) of PS II or of the core subunit CP47 leads to the absence of all PS II subunits from the thylakoid membrane (de Vitry et al., 1989). By contrast, D2 shows significant accumulation in the absence of D1 in *Synechocystis*, while RC II can be detected in strains lacking CP47 (Komenda et al., 2004). Furthermore, in cases where minor subunits were not expressed, the bulk of the protein complex still accumulated close to wild-type levels, in striking contrast to what had been reported for *Chlamydomonas*. A mutant of *Synechocystis* 6803 lacking the *psbI* gene retained 70–75% of wild-type levels of oxygen evolution capability and PS II accumulation (Ikeuchi et al., 1995), while in *Chlamydomonas* inactivation of the *psbI* gene decreased oxygen evolution activity and D1 content almost 10-fold (Kunstner et al., 1995). *psbH* and *psbK* deletion strains of *Synechocystis* were still capable of photosynthetic growth even if they showed light sensitivity (Ikeuchi et al., 1991; Mayers et al., 1993; Komenda et al., 2002), while the corresponding mutant strains of *Chlamydomonas* failed to assemble PS II and thus were not photosynthetic (Takahashi et al., 1994; Summer et al., 1997).

These distinct features of protein accumulation in chloroplasts and cyanobacteria point to an increased quality control of protein assembly in a eukaryotic context. In the latter, functional assembly of photosynthesis proteins in the chloroplast results from intricate biogenesis processes that can be temporarily desynchronized because they are spatially separated between the organelle and the nucleus-cytosol.

D. Proteolytic degradation of unassembled subunits in the chloroplast

Two major regulatory processes which together achieve the concerted accumulation of protein subunits have been identified in organelles. First, the chloroplast is the site of a proteolytic disposal of unassembled, partially assembled or misassembled subunits (see Chapter 19). Pulse-chase or immunochase experiments performed on *petA*, *petB*, or *petD* deletion strains have shown that the half-lives of subunit IV and cytochrome b_6, but not their initial synthesis rates, were strongly decreased when another major subunit from the complex was missing. For example, the half-life of subunit IV drops from almost 2 hours in the wild type to 45 minutes and 15 minutes in strains lacking cytochrome *f* or cytochrome b_6, respectively (Kuras and Wollman, 1994). Unassembled subunit IV is thus rapidly degraded, whereas it is converted upon assembly to a more stable form, probably because the motifs recognized by proteases are shielded by the other subunits. The protease responsible for the degradation of unassembled subunit IV has not been identified yet. The ClpP protease does not seem to be involved as translational attenuation of the *clpP* gene did not enhance the accumulation of unassembled subunits in deletion strains (Majeran et al., 2000). Similarly, the turnover of the D2 polypeptide from the PS II RC is strongly increased in mutant strains defective for the expression of D1 (de Vitry et al., 1989; Preiss et al., 2001; Minai et al., 2006) or CP43 (de Vitry et al., 1989). The latter situation illustrates the more active degradation of unassembled subunits in the chloroplast than in cyanobacteria, which do accumulate D2 in the absence of D1 (Komenda et al., 2004). Indeed, the chloroplast houses a particularly sophisticated equipment of proteolytic enzymes with some protein families having more members than in the cyanobacteria, like the Deg (6 vs. 3) or FtsH (12 vs. 4) families (Sokolenko et al., 2002).

III. CES, A MAJOR CONTROL STEP IN THE BIOGENESIS OF PHOTOSYNTHETIC PROTEINS

A. Overview

The chloroplast also displays a unique regulatory process for protein assembly, the CES process (for control by epistasy of synthesis) that consists in an assembly dependent regulation of translation. The CES concept is derived from studies that consistently reported that the absence of some core subunits in certain photosynthesis mutants of *Chlamydomonas* caused a decrease in the rate of synthesis of another chloroplast-encoded subunit from the same protein complex, which was then designated as a CES subunit (Wollman et al., 1999; Choquet and Vallon, 2000; Choquet and Wollman, 2002). This regulatory process probably arose with endosymbiosis since it had not been observed in cyanobacteria for bioenergetic complexes. Its contribution to the stoichiometric assembly of the subunits from photosynthetic proteins is well documented in *Chlamydomonas*, as will be described below. Several lines of evidence indicate that it also occurs in higher plant chloroplasts and in the mitochondria of the yeast *Saccharomyces cerevisiae*.

B. Generality of CES in the chloroplast of *Chlamydomonas*

Cytochrome *f* from the cytochrome b_6f complex is a typical CES protein whose rate of synthesis decreases in the absence of cytochrome b_6 or subunit IV, to about 10% of that observed in the wild type (Lemaire et al., 1986; Kuras and Wollman, 1994). Since the abundance of the *petA* mRNA remains unchanged upon defective assembly, this regulation likely occurs at the translational level. Thus, the expression of cytochrome b_6f subunits is seemingly hierarchical with the presence of cytochrome b_6 and subunit IV being required for sustained expression of the CES subunit cytochrome *f*.

Studies of *Chlamydomonas* mutants have uncovered similar hierarchies in the synthesis of chloroplast-encoded subunits from all major photosynthetic proteins, as illustrated in Figure 29.1. PS I mutants defective for PsaB expression show no detectable synthesis of the CES subunit PsaA, whether failure to synthesize PsaB results from mutations in the chloroplast *psaB* gene itself (Girard-Bascou et al., 1987) or in the *TAB1* or *TAB2* nuclear genes, whose products are specifically required for the translation of *psaB* mRNA (Stampacchia et al., 1997; Dauvillée et al., 2003). In contrast, mutants affected in *psaA* gene expression show wild-type levels of *psaB* translation (Girard-Bascou, 1987; Goldschmidt-Clermont et al., 1990). The requirement for the TAB1 factor to translate *psaB* mRNA can be bypassed in a suppressor strain that carries a chloroplast mutation in the *psaB* 5′ untranslated region (UTR). Restored PsaB synthesis in this suppressor strain is correlated with that of PsaA, suggesting that the presence

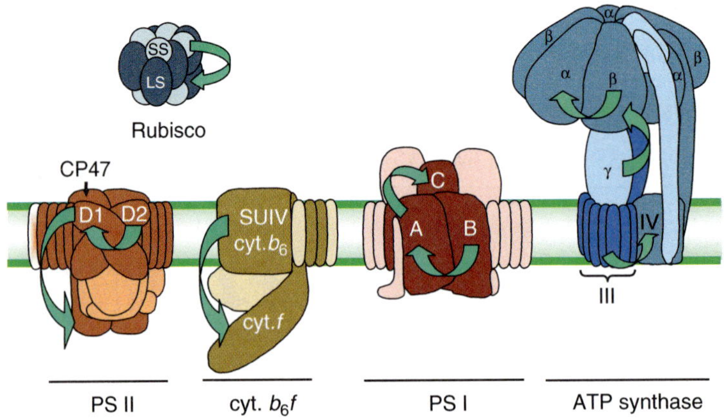

FIGURE 29.1 *CES contributes to the biogenesis of all major photosynthetic protein complexes in Chlamydomonas chloroplasts. Schematic representation of the major photosynthetic proteins. Subunits encoded by the chloroplast genome are shown in dark colors, while those encoded in the nucleus are pale colors. Green arrows indicate epistatic relationships between the subunits of a photosynthetic protein complex and connect an assembly partner whose presence is required for efficient synthesis of a CES subunit, to this CES subunit, emphasized by the arrowhead.*

of PsaB is required for efficient translation of the *psaA* mRNA (Stampacchia et al., 1997). In a similar way, *psaA* mutants show reduced synthesis of the CES subunit PsaC (Takahashi et al., 1991), while PsaB and PsaA are synthesized at wild-type rates – but rapidly degraded – in *psaC* deletion strains (Takahashi et al., 1991). These observations define a "CES cascade," where PsaB is required for sustained expression of PsaA, which, in turn, is required for significant production of PSAC (Wostrikoff et al., 2004).

An analogous CES cascade participates in *Chlamydomonas* PS II biogenesis. While mutants lacking D1 show a greatly reduced level of CP47 synthesis but maintain synthesis of D2 (Bennoun et al., 1986; Jensen et al., 1986; de Vitry et al., 1989), the rate of both D1 and CP47 synthesis is strongly decreased in mutants lacking expression of D2 (Erickson et al., 1986; de Vitry et al., 1989). Thus, the presence of the D2 subunit is required for efficient synthesis of the D1 subunit, while the latter is required for sustained translation of the *psbB* mRNA encoding the CP47 core antenna. In this view, both CP47 and D1 are CES subunits in PS II biogenesis.

CES also contributes to the biogenesis of both CF_0 and CF_1 moieties of the ATP synthase complex. CF_1, the soluble subcomplex which catalyzes ATP synthesis, is composed of five subunits α, β, γ, δ, and ε, assembled in the stoichiometry $\alpha_3\beta_3\gamma_1\delta_1\varepsilon_1$ (see Chapter 18). Subunits β and α can both be regarded as CES subunits for CF_1 biogenesis: mutant strains defective for the expression of the nucleus-encoded subunit γ show reduced synthesis of subunit β (Drapier et al., 2007), the product of the chloroplast *atpB* gene, while the rate of translation of *atpA* mRNA, encoding subunit α, is strongly reduced in mutant strains defective for subunit β expression (Lemaire and Wollman, 1989b; Drapier et al., 1992). The membrane-embedded CF_0 portion of the ATP synthase is a selective proton channel that comprises subunits I, II, III, and IV, assembled in a 1:1:(13–14):1 stoichiometry (Seelert et al., 2003; Meyer zu Tittingdorf et al., 2004). A mutant strain lacking synthesis of CF_0 subunit III because it lacks stable accumulation of the *atpH* mRNA (Majeran et al., 2001) shows a pleiotropic loss of all ATP synthase subunits, but a specific reduction in the rate of synthesis of subunit IV (Lemaire and Wollman, 1989b).

CES is not restricted to the integral thylakoid membrane proteins: the large subunit of Rubisco is also a CES protein, whose rate of synthesis is highly reduced in deletion strains lacking expression of the small subunit (Khrebtukova and Spreitzer, 1996).

CES thus appears to be a major control step during the biogenesis of the photosynthetic apparatus in *Chlamydomonas*, as almost all major chloroplast-encoded polypeptides – those polypeptides whose rate of synthesis can be accurately measured by pulse-labeling experiments – participate in CES, being either CES subunits or assembly partners required for the expression of CES subunits from the same protein complex, as illustrated in Figure 29.1. One exception is the CP43 core antenna of PS II: its rate of synthesis does not depend on the presence of other PS II subunits, even if it is less stable in strains showing compromised PS II assembly. Conversely, its absence does not affect the rate of synthesis of other PS II subunits, even though they are more rapidly degraded in the absence of CP43 (de Vitry et al., 1989). In contrast, it remains to be determined whether small chloroplast-encoded subunits participate in CES cascades because being poorly resolved by SDS-PAGE, their actual rates of translation are difficult to assess by pulse-labeling experiments with assembly-defective strains.

C. CES as a regulation of translation initiation

The reduced synthesis of CES proteins, when their assembly is compromised, might rely on a variety of mechanisms that could differ from one CES subunit to another. In all instances reported so far, changes in protein synthesis were not correlated with changes in mRNA levels (Drapier et al., 1992, 2007; Khrebtukova and Spreitzer, 1996; Choquet et al., 1998; Wostrikoff et al., 2004; Minai et al., 2006), suggesting that the regulation operates at the (post)translational level. These observations, based on short pulse-labeling experiments, were equally consistent with an actual regulation of translation or an early degradation process – possibly co-translational – that would be responsible for the diminished labeling of the CES subunits when their assembly was prevented.

The actual demonstration that the CES process corresponded to an *assembly dependent regulation of translation* came with chloroplast transformation experiments based on the use of chimeric reporter genes expressed under the control of the 5′ UTR from the mRNA encoding the CES protein under study. For the *petA* gene product, cytochrome *f*, Choquet et al. (1998, 2003) demonstrated that the *petA* 5′ UTR was sufficient to confer decreased synthesis to reporter proteins when they were expressed in the absence of cytochrome *f* assembly partners. Thus, protein reporters mimicked the behavior of the CES protein cytochrome *f*, provided they were expressed under the same *cis*-acting signal for translation: the CES process operated at an early control step of chloroplast mRNA translation. Furthermore, cytochrome *f* no longer behaved as a CES subunit when its coding sequence was translated under the control of an unrelated 5′ UTR: its rate of synthesis and its accumulation remained constant, irrespective of the presence of its assembly partners (Choquet et al., 1998). These experiments excluded any cotranslational degradation of unassembled cytochrome *f*.

Similar experiments performed with the other CES genes consistently led to the same observations. Reporter genes translated under the control of the 5′ UTR of a CES protein were consistently less expressed upon compromised assembly, despite the fact that the reporter gene product was unrelated to the protein complex whose assembly was investigated. The reduced synthesis of the CES subunits from PS I (PsaA and PsaC: Wostrikoff et al., 2004), PS II (D1 and CP47; Minai et al., 2006) or from the ATP synthase CF_1 (subunits α and β; Drapier et al., 2007) resulted from regulation at some early step in the translation initiation process, either an activation step prior to initiation, translation initiation itself, or a transitional step between initiation and elongation of translation.

D. Unassembled cytochrome *f* negatively regulates its own synthesis

Further clues to the assembly-dependent regulation of cytochrome *f* synthesis came from the study of cytochrome *f* variants generated by site-directed mutagenesis. Cytochrome *f* exposes a large N-terminal domain in the lumen of the thylakoid, to which the *c*-type heme is covalently attached by thioester bonds. This globular domain is tethered to the membrane by a C-terminal domain comprising a 20 residue-long transmembrane helix that terminates in the stroma with a stretch of 15 residues (see Chapter 17). In one mutant strain, the cysteinyl residues liganding the heme were substituted, leading to the synthesis of a highly unstable apocytochrome *f*, barely detectable in the thylakoid membrane (Kuras et al., 1995a). Another mutant strain expressed the N-terminal domain only, that is a soluble form of the protein, which shows stable accumulation in the thylakoid lumen (Kuras et al., 1995b). Despite large differences in their stabilities, both mutated cytochrome *f* forms presented the same behavior: a threefold stimulated synthesis when compared to the wild-type protein, independent of the presence or absence of subunit IV, an assembly partner of cytochrome *f* (Kuras et al., 1995a; Choquet et al., 1998).

The common feature of the two strains was the absence of the C-terminal domain, either because it was not synthesized in the strain expressing soluble cytochrome *f*, or because it failed to accumulate to any significant level in the strain expressing unstable apocytochrome *f*. These studies strongly suggested that accumulation of the C-terminal domain of cytochrome *f* at the membrane surface facing the stroma was required to down-regulate translation of the *petA* mRNA. In this view, the C-terminal repressor domain would be shielded upon assembly with other cytochrome b_6f subunits, allowing cytochrome *f* synthesis to resume.

This hypothesis was confirmed upon further characterization of the putative translational repressor domain by site-directed mutagenesis. A critical tetrapeptide cluster $K_{305}QFE_{308}$ was identified in the stromal extension of cytochrome *f*, together with two upstream residues, K_{302}, which lies at the interface between the membrane and the stroma and Q_{297}, buried in the thylakoid membrane (Figure 29.2). Substitutions of these critical residues led to enhanced translation of cytochrome *f*, which became independent of, or poorly dependent on, the presence or absence of its assembly partners. In contrast, substitutions of other residues within the C-terminal domain had no impact on the regulation of cytochrome *f* translation

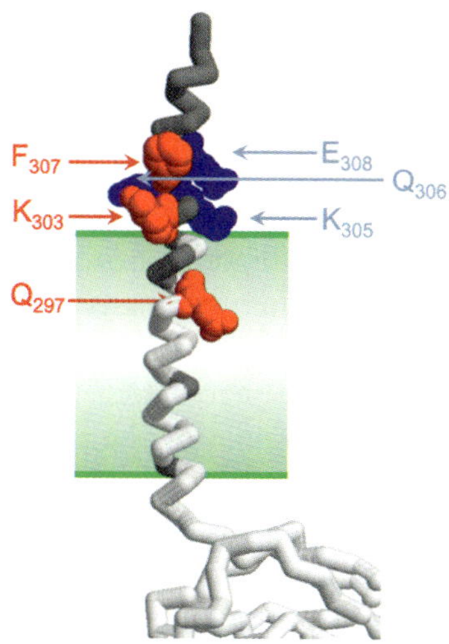

FIGURE 29.2 *The translation repressor domain from cytochrome f. Schematic representation of the cytochrome f C-terminal domain (drawn from the structure of Stroebel et al., 2003). Residues shown in spacefill contribute to the repressor domain of cytochrome f. They are drawn in red when substitutions completely abolish the CES process, or in blue if substitutions result in significant, but not total, decrease in the efficiency of the CES process. Substitution of the residues represented as backbone have no effect on the CES control of cytochrome f expression (shown in dark grey) or have not been tested (shown in pale grey).*

(Choquet et al., 2003). Together, these experiments demonstrated that unassembled cytochrome *f* exerts a negative feedback on its own synthesis.

E. Synthesis of most CES proteins is autoregulated

Because the other CES subunits of *Chlamydomonas* chloroplasts are polytopic membrane proteins, the identification of putative repressor motifs in their primary sequences was not straighforward. Nevertheless, the CES mechanisms could still be investigated. Basically, two different hypotheses may account for the reduced synthesis of a CES protein in the absence of its assembly partner (Figure 29.3, upper panel). As depicted above for cytochrome *f*, the unassembled CES subunit could exert a negative feedback on its own translation. Alternatively, the assembly partner could be required – directly or indirectly – to stimulate synthesis of the CES subunit, in a transactivation mechanism. As reporter genes driven by the 5′ UTR of the CES subunit mRNA recapitulate the behavior of the CES protein itself, that is a reduced expression in the absence of assembly partners, it became possible to discriminate between the two hypotheses by assessing the expression of the reporter genes in the combined absence of the CES subunit and its assembly partner (Figure 29.3, lower panel). In the case of a transactivation mechanism, translation of the chimeric gene should remain low in the absence of the assembly partner, irrespective of the presence of the CES protein. The opposite behavior should be observed in case of a negative feedback mechanism, with the reporter gene being highly expressed when the CES protein carrying out the autoregulation is missing, whether the assembly partner is present or not.

A pitfall in these experiments lies in the possible titration of limiting amounts of translation factors by the concentration of their 5′ UTR targets, in which case expression of the reporter gene together with the original CES gene may be weaker than when expressed alone. Such a competition has indeed been observed for the accumulation of 5′*atpA*-driven endogenous and chimeric transcripts (Drapier et al., 2007) (Figure 29.4). To avoid this flaw, it is important, when comparing the expression of a reporter gene in different strains, to keep a constant number of CES 5′ UTRs. Thus, truncated versions of several CES proteins were used, which had lost their ability to assemble in a protein complex and failed to accumulate to substantial amount in the thylakoid membrane because they were rapidly degraded *in vivo* (Choquet et al., 1998; Wostrikoff et al., 2004; Minai et al., 2006; Drapier et al., 2007). In strains expressing these truncated versions, the CES proteins did not accumulate, but their mRNAs still accumulated and were translated, as in strains lacking expression of the sole assembly partner. With these tools in hand, one could safely compare the expression of a reporter construct in single mutants that allowed or prevented expression of either the assembly partner or the CES protein – because of its

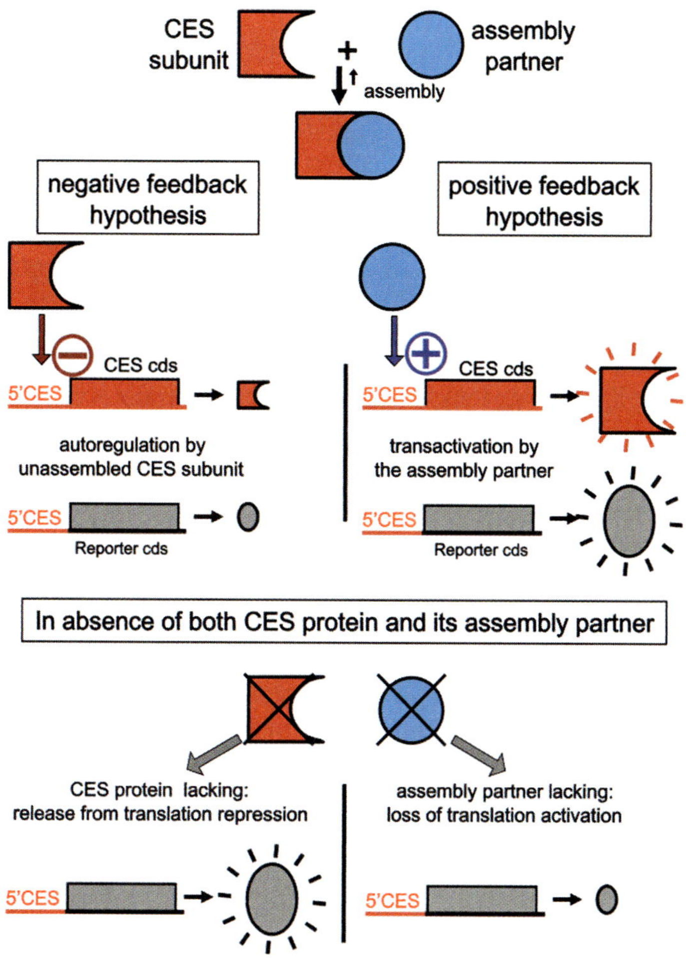

FIGURE 29.3 *The two models accounting for the repression of CES mRNA translation in the absence of its assembly partner. Top panel: the autoregulation hypothesis is detailed in the left part of the figure while trans-activation is shown at the right. Expression of 5′-CES-UTR-driven coding sequences, symbolized by filled boxes, either that of the endogenous CES protein or that of a reporter gene (shown in grey because its gene product depicted as a grey ellipse, is unrelated to the protein complex whose biogenesis is under study), are shown for each hypothesis. Lower panel: one can discriminate between these two models since they lead to opposite predictions concerning the level of expression of a 5′ CES UTR-driven reporter gene in the combined absence of the CES subunit and of its assembly partner.*

truncation – or in double mutants lacking both subunits. When reporter genes translated under the control of the 5′ UTRs for CES subunits were expressed together with truncated versions of CES subunits for PS II (D1 or CP47) or PS I (PsaA and PsaC), they were consistently highly expressed, whether the assembly partners (D2, D1, PsaB, and PsaA, respectively) were present or not. Thus, as was the case for cytochrome *f*, synthesis of CES subunits from PS I or PS II is not transactivated by their assembly partners,

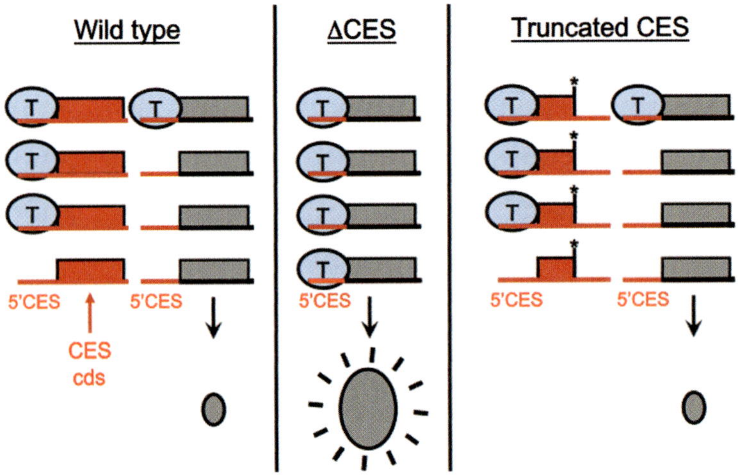

FIGURE 29.4 *Possible artifacts when assessing the expression of a 5′ CES UTR-driven reporter gene. 5′ CES UTRs of endogenous and chimeric reporter genes may be in competition for interaction with nucleus-encoded expression factors in the wild type (left column, where for sake of clarity only a translation factor is represented as a pale blue ellipse). In a strain lacking expression of the CES protein because of a deletion (middle column), all nucleus-encoded factors are free to interact with the reporter transcript, whose expression may therefore be increased for reasons unrelated to the CES process under study. By contrast, in strains synthesizing a truncated version of the CES subunit (right column), nucleus-encoded factors should be distributed as in the wild type, and changes in the expression of the reporter will only originate from the CES regulation.*

but rather autoregulated when CES subunits accumulate in an unassembled state.

IV. CES CASCADES AND ASSEMBLY OF MULTIMERIC PROTEIN COMPLEXES

A. CES plays a major role in the sequential assembly of oligomeric complexes

Translational autoregulation thus appears to be a common control mechanism in *Chlamydomonas* chloroplasts, for CES subunits from cytochrome b_6f (cytochrome *f*), PS I (PsaA and PsaC), and PS II (D1 and CP47). These oligomeric complexes have a core comprised of integral membrane subunits assembled in a 1:1 stoichiometry. Comparisons between strains defective in the production of specific core subunits have shown that assembly is a stepwise process, with a progressive increase in the stability of assembly intermediates as more subunits are added (reviewed in Choquet and Vallon, 2000). For PS I biogenesis (Figure 29.5A, free PsaA negatively regulates *psaA* mRNA translation, unless it assembles with PsaB to form the CPI core complex. Once formed, CPI can recruit PsaC in a subsequent assembly step, releasing the negative feedback exerted by unassembled PsaC on

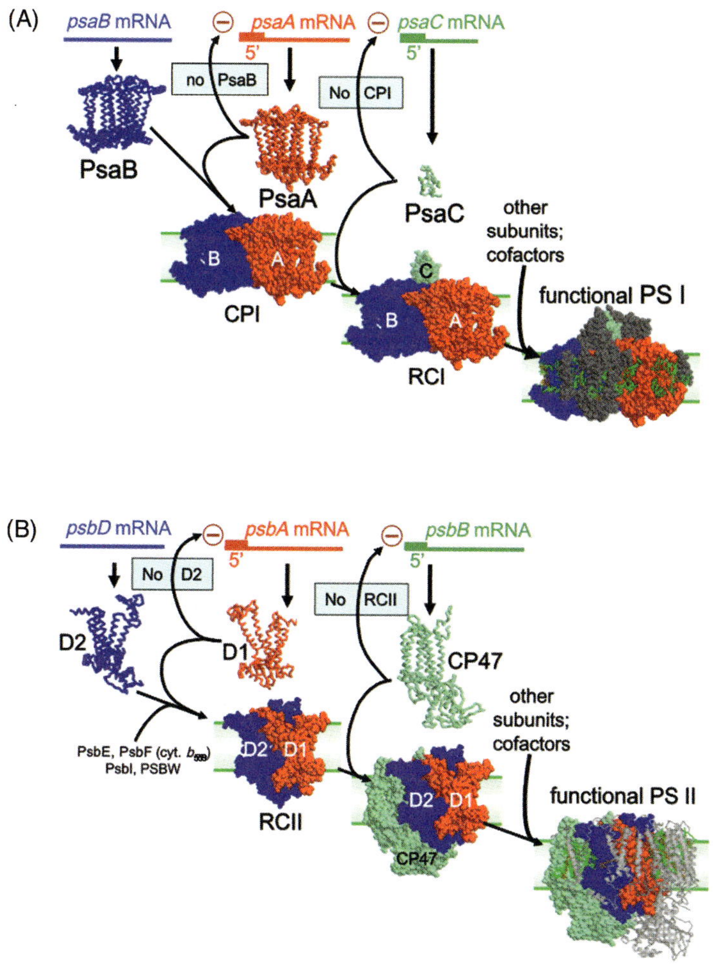

FIGURE 29.5 *Epistatic cascades contributing to the biogenesis of PS I (A) and PS II (B). Assembly pathways of PS I and PS II are schematically depicted. Structures of PS I and PS II are drawn from Jordan et al. (2001) and Loll et al. (2005), respectively. Most upstream assembly partners are drawn in blue, while CES subunits are shown in red and green. When newly-synthesized CES polypeptides, represented as helical backbones cannot assemble, they repress the translation initiation of their encoding mRNAs, as indicated by the Θ symbol in red.*

its own synthesis. Most similarly, in the course of PS II core assembly, unassembled D1 subunit represses its own synthesis unless D2 production sequesters it, whereas unassembled CP47 prevents further translation of its mRNA as long as it cannot assemble with the D1/D2 RC (Figure 29.5B). Thus, a cascade of autoregulation by unassembled subunits couples assembly of these complexes to the expression of their individual subunits.

Feedback loops working in series prevent excessive production of a CES subunit until the substrate for its assembly is available, when all upstream steps in the sequential assembly pathway have been completed.

Consequently, a high expression of CES subunits requires the presence of all subunits that participate in upstream steps in the assembly pathway, but not that of subunits that assemble in downstream steps. In other words, the CES hierarchy reflects necessarily the sequence of the successive steps in protein complex assembly.

Efficient translational autoregulation requires that unassembled CES proteins accumulate at least to some extent in the absence of their assembly partners. While we lack experimental data on the behavior of unassembled PsaA and PaC subunits in strains devoid of PsaB and CPI, respectively, the other CES proteins proved quite long-lived in their fully unassembled forms. D1 and CP47, although synthesized in markedly reduced amounts, resisted proteolytic degradation in absence of D2 and D1, respectively (de Vitry et al., 1989), as did unassembled cytochrome *f* in the absence of cytochrome b_6 or subunit IV (Kuras and Wollman, 1994; Choquet et al., 2003). Conversely, one could argue that the stability of unassembled CES subunits *in vivo* requires a tight control of their rate of synthesis, to prevent their overaccumulation. It is of note that when assembled or partially assembled in protein complexes, CES subunits may become more susceptible to degradation than when fully unassembled. The CES subunits D1 and CP47 are prone to proteolytic degradation in mutant strains deficient for the synthesis of CP43, where they accumulate in a D2/D1/CP47 subcomplex. Similarly, once assembled within cytochrome b_6f complexes cytochrome *f* becomes a target for proteases, either upon nitrogen starvation (Bulte and Wollman, 1992) or in strains expressing a mutated form of the Rieske protein (Majeran et al., 2000).

It should be stressed that the CES process, although readily evident in assembly-defective mutants, also operates in a wild-type context, as CES protein synthesis is fine-tuned to the availability of their assembly partners. This can be seen by the threefold overexpression of cytochrome *f* variants lacking the C-terminal repressor domain of the protein, when compared with the level of cytochrome *f* synthesis in the wild type (Kuras et al., 1995b; Choquet et al., 1998, 2003), or from the higher expression of 5′-CES-UTR-driven reporter genes in the absence of the CES subunits regardless of the presence of their assembly partners (Wostrikoff et al., 2004; Minai et al., 2006; Drapier et al., 2007). In the latter case, the strains lack any accumulation of unassembled CES subunits to repress expression of the reporter gene. As an exception, the expression of 5′-*psaA*-UTR-reporter genes was similar in strains expressing or lacking PsaA. Most probably the concentration of unassembled PaA, which can bind up to 40 chlorophyll molecules, has to be kept very low in the wild type to avoid the production of damaging radicals.

In contrast to CES subunits, the most upstream assembly partners of CES cascades are quite unstable when unassembled, as discussed in sections II.B and II.C. Assembly thus must necessarily be more efficient than the competing degradation process. The CES process would optimize the rate of assembly of these partner subunits, which would encounter upon thylakoid insertion low, but stable, levels of unassembled CES subunits, available for a rapid interaction that would protect them from degradation. This hypothesized protective effect of CES subunits should be very efficient provided the biogenesis of photosynthetic proteins is not delocalized throughout the thylakoid membrane, but instead restricted and thus concentrated in specialized membrane compartments, as has been suggested (Zerges and Rochaix, 1998).

Thus, the particular properties of CES subunits offer unique means to optimize a sequential and multistep process for protein assembly. Strikingly, CES is apparently restricted to the *de novo* biogenesis of photosynthetic proteins, since CES plays no role in the regulation of D1 synthesis during the D1 repair cycle. Under bright illumination, PS II charge separation damages the D1 subunit, which should then be degraded and replaced by a newly synthesized D1 polypeptide (reviewed in Ohad et al., 1994). D1 synthesis is strongly increased (Kettunen et al., 1997) in response to D1 degradation and to the formation of D1-less PS II complexes (Adir et al., 1990; Zer et al., 1994; Bailey et al., 2002). This could be consistent with an autoregulated synthesis of the CES protein D1, but the expression of a 5′-*psbA*-UTR-driven reporter gene, which shows CES behavior when assembly of PS II cores is prevented, is not increased upon activation of the D1 repair cycle. Thus, *psbA* gene expression is controlled by different mechanisms during PSII biogenesis, which involves CES regulation of translation initiation, and repair, where elongation is the key control step (Kettunen et al., 1997).

B. Atypical features of CES in the assembly of ATP synthase

The assembly of the ATP synthase complex requires achieving uneven stoichiometries since the five CF_1 subunits, α, β, γ, δ, and ε, are present in a 3:3:1:1:1 stoichiometry. Among these, subunits α and β are both CES proteins whose rates of translation respectively depend on the presence of subunit β and subunit γ. As observed for the other CES subunits, down-regulation of their synthesis in the absence of assembly relies on control of translation initiation mediated by the 5′ UTRs of *atpA* and *atpB* mRNAs, respectively (Drapier et al., 2007).

In contrast to all other CES subunits studied so far, α subunit synthesis is not autoregulated, but *trans*-activated by its assembly partner, subunit β. Indeed, a reporter gene translated under the control of the *atpA* 5′ UTR remains poorly expressed in the combined absence of both α and β subunits. This unique regulatory mechanism probably originates in the propensity of unassembled α subunit to self-aggregate, becoming thereby unavailable for autoregulation (Ketchner et al., 1995; Du and Gromet-Elhanan, 1999; Drapier et al., 2007).

As another unusual characteristic, translation of *atpB* mRNA is not merely autoregulated. Fully unassembled β subunit, in the absence of both subunits α and γ, is unable to down-regulate its own synthesis: despite a significant accumulation of unassembled β subunit, translation of the *atpB* mRNA (or of a reporter gene under control of the *atpB* 5′ UTR) remains high in double mutant strains lacking expression of both subunits α and γ (Drapier et al., 2007). A negative feedback on the synthesis of subunit β is observed only when a CF_1 assembly intermediate comprising both α and β subunits (either the catalytic hexamer $\alpha_3\beta_3$ or α/β dimers in equilibrium with the hexamer) accumulates. This regulatory mechanism sets the accumulation of subunits α and β to a level three times higher than that of their assembly partners in CF_1, since the negative feedback exerted by α/β assembly intermediates on the translation of subunit β is only released upon assembly of subunit γ with α_3 β_3 hexamers.

These intertwined translational feedback loops (see Figure 18.2) are well suited to control the uneven accumulation of CF_1 subunits. They ensure a tight control of the nucleo-cytoplasmic compartment on the biogenesis of CF_1 solely through the regulation of the nucleus-encoded γ subunit. They also allow a tight control of the relative stoichiometry between subunits α and β as an excess of subunit β will *trans*-stimulate translation of subunit α until stoichiometry is reached again, while reduced accumulation of subunit β will prevent new synthesis of subunit α and deplete the pool of α/β assembly intermediates, increasing the rate of synthesis of subunit β.

CES is thus a versatile regulatory process, the precise molecular mechanism of which can be adapted to the specific features of the biogenesis of a given photosynthetic oligomeric protein. It seems evident that characterization of the CES regulatory loop that sets the synthesis of subunit IV to the presence of subunit III (Lemaire and Wollman, 1989b) will shed light on the mechanisms responsible for the specific 14:1 stoichiometry between subunit III and the other CF_0 subunits.

V. IS CES CENTRAL TO ORGANELLE PROTEIN EXPRESSION?

A. *Chlamydomonas* and vascular plant chloroplasts

Data presented above establish CES as a major regulatory process for photosynthetic protein assembly in *Chlamydomonas*. By contrast, the lack of plant mutant lines specifically defective for the expression of a single chloroplast-encoded subunit, together with technical limitations in the determination of the rate of synthesis of the remaining chloroplast-encoded polypeptides in higher plants, have long hampered a precise assessment of the prevalence of CES in the biogenesis of chloroplast proteins. Only in 2007 did an elegant study show that the Rubisco large subunit (LS) is a typical CES protein in higher plant chloroplasts (Wostrikoff and Stern, 2007).

LS, the product of the chloroplast gene *rbcL*, associates with the Rubisco small subunit (SS), encoded by a small family of nuclear genes, to form the hexadecameric functional enzyme L_8S_8 (see also Chapter 9). LS shows reduced synthesis in RNAi or antisense tobacco plants with decreased SS expression (Rodermel et al., 1996; Wostrikoff and Stern, 2007), because *rbcL* mRNA, although accumulating to the same level as in wild-type plants, is less associated with polysomes, suggesting a regulation of translation initiation. This establishes LS as a CES protein in higher plants, as it is in *Chlamydomonas* (Khrebtukova and Spreitzer, 1996), since its rate of translation would decrease in the absence of assembly with SS. Insight into the mode of CES regulation for LS in higher plants – autoregulation versus *trans*-activation – came with the demonstration that in material showing reduced expression of SS, the rate of synthesis of LS and the polysomal association of *rbcL* mRNA remained high when stable accumulation of LS was compromised (Wostrikoff and Stern, 2007). This was observed either in transplastomic plants expressing a truncated and unstable version of LS or in transgenic lines showing reduced expression of BSD2, an LS-specific chaperone required for proper folding and stable accumulation of neosynthesised LS polypeptide (Brutnell et al., 1999). Together these experiments exclude the possibility that SS could be required to *trans*-activate synthesis of LS and demonstrate that accumulation of unassembled LS down-regulates its own translation.

As mentioned by Wostrikoff and Stern (2007), CES may also account for the contrasting expression of LS in mesophyll and bundle sheath cells of C4 plants. While LS and SS are normally expressed in bundle sheath cells, mesophyll cells lack accumulation and expression of *RBCS* mRNAs and show reduced accumulation of *rbcL* mRNA which, in addition is not significantly associated with polysomes. This is consistent with a strong negative feedback of LS on translation of *rbcL* mRNA in the absence of the small subunit of Rubisco. A further indication that CES could be central in this differential expression comes from the expression properties of LS in the maize *bsd2* mutant, defective for the expression of the assembly chaperone. In this case, bundle sheath cells fail to accumulate Rubisco because of the improper folding of LS (Brutnell et al., 1999). Surprisingly, mesophyll cells from *bsd2* lines show increased accumulation and polysome association of *rbcL* mRNA as compared to wild-type controls. A reasonable explanation is that autoregulation of LS synthesis is lost in *bsd2* mesophyll cells, as it is in tobacco plants expressing reduced levels of the chaperone (Wostrikoff and Stern, 2007), because misfolded LS can not properly repress its own synthesis, resulting in active translation of *rbcL* mRNA.

CES may also contribute to another control step of Rubisco biogenesis, the light regulation of *rbcL* mRNA translation. Following the transfer of *Chlamydomonas* cells from low to high light, the rate of synthesis of LS decreases dramatically (Shapira et al., 1997). Accumulation of *rbcL* mRNA is not affected, but after sedimentation on sucrose gradients, most transcripts are found in low density fractions that contain monosomes and small polysomes (Cohen et al., 2005). This decrease in LS synthesis likely results from oxidative stress, which transiently generates reactive oxygen species and shifts the glutathione pool towards its oxidized form. Indeed *rbcL* mRNA translation decreases, even in low light conditions, in *Chlamydomonas* or higher plant cells treated with methyl viologen but not in *Chlamydomonas* cells transferred to high light if incubated with a reducing agent such as ascorbate (Irihimovitch and Shapira, 2000). Under these oxidative conditions, assembly of new Rubisco is arrested and newly synthesized SS is rapidly degraded (Cohen et al., 2005). LS, although synthesized in reduced amount, is more stable and thus would become available to repress its own synthesis. Most strikingly, oxidizing conditions weaken the interaction of the *rbcL* 5′ UTR with specific RNA-binding proteins and concomitantly increase the binding of a 55-kD protein, which is probably LS, to the *rbcL* 5′ UTR (Yosef et al., 2004). The N-terminal domain of Rubisco shows structural similarities with "ferredoxin-like" RNA-binding domains (Yosef et al., 2004). This LS N-terminal domain, highly conserved in photosynthetic organisms from purple bacteria to higher plants (Cohen et al., 2006), is normally buried within assembled Rubisco and only becomes exposed when unassembled LS experiences a conformational change under oxidative conditions (Yosef et al., 2004). However, it is still unclear whether this RNA-binding domain participates in the autoregulation of LS synthesis, as RNA binding is not sequence-specific (Yosef et al., 2004), a requirement for its contribution to the CES process.

In the case of PS II biogenesis in higher plant chloroplasts, a possible CES contribution is suggested by the phenotype of the vir^{115} mutant of barley, which lacks PS II activity. This mutant is primarily affected, in a developmentally regulated manner, in the translation of D1 (Kim et al., 1994), but shows concomitant down-regulation of CP47 expression (Gamble and Mullet, 1989) which is reminiscent of the CES for PS II biogenesis in *Chlamydomonas*. The molecular mechanism of the translational coupling between these two PS II subunits in higher plants remains, however, to be studied.

Whether cytochrome *f* is a CES protein in the chloroplast of higher plants has also been investigated, using transplastomic tobacco lines harboring deletions of *petD* or *petB* genes. These plants fail to accumulate the whole set of cytochrome b_6f subunits and produce cytochrome *f* at very low levels. The *petA* coding sequence being the last one on a ~4-kb polycistronic transcript that also contains *psaI*, *ycf4*, and *cemA*, it was expected that polysome binding of the tetracistronic mRNA would be largely preserved in the mutants. Still in those strains, the *petA* mRNA was shifted to polysomes of smaller size when compared to wild-type plants (Monde et al., 2000) but these polysomes were larger than those observed in the *crp1* maize mutant, which is fully defective for *petA* mRNA translation (Barkan et al., 1994). A definitive conclusion regarding the CES status of cytochrome *f* in higher plants might well be possible with the characterization of transplastomic lines that express a cytochrome *f* variant insensitive to the CES process because of a mutation in its repressor domain, which is fully conserved between *Chlamydomonas* and higher plant cytochrome *f*.

B. CES and biogenesis of respiratory proteins in yeast mitochondria

The CES process also contributes to the biogenesis of respiratory complexes in yeast mitochondria, as evidenced by studies of the expression of Cox1, a core subunit of the cytochrome oxidase complex (COX) encoded by the mitochondrial gene *cox1*. Cox1 is a *bona fide* CES subunit since its rate of synthesis is strongly reduced in yeast strains deficient in other COX subunits (Cabral et al., 1978; Cabral and Schatz, 1978; Poutre and Fox, 1987; Calder and McEwen, 1991; Nakai et al., 1995; Barrientos et al., 2004), or expressing the full complement of COX subunits but lacking COX assembly factors (Barrientos et al., 2004). Furthermore, Cox1, although synthesized in markedly reduced amount, is stable when COX assembly is compromised, while Cox2 and Cox3 are synthesized at wild-type levels but rapidly degraded (Calder and McEwen, 1991; Nakai et al., 1995; Barrientos et al., 2004). Strinkingly, newly synthesized Cox1 interacts physically with Mss51, one of the translational activators required for expression of *cox1* mRNA and with Cox14, a COX assembly factor that probably stabilizes the interaction between Mss51 and Cox1 (Perez-Martinez et al., 2003; Barrientos et al., 2004). Once sequestered in this ternary complex, as depicted in Figure 29.6A, Mss51 is prevented from initiating new rounds of Cox1 synthesis until it is released, which occurs when Cox1 interacts with other COX subunits in further steps of COX assembly catalyzed by Shy1, another COX assembly factor (Mick et al., 2007). Mss51, responsible for the tight coordination of Cox1 synthesis with downstream steps of COX assembly, has a dual role in *cox1* mRNA translation, being required for both initiation and elongation steps (Perez-Martinez et al., 2003; Zambrano et al., 2007). Upon compromised COX assembly, an aberrant *cox1* translation product of about 15 kD is synthesized in strains harboring the full complement of *cox1* introns, but not in strains expressing an intronless version of the gene (Zambrano et al., 2007). This aberrant translation, probably originating from an alternative initiation codon, suggests that assembly-dependent control of translation may also operate on another target step, translation elongation. This latter regulatory step was probably

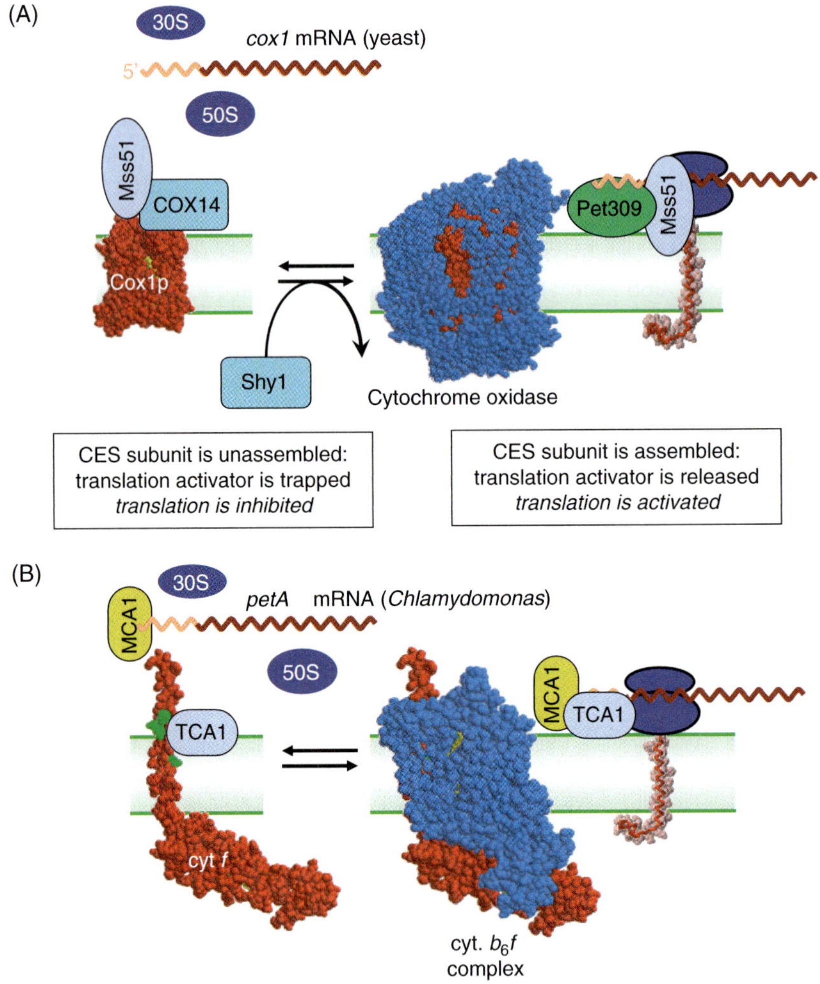

FIGURE 29.6 *Molecular mechanism of the CES autoregulation. (A) Regulation of COX1 synthesis in yeast mitochondria. cox1-specific translational activators (Mss51 and Pet309) are shown in pale blue and green, respectively, while COX assembly factors are represented in cyan. Unassembled Cox1 subunit traps Mss51, which is prevented from initiating new rounds of translation. Upon assembly with other COX subunits, catalyzed by Shy1, Mss51 becomes free to activate translation, together with Pet309. For the sake of simplicity, the effect of Mss51 on translation elongation has not been represented in this figure. (B) Model for the autoregulation of petA mRNA translation in Chlamydomonas. An epistatic effector acting as a translation factor, most probably TCA1 (drawn in pale blue), is sequestered by unassembled cytochrome f and is released when the latter is assembled within cytochrome b_6f complex. The residues from cytochrome f that contribute to the translational repressor domain are shown in green; note that they are shielded in the assembled cytochrome b_6f complex by the other subunits, represented in blue. A possible interaction of TCA1 with MCA1, the petA mRNA-specific stabilization factor that could also contribute to the CES effector, is suggested.*

evolutionarily selected to maintain the correct splicing of *cox1* mRNA, which requires translation of intron-encoded maturases.

Despite its limited coding capacity, the mitochondrial genome of yeast probably encodes a second CES protein, as mutants lacking expression of Atp9, the mitochondrial counterpart of chloroplast CF_0 subunit III, show reduced synthesis of the mitochondrially-encoded subunits Atp6 and Atp8, the latter of which corresponds to CF_0 subunit IV (Jean-Francois et al., 1986; Ooi et al., 1987; Payne et al., 1991). This suggests a conservation of this particular CES process through evolution.

C. CES and endosymbiosis: Cyanobacteria do not exhibit CES processes

CES thus appears to contribute to the biogenesis of protein complexes in organelles from various eukaryotes. By contrast, there is no evidence for a CES process in cyanobacteria. For example, mutations preventing PS II assembly because of defective expression of one core subunit had no apparent effect on the rate of transcription or translation of the remaining subunits in cyanobacteria (Komenda et al., 2004; Yu and Vermaas, 1990). Evidence includes a *Synechocystis* deletion strain lacking expression of D2, which still showed significant synthesis of D1, and mutations preventing *psb*A expression, which did not down-regulate expression of *psbB* (Komenda et al., 2004).

Cytochrome b_6f complexes being essential in cyanobacteria, deletion strains are not available to assess the participation of CES in their biogenesis. However, in contrast to what is observed in *Chlamydomonas*, unassembled and unprocessed pre-apo-cytochrome *f* accumulates to a high level in the thylakoid of an *S. elongatus* mutant which shows decreased heme-binding to apocytocrome *f*, with no indication of autocontrol of synthesis (Tichy and Vermaas, 1999). It is also noteworthy that one critical residue of the repressor domain for the CES regulation of cytochrome *f* translation in *Chlamydomonas*, the aromatic F_{307} residue from the regulatory tetrapeptide cluster, is a conserved signature of cytochrome *f* in chloroplasts, while aliphatic residues (V or I) are found at a similar position in cyanobacterial sequences (Choquet et al., 2003). This supports the view that cytochrome *f* acquired its CES regulatory properties following the initial endosymbiosis event.

VI. CES AND THE NUCLEAR CONTROL OF CHLOROPLAST GENE EXPRESSION

A. Specificity of CES autoregulation

In most instances apart from the CES process being discussed here, assembly-controlled autoregulation of translation involves proteins with a known RNA-binding activity such as ribosomal subunits (Yates et al., 1980; Yates and Nomura, 1980; Portier et al., 1990; Li et al., 1996; Hendrick et al., 2001; Bachand et al., 2006), translation initiation factor IF-3 (Butler et al., 1986), the β subunit of RNA polymerase (Peacock et al., 1983), the poly(A) binding protein (Wu and Bag, 1998), polynucleotide phosphorylase (Robert-Le Meur and Portier, 1992), an RNA chaperone (Vecerek et al., 2005), a yeast RNA helicase (Barta and Iggo, 1995), *Escherichia coli* SecA (Schmidt and Oliver, 1989), the tumour suppressor protein p53 (Mosner et al., 1995), and various viral proteins (Russel et al., 1976; Liang et al., 1988; Tuerk et al., 1990; reviewed in Kozak, 2005). In their unassembled state these proteins most often bind a structural element of their own 5′ UTR that mimics their target RNA, thereby preventing the formation of the translation initiation complex.

With the exception of the Rubisco large subunit (see above), chloroplast CES subunits have no known RNA-binding properties. Their sequences, constrained by their function in light capture, electron transfer and energy conversion, are highly conserved from cyanobacteria lacking CES processes to unicellular algae and higher plants. The regulatory motif that has been identified in the case of cytochrome *f* (Choquet et al., 2003), is located within a single α-helix extending from the membrane to the stroma (Stroebel et al., 2003). It does not have any feature typical of an RNA-binding protein and encompasses a critical residue, Q_{297}, that is unlikely to interact directly with the *petA* 5′ UTR, because it is embedded in the membrane bilayer. One should also note that the nucleotide sequences of the CES 5′ UTRs show considerable variation between organisms, arguing against their specific recognition by the conserved CES proteins.

There are only a few examples of protein assembly control of translation that do not merely result from the known RNA-binding function of the autoregulated protein. A first example is that of α- and β-tubulins, two proteins showing assembly-dependent autoregulation of gene expression in animal cells. Whereas expression of β-tubulin is regulated at a pre-translational level, since unassembled β-tubulin induces the degradation of polysome-bound tubulin mRNA (Cleveland, 1989; Bachurski et al., 1994), α-tubulin shows genuine assembly controlled regulation of translation in Chinese hamster ovary cells (Gonzalez-Garay and Cabral, 1996). Accordingly, overexpression of β-tubulin, its assembly partner, stimulates the synthesis of α-tubulin by releasing the negative feedback mediated by unassembled α-tubulin (Gonzalez-Garay and Cabral, 1995).

The complex process of the bacterial flagellum biogenesis is another example of assembly feedback on the expression of the constituent subunits reviewed by (Chilcott and Hughes, 2000; Aldridge and Hughes, 2002). A complex multi-layered regulatory network finely tunes the hierarchical expression of flagellar subunits to facilitate their delivery in a timely manner during the assembly process. Most of the regulation develops at the transcriptional level through the sequential expression of transcription and anti-transcription factors, whose expression is controlled at the level of translation and secretion, through interlaced regulatory loops modulated by the state of flagella assembly (Karlinsey et al., 1998, 2000; Aldridge et al., 2006; Rosu et al., 2006).

B. Role of nuclear-encoded and gene-specific *trans*-acting factors in CES

In the case of the organelle-encoded CES proteins, it is of note that a variety of gene-specific translation activators play a major role in organelle gene expression (see Chapter 28, also Grivell, 1995; Barkan and Goldschmidt-Clermont, 2000; Fox, 1996; Zerges, 2000). As is the case for CES, these factors have probably evolved after endosymbiosis to coordinate the expression of organellar and nuclear genomes since they rarely have a counterpart in cyanobacteria, where gene expression is mainly – although not exclusively – regulated at the transcriptional level. These factors show sequence-specific interactions (either direct or indirect) with the 5′ UTR of their target transcripts. Some of them could have been later recruited for the CES process or they could have coevolved with the CES process, acquiring an additional affinity for unassembled CES subunits. This model would set efficiency of translation initiation, the step that requires the participation of these nucleus-coded translational activators, to the rate of assembly of a CES subunit within a (sub)complex along the biogenesis pathway of a particular photosynthesis protein complex.

In a typical chloroplast, the accumulation of about $1 - 2 \times 10^6$ copies of each thylakoid-embedded photosynthetic protein is in large excess over that of the corresponding mRNAs, which range from ~5000 mRNA molecules for *psbA* mRNA to about 500 copies of the *atpB* mRNA (Rapp et al., 1992). LS accumulation was estimated to about 10^7 copies versus ~5500 molecules of the *rbcL* mRNA (Dean and Leech, 1982). Thus, CES subunits, even when they accumulate to low levels in the absence of their assembly partners, remain abundant enough to trap their translational activators, whose accumulation is expected to be in the same range – or lower – than that of their target mRNA.

Mss51, whose action on yeast Cox1 was described above, provides the best example of such a molecular mechanism. An analogous process is presented for cytochrome *f* in Figure 29.6B. Upon interaction with the repressor domain of unassembled cytochrome *f*, a ternary CES effector capable of translational activation is prevented from promoting new rounds of cytochrome *f* synthesis until it is released, which occurs when cytochrome *f* assembles within a b_6f complex. Thus, depending on its binding to unassembled cytochrome *f*, the ternary effector should modulate translation of the *petA* transcript. Extensive genetic and molecular analyses of a number of independent mutant strains specifically defective for cytochrome *f* expression have shown the involvement of two, and probably only two, specific nucleus-encoded factors which constitute the *petA* gene expression system. MCA1 primarily controls the accumulation of the transcript (Girard-Bascou et al., 1995; Gumpel et al., 1995; Lown et al., 2001) while TCA1 is required for its translation (Wostrikoff et al., 2001; Raynaud et al., 2007). The most likely candidates for the epistatic ternary effector are TCA1, or a complex formed by TCA1 and MCA1. The characterization of a partial CES process in suppressor strains with altered TCA1 activity (Wostrikoff et al., 2001) is in agreement with this hypothesis. Furthermore, while TCA1 was found to be rate-limiting for cytochrome *f* synthesis when expressed at low levels, cytochrome *f* expression became rather insensitive to changes in the amount of TCA1 in strains expressing higher levels of this factor (Raynaud et al., 2007). This particular concentration dependency is consistent with the amount of TCA1 available for translation being itself regulated by cytochrome b_6f assembly, as predicted by CES for a negative feedback mechanism. Still, the actual participation of TCA1 (/MCA1) in the CES process awaits a conclusive demonstration.

C. Physiology of CES

CES represents an optimal strategy to control tightly the rate of production of chloroplast-encoded subunits in order to facilitate their assembly within a functional complex. Depending on light quality and physiological conditions, the relative cellular demand for *de novo* biogenesis of photosynthetic proteins should vary (Chow et al., 1990). With the regulation mechanism we have described here, the nucleus needs only to control the rate of expression of the most upstream assembly partner in a biogenesis pathway, with CES cascades then ensuring the stoichiometric production of the other subunits. For example, the light-induced increase in the expression of *psbD*, encoding the RC II subunit D2 (Herrin et al., 1986) should trigger synthesis of the other RC II subunit D1 and of the PS II core subunit CP47 *via* the hierarchical control of translation of *psbD* over *psbA*, and *psbB* mRNAs. In that respect, it is worth noting that a site-directed mutation of the *psbD* 5′ UTR that destabilizes a secondary structure sequestering the initiation codon, causes an increase in the rate of synthesis of D2, accompanied by its increased accumulation. Since unassembled D2 is rapidly degraded *in vivo* (de Vitry et al., 1989), its increased accumulation in the mutant indicates it should be protected from degradation by assembly, as suggested by the moderate, but significant, increase in the photoautotrophic growth rate of that strain (Klinkert et al., 2006). Similarly, mutant strains that accumulate excess *petD* mRNA because of the insertion of a polyguanosine tract within the *petD* 5′ UTR, display increased accumulation of subunit IV (Drager et al., 1999). The fates of the other subunits was not investigated in that study. However, subunit IV being highly unstable when unassembled (Kuras and Wollman, 1994), its increased level in that strain suggests an increased accumulation of the whole b_6f complex. By contrast, a strain that overaccumulates cytochrome *f*, the downstream CES subunit in the biogenesis pathway, does not show an increased accumulation of subunit IV (Choquet et al., 2003).

VII. CONCLUSION

This description of CES further confirms the prevalent role of translation as a major regulatory step in the expression of chloroplast genes, as was previously proposed (Eberhard et al., 2002). CES probably relies extensively on nucleus-encoded gene-specific factors. Their participation in CES remains to be further substantiated in most cases, which represents a major research goal for the future. The actual regulatory function of some of these nucleus-encoded factors has been demonstrated only recently (Raynaud et al., 2007; see also Chapter 28). They should turn out to be critical for the assembly-controlled translational autoregulation in organelles.

Finally, one should consider CES as reflecting one of the modes of nuclear control of chloroplast gene expression. Not only would it have co-evolved with the production of nuclear *trans*-acting factors in organelle gene expression, but it also couples the expression of chloroplast genes to that of nucleus-encoded subunits. This is most obvious for the biogenesis of Rubisco and of CF_1 from the ATP synthase, where delivery of the nucleus-encoded subunits – SS and subunit γ respectively – turns out to act as rate-limiting in the expression of the other chloroplast-encoded subunits of the same protein complex.

ACKNOWLEDGMENTS

Our deepest thanks to those that embarked with us for the CES adventure: D. Drapier, J. Girard-Bascou, R. Kuras, C. Loiselay, L. Minai, C. Raynaud, B. Rimbault, D. B. Stern, O. Vallon, K. Wostrikoff, and F. Zito. Work performed in the authors' laboratory "Molecular and Membrane Physiology of the Chloroplast" was supported by CNRS, Collège de France and Université Pierre et Marie Curie-Paris 6.

REFERENCES

Adir, N., Shochat, S., and Ohad, I. (1990). Light-dependent D1 protein synthesis and translocation is regulated by reaction center II. Reaction center II serves as an acceptor for the D1 precursor. *J. Biol. Chem.* **265**, 12563–12568.

Aldridge, P. and Hughes, K.T. (2002). Regulation of flagellar assembly. *Curr. Opin. Microbiol.* **5**, 160–165.

Aldridge, P., Karlinsey, J.E., Becker, E., Chevance, F.F., and Hughes, K.T. (2006). Flk prevents premature secretion of the anti-sigma factor FlgM into the periplasm. *Mol. Microbiol.* **60**, 630–643.

Bachand, F., Lackner, D.H., Bahler, J., and Silver, P.A. (2006). Autoregulation of ribosome biosynthesis by a translational response in fission yeast. *Mol. Cell Biol.* **26**, 1731–1742.

Bachurski, C.J., Theodorakis, N.G., Coulson, R.M., and Cleveland, D.W. (1994). An amino-terminal tetrapeptide specifies cotranslational degradation of beta-tubulin but not alpha-tubulin mRNAs. *Mol. Cell Biol.* **14**, 4076–4086.

Bailey, S., Thompson, E., Nixon, P.J., Horton, P., Mullineaux, C.W., Robinson, C., and Mann, N.H. (2002). A critical role for the Var2 FtsH homologue of *Arabidopsis thaliana* in the photosystem II repair cycle *in vivo*. *J. Biol. Chem.* **277**, 2006–2011.

Barkan, A. and Goldschmidt-Clermont, M. (2000). Participation of nuclear genes in chloroplast gene expression. *Biochimie* **82**, 559–572.

Barkan, A., Walker, M., Nolasco, M., and Johnson, D. (1994). A nuclear mutation in maize blocks the processing and translation of several chloroplast mRNAs and provides evidence for the differential translation of alternative mRNA forms. *EMBO J.* **13**, 3170–3181.

Barrientos, A., Zambrano, A., and Tzagoloff, A. (2004). Mss51p and Cox14p jointly regulate mitochondrial Cox1p expression in *Saccharomyces cerevisiae*. *EMBO J.* **23**, 3472–3482.

Barta, I. and Iggo, R. (1995). Autoregulation of expression of the yeast Dbp2p "DEAD-box" protein is mediated by sequences in the conserved *DBP2* intron. *EMBO J.* **14**, 3800–3808.

Bennoun, P., Girard, J., and Chua, N.-H. (1977). A uniparental mutant of *Chlamydomonas reinhardtii* deficient in the chlorophyll-protein complex CP1. *Mol. Gen. Genet.* **153**, 343–348.

Bennoun, P., Spierer-Herz, M., Erickson, J., Girard-Bascou, J., Pierre, Y., Delosme, M., and Rochaix, J.D. (1986). Characterization of photosystem II mutants of *Chlamydomonas reinhardii* lacking the *psbA* gene. *Plant Mol. Biol.* **6**, 151–160.

Berthold, D.A., Schmidt, C.L., and Malkin, R. (1995). The deletion of *petG* in *Chlamydomonas reinhardtii* disrupts the cytochrome *bf* complex. *J. Biol. Chem.* **270**, 29293–29298.

Brutnell, T.P., Sawers, R.J., Mant, A., and Langdale, J.A. (1999). BUNDLE SHEATH DEFECTIVE2, a novel protein required for post-translational regulation of the *rbcL* gene of maize. *Plant Cell* **11**, 849–864.

Bulté, L. and Wollman, F.A. (1992). Evidence for a selective destabilization of an integral membrane protein, the cytochrome b_6f complex, during gametogenesis in *Chlamydomonas reinhardtii*. *Eur. J. Biochem.* **204**, 327–336.

Butler, J.S., Springer, M., Dondon, J., Graffe, M., and Grunberg-Manago, M. (1986). *Escherichia coli* protein synthesis initiation factor IF3 controls its own gene expression at the translational level *in vivo*. *J. Mol. Biol.* **192**, 767–780.

Cabral, F. and Schatz, G. (1978). Identification of cytochrome c oxidase subunits in nuclear yeast mutants lacking the functional enzyme. *J. Biol. Chem.* **253**, 4396–4401.

Cabral, F., Solioz, M., Rudin, Y., Schatz, G., Clavilier, L., and Slonimski, P.P. (1978). Identification of the structural gene for yeast cytochrome *c* oxidase subunit II on mitochondrial DNA. *J. Biol. Chem.* **253**, 297–304.

Calder, K.M. and McEwen, J.E. (1991). Deletion of the *COX7* gene in *Saccharomyces cerevisiae* reveals a role for cytochrome *c* oxidase subunit VII in assembly of remaining subunits. *Mol. Microbiol.* **5**, 1769–1777.

Chilcott, G.S. and Hughes, K.T. (2000). Coupling of flagellar gene expression to flagellar assembly in *Salmonella enterica* serovar typhimurium and *Escherichia coli*. *Microbiol. Mol. Biol. Rev.* **64**, 694–708.

Chitnis, P.R., Reilly, P.A., and Nelson, N. (1989). Insertional inactivation of the gene encoding subunit II of photosystem I from the cyanobacterium *Synechocystis* sp. PCC 6803. *J. Biol. Chem.* **264**, 18381–18385.

Choquet, Y. and Vallon, O. (2000). Synthesis, assembly and degradation of thylakoid membrane proteins. *Biochimie* **82**, 615–634.

Choquet, Y. and Wollman, F.A. (2002). Translational regulations as specific traits of chloroplast gene expression. *FEBS Lett.* **529**, 39–42.

Choquet, Y., Goldschmidt-Clermont, M., Girard-Bascou, J., Kuck, U., Bennoun, P., and Rochaix, J.D. (1988). Mutant phenotypes support a *trans*-splicing mechanism for the expression of the tripartite *psaA* gene in the *C. reinhardtii* chloroplast. *Cell* **52**, 903–913.

Choquet, Y., Stern, D.B., Wostrikoff, K., Kuras, R., Girard-Bascou, J., and Wollman, F.A. (1998). Translation of cytochrome *f* is autoregulated through the 5′ untranslated region of *petA* mRNA in *Chlamydomonas* chloroplasts. *Proc. Natl. Acad. Sci. U. S. A.* **95**, 4380–4385.

Choquet, Y., Zito, F., Wostrikoff, K., and Wollman, F.A. (2003). Cytochrome *f* translation in *Chlamydomonas* chloroplast is autoregulated by its carboxyl-terminal domain. *Plant Cell* **15**, 1443–1454.

Chow, W.S., Melis, A., and Anderson, J.M. (1990). Adjustments of photosystem stoichiometry in chloroplasts improve the quantum efficiency of photosynthesis. *Proc. Natl. Acad. Sci. U. S. A.* **87**, 7502–7506.

Cleveland, D.W. (1989). Autoregulated control of tubulin synthesis in animal cells. *Curr. Opin. Cell Biol.* **1**, 10–14.

Cohen, I., Knopf, J.A., Irihimovitch, V., and Shapira, M. (2005). A proposed mechanism for the inhibitory effects of oxidative stress on Rubisco assembly and its subunit expression. *Plant Physiol.* **137**, 738–746.

Cohen, I., Sapir, Y., and Shapira, M. (2006). A conserved mechanism controls translation of Rubisco large subunit in different photosynthetic organisms. *Plant Physiol.* **141**, 1089–1097.

Dauvillée, D., Stampacchia, O., Girard-Bascou, J., and Rochaix, J.D. (2003). Tab2 is a novel conserved RNA binding protein required for translation of the chloroplast *psaB* mRNA. *EMBO J.* **22**, 6378–6388.

Dean, C. and Leech, R.M. (1982). Genome expression during normal leaf development: I. Cellular and chloroplast numbers and DNA, RNA, and protein levels in tissues of different ages within a seven-day-old wheat leaf. *Plant Physiol.* **69**, 904–910.

de Vitry, C., Olive, J., Drapier, D., Recouvreur, M., and Wollman, F.A. (1989). Posttranslational events leading to the assembly of photosystem II protein complex: A study using photosynthesis mutants from *Chlamydomonas reinhardtii*. *J. Cell Biol.* **109**, 991–1006.

Diner, B.A. and Wollman, F.A. (1980). Isolation of highly active photosystem II particles from a mutant of *Chlamydomonas reinhardtii*. *Eur. J. Biochem.* **110**, 521–526.

Drager, R.G., Higgs, D.C., Kindle, K.L., and Stern, D.B. (1999). 5′ to 3′ exoribonucleolytic activity is a normal component of chloroplast mRNA decay pathways. *Plant J.* **19**, 521–531.

Drapier, D., Girard-Bascou, J., and Wollman, F.-A. (1992). Evidence for nuclear control of the expression of the *atpA* and *atpB* chloroplast genes in *Chlamydomonas*. *Plant Cell* **4**, 283–295.

Drapier, D., Suzuki, H., Levy, H., Rimbault, B., Kindle, K.L., Stern, D.B., and Wollman, F.A. (1998). The chloroplast *atpA* gene cluster in *Chlamydomonas reinhardtii.* Functional analysis of a polycistronic transcription unit. *Plant Physiol.* **117**, 629–641.

Drapier, D., Rimbault, B., Vallon, O., Wollman, F.A., and Choquet, Y. (2007). Intertwined translational regulations set uneven stoichiometry of chloroplast ATP synthase subunits. *EMBO J.* **26**, 3581–3591.

Du, Z. and Gromet-Elhanan, Z. (1999). Refolding of recombinant a and b subunits of the *Rhodospirillum rubrum* F_0F_1 ATP synthase into functional monomers that reconstitute an active a_1b_1-dimer. *Eur. J. Biochem.* **263**, 430–437.

Eberhard, S., Drapier, D., and Wollman, F.A. (2002). Searching limiting steps in the expression of chloroplast-encoded proteins: Relations between gene copy number, transcription, transcript abundance and translation rate in the chloroplast of *Chlamydomonas reinhardtii*. *Plant J.* **31**, 149–160.

Erickson, J.M., Rahire, M., Malnoe, P., Girard-Bascou, J., Pierre, Y., Bennoun, P., and Rochaix, J.-D. (1986). Lack of the D2 protein in a *Chlamydomonas reinhardtii psbD* mutant affects photosystem II stability and D1 expression. *EMBO J.* **5**, 1745–1754.

Ferreira, K.N., Iverson, T.M., Maghlaoui, K., Barber, J., and Iwata, S. (2004). Architecture of the photosynthetic oxygen-evolving center. *Science* **303**, 1831–1838.

Fox, T.D. (1996). *Genetics of Mitochondrial Translation*. Cold Spring Harbor Laboratory Press, Cold Spring Harbor, New York.

Gamble, P.E. and Mullet, J.E. (1989). Translation and stability of proteins encoded by the plastid *psbA* and *psbB* genes are regulated by a nuclear gene during light-induced chloroplast development in barley. *J. Biol. Chem.* **264**, 7236–7243.

Girard, J., Chua, N.-H., Bennoun, P., Schmid, G.H., and Delosme, M. (1980). Studies of mutants deficient in the photosystem I reaction centers in *Chlamydomonas reinhardtii*. *Curr. Genet.* **2**, 215–221.

Girard-Bascou, J. (1987). Mutations in four chloroplast loci of *Chlamydomonas reinhardtii* affecting the photosystem I reaction centers. *Curr. Genet.* **12**, 483–488.

Girard-Bascou, J., Choquet, Y., Schneider, M., Delosme, M., and Dron, M. (1987). Characterization of a chloroplast mutation in the *psaA2* gene of *Chlamydomonas reinhardtii*. *Curr. Genet.* **12**, 489–495.

Girard-Bascou, J., Choquet, Y., Gumpel, N.J., Culler, D., Purton, S., Merchant, S., Laquerriere, F., and Wollman, F.-A. (1995). Nuclear control of the expression of the chloroplast *pet* genes in *Chlamydomonas reinhardtii*. In: *Photosynthesis: From Light to Biosphere*, Vol. III (P. Mathis, Ed.), pp. 683–686. Kluwer Academic Publishers, Dordrecht.

Goldschmidt-Clermont, M., Girard-Bascou, J., Choquet, Y., and Rochaix, J.D. (1990). *Trans*-splicing mutants of *Chlamydomonas reinhardtii*. *Mol. Gen. Genet.* **223**, 417–425.

Gonzalez-Garay, M.L. and Cabral, F. (1995). Overexpression of an epitope-tagged beta-tubulin in Chinese hamster ovary cells causes an increase in endogenous alpha-tubulin synthesis. *Cell Motil. Cytoskeleton* **31**, 259–272.

Gonzalez-Garay, M.L. and Cabral, F. (1996). Alpha-tubulin limits its own synthesis: Evidence for a mechanism involving translational repression. *J. Cell Biol.* **135**, 1525–1534.

Grivell, L.A. (1995). Nucleo-mitochondrial interactions in mitochondrial gene expression. *Crit. Rev. Biochem. Mol. Biol.* **30**, 121–164.

Gumpel, N.J., Ralley, L., Girard-Bascou, J., Wollman, F.A., Nugent, J.H., and Purton, S. (1995). Nuclear mutants of *Chlamydomonas reinhardtii* defective in the biogenesis of the cytochrome b_6f complex. *Plant Mol. Biol.* **29**, 921–932.

Hager, M., Biehler, K., Illerhaus, J., Ruf, S., and Bock, R. (1999). Targeted inactivation of the smallest plastid genome-encoded open reading frame reveals a novel and essential subunit of the cytochrome b_6f complex. *EMBO J.* **18**, 5834–5842.

Hansson, A., Amann, K., Zygadlo, A., Meurer, J., Scheller, H.V., and Jensen, P.E. (2007). Knock-out of the chloroplast-encoded PSI-J subunit of photosystem I in *Nicotiana tabacum*. *FEBS J.* **274**, 1734–1746.

Hendrick, J.L., Wilson, P.G., Edelman, II., Sandbaken, M.G., Ursic, D., and Culbertson, M.R. (2001). Yeast frameshift suppressor mutations in the genes coding for transcription factor Mbf1p and ribosomal protein S3: evidence for autoregulation of S3 synthesis. *Genetics* **157**, 1141–1158.

Herrin, D.L., Michaels, A.S., and Paul, A.L. (1986). Regulation of genes encoding the large subunit of ribulose-1,5-bisphosphate carboxylase and the photosystem II polypeptides D-1 and D-2 during the cell cycle of *Chlamydomonas reinhardtii*. *J. Cell Biol.* **103**, 1837–1845.

Ihnatowicz, A., Pesaresi, P., Varotto, C., Richly, E., Schneider, A., Jahns, P., Salamini, F., and Leister, D. (2004). Mutants for photosystem I subunit D of *Arabidopsis thaliana*: Effects on photosynthesis, photosystem I stability and expression of nuclear genes for chloroplast functions. *Plant J.* **37**, 839–852.

Ihnatowicz, A., Pesaresi, P., and Leister, D. (2007). The E subunit of photosystem I is not essential for linear electron flow and photoautotrophic growth in *Arabidopsis thaliana*. *Planta* **226**, 889–895.

Ikeuchi, M., Eggers, B., Shen, G.Z., Webber, A., Yu, J.J., Hirano, A., Inoue, Y., and Vermaas, W. (1991). Cloning of the *psbK* gene from *Synechocystis* sp. PCC 6803 and characterization of photosystem II in mutants lacking PSII-K. *J. Biol. Chem.* **266**, 11111–11115.

Ikeuchi, M., Shukla, V.K., Pakrasi, H.B., and Inoue, Y. (1995). Directed inactivation of the *psbI* gene does not affect photosystem II in the cyanobacterium *Synechocystis* sp. PCC 6803. *Mol. Gen. Genet.* **249**, 622–628.

Irihimovitch, V. and Shapira, M. (2000). Glutathione redox potential modulated by reactive oxygen species regulates translation of Rubisco large subunit in the chloroplast. *J. Biol. Chem.* **275**, 16289–16295.

Jean-Francois, M.J., Lukins, H.B., and Marzuki, S. (1986). Post-transcriptional defects in the synthesis of the mitochondrial H+-ATPase subunit 6 in yeast mutants with lesions in the subunit 9 structural gene. *Biochim. Biophys. Acta* **868**, 178–182.

Jensen, K.H., Herrin, D.L., Plumley, F.G., and Schmidt, G.W. (1986). Biogenesis of photosystem II complexes: Transcriptional, translational, and posttranslational regulation. *J. Cell Biol.* **103**, 1315–1325.

Jordan, P., Fromme, P., Witt, H.T., Klukas, O., Saenger, W., and Krauss, N. (2001). Three-dimensional structure of cyanobacterial photosystem I at 2.5 Å resolution. *Nature* **411**, 909–917.

Karlinsey, J.E., Tsui, H.C., Winkler, M.E., and Hughes, K.T. (1998). Flk couples flgM translation to flagellar ring assembly in *Salmonella typhimurium*. *J. Bacteriol.* **180**, 5384–5397.

Karlinsey, J.E., Lonner, J., Brown, K.L., and Hughes, K.T. (2000). Translation/secretion coupling by type III secretion systems. *Cell* **102**, 487–497.

Ketchner, S.L., Drapier, D., Olive, J., Gaudriault, S., Girard-Bascou, J., and Wollman, F.A. (1995). Chloroplasts can accommodate inclusion bodies. Evidence from a mutant of *Chlamydomonas reinhardtii* defective in the assembly of the chloroplast ATP synthase. *J. Biol. Chem.* **270**, 15299–15306.

Kettunen, R., Pursiheimo, S., Rintamaki, E., van Wijk, K.J., and Aro, E.M. (1997). Transcriptional and translational adjustments of *psbA* gene expression in mature chloroplasts during photoinhibition and subsequent repair of photosystem II. *Eur. J. Biochem.* **247**, 441–448.

Khrebtukova, I. and Spreitzer, R.J. (1996). Elimination of the *Chlamydomonas* gene family that encodes the small subunit of ribulose-1,5-bisphosphate carboxylase/oxygenase. *Proc. Natl. Acad. Sci. U. S. A.* **93**, 13689–13693.

Kim, J., Klein, P.G., and Mullet, J.E. (1994). Vir-115 gene product is required to stabilize D1 translation intermediates in chloroplasts. *Plant Mol. Biol.* **25**, 459–467.

Klinkert, B., Elles, I., and Nickelsen, J. (2006). Translation of chloroplast *psbD* mRNA in *Chlamydomonas* is controlled by a secondary RNA structure blocking the AUG start codon. *Nucl. Acids. Res.* **34**, 386–394.

Komenda, J., Lupinkova, L., and Kopecky, J. (2002). Absence of the *psbH* gene product destabilizes photosystem II complex and bicarbonate binding on its acceptor side in *Synechocystis* PCC 6803. *Eur. J. Biochem.* **269**, 610–619.

Komenda, J., Reisinger, V., Muller, B.C., Dobakova, M., Granvogl, B., and Eichacker, L.A. (2004). Accumulation of the D2 protein is a key regulatory step for assembly of the photosystem II reaction center complex in *Synechocystis* PCC 6803. *J. Biol. Chem.* **279**, 48620–48629.

Kozak, M. (2005). Regulation of translation via mRNA structure in prokaryotes and eukaryotes. *Gene* **361**, 13–37.

Kunstner, P., Guardiola, A., Takahashi, Y., and Rochaix, J.D. (1995). A mutant strain of *Chlamydomonas reinhardtii* lacking the chloroplast photosystem II *psbI* gene grows photoautotrophically. *J. Biol. Chem.* **270**, 9651–9654.

Kuras, R. and Wollman, F.A. (1994). The assembly of cytochrome b_6f complexes: An approach using genetic transformation of the green alga *Chlamydomonas reinhardtii*. *EMBO J.* **13**, 1019–1027.

Kuras, R., Buschlen, S., and Wollman, F.A. (1995a). Maturation of pre-apocytochrome *f in vivo*. A site-directed mutagenesis study in *Chlamydomonas reinhardtii*. *J. Biol. Chem.* **270**, 27797–27803.

Kuras, R., Wollman, F.A., and Joliot, P. (1995b). Conversion of cytochrome *f* to a soluble form *in vivo* in *Chlamydomonas reinhardtii*. *Biochemistry* **34**, 7468–7475.

Lemaire, C. and Wollman, F.A. (1989a). The chloroplast ATP synthase in *Chlamydomonas reinhardtii*. I. Characterization of its nine constitutive subunits. *J. Biol. Chem.* **264**, 10228–10234.

Lemaire, C. and Wollman, F.A. (1989b). The chloroplast ATP synthase in *Chlamydomonas reinhardtii*. II. Biochemical studies on its biogenesis using mutants defective in photophosphorylation. *J. Biol. Chem.* **264**, 10235–10242.

Lemaire, C., Girard-Bascou, J., Wollman, F.-A., and Bennoun, P. (1986). Studies on the cytochrome b_6f complex. I. Characterization of the complex subunits in Chlamydomonas reinhardtii. Biochim. Biophys. Acta 851, 229–238.

Lezhneva, L. and Meurer, J. (2004). The nuclear factor HCF145 affects chloroplast *psaA-psaB-rps14* transcript abundance in *Arabidopsis thaliana*. *Plant J.* **38**, 740–753.

Li, X., Lindahl, L., and Zengel, J.M. (1996). Ribosomal protein L4 from *Escherichia coli* utilizes nonidentical determinants for its structural and regulatory functions. *RNA* **2**, 24–37.

Liang, Y.M., Wei, R.X., Hsu, T., Alford, C., Dawson, M., and Karam, J. (1988). Autogenous regulation of the *regA* gene of bacteriophage T4: Derepression of translation. *Genetics* **119**, 743–749.

Loll, B., Kern, J., Saenger, W., Zouni, A., and Biesiadka, J. (2005). Towards complete cofactor arrangement in the 3.0 Å resolution structure of photosystem II. *Nature* **438**, 1040–1044.

Lown, F.J., Watson, A.T., and Purton, S. (2001). *Chlamydomonas* nuclear mutants that fail to assemble respiratory or photosynthetic electron transfer complexes. *Biochem. Soc. Trans.* **29**, 452–455.

Maiwald, D., Dietzmann, A., Jahns, P., Pesaresi, P., Joliot, P., Joliot, A., Levin, J.Z., Salamini, F., and Leister, D. (2003). Knock-out of the genes coding for the Rieske protein and the ATP-synthase delta-subunit of *Arabidopsis*. Effects on photosynthesis, thylakoid protein composition, and nuclear chloroplast gene expression. *Plant Physiol.* **133**, 191–202.

Majeran, W., Wollman, F.A., and Vallon, O. (2000). Evidence for a role of ClpP in the degradation of the chloroplast cytochrome b_6f complex. *Plant Cell* **12**, 137–150.

Majeran, W., Olive, J., Drapier, D., Vallon, O., and Wollman, F.A. (2001). The light sensitivity of ATP synthase mutants of *Chlamydomonas reinhardtii*. *Plant Physiol.* **126**, 421–433.

Mannan, R.M., Whitmarsh, J., Nyman, P., and Pakrasi, H.B. (1991). Directed mutagenesis of an iron-sulfur protein of the photosystem I complex in the filamentous cyanobacterium *Anabaena variabilis* ATCC 29413. *Proc. Natl. Acad. Sci. U. S. A.* **88**, 10168–10172.

Mayers, S.R., Dubbs, J.M., Vass, I., Hideg, E., Nagy, L., and Barber, J. (1993). Further characterization of the *psbH* locus of *Synechocystis* sp. PCC 6803: inactivation of *psbH* impairs Q_A to Q_B electron transport in photosystem 2. *Biochemistry* **32**, 1454–1465.

Meyer zu Tittingdorf, J.M., Rexroth, S., Schafer, E., Schlichting, R., Giersch, C., Dencher, N.A., and Seelert, H. (2004). The stoichiometry of the chloroplast ATP synthase oligomer III in *Chlamydomonas reinhardtii* is not affected by the metabolic state. *Biochim. Biophys. Acta* **1659**, 92–99.

Mick, D.U., Wagner, K., van der Laan, M., Frazier, A.E., Perschil, I., Pawlas, M., Meyer, H.E., Warscheid, B., and Rehling, P. (2007). *Shy1* couples *Cox1* translational regulation to cytochrome *c* oxidase assembly. *EMBO J.* **26**, 4347–4358.

Minai, L., Wostrikoff, K., Wollman, F.A., and Choquet, Y. (2006). Chloroplast biogenesis of photosystem II cores involves a series of assembly-controlled steps that regulate translation. *Plant Cell* **18**, 159–175.

Monde, R.A., Zito, F., Olive, J., Wollman, F.A., and Stern, D.B. (2000). Post-transcriptional defects in tobacco chloroplast mutants lacking the cytochrome b_6f complex. *Plant J.* **21**, 61–72.

Mosner, J., Mummenbrauer, T., Bauer, C., Sczakiel, G., Grosse, F., and Deppert, W. (1995). Negative feedback regulation of wild-type p53 biosynthesis. *EMBO J.* **14**, 4442–4449.

Nakai, T., Yasuhara, T., Fujiki, Y., and Ohashi, A. (1995). Multiple genes, including a member of the AAA family, are essential for degradation of unassembled subunit 2 of cytochrome c oxidase in yeast mitochondria. *Mol. Cell Biol.* **15**, 4441–4452.

Nyhus, K.J., Thiel, T., and Pakrasi, H.B. (1993). Targeted interruption of the *psaA* and *psaB* genes encoding the reaction-centre proteins of photosystem I in the filamentous cyanobacterium *Anabaena variabilis* ATCC 29413. *Mol. Microbiol.* **9**, 979–988.

Ohad, I., Keren, N., Zer, H., Gong, H., Mor, T.S., Gal, A., Tal, S., and Domovich, Y. (1994). Light-induced degradation of the photosystem II reaction centre D1 protein *in vivo*: An integrative approach. In: *Photoinhibition from the Molecule to the Field* (N. Baker, Ed.), pp. 161–177. Bios Scientific Publishers, Oxford, UK.

Ooi, B.G., Lukins, H.B., Linnane, A.W., and Nagley, P. (1987). Biogenesis of mitochondria: A mutation in the 5′-untranslated region of yeast mitochondrial *oli1* mRNA leading to impairment in translation of subunit 9 of the mitochondrial ATPase complex. *Nucl. Acids Res.* **15**, 1965–1977.

Payne, M.J., Schweizer, E., and Lukins, H.B. (1991). Properties of two nuclear *pet* mutants affecting expression of the mitochondrial *oli1* gene of *Saccharomyces cerevisiae*. *Curr. Genet.* **19**, 343–351.

Peacock, S., Brot, N., and Weissbach, H. (1983). Translational control of the expression of the β subunit gene of *E. coli* RNA polymerase. *Biochem. Biophys. Res. Commun.* **113**, 1018–1025.

Perez-Martinez, X., Broadley, S.A., and Fox, T.D. (2003). Mss51p promotes mitochondrial Cox1p synthesis and interacts with newly synthesized Cox1p. *EMBO J.* **22**, 5951–5961.

Portier, C., Dondon, L., and Grunberg-Manago, M. (1990). Translational autocontrol of the *Escherichia coli* ribosomal protein S15. *J. Mol. Biol.* **211**, 407–414.

Poutre, C.G. and Fox, T.D. (1987). *PET111*, a *Saccharomyces cerevisiae* nuclear gene required for translation of the mitochondrial mRNA encoding cytochrome *c* oxidase subunit II. *Genetics* **115**, 637–647.

Preiss, S., Schrader, S., and Johanningmeier, U. (2001). Rapid, ATP-dependent degradation of a truncated D1 protein in the chloroplast. *Eur. J. Biochem.* **268**, 4562–4569.

Rapp, J.C., Baumgartner, B.J., and Mullet, J. (1992). Quantitative analysis of transcription and RNA levels of 15 barley chloroplast genes. Transcription rates and mRNA levels vary over 300-fold; predicted mRNA stabilities vary 30-fold. *J. Biol. Chem.* **267**, 21404–21411.

Raynaud, C., Loiselay, C., Wostrikoff, K., Kuras, R., Girard-Bascou, J., Wollman, F.A., and Choquet, Y. (2007). Evidence for regulatory function of nucleus-encoded factors on mRNA stabilization and translation in the chloroplast. *Proc. Natl. Acad. Sci. U. S. A.* **104**, 9093–9098.

Robert-Le Meur, M. and Portier, C. (1992). *E. coli* polynucleotide phosphorylase expression is autoregulated through an RNAse III-dependent mechanism. *EMBO J.* **11**, 2633–2641.

Robertson, D., Boynton, J.E., and Gillham, N.W. (1990). Cotranscription of the wild-type chloroplast *atpE* gene encoding the CF1/CF0 epsilon subunit with the 3′ half of the *rps7* gene in *Chlamydomonas reinhardtii* and characterization of frameshift mutations in *atpE*. *Mol. Gen. Genet.* **221**, 155–163.

Rodermel, S., Haley, J., Jiang, C.Z., Tsai, C.H., and Bogorad, L. (1996). A mechanism for intergenomic integration: Abundance of ribulose bisphosphate carboxylase small-subunit protein influences the translation of the large-subunit mRNA. *Proc. Natl. Acad. Sci. U. S. A.* **93**, 3881–3885.

Rosu, V., Chevance, F.F., Karlinsey, J.E., Hirano, T., and Hughes, K.T. (2006). Translation inhibition of the *Salmonella fliC* gene by the *fliC* 5′ untranslated region, *fliC* coding sequences, and FlgM. *J. Bacteriol.* **188**, 4497–4507.

Russel, M., Gold, L., Morrissett, H., and O'Farrell, P.Z. (1976). Translational, autogenous regulation of gene 32 expression during bacteriophage T4 infection. *J. Biol. Chem.* **251**, 7263–7270.

Rymarquis, L.A., Higgs, D.C., and Stern, D.B. (2006). Nuclear suppressors define three factors that participate in both 5′ and 3′ end processing of mRNAs in *Chlamydomonas* chloroplasts. *Plant J.* **46**, 448–461.

Schmidt, M.G. and Oliver, D.B. (1989). SecA protein autogenously represses its own translation during normal protein secretion in *Escherichia coli*. *J. Bacteriol.* **171**, 643–649.

Schottler, M.A., Flugel, C., Thiele, W., and Bock, R. (2007). Knock-out of the plastid-encoded PetL subunit results in reduced stability and accelerated leaf age-dependent loss of the cytochrome b_6f complex. *J. Biol. Chem.* **282**, 976–985.

Schwenkert, S., Legen, J., Takami, T., Shikanai, T., Herrmann, R.G., and Meurer, J. (2007). Role of the low-molecular-weight subunits PetL, PetG, and PetN in assembly, stability, and dimerisation of the cytochrome b_6f complex in tobacco. *Plant Physiol.* **144**, 1924–1935.

Seelert, H., Dencher, N.A., and Muller, D.J. (2003). Fourteen protomers compose the oligomer III of the proton-rotor in spinach chloroplast ATP synthase. *J. Mol. Biol.* **333**, 337–344.

Shapira, M., Lers, A., Heifetz, P.B., Irihimovitz, V., Osmond, C.B., Gillham, N.W., and Boynton, J.E. (1997). Differential regulation of chloroplast gene expression in *Chlamydomonas reinhardtii* during photoacclimation: Light stress transiently suppresses synthesis of the Rubisco LSU protein while enhancing synthesis of the PS II D1 protein. *Plant Mol. Biol.* **33**, 1001–1011.

Smart, L.B. and McIntosh, L. (1993). Genetic inactivation of the *psaB* gene in *Synechocystis* sp. PCC 6803 disrupts assembly of photosystem I. *Plant Mol. Biol.* **21**, 177–180.

Smart, L.B., Anderson, S.L., and McIntosh, L. (1991). Targeted genetic inactivation of the photosystem I reaction center in the cyanobacterium *Synechocystis* sp. PCC 6803. *EMBO J.* **10**, 3289–3296.

Sokolenko, A., Pojidaeva, E., Zinchenko, V., Panichkin, V., Glaser, V.M., Herrmann, R.G., and Shestakov, S.V. (2002). The gene complement for proteolysis in the cyanobacterium *Synechocystis* sp. PCC 6803 and *Arabidopsis thaliana* chloroplasts. *Curr. Genet.* **41**, 291–310.

Stampacchia, O., Girard-Bascou, J., Zanasco, J.L., Zerges, W., Bennoun, P., and Rochaix, J.D. (1997). A nuclear-encoded function essential for translation of the chloroplast *psaB* mRNA in *Chlamydomonas*. *Plant Cell* **9**, 773–782.

Stroebel, D., Choquet, Y., Popot, J.L., and Picot, D. (2003). An atypical haem in the cytochrome b_6f complex. *Nature* **426**, 413–418.

Sturm, N.R., Kuras, R., Buschlen, S., Sakamoto, W., Kindle, K.L., Stern, D.B., and Wollman, F.A. (1994). The *petD* gene is transcribed by functionally redundant promoters in *Chlamydomonas reinhardtii* chloroplasts. *Mol. Cell Biol.* **14**, 6171–6179.

Summer, E.J., Schmid, V.H., Bruns, B.U., and Schmidt, G.W. (1997). Requirement for the H phosphoprotein in photosystem II of *Chlamydomonas reinhardtii*. *Plant Physiol.* **113**, 1359–1368.

Takahashi, Y., Goldschmidt-Clermont, M., Soen, S.Y., Franzen, L.G., and Rochaix, J.D. (1991). Directed chloroplast transformation in *Chlamydomonas reinhardtii*: Insertional inactivation of the *psaC* gene encoding the iron sulfur protein destabilizes photosystem I. *EMBO J.* **10**, 2033–2040.

Takahashi, Y., Matsumoto, H., Goldschmidt-Clermont, M., and Rochaix, J.D. (1994). Directed disruption of the *Chlamydomonas* chloroplast *psbK* gene destabilizes the photosystem II reaction center complex. *Plant Mol. Biol.* **24**, 779–788.

Takahashi, Y., Rahire, M., Breyton, C., Popot, J.L., Joliot, P., and Rochaix, J.D. (1996). The chloroplast *ycf7* (*petL*) open reading frame of *Chlamydomonas reinhardtii* encodes a small functionally important subunit of the cytochrome b_6f complex. *EMBO J.* **15**, 3498–3506.

Tichy, M. and Vermaas, W. (1999). Accumulation of pre-apocytochrome *f* in a *Synechocystis* sp. PCC 6803 mutant impaired in cytochrome c maturation. *J. Biol. Chem.* **274**, 32396–32401.

Tuerk, C., Eddy, S., Parma, D., and Gold, L. (1990). Autogenous translational operator recognized by bacteriophage T4 DNA polymerase. *J. Mol. Biol.* **213**, 749–761.

Turmel, M. and Otis, C. (1994). The chloroplast gene cluster containing *psbF, psbL, petG* and *rps3* is conserved in *Chlamydomonas*. *Curr. Genet.* **27**, 54–61.

Varotto, C., Pesaresi, P., Jahns, P., Lessnick, A., Tizzano, M., Schiavon, F., Salamini, F., and Leister, D. (2002). Single and double knockouts of the genes for photosystem I subunits G, K, and H of *Arabidopsis*. Effects on photosystem I composition, photosynthetic electron flow, and state transitions. *Plant Physiol.* **129**, 616–624.

Vecerek, B., Moll, I., and Blasi, U. (2005). Translational autocontrol of the *Escherichia coli hfq* RNA chaperone gene. *RNA* **11**, 976–984.

Wollman, F.A., Olive, J., Bennoun, P., and Recouvreur, M. (1980). Organization of the photosystem II centers and their associated antennae in the thylakoid membranes: A comparative ultrastructural, biochemical, and biophysical study of *Chlamydomonas* wild type and mutants lacking in photosystem II reaction centers. *J. Cell Biol.* **87**, 728–735.

Wollman, F.A., Minai, L., and Nechushtai, R. (1999). The biogenesis and assembly of photosynthetic proteins in thylakoid membranes. *Biochim. Biophys. Acta* **1411**, 21–85.

Wostrikoff, K. and Stern, D. (2007). Rubisco large-subunit translation is autoregulated in response to its assembly state in tobacco chloroplasts. *Proc. Natl. Acad. Sci. U. S. A.* **104**, 6466–6471.

Wostrikoff, K., Choquet, Y., Wollman, F.A., and Girard-Bascou, J. (2001). TCA1, a single nuclear-encoded translational activator specific for *petA* mRNA in *Chlamydomonas reinhardtii* chloroplast. *Genetics* **159**, 119–132.

Wostrikoff, K., Girard-Bascou, J., Wollman, F.A., and Choquet, Y. (2004). Biogenesis of PSI involves a cascade of translational autoregulation in the chloroplast of *Chlamydomonas*. *EMBO J.* **23**, 2696–2705.

Wu, J. and Bag, J. (1998). Negative control of the poly(A)-binding protein mRNA translation is mediated by the adenine-rich region of its 5′-untranslated region. *J. Biol. Chem.* **273**, 34535–34542.

Xu, Q., Jung, Y.S., Chitnis, V.P., Guikema, J.A., Golbeck, J.H., and Chitnis, P.R. (1994). Mutational analysis of photosystem I polypeptides in *Synechocystis* sp. PCC 6803. Subunit requirements for reduction of NADP+ mediated by ferredoxin and flavodoxin. *J. Biol. Chem.* **269**, 21512–21518.

Yates, J.L. and Nomura, M. (1980). *E. coli* ribosomal protein L4 is a feedback regulatory protein. *Cell* **21**, 517–522.

Yates, J.L., Arfsten, A.E., and Nomura, M. (1980). *In vitro* expression of *Escherichia coli* ribosomal protein genes: Autogenous inhibition of translation. *Proc. Natl. Acad. Sci. U. S. A.* **77**, 1837–1841.

Yosef, I., Irihimovitch, V., Knopf, J.A., Cohen, I., Orr-Dahan, I., Nahum, E., Keasar, C., and Shapira, M. (2004). RNA binding activity of the ribulose-1,5-bisphosphate carboxylase/oxygenase large subunit from *Chlamydomonas reinhardtii*. *J. Biol. Chem.* **279**, 10148–10156.

Yu, J. and Vermaas, W. (1990). Transcript levels and synthesis of Photosystem II components in cyanobacterial mutants with inactivated Photosystem II genes. *Plant Cell* **2**, 315–322.

Yu, J., Smart, L.B., Jung, Y.S., Golbeck, J., and McIntosh, L. (1995). Absence of PsaC subunit allows assembly of photosystem I core but prevents the binding of PsaD and PsaE in *Synechocystis* sp. PCC6803. *Plant Mol. Biol.* **29**, 331–342.

Zambrano, A., Fontanesi, F., Solans, A., de Oliveira, R.L., Fox, T.D., Tzagoloff, A., and Barrientos, A. (2007). Aberrant translation of cytochrome *c* oxidase subunit 1 mRNA species in the absence of Mss51p in the yeast *Saccharomyces cerevisiae*. *Mol. Biol. Cell* **18**, 523–535.

Zer, H., Prasil, O., and Ohad, I. (1994). Role of plastoquinol oxidoreduction in regulation of photochemical reaction center II D1 protein turnover *in vivo*. *J. Biol. Chem.* **269**, 17670–17676.

Zerges, W. (2000). Translation of mRNAs encoded by chloroplast genomes. *Biochimie* **82**, 583–601.

Zerges, W. and Rochaix, J.D. (1998). Low density membranes are associated with RNA-binding proteins and thylakoids in the chloroplast of *Chlamydomonas reinhardtii*. *J. Cell Biol.* **140**, 101–110.

Zouni, A., Witt, H.T., Kern, J., Fromme, P., Krauss, N., Saenger, W., and Orth, P. (2001). Crystal structure of photosystem II from *Synechococcus elongatus* at 3.8 Å resolution. *Nature* **409**, 739–743.

Index

Note: Page numbers in italics indicate tables or figures.

Q

R

S

T

U

V

W

X

Y

Z